Digital Logic and Computer Design

Thomas Richard McCalla

University of Science and Arts of Oklahoma

Merrill, an imprint of
Macmillan Publishing Company
New York

Maxwell Macmillan Canada
Toronto

Maxwell Macmillan International
New York Oxford Singapore Sydney

Editor: Dave Garza
Developmental Editor: Carol Hinklin Robison
Production Editor: Constantina Geldis
Text Designer: Cynthia M. Brunk
Cover Designer: Robert Vega
Production Buyer: Pamela D. Bennett

This book was set in Times Roman and was printed and bound by Book Press, Inc., a Quebecor America Book Group Company. The cover was printed by Lehigh Press, Inc.

Printed in the United States of America

Macmillan Publishing Company
866 Third Avenue
New York, NY 10022

Macmillan Publishing Company is part of the
Maxwell Communication Group of Companies.

Maxwell Macmillan Canada, Inc.
1200 Eglinton Avenue East, Suite 200
Don Mills, Ontario M3C 3N1

Library of Congress Cataloging-in-Publication Data
McCalla, Thomas Richard.
Digital logic and computer design / Thomas Richard McCalla.
p. cm.
Includes index.
ISBN 0-675-21170-0
1. Logic design. 2. Logic circuits. 3. Electronic digital computers—Circuits. I. Title.
TK7888.4.M42 1992
621.39'5—dc20 91-17012
CIP

Printing: 1 2 3 4 5 6 7 8 9 Year: 2 3 4 5

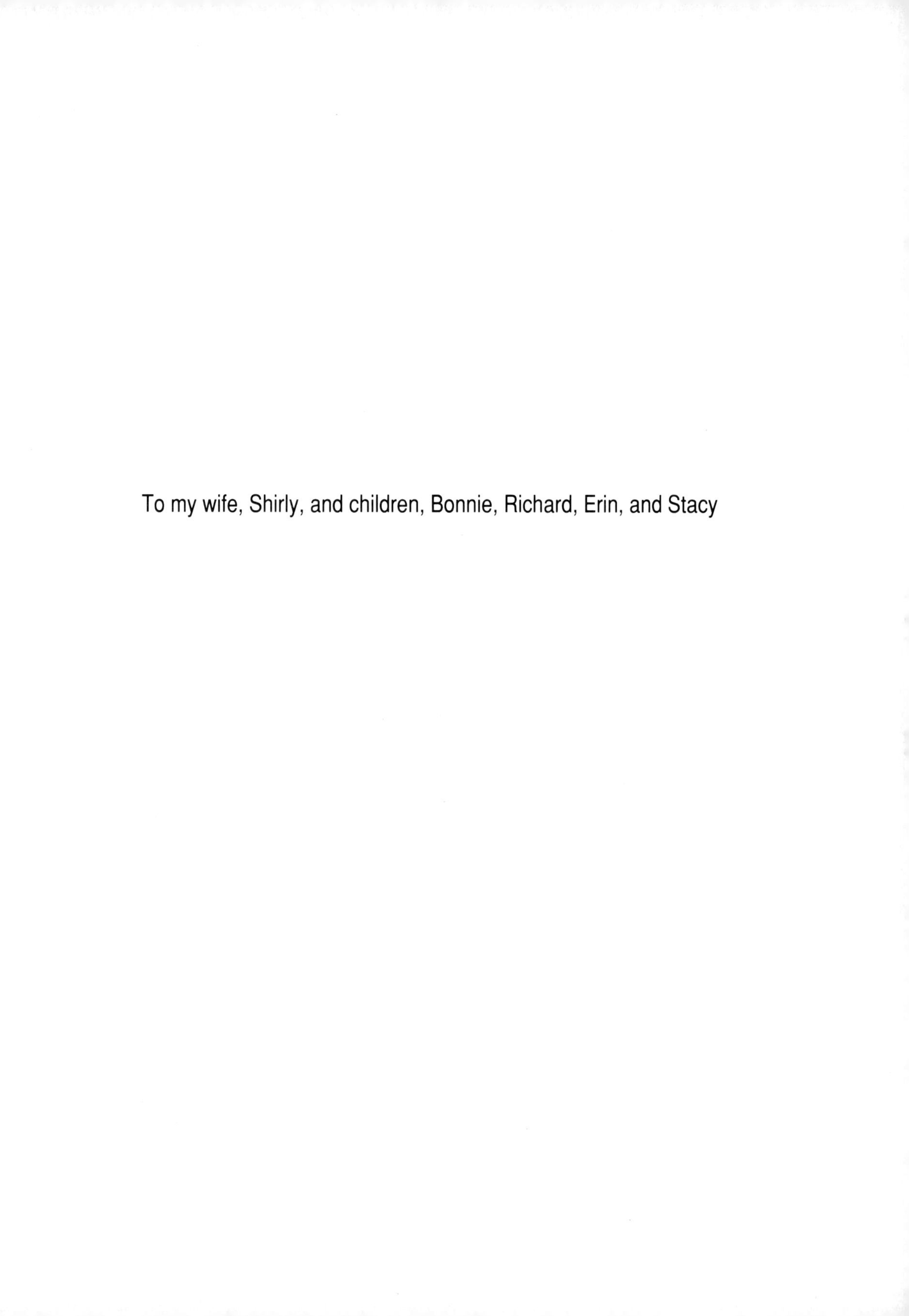

To my wife, Shirly, and children, Bonnie, Richard, Erin, and Stacy

Merrill's International Series in Engineering Technology

ADAMSON	*Applied Pascal for Technology*, 0-675-20771-1
	The Electronic Dictionary for Technicians, 0-02-300820-2
	Microcomputer Repair, 0-02-300825-3
	Structured BASIC Applied to Technology, 0-675-20772-X
	Structured C for Technology, 0-675-20993-5
	Structured C for Technology (w/disks), 0-675-21289-8
ANTONAKOS	*The 68000 Microprocessor: Hardware and Software Principles and Applications*, 0-675-21043-7
ASSER/STIGLIANO/ BAHRENBURG	*Microcomputer Servicing: Practical Systems and Troubleshooting*, 0-675-20907-2
	Microcomputer Theory and Servicing, 0-675-20659-6
	Lab Manual to accompany Microcomputer Theory and Servicing, 0-675-21109-3
ASTON	*Principles of Biomedical Instrumentation and Measurement*, 0-675-20943-9
BATESON	*Introduction to Control System Technology, Third Edition*, 0-675-21010-0
BEACH/JUSTICE	*DC/AC Circuit Essentials*, 0-675-20193-4
BERLIN	*Experiments in Electronic Device to accompany Floyd's Electronic Devices and Electronic Devices: Electron Flow Version, Third Edition*, 0-02-308422-7
	The Illustrated Electronics Dictionary, 0-675-20451-8

BERLIN/GETZ	*Experiments in Instrumentation and Measurement*, 0-675-20450-X
	Fundamentals of Operational Amplifiers and Linear Integrated Circuits, 0-675-21002-X
	Principles of Electronic Instrumentation and Measurement, 0-675-20449-6
BERUBE	*Electronic Devices and Circuits Using MICRO-CAP II*, 0-02-309160-6
BOGART	*Electronic Devices and Circuits, Second Edition*, 0-675-21150-6
BOGART/BROWN	*Experiments in Electronic Devices and Circuits, Second Edition*, 0-675-21151-4
BOYLESTAD	*DC/AC: The Basics*, 0-675-20918-8
	Introductory Circuit Analysis, Sixth Edition, 0-675-21181-6
BOYLESTAD/ KOUSOUROU	*Experiments in Circuit Analysis, Sixth Edition*, 0-675-21182-4
	Experiments in DC/AC Basics, 0-675-21131-X
BREY	*Microprocessors and Peripherals: Hardware, Software, Interfacing and Applications, Second Edition*, 0-675-20884-X
	The Intel Microprocessors—8086/8088, 80186, 80286, 80386, and 80486—Architecture, Programming, and Interfacing, Second Edition, 0-675-21309-6
BROBERG	*Lab Manual to accompany Electronic Communication Techniques, Second Edition*, 0-675-21257-X
BUCHLA	*Digital Experiments: Emphasizing Systems and Design, Second Edition*, 0-675-21180-8
	Experiments in Electric Circuits Fundamentals, Second Edition, 0-675-21409-2
	Experiments in Electronics Fundamentals: Circuits, Devices and Applications, Second Edition, 0-675-21407-6
BUCHLA/McLACHLAN	*Applied Electronic Instrumentation and Measurement*, 0-675-21162-X
CICCARELLI	*Circuit Modeling: Exercises and Software, Second Edition*, 0-675-21152-2
COOPER	*Introduction to VersaCAD*, 0-675-21164-6
COX	*Digital Experiments: Emphasizing Troubleshooting, Second Edition*, 0-675-21196-4
CROFT	*Getting a Job: Resume Writing, Job Application Letters, and Interview Strategies*, 0-675-20917-X

DAVIS	*Technical Mathematics*, 0-675-20338-4
	Technical Mathematics with Calculus, 0-675-20965-X
	Study Guide to accompany Technical Mathematics, 0-675-20966-8
	Study Guide to accompany Technical Mathematics with Calculus, 0-675-20964-1
DELKER	*Experiments in 8085 Microprocessor Programming and Interfacing*, 0-675-20663-4
FLOYD	*Digital Fundamentals, Fourth Edition*, 0-675-21217-0
	Electric Circuits Fundamentals, Second Edition, 0-675-21408-4
	Electronic Devices, Third Edition, 0-675-22170-6
	Electronic Devices: Electron Flow Version, 0-02-338540-5
	Electronics Fundamentals: Circuits, Devices, and Applications, Second Edition, 0-675-21310-X
	Fundamentals of Linear Circuits, 0-02-338481-6
	Principles of Electric Circuits, Electron Flow Version, Second Edition, 0-675-21292-8
	Principles of Electric Circuits, Third Edition, 0-675-21062-3
FULLER	*Robotics: Introduction, Programming, and Projects*, 0-675-21078-X
GAONKAR	*Microprocessor Architecture, Programming, and Applications with the 8085/8080A, Second Edition*, 0-675-20675-8
	The Z80 Microprocessor: Architecture, Interfacing, Programming, and Design, 0-675-20540-9
GILLIES	*Instrumentation and Measurements for Electronic Technicians*, 0-675-20432-1
GOETSCH	*Industrial Supervision: In the Age of High Technology*, 0-675-22137-4
GOETSCH/RICKMAN	*Computer-Aided Drafting with AutoCAD*, 0-675-20915-3
GOODY	*Programming and Interfacing the 8086/8088 Microprocessor*, 0-675-21312-6
HUBERT	*Electric Machines: Theory, Operation, Applications, Adjustment, and Control*, 0-675-21136-0
HUMPHRIES	*Motors and Controls*, 0-675-20235-3
HUTCHINS	*Introduction to Quality: Management, Assurance and Control*, 0-675-20896-3
KEOWN	*PSpice and Circuit Analysis*, 0-675-22135-8

KEYSER	*Materials Science in Engineering, Fourth Edition*, 0-675-20401-1
KIRKPATRICK	*The AutoCAD Book: Drawing, Modeling and Applications, Second Edition*, 0-675-22288-5
	Industrial Blueprint Reading and Sketching, 0-675-20617-0
KRAUT	*Fluid Mechanics for Technicians*, 0-675-21330-4
KULATHINAL	*Transform Analysis and Electronic Networks with Applications*, 0-675-20765-7
LAMIT/LLOYD	*Drafting for Electronics*, 0-675-20200-0
LAMIT/WAHLER/HIGGINS	*Workbook in Drafting for Electronics*, 0-675-20417-8
LAMIT/PAIGE	*Computer-Aided Design and Drafting*, 0-675-20475-5
LAVIANA	*Basic Computer Numerical Control Programming, Second Edition*, 0-675-21298-7
MacKENZIE	*The 8051 Microcontroller*, 0-02-373650-X
MARUGGI	*Technical Graphics: Electronics Worktext, Second Edition*, 0-675-21378-9
	The Technology of Drafting, 0-675-20762-2
	Workbook for the Technology of Drafting, 0-675-21234-0
McCALLA	*Digital Logic and Computer Design*, 0-675-21170-0
McINTYRE	*Study Guide to accompany Electronic Devices and Electronic Devices: Electron Flow Version, Third Edition*, 0-02-379296-5
	Study Guide to accompany Electronics Fundamentals, Second Edition, 0-675-21406-8
MILLER	*The 68000 Microprocessor Family: Architecture, Programming, and Applications, Second Edition*, 0-02-381560-4
MONACO	*Essential Mathematics for Electronics Technicians*, 0-675-21172-7
	Introduction to Microwave Technology, 0-675-21030-5
	Laboratory Activities in Microwave Technology, 0-675-21031-3
	Preparing for the FCC General Radiotelephone Operator's License Examination, 0-675-21313-4
	Student Resource Manual to accompany Essential Mathematics for Electronics Technicians, 0-675-21173-5
MONSEEN	*PSPICE with Circuit Analysis*, 0-675-21376-2
MOTT	*Applied Fluid Mechanics, Third Edition*, 0-675-21026-7

	Machine Elements in Mechanical Design, Second Edition, 0-675-22289-3
NASHELSKY/ BOYLESTAD	*BASIC Applied to Circuit Analysis*, 0-675-20161-6
PANARES	*A Handbook of English for Technical Students*, 0-675-20650-2
PFEIFFER	*Proposal Writing: The Art of Friendly Persuasion*, 0-675-20988-9
	Technical Writing: A Practical Approach, 0-675-21221-9
POND	*Introduction to Engineering Technology*, 0-675-21003-8
QUINN	*The 6800 Microprocessor*, 0-675-20515-8
REIS	*Digital Electronics Through Project Analysis*, 0-675-21141-7
	Electronic Project Design and Fabrication, Second Edition, 0-02-399230-1
	Laboratory Manual for Digital Electronics Through Project Analysis, 0-675-21254-5
ROLLE	*Thermodynamics and Heat Power, Third Edition*, 0-675-21016-X
ROSENBLATT/ FRIEDMAN	*Direct and Alternating Current Machinery, Second Edition*, 0-675-20160-8
ROZE	*Technical Communication: The Practical Craft*, 0-675-20641-3
SCHOENBECK	*Electronic Communications: Modulation and Transmission, Second Edition*, 0-675-21311-8
SCHWARTZ	*Survey of Electronics, Third Edition*, 0-675-20162-4
SELL	*Basic Technical Drawing*, 0-675-21001-1
SMITH	*Statistical Process Control and Quality Improvement*, 0-675-21160-3
SORAK	*Linear Integrated Circuits: Laboratory Experiments*, 0-675-20661-8
SPIEGEL/ LIMBRUNNER	*Applied Statics and Strength of Materials*, 0-675-21123-9
STANLEY, B.H.	*Experiments in Electric Circuits, Third Edition*, 0-675-21088-7
STANLEY, W.D.	*Operational Amplifiers with Linear Integrated Circuits, Second Edition*, 0-675-20660-X
SUBBARAO	*16/32-Bit Microprocessors: 68000/68010/68020 Software, Hardware, and Design Applications*, 0-675-21119-0
TOCCI	*Electronic Devices: Conventional Flow Version, Third Edition*, 0-675-20063-6
	Fundamentals of Pulse and Digital Circuits, Third Edition, 0-675-20033-4

	Introduction to Electric Circuit Analysis, Second Edition, 0-675-20002-4
TOCCI/OLIVER	*Fundamentals of Electronic Devices, Fourth Edition*, 0-675-21259-6
WEBB	*Programmable Logic Controllers: Principles and Applications, Second Edition*, 0-02-424970-X
WEBB/GRESHOCK	*Industrial Control Electronics*, 0-675-20897-1
WEISMAN	*Basic Technical Writing, Sixth Edition*, 0-675-21256-1
WOLANSKY/AKERS	*Modern Hydraulics: The Basics at Work*, 0-675-20987-0
WOLF	*Statics and Strength of Materials: A Parallel Approach*, 0-675-20622-7

Preface

Digital Logic and Computer Design is intended for use in first courses in digital logic design. The text covers the material for subject area 6 (logic design) of the IEEE recommended curriculum. Material on elementary computer design is also included to illustrate the hierarchical nature of digital systems and to provide a meaningful, attainable goal for the reader.

The text is suitable for teaching combinational and sequential logic design in university computer science and electrical and computer engineering programs. It may also be used for logic design courses in electronics or computer technology programs in technical institutes and community colleges.

Organization

The text material is hierarchically organized in parts that relate to areas of study necessary to the logic designer:

Part	Area of Study
One, Two	Combinational Logic
Thee	Sequential Logic
Four	Digital systems
Five	Programmable digital systems (computers)

Further, the text integrates digital design methods (hierarchical organization, divide and conquer, evolutionary modular design) with computer organization concepts (data flow, control, memory, and input/output).

Philosophy

The hierarchical organization of the text material and a standardized design procedure (with appropriate conceptual models and design tools) are coupled with the use of logic simulation programs to provide an interactive environment that enables students to effectively learn and apply digital logic concepts to the design of a wide range of digital devices and systems (including computers).

An integrated approach to the study of sequential logic, digital systems, and programmable systems is made possible by the use of an evolutionary set of conceptual models to represent processes:

Finite state machines (FSM)—sequential processes

Algorithmic state machines (ASM)—controlled processes

Programmable ASM (PASM)—multifunction processes

This evolutionary approach within each area of study is illustrated by the design evolution from a 1-bit adder to an 8-bit arithmetic unit, from rudimentary-control sequential machines to sophisticated ASMs, and from simple stored-program computers to more complex computers.

Understanding of each area of study (and how these areas are related) is facilitated by the theoretical development of conceptual models and the application of these models in the design of a variety of digital devices and systems.

Evolutionary Hierarchical Building-Block Approach

The text emphasizes a hierarchical approach to digital logic and computer design that encourages system-to-circuits (top-down) thinking prior to circuits-to-systems thinking. A diagram called *Evolutionary Hierarchical Building-Block Approach* (EHBBA) appears in each of the five part openers. The EHBBA diagram provides a road map for students that enables them to relate each subject area to the overall study of digital logic and computer design.

A standard 4-step problem-solving procedure (PSP) (with appropriate conceptual models and design tools for combinational logic, sequential logic, digital systems, and stored-program computers) is used throughout the text to integrate the areas of study and provide a unified approach to digital design. The 4-step PSP is used to design a number of common combinational logic function modules (such as multiplexers, demultiplexers, encoders, decoders, parity circuits, and shifters) and sequential logic function modules (such as counters, shift registers, and shift-register counters). These function modules are then used as building blocks for implementing higher-level devices and systems.

The four steps (problem statement, conceptualization, solution/simplification, and realization), first described in Chapter 2, are applied to circuit design examples in Chapters 3–9 and systems/computer design examples in Chapters 10–14. The following table shows how the 4-step problem-solving procedure is used in the design of combinational logic circuits (Part Two), sequential logic circuits (Part Three), digital systems (Part Four), and computers (Part Five).

Part	Problem Statement (Description of)	Conceptualization (Design Tools)	Solution/Simplification	Realization
2	Output as function of input combinations	Combinational logic function: truth table	Canonical expression Simplified expression	Combinational circuit
3	Sequential process I/O sequence relation	Sequential machine: state table and diagram	State reduction State assignment Excitation and output equations	Sequential circuit
4	Process and its control algorithm	Algorithmic state machine (ASM): structure diagram and control-flow diagram	Data path design; controller next-state generator and output decoder design	Digital system
5	Computer instruction set (architecture)	Programmable ASM (memory, CPU, I/O): structure diagram and control-flow diagram	CPU data path and control unit design; Memory, CPU, and I/O integration	Digital computer

A number of the more complex design problems in the text are presented as interactive design applications (IDAs) in which the 4-step PSP uses appropriate conceptual-level design tools, such as structure diagrams (for control/process partitioning), control-flow diagrams, and register transfer notation. This problem-solving procedure provides a framework to guide the reader through the process of designing and implementing a number of increasingly sophisticated, yet conceptually simple, digital devices and systems (including computers).

The emphasis throughout is on learning digital logic concepts and applying these concepts in the analysis or design of physically realizable circuits and systems. In the process, the text

- Develops conceptual models for combinational functions, sequential processes, controlled processes, and stored-program computers
- Develops algorithms to replace manual techniques (such as Karnaugh maps) and trial-and-error methods (such as translating design specifications into state diagrams)
- Develops geometric models that permit visualization and interpretation of procedures and processes
- Uses an evolutionary, hierarchical, building-block approach to digital design, using logic gates to form function modules that are combined to form higher-level devices that, in turn, are used as larger building blocks for the design and implementation of a number of increasingly sophisticated devices and systems
- Uses a structured algorithmic approach to the design of digital systems for realizing processes controlled by a set of rules (for example, a dice game, a blackjack dealer, a serial adder, a serial multiplier, a serial divider)
- Promotes interactive design through the use of logic simulation programs
- Develops problem-solving skills, using digital design tools, techniques, and functional building blocks to design and implement a variety of digital devices and systems

The text is highly interactive, using logic simulation programs that allow students to simulate digital logic devices and systems that they have designed. In Chapters 2 through 14 a logic simulation program (Designworks by Capilano Computing) is used on a Macintosh computer to simulate a wide range of digital logic devices and systems, starting with a 1-bit adder and culminating in a variety of functional 8-bit digital computers.

Throughout the text, design theory and its application are described in detail, starting with a design problem statement and ending with the simulation/implementation of a digital device or system.

Coverage

Chapter 1 describes number systems and computer codes, develops algorithms for converting integers and fractions from binary to decimal and decimal to binary, and develops algorithms for generating various computer codes.

Chapter 2 develops logic concepts, switching theory, and Boolean algebra in a context independent of hardware implementation. The application of Boolean algebra to digital-logic circuits is then described using compatible fixed-logic and mixed-logic systems. The appropriate use of mixed-logic methodology enhances the student's ability to analyze logic diagrams and visualize equations representing a circuit.

Chapter 3 presents traditional map techniques (Karnaugh maps) and algorithmic methods (Quine–McCluskey) for simplifying logic expressions. An integrated mapping/algorithmic approach is developed for simplifying logic expressions in more than four variables using the McCalla minterm-ring map/algorithm; this method provides both visual and algorithmic identification of groupings of logically adjacent minterms (prime implicants and essential prime implicants) used to produce a minimal sum-of-products expression. The minterm-ring map/algorithm can be used to replace the traditional map techniques and algorithmic methods.

Chapter 4 uses the 4-step PSP to design a number of common combinational-logic function modules, such as multiplexers, demultiplexers, encoders, decoders, parity circuits, and shifters.

Chapter 5 develops alternative designs of multibit adders (cascaded, direct, and carry look-ahead) and describes the design evolution of combinational-logic computing devices, ranging from a 1-bit adder to an 8-bit arithmetic unit.

Chapter 6 describes alternative approaches for designing higher-level combinational-logic computing devices. A bit-slice approach is used to develop an arithmetic logic unit (ALU), and a top-down design approach is used to design a BCD arithmetic unit.

Chapter 7 introduces sequential (finite state) machine concepts and uses mixed-logic methodology to describe the design of basic sequential-logic function modules (latches and flip-flops). The 4-step PSP is used to design specialized synchronous sequential-logic circuits such as counters, registers, and register counters.

Chapter 8 describes and illustrates procedures for analyzing general sequential-logic circuits using classical design tools (state diagrams and tables and transition maps), develops algorithms for translating design specifications into state diagrams, and describes procedures for state table reduction. The 4-step PSP is used to design a variety of general synchronous sequential-logic circuits.

Chapter 9 describes basic concepts of fundamental-mode asynchronous sequential-logic circuits and illustrates analysis procedures to determine the presence of hazards or race conditions. It also covers state reduction techniques and develops an algorithm for designing race-free asynchronous circuits. The 4-step PSP is used to design a variety of fundamental-mode asynchronous circuits.

Chapter 10 presents an algorithmic approach for designing digital logic systems. Finite-state machine concepts are used to describe iterative processes, and 4-step PSP designs are presented for rudimentary control machines (serial parity checker, 2s complementer, and serial adder). Algorithmic state machines (ASMs) are used to model general processes that are governed by a set of rules. ASMs are designed for a serial adder, a dice game, and a blackjack dealer using the 4-step PSP.

Chapter 11 uses an algorithmic approach and the 4-step PSP to describe the design of realistic digital systems; these include a serial multiplier, serial divider, and a combined serial multiplier/divider. Techniques for shared data path design and combined control unit design are discussed in the development of the combined multiplier/divider.

Chapter 12 uses a building-block design approach to describe the evolution from a 4-bit adder to a functional stored-program computer processor in 12 steps. Each step describes a functional building block and its integration with the preceding components of its subsystem.

Chapter 13 uses a systems-design approach to describe the design of simple digital computers. In a systems context, a computer processor is a programmable algorithmic state machine (PASM) composed of a memory and central processing unit (CPU), consisting of a control unit and a process data path. Chapter 13 describes in detail general procedures for the design of simple stored-program digital computers. These procedures are illustrated by describing the design evolution of a hierarchy of four 8-bit mux-oriented computers (MC-8X). Each succeeding computer is designed to implement an increasingly complex instruction set (architecture).

Chapter 14 describes the design evolution of a hierarchy of 8-bit bus-oriented computers (BC-8X) with instruction sets corresponding to those of the computers designed in Chapter 13.

Each computer designed in Part Five is accompanied by a complete logic diagram and a description of its assembly-language instruction set. The reader can use the logic diagram to implement the computer in hardware or simulate it using a logic simulation program such as Designworks. Each assembly language set allows the reader to program the computer and debug both the computer and the program using Designworks. The MC-8D and BC-8D assembly language instructions can implement a wide range of algorithms, which include

List processing (sorting, merging, and comparing lists of data)

Statistics (computing mean, variance, regression, correlation)

Numerical methods (interpolation, curve fitting, numerical integration, and the solution of differential equations) *

Each chapter contains a set of relevant exercises. In addition to the set of general exercises related to concepts in the chapter, Chapters 2–14 each contain a set of design/implementation exercises to allow students to further develop their problem-solving skills.

*Knowledge of numerical methods is not required.

In addition to providing complete and understandable coverage of the fundamentals of digital logic and computer design, it is my hope that students will find the material in this text presented in a manner that adds clarification to complex topics, provides visual representations of abstract concepts, and generally enhances their learning efforts. If this text effectively raises the level of learning that the students experiences while eliminating some of the confusion and frustration, which can commonly occur, it has successfully achieved its goal.

I appreciate the helpful comments and suggestions of the following reviewers: Lyle McCurdy, California Polytechnic State University—Pomona; Warren Foxwell, DeVry Institute of Technology—Lombard; Albert McHenry, Arizona State University; Ronald Lessard, Norwich University; Dale Pollack, Northern Illinois University; Ed Geckler, Hocking Technical College; Nazar Karzay, Indiana Vocational Technical College; Madjid Mousavi, Greenville Technical College; Martin Taylor, University of California—Santa Cruz; Glen Langdon, University of California—Santa Cruz; Sam Bell, University of Louisville; Richard Cockrum, California Polytechnic State University—Pomona; Richard Adamec, Penn State University—Mont Alto Campus; Lee Rosenthal, Fairleigh Dickinson University; Gary Johnsey, University of Southern Mississippi; and John Yarbrough, Oregon Institute of Technology.

My heartfelt thanks also to all who contributed to the development of this book: Dr. Glen Langdon for introducing me to the subject of computer design; Kay Mote, Vanessa Vann, Donna Jo McCalla, Ron Kemper, and Cheryl Sharp for assistance in typing and/or printing early versions of the manuscript; my students who have suffered through various manuscript versions and offered encouragement and support; Chris Dewhurst of Capilano Computing and Steve Narmontas of Clarity Software for developing logic simulation programs (Designworks and CCW, respectively) to facilitate the interactive design of digital circuits, systems, and computers; Alan Rossman of Radius Corporation for the loan of a Radius portrait-size display which facilitated the development of the larger logic diagrams in the text; Steve Helba and Dave Garza at Merrill, an imprint of Macmillan Publishing, for believing in and supporting the development of an integrated treatment of digital logic and computer design; my developmental editor, Carol Robison, who patiently guided the manuscript through the reviews and rewrites and helped sharpen the book's focus; Betty O'Bryant who diligently edited the final manuscript; and Connie Geldis and Sharon Rudd for their work in guiding the manuscript through the production process.

Special thanks to my daughter, Stacy, for sketching the "mouse" diagrams in Chapter 12, and to my wife, Shirly, for her unwavering love and support through the years.

I wish my readers well in their pursuit of knowledge in this exciting field of digital logic and computer design.

Thomas Richard McCalla

Contents

PART ONE
Number Systems, Computer Codes, and Combinational-Logic Design Tools — 2

1
Number Systems and Computer Codes — 5

1.1 Introduction — 6
1.2 Number Systems — 8
1.3 The Decimal (Base-10) Number System — 9
1.4 The Binary (Base-2) Number System — 16
1.5 Converting Numbers Between Base 10 and Base 2 — 23
1.6 Other Number Systems — 27
1.7 Computer Codes — 31
1.8 Summary — 37

2
Logic, Switching Circuits, Boolean Algebra, and Logic Gates — 41

2.1 Introduction — 42
2.2 Elements of Problem Solving — 42
2.3 Conceptualization in Logic Design—Logic and Logic Functions — 44
2.4 Realization of Logic Operations by Switches—Switch Network Algebra — 50
2.5 Boolean Switching Algebra — 54
2.6 Realization of Logic Equations Using Gates as Logic Functions — 67
2.7 Realization of Application Function Modules and Devices — 81
2.8 Summary — 85

3

Methods for Simplifying Boolean Expressions 91

3.1 Introduction 92
3.2 Karnaugh Maps in 2 Variables 92
3.3 Karnaugh Maps in 3 Variables 95
3.4 Karnaugh Maps in 4 Variables 101
3.5 Minterm-Ring Maps and Karnaugh Maps in 5 and 6 Variables 119
3.6 The Quine–McCluskey Method for Simplifying Boolean Expressions 135
3.7 Variable-Entered Maps 140
3.8 Summary 144

PART TWO
Evolution of Combinational-Logic Computing Devices 150

4

Design of Combinational-Logic Function Modules 153

4.1 Introduction 154
4.2 Design Procedure for a Combinational Logic Circuit 160
4.3 Multiplexers, Demultiplexers, and Buses 163
4.4 Encoders, Decoders, and Code Converters 168
4.5 Parity Circuits—Generators and Checkers 177
4.6 Shifters 180
4.7 Implementation of an Arbitrary Boolean Function Using MSI ICs 181
4.8 Building-Block Implementation of Higher-Level Devices 183
4.9 Programmable Logic Devices 188
4.10 Summary 191

5

Evolution from a 1-Bit Adder to an 8-Bit Arithmetic Unit 195

5.1 Introduction 196
5.2 Design of 2-Bit Binary Adders 197
5.3 Design of 4-Bit Binary Adders 205
5.4 Design of a 4-Bit Adder/Subtracter 210
5.5 Design of a 4-Bit 8-Function Arithmetic Unit 214
5.6 Use of 4-Bit MSI ICs to Implement 8-Bit Computing Devices 217
5.7 Summary 218

6

Logical and Decimal Operations—Hierarchical Development 221

6.1 Introduction 222
6.2 Design of Binary Arithmetic Logic Units 223

6.3 BCD Arithmetic Unit—A Top-Down Design 234
6.4 Summary 254

PART THREE
Sequential Processes and Machines and Sequential Logic Circuits 258

7

Latches, Flip-Flops, Counters, and Registers 261

7.1 Introduction 262
7.2 Gated Latches and Level-Sensitive Flip-Flops 275
7.3 Pulse-Triggered and Edge-Triggered Flip-Flops 283
7.4 Counters 292
7.5 Registers, Shift Registers, and Shift-Register Counters 312
7.6 Summary 325

8

Sequential Processes and Machines and Synchronous Sequential Circuits 331

8.1 Introduction 332
8.2 Analysis of General Synchronous Sequential-Logic Circuits 335
8.3 Design of General Synchronous Sequential-Logic Circuits 348
8.4 Design of Various Types of Synchronous Sequential-Logic Circuits 374
8.5 Summary 396

9

Asynchronous Sequential Circuits 401

9.1 Introduction 402
9.2 Basic Concepts of Fundamental-Mode Circuits 403
9.3 Analysis of Fundamental-Mode Circuits 409
9.4 Design of Fundamental-Mode Circuits 418
9.5 Design of Representative Fundamental-Mode Circuits 438
9.6 Summary 455

PART FOUR
Digital Systems Design—An Algorithmic Approach 462

10

Digital Systems Design 1—State Machines and Systems Design 465

10.1 Introduction 466
10.2 One-Dimensional Iterative Processes 467
10.3 Sequential Machines with Rudimentary Control 471

10.4 Algorithmic State Machines and Systems-Level Design Procedures 481
10.5 Summary 509

11

Digital Systems Design 2—Multipliers, Dividers, and Control Units 515

11.1 Introduction—An Algorithmic Approach to Design 516
11.2 Design and Implementation of a Serial Multiplication ASM 517
11.3 Design and Implementation of a Serial Division ASM 528
11.4 Design and Implementation of a Combined Serial Multiplier/Divider 548
11.5 BCD-to-Binary and Binary-to-BCD Conversion 560
11.6 Summary 561

PART FIVE
Computer Design, Simulation/Implementation, and Programming 564

12

Building-Block Evolution of a Simple Computer 567

12.1 Introduction 568
12.2 Design and Implementation of a Data Path Subsystem 577
12.3 Design and Implementation of a Control Unit Subsystem 581
12.4 Design and Implementation of a Memory Subsystem 591
12.5 Summary 604

13

Design of Mux-Oriented 8-Bit Computers 611

13.1 Introduction 612
13.2 A Programmer's View of a Stored-Program Computer 612
13.3 Direct Address Mode, Register Reference, I/O, and Branch Instructions 614
13.4 Indirect Address Mode, Compare, and Conditional Branch Instructions 630
13.5 Subroutine Jump, Jump Indirect, Multiply, Divide, and Logic Instructions 646
13.6 Indexed Address Mode, Compare, Swap, and Skip Instructions 659
13.7 Summary 662

14

Design of Bus-Oriented 8-Bit Computers 665

14.1 Introduction 666
14.2 Direct Address Mode, Register Reference, I/O, and Branch Instructions 668
14.3 Indirect Address Mode, Compare, and Conditional Branch Instructions 680
14.4 Subroutine Jump, Jump Indirect, Multiply, Divide, and Logic Instructions 691
14.5 Summary 699

Appendices 705

A Integrated Circuit List 707
B ANSI/IEEE Standard 91–1984 for Logic Functions 709
C TTL User's Guide 713
D Introduction to Signetics Programmable Logic 721

Solutions to Odd-Numbered Exercises 733

Index 779

PART ONE

Number Systems, Computer Codes, and Combinational-Logic Design Tools

1 **Number Systems and Computer Codes**
2 **Logic, Switching Circuits, Boolean Algebra, and Logic Gates**
3 **Methods for Simplifying Boolean Expressions**

This text uses an evolutionary hierarchical building-block approach (**EHBBA**) to describe the design and implementation of a variety of digital logic devices and systems, ranging from simple devices to complex systems, such as controllers and computers. The EHBBA diagram at the beginning of this part provides a roadmap to guide you in the study of digital logic. The relevant sections of the EHBBA diagram will be highlighted in subsequent parts of this text.

Part I (Chapters 1 through 3) describes number systems and computer codes; develops fundamental elements of logic, switching theory, and Boolean algebra; describes methods for simplifying logic expressions; and illustrates the application of Boolean algebra in the design of combinational logic circuits.

Chapter 1 presents a brief overview of systems in general and digital systems in particular. Because humans usually count and compute with decimal numbers and most computers use binary-related systems, decimal data is generally converted to binary form for input and processing. The binary results of processing are generally converted back to decimal output to facilitate understanding by humans. For this reason, it is important to understand both the decimal and the binary number systems and the relationship between the two.

Chapter 2 describes a 4-step problem-solving procedure for solving a range of problems in different application areas. Basic logic concepts and switching theory are developed, along with Boolean algebra and its application to digital circuits using fixed-logic and mixed-logic systems. Chapter 2 also describes the logic gates that realize the principal switching functions and illustrates their use in implementing combinational logic devices.

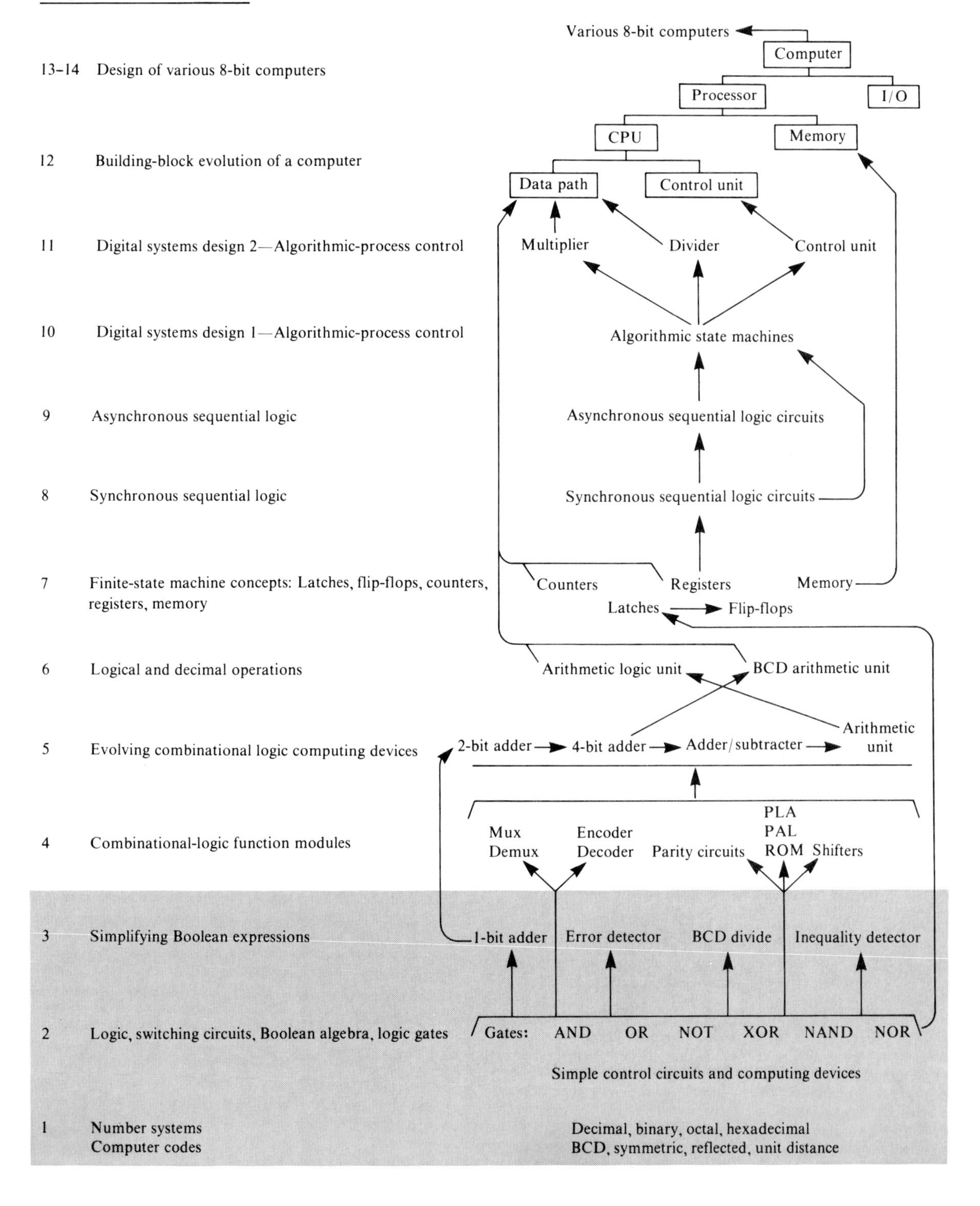

Chapters
Coverage
13–14 Design of various 8-bit computers
12 Building-block evolution of a computer
11 Digital systems design 2—Algorithmic-process control
10 Digital systems design 1—Algorithmic-process control
9 Asynchronous sequential logic
8 Synchronous sequential logic
7 Finite-state machine concepts: Latches, flip-flops, counters, registers, memory
6 Logical and decimal operations
5 Evolving combinational logic computing devices
4 Combinational-logic function modules
3 Simplifying Boolean expressions
2 Logic, switching circuits, Boolean algebra, logic gates
1 Number systems
Computer codes
Various 8-bit computers
Computer
Processor
I/O
CPU
Memory
Data path
Control unit
Multiplier
Divider
Control unit
Algorithmic state machines
Asynchronous sequential logic circuits
Synchronous sequential logic circuits
Counters
Registers
Memory
Latches
Flip-flops
Arithmetic logic unit
BCD arithmetic unit
Arithmetic unit
2-bit adder
4-bit adder
Adder/subtracter
Mux
Demux
Encoder
Decoder
Parity circuits
PLA
PAL
ROM
Shifters
1-bit adder
Error detector
BCD divide
Inequality detector
Gates: AND OR NOT XOR NAND NOR
Simple control circuits and computing devices
Decimal, binary, octal, hexadecimal
BCD, symmetric, reflected, unit distance

Chapter 3 describes methods for simplifying Boolean algebra expressions. In this chapter, the focus is on visual techniques and algorithmic methods. A new minterm-ring method is described that combines map techniques and an algorithmic method for the systematic reduction of logic expressions in 5 or more variables. Since most logic minimization currently is accomplished with software tools, the minterm-ring method is important because it provides a contemporary algorithmic method for simplifying expressions and provides a three-dimensional perspective for visual identification of logically adjacent minterms of switching functions.

1
Number Systems and Computer Codes

OBJECTIVES

After you complete this chapter, you will be able

- ☐ To use positional notation number systems
- ☐ To add and subtract in the decimal and binary number systems
- ☐ To convert decimal integers and fractions to binary integers and fractions
- ☐ To convert binary integers and fractions to decimal integers and fractions
- ☐ To represent binary integers in three forms: sign magnitude, 1s complement, and 2s complement
- ☐ To perform binary subtraction through binary addition, as is done in the central processing unit (CPU) of most computers
- ☐ To express and interpret numbers in octal and in hexadecimal systems and to convert these numbers to and from decimal and binary
- ☐ To represent numbers in floating-point form
- ☐ To recognize the characteristics and the applications of various computer codes

1.1 INTRODUCTION

In this chapter, we present a brief overview of systems in general and digital systems in particular. Then, we describe in detail the number and coding systems used for man–machine communication.

Humans usually count and compute with decimal numbers, while most computers use binary-related systems. Hence, decimal data are generally converted to binary form for input and processing, and the binary results of the computer's processing are usually converted back to decimal form to facilitate understanding by humans. For this reason, it is important to understand both the decimal and binary number systems and the relationship between the two. The octal (base-8) and hexadecimal (base-16) number systems, which we will cover in Section 1.6, are compact representations of binary numbers that are used for the convenience of humans.

Digital systems use binary numbers to represent logic values and signal voltage levels, memory addresses, and program instructions and data. Most computer memories are composed of 2-state storage elements (cell), each of which can store a *binary digit* (bit). A bit is either a 0 or a 1. Each addressable memory location is a *word* consisting of N bits. Therefore, the binary number system and binary coding schemes are used to store program instructions and data.

Alphanumeric computer codes, such as ASCII (American Standard Code for Information Interchange) and EBCDIC (Extended BCD Interchange Code), are binary coding schemes used to represent keyboard characters such as decimal digits, letters of the alphabet, and special symbols. *Numeric codes*, such as binary-coded decimal (BCD), are used in digital systems to represent decimal numbers in binary form.

An Overview of Systems

The study of digital logic encompasses the analysis and design of digital devices and systems, ranging from devices that perform simple logic functions to complex systems such as controllers and digital computers. A **system** is (1) a group of things or parts working together or connected in some way so as to form a whole (e.g., the solar system, a school system), or (2) a set of facts, rules, ideas, and so forth that make up an orderly plan (e.g., a democratic system of government).

We are all familiar with systems. We live in the solar system, we are educated in school systems, and our lives are governed by political and economic systems. Our daily class activities are scheduled by computer systems. Accounting systems maintain records of our bills and the payment or nonpayment of those bills.

An electronic system consists of a number of electrically powered devices, each of which is made up of physical components such as resistors, capacitors, and inductors. Electronic systems can be categorized as either analog, digital, or hybrid (a combination analog-digital system). The word **analog** derives from analogous (alike in some way), while the word **digital** derives from the Latin word *digitus* (finger). An analog computer deals directly with physical quantities (weights, voltages, lengths, and so on) rather than with a numerical code. A digital computer uses numbers expressed as digits of a number system (usually a base-2 or binary number system). For example,

- An automobile ignition system, from an overall viewpoint, is a digital system; the ignition is either on or off.
- An automobile gas gauge is a simple analog system; the position of the indicator is an analog of the level of gasoline in the tank.

A system can be described verbally or can be represented by a model. A *model* is any convenient representation of a system. For example,

- A diagram that graphically describes the organization of a system.
- A set of equations that defines the operation of a system.

Digital Systems

A digital system may be divided into interconnected subsystems that together accomplish the overall system function. Each subsystem may in turn be divided into functional devices or circuits that, when interconnected, accomplish the subsystem function. A **hierarchical model** of a digital system is presented in Figure 1.1.

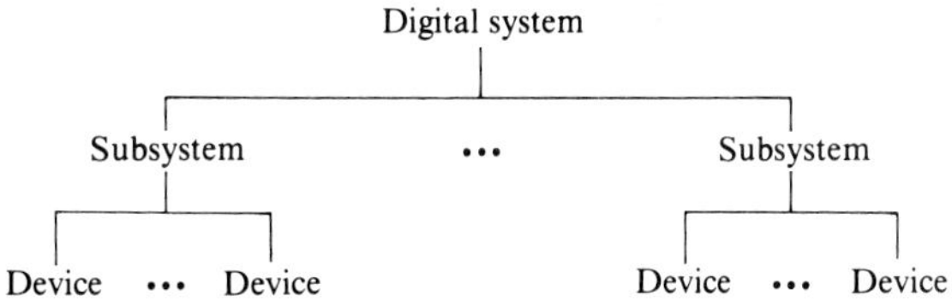

FIGURE 1.1
A hierarchical model of a digital system

Each functional device or circuit can be realized using basic logic elements, such as gates and memory elements, that are formed from resistors, diodes, transistors, and capacitors. At the most elemental level, a digital circuit may be interpreted as an electrical network of diodes, transistors, resistors, capacitors, inductors, and energy sources. The voltages and currents in the network are called **signals**. Thus, a *digital system* may be interpreted as a collection of interconnected digital circuits that process various signals.

An example of a digital system is the simple weather-monitoring system illustrated in Figure 1.2. The system has three subsystems: temperature, wind direction/speed, and barometric pressure. Each subsystem may be further subdivided into functional devices. For example, the temperature subsystem may include a thermocouple, an analog-to-digital converter *(ADC)*, and a digital readout (which displays temperature in either degrees centigrade or degrees Fahrenheit).

FIGURE 1.2
A simple weather-monitoring system

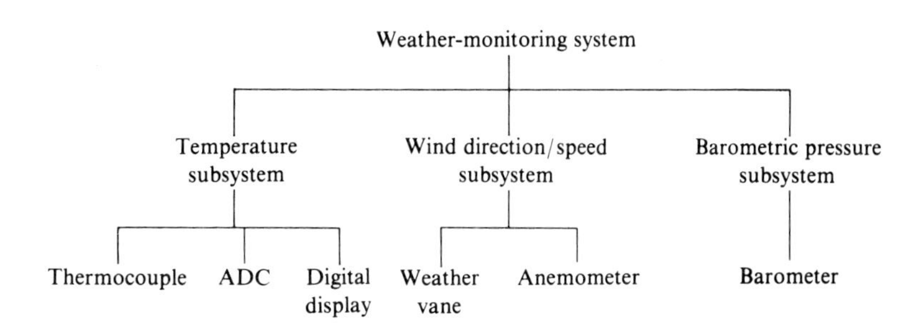

A thermocouple is a transducer (converter) whose output voltage is proportional to the ambient temperature. The voltage signal of the thermocouple is an analog of temperature and a continuous function of time. At discrete time increments, the analog signal can be quantized and converted to digital form using an ADC. Since digital signals are usually represented in some form of binary code and humans are most familiar with the decimal number system, the temperature in binary is usually converted to BCD form for display. BCD numbers are also used in most calculators and in some computers.

Systems vary in type, size, and complexity. For example, contrast the simple digital weather-monitoring system, illustrated in Figure 1.2, with a complex worldwide weather-predicting system. As another example, a simple accounting system may consist of a paycheck, a checkbook, and a set of rules for keeping track of the bank balance. In contrast, a computerized accounting system usually consists of a collection of related files in a data base and a number of programs for maintaining those files according to the rules of accounting.

1.2 NUMBER SYSTEMS

Numbers provide a means to quantitatively describe systems and their operation. In order to be useful, a number system must be easy to use and have characteristics that are compatible to the particular system.

For example, the decimal number system is most familiar to humans and provides an easy-to-use system for counting and computing. On the other hand, the binary number system is ideally suited for digital systems. Both the decimal and binary number systems use **positional notation**, in which each position is assigned a weighted value.

Since the beginning of recorded time, a variety of number systems have been developed to provide humans with a means to count and compute. The early Egyptian hieroglyphic system, dating from about 3000 B.C., was a nonpositional decimal system that used different symbols to represent powers of 10. Each symbol represented a particular numeric value (1, 10, 100, . . .), irrespective of the symbol's position in a composite number. The early Babylonian cuneiform system, dating from about 1800 B.C., was a sexagesimal (base-60) positional number system

The Roman numeral system is a nonpositional number system that uses strings of symbols (I = 1, V = 5, X = 10, L = 50, C = 100, M = 1000) to represent integers. For example, the number 1987 is represented in Roman numerals as MCMLXXXVII. The **decimal number system**, also referred to as the Hindu–Arabic number system, is a positional notation system that facilitates counting and computing. Starting with the number 0, it is easy to count 0, 1, 2, . . . , 9, 10, 11, 12, . . . , 19, 20, 21, . . . , 99, 100, 101, . . . , 999, 1000, 1001, ... Once the addition and multiplication tables are learned, computations can be performed easily using the decimal number system.

If powers of a number r are used as weights in a positional number system, then the number r is called the **radix** (or base) of that number system. The binary, octal, decimal, and hexadecimal number systems are positional-notation number systems with base (radix) 2, 8, 10, and 16, respectively. An integer N can be represented in a base-r positional number system by an expression of the form

$$N = a_n(r^n) + a_{n-1}(r^{n-1}) + \cdots + a_1(r^1) + a_0(r^0)$$

where $r^0 = 1$, and each coefficient a_k is an integer in the set $\{0, r-1\}$.

The numbers above the line are exponents (powers of the base r). The numbers below the line are subscripts (index numbers) that indicate the zeroth, first, second, and so forth, coefficients of N.

A fraction F can be represented in a base-r positional number system by an expression of the form

$$F = a_{-1}(r^{-1}) + a_{-2}(r^{-2}) + \cdots$$

Historical Note. In *A History of Mathematics*, Boyer notes "A study of several hundred tribes among the American Indians showed that almost one third used a decimal base and about another third had adopted a quinary or a quinary-decimal system; fewer than a third had a binary scheme, and those using a ternary system constituted less than 1 percent of the group. The vigesimal system, with twenty as a base, occurred in about 10 percent of the tribes."*

"Mohammed ibn-Musa al-Khowarizmi (about A.D. 850), an Arab mathematician and astronomer, left such a complete account of the Hindu numerals that he is probably responsible for the widespread, but false, impression that our system of numeration is Arabic in origin. The new notation came to be known as that of al-Khowarizmi, or more carelessly, algorismi. Ultimately, the scheme of numeration making use of the Hindu numerals came to be simply algorism, or *algorithm*, a word that now means, more generally, any peculiar rule of procedure or operation, such as the Euclidean method for finding the greatest common divisor."

"Through his arithmetic," Boyer points out, "al-Khowarizmi's name has become a common English word; through the title of his most important book, *Al-jabr wa'l mugabalah*, he has supplied us with an even more popular household term. From this title has come the word *algebra*, for it is from this book that Europe later learned the branch of mathematics bearing this name."

1.3 THE DECIMAL (BASE-10) NUMBER SYSTEM

Since you are familiar with the decimal number system, we will use this system to explain the concept of positional notation and its applications. The use of positional notation in any number system facilitates* arithmetic operations *in that system and the conversion of numbers from one system to another.

An integer can be represented in the decimal number system by juxtaposing appropriate decimal digits 0, 1, 2, . . . , 9 in a symbol string, such as 1989. The decimal number 1989 represents the sum of the following:

$$1000 + 900 + 80 + 9$$

---

*Carl Boyer, *A History of Mathematics* (New York: Wiley, 1968), p. 3.

This number also can be written as the sum of the following:

$$1(1000) + 9(100) + 8(10) + 9(1)$$

The decimal number system is a positional notation system because each digit in the number 1989 has a weight of 1, 10, 100, and 1000 that depends on its position (units, tens, hundreds, thousands) in the composite number. Each term of the sum is a product formed by multiplying a power of 10 by a decimal digit (coefficient) that indicates how many units, tens, hundreds, and thousands make up the overall number. Therefore, the number 1989 is also obtained by adding the products of each coefficient and its corresponding positional weight. That is,

$$1989 = 1(10^3) + 9(10^2) + 8(10^1) + 9(10^0)$$

where 10^k = the kth power of 10, and $10^0 = 1$ by definition.

Base-10 Representation of Nonnegative Integers

Any nonnegative integer can be represented in positional notation in the decimal number system. For example, the decimal number 367 = 300 + 60 + 7 is represented simply by juxtaposing the coefficients 3, 6, and 7 to form the decimal number 367. In general, a nonnegative integer

$$A = a_2(10^2) + a_1(10^1) + a_0(10^0)$$

can be expressed in decimal notation by simply juxtaposing the coefficients a_2, a_1, and a_0 and writing $a_2a_1a_0$. Similarly, a nonnegative integer

$$B = b_1(10^1) + b_0(10^0)$$

can be written simply as b_1b_0.

Note that the numbers A and B used in the foregoing examples were chosen for illustration purposes and just happen to be a 3-digit decimal number A and a 2-digit decimal number B.

Base-10 Representation of Fractions

Fractional numbers can also be represented easily in the decimal number system. The fraction ⅛ can be represented as 0.125, obtained by dividing 1 by 8. This decimal fraction can be represented in full positional notation as

$$0.125 = 1(10^{-1}) + 2(10^{-2}) + 5(10^{-3})$$

Any number in the decimal number system can, therefore, be represented in positional notation by combining the integer and fractional parts of the number. For example,

$$367\ 1/8 = 367.125$$

Base-10 Addition and Subtraction of Nonnegative Integers

The nonnegative integers A and B, used in our earlier example, can be added by writing

$$A + B = [a_2(10^2) + a_1(10^1) + a_0(10^0)] + [b_1(10^1) + b_0(10^0)]$$

extracting the common factors,

$$A + B = a_2(10^2) + (a_1 + b_1)(10^1) + (a_0 + b_0)(10^0)$$

and, finally, adding the coefficients of like powers of 10. For example, let $A = 367$ and $B = 31$. The sum is computed as follows:

$$\begin{aligned} 367 + 31 &= 3(10^2) + 6(10^1) + 7(10^0) + 3(10^1) + 1(10^0) \\ &= 3(10^2) + (6 + 3)(10^1) + (7 + 1)(10^0) \\ &= 3(10^2) + 9(10^1) + 8(10^0) \\ &= 300 + 90 + 8 = 398 \end{aligned}$$

Note that, if the sum $a_i + b_i$ of any power is 10 or greater, the excess is carried over to the next higher power.

The arithmetic operation of subtracting nonnegative integers is also easy to perform in the decimal number system. For example,

$$\begin{aligned} A - B &= [a_2(10^2) + a_1(10^1) + a_0(10^0)] - [b_1(10^1) + b_0(10^0)] \\ &= a_2(10^2) + (a_1 - b_1)(10^1) + (a_0 - b_0)(10^0) \end{aligned}$$

Each b coefficient is subtracted from the corresponding a coefficient. If any coefficient $b_i > a_i$, a borrow from the next higher power of A is required; we compute $a_i + 10 - b_i$ and reduce a_{i+1} by 1 to reflect the borrow. For example, the subtraction $832 - 651 = 181$ is performed as follows:

$$\begin{aligned} 832 - 651 &= [8(10^2) + 3(10^1) + 2(10^0)] - [6(10^2) + 5(10^1) + 1(10^0)] \\ &= (8 - 6)(10^2) + (3 - 5)(10^1) + (2 - 1)(10^0) \end{aligned}$$

Since the coefficient $(3 - 5)$ is negative, 1 is borrowed from the coefficient of the next higher power and 10 is added to $(3 - 5)$ to get

$$\begin{aligned} 832 - 651 &= (2 - 1)(10^2) + [10 + (3 - 5)](10^1) + 1(10^0) \\ &= 1(10^2) + 8(10^1) + 1(10^0) \\ &= 100 + 80 + 1 = 181 \end{aligned}$$

Multiplication and division are more difficult operations to perform in any number system, including the decimal number system. Such complications have motivated the development of computing devices ranging from the ancient abacus to the modern computer. Early computing devices, such as Pascal's calculator (1642), used decimal numbers represented by a succession of 10-position wheels or rotors.

Base-10 Representation of Signed Integers

Let A and B be two nonnegative integers. The difference D obtained by subtracting B from A can be expressed as

$$D = A - B$$

This difference can also be written in the form

$$D = A + (-B)$$

by changing the sign of B and performing addition instead of subtraction. For example, if $A = 7$ and $B = 4$, then

$$\begin{aligned} D &= A - B = 7 - 4 \\ &= A + (-B) = 7 + (-4) = 3 \end{aligned}$$

9s Complement Method of Subtraction

In the decimal number system, the **9s complement** of a decimal digit d is defined as a number d_c for which $d + d_c = 9$, as seen in Table 1.1.

TABLE 1.1
Table of 9s complements

d	d_c
0	9
1	8
2	7
3	6
4	5
5	4
6	3
7	2
8	1
9	0

Subtraction can be performed by first adding the complement of the subtrahend to the minuend and then adding the **end-around carry** (denoted EAC) to the result. The procedure is illustrated in the following examples:

Subtract	*Add 9s Complement*
$\begin{array}{r} 7 \\ -\,4 \\ \hline 3 \end{array}$	$7 + 4_c = \begin{array}{r} 7 \\ +\,5 \\ \hline 12 \end{array}$
	EAC → $\begin{array}{r} +\,1 \\ \hline 3 \end{array}$

$$\begin{array}{r} 8 \\ -\ 6 \\ \hline 2 \end{array} \qquad 8 + 6_c = \begin{array}{r} 8 \\ +\ 3 \\ \hline 11 \\ \text{EAC} \rightarrow +\ 1 \\ \hline 2 \end{array}$$

In a multidigit integer B, B_c is formed by replacing each individual digit of B by the complement of that single digit. This is illustrated by the following examples:

Subtract *Add 9s Complement*

$$\begin{array}{r} 48 \\ -\ 23 \\ \hline 25 \end{array} \qquad 48 + 23_c = \begin{array}{r} 48 \\ +\ 76 \\ \hline 124 \\ \text{EAC} \rightarrow +\ 1 \\ \hline 25 \end{array}$$

$$\begin{array}{r} 136 \\ -\ 078 \\ \hline 58 \end{array} \qquad 136 + 078_c = \begin{array}{r} 136 \\ +\ 921 \\ \hline 1057 \\ \text{EAC} \rightarrow +\ 1 \\ \hline 58 \end{array}$$

Thus, the subtraction of a decimal integer B from a decimal integer A can be accomplished by adding the 9s complement of B to A and then adding the EAC to the result.

10s Complement Method of Subtraction

The principal disadvantage of performing decimal subtraction using the 9s complement method is that the EAC must be added to the addition result. This situation can be remedied by using a **10s complement** B', defined by the relation

$$B' = B_c + 1$$

where B_c is the 9s complement of B. That is, the 10s complement of a number is obtained by adding 1 to the 9s complement of the number, as shown in Table 1.2.

TABLE 1. 2
Table of 9s and 10s complements

Integer B		9s Complement B_c		10s Complement B'
4		5		6
3		6		7
23	→	76	→	77
78		21		22

The following examples illustrate the subtraction of *B* from *A* by adding the 10s complement of *B* to integer *A*:

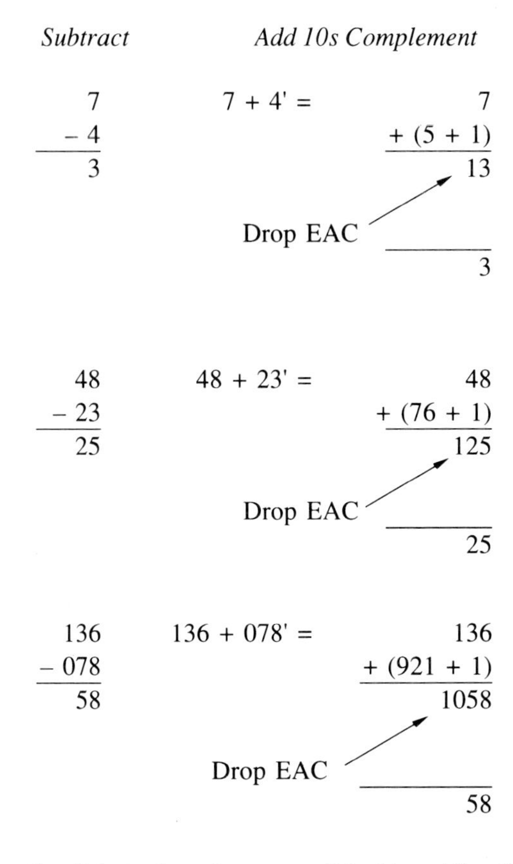

Consequently, subtracting *B* from *A* can be accomplished by adding the 10s complement of *B* to *A* and dropping the EAC from the resulting sum.

Base-10 Representation of Floating-Point Numbers

Extremely large or small numbers can be represented in a compact form called **floating-point notation** (historically referred to as **scientific notation**). For example, the decimal numbers 179,000,000 and 0.0000762 can be written as 179 E+6 and 0.762 E–4, respectively, where E+6 and E–4 denote the base-10 exponents, representing 10^6 and 10^{-4}.

Floating-point numbers are used in computers to extend the range of numbers that can be stored in a word that contains a fixed number of bits. For example, if a signed number is stored in a 16-bit word in integer form (using 1 bit for the sign and 15 bits for the magnitude), the number range is from $-(2^{15} - 1)$ to $+(2^{15} - 1)$.

A floating-point number *N* can be represented in the form

$$a(b^e)$$

where the coefficient a is any number in base b, b is the number system base, and e is an exponent. For example, the decimal number 245.3 can be expressed in floating-point form as $245.3(10^0)$, $24.53(10^1)$, $2.453(10^2)$, or $0.2453(10^3)$.

If the coefficient a is a proper fraction F in the base system such that $1/b \leq |F| < 1$, then the number $F(b^e)$ is called a normalized floating-point number. For example, in the decimal number system (base 10), the number $0.2453(10^3)$ is a normalized floating-point number, while other floating-point representations of 245.3 are not normalized. The fraction F is also referred to as the **mantissa**.

The rules for adding, subtracting, multiplying, and dividing two floating-point numbers $N_1 = a_1 \cdot b^{e_1}$ and $N_2 = a_2 \cdot b^{e_2}$, where $e_1 \geq e_2$, are

$$N_1 + N_2 = (a_1 \cdot b^{e_1}) + (a_2 \cdot b^{e_2}) = (a_1 + a_2 \cdot b^{e_2-e_1}) \cdot b^{e_1}$$
$$N_1 - N_2 = (a_1 \cdot b^{e_1}) - (a_2 \cdot b^{e_2}) = (a_1 - a_2 \cdot b^{e_2-e_1}) \cdot b^{e_1}$$
$$N_1 \cdot N_2 = (a_1 \cdot b^{e_1}) \cdot (a_2 \cdot b^{e_2}) = (a_1 \cdot a_2) \cdot b^{e_1+e_2}$$
$$\frac{N_1}{N_2} = \frac{a_1 \cdot b^{e_1}}{a_2 \cdot b^{e_2}} = \frac{a_1}{a_2} \cdot b^{e_1-e_2}$$

Let the numbers X and Y have the normalized binary floating-point representations $X = F_x \cdot 10^{e_x}$ and $Y = F_y \cdot 10^{e_y}$, so that $0.1 \leq |F_x| < 1$ and $0.1 \leq |F_y| < 1$. Then the numbers X and Y can be combined by the arithmetic operations $+$, $-$, $\times$, and $/$, according to the foregoing rules for combining numbers of the form $N = a \times b^e$.

If $Z = X \bigcirc Y$, where $\bigcirc$ is any one of the operations $+, -, \times, /$, then the normalized binary floating-point representation $Z = F_z \cdot 10^{e_z}$, with $0.1 \leq |F_z| < 1$, can be summarized as follows:

1. Multiplication:

$$Z = XY = (F_x \cdot 10^{e_x})(F_y \cdot 10^{e_y}) = F_xF_y \cdot 10^{e_x+e_y}$$

a. If $|F_xF_y| < 0.1$, $F_z = 10F_xF_y$, $e_z = e_x + e_y - 1$.
b. If $0.1 \leq |F_xF_y| < 1$, $F_z = F_xF_y$, $e_z = e_x + e_y$.

2. Division:

$$Z = \frac{X}{Y} = \frac{F_x \cdot 10^{e_x}}{F_y \cdot 10^{e_y}} = \frac{F_x}{F_y} \cdot 10^{e_x-e_y}$$

a. If $0.1 \leq \left|\frac{F_x}{F_y}\right| < 1$, $F_z = \frac{F_x}{F_y}$, $e_z = e_x - e_y$.

b. If $\left|\frac{F_x}{F_y}\right| > 1$, $F_z = \frac{F_x/F_y}{10}$, $e_z = e_x - e_y + 1$.

3. Addition (assume that $e_x \geq e_y$):

$$Z = X + Y = (F_x \cdot 10^{e_x}) + (F_y \cdot 10^{e_y}) = (F_x + F_y \cdot 10^{e_y-e_x}) \cdot 10^{e_x}$$

Let $f_y = F_y \cdot 10^{e_y-e_x}$

a. If $|F_x + f_y| < 0.1$, $F_z = 10^m (F_x+f_y)$, $e_z = e_x - m$, where m is such that $0.1 \leq |F_z| < 1$.

b. If $0.1 < |F_x + f_y| < 1$, $F_z = F_x + f_y$, $e_z = e_x$.

c. If $|F_x + f_y| > 1$, $F_z = \dfrac{F_x + f_y}{10}$, $e_z = e_x + 1$.

4. Subtraction: Replace + with – in the addition mode.

1.4 THE BINARY (BASE-2) NUMBER SYSTEM

Because it is cost effective, the binary number system is used in most computers. Binary numbers are used to represent logic values and signal voltage levels, memory addresses, and program instructions and data. Most computer memories are composed of 2-state storage elements, each of which can store either 0 or 1. Therefore, the **binary number system** ***and binary-coding schemes are used to encode numeric and alphanumeric data (letters, numbers, and symbols) and to represent program instructions.***

Base-2 Representation of Nonnegative Integers

Any nonnegative integer can be represented in binary (base-2) positional notation. The notation is analogous to that used in the decimal number system except that the base-2 system uses only 0 and 1 as coefficients.

Recall from Section 1.2 that an integer N in a base r positional number system can be represented as

$$N = a_n(r^n) + \cdots + a_3(r^3) + a_2(r^2) + a_1(r^1) + a_0(r^0)$$

For example, the decimal integer $31 = 3(10^1) + 1(10^0)$. The corresponding binary integer $11111 = 1(2^4) + 1(2^3) + 1(2^2) + 1(2^1) + 1(2^0)$. The powers of 2 ($2^0, 2^1, 2^2, 2^3, 2^4, \ldots$) represent the weights 1, 2, 4, 8, 16, and so on. Table 1.3 illustrates the relation between corresponding integers in the decimal and binary systems.

Note that decimal 15 equals binary $1111 = 1(2^3) + 1(2^2) + 1(2^1) + 1(2^0) = 8 + 4 + 2 + 1$, which is the largest unsigned number in 4 bits, and that decimal 31 equals binary $11111 = 1(2^4) + 1(2^3) + 1(2^2) + 1(2^1) + 1(2^0) = 16 + 8 + 4 + 2 + 1$, which is the largest unsigned number in 5 bits.

Base-2 Representation of Fractions

Fractional numbers can also be represented easily in the binary number system. The fraction ⅛ can be represented in the binary number system as .001. This binary fraction can be represented in full positional notation as

$$.001 = 0(2^{-1}) + 0(2^{-2}) + 1(2^{-3})$$

Any number composed of a binary integer part and a binary fractional part can be represented in positional notation by combining the integer and fractional parts of the number. For example,

$$5\ 1/8 = 101.001$$

TABLE 1.3
Table of corresponding decimal and binary integers

Decimal 10 1	Binary 16 8 4 2 1	Binary Positional Notation
0	0 0 0 0	$= 0 \times 2^3 + 0 \times 2^2 + 0 \times 2^1 + 0 \times 2^0$
1	0 0 0 1	$= 0 \times 2^3 + 0 \times 2^2 + 0 \times 2^1 + 1 \times 2^0$
2	0 0 1 0	$= 0 \times 2^3 + 0 \times 2^2 + 1 \times 2^1 + 0 \times 2^0$
3	0 0 1 1	$= 0 \times 2^3 + 0 \times 2^2 + 1 \times 2^1 + 1 \times 2^0$
4	0 1 0 0	$= 0 \times 2^3 + 1 \times 2^2 + 0 \times 2^1 + 0 \times 2^0$
5	0 1 0 1	$= 0 \times 2^3 + 1 \times 2^2 + 0 \times 2^1 + 1 \times 2^0$
6	0 1 1 0	$= 0 \times 2^3 + 1 \times 2^2 + 1 \times 2^1 + 0 \times 2^0$
7	0 1 1 1	$= 0 \times 2^3 + 1 \times 2^2 + 1 \times 2^1 + 1 \times 2^0$
8	1 0 0 0	$= 1 \times 2^3 + 0 \times 2^2 + 0 \times 2^1 + 0 \times 2^0$
9	1 0 0 1	$= 1 \times 2^3 + 0 \times 2^2 + 0 \times 2^1 + 1 \times 2^0$
1 0	1 0 1 0	$= 1 \times 2^3 + 0 \times 2^2 + 1 \times 2^1 + 0 \times 2^0$
1 1	1 0 1 1	$= 1 \times 2^3 + 0 \times 2^2 + 1 \times 2^1 + 1 \times 2^0$
1 2	1 1 0 0	$= 1 \times 2^3 + 1 \times 2^2 + 0 \times 2^1 + 0 \times 2^0$
1 3	1 1 0 1	$= 1 \times 2^3 + 1 \times 2^2 + 0 \times 2^1 + 1 \times 2^0$
1 4	1 1 1 0	$= 1 \times 2^3 + 1 \times 2^2 + 1 \times 2^1 + 0 \times 2^0$
1 5	1 1 1 1	$= 1 \times 2^3 + 1 \times 2^2 + 1 \times 2^1 + 1 \times 2^0$
1 6	1 0 0 0 0	
⋮	⋮	⋮
2 0	1 0 1 0 0	
⋮	⋮	
3 1	1 1 1 1 1	
3 2	1 0 0 0 0 0	
⋮	⋮	

Base-2 Addition of Nonnegative Integers

The addition operation for 1-bit binary numbers is defined by the binary addition table (Table 1.4).

TABLE 1.4
Binary addition table

+	0	1
0	0	1
1	1	10

or

$$\begin{array}{r}0\\+0\\\hline 0\end{array}\qquad\begin{array}{r}0\\+1\\\hline 1\end{array}\qquad\begin{array}{r}1\\+0\\\hline 1\end{array}\qquad\begin{array}{r}1\\+1\\\hline 10\end{array}$$

The addition of multibit binary numbers is accomplished by a method analogous to the addition of multidigit decimal numbers, as illustrated in the following examples:

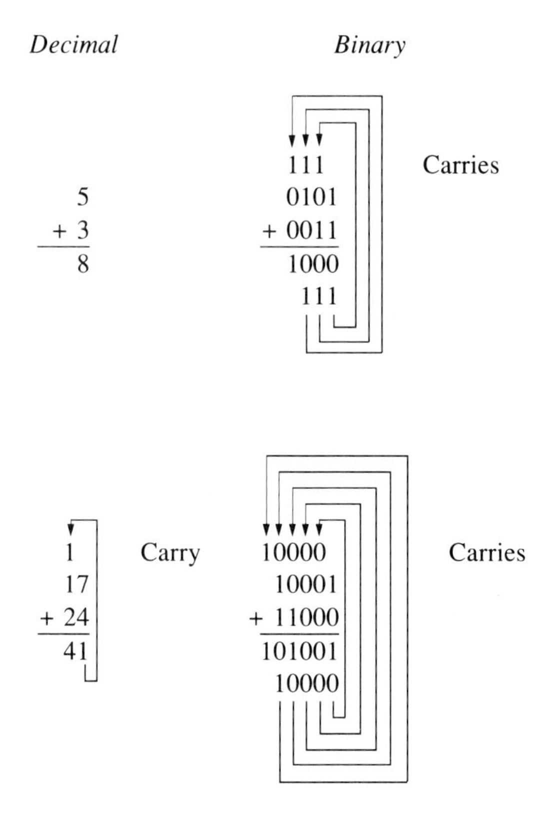

Base-2 Representation of Signed Integers

Up to this point, we have considered only arithmetic operations with nonnegative integers represented in base 2. Operations with negative integers, as well as nonnegative integers, must now be considered. However, it is first necessary to discuss alternative methods of representing any integer (whether negative, zero, or positive) in some binary form.

For any integer, both the sign and magnitude of the number must be represented. For the present, consider only integers whose magnitude can be represented in 3 bits. A fourth bit will be used for the sign of the number: 0 = positive, 1 = negative. This **sign-magnitude form** of representing signed integers using 4-bit words is illustrated in Table 1.5. An obvious disadvantage of the sign-magnitude method is that there are two representations for zero.

The subtraction of a binary integer B from a binary integer A can be accomplished in a manner analogous to the process using decimal integers and their complements.

1s Complement Method of Subtraction

Suppose that positive numbers are represented in sign-magnitude form and negative numbers are represented in **1s complement form**. The 1s complement is obtained by

TABLE 1.5
Sign-magnitude representation

Positive Integers		Negative Integers	
Decimal	Binary	Decimal	Binary
	Sign bit ↓		Sign bit ↓
+7	0111	–7	1111
+6	0110	–6	1110
+5	0101	–5	1101
+4	0100	–4	1100
+3	0011	–3	1011
+2	0010	–2	1010
+1	0001	–1	1001
+0	0000	–0	1000

TABLE 1.6
The 1s complement representation

Positive Integers		Negative Integers	
Decimal	Binary	Decimal	Binary
	Sign bit ↓		Sign bit ↓
+7	0111	–7	1000
+6	0110	–6	1001
+5	0101	–5	1010
+4	0100	–4	1011
+3	0011	–3	1100
+2	0010	–2	1101
+1	0001	–1	1110
+0	0000	–0	1111

inverting (complementing) each bit of the number, including the sign bit, as shown in Table 1.6.

The 1s complement method has the same disadvantage as the sign-magnitude method: There are two distinct representations for zero.

The use of the 1s complement can be illustrated by considering an n-bit machine that uses words containing n bits. For example, in a 4-bit machine, decimal 3 = 0011, and the 1s complement of 3 is 1100. In an 8-bit machine, decimal 3 = 00000011, and the 1s complement of 3 is 11111100.

Let B_c denote the 1s complement of an integer B in the binary number system. The process of performing the subtraction $A - B$ by adding B_c to A can be illustrated using a 4-bit word size as follows:

Subtract			*Add 1s Complement*	
Decimal	*Binary*			
			1110	Carries
7	0111	$7 + 5_c$	= 0111	
− 5	− 0101		+ 1010	(Since 5 = 0101)
2	0010		10001	
		EAC	+ 1	
			0010	= 2

Note that the 1s complement method for subtraction suffers from the same disadvantage as the 9s complement method for decimal numbers. That is, the carry-out of the **most significant bit** (MSB), the leftmost bit, must be end-around carried and added to produce the correct result.

2s Complement Method of Subtraction

If negative integers are represented in **2s complement** form (obtained by adding 1 to the 1s complement of the number), Table 1.7 results.

TABLE 1.7
The 2s complement representation

Positive Integers		Negative Integers	
Decimal	**Binary**	**Decimal**	**Binary**
			Sign bit ↓
	Sign bit ↓	−8	1000
+7	0111	−7	1001
+6	0110	−6	1010
+5	0101	−5	1011
+4	0100	−4	1100
+3	0011	−3	1101
+2	0010	−2	1110
+1	0001	−1	1111
+0	0000	−0	Not represented

An advantage of the 2s complement method is that there is a unique representation for zero. However, the range of positive integers differs from the range for negative integers, that is, +0 to +7 and −1 to −8.

The 2s complement B' of a multibit binary integer B is obtained by adding 1 to the 1s complement B_c, where B_c is obtained by complementing each bit of B. The process of accomplishing binary subtraction using the 2s complement method is illustrated by the following 4-bit example.

Subtract	
Decimal	*Binary*
7	0111
− 5	− 0101
2	0010

Add 2s Complement

7 + (5)' = 0111 + (1010 + 1)

```
  1111     Carries
  0111
+ 1011
 10010     = 2
Drop EAC
  1111
```

Addition and Subtraction Using 2s Complement

Addition and subtraction with 2s complement is best illustrated in the following examples. Assume that signed integers A and B are represented in 2s complement form; note that, for addition, both A and B are used without conversion and that, for subtraction, B is converted to 2s complement form before adding B' to A to accomplish the subtraction $A - B$. In the following examples, a and b represent the magnitudes of A and B, respectively.

Examples	*Addition*		*Subtraction*		
$A \geq 0, B \geq 0, a \geq b$	5	0101	5	0101	
	+ 2	+ 0010	− 2	+ 1110	= 2'
	7	0111	3	10011	
			Drop EAC		
				0011	
$A \geq 0, B \geq 0, a < b$	2	0010	2	0010	
	+ 3	+ 0011	− 3	+ 1101	= 3'
	5	0101	−1	1111	
$A \geq 0, B < 0, a \geq b$	3	0011	3	0011	
	+ (−2)	+ 1110	− (−2)	+ 0010	= (−2)'
	1	10001	5	0101	
	Drop EAC				
		0001			
$A \geq 0, B < 0, a < b$	2	0010	2	0010	
	+ (−4)	+ 1100	− (−4)	+ 0100	= (−4)'
	−2	1110	6	0110	

```
A < 0, B ≥ 0, a ≥ b        –5      1011              –5       1011
                          + 2    + 0010             – 2     + 1110    = 2'
                          ---    ------             ---     ------
                          –3      1101              –7      11001
                                              Drop EAC ↗
                                                            ------
                                                             1001

A < 0, B ≥ 0, a < b        –1      1111              –1       1111
                          + 6    + 0110             – 6     + 1010    = 6'
                          ---    ------             ---     ------
                            5     10101             –7      11001
                  Drop EAC ↗                  Drop EAC ↗
                                 ------                     ------
                                  0101                       1001

A < 0, B < 0, a ≥ b        –4      1100              –4       1100
                       + (–2)    + 1110          – (–2)     + 0010    = (–2)'
                       ------    ------          ------     ------
                           –6     11010              –2       1110
                  Drop EAC ↗
                                 ------
                                  1010

A < 0, B < 0, a < b        –1      1111              –1       1111
                       + (–5)    + 1011          – (–5)     + 0101    = (–5)'
                       ------    ------          ------     ------
                           –6  =  11010              +4      10100
                  Drop EAC ↗                  Drop EAC ↗
                                 ------                     ------
                                  1010                       0100
```

Base-2 Representation of Floating-Point Numbers

A number N represented in the form $a(b^e)$ is referred to as a floating-point number. For example, the binary number 101.11 can be expressed in floating-point form as $101.11(2^0)$, $10.111(2^1)$, $1.0111(2^2)$, or $0.10111(2^3)$.

If the coefficient a is a proper fraction F in the base system such that $1/b \leq |F| < 1$, then the number $F(b^e)$ is called a normalized floating-point number.

Let the numbers X and Y have the normalized binary floating-point representations $X = F_x(2^{e_x})$ and $Y = F_y(2^{e_y})$, so that $0.1 \leq |F_x| < 1$ and $0.1 \leq |F_y| < 1$. Then the numbers X and Y can be combined by the arithmetic operations +, –, ×, and /, according to the foregoing rules for combining numbers of the form $N = a(b^e)$.

If $Z = X \bigcirc Y$, where $\bigcirc$ is any one of the operations +, –, ×, /, then the normalized binary floating-point representation $Z = F_z(2^{e_z})$, with $0.1 \leq |F_z| < 1$, can be summarized as follows:

1. Multiplication:

$$Z = XY = (F_x \cdot 2^{e_x})(F_y \cdot 2^{e_y}) = F_xF_y \cdot 2^{e_x+e_y}$$

   a. If $|F_xF_y| < 0.1$, $F_z = 2F_xF_y$, $e_z = e_x + e_y - 1$.
   b. If $0.1 \leq |F_xF_y| < 1$, $F_z = F_xF_y$, $e_z = e_x + e_y$.

2. Division:

$$Z = \frac{X}{Y} = \frac{F_x \cdot 2^{e_x}}{F_y \cdot 2^{e_y}} = \frac{F_x}{F_y} \cdot 2^{e_x-e_y}$$

a. If $0.1 \le \left| \dfrac{F_x}{F_y} \right| < 1$, $F_z = \dfrac{F_x}{F_y}$, $e_z = e_x - e_y$.

b. If $\left| \dfrac{F_x}{F_y} \right| > 1$, $F_z = \dfrac{F_x/F_y}{2}$, $e_z = e_x - e_y + 1$.

3. Addition (assume that $e_x \ge e_y$):

$$Z = X + Y = (F_x \cdot 2^{e_x}) + (F_y \cdot 2^{e_y}) = (F_x + F_y \cdot 2^{e_y - e_x}) \cdot 2^{e_x}$$

Let $f_y = F_y \cdot 2^{e_y - e_x}$

a. If $|F_x + f_y| < 0.1$, $F_z = 2^m(F_x + f_y)$, $e_z = e_x - m$, where m is such that $0.1 \le |F_z| < 1$

b. If $0.1 < |F_x + f_y| < 1$, $F_z = F_x + f_y$, $e_z = e_x$.

c. If $|F_x + f_y| > 1$, $F_z = (F_x + f_y)/2$, $e_z = e_x + 1$.

4. Subtraction: Replace + with – in the addition mode.

1.5 CONVERTING NUMBERS BETWEEN BASE 10 AND BASE 2

Up to this point we have reviewed base-10 operations and developed our skills in base-2 operations. As stated earlier, humans usually count and compute with decimal numbers, and most computers use some binary number form. For this reason, it is important to understand how decimal numbers are converted to binary form and vice versa.

Converting Integers from Decimal to Binary

Any integer in the decimal number system can be converted to an equivalent representation in the binary number system. For example, the decimal integer 87 can be converted to an equivalent binary integer by subtracting powers of 2 until nothing remains. Note that the kth power of 2 can be written as 2^k. In each step, subtract 2^k, where 2^k does not exceed the number remaining at that time.

$$\begin{aligned}
87 - 2^6 &= 87 - 64 = 23 \\
23 - 2^4 &= 23 - 16 = 7 \\
7 - 2^2 &= 7 - 4 = 3 \\
3 - 2^1 &= 3 - 2 = 1 \\
1 - 2^0 &= 1 - 1 = 0
\end{aligned}$$

Therefore

$$\begin{aligned}
87 &= 1(2^6) + 0(2^5) + 1(2^4) + 0(2^3) + 1(2^2) + 1(2^1) + 1(2^0) \\
&= 1010111 \text{ in binary positional notation}
\end{aligned}$$

```
| | |||
6543210
```

where 6, 5, 4, . . . , 0 = the corresponding power of 2 for the position.

An alternative, and perhaps more popular, method of converting from decimal to binary is accomplished by dividing the decimal integer successively by 2 and noting the remainder at each stage:

87/2 = 43	Remainder 1	→ 1010111
43/2 = 21	1	
21/2 = 10	1	
10/2 = 5	0	
5/2 = 2	1	
2/2 = 1	0	
1/2 = 0	1	

The binary integer is formed by reading up the remainder column to obtain 1010111 (base-2) = 87 (base-10).

Therefore, any decimal integer can be converted to an equivalent binary integer. In general, the binary integer can be represented in binary positional notation in the form

$$a_n(2^n) + \cdots + a_i(2^i) + \cdots + a_2(2^2) + a_1(2^1) + a_0(2^0)$$

where the binary coefficients $a_i = 0$ or 1, $(i = 0,1, \ldots, n)$.

Converting Integers from Binary to Decimal

The conversion of a binary integer to its decimal equivalent can be accomplished by adding the products obtained by multiplying each coefficient by the corresponding power of 2. For instance, to convert the binary integer $A = 1010111$ to an equivalent decimal integer, start on the right with the smallest power of A:

$$\begin{aligned}
1 \times 2^0 &= 1 \times 1 = 1 \\
1 \times 2^1 &= 1 \times 2 = 2 \\
1 \times 2^2 &= 1 \times 4 = 4 \\
0 \times 2^3 &= 0 \times 8 = 0 \\
1 \times 2^4 &= 1 \times 16 = 16 \\
0 \times 2^5 &= 0 \times 32 = 0 \\
1 \times 2^6 &= 1 \times 64 = \underline{64} \\
&= 87 \text{ decimal}
\end{aligned}$$

Converting Fractions from Decimal to Binary

Some decimal fractions can be converted to binary fractions that are exact equivalents, as indicated in Table 1.8.

TABLE 1.8
Table of corresponding decimal and binary fractions

Decimal Fraction	Binary Fraction
1/2 = 0.5	= .1 = 2^{-1}
1/4 = 0.25	= .01 = 2^{-2}
1/8 = 0.125	= .001 = 2^{-3}
1/16 = 0.0625	= .0001 = 2^{-4}
1/32 = 0.03125	= .00001 = 2^{-5}
1/64 = 0.015625	= .000001 = 2^{-6}
1/128 = 0.0078125	= .0000001 = 2^{-7}
⋮ ⋮	⋮ ⋮

A decimal fraction F that is not exactly equivalent to $1/(2^n)$ can be converted to an equivalent base-2 representation, provided that F is equal to some sum of binary fractions listed in Table 1.8. For example,

$$\begin{aligned} 0.75 &= 0.50 + 0.25 \\ &= 1/2 + 1/4 \\ &= 1(2^{-1}) + 1(2^{-2}) = .11 \end{aligned}$$

and

$$\begin{aligned} 0.375 &= 0.25 + 0.125 \\ &= 1/4 + 1/8 \\ &= 0(2^{-1}) + 1(2^{-2}) + 1(2^{-3}) = .011 \end{aligned}$$

However, not all decimal fractions have an exact representation in base 2. For example,

$$\begin{aligned} 0.66 &= 0.5 + 0.125 + 0.03125 + \cdots \\ &= 1/2 + 1/8 + 1/32 + \cdots \\ &= 1(2^{-1}) + 1(2^{-3}) + 1(2^{-5}) + \cdots \\ &= .10101 \cdots \end{aligned}$$

Consequently, the degree of accuracy of representing an arbitrary decimal fraction in base 2 depends on the number of bits used in the base-2 representation of that decimal fraction.

Since a decimal integer can be converted to an equivalent binary integer by a division process, it seems logical that a decimal fraction may be converted to an equivalent binary fraction by a multiplication process. In fact, the process of converting a decimal fraction to an equivalent representation in binary can be accomplished by repeatedly multiplying successive decimal fractions by 2. If the process results in a zero decimal fraction at some point, the original decimal fraction has an exact base-2 representation. We will illustrate with the following example, where we convert 0.375 to an equivalent binary fraction (I equals integer and F equals fraction).

$$
\begin{array}{rrrr}
 & 0.375 & & F \\
 & \times 2 & & \times 2 \\
= & 0.750 & = & I_1 + F_1 \\
 & \times 2 & & \times 2 \\
= & 1.500 & = & I_2 + F_2 \\
 & \times 2 & & \times 2 \\
= & 1.000 & = & I_3 + F_3
\end{array}
$$

.0 1 1

The binary fraction is obtained by writing the integer that results at each stage; that is, 0.375 = .011 in binary. Note that only the decimal fraction is multiplied by 2 in each stage!

If the decimal-to-binary conversion does not result in a zero decimal fraction at some point, the original decimal fraction has no exact base-2 representation, as illustrated in the following example:

$$
\begin{array}{rrrr}
 & 0.66 & & I_1 + F_1 \\
 & \times 2 & & \times 2 \\
= & 1.32 & = & I_2 + F_2 \\
 & \times 2 & & \times 2 \\
= & 0.64 & = & I_3 + F_3 \\
 & \times 2 & & \times 2 \\
= & 1.28 & = & I_4 + F_4 \\
 & \times 2 & & \times 2 \\
= & 0.56 & = & I_5 + F_5 \\
 & \times 2 & & \times 2 \\
= & 1.12 & = & I_6 + F_6 \\
 & \vdots & & \vdots
\end{array}
$$

.1 0 1 0 1 . . .

$$0.66 = 0.5 + 0(0.25) + 0.125 + 0(0.0625) + 0.03125 + \cdots$$

The process of converting a decimal fraction to a base-2 representation can be shown in algebraic form as follows: Let F denote the decimal fraction and assume that F has an equivalent base-2 representation. Thus,

$$F = b_{-1}(2^{-1}) + b_{-2}(2^{-2}) + b_{-3}(2^{-3}) + b_{-4}(2^{-4}) + \cdots$$

If we multiply both sides of the equation by 2, then

$$2F = I_1 + F_1 = b_{-1} + [b_{-2}(2^{-1}) + b_{-3}(2^{-2}) + b_{-4}(2^{-3}) + \cdots]$$

where I_1 is an integer and F_1 is a decimal fraction.

Since $F < 1$, it follows that $2F = I_1 + F_1 < 2$, so that integer $I_1 = 0$ or 1 and fraction $F_1 < 1$.

Now, the integral part of the left side equals the integral part of the right side, and the fractional part on the left equals the fractional part on the right. Therefore,

$$I_1 = b_{-1}$$

and

$$F_1 = b_{-2}(2^{-1}) + b_{-3}(2^{-2}) + b_{-4}(2^{-3}) + \cdots$$

We repeat the process of multiplying the decimal fraction by 2:

$$2F_1 = I_2 + F_2 = b_{-2} + [b_{-3}(2^{-1}) + b_{-4}(2^{-2}) + \cdots]$$

Equating integral left part to integral right part and equating fractional left part to fractional right part, we have

$$I_2 = b_{-2}$$

and

$$F_2 = b_{-3}(2^{-1}) + b_{-4}(2^{-2}) + \cdots$$

The process is repeated until the decimal fraction F_i on the left either becomes zero or is sufficiently close to zero, depending on the required degree of accuracy of the base-2 representation of the decimal fraction.

Converting Fractions from Binary to Decimal

A binary fraction can be converted to an equivalent decimal fraction by forming the sum of the products obtained by multiplying each coefficient by the corresponding (negative) power of 2. To illustrate, the binary fraction

```
.011
 |||
 123
```

where 1, 2, 3 are the positional negative powers of 2, can be converted to an equivalent decimal fraction as follows:

$$\begin{aligned} 0 \times 2^{-1} &= 0 \times 0.5 &&= 0.0 \\ 1 \times 2^{-2} &= 1 \times 0.25 &&= 0.25 \\ + \; 1 \times 2^{-3} &= 1 \times 0.125 &&= \underline{0.125} \\ & &&= 0.375 \text{ decimal} \end{aligned}$$

1.6 OTHER NUMBER SYSTEMS

The decimal and binary number systems and their relationship have been described in the preceding sections. Now we look briefly at the octal (base-8) and hexadecimal (base-16) number systems. The octal and hexadecimal numbers are compact representations of binary numbers, devised for human convenience.

The Octal (Base-8) Number System

The **octal number system** is a base-8 positional notation system that uses the decimal digits 0 through 7 as coefficients for each position that represents a power of 8. The octal number system is useful in applications in which 3 binary digits are grouped to form an octal digit. Early computers used "octal dumps" to provide a more readable printout of data because each 3 bits are represented by a corresponding octal digit (e.g., $000 \rightarrow 0_8$, $111 \rightarrow 7_8$).

An integer number can be represented in the octal system in the form

$$a_n(8^n) + \cdots + a_i(8^i) + \cdots + a_2(8^2) + a_1(8^1) + a_0(8^0)$$

where any coefficient $a_i = 0, 1, \ldots, 7$. The relationships between numbers in the decimal, octal, and binary systems are illustrated in Table 1.9.

TABLE 1.9
Table of decimal, octal, and binary equivalents

Decimal 10 1	Octal 8 1	Binary 8 4 2 1
0	0	0 0 0
1	1	0 0 1
2	2	0 1 0
3	3	0 1 1
4	4	1 0 0
5	5	1 0 1
6	6	1 1 0
7	7	1 1 1
8	1 0	1 0 0 0
9	1 1	1 0 0 1
1 0	1 2	1 0 1 0
1 1	1 3	1 0 1 1
1 2	1 4	1 1 0 0
1 3	1 5	1 1 0 1
1 4	1 6	1 1 1 0
1 5	1 7	1 1 1 1
1 6	2 0	1 0 0 0 0

Octal and hexadecimal numbers have a direct relationship to binary numbers. That is, each octal (or hexadecimal) character represents 3 (or 4) binary digits. Decimal numbers do not have a direct character relation to groups of binary digits.

The Hexadecimal (Base-16) Number System

The **hexadecimal number system** (hexa = 6; deci = 10) is a base-16 positional notation system that uses the 10 decimal digits 0 through 9 and the 6 letters A, B, C, D, E, and F as coefficients for each position representing powers of 16. The hexadecimal system is useful in applications in which 4 binary digits are grouped to form a hexadecimal digit. For example, hexadecimal digits are useful in representing memory addresses and in representing machine-language instructions.

An integer number can be represented in the hexadecimal system in the form

$$a_n(16^n) + \cdots + a_i(16^i) + \cdots + a_2(16^2) + a_1(16^1) + a_0(16^0)$$

where any coefficient $a_i = 0, 1, \ldots, 9, A, \ldots, F$. The relationships between the decimal, hexadecimal, and binary systems are illustrated in Table 1.10.

TABLE 1.10
Decimal, hexadecimal, and binary equivalents

Decimal 10 1	Hexadecimal 16 1	Binary 16 8 4 2 1
0	0	0 0 0
1	1	0 0 1
2	2	0 1 0
3	3	0 1 1
4	4	1 0 0
5	5	1 0 1
6	6	1 1 0
7	7	1 1 1
8	8	1 0 0 0
9	9	1 0 0 1
1 0	A	1 0 1 0
1 1	B	1 0 1 1
1 2	C	1 1 0 0
1 3	D	1 1 0 1
1 4	E	1 1 1 0
1 5	F	1 1 1 1
1 6	1 0	1 0 0 0 0
1 7	1 1	1 0 0 0 1
1 8	1 2	1 0 0 1 0
1 9	1 3	1 0 0 1 1
2 0	1 4	1 0 1 0 0
⋮	⋮	⋮
2 6	1 A	1 1 0 1 0
⋮	⋮	⋮
3 1	1 F	1 1 1 1 1
3 2	2 0	1 0 0 0 0 0
⋮	⋮	⋮

Note the following relationships:

$$16^0 = 2^0 = 1$$
$$16^1 = 2^4 = 16$$
$$16^2 = 2^8 = 256$$
$$16^3 = 2^{12} = 4096$$
$$16^4 = 2^{16} = 65636$$

A hexadecimal digit can represent 4 binary digits and, therefore, 4 hexadecimal digits (representing 16 binary digits) simplifies memory addressing and machine instructions by representing more information in less digits. The use of the hexadecimal system to simplify memory addressing can be illustrated by the following example.

Suppose the addresses in a 64K memory have the following representations in hexadecimal, binary, and decimal:

Hexadecimal	Binary	Decimal
0 0 0 0	0000 0000 0000 0000	0
0 0 0 F	0000 0000 0000 1111	15
0 0 1 F	0000 0000 0001 1111	31
0 0 2 0	0000 0000 0010 0000	32
0 0 F F	0000 0000 1111 1111	255
0 F F F	0000 1111 1111 1111	4095
F F F F	1111 1111 1111 1111	65635

We can see that the largest unsigned number that can be represented in 16 binary digits is $2^{16} - 1 = 1111\ 1111\ 1111\ 1111 = 65635$ and the largest unsigned number that can be represented in 4 hexadecimal digits is $16^4 - 1 = \text{FFFF} = 65635$.

The use of the hexadecimal system to simplify **machine language instructions** can be further illustrated by showing the relationship between a computer's instructions in mnemonic code (three-letter codes such as LDA and ADD), binary code, and hexadecimal code. Suppose that a computer uses a 3-bit operation code (**opcode**) and a 5-bit data **address field** to form an 8-bit instruction. The opcodes are as follows:

Mnemonic Opcode	Binary Opcode
LDA	000
ADD	001
STA	010
OUT	011

The following table is a comparison of mnemonic, binary (which consists of a 3-bit opcode and a 5-bit address field), and 2-digit hexadecimal instructions. These examples are illustrations only. Detailed examples of binary and hexadecimal computer instructions will be presented in Chapters 12–14.

Mnemonic Opcode	Mnemonic Address*	Binary Opcode	Binary Address	Hexadecimal
LDA	o0	000	00000	0 0
LDA	oA	000	01010	0 A
ADD	o1	001	00001	2 1
ADD	oC	001	01100	2 C
STA	1F	010	11111	5 F
OUT	o0	011	00000	6 0

*In the mnemonic address the symbol o represents a 1-bit zero, and the symbol 0 represents 4 binary zeros.

1.7 COMPUTER CODES

Most calculators and computers use groups of 2-state memory cells for storing encoded characters. Special names have been given to groups of 4 and 8 bits: A group of 4 bits is called a* nibble*, and a group of 8 bits is called a* byte*. Thus, a byte consists of 2 nibbles.

Decimal (numeric) data input to a computer must be converted to some binary form to be stored in memory and processed by the central processing unit (CPU). Alphanumeric (text) data and keyboard characters (letters, digits, and symbols) used in program instructions must also be converted to some binary form to be stored in memory. Various computer codes are used to represent information, including both data and programs, in binary form for purposes of storage, processing, and communication. There are four basic types of **computer codes**: decimal, Gray, alphanumeric, and error codes. We will discuss the first three codes in this section. In Chapter 4, we will describe the use of a parity bit to detect single errors, but error codes, per se, are beyond the scope of this text.

Decimal Codes

A **decimal code** is used to represent the decimal digits 0 through 9. The overall category of decimal codes can be divided into two subcategories: weighted and nonweighted. Each of these subcategories can be further subdivided into symmetric, nonsymmetric, unit distance, and reflected, as shown in Figure 1.3.

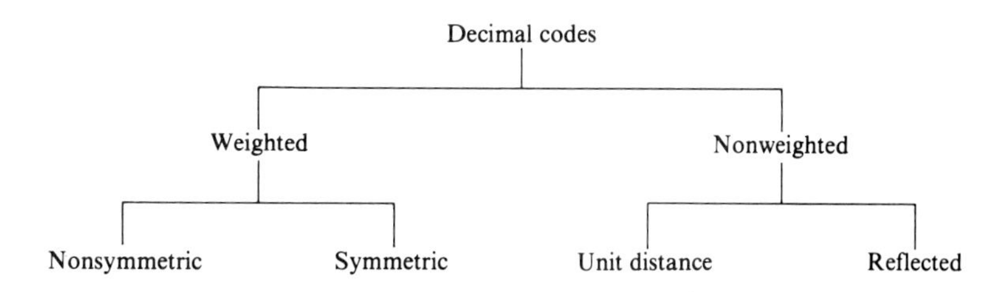

FIGURE 1.3
Organization of decimal codes

The codes in each subcategory and sub-subcategory possess certain properties.

1. All weighted decimal codes possess the property that the encoded decimal digit is the sum of the weights of positions with 1s.
2. A symmetric weighted decimal code is self-complementing; that is, the arithmetic complement $(9 - X)$ of code word X is the same as the logical (bitwise) complement of X formed by interchanging 0s and 1s in code word X.
3. In unit distance (cyclic) code, the adjacent code words differ in only one bit position.

The selection of a particular code for a given application is based on the properties of that code as it satisfies certain requirements of the application. For instance,

- Unit distance codes are useful in such applications as shaft-position encoders, where other binary codes are vulnerable to errors in transitions in which adjacent code words differ in more than one bit position.
- Self-complementing codes are useful in applications requiring the arithmetic complement of a number.
- Weighted codes are useful in coding data to be used in computations.

Weighted Nonsymmetric Decimal Codes

The most widely used decimal code is the weighted nonsymmetric 8421 BCD code, which is also referred to as NBCD (natural binary-coded decimal). The 8421 code was used in the 4004 CPU of the first microprocessor system, the Intel MCS. Today, the 8421 code is used in many calculators and in the ASCII and EBCDIC alphanumeric codes used in computers. Table 1.11 shows a number of **weighted nonsymmetric decimal codes**. Other weighted codes can also be constructed.

TABLE 1.11
Weighted nonsymmetric decimal codes

Decimal	NBCD 8421	7421	5421	5311
0	0000	0000	0000	0000
1	0001	0001	0001	0001
2	0010	0010	0010	0011
3	0011	0011	0011	0100
4	0100	0100	0100	0101
5	0101	0101	0101	1000
6	0110	0110	0110	1001
7	0111	1000	0111	1011
8	1000	1001	1011	1100
9	1001	1010	1100	1101

Weighted Symmetric Decimal Codes

Weighted symmetric decimal codes possess a line of symmetry between the first five lines and the last five lines of the table. That is, a code below the dashed line in the following table is the complement of the code in the corresponding position above the line. For example, the last line (code for 9) is the **logical (1s) complement** of the first line (code for 0).

Decimal	631(−1)	2421	84(−2)(−1)
0	0011	0000	0000
1	0010	0001	0111
2	0101	0010	0110
3	0111	0011	0101
4	0110	0100	0100
5	1001	1011	1011
6	1000	1100	1010
7	1010	1101	1001
8	1101	1110	1000
9	1100	1111	1111

Consequently, symmetric decimal codes are self-complementing. That is, the **arithmetic complement** $(9 - X)$ of a code word X is the same as the logical (1s) complement of X, as is shown in the following weighted 631(−1) examples:

	X	X'	$= 9 - X$
Decimal	631(−1)	631(−1)	Decimal
2	0101	1010	7
8	1101	0010	1

Nonweighted Unit Distance Codes

A **nonweighted unit distance code** is a code in which adjacent (successive) code words differ in only one bit position. Unit distance codes are useful in such applications as shaft-position encoders, as indicated earlier.

A unit distance code can be generated using the following procedure:

1. Draw an $X = x_1x_0$ axis, with adjacent points differing by 1 bit. Draw a $Y = y_1y_0$ axis, with adjacent points differing by 1 bit. Construct a matrix such that the Y axis is perpendicular to the X axis, as in the following table:

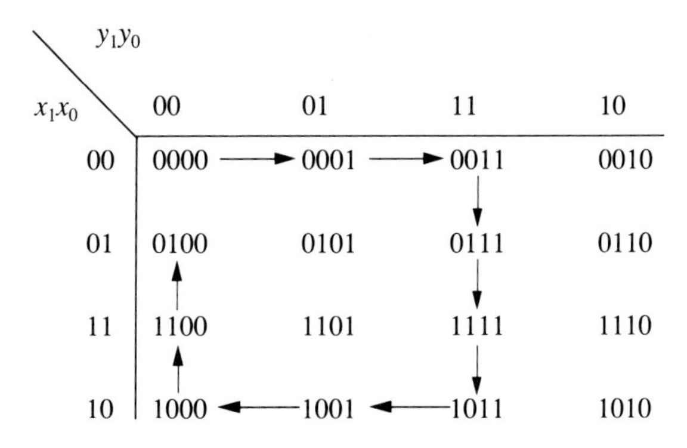

2. To generate a unit distance decimal code, start at any point in the matrix and follow any path that covers 10 codes and that ends at a point adjacent to the starting point, such that no code word is used more than once in the code table. For example, the path outlined above is

0000 0001 0011 0111 1111 1011 1001 1000 1100 0100

Note that any two adjacent codes, including the first and last, differ in only one bit position.

Nonweighted Reflected Codes

A **nonweighted reflected code** consists of pairs of code words in which pair members (1) occupy corresponding positions above and below a midline of a table and (2) differ in one bit position. An n-bit reflected code can be generated by the following procedure:

1. Write the same sequence of $(n - 1)$ bit numbers above and below a horizontal line. For example, for a 4-bit reflected code, write the sequences:

```
000     001     100     101     110
-----------------------------------
000     001     100     101     110
```

2. For each 3-bit number, append on either end, or insert between any 2 bits, a 0 above the line and a 1 below the line. For example, append on the left end of each 3-bit number a 0 above the line and a 1 below the line:

```
0000    0001    0100    0101    0110
------------------------------------
1000    1001    1100    1101    1110
```

3. Let the resulting 4-bit words represent the (reflected) codes for the decimal digits 0 through 9, counting clockwise from the left end of the upper line.

```
0       1       2       3       4
0000    0001    0100    0101    0110
------------------------------------
1000    1001    1100    1101    1110
9       8       7       6       5
```

The resulting code is a reflected code since each vertical pair of 4-bit words consists of members that differ in only one bit position.

Gray Codes

The Gray code is a reflected, unit distance code. The following 1-bit Gray code

Decimal	Gray Code
0	0
1	1

satisfies the unit distance property (adjacent words differ in 1 bit) and reflected property (pair members differ in 1 bit).

A 2-bit Gray code can be generated from a 1-bit Gray code as follows:

1. Duplicate the 1-bit Gray code.
2. Draw a line below the last 1-bit Gray code and reflect the 1-bit Gray code about this line to produce the **least significant bit** (LSB) of a 2-bit Gray code.
3. Prefix each LSB with 0 above the line and 1 below the line, thereby producing the most significant bit (MSB) of the 2-bit Gray code.

2 Bit	1 Bit
0 0	0
0 1	1
----	
1 1	
1 0	

MSB LSB

An n-bit Gray code can be generated from an $(n-1)$-bit Gray code as follows:

1. Duplicate the $(n-1)$-bit Gray code.
2. Draw a line below the last $(n-1)$-bit Gray code, and reflect the Gray code about this line to produce the $(n-1)$-LSB of an n-bit Gray code.
3. Prefix the LSBs with 0 above the line and 1 below the line, thereby producing the MSB of the n-bit Gray code.

3 Bit	2 Bit	1 Bit
0 00	0 0	0
0 01	0 1	1
0 11	1 1	
0 10	1 0	
------		
1 10		
1 11		
1 01		
1 00		

MSB LSB

In the resulting n-bit Gray code, upper and lower pair members are identical in $(n-1)$ bits, with adjacent numbers differing in only 1 bit (satisfying the unit distance property). Appending 0 above and 1 below to each $(n-1)$-bit number produces the reflected property (pair members differ by 1 bit). Thus, the resulting n-bit Gray code possesses both the unit distance and reflected properties.

Alphanumeric Codes

Alphanumeric computer codes are used for encoding numeric data, alphanumeric (text) data, and program instructions. Consequently, alphanumeric codes are used to represent characters, such as letters, decimal digits, and special symbols, of character sets used by various programming languages. Early alphanumeric codes include the Hollerith (card) code and the 6-bit flexowriter code.

The two most widely used alphanumeric codes are the 7-bit ASCII and the 8-bit EBCDIC codes. An extended 8-bit ASCII code is used in most modern computers. The additional 128 codes resulting from the extended ASCII code are mostly used to represent graphic characters. The 7-bit ASCII code is presented in Table 1.12, where the code word is denoted by $b_6b_5b_4b_3b_2b_1b_0$.

TABLE 1.12
ASCII code table

	MSB = $b_6b_5b_4$							
LSB = $b_3b_2b_1b_0$	000	001	010	011	100	101	110	111
0000	NUL	DLE	SP	0	@	P		P
0001	SOH	DC1	!	1	A	Q	a	q
0010	STX	DC2	"	2	B	R	b	r
0011	ETX	DC3	#	3	C	S	c	s
0100	EOT	DC4	$	4	D	T	d	t
0101	ENQ	NAK	%	5	E	U	e	u
0110	ACK	SYN	&	6	F	V	f	v
0111	BEL	ETB	'	7	G	W	g	w
1000	BS	CAN	(	8	H	X	h	x
1001	HT	EM	)	9	I	Y	i	y
1010	LF	SUB	*	:	J	Z	j	z
1011	VI	ESC	+	;	K	[	k	{
1100	FF	FS	,	<	L	\	l	|
1101	CR	GS	-	=	M	]	m	}
1110	SO	RS	.	>	N		n	
1111	SI	US	/	?	O	<-	o	DEL

Each decimal digit is composed of a 011 followed by the BCD representation of a decimal digit. For example, decimal 0 = 0110000, . . . , decimal 9 = 0111001. The first 15 capital letters are represented by 1000001 through 1001111, while the remaining 11 capital letters are 1010000 through 1011010. The first two columns of the ASCII table are coded representations of various control characters. For example, ACK is acknowledge, BEL is bell, NUL is null, DLE is data link escape, and SUB is substitute.

1.8 SUMMARY

This chapter has dealt with how numbers are represented in the decimal, binary, octal, and hexadecimal positional notation systems and in floating-point form. We looked at algorithms for converting decimal integers to binary integers, binary integers to decimal integers, decimal fractions to binary fractions, and binary fractions to decimal fractions.

We saw that there are alternative methods for representing binary numbers: the sign-magnitude method, the 1s complement method, and the 2s complement method. We then noted the utility of the 2s complement form for performing subtraction by complementing the subtrahend and adding the result to the minuend. The 2s complement method is used in the central processing unit of most computers to combine an add and a subtract capability into an integrated adder/subtracter.

We saw examples of the use of the hexadecimal system to represent memory addresses and machine-language instructions. Finally, we dealt with some computer codes (decimal, Gray, and alphanumeric) and their applications.

KEY TERMS

EHBBA
Alphanumeric computer codes
System
Analog
Digital
Hierarchical model
Signals
Positional notation
Decimal number system
Radix
Arithmetic operations
9s complement
End-around carry
10s complement
Floating-point notation
Scientific notation
Mantissa
Binary number system
Sign-magnitude form
1s complement
Most significant bit (MSB)
2s complement
Octal number system
Hexadecimal number system
Machine language instructions
Opcode
Address field
Computer codes
Decimal code
Weighted nonsymmetric decimal code
Weighted symmetric decimal code
Logical 1s complement
Arithmetic complement
Nonweighted unit distance code
Nonweighted reflected code
Gray code
Least significant bit (LSB)

EXERCISES

General

1. Convert the following numbers from decimal to binary.

a. 43 **b.** 352 **c.** 76 **d.** 18

2. Convert the following fractions from decimal to binary.

a. 7/8 **b.** 5/32 **c.** 0.625 **d.** 0.875

3. Convert the following unsigned numbers from binary to decimal.

a. 11001 **b.** 101010 **c.** 110110 **d.** 11101

4. Write the equivalent octal numbers for the following decimal numbers.

a. 7 **b.** 15 **c.** 16

5. Write the equivalent hexadecimal numbers for the following decimal numbers.

a. 10 **b.** 15 **c.** 16 **d.** 31 **e.** 32

6. Perform the following computations (use 2s complement form).

a. Add $A = +5$, $B = -2$ **b.** Subtract $A = +6$, $B = +3$

7. Write the following decimal integers in binary and in binary positional notation.

a. 1 **b.** 10 **c.** 18 **d.** 34

8. Convert the following fractions to decimal fractions and binary fractions.

a. 1/2 **b.** 1/8

9. Find the sum and its decimal equivalent of these unsigned binary integers.

a. $\begin{array}{r} 10001 \\ +\ 11011 \\ \hline \end{array}$ **b.** $\begin{array}{r} 101 \\ +\ 010 \\ \hline \end{array}$

10. Using positional notation, multiply and divide each of these pairs of integers.

a. 836, 20 **b.** 455, 15

11. Write the following binary integers in base-2 positional notation.

a. 1110 **b.** 10111

12. Add the following unsigned binary integers.

a. $\begin{array}{r} 1011 \\ +\ 1100 \\ \hline \end{array}$ **b.** $\begin{array}{r} 1111 \\ +\ 10110 \\ \hline \end{array}$ **c.** $\begin{array}{r} 110101 \\ +\ 10111 \\ \hline \end{array}$

13. Using integer subtraction, 9s complement subtraction, and 10s complement subtraction, do the following problems.

a. $\begin{array}{r} 76 \\ -\ 55 \\ \hline \end{array}$ **b.** $\begin{array}{r} 5 \\ -\ 4 \\ \hline \end{array}$ **c.** $\begin{array}{r} 7 \\ -\ 5 \\ \hline \end{array}$ **d.** $\begin{array}{r} 171 \\ -\ 28 \\ \hline \end{array}$

14. Do the following subtractions in 1s complement and 2s complement form. (Convert to binary first.)

a. $\begin{array}{r} 8 \\ -3 \\ \hline \end{array}$ **b.** $\begin{array}{r} 16 \\ -7 \\ \hline \end{array}$ **c.** $\begin{array}{r} 25 \\ -10 \\ \hline \end{array}$

15. Given the following decimal integers and their binary equivalent in sign-magnitude form, write the corresponding 1s complement and 2s complement representation.

Decimal	Binary
4	0100
1	0001
6	0110
7	0111

16. What relationship exists between the ASCII codes for successive letters of the alphabet? How could this relationship be used in generating a telephone book?

17. Generate a unit distance code starting at point 0101 in the unit distance generation matrix.

18. Generate a 4-bit reflected code for the decimal digits 0 through 9.

2

Logic, Switching Circuits, Boolean Algebra, and Logic Gates

OBJECTIVES

After you complete this chapter, you will be able

- To translate control problems and computation problems into a common logic model
- To conceptualize a device's input-output relations as a logic truth table
- To understand logic concepts concerning constants, variables, operations, and functions
- To use statements with AND, OR, and NOT connectors to form logic expressions for a device output as a function of its inputs
- To realize the logic operations AND, OR, and NOT with switches
- To develop concepts of duality and positive and negative logic systems
- To understand Boolean (switching) algebra axioms and theorems
- To understand switching algebra concepts as they relate to switching functions, canonical forms of expressions, functional completeness, universal operators, equivalence of expressions, function dual and complement
- To convert logic expressions from one form to another and simplify expressions using algebraic techniques
- To apply Boolean algebra in digital logic analysis and design
- To realize switching functions with gates
- To use a gate as either an AND or OR function
- To use mixed-logic methodology to implement logic equations
- To realize application function modules using logic gates as building blocks
- To realize higher-level devices using function modules and logic gates as building blocks
- To use software logic simulators for simulating digital logic circuits

2.1 INTRODUCTION

This chapter opens with a description of the principal elements of problem solving, elements that can be used to solve a range of problems in different applications areas, such as digital logic, physics, mechanics, and so forth. These elements are combined to form a 4-step problem-solving procedure that is used throughout this text. Then, we see how these elements can be used to develop basic concepts and tools for analyzing and designing digital logic circuits.

In this chapter, we also develop the basic concepts of logic and logic functions, an algebra of switch networks, and the more general Boolean (switching) algebra—the principal theoretical tool used in digital logic analysis and design. Then, we show how Boolean algebra is used to solve combinational logic design problems and how the design solution is implemented using physical logic gates to realize logic functions. We also describe fixed-logic and mixed-logic systems and the dual AND, OR use of gates in a mixed-logic system.

Finally, we describe a building-block approach for realizing function modules using logic gates. We then realize higher-level devices using the function modules and gates as building blocks.

Several logic simulation programs are available to generate the physical digital logic circuits that are used throughout this text. The logic diagrams in this text are generated by a Macintosh using Capilano Computing's logic simulation program called Designworks.®

2.2 ELEMENTS OF PROBLEM SOLVING

Various types of problems in different applications areas (such as digital logic, physics, and mechanics) can be solved using a simple 4-step problem-solving procedure (PSP).

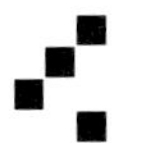

PSP

Step 1: Problem Statement. State the problem in terms of the particular application area.

Step 2: Conceptualization. Translate the problem into a conceptual model such as a drawing, an engineer's "black box," logic equations, or mathematical equations (math model).

Step 3: Solution/Simplification. Determine a solution for the conceptual model using appropriate methods. For example, the solution of two linear equations in x, y can be determined by Cramer's rule.

Step 4: Realization. The solution of the original problem can be realized by translating the result of Step 3 into the original application area. Alternatively, the solution may be realized using an appropriate simulation, such as the logic-simulation program we use in this text. ■

Consider the problem-solving procedure's use in solving the following problem.

Step 1: Problem Statement. Given masses m and M (or distances d and D), determine the conditions required for balancing the beam shown in Figure 2.1.

®Registered trademark of Capilano Computing, Vancouver, B.C.

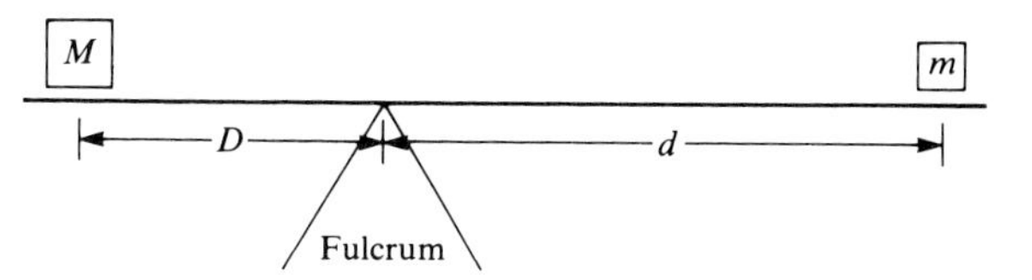

FIGURE 2.1
Illustration of the balance-beam problem

a. If masses M and m are given, determine the (relative) values of distances D and d so that the beam is balanced.
b. If distances D and d are given, determine the (relative) values of masses M and m so that the beam is balanced.

Step 2: Conceptualization. A mathematical model of the balance-beam problem can be developed by applying basic principles of physics: Assuming a weightless beam, the beam is balanced if the torques $M \times D$ and $m \times d$, resulting from gravitational forces, are equal. Therefore, the beam is balanced (in a state of equilibrium) if

$$M \times D - m \times d = 0 \tag{2.1}$$

Equation (2.1) is a mathematical (conceptual) model of the balance-beam problem.

Step 3: Solution/Simplification.

a. If the masses M and m are given constants, equation (2.1) can be rewritten as

$$d = \frac{M}{m} \times D \quad \text{or} \quad D = \frac{m}{M} \times d \tag{2.2}$$

Equation (2.2) can be solved for d (or D) by assuming a value for D (or d) and evaluating the expression on the right-hand side.

b. If the distances D and d are given constants, equation (2.1) can be rewritten as

$$m = \frac{D}{d} \times M \quad \text{or} \quad M = \frac{d}{D} \times m \tag{2.3}$$

Equation (2.3) can be solved for m (or M) by assuming a value for M (or m) and evaluating the expression on the right-hand side.

Step 4: Realization. We can use equation (2.2) or (2.3) to realize the solution of the original balance-beam problem (in the physical realm). The solution can be realized by either of the following two methods:

a. Physical implementation using an actual balance beam
b. Simulation of a balance beam using an analog device with a known mapping to the model

A range of problems in different application areas can be systematically solved using this same 4-step problem-solving procedure. The principal elements of problem solving are illustrated in the model shown in Figure 2.2.

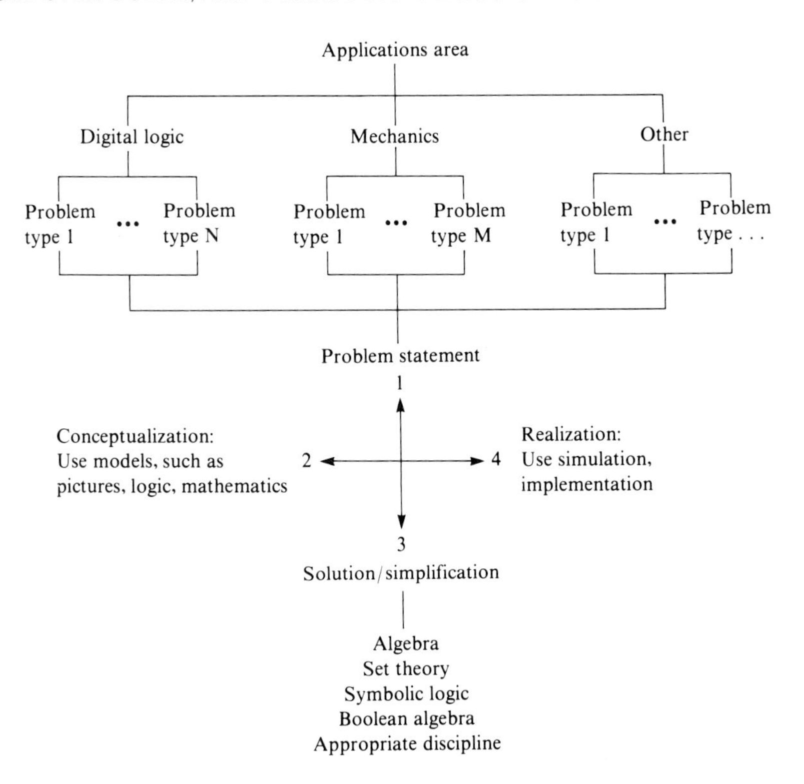

FIGURE 2.2
A model for a 4-step problem-solving procedure

2.3 CONCEPTUALIZATION IN LOGIC DESIGN—LOGIC AND LOGIC FUNCTIONS

Digital logic devices and systems can be designed to solve a variety of problems in diverse application areas. In this section, we will show how computation problems and control problems can be translated into a common framework or conceptual model. This conceptual model may take the form of a truth table or a logic expression.

Translating Computation and Control Problems into Logic

Two simple design problems are presented here to illustrate how the solution of different types of application problems can be represented by a single **conceptual model** called a **logic truth table**.

Design of a Half Adder

Step 1: Problem Statement. Design a computing device, called a **half adder** (HA), that will add two 1-bit numbers x and y to produce a 2-bit sum CH defined by the following equation:

$$x \times 2^0 + y \times 2^0 = C \times 2^1 + H \times 2^0 \tag{2.4}$$

Step 2: Conceptualization. A binary addition table (Table 2.1) can be used to list the sum bit H and the carry bit C for each possible combination of x and y. Then, the binary addition table can be converted to a logic truth table (Table 2.2) that lists the truth values of the statements $C = 1$ and $H = 1$ for each combination of truth values of the statements $x = 1$ and $y = 1$.

TABLE 2.1
Binary addition table

x	y	C	H
0	0	0	0
0	1	0	1
1	0	0	1
1	1	1	0

TABLE 2.2
Logic truth table

$x = 1$	$y = 1$	$C = 1$	$H = 1$
F	F	F	F
F	T	F	T
T	F	F	T
T	T	T	F

Design of an Alarm Circuit

Step 1: Problem Statement. Design a control circuit that activates alarms F_1 and F_2 under the following conditions:

Alarm F_1 is ON if temperature $T > 32°F$ or humidity $H > 50\%$.
Alarm F_2 is ON if temperature $T > 32°F$ and humidity $H > 50\%$.

Step 2: Conceptualization. The state (ON, OFF) of each alarm at a given time depends on the combination of values of temperature and humidity. The operation of each alarm can be described by a physical function table (Table 2.3) that lists the state of alarms F_1 and F_2. Since there are two temperature ranges (≤32 and >32) and two humidity ranges (≤50% and >50%) of interest, there are four possible combinations of values.

The physical function table can be converted into a logic truth table (Table 2.4) that lists the truth values of the statements F_1 is ON and F_2 is ON for each combination of truth values of statements $T > 32°F$ and $H > 50\%$.

TABLE 2.3
Physical function table

		Alarms	
Temperature	Humidity	F_1	F_2
≤ 32	≤ 50%	OFF	OFF
≤ 32	> 50%	ON	OFF
> 32	≤ 50%	ON	OFF
> 32	> 50%	ON	ON

TABLE 2.4
Logic truth table

$T > 32°$	$H > 50\%$	$F_1 = \text{ON}$	$F_2 = \text{ON}$
F	F	F	F
F	T	T	F
T	F	T	F
T	T	T	T

Note that the function tables for the computing device and the control circuit are each expressed in terms peculiar to the application. However, by expressing each problem in logic terms, we can conceptualize different functions in a common form, the logic truth table. Since a truth table is one form of a logic model, we can see the usefulness of logic as a single unifying concept for representing different types of physical systems.

Only the most basic elements of logic are required to understand how logic models can be used to conceptualize physical systems:

1. A simple statement (proposition) is a declarative sentence that is either true (T) or false (F).
2. Compound statements are formed from simple statements and other compound statements using logic connectives such as AND, OR, NOT, and IF . . . THEN.
3. The compound statements *X* AND *Y*, *X* OR *Y*, and NOT *X* are defined by truth tables as follows:

X	*Y*	*X* AND *Y*
F	F	F
F	T	F
T	F	F
T	T	T

X	*Y*	*X* OR *Y*
F	F	F
F	T	T
T	F	T
T	T	T

X	NOT *X*
F	T
T	F

For example, a compound statement *X* AND *Y* is true if statement *X* is true *and* statement *Y* is true. A compound statement *X* OR *Y* is true if *X* is true *or* *Y* is true *or* both are true. A compound statement NOT *X* is true if statement *X* is false, and NOT *X* is false if statement *X* is true.

4. Truth tables can be used to determine the truth values of compound statements such as *X* OR (*Y* AND *Z*), (*X* AND *Y*) OR (*X* AND NOT *Z*).

If statements *X*, *Y*, and *Z* are represented by 2-state variables *X*, *Y*, and *Z*, and if the connectives AND, OR, and NOT are represented by symbols ·, +, ', then compound statements can be represented symbolically in algebra-like logic expressions such as $X \cdot Y + X' \cdot Z + Y \cdot Z'$. A set of rules that govern the formation and manipulation of such expressions is the basis for a discipline called **symbolic logic**.

Symbolic Logic and Logic Functions

In 1854 George Boole laid the foundation for symbolic logic (and Boolean algebra) in a book entitled *An Investigation of the Laws of Thought*. Basically, symbolic logic is a discipline in which statements and the connectives AND, OR, and NOT are represented by symbols. The essential elements of symbolic logic are

Logic constants 0 and 1, which are used to represent false and true, respectively

Logic variables such as *X*, which is a two-valued variable that can be either 0 or 1; a logic variable is used to represent a simple statement

Logic operations AND, OR, and NOT, represented by the symbols ·, +, and ' (or overbar), which are used to combine operands (logic constants and logic variables) to form logic expressions.

We can write, for example,

$$X, \quad X \cdot Y, \quad X + (Y \cdot Z), \quad (X + Y) \cdot (X + Z')$$

The symbol · (AND) is used to combine two operands to form a **logic product** such as $Y \cdot Z$. The symbol + (OR) is used to combine two operands to form a **logic sum** such as $X + (Y \cdot Z)$. The symbol for NOT (' or ¯) is used to represent the complement operation.

The logic operations AND, OR, and NOT are defined by truth tables expressed in terms of logic values 0 and 1, as shown in Table 2.5.

TABLE 2.5
Truth tables for logic operations AND, OR, and NOT

AND			OR			NOT	
X	Y	$X \cdot Y$	X	Y	$X + Y$	X	X'
0	0	0	0	0	0	0	1
0	1	0	0	1	1	1	0
1	0	0	1	0	1		
1	1	1	1	1	1		

For example, if X has value 1, a complementary variable X' has value 0. If X has value 0, a complementary variable X' has value 1.

The symbolic logic operations · and + have the properties of commutativity (X × Y = Y × X) and associativity [X × (Y × Z) = (X × Y) × Z], where × denotes operator · or +.

A logic function $F(X, Y)$ assigns a logic value 0 or 1 to each combination of (X, Y) values. A binary logic operator combines two operands X, Y, and the operation results in a logic value 0 or 1. Consequently, the terms **logic function** and **logic operation** are often used interchangeably, although each has a different interpretation.

The logic operations AND, OR, and NOT may be interpreted as logic functions since each of these operations assigns a value 0 or 1 to each combination of (X, Y) values:

$$X \text{ AND } Y = F(X, Y)$$

where $F(0, 0) = 0$, $F(0, 1) = 0$, $F(1, 0) = 0$, $F(1, 1) = 1$.

$$X \text{ OR } Y = G(X, Y)$$

where $G(0, 0) = 0$, $G(0, 1) = 1$, $G(1, 0) = 1$, $G(1, 1) = 1$.

$$\text{NOT } X = H(X, Y)$$

where $H(0, 0) = 1$, $H(0, 1) = 1$, $H(1, 0) = 0$, $H(1, 1) = 0$.

An application of symbolic logic is illustrated by solving the following example problem using the 4-step problem-solving procedure:

The Corn, the Goat, and the Wolf

Step 1: Problem Statement. A farmer has corn stored in an open crib in a barn. He also has a goat and an absent-minded hired hand who often leaves the barn door open, and there is a wolf lurking in the woods near the barn. The farmer wants to devise a system that will activate an alarm if (1) the goat is in danger of being eaten by the wolf and/or (2) the corn is in danger of being eaten by the goat.

Step 2: Conceptualization. The operation of the alarm system can be described by the following physical function table:

Wolf Is in Sight	Goat Is in Sight	Barn Door Is Open	Danger Exists
No	No	No	No
No	No	Yes	No
No	Yes	No	No
No	Yes	Yes	Yes
Yes	No	No	No
Yes	No	Yes	No
Yes	Yes	No	Yes
Yes	Yes	Yes	Yes

The physical function table can be transformed into a logic truth table by substituting the logic variables W, G, B, and D for wolf is in sight, goat is in sight, barn door is open, and danger exists, respectively. The truth table can be written in either of two forms, as shown below. (In the table on the right, the entries W and W' denote $W = 1$ and $W' = 1$, where $W' =$ NOT W.)

W	G	B	D
0	0	0	0
0	0	1	0
0	1	0	0
0	1	1	1
1	0	0	0
1	0	1	0
1	1	0	1
1	1	1	1

W	G	B	D
W'	G'	B'	0
W'	G'	B	0
W'	G	B'	0
W'	G	B	1
W	G'	B'	0
W	G'	B	0
W	G	B'	1
W	G	B	1

Each line of the truth table represents a logic product, such as $W' \cdot G' \cdot B'$ or $W' \cdot G' \cdot B$. The logic function $D(W, G, B)$ defined by the truth table can be represented as a logic expression formed as a sum of the logic products for which the corresponding function value is 1:

$$D(W, G, B) = W' \cdot G \cdot B + W \cdot G \cdot B' + W \cdot G \cdot B$$

Thus, a logic expression and a truth table are alternative forms of a logic model used to conceptualize a problem statement that describes a relation between a functional device's inputs and outputs.

Functions in Ordinary Algebra and in Logic

A fundamental structure of mathematics is a set or collection of elements. A set S consisting of elements a, b, and c is denoted $S = \{a, b, c\}$. Any element a in set S, denoted by $a \in S$, is a member of set S.

Let I denote the set of integers $\{\ldots, -2, -1, 0, 1, 2, \ldots\}$. In ordinary algebra, an integer function $f(x, y)$, where $x \in I$ and $y \in I$, is a mapping of each point (x, y) into an integer number. That is, each point (x, y), $x \in I$ and $y \in I$, is assigned a function value.

If $f(x, y) = x \times y$, the function assigns a value $(x \times y)$ to each point (x, y), where $x \in I$ and $y \in I$. The function can be represented by either tabulating or plotting function values for a specified set of (x, y) values, as shown in Figures 2.3 and 2.4.

FIGURE 2.3
Values for $f(x, y)$ tabulated for $(-1 \le x \le 1, -1 \le y \le 1)$

x	y	$f(x, y) = x \times y$
−1	−1	1
−1	0	0
−1	1	−1
⋮	⋮	⋮
1	1	1

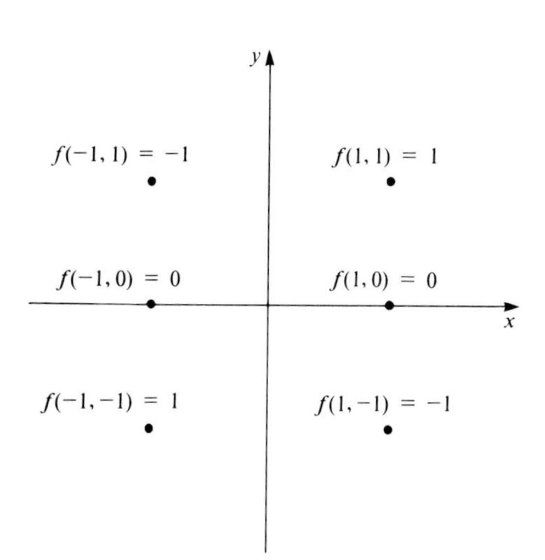

FIGURE 2.4
Values of $f(x, y)$ plotted in cartesian coordinates

In symbolic logic, a variable has value 0 or 1. The corresponding binary set $S = \{0, 1\}$ contains the elements 0 and 1.

The **cartesian product** $A \times B$ is defined as the set $\{a_i, b_i\}$ that consists of all ordered pairs (a_i, b_i), where $a_i \in A$ and $b_i \in B$. If $S = \{0, 1\}$,

$$S \times S = \{(0, 0), (0, 1), (1, 0), (1, 1)\}$$

A logic function of two variables A and B, denoted $f(A, B)$, assigns a value (0 or 1) to each of the ordered pairs: (0, 0), (0, 1), (1, 0), (1, 1). For example,

$$f(0, 0) = 1, f(0, 1) = 0, f(1, 0) = 1, f(1, 1) = 1$$

Thus, a logic function $f(A, B)$ is a mapping of points of $S \times S$ into S.

If set A = set B = set $S = \{0, 1\}$, a logic function $f(A, B)$ is a mapping of points (a_i, b_i), $a_i \in A$, $b_i \in B$, into set $S = \{0, 1\}$. The function $f(A, B)$ can be represented in either of two forms:

A truth table, which can be constructed in alternative forms:

A	B	$f(A, B)$
0	0	1
0	1	0
1	0	1
1	1	0

A	B	$f(A, B)$
A'	B'	1
A'	B	0
A	B'	1
A	B	0

A logic equation such as

$$f(A, B) = A' \cdot B' + A \cdot B'$$

2.4 REALIZATION OF LOGIC OPERATIONS BY SWITCHES—SWITCH NETWORK ALGEBRA

Section 2.3 showed how different application functions, each defined by an application-specific function table, can be translated into a logic model. This logic model may be either a logic truth table or a logic equation. If the logic operations AND, OR, and NOT are realized by physical devices, then a logic equation can be physically realized. This section describes how the logic operations AND, OR, and NOT can be realized by electrical 2-state switches.

Switches in Series or Parallel

A single-pole single-throw switch is a 2-state device, with states "open" and "closed." A switch must be either open or closed, so open = not closed and closed = not open. For any switch A, there exists a complement A'. If switch A is open, then switch A' is closed; conversely, when switch A is closed, switch A' is open.

A light in a simple loop circuit can be controlled by connecting two independent 2-state (open, closed) switches *A* and *B* in **series** or in **parallel**, as shown in Figures 2.5a and b.

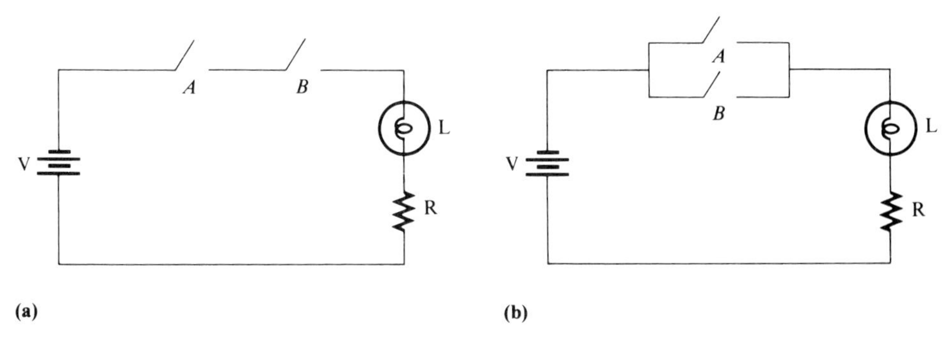

FIGURE 2.5
Loop circuits with switches. (a) Series, (b) Parallel

A loop circuit is closed if a path exists for electrons to flow from the voltage source (battery) V to the light L. A loop circuit is open if no path exists for electrons to flow from V to L. At a given time, the state (open, closed) of the loop with switches in series (or parallel) depends on the combination of the setting of switches *A* and *B*, as shown in Table 2.6.

TABLE 2.6
Physical function tables for loop circuits

Switches *A*, *B* in Series			Switches *A*, *B* in Parallel		
A	***B***	**Loop *S*(*A*, *B*)**	***A***	***B***	**Loop *P*(*A*, *B*)**
Open	Open	Open	Open	Open	Open
Open	Closed	Open	Open	Closed	Closed
Closed	Open	Open	Closed	Open	Closed
Closed	Closed	Closed	Closed	Closed	Closed

We can infer from Table 2.6 certain conditions for a closed or open loop. In series,

- The loop is closed if *A* is closed and *B* is closed.
- The loop is open for all other *A*, *B* value combinations.

In parallel,

- The loop is closed for *A*, *B* combinations: (open, closed), (closed, open), (closed, closed).
- The loop is open if *A* is open and *B* is open.

A switch can be represented by a binary variable. Since switch *A* is independent of switch *B*, the variables *A* and *B* are independent variables. The loop state can be considered as a function of variables *A* and *B*. The physical function table is transformed into truth table form by substituting logic variables A, B, and S (or P or any other variable name) for the state of switches A, B, and the loop state.

Closed (Positive) Logic and Open (Negative) Logic Systems

The function table for the circuit in Figure 2.5a (series) can be translated into either of two logic truth tables by assigning value "closed" to logic value 1 or value "open" to logic value 1, as shown in Table 2.7.

TABLE 2.7
Loop circuit with switches in series

Truth Tables						Function Table		
Closed = 1			Open = 1					
A	B	$S(A, B)$	A	B	$S(A, B)$	A	B	S
0	0	0	1	1	1	o	o	o
0	1	0	1	0	1	o	c	o
1	0	0	0	1	1	c	o	o
1	1	1	0	0	0	c	c	c

That is, if value closed = logic 1, $S(A, B) = A$ AND B; if value open = logic 1, $S(A, B) = A$ OR B. Switches in series perform logic operation AND if closed = 1 and operation OR if open = 1.

The function table for the circuit in Figure 2.5b (parallel) can be translated into either of two logic truth tables by assigning value "closed" to logic value 1 or assigning value "open" to logic value 1, as shown in Table 2.8.

TABLE 2.8
Loop circuit with switches in parallel

Truth Tables						Function Table		
Closed = 1			Open = 1					
A	B	$P(A, B)$	A	B	$P(A, B)$	A	B	P
0	0	0	1	1	1	o	o	o
0	1	1	1	0	0	o	c	c
1	0	1	0	1	0	c	o	c
1	1	1	0	0	0	c	c	c

That is, if value closed = logic 1, $P(A, B) = A$ OR B; if value open = logic 1, $P(A, B) = A$ AND B. Switches in parallel perform logic operation OR if closed = 1 and operation AND if open = 1.

Duality of Functions S(A, B) and P(A, B)

The principle of duality allows us to simplify more complicated circuits, as we will see when we study Boolean algebra and mixed logic. The following are definitions of duality.

DEFINITION: If a **logic expression** for function F is transformed into a logic expression for function G by interchanging $\cdot$ with + and 1 with 0, then G is a **dual** of F.

DEFINITION: A closed (**positive**) **logic system** is a system in which value closed = logic value 1.

$$\text{If value closed} = \text{logic } 1,\ S(A, B) = A \text{ AND } B = A \cdot B$$

$$\text{If value closed} = \text{logic } 1,\ P(A, B) = A \text{ OR } B = A + B$$

The expression $A \cdot B$ is transformed into the expression $A + B$ by the interchanges of + with $\cdot$ and 1 with 0. Therefore, in a closed logic system, $S(A, B) = A \cdot B$ is a dual of $P(A, B) = A + B$.

DEFINITION: An open (**negative**) **logic system** is a system in which value open = logic value 1.

$$\text{If value open} = \text{logic } 1,\ S(A, B) = A \text{ OR } B = A + B$$

$$\text{If value open} = \text{logic } 1,\ P(A, B) = A \text{ AND } B = A \cdot B$$

The expression $A + B$ is transformed into the expression $A \cdot B$ by the interchanges + with $\cdot$ and 1 with 0. Therefore, in an open logic system, $S(A, B) = A + B$ is a dual of $P(A, B) = A \cdot B$.

The convention that logic 1 represents the *asserted* (active) state is used throughout the rest of this text.

An Algebra of Switch Networks

Consider a network of switches between points 1 and 2 in a simple loop circuit, as shown in Figure 2.6. Any switch can be connected in series or in parallel to another switch, or it may be connected to a series or parallel combination of other switches. The way in which switches are interconnected in a given network is called the *configuration* of the switches in that network.

FIGURE 2.6
Switches in series–parallel combinations

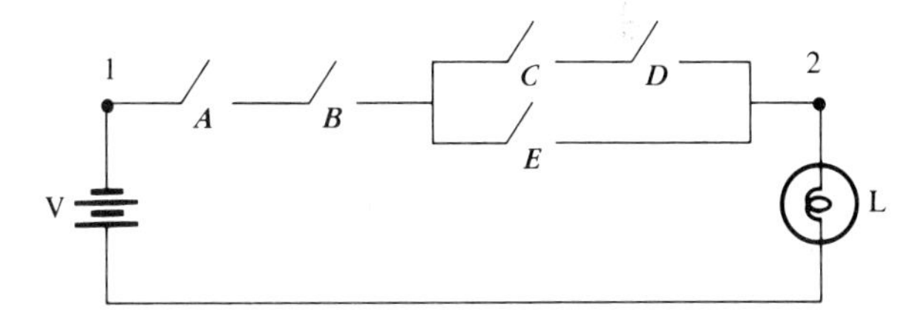

A particular configuration of switches A, B, C, D, and E, in a network, like that of Figure 2.6, can be represented by a logic expression such as $A \cdot B \cdot [E + (C \cdot D)]$. The loop state is either open or closed at any given time, depending on the configuration of switches and the combination of state values (open, closed) of the switches in the network.

A set of rules can be developed to form an algebra of switch networks. Then, any logic expression representing a given configuration of switches can be algebraically formed and manipulated according to the rules of the algebra. If a logic expression can be reduced to a simpler equivalent expression, then the original configuration of switches can be replaced by a simpler, functionally equivalent configuration.

Such an algebra of switch networks satisfies the properties of a Boolean algebra that will be described in the next section and provides a theoretical basis whereby a network of switches can be used to implement any given logic expression representing a physical digital logic device. However, there are two basic drawbacks in using an algebra of switch networks and switch networks in digital logic design:

1. The elements and operations in the algebra of switch networks are defined in specific terms of switches and series and parallel configurations of switches. An analogous, general Boolean algebra is based on the same five properties (called axioms), but its elements and operations are not limited by being switch-specific. Therefore, this more general Boolean switching algebra is the principal theoretical tool used in the analysis and design of digital logic circuits.
2. Today integrated-circuit devices, called **logic gates** (to be described in Section 2.6), are used (instead of switches connected in series or parallel) to implement logic functions in digital logic circuits.

Before we move on to Boolean switching algebra, be sure you understand the concepts presented in this section:

Realization of logic operations (functions) using physical devices

Closed (positive) logic and open (negative) logic systems

Duality of logic expressions

Implementation of a logic expression using physical devices

2.5 BOOLEAN SWITCHING ALGEBRA

Switching algebra, which deals only with two-valued variables, is an important theoretical tool used in the analysis and design of digital logic circuits. Switching algebra is a special case of a general Boolean algebra, which deals with **n*-valued variables.***

A math structure, such as Boolean algebra, can be defined by a set of axioms or postulates. By definition, an **axiom** is a true statement about the properties of the structure.

Axiomatic Definition of Boolean Switching Algebra

Let $S = \{0, 1\}$ be a set, with distinct elements 0 and 1 and with two binary operations ($\cdot$, +) and a unary operation (') defined as

A	B	A · B
0	0	0
0	1	0
1	0	0
1	1	1

A	B	A + B
0	0	0
0	1	1
1	0	1
1	1	1

A	A'
0	1
1	0

The structure $[S, \cdot, +, ', 0, 1]$ is a Boolean algebra, called **switching algebra**, that satisfies certain axioms for all elements A, B, and C in S. (Any element A, B, C has value 0 or 1.) These axioms are given here as axioms 1 through 5 (and their duals as 1d through 5d).

- Closure of set S with respect to $\cdot$ and $+$:
  - AXIOM 1 $A \cdot B \in S$
  - AXIOM 1d $A + B \in S$
- Existence of identities:
  - AXIOM 2 There exists an identity 0 for + such that $A + 0 = A$
  - AXIOM 2d There exists an identity 1 for $\cdot$ such that $A \cdot 1 = A$
- Commutative laws:
  - AXIOM 3 $A + B = B + A$
  - AXIOM 3d $A \cdot B = B \cdot A$
- Distributive laws:
  - AXIOM 4 $A \cdot (B + C) = A \cdot B + A \cdot C$
  - AXIOM 4d $A + (B \cdot C) = (A + B) \cdot (A + C)$
- Existence of a complement. For every element $A \in S$, there exists a complement $A' \in S$ such that
  - AXIOM 5 $A + A' = 1$
  - AXIOM 5d $A \cdot A' = 0$

In switching algebra a certain precedence of operations must be observed: (1) An expression in parentheses must be evaluated before any other operation. (2) Then, the unary operation ' is carried out, followed by the binary operation $\cdot$, and finally, the binary operation +.

The dual of any statement in a Boolean algebra is the statement obtained by interchanging the operations + and $\cdot$ and the identity elements 0 and 1 in the original statement. It should be noted that Axiom 1d is the dual of Axiom 1, Axiom 2d is the dual of Axiom 2, and so on. Note also the symmetry in the axioms of a Boolean algebra: The dual of the set of original axioms is the same as the original set of axioms.

Theorems in Switching Algebra

A **theorem** is a mathematical rule that is not assumed in the axioms on which the math structure is based. A proof of a theorem is often presented as a sequence of assertions (true statements) such that each assertion is either: an axiom, a previous theorem, or a logical inference from previous steps of the proof.

The switching algebra theorems below can be proved using the axioms of switching algebra (A, B, and C are any elements in $S = \{0, 1\}$).

- Idempotent laws:
  - THEOREM 1 $A + A = A$
  - THEOREM 1d $A \cdot A = A$
- DeMorgan's theorems:
  - THEOREM 2 $(A + B)' = A' \cdot B'$
  - THEOREM 2d $(A \cdot B)' = A' + B'$

- Boundness laws:

  THEOREM 3 $A + 1 = 1$

  THEOREM 3d $A \cdot 0 = 0$

- Absorption laws:

  THEOREM 4 $A + (A \cdot B) = A$

  THEOREM 4d $A \cdot (A + B) = A$

- Elimination laws:

  THEOREM 5 $A + (A' \cdot B) = A + B$

  THEOREM 5d $A \cdot (A' + B) = A \cdot B$

- Unique complement theorem:

  THEOREM 6 If $A + X = 1$ and $A \cdot X = 0$, then $X = A'$.

- Involution theorem:

  THEOREM 7 $(A')' = A$

  THEOREM 7d $0' = 1$

- Associative properties:

  THEOREM 8 $A + (B + C) = (A + B) + C$

  THEOREM 8d $A \cdot (B \cdot C) = (A \cdot B) \cdot C$

- Consensus theorem:

  THEOREM 9 $\mathrm{A \cdot B + A' \cdot C + B \cdot C = A \cdot B + A' \cdot C}$

  THEOREM 9d $(A + B) \cdot (A' + C) \cdot (B + C) = (A + B) \cdot (A' + C)$

Switching algebra is also based on **DeMorgan's generalized law**, which states that the complement of a logical sum is equal to the product of the complements of the terms of the sum. DeMorgan's theorems can be extended to any number of variables. In three variables, we have

$$(A + B + C)' = A' \cdot B' \cdot C'$$
$$(A \cdot B \cdot C)' = A' + B' + C'$$

In four variables, we have

$$(A + B + C + D)' = A' \cdot B' \cdot C' \cdot D'$$
$$(A \cdot B \cdot C \cdot D)' = A' + B' + C' + D'$$

Boolean Variables, Expressions, and Switching Functions

In Boolean algebra, logic expressions can be formed by combining logic constants and variables using logic operator symbols in much the same way that standard algebraic expressions are formed by combining numeric constants and variables using algebraic operator symbols (+, −, ×, and /). Similarly, logic functions can be defined using logic expressions just as standard algebraic functions are defined using algebraic expressions.

- In the two-element Boolean (switching) algebra, a variable x can have only the value 0 or 1.
- A Boolean expression E is formed by combining the logic constants (0 and 1) and variables (x, y, and z) using the logic operators (+, ·, '). For example, if $x \in S$ and $y \in S$, then $E = x' \cdot y + x \cdot y'$ is an expression.

- □ An ordered *n-tuple* is of the form $(x_1, \ldots, x_n)$ where order is important.
- □ A cartesian product of n sets A_i, $(i = 1, n)$, consists of all ordered n-tuples $(x_1, \ldots, x_n)$, $x_i \in A_i$, $(i = 1, n)$. If each set $A_i = S = \{0, 1\}$, the cartesian product $S \times S \cdots \times S$ is denoted S^n.

$$S \times S = \{(0, 0), (0, 1), (1, 0), (1, 1)\} = \{(x, y)\}$$
$$S \times S \times S = \{(0, 0, 0), (0, 0, 1), \ldots, (1, 1, 1)\} = \{(x, y, z)\}$$

- □ A combinational or **switching function** F: $S^n \rightarrow$ S, where $S = \{0, 1\}$, consists of

A nonempty set S^n, called the **domain** of the function

A nonempty set S, called the **range** of the function

A rule that assigns one and only one element of S to each element of S^n. (Conversely, each element of S^n is "mapped" to only one element of S.)

- □ A 2-variable switching function $F(x, y)$ is a function that maps each element of the domain $S \times S = \{(0, 0), (0, 1), (1, 0), (1, 1)\}$ to only one element of $S = \{0, 1\}$.

The function $F(x, y) = 0 \cdot x' \cdot y' + 1 \cdot x' \cdot y + 1 \cdot x \cdot y' + 1 \cdot x \cdot y$ has the mapping

$$S \times S = \left\{ \begin{array}{c} (x, y) \quad F \\ (0, 0) \rightarrow 0 \\ (0, 1) \rightarrow 1 \\ (1, 0) \rightarrow 1 \\ (1, 1) \rightarrow 1 \end{array} \right\} = \text{Elements of } S$$

Note that the coefficient (0 or 1) of product term $x^* \cdot y^*$ is the value assigned to $F(x^*, y^*)$, where * denotes specific x, y combinations $(x' \cdot y', x' \cdot y, x \cdot y', x \cdot y)$. Therefore, a truth table is a table that lists each product-term combination and its corresponding value F. The tables below illustrate the **mapping** F: $S \times S \rightarrow S$ and the corresponding truth table:

	x	y	F		x y	F
$S \times S =$	0	0	0	= Elements of S	$x' \cdot y'$	0
	0	1	1		$x' \cdot y$	1
	1	0	1		$x \cdot y'$	1
	1	1	1		$x \cdot y$	1

The four rows of a 2-variable truth table represent the products $x' \cdot y'$, $x' \cdot y$, $x \cdot y'$, and $x \cdot y$. Each product term that has a 1 in the output column is a minterm of the function defined in the truth table.

Canonical Forms of Switching Functions in N *Variables*

A **literal** denotes a variable with or without a prime. Therefore, x and x' are two literals that refer to the same variable. A product term of n-distinct literals is called a **minterm**, provided each literal appears once and only once. A sum term of n-distinct literals is called a

maxterm, provided each literal appears once and only once. Table 2.9 lists the possible minterms and maxterms for switching functions in 2 variables x and y:

TABLE 2.9
The 2-variable minterms and maxterms

i	x	y	**Minterm**	**Maxterm**
0	0	0	$m_0 = x' \cdot y'$	$M_0 = x + y$
1	0	1	$m_1 = x' \cdot y$	$M_1 = x + y'$
2	1	0	$m_2 = x \cdot y'$	$M_2 = x' + y$
3	1	1	$m_3 = x \cdot y$	$M_3 = x' + y'$

The relation between corresponding minterms and maxterms:

$$m_i = M_i' \text{ and } m_i' = M_i \; (i = 0, 1, 2, 3)$$

can be proved by DeMorgan's law. For example,

$$m_0' = (x' \cdot y')' = (x')' + (y')' = x + y = M_0.$$

Minterms satisfy the relations

$$\begin{aligned} m_i \cdot m_j &= 0 \quad \text{if } i \neq j \\ &= m_i \quad \text{if } i = j \end{aligned}$$

Maxterms satisfy the relations

$$\begin{aligned} M_i + M_j &= 1 \quad \text{if } i \neq j \\ &= M_i \quad \text{if } i = j \end{aligned}$$

Every switching function F: $S \times S \cdots \times S \rightarrow S$ can be written in either of the following forms:

- □ **Canonical sum-of-products (SOP)** form

$$F(x, y, z, \ldots) = \text{Sum } (a_k \cdot m_k), \text{ where } a_k = 0 \text{ or } 1$$

- □ **Canonical product-of-sums (POS)** form

$$F(x, y, z, \ldots) = \text{Product } (b_k + M_k), \text{ where } b_k = 0 \text{ or } 1$$

We can express this in tabular form as

i	x	y	Minterm	Maxterm	F
0	0	0	$m_0 = x' \cdot y'$	$M_0 = x + y$	0
1	0	1	$m_1 = x' \cdot y$	$M_1 = x + y'$	1
2	1	0	$m_2 = x \cdot y'$	$M_2 = x' + y$	1
3	1	1	$m_3 = x \cdot y$	$M_3 = x' + y'$	0

The canonical SOP expression for F is the logical sum-of-product terms with function value 1:

$$F = m_1 + m_2 = x' \cdot y + x \cdot y'$$

This expression can be written in abbreviated form using the minterm decimal numbers:

$$F(x, y) = \Sigma(1, 2) = m_1 + m_2$$

The canonical POS expression for F is the logical product-of-sum terms with function value 0:

$$F = M_0 \cdot M_3 = (x + y) \cdot (x' + y')$$

This expression can be written in abbreviated form using the maxterm decimal numbers:

$$F(x, y) = \pi(0, 3) = M_0 \cdot M_3$$

There are 16 possible switching functions of 2 variables, x and y, as defined in Table 2.10.

TABLE 2.10
Table of switching functions $F(x, y)$

		Functions $F: S \times S \to S$															
	(x, y)	**0**	$x \cdot y$	$x \cdot y'$	x	$x' \cdot y$	y	$x \oplus y$	$x + y$	**NOR**	$x \equiv y$	y'		x'		**NAND**	**1**
$S \times S =$	(0, 0)	0	0	0	0	0	0	0	0	1	1	1	1	1	1	1	1
	(0, 1)	0	0	0	0	1	1	1	1	0	0	0	0	1	1	1	1
	(1, 0)	0	0	1	1	0	0	1	1	0	0	1	1	0	0	1	1
	(1, 1)	0	1	0	1	0	1	0	1	0	1	0	1	0	1	0	1
	Decimal F:	0	1	2	3	4	5	6	7	8	9	10	11	12	13	14	15

Each column of the 16-function table of switching functions in x, y is a truth table column for one of the 16 distinct functions. If a, b, c, and d are in $\{0, 1\}$, then the column (a, b, c, d) is one of 16 functions in 2 variables, and $a \cdot x' \cdot y' + b \cdot x' \cdot y + c \cdot x \cdot y' + d \cdot x \cdot y$ is the corresponding canonical SOP expression. Therefore, any function can be expressed, using only the operations $\cdot$, $+$, $'$ in canonical SOP form as the logical sum of the minterms of the function. For example, the canonical SOP expression for the function $F(x, y)$ with column (0, 1, 1, 1) is

$$0 \cdot x' \cdot y' + 1 \cdot x' \cdot y + 1 \cdot x \cdot y' + 1 \cdot x \cdot y = x' \cdot y + x \cdot y' + x \cdot y$$

Switching Functions in 2 or More Variables

There are 6 functions $F(x, y)$ in 2 variables that are of primary interest in digital logic:

1	$F(x, y) = x \cdot y$	AND function
7	$F(x, y) = x + y$	OR function
14	$F(x, y) = (x \cdot y)'$	NAND function
8	$F(x, y) = (x + y)'$	NOR function
6	$F(x, y) = x \oplus y$	Exclusive-OR function
9	$F(x, y) = (x = y)$	Equivalence function

Switching Functions in 3 or More Variables

There are 6 functions in 3 or more variables that are of primary interest in digital logic:

$F(x, y, z) = x \cdot y \cdot z$	AND function
$F(x, y, z) = x + y + z$	OR function
$F(x, y, z) = (x \cdot y \cdot z)'$	NAND function
$F(x, y, z) = (x + y + z)'$	NOR function
$F(x, y, z) = x \oplus y \oplus z$	Exclusive-OR function
$F(x, y, z) = (x \oplus y \oplus z)'$	Exclusive-NOR function

The NAND and NOR functions in 3 variables are defined as the AND-NOT and OR-NOT functions, respectively, because the NAND and NOR operations are not associative.

Logic Conditions for Assertion of Logic Variables and Operations

Any switching function can be represented by an expression formed by combining variables using the logic operations AND, OR, and NOT. To determine when the switching function is asserted (true), it is necessary to know under what logic conditions the operations AND, OR, and NOT are true, independent of any physical system in which the function is to be implemented.

A variable X is **asserted** when it has a value of 1; X is not asserted when it has a value of 0. A logic operation is asserted (has value 1), or not asserted (has value 0), depending on the combination of inputs A and B (see Table 2.11):

- **AND** is asserted when all inputs are asserted.
- **OR** is asserted when any input is asserted.

The NOT operation performs a **complement operation** in a fixed-logic system ($A \rightarrow A'$, or $A' \rightarrow A$). The NOT operation is defined by the following truth table:

A	NOT A
0	1
1	0

TABLE 2.11
Assertion tables for AND and OR

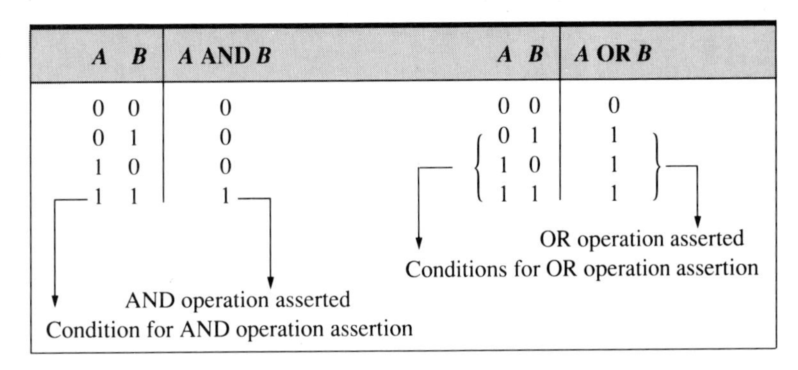

A	B	A AND B
0	0	0
0	1	0
1	0	0
1	1	1

AND operation asserted
Condition for AND operation assertion

A	B	A OR B
0	0	0
0	1	1
1	0	1
1	1	1

OR operation asserted
Conditions for OR operation assertion

A switching function F, represented by a logic expression, is asserted when a condition exists for that expression's assertion. For example, a switching function $F(C, W, D) = C \cdot (W + D)$ is asserted when $(C = 1)$ and $(W = 1$ or $D = 1)$.

Application of Switching Algebra to Digital Circuits

The application of switching algebra in the design of digital logic circuits can be illustrated by the following problem.

Design of an Alarm System

Step 1: Problem Statement. Design a logic circuit that will activate an alarm if a door or window is opened during nonbusiness hours.

Step 2: Conceptualization. Activation of an alarm F depends on three independent logic variables:

Clock	$C = 0$ (business hours) or 1 (nonbusiness hours)
Door	$D = 0$ (closed) or 1 (open)
Window	$W = 0$ (closed) or 1 (open)

The operation of the alarm circuit is described by the function table below, which can be transformed into a logic truth table by substituting the logic variables, C, D, W, and F for clock, door, window, and alarm, respectively.

Alarm Circuit Function Table

Clock	Door	Window	Alarm
Business	Closed	Closed	Off
Business	Closed	Open	Off
Business	Open	Closed	Off
Business	Open	Open	Off
Nonbusiness	Closed	Closed	Off
Nonbusiness	Closed	Open	On
Nonbusiness	Open	Closed	On
Nonbusiness	Open	Open	On

Truth Table

C	D	W	F
0	0	0	0
0	0	1	0
0	1	0	0
0	1	1	0
1	0	0	0
1	0	1	1
1	1	0	1
1	1	1	1

The canonical SOP expression for F can be obtained directly from the truth table:

$$F = C \cdot D' \cdot W + C \cdot D \cdot W' + C \cdot D \cdot W$$

Step 3: Solution/Simplification. The canonical SOP expression can be reduced to a simpler equivalent expression using the laws of Boolean algebra:

	Law Used
$F = C \cdot D' \cdot W + C \cdot D \cdot W' + C \cdot D \cdot W + C \cdot D \cdot W$	Idempotent
$F = C \cdot D' \cdot W + C \cdot D \cdot W + C \cdot D \cdot W' + C \cdot D \cdot W$	Commutative
$F = C \cdot W \cdot D' + C \cdot W \cdot D + C \cdot D \cdot W' + C \cdot D \cdot W$	Commutative
$F = C \cdot W \cdot (D' + D) + C \cdot D \cdot (W' + W)$	Distributive
$F = C \cdot W \cdot 1 + C \cdot D \cdot 1$	$A' + A = 1$
$F = C \cdot W + C \cdot D$	$A \cdot 1 = A$
$F = C \cdot (W + D)$	Distributive

It should be noted that the Boolean expressions representing a circuit's design specification are independent of physical quantities (voltages) and physical gates used to implement the circuit.

Step 4: Realization. The original application function (alarm system) is realized by implementing the theoretical solution $F(C, D, W) = C \cdot (W + D)$ as a digital logic circuit. The implementation of a logic equation requires (1) implementing (realizing) the logic functions AND, OR, and NOT as voltage-operated devices and (2) implementing the logic variables as voltage-implemented signals.

A logic equation can be represented as a logic diagram by

- Replacing each logic variable with a corresponding signal name
- Replacing each logic operator symbol (·, +, ') with the corresponding diagram symbol shown in Figure 2.7
- Connecting the signals to inputs of initial-stage gate symbols whose outputs are then connected to the inputs of the next stage, and so on, until the output of the final state is connected to the function signal line.

FIGURE 2.7
Illustration of logic symbols

The AND logic operator (·) is replaced with symbol

The OR logic operator (+) is replaced with symbol

The NOT logic operator (') is replaced with symbol

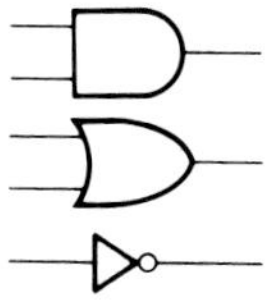

The Boolean equation $F = C \cdot (W + D)$ can then be represented by the logic diagram shown in Figure 2.8.

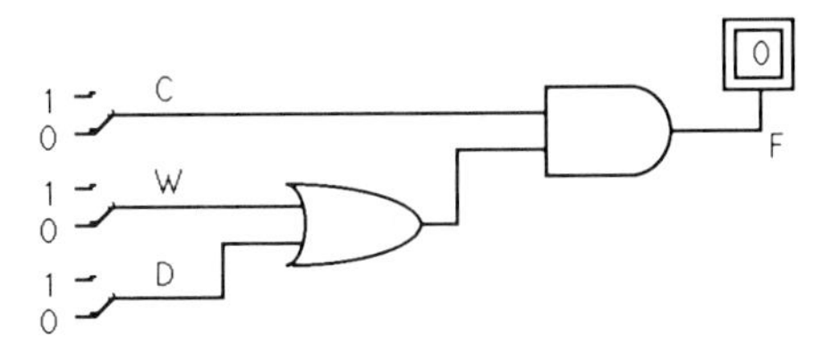

FIGURE 2.8
Hardware-independent logic diagram for alarm circuit

In Section 2.6 we will look at the hardware implementation of logic equations in fixed-logic (positive-logic, negative-logic) and mixed-logic systems.

Properties of Boolean Switching Algebra

Functional Completeness

Early in this section we saw that any switching function can be expressed using only the operations $\cdot$, $+$, $'$, in canonical SOP form as the logical sum of the minterms of the function. Since any switching function can be represented by combining elements x, y of S, and the logic constants 0, 1 using only the operations $\cdot$, $+$, $'$, the set of operations $\{\cdot, +, '\}$ is said to be functionally complete.

By DeMorgan's theorem, the logic sum $x' + y' = (x \cdot y)'$. Therefore, an + operation can be accomplished using the set of operations $\{\cdot, '\}$, and any switching function can be formed using only the set of operations $\{\cdot, '\}$. Consequently, the set $\{\cdot, '\}$ is **functionally complete**. Similarly, it can be shown that the $\{+, '\}$ is functionally complete.

Any operation is functionally complete (universal) if it can accomplish every operation of a functionally complete set of operations, such as $\{\cdot, '\}$ or $\{+, '\}$.

The NAND operation (denoted by $\uparrow$) is a functionally complete (universal) operation because it can accomplish both the $\cdot$ and $'$ operations.

PROOF:
$$
\begin{aligned}
(x \uparrow y) \uparrow (x \uparrow y) &= [(x \cdot y)' \cdot (x \cdot y)']' \\
&= [(x \cdot y)']' && \text{Idempotent} \\
&= x \cdot y && (A')' = A) \\
(x \uparrow x) = (x \cdot x)' &= (x)' = x'
\end{aligned}
$$

The NOR operation (denoted by $\downarrow$) is a functionally complete (universal) operation because it can accomplish both the + and $'$ operations.

PROOF:
$$
\begin{aligned}
(x \downarrow y) \downarrow (x \downarrow y) &= [(x + y)' + (x + y)']' \\
&= [(x + y)']' && \text{Idempotent} \\
&= x + y && (A')' = A \\
(x \downarrow x) = (x + x)' &= (x)' = x'
\end{aligned}
$$

Equivalence of Boolean Expressions

Two Boolean expressions F and G in variables x and y are logically equivalent, written $F = G$, if the truth table for F is identical to the truth table for G. For example, suppose $F = x' \cdot y + x \cdot y' + x \cdot y$ and $G = x + y$.

$x\,y$	$x' \cdot y$	$x \cdot y'$	$x \cdot y$	F	x	y	G
$x'y'$	0	0	0	0	0	0	0
$x'y$	1	0	0	1	0	1	1
$x\,y'$	0	1	0	1	1	0	1
$x\,y$	0	0	1	1	1	1	1

Since the truth tables for F and G are identical, expression F is *logically equivalent* to expression G. A proof accomplished by the use of truth tables is referred to as a proof by total induction.

For a given truth table there exists a unique function $F(x, y)$. However, a given function $F(x, y)$ may be represented by two or more different logically equivalent expressions. For example, the unique function $F(x, y) = x + y$ can also be represented by the expression $x' \cdot y + x \cdot y' + x \cdot y$. Therefore, two Boolean expressions E_1 and E_2 represent the same function if E_1 can be transformed into E_2 using switching algebra laws. For example, given $E_1 = x' \cdot y + x \cdot y' + x \cdot y$ and $E_2 = x + y$,

$$
\begin{aligned}
x' \cdot y + x \cdot y' + x \cdot y &= x' \cdot y + x \cdot y' + x \cdot y + x \cdot y && \text{Idempotent} \\
&= x' \cdot y + x \cdot y + x \cdot y' + x \cdot y && \text{Commutative} \\
&= (x' + x) \cdot y + x \cdot (y' + y) && \text{Distributive} \\
&= 1 \cdot y + x \cdot 1 && \text{Complement} \\
&= y + x && \text{Identity} \\
&= x + y && \text{Commutative}
\end{aligned}
$$

Therefore, $E_1 = E_2$, and the unique function $F(x, y) = x + y$ can be represented by any of the foregoing expressions.

Dual of a Boolean Expression

The principle of duality states that the dual of any theorem in a Boolean algebra is also a theorem. It also implies that if two expressions can be shown to be logically equivalent using a sequence of axioms and theorems, then the duals of the two expressions can be shown to be logically equivalent using a sequence of dual axioms and theorems.

The dual of a Boolean expression is the expression obtained by interchanging the operations (+ and ·) and the identity elements (0 and 1) in the original expression. For example, the dual of the expression $x \cdot y + x' . z$ is the expression $(x + y) \cdot (x' + z)$.

Complement of a Boolean Expression

If $G(x, y) = F'(x, y)$ for all $(x, y) \in S \times S$, then function G is the *complement* of function F, written $G = F'$, and $F + G = F + F' = 1$. DeMorgan's theorem can be used to show that the canonical SOP expression and the canonical POS expression for the complement of the function satisfy the following relation:

$$\text{If } F = \text{SOP } m_i, \text{ then } F' = \text{POS } M_i$$

For example,

$$\text{If } F = m_1 + m_2, \text{ then } F' = (m_1 + m_2)' = m_1' \cdot m_2' \qquad \text{DeMorgan}$$
$$= M_1 \cdot M_2 = G$$

The complement F' of a function F can be obtained by interchanging the 0s and 1s in the truth table function values of F. For example, given a function $F(x, y) = x \cdot y$ and its complement function $F'(x, y) = (x \cdot y)'$,

x	y	F	F'	$F + F'$
0	0	0	1	1
0	1	0	1	1
1	0	0	1	1
1	1	1	0	1

The complement of a function F can also be obtained by taking the **dual function** of F and then complementing each literal in the expression of the dual function. For example, given the following function:

$$F(x, y) = x' \cdot y + x \cdot y'$$

the dual obtained by interchanging $\cdot$ with + and 0 with 1 is

$$(x' + y) \cdot (x + y')$$

Then

$$F' = (x + y') \cdot (x' + y) = x \cdot y + x' \cdot y'$$

Converting Logic Expressions

The conversion of a logic expression from one form to another, such as from AND-OR form to NAND-NAND form, can be accomplished by application of DeMorgan's theorem. However, repeated application of DeMorgan's theorem to convert an expression from one form to another can easily result in errors, especially if the expression is complicated.

Consider some examples of converting logic expressions, starting in AND-OR form:

$$F = A + B \cdot D + C' \cdot D \qquad \text{AND-OR}$$
$$F = (F')' = [(A + B \cdot D + C' \cdot D)']'$$
$$= [A' \cdot (B \cdot D)' \cdot (C' \cdot D)']' \qquad \text{NAND-NAND}$$
$$= [A' \cdot (B' + D') \cdot (C + D')]' \qquad \text{OR-NAND}$$
$$= A + (B' + D')' + (C + D')' \qquad \text{NOR-OR}$$

Figure 2.9 (p. 66) presents the logic diagrams of an example AND-OR circuit and its equivalent in NAND-NAND form.

FIGURE 2.9
An AND-OR circuit and its equivalent in NAND-NAND form. (a) $F = A \cdot B + B' \cdot C$ (b) $F = [(A \cdot B)' \cdot (B' \cdot C)']'$

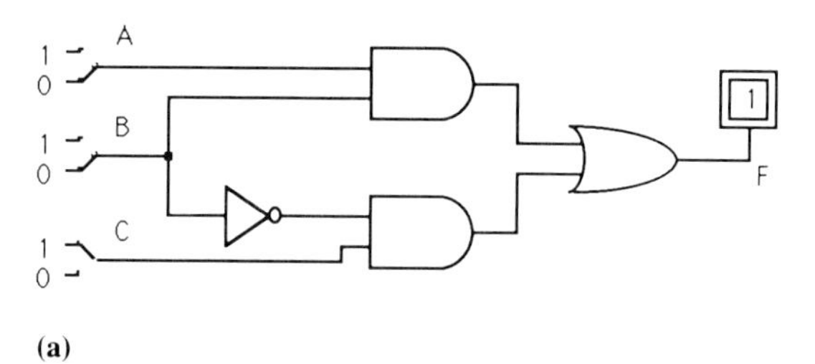

(a)

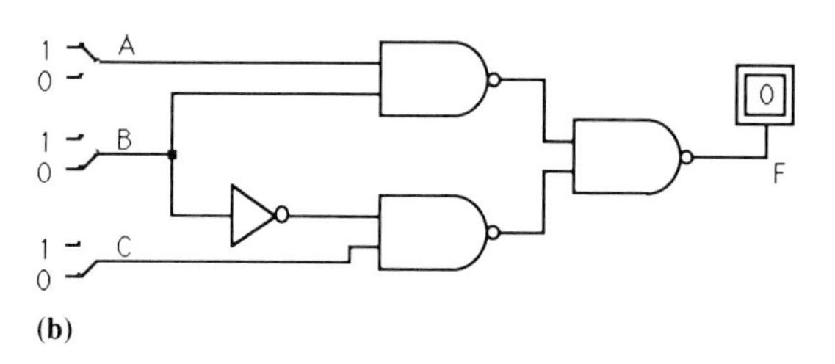

(b)

Now consider some examples of converting logic expressions, starting in OR-AND form:

$G = A \cdot (B + D) \cdot (C' + D)$	OR-AND
$G = (G')' = \{[A \cdot (B + D) \cdot (C' + D)]'\}'$	
$= [A' + (B + D)' + (C' + D)']'$	NOR-NOR
$= [A' + (B' \cdot D') + (C \cdot D')]'$	AND-NOR
$= A \cdot (B' \cdot D')' \cdot (C \cdot D')'$	NAND-AND

Algebraic Techniques for Simplifying Boolean Expressions

In addition to solving equations using appropriate algorithms, mathematicians employ a number of techniques to simplify equations and/or expressions in ordinary algebra. These techniques include

- □ Adding the same quantity to both sides of an equation, which can include adding a quantity equal to 0
- □ Multiplying both sides of an equation by the same quantity, which can include multiplying by a quantity equal to 1
- □ Cancelling a factor common to both sides of an equation

It should be noted that the cancellation law of ordinary algebra does not hold in Boolean algebra.

Similar techniques can be used to simplify expressions in Boolean algebra, since a Boolean expression is invariant (that is, it does not change) under the following operations:

- □ $X \cdot X'$ $(= 0)$ can be ORed (added) to an expression E since $E + 0 = E$.
- □ An expression E can be ANDed (multiplied) by $X + X'$ $(= 1)$ since $E \cdot 1 = E$.

In addition, Boolean expressions can be simplified using various axioms and theorems. The methods that are used most frequently include

- The **logical adjacency theorem** ($X \cdot Y + X \cdot Y' = X$), which can be used to eliminate a variable by combining logically adjacent terms. It should be noted that either X or Y can be product terms. For example,

$$\text{If } X = A \cdot B \text{ and } Y = C, \text{ then } (A \cdot B) \cdot C + (A \cdot B) \cdot C' = A \cdot B$$

- The absorption law ($X + X \cdot Y = X$), which can be used to eliminate term $X \cdot Y$.
- The elimination law ($X + X' \cdot Y = X + Y$), which can be used to eliminate the literal X'.
- The idempotent law ($X + X = X$), which can be used to include an existing term more than once to provide additional terms, which will then be grouped with existing terms for the purpose of factoring. For example,

$$\begin{aligned} C &= c' \cdot x \cdot y + c \cdot x' \cdot y + c \cdot x \cdot y' + c \cdot x \cdot y \\ &= c' \cdot x \cdot y + c \cdot x' \cdot y + c \cdot x \cdot y' + c \cdot x \cdot y + c \cdot x \cdot y + c \cdot x \cdot y \\ &= x \cdot y \cdot (c' + c) + c \cdot y \cdot (x' + x) + c \cdot x \cdot (y' + y) \\ &= x \cdot y + c \cdot y + c \cdot x \end{aligned}$$

- The consensus theorem ($X \cdot Y + X' \cdot Z = X \cdot Y + X' \cdot Z + Y \cdot Z$), which states that the addition of the (redundant) consensus term $Y \cdot Z$ to the expression $X \cdot Y + X' \cdot Z$ does not change the value of the expression. Conversely, it allows elimination of the term $Y \cdot Z$ from the expression $X \cdot Y + X' \cdot Z + Y \cdot Z$.

If two Boolean expressions, E_1 and E_2, are logically equivalent, a Boolean equation can be formed by equating the two expressions, ($E_1 = E_2$). If the validity of a Boolean equation is in doubt, the validity of that equation can be determined by one of the following methods:

- Use **total induction**. Construct two truth tables, one for the expression on each side of the equal sign. Then evaluate both truth tables. If the truth table outputs are identical, the expressions are logically equivalent and the equation is valid.
- Reduce the expression on the left side and the expression on the right side of the equal sign independently. If each of the expressions, for example E_1 and E_2, reduce to the same expression E, then the equation is valid.
- If an expression on one side of the equal sign can be reduced to the expression on the other side, then the equation is valid.

2.6 REALIZATION OF LOGIC EQUATIONS USING GATES AS LOGIC FUNCTIONS

***A logic design can be implemented using physical electronic devices called gates, provided that the output at any given time depends only on the combination of values of the inputs. Consequently, such circuits are referred to as* combinational logic circuits.**

The actual electronic devices (AND gates, OR gates, and inverters) that realize the logic AND, OR, and NOT functions are operated by two-level DC (direct current) voltages. For example, transistor-transistor logic (TTL) devices use 0 and +5 volts, represented by the symbols L and H (or LV and HV), respectively.

Fixed-Logic Systems—Positive Logic, Negative Logic

A positive-logic (active-high) system assigns the high voltage level to logic value 1 and the low voltage level to logic value 0. A negative-logic (active-low) system assigns the low voltage level to logic value 1 and the high voltage level to logic value 0. (See Table 2.12).

TABLE 2.12
Positive-logic and negative-logic assignment table

Positive-Logic System		Negative-Logic System	
Level of Physical Quantity	External Logic State	Level of Physical Quantity	External Logic State
H (high)	1	H (high)	0
L (low)	0	L (low)	1

In Boolean algebra a logic variable is defined independently of any physical system. When a logic equation is implemented using a physical voltage-referenced system, each logic variable is implemented as a voltage-referenced signal.

An input signal is asserted (true) when the corresponding logic variable has value 1. An output signal F is asserted (true) when a condition exists to cause the associated function to have logic value 1. From the example in Section 2.5, the alarm-system output signal $F = C \cdot (W + D)$ is asserted when ($C = 1$) and ($W = 1$ or $D = 1$).

The voltage value (H or L) that a signal has at any time depends on the following:

The logic value (0 or 1) that the associated logic variable has at that time

The system (positive logic or negative logic) in which the logic equation is implemented

Consequently, each signal in a fixed-logic system (whether positive logic or negative logic) must be identified by its associated variable name and a polarity indicator denoting the system in which the expression is implemented. The polarity indicator denotes the voltage level assigned to logic value 1 (for the associated logic variable or identifier). In this way, a conceptual logic entity and its physical implementation are combined to form a signal name. The following convention will be used to identify signal names:

Signal name = Logic identifier.Voltage polarity

If a logic expression is implemented in a positive-logic system ($H = 1$), a signal $A.H$ is asserted when its voltage value is H since the associated logic variable A has value 1 (asserted). If a logic expression is implemented in a negative-logic system ($L = 1$), a signal $B.L$ is asserted when its voltage value is L since the associated logic variable B has value 1 (asserted).

■ **EXAMPLE 2.1**

If a logic expression $A \cdot B + B' \cdot C$ is implemented in a positive-logic system ($H = 1$), the logic variables A, B, and C are implemented as signals $A.H$, $B.H$, and $C.H$, respectively.

The suffix .H is the signal's polarity, indicating that the associated logic variable is implemented in a positive-logic system. If a logic expression is implemented in a negative-logic system, each signal is assigned a polarity L, denoted by the suffix $.L$ appended to the associated logic variable name. ■

The signal name $A.H$ does not mean that the signal voltage is high. It means that when logic identifier A has logic value 1, the signal voltage is high, and, when logic identifier A has logic value 0, the signal voltage is low.

Mixed-Logic System

A digital logic design may require the integration of positive-logic and negative-logic subsystems into a combined **mixed-logic system**, as illustrated in Figure 2.10.

FIGURE 2.10
A mixed-logic system with fixed-logic subsystems

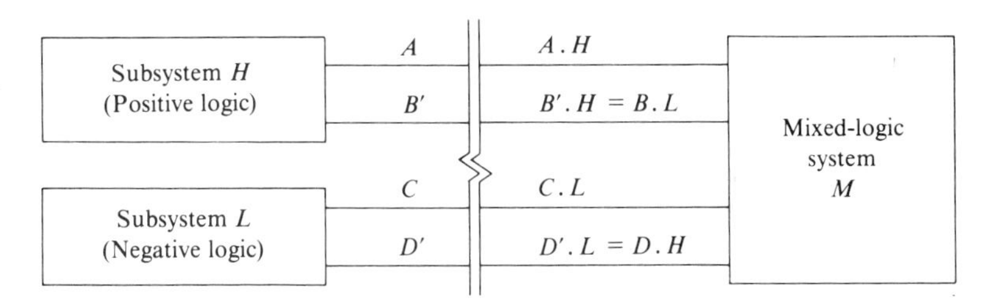

In a subsystem that uses only positive logic (or only negative logic), signal names are often written without a polarity indicator. For example, in a positive-logic system, $A.H$ and $B'.H$ are written as A and B', respectively. In a mixed-logic system, it is essential that each gate input signal and output signal be identified by both a **logic identifier** and a **polarity indicator** for that logic identifier. Again the polarity indicator denotes the voltage level assigned to logic value 1 for the associated logic identifier. In this way, a conceptual logic entity and its physical implementation are combined to form a signal name, and the same convention as used with fixed-logic systems will be used to identify signal names:

$$\text{Signal name} = \text{Logic identifier.Voltage polarity}$$

■ **EXAMPLE 2.2**

Signal $A.H$ = (logic identifier A).(polarity H) denotes that for identifier A, logic value 1 = voltage H. ■

■ **EXAMPLE 2.3**

Signal $B.L$ = (logic identifier B).(polarity L) denotes that for identifier B, logic value 1 = voltage L. ■

■ **EXAMPLE 2.4**

Signal $(A + B).H$ = (logic identifier $A + B$).(polarity H) denotes that for identifier $A + B$, logic value 1 = voltage H. ■

In a mixed-logic system, as in a fixed-logic system, the signal name $A.H$ does not mean that the signal voltage is high. It means that when logic identifier A has logic value 1 the signal voltage is high, and when logic identifier A has logic value 0 the signal voltage is low.

Gate Realization of Logic Functions

Digital logic gates, usually **integrated circuits** (ICs), are physical realizations of conceptual logic functions. Manufacturers' data books describe each logic gate using a function table in which the gate's inputs and output are listed in terms of voltage. Note that there are 4 AND gates with 2 inputs each on the 7408 IC; hence, the 7408 is called a quad 2-input AND. Figure 2.11 provides an illustration of a manufacturer's description.

PIN CONFIGURATION

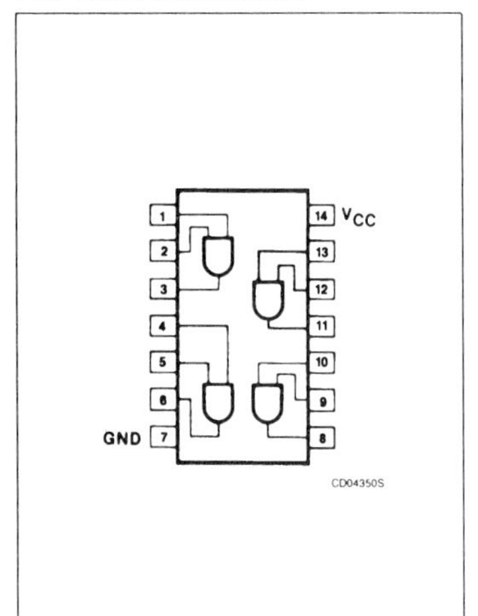

LOGIC SYMBOL

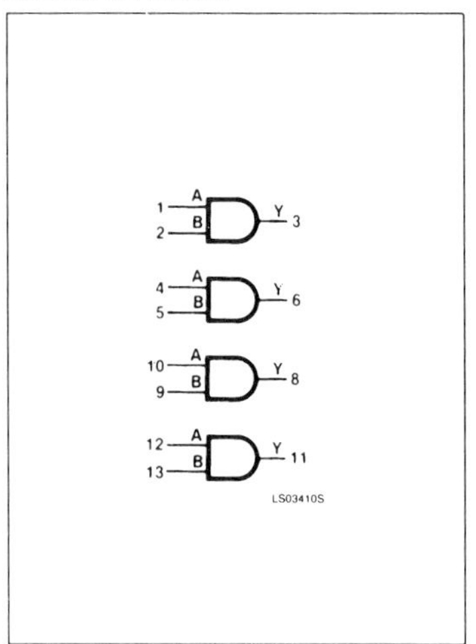

LOGIC SYMBOL (IEEE/IEC)

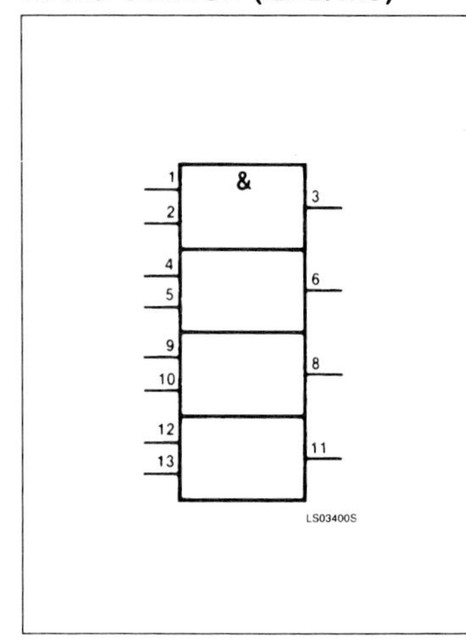

FIGURE 2.11
7408 IC pin configuration, logic symbol, and IEEE logic symbol (Reprinted by courtesy of Philips Components/Signetics)

Of the 16 possible logic functions $F(x, y)$ listed in Table 2.10, the 6 that are of primary interest in digital logic are listed in Table 2.13, together with IC gates that realize the functions.

TABLE 2.13
Important logic functions $F(x, y)$ and TTL gates

Logic Function	Function Name	TTL Gate
$F(x, y) = x \cdot y$	AND	7408
$F(x, y) = x + y$	OR	7432
$F(x, y) = (x \cdot y)'$	NAND	7400
$F(x, y) = (x + y)'$	NOR	7402
$F(x, y) = x \oplus y$	Exclusive-OR	7486
$F(x, y) = (x = y)$	Equivalence (XNOR)	7486 followed by an inverter

The following examples illustrate gate realizations of the logic functions listed in Table 2.13.

EXAMPLE 2.5: 7408 IC (Quad 2-Input AND)

Figure 2.12 shows that a 7408 gate can be used either as an AND function with active-high inputs and outputs (voltage H = logic 1) or as an OR function with active-low inputs and outputs (voltage L = logic 1). It should be noted that any gate that exhibits an AND characteristic also exhibits an OR characteristic.

7408 Gate Function Table

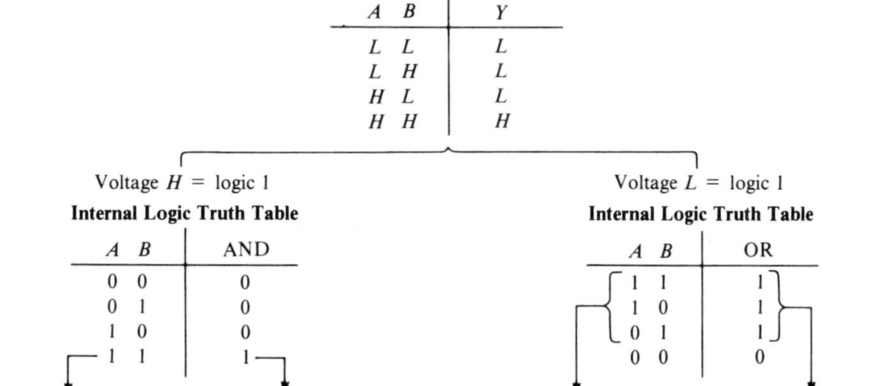

Inputs		Output
A	B	Y
L	L	L
L	H	L
H	L	L
H	H	H

Voltage H = logic 1

Internal Logic Truth Table

A	B	AND
0	0	0
0	1	0
1	0	0
1	1	1

Condition for AND — AND asserted

Voltage L = logic 1

Internal Logic Truth Table

A	B	OR
1	1	1
1	0	1
0	1	1
0	0	0

Conditions for OR — OR asserted

Logic 1 is always used as the asserted value for function inputs.
Logic 1 is always used as the asserted value of the logic function.

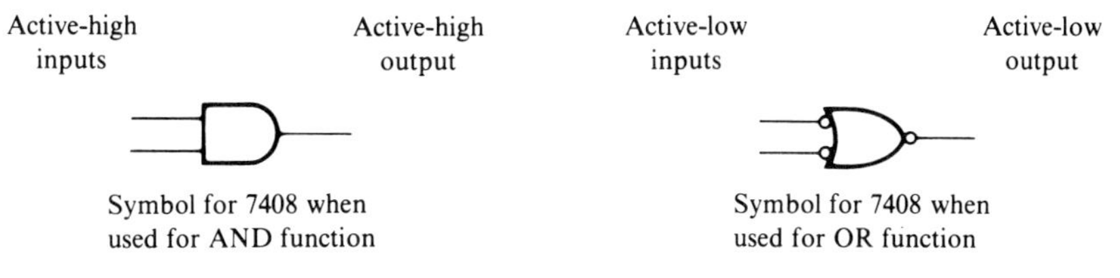

FIGURE 2.12
Illustration of dual use of the 7408 gate

EXAMPLE 2.6: 7432 IC (Quad 2-Input OR)

Figure 2.13 (p. 72) shows that a 7432 gate can be used either as an OR function with active-high inputs and output (voltage H = logic 1) or as an AND function with active-low inputs and output (voltage L = logic 1).

EXAMPLE 2.7: 7400 IC (Quad 2-Input NAND)

It is easier to analyze and design digital logic circuits by thinking in terms of AND, OR, and NOT than in terms of NAND and NOR. The 7400 gate is really a mixed-logic device; that

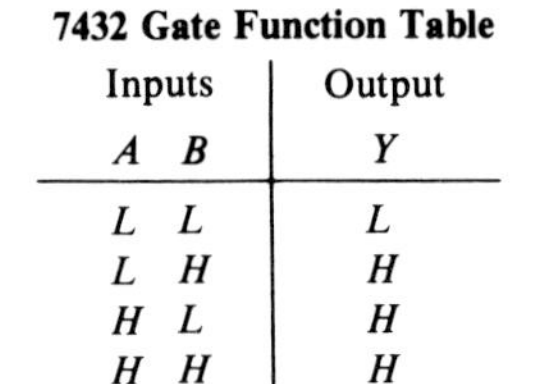

7432 Gate Function Table

Inputs A	Inputs B	Output Y
L	L	L
L	H	H
H	L	H
H	H	H

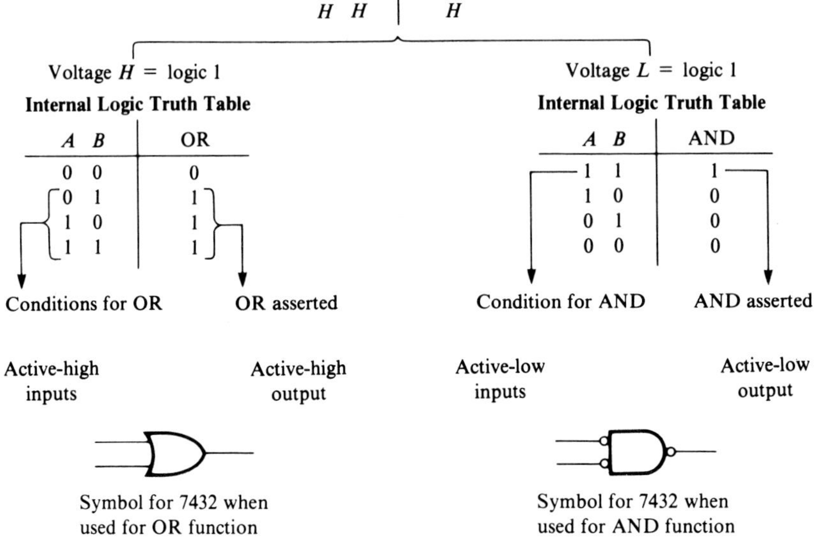

Voltage H = logic 1

Internal Logic Truth Table

A	B	OR
0	0	0
0	1	1
1	0	1
1	1	1

Voltage L = logic 1

Internal Logic Truth Table

A	B	AND
1	1	1
1	0	0
0	1	0
0	0	0

FIGURE 2.13
Illustration of dual use of the 7432 gate

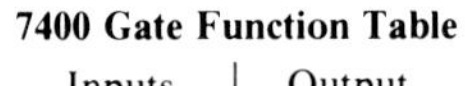

7400 Gate Function Table

Inputs A	Inputs B	Output Y
L	L	H
L	H	H
H	L	H
H	H	L

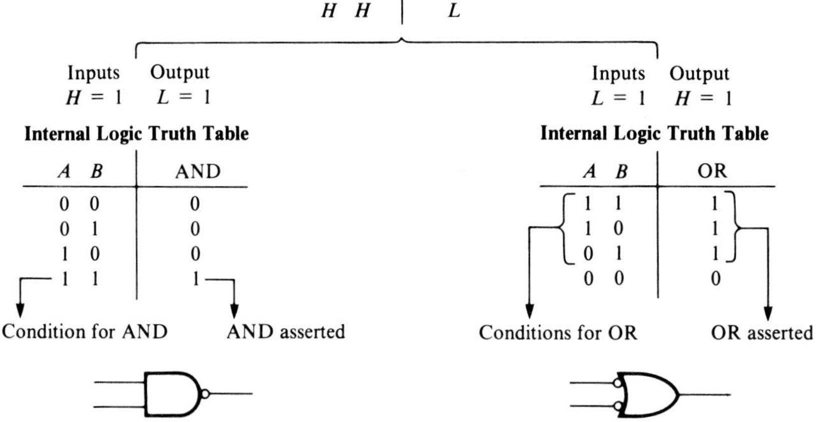

Inputs $H = 1$ Output $L = 1$

Internal Logic Truth Table

A	B	AND
0	0	0
0	1	0
1	0	0
1	1	1

Inputs $L = 1$ Output $H = 1$

Internal Logic Truth Table

A	B	OR
1	1	1
1	0	1
0	1	1
0	0	0

FIGURE 2.14
Dual use of the 7400 gate as AND or OR function

is, the asserted level of its inputs is opposite to the active level of its output. The mixed-logic nature of the 7400 gate, illustrated in Figure 2.14, allows us to think in terms of AND, OR, and NOT logic. It also eliminates the need to use DeMorgan's theorem in converting a circuit from one form to another, such as from AND-OR form to NAND-NAND form.

Figure 2.14 shows that a 7400 gate can be used either as an AND function with active-high inputs, active-low output or as an OR function with active-low inputs, active-high output. ■

■ **EXAMPLE 2.8: 7402 IC (Quad 2-Input NOR)**

The 7402 gate, like the 7400, is a mixed-logic device; that is, the asserted level of its inputs is opposite to the active level of its output. The mixed-logic nature of the 7402 gate, illustrated in Figure 2.15, allows us to think in terms of AND, OR, and NOT logic. It also eliminates the need to use DeMorgan's theorem in converting a circuit from one form to another, such as from OR-AND form to NOR-NOR form.

Figure 2.15 shows that a 7402 gate can be used either as an OR function with active-high inputs, active-low output or as an AND function with active-low inputs, active high output.

7402 Gate Function Table

Inputs		Output
A	*B*	*Y*
L	*L*	*H*
L	*H*	*L*
H	*L*	*L*
H	*H*	*L*

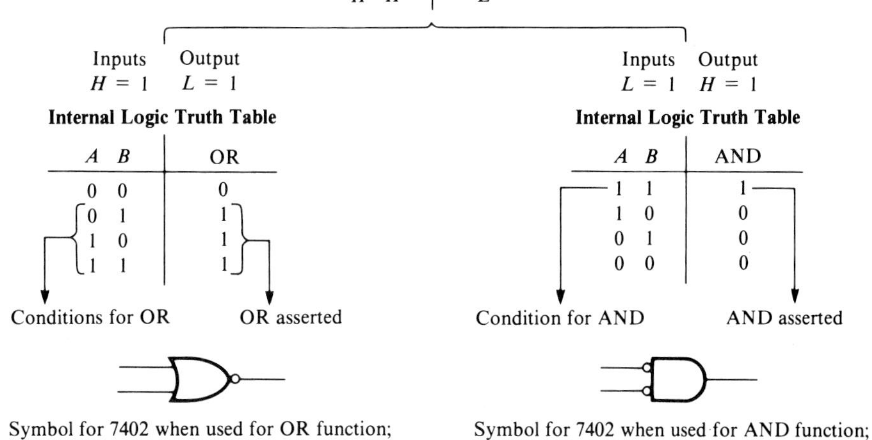

FIGURE 2.15
Dual use of the 7402 gate as OR or AND function ■

■ **EXAMPLE 2.9: 7486 IC (Quad 2-Input XOR)**

Figure 2.16 (p. 74) shows that a 7486 gate can be used either as an XOR function with active-high inputs and output (voltage H = logic 1) or as an inclusive-AND function with active-low inputs and output (voltage L = logic 1).

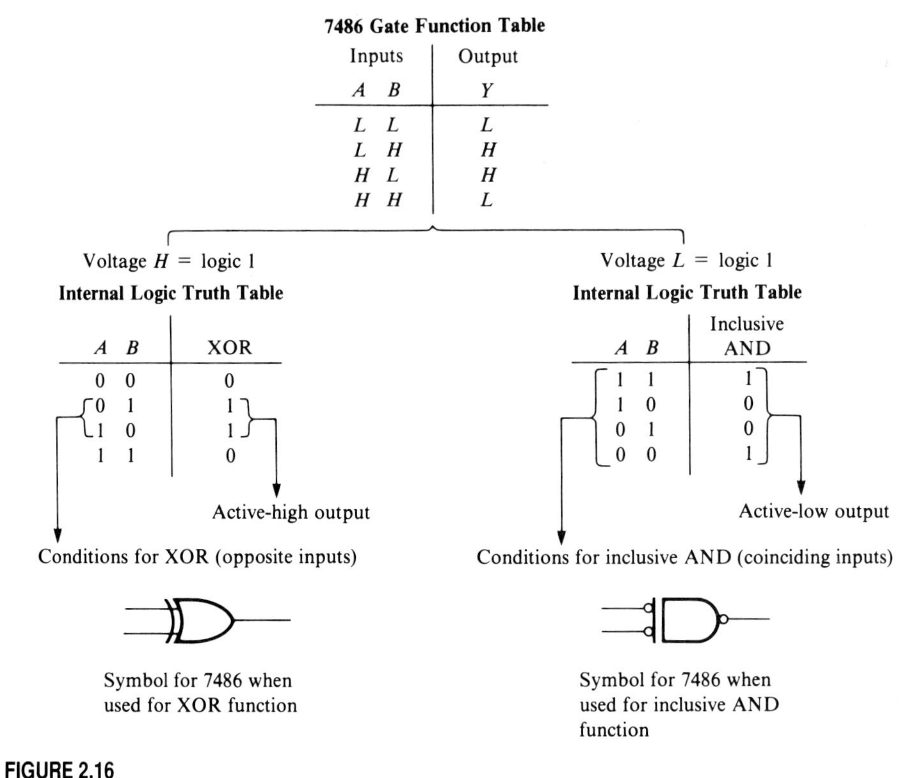

FIGURE 2.16
Dual use of the 7486 gate as XOR or Inclusive-AND function ■

■ **EXAMPLE 2.10: 7404 IC (Hex Inverter)**

Figure 2.17 (p. 75) presents the function table for an inverter, along with logic truth tables that show that in mixed logic the logic variable is unchanged by the inverter, while the assigned voltage level for assertion is reversed. In a mixed-logic system, an inverter (7404) reverses the assignment of the asserted voltage level for the associated logic variable:

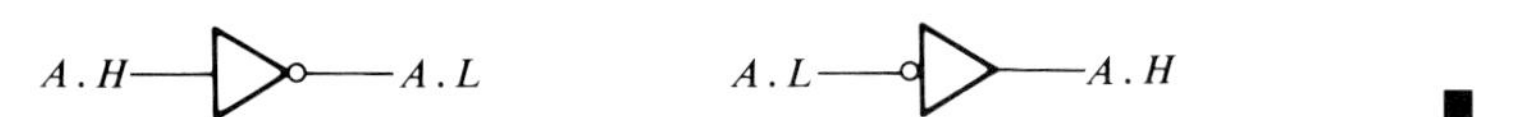

■

Other Useful Relations in Mixed Logic

Several useful mixed-logic relations are proved in the following truth tables.

1.

$A.H$	A		A'	$A'.L$
L	0	$0' = 1$	1	L
H	1	$1' = 0$	0	H

$A.H = A'.L$

2.

$A.H$	A	$A.L$
L	0	H
H	1	L

$(A.H)' = A.L$

3.

$A.H$	A		A'	$A'.H$
L	0	$0' = 1$	1	H
H	1	$1' = 0$	0	L

$(A.H)' = A'.H$

7404 Inverter Function Table

Input A	Output Y
L	H
H	L

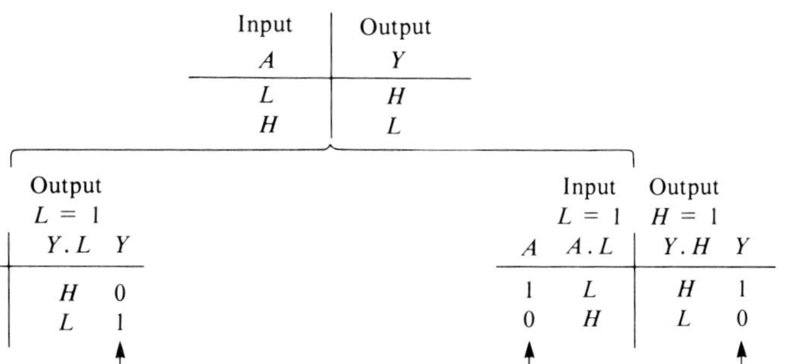

	Input $H = 1$	Output $L = 1$	
A	$A.H$	$Y.L$	Y
0	L	H	0
1	H	L	1

$A = Y$

Input $A.H$ = asserted high
Output $Y.L$ = asserted low

	Input $L = 1$	Output $H = 1$	
A	$A.L$	$Y.H$	Y
1	L	H	1
0	H	L	0

$A = Y$

Input $A.L$ = asserted low
Output $Y.H$ = asserted high

FIGURE 2.17
Inverter function table and mixed-logic truth tables for the 7404 IC

Corresponding truth tables for $A.L$ also can be constructed. A summary of these useful mixed-logic relations is provided in Table 2.14.

TABLE 2.14
Useful mixed-logic relations

1a.	$A.H = A'.L$	1b.	$A.L = A'.H$
2a.	$(A.H)' = A.L$	2b.	$(A.L)' = A.H$
3a.	$(A.H)' = A'.H$	3b.	$(A.L)' = A'.L$

Implementing Logic Equations in Mixed Logic

Four pairs of dual AND, OR symbols are used in mixed-logic diagrams, as shown in Table 2.15. Note that the dual of a given symbol is opposite in shape to the original symbol and its level indicators on all leads are reversed.

TABLE 2.15
Mixed-logic symbols

Function			
AND	OR	Gate	Type
		7408	AND
		7432	OR
		7400	NAND
		7402	NOR

In a mixed-logic system, an inverter (7404) reverses the assignment of the asserted voltage level for the associated logic variable. However, the 7404 performs no logic function in a mixed-logic system since the output A is the same as the input A.

Analysis of a Mixed-Logic Gate Network

In mixed logic, a **bubble** (o) on a gate input or output line indicates that the line is asserted low, while the absence of the bubble indicates that the line is asserted high. In other words, the bubble is used as an active-low ($L = 1$) level indicator. Remember that a bubble denotes logic inversion (complementation) in positive logic.

A logic equation for which the polarities of the inputs and output are specified can be implemented in a mixed-logic system (using the relations of Table 2.14 and the gate symbols of Table 2.15). An example of a gate network is presented in Example 2.11 to illustrate how analysis is accomplished using mixed-logic methodology.

■ **EXAMPLE 2.11**

A switching function represented by the logic equation

$$F = (A \cdot B + C') \cdot D' \cdot E$$

can be implemented in mixed logic, as shown in Figure 2.18. Each input A, B, C, D, and E has polarity H, and F has polarity L. The network gate levels are indicated by the numbers 1, 2, and 3 above the gate symbols, with the output gate being gate level 1. Note that the gate output levels are in the reverse order in which the operations are performed in evaluating the expression.

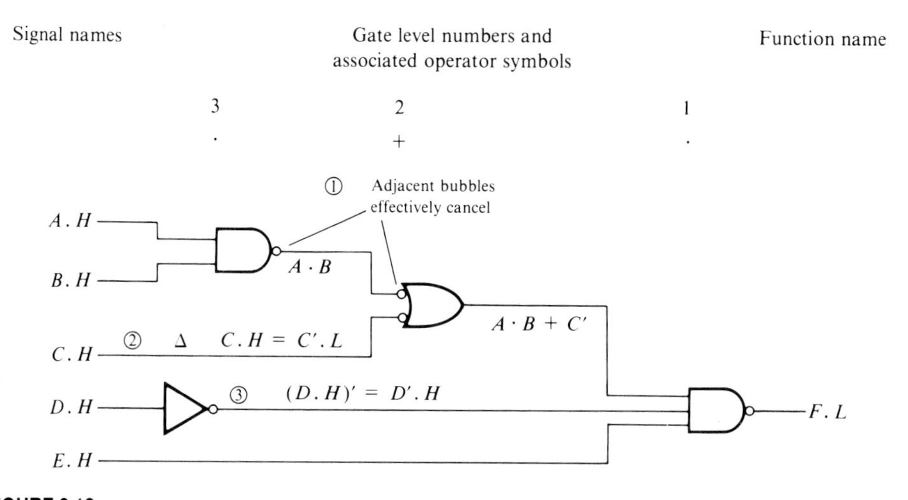

FIGURE 2.18
Mixed-logic diagram for circuit of Example 2.11 ■

The principal features of mixed-logic methodology are illustrated in Figure 2.18 as follows:

1. Adjacent bubbles (low-level indicators) effectively cancel.
2. A level mismatch (difference denoted by Δ, the Greek letter delta) between a signal polarity and its associated **gate asserted input level** (GAIL), results in a complemented variable C'.
3. An inverter, with no level mismatch between a signal polarity and its associated GAIL, results in a complemented variable D'.

We can use the features just described to develop an algorithm for designing mixed-logic gate networks. The resulting network can be implemented using any of the gates in Table 2.15, or using only NAND/NOR/NOT gates if specified.

The design of a gate network that realizes a specified switching function can be accomplished using the following 4-step procedure, which we will refer to as the **mixed-logic gate network design algorithm**:

1. Write the expression with explicit operator ($\cdot$, +, ') symbols and number the symbols in reverse order of operation (last operation = gate level 1). In complicated expressions, it may be necessary to use parentheses in the expression to indicate the order in which operations are performed.
2. Construct a framework for the logic diagram as follows:
   - **a.** Write the input signal names (variable.polarity) in a column on the left side and write the function name (variable.polarity) on the right side.
   - **b.** Write the gate level numbers in right-to-left order, with the operator symbols ($\cdot$ or +) directly below the associated number.
3. For each level number, select the appropriate gate(s) from Table 2.15 as follows:
   - **a.** Start with level $n = 1$. Use level 1 operator and the polarity of F to select the level-1 gate: **gate asserted output level** (GAOL) = polarity of F; gate function = OR for +, AND for $\cdot$.
   - **b.** Select level $n + 1$ gates:
     GAOL = gate-n GAIL; gate function = OR for +, AND for $\cdot$.
     Write the gate expression under each gate output line, and
     (1) If a level $n + 1$ gate output expression is complemented, reverse the GAOL of the level $n + 1$ gate.
     (2) If a level $n + 1$ gate input variable is complemented, use a symbol ' to flag the corresponding gate input.
     Connect level $n + 1$ gate output to inputs of level n gate. Increment n and repeat Step 3b until circuit inputs are connected to the leftmost gates of the circuit.
4. Match each circuit input signal polarity with its associated GAIL, and
   - **a.** If the circuit input signal polarity differs from the associated GAIL, flag the variable with symbol Δ (delta).
   - **b.** Insert an inverter in any signal line if that variable is flagged with either a delta or a prime, but not both.
   - **c.** Connect each circuit input to the associated gate input (or inverter).

We will use the mixed-logic gate network design algorithm in Examples 2.12 through 2.15 to illustrate the design of a number of circuits, which are implemented in the corresponding logic diagrams.

■ **EXAMPLE 2.12**

Implement the logic equation

$$F = (A \cdot B') + (C + D)$$

where each input A, B, C has polarity H, input D has polarity L, and F has polarity L. Use NAND or NOR gates.

1. Write the expression and number the operation symbols

$$F = (A \underset{2}{\cdot} B') \underset{1}{+} (C \underset{2}{+} D)$$

2. Construct a framework for the logic diagram as shown in Figure 2.19.

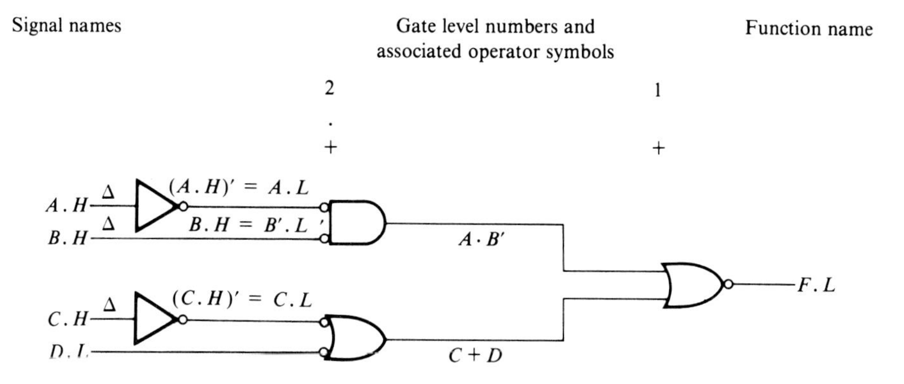

FIGURE 2.19
Mixed-logic diagram for circuit of Example 2.12

3. Select appropriate gate(s) for each level and connect level $n + 1$ gate outputs to the associated level n gate input.
4. Match each circuit input signal polarity with the polarity of its gate input. Connect the associated leads. ■

■ **EXAMPLE 2.13**

Implement the logic equation

$$F = [(A \cdot B)' \cdot (C + D)] + E$$

where each input A, B, C, D, and E has polarity H, and F has polarity H. Use NAND or NOR gates.

1. Write the expression and number the operation symbols

$$F = [(A \underset{3}{\cdot} B)' \underset{2}{\cdot} (C \underset{3}{+} D)] \underset{1}{+} E$$

2. Construct a framework for the logic diagram as shown in Figure 2.20.

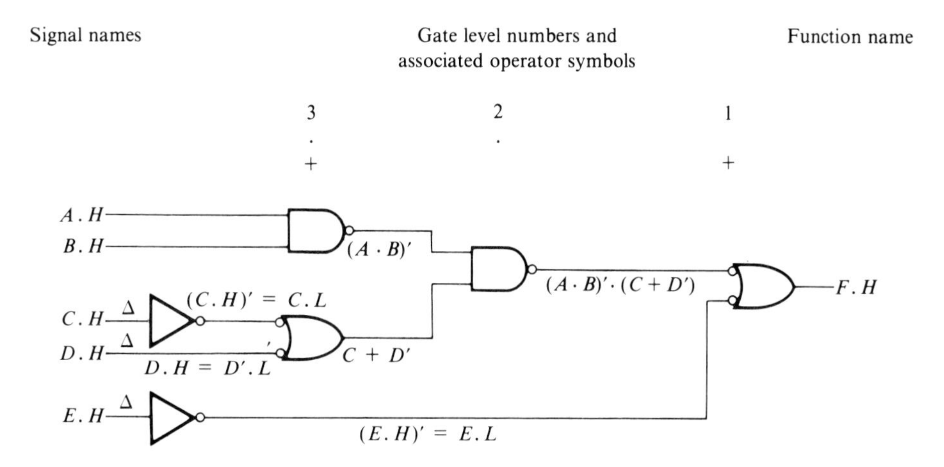

FIGURE 2.20
Mixed-logic diagram for circuit of Example 2.13

3. Select appropriate gate(s) for each level and connect level $n + 1$ gate outputs to the associated level n gate input.

4. Match each circuit input signal polarity with the polarity of its gate input. Connect the associated leads. ■

■ **EXAMPLE 2.14**

Implement the logic equation

$$F = (A + B \cdot C') \cdot (D + E)$$

where each input A, C, E has polarity H, each input B and D has polarity L, and F has polarity H. Use NAND or NOR gates.

1. Write the expression and number the operation symbols

$$F = (A + B \cdot C') \cdot (D + E)$$
$$\quad 2 \quad 3 \quad 1 \quad 2$$

2. Construct a framework for the logic diagram as shown in Figure 2.21.

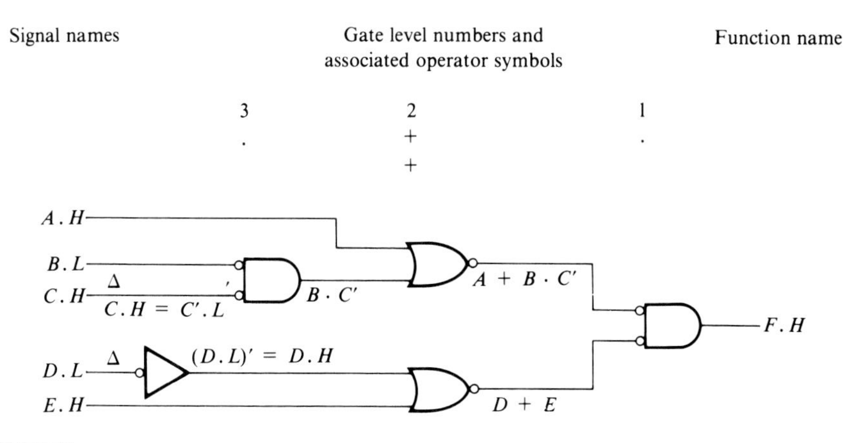

FIGURE 2.21
Mixed-logic diagram for circuit of Example 2.14

3. Select appropriate gate(s) for each level and connect level $n + 1$ gate outputs to associated level n gate input.
4. Match each circuit input signal polarity with the polarity of its gate input. Connect the associated leads. ■

■ **EXAMPLE 2.15**

Implement the logic equation

$$F = [(D + E') \cdot C + B'] \cdot A$$

where each input $A, B, C, D,$ and E has polarity H, and F has polarity H. Use NAND or NOR gates.

1. Write the expression and number the operation symbols in reverse order (last operation = gate level 1).

$$F = [(D + E') \cdot C + B'] \cdot A$$
$$\quad 4 \quad 3 \quad 2 \quad 1$$

2. Construct a framework for the logic diagram as shown in Figure 2.22.

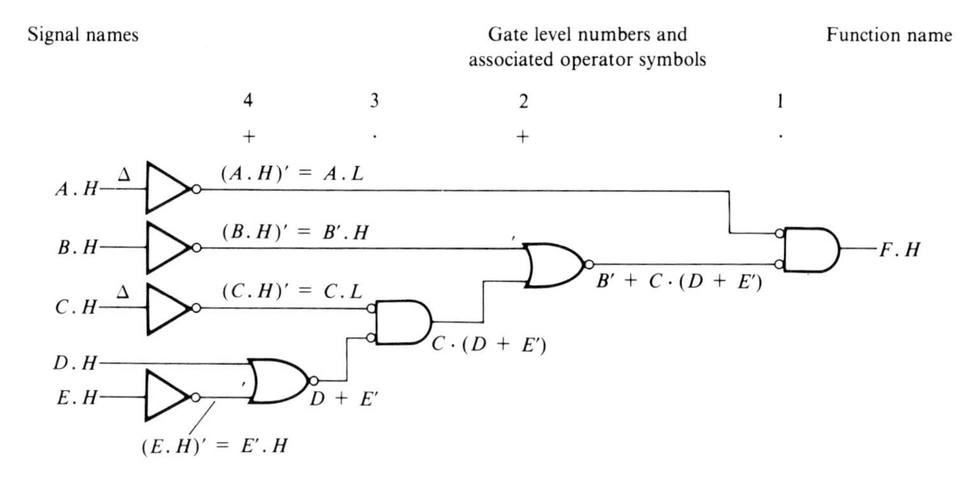

FIGURE 2.22
Mixed-logic diagram for circuit of Example 2.15

3. Select appropriate gate(s) for each level and connect level $n + 1$ gate outputs to associated level n gate input.
4. Match each circuit input signal polarity with the polarity of its gate input. Connect the associated leads. ■

It should be noted that the mixed-logic symbols in Table 2.15 can also be used in implementing logic equations in a fixed-logic system, that is, in a system in which the polarity is the same for all of the inputs and the output.

2.7 REALIZATION OF APPLICATION FUNCTION MODULES AND DEVICES

One of the simplest application functions in digital logic is the addition of two 1-bit numbers. In this section we describe how this function can be realized using logic gates and how this function can be used as a building block to realize the application function for adding three 1-bit numbers.

Realization of Application Function Modules Using Gates

In Section 2.3, the problem of adding two 1-bit binary numbers x and y was used to illustrate how the design of a simple computing device can be translated into a logic model. The procedure for solving digital-logic design problems is illustrated by the design of a device that adds two 1-bit numbers x and y.

Design of a Half Adder

Step 1: Problem Statement. Design a combinational logic circuit that realizes the addition of two 1-bit numbers x and y represented by the equation

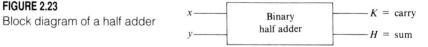

$$x \times 2^0 + y \times 2^0 = K \times 2^1 + H \times 2^0$$

The resulting circuit is a digital logic device called a *half adder* (HA).

Step 2: Conceptualization. The process of adding two 1-bit numbers x, y produces a carry K and sum H defined by the following arithmetic addition table:

x		y		K	H
0	+	0	=	0	0
0	+	1	=	0	1
1	+	0	=	0	1
1	+	1	=	1	0
				Carry K	Sum H

The arithmetic addition table can be converted to a logic truth table by replacing the numeric variables x, y, K, and H with corresponding logic variables:

x	y	K	H
0	0	0	0
0	1	0	1
1	0	0	1
1	1	1	0

The half-adder carry and sum functions can be represented by the logic equations:

$$K = x \cdot y$$
$$H = x' \cdot y + x \cdot y'$$

The binary half adder has two 1-bit inputs x and y and two 1-bit outputs K and H. The binary half adder can be represented by a black-box diagram, as illustrated in Figure 2.23.

FIGURE 2.23
Block diagram of a half adder

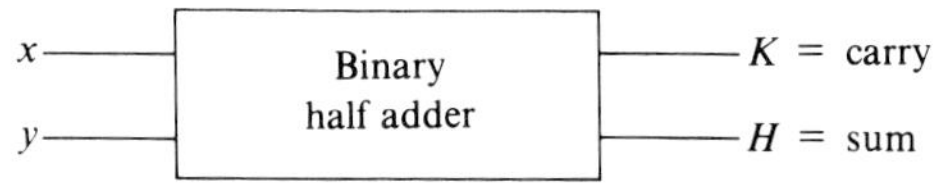

Step 3: Solution/Simplification. Simplify the Boolean equations for K and H. The expression $x \cdot y$, representing the carry function, is already in its simplest form. The expression $x' \cdot y + x \cdot y'$, representing the sum function, can be replaced by $x \oplus y$:

$$K = x \cdot y$$
$$H = x \oplus y$$

Step 4: Realization. The half-adder sum and carry functions can be realized using logic gates, as shown in Figure 2.24. Use a logic simulator to realize the half adder shown in the figure. Verify that the device operates in accordance with the design specifications described in the Problem Statement. This can be accomplished by testing the half adder for all possible combinations of the values for the inputs x and y, and verifying that the outputs K and H are correct for each input combination.

FIGURE 2.24
Logic diagram of half adder

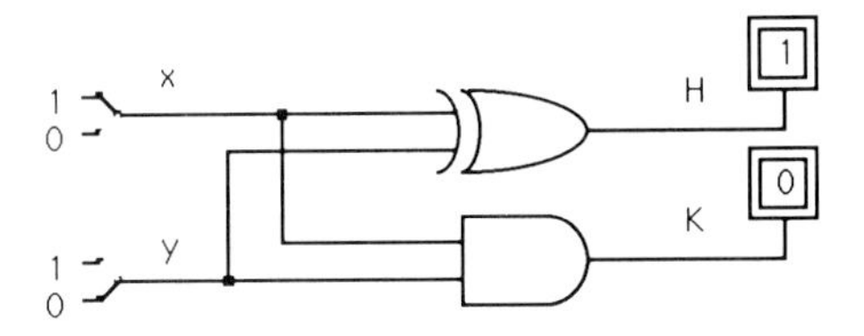

Realization of Higher-Level Devices Using Function Modules

A computer must be able to perform the arithmetic operations of addition, subtraction, multiplication, and division. These operations can be accomplished by higher-level digital logic devices such as the arithmetic unit to be described in Chapter 4.

Basic application **function modules** may be used as **building blocks** to implement higher-level digital logic devices. For example, the binary half adder can be used as a building block to implement a binary **full adder** (FA), as described in the following example design problem.

Design of a Full Adder

Step 1: Problem Statement. Design a combinational logic device, called a *1-bit full adder*, that adds two 1-bit binary numbers x and y and a 1-bit carry-in c. A 1-bit full adder can be used to compute the sum and carry of each stage of addition of two multibit numbers X and Y. The following example illustrates the binary addition of $X = 0111$ and $Y = 0110$.

	3210		Stage of adding 4-bit numbers
$c_3c_2c_1c_0$	1100		Carries
$X = x_3x_2x_1x_0$	0111	= 7	
$Y = y_3y_2y_1y_0$	0110	= 6	
$S_3S_2S_1S_0$	1101	=13	Sum
$C_4C_3C_2C_1$	0110		Carry

Each stage of the addition has three 1-bit inputs (c, x, and y) and two 1-bit outputs C and S.

Step 2: Conceptualization. A binary addition table can be constructed to illustrate the eight possible combinations of input values of the carry bit c and the x and y bits at stage i of the binary addition process. Addition of the 3 inputs (for any stage of the process) results in a sum S and a carry C. The binary addition table (in numeric variables) is as follows:

Inputs			Carry	Sum
c	x	y	C	S
0	0	0	0	0
0	0	1	0	1
0	1	0	0	1
0	1	1	1	0
1	0	0	0	1
1	0	1	1	0
1	1	0	1	0
1	1	1	1	1

The arithmetic addition table can be interpreted as a logic truth table by replacing the numeric variables c, x, y, C, and S with corresponding logic variables.

A 1-bit full adder can be represented by a black box that has 3 inputs—c, x, y—and 2 outputs—C and S—as illustrated in Figure 2.25.

FIGURE 2.25
Black-box diagram of 1-bit full adder

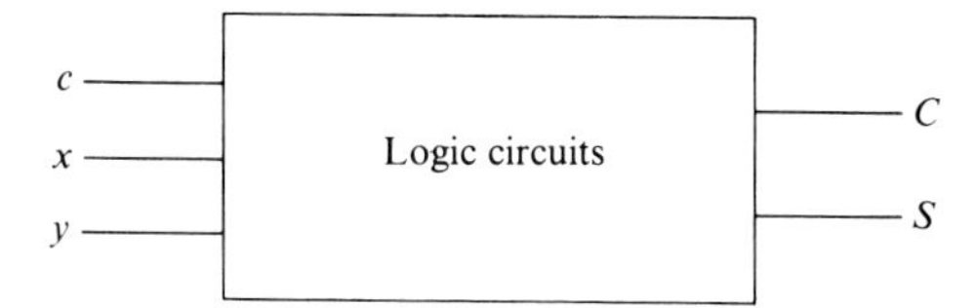

Step 3: Solution/Simplification. From the truth table, the following canonical SOP equations are obtained for the carry and sum functions of the binary 1-bit full adder. (The AND operation symbol (·) can be omitted; juxtaposition of variables then indicates the AND operation.)

$$C = c'xy + cx'y + cxy' + cxy$$
$$S = c'x'y + c'xy' + cx'y' + cxy$$

Without any simplification of these SOP expressions, the corresponding circuits could be implemented using basic logic gates. However, it would be advantageous to implement these functions using simpler, yet equivalent, circuits. Hence we will simplify the Boolean equation for C and S.

The expression for C can be simplified as follows:

$$\begin{aligned} C &= cxy + c'xy + cxy' + cx'y \\ &= (c + c')xy + c(xy' + x'y) \\ &= xy + c(x \oplus y) \end{aligned}$$

The expression for S can be simplified as follows:

$$\begin{aligned} S &= c'(x'y + xy') + c(x'y' + xy) \\ &= c'(x \oplus y) + c(x \oplus y)' \\ &= c \oplus (x \oplus y) \end{aligned}$$

Step 4: Realization. The half adder can be used as a building block to implement a 1-bit full adder. To do so, we substitute the half-adder functions, $K = x \cdot y$ and $H = x \oplus y$, into the equations for C and S:

$$C = x \cdot y + c \cdot (x \oplus y) = K + c \cdot H$$
$$S = c \oplus (x \oplus y) = c \oplus H$$

The resulting equations

$$C = K + c \cdot H$$
$$S = c \oplus H$$

define the full-adder carry and sum functions in terms of the half-adder carry and sum functions. Therefore, the full adder can be implemented using half adders as building blocks, as illustrated in Figure 2.26.

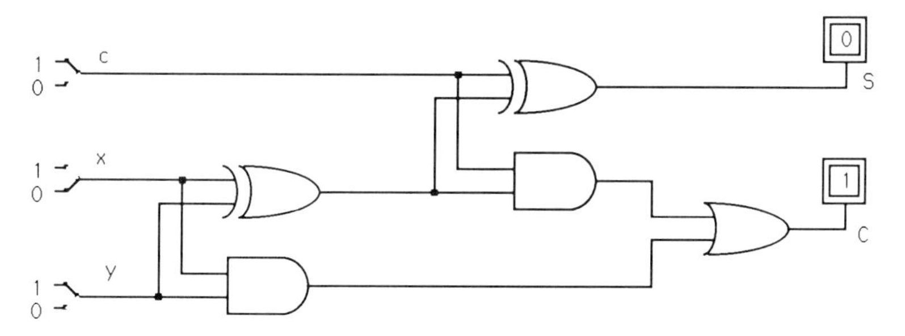

FIGURE 2.26
Using half adders to implement a full adder

TTL (transistor-transistor logic) ICs are used extensively in a wide range of applications. TTL ICs are commercially available in the 74xx series devices, which include a variety of small-scale integration (SSI) and medium-scale integration (MSI) ICs. A TTL User's Guide is presented in Appendix C.

2.8 SUMMARY

This chapter began with a 4-step problem-solving procedure (PSP) that can be used to solve a wide range of problems in different applications areas, such as digital logic, physics, mechanics, and so forth. The problem-solving procedure consists of

Step 1: Problem statement
Step 2: Conceptualization (modeling)
Step 3: Solution/simplification
Step 4: Realization of the solution by either simulation or physical implementation

This 4-step procedure will be used in this text to design combinational logic circuits, sequential logic circuits, digital systems, and computers.

In Section 2.3, we described how different types of digital-logic design problems could be modeled and presented the basic concepts of logic and logic functions. In Section 2.4, we extended the realization of logic functions using switching circuits and developed an algebra of switching networks.

In Sections 2.5 and 2.6, we developed Boolean (switching) algebra, the principal theoretical tool used in digital logic analysis and design. Then, we showed how Boolean algebra is applied to digital logic circuits and described the realization of logic functions in fixed (positive or negative) logic and mixed logic using physical logic gates. We then developed an algorithm to design mixed-logic gate networks and used the algorithm to illustrate the design of a number of representative circuits.

Finally, in Section 2.7, we described an approach for realizing function modules using logic gates as building blocks. When function modules and gates are used as building blocks, higher-level devices can be realized.

In this chapter, we also began to use a logic simulation program to realize physical digital logic circuits. For example, Figures 2.24 and 2.26 were generated by Capilano Computing's logic simulation program, Designworks.

KEY TERMS

Conceptual model
Half adder (HA)
Logic truth table
Symbolic logic
Logic product
Logic sum
Logic function
Logic operation
Cartesian product
Series
Parallel
Logic expression
Positive logic system
Negative logic system
Logic gates
Axiom
Switching algebra
Theorem
DeMorgan's generalized law
Switching function
Domain
Range
Literal
Minterm
Maxterm
Canonical sum-of-products (SOP)
Canonical product-of-sums (POS)
Asserted
Complement operation
Functionally complete
Dual function
Logical adjacency theorem
Total induction
Combinational logic circuits

Mixed logic system
Logic identifier
Polarity indicator
Digital logic gates
Integrated circuits (ICs)
Bubbles
Gate asserted input level (GAIL)
Mixed-logic gate network design algorithm
Gate asserted output level (GAOL)
Function modules
Building blocks
Full adder (FA)

EXERCISES

General

1. Given the Boolean function

$$F = A'B'C + A'BC' + A'BC + ABC' + ABC$$

determine a minimal sum-of-products expression for F.

2. Simplify $F = \Sigma(3, 5, 6, 7)$.

3. Find a simplified expression for

$$F = A'B'C + A'BC + ABC + ABC'$$

4. Convert the following SOP expression for

$$F = A'B + B'C + AC$$

to a canonical SOP form.

5. Convert the following SOP expression for function

$$F = A + BC$$

to a canonical SOP form.

6. Given a Boolean function

$$F = (A + C) \cdot (B + C)$$

convert the expression to a canonical POS form.

7. Convert the SOP expression for

$$F(A, B, C) = A'B' + A'C$$

to a canonical SOP form.

8. Convert the POS expression for

$$F = (A + B') \cdot (A + C)$$

to a canonical POS form.

9. Given $F = A \cdot B' + B \cdot C$, find the dual function G.

10. Given $F = (A + B) \cdot (B' + C)$, find the dual function G.

11. Given $F = A' \cdot B + B' \cdot C$, find an expression for its complement F' by taking the dual of F and then complementing each literal in the resulting dual expression.

12. Find the complement of function

$$F = A'B'C + A'BC + ABC + ABC'$$

by interchanging the 0s and 1s in the truth table function values of F.

13. Use DeMorgan's theorem to convert $F = A + BC + CD'$ to NAND-NAND form.

14. Use DeMorgan's theorem to convert $F = A \cdot (B + C) \cdot (C + D')$ to NOR-NOR form.

15. Prove the Associative law of Boolean algebra using the method of total induction (by truth tables).

16. Prove the Idempotent theorem: $A + A = A$.

17. Prove the consensus theorem: $AB + A'C + BC = AB + A'C$.

Design/Implementation

1. Use AND and OR gates to implement switching function

$$F = AC' + BC$$

where A, B, and C are active-high inputs, and F is an active-high output.

2. Use NAND/NOR gates to implement switching function

$$F = A'C + AB'$$

where A, B, and C are active-high inputs, and F is an active-low output.

3. Use NAND/NOR gates to implement the switching function

$$F = AB + BC$$

where A, B, and C are active-high inputs, and F is an active-high output.

4. Use NAND/NOR gates to implement the switching function

$$F = (A + B) \cdot (B + C)$$

where A, B, and C are active-high inputs, and F is an active-high output.

5. Use NAND/NOR gates to implement the switching function

$$F = (AC)' + B$$

where A, B, and C are active-high inputs, and F is an active-low output.

6. Use NAND/NOR gates to implement the switching function

$$F = AC' + BC$$

where A, B, and C and active-high inputs, and F is an active-low output.

7. Use NOR gates to implement switching function

$$F = (A + B') \cdot (A' + C)$$

where A, B, and C are active-high inputs, and F is an active-high output.

8. Use any gates to implement switching function

$$F = [(A \cdot B)' \cdot (C + D)] + E$$

where A, B, C, D, and E are active-high inputs, and F is an active-low output.

9. Use NAND gates to implement the switching function

$$F = AB' + CD$$

where A, B, C, and D are active-high inputs, and F is an active-high output.

10. Use NAND gates to implement the switching function

$$F = A + BC' + A'D$$

where A, B, C, and D are active-high inputs, and F is an active-high output.

11. Three judges (A, B, and C) preside over a court. A person tried by the court can be convicted only by a majority vote of the judges. Find a Boolean function G that indicates a guilty verdict, assuming that no judge abstains on a vote.

12. Three judges (A, B, and C) preside over a court. A verdict can be rendered only if all three judges are in agreement, either for a guilty or for an innocent verdict.

- **a.** Find a Boolean function V that has value 1 when a verdict is reached. Use 3 switches and NAND gates to activate V when the three judges are in agreement.
- **b.** Use 3 switches and NAND gates to implement a decision circuit that activates I for an innocent verdict and G for a guilty verdict.

13. A warehouse has three accesses: a door, a window, and a skylight. The skylight can only be reached by climbing a pull-down ladder. Design a simple alarm circuit that will activate when a switch E is closed and one or more of the accesses is opened.

14. Three coins A, B, and C are tossed into a fountain.

- **a.** Write an expression for a Boolean function F that has value 1 when the toss results in an odd number of tails.
- **b.** Write an expression for a Boolean function G that has value 1 when the toss results in an even number of heads.

15. Design a combinational logic circuit with 4 inputs (A, B, C, and D) and 1 output F, where $F = 1$ when $A \neq C$ and $B = D$. Write F in canonical SOP form and in simplified SOP form.

16. Design a combinational logic circuit with 4 inputs (A, B, C, and D) and 1 output G, where $G = 1$ when an even number of inputs = 1. Write G in canonical SOP form and in simplified SOP form.

17. A farmer must ferry a fox, a rooster, and a bag of corn across the river in a boat. If he uses both hands to row the boat, how many trips must he make to ensure that the fox does not eat the rooster and the rooster does not eat the corn?

3 Methods for Simplifying Boolean Expressions

OBJECTIVES

After you complete this chapter, you will be able

- ☐ To graphically represent Boolean (switching) functions using Karnaugh maps, minterm ring maps, and variable-entered maps
- ☐ To determine implicants, prime implicants, and essential prime implicants of a switching function
- ☐ To realize simpler, less costly, logically equivalent circuits by simplifying the Boolean expression
- ☐ To use map techniques for simplifying Boolean expressions
- ☐ To use algorithmic methods for simplifying Boolean expressions in any number of variables

3.1 INTRODUCTION

The principal objectives to be observed in the design and implementation of any digital logic device include the following: (1) The device must function properly in accordance with the design specifications; (2) the design implementation should be cost effective, that is, the digital logic circuit should be as simple as possible and contain a minimum number of components while satisfying the design specifications; and (3) the operation of the device should be fast and reliable throughout the range of specified conditions.

Trade-offs in the design of a digital logic device are frequent, especially the trade-offs between speed and complexity. The canonical sum-of-products (SOP) form of a Boolean (logic) expression, obtained directly from a truth table, can usually be reduced to a minimal sum that contains fewer terms with fewer literals in many terms. In this chapter, we will look at combinational-logic simplification methods that can be used to obtain the simplest expression for a given function. The resulting expression can be used to implement a logically equivalent circuit that is simpler, hence, more economical, more reliable, and often faster than a circuit implemented using the original expression in canonical form.

In this chapter, we will become familiar with map techniques and algorithmic methods for simplifying logic expressions. These map techniques and algorithmic methods are mechanical and, hence, easier to use and less error prone than the algebraic simplification techniques described in Chapter 2. Map techniques are graphical representations of switching functions used as visual aids for simplifying logic expressions. In Sections 3.2 through 3.5, we describe the Karnaugh map method for expressions in 2 to 6 variables.

In Section 3.5, we also describe a minterm-ring method that combines map techniques and an algorithmic method for the systematic reduction of logic expressions in 5 or more variables. Since most logic minimization currently is accomplished with software tools, the minterm-ring method is important because it provides a contemporary algorithmic method for simplifying expressions and provides a three-dimensional perspective for visual identification of prime implicants and essential prime implicants of switching functions. The derivation of the minterm-ring algorithm also provides an excellent illustration of how algorithms are developed.

The Quine–McCluskey tabular method is a classical algorithmic method for simplifying Boolean expressions. This method is described in Section 3.6. Finally, in Section 3.7, we describe the use of variable-entered maps to plot an n-variable switching function in less than n dimensions.

In each section of this chapter, design tools and techniques are illustrated with examples of simplifying logic expressions and implementing practical combinational logic circuits. These practical examples include binary adders, error detectors, parity generators, and BCD divide-by-n circuits, which can be used as building blocks for implementing higher-level digital devices.

3.2 KARNAUGH MAPS IN 2 VARIABLES

The implementation of a simplified logic expression results in a simpler, more economical, functionally equivalent circuit. A logic expression often can be simplified by factoring or by substitution, in much the same way that standard algebraic expressions can be simplified. However, it is not always known in what order the laws

and theorems should be applied in the algebraic simplification of a Boolean expression. Also, it is not always clear when a minimal expression has been produced. For these reasons, design tools were developed to aid in the reduction of Boolean expressions. One of the principal design tools is the* Karnaugh map technique *(K-map) developed by Veitch and refined by Maurice Karnaugh.

A switching function $F: S \times S \rightarrow S$ maps each element of the domain $S \times S = \{(0, 0), (0, 1), (1, 0), (1, 1)\}$ to only one element of $S = \{0, 1\}$. For example, the function $F(A, B) = 0 \cdot A' \cdot B' + 1 \cdot A' \cdot B + 1 \cdot A \cdot B' + 1 \cdot A \cdot B$ has the following mapping:

$$S \times S = \left\{ \begin{array}{cc} (A, B) & F \\ (0, 0) \rightarrow & 0 \\ (0, 1) \rightarrow & 1 \\ (1, 0) \rightarrow & 1 \\ (1, 1) \rightarrow & 1 \end{array} \right\} = \text{Elements of } S$$

Note that the coefficient (0 or 1) of the product term $a \cdot b$ is the value assigned to $F(a, b)$. Therefore, a truth table is a table that lists each product-term combination and its corresponding function value F. The tables below illustrate the mapping $F: S \times S \rightarrow S$ and the corresponding truth table.

	Mapping				Truth Table	
	A	*B*	*F*		*A B*	*F*
$S \times S =$	0	0	0	= Elements of S	$A' \cdot B'$	0
	0	1	1		$A' \cdot B$	1
	1	0	1		$A \cdot B'$	1
	1	1	1		$A \cdot B$	1

The four rows of a 2-variable truth table represent the standard products: $A' \cdot B'$, $A' \cdot B$, $A \cdot B'$, and $A \cdot B$. Each product term that has a 1 in the function column F is a minterm of the function defined in the truth table. The function can, therefore, be expressed in canonical SOP form as the logical sum of the minterms of the function.

$$F(A, B) = A' \cdot B + A \cdot B' + A \cdot B$$

If variables A and B are considered as axes in two dimensions, then the function domain $S \times S = \{(0, 0), (0, 1), (1, 0), (1, 1)\}$ divides the 2-variable universe into 4 **cells** of a map, as shown in Figure 3.1. Cell (a, b) is the cell in row a and column b of the map.

FIGURE 3.1
Map of elements of function domain $S \times S$

$S \times S =$ Cartesian product
$A = S, B = S$
$S = \{0, 1\}$

$A \backslash B$	0	1
0	(0, 0)	(0, 1)
1	(1, 0)	(1, 1)

A switching function $F: S \times S \rightarrow S = F(A, B)$ maps each domain element (a, b) into a range element $F(a, b)$ in S, where $a \in A$ and $b \in B$. If the function is represented in map form, cell (a, b) contains the function value $F(a, b)$. Figure 3.2 is a K-map of the function

$$F(A, B) = 0 \cdot A' \cdot B' + 1 \cdot A' \cdot B + 1 \cdot A \cdot B' + 1 \cdot A \cdot B$$
$$= A' \cdot B + A \cdot B' + A \cdot B$$

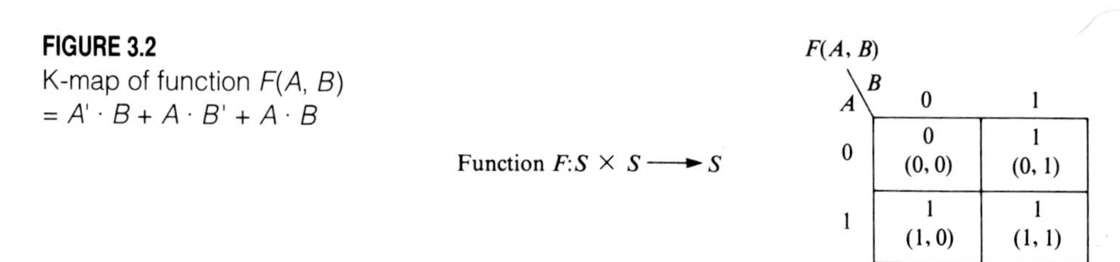

FIGURE 3.2
K-map of function $F(A, B) = A' \cdot B + A \cdot B' + A \cdot B$

The K-map is a special case of the Venn diagram used in the algebra of sets. In set theory terms, the entire K-map represents the cartesian product $S \times S = \{(0, 0), (0, 1), (1, 0), (1, 1)\}$, which is the **domain** of the function $F: S \times S \rightarrow S$. Domain element (a, b) is cell (a, b), located in row a and column b. Cell (a, b) contains the function value $F(a, b)$, the coefficient of product term $a \cdot b$ in the canonical SOP expression:

$$\text{Function } F(A, B) = \text{the sum of product terms with coefficient } 1$$

The map rows may be labeled A' and A, and the columns B' and B. The product term corresponding to each element of $S \times S$ is illustrated in Figure 3.3. The row labels A' and A represent the values 0 and 1 of variable A. The column labels B' and B represent the values 0 and 1 of variable B.

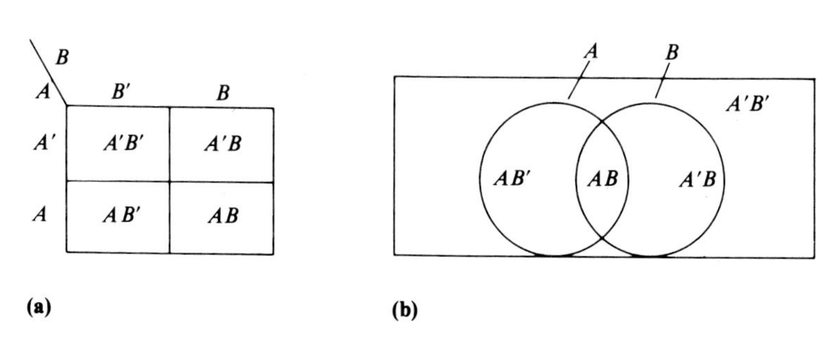

FIGURE 3.3
Product terms corresponding to each element of $S \times S$.
(a) K-map (b) Venn diagram

K-map cells that have a side in common are logically adjacent. The product terms of two logically adjacent cells differ in only 1 variable. Consequently, the Boolean algebra logical adjacency theorem ($X \cdot Y + X \cdot Y' = X$) can be used to replace the logical sum of 2 product terms by a single term in which 1 variable has been eliminated. For example, the logical sum of product terms $A' \cdot B$ and $A \cdot B$ is

$$A' \cdot B + A \cdot B = B$$

This is accomplished in the K-map by combining two logically adjacent cells into a 2-cell pair. Algebraically A is a common factor of each row-A product term ($A \cdot B'$, $A \cdot B$), and B is a common factor of each column-B product term ($A' \cdot B$, $A \cdot B$).

3.3 KARNAUGH MAPS IN 3 VARIABLES

The rows of a truth table are numbered in binary sequence. For a function of 3 variables, the rows are numbered 000, 001, 010, 011, 100, 101, 110, and 111. Notice that adjacent terms in the binary sequence sometimes differ in more than one bit position. To take advantage of the logical adjacency theorem ($XY + XY' = X$), the rows of a truth table should be mapped so that adjacent map cells will be occupied by 2 terms that differ in only 1 variable. Therefore, the columns are arranged in Gray-code order so that adjacent columns differ in only 1 variable.

The domain of $F: S \times S \times S \rightarrow S$ is $S \times S \times S = \{(0, 0, 0), (0, 0, 1), \ldots, (1, 1, 1)\}$. Each of the elements (A, B, C) in set $S \times S \times S$ can be represented by a K-map, as illustrated in Figure 3.4.

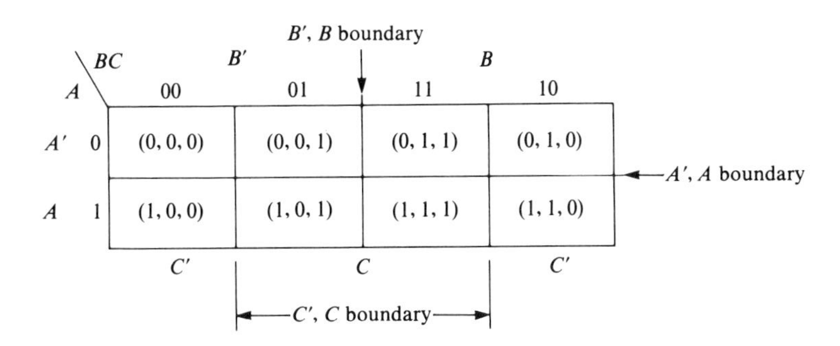

FIGURE 3.4
Cell locations of a K-map in 3 variables

Adjacent cells in the map in Figure 3.4 differ in only 1 variable. If the rectangular map is formed into a cylinder by joining the left and right ends, then the leftmost and rightmost cells in a row are physically adjacent and also differ only in 1 variable (B). Therefore, the map can be interpreted as a K-map in which physically adjacent cells are logically adjacent. In a 3-variable K-map each cell can be identified by a standard product term, as illustrated in Figure 3.5.

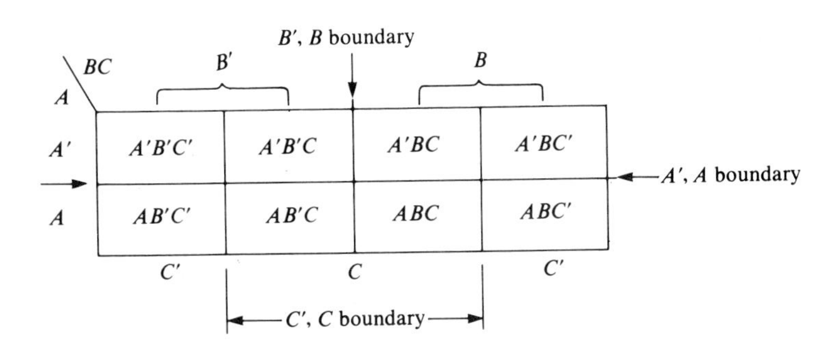

FIGURE 3.5
K-map with cells identified by product term

As in a 2-variable K-map, the Boolean property $XY + XY' = X$ can be used to combine 2 logically adjacent cells into a 2-cell block that produces a product term in which the boundary-crossing variable is eliminated. For example, if logically adjacent cells ABC and ABC' are combined into a 2-cell block, then that block is represented by the product term AB, as follows:

$$(AB)C + (AB)C' = AB$$

Each truth-table entry for which function $F = 1$ is called a **minterm**. A minterm is a logical product in which each input variable or its complement occurs once and only once. If we represent the states of a binary variable X by a logic 0 and 1, each of the cells can be identified by a 3-bit minterm number or its decimal equivalent, as illustrated in Figure 3.6. The upper half of the map (A') consists of all cells in row 0, while the lower half (A) consists of all cells in row 1. The columns are numbered 00, 01, 11, 10 to correspond to the values of the variables B, C.

FIGURE 3.6
K-map with cells identified by minterm number

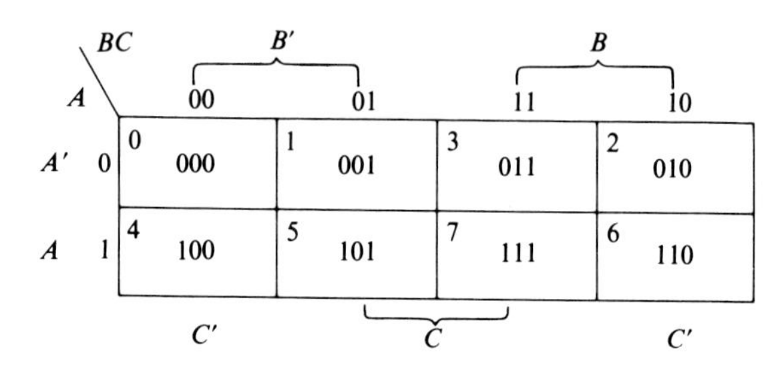

Note that the truth table inputs may be represented in binary, decimal, or as symbols:

$$0 = 000 = A'B'C',\ 1 = 001 = A'B'C,\ \ldots,\ 7 = 111 = ABC$$

■ **EXAMPLE 3.1**

Figure 3.7 illustrates the truth table and corresponding K-map of the function F, given $F = A'BC + AB'C + ABC + ABC'$.

Decimal	Inputs A	B	C	Output F
0	0	0	0	0
1	0	0	1	0
2	0	1	0	0
3	0	1	1	1
4	1	0	0	0
5	1	0	1	1
6	1	1	0	1
7	1	1	1	1

F: A \ BC	B'C'	B'C	BC	BC'
A'			A'BC 1	
A		AB'C 1	ABC 1	ABC' 1

FIGURE 3.7
Truth table and K-map for function $F(A, B, C)$ ■

Using K-Maps to Simplify SOP Expressions

A Boolean expression in canonical SOP form often can be reduced to a minimal sum that consists of fewer product terms, each of which may have fewer variables. The reduction can be accomplished by use of one or more Boolean algebra laws and theorems, including

- The idempotent law ($X + X = X$), which allows a product term to be used more than once in the same Boolean expression
- The logical adjacency theorem ($XY + XY' = X$), which eliminates 1 variable by forming the logical union of 2 logically adjacent terms
- The consensus theorem ($XY + X'Z + YZ = XY + X'Z$), which eliminates the redundant product term YZ created by the logical union of terms XY and $X'Z$
- The absorption and elimination laws: $A + AB = A$ and $A + A'B = A + B$

The following example shows how a K-map can be used as a visual aid for simplifying an SOP form Boolean expression.

EXAMPLE 3.2

Figure 3.8 illustrates the truth table and corresponding K-map of function F, given $F = A'BC + AB'C + ABC + ABC'$.

Inputs A	B	C	Output F
0	0	0	0
0	0	1	0
0	1	0	0
0	1	1	1
1	0	0	0
1	0	1	1
1	1	0	1
1	1	1	1

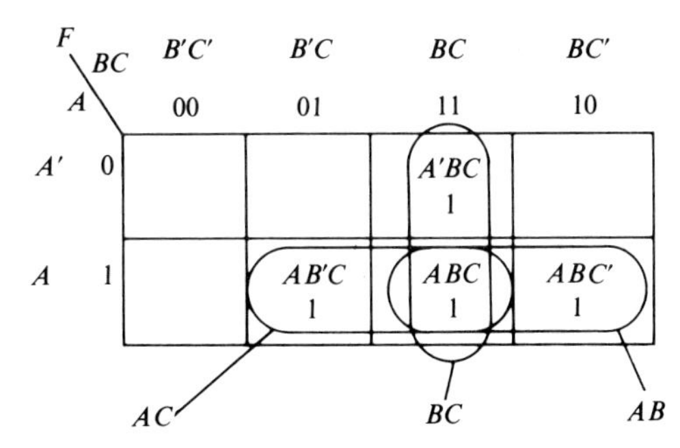

FIGURE 3.8
Truth table and K-map for function $F(A, B, C)$

The K-map shows that cell ABC is logically adjacent to each of the other 1 cells in the map. The idempotent law ($X + X = X$) allows use of a term more than once in a Boolean expression. In order to use term ABC in each of three 2-cell pairs, term ABC is "added" twice to the standard SOP expression:

$$F = A'BC + AB'C + ABC + ABC' + ABC + ABC$$

Using the commutative addition law, the terms in F can be reordered so that each pair of terms contains a variable and its complement:

$$F = (A'BC + ABC) + (AB'C + ABC) + (ABC' + ABC)$$

Finally, the logical adjacency theorem ($XY + XY' = X$) can be used to simplify the expression to a minimal SOP form

$$F = BC + AC + AB$$

Simplification of an expression by direct use of the K-map can be accomplished as follows: Each 2 logically adjacent cells are replaced by a 2-cell pair: ($A'BC$, ABC) by

BC, ($AB'C$, ABC) by AC, and (ABC', ABC) by AB. Hence, function $F = BC + AC + AB$.

If a digital logic circuit for function F were to be implemented using the original SOP expression, the circuit would require three inverters, four 3-input AND gates, and one 4-input OR gate. A much simpler circuit that contains fewer components can be implemented using the minimal SOP expression; this simplified circuit would require only three 2-input AND gates and one 3-input OR gate.

The primary objective of reducing Boolean expressions is to design digital circuits that are simpler, contain fewer components, and consequently are more cost effective. K-maps can be used as a design tool to develop many combinational-logic building blocks that are used in various digital systems. K-maps provide a visual representation of Boolean expressions, and they also provide techniques for the simplification of these expressions.

Consider a particular 3-input function defined by the following truth table:

	Inputs $A\ B\ C$	Output F
$A'B'C'$	0 0 0	0
$A'B'C$	0 0 1	0
$A'BC'$	0 1 0	1 *
$A'BC$	0 1 1	0
$AB'C'$	1 0 0	1 *
$AB'C$	1 0 1	1 *
ABC'	1 1 0	1 *
ABC	1 1 1	0

where * = minterms

Let F denote the Boolean function that has value 1 for each input combination that produces an output $F = 1$. The function F is the logical sum of the minterms (truth-table entries for which $F = 1$):

$$F = A'BC' + AB'C' + AB'C + ABC'$$

A K-map for the truth table is shown in Figure 3.9.

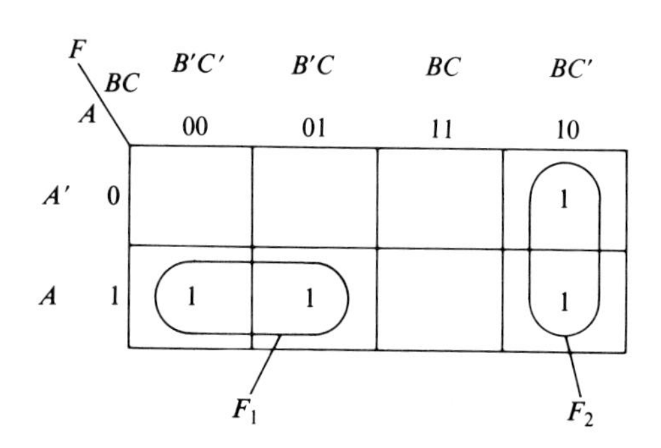

FIGURE 3.9
K-map of function with subfunctions

Each K-map cell that has value 1 is a minterm of function F. Adjacent terms in blocks of 2, 4, or 8 cells can be grouped together to compose subfunctions whose expressions can be simplified using the Boolean property $XY + XY' = X$. For example, the K-map in Figure 3.9 has subfunctions F_1 and F_2, where $F = F_1 + F_2$, and $F_1 = AB'C' + AB'C$ is composed of adjacent terms in row A and $F_2 = A'BC' + ABC'$ is composed of adjacent terms in column BC'.

If we move from any cell to an adjacent cell in the same row or column, it produces a change in the value of only 1 variable because each dimension of a K-map consists of terms ordered in Gray-code sequence.

The expression for subfunction F_1 can be simplified by combining cells $AB'C'$ and $AB'C$ into a 2-cell pair represented by product AB'. The expression for subfunction F_2 can be simplified by combining cells $A'BC'$ and ABC' into a 2-cell pair represented by product BC'. We can now write overall function F, where $F = F_1 + F_2$, in the following simplified form:

$$F = AB' + BC'$$

It should be emphasized that in adjacent cells of a K-map only 1 variable changes because of the unit distance property of the Gray code. For example, function $F_2 = A'BC' + ABC' = BC'(A' + A)$ has terms that differ only in variable A. Therefore, variable A and its complement A' are eliminated by combining 2 adjacent cells, one of which contains A and the other A'.

Simplification of a 1-Bit Full Adder

Section 2.7 described a 1-bit full adder with 3 inputs c, x, and y and 2 outputs C and S. The output functions can be represented by the following truth table:

Inputs $c\ x\ y$	Outputs $C\ S$
0 0 0	0 0
0 0 1	0 1
0 1 0	0 1
0 1 1	1 0
1 0 0	0 1
1 0 1	1 0
1 1 0	1 0
1 1 1	1 1

The Boolean expressions in canonical SOP form for C and S can be obtained directly from the truth table:

$$C = c'xy + cx'y + cxy' + cxy = \Sigma(3, 5, 6, 7)$$
$$S = c'x'y + c'xy' + cx'y' + cxy = \Sigma(1, 2, 4, 7)$$

These output functions can be represented by the K-maps shown in Figure 3.10 (p. 100).

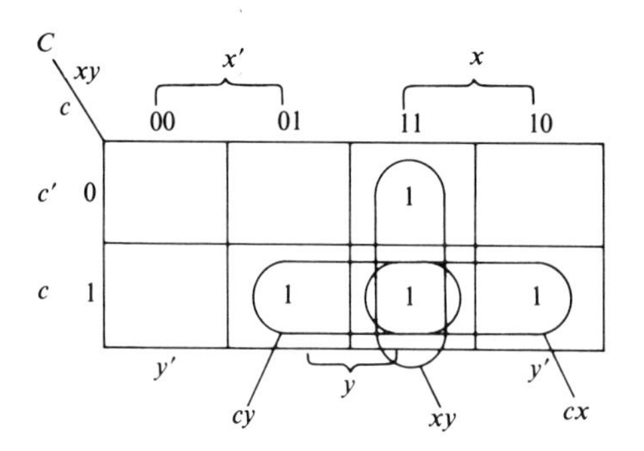

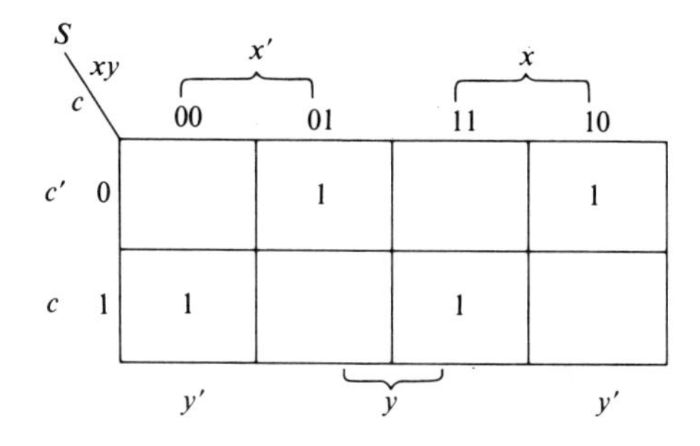

FIGURE 3.10
K-maps for full adder carry and sum

The expression for C can be simplified using the K-map by combining adjacent cell terms into product terms as follows:

$$C = xy + cx + cy$$

The expression for S can be simplified using the K-map by noting that row 0 terms contain the common factor c' while the row 1 terms contain the common factor c:

$$\begin{aligned} S &= c'(x'y + xy') + c(x'y' + xy) \\ &= c'z + cz' \end{aligned}$$

where $z = x'y + xy' = x \oplus y$.

Therefore,

$$\begin{aligned} S &= c \oplus z \\ &= c \oplus (x \oplus y) \end{aligned}$$

Using K-Maps to Produce a Minimal POS

A K-map can be used to obtain a minimal POS expression for function C'. This representation can then be converted to a minimal NOR-NOR form for C.

The full-adder carry function can be represented by the K-map shown in Figure 3.11.

FIGURE 3.11
Using blocks of 0 cells

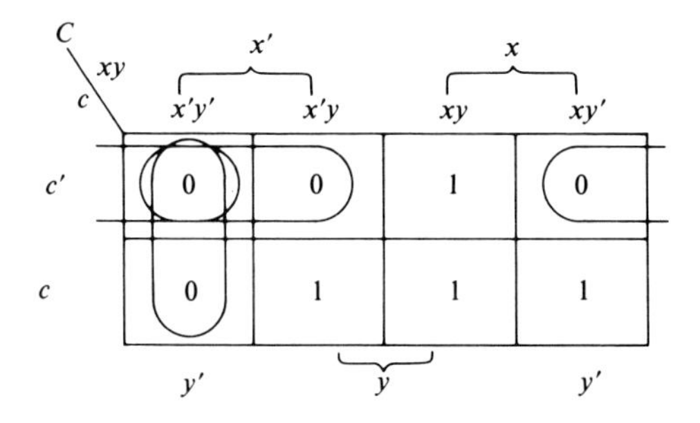

Since the carry function C is represented by the 1 cells in the map, the complement function C' is represented by the 0 cells:

$$C' = c'x'y' + c'x'y + c'xy' + cx'y'$$

Combining adjacent 0 cells into blocks, we can represent function C' in simplified form as follows:

$$C' = x'y' + c'x' + c'y'$$

Using the involution theorem and DeMorgan's theorem, we get

$$\begin{aligned} C = \{C'\}' &= \{x'y' + c'x' + c'y'\}' \\ &= (x'y')'(c'x')'(c'y')' \\ &= (x + y)(c + x)(c + y) \end{aligned}$$

This is a minimal POS expression for C. Hence, a POS expression can be read directly from the K-map.

Starting with this minimal POS form of C, we again use the involution theorem and DeMorgan's theorem to get

$$\begin{aligned} C = (C')' &= \{[(x + y) \cdot (c + x) \cdot (c + y)]'\}' \\ &= \{(x + y)' + (c + x)' + (c + y)'\}' \end{aligned}$$

This is a minimal NOR-NOR representation of the carry function C.

3.4 KARNAUGH MAPS IN 4 VARIABLES

The rows of a truth table are numbered in binary sequence. For a function of 4 variables, the rows are numbered 0000, 0001, 0010, 0011, 0100, 0101, 0110, 0111, . . . , 1111. Notice that adjacent terms in the binary sequence sometimes differ in more than one bit position. To take advantage of the logical adjacency theorem (XY + XY' = X), the rows of a truth table should be mapped so that adjacent map cells will be occupied by 2 terms that differ in only 1 variable.

For a switching function $F(A, B, C, D)$ in 4 variables, the domain of $F: S \times S \times S \times S \rightarrow S$ is $S \times S \times S \times S = \{(0, 0, 0, 0), (0, 0, 0, 1), \ldots, (1, 1, 1, 1)\}$. Each of the elements (A, B, C, D) in set $S \times S \times S \times S$ can be represented by a 4-variable K-map, as shown in Figure 3.12.

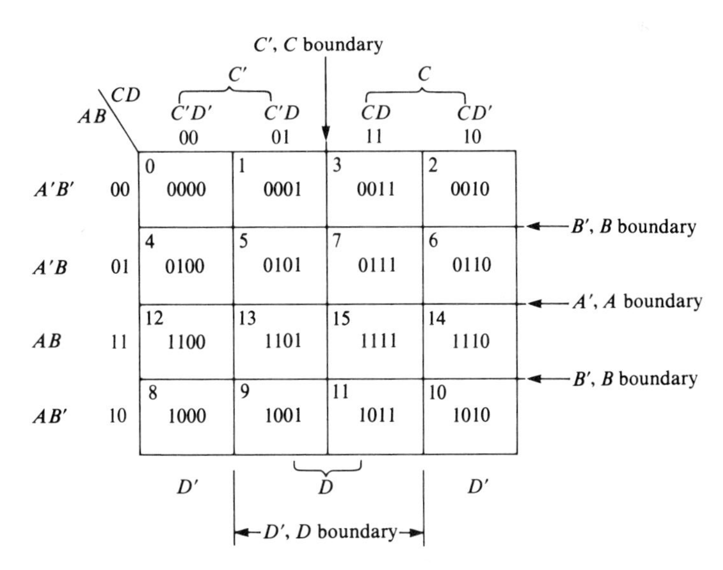

FIGURE 3.12
A 4-variable K-map

Again, the cells of a K-map are arranged to take advantage of the Boolean property $XY + XY' = X$. To accomplish this we arrange the cells in the map so that any 2 standard 4-variable product terms that differ in only 1 variable occupy cells that have a side in common. The 2-bit Gray code (00, 01, 11, 00) has the property that adjacent terms in the sequence differ in only 1 bit (variable). Each cell's location in the resulting K-map is determined by the concatenation of its row and column numbers. The numbering along each axis is in 2-bit Gray-code order (00, 01, 11, 10).

In any row, any 2 cells that have a side in common are logically adjacent because they differ in only 1 column variable (since the columns are numbered in Gray-code order). In any column, any 2 cells that have a side in common are logically adjacent because they differ in only 1 row variable (since the rows are numbered in Gray-code order). In any row, the rightmost cell is logically adjacent to the leftmost cell because columns 00 and 10 differ only in the C variable. In any column, the top cell is logically adjacent to the bottom cell because rows 00 and 10 differ only in the A variable. Therefore, in any 4-variable K-map, any 2 cells that have a side in common are logically adjacent.

The 4-variable K-map permits the combination of 2 cells that have a side in common, regardless of whether that side is a vertical or a horizontal or is at the top, bottom, extreme left or right side of the map. The 2-cell block (**pair**) formed by the union of 2 such logically adjacent cells corresponds to a 3-variable product term in which the boundary-crossing variable has been eliminated. Further, any two 2-cell pairs can be combined into a 4-cell block, called a **quartet**, provided that each cell in 1 pair is logically adjacent to a cell in the other pair.

Each cell in a 4-variable K-map can also be identified by the minterm corresponding to the cell's 4-bit number. For example, cell 0101 can also be identified by the minterm $A'BC'D$, as shown in Figure 3.13.

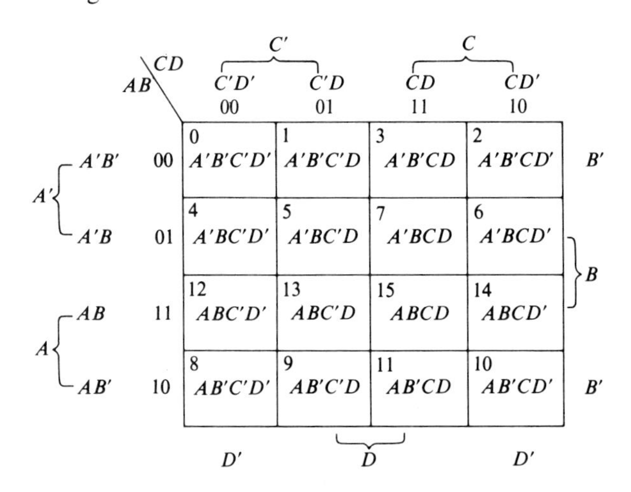

FIGURE 3.13
A 4-variable K-map identifying cells by minterm

A canonical SOP expression for a function $F(A, B, C, D)$ can be represented in any of the following alternative forms:

$$F = \sum \text{(decimal cell numbers)}$$
$$F = \sum \text{(binary cell numbers)}$$
$$F = \sum \text{(minterms)}$$

For example, we can have

$$F = \Sigma\ (1, 5, 14, 15)$$
$$F = \Sigma\ (0001, 0101, 1110, 1111)$$
$$F = A'B'C'D + A'BC'D + ABCD' + ABCD$$

Straightforward K-Map Simplification Techniques

The K-map has been constructed so that Boolean expressions can be simplified by combining logically adjacent 1 cells into a 2-cell pair using the Boolean property $XY + XY' = X$. Two 2-cell pairs can also be combined into a 4-cell block (quartet) provided that each cell in a pair is logically adjacent to a cell in the other pair. By repeated combination, blocks of 4, 8, . . . , 2^{n-1} cells can be combined into blocks in which 2, 3, . . . , $n - 1$ variables are eliminated.

■ **EXAMPLE 3.3**

Suppose we are given a function $F(A, B, C, D) = A'B'C'D + A'B'CD$, represented in the K-map by logically adjacent cells 1 and 3. These cells can be combined into a 2-cell pair that corresponds to the 3-variable product term $(A'B'D)$:

$$\begin{aligned} F &= A'B'C'D + A'B'CD \\ &= (A'B'D)C' + (A'B'D)C = A'B'D \end{aligned}$$

Note that the boundary-crossing variable $(C' \leftrightarrow C)$ has been eliminated by combining cells 1 and 3. ■

Combining Two 2-Cell Pairs into a 4-Cell Quartet

Let a 2-cell pair be denoted by P_i, with its component cells being denoted by c_{i1} and c_{i2}. If each cell of pair $P_1 = (c_{11}, c_{12})$ is logically adjacent to a cell in pair $P_2 = (c_{21}, c_{22})$, then the 2 pairs P_1 and P_2 can be combined into a 4-cell quartet corresponding to a 2-variable product term in which the 2 boundary-crossing variables have been eliminated.

■ **EXAMPLE 3.4**

Suppose we are given the following function:

$$Q = A'BCD + A'BCD' + ABCD + ABCD' = \Sigma(6, 7, 14, 15)$$

As seen in Figure 3.14 (p. 104), cells 6 and 7 can be combined into a 2-cell pair P_1, and cells 14 and 15 can be combined into pair P_2:

$$P_1 = (A'BC)D + (A'BC)D' = A'BC$$
$$P_2 = (ABC)D + (ABC)D' = ABC$$

Note that variable D has been eliminated in each of the pairs.

Since cell 6 (in P_1) is logically adjacent to cell 14 (in P_2), and cell 7 (in P_1) is logically adjacent to cell 15 in P_2, pairs P_1 and P_2 can be combined into a 4-cell quartet:

$$Q = P_1 + P_2 = (BC)A' + (BC)A = BC$$

Note that variables A and D have been eliminated in forming the quartet.

FIGURE 3.14
Forming a quartet of 4 cells

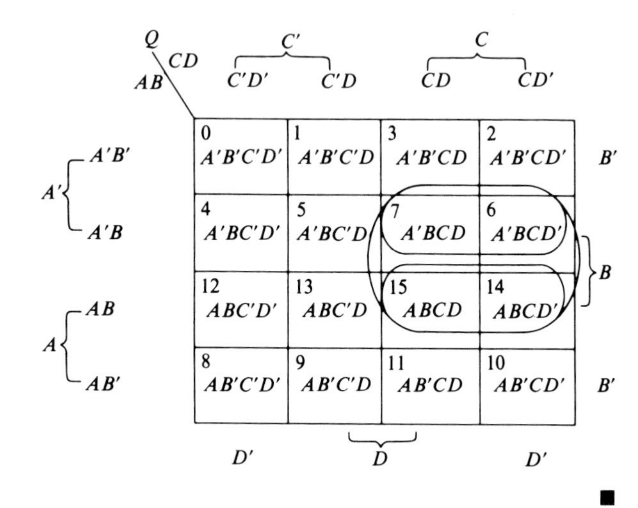

■

Figure 3.15 illustrates several examples of combining logically adjacent cells to form pairs, quartets, and octets, thus simplifying Boolean expressions.

- □ Figure 3.15a shows pairs (1, 0) = $A'B'C'$, (13, 5) = $BC'D$, (15, 11) = ACD, and (10, 8) = $AB'D'$ and redundant pair (15, 13) = ABD formed by the union of $BC'D$ and ACD.
- □ Figure 3.15b shows some quartets: (3, 2, 1, 0) = $A'B'$ = a row, (13, 9, 5, 1) = $C'D$ = a column, (15, 13, 7, 5) = BD = a square, and (14, 12, 10, 8) = AD' = a wraparound block.
- □ Figure 3.15c shows some octets: (14, 12, 10, 8, 6, 4, 2, 0) = D', (7, 6, 5, 4, 3, 2, 1, 0) = A', and (15, 14, 13, 12, 7, 6, 5, 4) = B.

Design of a 6311 Error Detector

A K-map block formed by combining 2, 4, or 8 rectangularly arrayed cells produces a product term in which 1, 2, or 3 variables, respectively, are eliminated. To illustrate the use of blocks of 2, 4, or 8 cells in the simplification of a Boolean expression, consider the following example.

Each of the decimal digits (0, 1, . . . , 9) can be represented by a 4-bit code with weights 6, 3, 1, 1. Since there are $2^4 = 16$ possible combinations to represent the 10 digits, there are 6 unneeded combinations. If a particular decimal digit can be represented by either of 2 combinations, then the first such combination is considered valid and the second combination is considered invalid.

A combinational logic circuit can be designed to detect an invalid 6311 input combination; this circuit has an output $F = 1$ when the input combination is not a valid 6311 code. A truth table and K-map for a 6311 error detector are presented in Figure 3.16.

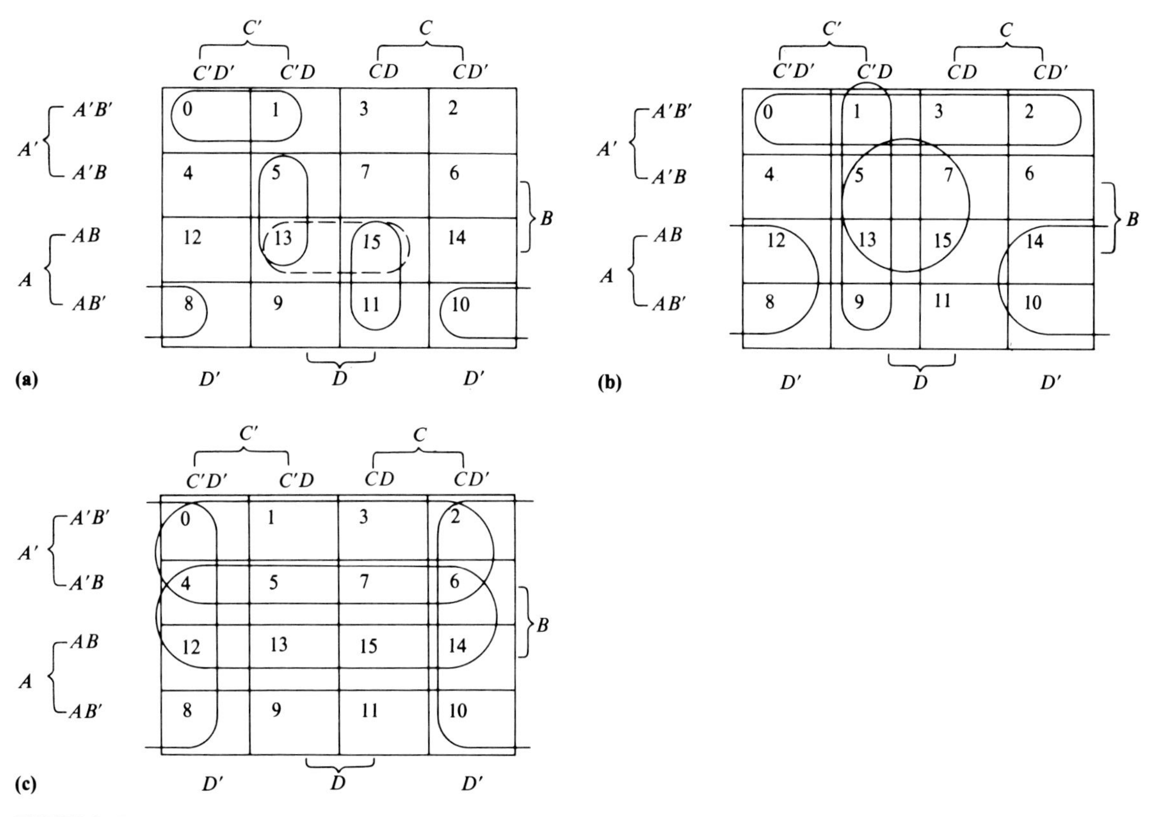

FIGURE 3.15
Examples of blocks of minterms. (a) Pairs (b) Quartets (c) Octets

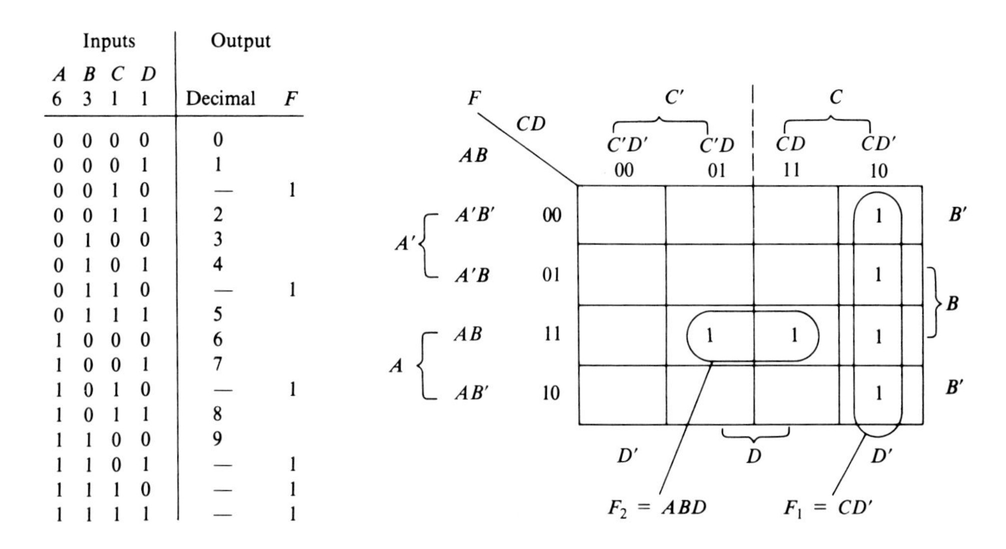

Inputs				Output	
A 6	B 3	C 1	D 1	Decimal	F
0	0	0	0	0	
0	0	0	1	1	
0	0	1	0	—	1
0	0	1	1	2	
0	1	0	0	3	
0	1	0	1	4	
0	1	1	0	—	1
0	1	1	1	5	
1	0	0	0	6	
1	0	0	1	7	
1	0	1	0	—	1
1	0	1	1	8	
1	1	0	0	9	
1	1	0	1	—	1
1	1	1	0	—	1
1	1	1	1	—	1

FIGURE 3.16
Truth table and K-map for 6311 error detector

Algebraically, the minterms of a column of a K-map have a common factor (denoted by the column label). Therefore, the right-hand column in Figure 3.16 can be algebraically simplified by the following steps:

$$\begin{aligned} F_1 &= CD'(A'B' + A'B + AB + AB') \\ &= CD'[(A'(B' + B) + A(B + B')] \\ &= CD'[A'(1) + A(1)] \\ &= CD'(1) = CD' \end{aligned}$$

Therefore, if all of the cells of a column equal 1, then the function F_1 can be written as the column label; that is, $F_1 = CD'$. The adjacent 1 cells in the third row ($ABC'D, ABCD$) can be combined to form a 2-cell block represented by the product $F_2 = ABD$. The overall function $F = F_1 + F_2$ can then be written in simplified form as $F = CD' + ABD$.

A Simple Algorithm for Finding a Minimal Cover for a K-Map

In many cases, it is obvious which isolated cells and/or blocks compose a minimal cover for a particular K-map. However, in some cases, it is not so obvious, and it is useful to have an algorithm for finding a minimal cover. Such an algorithm is as follows:

1. Circle each isolated 1 cell in the map.
2. Circle each block consisting of 8 rectangularly arrayed 1 cells that are not contained in a larger block.
3. Circle each block consisting of 4 rectangularly arrayed 1 cells that are not contained in a larger block.
4. Circle each block consisting of 2 adjacent 1 cells that are not contained in a larger block.
5. Starting with blocks of largest size, 2^k, use an asterisk (*) to mark any 1 cell that belongs to only 1 block. Check off all minterms of each block that contains a minterm with an asterisk. Repeat the process using blocks of size $2^{(k-1)}, \ldots, 2^1$.
6. A minimal cover consists of all essential blocks (those that contain a minterm with an asterisk), augmented by the smallest collection of other blocks that cover the remaining (unchecked) minterms.

■ **EXAMPLE 3.5**

Consider the K-map of Figure 3.17, where all the minterms are covered by the essential

FIGURE 3.17
K-map illustrating essential blocks

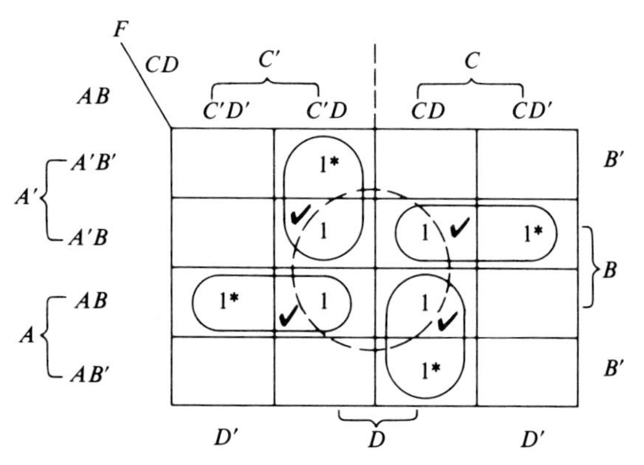

blocks. Consequently, the remaining block (BD) is not required. A minimal cover consists of $A'C'D$, $A'BC$, ACD, and ABC'. A minimal sum is

$$F = A'C'D + A'BC + ACD + ABC'$$

■

Design of a 2-Bit Inequality Detector

Let $N = AB$ and $n = ab$ represent two 2-bit numbers. An inequality detector, with output $I = 1$ when N is not equal to n, has the following truth table:

AB	ab	I
00	00	0
	01	1
	10	1
	11	1
01	00	1
	01	0
	10	1
	11	1
10	00	1
	01	1
	10	0
	11	1
11	00	1
	01	1
	10	1
	11	0

A 2-bit inequality function can be represented by the K-map shown in Figure 3.18. If we use the blocks indicated on the K-map, the 2-bit inequality function can be represented by the following Boolean equation:

$$I = Aa' + A'a + Bb' + B'b \qquad \text{Minimal sum}$$
$$= (A \oplus a) + (B \oplus b)$$

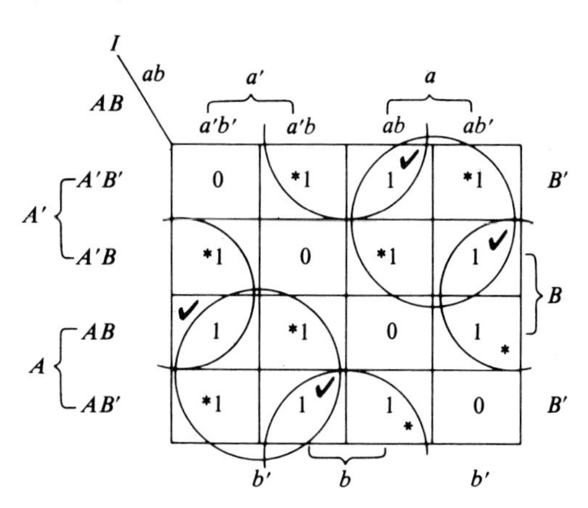

FIGURE 3.18
K-map for 2-bit inequality function

The intuitive solution can be expressed in this way: The number *AB* will not equal the number *ab* if *A* is not equal to *a* or if *B* is not equal to *b*. This implies that the inequality function is represented by the following Boolean equation:

$$I = (A \oplus a) + (B \oplus b)$$

Don't-Care Conditions in K-Maps

Consider a combinational logic circuit in which some combinations of inputs cannot occur. For example, if inputs *A*, *B*, *C*, and *D* represent only valid BCD digits, then the input combinations 1010, 1011, 1100, 1101, 1110, and 1111 cannot occur. Since the input combinations 1010 through 1111 cannot occur, we "don't care" whether any of these input combinations has function value 0 or 1; consequently, the function value for each of these **don't-care conditions** is represented by an *X* in the truth table and K-map of the function, as shown in Figure 3.19. (Note that $F = 1$ indicates valid BCD digits that are exactly divisible by 3.)

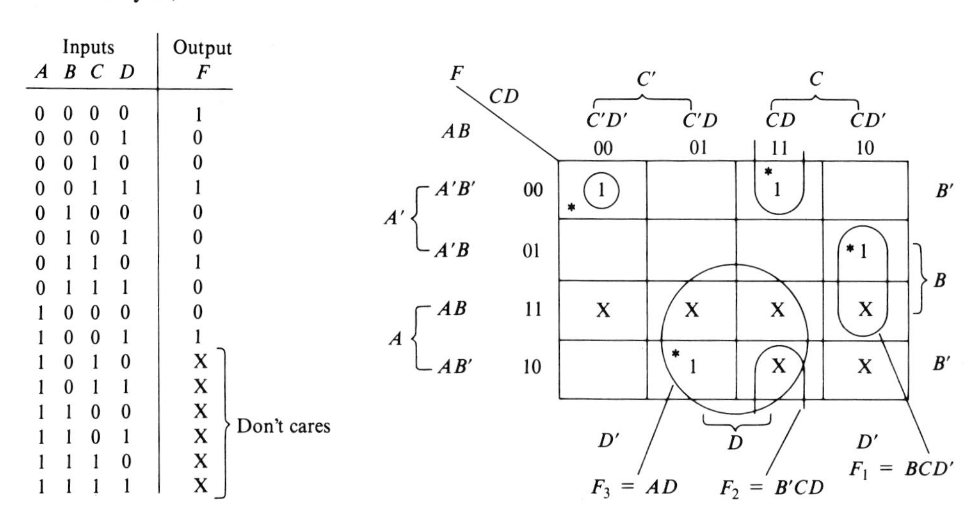

A	B	C	D	Output F
0	0	0	0	1
0	0	0	1	0
0	0	1	0	0
0	0	1	1	1
0	1	0	0	0
0	1	0	1	0
0	1	1	0	1
0	1	1	1	0
1	0	0	0	0
1	0	0	1	1
1	0	1	0	X
1	0	1	1	X
1	1	0	0	X
1	1	0	1	X
1	1	1	0	X
1	1	1	1	X

FIGURE 3.19
A BCD divide-by-3 function

Don't-care cells can be grouped with adjacent 1 cells to form blocks of 2^k cells represented by product terms that have eliminated *k* variables. From Figure 3.19, we see the following:

- □ The isolated 1 cell in the first row is minterm *A'B'C'D'*.
- □ The 2-cell block composes $F_1 = BCD'$.
- □ The 2-cell wraparound block composes $F_2 = B'CD$.
- □ The 4-cell block composes $F_3 = AD$.

Therefore, the function *F* can be represented by the two-level AND-OR expression:

$$F = A'B'C'D' + BCD' + B'CD + AD$$

The BCD divide-by-3 circuit can be implemented using the logic diagram shown in Figure 3.20.

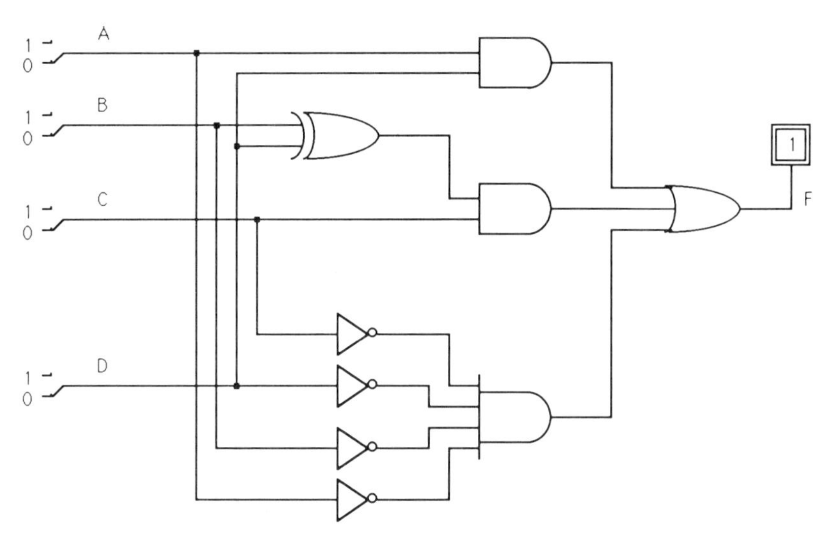

FIGURE 3.20
Logic diagram of a BCD divide-by-3 circuit

More-Sophisticated K-Map Techniques—Prime Implicants

Let F be a Boolean switching function. A 4-variable function $F(A, B, C, D)$ can be used to illustrate our definitions.

DEFINITION. For a Boolean function F represented in SOP form, a product term P is an **implicant** of function F if, and only if, $F = 1$ for every combination of input values for which $P = 1$.

For example, let

$$F(A, B, C, D) = A'C'D + A'B'D + A'BC' + ABC' + AB'CD'$$

Each of the product terms ($A'C'D$, $A'B'D$, $A'BC'$, ABC', and $AB'CD'$) is an implicant of F. In general, if F is written in SOP form, any product term in the sum is an implicant of F.

DEFINITION. An implicant of F is a **prime implicant** (PI) of F if it is not contained in a larger block of minterms that form an implicant of F.

For example, implicants $A'B'D$, $A'C'D$, and $AB'CD'$ are PIs of F because they are not contained in a larger block of minterms that form an implicant of F. Implicants $A'BC'$ and ABC' are not PIs of F because they are contained in a larger implicant $BC' = A'BC' + ABC'$. The product term BC' is a PI of F. Refer to Figure 3.21 to visualize this definition.

DEFINITION. A prime implicant P of F is an **essential prime implicant** (EPI) of F if P contains a minterm that is not contained in any other prime implicant of F.

For example, prime implicant $A'B'D$ is an EPI of F because it contains a minterm ($A'B'CD$) that is not contained in any other PI of F. Prime implicant BC' is also an EPI of F because it contains minterms (4, 12, 13) not contained in any other PI of F. Minterm $AB'CD'$ is an EPI because it is an isolated minterm of F. See also Figure 3.21.

Prime implicant $A'C'D$ is not an EPI because each minterm contained in it is contained in another PI of F. ($A'C'D$ is the consensus term created by the union of $A'BC'$ and $A'B'D$.)

DEFINITION. Any isolated minterm of F or any EPI minterm not contained in any other prime implicant of F is a **distinguished minterm** (denoted by *) of F.

For example, distinguished minterms of F are cells 3, 4, 10, 12, and 13 in Figure 3.21.

DEFINITION. A minimal cover of a Boolean function F consists of all EPIs and a collection of the fewest, nonredundant, **nonessential prime implicants** (non-EPIs) that cover all of the minterms that are not covered by the EPIs of F.

DEFINITION. A minimal sum for a Boolean function F consists of the logical sum of all EPIs and the smallest collection of non-EPIs that cover the minterms not covered by the EPIs of F.

Note that a minimal cover (sum) of a given function F is not always unique. Examples of this are given in the section discussing prime implicant tables and reduced prime implicant tables.

Using a K-Map to Find Implicants of a Boolean Function

A K-map provides a visual method for identifying implicants, PIs, and EPIs of a function.

In K-map terms,

1. An implicant is a rectangular block of 2^k adjacent 1 cells. An isolated minterm, block of 2^0 1 cells, is an implicant.
2. A prime implicant (PI) is a rectangular block of adjacent 1 cells not contained in a larger block.
3. An essential prime implicant (EPI) is a prime implicant that contains a distinguished minterm (*).

Figure 3.21 illustrates the use of a K-map to identify implicants, prime implicants, and essential prime implicants of F. In Figure 3.21,

$$F(A, B, C, D) = A'C'D + A'B'D + A'BC' + ABC' + AB'CD'$$

The circled implicants are $A'C'D$, $A'B'D$, $A'BC'$, ABC', and $AB'CD'$, and the prime implicants are $A'C'D$, $A'B'D$, BC', $AB'CD'$. Note that implicants $A'BC'$ and ABC' are contained in PI = BC'. Essential prime implicants are $AB'CD'$, $A'B'D$, and BC'.

A More-Efficient Algorithm for Finding a Minimal Cover

In many cases it is obvious which collection of PIs must be used to cover the remaining minterms not covered by the EPIs of a given function. However, in some cases, it is not so

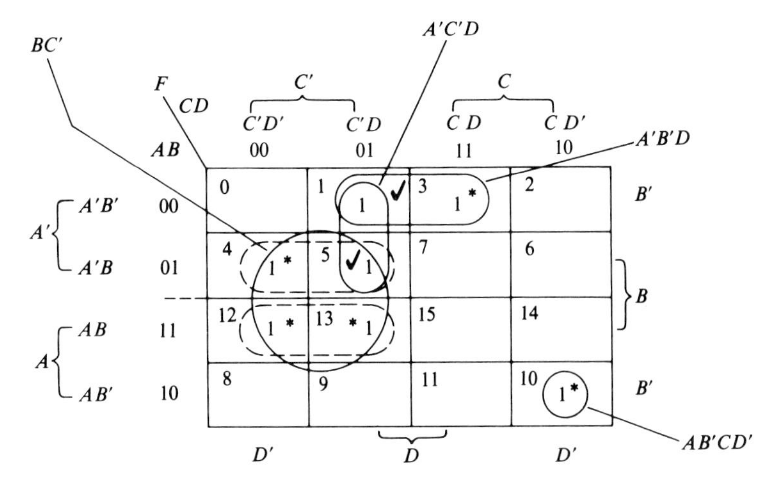

FIGURE 3.21
K-map illustrating implicants of F

obvious, and it is useful to have an algorithm for finding a minimal cover using a K-map. Such an algorithm is the K-map reduction algorithm, outlined as follows:

1. Find the PIs.
   a. Circle isolated minterms.
   b. Circle each octet of eight logically adjacent 1 cells that are not contained in a block of 16 1-cells.
   c. Circle each quartet of four logically adjacent 1 cells that are not contained in a larger block of 8 or 16 1-cells.
   d. Circle each pair of logically adjacent 1 cells that are not contained in a larger block of 4, 8, or 16 1-cells.

2. Find the EPIs. Use an asterisk (*) to mark any distinguished minterm. A PI containing a distinguished minterm is an EPI. Check off all minterms of each EPI.

3. Determine a minimal cover and corresponding minimal sum for F.
   a. If all 1 cells are covered by EPIs, the EPIs comprise a minimal cover of F, and a minimal sum is the logical sum of the EPI terms.
   b. If all 1 cells are not covered by EPIs, find the smallest collection of PIs that cover the remaining minterms (unchecked and having no *).
      (1) $PI_1 = PI_2$ if they cover the same set of unchecked minterms. If two equal PIs have the same size, select one of the PIs and check off its minterms. If two equal PIs differ in size, select the larger PI and check off its minterms.
      (2) PI_1 dominates PI_2, denoted $PI_1 > PI_2$, if they cover the same unchecked minterms and PI_1 covers at least one additional unchecked term. If $PI_1 > PI_2$, select PI_1 and check off its minterms.
      (3) A minimal cover consists of all EPIs augmented by the PIs selected in Step 3b. A minimal sum is the logical sum of all EPI product terms and the product terms of the PIs selected in step 3b.

The following examples illustrate the use of the K-map for finding a minimal cover for a given Boolean function.

EXAMPLE 3.6

A given function F is represented by the K-map shown in Figure 3.22.

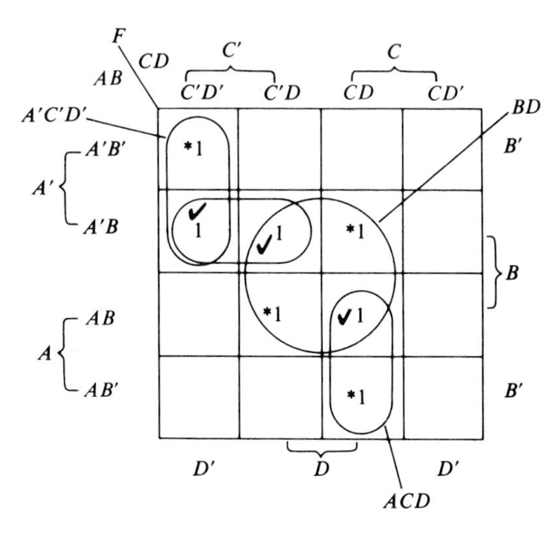

FIGURE 3.22
K-map for function F

A minimal cover for the function F in Figure 3.22 consists of the EPIs:

$$F = A'C'D' + BD + ACD$$

EXAMPLE 3.7

A given function F is represented by the K-map shown in Figure 3.23.

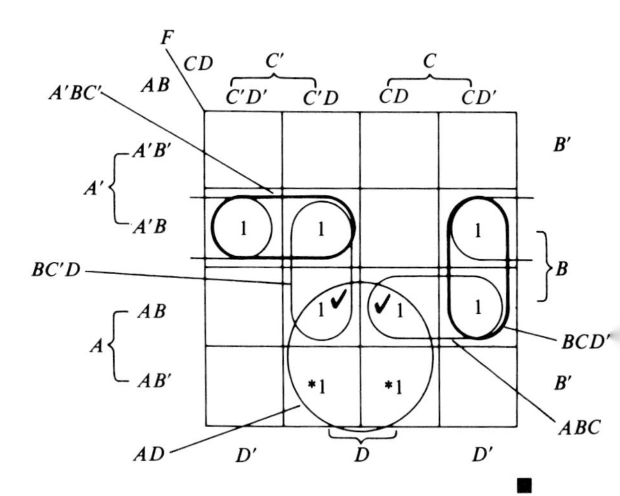

FIGURE 3.23
K-map for function F

PI_1 **dominates** PI_2, denoted $PI_1 > PI_2$, if they cover the same unchecked minterms and PI_1 covers at least one additional unchecked term. From Figure 3.23, it can be determined that *A'BC'* dominates *BC'D*, and *BCD'* dominates *ABC*. A minimal cover for the function in Figure 3.23 consists of the EPI *AD* and the dominant PIs *A'BC'* and *BCD'* that cover all unchecked 1 cells. Therefore, a minimal sum for *F* is

$$F = AD + A'BC' + BCD'$$

A K-map may appear less cluttered if the PI members are indicated by links, as shown in Figure 3.24, instead of using circles or ovals to enclose the members of each group. The following rules are used to identify PIs and EPIs by counting the number of links of each minterm of function *F*.

In terms of minterm links to logically adjacent minterms:

1. An isolated minterm has no links.
2. Each member of a PI pair has at least one link. A pair member with only one link is a distinguished minterm (*) because its only link is to the other member of that pair. A PI pair is an EPI if a pair member has only 1 link.
3. Each member of a PI quartet has at least two links. A quartet member with only two links is a distinguished minterm (*) because its only links are to other members of that quartet. A PI quartet is an EPI if a member has only two links.
4. Each member of a PI octet has at least three links. An octet member with only three links is a distinguished minterm (*) because its only links are to other members of that octet. A PI octet is an EPI if a member has only three links.

Figure 3.24 illustrates the use of minterm links to identify the implicants, PIs, and EPIs of a function $F = \Sigma(0, 4, 5, 7, 11, 13, 15)$. All the minterms of *F* are covered by the EPIs: *A'C'D'*, *BD*, and *ACD*. Therefore, *F* can be represented by the minimal sum

$$F = A'C'D' + BD + ACD$$

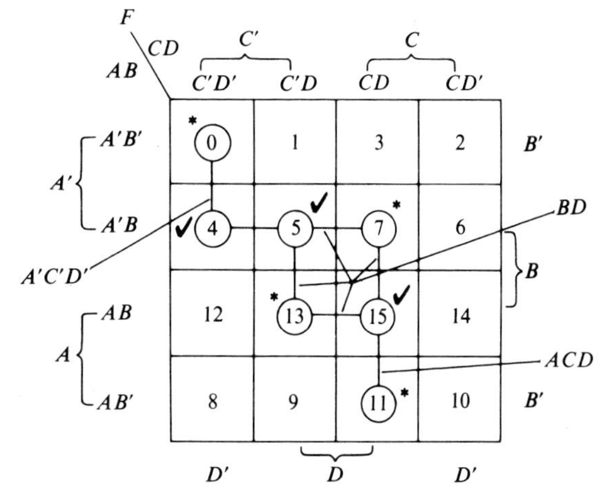

FIGURE 3.24
Using links to identify implicants

An example of a function that contains dominant PIs was illustrated in Figure 3.23. The same function, $F = \Sigma(4, 5, 6, 9, 11, 13, 14, 15)$, is illustrated in Figure 3.25 to demonstrate how links can be used to identify dominant PIs.

FIGURE 3.25
K-map for function F

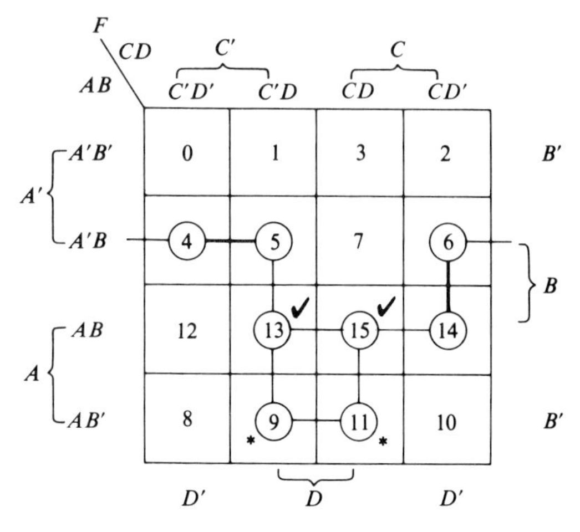

In Figure 3.25, linked quartet 15–13–9–11 is an EPI since minterm 9 has only two links and minterm 11 has only two links. Similarly, linked pairs 4–5, 4–6, 5–13, 6–14, and 14–15 are PIs since each member of each pair has two links.

Since terms 9, 11, 13, and 15 are covered by an EPI, the remaining uncovered terms are 4, 5, 6, and 14. Linked pair 4–5 dominates pair 5–13 because 4–5 covers two unchecked terms, while 5–13 covers only one unchecked term ($A'BC'$ dominates $BC'D$). Linked pair 6–14 dominates pair 14–15 because 6–14 covers two unchecked terms, while 14–15 covers only one unchecked term (BCD' dominates ABC). Thus, a minimal cover consists of EPI AD, and the dominant PIs, $A'BC'$ and BCD'. A minimal sum of F is

$$F = AD + A'BC' + BCD'$$

Boolean Functions with a Nonunique Minimal Cover

The canonical SOP form of a Boolean expression, derived directly from a truth table, can usually be reduced to a minimal sum that contains fewer terms with fewer literals in many terms. However, a minimal cover (sum) of a given function F is not always unique. That is, the remaining (unchecked) terms, those not covered by EPIs, may be covered in more than one way. A nonunique-minimal-cover condition occurs if minterms of non-EPIs belong to more than one PI. An example of a function, $F = \Sigma(0, 1, 3, 4, 11)$, that does not have a unique minimal cover is illustrated in Figure 3.26.

The implicants, PIs, and EPIs of F are indicated by the linked minterms. The function has four PI pairs of F: (11, 3), (4, 0), (3, 1), and (1, 0). Since minterm 11 has only one link and minterm 4 has only one link, F has two EPIs: (11*, 3) and (4*, 0).

Minterm 1 is the only term not covered by an EPI. Since minterm 1 is covered by PI pairs (3, 1) and (1, 0), the function F does not have a unique minimal cover. Consequently, F can be represented by either of the following SOP expressions:

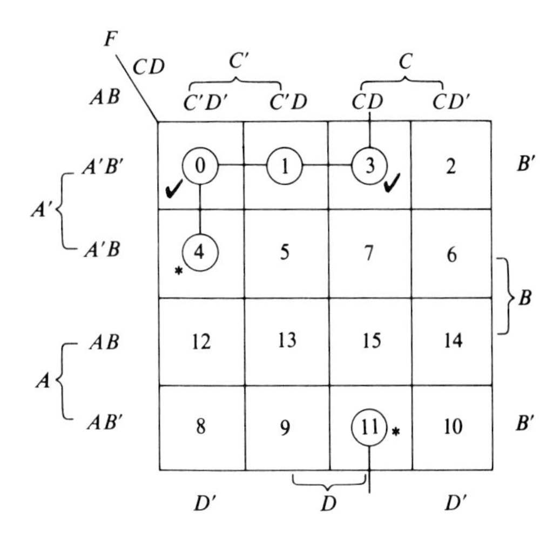

FIGURE 3.26
K-map illustrating nonunique minimal cover

$$F = A'C'D' + B'CD + A'B'C' \leftrightarrow (4, 0), (11, 3), (1, 0)$$

or

$$F = A'C'D' + B'CD + A'B'D \leftrightarrow (4, 0), (11, 3), (3, 1)$$

Figure 3.27 illustrates another function, $F = \Sigma(0, 2, 4, 5, 7, 8, 10, 12, 13, 15)$, that has no unique minimal cover. From Figure 3.27, we see

$$F = BD + B'D' + \begin{cases} BC' \leftrightarrow (13, 12, 5, 4) \\ C'D' \leftrightarrow (12, 8, 4, 0) \end{cases}$$

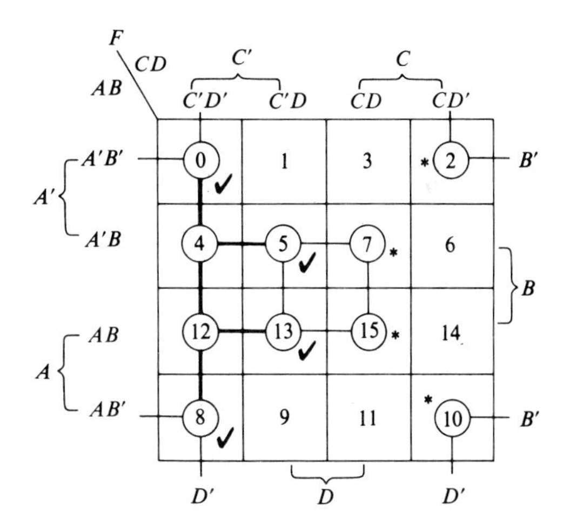

FIGURE 3.27
K-map illustrating nonunique minimal cover

Minterm Rings in a K-Map

For Boolean functions of 4 or fewer variables, the PIs of the function can usually be identified using a standard K-map. If the configuration of 1 cells is fairly simple, identification of the PIs and EPIs can be accomplished visually. If the configuration of 1 cells is more complicated, use of the K-map reduction algorithm may be required for identification of the PIs and EPIs.

However, there are some configurations of 1 cells that lead to a simplification of the Boolean expression even though there may be no logically adjacent 1 cells in the K-map. The configuration in Figure 3.28 illustrates such a case.

The function $F = \Sigma(1, 2, 4, 7, 8, 11, 13, 14)$, illustrated in Figure 3.28, can be algebraically simplified by extracting the common factor in each row and defining variables $Z = A'B + AB'$ and $W = C'D + CD'$. With these substitutions, F can be expressed in terms of exclusive-ORs:

$$F = (A \oplus B) \oplus (C \oplus D)$$

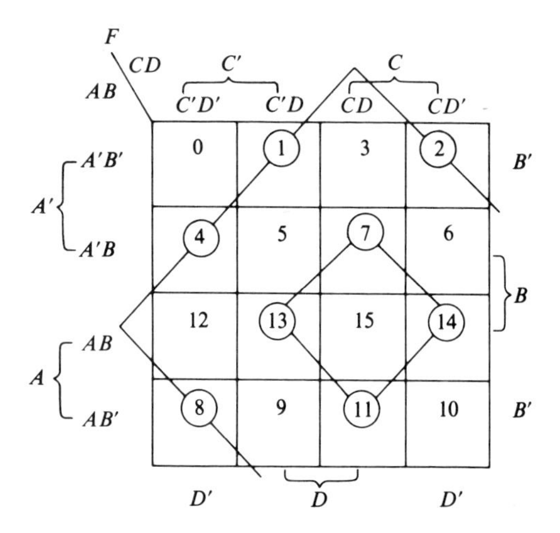

FIGURE 3.28
Minterm rings in a K-map

The function contains minterms 1, 2, 4, and 8 (0001, 0010, 0100, 1000), and the minterms 7, 11, 13, and 14 (0111, 1011, 1101, 1110). The binary cell number of each number in the first group contains one 1 while each binary cell number of the second group contains three 1s. By connecting the cells of the first group (1, 2, 4, 8) we find a *ring* of cells in which each cell number contains one 1. By connecting each binary cell number in the second group (7, 11, 13, 14) we find a ring of cells in which each cell number contains three 1s. This function, which contains 1 cells that appear in rings rather than in rectangular blocks (called **minterm rings**), can be simplified using exclusive–ORs.

INTERACTIVE DESIGN APPLICATION

Design of a 4-Bit Even-Parity Checker

The simplification of a Boolean function using common factors can be illustrated by the following design problem. We will use the 4-step problem-solving procedure introduced in Chapter 2 to obtain a solution.

Step 1: Problem Statement. Design a combinational logic circuit that has output function $F = 1$ if, and only if, the number of active-high (ON = 1) inputs A, B, C, D is odd. Bits A, B, and C represent a 3-bit message, and bit D represents an even-parity bit. The output F

represents an error function that has value 1 if the input word (message + parity bit) contains an odd number of 1s.

Step 2: Conceptualization. The function F can be represented by the truth table and its corresponding K-map shown in Figure 3.29.

FIGURE 3.29
Truth table and K-map for function F

A	B	C	D	F
0	0	0	0	0
0	0	0	1	1
0	0	1	0	1
0	0	1	1	0
0	1	0	0	1
0	1	0	1	0
0	1	1	0	0
0	1	1	1	1
1	0	0	0	1
1	0	0	1	0
1	0	1	0	0
1	0	1	1	1
1	1	0	0	0
1	1	0	1	1
1	1	1	0	1
1	1	1	1	0

F: AB \ CD	C'D' 00	C'D 01	CD 11	CD' 10
A'B' 00		1		1
A'B 01	1		1	
AB 11		1		1
AB' 10	1		1	

Step 3: Solution/Simplification. If the minterms of F are written in order, that is, first row, second row, and so forth, then

$$F = (A'B'C'D + A'B'CD') + (A'BC'D' + A'BCD) + (ABC'D + ABCD') + (AB'C'D' + AB'CD)$$

Note that $A'B'$ is a common factor of K-map row 00, $A'B$ is a common factor of K-map row 01, AB is a common factor of K-map row 11, and AB' is a common factor of K-map row 10. The expression for F can be simplified by extracting common factors, as follows:

$$F = A'B'\ (C'D + CD') + A'B(C'D' + CD) + AB\ (C'D + CD') + AB'(C'D' + CD)$$

The expression can be further simplified by factoring out the common factors to obtain

$$F = (A'B' + AB)(C'D + CD') + (A'B + AB')(C'D' + CD)$$

Substituting variables

$$X = A'B + AB' = A \oplus B$$

and

$$Y = C'D + CD' = C \oplus D$$

we find

$$F = X'Y + XY' = X \oplus Y = (A \oplus B) \oplus (C \oplus D)$$

Step 4: Realization. A logic diagram of the 4-bit even-parity checker can be constructed using the simplified expression for F. The logic diagram is presented in Figure 3.30.

FIGURE 3.30
Logic diagram for 4-bit even-parity checker

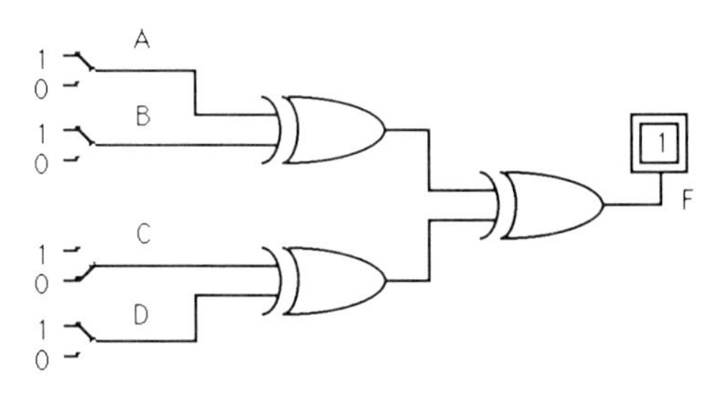

Using Minterm Rings to Determine Logical Adjacency

The minterm rings can also be used in simplifying functions that have a minimal AND-OR form. An examination of the binary cell numbers of each cell in a 4-variable K-map reveals a concentric arrangement of 5 minterm rings. Each ring contains all K-map cells containing the same number of 1s in the binary cell number, as illustrated in Figure 3.31.

FIGURE 3.31
Minterm rings in a K-map

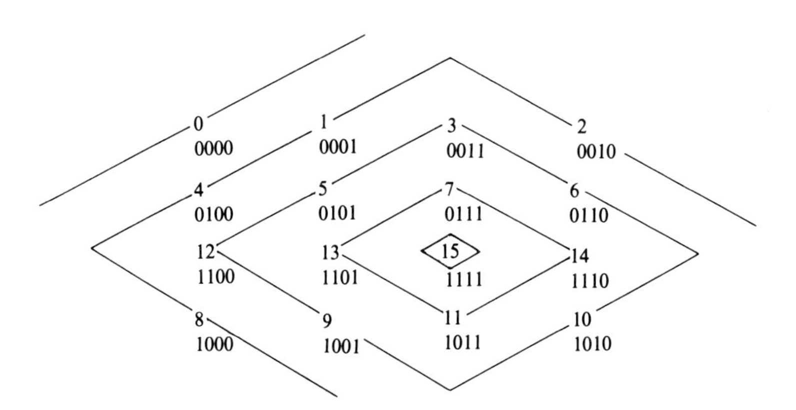

A table of the minterms in each ring, together with the logically adjacent minterms in adjacent rings, is illustrated in Table 3.1.

Since any cell of a given ring differs in 1 variable with any cell in an adjacent ring, any cell in 1 ring is logically adjacent to any cell in an adjacent ring *if* that cell is in the same row or column as the first cell (so that the two cells are physically adjacent). Therefore, the adjacent-ring minterms that can be paired with a given minterm can be found by looking left, right, up, and down (from the given cell) on the K-map, as exemplified in Figure 3.32.

This minterm-ring arrangement is very useful in dealing with Boolean functions in 5, 6, 7, and 8 variables. Minterm-ring maps and a **minterm-ring algorithm** for determining PIs and EPIs provide a powerful method for the systematic reduction of Boolean expressions in 5 or more variables. These concepts and methods will be described in the next section.

TABLE 3.1
Minterm-ring table for a 4-variable K-map

Ring Number	Minterms		Pair Members: Logically Adjacent Terms in	
	Decimal	Binary	Next Lower Ring	Next Higher Ring
0	0	0000	—	1, 2, 4, 8
1	1	0001	0	3, 5, 9
	2	0010	0	3, 6, 10
	4	0100	0	5, 6, 12
	8	1000	0	9, 10, 12
2	3	0011	1, 2	7, 11
	5	0101	1, 4	7, 13
	6	0110	2, 4	7, 14
	9	1001	1, 8	11, 13
	10	1010	2, 8	11, 14
	12	1100	4, 8	13, 14
3	7	0111	3, 5, 6	15
	11	1011	3, 9, 10	15
	13	1101	5, 9, 12	15
	14	1110	6, 10, 12	15
4	15	1111	7, 11, 13, 14	—

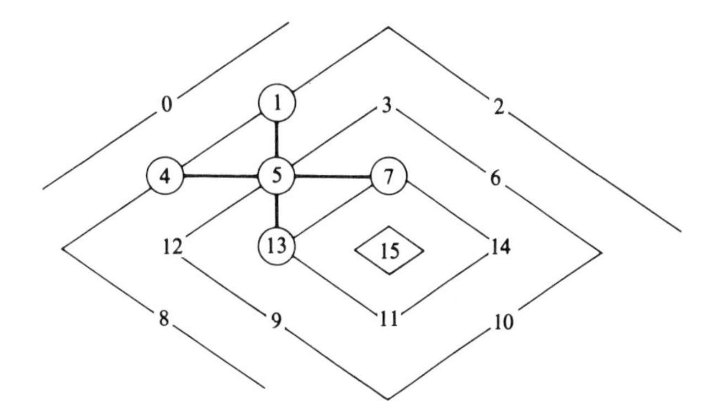

FIGURE 3.32
Logically adjacent terms in a minterm ring map

3.5 MINTERM-RING MAPS AND KARNAUGH MAPS IN 5 AND 6 VARIABLES

A switching function f in 5 variables can be plotted using two 4-variable K-maps that are arranged side by side or one above the other, as shown in Figure 3.33. A minimal cover of f can then be determined by the following procedure:

1. **Link** logically adjacent minterms in each 4-variable map to form linked pairs, quartets, octets, and so forth.

2. Mentally superimpose one map over the other to determine if any block of 2^n minterms in map $E = 0$ coincides with the same block in map $E = 1$; if so, the two blocks are combined to form a block of $2^{(n+1)}$ minterms, in which variable E is eliminated.
3. Find minimal cover by counting links to determine the EPIs and PIs of the function f.

The following example illustrates the use of this procedure in finding a minimal sum of a switching function in 5 variables.

EXAMPLE 3.8

Suppose we are given

$$f(E, A, B, C, D) = \Sigma(1, 3, 6, 8, 9, 12, 13, 18, 19, 20, 21, 24, 25, 27, 28, 29)$$

1. We link logically adjacent minterms in each 4-variable map to form pairs $EA'B'C$, $E'A'B'D$, and quartets EBC', EAC', $E'AC'$. (See Figure 3.33.)
2. We mentally superimpose one map over the other and find that the quartets EAC' and $E'AC'$ can be combined into an octet AC'.
3. By counting links, we determine the EPIs and PIs that compose a minimal cover of the function f. The function f can then be represented by the minimal sum

$$f(E, A, B, C, D) = AC' + EBC' + EA'B'C + E'A'B'D + E'A'BCD' + \begin{cases} EAB'D \\ EB'CD \end{cases}$$

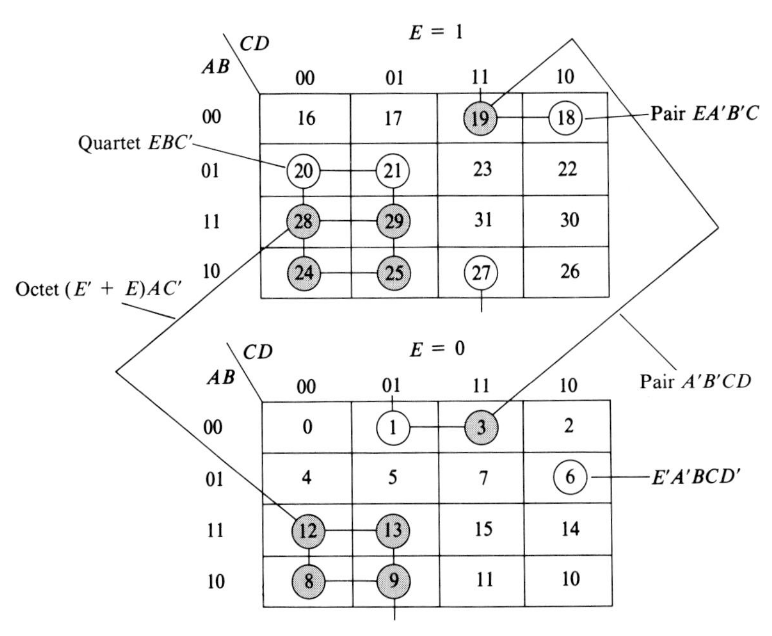

FIGURE 3.33
Using two 4-variable K-maps to plot a 5-variable function

Minterm-Ring Maps in 5 Variables

A switching function $f(E, A, B, C, D)$ in 5 variables has domain $S \times S \times S \times S \times S = \{(0,0,0,0,0), \ldots, (1,1,1,1,1)\}$ corresponding to the 32 possible combinations of the values of the 5 variables. Thus, a map of the cartesian product $S \times S \times S \times S \times S$ contains 32 cells. Picture the cells in two 16-cell maps. Cells in the lower map are numbered 0 (00000) through 15 (01111) corresponding to the combinations $E'A'B'C'D'$ through $E'ABCD$. Cells in the upper map are numbered 16 (10000) through 31 (11111) corresponding to the combinations $EA'B'C'D'$ through $EABCD$. The two maps, each with its minterm rings, are shown in Figure 3.34.

FIGURE 3.34
Diagram of two 16-cell K-maps

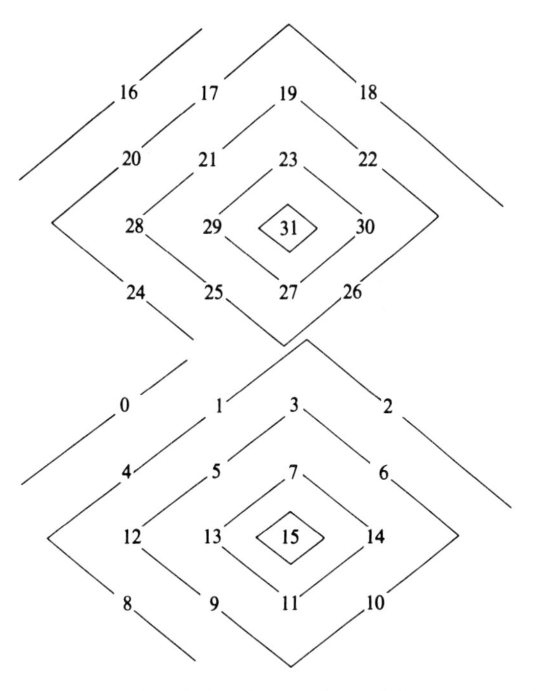

The number of each cell in the upper map is 16 plus the number of the corresponding cell in the lower map. In binary, the corresponding cells differ only in the variable E; therefore, each cell in the upper map is logically adjacent to the corresponding cell in the lower map. Consequently, the corresponding upper and lower cells can be combined into a 2-cell pair represented by a product term in which variable E has been eliminated.

Now we begin to see the usefulness of the minterm rings introduced in Section 3.4. Each cell in the lower map belongs to a minterm ring that contains one less 1 than the ring of the corresponding cell in the upper map. For example, lower-map cell number 0 has zero 1s while its corresponding upper-map cell number (16) contains one 1; lower-map cell number 1 has one 1 while its corresponding upper-map cell number (17) contains two 1s, and so on.

A double-layer map can be constructed by superimposing the upper map on the lower map, as illustrated in Figure 3.35 (p. 122).

FIGURE 3.35
Minterm ring map in 5 variables, *E*, *A*, *B*, *C*, and *D*

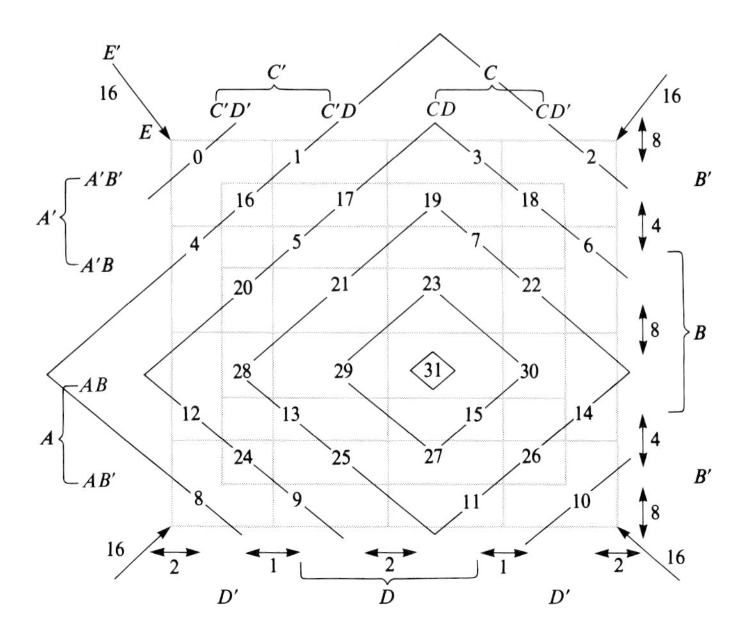

Lower-layer cell numbers containing N 1s are in the same ring as upper-layer cells with N 1s. For example, ring 1 contains lower-layer cells 1, 2, 4, and 8 and upper-layer cell 16. In the double-layer map, lower-layer cell numbers and corresponding upper-layer cell numbers are displaced from one another so that both are visible and so that each cell number lies in the ring having the same number of 1s as the cell number. Note that each lower-layer cell is logically adjacent to the corresponding upper-layer cell since they differ only in variable E.

Table 3.2 lists the "composite" minterm rings formed by superimposing the two 16-cell maps. The left column lists the number of 1s in the binary number of each cell in that ring (row).

TABLE 3.2
A 5-variable minterm-ring table

Number of 1s in Ring	Lower-Layer Cells	Upper-Layer Cells
0	0	—
1	1, 2, 4, 8	16
2	3, 5, 6, 9, 10, 12	17, 18, 20, 24
3	7, 11, 13, 14	19, 21, 22, 25, 26, 28
4	15	23, 27, 29, 30
5	—	31

Minterm Links—The Logical Connection

Minterm-ring maps allow us to visualize implicants, PIs, and EPIs for functions of 5, 6, 7, and 8 variables. Let M denote the highest numbered minterm in a block of size 2^n in f. The following definitions describe the terms used in minterm-ring maps.

DEFINITION. Two minterms are logically adjacent if their binary cell numbers differ in only 1 variable; this boundary-crossing variable appears in 1 minterm in uncomplemented form and in the other minterm in complemented form. In a minterm-ring map, a minterm M is logically adjacent to any adjacent ring minterm that is located to the left, right, up, down, or diagonally inward from M.

Consider Example 1 in Figure 3.36. Cell 5 is logically adjacent to each of the adjacent ring cells that are to the left, right, up, down, or diagonally inward of cell 5; the cells logically adjacent to cell 5 are 4, 7, 1, 13, and 21.

DEFINITION. An isolated minterm is any minterm that has no logically adjacent minterm.

DEFINITION. A pair (M, N) consists of 2 logically adjacent minterms, M and N.

DEFINITION. In a pair (M, N) the logical link is the difference $x = M - N$. The variable with value x is eliminated in the product formed by combining minterms M and N into a pair. For example, consider the pair $20 - 16 = 4$: In literal form,

$$EA'BC'D' + EA'B'C'D' = EA'C'D'(B + B') = EA'C'D'$$

DEFINITION. A quartet of f consists of 4 minterms belonging to 4 pairs where each member of any pair is logically adjacent to a member of any other pair that does not contain that minterm. If pairs of the quartet are formed by crossing boundaries x_1 and x_2, then x_1, x_2 are links of the quartet with minterm members: $M, M - x_1, M - x_2, M - x_1 - x_2$. The variables with values x_1 and x_2 are eliminated in the product produced in forming the quartet.

Consider Example 2 in Figure 3.36. Minterms 31, 30, 23, and 22 compose a quartet Q of f since a member of each of the pairs (31, 30), (23, 22), (31, 23), and (30,

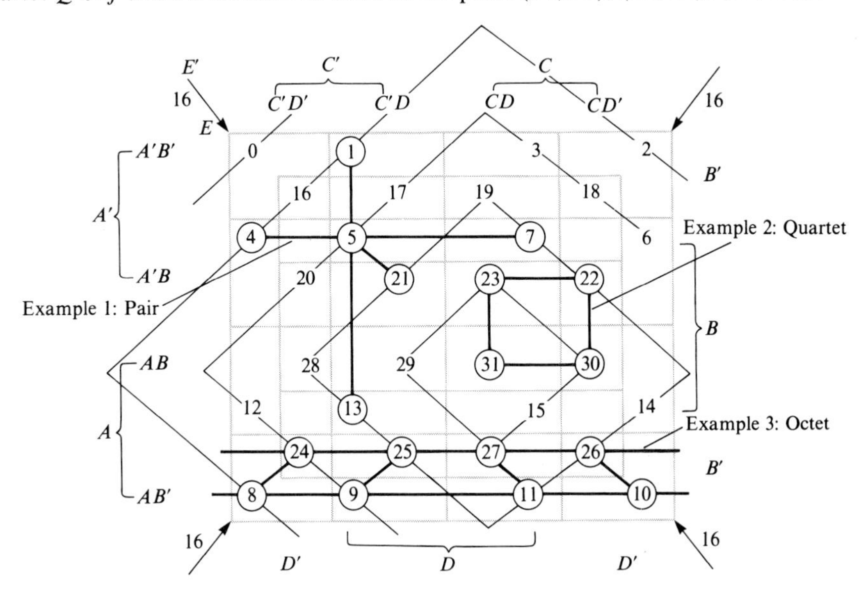

FIGURE 3.36
Illustration of pair, quartet, and octet

22) is logically adjacent to a member of any other pair (which does not contain that minterm) of the quartet. For example, 31 in pair (31, 30) is logically adjacent to 23 in pair (23, 22) and 30 in pair (30, 22).

DEFINITION. An **octet** of f consists of 8 minterms belonging to 6 quartets of f in which each member of 1 quartet is logically adjacent to a member of any other quartet that does not contain that minterm. If pairs within the octet are formed by crossing boundaries x_1, x_2, x_3, then x_1, x_2, x_3 are links of the octet with minterm members: M, $M - x_1$, $M - x_2$, $M - x_3$, $M - x_1 - x_2$, $M - x_1 - x_3$, $M - x_2 - x_3$, $M - x_1 - x_2 - x_3$.

Consider Example 3 of Figure 3.36. Minterms 27, 26, 25, 24, 11, 10, 9, and 8 compose an octet O of f since 6 quartets (each with a member that is logically adjacent to a minterm of any quartet not containing it) can be formed with the 8 points.

The product term produced by combining logically adjacent minterms M and N into a pair, where $M > N$, can be expressed in any one of the following alternative forms:

Highest decimal number cell (link); for example, 7(2)

Highest binary number cell (link); for example, 00111(2)

Literal form; for example, $E'A'BCD = E'A'BD$

The link variable is eliminated.

The product term produced by forming a quartet can be expressed in any one of the following forms:

Highest decimal number cell (links); for example, 31(8,1)

Highest binary number cell (links); for example, 11111(8,1)

Literal form; for example, $EABCD = EBC$

The link variables are eliminated.

The product term produced by forming an octet can be expressed in any one of the following forms:

Highest decimal number cell (links); for example, 27(16,2,1)

Highest binary number cell (links); for example, 11011(16,2,1)

Literal form; for example, $EAB'CD = AB'$

The link variables are eliminated.

In the literal form, each variable corresponding to a link is eliminated.

The implicant definitions presented in Section 3.4 can be rephrased in terms of minterm rings.

DEFINITION. An implicant of f is the product term produced by combining members of a pair, quartet, octet, or larger block of minterms.

DEFINITION. An implicant of f is a PI of f if the implicant group is not contained in a larger block of minterms of f.

DEFINITION. A PI of f is an EPI of f if the implicant group contains a minterm not contained in any other prime implicant group of f.

In terms of minterm links,

1. A minterm is an isolated minterm (hence, an EPI) if it has 0 links.
2. A PI pair is an EPI if either member has only 1 link. A pair member that has only 1 link is a distinguished minterm (*).
3. A PI quartet is an EPI if any member has only 2 links. Any quartet member that has only 2 links is a distinguished minterm (*).
4. A PI octet is an EPI if any member has only 3 links. Any octet member which has only 3 links is a distinguished minterm (*).

Missing-Link Theorems

A quartet (or octet) is a group of 4 (or 8) minterms that are pairwise logically adjacent. If a function f has 3 (or 7) logically adjacent minterms, but *not* the fourth (or eighth) minterms that would complete a quartet (or octet), then we say that f has **missing links**.

QUARTET MISSING-LINK THEOREM. Given a minterm M that has at least 2 links x_i, x_j (so that $M - x_i \in \Sigma$, $M - x_j \in \Sigma, \ldots$). If for all x_i, x_j combinations, $M - x_i - x_j \notin \Sigma$, then M cannot spawn a quartet of function f, and all pairs $(M, M - x_i)$ and $(M, M - x_j, \ldots)$ are PIs of f.

Example 1 in Figure 3.37 illustrates the quartet missing-link theorem.

OCTET MISSING-LINK THEOREM. Given a minterm M that has at least 3 links x_i, x_j, x_k (so that $M - x_i \in \Sigma$, $M - x_j \in \Sigma$, $M - x_k \in \Sigma, \ldots$). If for all x_i, x_j, x_k combinations, $M - x_i - x_j - x_k \notin \Sigma$, then M cannot spawn an octet of function f, and quartets $(M, M - x_i,$

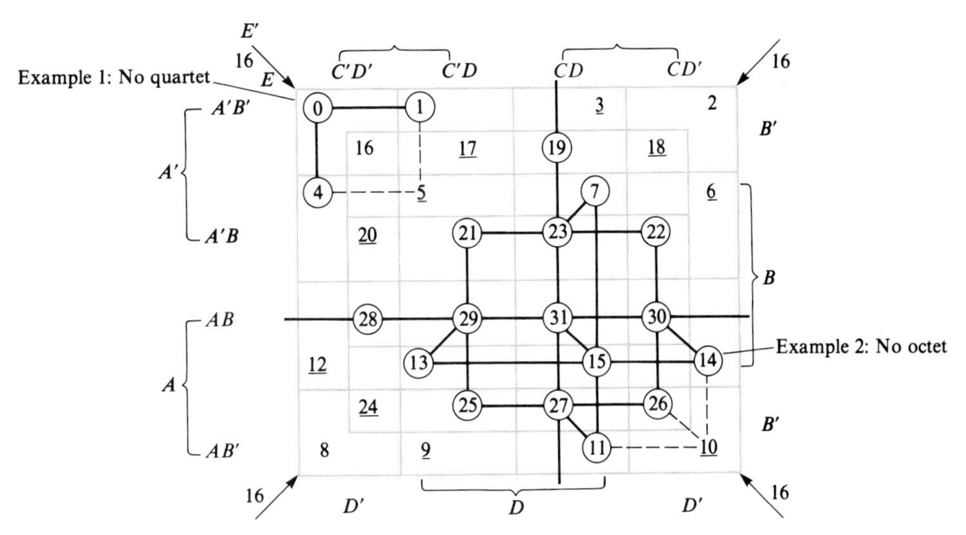

FIGURE 3.37
Illustration of missing-link theorems

$M - x_j$, $M - x_i - x_j$), (M, $M - x_i$, $M - x_k$, $M - x_i - x_k$), (M, $M - x_j$, $M - x_k$, $M - x_j - x_k$), . . . are PIs of f.

Example 2 in Figure 3.37 illustrates the octet missing-link theorem. Minterm M = 31 has 5 links (x_i = 1, 2, 4, 8, 16) and spawns 10 quartets (M, $M - x_i$, $M - x_j$, $M - x_i - x_j$). There are no octets in f since $M - x_i - x_j - x_k \notin \Sigma$ for all x_i, x_j, x_k combinations. If any point R were in Σ so that $M - x_i - x_j - x_k \in \Sigma$, then an octet would exist (see underlined points: 17, 3, 18, 6, . . .).

Use of Minterm-Ring Maps in Simplifying Boolean Expressions

A minterm-ring map facilitates a systematic reduction of Boolean expressions in 5, 6, 7, and 8 variables. The minterm-ring map is easier to use than vertical or side-by-side K-maps for several reasons:

1. A straightforward minterm-ring algorithm provides a systematic means for forming implicants (pairs, quartets, octets, and so forth) of f.
2. Identification of EPIs at each stage simplifies formation of larger blocks of minterms in the next stage of the process. A reduced PI table is directly produced by identifying the EPIs of f.
3. The three-dimensional perspective of the minterm-ring map, together with the links connecting implicant group members, simplifies the visual identification of PI and EPI pairs, quartets, octets, and so on.

The McCalla Minterm-Ring Algorithm

Since most logic minimization currently is accomplished with software tools, the minterm-ring method is important because it combines a contemporary algorithmic method for simplifying Boolean expressions with a three-dimensional mapping technique that facilitates visual identification of PIs and EPIs. The derivation of the minterm-ring algorithm also provides an excellent illustration of how algorithms are developed.

The McCalla minterm-ring algorithm† and the Quine–McCluskey tabulation method (Section 3.6) can be programmed on a computer to simplify Boolean expressions in any number of variables. The minterm-ring algorithm is more powerful than the Quine–McCluskey method since it automatically determines EPIs and directly produces a reduced PI table. The minterm-ring algorithm determines a function's PIs and EPIs by counting the number of links of each minterm. A distinguished member of an EPI is labeled with an asterisk (*). EPIs are determined as follows:

A minterm is an isolated minterm (hence, an EPI) if it has no links. An isolated minterm is a distinguished minterm (*).

A PI pair is an EPI if either member has only 1 link. A pair member with only 1 link is a distinguished minterm (*).

A PI quartet is an EPI if any member has only 2 links. Any quartet member with only 2 links is a distinguished minterm (*).

---

†"A Minterm-Ring Algorithm for Simplifying Boolean Expressions," IEEE Thirty-Third Symposium on Circuits and Systems, Calgary, Alberta, Canada, August 1990.

A PI octet is an EPI if any member has only 3 links. Any octet member with only 3 links is a distinguished minterm (*).

The McCalla minterm-ring algorithm can be used to find a minimal sum of a switching function f of 5 or more variables. The development of this algorithm is described in Example 3.9. The map accompanying this discussion (Figure 3.38) shows rings of minterms by the number of 1s. A procedure guide or template for Example 3.9 is shown on page 129.

EXAMPLE 3.9

Use the minterm-ring algorithm to determine the PIs and EPIs of the switching function $f(E, A, B, C, D) = \Sigma(1, 3, 6, 8, 9, 12, 13, 18, 19, 20, 21, 24, 25, 27, 28, 29)$.

Note that each variable has an associated weight: $E = 16, A = 8, B = 4, C = 2, D = 1$. The switching function f is plotted in the minterm-ring map in Figure 3.38 (p. 128) to show how to visualize the procedure the minterm-ring algorithm uses to determine PIs and EPIs of f. ■

The McCalla minterm-ring algorithm determines the PIs and EPIs of a switching function and directly produces a reduced PI table using the following steps:

Step 0. Circle each minterm of the specified function f on the accompanying minterm-ring map (Figure 3.38).

Step 1. Find the PIs and EPIs.

a. Find the *pairs*. Start with the highest-numbered minterm M and connect (link) adjacent-ring logically adjacent minterms by looking left, right, up, down, and diagonally from M. For example, in Figure 3.38, we see that minterm 6* is an EPI since it is an isolated minterm.

$$6^* = 00110 = E'A'BCD'$$

Pair (19, 18*) is an EPI since it has a distinguished minterm member (a member with only 1 link).

$$(19,\ 18^*) = 10011(1) = EA'B'C\not{D} = EA'B'C$$

Note that the link variable ($D = 1$) is eliminated.

b. Find the *quartets* (from non-EPI pairs) of 4 pairwise logically adjacent points. From Figure 3.38, quartet (29, 28, 21*, 20*) is an EPI since it has 2 distinguished minterm members (those with only 2 links).

$$(29,\ 28,\ 21^*,\ 20^*) = 11101(8,1) = E\not{A}BC'\not{D} = EBC'$$

The link variables ($A = 8, D = 1$) are eliminated.

c. Find the *octets* (from non-EPI quartets) of 8 pairwise logically adjacent points. From Figure 3.38, octet (29, 28, 25, 24*, 13*, 12*, 9, 8*) is an EPI since it has distinguished minterm members (those with only 3 links).

$$(29,28,25,24^*,13^*,12^*,9,8^*) = 11101(16,4,1) = \not{E}A\not{B}C'\not{D} = AC'$$

The link variables ($E = 16, B = 4, D = 1$) are eliminated.

FIGURE 3.38
Minterm ring map of function $f(E, A, B, C, D)$

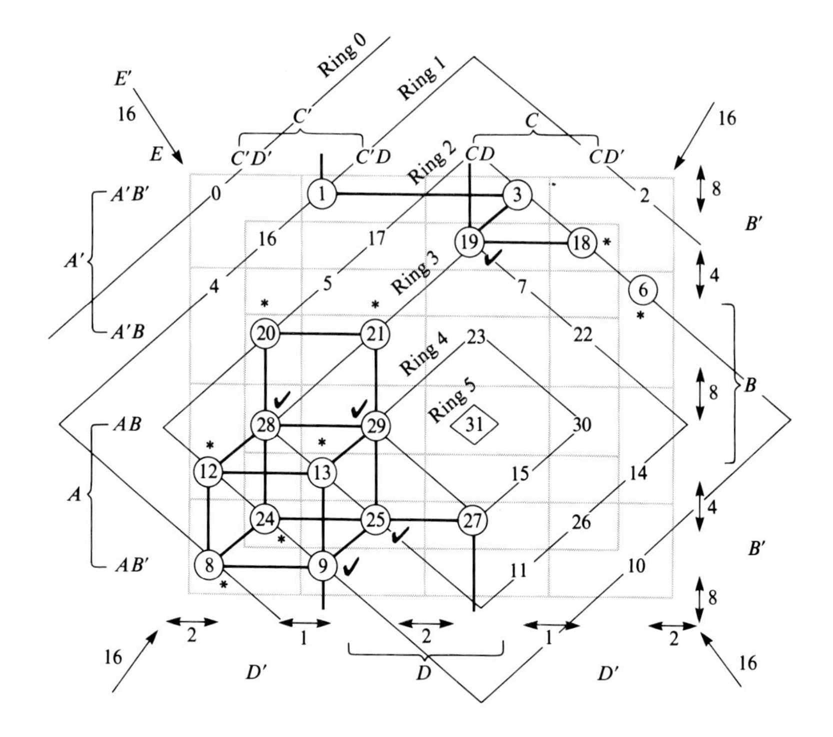

d. Find groups of 16 (from non-EPI octets) of 16 pairwise logically adjacent points. For our example, there are no 16-groups in Figure 3.38.

We can now check off the minterm members of the EPIs determined in Steps 1a, b, c, and d. The remaining (unchecked) minterms are 1, 3, and 27. Thus, Step 1 of the algorithm produces a reduced PI table as follows:

	Unchecked Minterms		
Non-EPI PIs	1	3	27
(3, 1)	x	x	
(9, 1)	x		
(19, 3)		x	
(27, 25)			x
(27, 19)			x

in which each column is an unchecked minterm and each row is a non-EPI PI that covers one or more unchecked minterms. In this example, the unchecked minterms are covered by the following non-EIP pairs:

	(3, 1),	(27, 25),	(27, 19),	(9, 1),	(19, 3)
Decimal (link):	3(2),	27(2),	27(8),	9(8),	19(16)
Binary (link):	00011(2),	11011(2),	11011(8),	01001(8),	10011(16)
Literal:	E'A'B'~~C'~~D,	EAB'~~C'~~D,	E~~A~~B'CD,	E'~~A'~~B'C'D,	~~E~~A'B'CD
	E'A'B'D,	EAB'D,	EB'CD,	E'B'C'D,	A'B'CD

Find the PIs and EPIs for the switching function
f(E, A, B, C, D) = Σ (1, 3, 6, 8, 9, 12, 13, 18, 19, 20, 21, 24, 25, 27, 28, 29, , , , ,)
* * ✓ * * * ✓ * * * ✓ ✓ ✓

Step 0. Circle minterms on minterm ring map.

Step 1. Find PIs and EPIs. (Start with highest # M in Steps a, b, c, and d)

a. Find the PAIRS:

For each M
For each x = {1, 2, 4, 8, 16, 32}
If M − x ∈ Σ, store M − x & increment M's CTd.
If M + x ∈ Σ, store M + x & increment M's CTu.
CTa = CTd + CTu.

If CTa = 0, M is D.M.* 6*

If CTa = 1, M is D.M.* and (M, M ± x) is EPI pair. {18*, 19}

b. Find QUARTETS (among non-EPI pairs):

For each unstarred M with CTd ⩾ 2
For each x, y combination, check 2 pairs (M, M − x), (M, M − y)
If M − x − y ∈ Σ, store M − x − y & increment M's CTq:
Q = {M, M − x, M − y, M − x − y} is Quartet.
{29, 28, 21*, 20*}
If any Q member q has CTa = 2, q is D.M.* and Q is EPI quartet.

c. Find OCTETS (among non-EPI quartets):

For each unstarred M with CTd ⩾ 3 & CTq ⩾ 3
For each x, y, z combination

If M − x − y − z ∈ Σ, store M − x − y − z & increment M's CTo.

O = {M, M − x, M − y, M − z, M − x − y, M − x − z, M − y − z,
{29, 28, 25, 13*, 24*, 12*, 9,
M − x − y − z} is octet.
8*}

If any member o has CTa = 3, o is D.M.* and O is EPI octet.

d. Find 16-GROUP (among non-EPI octets):

Check off the minterm members of the EPIs determined in Steps a, b, c, and d.
Unchecked minterms: 1, 3, 27

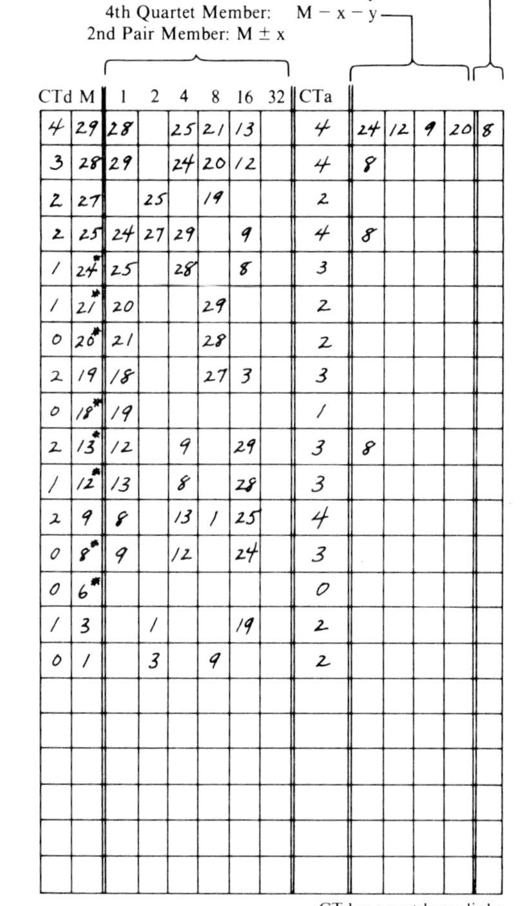

8th Octet Member: M − x − y − z
4th Quartet Member: M − x − y
2nd Pair Member: M ± x

CTd	M	1	2	4	8	16	32	CTa					
4	29	28		25	21	13		4	24	12	9	20	8
3	28	29		24	20	12		4	8				
2	27		25		19			2					
2	25	24	27	29		9		4	8				
1	24*	25		28		8		3					
1	21*	20			29			2					
0	20*	21			28			2					
2	19	18			27	3		3					
0	18*	19						1					
2	13*	12		9		29		3	8				
1	12*	13		8		28		3					
2	9	8		13	1	25		4					
0	8*	9		12		24		3					
0	6*							0					
1	3		1			19		2					
0	1		3		9			2					

CTd = countdown links
CTu = countup links
CTa = countall links

Step 2. List the remaining (non-EPI) pairs, quartets, octets:

Non-EPI pairs: (3, 1), (27, 25), (27, 19), (9, 1), (19, 3)

Non-EPI quartets: (, , ,), (, , ,), (, , ,)

Non-EPI octets: (, , , , , , ,), (, , , , , , ,)

The remaining (non-EPI) pairs, quartets, and octets compose the reduced PI table.

Step 2. Determine the smallest number of non-EPI PIs that cover the unchecked minterms.

	Non-EPI PIs	Unchecked Minterms 1	3	27
	(3, 1)	x	x	
Dominated by (3, 1)	(9, 1)	x		
Dominated by (3, 1)	(19, 3)		x	
Choice {	(27, 25)			x
	(27, 19)			x

Having found the EPIs and the smallest number of non-EPIs that form a minimal cover, we can write a minimal sum for f.

$$f = AC' + EBC' + EA'B'C + E'A'BCD' + E'A'B'D + \begin{cases} EAB'D \\ EB'CD \end{cases}$$

Note that f is the same function plotted in Example 3.8 using two 4-variable K-maps.

Minterm-Ring Algorithm Procedure Guide

The manual use of the minterm-ring algorithm is facilitated by a procedure guide or template such as the one shown on page 129. We will illustrate the use of this guide for 6 variables on page 133.

Minterm-Ring Maps and K-Maps in 6 Variables

A 6-variable K-map is sometimes pictured as four 4-variable K-maps, which are arranged in a four-plane stack or in the configuration shown in Figure 3.39. Studying the figure, you can see that the cell number of a given cell in the lower-left map differs by 16, 32, and 48, respectively, from the corresponding cell in the upper-left, lower-right, and upper-right maps.

Each lower-left map cell is logically adjacent to the corresponding upper-left map cell, differing in variable E. Each lower-left map cell is logically adjacent to the corresponding lower-right map cell, differing in variable F. Each upper-left map cell is logically adjacent to the corresponding upper-right map cell, differing in variable F. Each lower-right map cell is logically adjacent to the corresponding upper-right map cell, differing in variable E.

The function

$$f = \Sigma(0, 1, 4, 15, 16, 17, 19, 20, 21, 22, 23, 34, 40, 44, 48, 49, 50, 51, 52, 53, 55, 60, 61)$$

is plotted in Figure 3.39 to illustrate the use of four 16-cell K-maps. The same function is plotted in Figure 3.40 so that we see how the three-dimensional perspective and link counting of the minterm-ring map facilitates identification of the PIs and EPIs of a function.

A 6-variable minterm-ring map can be formed by superimposing four 4-variable K-

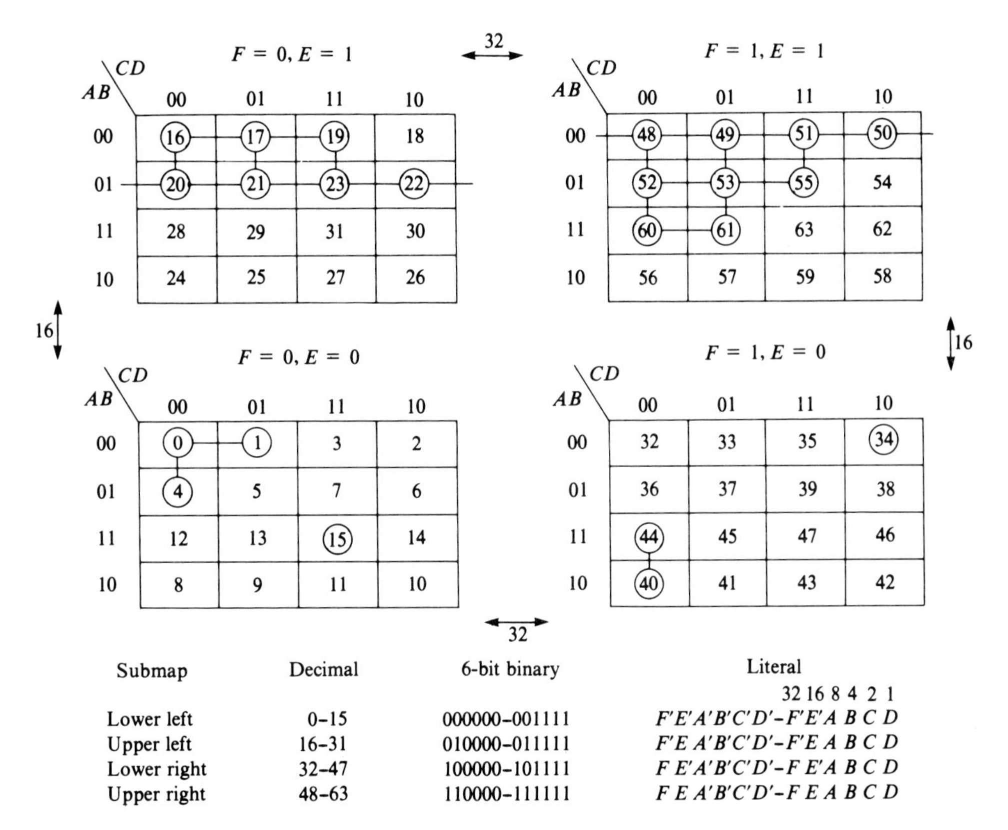

Submap	Decimal	6-bit binary	Literal (32 16 8 4 2 1)
Lower left	0–15	000000–001111	$F'E'A'B'C'D'$–$F'E'ABCD$
Upper left	16–31	010000–011111	$F'EA'B'C'D'$–$F'EABCD$
Lower right	32–47	100000–101111	$FE'A'B'C'D'$–$FE'ABCD$
Upper right	48–63	110000–111111	$FEA'B'C'D'$–$FEABCD$

FIGURE 3.39
Diagram of four 16-cell K-maps

maps, as illustrated in Figure 3.40. In this map, a given cell number is logically adjacent to any adjacent ring cell that is to the left, right, up, down, or on an inward diagonal (16 or 32). To find a minimal cover and corresponding minimal sum for function f we will use the minterm-ring algorithm that we used for the 5-variable case.

Minterm-Ring Algorithm for 6 Variables

The minterm-ring algorithm can be used to find a minimal cover of a Boolean function f with up to 8 variables (the accompanying map shows rings of minterms by number of 1s). The following function f is used for illustration:

$$f = \Sigma(0,1,4,15,16,17,19,20,21,22,23,34,40,44,48,49,50,51,52,53,55,60,61)$$

A detailed description of the application of the minterm-ring algorithm for the function illustrated in Figure 3.40 (p. 132) follows:

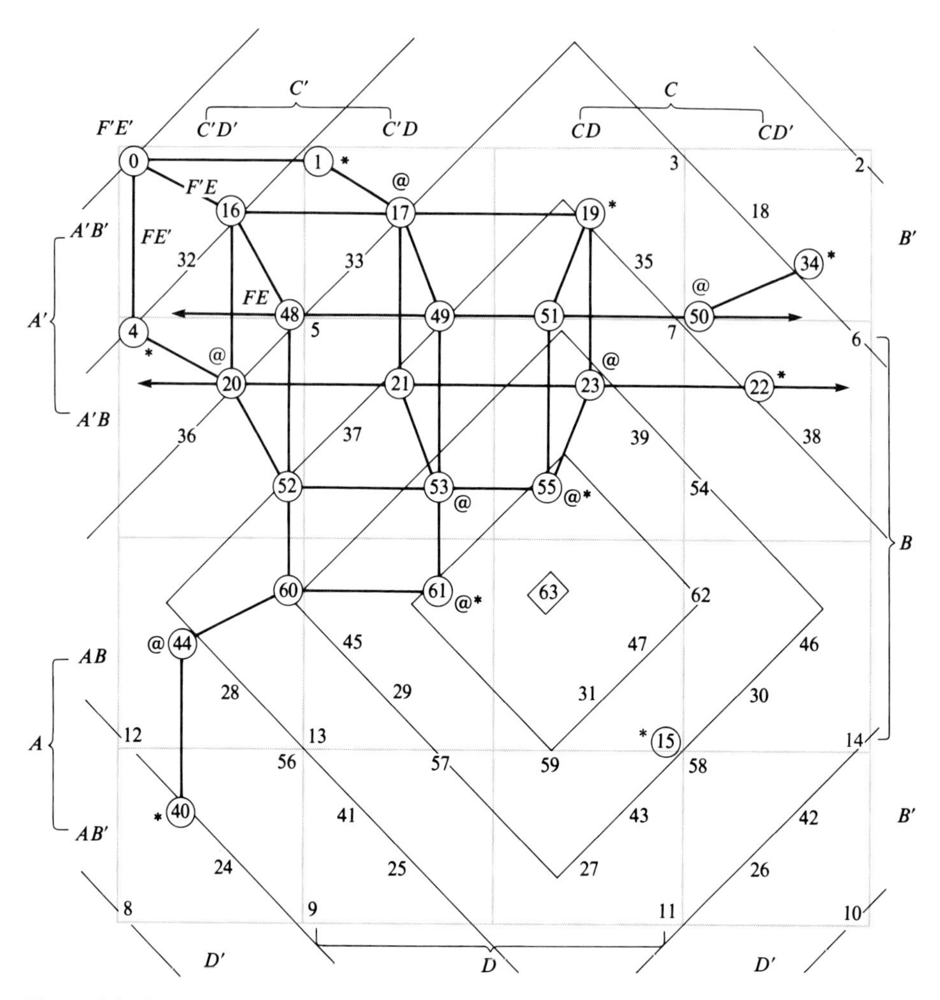

FIGURE 3.40
A 6-variable minterm ring map

Step 0 Circle each minterm of specified function f on the map.

Step 1. Find the PIs and EPIs.

a. Find pairs. Start with the highest-numbered minterm M and connect adjacent-ring logically adjacent minterms by looking left, right, up, down, and diagonally inward from M.

EPI minterm* has no links:

$$15^* = 001111 = F'E'ABCD$$

Find the PIs and EPIs for the switching function

f(F, E, A, B, C, D) = Σ (0 , 1 , 4 , 15 , 16 , 17 , 19 , 20 , 21 , 22 , 23 , 34 , 40 , 44 , 48 , 49 , 50 , 51 , 52 , 53 , 55 , 60 , 61).

✓ * * * ✓ ✓ * ✓ ✓ * ✓ * * ✓ ✓ ✓ ✓ ✓ ✓ * ✓ *

Step 0. Circle minterms on minterm ring map.

Step 1. Find PIs and EPIs. (Start with highest # M in Steps a, b, c, and d)

a. Find the PAIRS:

For each M
For each x = 1, 2, 4, 8, 16, 32
If M − x ∈ Σ, store M − x & increment M's CTd.
If M + x ∈ Σ, store M + x & increment M's CTu.
CTa = CTd + CTu.

If CTa = 0, M is D.M.* 15*

If CTa = 1, M is D.M.* and (M, M ± x) is EPI pair. {34*50}{40,44}

b. Find QUARTETS (among non-EPI pairs):

For each unstarred M with CTd ⩾ 2
For each x, y combination, check 2 pairs (M, M − x), (M, M − y)
If M − x − y ∈ Σ, store M − x − y & increment M's CTq:
Q = {M, M − x, M − y, M − x − y} is Quartet.
{61*, 60, 53, 52}, {20, 16, 4*, 0}, {23, 22*, 21, 20}, {17, 16, 1*, 0}

If any Q member q has CTa = 2, q is D.M.* and Q is EPI quartet.

c. Find OCTETS (among non-EPI quartets):

For each unstarred M with CTd ⩾ 3 & CTq ⩾ 3
For each x, y, z combination

If M − x − y − z ∈ Σ, store M − x − y − z & increment M's CTo.

O = {M, M − x, M − y, M − z, M − x − y, M − x − z, M − y − z, M − x − y − z} is octet.
{55*, 53, 51, 23, 49, 21, 19*, 17}

If any member o has CTa = 3, o is D.M.* and O is EPI octet.

d. Find 16-GROUP (among non-EPI octets):

8th Octet Member: M − x − y − z
4th Quartet Member: M − x − y
2nd Pair Member: M ± x

CTd	M	1	2	4	8	16	32	CTa					
2	61*	60			53			2	52				
2	60	61			52	44		3					
3	55*		53	51			23	3	49	21	19		17
3	53	52	55	49	61		21	5	48	20	17		16
2	52	53		48	60		20	4	16				
3	51	50	49	55			19	4	48	17			
2	50	51	48			34		3					
2	49	48	51	53			17	4	16				
1	48	49	50	52			16	4					
1	44			40		60		2					
0	40*			44				1					
0	34*					50		1					
3	23	22	21	19			55	4	20	17			
1	22*	23	20					2					
2	21	20	23	17			53	4	16				
2	20	21	22	16		4	52	5	0				
1	19*		17	23			51	3					
2	17	16	19	21		1	49	5	0				
1	16	17		20		0	48	4					
0	15*							0					
1	4*			0		20		2					
1	1*	0				17		2					

CTd = countdown links
CTu = countup links
CTa = countall links

Check off the minterm members of the EPIs determined in Steps a, b, c, and d.
Unchecked minterms: 48

Step 2. List the remaining (non-EPI) pairs, quartets, octets:

Non-EPI pairs: (50, 48), (,), (,), (,), (,), (,)

Non-EPI quartets: (51, 50, 49, 48), (, , ,), (, , ,), (, , ,), (, , ,)

Non-EPI octets: (53, 52, 49, 48, 21, 20, 17, 16), (, , , , , , ,), (, , , , , , ,)

The remaining (non-EPI) pairs, quartets, and octets compose the reduced PI table.

EPI pair has member* with only one link:

(44,40*) = 101100(4) = $FE'AC'D'$
(50,34*) = 110010(16) = $FA'B'CD'$

b. Find the quartets of 4 connected points (from non-EPI pairs):
EPI quartet has member* with only two links:

(61*,60,53,52) = 111101(8,1) = $FEBC'$
(23,22*,21,20) = 010111(2,1) = $F'EA'B$
(20,16,4*,0) = 010100(16,4) = $F'A'C'D'$
(17,16,1*,0) = 010001(16,1) = $F'A'B'C'$

c. Find the octets of 8 connected points (from non-EPI quartets):
EPI octet has member* with only three links:

(55*,53,51,49,23,21,19*,17) = 110111(32,4,2) = $EA'D$

d. Find groups of 16 connected points (from non-EPI octets):

In Steps 1a, b, c, and d, check minterms of f covered by EPIs*.

Step 2. Find the smallest number of non-EPIs to cover unchecked minterms:

Non-EPI pairs: (50,48)
Decimal (link) 50(2)
Binary (link) 110010(2)
Literal $FEA'B'D'$

Non-EPI quartets: (51,50,49,48)
Decimal (links) 51(2, 1),
Binary (links) 110011(2, 1)
Literal $FEA'B'$

Non-EPI octets: (53, 52, 49, 48, 21, 20, 17, 16)
Decimal (links) 53(32, 4, 1)
Binary (links) 110101(32, 4, 1)
Literal $EA'C'$

Note that the minterm-ring algorithm directly produces a reduced PI table:

Non-EPIs		Unchecked Minterms 48
$FEA'B'D'$	(50, 48)	x
$FEA'B'$	(51,50,49,48)	x
$EA'C'$	(53,52,49,48,21,20,17,16)	x

Having found the PIs, EPIs, and the smallest number of non-EPIs that form a minimal cover (1 and 2), we can write a minimal sum for f.

$$f = EA'D + FEBC' + F'EA'B + F'A'C'D' + F'A'B'C' + FA'B'CD' + FE'AC'D' + F'E'ABCD + EA'C'$$

Advantages of the Minterm-Ring Algorithm

The advantages of using links to connect logically adjacent minterms (in Karnaugh and minterm-ring maps) include

Reducing the clutter caused by enclosing PIs with loops

Facilitating identification of dominant PIs

Facilitating the determination of a nonunique minimal sum and, thereby, simplifying the formation of alternative minimal sums

Manual use of the minterm-ring algorithm is facilitated by a procedure that

Simplifies the tabulation of minterm members of implicant pairs, quartets, octets, and so on

Identifies distinguished minterms (and, hence, EPIs) by counting the links of each minterm

Directly produces a reduced PI table by identifying EPIs and checking off their minterms

Since most logic minimization currently is accomplished with software tools, the minterm-ring method is important because it combines a contemporary algorithmic method and a three-dimensional mapping technique for simplifying Boolean expressions and identifying PIs and EPIs. The derivation of the minterm-ring algorithm also provides an excellent illustration of how algorithms are developed.

3.6 THE QUINE–MCCLUSKEY METHOD FOR SIMPLIFYING BOOLEAN EXPRESSIONS

The* Quine–McCluskey method *is a classical tabular method that can be used to simplify SOP Boolean expressions.

Briefly, the procedure can be organized as follows:

Step 1. Find the PIs.

- **a.** Arrange the minterms of function f in groups according to the number of 1s in the binary cell number.
- **b.** For minterms in adjacent groups, pair those minterms that differ in only 1 variable.
- **c.** For pairs in adjacent groups, group those pairs that differ in only 1 variable.
- **d.** For quartets in adjacent groups, group those quartets that differ in only 1 variable.
- **e.** Continue this process of forming blocks of size 2^k, until no larger rectangular blocks of minterms can be found.

Step 2. Find the EPIs.

a. Form a PI table in which each minterm is a column heading and each PI is a row heading; place an X in each column of a row if that minterm is contained in the row's PI.

b. If any column in the PI table contains only one X, that minterm is a distinguished minterm (*), and the corresponding row is an EPI. Check off all columns that are covered by an EPI.

c. If all minterms of *f* are covered by the EPIs, the EPIs form a minimal cover; if not, continue to Step 3.

Step 3. Find the smallest collection of non-EPIs that cover the unchecked minterms.

a. Form a reduced PI table in which each remaining minterm is a column heading and each non-EPI PI is a row heading; place an X in each column of a row if that minterm is contained in the row's PI.

b. Find dominated rows and dominating columns.

(1) Row i dominates row j if row i covers at least each minterm (column) covered by row j. A row dominated by another row can be eliminated if the dominated row PI has as many or more literals than the dominating row PI.

(2) Column m dominates column n if column m contains an X in at least each row in which column n contains an X. The PIs that cover a dominated column also cover the dominating column. Since a column need be covered only once, the dominating column can be eliminated.

(3) Remove each dominated row and each dominating column; the remaining rows and columns comprise a final reduced PI table.

If any column in the final reduced PI table contains a single X, that minterm is a **secondary distinguished minterm** +, and the corresponding row is a secondary EPI(+). Check off all columns covered by the secondary EPI(s). If all minterms are checked off, the EPIs and secondary EPIs form a minimal cover for *f*. If unchecked minterms remain, begin Step 3 again by forming a new reduced PI table.

Table 3.3 illustrates the Quine–McCluskey tabular method for simplifying the following Boolean expression:

$$f = \Sigma(0,1,4,7,8,9,10,11,12,14,15,16,17,20,22,26,27,29,30,31)$$

Table 3.3 enumerates the implicants (pairs, quartets, and octets) formed by combining logically adjacent minterms of *f*. The PIs of *f* are those implicants in Table 3.3 that are not contained in a larger implicant group.

A PI table is constructed by listing the minterms as columns and prime implicants as rows, as illustrated in Table 3.4. After the PI table is constructed, we must examine each column to determine if the corresponding minterm belongs to only one PI. Any minterm that belongs to only one PI is a distinguished minterm (denoted by *), and that PI is an EPI.

TABLE 3.3
Implicants determined using Quine–McCluskey method

Minterm	Pairs	Quartets	Octets
31			
30	31, 30		
29	31, 29		
27	31, 27		
15	31, 15		
26	30, 26 27, 26	31, 30, 27, 26	
22	30, 22		
14	30, 14 15, 14	31, 30, 15, 14	
11	27, 11 15, 11	31, 27, 15, 11	
7	15, 7		
20	22, 20		
17			
12	14, 12		
10	26, 10 14, 10 11, 10	27, 26, 11, 10	
9	11, 9	15, 14, 11, 10	31, 30, 27, 26, 15, 14, 11, 10
16	20, 16 17, 16		
8	12, 8 10, 8 9, 8	14, 12, 10, 8	
4	20, 4 12, 4	11, 10, 9, 8	
1	17, 1 9, 1		
0	16, 0 8, 0 4, 0 1, 0	20, 16, 4, 0	
		17, 16, 1, 0	
		12, 8, 4, 0	
		9, 8, 1, 0	

TABLE 3.4
PI table for function *f*

	✔ 0	✔ 1	4	* 7	8	9	✔ 10	✔ 11	12	✔ 14	✔ 15	✔ 16	* 17	20	22	* 26	* 27	* 29	✔ 30	✔ 31
31, 30, 27*, 26*, 15, 14, 11, 10							X	X		X	X					X	X		X	X
14, 12, 10, 8					X		X		X	X										
11, 10, 9, 8					X	X	X	X												
20, 16, 4, 0	X		X									X		X						
17*, 16, 1, 0	X	X										X	X							
9, 8, 1, 0	X	X			X	X														
12, 8, 4, 0	X		X		X				X											
31, 29*																		X		X
30, 22															X				X	
15, 7*				X							X									
22, 20														X	X					

*denotes a distinguished minterm

Check off all columns for the minterms covered by each EPI. After all the EPIs have been accounted for, the unchecked columns (minterms) and non-EPIs are used to form a reduced PI table, as shown in Table 3.5.

TABLE 3.5
Reduced PI table for *f*

	4	8	9	12	20	22	
20, 16, 4, 0	X				X		
12, 8, 4, 0	X	X		X			
14, 12, 10, 8		X		X			Dominated by 12, 8, 4, 0
11, 10, 9, 8		X	X				Choice
9, 8, 1, 0		X	X				Choice
22, 20					X	X	
30, 22						X	Dominated by 22, 20

The unchecked minterms (4, 8, 9, 12, 20, 22) are covered by the quartets (12, 8, 4, 0), the pair (22, 20), and either quartet (11, 10, 9, 8) or (9, 8, 1, 0). These non-EPIs and the EPIs make up a minimal cover for the function *f*. Therefore, the function *f* can be represented by either of the following minimal sums:

$$f = AC + A'B'C' + EABD + E'BCD + EC'D' + EA'BD' + \begin{cases} E'AB' \\ E'B'C' \end{cases}$$

A comparison of the tabular Quine–McCluskey method and the K-map technique can be accomplished by using two 4-variable K-maps, shown in Figure 3.41, to simplify the same expression that was reduced using the Quine–McCluskey method.

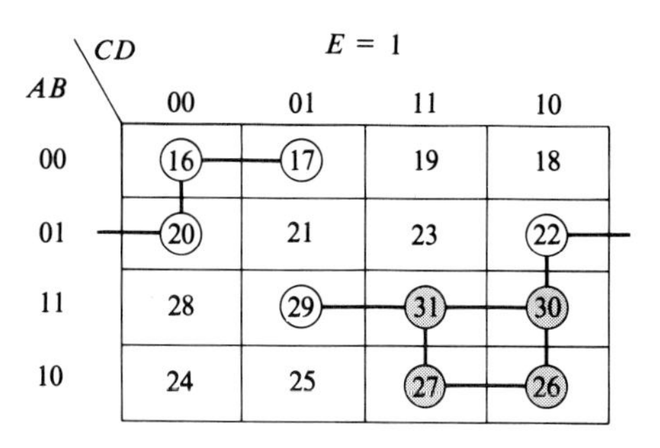

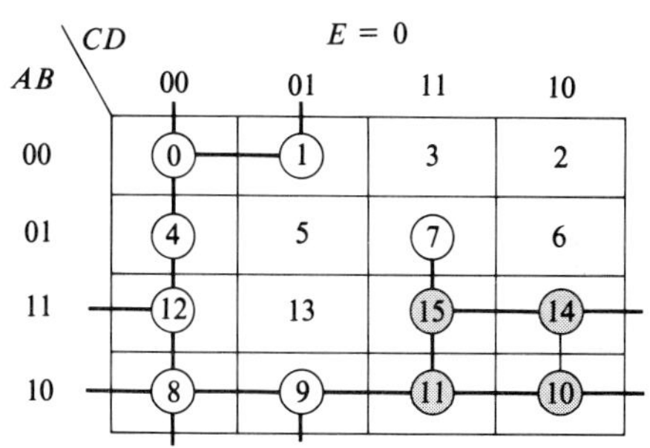

FIGURE 3.41
Using two 4-variable K-maps to plot a 5-variable function

A comparison of the tabular Quine–McCluskey method and the minterm-ring algorithm can be accomplished by using a 5-variable minterm ring map, shown in Figure 3.42, to simplify the same expression that was reduced using the Quine–McCluskey method.

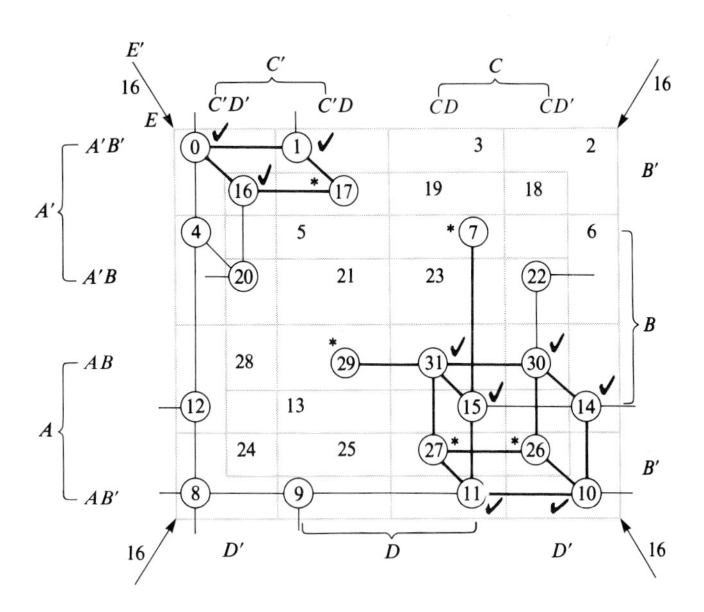

FIGURE 3.42
Minterm ring map of function $f(E, A, B, C, D)$

Manual use of the Quine–McCluskey method is cumbersome for 4 or more variables because

- Identification of implicant pairs, quartets, and octets requires comparison of binary representations of minterms, pairs, and so on.
- It requires generation of a complete PI table, containing both EPIs and non-EPIs.
- It requires reduction of the PI table by identifying EPIs by locating minterms that belong to only one PI.

It can be seen from the preceding sections that the McCalla minterm-ring algorithm, with accompanying minterm-ring maps for visual identification of PIs and EPIs, is a better method for simplifying Boolean expressions than the Quine–McCluskey method.

3.7 VARIABLE-ENTERED MAPS

Variable-entered mapping is a method by which a Boolean function in **n** ***variables can be plotted in fewer than*** **n** ***dimensions. In this method, one or more variables appear within the map and, consequently, are referred to as*** **map-entered variables.** ***We will describe this method first for a switching function in 3 variables.***

Variable-Entered Mapping of Functions in 3 Variables

Let $f(A, B, C)$ be a switching function written in canonical SOP form. If the terms containing A' are grouped first, followed by the terms containing A, then

$$f(A, B, C) = A' \cdot (\text{some terms in } B, C) + A \cdot (\text{some terms in } B, C)$$

■ **EXAMPLE 3.10**

Suppose we are given the following function:

$$\begin{aligned} f(A, B, C) &= A'B'C' + A'B'C + AB'C' + ABC \\ &= A'(B'C' + B'C) + A(B'C' + BC) \end{aligned}$$

The terms in B, C associated with A' compose a function $G(B, C)$, and the terms in B, C associated with A compose a function $H(B, C)$. Therefore,

$$f(A, B, C) = A' \cdot G(B, C) + A \cdot H(B, C) \qquad \textbf{(3.1)}$$

where

$$G(B, C) = g_0 \cdot B' \cdot C' + g_1 \cdot B' \cdot C + g_2 \cdot B \cdot C' + g_3 \cdot B \cdot C \qquad \textbf{(3.2a)}$$

$$H(B, C) = h_0 \cdot B' \cdot C' + h_1 \cdot B' \cdot C + h_2 \cdot B \cdot C' + h_3 \cdot B \cdot C \qquad \textbf{(3.2b)}$$

with g_i in $\{0, 1\}$ and h_i in $\{0, 1\}$.

Thus, a function $f(A, B, C)$ in 3 variables can be expressed in terms of functions $G(B, C)$ and $H(B, C)$ in 2 variables. ■

If Equations (3.2a) and (3.2b) are ANDed by A' and A, respectively, it follows that

$$\begin{aligned} A' \cdot G(B, C) &= A'g_0\, B'C' + A'g_1\, B'C + A'g_2\, BC' + A'g_3\, BC \\ A \cdot H(B, C) &= Ah_0\, B'C' + Ah_1\, B'C + Ah_2\, BC' + Ah_3\, BC \end{aligned} \quad \textbf{(3.3)}$$

Substituting Equation (3.3) into Equation (3.1) and collecting terms, we find

$$\begin{aligned} f(A, B, C) &= (A'g_0 + Ah_0)B'C' + (A'g_1 + Ah_1)B'C + (A'g_2 + Ah_2)BC' + (A'g_3 + Ah_3)BC \\ &= f_0\, B'C' + f_1\, B'C + f_2\, BC' + f_3\, BC \end{aligned}$$

Therefore,

$$\begin{aligned} f_0 &= A'g_0 + Ah_0 \\ f_1 &= A'g_1 + Ah_1 \\ f_2 &= A'g_2 + Ah_2 \\ f_3 &= A'g_3 + Ah_3 \end{aligned}$$

The function $f(A, B, C)$ in 3 variables can be represented by a variable-entered map in 2 variables, as shown in Figure 3.43.

FIGURE 3.43
Illustration of variable-entered mapping

f B \ C	0	1
0	$f_0 = A'g_0 + Ah_0$	$f_1 = A'g_1 + Ah_1$
1	$f_2 = A'g_2 + Ah_2$	$f_3 = A'g_3 + Ah_3$

The function $f = A' \cdot G + A \cdot H$ is called a **map-entered function**. It should be noted that each cell of the variable-entered map is a function of the literals A' and A, and A is a map-entered variable.

Since g_i, h_i each has value 0 or 1, any g_i, h_i pair may have one of four possible combinations of values:

If g_i, h_i = 0, 0, then cell value = 0, independent of A
0, 1 = A
1, 0 = A'
1, 1 = $A' + A = 1$

EXAMPLE 3.11

Consider variable-entered mapping in this way. Suppose

$$f(A, B, C) = A'B'C' + A'B'C + AB'C' + ABC$$

The function f can be represented by the following truth table:

	A B C	f
A'	0 0 0	$1 = g_0$
	0 0 1	$1 = g_1$
	0 1 0	$0 = g_2$
	0 1 1	$0 = g_3$

	A B C	f
A	1 0 0	$1 = h_0$
	1 0 1	$0 = h_1$
	1 1 0	$0 = h_2$
	1 1 1	$1 = h_3$

$$f_0 = A'g_0 + Ah_0 = A' \cdot 1 + A \cdot 1 = A' + A = 1$$
$$f_1 = A'g_1 + Ah_1 = A' \cdot 1 + A \cdot 0 = A'$$
$$f_2 = A'g_2 + Ah_2 = A' \cdot 0 + A \cdot 0 = 0$$
$$f_3 = A'g_3 + Ah_3 = A' \cdot 0 + A \cdot 1 = A$$

Figure 3.44a presents the function $f(A, B, C)$ mapped in a 3-variable K-map and Figure 3.44b presents the same function in a variable-entered map for comparison of the two maps for the same function.

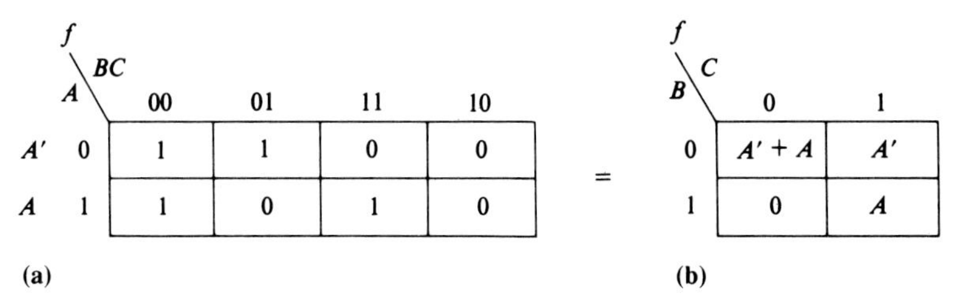

FIGURE 3.44
Comparison of two map techniques. (a) Standard K-map (b) Variable-entered map ■

Variable-Entered Mapping of Functions in 4 Variables

Let $f(A, B, C, D)$ be a switching function written in canonical SOP form. If the terms containing A' are grouped first, followed by the terms containing A, then

$$f(A, B, C, D) = A' \cdot G(B, C, D) + A \cdot H(B, C, D)$$

where

$$G(B, C, D) = g_0\, B'C'D' + g_1\, B'C'D + \cdots + g_7\, BCD$$
$$H(B, C, D) = h_0\, B'C'D' + h_1\, B'C'D + \cdots + h_7\, BCD$$

with g_i in $\{0, 1\}$ and h_i in $\{0, 1\}$. It follows that

$$\begin{aligned} f(A, B, C, D) &= (A'g_0 + Ah_0)B'C'D' + (A'g_1 + Ah_1)B'C'D \\ &\quad + (A'g_2 + Ah_2)B'C\,D' + \cdots + (A'g_7 + Ah_7)BCD \\ &= f_0\, B'C'D' + f_1\, B'C'D + f_2\, B'CD' + \cdots + f_7\, BCD \end{aligned}$$

The following example illustrates how a function $f(A, B, C, D)$ in 4 variables can be represented by a variable-entered map in 3 variables.

■ **EXAMPLE 3.12**

Consider the switching function

$$f(A, B, C, D) = \Sigma(0,1,3,5,7,11,12,15)$$

The 4-variable K-map in Figure 3.45 shows that the function $f(A, B, C, D)$ can be represented in SOP form as

$$f(A, B, C, D) = A'D + CD + A'B'C' + ABC'D'$$ ■

Decimal	Inputs A	B	C	D	Output f
0	0	0	0	0	$1 = g_0$
1	0	0	0	1	$1 = g_1$
2	0	0	1	0	$0 = g_2$
3	0	0	1	1	$1 = g_3$
4	0	1	0	0	$0 = g_4$
5	0	1	0	1	$1 = g_5$
6	0	1	1	0	$0 = g_6$
7	0	1	1	1	$1 = g_7$
8	1	0	0	0	$0 = h_0$
9	1	0	0	1	$0 = h_1$
10	1	0	1	0	$0 = h_2$
11	1	0	1	1	$1 = h_3$
12	1	1	0	0	$1 = h_4$
13	1	1	0	1	$0 = h_5$
14	1	1	1	0	$0 = h_6$
15	1	1	1	1	$1 = h_7$

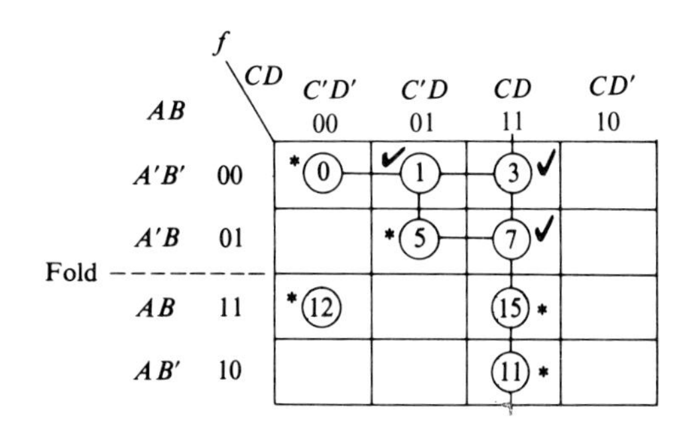

FIGURE 3.45
Truth table and K-map for $f(A, B, C, D)$

A function $f(A, B, C, D)$ can be plotted in a variable-entered map as follows:

1. Fold the 4-variable K-map on the $A'A$ boundary to form a folded 4-variable K-map. Circle the decimal number of each minterm of function f as shown in Figure 3.46.

FIGURE 3.46
Folded 4-variable K-map

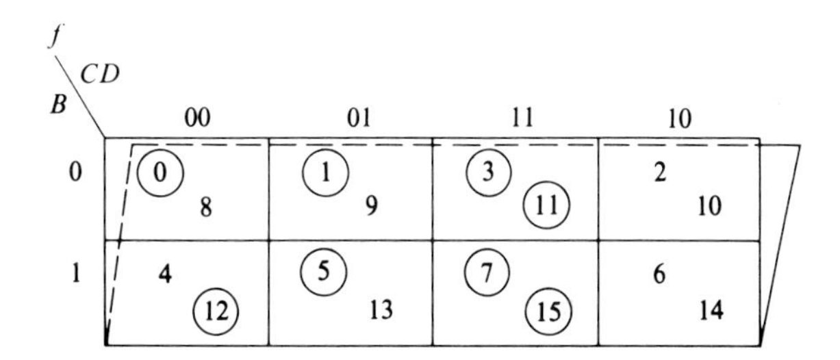

2. Convert the folded 4-variable K-map to a variable-entered map as shown in Figure 3.47.
   a. If circled number is ≤ 7, write A' in corresponding cell
   b. If circled number is ≥ 8, write A in corresponding cell
   c. If both A' and A appear in the same cell, enter 1 in that cell

FIGURE 3.47
Variable-entered map

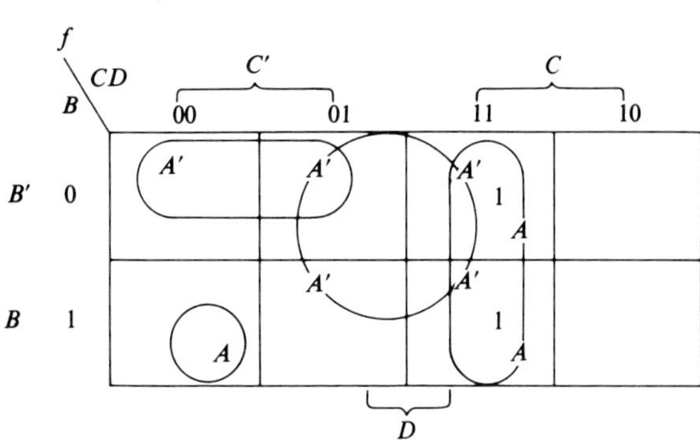

3. Combine like terms to form groups in the variable-entered map:
   a. Circle any lone literal that will not group with an adjacent corresponding literal;
   b. Circle any lone 1 or any pair of adjacent matching literals (at least one being a lone literal) that will not group in a larger group;
   c. Circle pairs of adjacent 1s or quartets of adjacent matching literals.

The variable-entered map can be used to determine SOP expression for f as follows:

$$f = A'D + CD + A'B'C' + ABC'D'$$

3.8 SUMMARY

In this chapter you have learned how to use map techniques and algorithmic methods to simplify Boolean expressions. The resulting expressions can be used to implement logically equivalent circuits that are simpler and, hence, more economical, more reliable, and often faster than a circuit implemented using the original expression in canonical form.

Map techniques and algorithmic methods for simplifying Boolean expressions are mechanical and, therefore, easier to use and less error prone than algebraic simplification techniques described in Chapter 2. Map techniques include K-maps and minterm-ring maps, both of which provide visual aids for identifying implicants. Algorithmic methods include the McCalla minterm-ring algorithm and the Quine–McCluskey method.

K-maps and minterm ring maps both have the property that physically adjacent minterms are logically adjacent. A Boolean expression can be simplified by map techniques that include

- Combining pairwise logically adjacent minterms into PIs of size 2^k to eliminate k variables
- Visual identification of PIs and EPIs
- Use of don't-care conditions to take advantage of input combinations that cannot occur

Map techniques can also be used to produce a simplified POS form of a Boolean expression. This approach is useful when a given function is to be implemented in NOR-NOR form.

Minterm-ring maps provide a three-dimensional perspective that facilitates identification of PIs and EPIs; hence, they are superior to K-maps for functions of 5 and 6 variables and can be extended to 7 or 8 variables.

The minterm-ring algorithm and the Quine–McCluskey method can be programmed on a computer to simplify Boolean expressions in any number of variables. The minterm-ring algorithm is more powerful than the Quine–McCluskey method since it automatically determines EPIs and directly produces a reduced PI table.

Variable-entered maps can be used to plot n-variable functions in less than n dimensions. As we will see in Chapter 10, variable-entered maps are especially useful in plotting control functions, with control variables as the map-entered variables.

You have also learned how to design a number of additional combinational logic circuits, such as inequality detectors, error detectors, parity generators, and divide-by-n circuits. In later chapters, we will use these and other circuits as functional modules to design and implement a variety of digital devices and systems, including computers.

Chapter 3 concludes Part I, which has developed the basic combinational logic design tools and hardware components (logic gates and integrated circuits) that you will use throughout the text. In Part II (Chapters 4, 5, and 6) you will learn how to design and implement a variety of combinational-logic computing devices, evolving from a 1-bit adder to an 8-bit arithmetic unit.

KEY TERMS

Karnaugh map technique (K-map)
Domain
Minterm
Pair
Quartet
Don't-care conditions
Implicant
Prime implicant (PI)
Essential prime implicant (EPI)
Distinguished minterm
Nonessential prime implicant (non-EPI)
Dominate
Minterm rings
Minterm-ring algorithm
Link
Octet
Missing links
Secondary distinguished minterm
Map-entered variables
Map-entered function
Quine–McCluskey method

EXERCISES

General

1. Given a Boolean function $f = A'B'C + A'BC + AB'C' + ABC'$, find a minimal SOP expression for f, using
   a. K-map method
   b. Simplification by algebraic reduction
2. Given a Boolean function $f = A'B'C + A'BC + AB'C' + AB'C$, find a minimal POS expression for f.
3. Given $f = A'B'C' + A'B'C + AB'C + ABC$, find a minimal SOP expression for f.
4. Given $f = A'B' + B'C' + AC + AB + BC'$, find a minimal SOP expression for f.
5. Convert $f = AB + C'$ to canonical SOP form, using a K-map as a visual aid.
6. Convert $f = AC + AB + B'$ to a canonical SOP form.
7. Convert $f = A \cdot (B + C)$ to a canonical POS form.
8. Convert $f = (A + B) \cdot (B' + C)$ to a canonical POS form.

9. Write a simplified expression for the Boolean function f defined by the following K-map:

f: A \ BC	00	01	11	10
0	0	1	0	0
1	0	1	1	1

10. Write a simplified expression for the Boolean function f defined by the following K-map:

f: A \ BC	00	01	11	10
0	1	1	1	1
1	0	0	0	1

11. Use a K-map to illustrate the Boolean function

$$F = B \cdot D + B' \cdot D'$$

12. Write a simplified expression for the Boolean function f defined by the following K-map:

f: AB \ CD	00	01	11	10
00	0	1	1	1
01	1	0	1	1
11	1	1	0	1
10	1	1	1	0

13. Determine the PIs and EPIs of f defined by the following K-map:

f: AB \ CD	00	01	11	10
00	0	1	1	0
01	0	0	1	1
11	1	0	1	1
10	0	0	0	0

14. Write a minimal SOP expression for the Boolean function defined by the following K-map:

f

AB \ CD	00	01	11	10
00	1	1	0	1
01	1	1	1	0
11	0	1	0	1
10	1	1	0	0

15. Write a simplified expression for the Boolean function f defined by the following K-map:

f

AB \ CD	00	01	11	10
00	0	0	1	1
01	0	0	0	0
11	X	X	X	X
10	1	1	X	X

16. Write a minimal SOP expression for the function defined by the following K-map:

f

AB \ CD	00	01	11	10
00	0	0	0	0
01	1	0	1	1
11	1	1	1	0
10	1	1	1	0

17. Determine if a unique, minimal SOP expression exists for the function f defined by the following K-map:

f

AB \ CD	00	01	11	10
00	1	0	1	0
01	0	1	1	1
11	0	1	0	1
10	1	1	0	0

18. Given $f = \Sigma(5, 7, 8, 11, 12, 13, 15)$, find the EPIs of f.

19. Given $f = \Sigma(3, 4, 5, 7, 9, 11, 12, 13)$, find the EPIs of f.

20. Given $f = \Sigma(0, 1, 2, 3, 5, 7, 8, 10, 13, 15)$, find
a. The EPIs of f
b. Equal PIs of f (that is, PIs that cover the same number of unchecked minterms)

21. Plot the function

$$f(E, A, B, C, D) = \Sigma(1, 3, 6, 8, 9, 12, 13, 18, 19, 20, 21, 24, 25, 27, 28, 29)$$

using two 4-variable K-maps.

22. Plot the function of Exercise 21 using a 5-variable minterm ring map.

23. Plot the function of Exercise 21 using a variable-entered map, with E as the map-entered variable.

24. Plot the function

$$f = \Sigma(0, 4, 6, 7, 12, 14, 15, 16, 17, 18, 22, 23, 28, 30, 31)$$

using a 5-variable minterm-ring map. Use the minterm ring algorithm to determine the EPIs of f.

25. Use the Quine–McCluskey method to determine the EPIs and PIs of the function in Exercise 24.

Design/Implementation

1. Design a combinational logic circuit with 3 inputs A, B, and C and output f, where $f = 1$ if, and only if, all 3 inputs are equal.

2. Design a 5-input minority circuit.

3. Design a 3-input combinational-logic circuit that has as its output a Boolean function f that has value 1 when a minority of its inputs (A, B, C) have value 1.
a. Write a truth table and canonical SOP expression for f.
b. Use a K-map to simplify the canonical SOP expression for f.
c. Implement the minority circuit using NAND gates only.

PART TWO

Evolution of Combinational-Logic Computing Devices

4 Design of Combinational-Logic Function Modules
5 Evolution from a 1-Bit Adder to an 8-Bit Arithmetic Unit
6 Logical and Decimal Operations—Hierarchical Development

A digital system is an integrated collection of functional components that together perform overall system functions. A system may be represented by a hierarchical model in which the system is divided into subsystems and the subsystems are further divided into sub-subsystems. The lowest level of the structure consists of functional modules, each of which performs a particular function. For example, a simple, stored program computer can be viewed as a digital system represented by a hierarchical (top-down) model as shown in Figure A (p. 152).

The design of digital systems is made easier by using function modules as building blocks to create higher-level devices. There are two types of function modules: combinational and sequential.

1. Combinational-logic function modules include devices such as adders, multiplexers, demultiplexers, encoders, decoders, code converters, parity generators, parity checkers, shifters, and programmable logic devices.
2. Sequential-logic function modules include devices such as counters, registers, shift registers, sequence generators, sequence checkers, and memories.

Part II, consisting of Chapters 4, 5, and 6, describes the design of common combinational-logic function modules and illustrates their use as building blocks of higher-level digital logic devices, such as arithmetic units and arithmetic logic units (ALU). Chapter 4 uses the 4-step problem-solving procedure to design

Chapters	Coverage
13–14	Design of various 8-bit computers
12	Building-block evolution of a computer
11	Digital systems design 2—Algorithmic-process control
10	Digital systems design 1—Algorithmic-process control
9	Asynchronous sequential logic
8	Synchronous sequential logic
7	Finite-state machine concepts: Latches, flip-flops, counters, registers, memory
6	Logical and decimal operations
5	Evolving combinational logic computing devices
4	Combinational-logic function modules
3	Simplifying Boolean expressions
2	Logic, switching circuits, Boolean algebra, logic gates
1	Number systems Computer codes

Various 8-bit computers
Computer
Processor
I/O
CPU
Memory
Data path
Control unit
Multiplier
Divider
Control unit
Algorithmic state machines
Asynchronous sequential logic circuits
Synchronous sequential logic circuits
Counters
Registers
Memory
Latches → Flip-flops
Arithmetic logic unit
BCD arithmetic unit
2-bit adder → 4-bit adder → Adder/subtracter → Arithmetic unit
Mux
Demux
Encoder
Decoder
Parity circuits
PLA
PAL
ROM
Shifters
1-bit adder
Error detector
BCD divide
Inequality detector
Gates: AND OR NOT XOR NAND NOR
Simple control circuits and computing devices

Decimal, binary, octal, hexadecimal
BCD, symmetric, reflected, unit distance

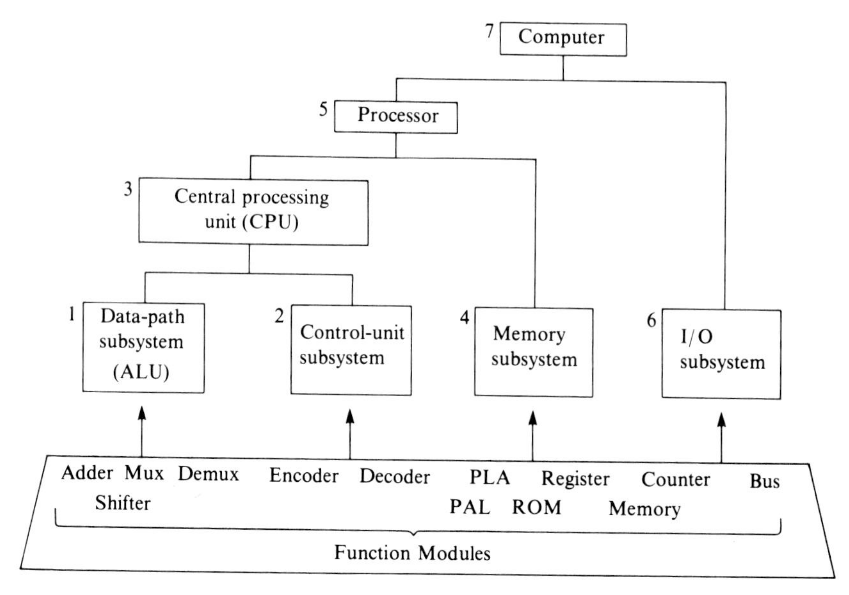

FIGURE A
A hierarchical model of a simple computer

and implement common combinational-logic function modules (multiplexers, demultiplexers, encoders, decoders, parity circuits, and shifters) that are used in digital systems. Chapter 4 also shows how arbitrary switching functions can be implemented using medium-scale integration (MSI) integrated circuits.

Chapter 5 uses an evolutionary hierarchical building-block approach (EHBBA) to describe the design and implementation of a variety of combinational-logic computing devices, ranging from a 1-bit adder to an 8-bit arithmetic unit. Chapter 6 describes a bit-slice design of a binary ALU and a top-down design of a BCD arithmetic unit.

The EHBBA diagram at the beginning of this part provides a roadmap to guide you in the design evolution of combinational-logic computing devices.

4 Design of Combinational-Logic Function Modules

OBJECTIVES

After you complete this chapter, you will be able

- □ To analyze combinational logic circuits using mixed-logic methods
- □ To convert combinational logic circuits from one form to another; for example, convert AND-OR form to NAND-NAND to OR-NAND to NOR-OR
- □ To identify and eliminate hazards in combinational logic circuits
- □ To design and implement specified combinational logic circuits, using the 4-step problem-solving procedure
- □ To design and implement common combinational-logic function modules that can be used as building blocks in various digital systems
- □ To read and interpret manufacturers' data books to select and use appropriate integrated-circuit function modules
- □ To implement arbitrary Boolean functions using medium-scale integration (MSI) integrated circuits (ICs) and programmable logic devices (PLDs)
- □ To develop proficiency in using digital-logic design tools such as Boolean algebra and K-maps
- □ To improve your problem-solving skills while designing a variety of combinational-logic function modules (multiplexers, demultiplexers, encoders, decoders, code converters, parity generators/checkers, among others)

4.1 INTRODUCTION

A variety of combinational-logic function modules are used as building blocks in the implementation of digital systems. This is somewhat analogous to building different types of motor vehicles (cars, trucks, tractors, and so forth) using basic functional components such as chassis, engine, transmission, drive train, axles, wheels, and so on.

A number of common function modules are designed in this chapter. We have two primary objectives for presenting these design examples:

1. To develop a set of common combinational-logic function modules to use as building blocks in the implementation of higher-level devices and various digital systems.
2. To provide examples of practical digital-logic design problems while illustrating the use of Boolean algebra, K-maps, and minterm-ring maps in the design of combinational logic circuits.

We will look at the following combinational-logic function modules:

Multiplexers, which select and transfer one of N inputs to a single output line

Demultiplexers, which transfer a signal on a single input line to a selected (1 of N) output line

Buses, which form common data paths between source and destination devices

Encoders, which translate characters into binary coded form

Decoders, which translate coded data back into characters

Code converters, which convert data from one code to another

Parity generators, which generate a parity bit for a given data word

Parity checkers, which check a combined data word plus parity bit to determine if received data contains a transmission error

Shifters, which produce output by shifting the input word

Many of these combinational logic functions exist in MSI form and can be used as modules in the implementation of higher-level digital-logic devices and digital systems such as controllers and computers.

Programmable logic devices (PLDs), a class of integrated circuits for implementing switching functions, provide a cost-effective method of reducing random logic requirements. The flexibility inherent in PLDs allows the designer to implement a range of switching functions using a single type of programmable logic device.

At the end of this chapter, the use of function modules to implement a higher-level device is illustrated by designing and implementing a tiny calculator that uses a variety of combinational-logic function modules as building blocks.

A Model for Combinational Logic Functions

A model is any convenient representation that describes the organization or operation of an application function or system. A model can take on any appropriate form, such as an engineer's black box, a set of descriptive statements, a truth table, or a set of logic equations.

A switching (combinational logic) function can be represented by a black-box model with 2-state inputs A, B, C, . . . and output(s) F_1 (F_2, . . .), as shown in Figure 4.1. Such a model is usually accompanied by one or more statements that describe a functional relation between each output and the set of inputs.

FIGURE 4.1
Black-box model of a combinational logic function

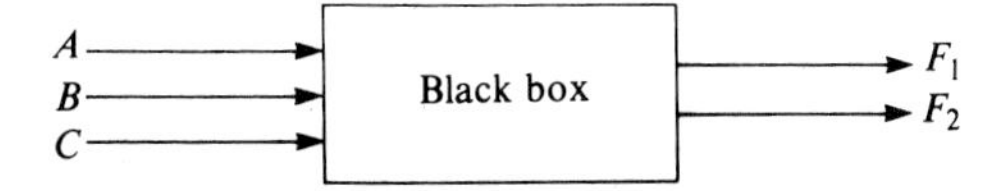

A combinational logic function can be conceptualized as a logic model in the form of a truth table that specifies the required output for each combination of the function's inputs.

A B C	F
0 0 0	0
0 0 1	0
0 1 0	0
0 1 1	1
1 0 0	0
1 0 1	1
1 1 0	1
1 1 1	1

A truth table defines a unique combinational logic function $F(A, B, C)$ that can be represented either in canonical SOP form or in canonical POS form. Such an expression can be simplified (reduced) to a logically equivalent minimal expression in any of the following ways:

- □ Algebraic reduction, using axioms and theorems of Boolean algebra. The principal theorems used are the logical adjacency theorem, the idempotent theorem, and the consensus theorem.
- □ Map methods, using K-maps or minterm-ring maps.
- □ Algorithmic methods, using the Quine–McCluskey method or the McCalla minterm-ring algorithm.

The minimal SOP expression in AND-OR form can then be converted to NAND-NAND, OR-NAND, and NOR-OR form by successive application of DeMorgan's theorem. A minimal POS expression in OR-AND form can likewise be converted to NOR-NOR, AND-NOR, and NAND-AND form by successive application of DeMorgan's theorem. Thus, a combinational logic (switching) function can be represented in any one of these eight forms.

Finally, a cost-effective realization of a logic function can be achieved by implementing a minimal SOP (or POS) expression as a combinational logic circuit. It should be noted that a combinational logic circuit's output at any given time depends only on the combination of values of its inputs.

The specification, conceptualization, solution/simplification, and realization of a combinational logic function are summarized in Figure 4.2 (p. 156).

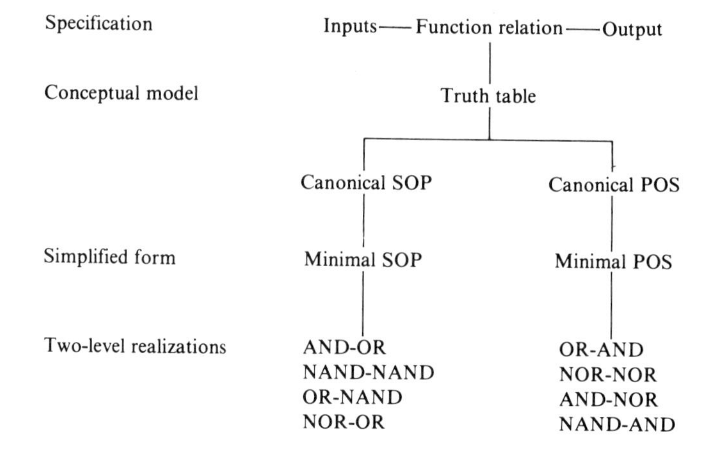

FIGURE 4.2
A combinational logic function from specification through realization

Analysis of Combinational Logic Circuits

The analysis of a combinational logic circuit is the process of analyzing the circuit's logic diagram and determining a functional relationship between the circuit's inputs and output(s). The output of a combinational logic circuit at a given time depends only on the combination of its inputs. That is, the output F is a *function* of the inputs; this function may be specified by a truth table, by logic equation, or by a verbal description.

The first step in the analysis process is to verify that the circuit is indeed a combinational logic circuit. That is, the circuit's logic diagram should be examined to verify that it does not contain any feedback paths or memory elements. A **feedback path** is a path from the output of one gate (B) to the input of another gate (A) whose output forms part of the input of gate B.

The analysis of a combinational logic circuit can be accomplished by the following procedure:

1. Label the inputs of the circuit using the letters $A, B, C, \ldots$. Since any input can change independently of any other input, the inputs are independent variables of the circuit.
2. Starting on the input side of the logic diagram, and proceeding to the output side, label each gate with a logic expression as follows:
   a. If the gate has an AND shape, the expression is formed by ANDing its inputs.
   b. If the gate has an OR shape, the expression is formed by ORing its inputs.
   c. If an input of gate G has an asserted-input level (GAIL) that does not match the asserted-output level (GAOL) on the connecting line, then that input is complemented in the logic expression for gate G.
3. Repeat Step 2 until a logic expression has been obtained for each circuit output.
4. Interpret the logic expression(s) obtained in Step 3 to determine a functional description of the circuit's operation. If necessary, convert the logic expression to a canonical form and construct a truth table for the circuit.

■ **EXAMPLE 4.1**

Analyze the combinational logic circuit shown in Figure 4.3. Note that bubbles on opposite ends of a connecting line effectively cancel one another.

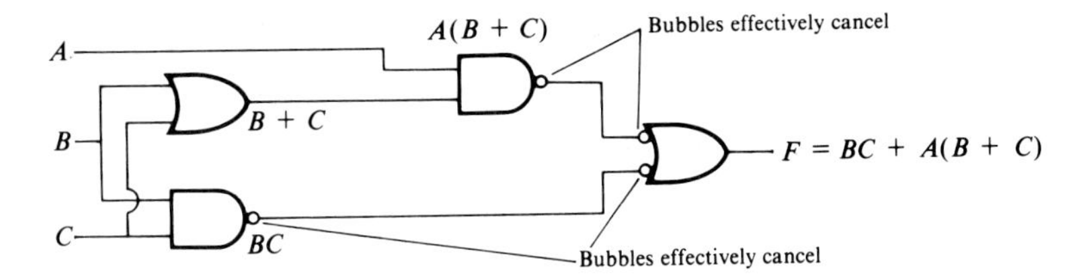

FIGURE 4.3
Logic diagram of a combinational logic circuit ■

By applying the distributive law of Boolean algebra, we can rewrite the logic equation for the circuit in Figure 4.3

$$F = BC + A(B + C)$$

as a 2-level AND-OR expression

$$F = BC + AB + AC$$

and then convert this to a canonical SOP expression by ANDing each product with $(X + X')$, where X represents the variable that does not appear in that product:

$$F = BC(A + A') + AB(C + C') + AC(B + B')$$

This expression can then be converted to an SOP form by a second application of the distributive law.

$$F = BCA + BCA' + ABC + ABC' + ACB + ACB'$$

The variables in each product term can be reordered using the commutative law.

$$F = ABC + A'BC + ABC + ABC' + ABC + AB'C$$

The redundant terms can then be eliminated, by using the idempotent theorem, to produce a canonical SOP expression as follows:

$$F = ABC + A'BC + ABC' + AB'C$$

Finally, a truth table can be constructed from the canonical SOP expression as follows:

A B C	F
0 0 0	0
0 0 1	0
0 1 0	0
0 1 1	1
1 0 0	0
1 0 1	1
1 1 0	1
1 1 1	1

An analysis of the truth table reveals that the output $F = 1$ if two or more of the inputs are 1. Therefore, the circuit shown in Figure 4.3 is a 3-input majority circuit.

■ **EXAMPLE 4.2**

Analyze the combinational logic circuit shown in Figure 4.4.

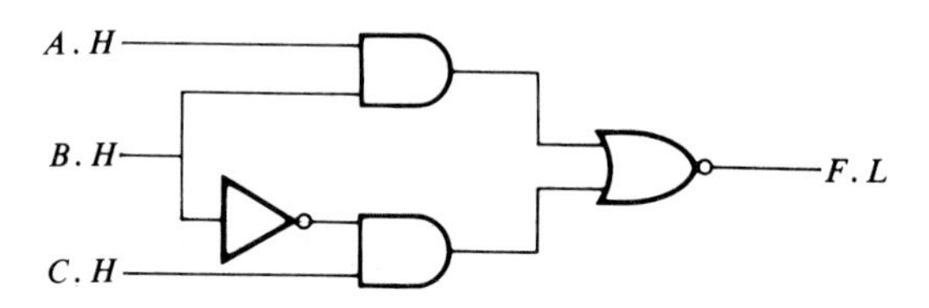

FIGURE 4.4
Logic diagram of a combinational logic circuit

The circuit output in Figure 4.4 is represented by the logic equation $F = AB + B'C$. The output $F.L$ is asserted (low) if A and B are asserted or if B' and C are asserted. ■

Hazards in Combinational Logic Circuits

A **hazard** is a condition in a circuit that can cause unreliable circuit operation. If different signal paths to a gate's inputs have different propagation delays, then undesirable output transients (temporary change in signal state) may occur after an input change. There are two types of hazards (**static** and **dynamic**) that may be present in a combinational logic circuit. These hazards must be eliminated in order to ensure reliable operation of a circuit.

The design of digital circuits involves a trade-off between component minimization and circuit reliability; component minimization may be achieved by eliminating redundant terms in logic expressions, while reliability of a circuit's operation is generally enhanced by the addition of redundant terms. The following sections describe methods for detecting and eliminating static and dynamic hazards.

Static Hazards

A static hazard causes a single transient in a combinational-logic output signal that should have remained unchanged in response to an input change. A circuit with a static hazard, caused by unequal propagation delays of signal B, is illustrated in Figure 4.5.

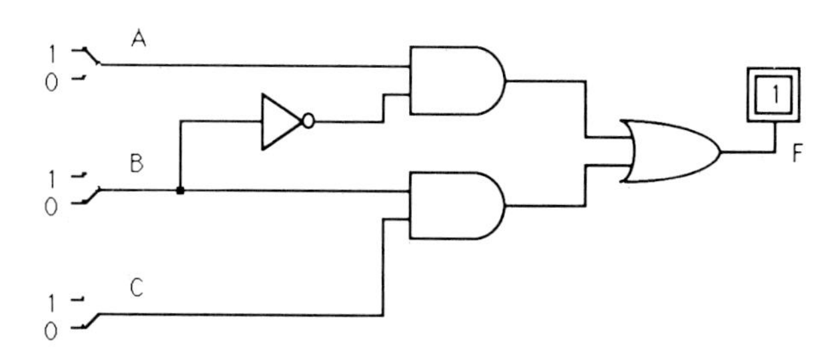

FIGURE 4.5
A combinational logic circuit with a static hazard

If $A, B, C = 1, 1, 1$, and input B changes from 1 to 0, the unequal propagation delays in signal B may momentarily cause both inputs of Gate 3 to be simultaneously 0, thereby causing F to momentarily change to 0 (see Figure 4.6b) when F should not have changed in value. The presence of a static hazard is indicated by the juxtaposition of rectangular groups of 1 cells (representing prime implicants) in the K-map, as illustrated in Figure 4.6a.

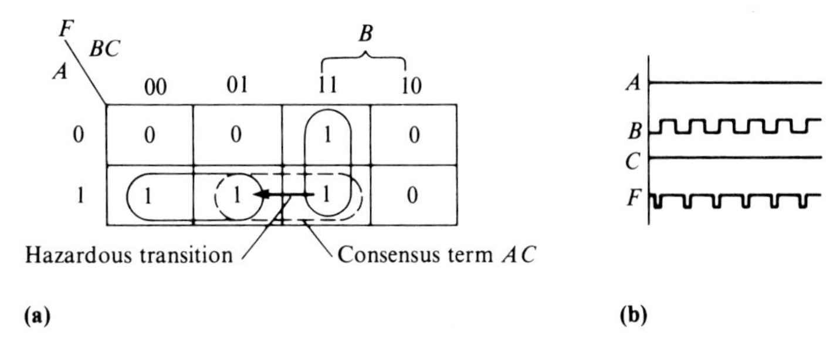

FIGURE 4.6
Illustration of a static hazard. (a) K-map (b) Timing diagram

The static hazard can be eliminated by adding the redundant consensus term (AC) to connect such rectangular groups. That is, the static hazard can be eliminated by replacing the original logic equation,

$$F = AB' + BC$$

by a logically equivalent equation,

$$F = AB' + BC + AC$$

that contains the consensus term AC.

Dynamic Hazards

A dynamic hazard occurs when the output of a combinational logic circuit is supposed to change once but instead changes 3 or more (odd) times. The presence of a dynamic hazard implies that the changing input variable must have three different paths to a single gate. A circuit with a dynamic hazard is illustrated in Figure 4.7. Note that there are three different signal paths from input A to the OR gate.

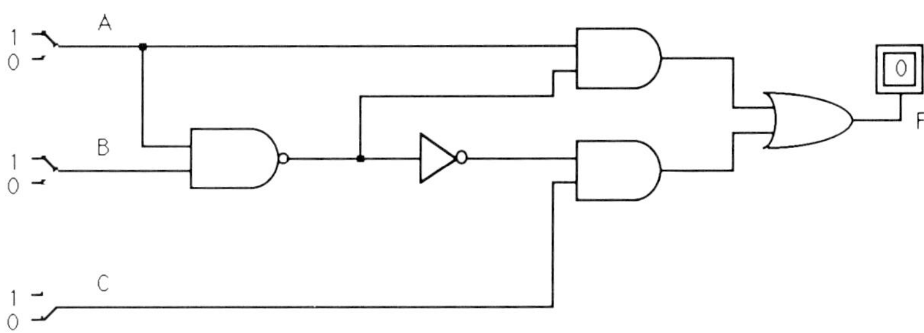

FIGURE 4.7
Logic diagram of a circuit with a dynamic hazard

The presence of a dynamic hazard can be detected by analyzing a timing diagram of the circuit. The presence of a dynamic hazard can also be detected by analyzing the following logic equation of the circuit:

$$F = A(A' + B') + ABC$$

The occurrence of input A in one term with A' and in another term without A' is indicative of the presence of a dynamic hazard. The hazard may be eliminated by the inclusion of a consensus term or by rewriting the logic equation so that an input variable does not appear in one product term with its complement and in another term without its complement. For example, the foregoing equation can be replaced by the following logically equivalent equation.

$$F = AB' + ABC$$

A circuit implemented using the preceding equation (see Figure 4.8) has no dynamic hazard.

FIGURE 4.8
Circuit with dynamic hazard eliminated

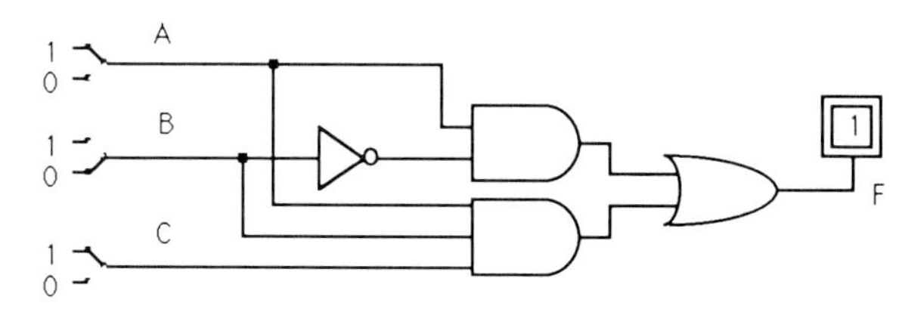

4.2 DESIGN PROCEDURE FOR A COMBINATIONAL LOGIC CIRCUIT

A particular combinational logic circuit can be designed using the 4-step problem-solving procedure (PSP) described in Chapter 2.

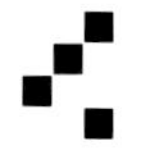

PSP

Step 1: Problem Statement. Describe the required circuit using a set of design specifications that include the circuit's inputs, output(s), and a functional relation between each output and the set of inputs. For mixed logic, indicate the polarity of each input and output.

Step 2: Conceptualization. Translate the design specifications into a function table and truth table.

Step 3: Solution/Simplification. Each output can be represented by a logic expression in canonical SOP (or POS) form. Simplify, if possible, the expression(s) using algebraic techniques, map methods (a K-map or a minterm-ring map), or algorithmic methods (the Quine–McCluskey tabular method or the McCalla minterm-ring algorithm).

Step 4: Realization. Draw a logic diagram of the circuit from the simplified expression for each output. Use the logic diagram to implement or simulate the circuit. Test the circuit for each possible combination of values of the inputs. Evaluate these results to determine if the circuit operates as required by the design specifications. ■

A particular circuit may be realized by one or more alternative implementations. In such cases, attempt to determine the "best" implementation by comparing the relative costs, circuit complexities, and operating speeds (the inverse of the maximum number of propagation delays) of the various implementations.

Boolean algebra is the principal theoretical tool used in the design of combinational logic circuits. The methods for simplifying Boolean expressions, described in Chapter 3, are essential to the design of cost-effective circuits.

A number of common, combinational-logic function-module circuits are designed in the remaining sections of this chapter. The design of each of these circuits is accomplished using the 4-step problem-solving procedure, outlined above, to the extent practicable. Off-the-shelf ICs will also be described, as applicable, for each type of function module.

■ **EXAMPLE 4.3: Design of a Combinational Logic Circuit**

Step 1: Problem Statement. Design a 3-input combinational logic circuit whose output equals 1 if two or more of its inputs are asserted. Such a circuit is referred to as a majority circuit.

Step 2: Conceptualization. The output F of a 3-input (A, B, C) majority function is defined by the following truth table:

A B C	F
0 0 0	0
0 0 1	0
0 1 0	0
0 1 1	1
1 0 0	0
1 0 1	1
1 1 0	1
1 1 1	1

Step 3: Solution/Simplification. The output F can be represented either in canonical SOP or POS form. A canonical SOP expression is obtained using the minterms (standard product terms for which $F = 1$):

$$F = \Sigma(3, 5, 6, 7) = A'BC + AB'C + ABC' + ABC$$

Since the complement F' is equal to the sum of standard product terms for which $F = 0$, F' is represented by the canonical SOP expression

$$F' = \Sigma(0, 1, 2, 4) = A'B'C' + A'B'C + A'BC' + AB'C'$$

Application of the involution theorem ($F = \{F'\}'$) and DeMorgan's theorem produces

$$\begin{aligned} F = \{F'\}' &= \{A'B'C' + A'B'C + A'BC' + AB'C'\}' \\ &= (A'B'C')'(A'B'C)'(A'BC')'(AB'C')' \\ &= (A + B + C)(A + B + C')(A + B' + C)(A' + B + C) \end{aligned}$$

which is a canonical POS expression for F.

The output F can be graphically represented by the K-map shown in Figure 4.9 (p. 162).

FIGURE 4.9
K-map of a majority circuit

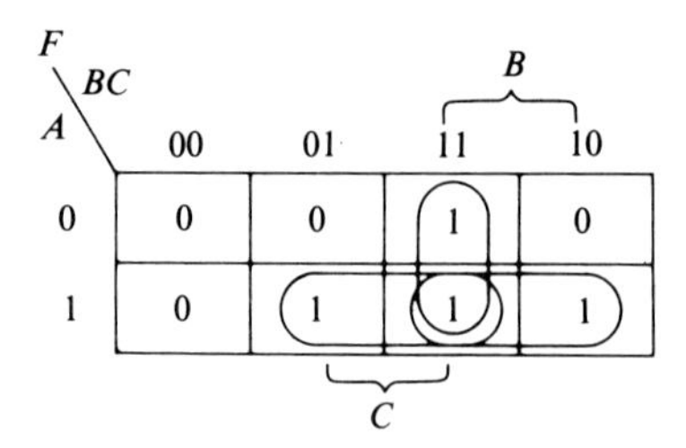

A simplified SOP expression for F can be obtained by combining the logically adjacent 1 cells:

$$F = BC + AB + AC$$

A simplified POS expression for F can be obtained by using the 0 cells of the K-map to determine F' and then applying DeMorgan's theorem:

$$F' = B'C' + A'B' + A'C'$$
$$F = (F')' = (B + C)(A + B)(A + C)$$

Step 4: Realization. The majority function may be realized in any of the following 2-level forms:

$F = BC + AB + AC$	AND-OR
$= [(BC)'(AB)'(AC)']'$	NAND-NAND
$= [(B' + C')(A' + B')(A' + C')]'$	OR-NAND
$= (B' + C')' + (A' + B')' + (A' + C')'$	NOR-OR

$F = (B + C)(A + B)(A + C)$	OR-AND
$= [(B + C)' + (A + B)' + (A + C)']'$	NOR-NOR
$= [(B'C') + (A'B') + (A'C')]'$	AND-NOR
$= (B'C')'(A'B')'(A'C')'$	NAND-AND

Other realizations are possible. For example, a 3-level AND-OR realization can be obtained by factoring:

$$F = BC + A(B + C)$$

A logic diagram of a 3-input majority circuit is presented in 2-level AND-OR form in Figure 4.10.

FIGURE 4.10
Logic diagram of a 3-input majority circuit

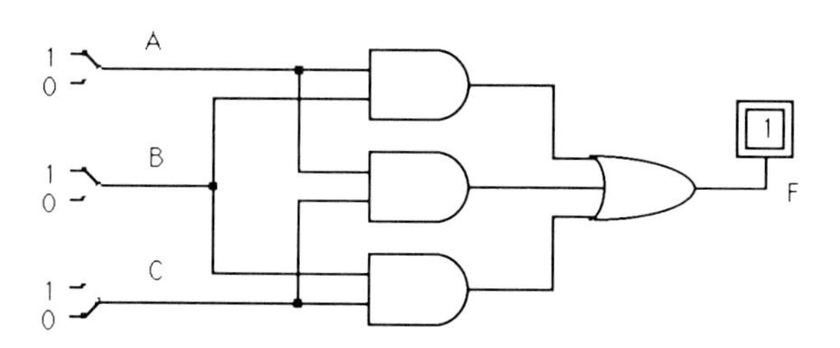

■

4.3 MULTIPLEXERS, DEMULTIPLEXERS, AND BUSES

A* multiplexer *(mux) is a multi-input, single-output device that selects one input and passes it to the output. Hence, a multiplexer is sometimes called a* data selector. *A mux that has 2^n inputs requires* n *select lines (with 2^n possible combinations of values) to choose which of the 2^n inputs will be passed to the output.

A mux can be used (1) as a data-selector function module in a digital logic system, (2) as a bus interface device, or (3) to implement an arbitrary switching function (see Section 4.7).

Multiplexer Design

Design of a 2-to-1 Multiplexer

Step 1: Problem Statement. Design a combinational logic circuit that has 2 input lines (i_0, i_1) and 1 output line (O). The circuit also has a control (select) line (S). The state of this control line determines which of the 2 input lines is to be electronically connected to the output line. The circuit is a 2-to-1 mux.

Step 2: Conceptualization. A 2-to-1 mux can be represented by the model shown in Figure 4.11.

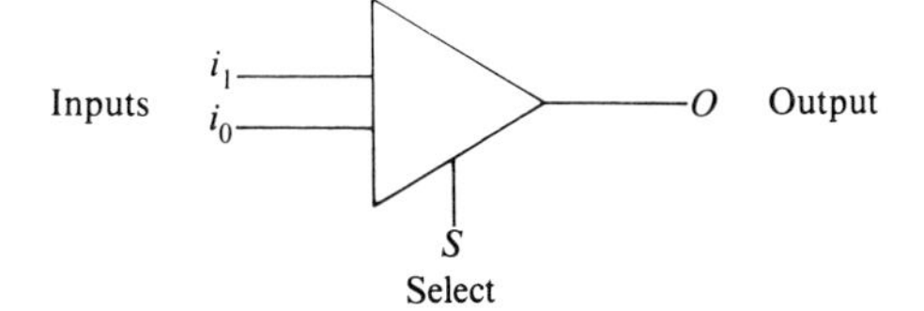

FIGURE 4.11
Symbolic representation of a 2-to-1 mux

Since there are $2 = 2^1$ inputs, one 2-state select line S is sufficient to select which input to pass to the output. The operation of the circuit can be described by the following function table:

Select S	Output O
0	i_0
1	i_1

In words, output O = input i_0 when $S = 0$, and output O = input i_1 when $S = 1$.

Step 3: Solution/Simplification. This longhand description of the output function can be expressed in the form of a Boolean equation

$$O = i_0S' + i_1S$$

which requires no simplification. Note that the mux output is an OR function, and the output is determined by the setting of the select line S.

Step 4: Realization. Draw a logic diagram of the 2-to-1 mux circuit, as shown in Figure 4.12, and use a logic simulator to realize the circuit. Test the circuit for each of the possible input combinations and evaluate the results to verify that the circuit operates as required by the design specifications.

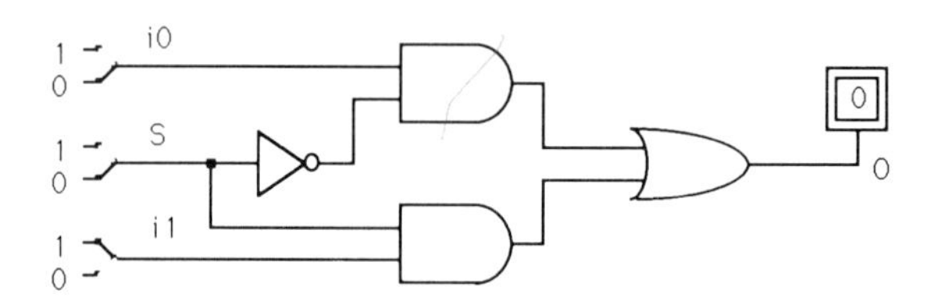

FIGURE 4.12
Logic diagram of a 2-to-1 mux

Design of a 4-to-1 Multiplexer

Step 1: Problem Statement. A 4-to-1 mux requires 2 select lines (with four possible combinations of values) to determine which one of the 4 inputs is to be passed to the output.

Step 2: Conceptualization. A 4-to-1 mux can be represented by the model shown in Figure 4.13.

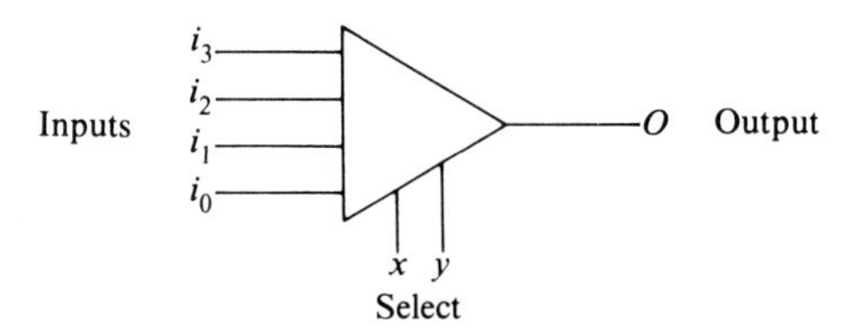

FIGURE 4.13
Symbolic representation of a 4-to-1 mux

The output O of a 4-to-1 mux is defined by the following function table:

Select x y	Output O
0 0	i_0
0 1	i_1
1 0	i_2
1 1	i_3

Step 3: Solution/Simplification. The output function O can, therefore, be defined by the following Boolean equation:

$$O = i_0\, x'y' + i_1\, x'y + i_2\, xy' + i_3\, xy$$

Note that the SOP expression contains all of the 2-variable standard products: $x'y'$, $x'y$, xy', and xy. The logic expression requires no simplification. The 4-to-1 mux output is an OR function, and the output is determined by the setting of the select lines x and y.

Step 4: Realization. Draw a logic diagram of the 4-to-1 mux (see Figure 4.14) and use a logic simulator to realize the circuit. Note that the off-the-shelf IC 74153 is a dual 4-line to 1-line mux.

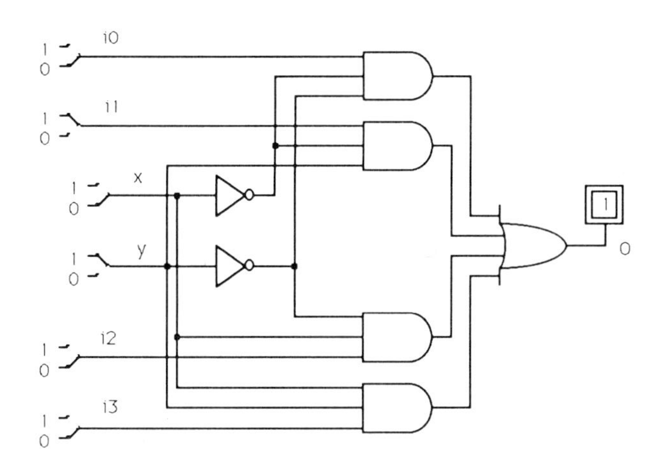

FIGURE 4.14
Logic diagram of a 4-to-1 mux

Demultiplexer Design

A **demultiplexer** (demux) is a 1-input, multi-output device that routes the single input to the output selected by the combination of values of the select lines. Hence, a demultiplexer is also referred to as a **data router**. A demux is usually implemented by a decoder with an enable input.

Design of a 1-to-4 Demultiplexer

Step 1: Problem Statement. Design a combinational logic circuit that passes a single input I to one of 4 output lines O_0, O_1, O_2, and O_3 as determined by the combination of values of the x, y select lines.

Step 2: Conceptualization. A 1-to-4 demux can be represented by the model shown in Figure 4.15.

FIGURE 4.15
Symbolic representation of a 1-to-4 demux

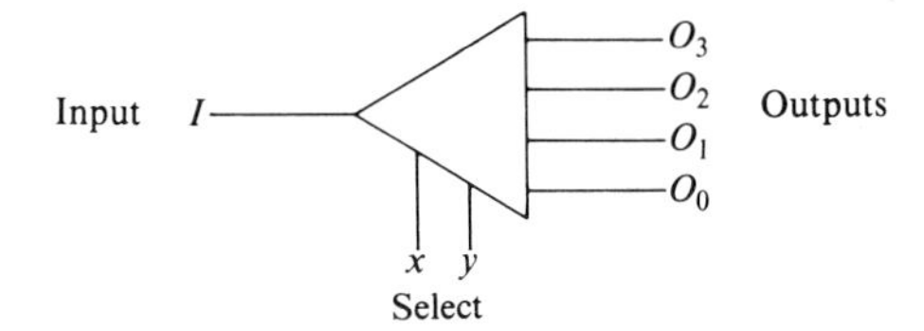

The operation of a 1-to-4 demultiplexer can be described by a function table with the 4 outputs O_0, O_1, O_2, and O_3 as functions of x and y as follows:

	Select	Outputs			
	x y	O_0	O_1	O_2	O_3
I	0 0	I	0	0	0
	0 1	0	I	0	0
	1 0	0	0	I	0
	1 1	0	0	0	I

Step 3: Solution/Simplification. From the function table, we see that the outputs can be expressed as the following Boolean functions:

$$O_0 = x'y'I$$
$$O_1 = x'yI$$
$$O_2 = xy'I$$
$$O_3 = xyI$$

These expressions require no simplification.

Step 4: Realization. Draw a logic diagram of the 1-to-4 demux (see Figure 4.16) and use a logic simulator to realize the circuit.

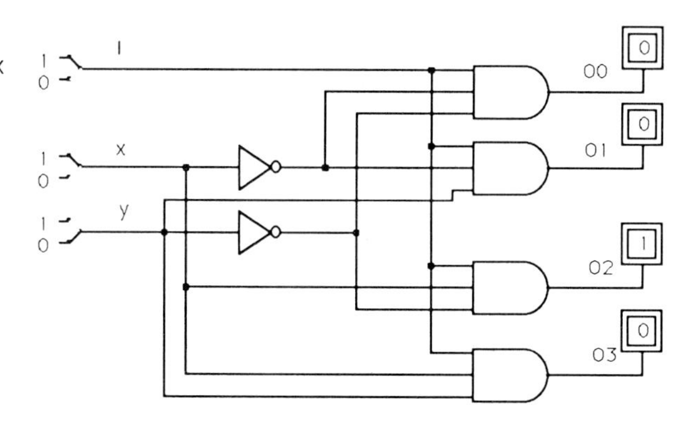

FIGURE 4.16
Logic diagram of a 1-to-4 demux

Busing and Wired Logic

A digital system may have a number of devices that are interconnected for the purpose of data transfer. A device from which data is transferred is referred to as a **source**; a device to which data is transferred is referred to as a **destination**. A **bus** is a common data path shared by all source and destination devices. If the source and destination devices are n-bit devices, then the bus must be an n-bit bus for parallel data transfer. Buses may be used for a variety of purposes, as their names imply—data bus, address bus, and control bus.

Design of a Digital Logic Bus

Step 1: Problem Statement. Design and implement a digital logic circuit (bus) that transfers any one of several sources to a single destination.

Step 2: Conceptualization. A bus can be conceptualized as an OR function of sources A, B, C, and D, wherein only one source is functionally connected to the bus at any given time. Figure 4.17 illustrates bit i of a bus model together with bit i of each source device.

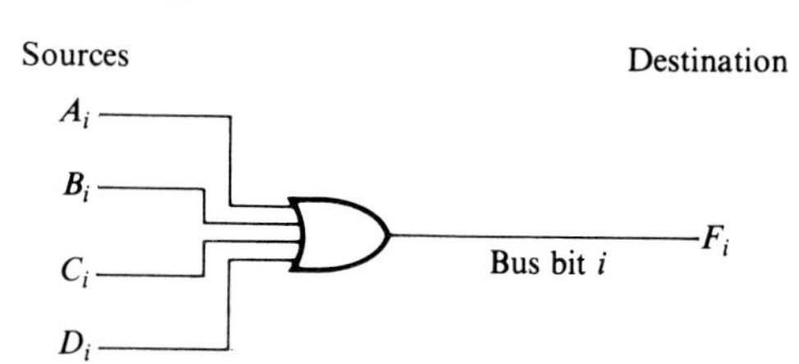

FIGURE 4.17
Bus conceptual model showing bit i of n bits

Step 3: Solution/Simplification. The bus function can be expressed as an OR function

$$F = c_0A + c_1B + c_2C + c_3D$$

where only one control c_i is to be asserted at any given time.

Step 4: Realization. A bus can be realized in any one of several ways. Alternative implementations include a centralized-access bus or a decentralized-access bus.

a. A centralized-access bus. Recall that a 4-to-1 mux, with inputs A, B, C, and D, output F, and select lines x and y, represented by the Boolean equation

$$F = x'y'A + x'yB + xy'C + xyD$$

is an OR function whose output depends on the setting of the x and y select lines. Therefore, the bus function F is equivalent to the mux output F if $c_0 = x'y'$, $c_1 = x'y$, $c_2 = xy'$, and $c_3 = xy$.

The source that is asserted on the bus is determined by the combination of x and y values selected at a given time. Consequently, access to the bus is centralized in the bus mechanism (mux).

b. A decentralized-access bus. This bus can have two implementations: (1) an implementation using a physical OR gate or (2) an implementation using a virtual (wired) OR gate.

1. The following bus function

$$F = c_0A + c_1B + c_2C + c_3D$$

can be implemented by a 2-level AND-OR circuit, where only one control c_i can be asserted at a given time, so that only one source is asserted on the bus at that time. In this implementation, bus access is decentralized, and access is determined by independent controls c_i. Figure 4.18 presents a logic diagram of a 2-level AND-OR implementation of the bus function.

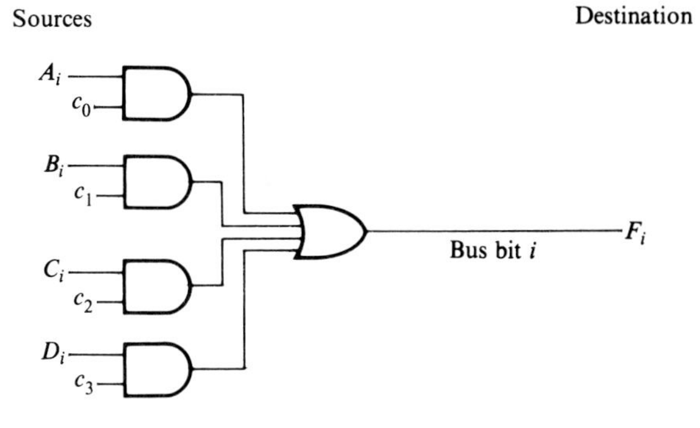

FIGURE 4.18
A 2-level AND-OR implementation of a bus

2. The bus function

$$F = c_0A + c_1B + c_2C + c_3D$$

can also be implemented using a virtual OR (wired OR), wherein the OR function is implemented by wiring together the outputs of open collector devices or 3-state devices.

Open-collector buffer/drivers, such as the 7407 IC, are designed for busing and their outputs can be tied together. A pull-up resistor is required when using open-collector buffers (see Figure 4.19). In this method of busing, only one source should be asserted at any one time. Consequently, access to the bus is decentralized in this method.

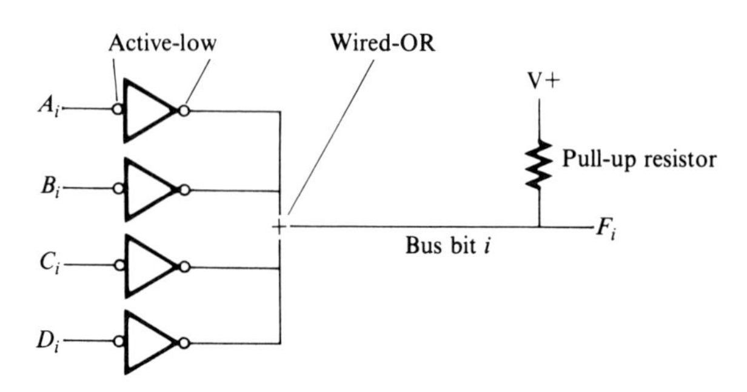

FIGURE 4.19
Implementation of a bus using a wired-OR

Buffer/drivers, such as the 74S244 IC, which have 3-state outputs are also designed for busing. In addition to the usual high- and low-voltage outputs, a 3-state output device has a high-impedance mode in which the output appears as if it is disconnected from its destination device(s). The 3-state outputs can be tied together, but they are not designed to be active simultaneously. Since each source has its own enable line, access to the bus is decentralized. When a 3-state device is enabled, it drives the bus actively high or low. When it is disabled, it appears as an open circuit. Figure 4.20 illustrates the wired-OR implementation using a 3-state buffer.

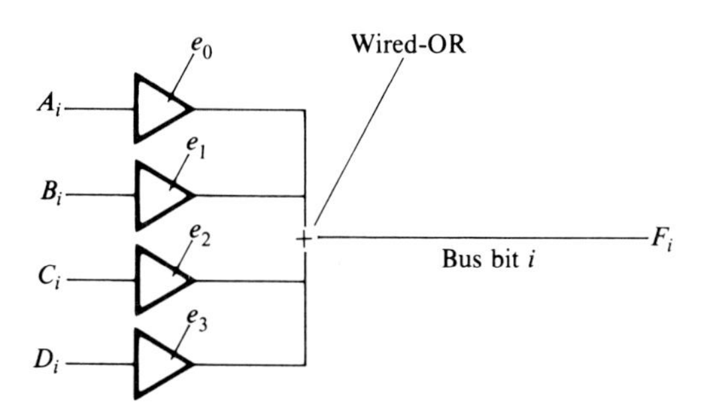

FIGURE 4.20
Implementation of a bus using 3-state buffers

4.4 ENCODERS, DECODERS, AND CODE CONVERTERS

Since the memory of most computers consists of elements that are 2-state devices, each character of an external character set (letters, digits, and symbols) must be encoded in some binary form to be stored in memory. Conversely, when the code word in memory is retrieved, it must be decoded in order to be output in its original form.

Consider a device that receives input from a simple alphabetic keyboard with 26 capital letters, 5 punctuation symbols, and a space bar—a total of 32 characters in the input character set. A 32-bit word could be used to store each character of the set. That is, set bit 1 for letter A, bit 2 for letter B, . . . , bit 26 for letter Z, and so on. However, this method of storing data would be very inefficient since it requires a memory in which each location (word) contains 32 bits.

A more efficient alternative is to encode each keyboard character, and the space character (represented by a □ symbol) in some combination of 1s and 0s. For example, the characters can be written in some order, with each character being assigned a particular combination of 1s and 0s, as illustrated in Table 4.1.

TABLE 4.1
A 32-character code table

Character	Decimal	Binary
□ = Space	0	00000
A	1	00001
B	2	00010
C	3	00011
D	4	00100
⋮	⋮	⋮
X	24	11000
Y	25	11001
Z	26	11010
⋮	⋮	⋮
?	31	11111

The simple binary code represented in Table 4.1 is a more efficient way to store the characters in memory since it requires only a 5-bit word to store any one of the 32 characters.

Then, we must consider something besides use and efficiency in a device's input and storage capacity; its output should be in a form that can be readily interpreted. A decoder is needed to decode any 5-bit binary code in the right-hand column of Table 4.1 back into the corresponding character in the left-hand column.

In practical applications, computer keyboards have 26 capital letters, 26 lowercase letters, 10 decimal digits, and a variety of punctuation symbols and special symbols. The resulting character set contains in excess of $64 = 2^6$ characters. Consequently, most schemes for encoding keyboard characters use more than 6 bits. For example, the ASCII code uses a 7-bit code as discussed in Chapter 1. Many modern computers use an 8-bit (extended) ASCII code.

Most high-level computer languages use an alphanumeric character set that includes the letters A–Z of the English alphabet, the digits 0–9 of the decimal number system, and various symbols. For example, a statement in Pascal, which computes the sum C of the numbers A and B, can be written C:=A+B; where the variables A, B, and C represent decimal numbers.

Since the memory of most computers consists of elements that are 2-state devices, each character of the external character set (A–Z, 0–9, and symbols) must be encoded in some internal binary code, usually ASCII or EBCDIC.

Numbers can also be stored in binary or packed decimal form. Computations are performed in the computer's CPU in a binary code, usually 2s complement or BCD. Numbers are sometimes stored in floating-point form, wherein the number is represented by a fraction and an exponent.

Numeric results generated by a computer program are usually output in decimal form while textual information is usually output in the character set (A–Z, 0–9, and symbols) of the high-level computer language.

The input conversion, internal storage, processing, and output conversion in a representative computer can be illustrated by the diagram shown in Figure 4.21.

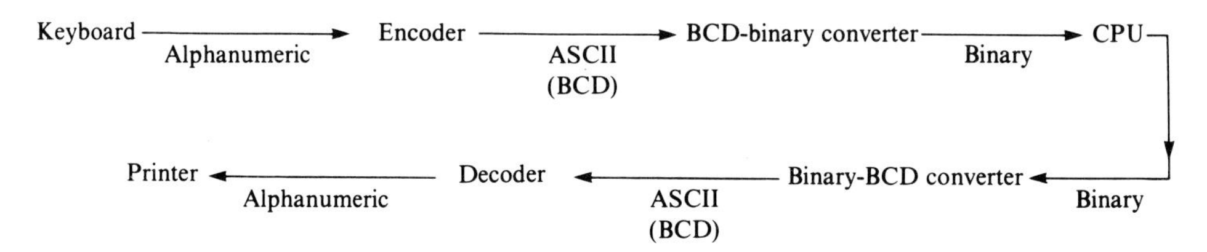

FIGURE 4.21
Input, binary processing, and output diagram

Many calculators and some computers process numbers in BCD form instead of straight binary form. In this case, Figure 4.21 can be simplified to the form shown in Figure 4.22 because BCD digits are represented directly as the rightmost 4 bits of an ASCII character. The ASCII code is an industry standard code that uses a 7-bit binary code to represent 128 possible distinct characters.

FIGURE 4.22
Input, BCD processing, and output diagram

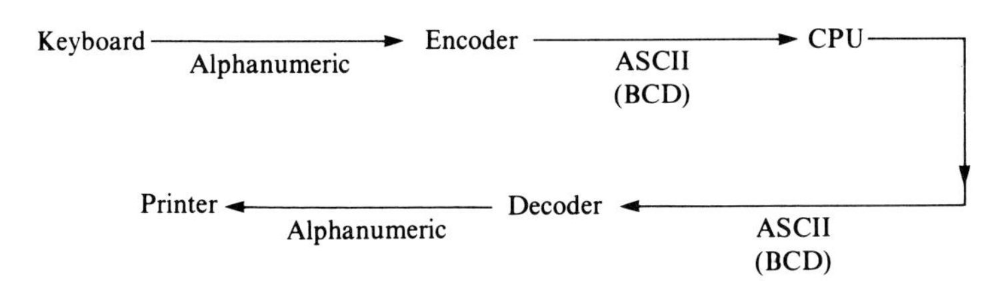

Encoder Design

Encoders in one form or another are used in most digital systems. The type of encoder used depends on the requirements of the particular application. An encoder function transforms each input character to a binary form. For example, a hexadecimal encoder converts each input character (0, 1, 2, 3, . . . , 9, A, B, C, D, E, F) to a binary code word (0000, 0001, 0010, 0011, . . . , 1001, 1010, 1011, 1100, 1101, 1110, 1111).

Consider the following problem, which can serve as an example to illustrate the design of a simple encoder.

Design of a 4-to-2 Encoder

Step 1: Problem Statement. A calculator has four function keys (add, subtract, multiply, divide). Only one of the function keys can be pressed at a time. When a particular key is pressed, that key is encoded according to the following function table:

Function	Code
Add	00
Sub	01
Mul	10
Div	11

Step 2: Conceptualization. Translate the design specifications into a truth table. Each of the function keys is an input of an encoder circuit. The inputs can be represented by 2-state variables as follows: A = add, S = subtract, M = multiply, D = divide. A 2-bit output, with combinations 00, 01, 10, 11, can be used to encode the 4 inputs. Therefore, the encoder outputs can be represented by 2-state variables x, y.

A truth table with 4 inputs has 16 possible combinations of values of these 4 inputs. However, since only one function key can be pressed at a time, only 4 of the 16 input combinations are needed, as illustrated in the following partial truth table:

Line	Inputs $A\ S\ M\ D$	Outputs $x\ y$
1	1 0 0 0	0 0
2	0 1 0 0	0 1
3	0 0 1 0	1 0
4	0 0 0 1	1 1

This encoding table is obtained by assigning one of the 4 x, y combinations (00, 01, 10, 11) to each of the combinations of settings of the function keys (1000, 0100, 0010, 0001).

Step 3: Solution/Simplification. From truth table lines 3 and 4, $x = 1$ if $A = 0$ and $S = 0$, independent of the values of M and D; therefore, $x = A'S'$. From truth table lines 2 and 4, $y = 1$ if $A = 0$ and $M = 0$, independent of the values of S and D; therefore, $y = A'M'$.

The outputs x and y are functions of the inputs A, S, M, and D and can be represented by the following Boolean equations:

$$x = A'S'$$

$$y = A'M'$$

Step 4: Realization. Draw a logic diagram of the 4-to-2 encoder circuit (see Figure 4.23) and use a logic simulator to simulate the circuit. Test the circuit for each of the possible input combinations and evaluate the results to verify that the circuit operates as required by the design specifications.

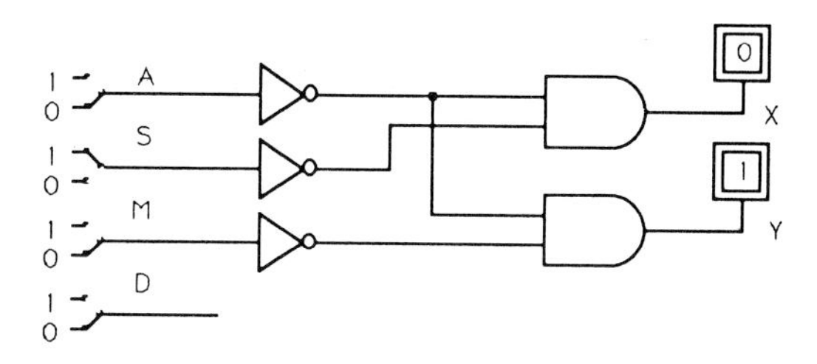

FIGURE 4.23
Logic diagram of a 4-to-2 encoder

Design of a 4-Input Priority Encoder

A priority encoder is a device that encodes the input line that has the highest priority.

Step 1: Problem Statement. Design a combinational logic circuit that has 4 prioritized inputs *A*, *B*, *C*, *D* (with *A* having the highest priority, *B* the next highest, and so on), encoded outputs *X* and *Y*, and a validity output *Z*, such that

If $Z = 0$, outputs *X* and *Y* are invalid

If $Z = 1$, outputs *X* and *Y* equal the encoded value of the highest priority input that is active

Step 2: Conceptualization. A 4-input priority encoder can be represented by the model shown in Figure 4.24.

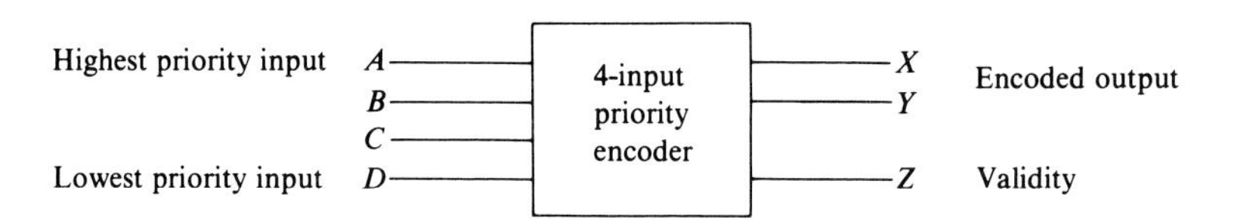

FIGURE 4.24
Black-box diagram of a 4-input priority encoder

The operation of the 4-input priority encoder is defined by the following function/truth table:

Inputs	Outputs
A B C D	*X Y Z*
0 0 0 0	– – 0
0 0 0 1	0 0 1
0 0 1 x	0 1 1
0 1 x x	1 0 1
1 x x x	1 1 1

where input value x denotes value 0 or 1; when all inputs are equal to 0, the output is undefined (– –) and $Z = 0$.

Step 3: Solution/Simplification. An input variable with value x, denoting either 0 or 1, is represented by (variable plus complement) in the following Boolean equations for the outputs:

$$X = A'B\ (C + C')(D + D') + A\ (B + B')(C + C')(D + D')$$
$$Y = A'B'C(D + D') + A(B + B')(C + C')(D + D')$$
$$Z = A + B + C + D$$

The expressions for *X*, *Y*, and *Z* can be simplified as follows:

$$X = A'B + A$$
$$= A + B$$
$$Y = A'B'C + A$$
$$= A + B'C$$
$$Z = A + B + C + D$$

Step 4: Realization. Draw a logic diagram of the 4-input priority encoder (see Figure 4.25) and use a logic simulator to simulate the circuit. Note that the 74148 IC is an off-the-shelf 8-input priority encoder. The 8 inputs and the 3 outputs are active-low.

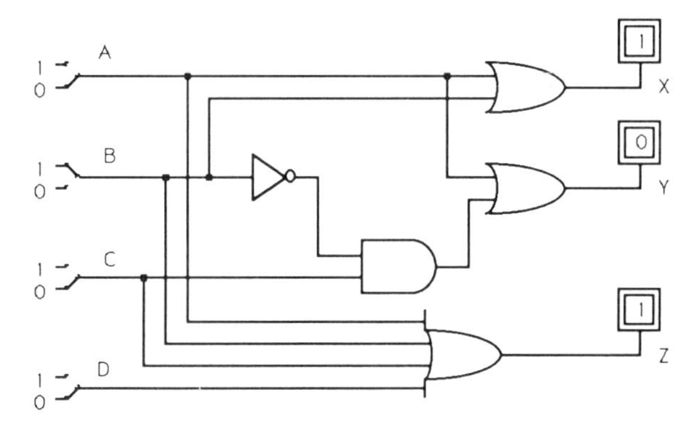

FIGURE 4.25
Logic diagram of a 4-input priority encoder

Decoder Design

An n-to-2^n decoder converts n bits of encoded data to 2^n bits of decoded data. The 2^n outputs are minterms of the n input variables. Decoders can be used for a number of functions, including (1) routing data to a selected output line, (2) addressing memory, (3) converting data from one code to another, (4) realizing an n-input truth table, and (5) performing a demultiplexer function.

An n-to-2^n decoder performs the reverse function of a 2^n-to-n encoder. The design of a decoder by a formal solution using truth tables is simpler than the design of the corresponding encoder because the decoder has n inputs while the encoder has 2^n inputs. Therefore, the truth table for the decoder has only 2^n rows while the truth table for the encoder has 2^{2^n} rows.

Design of a 2-to-4 Decoder

Step 1: Problem Statement. Design a 2-input, 4-output combinational logic circuit to decode the 2-bit output of the 4-to-2 encoder, designed in the preceding section. The circuit is a 2-to-4 decoder whose output determines which of the four function keys (A, S, M, D) was pressed at a given time.

Step 2: Conceptualization. A 2-to-4 decoder can be represented by the black-box diagram shown in Figure 4.26.

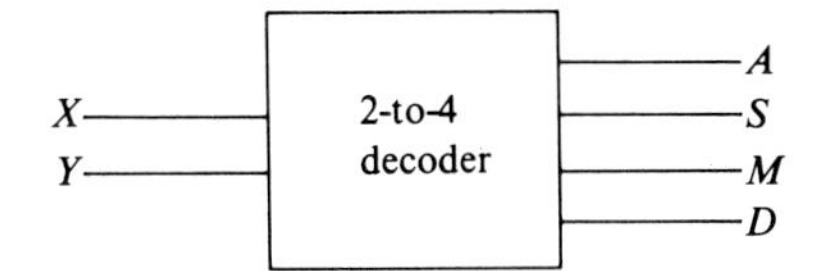

FIGURE 4.26
Black-box diagram of a 2-to-4 decoder

Let each of the 4 possible combinations (00, 01, 10, 11) of the 2 inputs (X and Y) represent one of the 4 keys (A, S, M, D), as listed in the following truth table:

Inputs	Outputs
X Y	A S M D
0 0	1 0 0 0
0 1	0 1 0 0
1 0	0 0 1 0
1 1	0 0 0 1

Step 3: Solution/Simplification. From the truth table, each of the outputs can be represented as a Boolean function of the inputs as follows:

$$A = X'Y'$$
$$S = X'Y$$
$$M = XY'$$
$$D = XY$$

Simplify the expressions for X and Y. Each of the 4 outputs of a 2-to-4 decoder is a standard product term of the inputs X and Y. Therefore, a decoder can be used to realize a truth table, and each combination of input values can be used to address a minterm (decoder output).

Step 4: Realization. Draw a logic diagram of the 2-to-4 decoder circuit (see Figure 4.27) and use a logic simulator to simulate the circuit. Test the circuit for each of the possible input combinations and evaluate the results to verify that the circuit operates as required by the design specifications. An example of an MSI decoder is the 74154, which is a 4-to-16 decoder, with two enable lines.

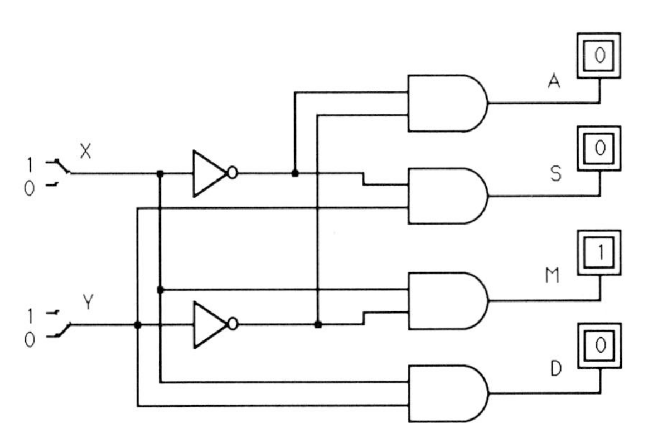

FIGURE 4.27
Logic diagram of a 2-to-4 decoder

Code Converter Design—The Gray Code

The Gray code, which is employed in K-maps, is useful in applications in which susceptibility to error increases with the number of bit changes between adjacent terms in a sequence. For example, the Gray code is used in shaft-position encoders to perform analog-to-digital conversion (where an analog quantity—position of a rotating shaft—is converted to digital form) as illustrated in Figure 4.28. In Figure 4.28, the shaft passes through the center of the wheel with the axis of the shaft perpendicular to the plane of the wheel. Note that any two adjacent sectors of the Gray-code wheel differ in only one bit position.

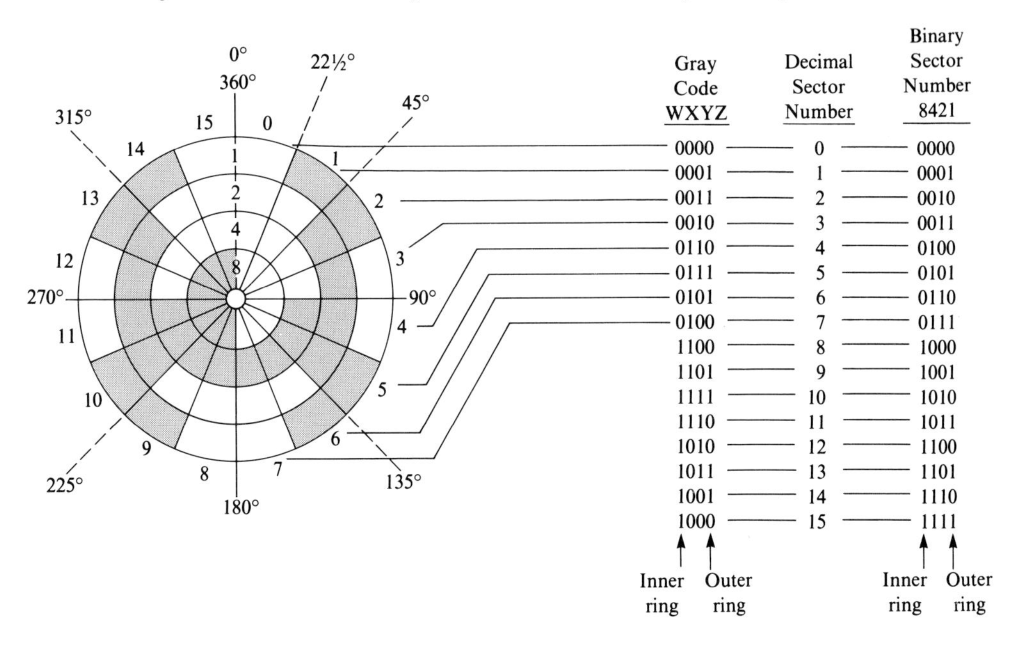

FIGURE 4.28
Diagram of a Gray-code wheel affixed to a rotating shaft

The advantage of the Gray code over straight binary code is due to the unit distance property of the Gray code, that is, that two consecutive numbers differ in only one bit position, as illustrated in the K-map shown in Figure 4.29.

FIGURE 4.29
Diagram of a 4-input K-map

AB \ CD	00	01	11	10
00	0	1	3	2
01	4	5	7	6
11	12	13	15	14
10	8	9	11	10

A number of other codes, such as 6311, 2'421, and excess-3 have been used in the past because of particular features specific to each code. For example, the 1s complement of an excess-3 code digit produces the 9s complement of the corresponding decimal digit. Similarly, the 1s complement of a 2'421 code digit produces the 9s complement of the corresponding digit. The 6311 code is a weighted code, while neither the excess-3 nor the 2'421 code is weighted.

Design of a Binary-to-Gray Code Converter

Step 1: Problem Statement. Design a binary-to-Gray code converter. This device accepts a 4-bit binary code as input and generates the corresponding Gray code as output.

Step 2: Conceptualization. The 4 inputs, labeled A, B, C, and D, represent weights 8, 4, 2, and 1 of the binary number. The 4 outputs, labeled W, X, Y, Z, represent the Gray code number. A truth table can be constructed by writing each Gray code number alongside its corresponding binary number, as shown in Table 4.2.

TABLE 4.2
Binary-to-Gray code table

Binary *A B C D*	Gray *W X Y Z*
0 0 0 0	0 0 0 0
0 0 0 1	0 0 0 1
0 0 1 0	0 0 1 1
0 0 1 1	0 0 1 0
0 1 0 0	0 1 1 0
0 1 0 1	0 1 1 1
0 1 1 0	0 1 0 1
0 1 1 1	0 1 0 0
1 0 0 0	1 1 0 0
1 0 0 1	1 1 0 1
1 0 1 0	1 1 1 1
1 0 1 1	1 1 1 0
1 1 0 0	1 0 1 0
1 1 0 1	1 0 1 1
1 1 1 0	1 0 0 1
1 1 1 1	1 0 0 0

Note that each minterm in A, B, C, and D serves as an address of the corresponding Gray code character. The truth table defines each Gray code bit W, X, Y, and Z as a function of the binary inputs A, B, C, and D.

Step 3: Solution/Simplification. Logic expressions for the output functions W, X, Y, and Z can be determined by an examination of Table 4.2, as follows:

- □ $W = 1$ when $A = 1$; therefore, $W = A$.
- □ $X = 1$ when ($A = 0$ AND $B = 1$) OR ($A = 1$ AND $B = 0$); therefore, $X = A'B + AB' = A \oplus B$.
- □ $Y = 1$ when ($B = 0$ AND $C = 1$) OR ($B = 1$ AND $C = 0$); therefore, $Y = B'C + BC' = B \oplus C$.
- □ $Z = 1$ when ($C = 0$ AND $D = 1$) OR ($C = 1$ AND $D = 0$); therefore, $Z = C'D + CD' = C \oplus D$.

Step 4: Realization. Draw a logic diagram of the binary-to-Gray code converter (see Figure 4.30). Use a logic simulator to simulate the circuit. Test the circuit for each of the possible input combinations and evaluate the results to verify that the circuit operates as required by the design specifications.

FIGURE 4.30
Logic diagram of a binary-to-Gray code converter

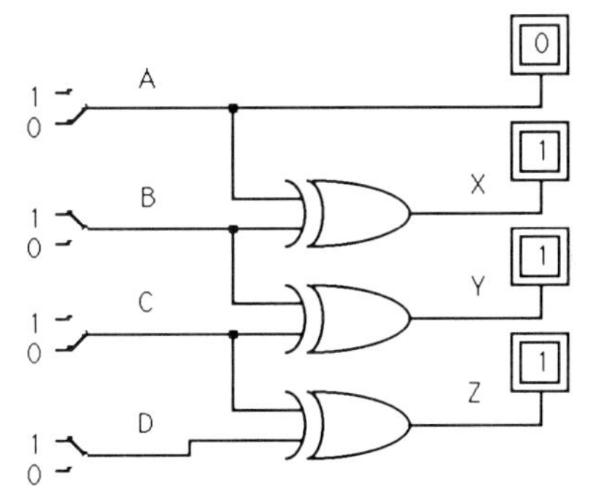

4.5 PARITY CIRCUITS—GENERATORS AND CHECKERS

Computers and other digital logic devices handle data in bit groups called* words. *A parity bit may be appended to a data word for error detection purposes. Parity can be even or odd. If a device uses even (or odd) parity, each data plus parity word handled by the device should contain an even (or odd) number of 1s.

A **parity generator** is used to append a parity bit to each data word so that the total number of 1s is even (or odd) in every data plus parity word. If a word is moved from one location or device to another, the received word is checked to determine if an error in transmission has occurred. If a single bit has changed, from 0 to 1 or vice versa, the total number of 1s changes, indicating an error that can be detected by a parity checker. If an even number of bits change, the number of 1s remains even (or odd) and the errors cannot be detected by a parity checker. There are special codes for error detection and error correction that may be used for the detection of two or more errors in a word.

Design of an Even-Parity Generator

Step 1: Problem Statement. Design a combinational logic circuit that has 4 inputs and 1 output, such that the output equals 1 if, and only if, the number of active-high (ON = 1) inputs is odd.

Step 2: Conceptualization. Let the input data bits be represented by variables A, B, C, and D. The output (the generated parity bit), denoted by P, can be represented by the truth table and K-map shown in Figure 4.31.

FIGURE 4.31
Truth table and K-map for even-parity generator

Inputs				Output
A	B	C	D	P
0	0	0	0	0
0	0	0	1	1
0	0	1	0	1
0	0	1	1	0
0	1	0	0	1
0	1	0	1	0
0	1	1	0	0
0	1	1	1	1
1	0	0	0	1
1	0	0	1	0
1	0	1	0	0
1	0	1	1	1
1	1	0	0	0
1	1	0	1	1
1	1	1	0	1
1	1	1	1	0

P — AB \\ CD	$C'D'$ 00	$C'D$ 01	CD 11	CD' 10
$A'B'$ 00		1		1
$A'B$ 01	1		1	
AB 11		1		1
AB' 10	1		1	

The minterms can be written in row order (1st row, 2nd row, and so on), so that

$$P = (A'B'C'D + A'B'CD') + (A'BC'D' + A'BCD) + (ABC'D + ABCD') + (AB'C'D' + AB'CD)$$

Step 3: Solution/Simplification. The expression for P can be simplified by extracting common factors, as follows:

$$P = A'B'(C'D + CD') + A'B(C'D' + CD) + AB(C'D + CD') + AB'(C'D' + CD)$$

Again, common factors can be extracted to obtain

$$P = (A'B' + AB)(C'D + CD') + (A'B + AB')(C'D' + CD)$$

Finally, the substitutions

$$X = A'B + AB' = A \oplus B$$

and

$$Y = C'D + CD' = C \oplus D$$

produce the final result:

$$P = X'Y + XY' = X \oplus Y = (A \oplus B) \oplus (C \oplus D)$$

Step 4: Realization. Construct a logic diagram of the even-parity generator (see Figure 4.32). Use a logic simulator to simulate the circuit.

FIGURE 4.32
Logic diagram for 4-bit even-parity generator

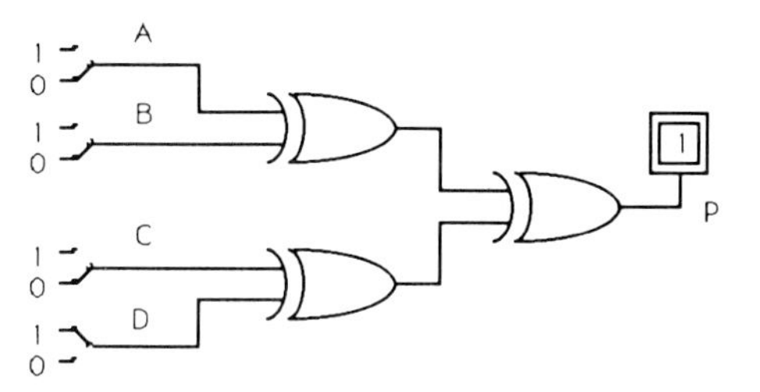

Design of an Even-Parity Checker

A **parity checker** is a device that tests a data plus parity word to determine if an error has been introduced into that word. Suppose that the 5-bit word *PABCD*, formed by appending the parity bit *P* to the data word *ABCD*, is transmitted to a receiving device. The parity of the received word *PABCD* can be checked by a 5-bit parity checker.

Since the originating device used an even-parity generator to generate parity bit *P*, each received 5-bit word should have an even number of 1s. If not, the parity checker output, denoted *F*, has value 1, indicating a single error in the word being checked. The parity-check error function *F* can be represented by a 5-input truth table, which has the K-maps shown in Figure 4.33.

FIGURE 4.33
K-maps for 5-bit even-parity checker

$P = 0$

AB \ CD	00	01	11	10
00		1		1
01	1		1	
11		1		1
10	1		1	

$P = 1$

AB \ CD	00	01	11	10
00	1		1	
01		1		1
11	1		1	
10		1		1

It can be shown that the function *F* can be represented by the Boolean equation:

$$F = P \oplus [(A \oplus B) \oplus (C \oplus D)]$$

It should be noted that the 5-bit parity checker, represented by the foregoing Boolean equation, can be used as a 4-bit even-parity generator by setting *P* equal to 0. A logic diagram for the 5-bit even-parity checker is presented in Figure 4.34.

An example of an off-the-shelf 9-bit parity generator/checker is the 74180 MSI IC.

FIGURE 4.34
Logic diagram for 5-bit even-parity checker

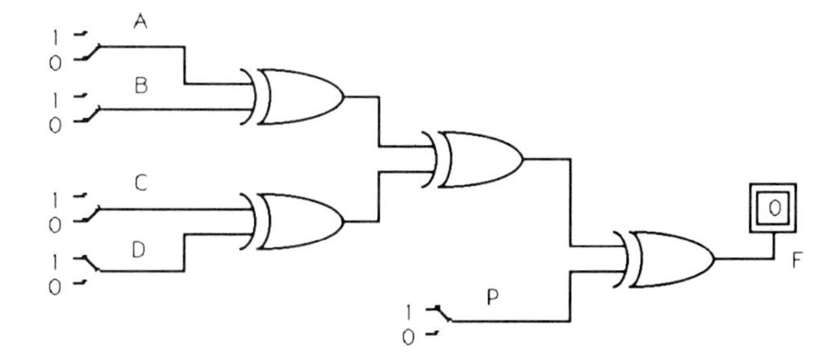

4.6 SHIFTERS

A combinational logic shifter is a device that produces an output obtained by shifting its input. For a right shift, the most significant bit (MSB) is called the* fill bit *and the least significant bit (LSB) is called the* spill (end off) bit. *For a left shift, the MSB is the spill bit and the LSB is the fill bit.

Shift operations can be categorized as follows:

Logical, where logic 0 is inserted in the fill position

Arithmetic, where the sign bit is extended in a right shift

End-around (or rotate)

For logical shifting, the MSB and LSB of a shifter input word A are filled in or spilled off depending on the direction of the shift performed, in accordance with the following table:

MSB	Function	LSB
Fill	Right shift	Spill
Spill	Left shift	Fill

For arithmetic shifts, where A is in 1s or 2s complement form, the magnitude bits are shifted while the sign bit is unaffected, as shown in the following table. The input word $A = a_n a_{n-1} a_{n-2} \cdots a_1 a_0$, where a_n is the sign bit.

Sign	Magnitude MSB	Function	Magnitude LSB
Same	Sign $\rightarrow$ Fill	Arithmetic shift right (ASR)	Spill bit lost
Same	Spill $\leftarrow a_{n-1}$	Arithmetic shift left (ASL)	Fill bit $\leftarrow 0$

For ASL, a_{n-1} is lost. An overflow occurs if $a_{n-1} \neq$ sign bit. For end-around (rotate) shifts, the data word bits form a continuous loop.

Design of a Logical Shifter

Step 1: Problem Statement. Design and implement a 4-bit logical shifter that has 4-bit input A, 4-bit output S, and 1-bit controls x and y, where

x y	S	Function
0 0	LS(A)	Left shift
0 1	A	Parallel load
1 0	RS(A)	Right shift
1 1	2RS(A)	Double right shift

Step 2: Conceptualization. A 4-bit shifter can be represented by the black-box model shown in Figure 4.35.

FIGURE 4.35
Black-box model of a 4-bit logical shifter

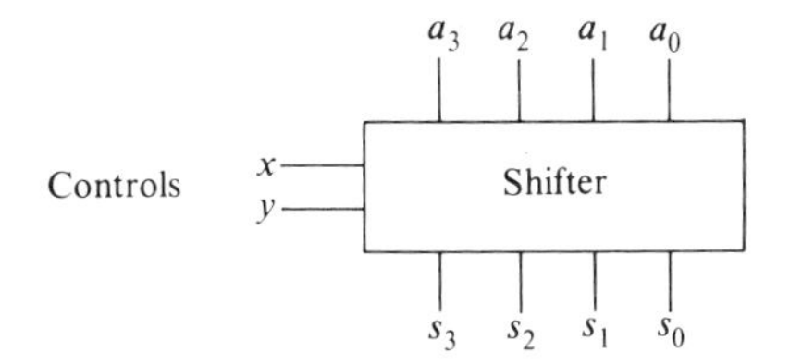

The output, as determined by the combination of x, y values, is as follows:

Result	s_3	s_2	s_1	s_0	Function
LS(A)	a_2	a_1	a_0	0	Left shift
A	a_3	a_2	a_1	a_0	Parallel load
RS(A)	0	a_3	a_2	a_1	Right shift
2RS(A)	0	0	a_3	a_2	Double right shift

Step 3: Solution/Simplification.

$$\begin{aligned} & \quad\;\; 2RS \qquad\; RS \qquad\;\; A \qquad\;\; LS \\ s_3 &= xy\ 0 + xy'\ 0 + x'y\ a_3 + x'y'a_2 \\ s_2 &= xy\ 0 + xy'\ a_3 + x'y\ a_2 + x'y'a_1 \\ s_1 &= xy\ a_3 + xy'\ a_2 + x'y\ a_1 + x'y'a_0 \\ s_0 &= xy\ a_2 + xy'\ a_1 + x'y\ a_0 + x'y'\ 0 \end{aligned}$$

Step 4: Realization. The shifter can be implemented using four 4-to-1 muxes configured as shown in Figure 4.36.

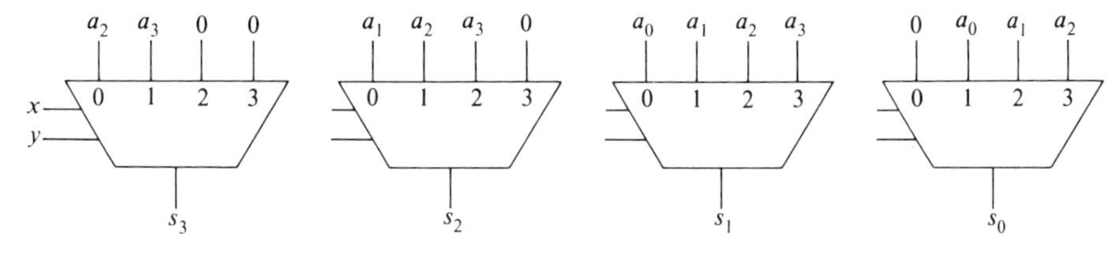

FIGURE 4.36
Implementation of 4-bit shifter using muxes

4.7 IMPLEMENTATION OF AN ARBITRARY BOOLEAN FUNCTION USING MSI ICS

An arbitrary Boolean (switching) function can also be implemented using MSI devices such as a multiplexer or a decoder, since each of these devices realizes all of the 2^n minterms of n-input variables.

Implementation of a Full Adder Using Multiplexers

This section describes the implementation of a 1-bit full adder function using 4-to-1 multiplexers (mux). You recall that the output O of a 4-to-1 mux can be expressed in the form

$$O = i_0\, x'y' + i_1\, x'y + i_2\, xy' + i_3\, xy$$

Note that the output function O is written in canonical SOP form, in which all possible combinations—minterms $x'y'$, $x'y$, xy', and xy—of variables x and y occur.

The sum and carry functions S and C for a full adder can be obtained from the following truth table:

Inputs	Carry	Sum
$c\ x\ y$	C	S
0 0 0	0	0
0 0 1	0	1
0 1 0	0	1
0 1 1	1	0
1 0 0	0	1
1 0 1	1	0
1 1 0	1	0
1 1 1	1	1

and expressed in canonical SOP form as

$$S = c'x'y + c'xy' + cx'y' + cxy = \Sigma(1, 2, 4, 7)$$
$$C = c'xy + cx'y + cxy' + cxy = \Sigma(3, 5, 6, 7)$$

Note that each of the functions C and S contain product terms that contain some or all of the factors $x'y'$, $x'y$, xy', and xy. Since the 4-to-1 mux output function O contains all of these factors, a 4-to-1 mux can be used to implement either S or C.

The relation between the mux output O and the full adder sum S is made apparent by reordering the terms of S so that the terms appear in the left to right order: $x'y'$, $x'y$, xy', xy. The reordered expression for S is written above the expression for the mux output function O:

$$S = cx'y' + c'x'y + c'xy' + cxy$$
$$O = i_0x'y' + i_1x'y + i_2xy' + i_3xy$$

Hence, if the mux inputs in O are replaced by the corresponding coefficient (c or c') in S:

$$i_0 = c, \qquad i_1 = c', \qquad i_2 = c', \qquad i_3 = c$$

then $O = S$, and S is, thereby, implemented using a 4-to-1 mux.

Implementation of a Full Adder Using a Decoder

This section describes the implementation of a 1-bit full adder function using a 3-to-8 decoder. You recall that the sum and carry functions of a full adder can be expressed in canonical SOP form as

$$S = c'x'y + c'xy' + cx'y' + cxy = \Sigma(1, 2, 4, 7)$$
$$C = c'xy + cx'y + cxy' + cxy = \Sigma(3, 5, 6, 7)$$

Since a 3-to-8 decoder realizes all 8 minterms of the input variables, c, x, y, the full adder outputs C and S can be implemented using a 3-to-8 decoder and 4-input NAND gates, as shown in Figure 4.37.

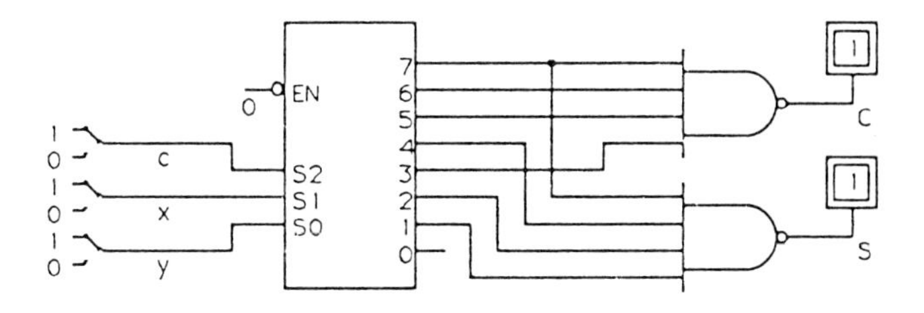

FIGURE 4.37
Implementation of a full adder using a 3-to-8 decoder

4.8 BUILDING-BLOCK IMPLEMENTATION OF HIGHER-LEVEL DEVICES

In this section we will see how combinational-logic function modules can be used as building blocks to design and implement higher-level devices.

As illustrated in the introduction to Part II, a simple computer can be organized in a hierarchical (top-down) structure. We review that structure here in Figure 4.38.

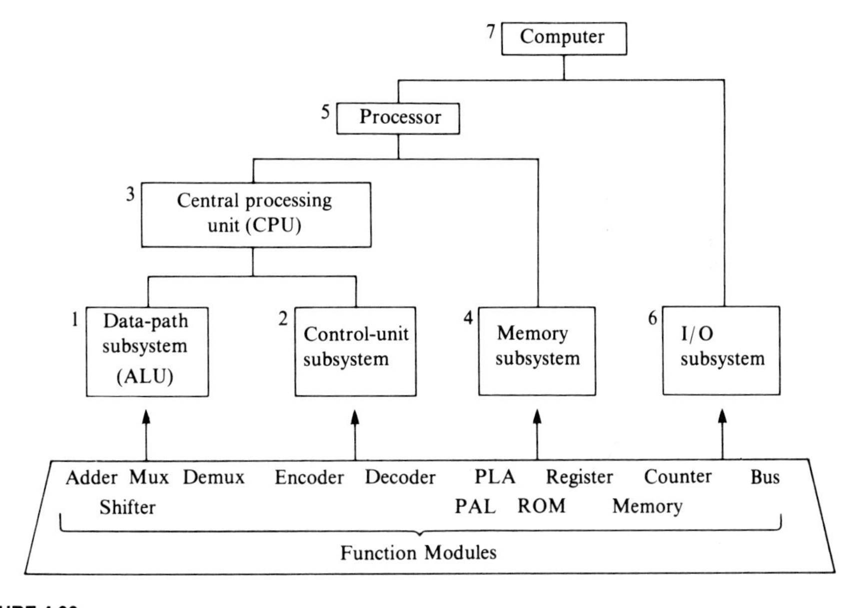

FIGURE 4.38
A hierarchical model of a simple computer

This hierarchically structured computer can be implemented in seven steps as follows:

1. Design a data-path subsystem (ALU) to perform arithmetic and logic operations on data.
2. Design a control-unit subsystem to control the operation of the data-path subsystem.
3. Integrate the (passive) data-path subsystem and the (active) control-unit subsystem to form a central processing unit (CPU).
4. Design a memory subsystem to store program instructions and data.
5. Integrate the CPU and memory subsystem to form a processor.
6. Design an input/output (I/O) subsystem to input data and output results.
7. Integrate the processor and the I/O subsystem to form a complete computer.

We will now apply this evolutionary hierarchical building-block approach to a simple project, the design of a tiny calculator.

Using Function Modules to Implement a Tiny Calculator

Most humans can interpret numbers represented in decimal (base = 10) form more easily than numbers in other bases. By contrast, computers and calculators usually store and process numbers in some binary form. Our objectives are to (1) design a simple device that handles the man–machine transfer of numerical data and performs simple computations, and (2) use a building-block approach to implement this device.

Step 1: Problem Statement. Design a device that accepts two decimal numbers X and Y (each limited to values 0, 1, 2, 3), computes the sum $S = X + Y$, and outputs S in decimal form.

Step 2: Conceptualization. Design and implement a functional, albeit tiny, calculator that

Accepts inputs (0, 1, 2, 3) in decimal form

Encodes decimal into binary form

Computes (adds) in binary

Decodes binary back to decimal

Outputs the result in decimal

Step 3: Solution/Simplification. Here we will use four combinational-logic function modules (a 4-to-2 encoder, a half adder, a 1-bit full adder, and a 3-to-8 decoder) to implement the tiny calculator.

First, the decimal numbers X and Y must be encoded into binary form. A 4-to-2 encoder, seen in Figure 4.39, can be used to translate from decimal to binary form.

A 2-bit number can be 00, 01, 10, or 11 with corresponding decimal equivalents 0, 1,

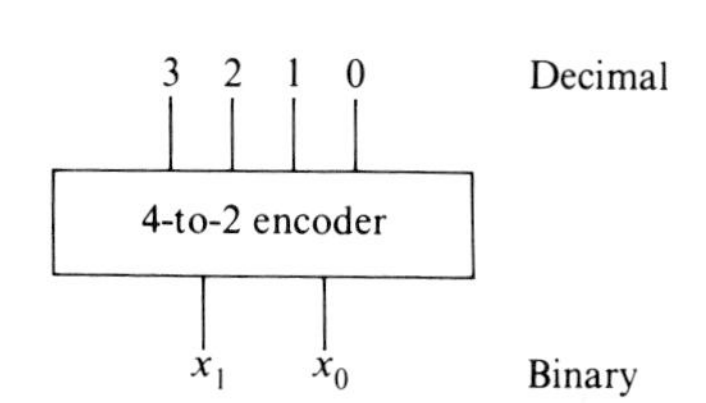

FIGURE 4.39
Symbolic diagram of 4-to-2 encoder

2, or 3. Therefore, the range of sums produced by adding two 2-bit numbers can be determined from the following decimal addition table:

Sum	0	1	2	3
0	0	1	2	3
1	1	2	3	4
2	2	3	4	5
3	3	4	5	6

The decimal sums range from 0 through 6, so that the corresponding binary sums range from 000 through 110. The encoded numbers, $X = x_1\, x_0$ and $Y = y_1\, y_0$, can be added using a half adder ($x_0 + y_0 \rightarrow C_1, S_O$) and a full adder ($C_1 + x_1 + y_1 \rightarrow C_2, S_1$).

At this point the outputs of two 4-to-2 encoders are connected to the inputs of the adders, as shown in Figure 4.40. Thus, function modules have been used to implement a higher-level device. This device accepts 2 decimal numbers X and Y, encodes each number into binary, and computes a 3-bit sum.

FIGURE 4.40
Diagram of 4-to-2 encoders and half and full adders

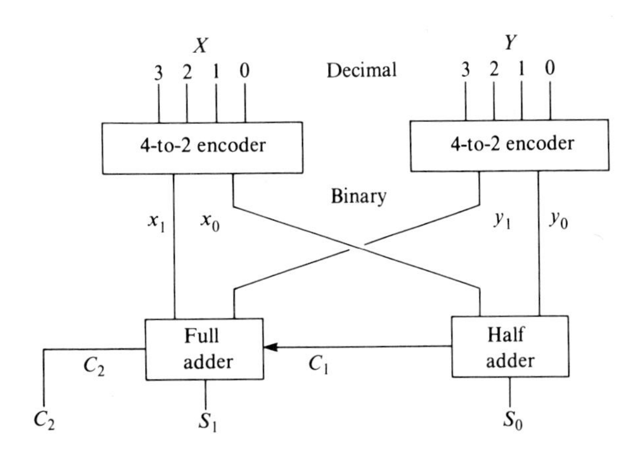

The final task is to translate (decode) this 3-bit sum from binary back to decimal so that the results can be easily interpreted. By connecting the 3 adder outputs to the inputs of the 3-to-8 decoder (and by connecting indicator lights to the 8 outputs of the 3-to-8 decoder), we complete the design of the tiny calculator.

Step 4: Realization. The tiny calculator can be implemented by interconnecting the combinational-logic function modules (a 4-to-2 encoder, a half adder, a 1-bit full adder, and a 3-to-8 decoder) as shown in Figure 4.41 (p. 186).

This tiny calculator doesn't do a lot. However, it does translate (encode), compute (add), and retranslate (decode), just as many larger computers do. Most importantly, the tiny calculator illustrates that higher-level devices can be designed and implemented by using function modules as building blocks.

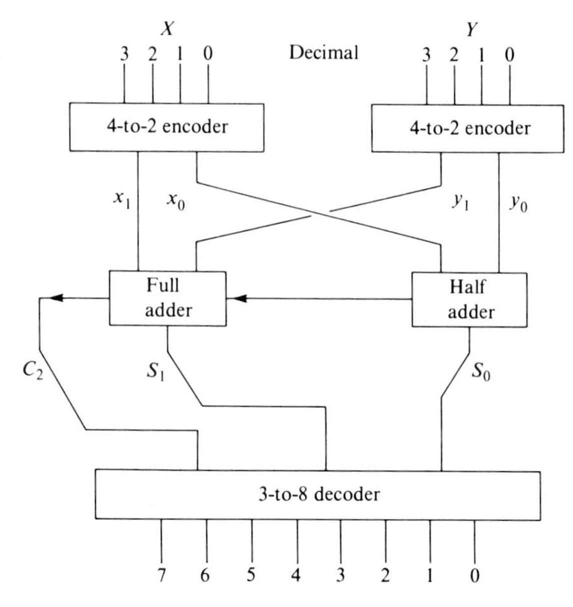

FIGURE 4.41
Block diagram of a tiny calculator

Using Function Modules to Implement a Pyramid-Dice Game Machine

A game called pyramid dice is played using two 4-faced dice shaped as pyramids. The faces on each die are numbered 0, 1, 2, 3. A player rolls the two dice, labeled X and Y. The outcome of a roll is one of 16 possibilities, as illustrated in Figure 4.42. Note that the 2-digit numbers in each block indicate the result of a roll; the left number is X, and the right number is Y.

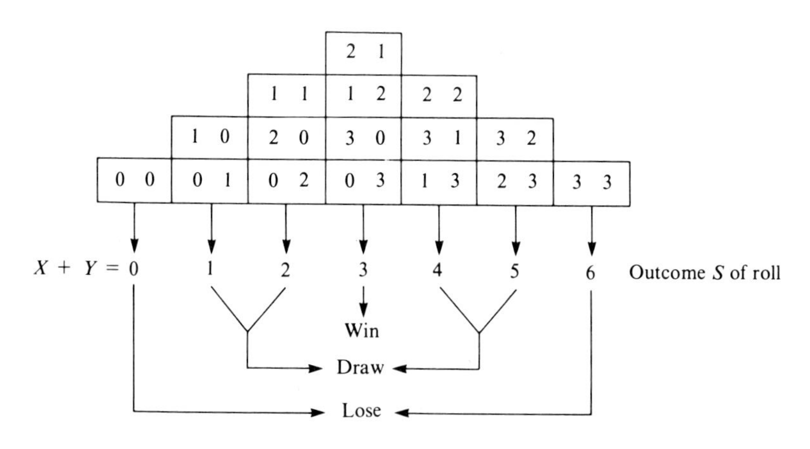

FIGURE 4.42
Table of possible outcomes for pyramid-dice game

From this figure, we see that a player can either

Win by rolling a 3

Lose by rolling a 0 or 6

Draw by rolling a 1, 2, 4, or 5

It should be noted that this game is a simplification of the game of dice played with two 6-faced dice. A pyramid-dice game machine can be designed using the 4-step problem-solving procedure as follows:

Step 1: Problem Statement. Design and implement a hardware device that will simulate the game of pyramid dice.

Step 2: Conceptualization. There are at least two approaches to automating the game. Each of these approaches can result in a different implementation of the game machine. In any event, we are to design a device that

Accepts decimal numbers X and Y, where X and Y result from rolling the pyramid dice

Computes the sum $S = X + Y$

Determines the game outcome: W if $S = 3$; L if $S = 0$ or 6; D if $S = 1, 2, 4,$ or 5

Displays the outcome (W, L, or D) of the game

Step 3a: Solution/Simplification. Using the first approach, the tiny calculator designed in the preceding section accepts 2 decimal numbers X and Y (0, 1, 2, 3), computes the sum $S = X + Y$, and outputs the decoded result. This calculator can be used as a building block in the construction of our pyramid-dice game machine.

Step 4a: Realization. Since the game outcomes are OR functions of possible outcomes of the sum S, OR gates can be used to determine whether a player wins, loses, or draws. The result is the pyramid-dice game machine illustrated in Figure 4.43.

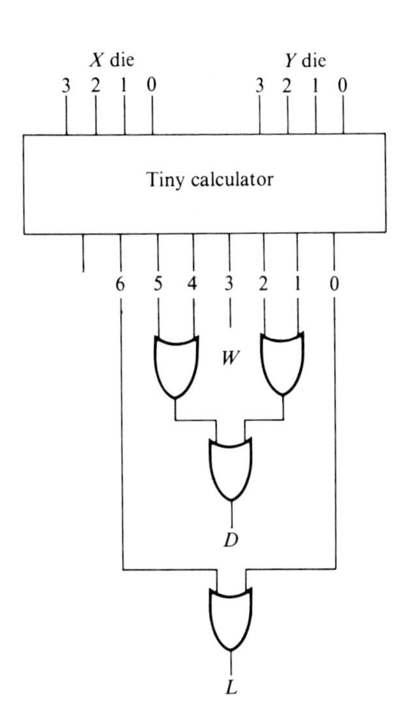

FIGURE 4.43
The pyramid-dice game machine—first implementation

Step 3b: Solution/Simplification. Using the second approach to implement a pyramid-dice game machine, we use the game outcomes

$$W = 1 \text{ if } S = 3$$
$$L = 1 \text{ if } S = 0 \text{ or } 6$$
$$D = 1 \text{ if } S = 1, 2, 4, \text{ or } 5$$

directly to develop logic equations for the machine's outputs W, L, and D. If a roll of the two dice is denoted by (X, Y), where $X = 0, 1, 2,$ or 3 and $Y = 0, 1, 2,$ or 3, these logic equations can be written as follows:

$$\begin{aligned} W &= (0 \cdot 3) + (3 \cdot 0) + (1 \cdot 2) + (2 \cdot 1) \\ L &= (0 \cdot 0) + (3 \cdot 3) \\ D &= (0 \cdot 1) + (1 \cdot 0) + (0 \cdot 2) + (2 \cdot 0) + (1 \cdot 1) \\ &\quad + (1 \cdot 3) + (3 \cdot 1) + (2 \cdot 2) + (2 \cdot 3) + (3 \cdot 2) \end{aligned}$$

Step 4b: Realization. Each of the logic equations in Step 3b can be implemented as a two-level AND-OR circuit.

4.9 PROGRAMMABLE LOGIC DEVICES

Any combinational logic function with one or more outputs can be represented by a multioutput truth table and the corresponding canonical SOP (or POS) expressions.

Programmable logic devices (PLDs) are a class of logic devices that can be used to realize (implement) any combinational logic function; they include the devices shown in Table 4.3.

TABLE 4.3
Programmable logic devices

Type of PLD	AND Array	OR Array
Programmable Logic Array (PLA)	Programmable	Programmable
Programmable AND-Array Logic (PAL)	Programmable	Fixed
Read-Only Memory (ROM)	Decodes all minterms	

Programmable Logic Array

A programmable logic array (PLA) is composed of a programmable AND array for realizing product terms and a programmable OR array for realizing sums of the product terms. A PLA implements only required product terms rather than the complete set of 2^n minterms for n-input variables. Therefore, they are useful for implementing switching functions in which the original canonical SOP expression has been replaced by a simpler equivalent minimal SOP expression. PLAs may also be used as a function module of a computer's control unit.

PLAs are available in mask-programmable or field-programmable form. In the mask-programmable PLA, the desired product terms (P_1, P_2, . . . , P_n) and sum terms are programmed when the device is manufactured. The field-programmable PLA has links that can be fused or blown to store the desired pattern in the AND and OR arrays. The PLA logic structure is shown in Figure 4.44. (P_i denotes the ith product term.)

FIGURE 4.44
Logic structure of a PLA

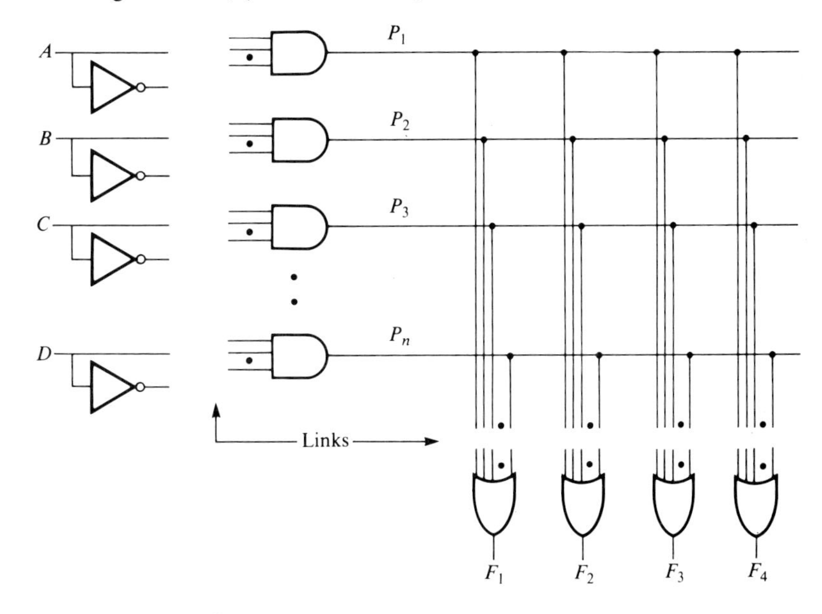

EXAMPLE 4.4: PLA Implementation of a 2-Bit Multiplier

If a 2-bit number AB is multiplied by a 2-bit number CD, the 4 bits f_3, f_2, f_1, f_0 of the 4-bit product are

$$f_0 = \Sigma(5, 7, 13, 15) = BD$$
$$f_1 = \Sigma(6, 7, 9, 11, 13, 14) = AB'D + AC'D + BCD' + A'BC$$
$$f_2 = \Sigma(10, 11, 14) = AB'C + ACD'$$
$$f_3 = \Sigma(15) = ABCD$$

The K-maps for the multiplier's outputs are shown in Figure 4.45. Since the 2-bit multiplier requires only 8 product terms, of which only 1 has 4 inputs, it can be implemented more economically with a PLA than with a ROM (to be described shortly).

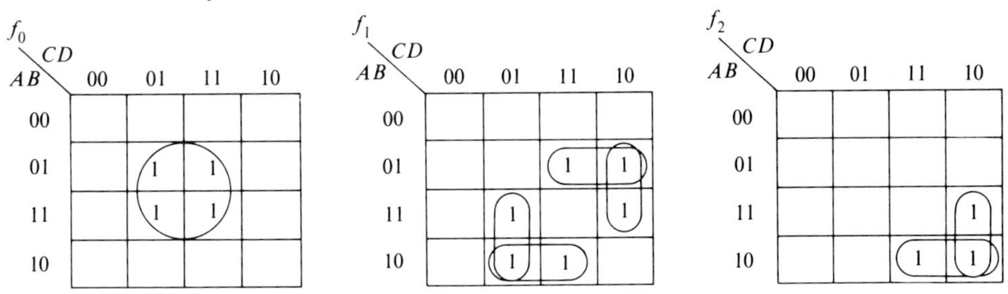

FIGURE 4.45
K-maps for 2-bit multiplier outputs

Programmable AND-Array Logic

Another type of programmable logic device is the programmable AND-array logic (PAL) device. This device has a programmable AND-array, as its name indicates, and a fixed OR array. Figure 4.46 shows the logic diagram for the PAL.

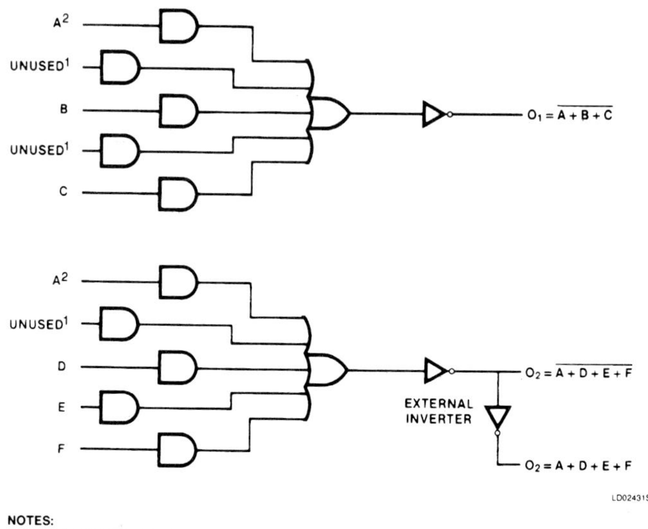

FIGURE 4.46
PAL logic diagram (Reprinted by courtesy of Philips Components/Signetics.)

Read-Only Memory Logic

A read-only memory (ROM) can be considered as a programmable logic device. The AND array of an n-by-m ROM realizes all of the 2^n minterms of the n-input variables. Each combination of input values determines a minterm, so the inputs of a ROM are referred to as address lines. Each OR gate output represents the output of an n-input switching function.

The logic structure of an n-by-m ROM consists of an AND array and an OR array, as shown in Figure 4.47.

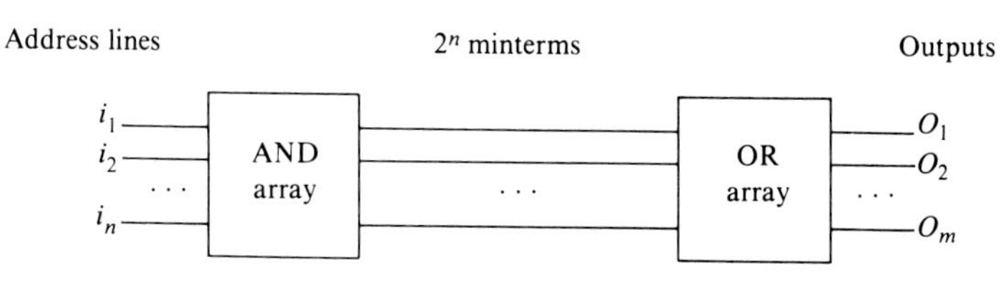

FIGURE 4.47
Logic structure of an n-by-m ROM

Therefore, a combinational logic function with n inputs and m outputs can be implemented using a n-by-m ROM. Since a ROM realizes an n-input truth table, ROMs are useful for implementing switching functions represented in canonical SOP form. For example, a binary-to-Gray code converter can be implemented using a 16-by-4 ROM.

Since a decoder realizes the 2^n minterms for n variables, the AND array of a ROM can be implemented using an n-to-2^n decoder. The outputs of a decoder are usually active-low, so the gates in the ROM's OR array can be implemented using NAND gates, as shown in Figure 4.48. Note that each output F_i is a canonical SOP expression.

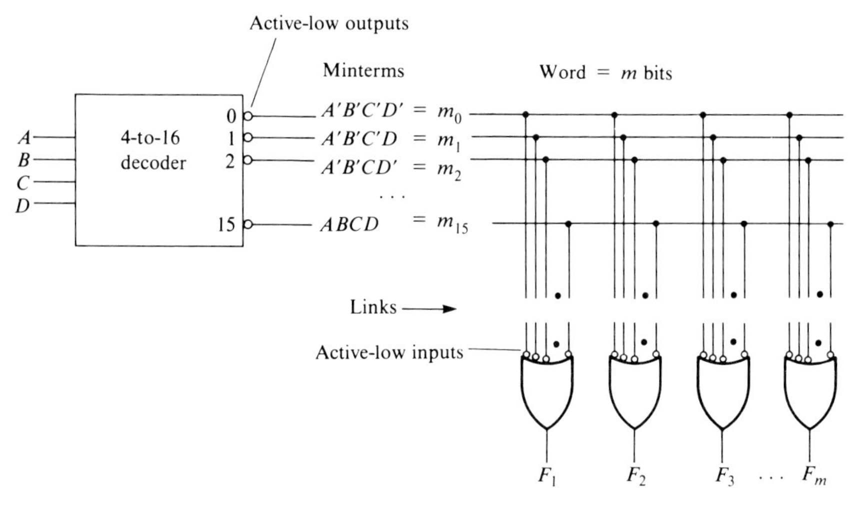

FIGURE 4.48
ROM implementation using a decoder and NAND gates

4.10 SUMMARY

The digital-logic design process involves the four principal elements of problem solving described in Chapter 2: problem statement, conceptualization, solution/simplification, and realization. Boolean algebra is the principal theoretical tool used in the design of combinational logic circuits. The methods for simplifying Boolean expressions, described in Chapter 3, are essential to the design of cost-effective circuits.

A minimal SOP expression in AND-OR form can be converted to NAND-NAND, OR-NAND, and NOR-OR form by successive application of DeMorgan's theorem. A minimal POS expression in OR-AND form can be converted to NOR-NOR, AND-NOR, and NAND-AND form by successive application of DeMorgan's theorem. A combinational logic function can be realized in any one of the following 2-level forms:

AND-OR	OR-AND
NAND-NAND	NOR-NOR
OR-NAND	AND-NOR
NOR-OR	NAND-AND

The 4-step problem-solving procedure was used in this chapter to design a number of common combinational-logic function modules, including multiplexers, demultiplexers, encoders, decoders, code converters, parity generators, parity checkers, and shifters. Many of these functions exist in medium-scale-integration (MSI) form and can be used as off-the-shelf building blocks to implement higher-level devices and systems, such as controllers and computers. The use of function modules to implement a higher-level device was illustrated by constructing a tiny calculator using a half adder, a full adder, 4-to-2 encoders, and a 3-to-8 decoder as building blocks.

These examples illustrated the use of Boolean algebra and K-maps in the design of combinational logic circuits. Other examples were presented to illustrate the use of MSI circuits to implement an arbitrary Boolean function.

Static and dynamic hazards in combinational logic circuits are caused by unequal signal path delays and may be eliminated in the initial design phase. A static hazard causes a single transient in an output signal that should have remained unchanged in response to an input change. A static hazard, indicated by adjacent 1-groups in the K-map, can be eliminated by adding a redundant consensus term.

A dynamic hazard occurs when an output circuit is supposed to change once but instead changes three or more (odd) times. A dynamic hazard, detected by analyzing a timing diagram or the circuit's logic equation, may be eliminated by the inclusion of a consensus term or by rewriting the logic equation so that an input variable does not appear in one product term with its complement and in another term without its complement.

A multiplexer can be used as (1) a data-selector function module in a digital logic system, (2) a bus interface device, or (3) to implement an arbitrary switching function.

A demultiplexer is a 1-input, multi-output device that routes the single input to the output selected by the combination of values of the select lines. A demultiplexer is generally implemented by a decoder with an enable input.

A bus can be conceptualized as an OR function of several sources, where only one source is functionally connected to the bus at any given time. A bus can be realized either as (1) a centralized-access bus, implemented with a multiplexer, or (2) a decentralized-access bus, implemented with a physical OR gate or a virtual (wired) OR gate.

A priority encoder is a device that encodes the input line with the highest priority. The 74148 IC is an example of an off-the-shelf 8-input priority encoder with active-low inputs and outputs.

Decoders can be used for a number of functions such as routing data to a selected output line, addressing memory, converting data from one code to another, realizing an n-input truth table, and performing a demultiplexer function.

A parity generator is used to append a parity bit to each data word so that the total number of 1s is even (or odd) in every data plus parity word. The parity of a data plus parity word can be checked by a parity checker.

A combinational logic shifter is a device that produces an output that is obtained by shifting its input. Types of shift operations include logical (logic 0 is inserted in vacated positions), arithmetic (sign bit is extended in a right shift), and end-around.

An arbitrary Boolean (switching) function can also be implemented using MSI devices such as a multiplexer or a decoder, since each of these devices realize all of the 2^n minterms of n-input variables.

Programmable logic devices (PLDs) are a class of logic devices that can be used to implement any combinational logic circuit. PLDs include programmable logic arrays (PLA), programmable AND-array logic (PAL), and read-only memory (ROM).

The next chapter describes the evolution of a variety of combinational-logic arithmetic circuits ranging from a 1-bit adder to an 8-bit arithmetic unit.

KEY TERMS

Feedback path
Hazard
Static hazard
Dynamic hazard
Multiplexer (mux)
Data selector
Demultiplexer (demux)
Data router
Source
Destination
Bus
Words
Parity generator
Parity checker
Fill bit
Spill (end off) bit
Programmable logic devices (PLDs)

EXERCISES

General

1. Construct a truth table for converting a 4-bit Gray code (A, B, C, D) to binary code (W, X, Y, Z), and use 4-input K-maps to simplify the resulting expressions.

2. Construct a truth table for a 5-input majority circuit with inputs E, A, B, C, and D.
   a. Construct a minterm-ring map for the 5-input majority function $f(E, A, B, C, D)$.
   b. Use the minterm-ring algorithm to simplify the expression for f.

3. Let $P = p_5p_4p_3p_2p_1p_0$ denote the product of two 3-bit integers $Y = FEA$ and $X = BCD$.
   a. Construct a minterm-ring map for each coefficient p_i, $(i = 0, 5)$.
   b. Use the minterm-ring algorithm to simplify the expression for each coefficient p_i, $(i = 0, 5)$.

4. Construct a function table for an 8-input (A, B, C, D, E, F, G, H) priority encoder, where A and H represent the highest and lowest priority inputs, respectively.

5. Show that a 4-to-16 decoder can be implemented using two 2-to-4 decoders and sixteen 2-input AND gates.

Design/Implementation

1. Use the 4-step problem-solving procedure to design a 3-to-8 decoder. Assume inputs A_2, A_1, A_0 are active-high and outputs $O_7, \ldots, O_0$ are active-low.

2. Use the 4-step problem-solving procedure to design a 1-to-4 demultiplexer using a 2-to-4 decoder with an enable input. *Hint:* When using a decoder as a demultiplexer, use the enable input as the data input and use the decoder inputs as select lines.

3. Design a Gray-to-binary code converter.

4. Implement a full-adder carry output C using a 4-to-1 mux.

5. Use a ROM to implement a BCD-to-7-segment display.

6. Use a ROM to implement a BCD-to-excess-3 code converter.

7. Use a ROM to implement a hexadecimal-to-7-segment display.

8. Use a mux to implement a true/complement/zero/one (TC01) function. The output $F = 0, 1, A$, or A', depending on the setting of select signals x and y.

9. Design a 2-input (A, B) combinational-logic circuit with output F, where $F = (A \cdot B), (A + B), (A \oplus B)$, or A', depending on the setting of select signals x and y.
a. Implement the circuit using a 4-to-1 multiplexer.
b. Implement the circuit using NAND gates.

10. Design and implement an 8-input priority encoder, with inputs $A, \ldots, H$, (A = highest priority, H = lowest priority). The 8 inputs and 3 outputs are active low.

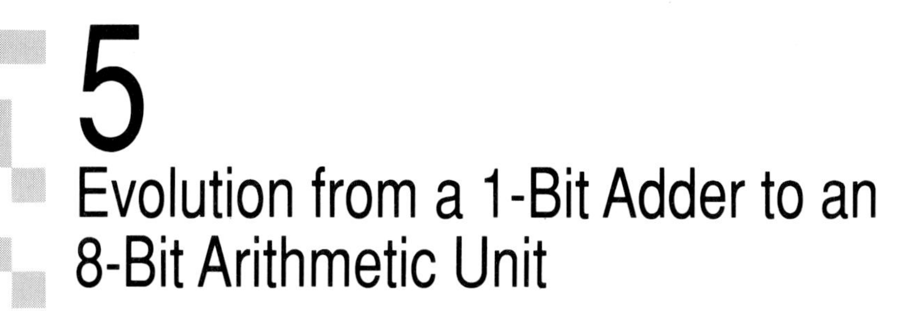

5 Evolution from a 1-Bit Adder to an 8-Bit Arithmetic Unit

OBJECTIVES

After you complete this chapter, you will be able

- □ To use alternative approaches for solving design problems, such as the intuitive approach and the formal solution approach
- □ To analyze a design problem with the idea in mind that there may be two or more alternative implementations
- □ To use a parallel adder to accomplish multibit addition
- □ To analyze alternative implementations to determine how they compare in operating speed and hardware organization
- □ To use an evolutionary approach to digital design
- □ To develop multifunction devices, such as an adder/subtracter and an 8-function arithmetic unit.
- □ To use medium-scale integration (MSI) integrated circuits to implement larger-scale digital devices.

5.1 INTRODUCTION

This chapter describes the evolutionary development of a simple, yet realistic, arithmetic unit for a digital computer. This arithmetic unit, like those of most conventional, general-purpose digital computers, is based on a multibit binary adder.

In this chapter, we will not always use the 4-step problem-solving procedure because we want to emphasize the evolution from a 1-bit adder to an 8-bit arithmetic unit.

We illustrate the design of small-scale combinational-logic computing devices (up to 8 bits) by describing the evolution of a device from a 1-bit adder to an 8-bit arithmetic unit. Our intent is to illustrate the design of computing devices with a limited number of bits so that each design can be presented in its entirety. Later you will see, from a conceptual viewpoint, how 16-bit and 32-bit devices can be implemented by extrapolating from the design of the 4-bit and 8-bit devices.

We begin with the design of simple computing devices such as 1-, 2-, and 4-bit adders. Then, we design a 4-bit adder/subtracter and a 4-bit, 8-function arithmetic unit. Finally, we look at the implementation of 8-bit adders, adder/subtracters, and arithmetic units using MSI 4-bit adders as functional building blocks.

There are a number of alternative approaches for the design and implementation of multibit adders. This chapter describes in detail the design and implementation of the following multibit binary adders:

A 4-bit adder implemented by cascading four 1-bit adders

A 4-bit adder implemented using carry look-ahead

An 8-bit adder implemented using two MSI 4-bit adders

The following are two reasons for illustrating alternative approaches to, and different implementations of, the same design problem:

- To see that a particular solution to a design problem may be arrived at by using either an intuitive approach or a formal solution.
- To see that a given design problem may yield alternative solutions (implementations) that represent trade-offs between circuit complexity and operating speed. That is, one implementation may be less complex (requiring less hardware), while another implementation may be faster (including fewer stages and less propagation delay). Yet another implementation may represent a compromise, being less complex than one and faster than another.

A number of 4-bit adders, arithmetic units, and ALUs exist in MSI IC form. These MSI ICs can be cascaded to implement 8- and 16-bit (and even more powerful) devices.

A practical computer must be capable of performing arithmetic operations on multibit numbers. For example, in an 8-bit computer, numeric data may be stored in single length form (8 bits) or in double length form (16 bits). For the time being, let us assume that numeric data is stored in 2s complement form in 8 bits, with the leftmost bit used for the sign and the remaining 7 bits used for the magnitude of the number.

5.2 DESIGN OF 2-BIT BINARY ADDERS

In Chapter 2 you learned how to design a 1-bit full-adder circuit that adds two 1-bit numbers* x *and* y *and a carry-in* c. *Now the question is how to approach the design of a digital logic circuit that accomplishes the addition of two 2-bit numbers with a carry-in.

The design of a 2-bit binary adder can be accomplished using the 4-step problem-solving procedure.

Step 1: Problem Statement. Design a 2-bit binary adder that adds two 2-bit numbers AB and CD and a carry-in E.

A 2-bit number can be 00, 01, 10, or 11, with corresponding decimal equivalents 0, 1, 2, or 3. Therefore, the range of sums produced by adding two 2-bit numbers and a carry-in E can be determined from the addition tables below. Here, we can see that since the decimal sums range from 0 through 7, the corresponding binary sums range from 000 through 111.

$E = 0$

Sum	0	1	2	3
0	0	1	2	3
1	1	2	3	4
2	2	3	4	5
3	3	4	5	6

$E = 1$

Sum	0	1	2	3
0	1	2	3	4
1	2	3	4	5
2	3	4	5	6
3	4	5	6	7

Step 2: Conceptualization. After we have determined the value ranges of the outputs, the next step is to develop a conceptual model of the process. In order to illustrate alternative approaches to the design of a 2-bit adder, we consider two representations of the 2-bit addition process: a symbolic form, as shown in Figure 5.1a, and a block diagram, as shown in Figure 5.1b.

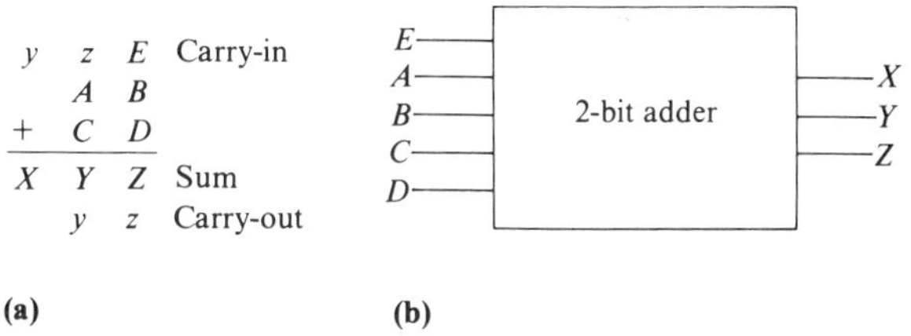

FIGURE 5.1
Alternative representations of 2-bit addition. (a) Symbolic form (b) Block diagram

The representation of the 2-bit addition process in both symbolic form and block diagram suggests alternative approaches for solving the design problem stated in Step 1:

1. The columnar form of Figure 5.1a suggests using a full adder to add the rightmost column bits (E, B, D) and then cascading the carry-out z to a second full adder that adds z, A, and C. This intuitive approach leads to a 2-bit **ripple-carry adder** implemented by cascading two 1-bit full adders.

2. A 5-input truth table, suggested by the block diagram form of Figure 5.1b, can be used to develop logic expressions for the outputs X, Y, and Z of a 2-bit direct adder (without cascading); this approach requires a formal solution using the truth table.

These alternative approaches to designing different versions of a 2-bit adder allow us to demonstrate that different approaches can result in different circuit implementations, which then must be compared with regard to simplicity of design, efficiency, and propagation delay.

The following sections describe Steps 3 and 4 of the design of alternative forms of a 2-bit adder using the two approaches described above.

Design of a 2-Bit Binary Adder by Cascading 1-Bit Adders

Step 3a: Solution/Simplification. We will begin with the first approach described in the preceding section. The addition of two 2-bit numbers AB, CD, and a carry-in E, can be accomplished using a standard pencil-and-paper procedure. The coefficients in the 2^0 position (LSBs E, B, and D) are added to form a sum Z and a carry-out z; then the coefficients in the 2^1 position (z, A, and C) are added to form a sum Y and carry-out y. The result is a 3-bit sum XYZ, where the 2^2 coefficient X equals the carry-out y:

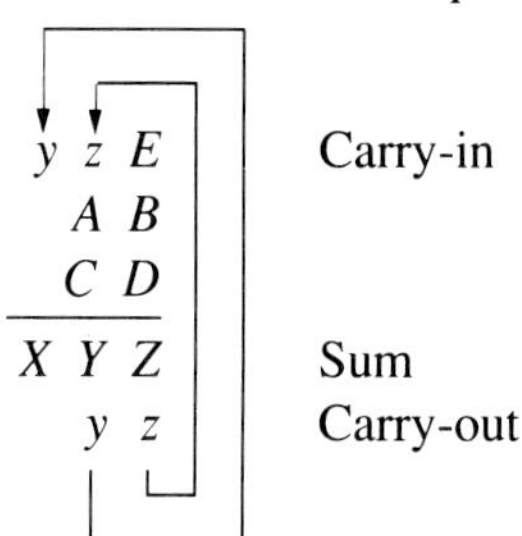

Intuitively, we can arrive at the sum Z and carry z in the right column as follows:

□ The sum $Z = 1$ if an odd number of inputs $E, B, D = 1$; hence,

$$Z = E'B'D + E'BD' + EB'D' + EBD = E \oplus (B \oplus D)$$

□ The carry-out $z = 1$ if two or more inputs $E, B, D = 1$; hence,

$$z = E'BD + EB'D + EBD' + EBD = EB + ED + BD$$

Similarly, the formulas for the second column sum and carry-out, Y and y, have the same form as the formulas for Z and z in the first column; that is,

$$Y = z'A'C + z'AC' + zA'C' + zAC = z \oplus (A \oplus C)$$

and

$$y = z'AC + zA'C + zAC' + zAC = zA + zC + AC$$

and the third column term $X = y$.

Step 4a: Realization. A 1-bit full adder, with inputs E, B, and D, produces the right column sum Z and carry-out z. A second full adder, cascaded with the first, adds the inputs z, A, and C to produce the second column sum Y and carry-out y. The carry-out y of the second full adder is the X coefficient of the 3-bit sum XYZ. A block diagram of a 2-bit ripple-carry adder, formed by cascading two 1-bit full adders, is shown in Figure 5.2.

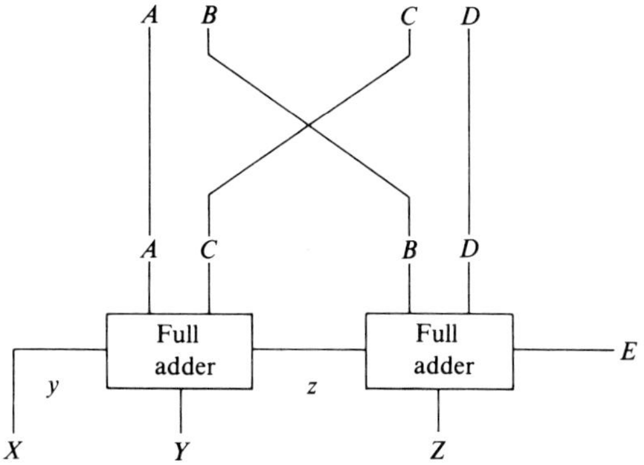

FIGURE 5.2
Block diagram of 2-bit ripple-carry adder

A logic diagram for a 2-bit ripple-carry adder is presented in Figure 5.3.

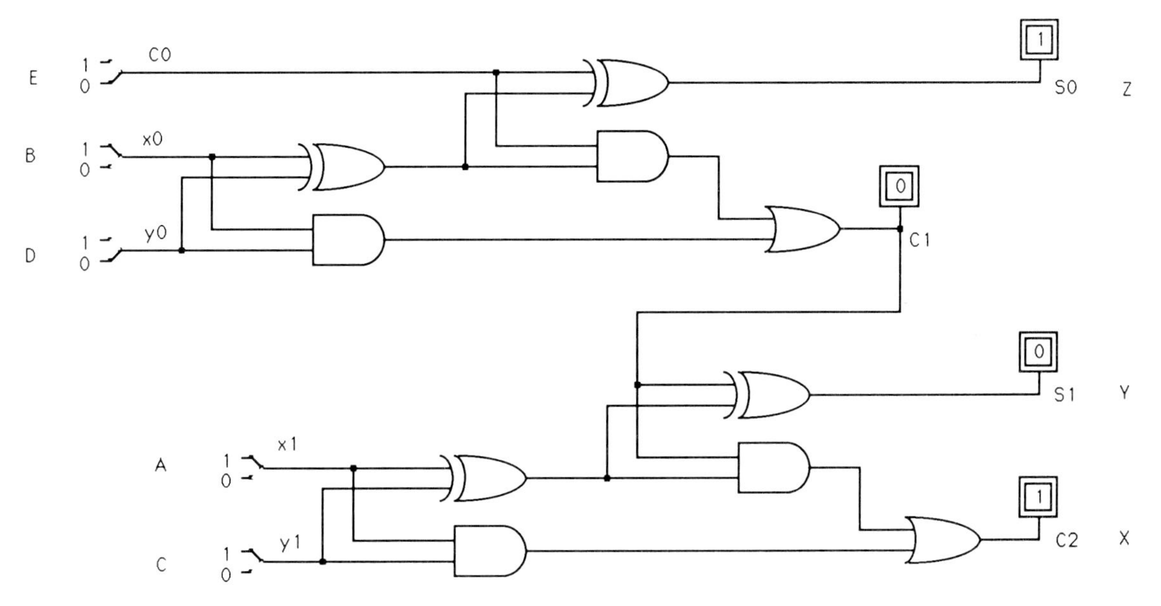

FIGURE 5.3
Logic diagram of 2-bit ripple-carry adder

Note that there are five stages of propagation delays in the longest path of a 2-bit ripple-carry adder formed by cascading two 1-bit full adders of the form shown in Figure 5.3. On the other hand, if each full adder is implemented in 2-level AND-OR form, where $C = xy + cx + cy$ and $S = c \oplus x \oplus y$, the 2-bit ripple-carry adder has four stages of propagation delays in the longest path.

Design of a 2-Bit Direct Binary Adder Without Cascading

Step 3b: Solution/Simplification. Now we will use the second approach to design a 2-bit adder. A 2-bit **direct adder** is defined as an adder, with two 2-bit inputs *AB* and *CD* and a carry-in *E*, that computes the outputs *X*, *Y*, and *Z* directly in terms of the inputs *A*, *B*, *C*, *D*, and *E*, without cascading the internal carry as in the ripple-carry adder. A symbolic diagram of a 2-bit direct binary adder is presented in Figure 5.4.

FIGURE 5.4
Symbolic diagram of 2-bit direct adder

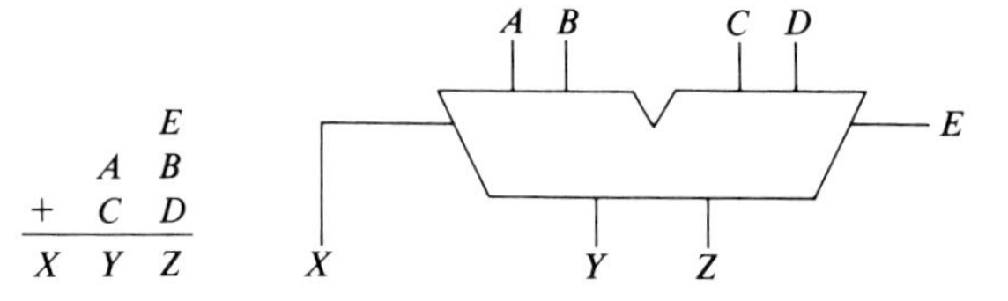

The outputs *X*, *Y*, and *Z* are functions of *A*, *B*, *C*, *D*, and *E*, as determined by the addition table shown in Table 5.1.

TABLE 5.1
Truth table for 2-bit addition

Decimal	*E*	*A*	*B*	*C*	*D*	*X*	*Y*	*Z*
0	0	0	0	0	0	0	0	0
1	0	0	0	0	1	0	0	1
2	0	0	0	1	0	0	1	0
3	0	0	0	1	1	0	1	1
4	0	0	1	0	0	0	0	1
5	0	0	1	0	1	0	1	0
6	0	0	1	1	0	0	1	1
7	0	0	1	1	1	1	0	0
8	0	1	0	0	0	0	1	0
9	0	1	0	0	1	0	1	1
10	0	1	0	1	0	1	0	0
11	0	1	0	1	1	1	0	1
12	0	1	1	0	0	0	1	1
13	0	1	1	0	1	1	0	0
14	0	1	1	1	0	1	0	1
15	0	1	1	1	1	1	1	0
16	1	0	0	0	0	0	0	1
17	1	0	0	0	1	0	1	0
18	1	0	0	1	0	0	1	1
19	1	0	0	1	1	1	0	0
20	1	0	1	0	0	0	1	0
21	1	0	1	0	1	0	1	1
22	1	0	1	1	0	1	0	0
23	1	0	1	1	1	1	0	1

TABLE 5.1
Continued

Decimal	E	A	B	C	D	X	Y	Z
24	1	1	0	0	0	0	1	1
25	1	1	0	0	1	1	0	0
26	1	1	0	1	0	1	0	1
27	1	1	0	1	1	1	1	0
28	1	1	1	0	0	1	0	0
29	1	1	1	0	1	1	0	1
30	1	1	1	1	0	1	1	0
31	1	1	1	1	1	1	1	1

Note the exponential increase in combinations; with 5 inputs, there are $2^5 = 32$ input combinations.

If the numeric variables E, A, B, C, D and X, Y, Z are replaced by corresponding logic variables, the binary addition table becomes a truth table. The output variables X, Y, and Z can be represented by the 5-variable minterm-ring maps shown in Figures 5.5, 5.6, and 5.7, respectively.

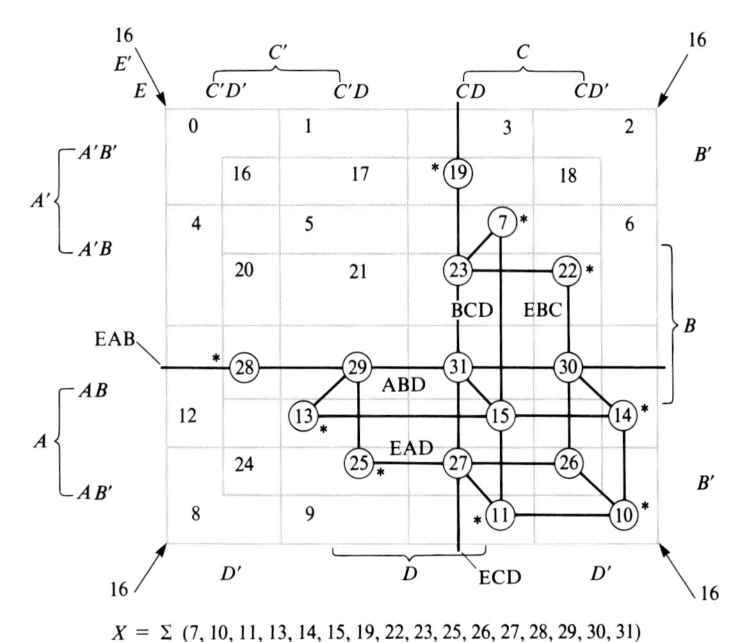

FIGURE 5.5
A 5-variable minterm ring map for X

In Figure 5.5 we see that a minimal cover of X is composed of 6 EPI quartets and 1 EPI octet, as follows:

Quartets

$$
\begin{aligned}
(31, 29, 15, 13) &= 31(16, 2) &&= 11111(16, 2) = \not{E}AB\not{C}D &&= ABD \\
(31, 30, 29, 28) &= 31(2, 1) &&= 11111(2, 1) = EAB\not{C}\not{D} &&= EAB \\
(31, 29, 27, 25) &= 31(4, 2) &&= 11111(4, 2) = EA\not{B}\not{C}D &&= EAD \\
(31, 23, 15, 7) &= 31(16, 8) &&= 11111(16, 8) = \not{E}\not{A}BCD &&= BCD \\
(31, 30, 23, 22) &= 31(8, 1) &&= 11111(8, 1) = E\not{A}BC\not{D} &&= EBC \\
(31, 27, 23, 19) &= 31(8, 4) &&= 11111(8, 4) = E\not{A}\not{B}CD &&= ECD
\end{aligned}
$$

Octet

$$(31, 30, 27, 26, 15, 14, 11, 10) = 31(16, 4, 1) = \not{E}A\not{B}C\not{D} = AC$$

A minimal sum of X in 2-level AND-OR form is

$$X = ABD + EAB + EAD + BCD + EBC + ECD + AC$$

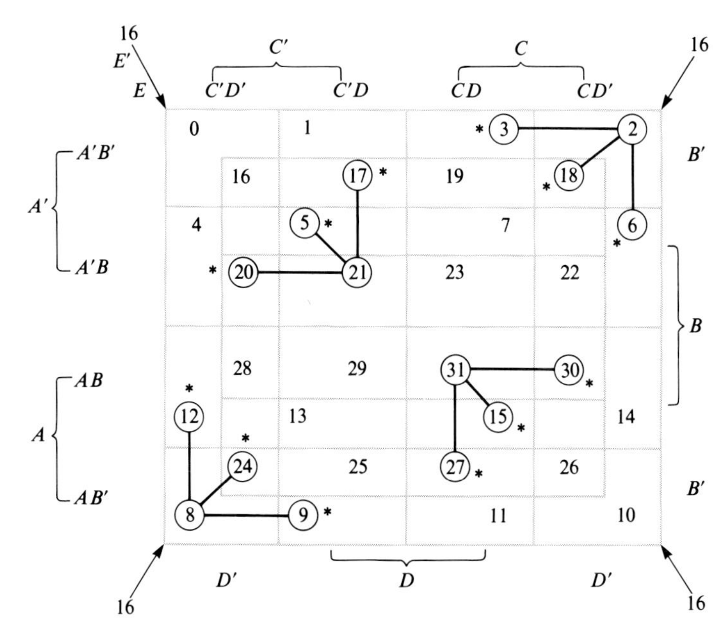

$Y = \Sigma\ (2, 3, 5, 6, 8, 9, 12, 15, 17, 18, 20, 21, 24, 27, 30, 31)$

FIGURE 5.6 A 5-variable minterm ring map for Y

Figure 5.6 shows that a minimal cover of Y is composed of 12 EPI pairs:

$$
\begin{aligned}
(3, 2) &= 00011(1) &&= E'A'B'C\not{D} = E'A'B'C \\
(6, 2) &= 00110(4) &&= E'A'\not{B}CD' = E'A'CD' \\
(18, 2) &= 10010(16) &&= \not{E}A'B'CD' = A'B'CD' \\
&\ \vdots
\end{aligned}
$$

A minimal sum of Y in 2-level AND-OR form is

$$\begin{aligned} Y = {} & E'A'B'C + E'A'CD' + E'AB'C' + E'AC'D' + EA'BC' + EA'C'D \\ & + EABC + EACD + A'B'CD' + AB'C'D' + A'BC'D + ABCD \end{aligned}$$

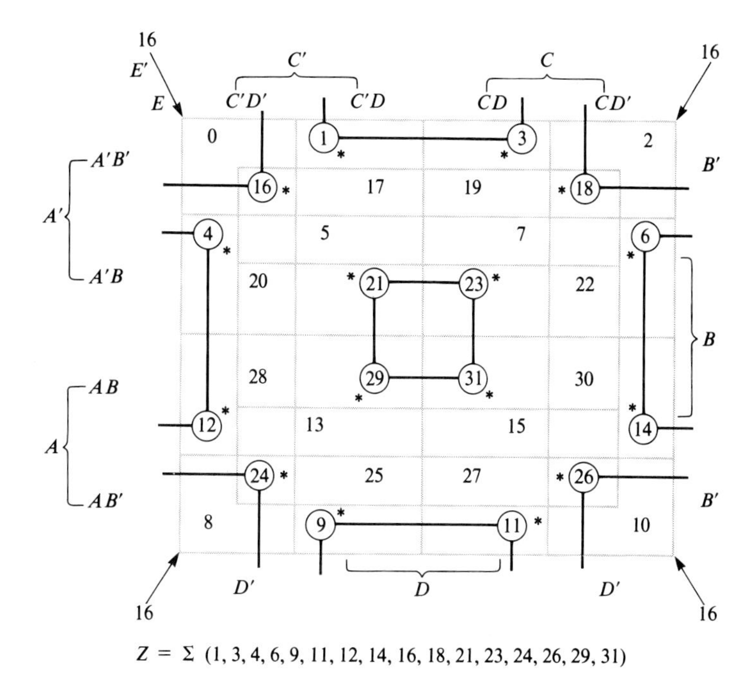

$Z = \Sigma\ (1, 3, 4, 6, 9, 11, 12, 14, 16, 18, 21, 23, 24, 26, 29, 31)$

FIGURE 5.7
A 5-variable minterm ring map for Z

Figure 5.7 shows that a minimal cover of Z is composed of 4 EPI quartets:

$$\begin{aligned}
(31, 29, 23, 21) &= 31(8, 2) = 11111(8, 2) = E\not{A}B\not{C}D = EBD \\
(26, 24, 18, 16) &= 26(8, 2) = 11010(8, 2) = E\not{A}B'\not{C}D' = EB'D' \\
(14, 12, 6, 4) &= 14(8, 2) = 01110(8, 2) = E'\not{A}B\not{C}D' = E'BD' \\
(11, 9, 3, 1) &= 11(8, 2) = 01011(8, 2) = E'\not{A}B'\not{C}D = E'B'D
\end{aligned}$$

A minimal sum of Z in 2-level AND-OR form is

$$Z = EBD + EB'D' + E'BD' + E'B'D$$

Step 4b: Realization. Figure 5.8 (p. 204) illustrates an implementation of a 2-bit direct adder using factored forms of the minimal SOP expressions for X, Y, and Z.

The speed and complexity trade-offs between the 2-level AND-OR implementation of a 2-bit direct adder and a 2-bit ripple-carry adder can be summarized as follows:

- □ The two-level AND-OR implementation of the 2-bit direct adder has 2 gate-propagation delays in each path.
- □ The longest path in the cascaded 2-bit ripple-carry adder has 5 gate-propagation delays, assuming each full adder is implemented using $S = c \oplus x \oplus y$, and $C = xy + c(x \oplus y)$.
- □ The 2-level AND-OR implementation of a 2-bit direct adder is more than twice as fast as the 2-bit ripple-carry adder because of these first two facts.

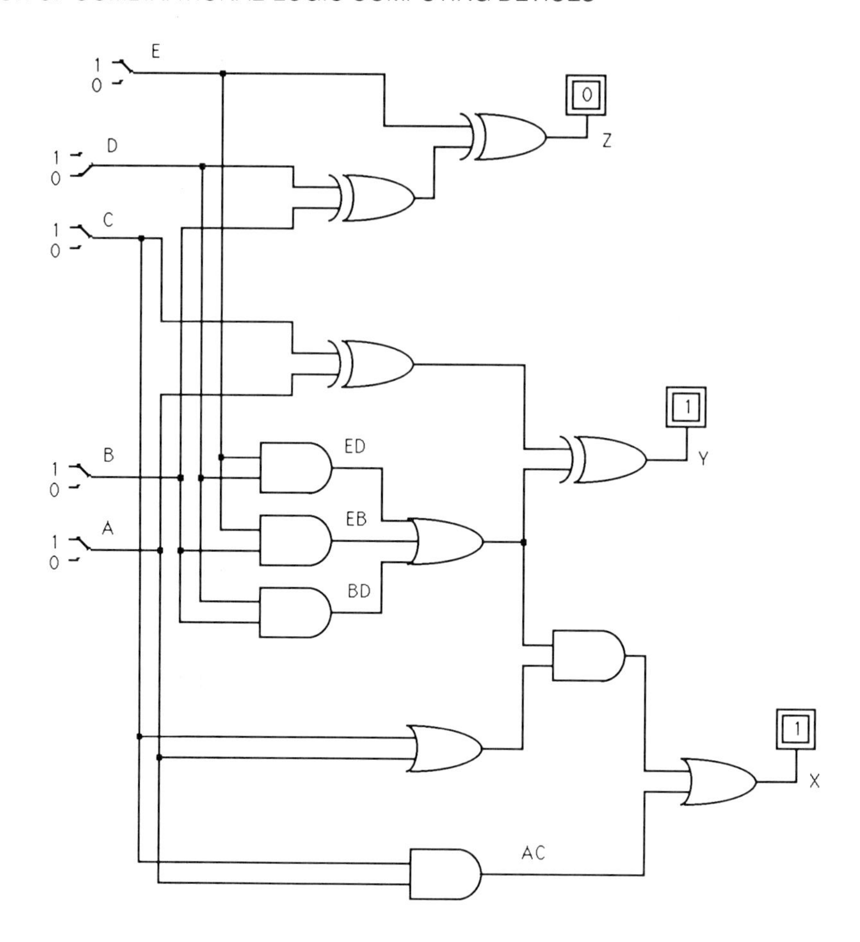

FIGURE 5.8
Implementation of a 2-bit direct adder

- The cascaded ripple-carry adder has an advantage over the direct adder, in that it uses the same circuit (full adder) twice. Hence, the ripple-carry adder is less complex than the direct adder.
- The maximum gate fan-in of the 2-level AND-OR implementation of a direct adder is 4, while the maximum gate fan-in for the ripple-carry adder is 2.

It can be shown that the 2-bit direct adder expressions for X, Y, and Z can be reduced to formulas logically equivalent to those used in the 2-bit ripple-carry adder as follows:

$$X = AC + A(BD + EB + ED) + C(BD + EB + ED)$$
$$= AC + (A + C)(BD + EB + ED)$$

$$Y = E'A'B'C + E'A'CD' + E'AB'C' + E'AC'D' + EA'BC' + EA'C'D$$
$$+ EABC + EACD + A'B'CD' + AB'C'D' + A'BC'D + ABCD$$

After algebraic reduction, Y can be represented in the following simplified form:

$$Y = (BD + EB + ED) \oplus (A \oplus C)$$

Note that the simplification requires use of the rule $E + E'D = E + D$.

Alternatively, this form for Y can be obtained using minterms in the same ring to determine exclusive-OR terms.

$$Z = EBD + EB'D' + E'BD' + E'B'D$$

By factoring and using the relation $B \oplus D = BD' + B'D$, we can represent Z in the form:

$$Z = E \oplus (B \oplus D)$$

Note that the sum $(BD + EB + ED)$ is the same as the first column carry-out term of the cascaded 2-bit ripple-carry adder.

5.3 DESIGN OF 4-BIT BINARY ADDERS

It is possible to design adders that are less complex than the two-level direct adder and faster than a cascaded ripple-carry adder. However, as the number of bits (n) increases, the complexity of a 2-level direct adder increases significantly while its speed remains basically unchanged. On the other hand, as n increases, the logic required for a ripple-carry adder does not increase (aside from an increased number of duplicate stages), but its operating speed is slower because of the additional number of propagation delays of the internal carries.

This section describes the design of 4-bit binary adders that add two 4-bit numbers and a carry-in. A number of alternate approaches can be used to design 4-bit adders; these alternatives include

Design of a 4-bit 2-level direct adder (without cascading)

Design of a 4-bit adder by cascading two 2-bit direct adders

Design of a 4-bit adder by cascading four 1-bit adders

Design of a 4-bit adder using carry look-ahead

The first two approaches, the 2-level direct adder and the cascaded ripple-carry adder, represent the extreme cases in the trade-off between complexity and speed. The design of a 4-bit direct adder would require a 9-input truth table with 512 possible combinations of input values. Therefore, this approach is ruled out because of its complexity. The second approach is reserved as a design exercise for the reader.

The third and fourth approaches, cascading four 1-bit adders and using carry look-ahead, will be used here to illustrate the differences in design simplicity and operating speed.

Design of a 4-Bit Adder by Cascading Four 1-Bit Full Adders

This section describes the design of a 4-bit adder using the intuitive approach used in Section 5.2 to design a 2-bit adder by cascading 1-bit adders.

The addition of two 4-bit numbers X and Y with a carry-in C_0, can be represented in the following form:

$$\begin{array}{rl} C_0 & \text{Carry-in} \\ x_3\ x_2\ x_1\ x_0 & \\ \underline{y_3\ y_2\ y_1\ y_0} & \\ S_3\ S_2\ S_1\ S_0 & \text{Sum} \\ C_4\ C_3\ C_2\ C_1 & \text{Carry-out} \end{array}$$

A 4-bit ripple-carry adder can be implemented by cascading four 1-bit adders, as illustrated in Figure 5.9. The addition of the LSB (C_0, x_0, y_0) is performed by the rightmost 1-bit full adder, and the sum S_0 and carry-out C_1 are represented by the following Boolean equations:

$$S_0 = C_0 \oplus (x_0 \oplus y_0)$$
$$C_1 = x_0\ y_0 + C_0\ (x_0 \oplus y_0)$$

The sum and carry-out functions of each column have the same form as these equations, where the carry-out of one stage is the carry-in of the next left stage:

$$S_i = C_i \oplus (x_i \oplus y_i)$$
$$C_{i+1} = x_i\ y_i + C_i\ (x_i \oplus y_i)$$

where $i = 1, 2, 3$.

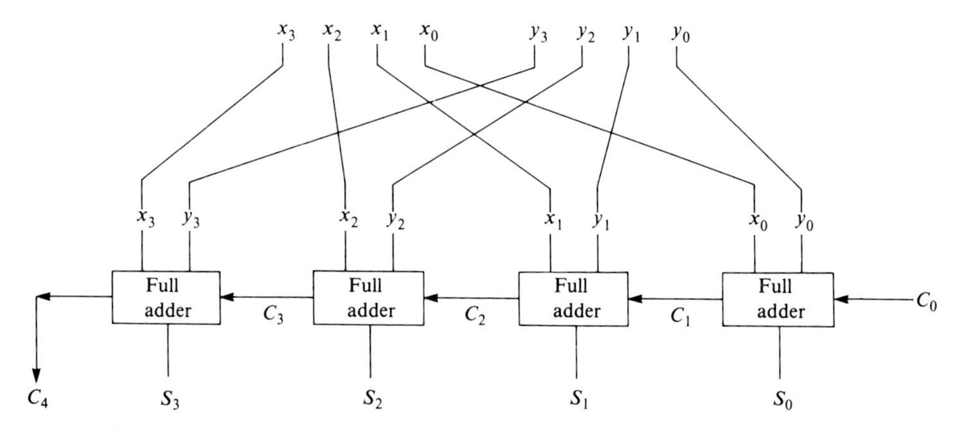

FIGURE 5.9
4-bit ripple-carry adder by cascading 1-bit adders

The 4-bit ripple-carry adder formed by cascading four 1-bit full adders is slow in operation because the carry-out of each stage is cascaded to the carry-in of the next left stage. An advantage of the ripple-carry adder is its logic simplicity, due to the use of the same building block (full adder) in each stage.

Design of a 4-Bit Adder Using Carry Look-Ahead

The design goals of any digital logic device include implementing a cost-effective device that has minimal logic complexity, a high operating speed, and reliability throughout the specified range of inputs. How can we design a multibit adder that is less complex than a 2-level direct adder and faster than a cascaded ripple-carry adder?

A relatively simple high-speed adder can be designed using a technique known as **carry look-ahead**. The resulting carry look-ahead adder represents a compromise between the complexity of a high-speed direct adder and the simplicity of a slow-speed ripple-carry adder.

Consider the truth table in Table 5.2 for adding two 1-bit numbers x and y and a carry-in c.

TABLE 5.2
Illustration of carry-generate and carry-propagate

	Inputs $c\ x\ y$	Outputs $C\ S$		
	0 0 0	0 0		
	0 0 1	0 1		
	0 1 0	0 1		
Carry-in = 0 →	0 1 1	1 0	← Carry-out = 1	(G)
	1 0 0	0 1		
Carry-in = 1 {	1 0 1	1 0	} Carry-out = 1	(P)
	1 1 0	1 0		
	1 1 1	1 1		

From Table 5.2, we see the following:

1. A carry-out is generated (G) if carry-in = 0 and carry-out = 1. A carry is generated if x_i, y_i = 1, 1, so $G_i = x_i y_i$.
2. A carry-out is propagated (P) if carry-in = 1 and carry-out = 1. A carry is propagated if x_i, y_i = 0, 1 or 1, 0 or 1, 1, so $P_i = x_i + y_i$.
3. Since $x_i y_i$ causes a carry-generate, the case x_i, y_i = 1, 1 can be omitted from the carry-propagate. Consequently, the carry-propagate can be accomplished using $P_i = x_i \oplus y_i$.

Since the carry-generate and carry-propagate are available 1 time delay after the inputs x_i and y_i are stable, the speed of an n-bit adder can be increased by using the carry-propagate and carry-generate terms instead of cascading the carries from one stage to the next.

A carry-out occurs when a carry is generated or when a carry is propagated. Therefore, the substitutions $G = xy$ and $P = x \oplus y$ in the expression $C = xy + c(x \oplus y)$ produces the following carry look-ahead equation:

$$C = G + cP$$

Each of the G_i, P_i can be generated independently and concurrently in one gate-delay time after the inputs X and Y are available at the inputs of the carry look-ahead adder. Thus, a 4-bit carry look-ahead adder can be implemented by using carry-generate G_i and carry-propagate P_i.

Using the recursion formula

$$C_{i+1} = G_i + C_iP_i$$

we see that each of the carry-out terms C_1, C_2, C_3, and C_4 can be simplified and expressed in terms of C_0. The carry-generate G_i and carry propagate P_i, for $i = 0$, are

$$C_1 = G_0 + C_0P_0$$

For $i = 1$,

$$\begin{aligned} C_2 &= G_1 + C_1P_1 \\ &= G_1 + (G_0 + C_0P_0)P_1 \\ &= G_1 + G_0P_1 + C_0P_0P_1 \end{aligned}$$

For $i = 2$,

$$\begin{aligned} C_3 &= G_2 + C_2P_2 \\ &= G_2 + (G_1 + G_0P_1 + C_0P_0P_1)P_2 \\ &= G_2 + G_1P_2 + G_0P_1P_2 + C_0P_0P_1P_2 \end{aligned}$$

For $i = 3$,

$$\begin{aligned} C_4 &= G_3 + C_3P_3 \\ &= G_3 + (G_2 + G_1P_2 + G_0P_1P_2 + C_0P_0P_1P_2)P_3 \\ &= G_3 + G_2P_3 + G_1P_2P_3 + G_0P_1P_2P_3 + C_0P_0P_1P_2P_3 \end{aligned}$$

The sum S_3, S_2, S_1, S_0 can be generated by the following Boolean equations:

$$\begin{aligned} S_0 &= C_0 \oplus x_0 \oplus y_0 \\ S_1 &= C_1 \oplus x_1 \oplus y_1 \\ S_2 &= C_2 \oplus x_2 \oplus y_2 \\ S_3 &= C_3 \oplus x_3 \oplus y_3 \end{aligned}$$

Note that each C_i is computed with only three propagation delays, and each S_i is computed with only one additional propagation delay. Therefore, the carry look-ahead adder has only 4 delays in any path, while a cascaded 4-bit adder, implemented in 2-level form ($S = c \oplus x \oplus y$, $C = xy + cx + cy$), has 8 delays in the longest path.

The speed advantage of computing S_i and C_i using carry look-ahead is somewhat offset by the requirement of using 3-, 4-, and 5-input gates in the carry look-ahead adder. A logic diagram for a 4-bit carry look-ahead adder is presented in Figure 5.10.

The carry look-ahead technique can be applied across 4-bit groups using the formulas for group-propagate $GP(3–0)$ and group-generate $GG(3–0)$, which are defined as follows:

$$\begin{aligned} GP(3–0) &= P_0P_1P_2P_3 \\ GG(3–0) &= G_3 + G_2P_3 + G_1P_2P_3 + G_0P_1P_2P_3 \end{aligned}$$

The group-propagate and group-generate can be combined to form higher-level carry look-ahead equations in the same way that the equations were developed for carry look-ahead C_1, C_2, C_3, and C_4.

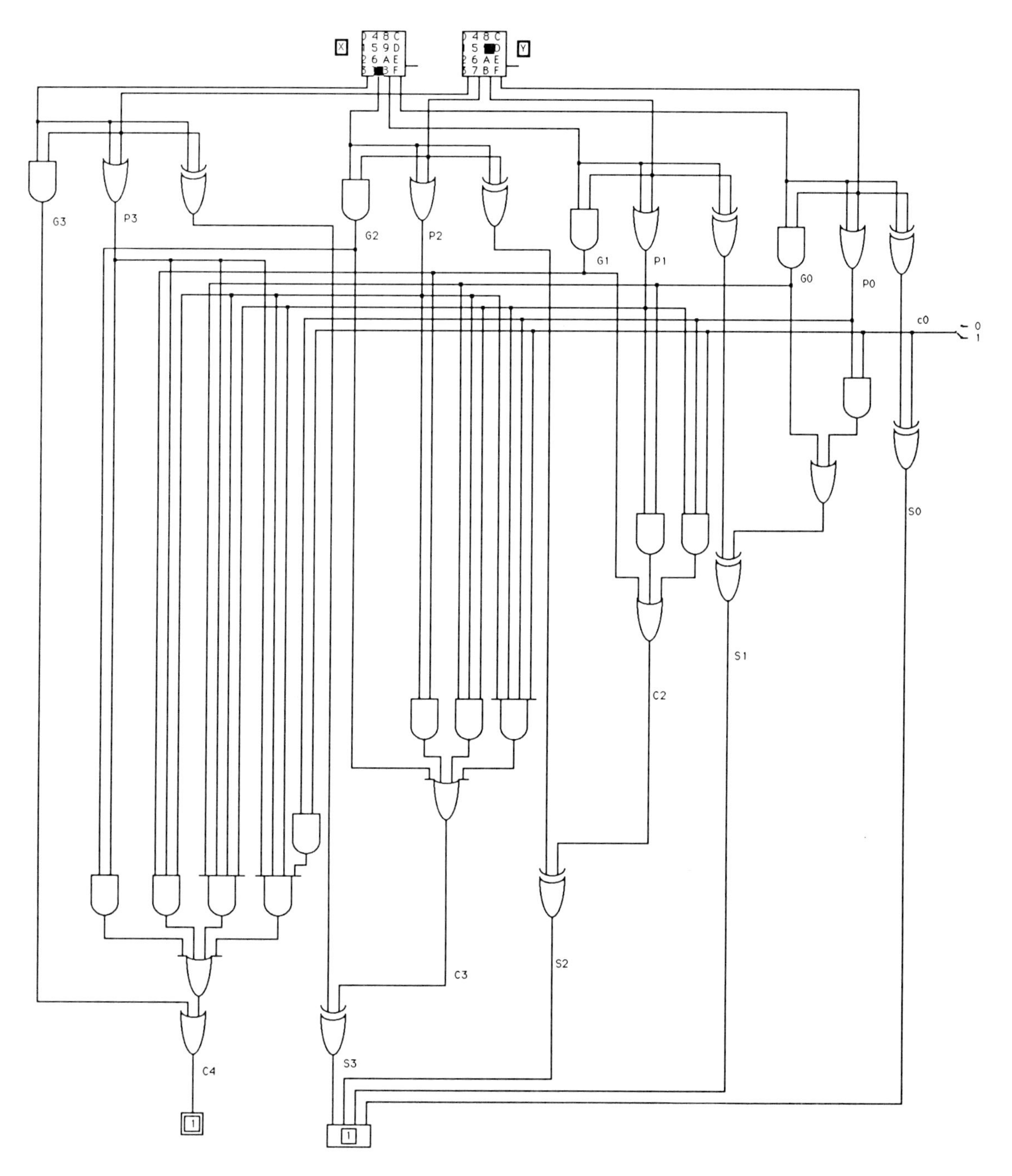

FIGURE 5.10
Logic diagram of a 4-bit carry look-ahead adder

The 74S182 carry look-ahead generator accepts up to 4 pairs of active-low carry-propagate and carry-generate signals and an active-high carry-in, and provides active-high anticipated carries across 4 groups of 4-bit binary adders. The 74S182 also has active-low carry-propagate and carry-generate outputs that may be used for further levels of carry look-ahead. The 74182 can be used in conjunction with four 74181 ICs to provide high-speed addition of two 16-bit inputs.

5.4 DESIGN OF A 4-BIT ADDER/SUBTRACTER

The implementation of an arithmetic unit can be simplified if the operations of addition and subtraction each can be performed by a combined adder/subtracter circuit, instead of by separate circuits (one for addition and another for subtraction).

A binary adder easily can be modified to perform subtraction. Chapter 1 showed that the subtraction $A - B$ can be accomplished by adding the 2s complement of B to A. Recall also that binary numbers are usually stored in 2s complement form in simple computers. The following examples of addition and subtraction are included here to refresh your memory.

Addition		*Subtraction*	
+5	0101	+5	0101
+ +2	0010	− +2	1110
+7	0111	+3	10011 (Drop EAC)
+6	0110	+4	0100
+ −3	1101	− −3	0011
+3	10011 (Drop EAC)	+7	0111

Remember that the 2s complement of a number can be produced by adding 1 to the 1s complement of that number, where the 1s complement is obtained by inverting each bit of the original number. Therefore, a binary adder can be modified to function as a binary subtracter by inverting each bit of the subtrahend (to produce the 1s complement) and adding 1 by setting carry-in $c = 1$ (to produce the 2s complement).

■ **EXAMPLE 5.1**

Given $A = +5 = 0101$ and $B = +2 = 0010$, perform binary addition or subtraction.

- □ For addition, retrieve A and B and use them without conversion.
- □ For subtraction, convert B to 2s complement form before adding B' to A to accomplish the subtraction $A - B$, as shown in Figure 5.11.

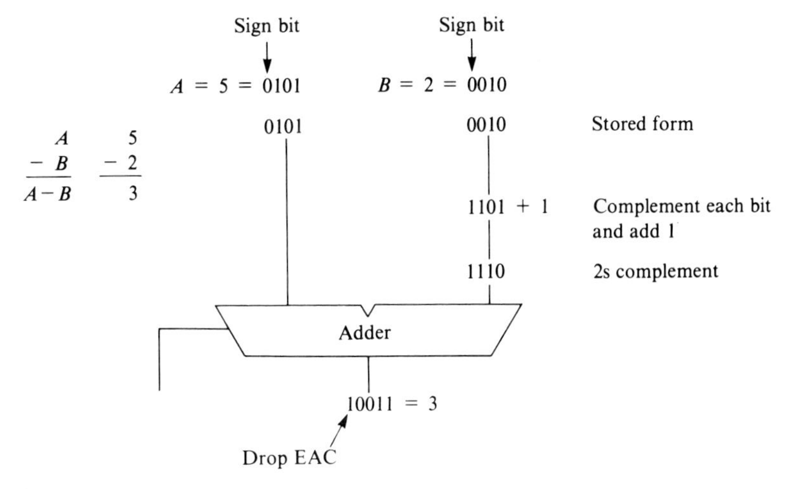

FIGURE 5.11
Illustration of subtraction process

Use of the Exclusive-OR as a Selective Complementer

In order to modify a binary adder to perform subtraction, we need to incorporate a circuit that produces a 2s complement of the subtrahend. This can be accomplished by producing a 1s complement of the subtrahend and adding 1 to the result.

Consider a combinational circuit that has a control input W and a data input y. If the control $W = 0$, data input y is passed unchanged to the output. If control $W = 1$, data input y is complemented before being passed to the output. The operation of this circuit, a selective complementer, is described by the following truth table:

Inputs			
Control W	Data y	Output F	
0	0	0	Pass y
0	1	1	
1	0	1	Complement y
1	1	0	

We see from the truth table that $F = W'y + Wy' = W \oplus y$, so that the selective complementer function can be implemented using an exclusive-OR (XOR) gate, as illustrated in Figure 5.12.

FIGURE 5.12
Using an XOR gate as a selective inverter

Now, a single control input W can be used with n data lines y_i to selectively complement (invert) all or none of the data inputs, as required in a complement circuit for an adder/subtracter. Figure 5.13 illustrates such a circuit.

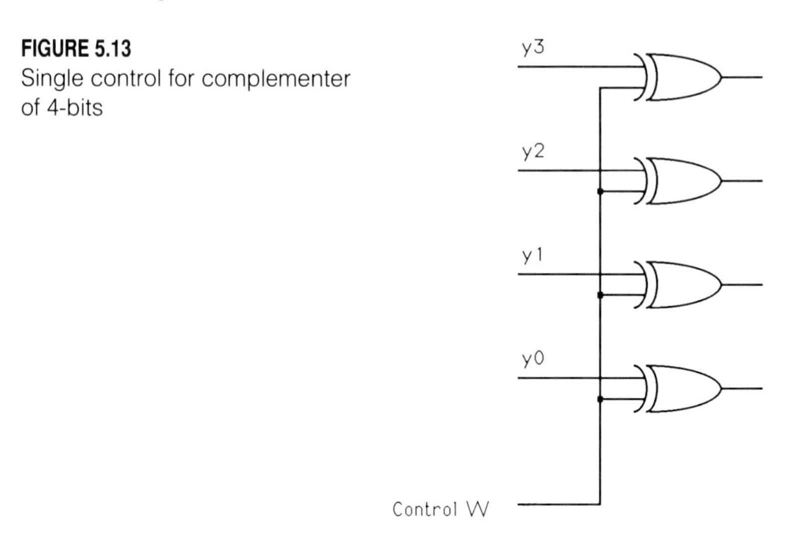

FIGURE 5.13
Single control for complementer of 4-bits

Implementation of a 4-Bit Binary Adder/Subtracter

The 1s complement function required in a binary adder/subtracter can be implemented using 4 XOR gates, as illustrated in Figure 5.14. Note that the 2s complement of Y is formed by inverting each y bit to form the 1s complement of Y and then adding 1 using the carry-in.

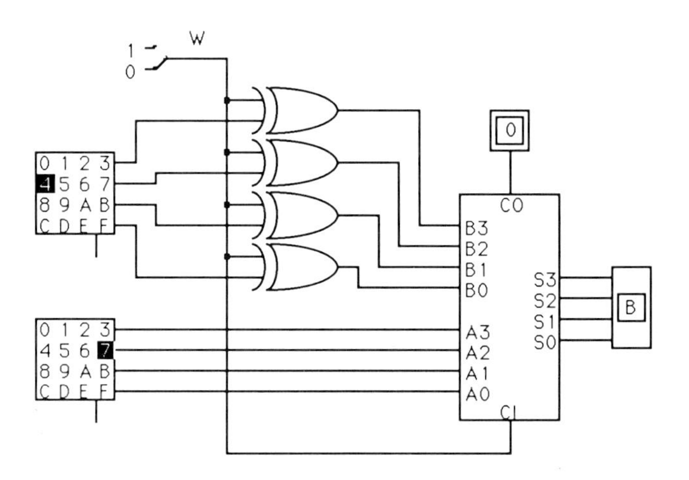

FIGURE 5.14
4-bit adder/subtracter

The 2s complement of the subtrahend Y can be accomplished by performing the 1s complement of Y (by inverting each bit of Y when $W = 1$) and then adding 1 using the carry-in. The resulting binary adder/subtracter can be represented in the block diagram shown in Figure 5.15.

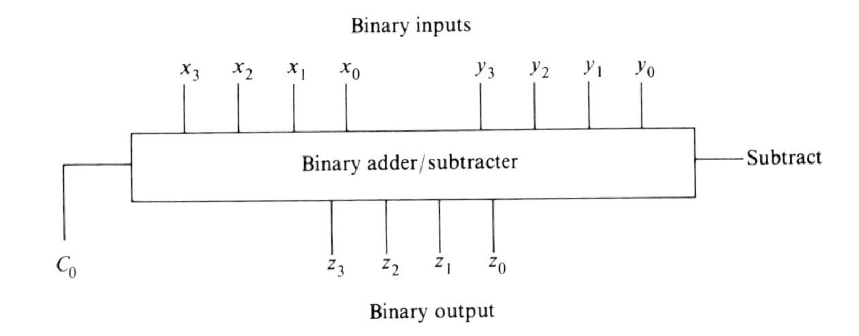

FIGURE 5.15
Block diagram of 4-bit adder/subtracter

Overflow Detection in Binary Addition and Subtraction

Overflow is a condition that occurs when the magnitude of a number exceeds the allowable maximum for the number of bits available to represent the number. In 2s complement arithmetic, the high order carry-out is ignored except for the overflow test: Overflow occurs if the sign-bit carry-in differs from the sign-bit carry-out, as shown in the following examples:

1. Overflow when two positive numbers are added:

```
          0100    Carries
  +4      0100
+ +4      0100
----      ----
  +8      1000    = −8 Erroneous
```

2. Overflow when two negative numbers are added:

```
          1000    Carries
  −5      1011
+ −4      1100
----      ----
  −9      0111    = +7 Erroneous
```

3. Overflow when numbers of opposite sign are subtracted:

```
          1001    Carries
  −7      1001
− +3      1101    (0011)' + 1 = 1101
----      ----
  −4      0110    = +6 Erroneous
```

An overflow detect can be implemented using the following formula:

$$\text{Overflow} = c_i \oplus c_o$$

where c_i and c_o denote the sign-bit carry-in and carry-out, respectively. An overflow-detect circuit sets a bit in a status register of an arithmetic unit to indicate that an overflow has occurred.

5.5 DESIGN OF A 4-BIT 8-FUNCTION ARITHMETIC UNIT

An arithmetic unit can be designed by incorporating some additional control circuitry in an adder/subtracter circuit. This section describes how an 8-function arithmetic unit can be designed to implement a given set of functions.

Design of a True/Complement/Zero/One Function

A true/complement/zero/one (TC01) function is a combinational logic function that has

One multibit input Y

Two 1-bit controls (E = enable, W = complement)

One multibit output F

The output F of the TC01 function is defined by the following physical function table:

Control W E	Output F
0 0	Zeros
0 1	Y
1 0	Ones
1 1	Y'

An examination of the TC01 function table reveals that the output F is enabled ($F = Y$ or Y') when $E = 1$. It should also be noted that if $E = 1$, then $F = Y$ when $W = 0$ and $F = Y'$ when $W = 1$.

A TC01 function can be designed by first examining an AND gate with an enable control E and a data input y. The following truth table illustrates how an AND gate can be used to enable or disable the gate output.

E y	$G = E \cdot y$	Output	Result
0 0	0	0	Disable gate
0 1	0	0	Disable gate
1 0	0	y	Enable gate
1 1	1	y	Enable gate

Thus, an AND gate with an enable control E enables the input y to pass to the gate output when $E = 1$ while the output is disabled (output = 0) when $E = 0$.

Now, consider a combinational logic function that has a control W, a data input y (which can be enabled or disabled), and an output F as described by the following truth table:

	$W\ E\ y$	$G = E \cdot y$	$F = W \oplus (E \cdot y)$	
Pass	0 0 0	0	0	0
	0 0 1	0	0	
	0 1 0	0	0	True
	0 1 1	1	1	
Complement	1 0 0	0	1	1
	1 0 1	0	1	
	1 1 0	0	1	Complement
	1 1 1	1	0	

Note that when control $W = 0$, enabled data $(E \cdot y)$ is passed and when control $W = 1$, enabled data $(E \cdot y)$ is complemented. This truth table function can be implemented then by the circuit illustrated in Figure 5.16.

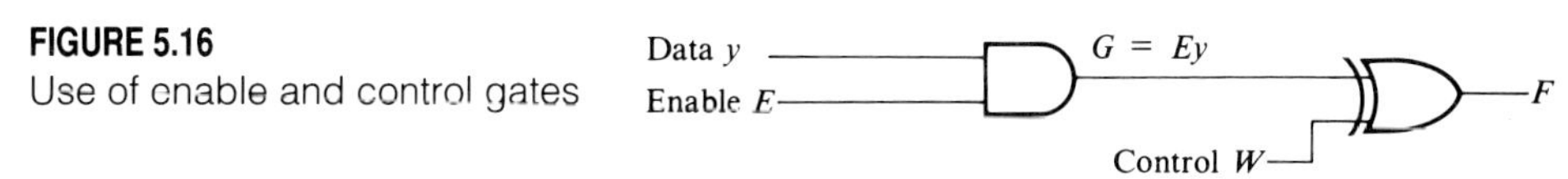

FIGURE 5.16
Use of enable and control gates

Now, a 1-bit enable E and a 1-bit control W can be used with multiple data lines y to form a TC01 function as shown in Figure 5.17.

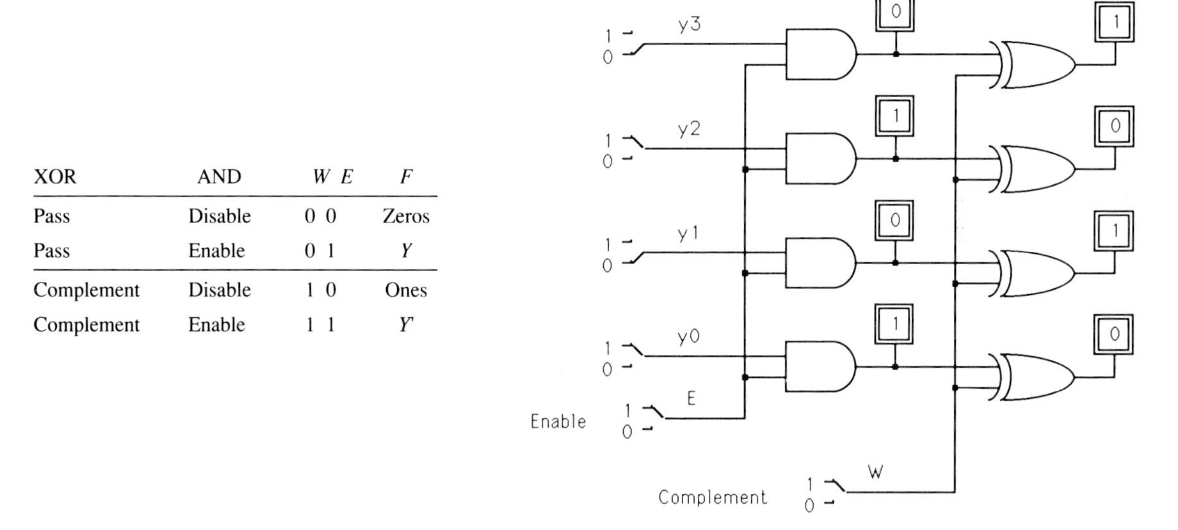

XOR	AND	W E	F
Pass	Disable	0 0	Zeros
Pass	Enable	0 1	Y
Complement	Disable	1 0	Ones
Complement	Enable	1 1	Y'

FIGURE 5.17
Use of enable and control gates with multiple inputs

Implementation of a 4-Bit 8-Function Arithmetic Unit

This section describes how an arithmetic unit can be designed, by modifying a binary adder, to accomplish any of the following functions: add, subtract, pass, decrement, or increment.

This simple arithmetic unit can be implemented by combining a TC01 function with a binary adder, as illustrated in Figure 5.18. The arithmetic unit accomplishes the functions indicated in the following truth table:

C_i W E	F	Z	Function
0 0 0	0s	X	Pass X
0 0 1	Y	$X + Y$	Add
0 1 0	1s	X + 1s	Decrement X
0 1 1	Y'	$X + Y'$	X + 1s complement
1 0 0	0s	$X + 1$	Increment X
1 0 1	Y	$X + Y + 1$	Add with carry-in
1 1 0	1s	X + 1s + 1	Pass X
1 1 1	Y'	$X + Y' + 1$	Subtract

where + signifies addition. We can illustrate decrement, increment, and pass with the following examples:

Decrement X	Increment X	Pass X
X = 0101 = 5	X = 0110 = 6	X = 0011 = 3
F = 1111	F = 0000	F = 1111
	C_i = 1	C_i = 1
Z = 10100 = 4	Z = 0111 = 7	Z = 10011 = 3

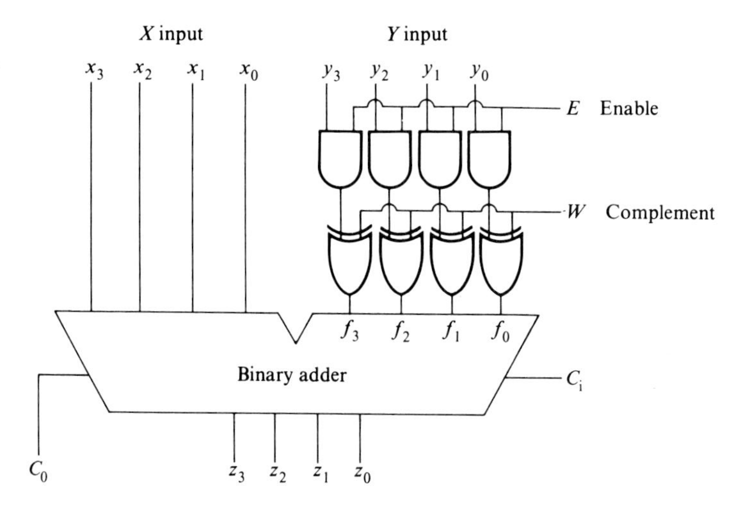

FIGURE 5.18
Combining AND and XOR gates with a binary adder

Finally, this simple binary arithmetic unit can be represented by the block diagram illustrated in Figure 5.19.

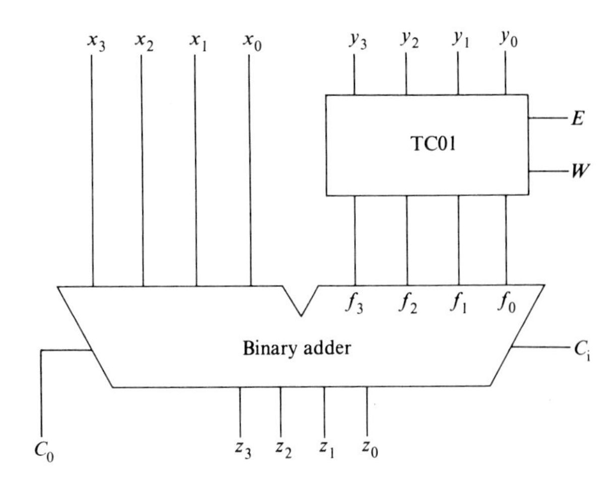

FIGURE 5.19
Block diagram of a simple binary arithmetic unit

5.6 USE OF 4-BIT MSI ICS TO IMPLEMENT 8-BIT COMPUTING DEVICES

The 7483 MSI IC is a 4-bit parallel adder in which the carry-out of each stage is cascaded to the carry-in of the next stage. Cascaded adders are also referred to as ripple-carry adders, as noted in Section 5.2. The 4-bit MSI adder 7483A uses carry look-ahead addition.

An 8-bit adder can be implemented by connecting the carry-out of one 7483A adder to the carry-in of a second 7483A adder. The first adder accepts the least significant nibbles of the 8-bit X and Y inputs, and the second adder accepts the most significant nibbles of the inputs, as illustrated in Figure 5.20. If additional speed is required, the 4-bit block-generate and block-propagate features of MSI adders can be used when cascading a number of such adders to form a multibit adder.

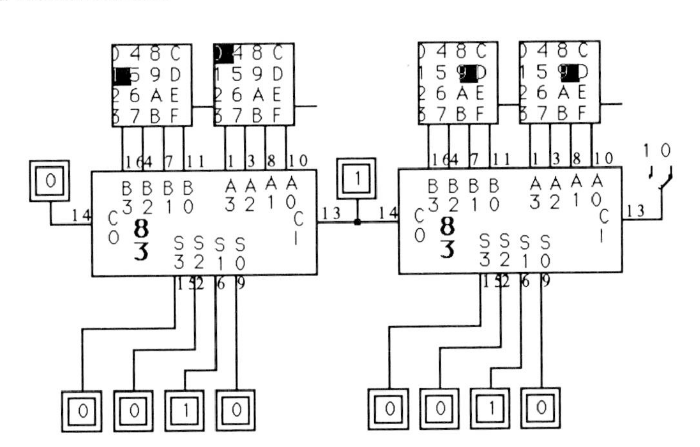

FIGURE 5.20
An 8-bit adder formed by cascading two 7483 ICs

An 8-bit adder/subtracter can be implemented by inserting an XOR gate in each y path (see Section 5.5) of the 8-bit adder and connecting the control W to the carry-in of the first, least significant nibble, of the 7483 IC. An 8-bit arithmetic unit can be implemented by inserting a TC01 in each y path of the 8-bit adder.

5.7 SUMMARY

Mathematicians use a bootstrap technique called the principle of finite induction to prove that a given formula holds for any positive integer k,

By showing that the formula holds for $k = 1$

By demonstrating that if the formula holds for any k, then it holds for $k + 1$

In this chapter we have used a similar approach (although less formal than the finite-induction method) to evolve from a 1-bit adder to an n-bit adder, where $n = 2, 4$, and 8 bits.

We have learned how to develop the design of simple computing devices using both intuitive and formal solution approaches. We also learned how to implement an 8-bit adder using 4-bit adders as building blocks. We could extend this building-block approach to the design of adders of 16 bits, 32 bits, and so on.

A given design problem may yield alternative realizations (implementations) that represent trade-offs between circuit complexity and operating speed. That is, one implementation may be less complex (require less hardware) while another implementation may be faster (include fewer stages and less propagation delay). Yet another implementation may be a compromise of others. You should be able to evaluate alternative implementations of a particular functional device by comparing the speed (inversely proportional to the number of propagation delays) and logic complexity (proportional to the component count) of the alternative implementations.

In this chapter, we also learned how to (1) design multifunction devices, such as a 4-bit adder/subtracter and a 4-bit 8-function arithmetic unit, and (2) implement 8-bit adders, adder/subtracters, and arithmetic units using MSI 4-bit adders as functional building blocks. As before, we have used a logic simulation program (DESIGNWORKS) to simulate the implementation of various digital logic devices.

Up to this point we have developed simple computing devices that input numbers in binary, perform addition and subtraction in binary, and output the results in binary. In the next chapter, we will develop an arithmetic-logic unit (ALU) and a BCD arithmetic unit.

KEY TERMS

Direct adder

Carry look-ahead

Overflow

EXERCISES

General

1. Given $A = +4$ and $B = +2$, perform the following:
   a. Add the two 4-bit numbers represented in 2s complement form.

b. Subtract $A - B$ by adding the 2s complement of B to A.

Addition		*Subtraction*	
+4	0100	+4	0100
+ +2	0010	− +2	1110

2. Given $A = +5$ and $B = -2$, perform the following:
a. Add the two 4-bit numbers represented in 2s complement form.
b. Subtract $A - B$ by adding the 2s complement of B to A.

Addition		*Subtraction*	
+5	0101	+5	0101
+ −2	1110	− −2	0010

3. Given $A = +5$ and $B = -3$, perform the following:
a. Add the two 4-bit numbers represented in 2s complement form.
b. Subtract $A - B$ by adding the 2s complement of B to A.

Addition		*Subtraction*	
+5	0101	+5	0101
+ −3	1101	− −3	0011

Does overflow exist in either of these problems? If so, indicate which one and show why.

4. Give an example of overflow resulting when two positive numbers are added.

5. Give an example of overflow resulting when two negative numbers are added.

6. Give an example of overflow resulting when numbers of opposite sign are subtracted.

7. An overflow detect can be implemented using the following formula:

$$\text{Overflow} = \_\_\_\_\_\_\_\_\_\_$$

where c_i and c_o denote the sign-bit carry-in and carry-out, respectively.

8. Summarize the speed and complexity trade-offs between the 2-level AND-OR implementation of a 2-bit direct adder and a 2-bit ripple-carry adder.

9. Show that the 2-bit direct adder expressions for X, Y, and Z can be reduced to logically equivalent formulas used in the 2-bit ripple-carry adder.

10. Cite an advantage of the 4-bit ripple-carry adder formed by cascading four 1-bit full adders.

11. Cite a disadvantage of the 4-bit ripple-carry adder formed by cascading four 1-bit full adders.

Design/Implementation

1. Draw a block diagram of a 4-bit adder formed by cascading two 2-bit direct adders.

2. Implement a 4-bit adder by cascading two 2-bit direct adders.

3. Use NAND gates to implement the carry look-ahead formulas for C_1 and C_2 of Section 5.3.

4. For 2s complement addition, overflow occurs if the sign-bit carry-in differs from the sign-bit carry-out. Modify the 4-bit ripple-carry adder (Figure 5.9) to incorporate an overflow detect: overflow = $C_3 \oplus C_4$.

5. Use AND gates, XOR gates, and a 7483 4-bit MSI adder to implement the simple arithmetic unit illustrated in Figure 5.18. Verify each of the 8 functions of this arithmetic unit.

6. Use two 7483 MSI 4-bit adders to implement the 8-bit adder illustrated in Figure 5.20.

6 Logical and Decimal Operations—Hierarchical Development

OBJECTIVES

After you complete this chapter, you will be able

- ☐ To perform arithmetic operations (such as addition and subtraction), logic operations (such as AND, OR, and XOR), and compare operations (such as less than, equal to, and greater than)
- ☐ To design different types of arithmetic logic units using 2s complement arithmetic and BCD arithmetic
- ☐ To use a bit-slice approach to design an arithmetic logic unit
- ☐ To organize a system into subsystems
- ☐ To design a system using top-down design

6.1 INTRODUCTION

A digital system, such as a computer, usually consists of a number of subsystems that are interconnected. Each subsystem may contain a number of functional modules that are also interconnected. For example, a computer system consists of a memory subsystem, a central processing unit (CPU), an input subsystem, and an output subsystem, as shown in Figure 6.1. The CPU is usually comprised of two functional modules: a control unit and an arithmetic logic unit (ALU).

FIGURE 6.1
Block diagram of a simple computer system

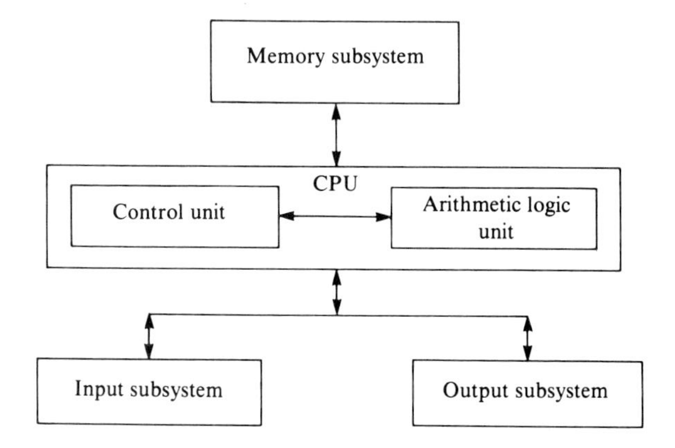

Program instructions and data to be processed by that program are stored in the computer's memory. As each instruction is retrieved from memory, it is decoded by the control unit and executed by the ALU. To execute a given instruction, the ALU, as directed by the control unit, performs a sequence of microoperations that together accomplish the execution of the program instruction. A given computer has its own set (repertoire) of instructions. The instructions in the repertoire determine the arithmetic, logical, and compare microoperations that the computer's ALU must be able to perform. Consequently, different computers usually contain different ALUs.

In 1965, the Digital Equipment Corporation marketed the world's first minicomputer, the PDP–8, whose CPU used a 2s complement ALU. In 1969, the Intel Corporation developed the first microcomputer, the MCS–4, whose 4004 CPU used a BCD arithmetic logic unit. This chapter describes the design of both a 2s complement ALU and a BCD arithmetic unit.

Recall that Chapter 5 described an evolutionary approach for developing a simple arithmetic unit. Starting with a 1-bit adder, we followed the evolutionary process through the design of 2-bit adders, 4-bit adders, and 4-bit adder/subtracters, finally achieving the design of a 4-bit 8-function arithmetic unit.

This chapter illustrates two other approaches to design:

A bit-slice approach that cascades 1-bit functional building blocks into an n-bit device

A top-down approach that examines an overall problem and uses a divide-and-conquer technique for designing subsystems (consisting of functional modules) of an overall system

A computer should have, in addition to arithmetic instructions, the capability to perform logic operations (AND, OR, and XOR) and to compare operations (greater than, equal to, and less than). Section 6.2 describes the design of a bit-slice ALU that performs arithmetic (2s complement), logic, and compare operations.

Humans communicate with one another using an alphanumeric character set (A–Z, 0–9, and symbols), while most computers handle data and instructions internally in some binary form. In Section 6.3, we describe how people communicate with computers and show how to perform calculations using BCD numbers. We also use a top-down approach to design a BCD arithmetic unit that consists of

A decimal-to-BCD encoder that encodes decimal digits into a BCD form the arithmetic unit can "read"

A BCD adder/subtracter that combines a 1-stage BCD adder (to add two 1-digit BCD numbers) and a BCD 9s complementer

A BCD-to-decimal decoder that decodes the BCD result into decimal form for output

All of these units are then integrated into a 1-stage BCD arithmetic unit. An n-stage BCD arithmetic unit can be implemented by cascading n 1-stage units.

6.2 DESIGN OF BINARY ARITHMETIC LOGIC UNITS

Prior to the mid-1960s two basic types of digital computers were marketed: (1) business computers (data processors) such as the IBM 1401, which was designed to perform relatively simple computations on large data sets or files, and (2) scientific computers (number crunchers) such as the IBM 7090, which was designed to perform relatively complex computations on smaller data sets. In 1964, IBM introduced the System 360, consisting of a family of upward-compatible computers that ranged from the small Model–20 data processor to the large Model–68 number cruncher.

What types of operations must a computer perform? To answer this question, consider three widely differing types of applications: file (or list) processing, which is common to data processors, and statistical computations and numerical methods, which are both common to number crunchers. Do these different types of applications have anything in common? Aside from the widely different amounts of data and the different complexity of the computations, all three of these applications involve the same types of operations on data sets $\{X(k)\}$, $\{Y(k)\}$. List processing involves adding, multiplying, comparing, sorting, and merging elements $x(k)$, $y(k)$. Statistical computations (computing the mean, variance, and covariance) and numerical methods (interpolation, curve fitting, numerical integration, and solution of differential equations) combine elements $x(k)$ and $y(k)$ using various arithmetic and logical operations.

In order to accomplish tasks such as list processing, statistical computations, and numerical methods, a computer must have the capability to

Input and store program instructions and data words

Read, interpret, and execute program instructions

Process data (both original data and data derived from processing)

Store the results of data processing

Output results

The focus of this section is to develop an ALU that performs the arithmetic, logic, and compare operations required to process numeric data in 2s complement form. In the following subsections, we will describe the design of 1- , 2- , and 4-bit comparators that determine whether one number is greater than, equal to, or less than another number. Then, we will develop the logic operations (AND, OR, and XOR) that are inherent in producing the arithmetic operations (sum and carry) in binary adders.

Finally, we will integrate a full adder function (with inherent logic operations), a true/complement/zero/one function (TC01) and a compare-equal function to form a 1-bit ALU. An n-bit bit-slice ALU is then realized by cascading n 1-bit ALUs. Then we will design a second binary ALU, realized by integrating a 4-bit adder and two TC01 function modules.

Compare Operations and Comparators

Two basic types of **compare operations** are used in applications programs: (1) comparing two numbers X and Y to determine if X is equal to, greater than, or less than Y; and (2) comparing two alphanumeric data words A and B to determine if A should precede or follow B in a sorted list, such as a telephone book. It should be noted that an alphanumeric data word may consist of a number of keyboard characters.

An n-bit comparator is a digital logic device that determines if an n-bit number X is equal to, greater than, or less than another n-bit number Y. Thus, an n-bit comparator can be used to solve the problem of comparing two numbers. The question is how to solve the problem of comparing two alphanumeric data words. The answer lies in the way that computer codes are constructed. In the ASCII and EBCDIC codes, letters in ascending alphabetic sequence are represented by binary codes in ascending numeric sequence. For example, in 7-bit ASCII A = 1000001, B = 1000010, C = 1000011, . . . , O = 1001111, P = 1010000, and so on. Consequently, a 7-bit comparator can be used to compare the ASCII codes of corresponding characters of two alphanumeric data words. Data words coded in 8-bit ASCII and EBCDIC can be handled in a similar fashion.

It is apparent, then, that a device that compares two numbers can also be used to compare two alphanumeric data words. In the following sections, we will describe the design and implementation of 1- , 2- , and 4-bit comparators.

The 1-Bit Comparator

The relation between two 1-bit numbers x and y can either be equal to (E), greater than (G), or less than (L) as described by the truth tables:

		E	G	L
	x y	$x = y$	$x > y$	$x < y$
$x'y'$	0 0	1	0	0
$x'y$	0 1	0	0	1
$x\,y'$	1 0	0	1	0
$x\,y$	1 1	1	0	0

Therefore, the E, G, and L functions can be represented by the following Boolean equations:

$$E = x'y' + xy = (x \oplus y)'$$
$$G = xy'$$
$$L = x'y$$

These E, G, and L functions can be combined into a 1-bit comparator function that has two inputs, x and y, and 3 outputs, E, G, and L, as illustrated in Figure 6.2a. This 1-bit comparator function can be implemented by the combinational circuit seen in Figure 6.2b.

FIGURE 6.2
A 1-bit comparator. (a) Block diagram (b) Logic diagram

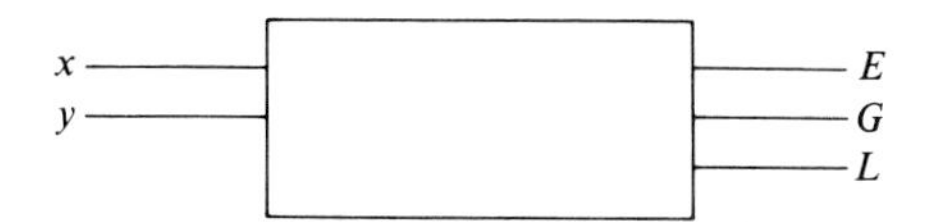

(a)

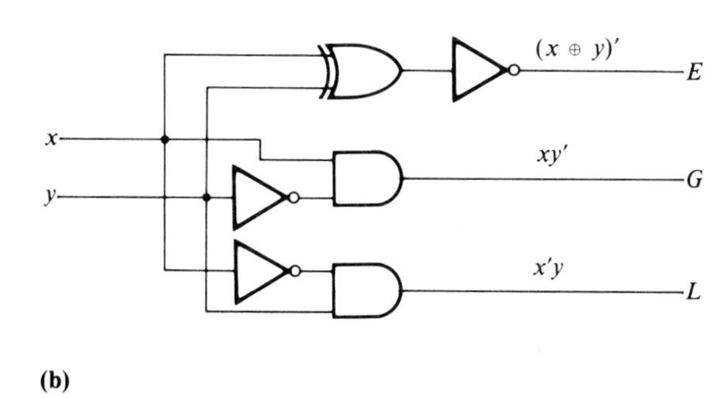

(b)

The 2-Bit Comparator

We set out now to design a combinational logic function that determines whether the 2-bit number x_1x_0 is equal to, greater than, or less than the 2-bit number y_1y_0. We can proceed intuitively as follows:

Let MSB denote the most significant bit of x_1x_0.

Let msb denote the most significant bit of y_1y_0.

Let LSB denote the least significant bit of x_1x_0.

Let lsb denote the least significant bit of y_1y_0.

The 2-bit numbers x_1x_0 and y_1y_0 are equal when $x_1 = y_1$ and $x_0 = y_0$. Therefore, the 1-bit equality functions $(x_1 \oplus y_1)'$ and $(x_0 \oplus y_0)'$ can be ANDed to produce the following 2-bit equality function

$$E = (x_1 \oplus y_1)' \cdot (x_0 \oplus y_0)'$$

The greater-than function can be deduced by noting that $x_1x_0 > y_1y_0$ when

$$\begin{pmatrix} \text{MSB} \\ \cdot \\ (\text{msb})' \end{pmatrix} \text{ OR } \begin{pmatrix} \text{MSB} & & \text{LSB} \\ = & \text{AND} & \cdot \\ \text{msb} & & (\text{lsb})' \end{pmatrix}$$

so that

$$G = x_1 y_1' + (x_1 \oplus y_1)' x_0 y_0'$$

The less-than function can be deduced by noting that $x_1 x_0 < y_1 y_0$ when

$$\begin{pmatrix} (\text{MSB})' \\ \cdot \\ \text{msb} \end{pmatrix} \text{ OR } \begin{pmatrix} \text{MSB} & & (\text{LSB})' \\ = & \text{AND} & \cdot \\ \text{msb} & & \text{lsb} \end{pmatrix}$$

so that

$$L = x_1' y_1 + (x_1 \oplus y_1)' \, x_0' y_0$$

Figure 6.3 presents a logic diagram for this 2-bit comparator.

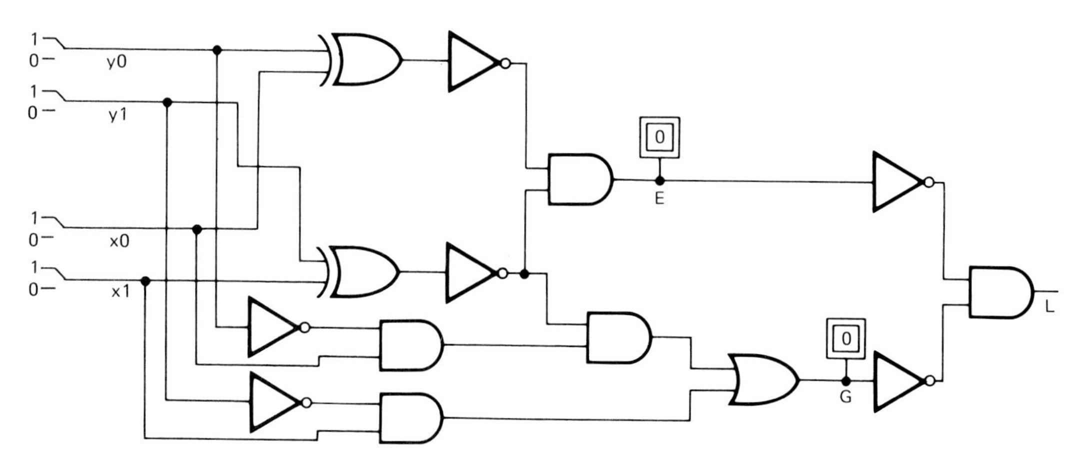

FIGURE 6.3
Logic diagram of 2-bit comparator

The 4-Bit Comparator

A 4-bit comparator can be designed by direct extension of the formulas for a 2-bit comparator. The resulting formulas for the 4-bit comparator are

$$\begin{aligned} E &= (x_3 \oplus y_3)'(x_2 \oplus y_2)'(x_1 \oplus y_1)'(x_0 \oplus y_0)' \\ G &= x_3 y_3' + (x_3 \oplus y_3)' x_2 y_2' + (x_3 \oplus y_3)'(x_2 \oplus y_2)' x_1 y_1' \\ &\quad + (x_3 \oplus y_3)'(x_2 \oplus y_2)'(x_1 \oplus y_1)' x_0 y_0' \\ L &= x_3' y_3 + (x_3 \oplus y_3)' x_2' y_2 + (x_3 \oplus y_3)'(x_2 \oplus y_2)' x_1' y_1 \\ &\quad + (x_3 \oplus y_3)'(x_2 \oplus y_2)'(x_1 \oplus y_1)' x_0' y_0 \end{aligned}$$

The MSI integrated circuit 7485 is an example of an off-the-shelf 4-bit comparator. Its logic diagram is presented in Figure 6.4

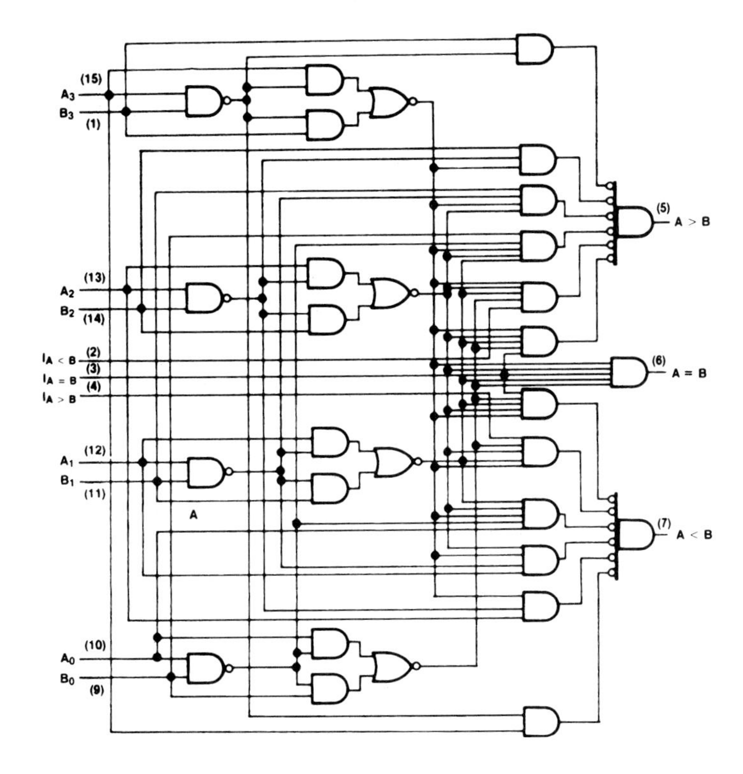

FIGURE 6.4
Logic diagram of 4-bit comparator 7485 (Reprinted by courtesy of Philips Components/Signetics)

Arithmetic and Logic Operations

In general, a simple ALU has the capability of performing the arithmetic operations addition and subtraction and the logic operations AND, OR, and XOR. Common applications of these logic operations are to selectively mask, merge, and complement individual bits of data words, as we now describe.

The AND operation can be used to mask (disable) part of a word and extract (enable) the remaining part of the word, as illustrated in Example 6.1.

EXAMPLE 6.1

An 8-bit mask is defined as 11110000. This mask is then ANDed with a data word abcdwxyz:

	abcdwxyz	Data word
AND	11110000	Mask
	abcd0000	Result

As a result, the left nibble (abcd) is extracted from the data word and the right nibble (wxyz) is masked off. ■

The OR operation can be used to combine (or merge) two words; the result of ORing two words is a third word that has a 1 in each bit position in which a 1 appears in either of the input words, as seen in Example 6.2.

EXAMPLE 6.2

Two data words are merged using the OR operation.

	00111100	Data word 1
OR	01010101	Data word 2
	01111101	Result

The XOR operation can be used to selectively pass or complement individual bits of a data word, as illustrated in Example 6.3.

EXAMPLE 6.3

An 8-bit control word is defined as 00001111. This control word is then XORed with a data word:

	11000101	Data word
XOR	00001111	Control word
	11001010	Result

As a result, the left nibble (1100) is passed from the data word without change, and each bit of the right nibble (0101) is complemented to produce 1010.

The required logic operations AND, OR, and XOR are inherent in a full adder function, as we will now show. A 1-bit full adder (FA) has 2 outputs (sum and carry) that can be represented by the following simplified Boolean equations

$$S = c \oplus (x \oplus y)$$
$$C = xy + c(x + y)$$

An examination of these equations shows that the logic operations AND $(x \cdot y)$, OR $(x + y)$, and XOR $(x \oplus y)$ are used to produce the arithmetic sum and carry operations. Figure 6.5 illustrates the arithmetic and logic functions produced in a 1-bit FA.

Design of a Cascaded (Bit-Slice) ALU

A variety of ALUs can be designed, each capable of performing a particular set of operations. What types of circuits are used to implement a particular ALU? How does the device know which operation is to be performed at a given time?

In this section, we will learn that (1) a 1-bit (bit-slice) ALU can be implemented by combining a full adder and a TC01 function and (2) an n-bit ALU can be implemented by cascading n 1-bit ALUs. We will use a TC01 function, with a selectable output (Y, Y', 0, or 1), in conjunction with the following function modules:

A compare equal function

A 1-bit adder with inherent logic functions

Since each of these modules performs an arithmetic, logic, or compare operation, we expect that the resulting device will be an ALU.

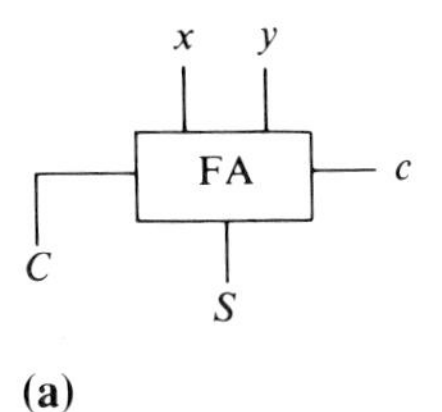

(a)

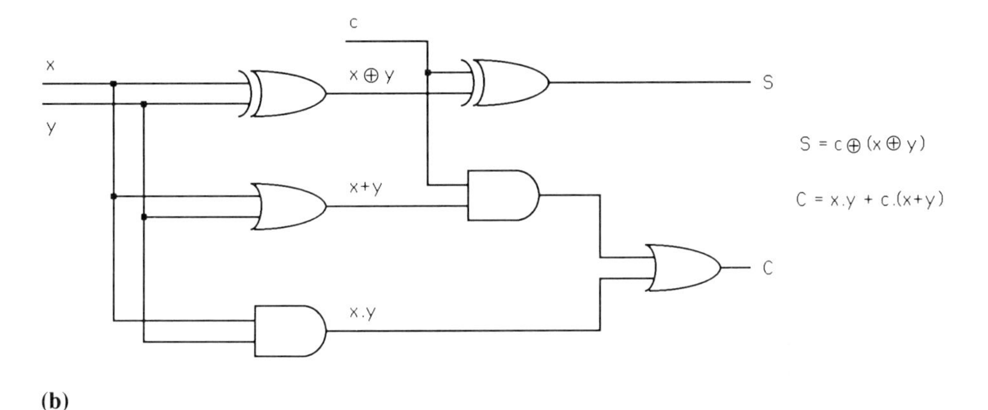

(b)

FIGURE 6.5
A 1-bit full adder with inherent logic functions. (a) Symbolic form (b) Detailed logic diagram

We first combine a full adder function and a TC01 function to form a 1-bit ALU and then determine what operations can be performed by the resulting device. Figures 6.6a and 6.6b (p. 230) present a block diagram and a logic diagram of a 1-bit ALU that is implemented by combining a full adder function and a TC01 function.

A number of logic operations can be implemented when a TC01 function is combined with a 1-bit adder. Table 6.1 lists the logic operations (at points 1, 2, and 3 in Figure 6.6b) for each combination of the enable (E) and complement (W) controls.

TABLE 6.1
Logic operations of a simple 1-bit ALU

	Logic Operations			Description		
$W\ E\ f$	AND $x \cdot f$	OR $x + f$	XOR $x \oplus f$	AND	OR	XOR
0 0 0	0	x	x	Clear	Pass x	Pass x
0 1 y	$x \cdot y$	$x + y$	$x \oplus y$	x AND y	x OR y	x XOR y
1 0 1	x	1	x'	Pass x	Set	Comp x
1 1 y'	$x \cdot y'$	$x + y'$	$x \oplus y'$	x AND y'	x OR y'	x XOR y'

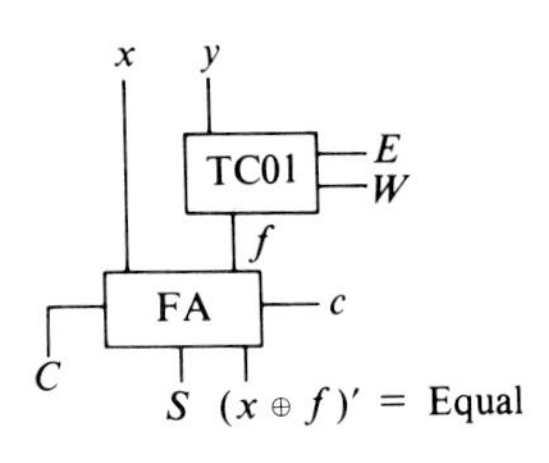

(a)

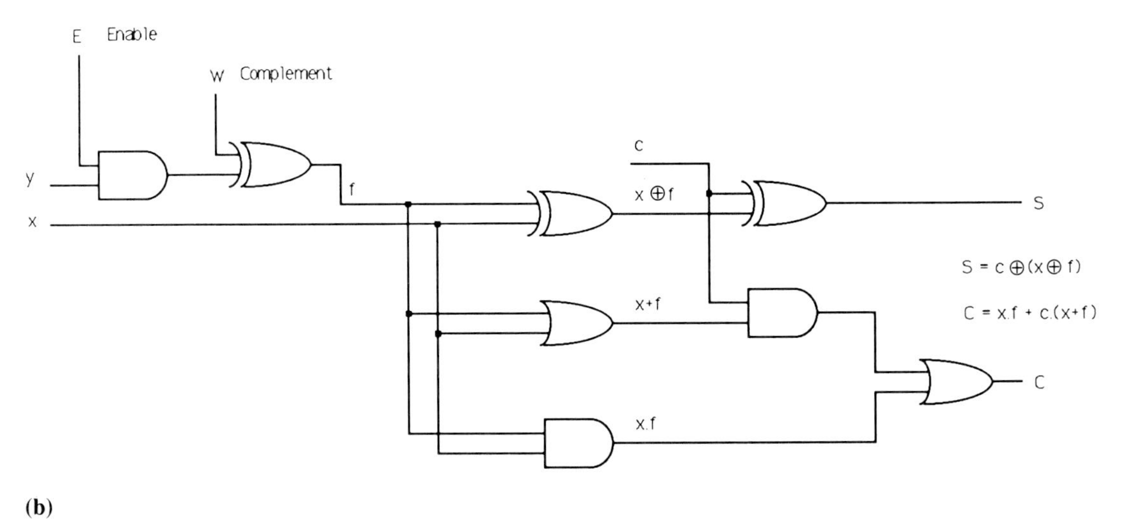

(b)

FIGURE 6.6
A simple 1-bit ALU. (a) Block diagram (b) Logic diagram

TABLE 6.2
Logic operations of a simple 4-bit ALU

	Logic Operations			Description		
$W\ E\ F$	AND $X \cdot F$	OR $X + F$	XOR $X \oplus F$	AND	OR	XOR
0 0 0	0	X	X	Clear	Pass X	Pass X
0 1 Y	$X \cdot Y$	$X + Y$	$X \oplus Y$	X AND Y	X OR Y	X XOR Y
1 0 1	X	1	X'	Pass X	Set	Comp X
1 1 Y'	$X \cdot Y'$	$X + Y'$	$X \oplus Y'$	X AND Y'	X OR Y'	X XOR Y'

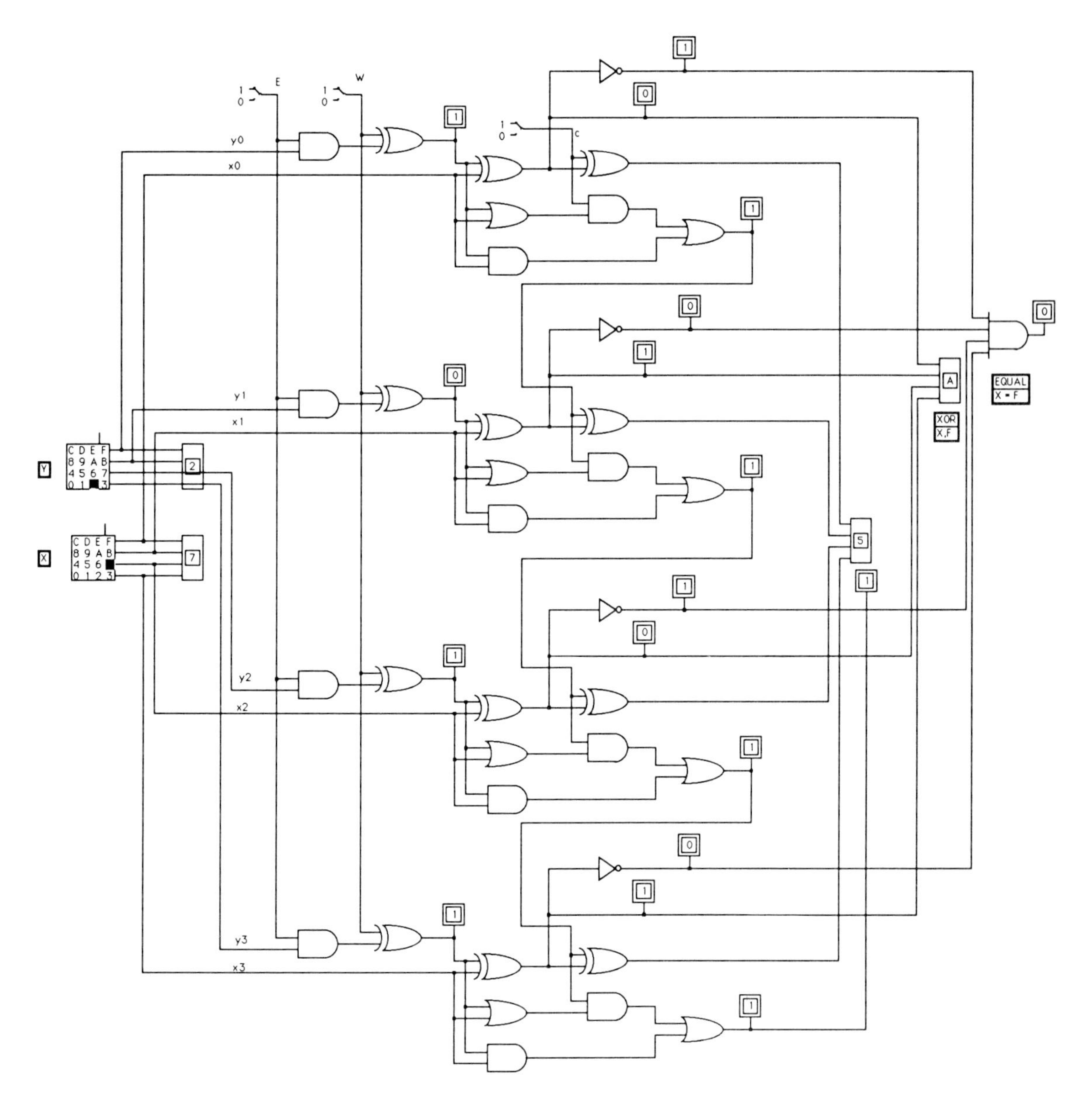

FIGURE 6.7
A 4-bit ALU formed by cascading four 1-bit ALUs

A 4-bit bit-slice ALU can be realized by cascading four 1-bit ALUs, as shown in Figure 6.7. Note that four 4-to-1 muxes could be used to select one of the four operations (AND, OR, XOR, or sum), using mux controls *a* and *b*. Table 6.2 lists the logic operations of the 4-bit ALU.

In addition to the logic operations listed, this simple 4-bit ALU can accomplish the arithmetic operations indicated in Table 6.3 (p. 232) where + signifies arithmetic addition.

TABLE 6.3
Arithmetic operations of a simple 4-bit ALU

c W E	F	$Z = X + F + c$	Function
0 0 0	Zeros	X	Pass X
0 0 1	Y	$X + Y$	Add
0 1 0	Ones	X + ones	Decrement X
0 1 1	Y'	$X + Y'$	X + 1s complement Y
1 0 0	Zeros	$X + 1$	Increment X
1 0 1	Y	$X + Y + 1$	Add with carry-in
1 1 0	Ones	X + Ones + 1	Pass X
1 1 1	Y'	$X + Y' + 1$	Subtract

We provide examples of these operations below.

	Decrement X			Increment X			Pass X	
$X =$	0101	= 5	$X =$	0110	= 6	$X =$	0011	= 3
$F =$	1111		$F =$	0000		$F =$	1111	
			$c_i =$	1		$c_i =$	1	
$Z =$	10100	= 4	$Z =$	0111	= 7	$Z =$	10011	= 3
	→ Drop EAC						→ Drop EAC	

Design of a Second Binary ALU

A myriad of different ALUs can be designed, depending on the mix of arithmetic, logic, and compare operations desired or specified. In the preceding section, a 1-bit FA (with inherent logic functions and arithmetic functions) was used to design a 1-bit ALU. Then, a 4-bit bit-slice ALU was implemented by cascading four 1-bit ALUs.

In this section, a 4-bit adder and two TC01 functions are used as building blocks to design and implement a second 4-bit ALU as illustrated in Figure 6.8.

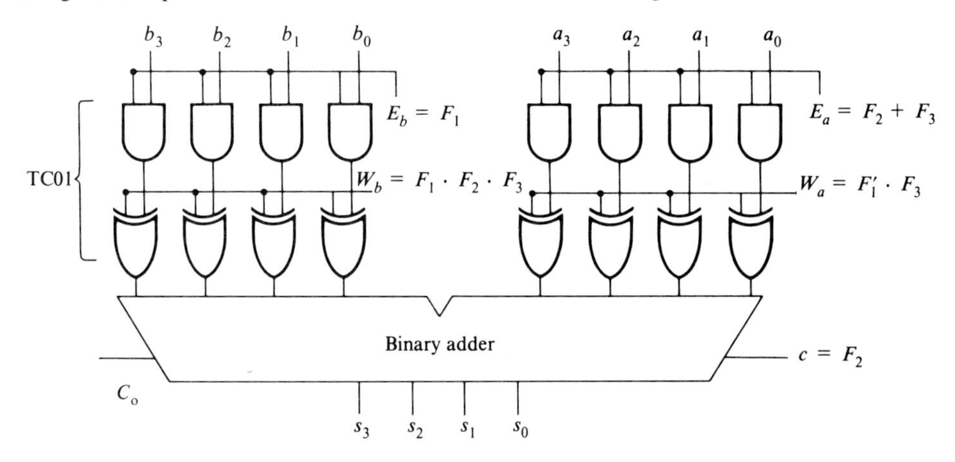

FIGURE 6.8
Block diagram of a binary ALU

If E_b, E_a, W_b, W_a, and c are assumed to be independent controls, then 32 operations can be defined for the ALU in Figure 6.8. For the time being, we will limit our design to ALUs with 8 operations, requiring 3 independent controls F_1, F_2, F_3.

F_1, F_2, and F_3 are 2-state controls, each of which can be set to 0 or 1. Each input combination of F_1, F_2, and F_3 determines which one of 8 possible operations is performed by this simple binary ALU. The input combinations and output functions are defined in Table 6.4, where + signifies arithmetic addition.

TABLE 6.4
Arithmetic operations of a second 4-bit ALU

Opcode F_1	F_2	F_3	Adder Output	Operation
0	0	0	$0 \rightarrow S$	Clear
0	0	1	$A' \rightarrow S$	Complement A
0	1	0	$A + 1 \rightarrow S$	Increment A
0	1	1	$A' + 1 \rightarrow S$	Negate A = 2s complement of A
1	0	0	$B \rightarrow S$	Transfer B
1	0	1	$A + B \rightarrow S$	Add without carry
1	1	0	$A + B + 1 \rightarrow S$	Add with carry
1	1	1	$A + B' + 1 \rightarrow S$	Subtract = $A - B$

Let the A and B inputs be two groups of 4 switches each. When a switch is open, the corresponding input has value 0; when a switch is closed, the corresponding input has value 1.

The operations listed in Table 6.4 can be verified by tracing the A and B inputs through the TC01 paths for each combination of F_1, F_2, and F_3. Value g = output of AND gate, and value f = output of XOR gate. The expanded table is shown in Table 6.5 where + signifies addition.

TABLE 6.5
Expanded form of operation table

Opcode F_1	F_2	F_3	g_b	f_b	g_a	f_a	c	Adder Output $Z = f_b + f_a + c$	Operation
0	0	0	0	0	0	0	0	$0 \rightarrow S$	Clear
0	0	1	0	0	A	A'	0	$A' \rightarrow S$	Complement A
0	1	0	0	0	A	A	1	$A + 1 \rightarrow S$	Increment A
0	1	1	0	0	A	A'	1	$A' + 1 \rightarrow S$	Negate A = 2s complement
1	0	0	B	B	0	0	0	$B \rightarrow S$	Transfer B
1	0	1	B	B	A	A	0	$A + B \rightarrow S$	Add without carry
1	1	0	B	B	A	A	1	$A + B + 1 \rightarrow S$	Add with carry
1	1	1	B	B'	A	A	1	$A + B' + 1 \rightarrow S$	Subtract = $A - B$

An examination of the F_1, F_2, and F_3 combinations that produce the operations listed in Table 6.5 shows that

- Either B or B' appears at the adder inputs, indicating $E_b = 1$ when $F_1 = 1$. Therefore, $E_b = F_1$.
- Either A or A' appears at the adder inputs, indicating $E_a = 1$ when $F_2 = 1$ or $F_3 = 1$. Therefore, $E_a = F_2 + F_3$.
- A' appears at the adder inputs, indicating $W_a = 1$ when $F_1 = 0$ and $F_3 = 1$. Therefore, $W_a = F_1' \cdot F_3$.
- B' appears at the adder inputs, indicating $W_b = 1$ when $F_1 = 1$ and $F_2 = 1$ and $F_3 = 1$. Therefore, $W_b = F_1 \cdot F_2 \cdot F_3$.
- $c = 1$ when $F_2 = 1$. Therefore, $c = F_2$.

The data-path control points E_b, W_b, E_a, W_a, and c are, thereby, controlled by a combinational logic circuit defined by the Boolean equations in controls F_1, F_2, and F_3:

$$\begin{aligned} E_b &= F_1 \\ W_b &= F_1 \cdot F_2 \cdot F_3 \\ E_a &= F_2 + F_3 \\ W_a &= F_1' \cdot F_3 \\ c &= F_2 \end{aligned}$$

A logic diagram of this ALU is presented in Figure 6.9.

The ALU designed in this section and illustrated in Figure 6.9 will serve as the basis for a simple computer to be described in Chapter 12.

6.3 BCD ARITHMETIC UNIT—A TOP-DOWN DESIGN

Top-down design is a common approach used in the design of systems. Suppose that you are given the task of designing a system to automate a particular process. A top-down approach divides the overall process into subprocesses that are divided into low-level functions, if possible. Devices are then designed to perform each function. These devices are then integrated into subsystems that are in turn integrated into a system that automates the overall process.

In this section, a top-down approach will be used to design a BCD arithmetic unit. You may ask yourself "Why do we need to perform computations in BCD if we can perform them in 2s complement arithmetic?" The answer to this question is explained in the next paragraph.

Consider an n-digit decimal number that consists of an integer part and a fractional part. The encoding of the n-digit decimal number in BCD results in an n-nibble word, in which each nibble is the BCD code for the corresponding decimal digit. Thus, the conversion from the external decimal form to the internal BCD form produces an exact representation of the n-digit decimal number. Contrast this with the conversion of the n-digit decimal number to an n-bit binary number; since 10 is not an integer power of 2, the

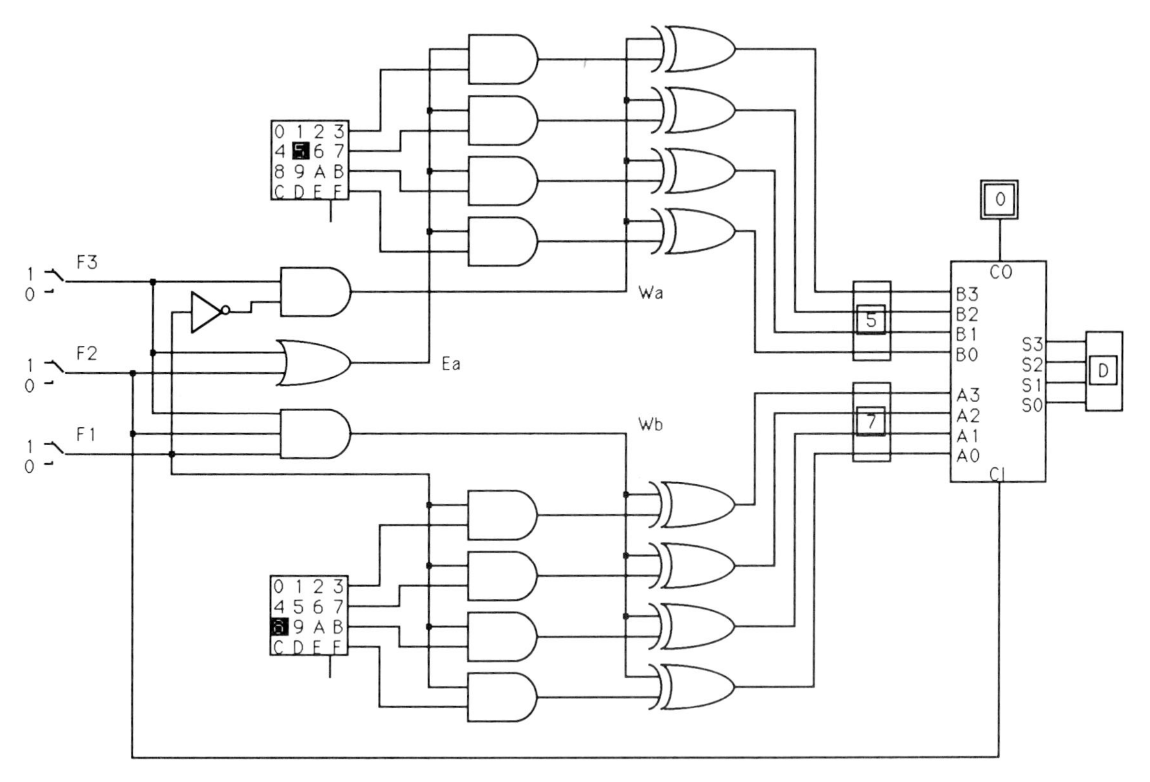

FIGURE 6.9
Logic diagram of a binary ALU

decimal-to-binary conversion can produce a binary number that is not an exact equivalent of the original decimal number.

Consequently, applications that require exact internal representation of external numeric data use BCD arithmetic rather than binary arithmetic. It is for this reason that some computers and many calculators perform computations in BCD.

Consider the addition of two 2-digit decimal numbers, as illustrated in the following examples:

00	Carry-in	11	Carry-in
31		45	
+ 25		+ 67	
056	Sum	112	Sum
000	Carry-out	011	Carry-out

Each digit of a multidigit decimal number must be encoded in BCD before these numbers can be added in BCD, as shown in the following examples:

	0100 001	Carry-in	11	1000 111	Carry-in
31	0011 0001		45	0100 0101	
+ 25	+ 0010 0101		+ 67	+ 0110 0111	
56 =	0101 0110	Sum	112 =	1010 1100	Sum
	0010 0001	Carry-out	11	0100 0111	Carry-out

Note that the addition in the second example produces nibbles 1010 = decimal 10 and 1100 = decimal 12. If each nibble is added using a 4-bit binary adder, the result must be corrected so that each nibble is a valid BCD digit.

Step 1: Problem Statement. Design a BCD arithmetic unit that inputs data in decimal form, converts the data to BCD form, performs computations (addition or subtraction) in BCD, and decodes the BCD results for output in decimal form. A BCD adder must adjust the result if an invalid BCD digit results, as we saw in the example above.

Step 2: Conceptualization. If two numbers (e.g., 45 and 67) are input in decimal, each digit of each decimal number is encoded in BCD: 45 = 0100 0101, 67 = 0110 0111. The rightmost nibbles can be added using a 1-stage BCD adder. The leftmost nibbles, with a carry-in, can be added using a second 1-stage BCD adder. The BCD result is then decoded back to decimal form for output.

The addition can be accomplished using a 2-digit BCD adder that computes a BCD sum using two cascaded stages of 1-digit BCD adders. A block diagram of a 2-stage BCD adder unit, with decimal-to-BCD encoders on the inputs, is shown in Figure 6.10.

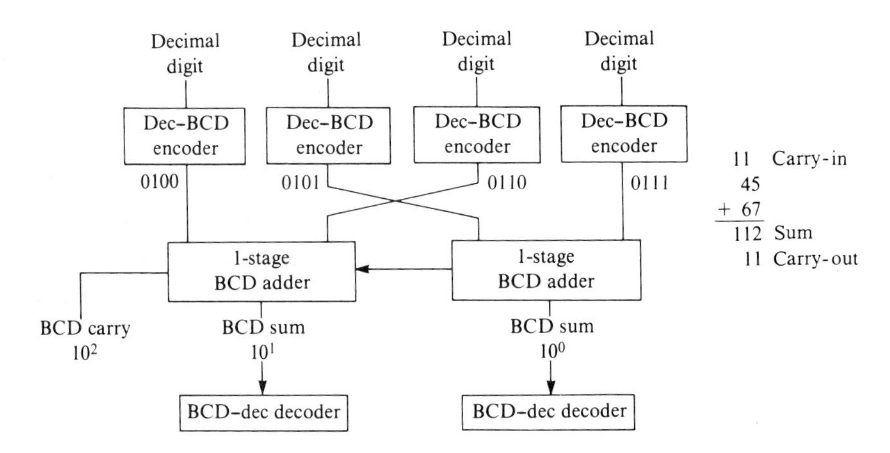

FIGURE 6.10
Block diagram of a 2-stage BCD adder

An analysis of the problem statement and the block diagram reveals the need for three functional modules, or building blocks, in the realization of our system:

A decimal-to-BCD encoder for input

A BCD adder/subtracter, consisting of a BCD adder and a BCD complementer (since BCD subtraction can be accomplished by adding the 9s complement of the subtrahend to the minuend)

Two BCD-to-decimal decoders for output: a BCD-to-decimal decoder and a BCD-to-7-segment decimal decoder

Figure 6.11 illustrates a top-down diagram for the design of this BCD arithmetic unit.

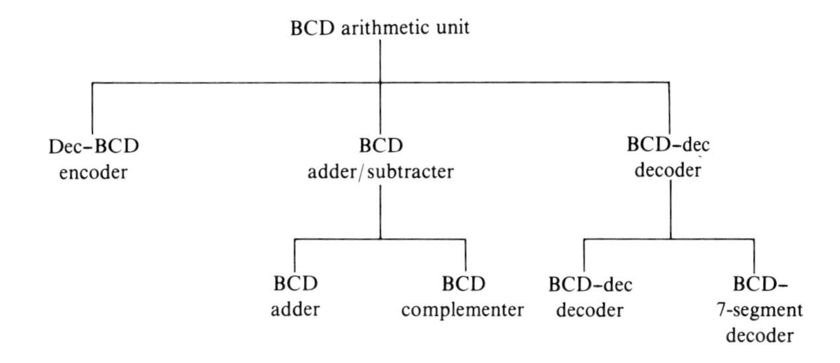

FIGURE 6.11
Top-down design of a BCD arithmetic unit

One stage of such a BCD arithmetic unit can be illustrated in block-diagram form, as indicated in Figure 6.12.

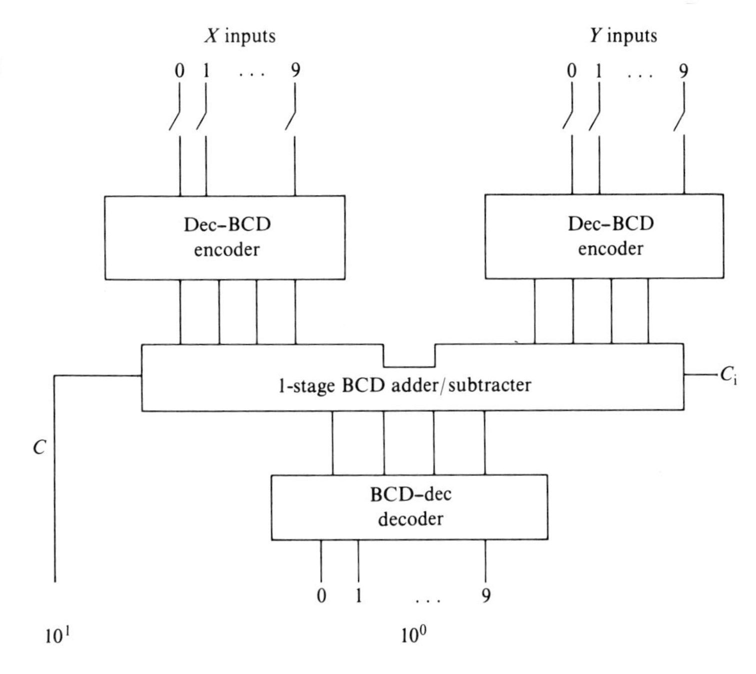

FIGURE 6.12
Block diagram of a 1-stage BCD arithmetic unit

This 1-stage BCD arithmetic unit accepts two 1-digit decimal numbers (0–9), encodes each in BCD, and then adds (or subtracts) the 2 BCD numbers and a carry-in C_i. Then, the BCD result is decoded into decimal form for output (using indicator lights). Since the result of 2 decimal digits can exceed 10, a possible carry-out C_o is included; this is output to a 10^1 light. An n-stage BCD arithmetic unit can be implemented by cascading n 1-stage units.

The design of the functional components (decimal-to-BCD encoder, BCD adder/subtracter, and BCD-to-decimal decoder) is described in the following sections.

Design of a Decimal-to-BCD Encoder for Input

A BCD arithmetic unit accepts input data in decimal form, but this data must then be encoded in BCD form to be processed. The design of a decimal-to-BCD encoder is accomplished using the 4-step problem-solving procedure as follows:

Step 1: Problem Statement. Design a decimal-to-BCD encoder that has 10 inputs (0, 1, . . . , 9) and 4 outputs (with weights 8, 4, 2, 1) representing the bits of a BCD number.

Step 2: Conceptualization. The correspondences between each decimal digit and its BCD equivalent, indicated in the following table, can be used to derive logic equations for the required combinational logic circuit.

Decimal	BCD abcd 8421
0	0000
1	0001
2	0010
3	0011
4	0100
5	0101
6	0110
7	0111
8	1000
9	1001

Step 3: Solution/Simplification. From the table above, we see that each output a, b, c, d of the decimal-to-BCD encoder is a logical sum of inputs, as defined in the following equations:

$$a = 8 + 9$$
$$b = 4 + 5 + 6 + 7$$
$$c = 2 + 3 + 6 + 7$$
$$d = 1 + 3 + 5 + 7 + 9$$

Step 4: Realization. Using the foregoing logical equations, we can implement a decimal-to-BCD encoder using OR gates, as illustrated in Figure 6.13.

The decimal digit to be encoded at any given time is selected by closing the switch for that digit. The weighted bits a, b, c, d of the output will be used as the BCD input to a BCD adder, which we will design next.

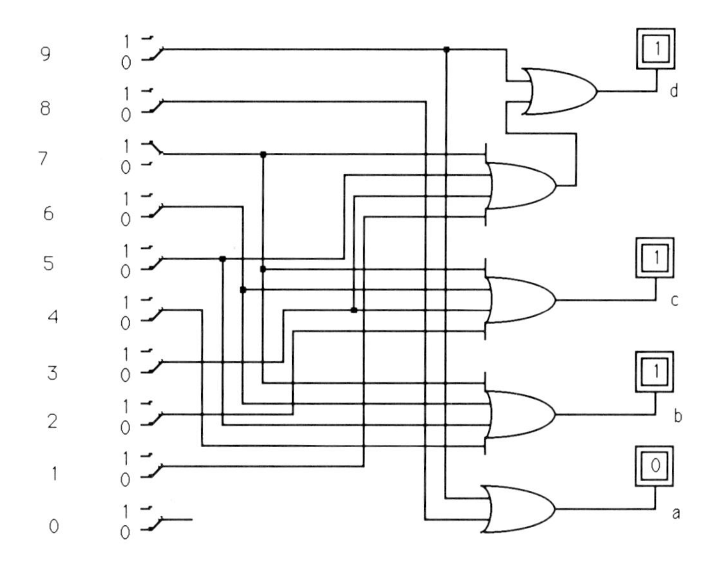

FIGURE 6.13
Logic diagram of a decimal-to-BCD encoder

Design of a BCD Adder/Subtracter

Design of a BCD Adder

A 1-stage BCD adder is a device that adds two 1-digit BCD numbers X and Y, with a possible carry-in C_i, as illustrated in the following example:

$$C_i = \{0, 1\}$$
$$X = x_3x_2x_1x_0 = \{0, 1, \ldots, 9\}$$
$$Y = y_3y_2y_1y_0 = \{0, 1, \ldots, 9\}$$

Note that subscripts 3, 2, 1, 0 denote powers of 2 weights of the positions in the BCD number. The addition of two BCD digits X and Y and a carry-in produces a sum that is in the range 0 to 19 decimal, as illustrated in Table 6.6.

TABLE 6.6
BCD addition table

Sum X \ Y ($C_i = 0$)	0	1	2	3	4	5	6	7	8	9
0	0	1	2	3	4	5	6	7	8	9
1	1	2	3	4	5	6	7	8	9	10
2	2	3	4	5	6	7	8	9	10	11
3	3	4	5	6	7	8	9	10	11	12
4	4	5	6	7	8	9	10	11	12	13
5	5	6	7	8	9	10	11	12	13	14
6	6	7	8	9	10	11	12	13	14	15
7	7	8	9	10	11	12	13	14	15	16
8	8	9	10	11	12	13	14	15	16	17
9	9	10	11	12	13	14	15	16	17	18

TABLE 6.6
Continued

Sum / X \ Y ($C_i = 1$)	0	1	2	3	4	5	6	7	8	9
0	1	2	3	4	5	6	7	8	9	10
1	2	3	4	5	6	7	8	9	10	11
2	3	4	5	6	7	8	9	10	11	12
3	4	5	6	7	8	9	10	11	12	13
4	5	6	7	8	9	10	11	12	13	14
5	6	7	8	9	10	11	12	13	14	15
6	7	8	9	10	11	12	13	14	15	16
7	8	9	10	11	12	13	14	15	16	17
8	9	10	11	12	13	14	15	16	17	18
9	10	11	12	13	14	15	16	17	18	19

The design of a 1-stage BCD adder can be accomplished using the 4-step problem-solving procedure as follows:

Step 1: Problem Statement. Design a 1-stage BCD adder that adds two 1-digit BCD numbers X, Y and a carry-in, as illustrated in Figure 6.14.

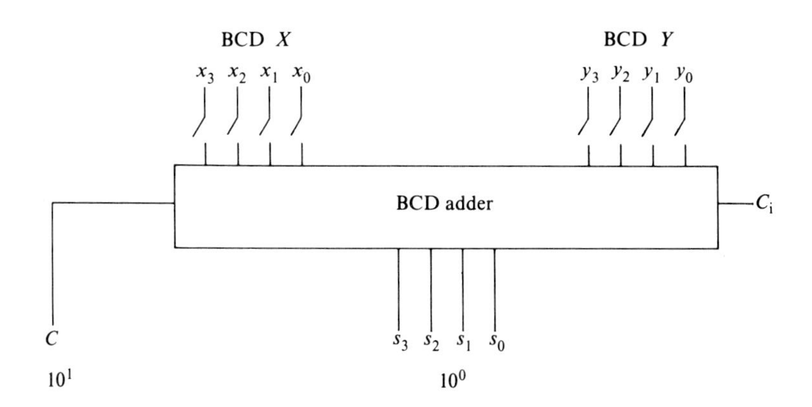

FIGURE 6.14
Block diagram of BCD adder

Step 2: Conceptualization. Since 4-bit binary adders were designed in Chapter 5, we can consider the possibility of modifying a binary adder to realize a BCD adder. A 4-bit binary adder with BCD inputs (0000 through 1001) for X and Y, and a carry-in C_i produces a sum Z and a carry-out K, as illustrated in Figure 6.15.

Step 3: Solution/Simplification. Examination of Table 6.7 shows that the resulting binary sum will produce the correct BCD result for decimal number sums in the range 0 to 9. For decimal number sums in the range 10 to 19, 6 must be added to the binary sum to produce the correct BCD sum.

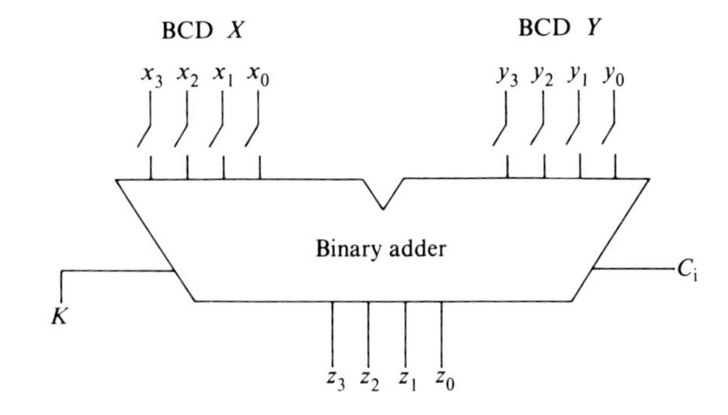

FIGURE 6.15
Block diagram of 4-bit binary adder

TABLE 6.7
Conversion of binary sums to BCD sums

Decimal	Binary Sum						BCD Sum					BCD Sum Weights	
Sum	K	z_3	z_2	z_1	z_0		C	s_3	s_2	s_1	s_0	10	1
0	0	0	0	0	0		0	0	0	0	0	0	0
1	0	0	0	0	1		0	0	0	0	1	0	1
2	0	0	0	1	0		0	0	0	1	0	0	2
3	0	0	0	1	1		0	0	0	1	1	0	3
4	0	0	1	0	0	+0 =	0	0	1	0	0	0	4
5	0	0	1	0	1		0	0	1	0	1	0	5
6	0	0	1	1	0		0	0	1	1	0	0	6
7	0	0	1	1	1		0	0	1	1	1	0	7
8	0	1	0	0	0		0	1	0	0	0	0	8
9	0	1	0	0	1		0	1	0	0	1	0	9
10	0	1	0	1	0		1	0	0	0	0	1	0
11	0	1	0	1	1		1	0	0	0	1	1	1
12	0	1	1	0	0		1	0	0	1	0	1	2
13	0	1	1	0	1		1	0	0	1	1	1	3
14	0	1	1	1	0	+6 =	1	0	1	0	0	1	4
15	0	1	1	1	1		1	0	1	0	1	1	5
16	1	0	0	0	0		1	0	1	1	0	1	6
17	1	0	0	0	1		1	0	1	1	1	1	7
18	1	0	0	1	0		1	1	0	0	0	1	8
19	1	0	0	1	1		1	1	0	0	1	1	9

Consider the following examples of converting binary sums to correct BCD sums:

EXAMPLE 6.4

Convert the following decimal sum in the range 0 to 9.

$$\begin{array}{rcll} 5 & = & 0101 & \\ +\ 3 & = & 0011 & \\ \hline 8 & = & 1000 & \text{Correct BCD sum} \end{array}$$

EXAMPLE 6.5

Convert the following decimal sum in the range 10 to 19.

$$\begin{array}{rcrl} 7 & = & 0111 & \\ +\ 5 & = & 0101 & \\ \hline 12 & = & 1100 & \text{In binary, note that this sum > decimal 9} \\ & + & 0110 & \text{Add 6, since sum > decimal 9} \\ \hline & & 1\ 0010 & = \text{12 correct BCD sum} \end{array}$$

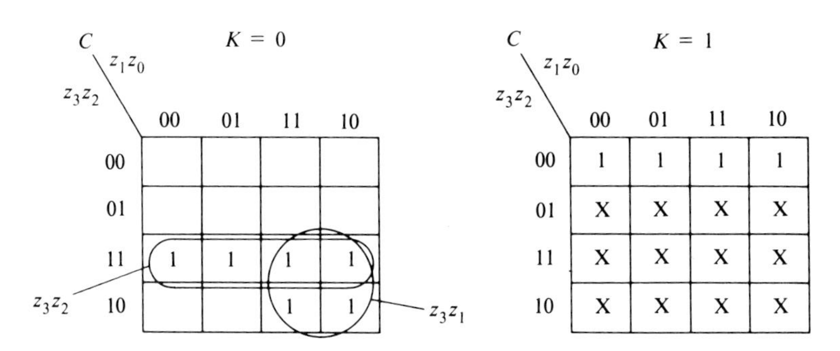

FIGURE 6.16
K-maps for BCD correction

Table 6.7 indicates that 6 must be added to the binary sum when $K = 1$ or $z_3z_1 = 1$ or $z_3z_2 = 1$. This conclusion can also be arrived at by using the K-maps shown in Figure 6.16. Thus, we see that

$$C = K + z_3z_1 + z_3z_2$$

Then, if $C = 1$, we must add 6 (binary 0110) to the binary sum to get the correct BCD sum.

Step 4: Realization. A 1-stage BCD adder can be implemented using two 4-bit binary adders and the circuitry required to correct any result that corresponds to an invalid BCD representation. This 1-stage BCD adder can be implemented as shown in Figure 6.17.

BCD Subtraction

In Chapter 1, we saw that the process of subtracting a decimal number Y from a decimal number X can be accomplished by adding the 9s complement of Y to the number X:

$$X - Y = X + Y'$$

where $Y' = 9 - Y$. Table 6.8 lists the 9s complement Y' for each Y in decimal and BCD form.

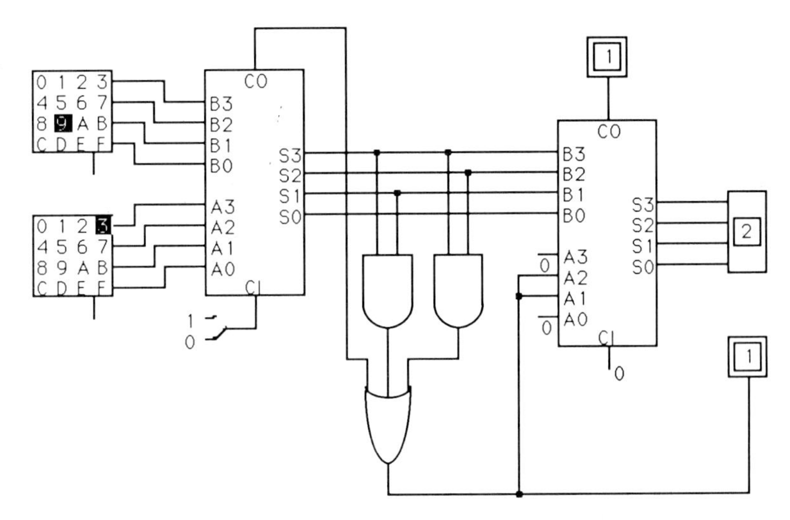

FIGURE 6.17
BCD adder realized using binary adders

TABLE 6.8
The 9s complements in decimal and BCD

Decimal		BCD	
Y	Y' = 9s Complement	Y *abcd*	Y' = 9s Complement *rstu*
0	9	0000	1001
1	8	0001	1000
2	7	0010	0111
3	6	0011	0110
4	5	0100	0101
5	4	0101	0100
6	3	0110	0011
7	2	0111	0010
8	1	1000	0001
9	0	1001	0000

The process of subtracting by complementing and adding is illustrated in the following examples:

■ **EXAMPLE 6.6**

For the case $X > Y$, subtract BCD $Y = 2$ from BCD $X = 5$.

$$
\begin{array}{rrl}
X = 5 = & 0101 & \\
Y = 2 \quad & \underline{0111} & \text{9s complement of 2} \\
X - Y = X + Y' = & 1100 & \textit{Note: } X + Y' = X + (9 - Y) \\
 & & \qquad\qquad = 9 + (X - Y) > 9 \\
 & \underline{+\ 0110} & \text{Add 6} \\
 & 10010 & \\
\text{EAC} \rightarrow & \underline{+\ 1} & \\
 = & 0011 & = 3
\end{array}
$$

■

EXAMPLE 6.7

For the case $X < Y$, subtract BCD $Y = 9$ from BCD $X = 5$.

$$
\begin{array}{rcll}
X = 5 & = & 0101 & \\
Y = 9 & & \underline{0000} & \text{9s complement of 9} \\
X - Y = X + Y' & = & 00101 &
\end{array}
$$

A zero carry MSB indicates a negative result, which necessitates taking the 9s complement to produce the correct answer; that is,

$$0100 = -4$$

The process of performing subtraction in BCD by adding the 9s complement of the subtrahend to the minuend requires a BCD complementer.

Design of a BCD Complementer Using a Binary Adder

Consider the realization of a BCD 9s complementer using a binary adder. Table 6.9 shows that the 9s complement of a BCD digit can be obtained by complementing the BCD number, adding decimal 10 (binary 1010 = 2s complement of 6), and then subtracting decimal 16 (binary 10000) by dropping the carry-out.

By mathematical justification, 16 plus the 9s complement N' is obtained from the equation

$$25 - N = 16 + (9 - N) = 16 + N'$$

and N' is obtained by dropping the carry-out (16).

TABLE 6.9
Generation of 9s complement of a BCD number

Decimal N	BCD abcd N	Bitwise Complement $15 - N$	+	Add 1010 10	=	Result $25 - N$ = $16 + N'$	Ignore carry-out = 9s complement N'
0	0000	1111		1010		11001	1001
1	0001	1110		1010		11000	1000
2	0010	1101		1010		10111	0111
3	0011	1100		1010		10110	0110
4	0100	1011		1010		10101	0101
5	0101	1010		1010		10100	0100
6	0110	1001		1010		10011	0011
7	0111	1000		1010		10010	0010
8	1000	0111		1010		10001	0001
9	1001	0110		1010		10000	0000

Consider the following examples:

BCD number	0000	0001	0010	0011
Complement	1111	1110	1101	1100
Add 1010	+ 1010	+ 1010	+ 1010	+ 1010
Result	11001	11000	10111	10110
9s complement	1001	1000	0111	0110

To summarize, the BCD 9s complement can be found as follows:

1. Complement each bit of the BCD digit.
2. Add 1010 (2s complement of 6).
3. Drop the carry-out to produce the BCD 9s complement.

A block diagram of the BCD 9s complementer is shown in Figure 6.18.

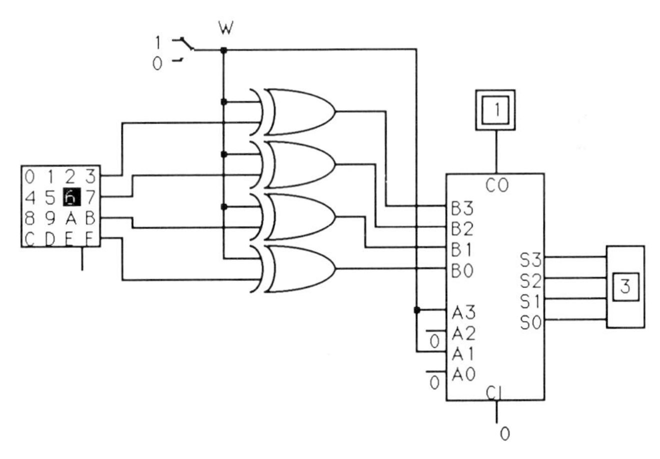

FIGURE 6.18
BCD 9s complementer using a binary adder

Combining the BCD Adder and Complementer

A BCD complementer can be combined with a BCD adder to form a BCD adder/subtracter that performs either addition or subtraction. A 1-bit control W is used to indicate addition or subtraction as follows:

- □ If $W = 0$, the adder/subtracter will perform addition.
- □ If $W = 1$, the adder/subtracter will perform subtraction.

An examination of the BCD 9s complementer (see Figure 6.18) reveals the following:

- □ If $W = 0$, the BCD input $Y = abcd$ will pass through the complementer unchanged, so that Y will be output from the complementer.
- □ If $W = 1$, the BCD input $Y = abcd$ will be complemented, so that Y' will be output from the complementer.

We can now insert a BCD complementer between the Y decimal-to-BCD encoder and the Y input to the BCD adder, as illustrated in Figure 6.19 (p. 246), to form a preliminary version of a BCD adder/subtracter.

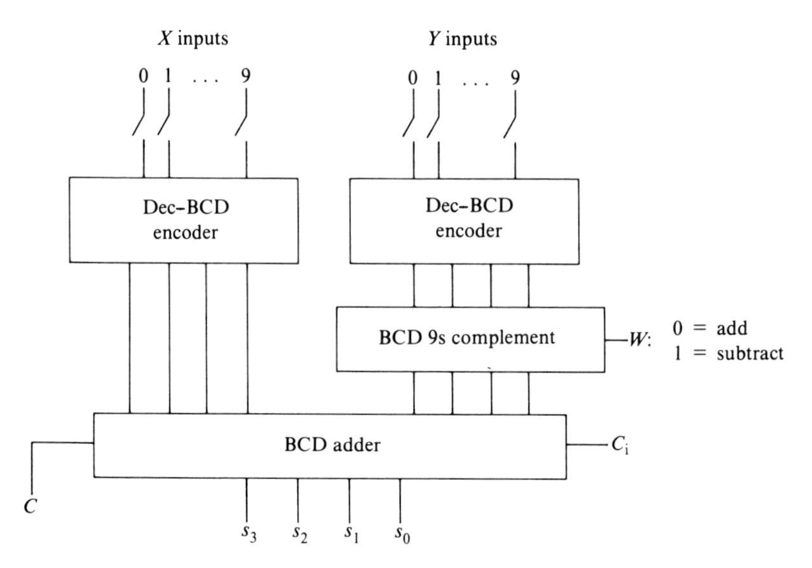

FIGURE 6.19
Preliminary block diagram of BCD adder/subtracter

We will analyze the preliminary BCD adder/subtracter by adding $X + Y + C_i$. When $W = 0$, input Y is passed through the complementer unchanged, and the preliminary BCD adder/subtracter illustrated in Figure 6.19 performs the required addition of the three inputs X, Y, and C_i.

Similarly, we will subtract $X - Y$. Let D denote the difference between X and Y, as defined in the equation $D = X - Y$. From the definition of the 9s complement, it follows that

$$\begin{aligned} D &= X + Y' \\ &= X + (9 - Y) \\ &= 9 + (X - Y) \end{aligned}$$

There are two separate cases to consider when performing the process of subtraction of Y from X:

1. If $X > Y$, then $X - Y > 0$ so that $D = X + Y' = 9 + (X - Y) =$ binary sum > 9. If the binary adder produces an output > 9, then 6 must be added to this result to produce a correct BCD sum; that is,

$$\begin{aligned} D + 6 &= [9 + (X - Y)] + 6 \\ &= 15 + (X - Y) \\ &= 15 + 1 + (X - Y - 1) \\ &= 16 + (X - Y - 1) \end{aligned}$$

   Hence, $D + 6$ produces a carry out (16) and a sum $= X - Y - 1$. If 1 is added to each side of the last equation, we get

$$\begin{aligned} D + 6 + 1 &= 16 + (X - Y - 1) + 1 \\ &= 16 + (X - Y) \end{aligned}$$

Therefore, if we add inputs X, Y', and 1 (by using a forced carry-in = 1), the BCD adder will produce a carry-out (16) and the desired difference $D = X - Y$.

$$C = 1 \text{ (ignore, since } D = X - Y < 16)$$
$$S = X - Y \text{ for the } 10^0 \text{ coefficient}$$

2. If $X < Y$, the difference $D = X - Y < 0$ so that $X + Y' = X + (9 - Y) = 9 + D < 9$. If the binary adder produces an output < 9, then the BCD adder produces:

   Output C, with value 0 for the 10^1 coefficient

   Output S, with value equal to the 10^0 coefficient

   With the substitution of a variable $Z = -D$, the equation

$$X + Y' = 9 + D$$

   can be written as $X + Y' = 9 - Z = Z'$.

   In this case, BCD adder output = Z', and Z' must be complemented to get $Z = -D$, the negative difference that occurs when Y is subtracted from X. Therefore, a 9s complementer is inserted after the S output of the BCD adder, and when $C = 0$, indicating a negative difference, the output S is complemented.

Figure 6.20 illustrates a block diagram of a complete BCD adder/subtracter. Note that if $C = 0$, subtraction produces a negative result, and the 9s complement of the result must be performed.

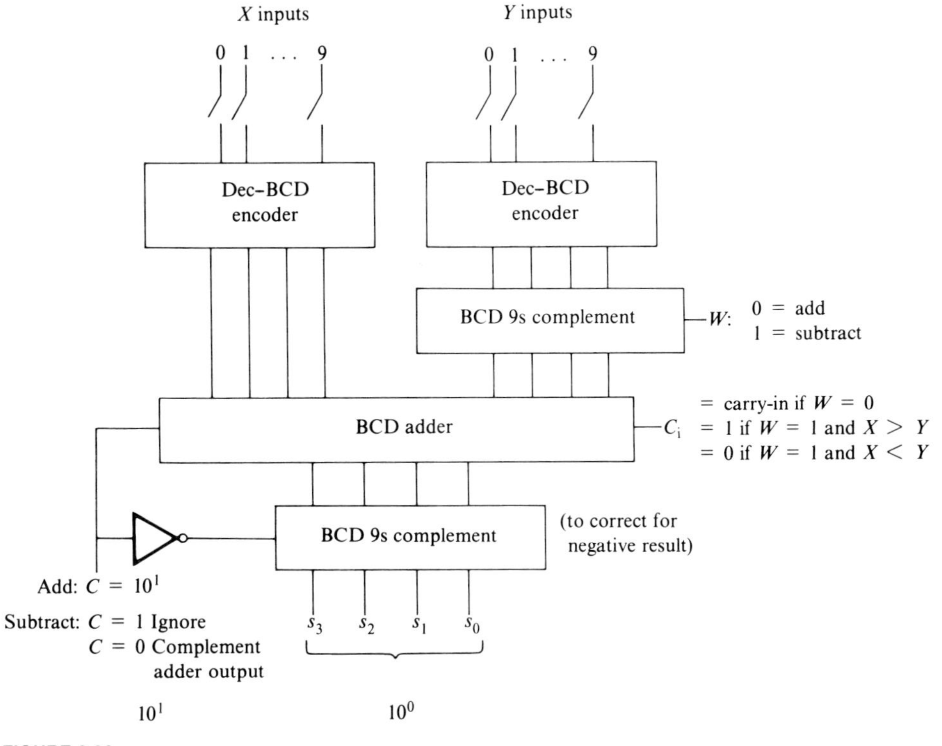

FIGURE 6.20
Block diagram of a complete BCD adder/subtracter

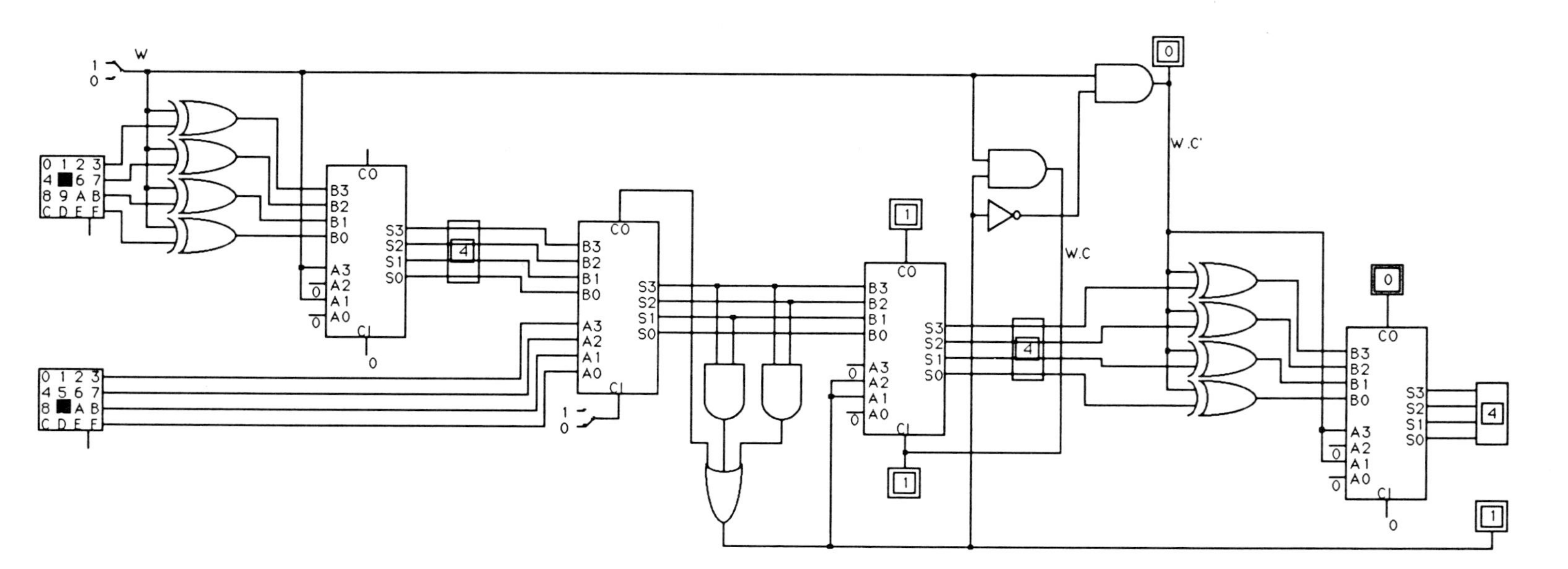

FIGURE 6.21
Implementation of a complete BCD adder/subtracter

Figure 6.21 shows an implementation of a complete BCD adder/subtracter, using 4-bit binary adders and logic gates.

Design of a BCD-to-Decimal Decoder for Output

Both the BCD adder and the BCD adder/subtracter produce BCD results, which should be decoded to decimal form for human consumption. To accomplish this, we must design a BCD-to-decimal decoder. Examine the following BCD-to-decimal correspondences.

	BCD *ABCD*	Decimal
	0000	0
	0001	1
	0010	2
	0011	3
	0100	4
	0101	5
	0110	6
	0111	7
	1000	8
	1001	9
Inputs that do not occur	1010	X
	1011	X
	1100	X
	1101	X
	1110	X
	1111	X

The functions 0, 1, . . . , 9 representing the decimal digits can be found from the K-map in Figure 6.22 (p. 250), which illustrates the grouping, wraparound, and don't-care features of the K-map. From Figure 6.22,

$$0 = A'B'C'D'$$
$$1 = A'B'C'D$$
$$2 = B'CD'$$
$$3 = B'CD$$
$$4 = BC'D'$$
$$5 = BC'D$$
$$6 = BCD'$$
$$7 = BCD$$
$$8 = AD'$$
$$9 = AD$$

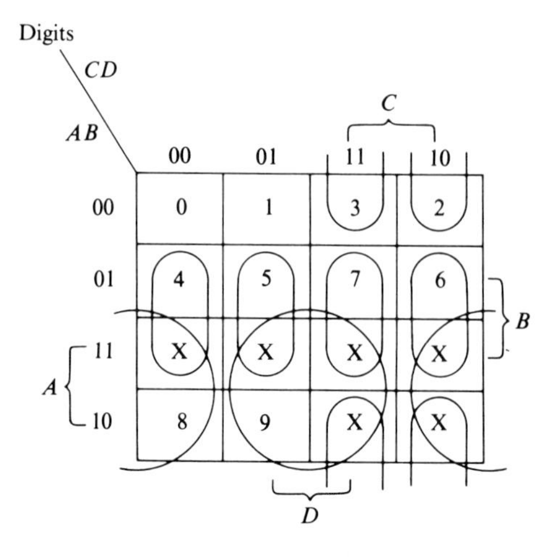

FIGURE 6.22
Decimal outputs of a BCD-to-decimal decoder

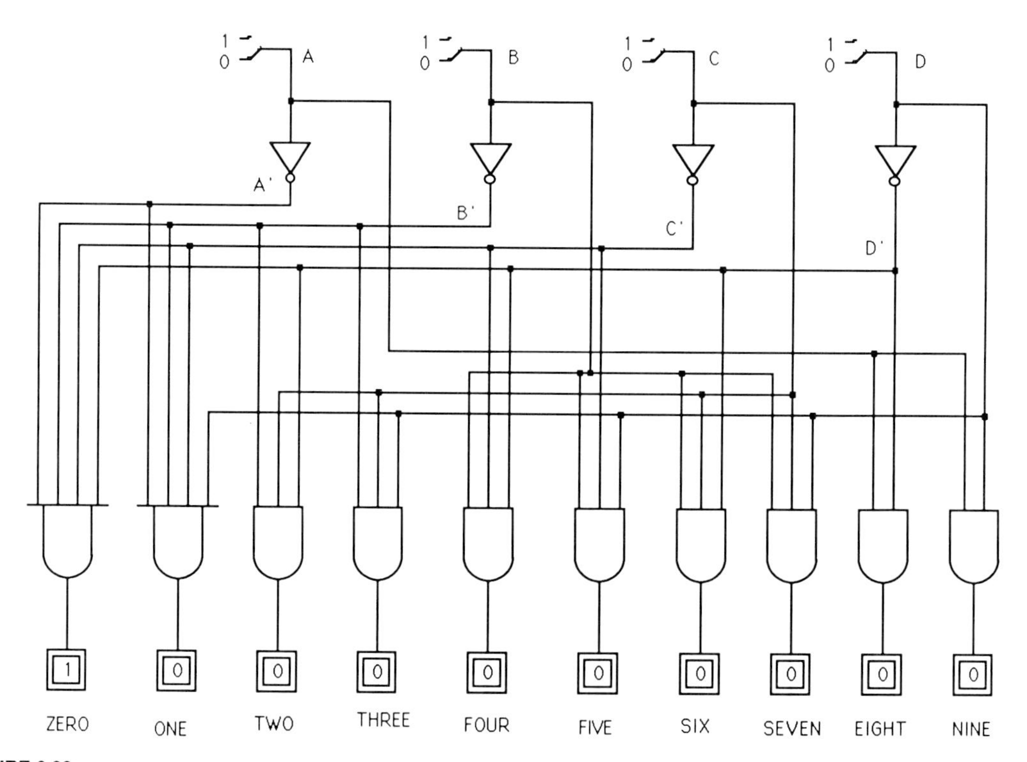

FIGURE 6.23
Logic diagram of BCD-to-decimal decoder

The Boolean functions for the decimal digits 0–9 are listed in the logic diagram for the BCD-to-decimal decoder shown in Figure 6.23.

This BCD-to-decimal decoder can then be combined with the BCD adder/subtracter and the decimal-to-BCD encoder, as illustrated in Figure 6.24.

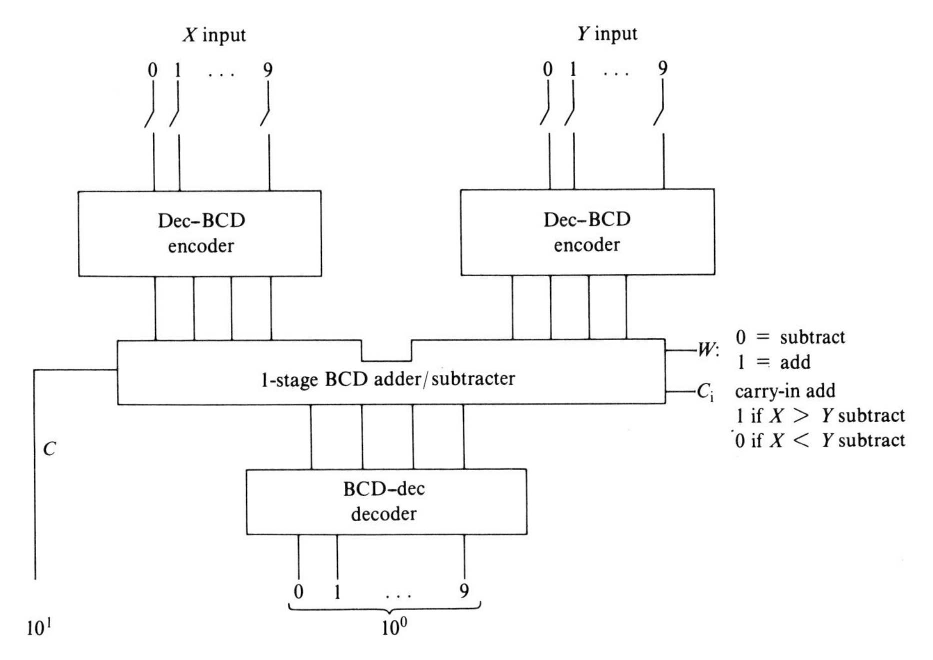

FIGURE 6.24
Block diagram of a 1-stage BCD arithmetic unit

This BCD arithmetic unit accepts as input two 1-digit decimal numbers (0–9), encodes each input in BCD, and then adds (or subtracts) the 2 BCD numbers and a carry-in C_i. The BCD result is then decoded into decimal form for output (using indicator lights). Since the result of 2 decimal digits can exceed 10, a possible carry-out C is included; this is output to a 10^1 light.

Design of a BCD-to-7-Segment Decoder for Display

We are now ready to design a combinational logic circuit that will decode each BCD digit and generate the corresponding decimal digit using a 7-segment display. Figure 6.25a and b illustrate the segments in a 7-segment display, together with a representation of each decimal digit.

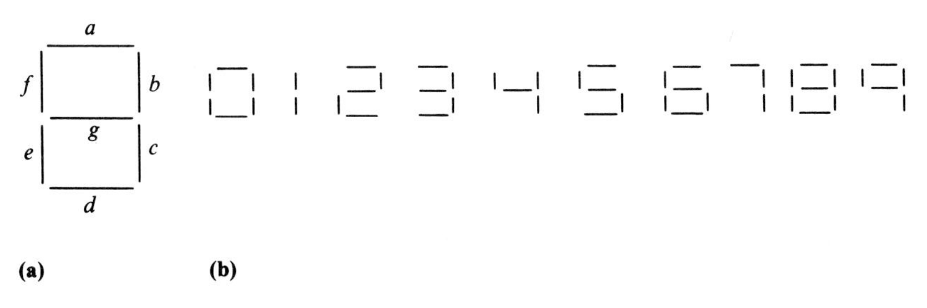

FIGURE 6.25
A 7-segment display. (a) Segments (b) Decimal digits

The operation of the BCD-to-7-segment display is described by Table 6.10.

TABLE 6.10
BCD-to-7-segment display

Decimal Digit	BCD $A\ B\ C\ D$	Segments $a\ b\ c\ d\ e\ f\ g$
0	0 0 0 0	1 1 1 1 1 1 0
1	0 0 0 1	0 1 1 0 0 0 0
2	0 0 1 0	1 1 0 1 1 0 1
3	0 0 1 1	1 1 1 1 0 0 1
4	0 1 0 0	0 1 1 0 0 1 1
5	0 1 0 1	1 0 1 1 0 1 1
6	0 1 1 0	1 0 1 1 1 1 1
7	0 1 1 1	1 1 1 0 0 0 0
8	1 0 0 0	1 1 1 1 1 1 1
9	1 0 0 1	1 1 1 0 0 1 1
—	1 0 1 0	Don't cares
—	1 0 1 1	Don't cares
—	1 1 0 0	Don't cares
—	1 1 0 1	Don't cares
—	1 1 1 0	Don't cares
—	1 1 1 1	Don't cares

Using K-maps, we can obtain the following simplified logic expressions for the segments of the 7-segment display:

$$a = A + C + B'D' + BD$$
$$b = B' + C'D' + CD$$
$$c = C' + D + B$$
$$d = B'D' + B'C + CD' + BC'D$$
$$e = B'D' + CD'$$
$$f = BD' + C'D' + BC' + A$$
$$g = A + BC' + B'C + CD'$$

Design of a BCD Error Detector

It may be necessary to verify that the number generated by the decimal-to-BCD encoder is indeed a valid BCD character. This verification function can be accomplished by a BCD error detector circuit, which we shall now design. This circuit should have 4 inputs A, B, C, and D and 1 output F, which equals 1 when any input combination is an invalid character, as shown in Table 6.11.

TABLE 6.11
BCD error detector

Inputs *ABCD*	Output *F*	
0000	0	
0001	0	
0010	0	
0011	0	
0100	0	
0101	0	
0110	0	
0111	0	
1000	0	
1001	0	
1010	1	Invalid BCD combinations
1011	1	
1100	1	
1101	1	
1110	1	
1111	1	

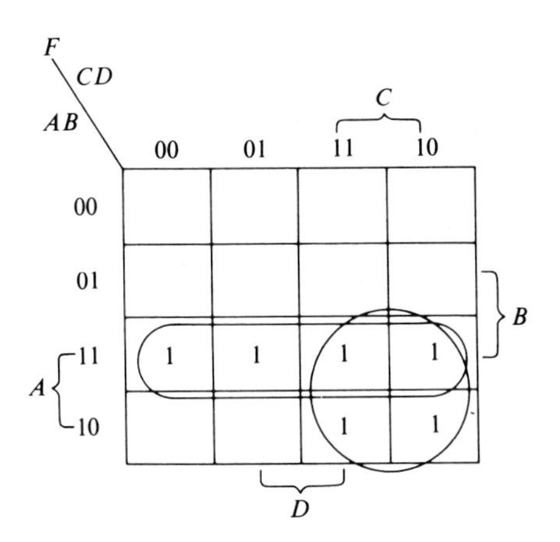

FIGURE 6.26
Output F of a BCD error detector

The output F can be represented by the K-map shown in Figure 6.26. From this K-map, we find that

$$F = AB + AC = A(B + C)$$

A logic diagram for the BCD error detector circuit is presented in Figure 6.27(p. 254).

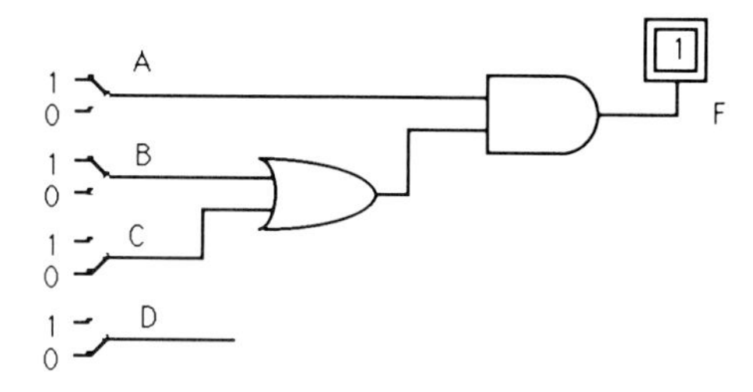

FIGURE 6.27
Logic diagram for a BCD error detector

Now consider a modification of this circuit in order to allow it to pass only valid BCD characters. That is, when $F = 0$, the 4 inputs $A, B, C,$ and D will be output as $W, X, Y,$ and Z. This can be accomplished by ANDing F' with each input A, B, C, or D to produce the corresponding output, as indicated in Figure 6.28.

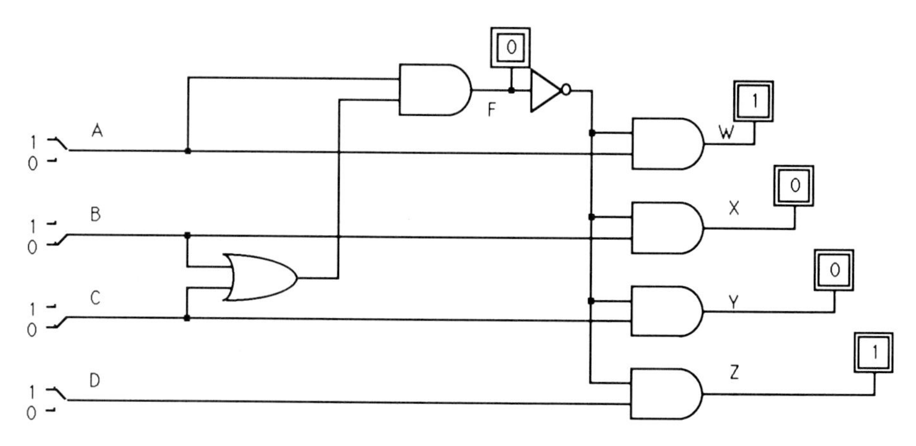

FIGURE 6.28
Pass valid BCD character circuit

6.4 SUMMARY

In this chapter we described how a digital system can be organized into subsystems, each of which may consist of a number of functional modules. For example, a computer system contains the following subsystems: memory, central processing unit (CPU), input, and output. In turn a CPU is composed of two functional units: a control unit and an arithmetic logic unit (ALU).

An ALU must perform a number of different types of operations so that the computer can accomplish tasks such as list processing, statistical computations, and numerical methods. We described the following basic categories of ALU operations:

Compare operations (less than, equal to, greater than)

Logic operations such as AND, OR, XOR

Arithmetic operations such as addition and subtraction

We described the design of 1-bit, 2-bit, and 4-bit comparators. We also described the basic arithmetic and logic operations that are inherent in a 1-bit full adder. We then showed how a 1-bit ALU can be implemented by integrating a true/complement/zero/one (TC01) unit and full adder. We then used a bit-slice design approach to implement a 4-bit (2s complement arithmetic) ALU by cascading four 1-bit ALUs. We also described the design of a second ALU by combining a 4-bit parallel adder and two TC01 function modules.

In a top-down design approach we use a divide-and-conquer technique to divide the process into subprocesses and divide subprocesses into functions that can be more easily conquered. We then design a device to accomplish each required function. Finally, we integrate the function devices into subsystems and the subsystems into a system that satisfies the overall specifications of the original design problem. In Section 6.3 we used a top-down approach to design a BCD arithmetic unit. After analyzing the requirements of a BCD arithmetic unit to determine the required subsystems and function modules, we described the design of decimal-to-BCD encoders, a BCD adder, a BCD 9s complementer, and BCD decoders. Finally, we integrated these function modules into subsystems and integrated the subsystems into a system (a BCD arithmetic unit).

This chapter concludes the study of digital systems that use only combinational logic circuits. In Part III (Chapters 7, 8, and 9) we will discuss sequential processes and learn how they can be represented by sequential (finite-state) machines and realized by sequential logic circuits.

EXERCISES

General

1. A 2-bit comparator, which compares two 2-bit numbers $A = a_1a_0$ and $B = b_1b_0$, has 3 outputs as follows:

$$E = 1 \text{ if } A = B$$
$$G = 1 \text{ if } A > B$$
$$L = 1 \text{ if } A < B$$

a. Construct a truth table for a 2-bit comparator.
b. Determine canonical SOP expressions for each of the outputs E, G, and L.
c. Use 4-input K-maps to simplify the expressions for E, G, and L.

2. A device converts a BCD input code (A, B, C, D) to an excess-3 output code (W, X, Y, Z). Each excess-3 codeword is obtained by adding 3 to the corresponding BCD codeword.
a. Construct a truth table for a BCD-to-excess-3 code converter.
b. Determine canonical SOP expressions for each of the outputs W, X, Y, and Z.
c. Use 4-input K-maps to simplify the expressions for W, X, Y, and Z.

3. A device converts a BCD input code (A, B, C, D) to a unit distance decimal code (W, X, Y, Z).
a. Construct a truth table for a BCD-to-unit distance decimal code converter.
b. Determine canonical SOP expressions for each of the outputs W, X, Y, and Z.
c. Use 4-input K-maps to simplify the expressions for W, X, Y, and Z. Note that the remaining six combinations (1010 through 1111) are don't-care conditions in the K-map.

4. A device complements a BCD input code (A, B, C, D) and generates a 9s complement output (W, X, Y, Z).
   a. Construct a truth table for a BCD 9s complementer.
   b. Determine canonical SOP expressions for each of the outputs W, X, Y, and Z.
   c. Use 4-input K-maps to simplify the expressions for W, X, Y, and Z.

Design/Implementation

Use the 4-step problem-solving procedure for the following design exercises.

1. Design a BCD-to-excess-3 converter that accepts valid BCD inputs A, B, C, and D and outputs the corresponding excess-3 code (W, X, Y, Z). Each excess-3 codeword is obtained by adding 3 to the corresponding BCD codeword.
2. Use 4-variable K-maps to design a BCD 9s complementer.
3. Use 4-variable K-maps to design a functional module that has the 4-bit number in a BCD decade after a right shift as input and outputs the corrected BCD number for that decade. Let $ABCD$ denote the 4-bit input; let $WXYZ$ denote the 4-bit output.

$$WXYZ = \text{don't cares for } (5, 6, 7, 13, 14, 15) \text{ combinations}$$
$$WXYZ = ABCD, \text{ if } A = 0$$
$$WXYZ = ABCD - 0011, \text{ if } A = 1$$

4. Design a 10-line-to-4-line priority encoder that accepts data from 9 active-low inputs (i_1' through i_9') and provides a binary representation on the 4 active-low outputs (a_0 through a_3). A priority is assigned to each input so that when 2 or more inputs are active at the same time, the input with the highest priority is output, with line i_9' having the highest priority. *Note:* The 74147 is an MSI 10-line-to-4 line priority encoder.
5. Implement the ALU illustrated in Figure 6.7. Verify each of the eight functions of this ALU.
6. Implement the decimal-to-BCD encoder illustrated in Figure 6.13. Verify that the encoder operates in accordance with design specifications.
7. Implement the BCD adder illustrated in Figure 6.17. Verify that the BCD adder operates in accordance with design specifications.
8. Implement the BCD 9s complementer illustrated in Figure 6.18. Verify that the complementer operates in accordance with design specifications.
9. Implement the BCD adder/subtracter illustrated in Figure 6.21. Verify that the adder/subtracter operates in accordance with design specifications.
10. Implement the BCD-to-decimal decoder illustrated in Figure 6.23. Verify that the decoder operates in accordance with design specifications.
11. Implement the BCD arithmetic unit illustrated in Figure 6.24. Verify that the BCD arithmetic unit operates in accordance with design specifications.
12. Design a BCD to unit-distance decimal code converter. (See General Exercise 3.)

PART THREE

Sequential Processes and Machines and Sequential Logic Circuits

7 Latches, Flip-Flops, Counters, and Registers
8 Sequential Processes and Machines and Synchronous Sequential Circuits
9 Asynchronous Sequential Circuits

Problem solving in digital logic involves the analysis of processes that accomplish some function(s), the design of conceptual models that represent the processes, and the implementation of digital circuits and systems that realize the process models.

A sequential process can be represented by a conceptual model called a **sequential (finite-state) machine**, which, in turn, can be realized by a sequential logic circuit. Part III (Chapters 7, 8, and 9) covers sequential processes and machines and sequential logic circuits.

A sequential logic circuit must have a memory capability and at least one feedback path. A sequential logic circuit may be classified as either asynchronous or synchronous, depending on the times at which its memory state is allowed to change. That is,

- Asynchronous sequential-logic circuits are basically combinational logic circuits with direct feedback paths. Memory state changes may occur at any time in response to changes in the circuit's inputs.
- Synchronous sequential-logic circuits use a clock signal to effect memory state changes at specific instants of time. These circuits are encountered more often in practice and are easier to design than asynchronous circuits.

Chapter 7 describes sequential processes and concepts of sequential (finite state) machines and covers 1-bit memory elements (latches and flip-flops). The chapter also describes the design of specialized, synchronous sequential-logic circuits such as counters, registers, and register counters.

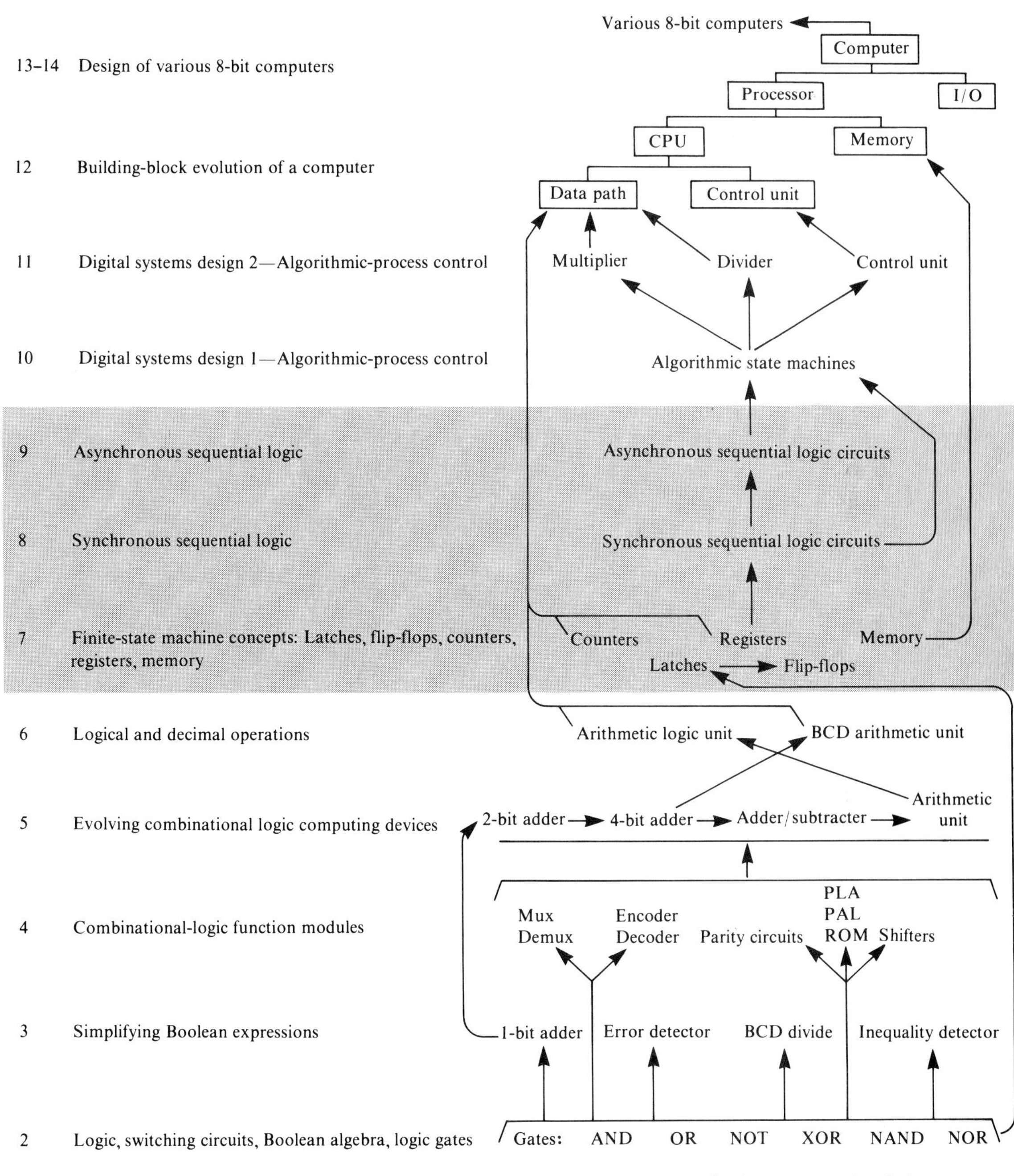

1	Number systems Computer codes	Decimal, binary, octal, hexadecimal BCD, symmetric, reflected, unit distance

Chapter 8 describes procedures for the analysis and design of general synchronous sequential-logic circuits. These procedures are illustrated using a variety of example circuits.

Chapter 9 describes procedures for the analysis and design of asynchronous sequential-logic circuits. These procedures are illustrated using a variety of example circuits.

A hierarchical diagram of sequential-logic circuit classifications is presented in Figure A.

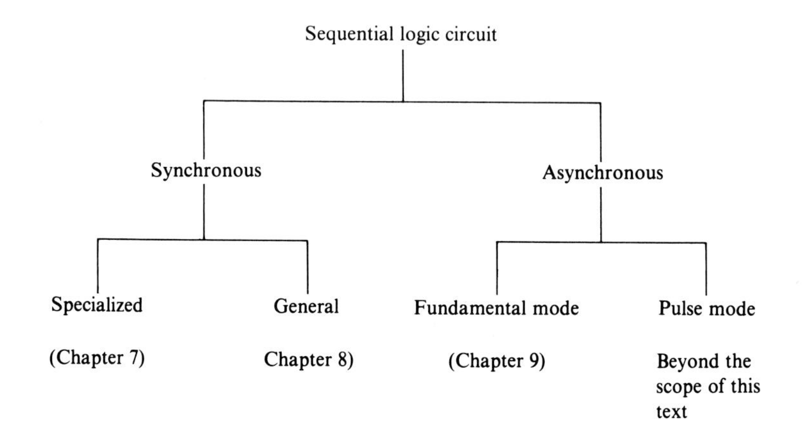

FIGURE A
Diagram of sequential logic circuit classifications

7 Latches, Flip-Flops, Counters, and Registers

OBJECTIVES

After you complete this chapter, you will be able

- ☐ To apply the basic concepts and terminology of sequential processes
- ☐ To understand the basic concepts of sequential (finite-state) machines
- ☐ To describe the characteristics of sequential logic circuits
- ☐ To understand the operation of a basic memory element (*SR* latch)
- ☐ To design level-sensitive memory elements
- ☐ To describe applications for level-sensitive memory elements
- ☐ To understand the operation of edge-sensitive flip-flops
- ☐ To describe applications for edge-sensitive flip-flops
- ☐ To convert flip-flops from one type to another
- ☐ To understand applications of registers (groupings of memory elements)
- ☐ To design various types of counters, shift registers, and shift-register counters

7.1 INTRODUCTION

Fundamental concepts of sequential processes are presented in this chapter, together with their representation by sequential (finite-state) machine models and realization by sequential logic circuits. Counting is an inherently sequential process. The natural counting numbers (the integers* 1,2, . . . , k, k + 1, . . .*) can be represented as a sequence of terms* t(1), t(2), . . . , t(k), t(k + 1), ... *If* t(k) *is the present term of the sequence,* t(k + 1) *is the next term. Starting with* t(1) = 1, *the sequence of integers can be generated using the recursion formula* t(k + 1) = t(k) + 1 (k = 1, infinity). *Thus, we see that counting is a special case of the process of sequence generation.

An odometer is a sequential counting/storage device that counts each mile traveled and registers (records or stores) the cumulative mileage. Starting with mileage 00000, the odometer counts each passing mile, successively storing terms of the integer sequence from 00000 to 99999. The odometer reading at a given time represents the counter's **present state**. With each mile traveled the odometer is incremented and the **next state** becomes the present state. Each odometer position represents a 10-state memory element that stores a digit (units, tens, hundreds, and so on) of a 5-digit decimal number.

An 8-bit binary odometer is a sequential counting/storage device, with a range from 00000000 to 11111111, in which each position represents a 2-state memory element that stores a bit of a multibit number. A binary 1-up counter is the digital logic counterpart of a binary odometer. The counter content at a given time represents the present state of the counter, and the content + 1 represents the next state. When the counter is incremented, the counter's next state becomes the present state.

This chapter also describes the design, operation, and application of a basic memory element and a number of specialized synchronous sequential-logic circuits such as gated latches, flip-flops, counters, shift registers, and shift-register counters. Many of these circuits are commercially available as MSI or LSI integrated circuits.

A 1-bit memory element is a **bistable device**, with stable states 0 and 1, that can store either a binary 0 or 1. The most basic 1-bit memory element is an asynchronous bistable device called an ***SR* latch**. A number of different types of 1-bit memory elements (gated latches and flip-flops) can be constructed using the basic memory cell as a logical building block.

An n-bit register consists of a group of n flip-flops. Since the register can store any n-bit number from 0 to 2^n–1, the register has 2^n distinct states. Shift registers can shift their contents in various ways to accomplish different arithmetic and logical operations. They are used in a variety of applications, such as serial-to-parallel and parallel-to-serial data conversion, time-delay circuits, and so on.

Shift-register counters can be designed to generate repetitions of specific sequences. For example, a 4-bit ring counter is a circular shift register that generates the sequence 1000, 0100, 0010, and 0001 over and over again.

Synchronous sequential-logic circuits are simpler to design than asynchronous circuits because the synchronization of memory state changes eliminates many of the problems caused by unequal propagation delays in different signal paths (critical races, essential hazards, and instability). State changes of a synchronous sequential-logic circuit are synchronized by a clock circuit that generates an alternating sequence of low and high DC voltage levels corresponding to logic 0s and 1s.

Sequential Processes, Machines, and Circuits

Digital logic applications involve a wide variety of processes such as adding, counting, comparing, encoding, decoding, sequence generation and detection, and so on. Some processes are strictly combinational in nature, in that the process output at a specific time depends only on the combination of the inputs. Examples of strictly combinational processes include encoding, decoding, multiplexing, demultiplexing, and code conversion.

Other processes are inherently sequential in nature, in that the output at a specific time depends on the process's present state as well as on the inputs. Examples of inherently sequential processes include counting, sequence generation, and sequence detection.

Still other processes, called **iterative processes**, exhibit both combinational and sequential characteristics; that is, a basic (combinational) operation is repeated a number of times (sequentially). Examples of one-dimensional and two-dimensional iterative processes include the addition and the multiplication of multibit numbers, respectively.

A **sequential process** is characterized by its state. The concept of state can be described using the following analogy. Consider the familiar process of boiling water in a tea kettle. Before we apply heat to the kettle, the water temperature is in an initial state, such as room temperature. When heat (input) is applied to the kettle, the water temperature passes through a sequence of states (lukewarm, hot, very hot) until it reaches the boiling state. If the water temperature is in a particular state, the application of a certain amount of heat at time t will cause a transition from the present state to the next state at some later time $t + dt$.

In Chapters 7, 8, and 9 we will be concerned with sequential processes that, at any given time, exist in one of a finite number of distinct states and (in response to a sequence of inputs) pass through a sequence of states and generate a sequence of outputs. If the next state and output are functions of the present state and input, then the process is a **functional sequential process**. Here's a simple verbal definition.

DEFINITION: A *functional sequential process* starts in an initial state s_0 and, in response to a sequence of inputs i, the process passes through a sequence of states s and generates a sequence of outputs z.

Functional Sequential Processes

We now develop fundamental concepts of sequential processes, define basic terminology, and illustrate tabular and graphic methods for representing such processes.

■ **EXAMPLE 7.1**

Consider a process that sequentially inputs bits of an n-bit word and determines if the word contains an even number of 1s. Before any bits are input, the process initial state is even, denoted by E, since zero 1s have been received and zero is an even number. The total number of 1s input, denoted *1s sum*, is initially zero.

As each bit is input, the 1s sum is incremented by 1 for a 1-input and 0 for a 0-input:

- If the new 1s sum is an even number, the new state is E and output $z = 0$.
- If the new 1s sum is an odd number, the new state is O (odd) and output $z = 1$.

If the final state (after bit n is processed) is E (even), then the n-bit word contains an even number of 1s. If the final state is O (odd), then the n-bit word contains an odd number of 1s. If the n-bit word consists of $n - 1$ data bits and an even parity bit, the output $z = 1$ for a final state O indicates a parity error in the n-bit word. ■

A sequential process can be defined by a **state table** that lists the next state and output for each combination of the present state and the input. In a state table,

- Each row represents a present state of the process.
- Each column represents an input.
- Each table entry pair represents the process next state and output corresponding to the row-column combination of present state and input.

Table 7.1 illustrates a state table of the process described in Example 7.1.

TABLE 7.1
State table for the Example 7.1 process

Present State	Next State, Input 0	Next State, Input 1	Output, Input 0	Output, Input 1
E	E	O	0	1
O	O	E	1	0

The operation of a sequential process can be graphically described by a **state diagram**. A state diagram is a directed graph that allows us to visualize the sequence of states that result from a sequence of inputs. For each possible combination of present state and input, the state diagram shows the next state and output of the process. In a state diagram, such as Figure 7.1 for Example 7.1,

- Each state of the process is represented by a node (circle) that contains the state's label.
- Each transition from state S_i to state S_j caused by input i, with output z, is represented by a directed branch (edge) from node S_i to node S_j, which is labeled with the input/output (I/O).

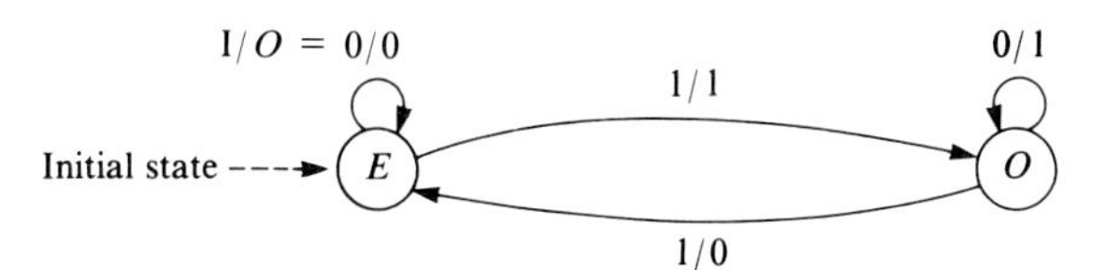

FIGURE 7.1
State diagram of the process in Example 7.1

This state diagram can be interpreted as follows:

1. If the process is in state E (even),
   a. A 0 input causes the process to remain in state E and produces a 0 output.
   b. A 1 input causes a transition to state O (odd) and produces a 1 output.
2. If the process is in state O (odd),
   a. A 0 input causes the process to remain in state O and produces a 1 output.
   b. A 1 input causes a transition to state E (even) and produces a 0 output.

Sequential (Finite-State) Machine Concepts

In mathematics each different class of problem can be represented by a mathematical model that facilitates the solution of any problem belonging to that class. For example, two nonhomogeneous linear equations in two unknowns x and y can be represented by the model

$$ax + by = c$$
$$dx + ey = f$$

The solution of any two specific nonhomogeneous linear equations in x and y, with known constants a, b, c, d, e, and f, can be solved using known methods, such as the elimination method or Cramer's rule.

In many cases, sequential processes can be represented (modeled) by recursion formulas that define a relation between successive terms (present term and next term) of a sequence. For example, consider the geometric sequence 1, 2, 4, 8, 16, . . . , with an initial term $t_1 = 1$, in which adjacent terms differ by a ratio of 2. Given a ratio r and an initial term t_1, the process of generating a geometric sequence can be represented by the following recursion formula:

$$t_{k+1} = rt_k, \quad (k = 1, 2, \ldots)$$

Similarly, a conceptual model exists in digital logic to represent any sequential process that at a given time exists in one of a finite number of distinct states and (in response to a sequence of inputs) passes through a sequence of states and generates a sequence of outputs.

This conceptual model, called a **sequential (finite-state) machine**, facilitates the analysis of sequential processes and the design and implementation of sequential logic circuits that realize these processes.

DEFINITION: A *sequential (finite-state) machine* is a model that has

A finite set (S) of states s

An initial state s_0 that is a member of set S

A finite set (\$) of inputs i, (\$ = alphabet)

A next-state function $d = d(s, i)$ that maps ordered pairs in $S \times \$$ into elements in S

An output function $f = f(s, i)$

If f is a function only of the internal state, the machine is called a **Moore machine**. If f is a function of both the inputs and the internal state the machine is called a **Mealy machine**.

If the machine is in state s and receives input i, then the machine enters into next state $d(s, i)$. If the machine is in state s and receives a string of inputs (word w), then the machine enters into state $d(s, w)$. State $d(s, w)$ is defined to be the state the machine is in after it has read the entire word w.

The operation of a sequential machine can be described by a state table or by a state diagram (see Figure 7.2), just as a sequential process is described. The state table (Figure 7.2a) lists the next state and output for each combination of present state and input.

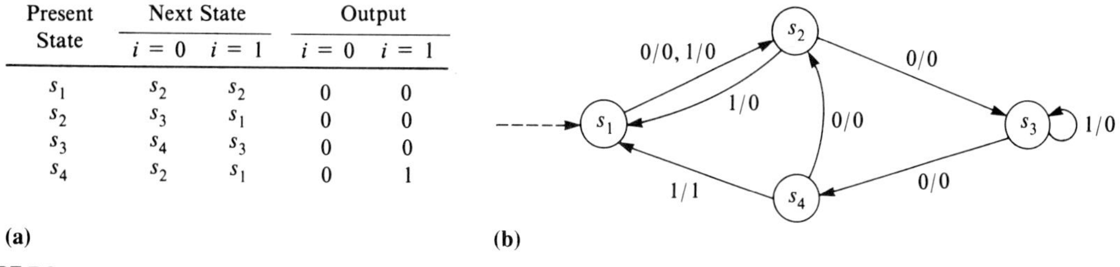

Present State	Next State $i = 0$	Next State $i = 1$	Output $i = 0$	Output $i = 1$
s_1	s_2	s_2	0	0
s_2	s_3	s_1	0	0
s_3	s_4	s_3	0	0
s_4	s_2	s_1	0	1

(a)

(b)

FIGURE 7.2
Simple finite-state machine. (a) State table (b) State diagram

In the state diagram (Figure 7.2b), each state of the sequential machine is represented by a circle (node) containing state s. If state $s' = d(s, i)$, then a directed arc from node s to node s' represents the machine's transition from state s to state s' when input i is received. The directed arc is labeled with the I/O. The initial state of the machine is indicated by a dashed arrow pointing to the node.

The underlying concepts of sequential processes can be illustrated further by the following example.

EXAMPLE 7.2

Consider the process of adding the unsigned binary numbers $X = 0101$ and $Y = 0011$ using the familiar pencil-and-paper method.

For any column	$i =$	3210	
Present-state carry	c	1110	Carry-in
Inputs	x	0101	(initial carry-in = 0)
	y	0011	
Output	$S =$	1000	Sum
Next-state carry	C	0111	Carry-out

For two arbitrary numbers X and Y, a column may contain any one of the eight possible combinations 000 through 111 for c, x, and y. The addition of any column (i) of three 1-bit numbers, c, x, y, produces a sum S and a carry-out C, as determined from the binary addition table in Table 7.2.

TABLE 7.2
Binary addition table

c	x	y	C	S
0	0	0	0	0
0	0	1	0	1
0	1	0	0	1
0	1	1	1	0
1	0	0	0	1
1	0	1	1	0
1	1	0	1	0
1	1	1	1	1

The sum S and carry-out C are Boolean functions of c, x, and y, which can be represented by logic equations in the following simplified form:

$$S = c \oplus x \oplus y$$
$$C = xy + c(x \oplus y)$$

By adding one column of Example 7.2 at a time, starting with the rightmost column, we see that the addition of two multibit numbers X and Y can be accomplished by a sequential process (serial addition) that has

- □ A set of inputs x and y
- □ An internal state (carry) that is one of a finite set of states $\{0, 1\}$, c = present state, and C = next state
- □ A next state C that is a Boolean function of the inputs x and y and the present state c
- □ An output S that is a Boolean function of the inputs x and y and the present state c

The process of serial addition can be described by a state table that lists the next state C and output S for each combination of the present state c and the inputs x and y. The state table and a corresponding state diagram for the process of serial addition are presented in Figure 7.3. For any input sequence, the state diagram can be used to determine the output sequence produced by the process (starting in a given state).

FIGURE 7.3
Serial addition process.
(a) State table (b) State diagram

Present	Next State C				Output S			
	Inputs xy				Inputs xy			
State c	00	01	11	10	00	01	11	10
0	0	0	1	0	0	1	0	1
1	0	1	1	1	1	0	1	0

(a)

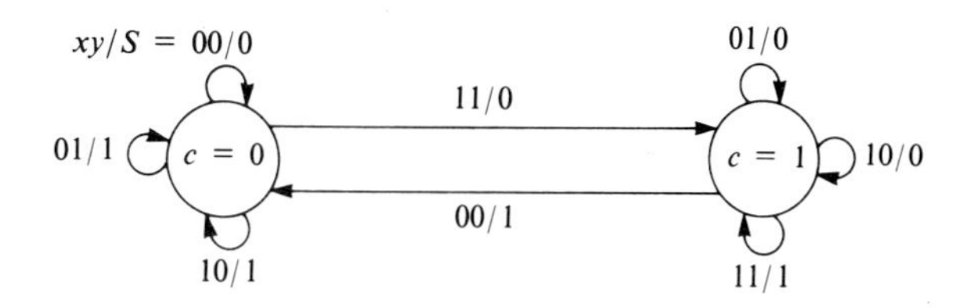

(b)

■

Realization of a Sequential Machine by a Sequential Circuit

A sequential machine can be realized by a sequential logic circuit. As an example, the serial addition process in Example 7.2 can be realized by a sequential logic circuit. The operation of this machine is analogous to the pencil-and-paper method of addition.

The sequential logic adder uses a full adder to add (at time t_i) the column i elements x_i and y_i and present-state carry c_i, producing sum s_i and next-state carry c_{i+1}. When time advances from t_i to t_{i+1}, the next-state carry c_{i+1}, is fed back as present-state carry c_i. The sequential logic adder has

- □ A set of external inputs $x_i = x(t_i)$, $y_i = y(t_i)$
- □ A memory (for carry) that has a finite set of states $\{0,1\}$, c_i = present state, and c_{i+1} = next state

- A next state c_{i+1} that is a Boolean function of the inputs x_i and y_i and the present state c_i
- An output s_i that is a Boolean function of the inputs x_i and y_i and the present state c_i.

The sequential-logic adder circuit is shown in detailed form in Figure 7.4. Since the memory consists of a single 1-bit element that holds the carry, the circuit state is represented by variable c.

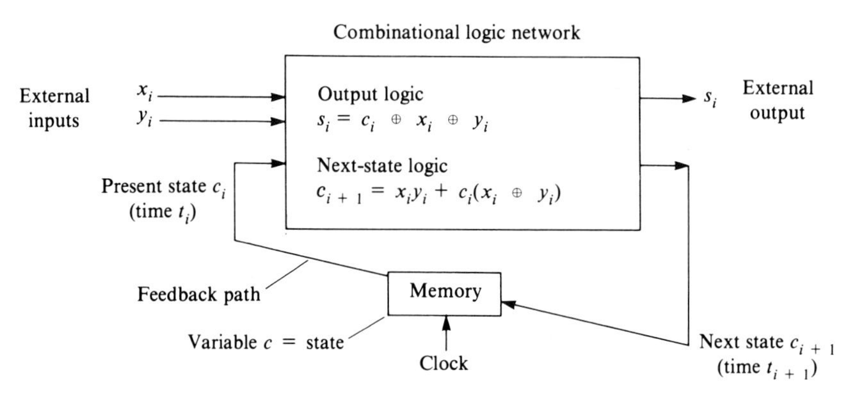

FIGURE 7.4
Sequential-logic adder circuit

Operation of a Basic Memory Element

This section describes the operation of a basic memory element (*SR* latch), which is used as the basis for constructing various types of 1-bit memory elements (gated latches and flip-flops) used as building blocks in counters, registers, memories, and other digital logic devices.

A basic memory element has a set input S, a reset input R, and outputs Q and P, as shown in Figure 7.5. The control inputs S and R of the memory element are allowed to change at any time, subject to the following constraints:

1. The memory element must be in a stable state prior to a change in either S or R.
2. Only one input, S or R, can be changed at a given time.
3. S and R cannot be asserted simultaneously.

FIGURE 7.5
Block diagram of a basic memory element

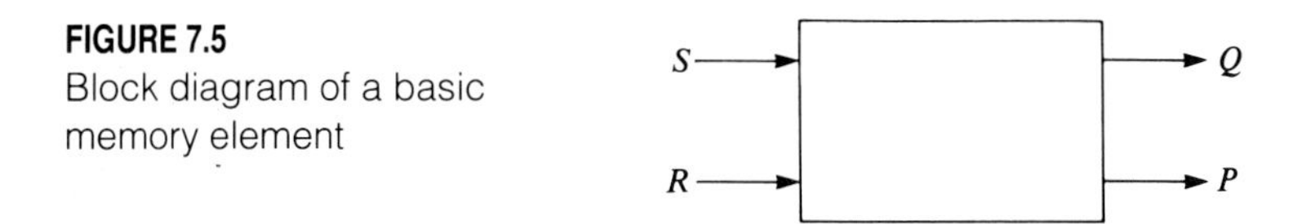

The basic memory element has the following properties:

1. The memory element outputs Q and P are complementary, that is, $P = Q'$.
2. The assertion of S will set the memory element (steady state $Q = 1$), and the output Q will remain at 1 when S is de-asserted.
3. The assertion of R will reset the memory element (steady state $Q = 0$) and the output Q will remain at 0 when R is de-asserted.

4. As long as neither S nor R is asserted ($S, R = 0, 0$), the memory element state remains unchanged.

The basic memory element is a special asynchronous sequential-logic device, the *SR* latch, that can store a logic 0 or 1. The *SR* latch uses feedback and the propagation delay of its gates to provide its memory capability. A basic memory element can be constructed by cross connecting 2 NOR gates or 2 NAND gates. A cross-connected NOR implementation is shown in Figure 7.6. The output of each NOR gate is fed back as an input to the other NOR gate.

FIGURE 7.6
Basic memory element implemented by NOR gates

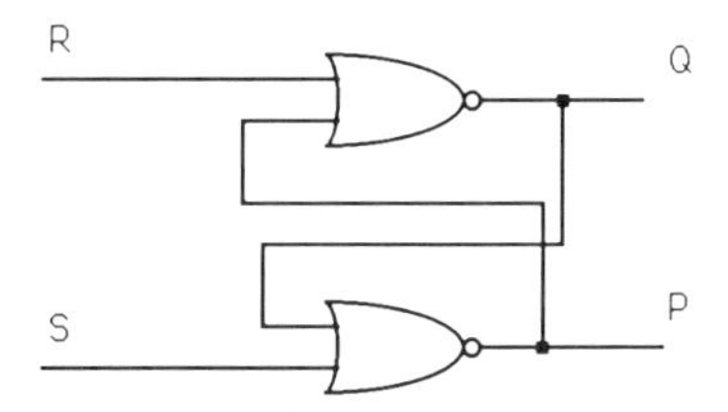

The NOR gate inversion properties, shown in Table 7.3, together with feedback and the gate propagation delays, are the basis of the memory capability of the cross-coupled NOR form of the *SR* latch. The inversion properties are

1. A NOR gate output is 0 if either input is 1.
2. A NOR gate output is 1 if both inputs are 0.

TABLE 7.3
NOR gate truth table

x	y	NOR	
0	0	1	2
0	1	0	
1	0	0	} 1
1	1	0	

The operation of the cross-coupled NOR-gate memory element can be described using properties 1 and 2 as follows:

- *Set operation* (S, R changes from 0, 0 to 1, 0). The assertion of S (= 1) drives the lower NOR output P to 0 (by property 1). Feedback P (= 0) drives an input of the upper NOR gate, and, since $R = 0$, Q goes to 1 (by property 2, since the upper NOR has inputs 0, 0).
- *Hold set operation* (S, R changes from 1, 0 back to 0, 0). When S is de-asserted, output $Q = 1$ holds P at 0 (by property 1). The 0, 0 inputs on the upper NOR hold Q at 1 (by property 2), and the cell remains set.
- *Reset operation* (S, R changes from 0, 0 to 0, 1). The assertion of R (= 1) drives the upper NOR output Q to 0 (by property 1). Feedback Q (= 0) drives an input of the lower NOR gate, and, since $S = 0$, P goes to 1 (by property 2, since the lower NOR has inputs 0, 0).
- *Hold reset operation* (S, R changes from 0, 1 back to 0, 0). When R is de-asserted, output $P = 1$ holds Q at 0 (by property 1). The 0, 0 inputs on the lower NOR hold P at 1 (by property 2), and the cell remains reset.

The asynchronous sequential-logic circuit shown in Figure 7.6 can be represented by a lumped-delay model by replacing the physical gates with idealized zero-delay gates, so that the total (lumped) time delay dt occurs in series in the feedback loop, as shown in Figure 7.7. The delay element represents the circuit's memory, since $Q(t) = q(t + dt)$.

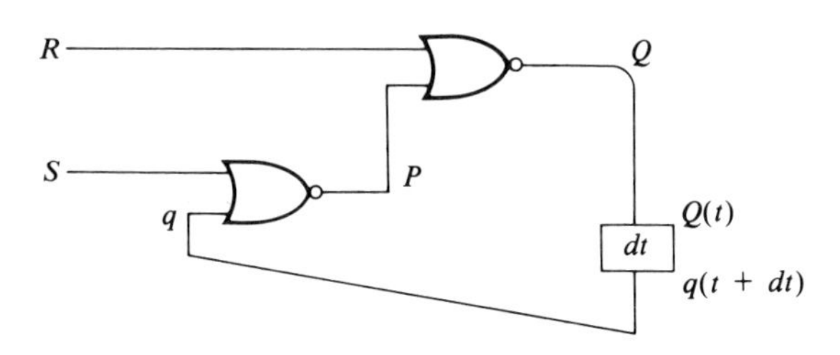

FIGURE 7.7
Lumped-delay model of a NOR gate memory cell

The operation of the basic memory element is defined by the excitation table shown in Table 7.4. If the circuit is in a stable state (where $q = Q$), a change in either S or R at time t produces an excitation Q, which is either equal to q or not equal to q. That is,

- □ If $Q = q$, the circuit is in a stable state.
- □ If $Q \neq q$, the circuit is in transition from one stable state to another stable state.

Thus, if an S, R, q combination at time t causes an excitation Q equal to the present state q, the circuit at time $t + dt$ is in a stable state $q(t + dt)$. Stable states ($q = Q$) are indicated by an * in the Q column of Table 7.4.

TABLE 7.4
Excitation table for the lumped-delay model

Function	S	R	q	P	Excitation Q	= Next State $q(t + dt)$
Hold	0	0	0	1	0 *	0
	0	0	1	0	1 *	1
Reset	0	1	0	1	0 *	0
	0	1	1	0	0	0
Set	1	0	0	0	1	1
	1	0	1	0	1 *	1
Not allowed	1	1	0	—	—	—
	1	1	1	—	—	—

The input combination $S, R = 1, 1$ is not allowed because the simultaneous assertion of S and R violates the requirement that steady state $P = Q'$.

The excitation table can be plotted in the form of an excitation map, as shown in Figure 7.8, where q represents the memory element present state, before an input change is applied, and Q represents the memory element next state that results from an input change. Stable states are circled in the excitation map.

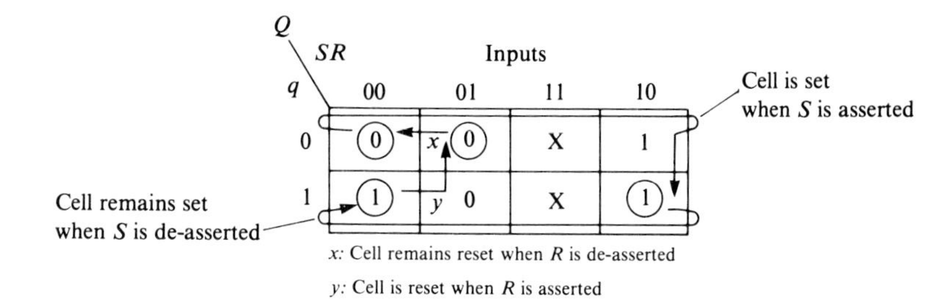

FIGURE 7.8
Excitation map with internal states 0, 1

The inputs S and R are called **primary variables**, and the feedback variable (q, Q) is referred to as a **secondary variable**. The total state (S, R, q) of the circuit consists of the combined values of the primary variables S and R and the feedback present state q. Each row of the excitation table with $Q = q$ represents an individual (primitive) stable total state of the circuit. It can be seen from Figure 7.8 that the memory element has four stable total states:

State a. Input $S, R = 0, 0$ holds element in state 0.

State b. Input $S, R = 0, 1$ resets element in state 0 to state 0.

State c. Input $S, R = 1, 0$ sets element in state 1 to state 1.

State d. Input $S, R = 0, 0$ holds element in state 1.

The excitation map can be plotted using the *primitive* stable total states a–d, as shown in Figure 7.9.

The operation of the basic memory element (*SR* latch) can be graphically represented by a primitive state diagram, as shown in Figure 7.10. Each stable total state (where $q = Q$) is a node of the state diagram. Each node is labeled inside with the state name. The

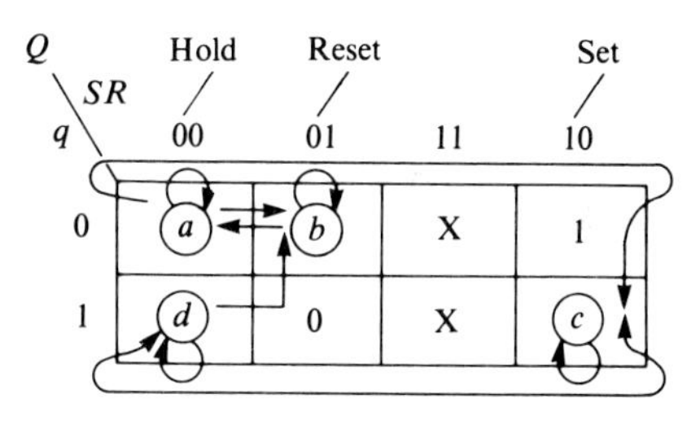

FIGURE 7.9
Excitation map with stable total states a–d

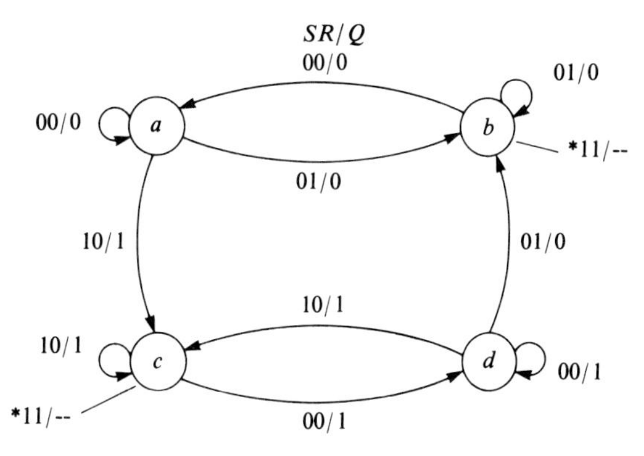

FIGURE 7.10
Primitive state diagram of an *SR* latch

directional edges connecting nodes are labeled by *SR/Q*, where the *SR* inputs cause the circuit to make a transition to next state *Q*.

The functions of an SR latch are summarized in Table 7.5.

TABLE 7.5
Function table of an *SR* latch

Function	*S*	*R*	*Q*	Description
Hold	0	0	*q*	Next state = present state
Reset	0	1	0	Next state = 0
Set	1	0	1	Next state = 1
Not allowed	1	1	X	Next state indeterminate

The excitation map in Figure 7.8 can be used to determine a simplified SOP expression for next state Q in terms of present state q and the control inputs S and R. The result is the characteristic equation

$$Q = S + R'q \tag{7.1}$$

Using the K-map zeros and Xs to determine $Q' = R + S'q'$, and complementing Q', a POS expression is obtained for Q. The result is the characteristic equation

$$Q = R'(S + q) \tag{7.2}$$

Using mixed-logic methodology, the latch can be implemented in one of the forms given in Table 7.6, depending on the polarities assigned to S, R, and Q.

TABLE 7.6
Alternate forms for implementing an *SR* latch

Polarities Assigned			
S, R	*Q*	Latch Form	Characteristic Equation
Low	Low	NAND	$Q = R'(S + q)$
Low	High	NAND	$Q = S + R'q$
High	Low	NOR	$Q = S + R'q$
High	High	NOR	$Q = R'(S + q)$

■ **EXAMPLE 7.3**

If the control inputs S, R are assigned polarity L and the next state Q is assigned polarity H, the latch takes the form of two cross-coupled NAND gates, as shown in Figure 7.11.

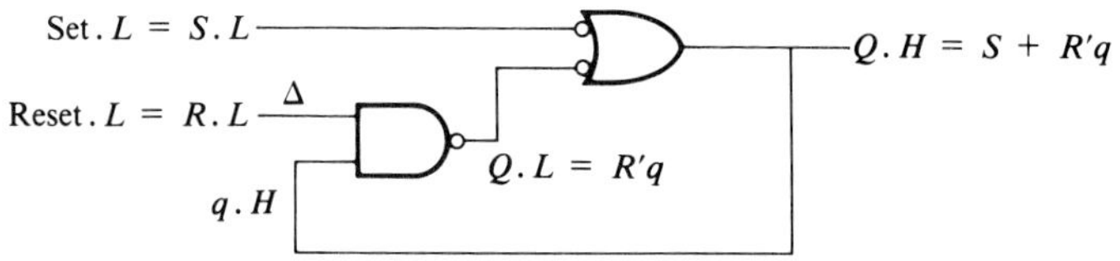

FIGURE 7.11
NAND form of the *SR* latch ■

Let q denote the present state of a 1-bit memory at some time t_n. A new input combination applied at time t_n produces the memory's next state Q at time t_{n+1}. The present-state-to-next-state transitions ($q \to Q$) are defined in Table 7.7. Note that a $0 \to 1$ transition is an alpha transition and a $1 \to 0$ transition is a beta transition. These state transition definitions are useful in analyzing and designing sequential logic circuits.

TABLE 7.7
Definition of state transitions

$q \to Q$	Transition type
$0 \to 0$	0
$0 \to 1$	alpha
$1 \to 0$	beta
$1 \to 1$	1

A **characteristic table** (Table 7.8) lists the next state Q for each input combination S, R, and q applied at time t.

TABLE 7.8
SR latch characteristic table

Function	S	R	$q \to Q$	Transition Type
Hold	0	0	$0 \to 0$	0
	0	0	$1 \to 1$	1
Reset	0	1	$0 \to 0$	0
	0	1	$1 \to 0$	beta
Set	1	0	$0 \to 1$	alpha
	1	0	$1 \to 1$	1
Not allowed	1	1	0 X	—
	1	1	1 X	—

An excitation table can be derived from the characteristic table by collecting like transitions. The resulting **excitation table** (Table 7.9) describes the inputs S and R that cause the corresponding state transition.

TABLE 7.9
SR latch excitation table

$q \to Q$	Transition	S	R
$0 \to 0$	0	0	X
$0 \to 1$	alpha	1	0
$1 \to 0$	beta	0	1
$1 \to 1$	1	X	0

The *SR* latch excitation table is a useful tool in designing gated latches and clocked flip-flops that use the *SR* latch as a basic building block. This table is also used in converting flip-flops from one type to another.

Memory Elements Derived from an *SR* Latch

A synchronous sequential-logic circuit is one whose state is allowed to change only at specific times. These times are usually determined by a signal referred to as a (system) **clock**. A clock is a periodic signal, in which each period is divided into a high level (ON) and a low level (OFF). The time durations of the high and low levels need not be equal. Clock signals are usually generated by a device called an **astable multivibrator**. A representative clock signal is shown in Figure 7.12.

FIGURE 7.12
Illustration of a clock signal

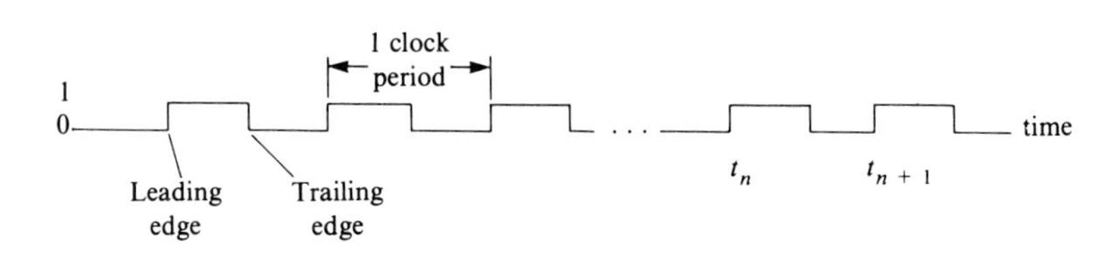

A transition between levels is an edge. Normally, the signal repeats itself regularly, and the time between like edges (leading or trailing) is the clock period. One of the edges is distinguished by calling it the **initial transition**. The initial transition starts a cycle and ends the previous cycle. The clock period is also called the **cycle time**.

The memory of a synchronous sequential-logic circuit consists of clocked, bistable memory elements, each having the capability of storing either a 0 or 1. The *SR* latch is a bistable device, which can be used as a building block to construct a variety of gated latches and clocked flip-flops, as shown in Figure 7.13. The numbers in parentheses are IC versions of these latches and flip-flops.

FIGURE 7.13
Memory elements derived from an *SR* latch

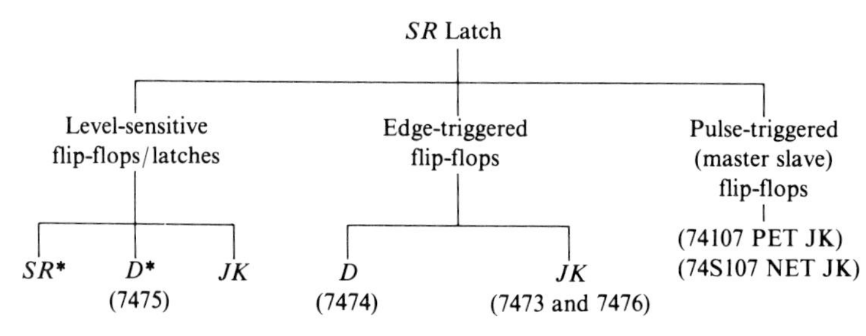

A **flip-flop** is a 1-bit memory element that is used as a basic cell in counters, registers, and other synchronous sequential-logic circuits. Flip-flops are categorized primarily by

1. How they are clocked: level-sensitive, pulse-width sensitive (master-slave), or edge-triggered.
2. How they are controlled: data type (*D*), set-reset (*SR*), or *JK* (usually edge-triggered or master-slave only).

Data type flip-flops are used primarily as elements of data registers. Set-reset and *JK* flip-flops are more suited for control logic implementations or remembering status or conditions.

7.2 GATED LATCHES AND LEVEL-SENSITIVE FLIP-FLOPS

A* gated latch *is a bistable device that has an enable (gate) input* G *in addition to the control input(s). When* G = 0 *(not asserted), the gate is disabled and the output* Q *is* latched *(is insensitive to changes in a control input), so* Q *remains as it was at the last instant* G *was asserted. When* G = 1*, the gate is enabled and the next state* Q *is determined by the combination of its present state and its control input(s).

If a clock signal is applied to the enable input of a gated latch, the gated latch operates as a clocked level-sensitive flip-flop.

A Gated *SR* Latch

The *SR* latch operates asynchronously. That is, its state may change at any time in response to a change in one of its control inputs, *S* and *R*. In this section, we will describe the design of a memory element with control inputs that can hold, set, or clear the memory state when the device is enabled, and retain its state when the device is disabled. This device is called a **gated *SR* latch**.

The gated *SR* latch is a useful modification of an *SR* latch that has an enable (gate) input *G* in addition to the control inputs *S* and *R*. When $G = 0$, the gate is disabled and the output *Q* is latched (is insensitive to changes in *S* or *R*), so *Q* remains as it was at the last instant *G* was asserted. When $G = 1$, the gate is enabled and the next state *Q* is defined for each *S*, *R* combination by the function table in Table 7.10.

TABLE 7.10
Function table of a gated *SR* latch

Function	*G*	*S*	*R*	*Q*	Description
Hold	1	0	0	*q*	Next state = present state
Reset	1	0	1	0	Next state = 0
Set	1	1	0	1	Next state = 1
Not Allowed	1	1	1	X	—
Disabled	0	X	X	*q*	Next state = present state

An *SR* latch can be converted to a gated *SR* latch by ANDing each of the external controls *S* and *R* with the enable (gate) input *G* to provide the inputs set = $G \cdot S$, and reset = $G \cdot R$ to the basic memory cell. A gated *SR* latch, represented by the symbol shown in Figure 7.14a (p. 276), can be implemented using a basic memory cell (*SR* latch) as a building block, as shown in Figure 7.14b.

FIGURE 7.14
Gated *SR* latch. (a) Symbol (b) Logic diagram

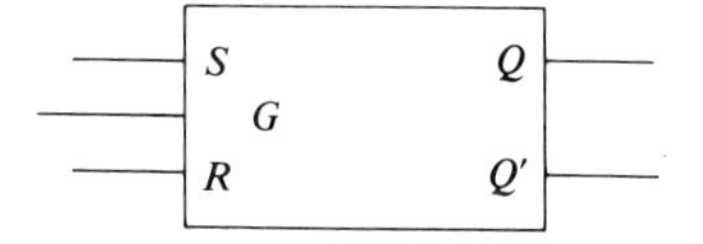

(a)

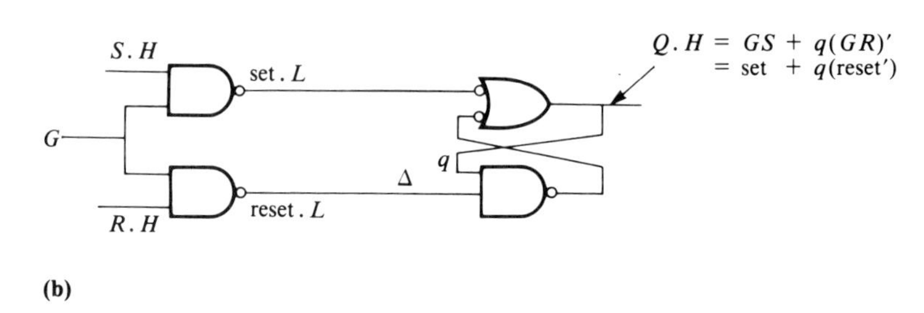

(b)

The requirement that the basic memory cell set and reset not be simultaneously asserted is satisfied by disallowing the simultaneous assertion of S and R.

When the gated *SR* latch is enabled it holds, sets, or clears its output depending on the combination of the S, R control lines. In the next section, we will describe a gated memory element that latches the data on its control line.

Design of a Gated *D* Latch

This section describes the design of a memory device, called a **gated *D* (data) latch** that captures incoming data when the device is enabled and retains its state when the device is disabled. A gated D latch (also called a *D-latch flip-flop*) is widely used in applications that require data storage in conjunction with an enabling signal. We will use the 4-step problem-solving procedure to design the gated D latch.

Step 1: Problem Statement. Use a basic memory cell as a building block to design and implement a memory element that has an enable (gate) input G and a data input D. This device is to operate as follows:

- When $G = 1$, the gate is enabled and next state Q = present input D.
- When $G = 0$, the gate is disabled and next state Q = present state q.

The operation of the gated D latch is described by the function table in Table 7.11.

TABLE 7.11
Function table of a gated *D* latch

Function	G	D	Q	Description
Data enabled	1	0	0	Next state = present input
Data enabled	1	1	1	Next state = present input
Data disabled	0	X	q	Next state = present state

Step 2: Conceptualization. The gated D latch next state Q is defined for each input combination G, D, and q in a characteristic table, which is derived from the function table and description given in Table 7.11. The characteristic table is given in Table 7.12.

TABLE 7.12
Characteristic table of a gated D latch

	G	D	q	Q	Transition
Gate disabled $G = 0 \Rightarrow Q = q$	0	0	0	0	0
	0	0	1	1	1
	0	1	0	0	0
	0	1	1	1	1
Gate enabled $G = 1 \Rightarrow Q = D$	1	0	0	0	0
	1	0	1	0	beta
	1	1	0	1	alpha
	1	1	1	1	1

A gated D latch can be designed using the block diagram shown in Figure 7.15.

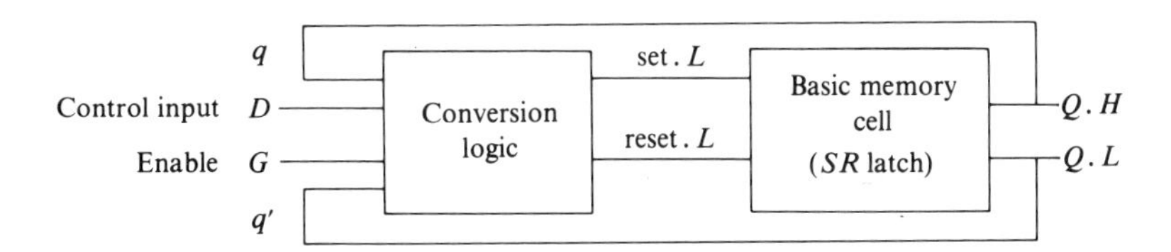

FIGURE 7.15
Gated D latch derived from an SR latch

Step 3: Solution/Simplification. The solution of the design problem requires the determination of the conversion logic outputs, set and reset, as functions of G, D, and q that satisfy the characteristic table. For each G, D, q combination, the corresponding set and reset values can be determined by using the D characteristic/transition table (Table 7.12) in conjunction with the excitation table of the SR latch (Table 7.9).

Each G, D, Q input combination to the conversion logic causes a $q \rightarrow Q$ transition in the D latch. These $q \rightarrow Q$ transitions of the gated D latch can be plotted in a transition map (Figure 7.16a, p. 278). For each cell GDq of the transition map, the SR latch excitation table (Table 7.9) is used to determine the set and reset values that cause the transition required for that cell; these set and reset values are stored in cell GDq in a set map and reset map, respectively. The basic memory cell controls, set and reset, are asserted or not asserted in accordance with the K-maps shown in Figure 7.16b.

The functions set and reset, as determined from the K-maps, are

$$\text{Set} = GD \quad \text{and} \quad \text{Reset} = GD'$$

Since set · reset = 0, the gated D latch satisfies the requirement that the basic memory cell inputs, set and reset, are not asserted simultaneously.

Step 4: Realization. A gated D latch can be realized by substituting

$$\text{Set} = GD \qquad \text{Reset} = GD' \qquad Q.H = \text{set} + q(\text{reset}')$$

into the block diagram of Figure 7.15. There are no feedback lines back to the device inputs since the functions for set and reset do not involve the variable q. A logic diagram of a gated D latch is shown in Figure 7.17 (p. 278), together with the gated D latch symbol.

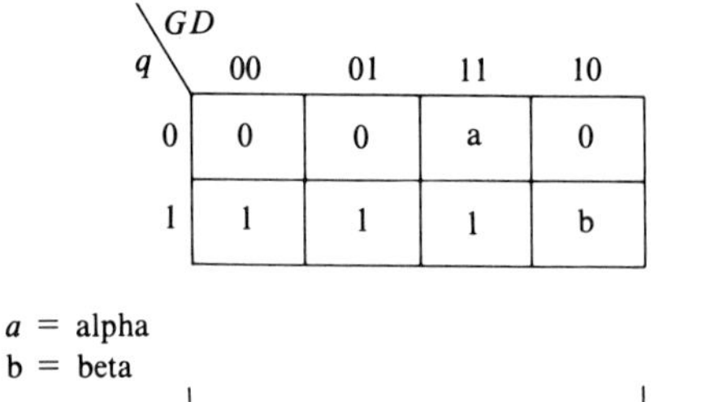

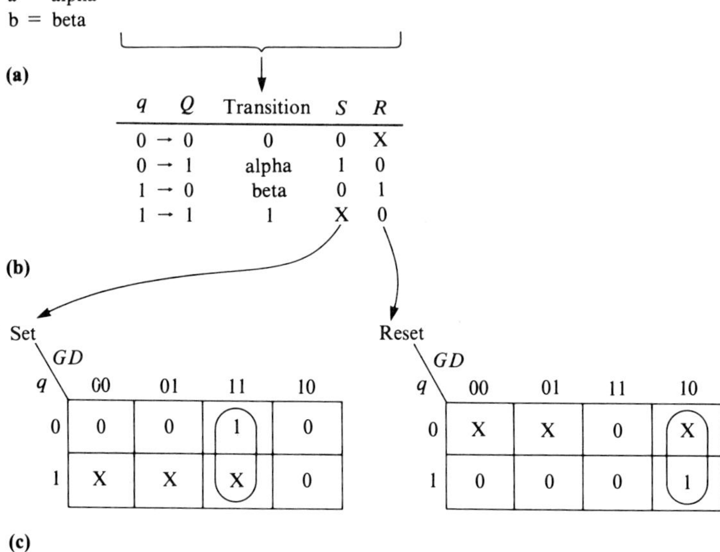

q	Q	Transition	S	R
0 →	0	0	0	X
0 →	1	alpha	1	0
1 →	0	beta	0	1
1 →	1	1	X	0

FIGURE 7.16
Gated *D* latch. (a) Transition map (b) Excitation table (c) K-maps for set and reset functions

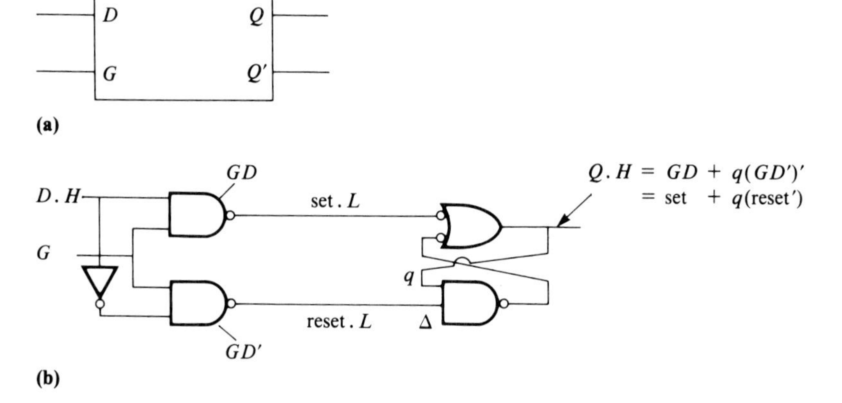

FIGURE 7.17
Gated *D* latch. (a) Symbol (b) Logic diagram

A next-state map of the gated D latch is presented in Figure 7.18. When $G = 1$ (asserted), the gate is enabled, and the gated D latch output Q follows the input D. Since the output "sees" the input when $G = 1$, the gated D latch is referred to as a **transparent device**.

The data must be stable on the latch input D some time prior to asserting the enable G. This interval, called the **setup time** t_s, is necessary to ensure that the correct data value is recognized by the latch when G is asserted. After G has been asserted, the data must be held constant for a period of time, called the **hold time** t_h, to ensure continued recognition of the data as it propagates through the latch. The setup and hold times are illustrated in Figure 7.19b. The integrated circuit 7475 is a commercial version of a gated D latch (see Figure 7.19).

FIGURE 7.18
Next-state map of a gated D latch

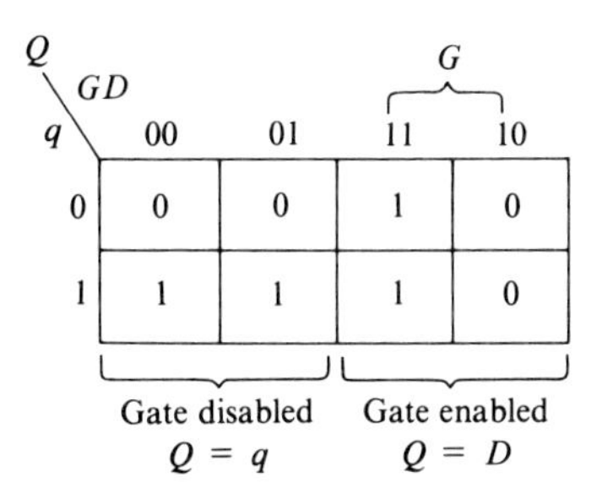

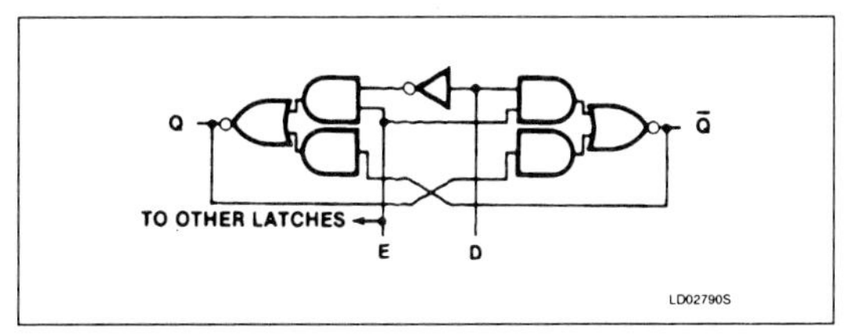

OPERATING MODE	INPUTS		OUTPUT	
	E	D	Q	$\bar{Q}$
Data enabled	H H	L H	L H	H L
Data latched	L	X	q	$\bar{q}$

H = HIGH voltage level
L = LOW voltage level
X = Don't care
q = Lower case letters indicate the state of referenced output one set-up time prior to the HIGH-to-LOW Enable transition.

(a)

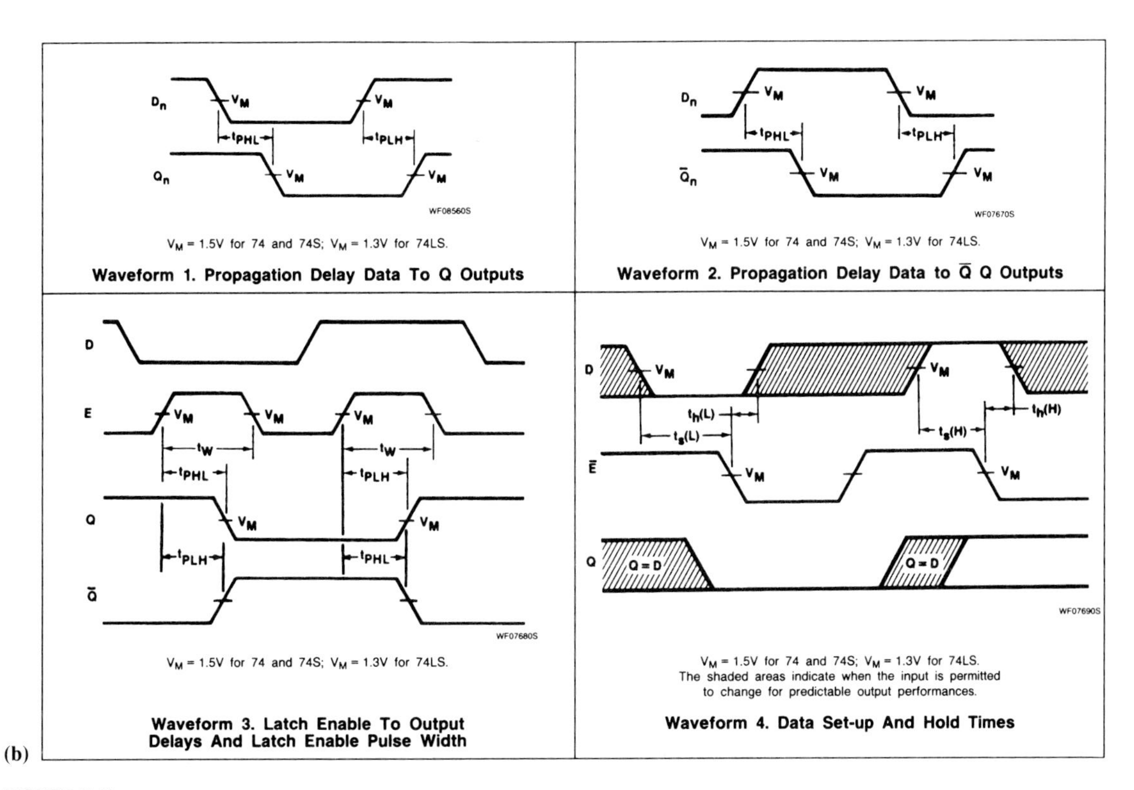

(b)

FIGURE 7.19
A 7475 D latch. (a) Logic diagram/function table (b) AC waveforms (Reprinted by courtesy of Philips Components/Signetics)

Design of a Clocked, Level-Sensitive, *JK* Flip-Flop

A limitation of the level-sensitive *SR* flip-flop is that the control inputs S and R cannot be simultaneously asserted. The *JK* flip-flop was developed to eliminate this restriction on the control inputs. In this section, we will design a clocked, level-sensitive, *JK* flip-flop that can hold, set, clear, or toggle its memory state when a clock input is asserted and retain its state when the clock input is not asserted.

Step 1: Problem Statement. Use a basic memory cell as a building block to design and implement a clocked *JK* flip-flop. The clock signal is used in conjunction with the J and K inputs to control the operation of the flip-flop. When the clock is asserted ($C = 1$), inputs J and K are enabled, and the *JK* combination determines the next state Q; when the clock is not asserted ($C = 0$), the next state Q equals the present state q. The clocked *JK* flip-flop is to operate in accordance with the function table in Table 7.13.

TABLE 7.13
Function table of a clocked *JK* flip-flop

Function	C	J	K	Q	Description
Hold	1	0	0	q	Next state = present state
Reset	1	0	1	0	Next state = 0
Set	1	1	0	1	Next state = 1
Toggle	1	1	1	q'	Next state = (present state)'
Disabled	0	X	X	q	Next state = present state

Step 2: Conceptualization. The next state Q is defined for each input combination C, J, K, q by the characteristic table in Table 7.14, which is derived from the function table and description contained in Table 7.13.

TABLE 7.14
Characteristic table of a *JK* flip-flop

Function		C	J	K	q	Q	Transition
Clock not asserted		0	0	0	0	0	0
		0	0	0	1	1	1
		0	0	1	0	0	0
		0	0	1	1	1	1
		0	1	0	0	0	0
		0	1	0	1	1	1
		0	1	1	0	0	0
		0	1	1	1	1	1
Clock asserted	Hold	1	0	0	0	0	0
		1	0	0	1	1	1
	Reset	1	0	1	0	0	0
		1	0	1	1	0	beta
	Set	1	1	0	0	1	alpha
		1	1	0	1	1	1
	Toggle	1	1	1	0	1	alpha
		1	1	1	1	0	beta

The *JK* flip-flop can be designed using the block diagram shown in Figure 7.20.

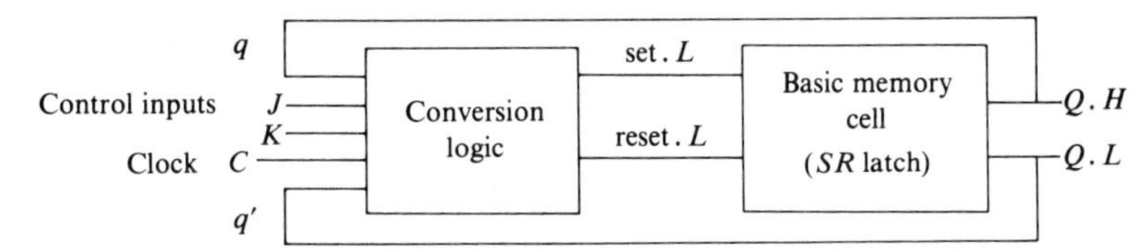

FIGURE 7.20
Clocked *JK* flip-flop derived from an *SR* latch

Step 3: Solution/Simplification. The solution of the design problem requires the determination of the conversion logic outputs, set and reset, as functions of C, J, K, and q that satisfy the characteristic table. For each C, J, K, q combination, the corresponding set and reset values can be determined by using the *JK* flip-flop characteristic/transition table in conjunction with the excitation table (Table 7.15) of the *SR* latch.

TABLE 7.15
Excitation table for an *SR* latch

q Q	Transition	Set	Reset
$0 \rightarrow 0$	0	0	X
$0 \rightarrow 1$	alpha	1	0
$1 \rightarrow 0$	beta	0	1
$1 \rightarrow 1$	1	X	0

Each C, J, K, q input combination to the conversion logic causes a $q \rightarrow Q$ transition in the *JK* flip-flop. These $q \rightarrow Q$ transitions of the *JK* flip-flop can be plotted in a *JK* flip-flop transition map (Figure 7.21).

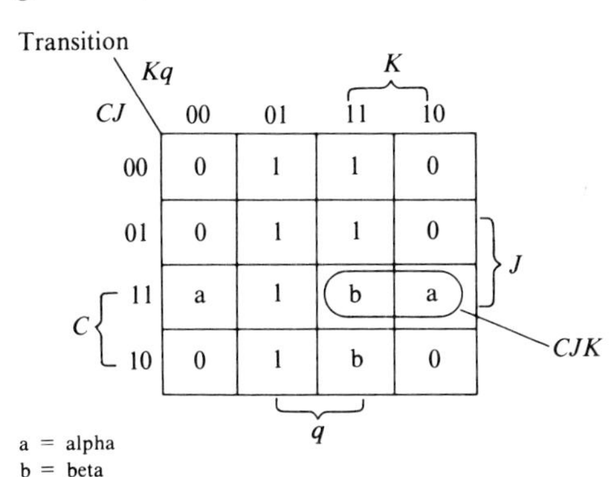

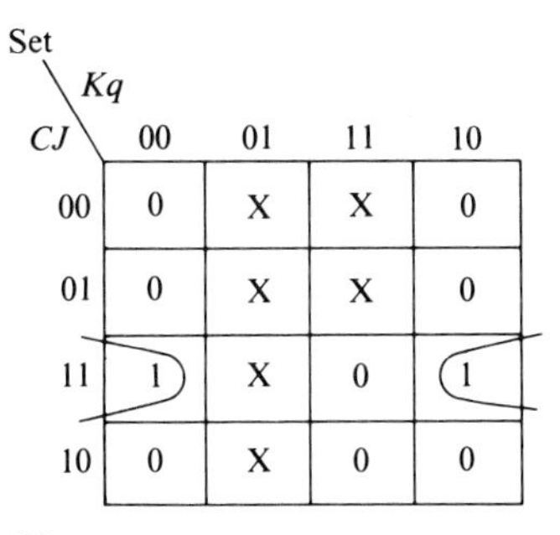

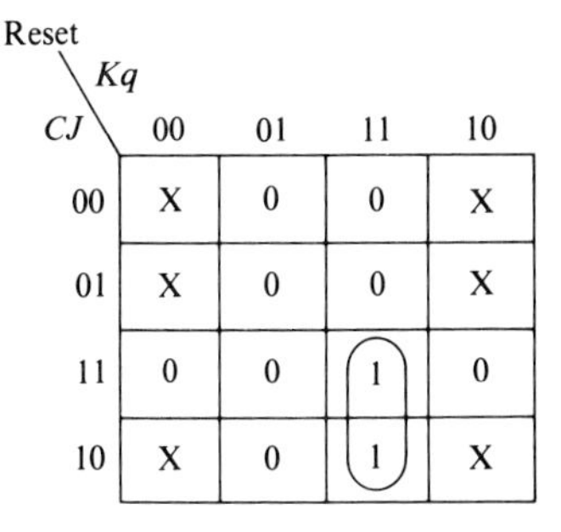

(b)

FIGURE 7.21
JK flip-flop. (a) Transition map (b) K-maps for set and reset functions

For each cell $CJKq$ of the transition map, the SR latch excitation table is used to determine the set and reset values that cause the transition required for that cell; these set and reset values are stored in cell $CJKq$ in a set map and reset map, respectively. The SR latch controls, set and reset, are asserted or not asserted in accordance with the K-maps shown in Figure 7.21b.

The functions set and reset, as determined from the K-maps, are

$$\text{Set} = CJq' \quad \text{and} \quad \text{Reset} = CKq$$

Step 4: Realization. A clocked JK flip-flop can be realized by substituting

$$\text{Set} = CJq' \qquad \text{Reset} = CKq \qquad Q.H = \text{set} + q(\text{reset}')$$

into the block diagram of Figure 7.20. A logic diagram of a clocked JK flip-flop is shown in Figure 7.22.

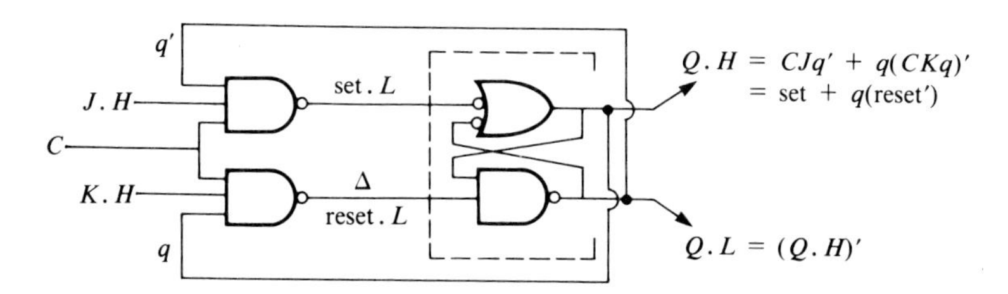

FIGURE 7.22
Logic diagram of a clocked *JK* flip-flop

Since the set and reset functions involve feedback variable q, the alpha and beta transitions must be checked to determine possible oscillations in the JK flip-flop. There are four cases in which $q \neq Q$, indicated by either an alpha or beta in the JK flip-flop transition map (Figure 7.21). Table 7.16 presents an analysis of these cases.

TABLE 7.16
$q \rightarrow Q$ transitions in a JK flip-flop

Case	C J K q	Q	Result	Function
1	1 1 0 0	1		Set: $Q = 1$
	1 1 0 1	1	Stabilizes	Stay set
2	1 0 1 1	0		Clear: $Q = 0$
	1 0 1 0	0	Stabilizes	Stay cleared
3	1 1 1 0	1		Toggle
	1 1 1 1	0		Toggle
	1 1 1 0	1		Toggle
	. . .		May oscillate	
4	1 1 1 1	0		Toggle
	1 1 1 0	1		Toggle
	1 1 1 1	0		Toggle
	. . .		May oscillate	

A disadvantage of a level-sensitive *JK* flip-flop is that oscillation may occur when $CJK = 1$ (toggle) either if the clock pulse is too long or if the propagation delays in the conversion logic gates are greater than the propagation delays in the basic memory cell gates. Another disadvantage is that the device is sensitive to noise on the control inputs *J* and *K* while the clock is asserted.

These disadvantages led to the development of edge-sensitive flip-flops that are sensitive to the levels of the control inputs only during a $0 \rightarrow 1$ (or $1 \rightarrow 0$) transition of the clock signal, referred to as the **state-changing clock transition.**

7.3 PULSE-TRIGGERED AND EDGE-TRIGGERED FLIP-FLOPS

Level-sensitive flip-flops are of limited use in practical application because they may oscillate if the clock pulse is too long and their outputs are sensitive to noise on their control lines while the clock is asserted. In the following sections we describe two types of edge-sensitive flip-flops (pulse-triggered and edge-triggered) that were developed to eliminate the disadvantages of level-sensitive flip-flops.

Pulse-Triggered (Master-Slave) Flip-Flops

The pulse-triggered (master-slave) flip-flop was an early attempt to design a bistable device that was insensitive to glitches in the control inputs while the clock is asserted. A master-slave *SR* flip-flop can be implemented by connecting 2 gated-*SR* latches so that the master latch is enabled while the clock is high and the slave latch is enabled while the clock is low. Thus, the output of the master latch becomes the output of the slave latch when the clock goes low. Because of the propagation delay of the inverter, the output of the master is transferred to the slave latch 1 gate-delay time after the trailing edge of the clock pulse. The *S*, *R* inputs must be stable prior to the clock's leading edge, but the output is postponed until the trailing edge of the clock. Consequently, the master-slave flip-flop is an edge-sensitive device. A block diagram of a master-slave *SR* flip-flop is presented in Figure 7.23a, and a timing diagram is presented in Figure 7.23b.

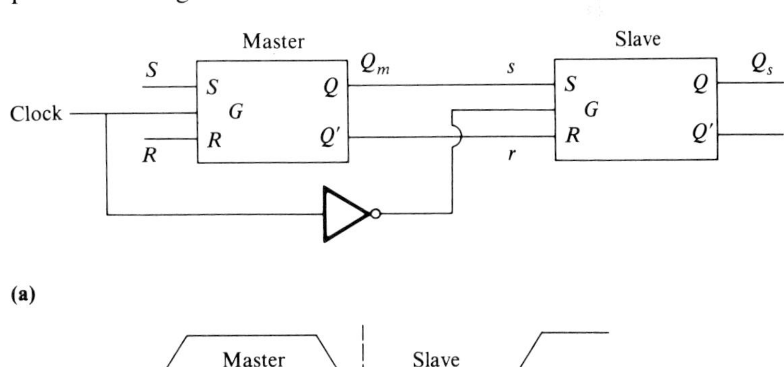

FIGURE 7.23
Master-slave *SR* flip-flop.
(a) Block diagram (b) Timing diagram

When the clock is high, the master latch is enabled; the output Q_m is set, reset, or held if SR = 10, 01, or 00; and the slave latch is disabled. When the clock is low, the slave latch is enabled; the output Q_s is set, reset, or held if sr = 10, 01, or 00; and the master latch is disabled so that any change in SR while the clock is low does not affect the output of the composite device. However, the control inputs S and R must be held constant while the clock is high. If a glitch occurs in S or R while the clock is high, the master latch will produce an improper output Q_m that is transferred to the slave latch when the clock goes low. Because of this "glitch catching" the master-slave flip-flop has been replaced in most applications by the edge-triggered flip-flop.

Edge-Triggered Flip-Flops

Edge-triggered flip-flops were designed to eliminate the oscillation and sensitivity-to-noise problems associated with level-sensitive flip-flops and to solve the glitch-catching problem inherent in master-slave flip-flops. An edge-triggered D flip-flop can be implemented by connecting 3 basic memory cells as shown in Figure 7.24. (The design of edge-triggered flip-flops is an asynchronous sequential-logic design problem and is discussed in detail in Chapter 9.)

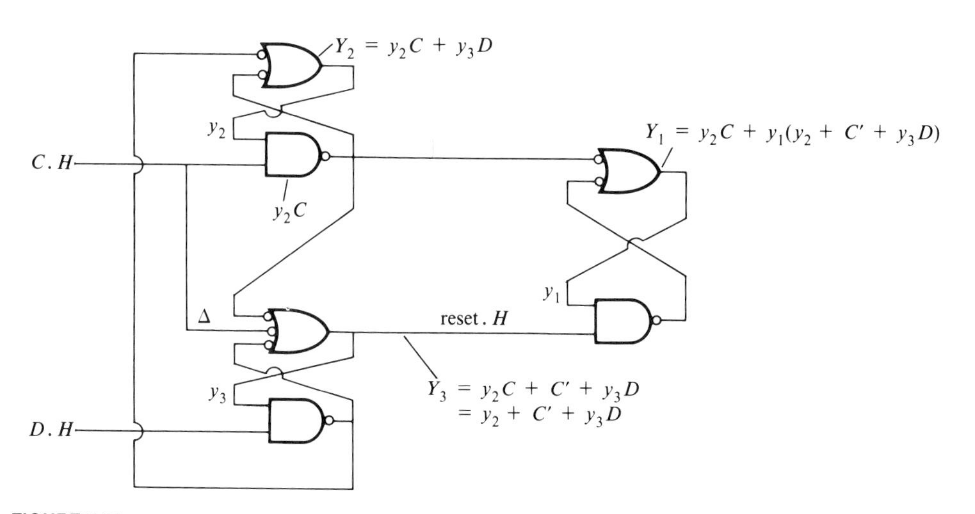

FIGURE 7.24
Logic diagram of an edge-triggered D flip-flop

An analysis of an edge-triggered D flip-flop can be accomplished by cutting the feedback lines in the driver latches. The resulting excitation equations for an edge-triggered D flip-flop are as follows:

$$Y_2 = y_2C + y_3D$$
$$Y_3 = y_2 + C' + y_3D$$
$$Y_1 = y_2C + y_1(y_2 + c^1 + y_3D)$$

The integrated circuit 7474 is a dual positive edge-triggered *D* flip-flop. The *D* inputs must be stable 1 set-up time prior to the low-to-high clock transition for predictable operation. Although the clock input is level-sensitive, the positive transition of the clock pulse between the 0.8V and 2.0V levels should be equal to or less than the clock-to-output delay time for reliable operation. A logic diagram of the 7474 IC is presented in Figure 7.25a, together with a mode-select function table and timing diagrams (Figure 7.25b and c).

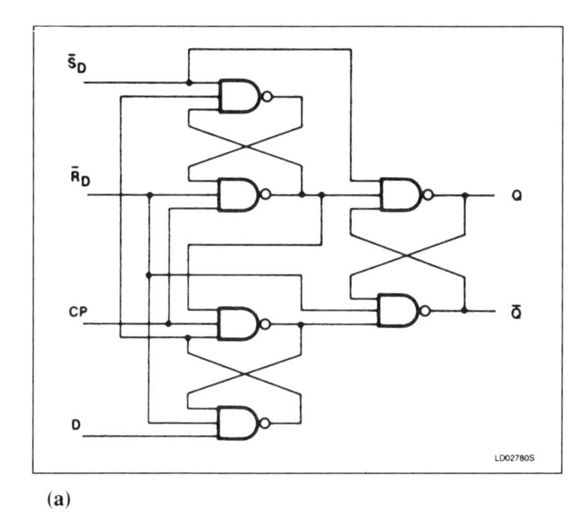

(a)

OPERATING MODE	INPUTS				OUTPUTS	
	$\bar{S}_D$	$\bar{R}_D$	CP	D	Q	$\bar{Q}$
Asynchronous Set	L	H	X	X	H	L
Asynchronous Reset (Clear)	H	L	X	X	L	H
Undetermined(1)	L	L	X	X	H	H
Load "1" (Set)	H	H	↑	h	H	L
Load "0" (Reset)	H	H	↑	l	L	H

H = HIGH voltage level steady state.
h = HIGH voltage level one set-up time prior to the LOW-to-HIGH clock transition.
L = LOW voltage level steady state.
l = LOW voltage level one set-up time prior to the LOW-to-HIGH clock transition.
X = Don't care.
↑ = LOW-to-HIGH clock transition.

NOTE:
(1) Both outputs will be HIGH while both $\bar{S}_D$ and $\bar{R}_D$ are LOW, but the output states are unpredictable if $\bar{S}_D$ and $\bar{R}_D$ go HIGH simultaneously.

(b)

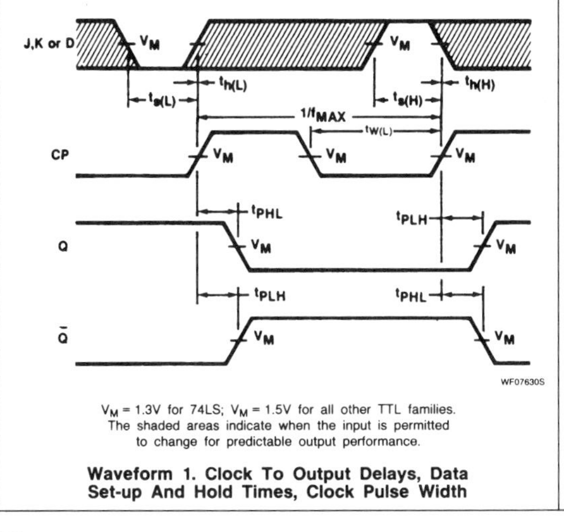

V_M = 1.3V for 74LS; V_M = 1.5V for all other TTL families.
The shaded areas indicate when the input is permitted to change for predictable output performance.

Waveform 1. Clock To Output Delays, Data Set-up And Hold Times, Clock Pulse Width

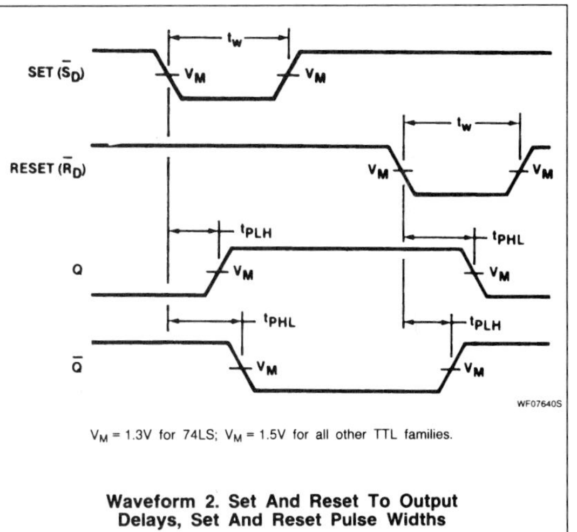

V_M = 1.3V for 74LS; V_M = 1.5V for all other TTL families.

Waveform 2. Set And Reset To Output Delays, Set And Reset Pulse Widths

(c)

FIGURE 7.25
A 7474 dual PET flip-flop. (a) Logic diagram (b) Mode-select function table (c) Timing diagrams (Reprinted by courtesy of Philips Components/Signetics)

Summary of Flip-Flop Characteristics and Excitations

The Edge-Triggered D Flip-Flop

The edge-triggered D flip-flop has a single control input that determines the state that the Q output assumes after a state-changing clock transition. If the control input $D = 1$, output Q will be set to 1. Conversely, if the control input $D = 0$, output Q will be reset to 0.

The D flip-flop operates in accordance with the characteristic table given in Table 7.17.

TABLE 7.17
D flip-flop characteristic table

D	$q \rightarrow Q$	Transition
0	$0 \rightarrow 0$	0
0	$1 \rightarrow 0$	beta
1	$0 \rightarrow 1$	alpha
1	$1 \rightarrow 1$	1

The characteristic table of a D flip-flop can be rewritten in the form of an excitation table (Table 7.18) that defines the input D that causes each transition.

TABLE 7.18
D flip-flop excitation table

$q \rightarrow Q$	Transition	D
$0 \rightarrow 0$	0	0
$0 \rightarrow 1$	alpha	1
$1 \rightarrow 0$	beta	0
$1 \rightarrow 1$	1	1

The D flip-flop has the characteristic equation

$$Q = D$$

since the D and Q columns of the excitation table are identical.

The general input equations can be derived from the excitation table as follows:

$$1_D = 1, \text{ alpha} \qquad 0_D = 0, \text{ beta}$$

In words, a 1 or an alpha transition is caused by a 1 signal on the D input of a D flip-flop, and a 0 or a beta transition is caused by a 0 signal on the D input of a D flip-flop.

The general input equations are useful in the design of sequential circuits. We will illustrate this in Section 7.4 in the design of a variety of counters.

The Edge-Triggered JK Flip-Flop

A JK flip-flop allows simultaneous assertion of the J and K inputs. If both $J = 1$ and $K = 1$ when the clock makes a state-changing transition, the flip-flop next state Q is the complement of its present state q. After a state-changing transition of the clock, the JK flip-flop next state Q is defined for each J, K, q combination by the characteristic table in Table 7.19.

TABLE 7.19
JK flip-flop characteristic table

Function	J	K	$q \to Q$	Transition
Hold	0	0	$0 \to 0$	0
	0	0	$1 \to 1$	1
Reset	0	1	$0 \to 0$	0
	0	1	$1 \to 0$	beta
Set	1	0	$0 \to 1$	alpha
	1	0	$1 \to 1$	1
Toggle	1	1	$0 \to 1$	alpha
	1	1	$1 \to 0$	beta

A K-map can be used to describe next state Q in terms of present state q and the control inputs J and K, as shown in Figure 7.26.

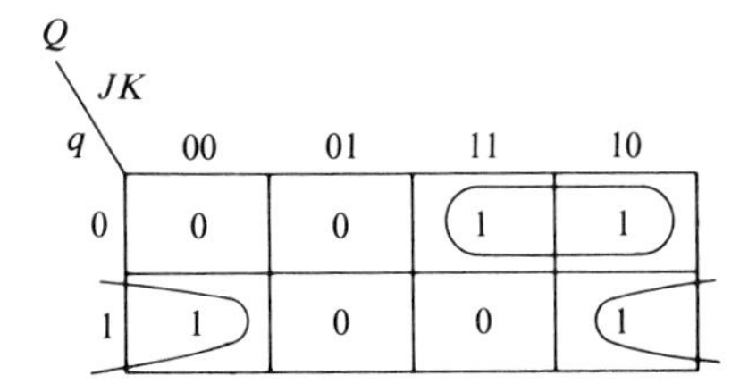

FIGURE 7.26
Next-state map of a *JK* flip-flop

The *JK* flip-flop has the following characteristic equation:

$$Q = Jq' + K'q$$

An excitation table can be derived from the characteristic table by collecting like transitions. The resulting excitation table (Table 7.20) defines the J and K inputs that cause the corresponding state transition, where input X indicates either 0 or 1.

TABLE 7.20
JK flip-flop excitation table

$q \to Q$	Transition	J	K
$0 \to 0$	0	0	X
$0 \to 1$	alpha	1	X
$1 \to 0$	beta	X	1
$1 \to 1$	1	X	0

The general input equations for a *JK* flip-flop, derived from the excitation table, are as follows:

1_J = alpha X_J = 1, beta $0_J = 0$

1_K = beta X_K = 0, alpha $0_K = 1$

The Edge-Triggered SR *Flip-Flop*

When the clock makes a state-changing transition, the *SR* flip-flop next state is defined for each S, R, q combination by the characteristic table shown in Table 7.21.

TABLE 7.21
SR flip-flop characteristic table

Function	S	R	$q \rightarrow Q$		Transition
Hold	0	0	0 →	0	0
	0	0	1 →	1	1
Reset	0	1	0 →	0	0
	0	1	1 →	0	beta
Set	1	0	0 →	1	alpha
	1	0	1 →	1	1
Not allowed	1	1	0	X	—
	1	1	1	X	—

A K-map can be used to describe next state Q in terms of present state q and the control inputs S and R, as shown in Figure 7.27. Note that the disallowed S, R input combination 1, 1 is shown as a don't care in the K-map.

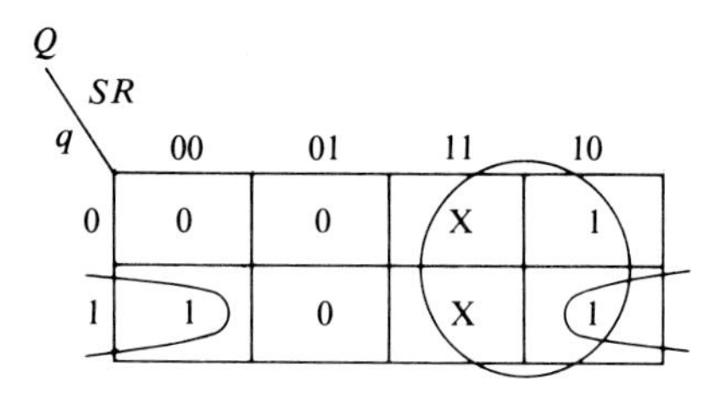

FIGURE 7.27
Next-state map of an *SR* flip-flop

The *SR* flip-flop has the following characteristic equation:

$$Q = S + R'q$$

The general input equations for an *SR* flip-flop, derived from the excitation table (Table 7.15), are as follows:

$$1_S = \text{alpha} \qquad X_S = 1 \qquad 0_S = 0, \text{beta}$$
$$1_R = \text{beta} \qquad X_R = 0 \qquad 0_R = 1, \text{alpha}$$

Converting Flip-Flops from One Type to Another

A particular type of flip-flop may be converted to another type of flip-flop by a procedure similar to that for designing a flip-flop using a basic memory cell. A procedure for converting one type of flip-flop to another type is illustrated in this section. An existing flip-flop of one type can be converted to a target flip-flop of another type using the block diagram shown in Figure 7.28.

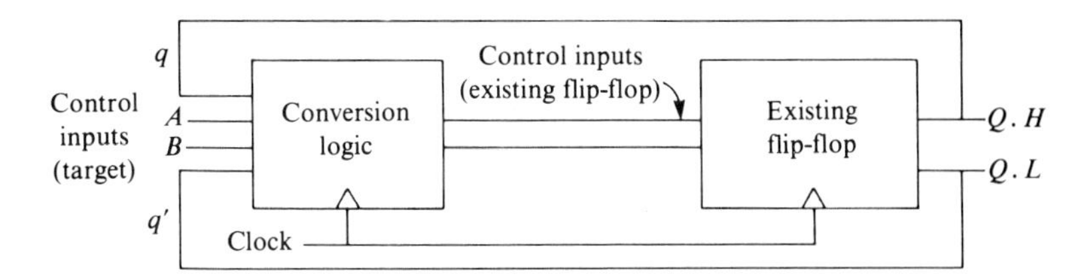

FIGURE 7.28
Diagram for converting one type of flip-flop to another

The resulting composite device operates in accordance with the function table of the target flip-flop. That is, the device state Q responds to the control inputs in the same manner as for the target flip-flop. Hence, the composite device appears to the "outside" world as the target flip-flop.

Converting a D Flip-Flop to a JK Flip-Flop

Step 1: Problem Statement. Convert a D flip-flop to a JK flip-flop.

Step 2: Conceptualization. An existing D flip-flop can be converted to a JK flip-flop (target) using the block diagram shown in Figure 7.29.

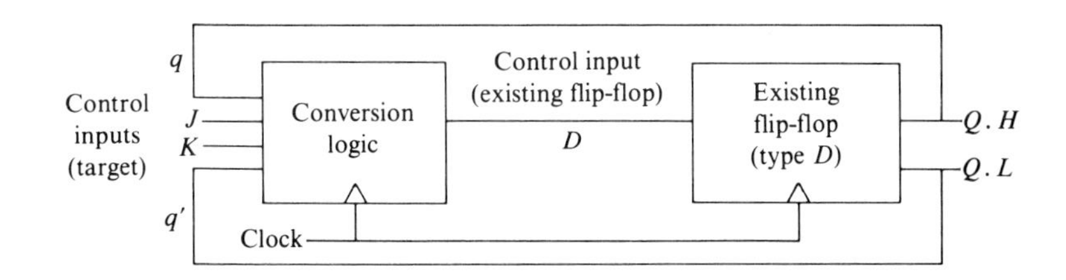

FIGURE 7.29
Diagram for converting a D flip-flop to a JK flip-flop

A D-to-JK conversion table can be constructed by forming a composite of the excitation tables for the D and the JK flip-flops, as shown in Table 7.22. Each row of the adjacent J, K, and q columns is formed by replacing the X value of J or K with 0 (or 1), together with the q and K or J value in the same row.

TABLE 7.22
Table for converting a D flip-flop to a JK flip-flop

		Existing Excitation			
***JK* Function**	**Transition**	***q Q***	***J K***	***J K q***	***D***
Hold	0	0 0	0 X	0 0 0	0
				0 1 0	
Set	alpha	0 1	1 X	1 0 0	1
				1 1 0	
Reset	beta	1 0	X 1	0 1 1	0
				1 1 1	
Toggle	1	1 1	X 0	0 0 1	1
				1 0 1	
		Target Excitation			

Step 3: Solution/Simplification. A D-to-JK conversion map (Figure 7.30) can be constructed using the JKq and D columns of Table 7.22.

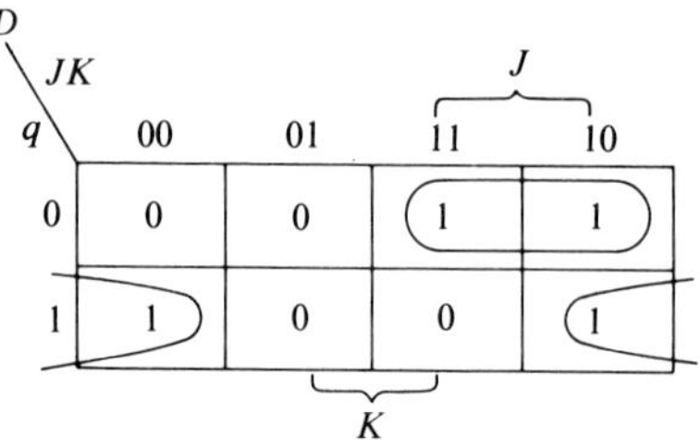

FIGURE 7.30
D-to-*JK* conversion map

The D-to-JK conversion map is then used to determine the control input (D) of the existing (D) flip-flop as a function of the target flip-flop controls (J, K) and variable q. The required function is the following logic equation:

$$D = J \cdot q' + K' \cdot q$$

Step 4: Realization. The existing D flip-flop is converted to the target JK flip-flop by the substitution of $D = J \cdot q' + K' \cdot$ q for the conversion logic in Figure 7.29. A block diagram for the converted D-to-JK flip-flop is presented in Figure 7.31.

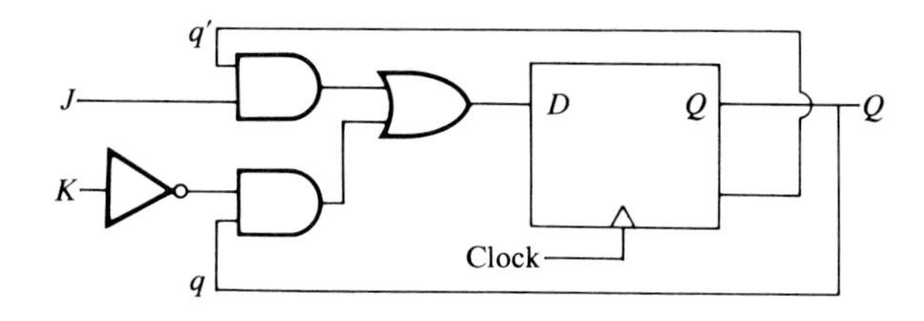

FIGURE 7.31
Conversion of a *D* flip-flop to a *JK* flip-flop

Converting an SR *Flip-Flop to a* JK *Flip-Flop*

Step 1: Problem Statement. Convert an SR flip-flop to a JK flip-flop.

Step 2: Conceptualization. An existing SR flip-flop can be converted to a JK flip-flop (target) using the block diagram shown in Figure 7.32.

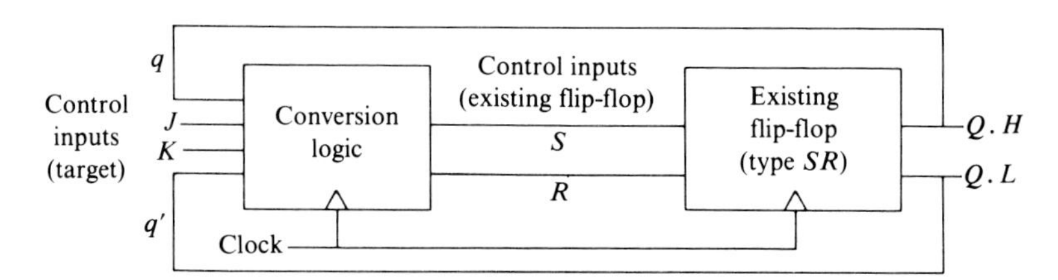

FIGURE 7.32
Diagram for converting an *SR* flip-flop to a *JK* flip-flop

An SR-to-JK conversion table can be constructed by forming a composite of the excitation tables for the SR and the JK flip-flops, as shown in Table 7.23. Each row of the adjacent JKq columns is formed by replacing the X value of J or K with 0 (and 1), together with the q and K or J value in the same row.

TABLE 7.23
Table for converting an *SR* flip-flop to a *JK* flip-flop

JK Function	Transition	*q Q*	*J K*	*J K q*	*S R*
			Existing Excitation		
Hold	0	0 0	0 X	0 0 0	0 X
				0 1 0	
Set	alpha	0 1	1 X	1 0 0	1 0
				1 1 0	
Reset	beta	1 0	X 1	0 1 1	0 1
				1 1 1	
Toggle	1	1 1	X 0	0 0 1	X 0
				1 0 1	
		Target Excitation			

Step 3: Solution/Simplification. An *SR*-to-*JK* conversion map (Figure 7.33) can be constructed using the *JKq* and *SR* columns of Table 7.23.

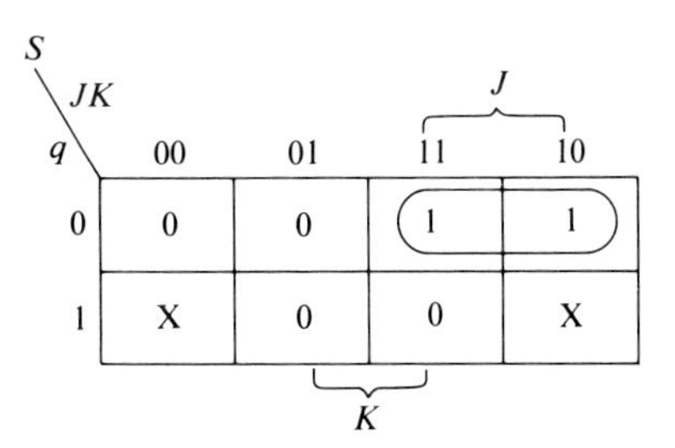

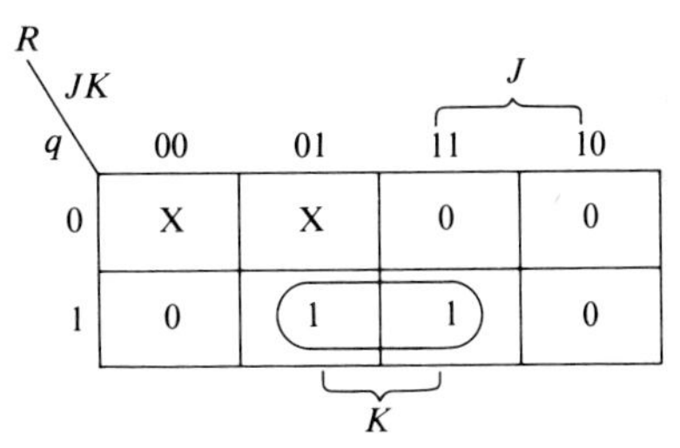

FIGURE 7.33
SR-to-*JK* conversion maps

The *SR*-to-*JK* conversion map is then used to determine the control inputs S, R of the existing *SR* flip-flop as a function of the target flip-flop controls J and K, and variable q. The required functions are the following logic equations:

$$S = J \cdot q' \text{ and } R = K \cdot q$$

Step 4: Realization. The existing *SR* flip-flop is converted to the target *JK* flip-flop by the substitutions $S = J \cdot q'$ and $R = K \cdot q$ for the conversion logic in Figure 7.32. A block diagram for the converted *SR*-to-*JK* flip-flop is presented in Figure 7.34 (p. 292).

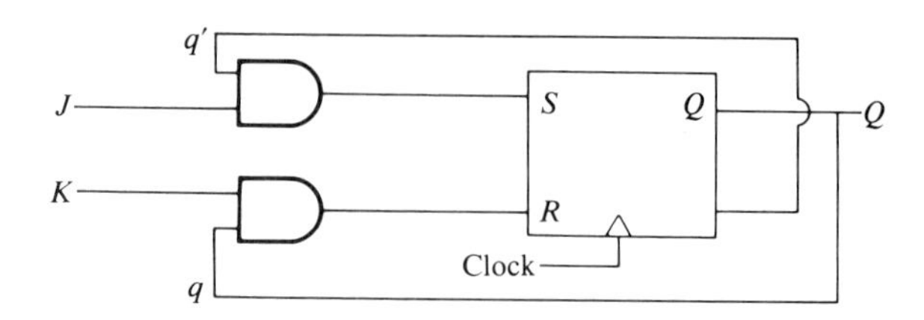

FIGURE 7.34
Conversion of an *SR* flip-flop to a *JK* flip-flop

7.4 COUNTERS

A* counter *is a specialized synchronous sequential-circuit that generates a predetermined number sequence over and over again. Digital systems use counters in a variety of applications, such as digital clocks, special sequence generators, time-delay circuits, pulse counters, control-state counters, program counters, and so on.

Counters are classified as either synchronous or ripple (asynchronous) depending on how they are clocked. For example,

- The counter is a synchronous counter if all the flip-flops of the counter are clocked by the same clock signal.
- The counter is a ripple (asynchronous) counter if individual flip-flops of the counter are clocked at different times.

Asynchronous counters are slower in operation than synchronous counters but their control circuitry is usually simpler.

Counters also can be classified according to a number of other characteristics:

- By the number of distinct states (**modulus**). A counter that has N distinct states is called a *modulus-N* (mod-N) *counter* or a divide-by-N counter. A mod-10 counter is also called a *decade* or *decimal counter*. If n is an integer, a mod-2^n counter is called an *n-bit counter* (or a natural-modulus counter).
- By the number of flip-flops contained in the counter. An n-bit counter contains n flip-flops.
- By the generated-code sequence.

Counters classified according to the generated-code sequence include

A binary up (or down) counter whose successive states form an increasing (or decreasing) binary count in normal sequence.

A binary up-down counter that counts up or down depending on the setting of a control parameter.

A Gray-code counter that generates a Gray-code sequence.

A BCD counter, which is a modulus-10 counter whose successive states form a normal BCD count sequence.

An n-bit ring counter that uses a circular shift to generate a sequence in which each term contains a single 1 and $n - 1$ 0s.

A twisted-ring counter that generates a sequence in which adjacent terms differ in only 1 bit.

An arbitrary-count counter whose successive states do not form a normal binary count sequence.

Counters are also referred to as single-mode or multimode counters. A binary up counter is an example of a single-mode counter. A binary up-down counter is an example of a multimode counter.

In general, if a mod-N counter is connected in cascade with a mod-M counter, the resulting counter is a mod-NM counter, with modulus $N \times M$. It should be noted that a mod-N counter (divide-by-N counter) repeats a particular count sequence in N periods of the clock.

Design of *N*-Bit Synchronous Binary Counters (Normal Sequence)

A formal approach and an intuitive approach are used in this section to design synchronous 2-bit and 4-bit binary counters. The design of 8-bit, 12-bit, and 16-bit counters can be accomplished by a direct extension of the techniques used in designing 2-bit and 4-bit counters.

Formal Design of a Synchronous 2-Bit Counter

This section uses a slightly modified 4-step problem solving procedure to design a 2-bit binary counter. Keep in mind that a counter is a special case of a sequence generator.

Step 1: Problem Statement. Design a device that generates the sequence of integers 0, 1, 2, and 3 over and over again.

Step 2: Conceptualization. Let t_k denote the kth term of the sequence. That is, $t_1 = 0$, $t_2 = 1$, $t_3 = 2$, $t_4 = 3$. We can generate the required sequence using the following recursion formula

$$t_{k+1} = t_k + 1, \; (k = 1, 2, 3) \qquad \textbf{(7.3)}$$

with a starter value $t_1 = 0$. Equation (7.3) is a mathematical model of the sequence generation problem.

A logic model for the sequence generation can be developed by converting the decimal integers to binary and listing the transition in each bit position between successive terms of the binary sequence (00, 01, 10, 11, and repeat), as shown in Figure 7.35a (p. 294). Recall that the transition definitions are as follows:

$q \rightarrow Q$	Transition
$0 \rightarrow 0$	0
$0 \rightarrow 1$	alpha
$1 \rightarrow 0$	beta
$1 \rightarrow 1$	1

A 2-bit binary number is denoted by AB, where A is the MSB and B is the LSB. A block diagram of a 2-bit counter is presented in Figure 7.35b.

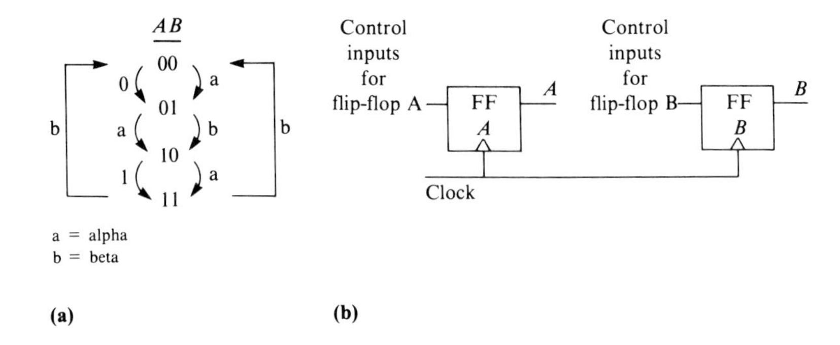

FIGURE 7.35
A 2-bit counter. (a) Transitions (b) Block diagram

We must design control circuits for the inputs of flip-flops A and B so that the device generates the specified sequence. Recall that a flip-flop excitation table (see Section 7.3) defines the input(s) required for each type of transition. If we know the required transition for bit A (and bit B) for each present state of the counter, then we can determine the input(s) for each flip-flop that will generate the next state. We, therefore, plot the counter transitions in transition maps, as shown in Figure 7.36.

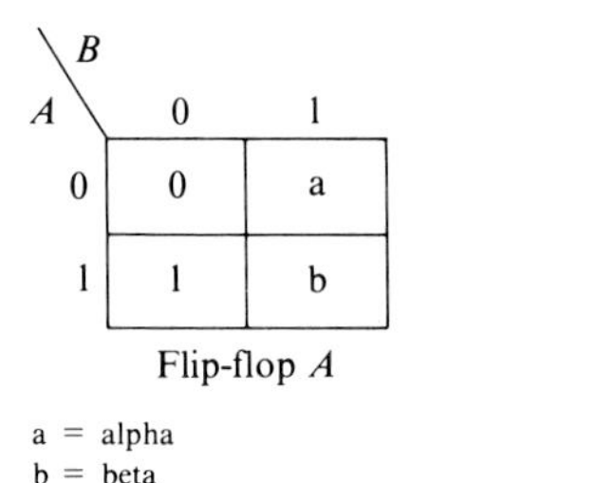

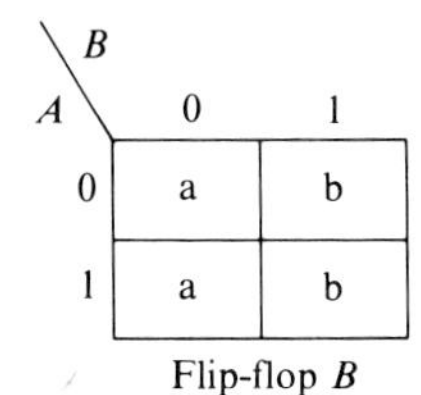

FIGURE 7.36
Transition maps for a 2-bit binary counter

It should be noted that the relation between successive terms of the sequence are defined in the transition maps independent of any hardware considerations. That is, the count sequence is uniquely defined by the transition maps. At this point, the design is independent of the type of flip-flops to be used in the counter implementation.

Step 3: Solution/Simplification. The next step is to select the type of flip-flops that are to be used in the counter. The selection of a particular type of flip-flop invokes the use of the general input equations for that flip-flop type. We will use JK flip-flops, which were designed primarily for control applications. The general input equations for a JK flip-flop, derived in Section 7.3, are as follows:

$$1_J = \text{alpha} \qquad X_J = 1, \text{beta} \qquad 0_J = 0$$
$$1_K = \text{beta} \qquad X_K = 0, \text{alpha} \qquad 0_K = 1$$

The JK flip-flop inputs required to generate the specified sequence are contained in the flip-flop input maps (Figure 7.37) obtained by substituting the flip-flop general input equations into the transition maps.

FIGURE 7.37
JK input maps for a 2-bit binary counter

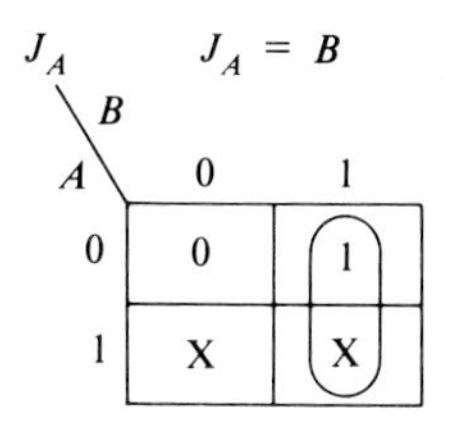

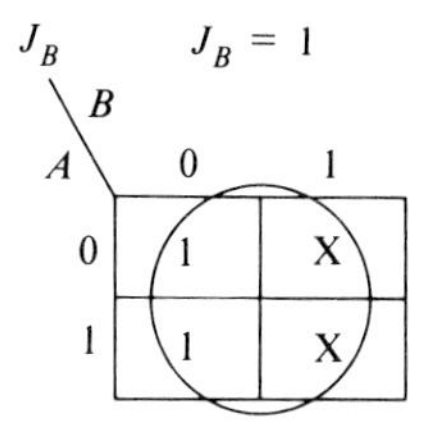

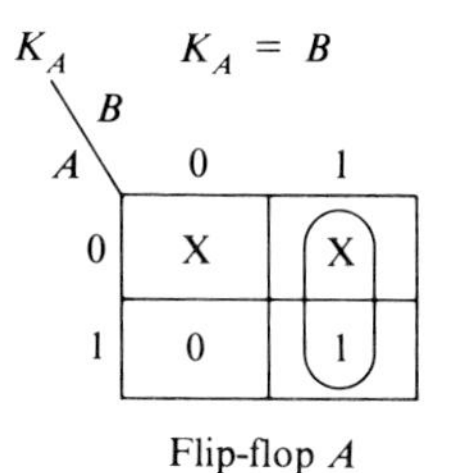

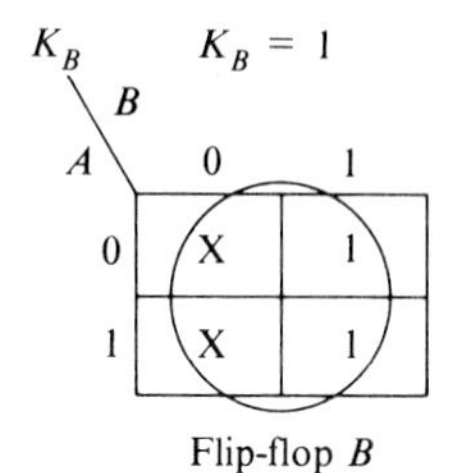

Application equations that define the counter inputs are then obtained using the *JK* input maps in Figure 7.37. It should be noted that these equations depend on the type of flip-flop that is to be used to realize the device. The application equations for the *J*, *K* inputs of the *A* and *B* flip-flops are

$$J_A = B \qquad J_B = 1$$
$$K_A = B \qquad K_B = 1$$

Step 4: Realization. We can now use the application equations to construct a logic diagram of the 2-bit binary counter implemented using *JK* flip-flops, as shown in Figure 7.38. Note that a synchronous counter is a counter in which all of the flip-flops are clocked by the same clock pulse.

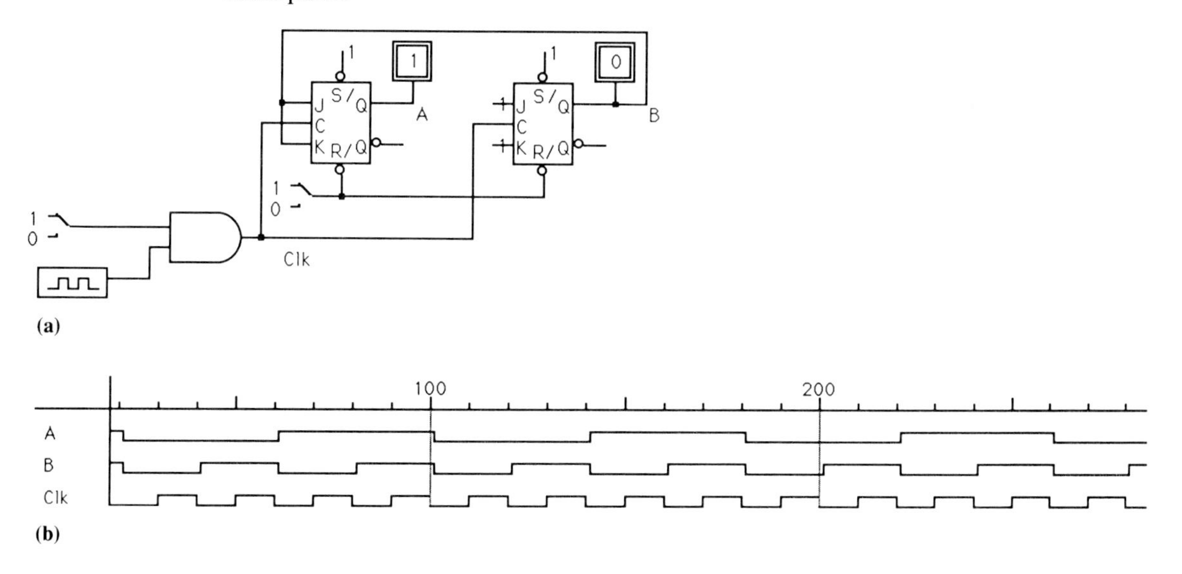

FIGURE 7.38
Synchronous 2-bit counter using *JK* flip-flops. (a) Logic diagram (b) Timing diagram

We can summarize the 4-step problem-solving procedure used to design counters.

Step 1: Problem Statement. Design a counter that generates a specified sequence of numbers.

Step 2: Conceptualization. Construct a count sequence/transition table. Then construct a transition map for each flip-flop.

Step 3: Solution/Simplification.

a. Select the type of flip-flops that are to be used and substitute the flip-flop general input equations into the transition maps to obtain flip-flop input maps.
For D flip-flops

$$1_D = 1, \text{ alpha} \qquad 0_D = 0, \text{ beta}$$

For JK flip-flops,

$$\begin{array}{lll} 1_J = \text{alpha}, & 0_J = 0, & X_J = 1, \text{ beta} \\ 1_K = \text{beta}, & 0_K = 1, & X_K = 0, \text{ alpha} \end{array}$$

b. Use the flip-flop input maps to determine the application equations for each flip-flop.

Step 4: Realization. Use the application equations to construct a logic diagram of the counter.

Shortcut Method for Counter Design Using JK Flip-Flops

Combined next-state/transition maps for a 2-bit binary counter are presented in Figure 7.39a, in which A and A^+ (B and B^+) denote the present state q and the next state Q, respectively. Using the JK excitation table (Figure 7.39b), a shortcut method can be devised for the design of counters implemented with JK flip-flops. Label the present state = 0 region of a flip-flop's transition map as the J input region since an alpha transition can occur only if the present state = 0 and 1_J = alpha, $0_J = 0$. Label the present state = 1 region as the K input region since a beta transition can occur only if the present state = 1 and 1_K = beta, $0_K = 1$. The transition maps (Figure 7.36) become the shortcut JK input maps, as shown in Figure 7.39c. Note that the variable A is eliminated (and its values are replaced by a dash, –) in the shortcut JK map for flip-flop A. Similarly, the variable B is eliminated in the shortcut JK map for flip-flop B.

We can then use these shortcut JK input maps to determine the application equations for the J, K inputs of the A and B flip-flops:

$$\begin{array}{ll} J_A = B & J_B = 1 \\ K_A = B & K_B = 1 \end{array}$$

Thus, the shortcut JK method produces the same result as obtained earlier using the standard JK method.

Intuitive Design of a Synchronous 2-Bit Binary Counter

An intuitive approach is developed in this section to design a 2-bit synchronous binary counter that counts in the sequence 00, 01, 10, 11, and then repeats this sequence over and

FIGURE 7.39
JK shortcut method.
(a) Combined next-state/transition maps (b) *JK* flip-flop excitation table (c) Shortcut *JK* input maps for a 2-bit binary counter

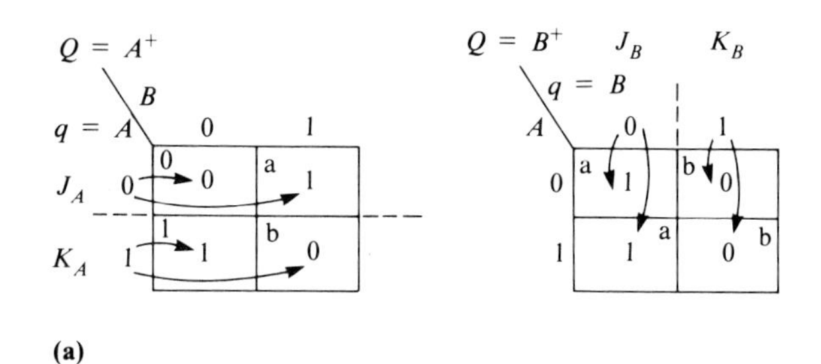

(a)

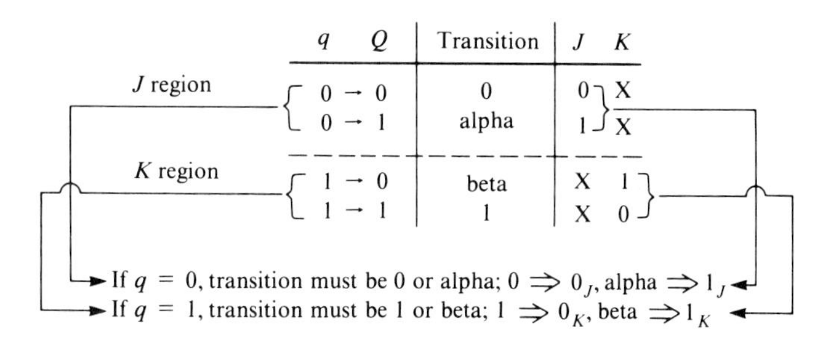

	q	Q	Transition	J	K
J region	0 → 0		0	0	X
	0 → 1		alpha	1	X
K region	1 → 0		beta	X	1
	1 → 1		1	X	0

If $q = 0$, transition must be 0 or alpha; $0 \Rightarrow 0_J$, alpha $\Rightarrow 1_J$
If $q = 1$, transition must be 1 or beta; $1 \Rightarrow 0_K$, beta $\Rightarrow 1_K$

(b)

	A \ B	0	1
J_A	—	0	1
K_A	—	0	1

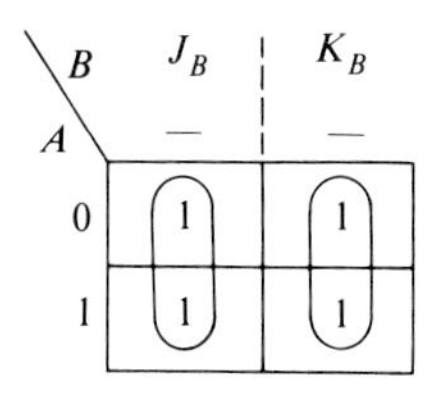

(c)

over again. A clock pulse is used to increment the counter; that is, each successive clock pulse causes the counter to advance to the next count in the sequence. A 2-bit binary number is denoted by *AB*, where *A* is the MSB and *B* is the LSB. The counter's count sequence is illustrated in Figure 7.40.

FIGURE 7.40
A 2-bit count sequence

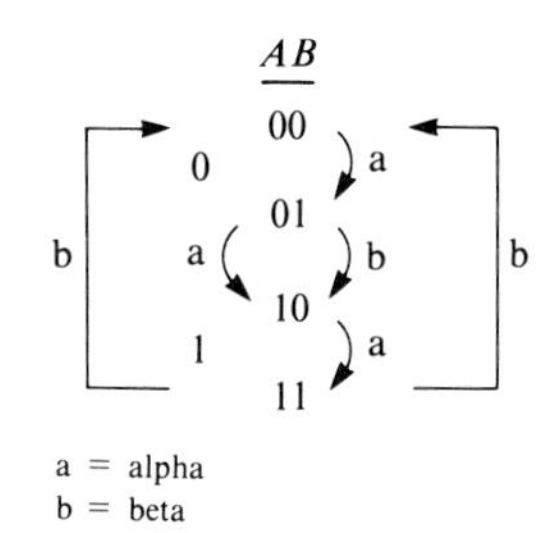

A flip-flop is used to hold each bit of the 2-bit number. Flip-flop *A* holds bit *A*, and flip-flop *B* holds bit *B*. The arrows in the count sequence table indicate when the individual flip-flops change state. With each clock pulse,

1. Flip-flop *B* toggles.
2. Flip-flop *A* operates as follows:
   a. When $B = 0$, *A* holds its state.
   b. When $B = 1$, *A* toggles.

An edge-triggered *JK* flip-flop, described in Section 7.3, has the function table shown in Table 7.24.

TABLE 7.24
JK flip-flop function table

Function	Inputs J K	Q
Hold	0 0	q
Reset	0 1	0
Set	1 0	1
Toggle	1 1	q'

The toggle property of the *JK* flip-flop is ideal for counter applications. Using a *JK* flip-flop for each bit of the 2-bit counter, we see that

1. Flip-flop *B* toggles with each clock pulse if both the *J* and *K* inputs of flip-flop *B* are set to 1.
2. Flip-flop *A* operates as follows:
   a. When $B = 0$, flip-flop *A* holds its state; this can be accomplished by *J*, *K* inputs 0, 0.
   b. When $B = 1$, flip-flop *A* toggles; this can be accomplished by *J*, *K* inputs 1, 1.

Conditions a and b can be accomplished by connecting output *B* to both the *J* and *K* inputs of flip-flop *A*. Therefore, the logic diagram for the 2-bit counter is the same as Figure 7.38a.

Analogous approaches are used in the next section to design a synchronous 4-bit binary counter. It will be shown that the design of a 4-bit counter is a direct extension of the techniques used in designing a 2-bit counter. You can then see that the design of multibit (8, 12, 16, . . .) counters can be accomplished by direct extension of the techniques used in designing 2-bit and 4-bit counters.

Intuitive Design of a Synchronous 4-Bit Binary Counter

An intuitive approach is used in this section to design a synchronous 4-bit binary counter that counts in the sequence 0000, 0001, 0010, . . . , 1111 and then repeats this sequence over and over again. A clock pulse is used to increment the counter; that is, each successive clock pulse causes the counter to advance to the next count in the sequence. Let a 4-bit binary number be denoted by *ABCD*, where *A* is the MSB and *D* is the LSB. The counter's count sequence is illustrated in Figure 7.41.

A flip-flop is used to hold each bit of the 4-bit number. Flip-flop *A* holds bit *A*, flip-flop *B* holds bit *B*, flip-flop *C* holds bit *C*, and flip-flop *D* holds bit *D*.

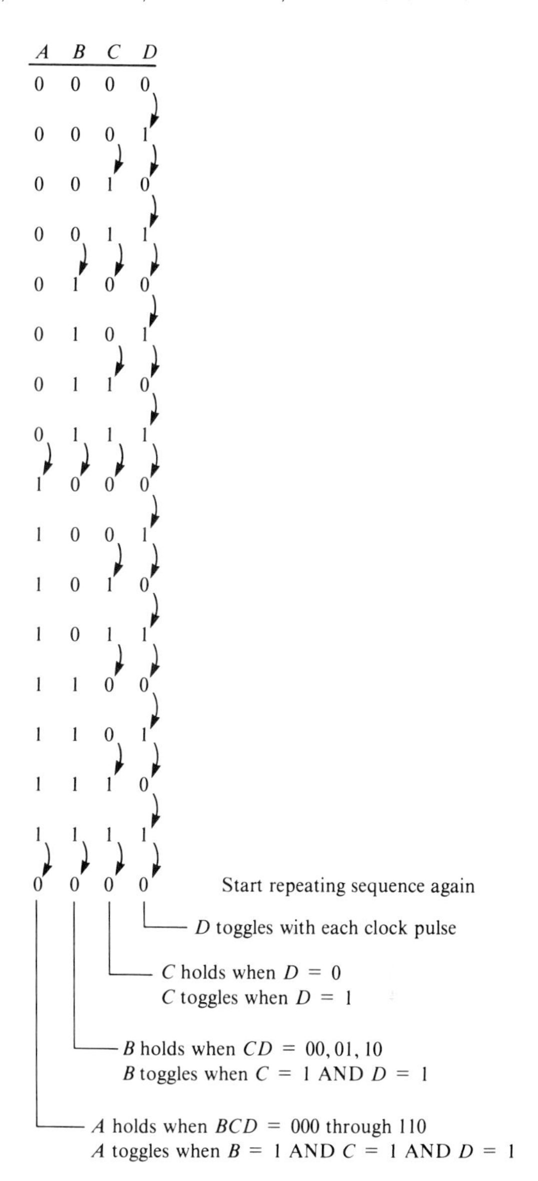

FIGURE 7.41
A 4-bit count sequence

Using the toggle property of the *JK* flip-flop, a synchronous 4-bit counter can be implemented as follows:

- Flip-flop *D*. Set both *J* and *K* inputs to 1 so that *D* toggles on each clock pulse.
- Flip-flop *C*. Connect both *J* and *K* inputs to *D*'s output so that flip-flop *C* holds when $D = 0$ and toggles when $D = 1$.
- Flip-flop *B*. Connect both *J* and *K* inputs to *C* AND *D* so that flip-flop *B* holds when $CD = 00, 01, 10$ and toggles when $CD = 11$.

□ Flip-flop *A*. Connect both *J* and *K* inputs to *B* AND *C* AND *D* so that flip-flop *A* holds when *BCD* = 000 through 110 and toggles when *BCD* = 111.

A synchronous 4-bit binary counter using *JK* flip-flops is illustrated in Figure 7.42. Note that the 4-bit counter operates just like a binary odometer.

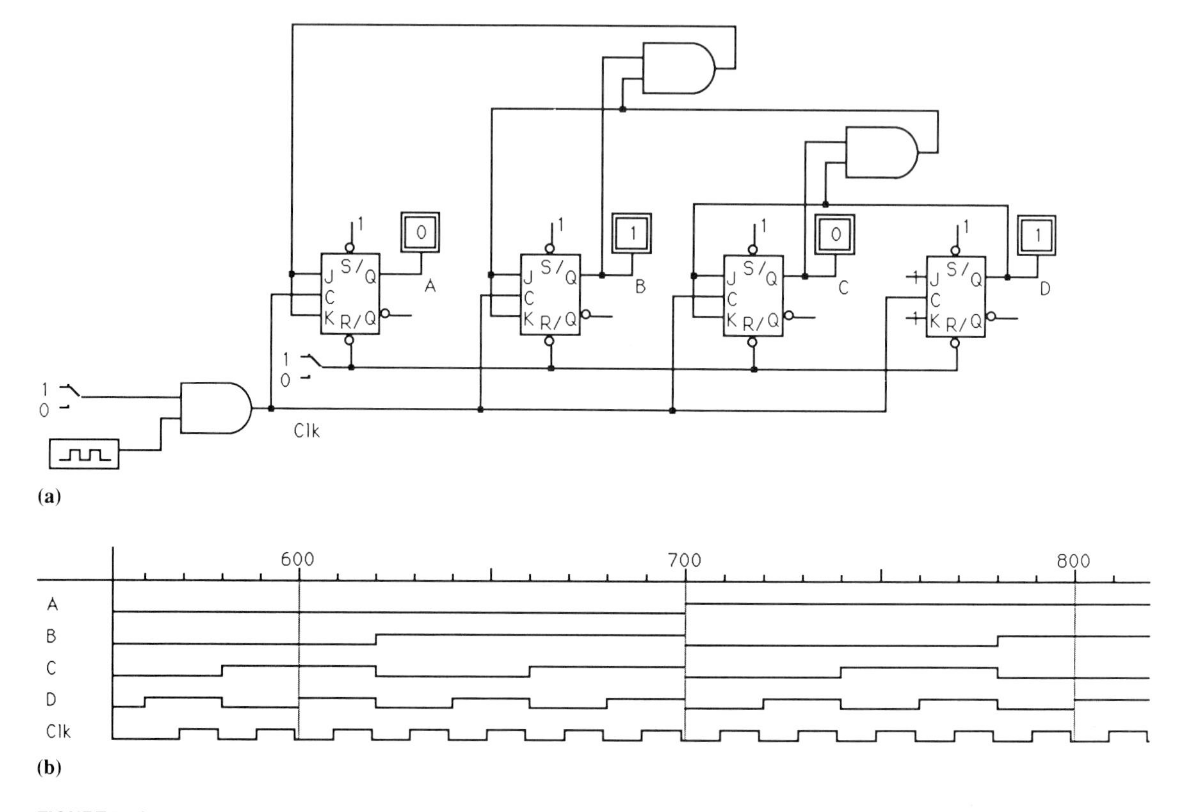

FIGURE 7.42
Logic diagram of a synchronous 4-bit binary counter

Formal Design of a Synchronous 4-Bit Binary Counter

A synchronous 4-bit binary counter is designed in this section using a formal solution. This solution is a direct extension of the formal solution used to design a 2-bit counter. Again, the following transition definitions are used.

$q \rightarrow Q$	Transition
$0 \rightarrow 0$	0
$0 \rightarrow 1$	alpha
$1 \rightarrow 0$	beta
$1 \rightarrow 1$	1

Step 1: Problem Statement. Design a synchronous 4-bit binary counter that generates the sequence 0000 through 1111 over and over again.

Step 2: Conceptualization. Examine the count sequence for a 4-bit binary counter and label the transitions for the outer columns (the reader may label the transitions for columns B and C) as shown in Figure 7.43.

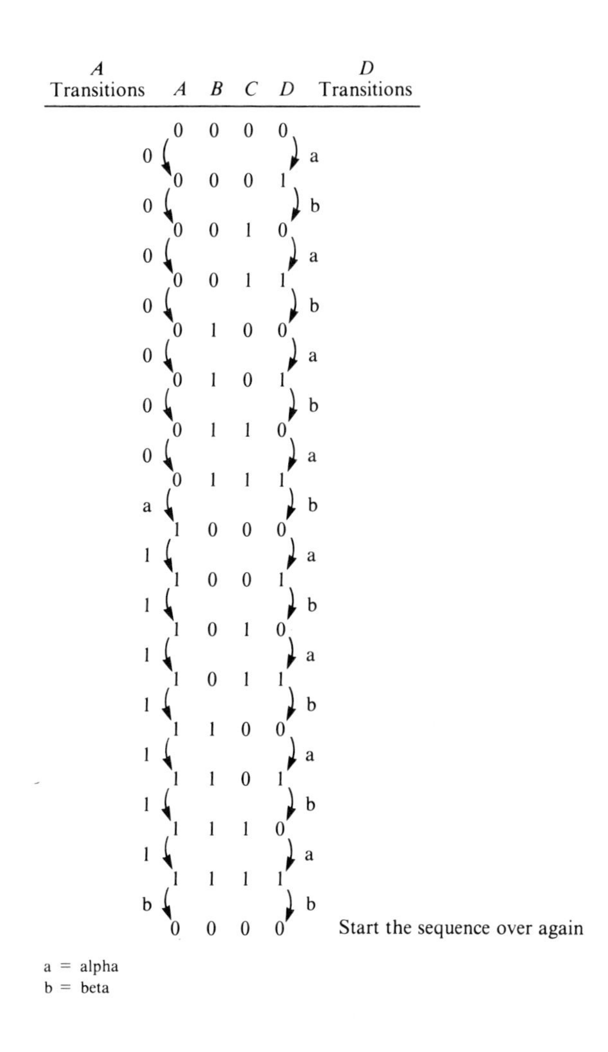

FIGURE 7.43
Transitions for a 4-bit binary counter

The transition maps can be constructed using the transitions taken from the count-sequence diagram, as shown in Figure 7.44 (p. 302).

The count sequence is uniquely defined by these transition maps. At this point the design is independent of the type of flip-flops to be used in the counter implementation.

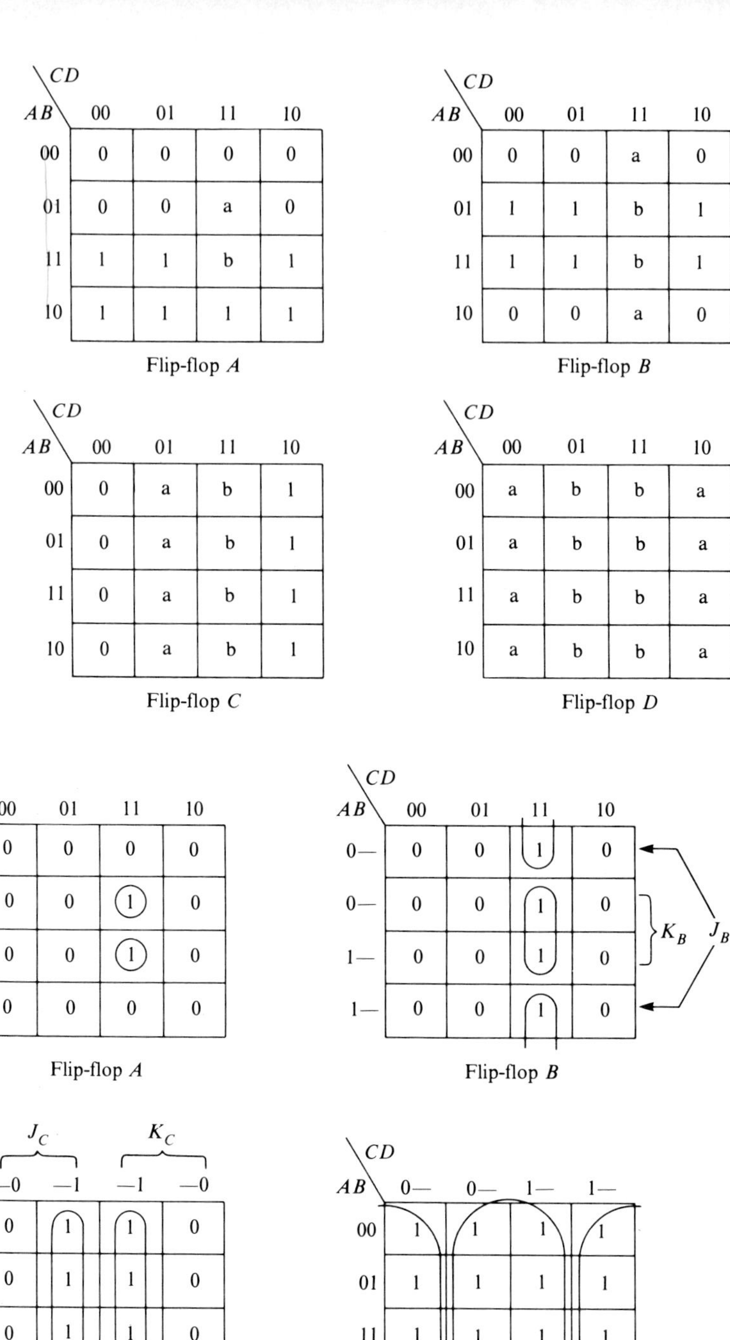

FIGURE 7.44
Transition maps for a 4-bit binary counter

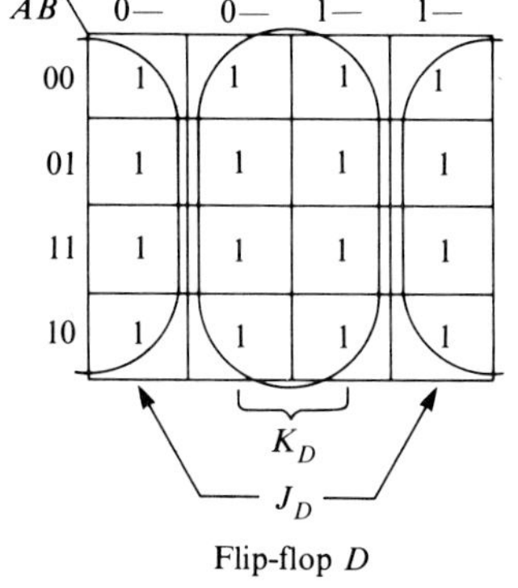

FIGURE 7.45
Shortcut *JK* input maps for a 4-bit binary counter

Step 3: Solution/Simplification. Selection of a particular type of flip-flop invokes the use of the general input equations for that type of flip-flop. Construct flip-flop input maps by substituting the general input equations into the transition maps.

If *JK* flip-flops are used to implement this 4-bit counter, the shortcut *JK* method may be used to design the counter. Label the present state = 0 region of a flip-flop's transition map as the *J* input region since an alpha transition can occur only if the present state = 0 and 1_J = alpha, $0_J = 0$. Label the present state = 1 region as the *K* input region since a beta transition can occur only if the present state = 1 and 1_K = beta, $0_K = 1$. The transition maps become shortcut *JK*-input maps, as shown in Figure 7.45.

The application equations are then obtained directly from the shortcut *JK*-input maps:

$$J_A = BCD \qquad J_B = CD \qquad J_C = D \qquad J_D = 1$$
$$K_A = BCD \qquad K_B = CD \qquad K_C = D \qquad K_D = 1$$

Step 4: Realization. As we can see, the formal solution and the intuitive solution produce the same application (control) equations for the flip-flops. The circuit diagram for the 4-bit counter is, therefore, the same as the diagram in Figure 7.42.

Design of *N*-Bit Synchronous Counters (Arbitrary Sequence)

The design of a binary counter that counts in an arbitrary sequence requires the use of transition maps to determine the application equations for each flip-flop of the counter. The design of an arbitrary sequence counter can be illustrated best by the following example.

Step 1: Problem Statement. Design a synchronous counter that has the following count sequence

0, 2, 4, 6, 8, 10, 12, 14, 15, 13, 11, 9, 7, 5, 3, 1, 0

and repeats the sequence over and over.

Step 2: Conceptualization. Convert the decimal numbers of the sequence to binary form and determine the transitions (for each bit position) between successive terms of the sequence, as shown in Figure 7.46.

The transition maps (Figure 7.47, p. 305) can be constructed using the transitions taken from the count sequence table shown in Figure 7.46.

Step 3: Solution/Simplification. If the counter is implemented using *JK* flip-flops, the transition maps can be converted to shortcut *JK*-input maps as shown in Figure 7.48 (p. 305).

The application equations for the counter can be found directly from the shortcut JK input maps:

$$J_A = BCD' \qquad K_A = B'C'D \qquad J_B = CD' + AC'D \qquad K_B = C'D + A'CD'$$
$$J_C = A + B + D' \qquad K_C = A' + B' + D \qquad J_D = ABC \qquad K_D = A'B'C'$$

FIGURE 7.46
Count sequence/transition table for an arbitrary sequence counter

State (Decimal)	State				Transition			
	A	*B*	*C*	*D*	*A*	*B*	*C*	*D*
0	0	0	0	0				
				↓	0	0	a	0
2	0	0	1	0				
				↓	0	a	b	0
4	0	1	0	0				
				↓	0	1	a	0
6	0	1	1	0				
				↓	a	b	b	0
8	1	0	0	0				
				↓	1	0	a	0
10	1	0	1	0				
				↓	1	a	b	0
12	1	1	0	0				
				↓	1	1	a	0
14	1	1	1	0				
				↓	1	1	1	a
15	1	1	1	1				
				↓	1	1	b	1
13	1	1	0	1				
				↓	1	b	a	1
11	1	0	1	1				
				↓	1	0	b	1
9	1	0	0	1				
				↓	b	a	a	1
7	0	1	1	1				
				↓	0	1	b	1
5	0	1	0	1				
				↓	0	b	a	1
3	0	0	1	1				
				↓	0	0	b	1
1	0	0	0	1				
				↓	0	0	0	b
0	0	0	0	0				

a = alpha
b = beta

Step 4: Realization. Use the application equations to draw a logic diagram for the counter (see Figure 7.49, p. 306).

Design of Synchronous Divide-by-*N* Counters

A counter that has N distinct states is called a divide-by-N (or modulus-N) counter. If $2^{(n-1)} < N < 2^n$, the divide-by-N counter does not use all of the possible n-bit number combinations. For example, a BCD counter is a divide-by-10 counter whose states are the binary numbers 0000 through 1001 corresponding to the decimal digits 0 through 9; the remaining six combinations, 1010 through 1111, are not used.

An arbitrary divide-by-N counter may use any N of the possible binary combinations to represent its N distinct states. However, a particular set of combinations may result in simpler flip-flop control circuitry than that implemented using a different set of combinations.

FIGURE 7.47
Transition maps for an arbitrary sequence counter

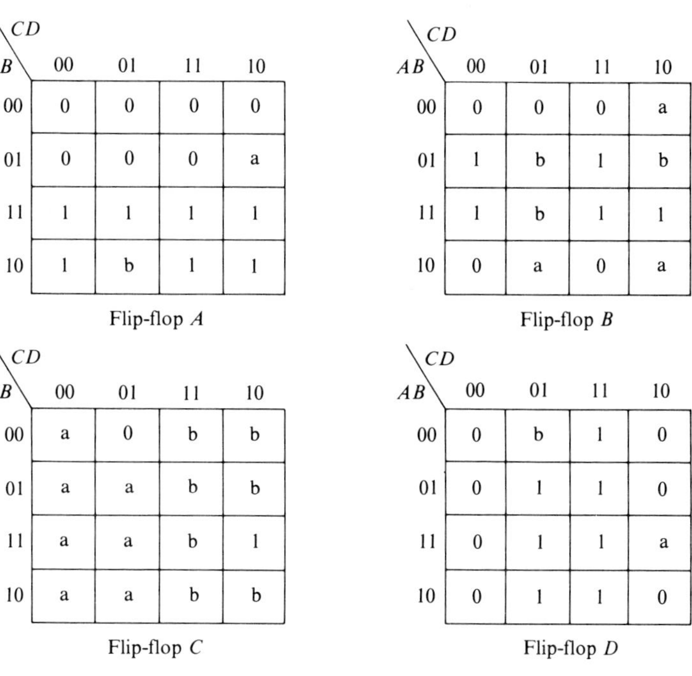

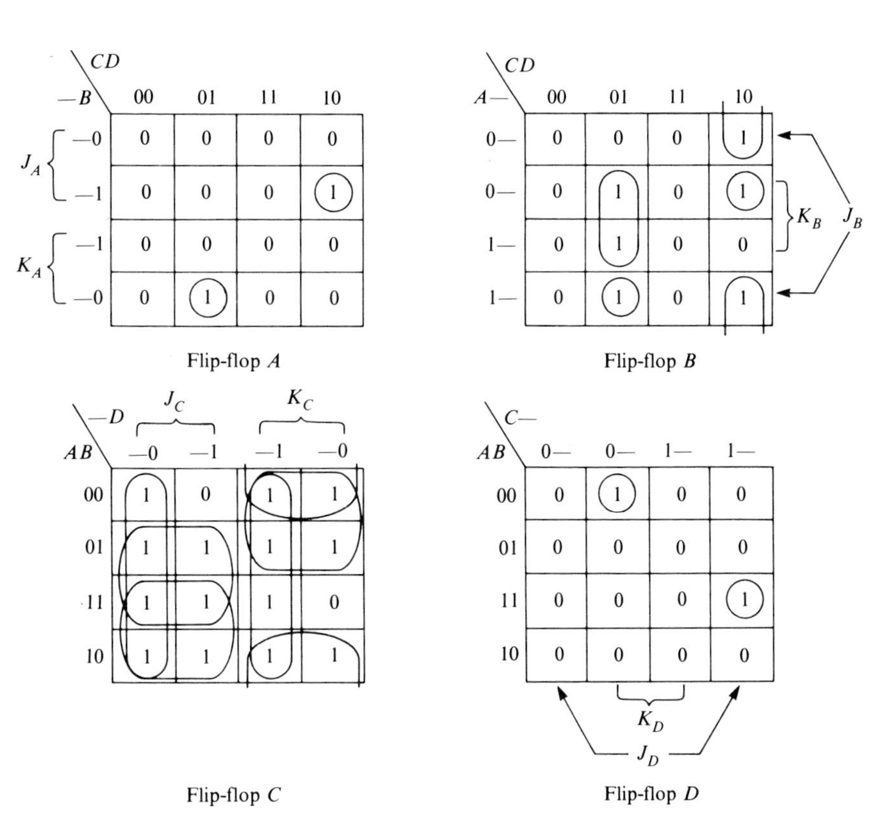

FIGURE 7.48
Shortcut *JK* input maps for an arbitrary-sequence counter

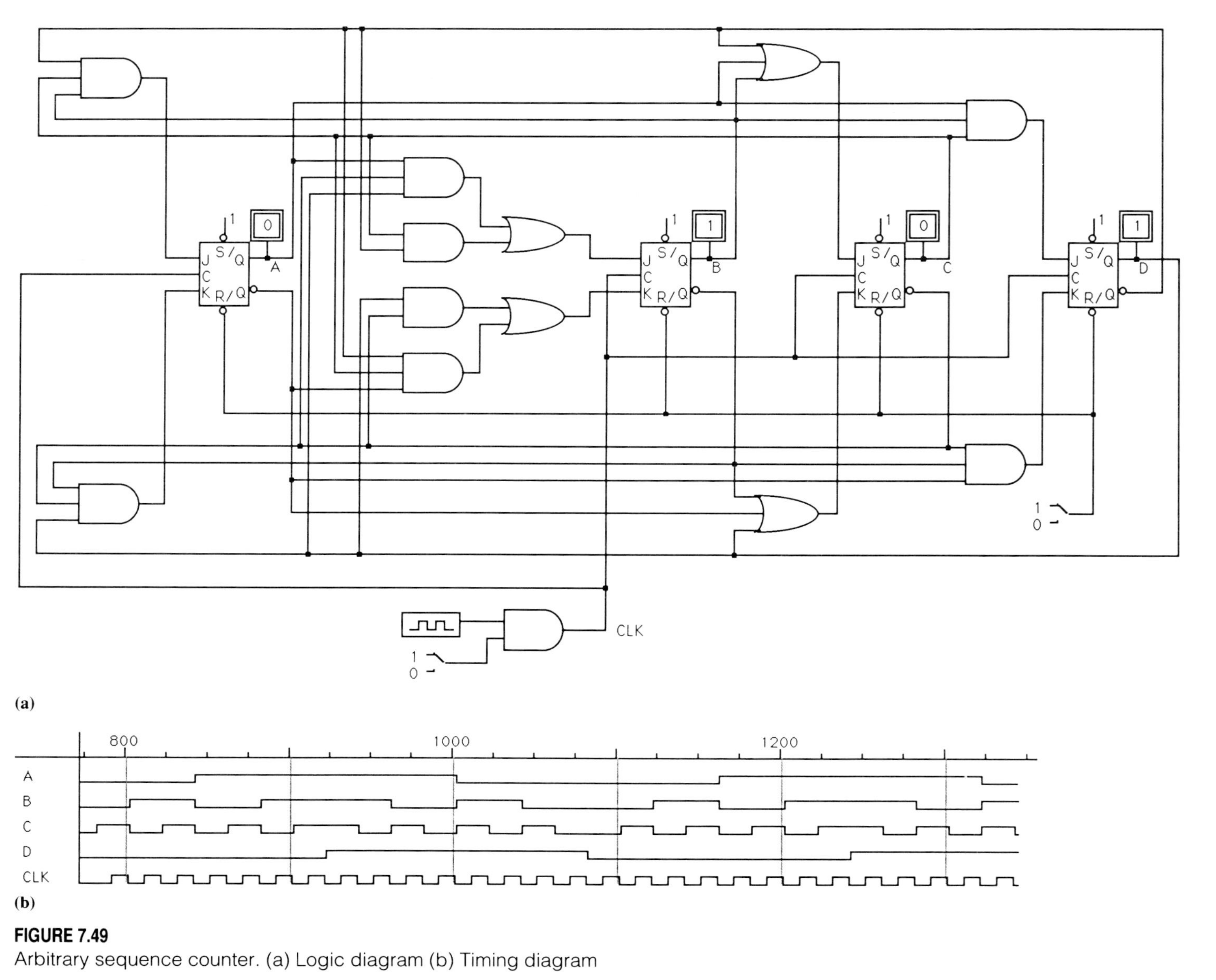

FIGURE 7.49
Arbitrary sequence counter. (a) Logic diagram (b) Timing diagram

Step 1: Problem Statement. Design a divide-by-5 counter that generates the sequence of decimal numbers 0, 2, 3, 5, and 7.

Step 2: Conceptualization. Convert the decimal numbers of the sequence to binary form and determine the transitions (for each bit position) between successive terms of the sequence, as shown in Figure 7.50.

FIGURE 7.50
Count sequence for a divide-by-5 counter

State			Transitions		
A	*B*	*C*	*A*	*B*	*C*
0	0	0			
			0	a	0
0	1	0			
			0	1	a
0	1	1			
			a	b	1
1	0	1			
			1	a	1
1	1	1			
			b	b	b
0	0	0			

a = alpha
b = beta

Use the transitions determined in Step 2 to construct the counter's transition maps shown in Figure 7.51.

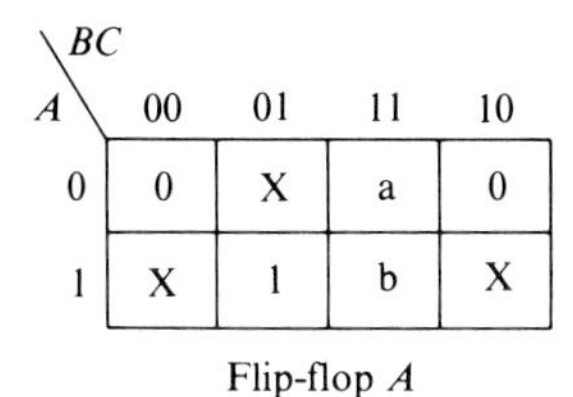

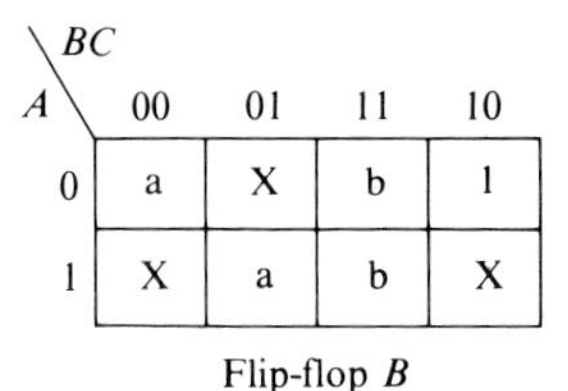

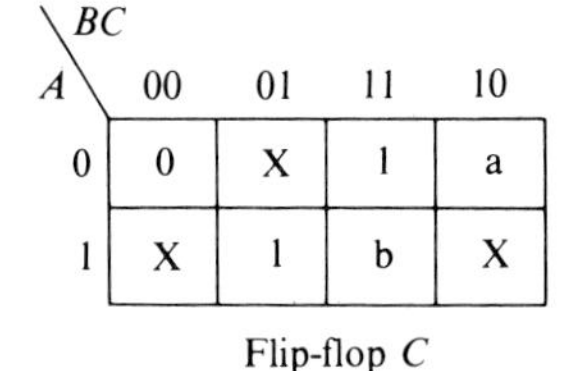

FIGURE 7.51
Transition maps for the divide-by-5 counter

Step 3: Solution/Simplification. If the counter is to be implemented using *D* flip-flops, the *D* input maps are constructed, as shown in Figure 7.52, by substituting the *D* flip-flop general input equations ($1_D = 1$, alpha; $0_D = 0$, beta) into the transition maps.

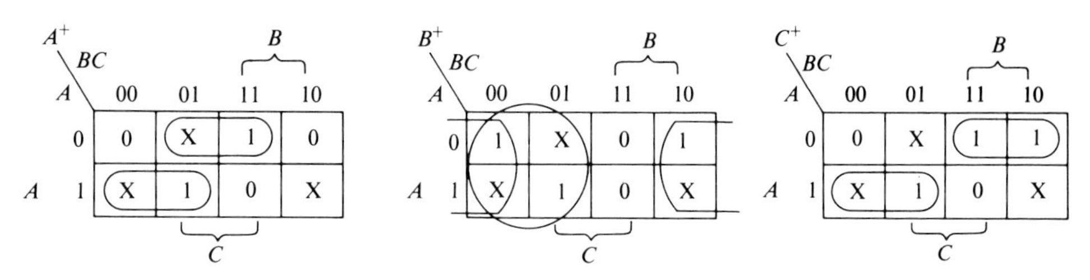

FIGURE 7.52
D flip-flop input maps for the divide-by-5 counter

The K-map 1s are used, together with the don't cares, to produce the D flip-flop application equations. The next state of a D flip-flop equals its present input D. Therefore, for flip-flop A, $A^+ = D_A$. A and A^+ denote the present state and next state, respectively, of the flip-flop A.

$$D_A = A'C + AB', \quad D_B = B' + C', \quad D_C = A'B + AB'$$

Step 4: Realization. These application equations are used to draw the logic diagram for the divide-by-5 counter implemented with D flip-flops, as shown in Figure 7.53. The present state of the counter is the combination ABC of the flip-flops and the counter next state is $A^+B^+C^+$.

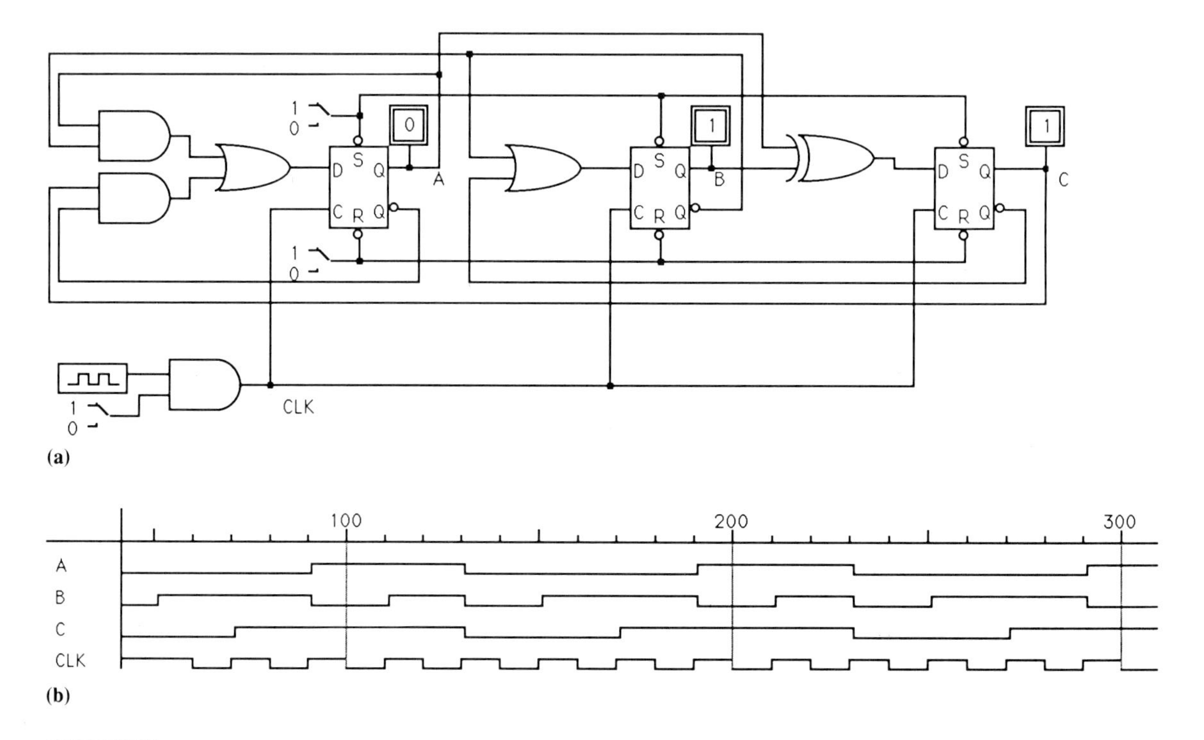

FIGURE 7.53
Divide-by-5 counter implemented by D flip-flops. (a) Logic diagram (b) Timing diagram

N-Bit Asynchronous Binary Counters

An asynchronous counter is one in which the flip-flops are not clocked at the same time. Usually only the flip-flop for the LSB is triggered by a clock pulse, while the remaining flip-flops are triggered by one or more outputs of the other stages. Asynchronous counters are also referred to as **ripple counters** because the triggering pulse of a given stage must ripple through the preceding (less significant) stages of the counter. For this reason, asynchronous counters are slower in operation than synchronous counters, but their control circuitry is usually simpler. A logic diagram of a 3-bit asynchronous counter is presented in Figure 7.54.

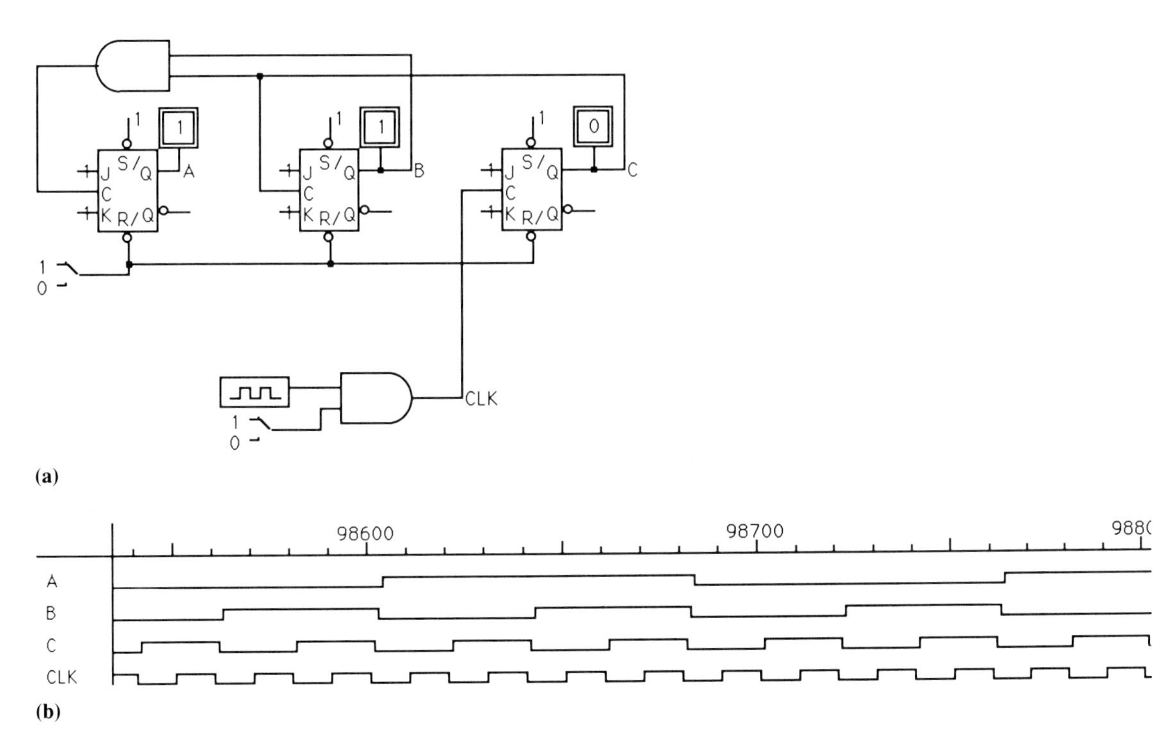

FIGURE 7.54
A 3-bit asynchronous counter. (a) Logic diagram (b) Timing diagram

Two or more counters can be cascaded to form a composite counter that has a higher modulus than the component counters. Recall that the modulus of a counter is the number of distinct states of the counter. A counter that has N distinct states is called a modulus-N (mod-N) counter or a divide-by-N counter. In general, if a mod-N counter is connected in cascade with a mod-M counter, the resulting counter is a mod-NM counter, with modulus $N \times M$. Figure 7.55 (p. 310) illustrates a cascaded counter formed by connecting a 2-bit asynchronous counter in cascade with a 3-bit asynchronous counter. Note that the output of the MSB of one counter is used to clock the LSB of the other counter.

A particular application may require that one or more outputs of a counter be decoded. For example, if a system requires generation of a control signal when a modulus-8 counter reaches a count of 6, we must decode the required count. Decoding a counter can be accomplished using decoders or logic gates. Figure 7.56 (p. 310) illustrates the decoding of a modulus-8 counter when the counter is at a count of 6; that is, when $Q_A = 1$, $Q_B = 1$, and $Q_C = 0$.

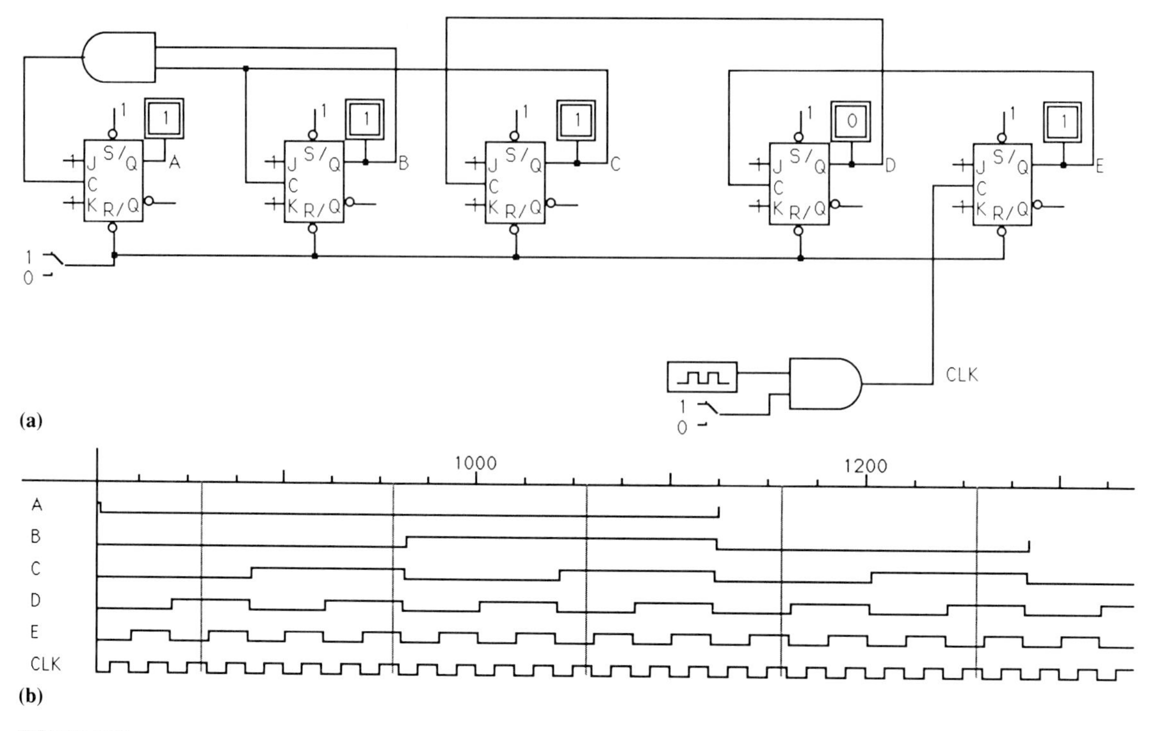

FIGURE 7.55
Cascaded counter. (a) Logic diagram (b) Timing diagram

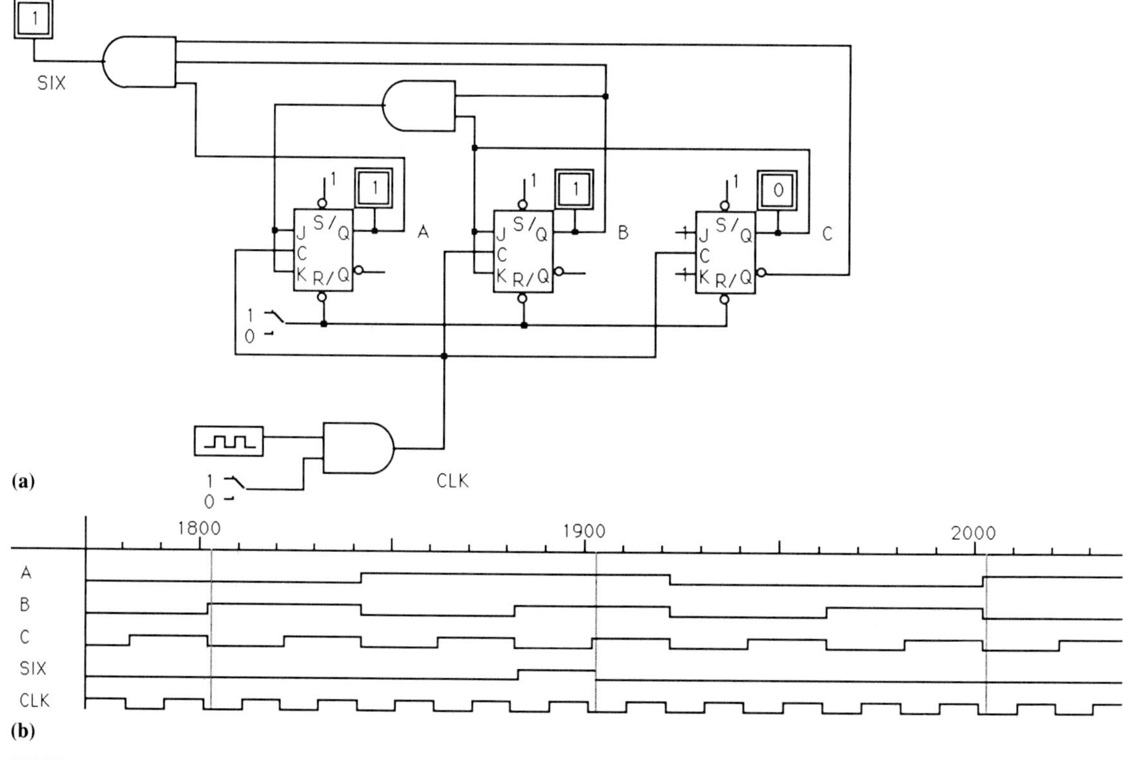

FIGURE 7.56
Decoding a counter using logic gates. (a) Logic diagram (b) Timing diagram

Integrated Circuit Counters

A variety of different types of counters are commercially available in IC form. For example, a BCD decade counter is available as either a 74160 or a 74162 IC, and a 4-bit binary counter is available either as a 74161 or a 74163 IC. A 4-bit up-down counter is available as either a 74192 or a 74193 IC. The 74192 counts in BCD mode, while the 74193 counts in binary mode. Logic diagrams of a 74LS163A 4-bit binary counter and a 74193 4-bit up-down binary counter are presented in Figures 7.57 and 7.58 (p. 312), respectively.

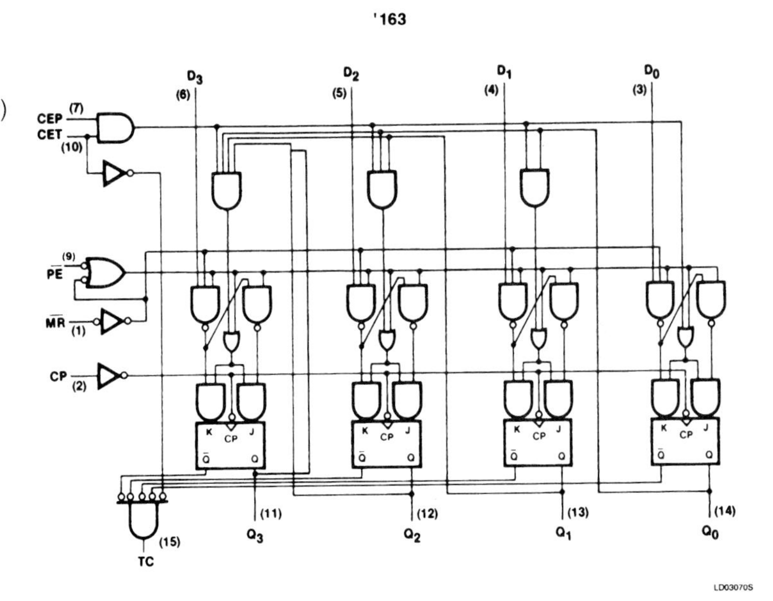

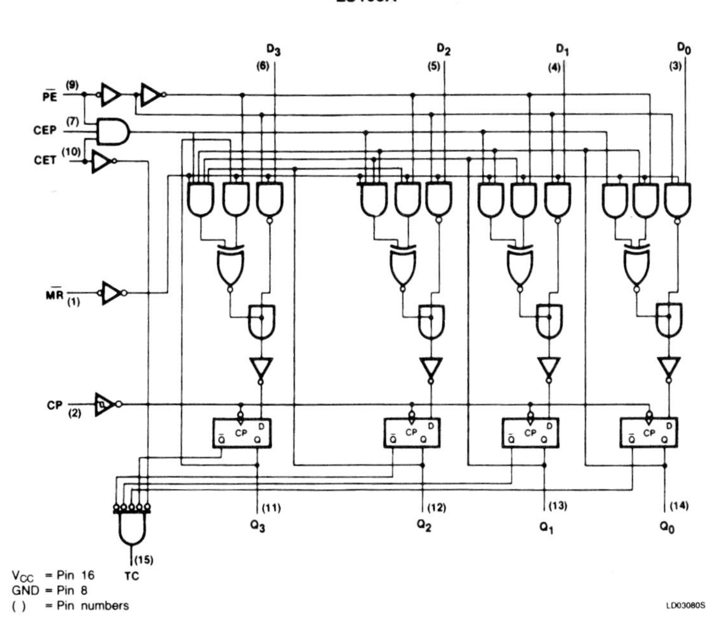

FIGURE 7.57
Logic diagram of a 74LS163A counter (Reprinted by courtesy of Philips Components/Signetics)

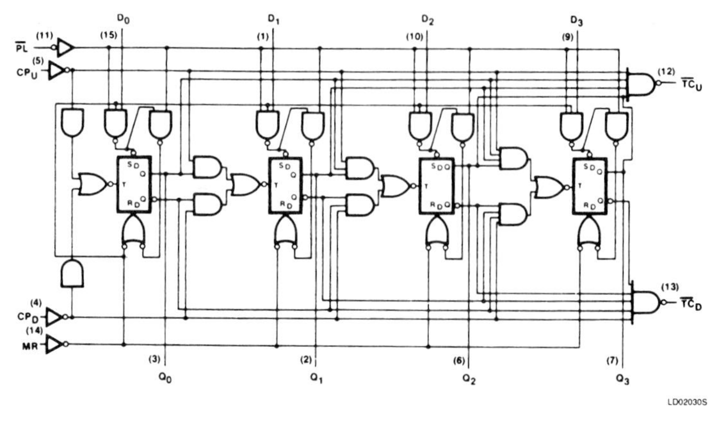

FIGURE 7.58
Logic diagram of a 74193 counter (Reprinted by courtesy of Philips Components/Signetics)

7.5 REGISTERS, SHIFT REGISTERS, AND SHIFT-REGISTER COUNTERS

A common requirement in digital systems is to input, process, store, and output* n*-bit alphanumeric data words. A given alphanumeric data word can represent numeric data or text data. A numeric data word may be in any one of several forms, such as fixed-point binary (usually 2s complement), floating-point binary, or binary-coded decimal (BCD). A text data word, representing an alphanumeric character, is usually encoded in either ASCII or EBCDIC. The system may also use data words that represent status or control information. Regardless of the type of data (numeric, text, status, or control), the data is stored as an* n*-bit word.

A data-storage register is a physical device that is capable of holding a data word. In digital devices, data-storage registers usually consist of a number of 1-bit memory elements, where each element can hold a binary 0 or 1. A 4-bit register can, therefore, hold any data word with a value between 0000 and 1111. A 4-bit register can be represented by the juxtaposition of four 1-bit memory elements (usually flip-flops) where each 1-bit memory element can hold a binary value 0 or 1, as shown in Figure 7.59.

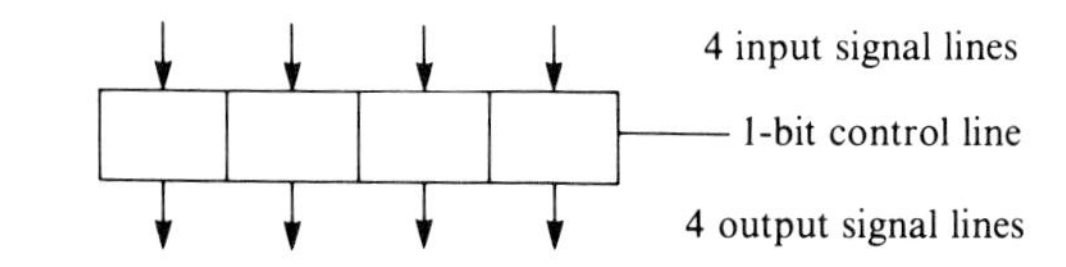

FIGURE 7.59
Symbolic representation of a 4-bit register

Functionally, a 4-bit data register has

A 4-bit data input

A 4-bit data output

A 1-bit control line (called **load register**)

A data register operates as follows: When the control line is strobed (goes from 0 to 1, remains at 1 long enough to be noticed by the register, then goes back to 0), the value on the

4-bit data input is stored in the register and becomes the new output value. Independent of new input values, the data register output is the last loaded value until the next input value is *loaded.*

An n-bit register $R = R < n - 1,0 >$ is composed of n flip-flops, labeled $R_{n-1}, \ldots, R_1, R_0$. The flip-flops of register R are ordered, with the LSB on the right designated as R_0, and the MSB on the left designated R_{n-1}. The flip-flops are usually controlled by a common clock that allows all bits to store data simultaneously. Most registers also have either synchronous or asynchronous set and clear inputs that allow all bits to be set or cleared (reset) simultaneously.

Registers are named using one or more capital letters. A register name may simply be a single letter (A, B, or C), or the name may indicate the function of the register; for example, *AC* (accumulator), *MD* (memory data), *MBR* (memory buffer register). In general, X and Y are used to represent n-bit register inputs, and Z denotes an n-bit register output.

Data Registers Using *D* Flip-Flops

The D flip-flop characteristic ($Q = D$), shown in Table 7.25, makes the D flip-flop ideally suited as a basic building block for constructing data registers.

TABLE 7.25
D flip-flop excitation table

State				
Present q	$\rightarrow$	Next Q	Transition	D
0	$\rightarrow$	0	0	0
0	$\rightarrow$	1	alpha	1
1	$\rightarrow$	0	beta	0
1	$\rightarrow$	1	1	1

Remember that the D flip-flop excitation table defines the input value that causes each of the basic transitions (0, alpha, beta, and 1). Since the second and last columns of the excitation table are identical, the D flip-flop characteristic is $Q = D$.

A common requirement of a digital system is to take n bits of data on parallel lines and store them simultaneously in a register composed of n flip-flops governed by a single clock. The parallel transfer of a 4-bit data word to a 4-bit data register N (for nibble) is illustrated in Figure 7.60. Note that the register has asynchronous set and reset inputs that allow the register to be set to 1111 or reset to 0000 independent of the operation of the clock.

The 4-bit data register, shown in Figure 7.60 (p. 314), can be loaded using the following procedure:

- Set each switch, S_1 through S_4, in accordance with the bit configuration of the data nibble to be loaded into the register.
- Strobe the clock input to load the register flip-flops.

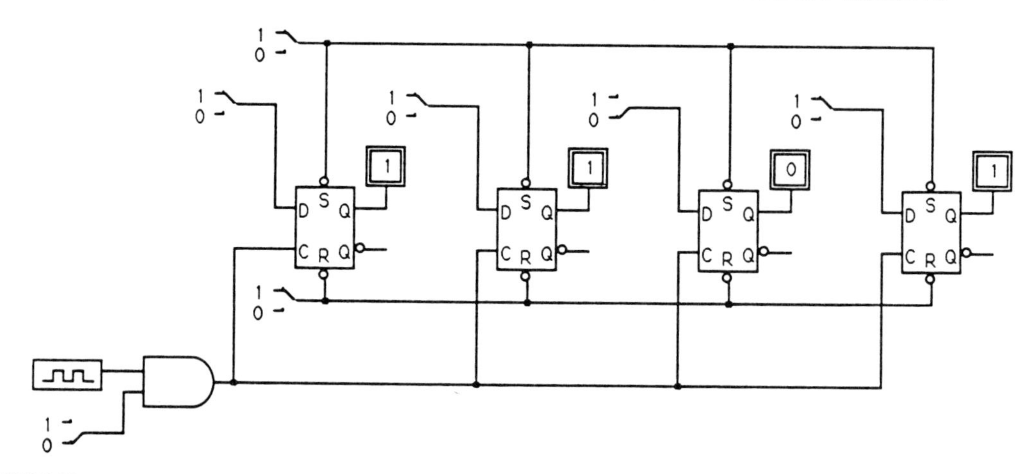

FIGURE 7.60
A 4-bit data register using D flip-flop

The Shift-Register Property

A shift register is a simple example of a digital system that illustrates the need for flip-flops that are insensitive to any changes in the level of the control inputs at times other than the state-changing transition of the clock signal. A 4-bit right-shift register is shown in Figure 7.61.

FIGURE 7.61
A 4-bit right-shift register

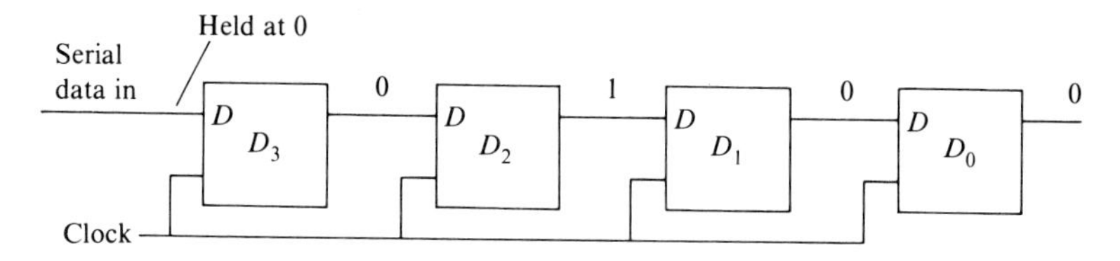

Let the initial output values of the flip-flops D_3 to D_0 = 0, 1, 0 0 (prior to clock pulse 0), as shown in Figure 7.61. If each flip-flop is clocked on the leading edge of the clock pulse, the shift register flip-flop outputs (prior to clock pulses 0, 1, 2, 3) should be shown in Table 7.26.

TABLE 7.26
Shift-register outputs

Clock Pulse	Outputs D_3	D_2	D_1	D_0
0	0	1	0	0
1	0	0	1	0
2	0	0	0	1
3	0	0	0	0

Consider flip-flop D_1 at clock pulse 0. It is "seeing" a 1 at its input, and this should be transferred to its output at the next $0 \rightarrow 1$ clock transition. However, one propagation delay later (the duration of time required to change the output of a flip-flop), D_1's data input goes to 0. In order for D_1 to behave properly, the flip-flop must not react to this input change prior to the next $0 \rightarrow 1$ clock transition. That is, the flip-flop must be insensitive to any change in the level of the control input D at any time other than the $0 \rightarrow 1$ transition of the clock. Thus, the flip-flop must not exhibit the transparency property of latches (and gated latches). This insensitivity is called the **shift register property**.

DEFINITION. A flip-flop has the *shift register property* if, following any transition on a control input that causes its output to change state, it immediately (or following a very short recovery time) becomes insensitive to any subsequent changes of its control input(s), such that its new output reflects its input state just prior to the state-changing transition. The flip-flop becomes aware of its new control input(s) following the next transition of its state-changing (clock) input.

Types of Shift Registers

A shift register is a specialized synchronous sequential-logic circuit that is widely used in applications that involve temporary storage and transfer of data within a digital system. A given shift register can be loaded serially or in parallel and has the capability to shift its contents left, right, or in either direction. There are four basic categories of shift registers, as shown in Table 7.27.

TABLE 7.27
Basic shift register categories

Type	Input	Output
PIPO	Parallel	Parallel
PISO	Parallel	Serial
SIPO	Serial	Parallel
SISO	Serial	Serial

A shift register in any one of these categories can be one of the following subcategories: shift left, shift right, shift bidirectionally (either left or right depending on the setting of a control line).

If you were assigned the task of designing and implementing a particular type of shift register (given a handful of positive-edge triggered D flip-flops), how would you do it? You know that a D flip-flop performs a 1-bit right shift. That is, if a D flip-flop has a 0 (or 1) on its input line at time t and the clock line is strobed, then a 0 (or 1) input appears on the flip-flop output line at some later time $t + dt$. Hence, the D flip-flop input is shifted to its output when the flip-flop clock line is strobed.

If you cascade four D flip-flops (D_3, D_2, D_1, D_0), as shown in Figure 7.62, you have a 4-bit right-shift register.

FIGURE 7.62
A 4-bit right-shift register

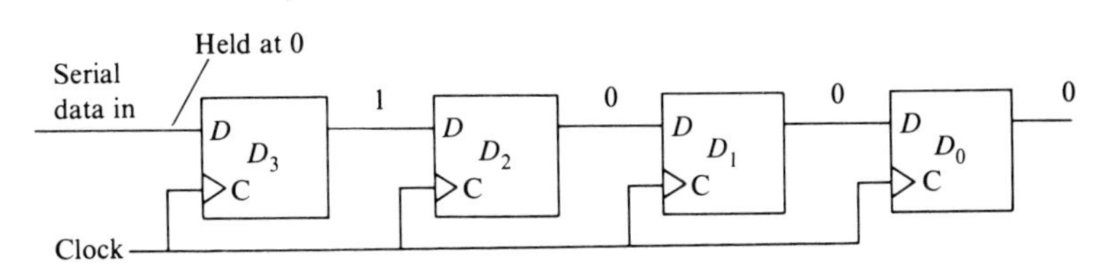

If each flip-flop is clocked on the leading edge of the clock pulse, the shift register flip-flop outputs D_3 to D_0 (initially 1000) will be 0100, 0010, 0001, and 0000 after the first, second, third, and fourth clock pulses. Since successive right shifts with a 0 on the serial-input line generates the sequence 1000, 0100, 0010, 0001, and 0000, a shift register performs a sequence generation function.

The 4-bit shift register in Figure 7.63 is loaded serially through the leftmost flip-flop D_3. After four clock pulses, the entire register is loaded and the output can be transferred in parallel to a destination register. Thus, this register is a serial in/parallel out (SIPO) right-shift register.

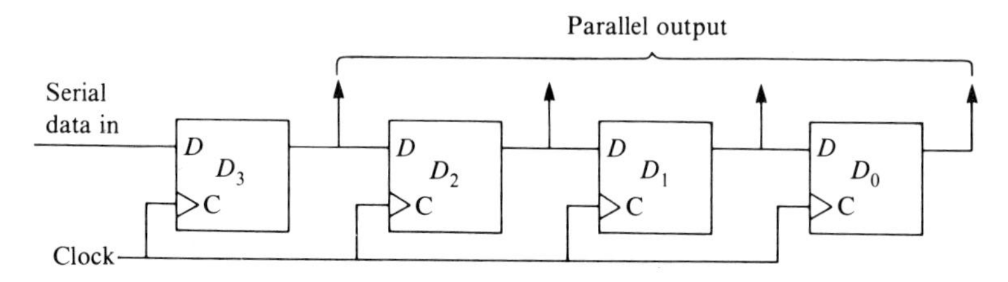

FIGURE 7.63
A 4-bit SIPO right-shift register

The 4-bit shift register in Figure 7.64 is loaded serially through the leftmost flip-flop D_3. After four clock pulses the entire register is loaded. The register contents can then be transferred in serial to a serial-input destination register by strobing the clock pulse four more times. Thus, this register is a serial in/serial out (SISO) right-shift register.

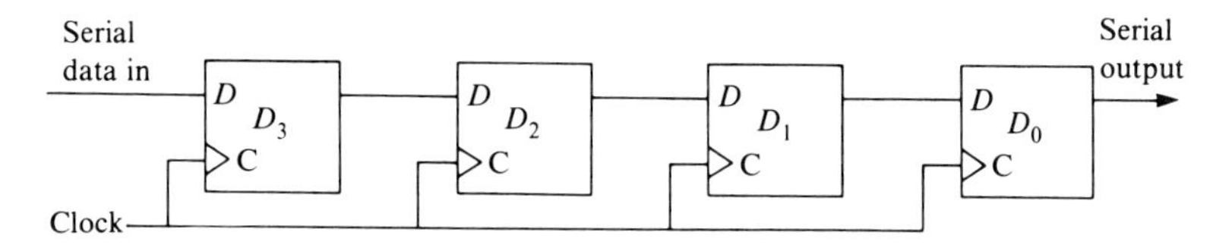

FIGURE 7.64
A 4-bit SISO right-shift register

Parallel in/serial out (PISO) and parallel in/parallel out (PIPO) shift registers can also be constructed using edge-triggered D flip-flops.

Shift registers can also be implemented using flip-flops other than the D type. Figure 7.65 illustrates a 4-bit shift register implemented with edge-triggered JK flip-flops.

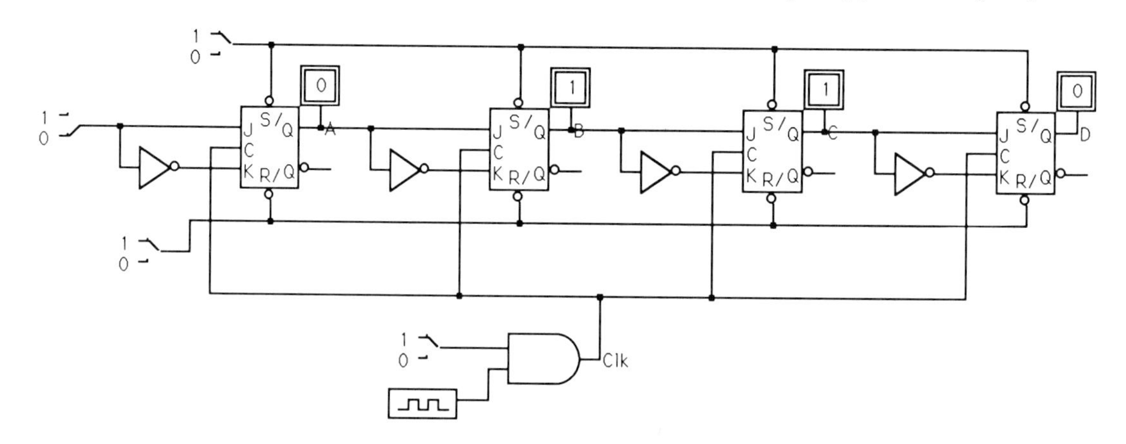

FIGURE 7.65
A 4-bit shift register using JK flip-flops

Registers and Register Transfer

To understand the role of registers in a computer, consider an adder-accumulator loop that processes data stored in a set of four 4-bit (nibble) registers, N_1, N_2, N_3, and N_4. These registers can be connected in any one of the three configurations shown in Figure 7.66.

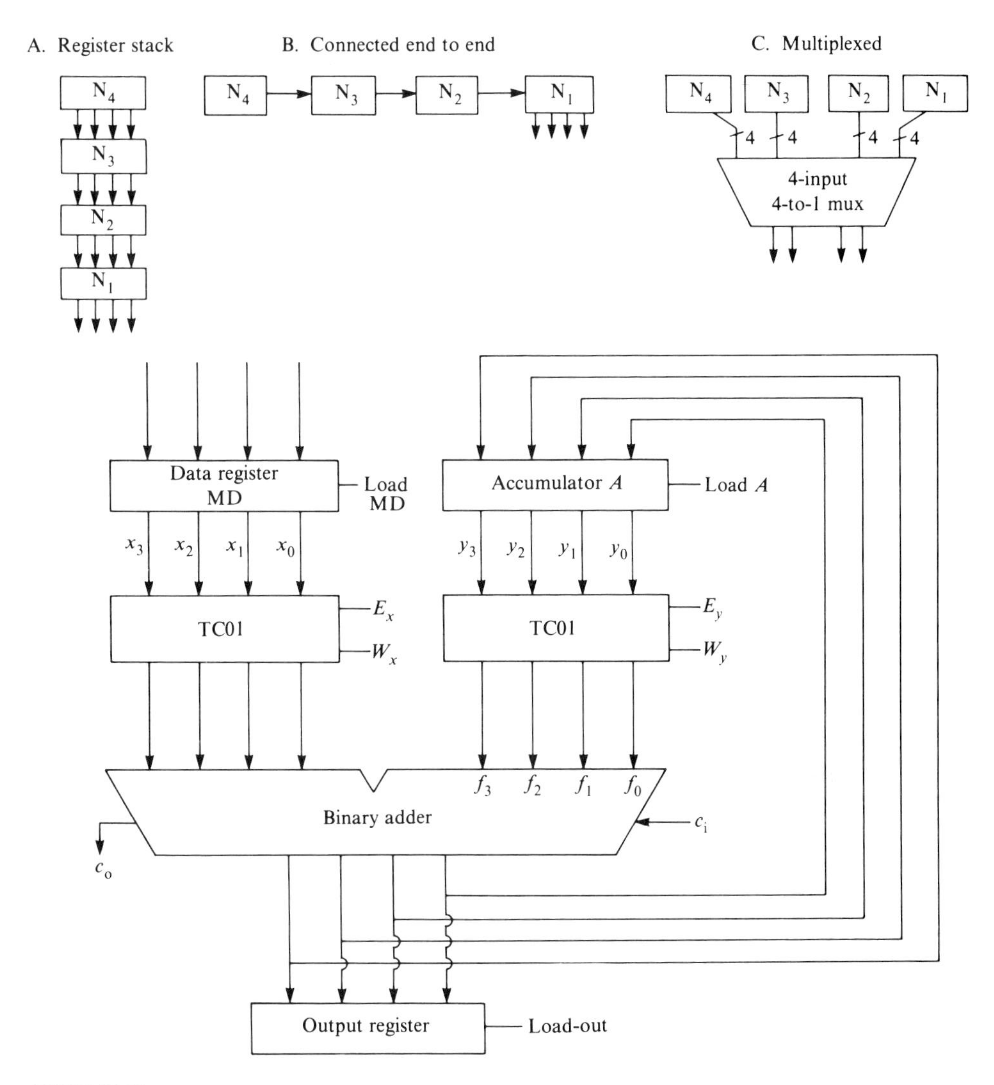

FIGURE 7.66
Block diagram of an adder-accumulator loop

In configuration C of Figure 7.66, the 4 output lines of each register are represented by —/—4, indicating a 4-line bus connecting that register to the mux. One of the register configurations (A, B, or C) will be selected to be connected to the inputs of the MD register. Data is loaded into registers N_1, N_2, N_3, and N_4 and processed using the manually controlled adder-accumulator loop. To load the registers and process data with this simple system requires a capability to load a data word into each of the registers N_1, N_2, N_3, and N_4.

1. To process the first data nibble (N_1) of a given configuration (A, B, or C), data must be transferred from that configuration's register N_1 to the MD register.
2. The second nibble (N_2) of a given configuration is made available for processing as follows:
   a. Using configuration A (register stack), data is transferred from N_2 to N_1 using a parallel transfer method.
   b. Using configuration B (registers connected end to end), data is transferred from N_2 to N_1 using a serial transfer method.
   c. Using configuration C (multiplexed registers), data is transferred via the mux from source register N_2 to destination register MD.

Thus, the processing of data stored in registers N_1 through N_4, using the adder-accumulator loop, requires the ability to

Load a register

Perform operations on registers (e.g., shifting for serial transfer)

Transfer data from register to register

A simple register transfer notation (RTN), summarized in Table 7.28, is used to describe basic types of register operations used in the digital devices and systems developed in the remainder of the text. The arrow in RTN statements indicates replacement.

TABLE 7.28
Register transfer operations

Operation	RTN Notation
Transfer register B contents to register A	$A \leftarrow B$
Load word X into register A	$A \leftarrow X$
Reset (clear) register A	$A \leftarrow 0$
Set (all bits of) register A	$A \leftarrow$ 1s
Increment a register's content	Inc(A)
Decrement a register's content	Dec(A)
Shift Register Operations	
Controls $S_1, S_0 =$ 0, 0	Hold
0, 1	Shift right
1, 0	Shift left
1, 1	Parallel load
Register Arithmetic Operations	
Addition	$C \leftarrow A + B$
Subtraction	$C \leftarrow A - B$
Multiplication	$C \leftarrow A \times B$
Division	$C \leftarrow A/B$
Register Logical Operations	
AND	$C \leftarrow A \wedge B\ (C_i = A_i \wedge B_i)$
OR	$C \leftarrow A \vee B\ (C_i = A_i \vee B_i)$
XOR	$C \leftarrow A \text{ XOR } B\ (C_i = A_i \text{ XOR } B_i)$
Complement	$C \leftarrow A'\ (C_i = A_i')$

$\wedge$ denotes AND; $\vee$ denotes OR

Design of Shift-Register Counters

Some applications require the generation of sequences other than a normal binary count sequence or an arbitrary count sequence. For example, a simple computer timing generator may require each of 4 timing pulses to be generated on successive clock pulses. These timing pulses can be generated by a 4-bit ring counter, which we will describe next. A ring counter will be used in various computers designed in Chapters 12 through 14.

For another example, a more sophisticated computer timing generator may require 8 timing pulses, each to be generated on successive clock pulses. A twisted-ring (Johnson or Moebius) counter generates 8 timing pulses in the sequence $A, B, C, D, A', B', C', D'$. We will also describe a twisted-ring counter.

INTERACTIVE DESIGN APPLICATION

Design of a 4-Bit Ring Counter

A 4-bit ring counter is a device that, starting at reset 0000, generates a sequence 1000, 0100, 0010, and 0001 over and over. An n-bit ring counter uses a circular shift to generate a sequence in which each term contains a single 1 and $n-1$ 0s. That is, a 1 "walks around" a ring. In this section, we will use the 4-step problem-solving procedure to describe the design of a ring counter.

Step 1: Problem Statement. Design a 4-bit ring counter that can be used, starting from reset position 0000, to generate timing pulses A, B, C, and D as indicated by the timing diagram shown in Figure 7.67.

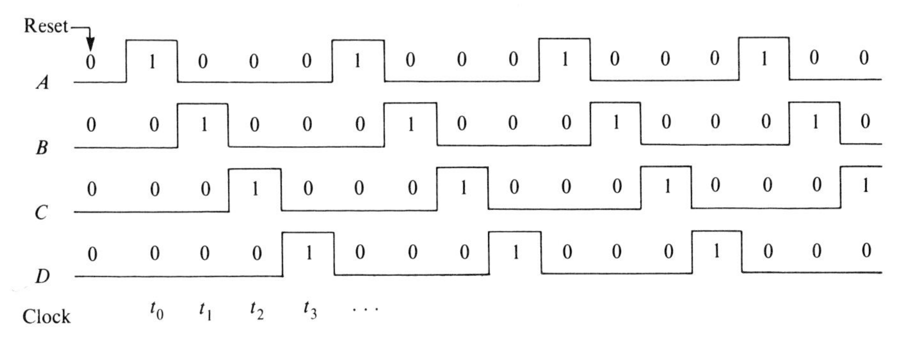

FIGURE 7.67
Timing diagram for a 4-bit ring counter

Step 2: Conceptualization. We know that a shift register has the capability to generate a sequence. Let's consider implementing a ring counter by modifying a shift register. Starting with reset position 0000, the ring counter generates the count sequence shown in Figure 7.68 (p. 320). (The transitions for the A bit and the D bit are indicated to the left and right, respectively, of the count sequence.)

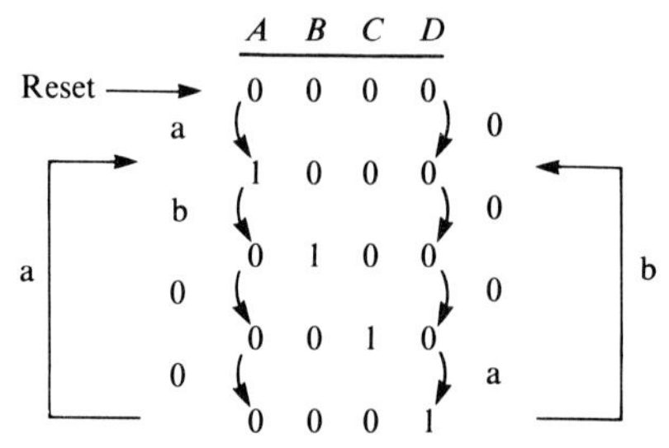

FIGURE 7.68
Transitions for a 4-bit ring counter

Step 3: Solution/Simplification. Plot the transitions for each bit of the ring counter, as shown in Figure 7.69. It should be noted that only 5 combinations (0000, 1000, 0100, 0010, and 0001) of the 16 possible combinations are used in this ring counter. The remaining 11 combinations are don't-care conditions in the transition maps shown in Figure 7.69.

FIGURE 7.69
Transition maps for a 4-bit ring counter

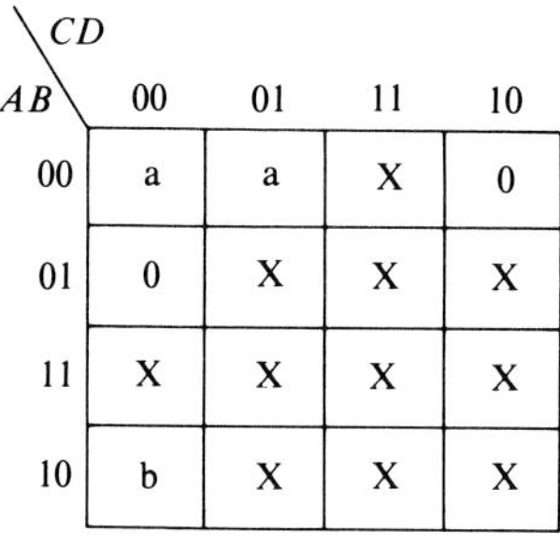

AB \ CD	00	01	11	10
00	a	a	X	0
01	0	X	X	X
11	X	X	X	X
10	b	X	X	X

Flip-flop A

AB \ CD	00	01	11	10
00	0	0	X	0
01	b	X	X	X
11	X	X	X	X
10	a	X	X	X

Flip-flop B

AB \ CD	00	01	11	10
00	0	0	X	b
01	a	X	X	X
11	X	X	X	X
10	0	X	X	X

Flip-flop C

AB \ CD	00	01	11	10
00	0	b	X	a
01	0	X	X	X
11	X	X	X	X
10	0	X	X	X

Flip-flop D

If D flip-flops are used to implement the ring counter, then the D general input equations

$$1_D = 1,\ \text{alpha} \qquad 0_D = 0,\ \text{beta}$$

are substituted into the transition maps to obtain the D input maps shown in Figure 7.70.

The application equations for the D inputs can be found from the input maps (Figure 7.70).

$$D_A = A'B'C' \qquad D_B = A \qquad D_C = B \qquad D_D = C$$

The output of a ring counter requires no decoding because only one flip-flop contains a 1 at a given time. Therefore, the ring counter can be used as a 4-phase clock.

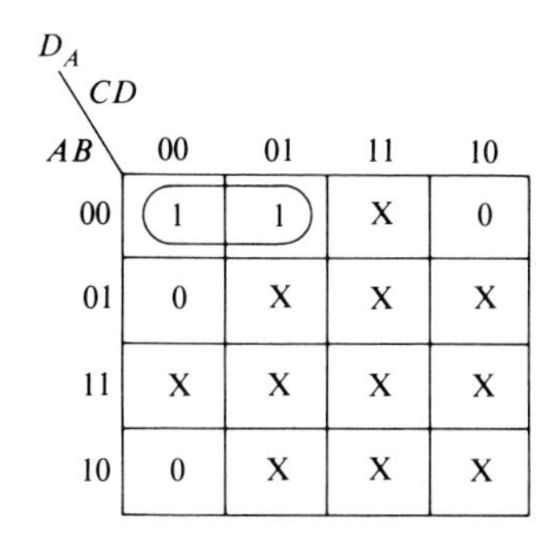

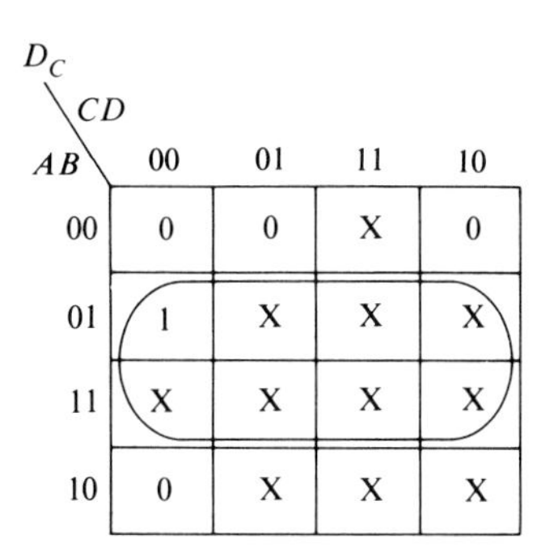

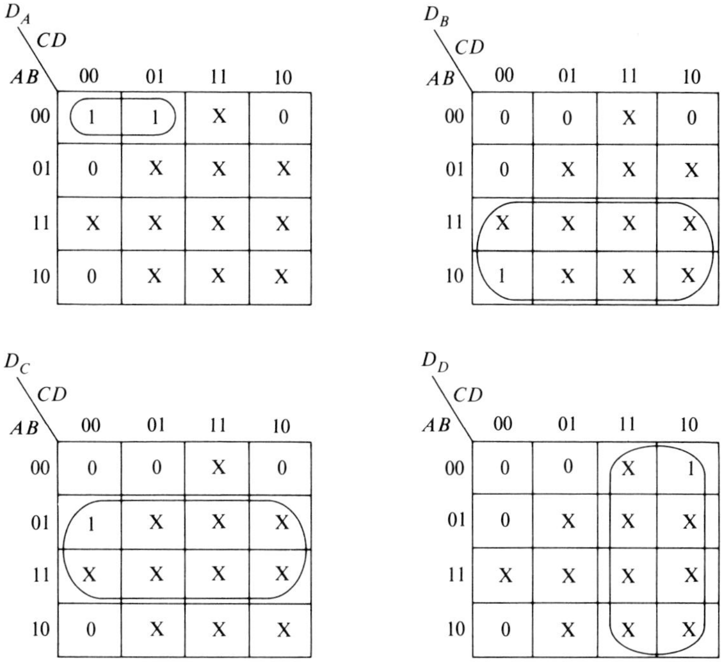

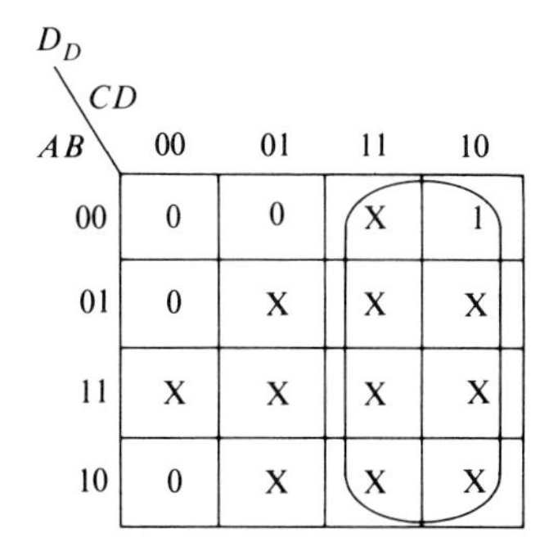

FIGURE 7.70
D input maps for a 4-bit ring counter

Step 4: Realization. Use the application equations to draw a logic diagram of a 4-bit ring counter, as shown in Figure 7.71a. A timing diagram for the ring counter is presented in Figure 7.71b.

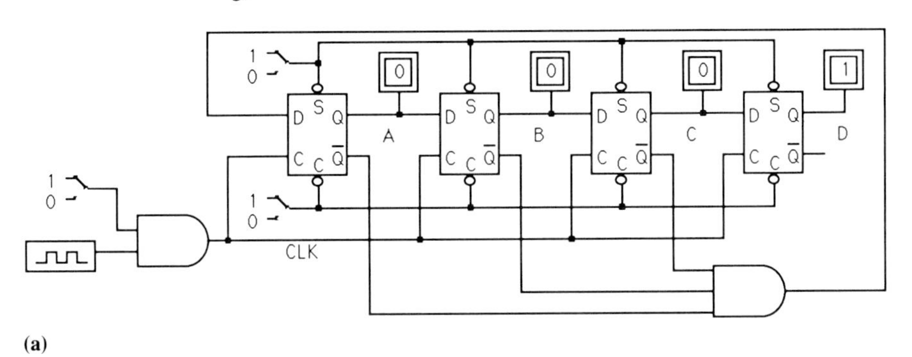

(a)

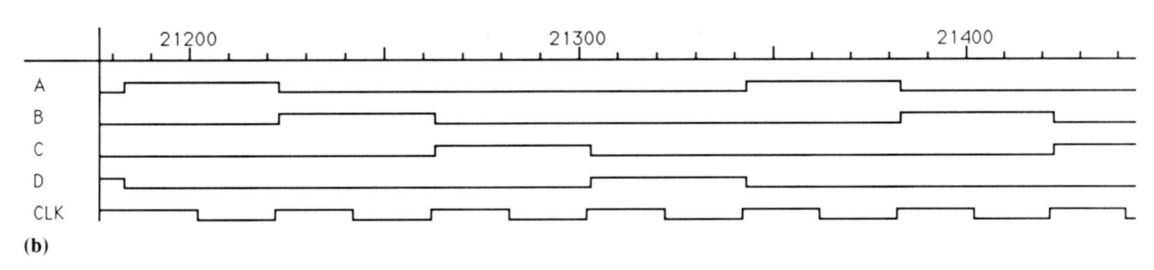

(b)

FIGURE 7.71
A 4-bit ring counter. (a) Logic diagram (b) Timing diagram

INTERACTIVE DESIGN APPLICATION

Design of a 4-Bit Twisted-Ring Counter

A twisted-ring counter, also called a Johnson or Moebius counter, is a device that generates a sequence in which adjacent terms differ in only 1 bit. This device can be used as an 8-phase clock.

Step 1: Problem Statement. Design a 4-bit twisted-ring counter that generates 8 outputs using only 4 flip-flops. The 0-to-1 transitions of the counter occur on successive clock times in the order A, B, C, D, A', B', C', D'. The twisted-ring counter has the timing diagram shown in Figure 7.72.

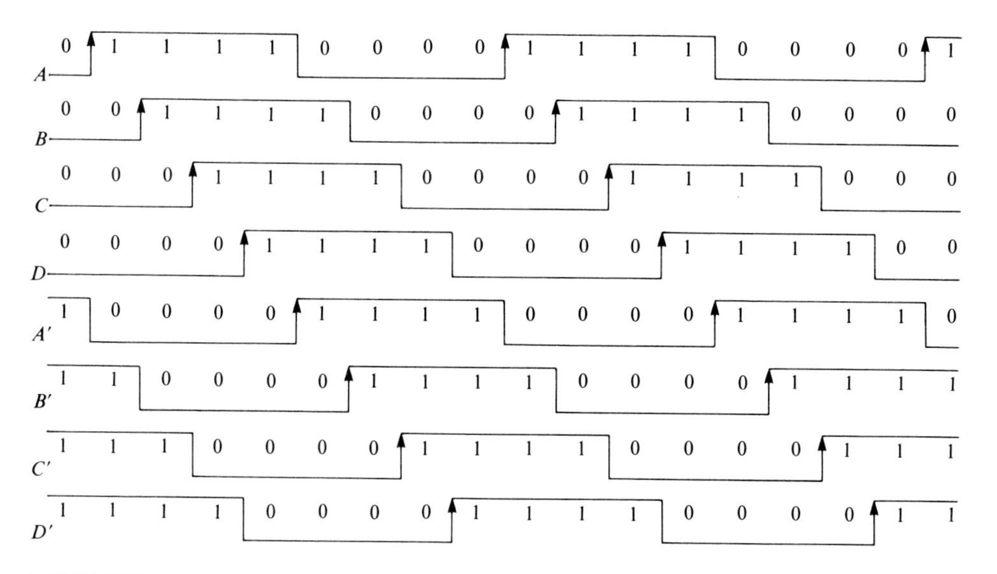

FIGURE 7.72
Timing diagram for a 4-bit twisted-ring counter

Step 2: Conceptualization. Let's consider implementing a twisted-ring counter by modifying a shift register. Starting with reset position 0000, the twisted-ring counter generates the count sequence shown in Figure 7.73. (The transitions for the *A* bit and the *D* bit are indicated to the left and right, respectively, of the count sequence.)

The outputs of a twisted-ring counter are decoded by ANDing the adjacent 10 terms, as indicated by the T_i column to the right of the count sequence.

Step 3: Solution/Simplification. Plot the transitions for each bit of the twisted-ring counter, as shown in Figure 7.74. It should be noted that only 8 combinations of the 16 possible are used in this counter. The remaining 8 combinations are don't-care conditions in the transition maps shown in Figure 7.74.

FIGURE 7.73
Transitions for a 4-bit twisted-ring counter

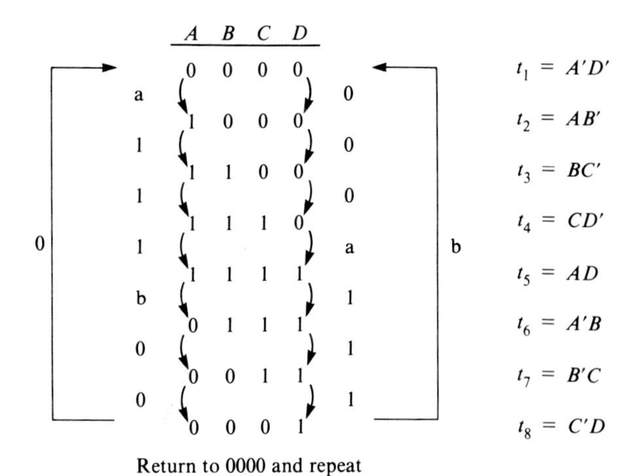

FIGURE 7.74
Transition maps for a 4-bit twisted-ring counter

AB \ CD	00	01	11	10
00	a	0	0	X
01	X	X	0	X
11	1	X	b	1
10	1	X	X	X

Flip-flop A

AB \ CD	00	01	11	10
00	0	0	0	X
01	X	X	b	X
11	1	X	1	1
10	a	X	X	X

Flip-flop B

AB \ CD	00	01	11	10
00	0	0	b	X
01	X	X	1	X
11	a	X	1	1
10	0	X	X	X

Flip-flop C

AB \ CD	00	01	11	10
00	0	b	1	X
01	X	X	1	X
11	0	X	1	a
10	0	X	X	X

Flip-flop D

If D flip-flops are used to implement the twisted-ring counter, then the D general input equations

$$1_D = 1, \text{alpha} \qquad 0_D = 0, \text{beta}$$

are substituted into the transition maps to obtain the D input maps shown in Figure 7.75.

The application equations for the D flip-flops in the twisted-ring counter can be obtained directly from the input maps (Figure 7.75, p. 324).

$$D_A = D' \qquad D_B = A \qquad D_C = B \qquad D_D = C$$

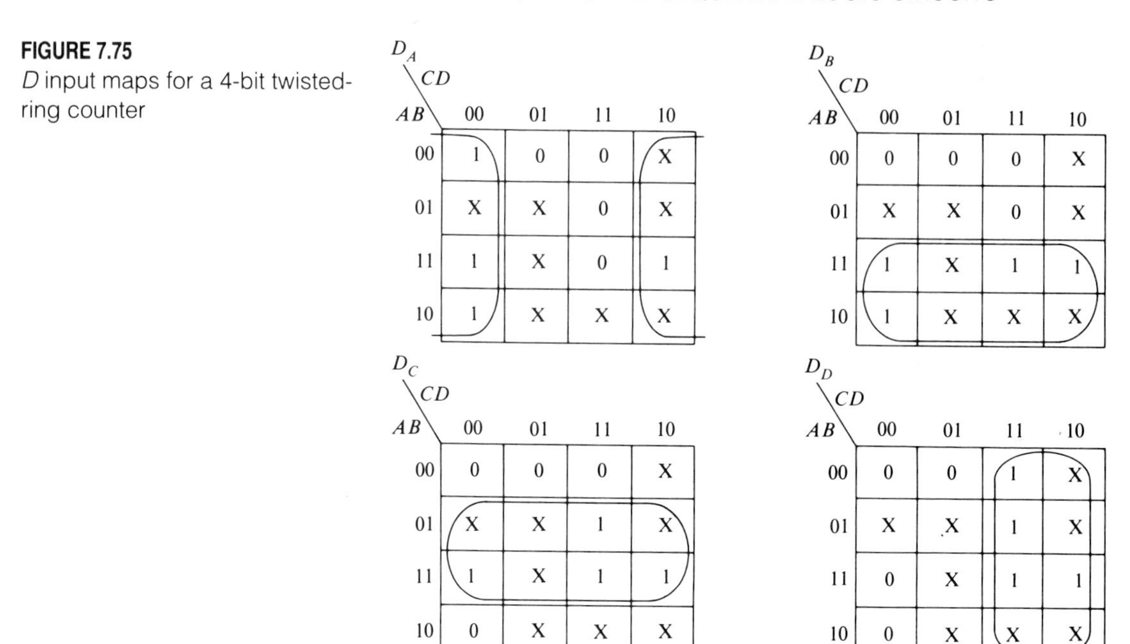

D_A

AB \ CD	00	01	11	10
00	1	0	0	X
01	X	X	0	X
11	1	X	0	1
10	1	X	X	X

D_B

AB \ CD	00	01	11	10
00	0	0	0	X
01	X	X	0	X
11	1	X	1	1
10	1	X	X	X

D_C

AB \ CD	00	01	11	10
00	0	0	0	X
01	X	X	1	X
11	1	X	1	1
10	0	X	X	X

D_D

AB \ CD	00	01	11	10
00	0	0	1	X
01	X	X	1	X
11	0	X	1	1
10	0	X	X	X

FIGURE 7.75
D input maps for a 4-bit twisted-ring counter

Step 4: Realization. Use these application equations to draw a logic diagram for a twisted-ring counter, as shown in Figure 7.76a. A timing diagram is presented in Figure 7.76b.

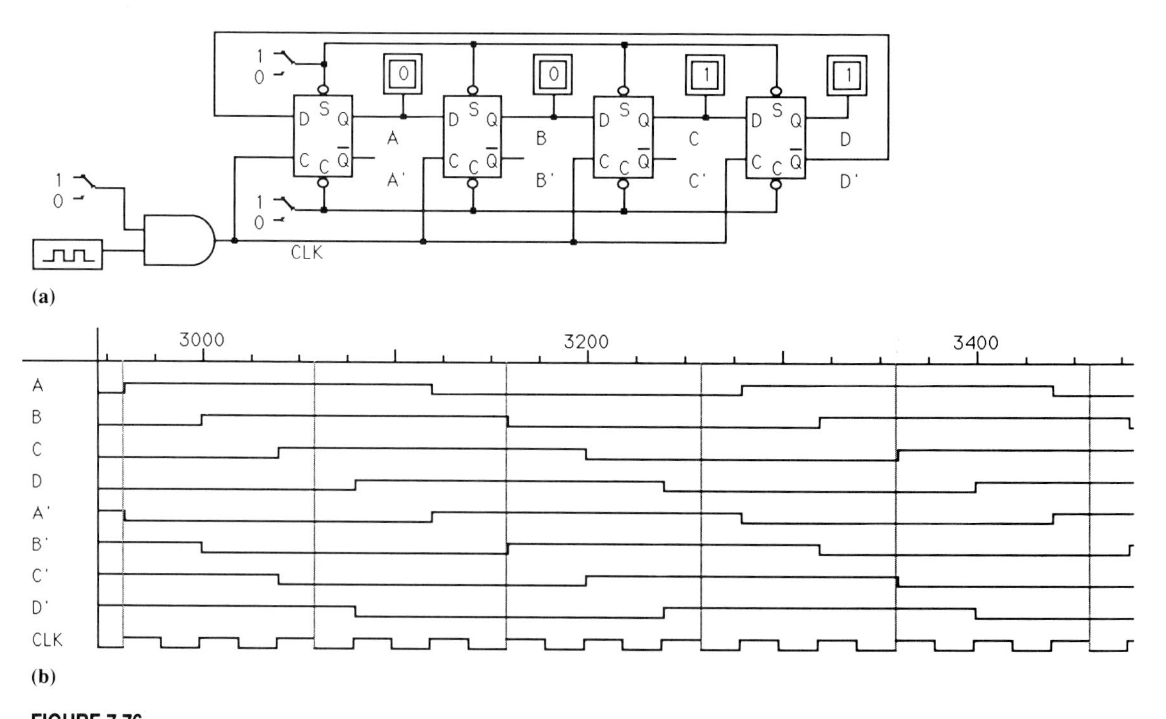

FIGURE 7.76
Twisted-ring counter. (a) Logic diagram (b) Timing diagram

7.6 SUMMARY

In this chapter, we described the fundamental concepts of sequential processes, their representation by sequential (finite-state) machines and realization as sequential logic circuits. Since sequential logic circuits are characterized by a memory capability, we described the design and operation of a basic memory cell (*SR* latch). The *SR* latch is a bistable device (with logic states 0 and 1) that operates asynchronously. That is, its state may change at any time in response to a change in one of its control inputs, *S* and *R*. The use of mixed-logic methodology facilitates the derivation of alternative implementations of the *SR* latch. We defined state transitions that are useful in analyzing and designing sequential logic circuits. We then showed how the *SR* latch can be used as a basic building block to construct a variety of gated latches and clocked flip-flops.

Latches and gated latches possess a transparency property. That is, a change in a control input affects the latch output directly, so that the output "sees" the device input. A gated latch is sensitive to changes in a control input if the enable input is asserted.

The memory of synchronous sequential-logic circuits consists of clocked bistable memory elements. We described procedures for the design of clocked (level-sensitive) flip-flops and for converting flip-flops from one type to another, and presented a summary of flip-flop characteristics and excitations.

The master-slave (pulse-triggered) flip-flop was an early attempt to design a bistable device that is insensitive to glitches in the control inputs while the clock is asserted. However, if a glitch occurs in *S* or *R* while the clock is high, the master latch will produce an improper output that is transferred to the slave latch when the clock goes low. Because of this "glitch catching," master-slave flip-flops have been replaced in most applications by edge-triggered flip-flops.

Edge-triggered flip-flops are designed to eliminate the oscillation and sensitivity-to-noise problems associated with level-sensitive flip-flops and to solve the glitch-catching problem inherent in master-slave flip-flops. An edge-triggered flip-flop is insensitive to input changes at any time other than during the state-changing clock transition. Since an input change does not directly affect its output, an edge-triggered flip-flop does not possess the transparency property. Consequently an edge-triggered flip-flop can "read in" a new state at the same time that its present state is being "read out"; this capability is the principal advantage of an edge-triggered flip-flop over a latch.

We used the 4-step problem-solving procedure to illustrate the design of counters, which are specialized sequential logic circuits. A variety of counters were designed, including synchronous and asynchronous *n*-bit binary counters and divide-by-*n* counters.

Since the operation of digital systems can be described in terms of register operations, we described data registers and shift registers and the basic concepts of register transfer. We then used the 4-step problem-solving procedure to describe the design of two types of shift-register counters (a ring counter and a twisted-ring counter) that can be used as timing circuits in digital systems.

KEY TERMS

Present state
Next state
Bistable device
SR latch

Iterative Processes
Sequential Process
Functional sequential process
State table
State diagram
Sequential (finite-state) machine
Moore machine
Mealy machine
Primary variables
Secondary variables
Characteristic table
Excitation table
Clock
Astable multivibrator
Initial transition
Cycle time
Flip-flop
Gated latch
Gated *SR* latch
Gated *D* (data) latch
Transparent device
Setup time
Hold time
State-changing clock transition
Counter
Modulus
Ripple counters
Load register
Shift register property

EXERCISES

General

1. For an *SR* latch, fill in column *Q*, and complete the description in the following table:

Function	*S*	*R*	*Q*	Description
Hold	0	0		Next state = ________
Reset	0	1		Next state = ________
Set	1	0		Next state = ________
Not allowed	1	1		Next state = ________

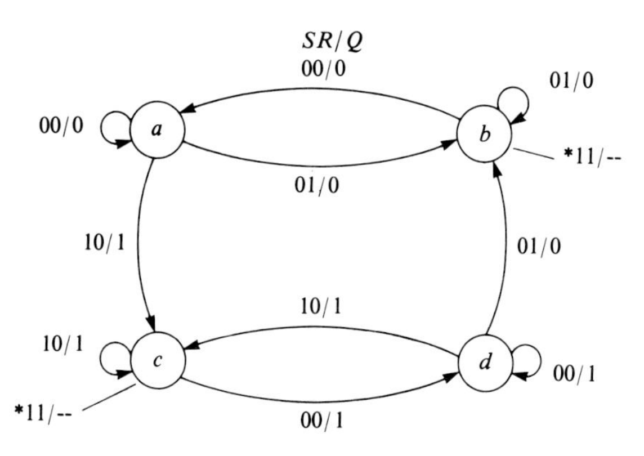

FIGURE 7.77
Primitive state diagram for Exercises 2 and 3

2. The operation of an *SR* latch can be described by the state diagram in Figure 7.77. Answer the following questions using this diagram.
   a. Each stable total state (S, R, q) is a _____ of the state diagram.
   b. Each edge of a state diagram is labeled by ___/___.

3. Use the primitive state diagram of Exercise 2 to complete the corresponding primitive flow table for an *SR* latch. Circle the stable states.

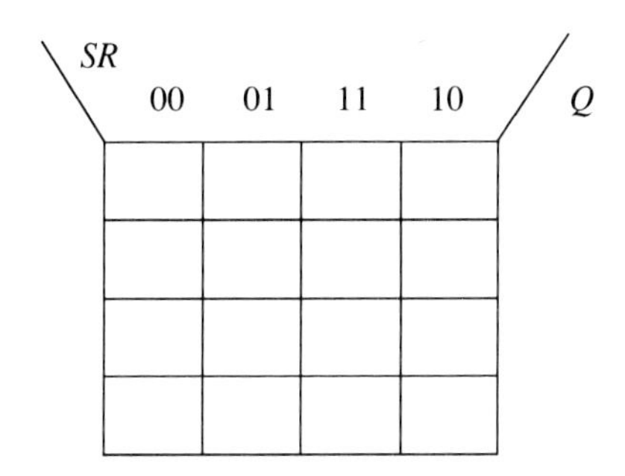

4. Given the following next-state map, determine a simplified SOP expression for Q.

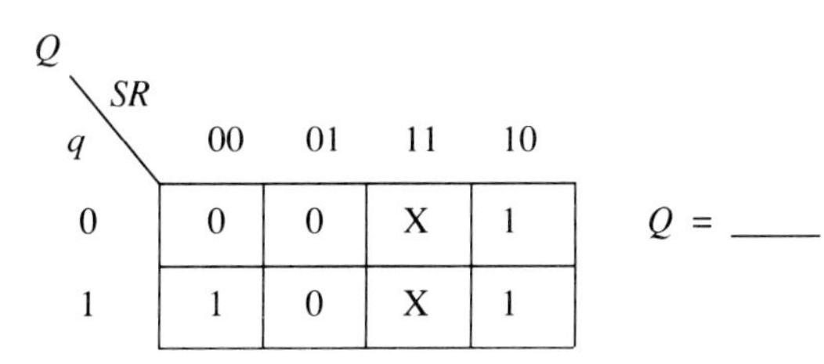

5. Label the timing diagram below with clock period, leading edge, trailing edge.

6. Fill in the following transition definition table:

$q \to Q$	Transition
$0 \to 0$	
$0 \to 1$	
$1 \to 0$	
$1 \to 1$	

7. Given a *JK* flip-flop characteristic table, fill in the transition column.

Function	J K $q \to Q$	Transition	Function	J K $q \to Q$	Transition
Hold	$0\ 0\ 0 \to 0$		Set	$1\ 0\ 0 \to 1$	
	$0\ 0\ 1 \to 1$			$1\ 0\ 1 \to 1$	
Reset	$0\ 1\ 0 \to 0$		Toggle	$1\ 1\ 0 \to 1$	
	$0\ 1\ 1 \to 0$			$1\ 1\ 1 \to 0$	

8. Given the *JK* characteristic table in Exercise 7, fill in the next-state map.

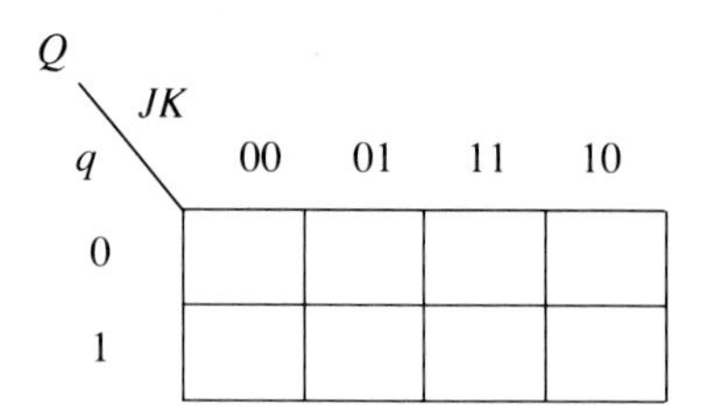

9. Use the completed map in Exercise 8 to write *JK* characteristic equation:

$$Q = \_\_\_\_\_$$

10. Fill in column Q and the transition column for a D flip-flop:

D	$q \rightarrow Q$	Transition
0	$0 \rightarrow$	
0	$1 \rightarrow$	
1	$0 \rightarrow$	
1	$1 \rightarrow$	

11. The D flip-flop has the following characteristic equation:

$$Q = \_\_\_\_\_$$

12. Given the count sequence shown below for a 2-bit counter AB, answer the following questions.

AB
00
01
10
11

a. Flip-flop B _____ with each clock pulse.
b. Flip-flop A operates on each clock pulse as follows:

A _____ when B = 0 A _____ when B = 1

13. A T flip-flop has the following characteristic table. Convert an SR flip-flop to a T flip-flop.

T	q	Q
0	0	0
0	1	1
1	0	1
1	1	0

14. Convert a JK flip-flop to a T flip-flop.

15. Draw a block diagram showing register transfer from either register A or B to register C using a 4-input 2-to-1 mux.

Design/Implementation.

Use the 4-step problem-solving procedure as appropriate.

1. Design a clocked T flip-flop using a basic memory cell as a building block.
2. Design a clocked SR flip-flop using a basic memory cell as a building block.
3. When an electrical switch, such as a single-pole double-throw (SPDT) switch, is opened or closed, the wiper (output) may bounce several times before settling in to its new position. A switch debounce circuit can be designed using NAND gates and pull-up resistors, as shown below. Write a logic equation for the debounced output $S.H$.

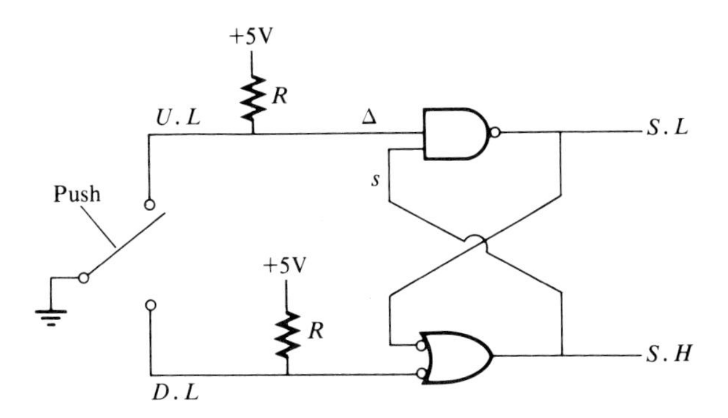

4. Design a synchronous 3-bit binary counter with JK flip-flops.
5. Design a synchronous 4-bit binary counter with D flip-flops.
6. Design a synchronous BCD counter with D flip-flops.
7. Design a synchronous 3-bit counter that generates the sequence 0, 2, 4, 6, 7, 5, 3, 1. Use JK flip-flops.
8. Design a divide-by-5 counter that generates the sequence (in decimal) 0, 1, 2, 3, 4. Implement the counter using edge-triggered JK flip-flops.
9. Design and implement a 3-bit Gray-code counter using edge-triggered D flip-flops.
10. Design a divide-by-3 counter that generates the sequence (in decimal) 0, 1, 2. Implement the counter using edge-triggered JK flip-flops.
11. Design a 3-bit counter that generates the sequence (in decimal) 0, 4, 6, 7, 3, 1, 5.
12. Design a 4-bit parallel in/parallel out register.
13. Design a parallel in/serial out right-shift register using AND and OR gates and edge-triggered D flip-flops.
14. Design a device that generates the sequence of integers 1, 2, 4, and 8 over and over again. Let t_k denote the kth term of the sequence. That is, $t_1 = 1$, $t_2 = 2$, $t_3 = 4$, $t_4 = 8$. Use JK flip-flops to implement the sequence generator.

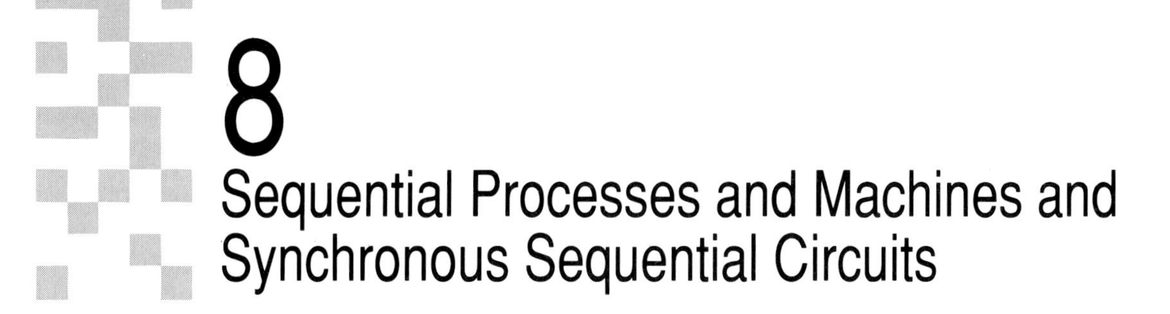

8
Sequential Processes and Machines and Synchronous Sequential Circuits

OBJECTIVES:

After you complete this chapter, you will be able

- ☐ To describe procedures for analyzing general synchronous sequential-logic circuits
- ☐ To describe procedures for designing general synchronous sequential-logic circuits that include translating design specifications into a state diagram and using state-table reduction methods and state assignment techniques
- ☐ To describe a number of general sequential processes
- ☐ To develop sequential (finite-state) machine models for specified processes
- ☐ To design synchronous sequential-logic circuit realizations for specified processes/machines

8.1 INTRODUCTION

Problem solving in the real world often involves analyzing the operation of a process and recognizing patterns in the sequences of states and outputs that are caused by different sequences of inputs. As we saw in Chapter 7, sequential processes can be represented (modeled) by recursion formulas. The recursion formula for a given process defines a relation between successive terms (present term and next term) of a sequence generated by that process. The solution of design problems in digital logic often involves the analysis of sequential processes, their representation by finite-state machine models, and the design of sequential logic circuits that realize these processes.

In this chapter, we are concerned with general sequential processes such as sequence generation, sequence detection, and filtering (elimination of a short-term sequence). We will investigate only those sequential processes that can be represented by finite-state machines, as defined in Chapter 7.

First, we describe two simple sequential processes to help visualize how such processes operate and how they are described using state tables and state diagrams. Then, we analyze a number of representative synchronous sequential-logic circuits that realize various sequential processes. Recall the following simple definition:

A **sequential process** starts in an initial state s_0 and, in response to a sequence of inputs i, the process passes through a sequence of states s and generates a sequence of outputs z.

Now, let's proceed with an example of a sequential process.

EXAMPLE 8.1

Consider a sequential process that performs a modulo-4 addition of a sequence of 2-bit numbers and generates a 1 output as long as the sum is a binary 11. This sequential process can be defined by a state table that lists the next state and output for each combination of the present state and input of the process.

Table 8.1 presents a state table of the modulo-4 addition process. The next state is the modulo-4 sum of two 2-bit numbers, AB and x_1x_2, which are the present state and input of the process. The present state represents the current sum and the next state represents the next sum. Since the process generates a 1 output only as long as the current sum is a binary 11, the output is a function only of the present state.

The state table can be used to draw a state diagram that graphically describes the modulo-4 addition process. In the state diagram, we let state 00 be denoted by label S_0, state

TABLE 8.1
State table for modulo-4 addition

	Next State A^+B^+				
	Input x_1x_2				
Present State AB	00	01	11	10	Output
00	00	01	11	10	0
01	01	10	00	11	0
11	11	00	10	01	1
10	10	11	01	00	0

01 by label S_1, state 10 by label S_2, and state 11 by label S_3. Recall that a state diagram is a directed graph that depicts the states of a process, together with the present-state/next-state transitions for each possible combination of present state and input:

- □ Each state of the process is represented by a node (circle) that contains the state's label and output.
- □ Each transition from state S_i to state S_j caused by input i is represented by a directed branch from node S_i to node S_j, which is labeled with the input.

A state diagram of the modulo-4 addition process is presented in Figure 8.1.

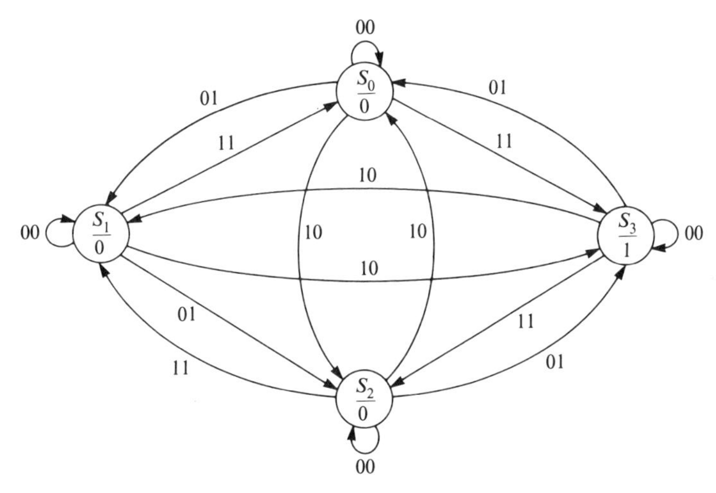

FIGURE 8.1
State diagram for the modulo-4 addition process

A sequential (finite-state) machine is a convenient model to represent a sequential process that has a finite number of states. The modulo-4 addition process has

A finite set of states: $S = \{00, 01, 10, 11\}$

An initial state 00

A finite set of inputs: $\$ = \{00, 01, 10, 11\}$

A next-state function $d(s, i) = s + i$ (modulo-4)

An output function $f(s, i) = 1$ if $s + i$ (modulo-4) $= 11$ ■

Sequence detection is another general sequential process. A simple example of this type of process is presented in the following example.

■ **EXAMPLE 8.2**

Consider a process that accepts an arbitrary sequence of 0s and 1s as input. At each state, the process generates an output as follows: If 3 or more consecutive 1s have been input, then the output z is 1; otherwise the output z is 0. A state table for a Moore 111 sequence detector is presented in Table 8.2 (p. 334).

TABLE 8.2
State table for a Moore 111 sequence detector

Present State	Next State $x = 0$	Next State $x = 1$	Output z
S_0	S_0	S_1	0
S_1	S_0	S_2	0
S_2	S_0	S_3	0
S_3	S_0	S_3	1

The process of detecting a 111 sequence can be graphically described by the state diagram shown in Figure 8.2.

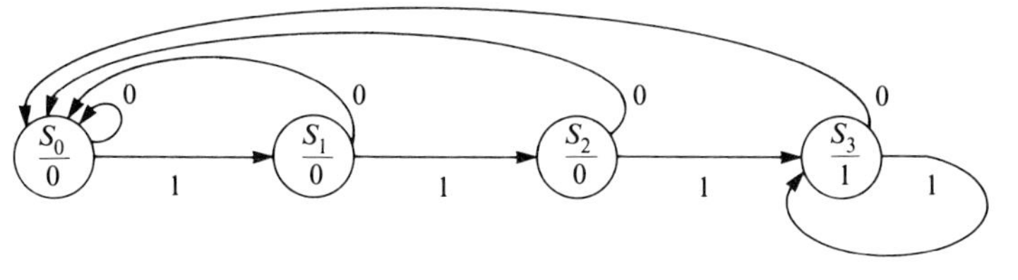

FIGURE 8.2
State diagram for a Moore 111 sequence detector ■

A general sequential process that has a finite set of states can be represented by a sequential (finite-state) machine and realized by a synchronous sequential-logic circuit of the form illustrated in Figure 8.3. Such a circuit includes the following components:

A set of inputs: $x_1, x_2, \ldots, x_n$

A memory section (p flip-flops) with outputs $y_1, \ldots, y_p$

A combinational logic network consisting of excitation (next state) logic and output logic

A set of outputs $z_1, z_2, \ldots, z_m$

A clock to synchronize the flip-flop state changes

FIGURE 8.3
Model for a general synchronous sequential-logic circuit

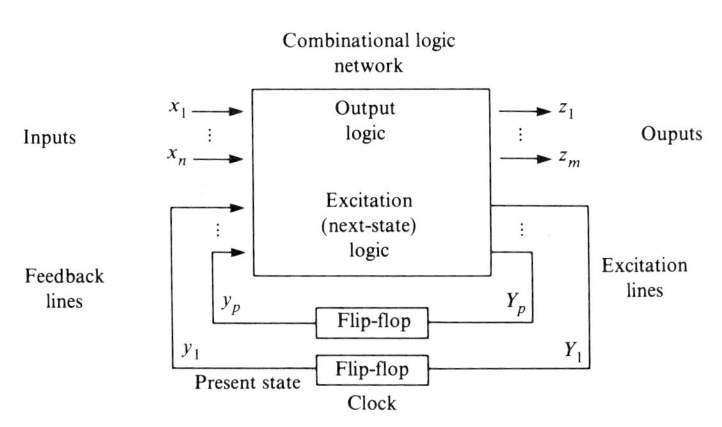

The behavior of a synchronous circuit is determined by its state and its external inputs at specific instants of time. In a synchronous sequential-logic circuit, the memory consists

of clocked flip-flops whose state changes are synchronized by the system clock.

When an input changes, the flip-flop controls respond after a propagation delay time [$t(pnsg)$] of the next-state generation logic and the flip-flop setup time [$t(su)$]. Thus, signals applied to flip-flop control inputs must be stable at least one setup time prior to the flip-flop state-changing clock transition.

In an alternate state transition method, the flip-flop state changes occur on the alternate clock transition to input signal changes. For a 50% duty cycle clock, this method allows one-half clock period between an input change and a state change; the minimum clock half period ($T/2$) is

$$T/2 = t(cd) + t(pnsg) + t(su)$$

where $t(cd)$ is the clock to data-stable time.

Synchronous sequential-logic circuits are simpler to design than asynchronous circuits because the synchronization of memory state changes eliminates many of the problems caused by unequal propagation delays in different signal paths. At any given time t, the memory is in a state ($y_1, \ldots, y_p$), called the **present state**, consisting of the p outputs of the flip-flops. The present state and the external inputs applied at time t determine the conditions on the excitation lines ($Y_1, \ldots, Y_p$), which in turn determine the memory next state after the next state-changing clock transition. Each flip-flop excitation is a Boolean function of the circuit's present state and inputs.

Each output is a Boolean function of the circuit's present state and the inputs. The sequential circuit is classified as either (1) a Moore circuit, if each output is a function only of the circuit's present state, or (2) a Mealy circuit, if any output is a function of the circuit's present state and the circuit inputs.

In the next section, we describe and illustrate procedures for analyzing synchronous sequential-logic circuits using a number of different types of circuits.

8.2 ANALYSIS OF GENERAL SYNCHRONOUS SEQUENTIAL-LOGIC CIRCUITS

The analysis of a synchronous sequential-logic circuit is the process of examining a detailed logic diagram to determine the circuit's function and operating characteristics. A step-by-step procedure for this analysis process is described in the following sections.

Analysis of a Moore Synchronous Sequential-Logic Circuit with *D* Flip-Flops

This section describes a step-by-step procedure for analyzing synchronous sequential-logic circuits, using a Moore circuit for illustration purposes.

■ **EXAMPLE 8.3**

Analyze the Moore sequential circuit shown in Figure 8.4 (p. 336) to determine its function. Begin by writing the excitation and output equations, then construct the excitation and next-state maps, and, finally, draw the state table and state diagram. Study the state diagram to determine the circuit's function.

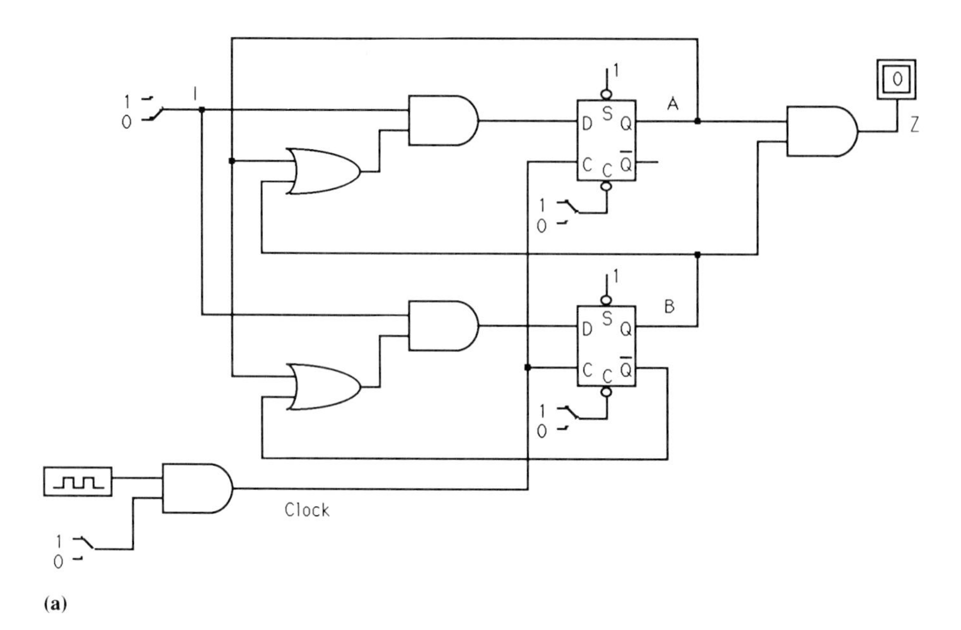

(a)

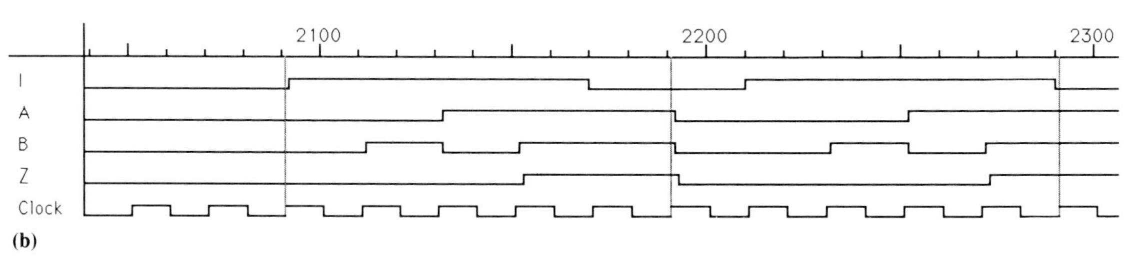

(b)

FIGURE 8.4
Logic diagram of a Moore sequential circuit

Step 1: Analyze the logic diagram in two parts.

a. Identify the following circuit elements: circuit input I; flip-flop excitations D_A, D_B; flip-flop outputs A, B; and circuit output Z.

b. Determine the circuit's excitation and output equations. Determine the excitation equation for each flip-flop by tracing each flip-flop input back to the circuit input.

$$\text{Excitation equations: } D_A = AI + BI \quad \text{and} \quad D_B = AI + B'I$$

Each logic expression specifies a flip-flop's excitation as a function of the circuit's input I and the circuit state A and B.

Determine the circuit's output equation by tracing the circuit output back to the flip-flop outputs.

$$\text{Output equation: } Z = AB$$

Since the output of this circuit is a function only of the circuit state, A and B, we verify that the circuit is a Moore sequential logic circuit.

Step 2: Construct an excitation map for each flip-flop. Each excitation map is a K-map representation of a flip-flop excitation as a function of the circuit's present state AB and input I. The excitation maps are shown in Figure 8.5.

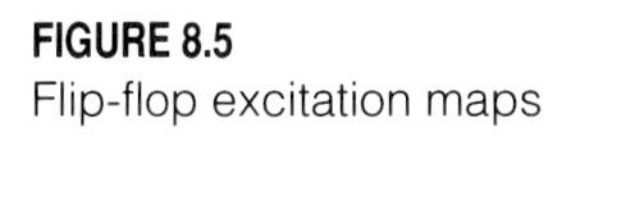

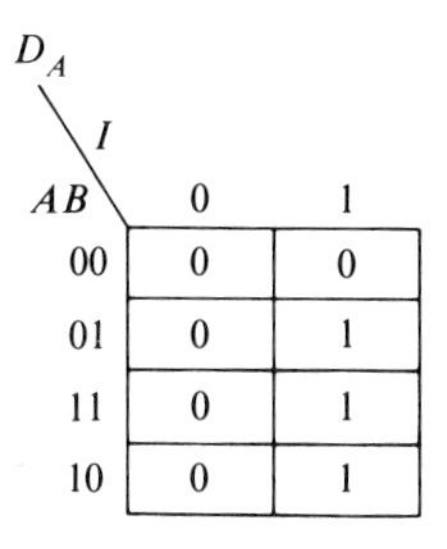

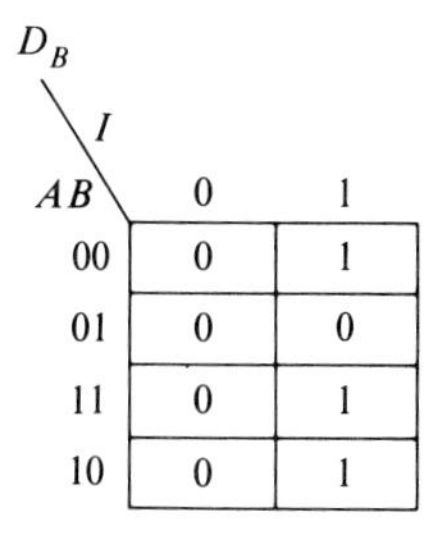

FIGURE 8.5
Flip-flop excitation maps

Step 3: Convert each flip-flop excitation map to a next-state map. Use the characteristic equation of the flip-flop type; that is, for a D flip-flop, $Q^+ = D$. Therefore, each flip-flop next-state map has the same form as the corresponding excitation map. The next-state map defines a flip-flop's next state as a function of the circuit's present state AB and input I, as shown in Figure 8.6.

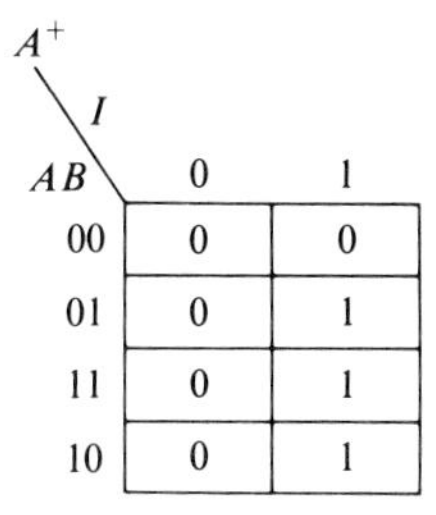

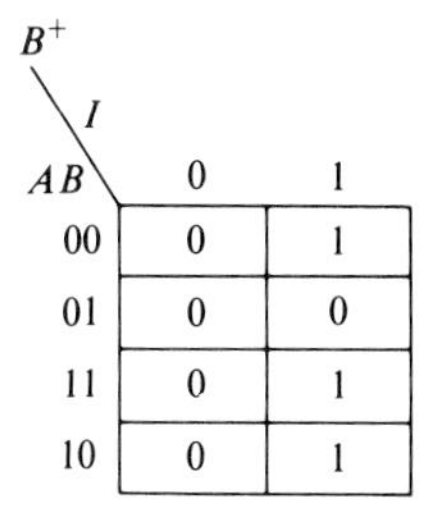

FIGURE 8.6
Flip-flop next-state maps

Step 4: Construct a circuit state table by combining the individual flip-flop next-state maps. Table 8.3 lists the circuit next states, A^+B^+, for each combination of circuit present state AB and input I. Note that the present state entries are written in the Gray-code order used in the K-maps. Since the circuit is a Moore circuit, the output is a function only of the present state.

TABLE 8.3
State table for circuit of Figure 8.4

Present State AB	Next State A^+B^+		Output
	$I = 0$	$I = 1$	
00	00	01	0
01	00	10	0
11	00	11	1
10	00	11	0

Each combination (00, 01, 10, 11) of values of state variables A and B is called a **state vector**. Each circuit input value causes a transition from circuit present state AB to circuit next state A^+B^+. As a general rule, we will use the letters $a, b, c, d, \ldots$ to denote the states of a sequential process, machine, or circuit.

Step 5: Construct a state diagram from the circuit's state table. If we assign a label to each state vector ($a = 00$, $b = 01$, $c = 10$, $d = 11$), we can write the state table shown in Table 8.4.

TABLE 8.4
State table for circuit of Figure 8.4

	Next State A^+B^+		
Present State AB	$I = 0$	$I = 1$	Output
a	a	b	0
b	a	c	0
d	a	d	1
c	a	d	0

The operation of a sequential logic circuit can be graphically described by a state diagram, as shown in Figure 8.7. Each node of the circuit's state diagram is a state of the circuit. Each edge of the state diagram is a directed arc from a present state node (origin) to that node's next state (destination); each edge is labeled with the circuit input value that causes the transition from the edge's origin node to the edge's destination node.

For a Moore circuit, the circuit state corresponding to a given node is written inside the circle for that node. Because the circuit output for a Moore circuit is a function only of the circuit's state, each circuit state has an associated output value. Therefore, for a Moore circuit, the circuit's output for a given state is listed inside the circle beneath the node's label.

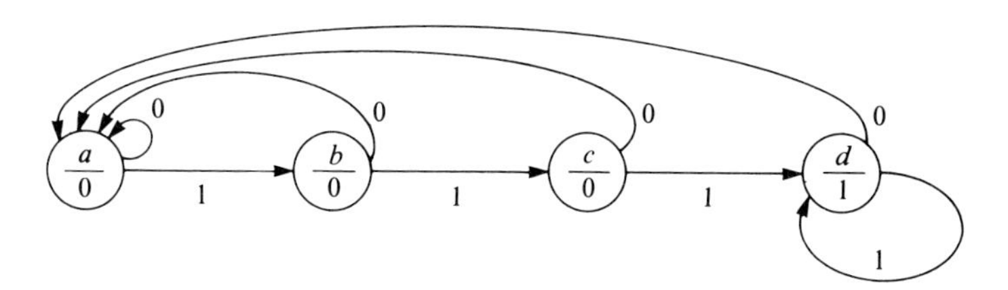

FIGURE 8.7
State diagram for a Moore 111 sequence detector

Step 6: Interpret the state diagram to determine the function of the circuit. Let a be the starting state of the circuit. The circuit remains in state a for each 0 input. A single 1 input causes a transition from state a to state b. If the circuit is in state b, a 0 input causes a transition back to state a, while a 1 input causes a transition to state c. If the circuit is in state c, a 0 input causes a transition back to state a, while a 1 input causes a transition to state d. If the circuit is in state d, a 0 input causes a transition back to state a, while a 1 input causes the circuit to remain in state d. ■

The circuit in this example can be interpreted as a sequence detector. When three or more consecutive 1 inputs occur, the circuit goes to state *d*, causing an output of 1. For a Moore circuit, an output change caused by a given input does not appear until after the circuit's transition to its next state because the output is a function only of the circuit state. Therefore, the output sequence is displaced in time relative to the input sequence.

Summary of Analysis Procedure for Synchronous Sequential-Logic Circuits

A general synchronous sequential-logic circuit, represented by a logic diagram, can be analyzed using the step-by-step procedure illustrated in the preceding example. The following is a summary of the analysis procedure.

Step 1: Analyze the logic diagram in two parts.

a. Identify the following circuit elements: circuit input(s), flip-flop excitations, flip-flop outputs, and circuit output(s).

b. Determine the circuit's excitation and output equations. Determine the excitation equation for each flip-flop by tracing each flip-flop input back to the circuit input. Determine the circuit output equation by tracing the circuit output back to the flip-flop outputs (and to the circuit inputs for a Mealy circuit).

Step 2: Construct an excitation map for each flip-flop.

Step 3: Convert each flip-flop excitation map to a next-state map. Use the characteristic equation of the flip-flop type to convert the excitation map to a next-state map.

Step 4: Construct a circuit state table by combining the individual flip-flop next-state maps.

Step 5: Construct a state diagram from the circuit's state table.

Step 6: Interpret the state diagram to determine the function of the circuit.

Several Moore and Mealy circuits, with either *D* or *JK* flip-flops and either 1 or 2 inputs, are analyzed in the following sections using this step-by-step analysis procedure.

Analysis of a Mealy Synchronous Sequential-Logic Circuit with *D* Flip-Flops

The procedure for analyzing synchronous sequential-logic circuits is illustrated in this section using a Mealy circuit. Remember that a Mealy circuit is a sequential logic circuit whose output is a function of both the present state and the inputs of the circuit. In general, a Mealy circuit requires fewer states than a corresponding Moore circuit.

■ **EXAMPLE 8.4**

Analyze the Mealy sequential circuit shown in Figure 8.8 (p. 340).

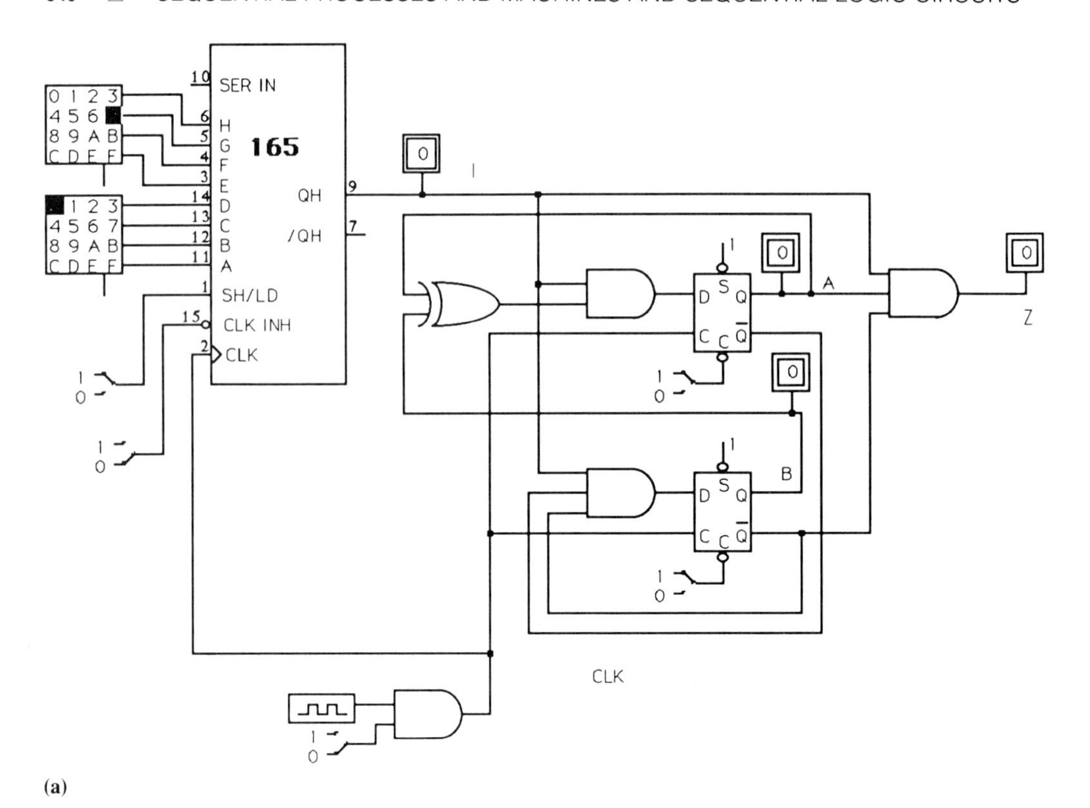

(a)

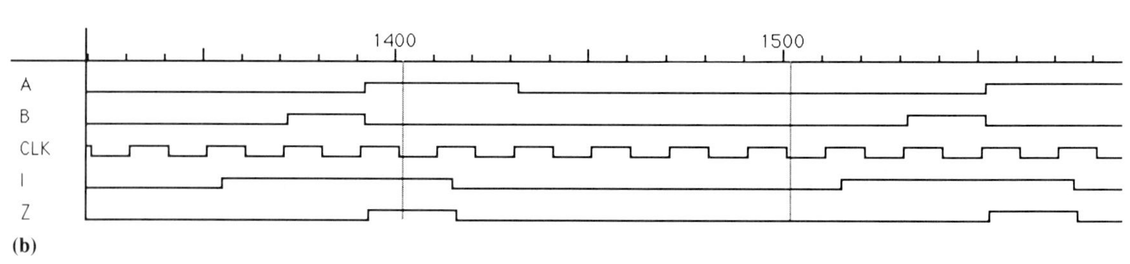

(b)

FIGURE 8.8
Logic diagram of a Mealy sequential logic circuit

Step 1: Analyze the logic diagram in two parts.

a. Identify the circuit elements: circuit input I; flip-flop excitations D_A,D_B; flip-flop outputs A and B; and circuit output Z.

b. Determine the circuit's excitation and output equations. Determine the excitation equation for each flip-flop by tracing each flip-flop input back to the circuit input.

$$\text{Excitation equations: } D_A = (A \oplus B)I \quad \text{and} \quad D_B = A'B'I$$

Each logic expression specifies the flip-flop's excitation as a function of the circuit's input, I, and state variables, A and B.

Determine the circuit output equation by tracing the circuit output back to the flip-flop outputs and the circuit input.

$$\text{Output equation: } Z = AB'I$$

Since the output of this circuit depends on both the input and state, the circuit is a Mealy synchronous sequential-logic circuit.

Step 2: Construct an excitation map for each flip-flop. The excitation maps for this circuit are shown in Figure 8.9.

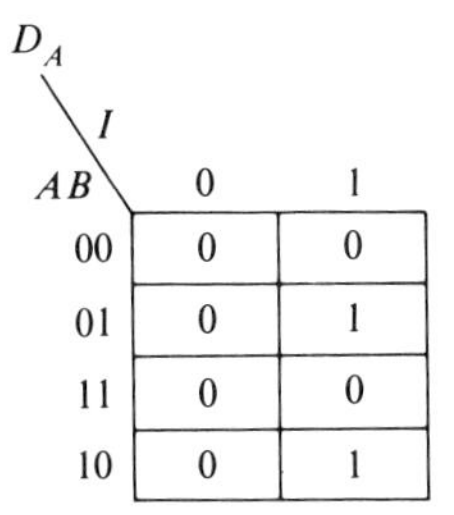

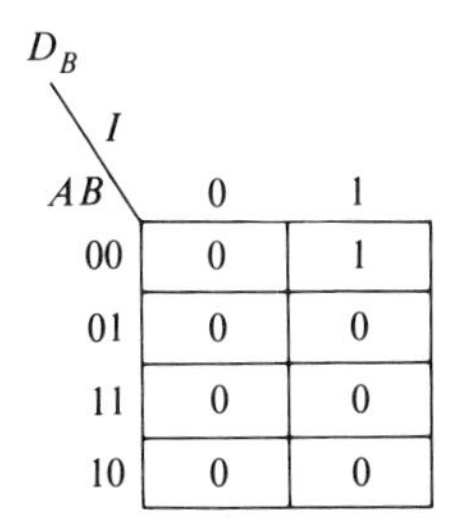

FIGURE 8.9
Flip-flop excitation maps for circuit in Figure 8.8

Step 3: Convert each flip-flop excitation map to a next-state map. Use the characteristic equation of the flip-flop type (for a D flip-flop, $Q^+ = D$), as shown in Figure 8.10

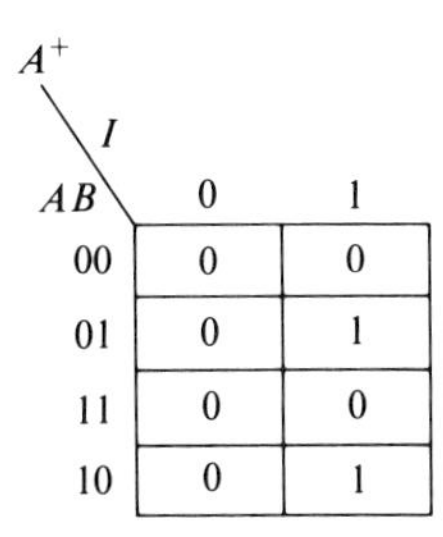

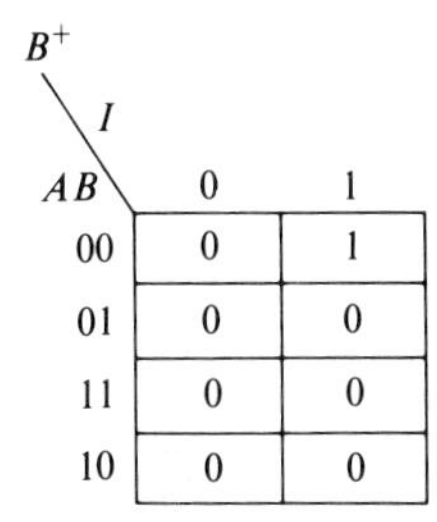

FIGURE 8.10
Flip-flop next-state maps for circuit in Figure 8.8

Step 4: Construct a circuit state table by combining the individual flip-flop next-state maps. Table 8.5 lists the circuit next-state, A^+B^+, for each combination of circuit present state AB and input I.

TABLE 8.5
State table for circuit of Figure 8.8

	Next State A^+B^+		Output	
Present State AB	$I = 0$	$I = 1$	$I = 0$	$I = 1$
00	00	01	0	0
01	00	10	0	0
11	00	00	0	0
10	00	10	0	1

By assigning a label to each state vector (a = 00, b = 01, c = 10, d = 11), we can write the state table shown in Table 8.6 (p. 342).

TABLE 8.6
State table for circuit of Figure 8.8

	Next State A^+B^+		Output	
Present State AB	$I = 0$	$I = 1$	$I = 0$	$I = 1$
a	*a*	*b*	0	0
b	*a*	*c*	0	0
d	*a*	*a*	0	0
c	*a*	*c*	0	1

Step 5: Draw a state diagram from the state table. For a Mealy circuit, the circuit output is a function of the circuit input and the state variables. Consequently each directed arc of the state diagram in Figure 8.11 is labeled with the input and output (separated by a /) associated with that present-state / next-state transition.

FIGURE 8.11
State diagram for a Mealy 111 sequence detector

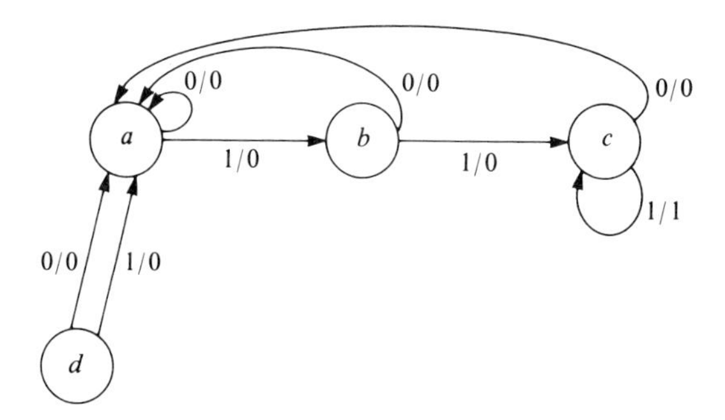

Step 6: Interpret the state diagram to determine the circuit function. Starting in state a, the circuit remains in a for input 0 or goes to state b for input 1. If the circuit is in state b, an input 0 causes a transition back to state a while an input 1 causes a transition to c. If the circuit is in state c, an input 0 causes a return to a while an input 1 causes an output pulse 1 with the circuit staying in c. If the circuit is in state d, an input 0 or 1 causes a return to a. Therefore,

a is the circuit's starting state.

b is the state corresponding to a single 1 preceded by one or more 0s.

c is the state corresponding to two or more consecutive 1s preceded by one or more 0s. Because the circuit outputs a 1 when an input 1 is received while the circuit is in state c, the circuit is a sequence detector for a sequence of three or more consecutive 1s. ■

To sum up our analysis, inputs in a synchronous sequential-logic circuit must be properly synchronized with the clock so that the input signal is stable prior to the state-changing clock transition. For proper operation input changes must occur so that the flip-flop inputs are stable no later than 1 setup time prior to the state-changing clock transition.

In a Mealy circuit, the output caused by a given input occurs immediately following the application of the input. Therefore, the output sequence of a Mealy circuit is not displaced in time relative to the input sequence.

The outputs of a Mealy circuit are associated with the transitions between states. Mealy circuit outputs are pulse signals, while Moore circuit outputs are level signals associated with the circuit states.

Analysis of a Mealy Synchronous Sequential-Logic Circuit With *JK* Flip-Flops

The procedure for analyzing synchronous sequential-logic circuits is illustrated in this section using an example of a Mealy circuit that is implemented using *JK* flip-flops.

EXAMPLE 8.5

Analyze a Mealy sequential-logic circuit using *JK* flip-flops. The logic diagram is given in Figure 8.12.

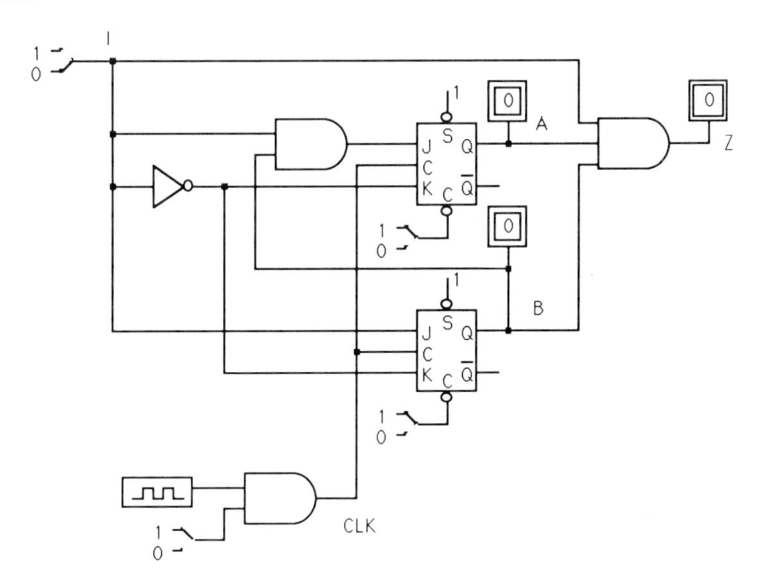

FIGURE 8.12
Logic diagram of a Mealy circuit with *JK* flip-flops

Step 1: Analyze the logic diagram in two parts.

a. Identify the circuit elements: circuit input I; flip-flop excitations J_A, K_A, J_B, K_B; flip-flop outputs A, B; and circuit output Z.

b. Determine the circuit's excitation and output equations. Determine the excitation equation for each flip-flop by tracing each flip-flop input back to the circuit input.

$$\text{Excitation equations: } J_A = BI' \qquad J_B = I \qquad K_A = 1 \qquad K_B = AI'$$

Determine the circuit output by tracing the circuit output back to the flip-flop outputs.

$$\text{Output equation: } Z = ABI$$

Step 2: Use the *JK* flip-flop characteristic equation

$$Q^+ = JQ' + K'Q$$

to convert the flip-flop excitation equations to transition equations:

$$A^+ = J_A A' + K_A'A = BI'A' + 0A = A'BI'$$
$$B^+ = J_B B' + K_B'B = IB' + (AI')'B = A'B + I$$

Step 3: Plot each flip-flop transition equation in a next-state map. The next-state maps are shown in Figure 8.13.

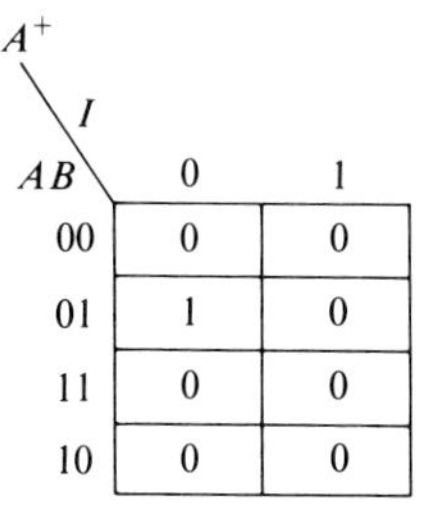

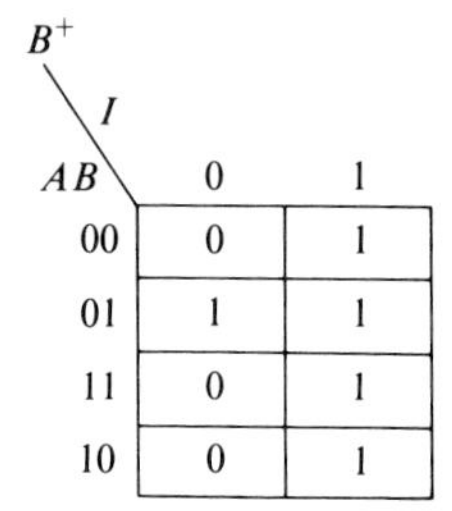

FIGURE 8.13
Flip-flop next-state maps for circuit in Figure 8.12

Step 4: Construct a circuit state table by combining the individual flip-flop next-state maps. Table 8.7 lists the circuit next states, A^+B^+, for each combination of circuit present state AB and input I.

TABLE 8.7
State table for circuit of Figure 8.12

	Next State A^+B^+		Output	
Present State AB	$I = 0$	$I = 1$	$I = 0$	$I = 1$
00	00	01	0	0
01	11	01	0	0
11	00	01	0	1
10	00	01	0	0

Step 5: Construct a state diagram from the circuit's state table. By assigning a label to each state vector, we can write the state table in the form shown in Table 8.8.

TABLE 8.8
State table for circuit of Figure 8.12

	Next State A^+B^+		Output	
Present State AB	$I = 0$	$I = 1$	$I = 0$	$I = 1$
a 00	*a* 00	*b* 01	0	0
b 01	*d* 11	*b* 01	0	0
d 11	*a* 00	*b* 01	0	1
c 10	*a* 00	*b* 01	0	0

The state diagram (Figure 8.14) can be constructed from this state table.

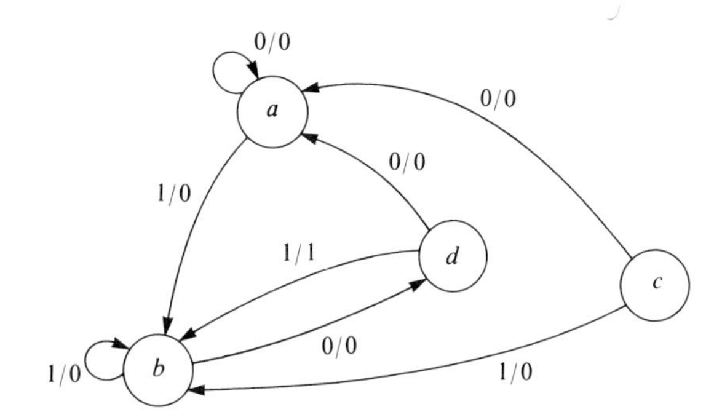

FIGURE 8.14
State diagram for a Mealy 101 sequence detector

Step 6: Interpret the state diagram to determine the function of the circuit. When the three most recent inputs are 101, the circuit goes to state b and generates an output = 1. Therefore, the circuit is a sequence detector for sequence 101. ■

Analysis of a Moore Synchronous Sequential-Logic Circuit with 2 Inputs

The procedure for analyzing synchronous sequential-logic circuits is illustrated in this section using an example of a Moore circuit with two inputs.

■ **EXAMPLE 8.6**

The analysis of a synchronous sequential-logic circuit with 2 inputs is described using the example circuit in Figure 8.15 (p. 346).

Step 1: Analyze the logic diagram in two parts.

a. Identify the circuit elements: circuit inputs x_1, x_2; flip-flop excitations D_A, D_B; flip-flop outputs A, B; and circuit output Z.

b. Determine the circuit's excitation and output equations. Determine the excitation equation for each flip-flop by tracing each flip-flop input back to the circuit inputs.

$$\text{Excitation equations: } D_A = B'(A \oplus x_1) + B(A \oplus (x_1 \oplus x_2))$$
$$D_B = B \oplus x_2$$

Determine the circuit output by tracing the circuit output back to the flip-flop outputs.

$$\text{Output equation: } Z = AB$$

Step 2: Construct flip-flop excitation maps from flip-flop excitation equations. The excitation maps are shown in Figure 8.16 (p. 346).

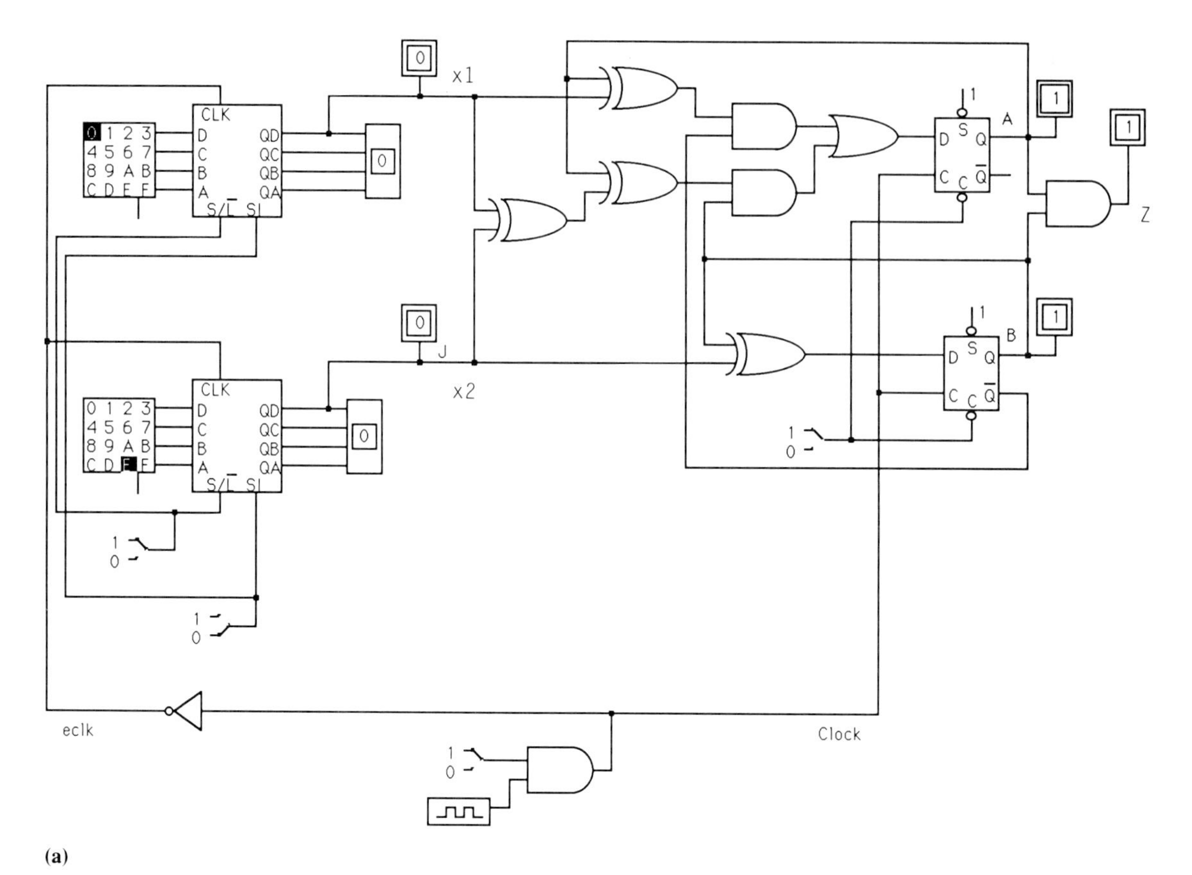

(a)

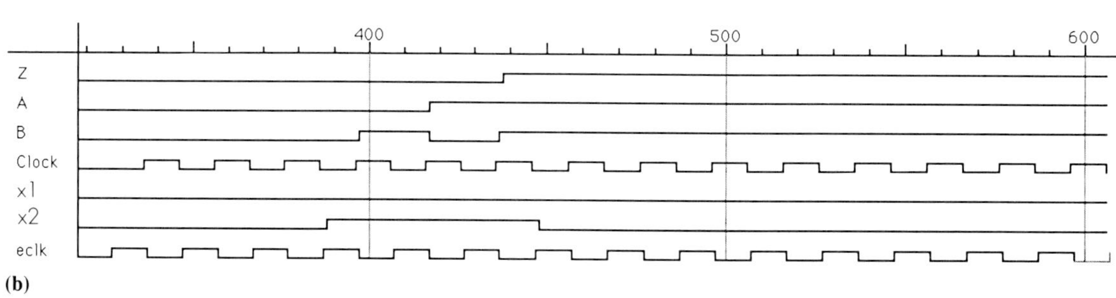

(b)

FIGURE 8.15
Logic diagram of a Moore circuit with 2 inputs

FIGURE 8.16
Flip-flop excitation maps for circuit in Figure 8.15

D_A

$AB \backslash x_1x_2$	00	01	11	10
00	0	0	1	1
01	0	1	0	1
11	1	0	1	0
10	1	1	0	0

D_B

$AB \backslash x_1x_2$	00	01	11	10
00	0	1	1	0
01	1	0	0	1
11	1	0	0	1
10	0	1	1	0

Step 3: Convert each flip-flop excitation map to a next-state map. For a D flip-flop, $Q^+ = D$. Therefore, each flip-flop next-state map (Figure 8.17) has the same form as the corresponding excitation map.

A^+

$AB \backslash x_1x_2$	00	01	11	10
00	0	0	1	1
01	0	1	0	1
11	1	0	1	0
10	1	1	0	0

B^+

$AB \backslash x_1x_2$	00	01	11	10
00	0	1	1	0
01	1	0	0	1
11	1	0	0	1
10	0	1	1	0

FIGURE 8.17
Flip-flop next-state maps for circuit in Figure 8.15

Step 4: Construct a circuit state table by combining the individual flip-flop next-state maps. Table 8.9 lists the circuit next states, A^+B^+, for each combination of circuit present state AB and input I.

TABLE 8.9
State table for circuit in Figure 8.15

Present State AB	Next State A^+B^+ (x_1x_2) 00	01	11	10	Output
00	00	01	11	10	0
01	01	10	00	11	0
11	11	00	10	01	1
10	10	11	01	00	0

Step 5: Construct a state diagram from the circuit's state table. Let state 00 be denoted by label S_0, state 01 by label S_1, state 10 by label S_2, and state 11 by label S_3. The state diagram for the circuit is presented in Figure 8.18.

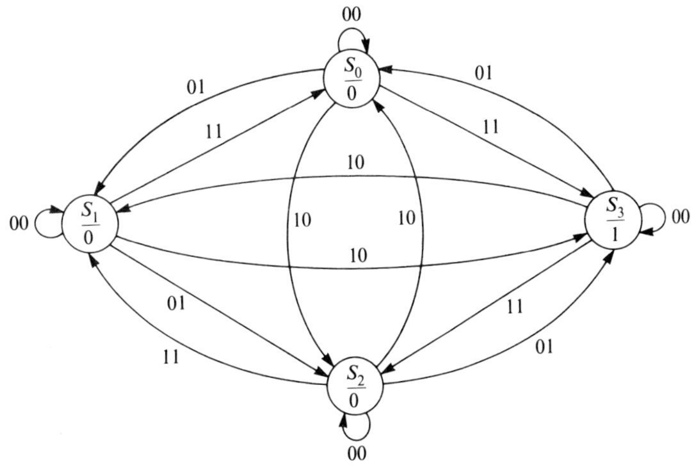

FIGURE 8.18
State diagram for a 2-input Moore circuit

Step 6: Interpret the state diagram to determine the function of the circuit. The circuit in Figure 8.15 performs a modulo-4 addition of the present state and the input. ■

8.3 DESIGN OF GENERAL SYNCHRONOUS SEQUENTIAL-LOGIC CIRCUITS

The design of a synchronous sequential-logic circuit is a process in which the functional requirements (design specifications) are used to develop a logic diagram of the circuit. The procedure for the design of a synchronous sequential circuit is essentially the reverse of the analysis procedure described in Section 8.2.

Design Procedure for a General Synchronous Sequential-Logic Circuit

The design of a synchronous sequential-logic circuit that realizes a specified finite-state machine can be accomplished using the 4-step problem-solving procedure introduced in Chapter 2, with some modifications.

PSP

Step 1: Problem Statement. Describe a set of design specifications for the required circuit.

Step 2: Conceptualization. Translate the design specifications into a machine state table and state diagram.

Step 3: Solution/Simplification.

a. Reduce the state table by eliminating any redundant machine states.
b. Assign binary codes to the machine states in such a way that the assignment achieves the simplest flip-flop excitation network.
c. Separate the circuit state table into flip-flop next-state maps.
d. Use the flip-flop characteristic equation to convert each flip-flop next-state map into a flip-flop excitation map.
e. Interpret each flip-flop excitation map to derive the corresponding flip-flop excitation equation(s). Use the state table to determine the circuit's output equation(s).

Step 4: Realization. Draw a circuit logic diagram from the flip-flop excitation equations and output equations. ■

This modified problem-solving procedure for the design of a synchronous sequential circuit will be illustrated by the following example.

■ **EXAMPLE 8.7**

Step 1: Problem Statement. Design a Moore synchronous sequential circuit to detect a string of three or more consecutive 1s in an arbitrary input string.

Step 2: Conceptualization. Translate the design specifications into a machine state table and state diagram.

The states required for a sequential machine that detects a sequence of three or more consecutive 1s include state *a*, the machine's initial state; state *b*, the state after a single 1; state *c*, the state after two consecutive 1s; state *d*, the state after three or more consecutive 1s. The dynamic operation of this finite-state machine can be graphically described by a state diagram, as shown in Figure 8.19.

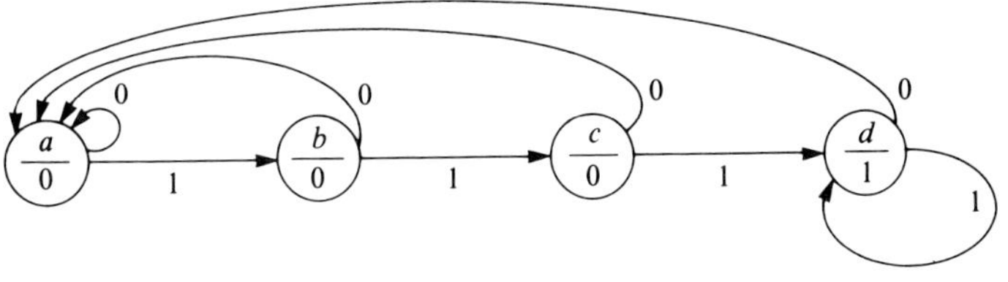

FIGURE 8.19
State diagram for the 111 sequence detector

If the machine is in state *a*, an input $I = 0$ will cause the machine to remain in state *a* while an input $I = 1$ will cause a transition to state *b*.

If the machine is in state *b*, an input $I = 0$ input causes a transition back to state *a* while an input $I = 1$ causes a transition to state *c*.

If the machine is in state *c*, an input $I = 0$ causes a transition back to state *a* while an input $I = 1$ causes a transition to state *d*.

If the machine is in state *d*, a 0 input causes a transition back to state *a* while a 1 input causes the circuit to remain in state *d*.

The design specifications for a finite-state machine that detects a sequence of three or more consecutive 1s can be represented in the form of the state table given in Table 8.10.

TABLE 8.10
State table for 111 sequence detector

Present State	Next State I = 0	Next State I = 1	Output
a	*a*	*b*	0
b	*a*	*c*	0
c	*a*	*d*	0
d	*a*	*d*	1

Step 3: Solution/Simplification.

a. Reduce the state table by eliminating any redundant machine states. Since no present state has the same next states and output, the state table has no redundant states.

b. Assign binary codes to the machine states in such a way that the simplest flip-flop excitation network is achieved. Since there are four distinct states, two flip-flops, denoted by *A* and *B*, are required to implement a sequential-logic circuit that realizes the machine. Each of the four machine states (*a*, *b*, *c*, *d*) is assigned one of the binary codes (00, 01, 10, 11). In this example we arbitrarily make the state assignment: $a = 00$, $b = 01$, $c = 11$, $d = 10$.

The circuit state at a given time is represented by the flip-flop state combination (AB = 00, 01, 10, 11) at that time. When the clock makes a state-changing transition, the circuit makes a transition from its present state AB to a next state A^+B^+, which depends on the input value at that time. The binary assignment is shown in Table 8.11.

TABLE 8.11
State table with binary codes

Present State	Next State I = 0	Next State I = 1	Output
a 00	00	01	0
b 01	00	11	0
c 11	00	10	0
d 10	00	10	1

c. Separate the circuit state table into flip-flop next-state maps. See Figure 8.20.

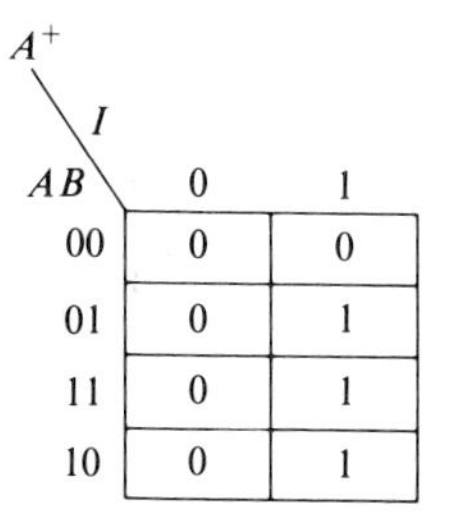

FIGURE 8.20
Next-state maps for 111 sequence detector

A^+

AB \ I	0	1
00	0	0
01	0	1
11	0	1
10	0	1

B^+

AB \ I	0	1
00	0	1
01	0	1
11	0	0
10	0	0

d. Use the flip-flop characteristic equation to convert each flip-flop next-state map into a flip-flop excitation map. Each excitation map defines that flip-flop's inputs for each row/column combination of present state and the input value. For D flip-flops, $Q^+ = D$. Therefore, the excitation map for each flip-flop has the same form as the corresponding next-state map, as shown in Figure 8.21.

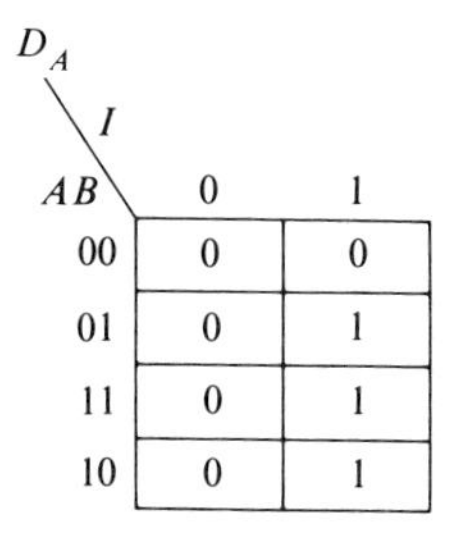

FIGURE 8.21
Excitation maps for 111 sequence detector

D_A

AB \ I	0	1
00	0	0
01	0	1
11	0	1
10	0	1

D_B

AB \ I	0	1
00	0	1
01	0	1
11	0	0
10	0	0

e. Interpret each flip-flop excitation map to derive that flip-flop's excitation equation. Each logic expression specifies the flip-flop's excitation as a function of the circuit's input I, and the circuit state A and B.

$$D_A = AI + BI \qquad D_B = A'I$$

An equation for the circuit's output can be derived directly from the state table, as follows:

$$Z = AB'$$

For a Moore circuit, the circuit's output is a function only of the circuit's states, A and B.

Step 4: Realization. Use the excitation and output equations to draw a logic diagram of the circuit (Figure 8.22). The logic diagram in Figure 8.22a uses 2 hexadecimal keypads and a 74165 IC. A shift register provides the input sequence.

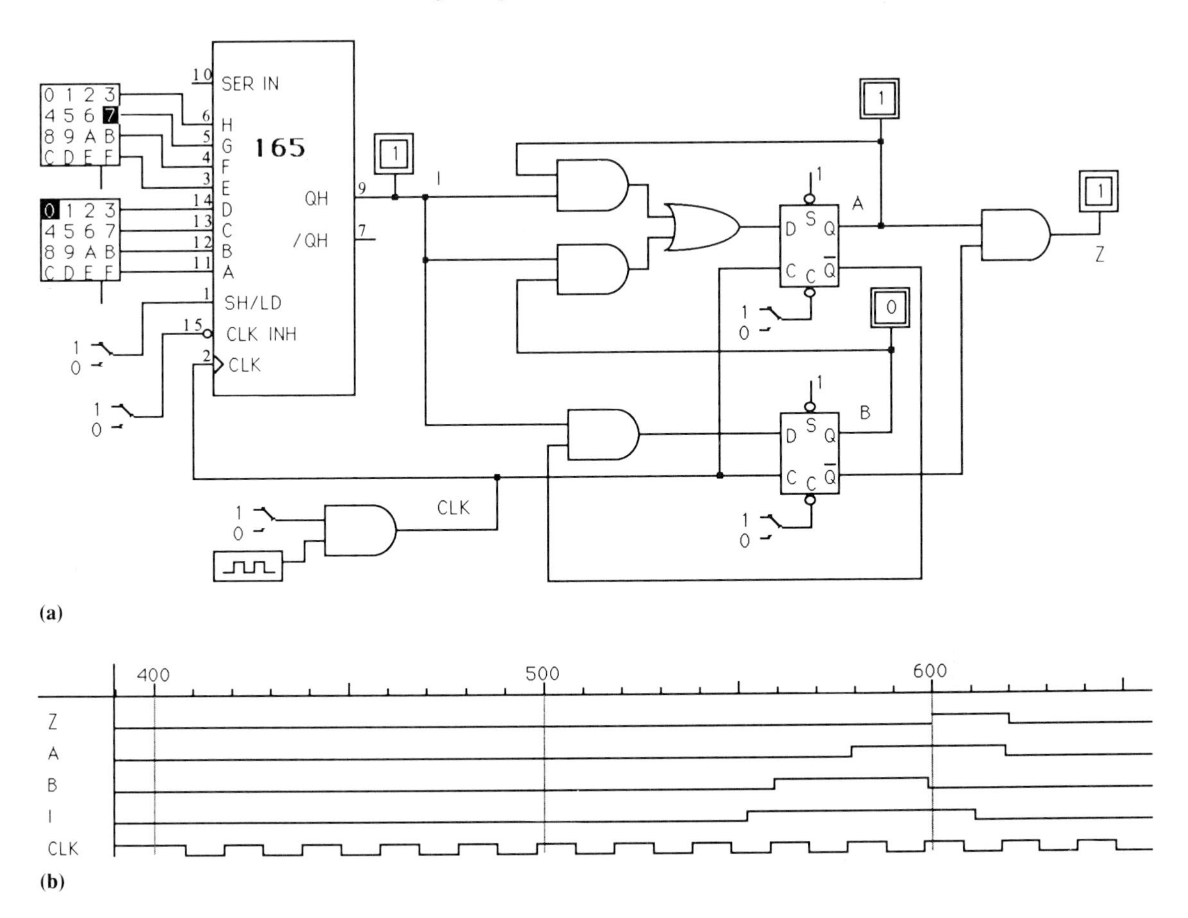

FIGURE 8.22
A Moore 111 sequence detector. (a) Logic diagram (b) Timing diagram ■

Translating Design Specifications into a State Diagram

The design specifications of a synchronous sequential-logic circuit that realizes a sequential process may be described by one or more statements of the interrelations between the input, state, and output sequences of the process. Given a set of design specifications, the translation of these specifications into a state diagram (or state table) is the first step in the design process.

A set of verbal specifications can be translated into a state diagram in a number of ways. For simple sequential processes, the verbal specifications can be translated into a state diagram using a heuristic approach as illustrated in the preceding section. For more complex processes, one of several systematic procedures can be used to translate the design specifications into a state diagram; these include (1) a binary-tree procedure, (1) a single-sequence-path algorithm, and (3) a multiple-sequence-path algorithm.

A binary tree has a root (initial state a) with a 0 branch (to state b) and a 1 branch (to state c). At each successive tree level, each state (at that level) has a 0 branch and a 1 branch to succession states in the next level. The binary tree procedure is described in detail in the following section.

Binary-Tree Procedure for Drawing State Diagrams

Let a denote the initial state. If the process is in any present state s, each of the possible input values (0 or 1) will cause the machine to make a transition to a next state $u = N(s, 0)$ or $v = N(s, 1)$ following the occurrence of a clock pulse. State $u = N(s, 0)$ is called the **0-successor of s**, and $v = N(s, 1)$ is called the **1-successor of s.**

Initial state a has a 0-successor (state b) and a 1-successor (state c). State b has states d and e as its 0-successor and 1-successor, respectively. State c has states f and g as its 0-successor and 1-successor, respectively. See Figure 8.23.

FIGURE 8.23
A binary-tree diagram

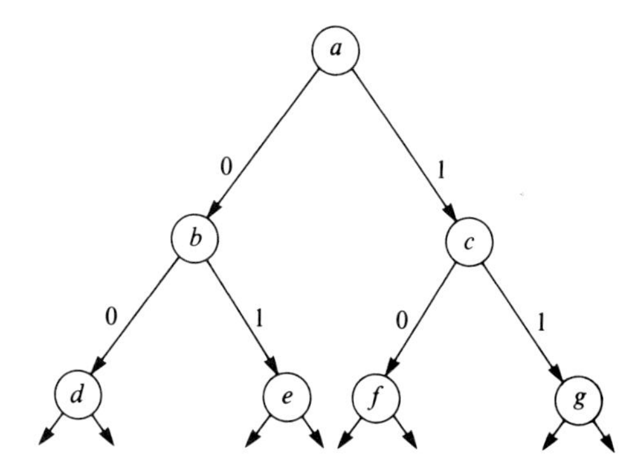

The resulting state diagram is a binary tree that has the initial state a as the root of the tree. A state diagram generated by the binary tree procedure contains all possible combinations of values of an input string of length n. Each of the 2^n combinations is represented by a path of n branches starting at state a and progressing through n levels to a destination state. For example, an input-value combination 01 is represented by path a–b–e.

The next example illustrates the use of the binary-tree procedure to design a Mealy sequence detector.

■ **EXAMPLE 8.8**

Step 1: Problem Statement. Design a Mealy synchronous sequential circuit to detect a sequence of three or more consecutive 1s in an arbitrary input string.

Step 2: Conceptualization. Translate the design specifications into a state diagram (Figure 8.24) using a binary-tree procedure. States d, e, f, and g can be labeled by the 2-bit combination corresponding to the two most recent input bits received by the circuit; that is, $d = 00$, $e = 01$, $f = 10$, and $g = 11$.

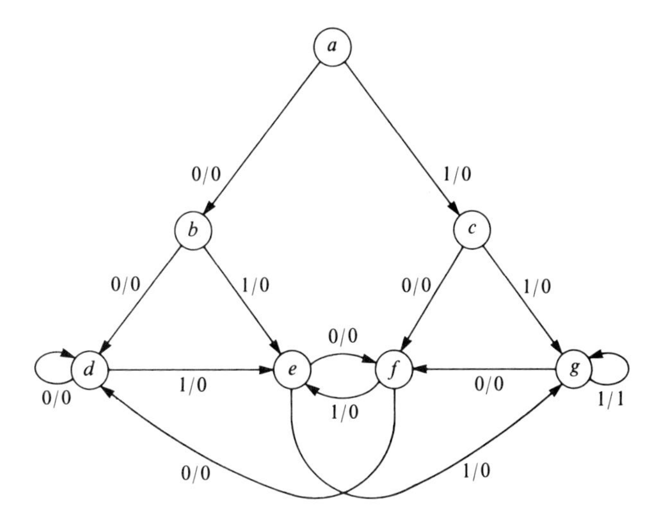

FIGURE 8.24
State diagram for a Mealy 111 sequence detector

The corresponding state table can then be constructed directly from the state diagram, as shown in Table 8.12.

TABLE 8.12
State table for a Mealy 111 sequence detector

Present State	Next State		Output	
	$I = 0$	$I = 1$	$I = 0$	$I = 1$
a	b	c	0	0
b	d	e	0	0
c	f	g	0	0
d	d	e	0	0
e	f	g	0	0
f	d	e	0	0
g	f	g	0	1

■

The binary-tree method for generating a state diagram is an exhaustive procedure that produces all 2^n possible combinations of input values for an input string of length n. As a result, the state diagram may contain one or more redundant states, which should be eliminated before the circuit is implemented. When we describe state-reduction techniques, we will continue this example.

Single-Sequence-Path Algorithm for Drawing State Diagrams

The exhaustive binary-tree procedure can be used to develop a state diagram for any synchronous sequential circuit. However, in some cases it may be more expedient to use an algorithmic approach based on a specified sequence or sequences. This section and the next describe two sequence-based algorithms for generating state diagrams.

The following method is the **single-sequence-path algorithm**, which is illustrated by the example sequence 0110.

1. Lay out a partial state diagram, starting from initial state *a*, with a set of states corresponding to the specified sequence. This is called the **main-path diagram** and is shown in Figure 8.25.

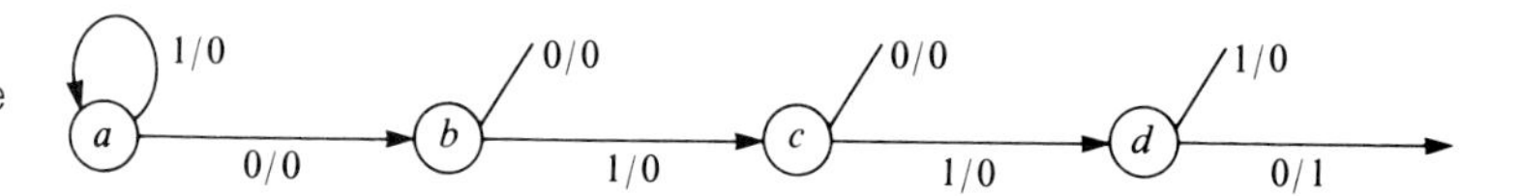

FIGURE 8.25
Main-path diagram for sequence 0110

2. Label the diagram as follows:
   a. At each main-path node (other than initial state *a*), label each unattached branch with a sequence number formed by appending the unattached branch input value (underlined) to the right of the main-path sequence that leads to that node. For example, the unattached 0 branch from node *b* is denoted 0*0*, formed by appending input *0* to the right of the main-path sequence (0) that leads to node *b*.
   b. Label each main-path node by the remaining bit sequence that will complete the overall specified sequence, shown in Figure 8.26. For example, node *b* is denoted by 110, which is the remaining bit sequence that will complete the overall specified sequence.

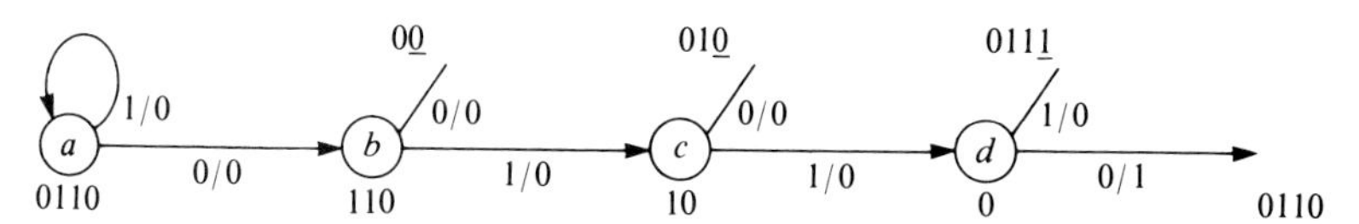

FIGURE 8.26
Main-path diagram with labeled branches and nodes

3. Construct a **connection matrix** (Figure 8.27) in which
   a. Each row is an unattached branch of the main-path diagram.
   b. Each column is a main-path node.
   c. Each cell contains the combined row-column labels separated by a dash. For example, the entry in row 00, column 110 is 00–110, formed by combining the row number (unattached branch) and the column number (main-path node). Note that any branch can return to the initial state *a* to begin a new sequence.

FIGURE 8.27
Connection matrix for a single-sequence-path detector

Branch Number \ Node Number	*a* 0110	*b* 110	*c* 10	*d* 0
00	↑	00–110	00–10	00–0
010	Any	010–110	010–10	010–0
0111		0111–110	0111–10	0111–0
0110	↓	0110–110	0110–10	0110–0

4. Examine each cell of the matrix to determine if it contains the specified sequence 0110.
   a. If the cell does not contain the specified sequence (including the dash), cross out the cell.
   b. If the cell contains the specified sequence, connect the unattached branch (row number) to the main-path node (column number).

The state diagram (Figure 8.28) produced by the single-sequence-path algorithm has no redundant states. Therefore, the corresponding state table is a minimal-state table for the 0110 sequence detector.

The state table for the 0110 sequence detector can be derived directly from the state diagram shown in Figure 8.28 and is given in Table 8.13.

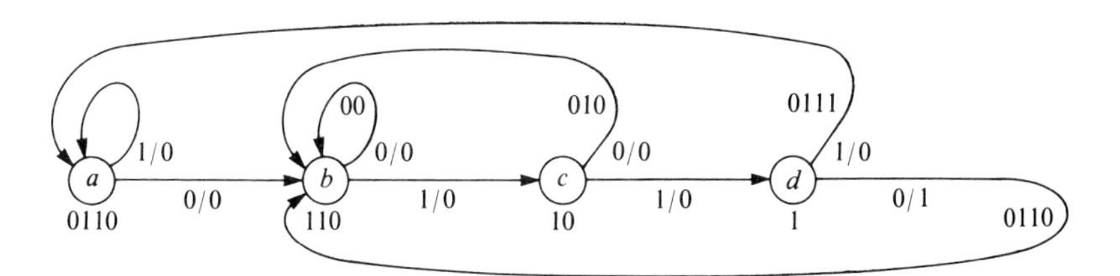

FIGURE 8.28
State diagram for a Mealy 0110 sequence detector

TABLE 8.13
Minimal-state table for 0110 sequence detector

Present State	Next State		Output	
	$I = 0$	$I = 1$	$I = 0$	$I = 1$
a	*b*	*a*	0	0
b	*b*	*c*	0	0
c	*b*	*d*	0	0
d	*b*	*a*	1	0

Multiple-Sequence-Path Algorithm for Drawing State Diagrams

A systematic procedure can also be developed for drawing a state diagram for synchronous sequential-logic circuits that recognize more than one specified sequence. The **multiple-sequence-path algorithm** is described in this section using the following example.

EXAMPLE 8.9

Design a Mealy synchronous sequential circuit that detects a sequence of at least four consecutive 1s or at least four consecutive 0s in an arbitrary input string.

1. Lay out a partial state diagram, starting from initial state *a*, with a set of states corresponding to the specified sequences. This is the main-path diagram given in Figure 8.29 (p. 356).
2. Label the diagram as follows:
   a. At each main-path node, label each unattached branch with a sequence number formed by appending the unattached branch input value (underlined) to the right of the main-path sequence that leads to that node.

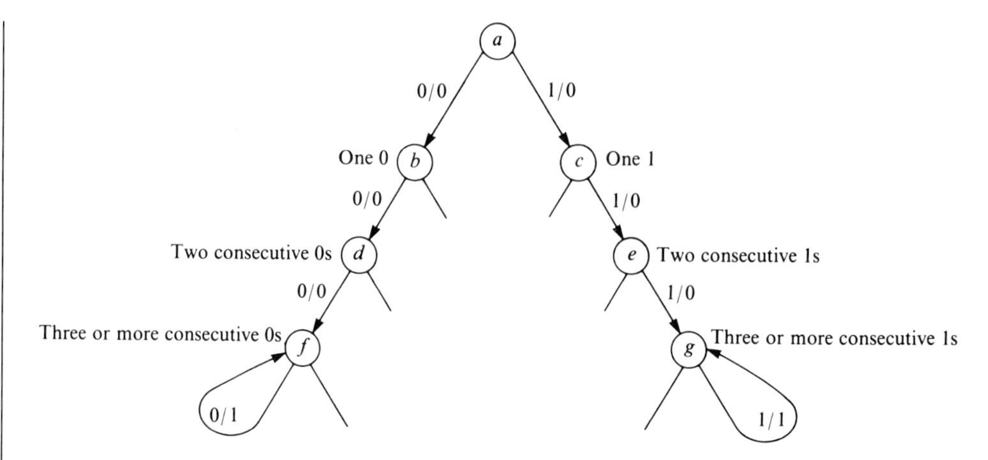

FIGURE 8.29
Main-path diagram for a multiple sequence 0000 or 1111 detector

b. Label each main-path node by the remaining sequence that will complete a specified sequence (Figure 8.30).

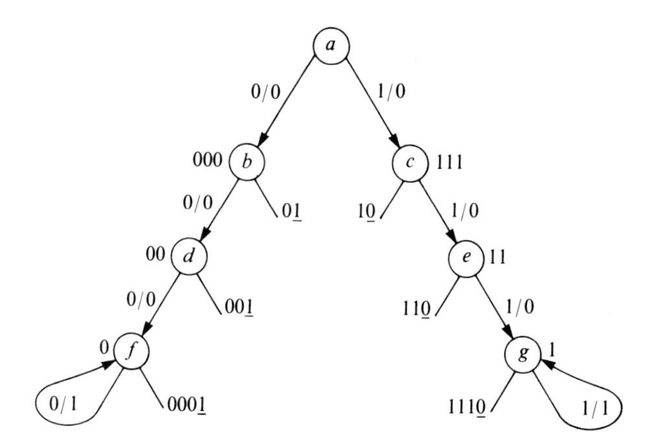

FIGURE 8.30
A multiple-sequence-path diagram

3. Construct a connection matrix (Figure 8.31) in which
   a. Each row is an unattached branch of the main-path diagram.
   b. Each column is a main-path node (other than node *a*).
   c. Each cell contains the juxtaposed row-column labels.
4. Examine each cell of the matrix to determine if it contains the specified sequence 0000 or 1111.
   a. If the cell does not contain the specified sequence (including the dash), cross out the cell.
   b. If the cell contains the specified sequence (including the dash), underline the sequence and connect the unattached branch (row number) to the main-path node (column number).

Branch Number \ Node Number	*b* 000	*d* 00	*f* 0	*c* 111	*e* 11	*g* 1
01	01–000	01–00	01–0	01–111	01–11	01–1
001	001–000	001–00	001–0	001–111	001–11	001–1
0001	0001–000	0001–00	0001–0	0001–111	0001–11	0001–1
10	10–000	10–00	10–0	10–111	10–11	10–1
110	110–000	110–00	110–0	110–111	110–11	110–1
1110	1110–000	1110–00	1110–0	1110–111	1110–11	1110–1

FIGURE 8.31
Connection matrix for a multiple sequence 0000 or 1111 detector

The state diagram (Figure 8.32) produced by the specified sequence path algorithm has no redundant states. Therefore, the corresponding state table is a minimal-state table for the sequence detector for either at least four consecutive 1s or for at least four consecutive 0s, as shown in Table 8.14.

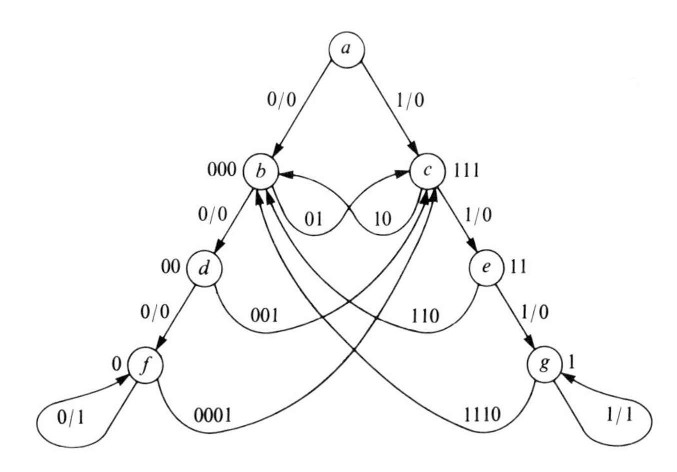

FIGURE 8.32
State diagram for a Mealy sequence detector 0000 or 1111

TABLE 8.14
Minimal-state table for 0000/1111 sequence detector

Present State	Next State $I = 0$	Next State $I = 1$	Output $I = 0$	Output $I = 1$
a	*b*	*c*	0	0
b	*d*	*c*	0	0
c	*b*	*e*	0	0
d	*f*	*c*	0	0
e	*b*	*g*	0	0
f	*f*	*c*	1	0
g	*b*	*g*	0	1

■

We will use this state table later to illustrate the design of a 2-person game.

State-Table Reduction Methods

A state table may contain one or more **redundant states**. The elimination of redundant states produces a reduced state table from which the circuit can be implemented. The flip-flop excitation network of the resulting circuit is usually simplified by using don't-care conditions introduced by the elimination of redundant states. If the number of states is reduced from greater than 2^n to 2^n or less, the resulting circuit requires fewer flip-flops in its implementation.

States p and q are distinguishable if there is at least one input sequence for which the circuit generates different output sequences depending on whether p or q is the starting state. But if state p is indistinguishable from any other state q, it is redundant.

The simplest case of redundant states is seen in a state table that contains two or more rows in which the next state and the output values are identical. In such a case, redundant states can be eliminated by the row-elimination method that we will describe next. Then we will present two additional reduction methods: the equivalence class partitioning and implication chart methods.

Row Elimination Method

Consider the state table (Table 8.15) for the Mealy 111 sequence detector given in Example 8.8.

TABLE 8.15
State table for 111 sequence detector

Present State	Next State		Output	
	$I = 0$	$I = 1$	$I = 0$	$I = 1$
a	b	c	0	0
b	d	e	0	0
c	f	g	0	0
d	d	e	0	0
e	f	g	0	0
f	d	e	0	0
g	f	g	0	1

Note in Table 8.15 that the next states and outputs for states b, d, and f are the same, and the next states and outputs for states c and e are the same. Therefore the group $\{b, d, f\}$ has 2 redundant states, and group $\{c, e\}$ has 1 redundant state.

If two states p and q have the same next state and the same outputs, then states p and q are equivalent and either p or q can be eliminated. The following **row elimination procedure** can be used to eliminate redundant state q:

1. Remove row q of the state table.
2. Replace q with p in any row that contains q as a next state.

We can now use the row elimination method to eliminate the redundant states shown in Table 8.15, by replacing *d* with *b*, *f* with *b*, and *e* with *c*, as shown in Table 8.16.

TABLE 8.16
Elimination of redundant states

	Next State		Output		
Present State	$I = 0$	$I = 1$	$I = 0$	$I = 1$	
a	*b*	*c*	0	0	
b	~~*d*~~*b*	~~*e*~~*c*	0	0	Replace *d* and *e* with *b* and *c*
c	~~*f*~~*b*	*g*	0	0	Replace *f* with *b*
d	~~*d*~~	~~*e*~~	~~0~~	~~0~~	Eliminate state *d* (= *b*)
e	~~*f*~~	~~*g*~~	~~0~~	~~0~~	Eliminate state *e* (= *c*)
f	~~*d*~~	~~*e*~~	~~0~~	~~0~~	Eliminate state *f* (= *b*)
g	~~*f*~~*b*	*g*	0	1	Replace *f* with *b*

The elimination of redundant states produces the reduced state table given in Table 8.17.

TABLE 8.17
Reduced state table

	Next State		Output	
Present State	$I = 0$	$I = 1$	$I = 0$	$I = 1$
a	*b*	*c*	0	0
b	*b*	*c*	0	0
c	*b*	*g*	0	0
g	*b*	*g*	0	1

States *a* and *b* now have the same next states and output. State *b* can be eliminated by the row elimination procedure as shown in Table 8.18.

TABLE 8.18
State table with redundant state

	Next State		Output		
Present State	$I = 0$	$I = 1$	$I = 0$	$I = 1$	
a	~~*b*~~*a*	*c*	0	0	Replace *b* with *a*
b	~~*b*~~	~~*c*~~	~~0~~	~~0~~	Eliminate state *b* (= *a*)
c	~~*b*~~*a*	*g*	0	0	Replace *b* with *a*
g	~~*b*~~*a*	*g*	0	1	Replace *b* with *a*

The resulting minimal state table for the finite-state machine is given in Table 8.19. This final reduced state table is a minimal state table since it has no redundant states.

TABLE 8.19
Minimal state table

	Next State		Output	
Present State	$I = 0$	$I = 1$	$I = 0$	$I = 1$
a	a	c	0	0
c	a	g	0	0
g	a	g	0	1

The row elimination method described in this section is the simplest method for eliminating redundant states of a state table. However, more complicated cases may exist for which it is not obvious that the state table contains one or more redundant states. In such cases, the redundant states may be eliminated by more complicated methods that determine groups of states wherein all states in any group are equivalent to one another. Next, we will describe the equivalence-class partitioning and implication chart methods for eliminating redundant states of a state table.

Equivalence-Class Partitioning Method

You recall from the preceding section that two states p and q are distinguishable if there is at least one input sequence for which the machine generates different output sequences depending on whether p or q is the starting state. But if state p is indistinguishable from any other state q, it is redundant. States p and q are **equivalent** if, and only if, for every input sequence, the machine generates the same output sequence starting from either p or q.

An **equivalence class** is a set of states in which all states are equivalent. If two states are in distinct equivalence classes, then those states are distinguishable. Conversely, distinguishable states must always be in different equivalence classes. Hence, the equivalence classes of a given machine are mutually exclusive. That is, the sets of equivalent states define a partition of the states of a finite-state machine M.

States p and q are k-equivalent if, for every input sequence of length k, the machine generates the same output sequence starting from either p or q. This definition leads to the following:

- □ Since no output is produced for an input sequence of length 0 (no input), all states of a machine are 0-equivalent.
- □ States that lead to the same output for an input sequence of length 1 (single input) are 1-equivalent.
- □ States that produce the same output sequence for an input sequence of length 2 are 2-equivalent.

The elimination of redundant states from an arbitrary state table can be accomplished by the **equivalence-class partitioning method**, which successively groups states into partitions of states that are 0-equivalent, 1-equivalent, 2-equivalent, and so on.

The equivalence-class partitioning method can be summarized as follows:

1. Partition $P(0)$. This consists of all of the states of machine M.
2. Partition $P(1)$. States (p, q) are grouped together in $P(1)$ if p and q produce identical output patterns for an input sequence of length 1.
3. Partition $P(k+1)$, $(k = 1, 2, \ldots)$. States (p, q) are grouped in $P(k+1)$ if, and only if,
   a. p and q are grouped in $P(k)$
   b. In each input column, the next states of p and q are in the same block of $P(k)$

The partitioning is repeated until $P(k+1) = P(k)$. Partition $P(k+1)$ is used to define the states of the minimal machine M', in which the redundant states in each equivalence class have been eliminated. For any input sequence, M' produces the same output sequence as M, assuming appropriate initial states.

We will now illustrate the equivalence-class partitioning method of state reduction using the example state table shown in Table 8.20.

TABLE 8.20
Partition $P(0)$ of state table

	Next State		Output	
Present State	$I = 0$	$I = 1$	$I = 0$	$I = 1$
a	*g*	*b*	0	0
b	*c*	*b*	0	0
c	*a*	*d*	0	0
[*d*	*e*	*b*	0	1]
e	*g*	*f*	0	0
[*f*	*h*	*b*	0	1]
g	*a*	*b*	0	0
h	*a*	*f*	0	0

Partition P(0): Since no output is produced for an input sequence of length 0, all states of a machine M are 0-equivalent and are denoted by partition $P(0)$ of the state table.

Partition P(1): The next step is to identify groups of states that lead to the same output for an input sequence of length 1 (1-equivalent states). For the single input $I = 1$, states d and f lead to an output 1 and the remaining states lead to an output 0. States d and f, indicated by [ ] in Table 8.20, are grouped into class S_1, and the remaining states are grouped into class S_2 as shown in Figure 8.33. Note that states c, e, and h, indicated by < and >, have successors in class S_1.

Partition P(2): Identify groups of states whose successors are in the same class. States c, e, and h, indicated by < > in Figure 8.33, have 1-successor in S_1 and are grouped into class S_{21}; the remaining states (with no successor in S_1) are grouped into class S_{22}, as shown in Figure 8.34. Note that state b, indicated by ( ), has a successor in Class S_{21}.

FIGURE 8.33
Partition $P(1)$. (a) State table (b) State diagram

Present Class	Present State	Next State $I = 0$	Next State $I = 1$	Next Class $I = 0$	Next Class $I = 1$
S_1	[*d*]	*e*	*b*	S_2	S_2
	[*f*]	*h*	*b*	S_2	S_2
S_2	*a*	*g*	*b*	S_2	S_2
	b	*c*	*b*	S_2	S_2
	<*c*	*a*	[*d*]	S_2	S_1 >
	<*e*	*g*	[*f*]	S_2	S_1 >
	g	*a*	*b*	S_2	S_2
	<*h*	*a*	[*f*]	S_2	S_1 >

(a)

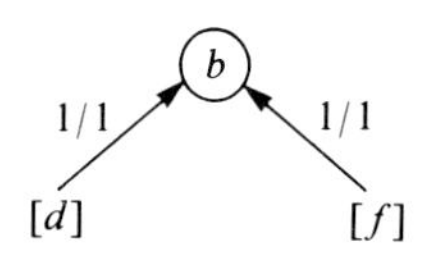

(b)

FIGURE 8.34
Partition $P(2)$. (a) State table (b) State diagram

Present Class	Present State	Next State $I = 0$	Next State $I = 1$	Next Class $I = 0$	Next Class $I = 1$
S_1	[*d*]	<*e*>	*b*	S_{21}	S_{22}
	[*f*]	<*h*>	*b*	S_{21}	S_{22}
S_{21}	<*c*>	*a*	[*d*]	S_{22}	S_1 >
	<*e*>	*g*	[*f*]	S_{22}	S_1 >
	<*h*>	*a*	[*f*]	S_{22}	S_1 >
S_{22}	*a*	*g*	*b*	S_{22}	S_{22}
	(*b*	<*c*>	*b*	S_{21})	S_{22}
	g	*a*	*b*	S_{22}	S_{22}

(a)

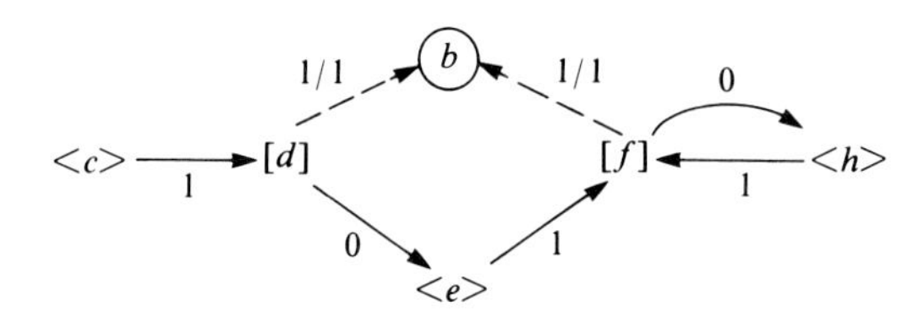

Key: Solid lines denote the most recent partition

(b)

Partition P(3): Identify groups of states whose successors are in the same class. State b has a 0-successor in $S_{21} = \{c, e, h\}$ and is placed in class S_{221}; the remaining states (with no 0-successor in S_{21}) are grouped into class S_{222}, as shown in Figure 8.35.

Partition P(4): Identify groups of states whose successors are in the same class. States a and g have 0-successors in the same class and 1-successors in the same class. Thus, partition $P(4)$ is the same as partition $P(3)$, and the process is complete. (See Figure 8.36.)

FIGURE 8.35
Partition P(3). (a) State table (b) State diagram

Present Class	Present State	Next State		Next Class	
		$I = 0$	$I = 1$	$I = 0$	$I = 1$
S_1	[d]	<e>	(b)	S_{21}	S_{221}
	[f]	<h>	(b)	S_{21}	S_{221}
S_{21}	<c>	a	[d]	S_{222}	S_1 >
	<e>	g	[f]	S_{222}	S_1 >
	<h>	a	[f]	S_{222}	S_1 >
S_{221}	(b	<c>	(b)	S_{21})	S_{221}
S_{222}	a	g	(b)	S_{222}	S_{221}
	g	a	(b)	S_{222}	S_{221}

(a)

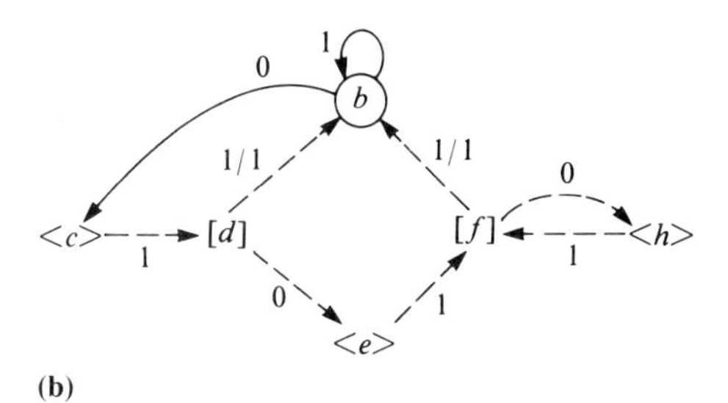

(b)

FIGURE 8.36
Equivalence-class partitioning. (a) State table (b) State diagram

Present Class	Present State	Next State		Next Class	
		$I = 0$	$I = 1$	$I = 0$	$I = 1$
S_1	[d]	<e>	(b)	S_{21}	S_{221}
	[f]	<h>	(b)	S_{21}	S_{221}
S_{21}	<c>	a	[d]	S_{222}	S_1 >
	<e>	g	[f]	S_{222}	S_1 >
	<h>	a	[f]	S_{222}	S_1 >
S_{221}	(b	<c>	(b)	S_{21})	S_{221}
S_{222}	a	g	(b)	S_{222}	S_{221}
	g	a	(b)	S_{222}	S_{221}

(a)

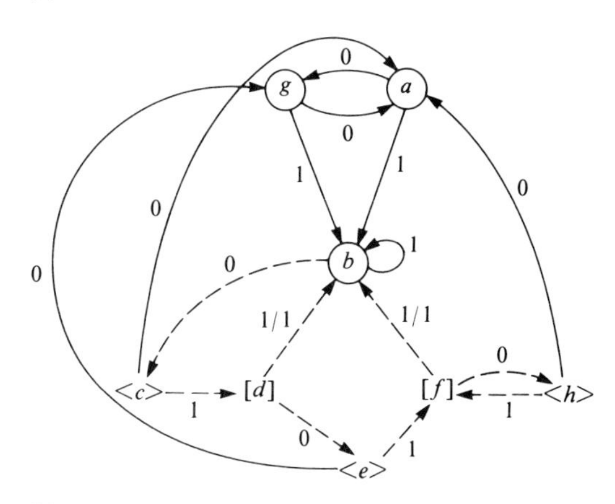

(b)

Note that the partitioning is discontinued when no group can be further subdivided, that is, when all states in each group have 0-successors in a single class and 1-successors in a single class.

States p and q are equivalent if, and only if, for every single input I the next states are equivalent and the outputs are the same. Therefore, the state table of machine M has been partitioned into the following equivalence classes: (d, f), (c, e, h), (b), and (a, g). The redundant states in each equivalence class can be eliminated to produce the minimal-state machine M'. The reduced state diagram is presented in Figure 8.37.

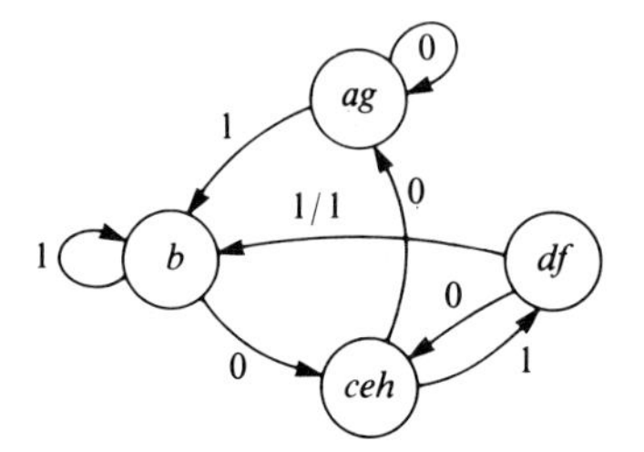

FIGURE 8.37
Reduced state diagram for 1011 sequence detector

Implication Chart Method

A *systematic* elimination of redundant states can be accomplished by the implication chart method described in this section. To illustrate the **implication chart method** for reducing a state table we will use the example state table shown in Table 8.21.

TABLE 8.21
Example state table

Present State	Next State		Output	
	$I = 0$	$I = 1$	$I = 0$	$I = 1$
a	b	a	0	0
b	b	c	0	0
c	d	e	0	0
d	f	g	0	1
e	h	a	0	0
f	b	g	0	0
g	d	a	0	0
h	f	g	0	0

An implication chart is a lower triangular matrix with $n - 1$ rows and $n - 1$ columns. The rows correspond to the second through the last state of the state table, and the columns correspond to the first through the next-to-last state. The cell in row R and column C contains either of the following:

1. An X if states R and C are not equivalent
2. A check mark if states R and C are unconditionally equivalent
3. The *next-state* pairs, $N(R, 0)$, $N(R, 1)$ and $N(C, 0)$, $N(C, 1)$, of states R and C if R and C are conditionally equivalent:

a. Left column of cell contains $N(C, 0)$ and $N(C, 1)$
b. Right column of cell contains $N(R, 0)$ and $N(R, 1)$

Therefore,

$$\text{cell } R, C = \begin{array}{ll} N(C, 0) & N(R, 0) \\ N(C, 1) & N(R, 1) \end{array}$$

We can construct an implication chart for Table 8.21 as shown in Table 8.22.

TABLE 8.22
Implication chart for Table 8.21

b	bb ac						
c	bd ae	bd ce					
d	X	X	X				
e	bh aa	bh ca	dh ea	X			
f	bb ag	bb cg	db eg	X	hb ag		
g	bd aa	bd ca	dd ea	X	hd aa	bd ga	
h	bf ag	bf cg	df eg	X	hf ag	bf gg	df ag
	a	b	c	d	e	f	g

From Table 8.21, we see state d has output 1 while the remaining states have output 0. Therefore, state d and each of the other states form a nonequivalent pair, and an X is placed in the initial implication chart.

The elimination of redundant states is accomplished by the following procedure (let cell R, C denote the cell in row R, column C):

1. Each non-X cell R, C of the implication chart is checked. If either row of cell R, C contains a pair of nonequivalent states, (initially $da, db, dc, de, df, dg, dh$) then cell R, C is crossed out (indicating that states R and C are not equivalent). In Table 8.23 (p. 366) we have drawn a diagonal line (/) through the cells crossed out in this pass.
2. Step 1 is repeated if any additional cells were checked off in the preceding pass. We have drawn a different diagonal line (\) through the cells crossed out in this pass.

The process is halted when no additional cells can be crossed out by application of Step 1. The non-X cells that have not been crossed out in the implication chart represent pairs of equivalent states.

TABLE 8.23
State reduction using an implication chart

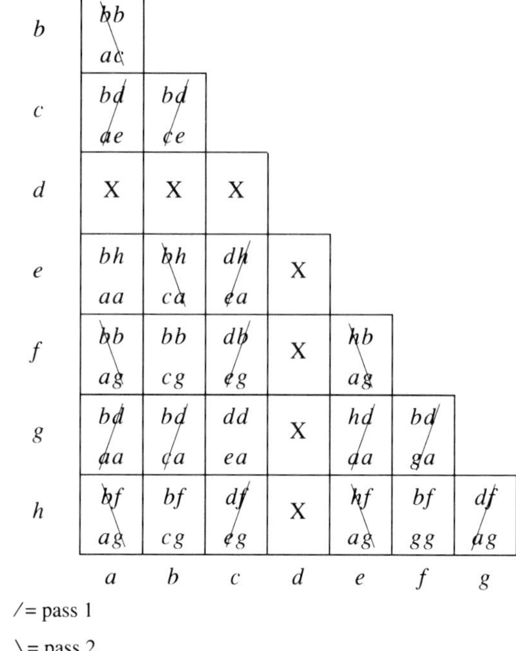

In this example, the following pairs represent equivalent states:

$$(a = e) \qquad (b = f), (b = h), (f = h) \qquad (c = g)$$

The state equivalence relation is transitive, so $b = f = h$. Therefore, the machine has the following equivalence classes:

$$\{a, e\} \qquad \{b, f, h\} \qquad \{c, g\} \qquad \{d\}$$

The redundant states in the equivalence classes are eliminated to produce the reduced state table and state diagram shown in Figure 8.38.

FIGURE 8.38
A 0101 detector. (a) Reduced state table (b) Reduced state diagram

Present State	Next State $I = 0$	Next State $I = 1$	Output $I = 0$	Output $I = 1$
a	b	a	0	0
b	b	c	0	0
c	d	a	0	0
d	b	c	0	1

(a)

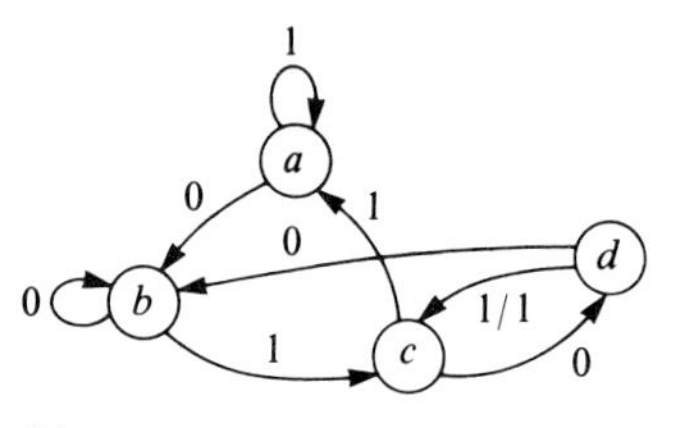

(b)

State Assignment

Starting with the original state tables and diagrams, and continuing through the state-reduction process, the machine states have been represented in symbolic form (a, b, c, and so on). If a sequential machine has K states, where $2^{n-1} < K \leq 2^n$ then n flip-flops with 2^n distinct states are required to implement a synchronous sequential-logic circuit that realizes the machine.

Since each circuit state is associated with a distinct binary code ($\leq 2^n$), each machine state must be assigned a binary code associated with a circuit state. If a minimal state table has three states (a, b, and c), the machine can be realized using any one of 24 possible assignments, as follows: For any state, say a, we have a choice of 4 codes: 00, 01, 10, 11. If we choose code 00 for state a, we have a choice of 3 codes for state b: 01, 10, 11. If we choose code 01 for state b, we have a choice of 2 codes for state c: 10 or 11. Thus, there are $4 \times 3 \times 2 = 24$ distinct code assignments for 3 states. If we always choose a particular code, say 00 for state a, there are only $1 \times 3 \times 2$ assignments to consider. In fact, it can be shown that there are only 3 nonequivalent assignments for 3 states (and for 4 states), as shown in Table 8.24.*

TABLE 8.24
Nonequivalent assignments for 3 and 4 states

State	3-State Assignments			4-State Assignments		
	1	2	3	1	2	3
a	00	00	00	00	00	00
b	01	01	11	01	01	11
c	10	11	01	10	11	01
d	—	—	—	11	10	10

The realization of a 5-state machine (a, b, c, d, and e) requires 3 flip-flops, denoted by A, B, and C. For any state, say a, we have a choice of 8 codes: 000, 001, 010, 011, 100, 101, 110, 111. If we choose code 000 for state a, we have a choice of 7 codes for state b (any code other than 000). If we choose code 001 for state b, we have a choice of 6 codes for state c (any code except 000 and 001), and so on. Thus, there are $8 \times 7 \times 6 \times 5 \times 4 = 6720$ distinct assignments for 5 states. It can be shown that of these there are 140 nonequivalent assignments. For 6, 7, and 8 states there are 420, 840, and 840 nonequivalent assignments.

The **state-assignment** problem is how to establish a 1-to-1 correspondence between each machine state and one of the 2^n possible circuit states. That is, how should the machine states be assigned binary codes? Can we arbitrarily assign states $a, b, c, \ldots$ the binary codes for decimal 0, 1, 2, ...? In a synchronous sequential-logic circuit the synchronization of memory state changes eliminates many of the problems caused by unequal propagation delays in different paths, so any arbitrary state assignment can be made provided that each machine state is assigned a unique code. However, each different state assignment usually results in a different combinational-logic excitation network, and a judicious assignment of states can significantly reduce the complexity of that network.

It should be noted that the state assignment in an asynchronous sequential-logic circuit must satisfy additional constraints to avoid critical races.

*See C. H. Roth, *Fundamentals of Logic Design* (St. Paul, Minn.: West, 1975), p. 261.

State Assignment Techniques

The problem of designing the 0101 sequence detector will be used to show how the application of state assignment guidelines can be used to reduce the complexity of a synchronous sequential-logic circuit that realizes a particular machine.

The original state table for the 0101 sequence detector in the preceding section has been reduced in Figure 8.38 to the form shown in Table 8.25. Since there are 4 distinct states in the reduced state table, 2 flip-flops, A and B, are required to implement the circuit.

TABLE 8.25
Reduced table for 0101 sequence detector

Present State	Next State		Output	
	$I = 0$	$I = 1$	$I = 0$	$I = 1$
a	b	a	0	0
b	b	c	0	0
c	d	a	0	0
d	b	c	0	1

The assignment of binary codes (00, 01, 10, 11) to machine states (a, b, c, d) should be made so that the combinational-logic circuits that provide excitation to flip-flops A and B are made as simple as possible. This can be accomplished by selecting a state assignment that produces flip-flop input (excitation) maps with the largest possible number of logically adjacent 1s, which then can be combined to simplify the flip-flop excitation network.

Thus, the binary codes should be assigned to the machine states in such a way as to satisfy as many logical adjacency conditions as possible. A set of state assignment guidelines that provide the greatest number of logical adjacency conditions are presented next.

The flip-flop excitation network for a circuit can be simplified by assigning each state a binary code so that as many 1 cells as possible are adjacent in the excitation maps of each flip-flop. The adjacent 1s can then be combined using K-map techniques to simplify the flip-flop excitation functions. The following rules are useful in selecting state assignment codes.

Rule 0: The circuit's initial state should be assigned to map cell 0.

Rule 1: If two states p and q *have* the same next state for a given input value, then p and q should be given adjacent state assignments.

Rule 2: If two states p and q *are* the next states of state s for adjacent input values, then p and q should be given adjacent state assignments.

Rule 3: Assign odd code numbers to majority states in the same input column ($I = 0$) and even code numbers to states in the input column ($I = 1$). This rule maximizes groups of adjacent 1s (and adjacent 0s) in the next-state map LSB (B flip-flop).

In selecting state assignment codes, preference should be given to those adjacency conditions that occur more than once (underlined) and to those imposed by rules 1 and 3.

Rules 1 through 3 impose the following adjacency conditions for the states in Table 8.26.

TABLE 8.26
State table

	Present State	Next State $I=0$	Next State $I=1$	Output $I=0$	Output $I=1$
	a	b	a	0	0
Rule 1	b	b	c	0	0
	c	d	a	0	0
	d	b	c	0	1
Rule 2					

By rule 1, a–b, a–d, $\underline{b\text{–}d}$, and a–c have the same next state and should be assigned adjacent codes.

By rule 2, a–b, a–d, and $\underline{b\text{–}c}$ are the next states of a given state and should be assigned adjacent codes.

By rule 3, assign odd-numbered codes to states b and d; assign even-numbered codes to states a and c.

Justification for rule 1: By assigning adjacent codes to a, b, d, and c, rule 1 adjacencies a–b, b–d, and a–c are satisfied as indicated by the reordered rows of the state table with next-state binary codes x_1x_0, where x denotes either a, b, c, or d. A state assignment matrix (Figure 8.39) can be used to assign binary codes to the machine states to satisfy the rule 1 adjacencies.

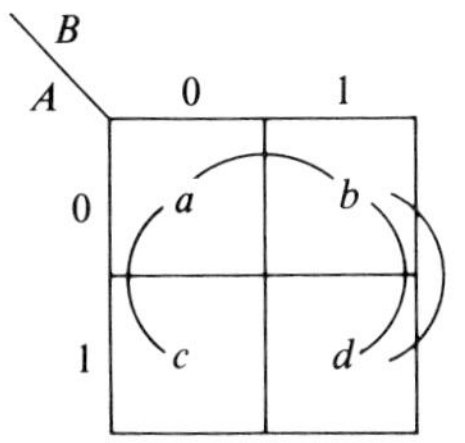

FIGURE 8.39
State assignment matrix

If states x and y have the same next state z for input column i, then by rule 1 x and y are assigned adjacent codes. This effectively places states x and y in adjacent rows so column i contains z in adjacent rows as shown in Table 8.27. Consequently equal MSBs ($z_1 = z_1$) and LSBs ($z_0 = z_0$) are grouped in *column i* of the table.

TABLE 8.27
Rule 1 grouping

(a) Present State	Next State $I=0$	Next State $I=1$	(b) Present State	Next State $I=0$	Next State $I=1$
a	b_1b_0	a_1a_0	a 00	b 01	a 00
b	b_1b_0	c_1c_0	b 01	b 01	c 10
d	b_1b_0	c_1c_0	d 11	b 01	c 10
c	d_1d_0	a_1a_0	c 10	d 11	a 00

The result of rule 1 (for the example) is as follows:

The assignment of adjacent codes to states *a* and *b* groups adjacent equal bits in next-state column $I = 0$ since *a* and *b* both have next state *b* for $I = 0$.

The assignment of adjacent codes to states *b* and *d* groups adjacent equal bits in next-state columns $I = 0, 1$ since *b* and *d* both have next state *b* for $I = 0$ and next state *c* for $I = 1$.

The assignment of adjacent codes to states *a* and *c* groups adjacent equal bits in next-state column $I = 1$ since *a* and *c* both have next state *a* for $I = 1$.

A state assignment matrix (Figure 8.40) can be used to assign binary codes to the machine states to satisfy the rule 2 adjacencies.

FIGURE 8.40
State assignment matrix

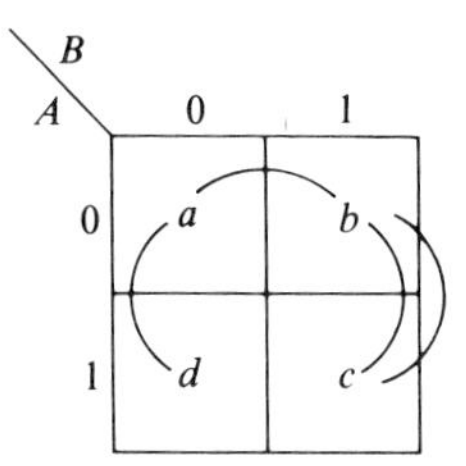

Justification for rule 2: By assigning adjacent codes to states *x*, *y*, the gray code implies that either $x_1 = y_1$ or $x_0 = y_0$, which groups equal bits in any next-state *row* for which *x* and *y are* the next state of *z*.

The result of rule 2 (without violating the rule 1 assignment) is as follows:

The assignment of adjacent codes to *x* and *y* (Table 8.28) groups adjacent equal bits in the next-state *row* for which *x* and *y are* the next state of *z* (either $x_1 = y_1$ or $x_0 = y_0$).

TABLE 8.28
Rule 2 grouping

(a) Present State	Next State $I = 0$	Next State $I = 1$
a_1a_0	b_1b_0	a_1a_0
b_1b_0	b_1b_0	c_1c_0
c_1c_0	d_1d_0	a_1a_0
d_1d_0	b_1b_0	c_1c_0

(b) Present State		Next State $I = 0$	Next State $I = 1$
a	00	*b* 01	*a* 00
b	01	*b* 01	*c* 11
c	11	*d* 10	*a* 00
d	10	*b* 01	*c* 11

In this example the state assignment using rule 2 violates some of the adjacencies satisfied by the assignment using rule 1. Therefore, the assignment should be made using rule 1 (as shown in Table 8.27).

Justification for rule 3: The assignment of odd codes to the majority next states in the same column (*b*, *d* for $I = 0$) and even codes to the majority states in the other column (*a*, *c* for $I = 1$) maximizes groups of adjacent 1s of *b*, *d* (and 0s of *a*, *c*) in LSB (*B* flip-flop).

The state assignment achieved, by applying rule 1, rule 2 (without violating rule 1), and, finally, rule 3 should result in a next-state map that provides the most groupings of adjacent equal code bits. That is,

Rule 1 groups equal MSBs and equal LSBs in a *column* of adjacent rows.

Rule 2 groups adjacent equal bits (either MSB or LSB) in the next-state *row* for which x and y *are* the next state of z.

Rule 3 maximizes groups of adjacent 1s of odd-numbered codes and groups of 0s of even-numbered codes in the LSB (B flip-flop).

With the code assignment made using rules 1, 2, and 3, the full state table appears as shown in Table 8.29.

TABLE 8.29
State assignment using Rules 1, 2, and 3

Present State	Next State $I=0$	Next State $I=1$	Output $I=0$	Output $I=1$	
a 00	*b* 01	*a* 00	0	0	
b 01	*b* 01	*c* 10	0	0	
d 11	*b* 01	*c* 10	0	1	$Z = ABI$
c 10	*d* 11	*a* 00	0	0	

From Table 8.29, the assignments are $a = 00, b = 01, d = 11, c = 10$. The state table can be presented in a K-map form (Figure 8.41) to show the adjacency conditions satisfied by the selected code assignment.

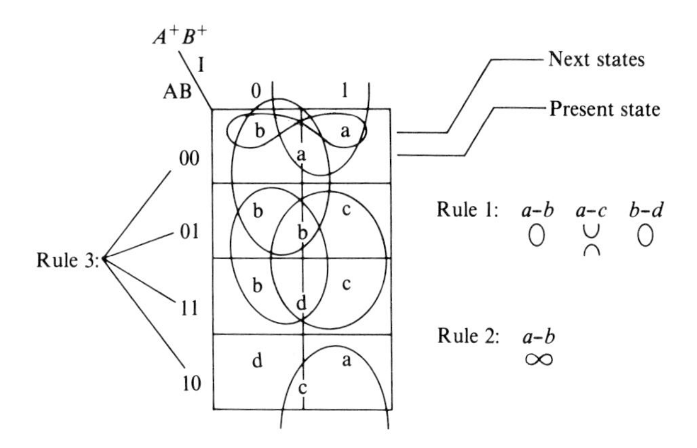

FIGURE 8.41
Illustration of adjacency conditions

State Assignment for State Tables with Unused States

This section describes procedures for designing synchronous sequential circuits for which the machine state table contains K distinguishable (nonredundant) states, where $2^{n-1} < K < 2^n$.

If a state table has K distinguishable states, where $2^{n-1} < K < 2^n$, then there are $2^n - K$ unused states and a corresponding set of unassigned n-bit binary codes. The unused states (codes) can be used as *don't cares* to simplify the flip-flop excitation network of the circuit. The rules for state assignment described in the preceding section can be used to select the binary codes for the K distinguishable states so that the flip-flop excitation network is made as simple as possible.

Consider a state diagram for a Mealy 000/111 sequence detector that has 5 distinguishable states. Figure 8.42 is a minimal state diagram for a Mealy circuit that detects either a sequence of at least three consecutive 1s or at least three consecutive 0s in an arbitrary bit string. The corresponding minimal state table is given in Table 8.30.

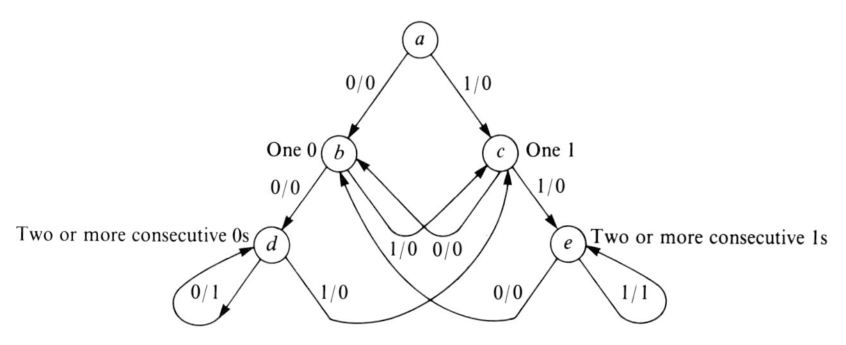

FIGURE 8.42
State diagram for a Mealy 000/111 sequence detector

TABLE 8.30
Minimal state table

	Present State	Next State $I = 0$	Next State $I = 1$	Output $I = 0$	Output $I = 1$
	a	b	c	0	0
Rule 1	b	d	c	0	0
	c	b	e	0	0
	d	d	c	1	0
	e	b	e	0	1
Rule 2					

We will use rules 0, 1, 2, and 3 presented in the preceding section to make the state assignments. Since there are 5 distinct states, 3 flip flops (with 8 possible states) will be required to implement the circuit. The complexity of the flip-flop excitation network depends on the selection of the binary codes for the distinguishable states of the circuit.

An examination of Table 8.30 indicates the following required state adjacencies:

By rule 1: a–c, a–e, $\underline{c\text{–}e}$ a–b, a–d, $\underline{b\text{–}d}$

By rule 2: b–c, $\underline{d\text{–}c}$, $\underline{b\text{–}e}$

By rule 3: Choose b and d as odd-numbered codes; choose c and e as even-numbered codes

The state assignments shown in Figure 8.43 satisfy the majority of the state adjacencies required by rules 1, 2 (without violating rule 1), and 3.

FIGURE 8.43
State assignment for 000/111 sequence detector

$AB \backslash C$	0	1
00	a	b
01	c	d
11	e	
10		

The resulting state table is given in Table 8.31.

TABLE 8.31
Circuit state table

Present State	Next State $I=0$	Next State $I=1$	Output $I=0$	Output $I=1$
a 000	b 001	c 010	0	0
b 001	d 011	c 010	0	0
d 011	d 011	c 010	1	0
c 010	b 001	e 110	0	0
e 110	b 001	e 110	0	1
111	xxx	xxx		
101	xxx	xxx		
100	xxx	xxx		

The state table for the circuit can be described in a K-map form that shows the adjacency conditions satisfied by the state assignment (Figure 8.44).

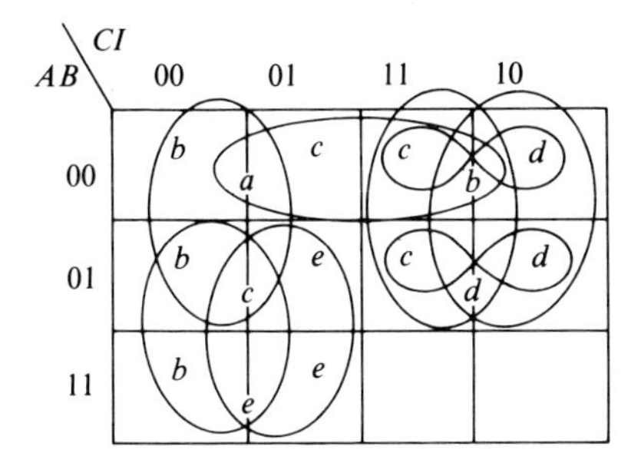

Rule 0: $a = 000$

Rule 1: a–b a–c b–d c–e

Rule 2: c–d

Rule 3: $b = 001, d = 011, c = 010, e = 110$

FIGURE 8.44
Illustration of adjacency conditions satisfied

A map-form state table that shows the state assignment codes is presented in Figure 8.45. Note that the rightmost bit of each next state is the same for all entries in a given column because of the restriction imposed by rule 3.

FIGURE 8.45
A map-form state table

AB \ CI	00	01	11	10
00	*b* 001	*c* 010	*c* 010	*d* 011
01	*b* 001	*e* 110	*c* 010	*d* 011
11	*b* 001	*e* 110	xxx	xxx
10	xxx	xxx	xxx	xxx

Present-state labels between columns: row 00: *a*, *b*; row 01: *c*, *d*; row 11: *e*, —; row 10: —, —.

8.4 DESIGN OF VARIOUS TYPES OF SYNCHRONOUS SEQUENTIAL-LOGIC CIRCUITS

In this section we consider a variety of sequential processes to illustrate the range of applications for the procedures described in this chapter. The following sections present the step-by-step design of synchronous sequential-logic circuits realizations of five different sequential processes:

A 3-bit up/down counter

A filter circuit

A 2-person game circuit

A combination lock

A Mealy 0101 sequence detector

The design of each of these circuits will be accomplished using the modified 4-step problem-solving procedure introduced in Section 8.3.

Design of a 3-Bit Up/Down Counter

Step 1: Problem Statement. Design a 3-bit up/down counter that has a control input U that determines the direction of the generated count sequence. When the control input $U = 1$, the counter counts up in the sequence 0, 1, 2, 3, 4, 5, 6, 7. When $U = 0$, the counter counts down in the sequence 0, 7, 6, 5, 4, 3, 2, 1. Implement a synchronous counter using edge-triggered JK flip-flops.

Step 2: Conceptualization. Translate the specifications into a state diagram (Figure 8.46) and a state table (Table 8.32).

Step 3: Solution/Simplification.

a. No state reduction is required since all the decimal numbers 0 through 7 are needed to represent the machine states. Each decimal-number state is assigned the corresponding binary code. A state diagram for the 3-bit up/down counter is presented in Figure 8.46.

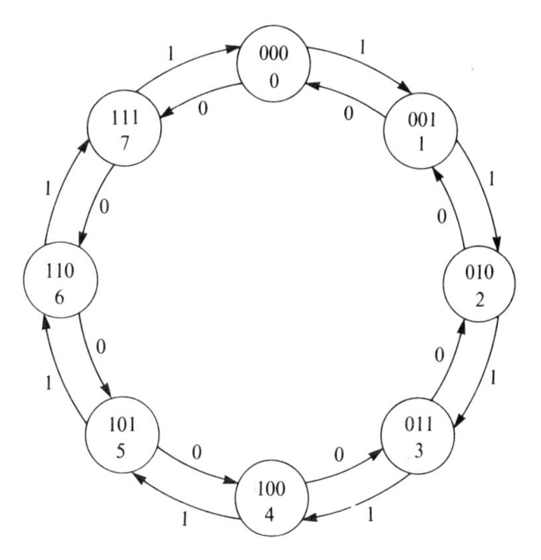

FIGURE 8.46 State diagram for a 3-bit up/down counter

b. The 3-bit counter has eight distinct machine states numbered 0 through 7 in decimal. Hence, three flip-flops, denoted by A, B, C, are required to implement the counter circuit. A state table for the 3-bit up/down counter can be constructed from the state diagram of Figure 8.46, as shown in Table 8.32.

TABLE 8.32 State table for a 3-bit up/down counter

		$A^+B^+C^+$	
Decimal	***ABC***	$U = 0$	$U = 1$
0	000	111	001
1	001	000	010
2	010	001	011
3	011	010	100
4	100	011	101
5	101	100	110
6	110	101	111
7	111	110	000

c. Separate the circuit state table into flip-flop next-state maps (Figure 8.47, p. 376).

A^+

AB \ CU	00	01	11	10
00	1	0	0	0
01	0	0	1	0
11	1	1	0	1
10	0	1	1	1

B^+

AB \ CU	00	01	11	10
00	1	0	1	0
01	0	1	0	1
11	0	1	0	1
10	1	0	1	0

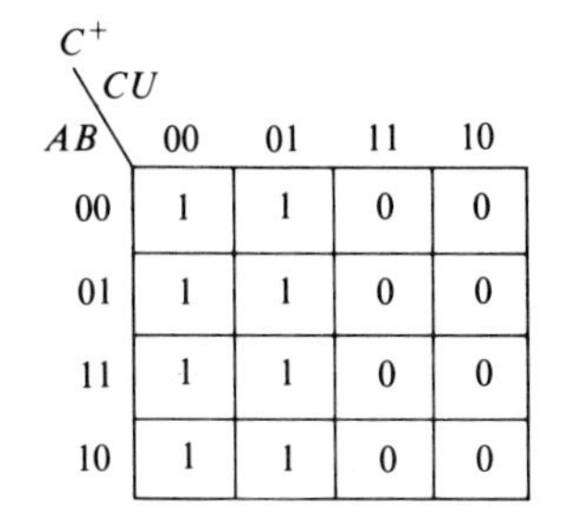

C^+

AB \ CU	00	01	11	10
00	1	1	0	0
01	1	1	0	0
11	1	1	0	0
10	1	1	0	0

FIGURE 8.47
Flip-flop next-state maps

d. For *JK* flip-flops convert each flip-flop next-state map to a transition map using the state transitions (a = alpha, b = beta), as shown in Figure 8.48(a). Then, convert the transition maps into shortcut *JK* excitation maps, as shown in Figure 8.48(b).

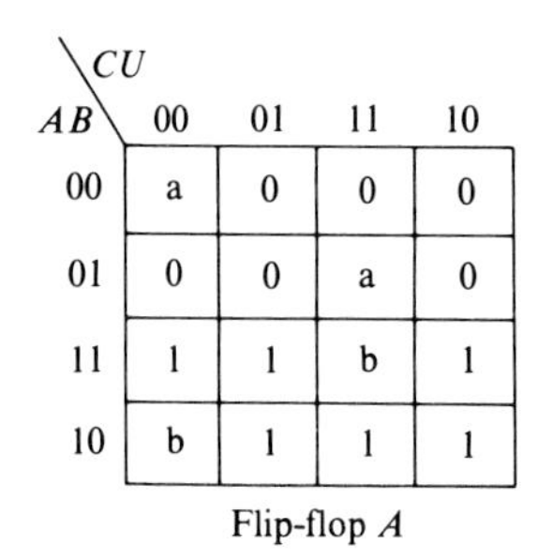

AB \ CU	00	01	11	10
00	a	0	0	0
01	0	0	a	0
11	1	1	b	1
10	b	1	1	1

Flip-flop *A*

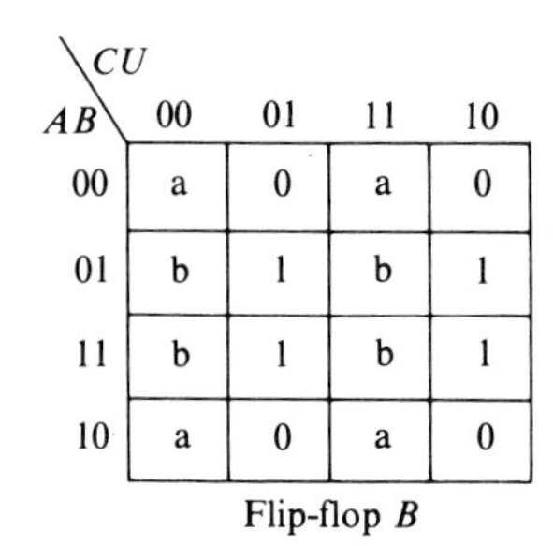

AB \ CU	00	01	11	10
00	a	0	a	0
01	b	1	b	1
11	b	1	b	1
10	a	0	a	0

Flip-flop *B*

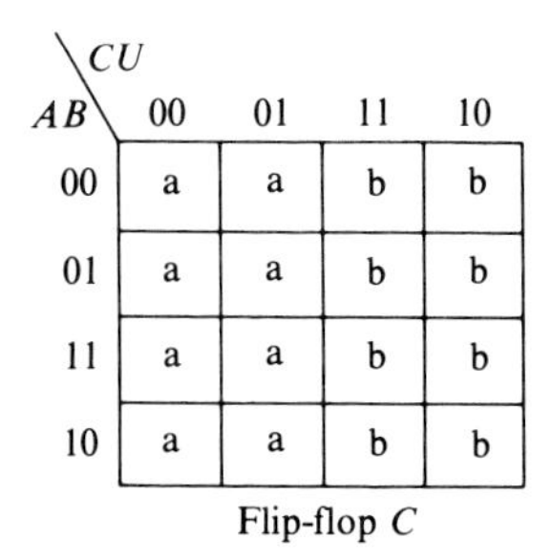

AB \ CU	00	01	11	10
00	a	a	b	b
01	a	a	b	b
11	a	a	b	b
10	a	a	b	b

Flip-flop *C*

(a)

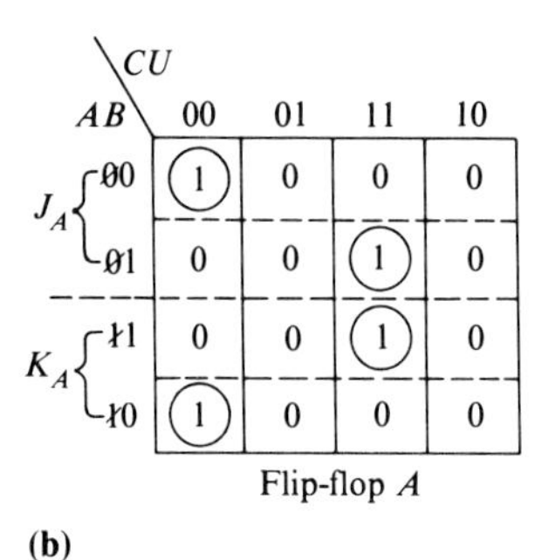

Flip-flop *A*

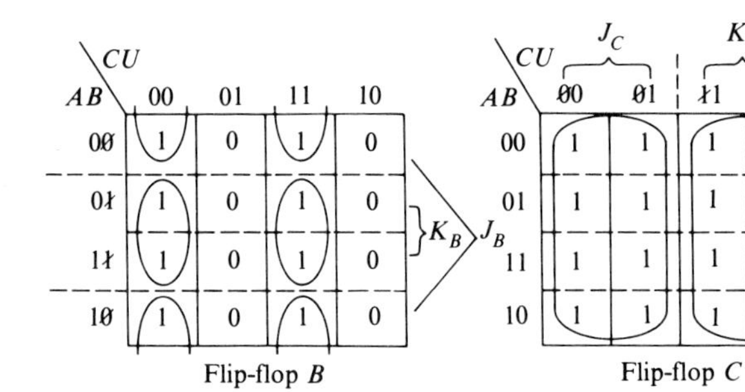

Flip-flop *B*

Flip-flop *C*

(b)

FIGURE 8.48
JK flip-flops. (a) Flip-flop transition maps (b) Shortcut *JK* excitation maps

e. Interpret each flip-flop excitation map to derive the corresponding flip-flop excitation equations:

$$J_A = B'C'U' + BCU \qquad J_B = C'U' + CU = (C \oplus U)' \qquad J_C = 1$$
$$K_A = B'C'U' + BCU \qquad K_B = C'U' + CU = (C \oplus U)' \qquad K_C = 1$$

Step 4: Realization. Use the flip-flop excitation equations to draw a logic diagram of the 3-bit up/down counter circuit, as shown in Figure 8.49.

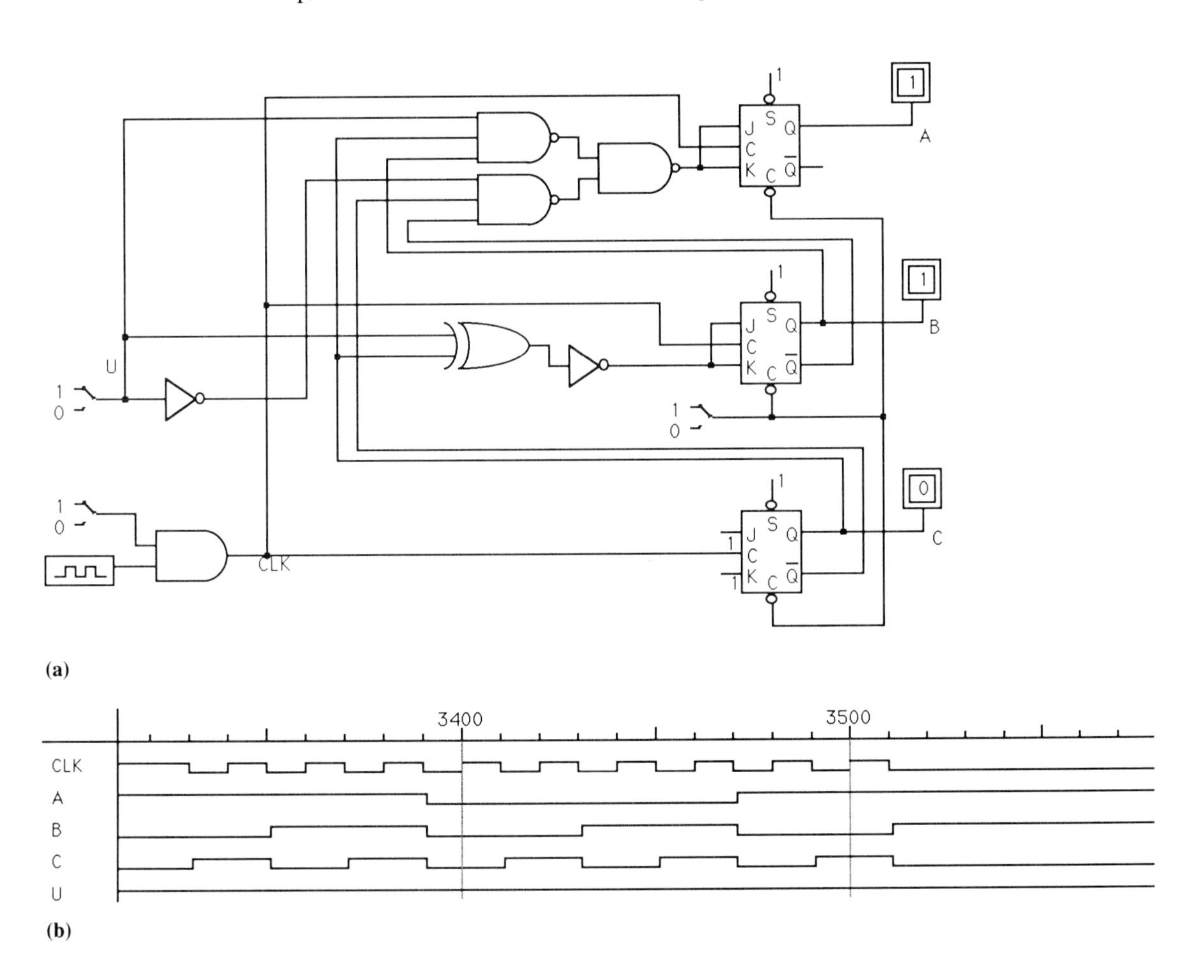

FIGURE 8.49
A 3-bit up/down counter. (a) Logic diagram (b) Timing diagram

Design of a Synchronous Sequential Filter Circuit

A filter is a device that accepts an arbitrary bit string as input and separates out short-term (1 or 2 bits) input sequences that differ from a preceding long-term input sequence.

Step 1: Problem Statement. Design a filter circuit according to the following specifications:

- Let *a* denote the initial state of the filter.
- Set the initial output of the filter equal to the initial input value.
- Let the filter output change value only when three successive inputs have the same value and that value is opposite to the current output.

Step 2: Conceptualization. Translate the design specifications into a state diagram (Figure 8.50).

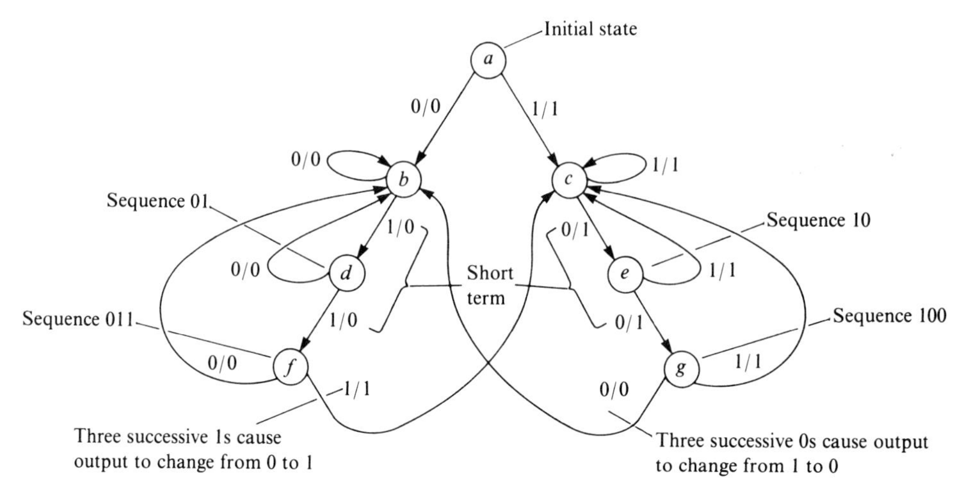

FIGURE 8.50
State diagram for a filter machine

The corresponding state table can then be constructed directly from the state diagram, as shown in Table 8.33.

TABLE 8.33
State table for a filter

	Next State		Output	
Present State	$I = 0$	$I = 1$	$I = 0$	$I = 1$
a	b	c	0	1
b	b	d	0	0
c	e	c	1	1
d	b	f	0	0
e	g	c	1	1
f	b	c	0	1
g	b	c	0	1

Step 3: Solution/Simplification.

a. Reduce the state table by eliminating any redundant machine states. An examination of the state table reveals that the next states and outputs in rows a, f, and g are identical. Therefore, states a, f, and g are equivalent. Consequently, redundant states f and g can be eliminated by the row-elimination method as shown in Table 8.34.

TABLE 8.34
Elimination of redundant states

	Next State		Output		
Present State	$I=0$	$I=1$	$I=0$	$I=1$	
a	b	c	0	1	
b	b	d	0	0	
c	e	c	1	1	
d	b	~~f~~ a	0	0	Replace f with a
e	~~g~~ a	c	1	1	Replace g with a
f	~~b~~	~~c~~	~~0~~	~~1~~	Eliminate row f
g	~~b~~	~~c~~	~~0~~	~~1~~	Eliminate row g

The elimination of the redundant states reduces the state table to the form shown in Table 8.35.

TABLE 8.35
Reduced state table for the filter

	Next State		Output	
Present State	$I=0$	$I=1$	$I=0$	$I=1$
a	b	c	0	1
b	b	d	0	0
c	e	c	1	1
d	b	a	0	0
e	a	c	1	1

After state reduction by the row elimination method, an implication chart can be used to determine if any redundant states remain. The implication chart for this state table is constructed as shown in Table 8.36.

TABLE 8.36
Implication table

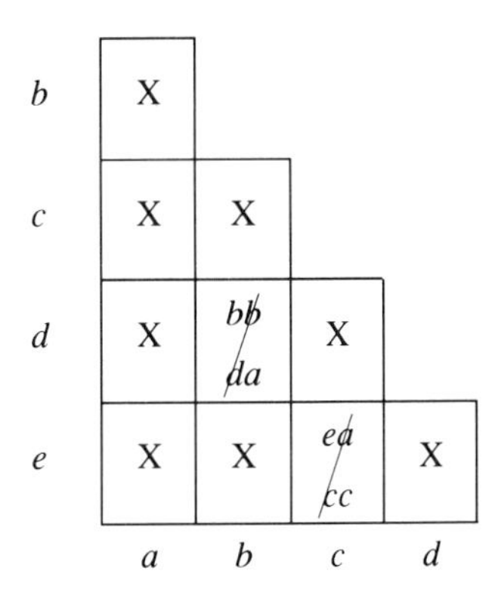

b	X			
c	X	X		
d	X	b~~b~~ ~~d~~a	X	
e	X	X	e~~a~~ ~~c~~c	X
	a	b	c	d

The only non-X cells can be crossed out because *e*, *a* and *d*, *a* comprise pairs of nonequivalent states. Therefore, there are no redundant states in the reduced state table. The final reduced state diagram has the form shown in Figure 8.51.

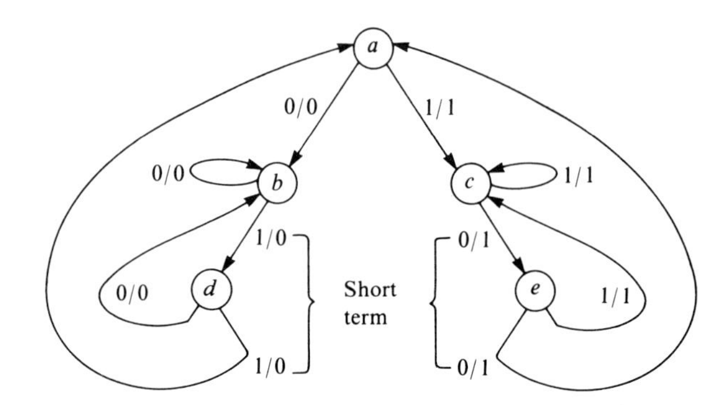

FIGURE 8.51
Reduced state diagram for a filter circuit

b. Assign states to simplify the flip-flop excitation network. Use the rules for state assignment (Section 8.3) to determine the optimum assignment of states for states of the reduced state table, as shown in Table 8.37.

TABLE 8.37
Reduced state table

	Present State	Next State $I = 0$	Next State $I = 1$	Output $I = 0$	Output $I = 1$
	a	*b*	*c*	0	1
	b	*b*	*d*	0	0
Rule 1	*c*	*e*	*c*	1	1
	d	*b*	*a*	0	0
	e	*a*	*c*	1	1
Rule 2					

By rule 0: Assign 000 to state *a*

By rule 1: $a–b,\ a–d,\ b–d,\ a–c,\ a–e,\ c–e$

By rule 2: $a–b, b–d, a–c, e–c, b–c$

By rule 3: Choose odd-numbered codes for states *b* and *e* and choose even-numbered codes for states *c* and *d*

The assignments are shown in Table 8.38.

The state table for the circuit can be described in a map form that shows which adjacency conditions are satisfied by the state assignment (Figure 8.52).

TABLE 8.38
State assignment table for the filter

ABC	$A^+B^+C^+$ $I=0$	$A^+B^+C^+$ $I=1$	Output $I=0$	Output $I=1$
a 000	*b* 001	*c* 010	0	1
b 001	*b* 001	*d* 100	0	0
e 011	*a* 000	*c* 010	1	1
c 010	*e* 011	*c* 010	1	1
110	XXX	XXX		
111	XXX	XXX		
101	XXX	XXX		
d 100	*b* 001	*a* 000	0	0

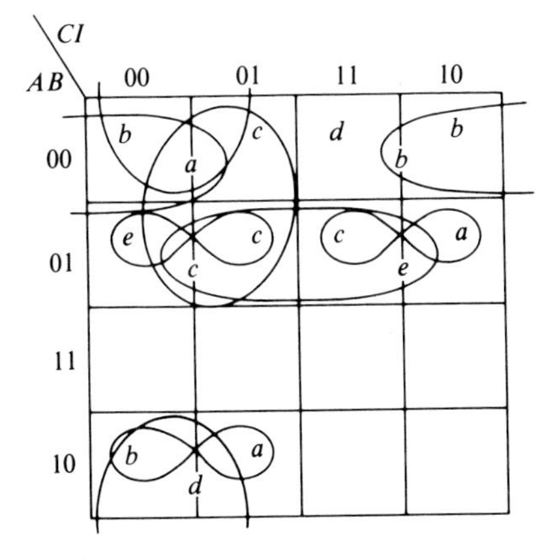

Rule 0: $a = 000$

Rule 1: a–b a–d c–e a–c

Rule 2: a–b a–c c–e

Rule 3: $b = 001, e = 011, c = 010, d = 100$

FIGURE 8.52
Illustration of adjacency conditions

A map-form state table, which shows the state assignment codes is presented in Figure 8.53. Note that the rightmost bit of each next state is the same for all entries (except when next state equals *a*) in a given column because of the restriction imposed by rule 3. The unused states can be used as don't cares in the state table, shown in Figure 8.53.

FIGURE 8.53
A map-form state table

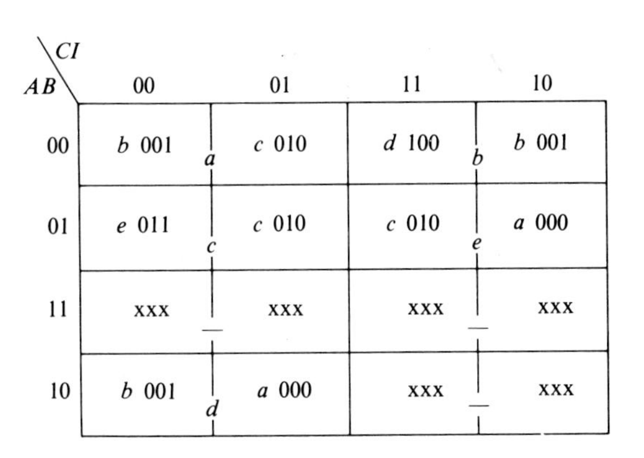

AB \ CI	00	01	11	10
00	*b* 001	*c* 010	*d* 100	*b* 001
01	*e* 011	*c* 010	*c* 010	*a* 000
11	XXX	XXX	XXX	XXX
10	*b* 001	*a* 000	XXX	XXX

c. Separate the circuit state table into flip-flop next-state maps, as shown in Figure 8.54.

d. Use the flip-flop characteristic equation to convert each flip-flop next-state map into a flip-flop excitation map. For D flip-flops, $Q^{+} = D$. Therefore, the excitation map for each flip-flop has the same form as the corresponding next-state map (Figure 8.54).

A^{+}

AB \ CI	00	01	11	10
00	0	0	1	0
01	0	0	0	0
11	X	X	X	X
10	0	0	X	X

B^{+}

AB \ CI	00	01	11	10
00	0	1	0	0
01	1	1	1	0
11	X	X	X	X
10	0	0	X	X

C^{+}

AB \ CI	00	01	11	10
00	1	0	0	1
01	1	0	0	0
11	X	X	X	X
10	1	0	X	X

FIGURE 8.54
Flip-flop next-state maps

To design a Mealy filter circuit using JK flip-flops construct flip-flop transition maps (Figure 8.55a) in terms of the state transitions (a = alpha, b = beta). Use the don't cares to make the largest possible 2^n groups of alphas and betas. Then convert the transition maps to shortcut JK excitation maps (Figure 8.55b).

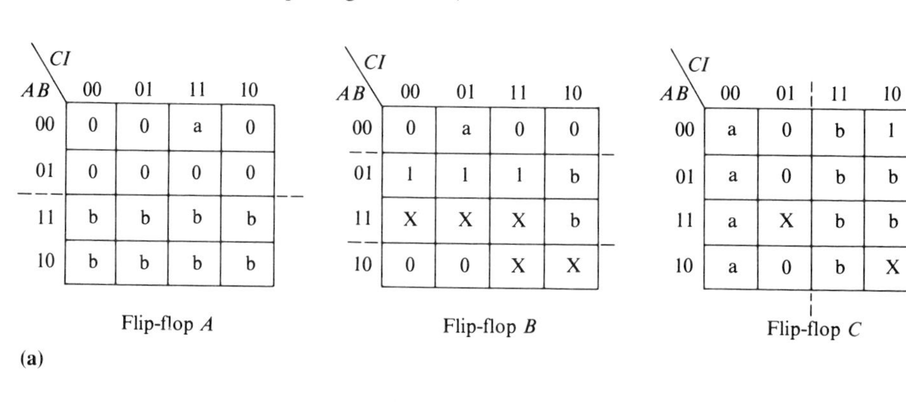

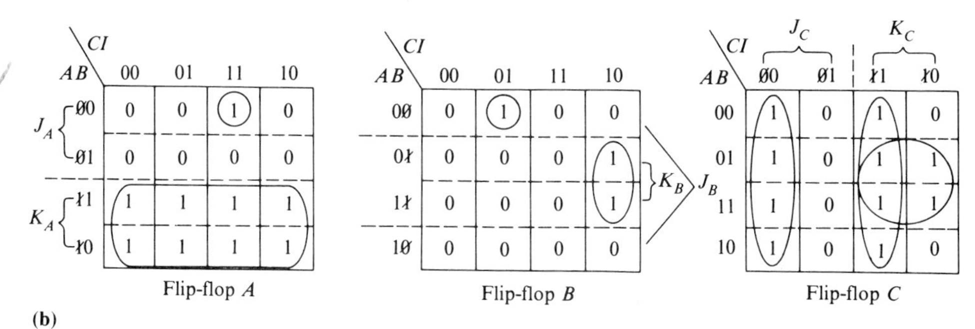

FIGURE 8.55
JK flip-flops. (a) Flip-flop transition maps (b) Shortcut *JK* excitation maps

e. Interpret the flip-flop excitation maps to derive flip-flop excitation equations: For *D* flip-flops,

$$D_A = B'CI \qquad D_B = BC' + BI + A'C'I \qquad D_C = C'I' + B'I'$$

For *JK* flip-flops,

$$J_A = B'CI \qquad J_B = A'C'I \qquad J_C = I'$$
$$K_A = 1 \qquad K_B = CI' \qquad K_C = B + I$$

Step 4: Realization. Use the flip-flop excitation equations and output equation to draw a logic diagram (Figure 8.56) of the filter circuit, $Z = A'(B + B'C'I)$

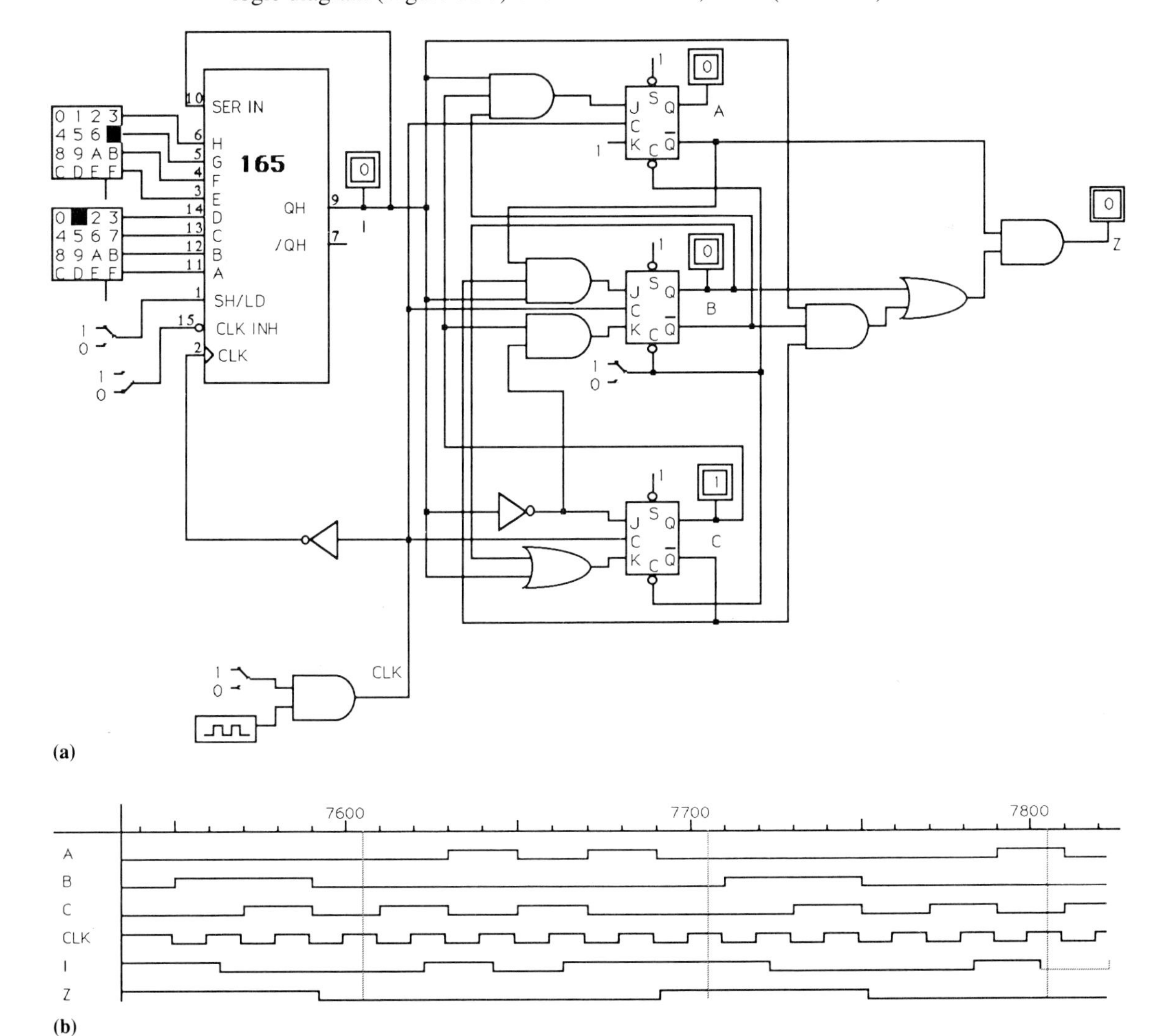

FIGURE 8.56
A Mealy filter circuit. (a) Logic diagram (b) Timing diagram

INTERACTIVE DESIGN APPLICATION

Design of a Two-Person Game

Step 1: Problem Statement. Design a 2-person game that has an output = 1 when the 2 inputs (x and y) match at least 4 times in succession or fail to match at least 4 times in succession. Let $I = (x \oplus y)'$ be the input to the sequential circuit.

Step 2: Conceptualization. Translate the specifications into a state diagram (Figure 8.57).

FIGURE 8.57
State diagram for a 2-person game circuit

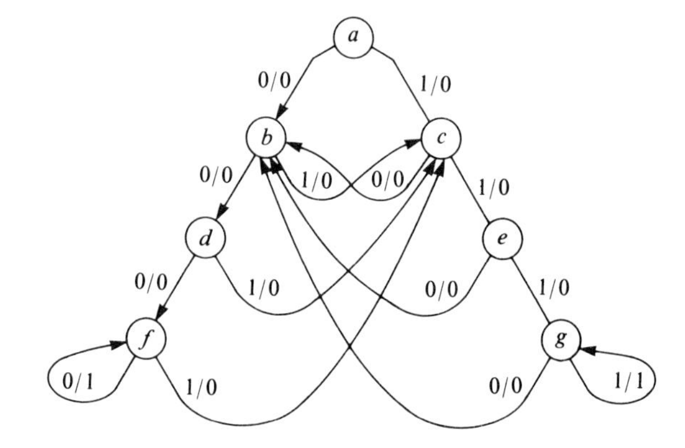

Step 3: Solution/Simplification.

a. The state diagram (Figure 8.57) produced by the multiple-sequence path algorithm has no redundant states. Therefore, the corresponding state table (Table 8.39) is a minimal-state table for the sequence detector for either at least four consecutive 1s or for at least four consecutive 0s.

TABLE 8.39
State table

Present State	Next State $I=0$	Next State $I=1$	Output $I=0$	Output $I=1$
a	b	c	0	0
b	d	c	0	0
c	b	e	0	0
d	f	c	0	0
e	b	g	0	0
f	f	c	1	0
g	b	g	0	1

(Rule 1 brackets group present states; Rule 2 brackets group next-state pairs.)

b. Make the state assignments in accordance with the following guidelines:

Rule 0: Assign 000 to initial state a

Rule 1: $a–b$, $a–c$, $a–d$, $a–e$, $a–f$, $a–g$, $b–d$, $b–f$, $c–e$, $c–g$, $d–f$, $e–g$

Rule 2: *b–c*, *b–e*, <u>*b–g*</u>, *d–c*, <u>*f–c*</u>

Rule 3: Assign odd-numbered codes to states *b*, *d*, and *f* and assign even-numbered codes to states *c*, *e*, and *g*

If we use the state assignment in Figure 8.58 we can construct the circuit state table shown in Table 8.40.

FIGURE 8.58
State assignment map

AB \ C	0	1
00	*a*	
01	*c*	*f*
11	*e*	*d*
10	*g*	*b*

TABLE 8.40
Circuit state table

	$A^+B^+C^+$		Output	
ABC	*I* = 0	*I* = 1	*I* = 0	*I* = 1
a 000	*b* 101	*c* 010	0	0
001	xxx	xxx		
f 011	*f* 011	*c* 010	1	0
c 010	*b* 101	*e* 110	0	0
e 110	*b* 101	*g* 100	0	0
d 111	*f* 011	*c* 010	0	0
b 101	*d* 111	*c* 010	0	0
g 100	*b* 101	*g* 100	0	1

The state table for the circuit can be described in a map form that shows the adjacency conditions that satisfy state assignment rules 0, 1, 2, and 3, as shown in Figure 8.59.

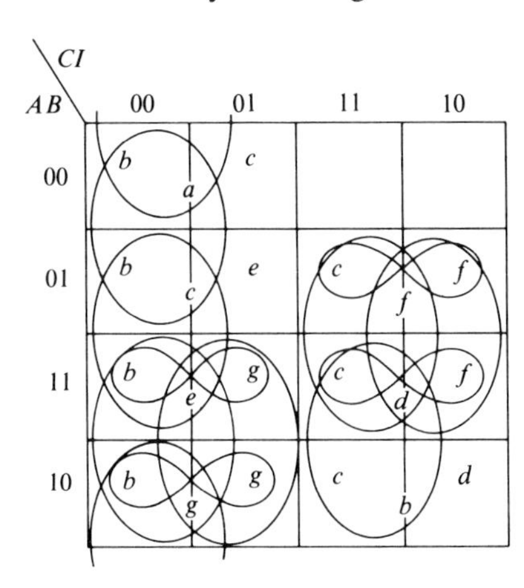

Rule 0: *a* = 000

Rule 1: *a–c* *a–g* *c–e* <u>*e–g*</u>
d–b <u>*f–d*</u>

Rule 2: <u>*c–f*</u> <u>*b–g*</u>

Rule 3: *c* = 010 *f* = 011
e = 110 *d* = 111
g = 100 *b* = 587

FIGURE 8.59
Illustration of adjacency conditions

A map-form state table that shows the state assignment codes is presented in Figure 8.60.

FIGURE 8.60
A map-form state table

AB \ CI	00	01	11	10
00	*b* 101 (*a*)	*c* 010	xxx	xxx
01	*b* 101 (*c*)	*e* 110	*c* 010 (*f*)	*f* 011
11	*b* 101 (*e*)	*g* 100	*c* 010 (*d*)	*f* 011
10	*b* 101 (*g*)	*g* 100	*c* 010 (*b*)	*d* 111

c. Separate the circuit state table into flip-flop next-state maps (Figure 8.61).

A^+

AB \ CI	00	01	11	10
00	1	0	X	X
01	1	1	0	0
11	1	1	0	0
10	1	1	0	1

B^+

AB \ CI	00	01	11	10
00	0	1	X	X
01	0	1	1	1
11	0	0	1	1
10	0	0	1	1

C^+

AB \ CI	00	01	11	10
00	1	0	X	X
01	1	0	0	1
11	1	0	0	1
10	1	0	0	1

FIGURE 8.61
Flip-flop next-state maps

d. For *D* flip-flops, the excitation maps have the same form as the next-state maps. For *JK* flip-flops rewrite the next-state maps as transition maps in terms of the state transitions (a = alpha, b = beta), as shown in Figure 8.62a. Then convert the transition maps to shortcut *JK* excitation maps as shown in Figure 8.62b.

AB \ CI	00	01	11	10
00	a	0	X	X
01	a	a	0	0
11	1	1	b	b
10	1	1	b	1

Flip-flip *A*

AB \ CI	00	01	11	10
00	0	a	a	a
01	b	1	1	1
11	b	b	1	1
10	0	0	a	a

Flip-flip *B*

AB \ CI	00	01	11	10
00	a	0	b	X
01	a	0	b	1
11	a	0	b	1
10	a	0	b	1

Flip-flip *C*

(a)

FIGURE 8.62
JK flip-flops. (a) Flip-flop transition maps (b) Shortcut *JK* excitation maps

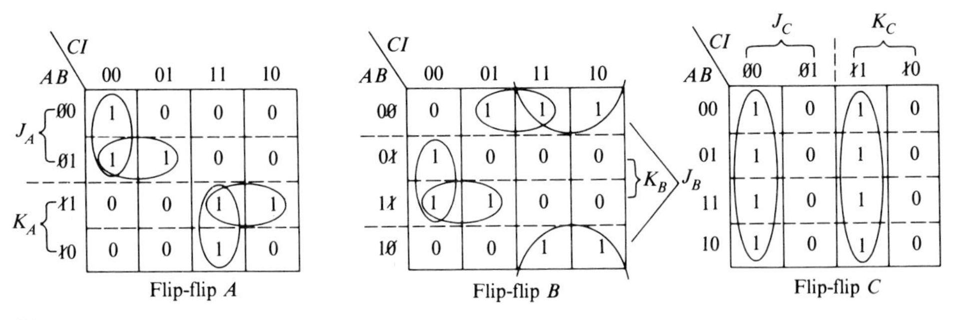

(b)

FIGURE 8.62
Continued

e. Interpret the flip-flop excitation maps or transition maps to determine the corresponding excitation equations:

$$D_A = B'I' + AC' + BC' \qquad D_B = C + A'I \qquad D_C = I'$$
$$J_A = C'I' + BC' \qquad J_B = A'I + C \qquad J_C = I'$$
$$K_A = BC + CI \qquad K_B = C'I' + AC' \qquad K_C = I$$

Step 4: Realization. Draw a logic diagram for the 2-person game circuit (Figure 8.63).

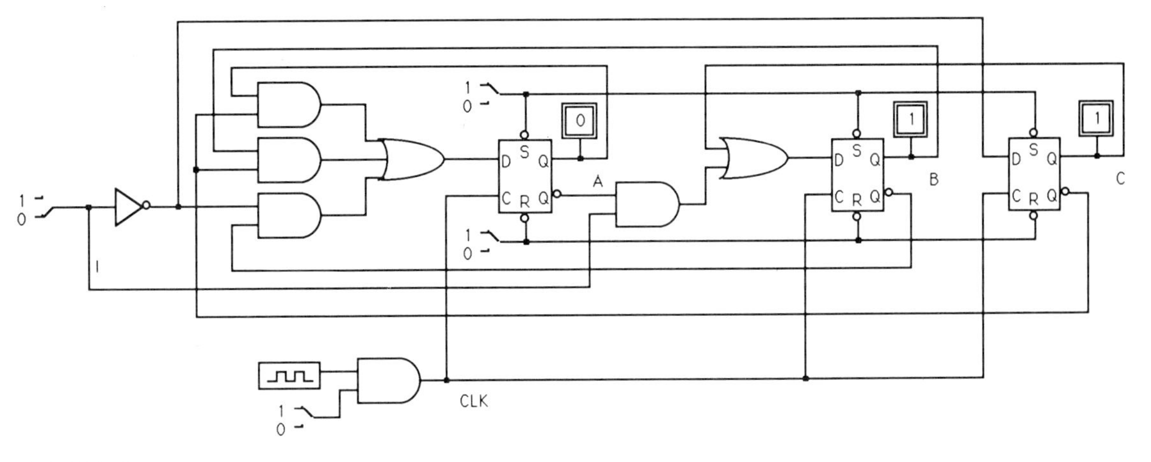

FIGURE 8.63
Logic diagram for 2-person game circuit

INTERACTIVE DESIGN APPLICATION

Design of a Sequential Logic Combination Lock

Step 1: Problem Statement. Design a digital combination lock that has a stored set of three BCD numbers as the "key" combination. Design the lock so that it opens when each of three consecutive BCD numbers input matches the corresponding stored BCD number. Let K_0, K_1, K_2 and X_0, X_1, X_2 denote the stored BCD key and input numbers, respectively.

Consider the specifications. We must design a combinational logic circuit that has output $I = 1$ when each bit of input X_i equals the corresponding bit of key K_i. (See Figure 8.64.) Therefore, we will design a Mealy synchronous circuit, with input I from the circuit seen in Figure 8.64, that detects a sequence of exactly three consecutive 1s.

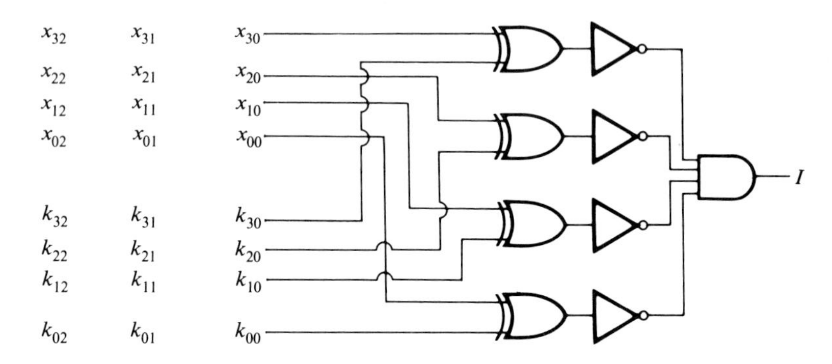

FIGURE 8.64
Combinational logic 4-bit compare-equal circuit

Step 2: Conceptualization. The state diagram can be constructed using a binary tree with reset after input of exactly 3 bits. The state diagram is shown in Figure 8.65.

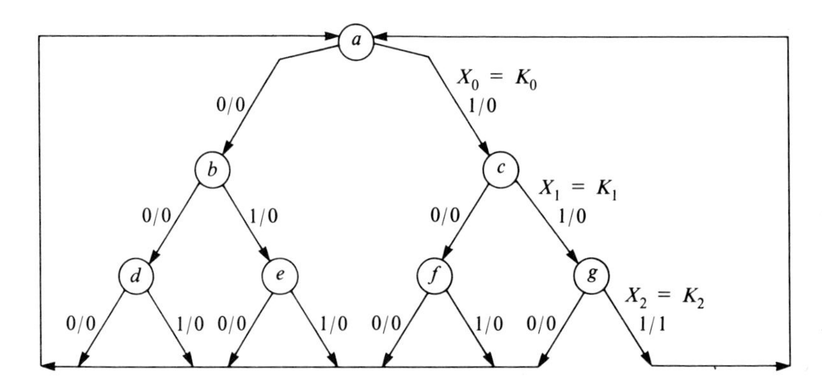

FIGURE 8.65
State diagram for a Mealy 111 detector with 3-bit reset

Step 3: Solution/Simplification.

a. Reduce the state table. Since the state table contains a number of rows with identical next states and outputs, the row-elimination method can be used to eliminate redundant states, as shown in Table 8.41.

TABLE 8.41
State table

Present State	Next State		Output	
	$I=0$	$I=1$	$I=0$	$I=1$
a	b	c	0	0
b	d	e	0	0
c	f	g	0	0
d	a	a	0	0 ← $d=e=f$
e	a	a	0	0 ←
f	a	a	0	0 ←
g	a	a	0	1

A second application of the row elimination method simplifies the state table to its final form, Table 8.42.

TABLE 8.42
Reduced state table

Present State	Next State		Output	
	$I=0$	$I=1$	$I=0$	$I=1$
a	b	c	0	0
Rule 1 [b	d	d	0	0
c	d	g	0	0
[d	a	a	0	0
g]	a	a	0	1
Rule 2				

b. Make the state assignments as follows (Figure 8.66):

Rule 0: Assign 000 to initial state a

Rule 1: b–c, $\underline{d\text{–}g}$

Rule 2: b–c, d–g

Rule 3: Assign odd-numbered codes to states b and d and assign even-numbered codes to states c and g

FIGURE 8.66
State assignment matrix

$AB \backslash C$	0	1
00	a	
01	c	b
11	g	d
10		

The state table has the form shown in Table 8.43 (p. 390).

TABLE 8.43
State assignment table

	$A^+B^+C^+$		Output	
ABC	*I* = 0	*I* = 1	*I* = 0	*I* = 1
a 000	*b* 011	*c* 010	0	0
001	xxx	xxx		
b 011	*d* 111	*d* 111	0	0
c 010	*d* 111	*g* 110	0	0
g 110	*a* 000	*a* 000	0	1
d 111	*a* 000	*a* 000	0	0
101	xxx	xxx		
100	xxx	xxx		

$$Z = ABC'I$$

The state table for the circuit can be described in a map form that shows the adjacency conditions satisfied by state assignment (Figure 8.67).

FIGURE 8.67
Map of adjacency requirements satisfied

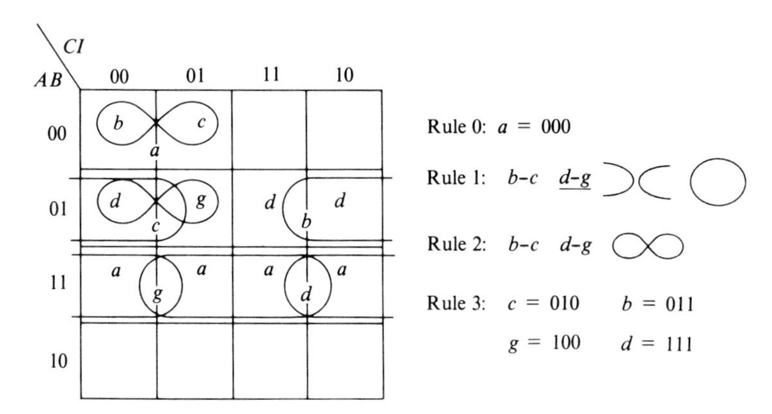

A map-form state table which shows the state assignment codes is presented in Figure 8.68.

FIGURE 8.68
A map-form circuit state table

AB \ CI	00		01	11		10
00	*b* 011	*a*	*c* 010	xxx		xxx
01	*d* 111	*c*	*g* 110	*d* 111	*b*	*d* 111
11	*a* 000	*g*	*a* 000	*a* 000	*d*	*a* 000
10	xxx		xxx	xxx		xxx

c. Separate the circuit state table into flip-flop next-state maps, as given in Figure 8.69.

A^+

AB \ CI	00	01	11	10
00	0	0	X	X
01	1	1	1	1
11	0	0	0	0
10	X	X	X	X

B^+

AB \ CI	00	01	11	10
00	1	1	X	X
01	1	1	1	1
11	0	0	0	0
10	X	X	X	X

C^+

AB \ CI	00	01	11	10
00	1	0	X	X
01	1	0	1	1
11	0	0	0	0
10	X	X	X	X

FIGURE 8.69
Flip-flop next-state maps

d. For D flip-flops the excitation maps have the same form as the next-state maps. For JK flip-flops (Figure 8.70) rewrite the next-state maps as transition maps in terms of the state transitions (a = alpha, b = beta). Then convert the transition maps to shortcut JK excitation maps.

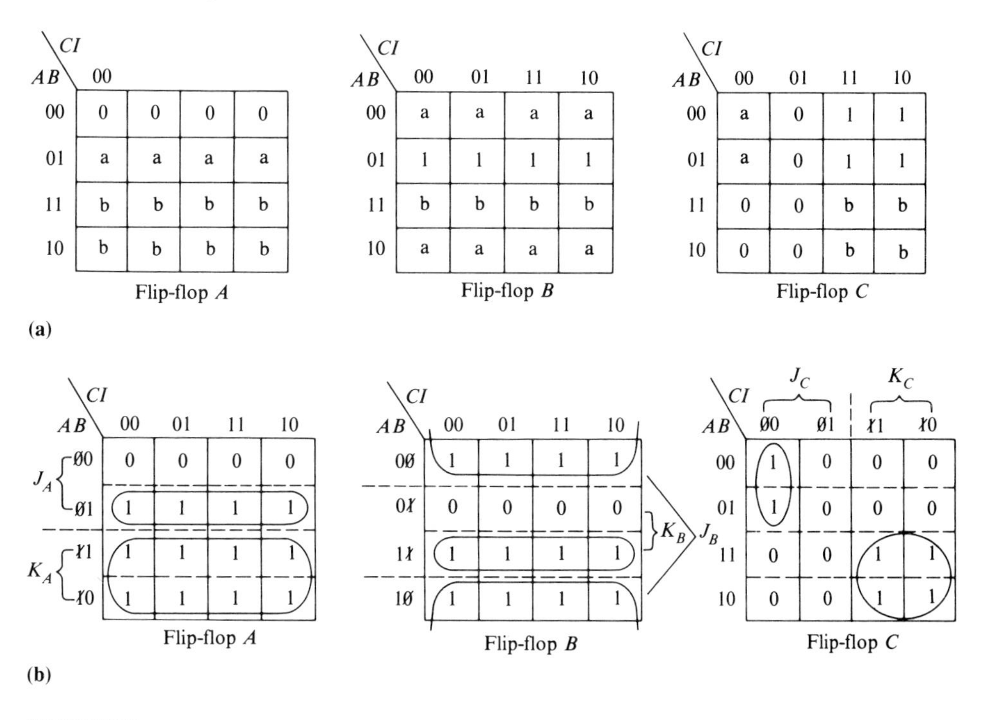

FIGURE 8.70
JK flip-flops. (a) Flip-flop transition maps (b) Shortcut *JK* excitation maps

e. Interpret each flip-flop excitation map to determine the corresponding excitation equations:

$$D_A = A'B \qquad D_B = A' \qquad D_C = A'I' + A'C$$
$$J_A = B \qquad J_B = 1 \qquad J_C = A'I'$$
$$K_A = 1 \qquad K_B = A \qquad K_C = A$$

Step 4: Realization. Draw a logic diagram (Figure 8.71)for the sequential-logic combination lock circuit.

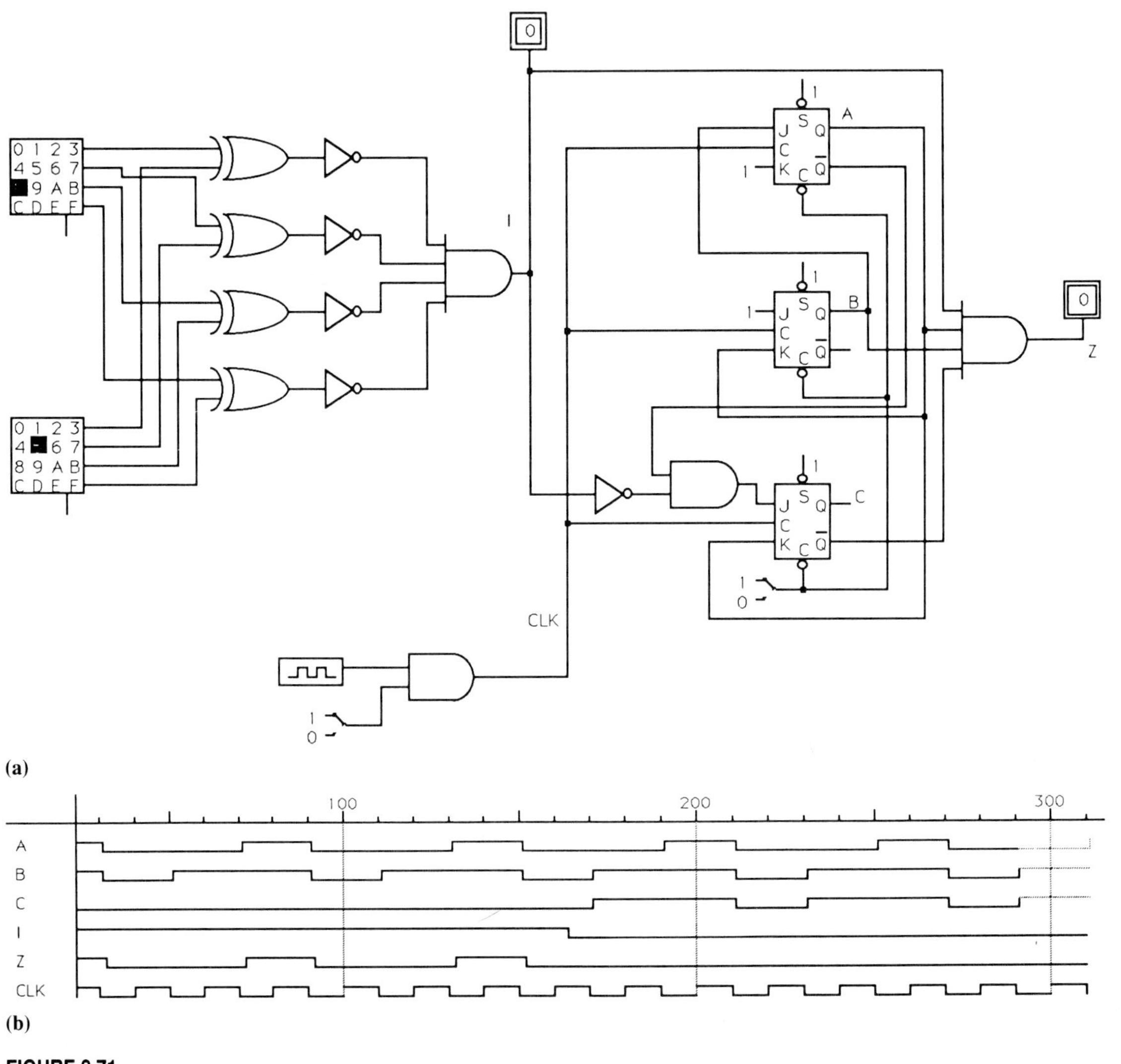

FIGURE 8.71
Combination lock using *JK* flip-flops. (a) Logic diagram (b) Timing diagram

INTERACTIVE DESIGN APPLICATION

Design of a Mealy 0101 Sequence Detector

Step 1: Problem Statement. Design a Mealy 0101 sequence detector.

Step 2: Conceptualization. Use the single sequence-path procedure to construct a state diagram, as shown in Figure 8.72a.

FIGURE 8.72
Mealy 0101 sequence detector. (a) State diagram (b) Connection matrix

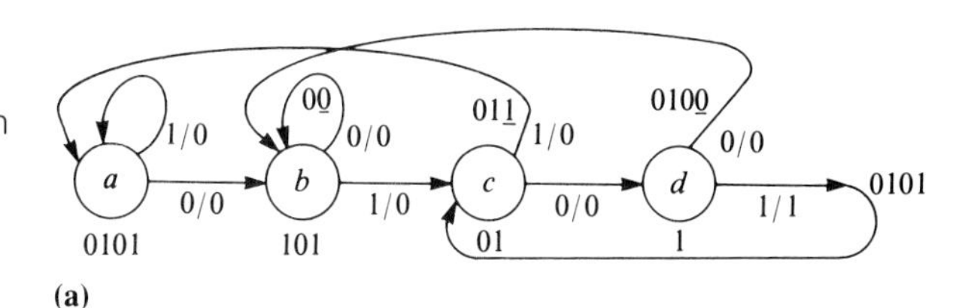

(a)

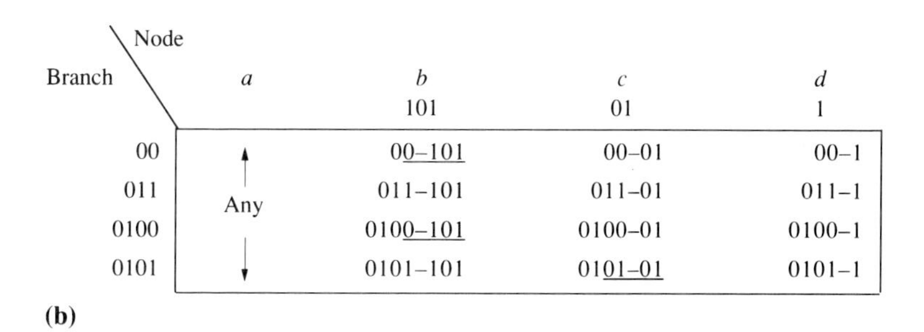

Branch \ Node	a	b 101	c 01	d 1
00	↑	00–101	00–01	00–1
011	Any	011–101	011–01	011–1
0100		0100–101	0100–01	0100–1
0101	↓	0101–101	0101–01	0101–1

(b)

A state table can then be constructed from the state diagram (Table 8.44).

TABLE 8.44
State table for a Mealy 0101 sequence detector

Present State	Next State $I = 0$	Next State $I = 1$	Output $I = 0$	Output $I = 1$
a	b	a	0	0
b	b	c	0	0
c	d	a	0	0
d	b	c	0	1

Step 3: Solution/Simplification.

a. State reduction. No reduction is required because the single-sequence-path procedure generates a minimal form state diagram and state table.

b. State assignment.

By rule 1: a–b, a–d, a–c, b–d

By rule 2: a–b, a–d, b–c

By rule 3: Assign odd-numbered codes to b, d; assign even-numbered code to c

The state assignment (a = 00, b = 01, d = 11, c = 10) that satisfies the majority of the required adjacencies is illustrated in Figure 8.73.

FIGURE 8.73
State assignment matrix

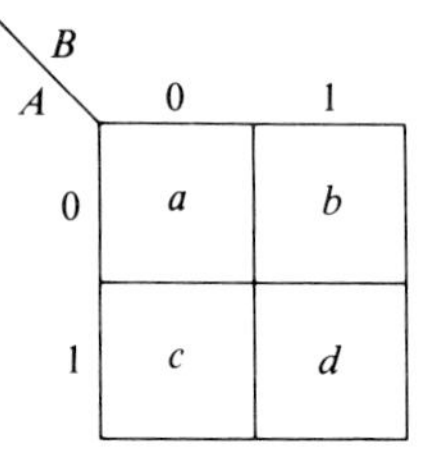

A map form circuit state table can be constructed as shown in Figure 8.74.

FIGURE 8.74
A map form circuit state table

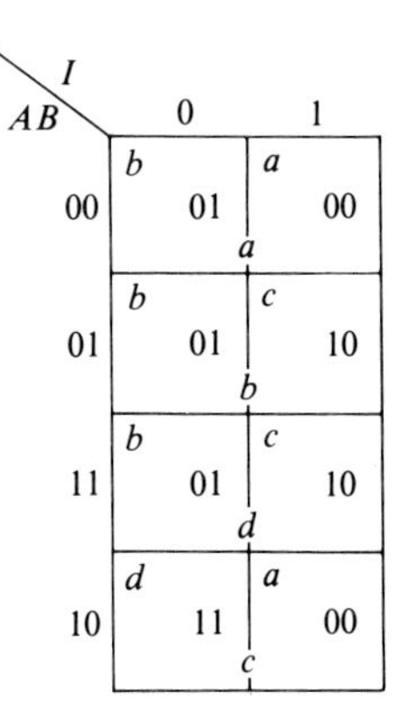

c. With the assignment complete, we can proceed to separate the circuit state table into flip-flop next-state maps (Figure 8.75).

FIGURE 8.75
Flip-flop next-state maps

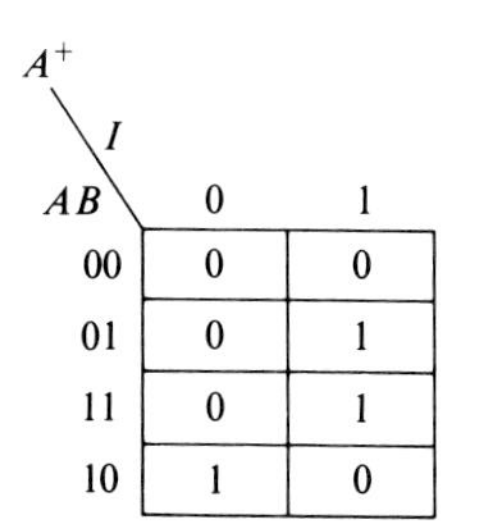

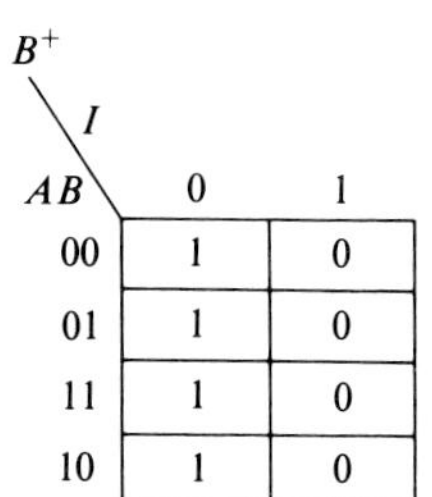

d. Then, we use the flip-flop characteristic equation to convert each next-state map into a flip-flop excitation map. For D flip-flops, $Q^+ = D$. Therefore, the excitation map for each flip-flop has the same form as the corresponding next-state map (Figure 8.76).

FIGURE 8.76
Excitation maps for D flip-flops

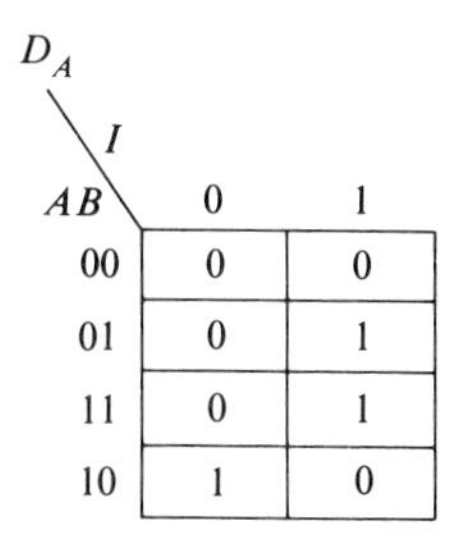

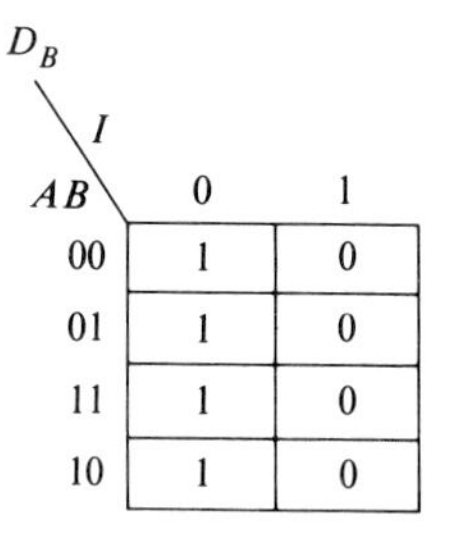

For JK flip-flops, $Q^+ = JQ' + K'Q$. The flip-flop next-state maps are rewritten as transition maps in terms of the state transitions (a = alpha, b = beta), as shown in Figure 8.77a. The transition maps can then be converted to shortcut JK excitation maps as shown in Figure 8.77b.

FIGURE 8.77
JK flip-flops. (a) Transition maps (b) Shortcut JK excitation maps

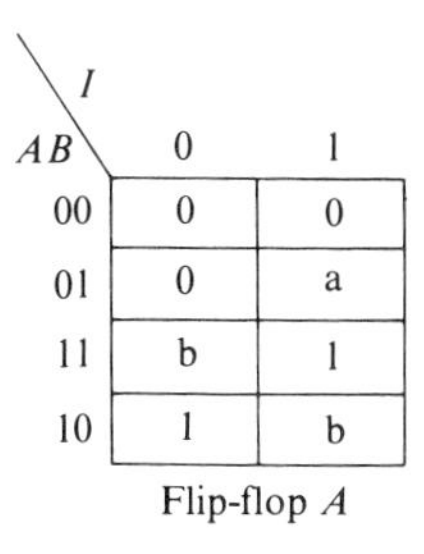

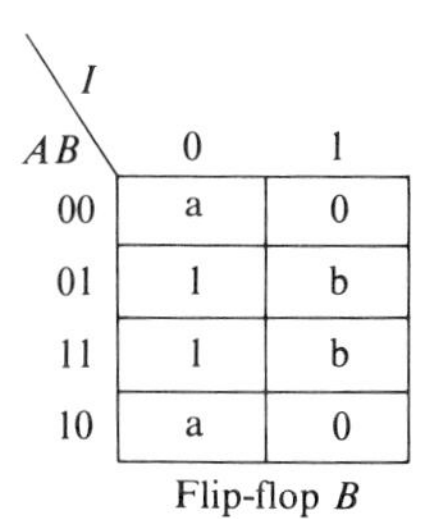

(a)

AB \ I	0	1
J_A: 00	0	0
J_A: 01	0	(1)
K_A: 11	(1)	0
K_A: 10	0	(1)

Flip-flop A

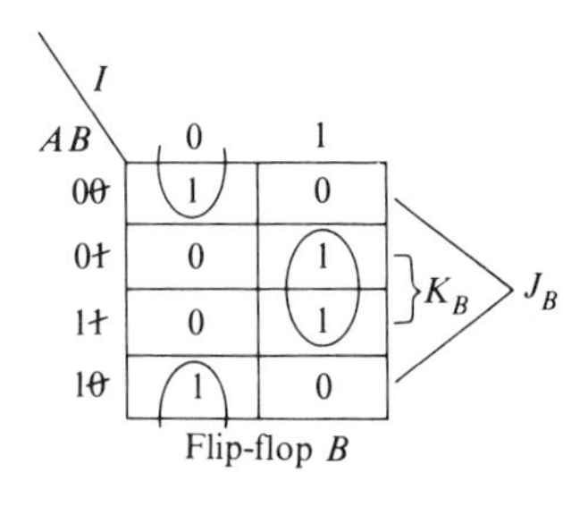

(b)

e. We can now interpret the flip-flop excitation maps to derive the flip-flop excitation equation(s):

$$D_A = AB'I' + BI \qquad D_B = I' \qquad Z = ABI$$

$$J_A = BI \qquad J_B = I'$$

$$K_A = BI' + B'I = B \oplus I \qquad K_B = I$$

Step 4: Realization. At this point, we can draw the logic diagram of the Mealy 0101 sequence detector, as shown in Figure 8.78a (p. 396). A timing diagram for the circuit is presented in Figure 8.78b.

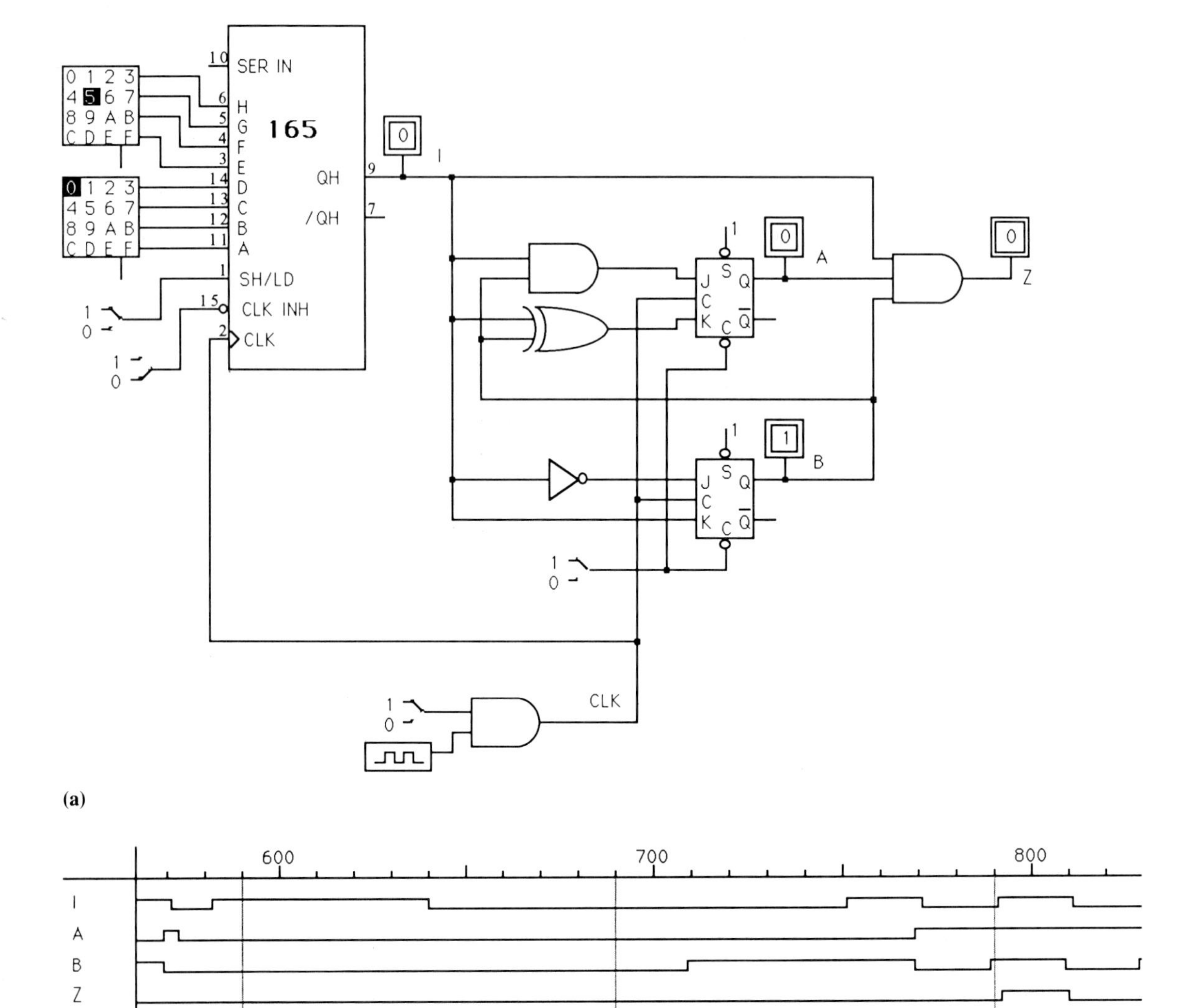

(b)

FIGURE 8.78
Mealy 0101 sequence detector. (a) Logic diagram (b) Timing diagram

8.5 SUMMARY

This chapter described a variety of general sequential processes (such as sequence generation, sequence detection, and filtering). It also showed how such processes can be represented by finite-state machine models that can be realized by sequential-logic circuits.

A step-by-step procedure for analyzing synchronous sequential circuits was described. This procedure was then illustrated in the analysis of a variety of Moore and Mealy sequential circuits. The output of a Moore circuit remains constant as long as the circuit state does not change. The output of a Mealy circuit remains constant as long as the circuit state and inputs do not change.

The 4-step problem-solving procedure was modified to accomplish the design of synchronous sequential-logic circuits. Five representative synchronous sequential circuits were then designed using this procedure:

A 3-bit up/down counter

A filter circuit

A two-person game

A combination lock

A Mealy 0101 sequence detector

KEY TERMS

Sequential process

Present state

0-successor of *s*

1-successor of *s*

Single-sequence-path algorithm

Main-path diagram

Connection matrix

Multiple-sequence-path algorithm

Redundant states

Row-elimination procedure

Equivalent

Equivalence class

Equivalence-class partitioning method

Implication chart method

State assignment

EXERCISES

General

1. Use the binary tree procedure to construct a state diagram for a Mealy sequence detector that detects three or more consecutive 0s. Construct the corresponding state table.
2. Use the single-sequence-path procedure to construct a state diagram for a Mealy sequence detector for the sequence 1011.
3. Use the single-sequence-path procedure to construct a state diagram for a Mealy sequence detector for the sequence 1010.
4. Use the multiple-sequence-path procedure to construct a state diagram for a Mealy sequence detector for the sequences 000 and 111.
5. Given the state table shown below, use the implication chart method to identify equivalent states in the table. Construct the minimal state table.

Present State	Next State $I = 0$	Next State $I = 1$	Output $I = 0$	Output $I = 1$
a	b	c	0	0
b	d	e	0	0
c	d	a	0	0
d	f	g	0	1
e	b	c	0	0
f	g	h	0	0
g	d	a	0	0
h	d	e	0	0

6. Given the state table shown in Exercise 5, use the row-elimination method to eliminate redundant states in the table.

7. Given the following state table, use the row elimination method to eliminate redundant states in the table.

Present State	Next State $I = 0$	Next State $I = 1$	Output $I = 0$	Output $I = 1$
a	b	c	0	0
b	c	a	0	0
c	g	d	0	0
d	f	g	0	1
e	b	c	0	0
f	g	h	0	0
g	c	a	0	0
h	a	e	0	0

8. Given the state table shown in Exercise 7, use the implication chart method to identify equivalent states in the table. Construct the reduced state table.

9. Given the following state table, use the equivalence-class partitioning method to determine the equivalence classes in the table.

Present State	Next State $I = 0$	Next State $I = 1$	Output $I = 0$	Output $I = 1$
a	b	a	0	0
b	b	c	0	0
c	d	e	0	0
d	f	g	0	1
e	h	a	0	0
f	b	g	0	0
g	d	a	0	0
h	f	g	0	0

10. Given the following state table, use the equivalence-class partitioning method to determine the equivalence classes in the table.

	Next State		Output	
Present State	$I = 0$	$I = 1$	$I = 0$	$I = 1$
a	d	b	0	1
b	f	f	0	0
c	h	f	1	0
d	a	h	1	0
e	c	d	0	1
f	g	d	0	1
g	b	e	1	0
h	f	e	0	0

11. Given the following state table, use the equivalence-class partitioning method to determine the equivalence classes in the table.

	Next State		Output	
Present State	$I = 0$	$I = 1$	$I = 0$	$I = 1$
a	c	b	0	0
b	c	d	0	0
c	e	f	0	0
d	e	a	1	0
e	a	b	0	0
f	e	d	0	0

12. Given the state table shown in Exercise 11, use the implication chart method to identify equivalent states in the table. Construct the reduced state table.

Design/Implementation

Use the modified 4-step problem-solving procedure in the following exercises.

1. Design a 2-bit up/down counter using *JK* flip-flops.

2. Design a Moore sequence detector that detects three or more consecutive 1s in an arbitrary input string.

3. Design a Mealy sequence detector for the sequence 110.

4. Design a Mealy sequence detector for the sequence 1010.

5. Design a Mealy sequence detector for the sequence 1011.

6. Design a Mealy sequence detector for the sequences 101 and 011.

7. Design a Mealy sequence detector for the sequences 001 and 101.

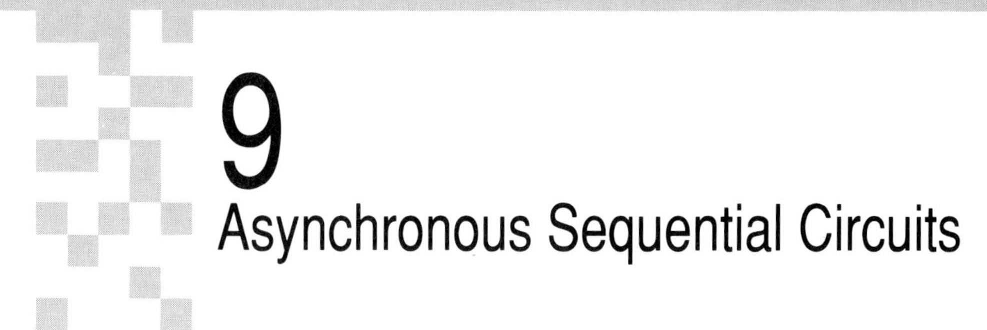

9 Asynchronous Sequential Circuits

OBJECTIVES

After you complete this chapter, you will be able

- To understand the basic concepts of fundamental-mode asynchronous circuits
- To describe the operation of a particular sequential process using a primitive flow table and/or a primitive state diagram
- To analyze an asynchronous sequential-logic circuit to determine its function and operating characteristics
- To identify and eliminate static, dynamic, and essential hazards that are caused by unequal propagation delays in different signal paths
- To design race-free fundamental-mode circuits that realize sequential processes using a 4-step problem-solving procedure

9.1 INTRODUCTION

What differentiates an asynchronous process from a synchronous process? Both have 0–1 level signal input sequences. However, an asynchronous process assumes a sequence of changing input combinations, while a synchronous process allows a sequence with repeated level values on an input line such as 00010011. The input levels of an asynchronous process can change at arbitrary times, while the input levels of a synchronous process can change only at specific (periodic) instants of time. Both types of processes can be represented by sequential (finite state) machines, as defined in Chapter 7. Asynchronous process machines are realized by asynchronous sequential-logic circuits, while synchronous process machines are realized by synchronous sequential-logic circuits, which were described in Chapters 7 and 8.

Asynchronous sequential-logic circuits are basically combinational logic circuits with direct feedback paths. Memory elements in an asynchronous circuit are either delay elements (which may be simply the propagation delay in the combinational logic circuit) or asynchronous latches. The behavior of an asynchronous circuit depends on the order in which its inputs change. Since its inputs can change at any arbitrary instant of time, memory state changes may occur at any time in response to these input changes.

By contrast, synchronous sequential-logic circuits use a clock signal to effect memory state changes at periodic times. Instead of delay elements or asynchronous latches, memory elements in a synchronous circuit are clocked (usually edge-triggered) flip-flops.

An asynchronous sequential-logic circuit can be represented by a model of the form shown in Figure 9.1. Such a circuit includes the following components:

- A set of n external inputs, $x_1, \ldots, x_n$
- A combinational logic network consisting of a next-state logic section and an output logic section
- A set of m external outputs, $z_1, \ldots, z_m$
- A memory section that stores system state $y_1, \ldots, y_p$
- A set of feedback lines

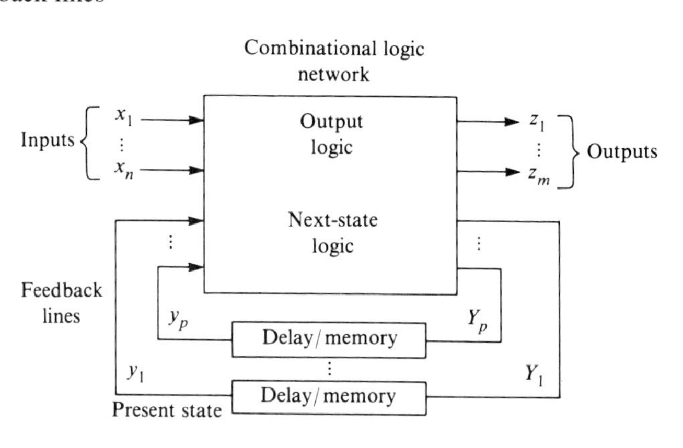

FIGURE 9.1
Model for an asynchronous sequential-logic circuit

An asynchronous sequential-logic circuit is basically a combinational logic circuit with direct feedback paths from the circuit's outputs to its inputs. If a new set of input values are applied at time t, the input changes filter through the combinational logic network along different paths with different propagation delays. The propagation delay time dt_i on the ith path can be represented by delay element i. Since the output $y_i(t + dt_i)$ of delay element i equals the input $Y_i(t)$, delay element i can be represented as a memory element in feedback path i.

Since an asynchronous circuit has no clock to synchronize the changes in the state variables, an asynchronous circuit is subject to a number of problems caused by unequal propagation delays in different signal paths; these problems (such as critical races and various types of hazards) are discussed in this chapter.

At any specific time t, the memory is in a state $(y_1, \ldots, y_p)$, called the present state, consisting of the p outputs of the memory section. The present state and the external inputs applied at time t determine the memory next state at time $t + d$, where d is greater than any dt_i, $(i = 1, p)$. Thus, the circuit's next state is a Boolean function of the circuit's present state and inputs.

A sequential logic circuit is classified as

- A Moore circuit if each output is a function only of the circuit's present state
- A Mealy circuit if any output is a function of the circuit's present state and circuit inputs.

Section 9.2 develops the basic concepts of fundamental-mode circuits (excitation tables and maps, feedback, primitive flow table, and state diagram).

Section 9.3 describes the analysis of fundamental-mode circuits and discusses problems (such as static, dynamic, and essential hazards and race conditions) caused by unequal propagation delays in different signal paths.

Section 9.4 describes the design of fundamental-mode circuits using a 4-step problem-solving procedure (PSP).

Section 9.5 describes the design of a number of representative fundamental-mode circuits, using the 4-step problem-solving procedure for design.

9.2 BASIC CONCEPTS OF FUNDAMENTAL-MODE CIRCUITS

An asynchronous sequential-logic circuit can be synthesized from a combinational logic network by connecting 1 or more outputs to the circuit's inputs; each such line is a feedback path that feeds an output signal back to an input of the combinational logic network. An asynchronous sequential-logic circuit can be represented by a black-box model, as shown in Figure 9.2.

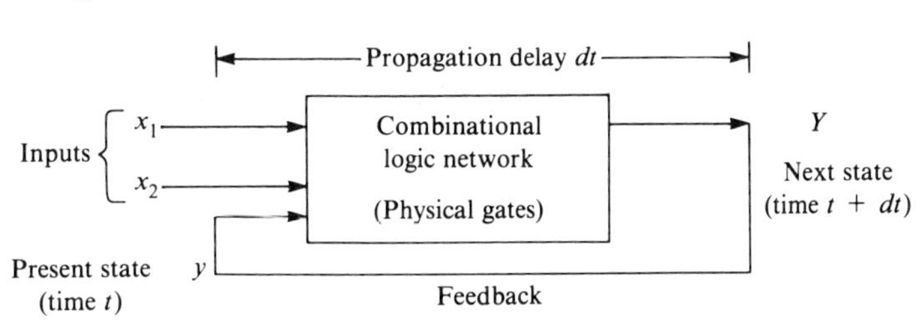

FIGURE 9.2
Black-box model of an asynchronous sequential-logic circuit

Lumped-Delay Model of an Asynchronous Sequential-Logic Circuit

The asynchronous sequential-logic circuit shown in Figure 9.2 can be represented by a **lumped-delay model** by replacing the physical gates with idealized zero-delay gates that have the total (lumped) time delay *dt* in series in the feedback loop, as shown in Figure 9.3.

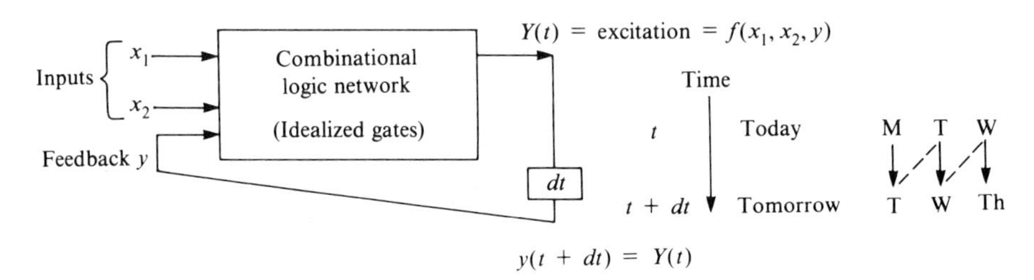

FIGURE 9.3
Lumped-delay model of the circuit of Figure 9.2

Let y denote the feedback variable. If either input, x_1 or x_2, is changed at time t, the combinational logic network, with gate inputs x_1, x_2, and y, produces an excitation Y. The excitation $Y(t)$ is fed back through a delay element with delay time dt, so $Y(t)$ = feedback $y(t + dt)$. Thus, when time advances from t to $t + dt$, the old excitation Y is the new feedback y.

Suppose that $dt = 24$ hours (t = today and $t + dt$ = tomorrow), so that only 1 input can change in any day (M = Monday, T = Tuesday, . . .). If today (t) is Monday, then tomorrow ($t + dt$) is Tuesday. When time advances from t to $t + dt$, old excitation Y = new feedback y. For example, Monday's excitation $Y(t)$ = Tuesday's feedback $y(t + dt)$.

Remember that the excitation $Y(t)$ is fed back as $y(t + dt)$. If this new feedback $y(t + dt)$ equals the old feedback $y(t)$, then $y(t)$ is a stable state of the circuit. Consequently, for a state to be stable, the combinational logic circuit must generate an excitation that does not cause a change in the feedback state. In short, if the excitation $Y(t)$ equals the feedback $y(t)$, the circuit is in a stable state.

For a given combination of x_1, x_2, and y values, the excitation Y is either equal to y or not equal to y. Therefore,

- If $Y = y$, the circuit is in a stable state.
- If $Y \neq y$, the circuit is in transition from one stable state to another stable state.

The circuit is asynchronous because state changes can occur at any time in response to changes at arbitrary times in the inputs x_1 and x_2. If an asynchronous circuit's inputs are constrained so that only 1 input can change at a time and only when the circuit is in a stable state, the circuit is a **fundamental-mode circuit**. We define a fundamental-mode input sequence as one in which successive input value combinations differ in only 1 variable (satisfy the unit-distance property). For example, the sequence 00, 01, 11, 10 is a fundamental-mode input sequence.

The circuit is sequential because a fundamental-mode sequence of inputs x_1x_2 at arbitrary times causes the excitation Y to take on a sequence of values, where each

$Y = f(x_1, x_2, y)$. At the end of each propagation delay dt, Y is fed back as y. Consequently, feedback is a "bridge" that causes a combinational logic circuit to become a sequential logic circuit.

This section describes how the lumped-delay model can be used to derive an excitation equation that defines the fundamental-mode operation of an asynchronous circuit. We will use the following example circuit for illustration.

EXAMPLE 9.1

A logic diagram of an asynchronous sequential-logic circuit with 2 inputs and 1 feedback line is presented in Figure 9.4.

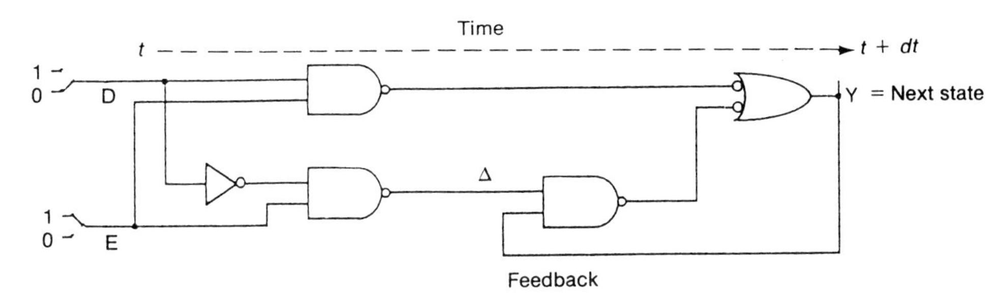

FIGURE 9.4
Logic diagram of an asynchronous sequential-logic circuit

The circuit in Figure 9.4 can be represented by a lumped-delay model, as shown in Figure 9.5, in which the physical gates are replaced with idealized (zero-delay) gates and the sum of the gate propagation delays is replaced by a lumped delay in series with the feedback loop.

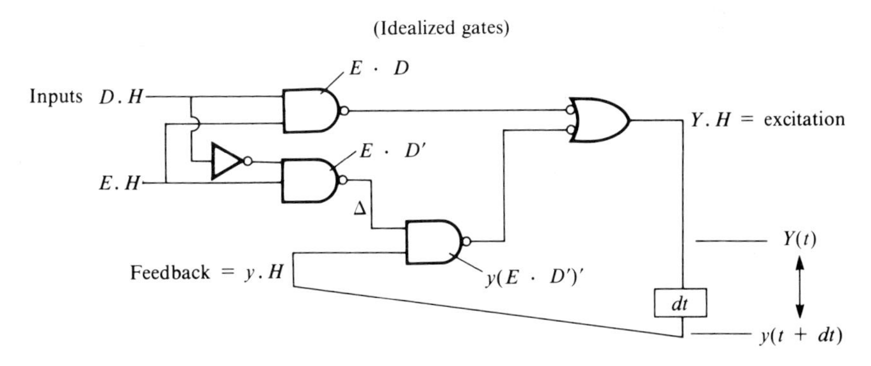

FIGURE 9.5
Lumped-delay model of the circuit of Figure 9.4

The circuit's external inputs E and D are called **primary variables**, and the feedback variable y is referred to as a **secondary variable**.

The underlying (characteristics) combinational logic circuit, with inputs E, D, and y (feedback), produces an excitation Y that is defined by the following excitation equation:

$$Y = Y(E, D, y) = E \cdot D + y \cdot (E' + D) \qquad \textbf{(9.1)}$$

This **excitation equation** is a logic model that defines the operation of a fundamental-mode circuit. For a given combination of E, D, and y values, the resulting excitation Y is either equal to y or not equal to y. Therefore,

- If $Y = y$, the circuit is in a stable state.
- If $Y \neq y$, the circuit is in transition from one stable state to another stable state. ■

Operation of a Fundamental-Mode Circuit

The operation of a fundamental-mode circuit can also be defined by an **excitation table**, which is obtained by evaluating the excitation equation (9.1) for each combination of E, D, and y values. The excitation table lists each possible E, D, and y combination and the corresponding excitation Y, as shown in Table 9.1. The **total state** (E, D, y) of the circuit consists of the combined values of primary variables, E and D and feedback variable y. The **internal state** of the circuit is the value of the feedback variable y.

TABLE 9.1
Excitation table for $Y = E \cdot D + y(E' + D)$

		Present Total State	Present Excitation = Next Feedback State
		E D y	Y
a		0 0 0	0
	d	0 0 1	1
b		0 1 0	0
	e	0 1 1	1
c		1 0 0	0
	–	1 0 1	0
	–	1 1 0	1
	f	1 1 1	1

Each row of the excitation table with $Y = y$ represents a primitive (individual) stable state of the circuit. The primitive stable total states are represented by the letters, **a** through **f**, in the column to the left of the excitation table. First, the stable states with $Y = 0$ are labeled (**a**, **b**, **c**). Then the stable states with $Y = 1$ are labeled (**d**, **e**, **f**). For each row with a dash, where $Y \neq y$, the circuit is in transition from one stable state to a different stable state.

Remember that if an asynchronous sequential-logic circuit's inputs are constrained so that only 1 input can change at a time and only when the circuit is in a stable state, the circuit is a fundamental-mode circuit.

Primitive Flow Table

A **primitive flow table** lists the transitions, caused by a valid fundamental-mode input change, from each starting stable state to each ending stable state. For a given starting state, a primitive flow table dynamically shows the sequence of states caused by a particular input sequence.

A skeletal primitive flow table, as shown in Table 9.2, is constructed by the following procedure:

- □ Create a column for each possible combination of input values. Arrange the columns in Gray-code order 00, 01, 11, 10, corresponding to a valid fundamental-mode input sequence.
- □ Create a row for each stable state of the circuit. On the left, label each row with the letter assigned in Table 9.1 and the present state y. On the right, label each row with the excitation (next state) Y.
- □ For each row, insert the stable state label (circled) in the column corresponding to the input combination for which that state is stable. For example, the circuit remains in state **a** ($ED\ y$ = 00 0) so long as the inputs E, D = 0, 0.

TABLE 9.2
Skeletal primitive flow table

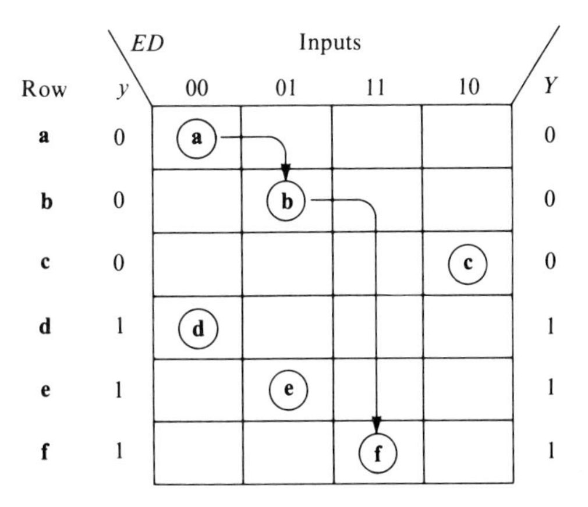

Row	*y* \ *ED* Inputs	00	01	11	10	*Y*
a	0	(a)				0
b	0		(b)			0
c	0				(c)	0
d	1	(d)				1
e	1		(e)			1
f	1			(f)		1

Let's trace another input through the flow table, Table 9.2. If the circuit is initially in stable state **a** ($ED\ y$ = 00 0), an input ED = 01 at time t will cause the circuit to pass to stable state **b** ($ED\ y$ = 01 0) at time $t + dt$. Then, if the circuit is in stable state **b** ($ED\ y$ = 01 0) an input ED = 11 will cause the circuit to pass through an intermediate unstable state before it reaches stable state **f** ($ED\ y$ = 11 1) with $Y = 1$.

For a given starting state, we see that the primitive flow table dynamically shows the sequence of states corresponding to a particular input sequence. For example, if the circuit is in stable state **a** ($ED\ y$ = 00 0), an input sequence ED = 01–11–10 causes the circuit to take on the sequence of states **b–f–c**.

Now we will complete the primitive flow table. Let SS denote stable state. The skeletal primitive flow table is then filled in for all SS-to-SS paths for each valid fundamental-mode input. For each SS-to-SS path, place the label of end-SS in the cell with row = start-SS and column = end-SS. For example, if the circuit is in SS **a** (start-SS) and receives input 01, end-SS is **b**. Therefore, we put label **b** in the row of start-SS and column of end-SS. Since $Y = y$ for each stable state (row), we can omit the y value on the left of the table. A dash is inserted in a column whose input is not a unit distance from the stable state column. The completed primitive flow table is shown in Table 9.3. (Paths are indicated by vectors for the input sequence 01–11–10.) Note that a primitive flow table has exactly one row for each stable state. In each row, the stable state is that state for which the next state is the same as the present state. Each unstable (transitional) state in a row has the same designation (label) as the destination state for that transition and is in the same column as the destination state.

TABLE 9.3
Completed primitive flow table

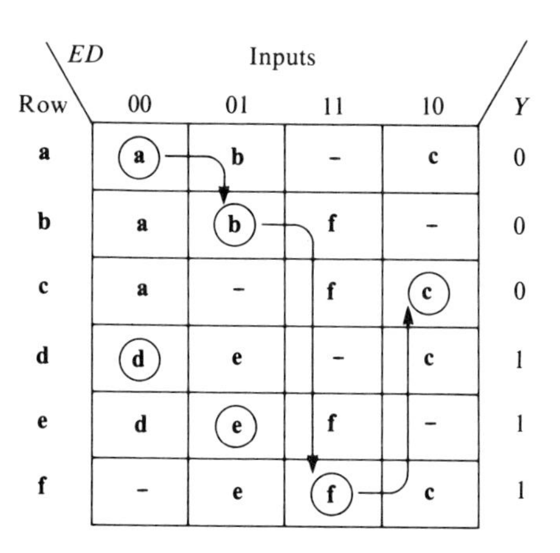

Row \ *ED* Inputs	00	01	11	10	*Y*
a	(a)	b	-	c	0
b	a	(b)	f	-	0
c	a	-	f	(c)	0
d	(d)	e	-	c	1
e	d	(e)	f	-	1
f	-	e	(f)	c	1

A primitive flow table is a logic model that completely defines the operation of the sequential machine. The primitive flow table is particularly useful for the state reduction process in the design of an optimal machine.

An alternative logic model is a primitive state diagram that contains the same information as a primitive flow table and *graphically* describes the operation of the sequential machine. Primitive state diagrams are described in the next section.

Primitive State Diagram

A primitive state diagram is a directed graph with the circuit's stable states as nodes and the SS-to-SS transitions as edges. For a given starting state, a primitive state diagram graphically shows the sequence of states caused by a particular input sequence. Thus, the operation of a fundamental-mode circuit may also be represented by a primitive state diagram in which each individual (primitive) stable state is a node. If the circuit is in a stable total state (start node), a valid fundamental-mode change (00–01–11–10–00) of one of the primary inputs E, D causes the circuit to make a transition to another stable state (end node). The start node is connected to the end node by an edge that is labeled by the input combination that causes that transition. A diagram of the fundamental-mode paths between stable states is called a primitive state diagram, as illustrated in Figure 9.6. Since the output of a Moore circuit is a function only of the present state, the circuit output is placed inside the present state node below the state label; a horizontal line inside the node separates the state label and the circuit output.

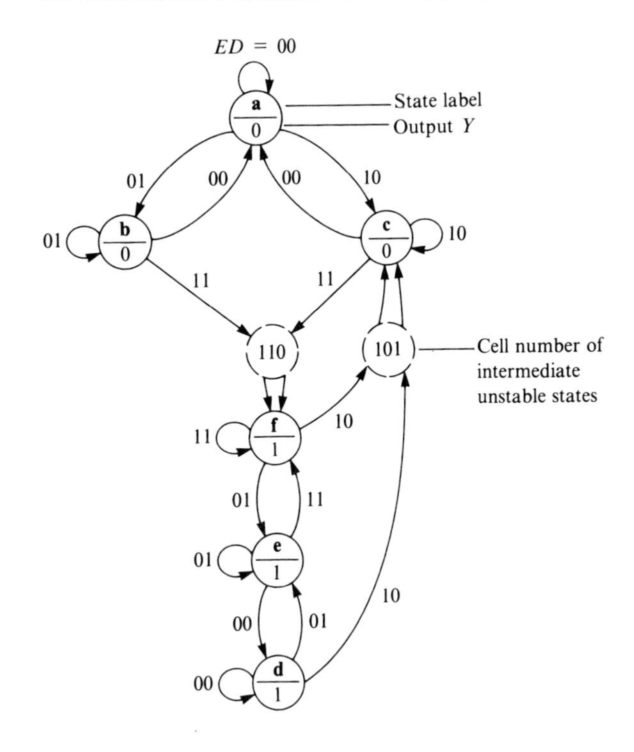

FIGURE 9.6
Moore primitive state diagram for circuit of Figure 9.4

Thus, the primitive state diagram provides a graphic description of the operation of a fundamental-mode circuit. It shows the transition between primitive (individual) total stable states for each valid fundamental-mode input.

A simple method of analyzing fundamental-mode circuits is described in the next section.

9.3 ANALYSIS OF FUNDAMENTAL-MODE CIRCUITS

The analysis of an asynchronous sequential-logic circuit is the process of examining a detailed logic diagram to determine the function and operating characteristics of the circuit. A procedure for analyzing asynchronous circuits, described in this section, is used to illustrate hazards and race conditions that may exist in asynchronous sequential-logic circuits.

The excitation equation (9.1), for the circuit in Example 9.1, can be plotted as an excitation map (Figure 9.7, p. 410). The form of the excitation map facilitates the comparison of y and Y for each combination of input and feedback conditions to determine whether or not the circuit is stable for a given input-feedback combination.

The combined values of the primary variables E and D and the feedback variable y comprise the *total state* ($ED\ y$) of the circuit. Thus, a total state is represented by a map cell at column ED and row y. For a given total state ($ED\ y$), the cell entry is the circuit excitation Y.

If inputs E and D can change only one at a time and only when the circuit is in a stable state, the circuit is a fundamental-mode circuit. The circuit operation is described by an excitation map as shown in Figure 9.7. If the circuit is in a stable total state, a valid fundamental-mode input change (only 1 input changes at a time) produces a y-to-Y transition as follows:

1. If $Y = y$, the transition is to a stable state in the same row. For example, if the circuit is in a stable total state $(ED\ y) = (00\ 0)$ a change in input D from 0 to 1 will result in the stable total state (01 0). Since $Y = y$, the transition remains in the same row.
2. If $Y \neq y$, the circuit transfers from the initial stable state, via an unstable state, to a stable state in a different row. For example, if the circuit is in a stable total state $(ED\ y) = (01\ 0)$, a change in input E from 0 to 1 will cause the circuit to momentarily make a transition through state (11 0) before it stabilizes in state (11 1) where $Y = y$. (Stable states $Y = y$ are circled.)

FIGURE 9.7
Excitation map

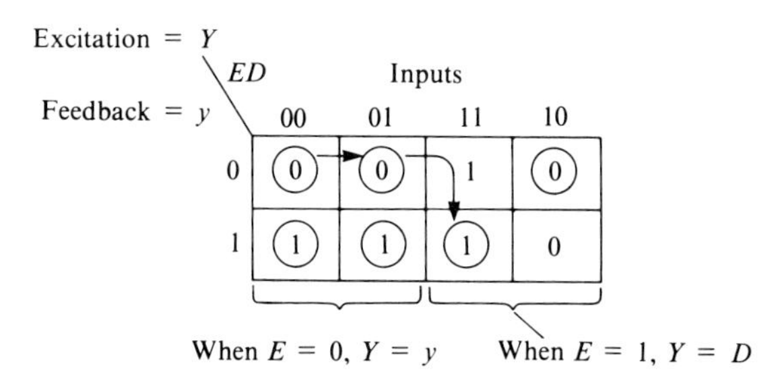

The example illustrated by Figure 9.7 represents the circuit given in Figure 9.4. The *function* of the circuit given in Figure 9.4 is revealed by an examination of the excitation map in Figure 9.7 as follows:

1. When $E = 0$ the excitation Y is stable, regardless of the value of D. Thus, when E is not asserted, excitation Y = feedback y. Input E can be described as an enable input.
2. When $E = 1$ the excitation Y is equal to the value of D. Thus, when enable E is asserted, excitation Y = input D.

Therefore, the circuit in Figure 9.4 is an enabled D latch, with data input D and an enable input E. When the enable E is not asserted, the latch holds its present state, regardless of the D input. When the enable E is asserted, the latch captures the D input.

Hazards in a Fundamental-Mode Circuit

A **hazard** is a condition in a circuit that can cause erratic circuit operation. It is usually manifested as a **transient** (temporary change in signal state) at a gate output caused by unequal propagation delays in different signal paths to the gate inputs. Hazards should be eliminated in order to ensure reliable operation of a circuit. The design of digital circuits usually involves a trade-off between component minimization and circuit reliability. While component minimization may be achieved by eliminating redundant terms in logic expressions, reliability is generally enhanced by the addition of redundant (consensus) terms. We will illustrate how hazards can be eliminated in Examples 9.2 through 9.4.

There are three types of hazards that may be present in a fundamental-mode circuit: static, dynamic, and essential. For circuits with static or dynamic hazards, the steady-state output is correct, but undesirable output transients may occur after an input change. Static and dynamic hazards, if present, are caused by the combinational logic part of the circuit and may be eliminated in the initial design phase. Essential hazards exist only in sequential circuits that have 2 or more feedback paths.

Static Hazards

A **static hazard** causes a single transient in a combinational-logic output signal that should have remained unchanged in response to a change in a single input.

EXAMPLE 9.2

A circuit with a static hazard, caused by unequal propagation delays of signal x_2, is illustrated in Figure 9.8a.

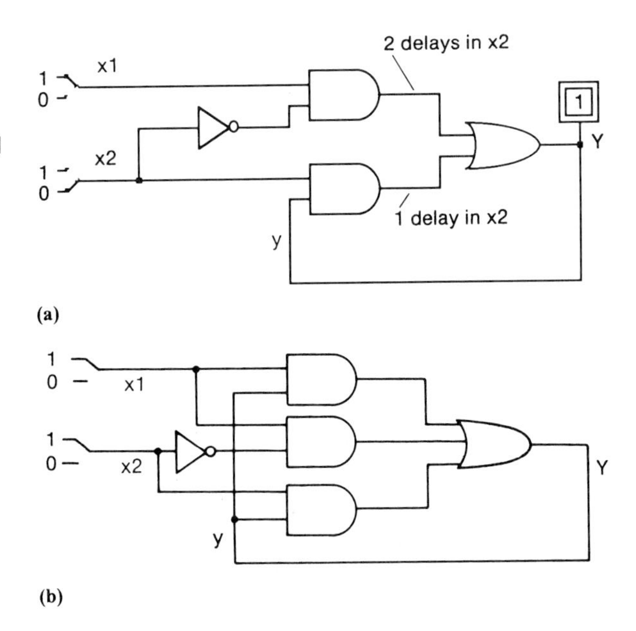

FIGURE 9.8
Fundamental-mode circuit.
(a) With a static hazard
(b) With static hazard eliminated

If the circuit is in stable total state $x_1x_2\ y = 11\ 1$, the circuit output (output of the OR gate that has inputs x_1x_2' and yx_2) should not change value when input x_2 changes from 1 to 0 (see Figure 9.9). However, the different propagation delays in the signal paths from x_2 to the OR gate inputs may cause Y to momentarily go to 0 and back to 1 (Circuit 9.8a in Table 9.4, p. 412) when it should have remained unchanged at 1. The hazardous transition is denoted by a in Figure 9.9.

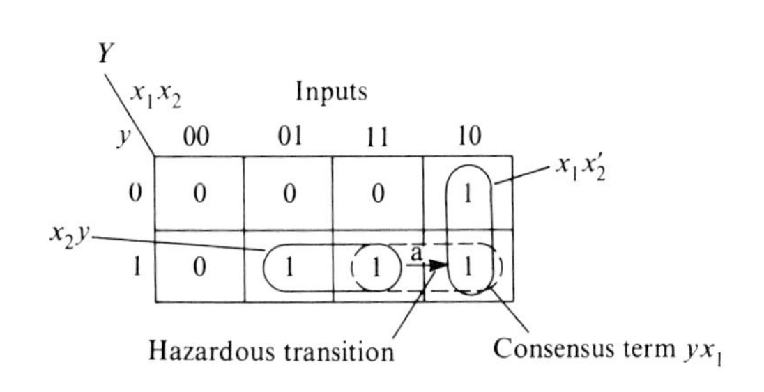

FIGURE 9.9
Excitation map illustrating a static hazard

TABLE 9.4
Circuit output Y after 1-to-0 change in x_2

	$Y(t)$	$Y(t + dt)$	$Y(t + 2dt)$
Circuit 9.8a: $Y = yx_2 + x_1x_2'$	1	0	1
Circuit 9.8b: $Y = yx_2 + x_1x_2' + yx_1$	1	1	1

The static hazard can be eliminated by adding the redundant consensus term yx_1 as shown in Figure 9.8b. That is, the static hazard can be eliminated by replacing the original excitation equation

$$Y = yx_2 + x_1x_2'$$

with the logically equivalent equation

$$Y = yx_2 + x_1x_2' + yx_1$$

that contains the consensus term yx_1.

Note that with the consensus term, the circuit output Y (see Circuit 9.8b in Table 9.4) has constant value 1 at times t, $t + dt$, and $t + 2dt$, so Y remains unchanged (as it should) in response to the 1-to-0 change in x_2.

The presence of a static hazard is also indicated by the juxtaposition of rectangular groups of 1 cells (representing prime implicants) in the excitation map illustrated in Figure 9.9.

The static hazard in this circuit could also be revealed by a close examination of the logic diagram in Figure 9.8a. That is, the lower input of the OR gate from input x_2 has 1 propagation delay while the upper input has 2 propagation delays. ■

Dynamic Hazards

A **dynamic hazard** occurs when the output of a combinational logic circuit is supposed to change once but instead changes 3 or more (odd) times. The presence of a dynamic hazard implies that the changing input variable must have 3 different paths to a single gate.

■ **EXAMPLE 9.3**

A circuit with a dynamic hazard is illustrated in the logic diagram shown in Figure 9.10a. Note that there are 3 different signal paths from input x_1 to the OR gate.

The presence of a dynamic hazard can be detected by analyzing a timing diagram of the circuit. The timing diagram will show that a change in input x_1 causes the excitation Y to change 3 times when the circuit is supposed to change only once.

The presence of a dynamic hazard can also be detected by analyzing the excitation equation of the combinational logic part of the circuit:

$$Y = x_1(x'_1 + x'_2) + x_1x_2y$$

The occurrence of input x_1 in one term with x'_1 and in another term without x'_1 is indicative of the presence of a dynamic hazard. The dynamic hazard can be eliminated by the inclusion of a consensus term or by rewriting the excitation equation so that an input variable does not appear in one product term with its complement and in another term

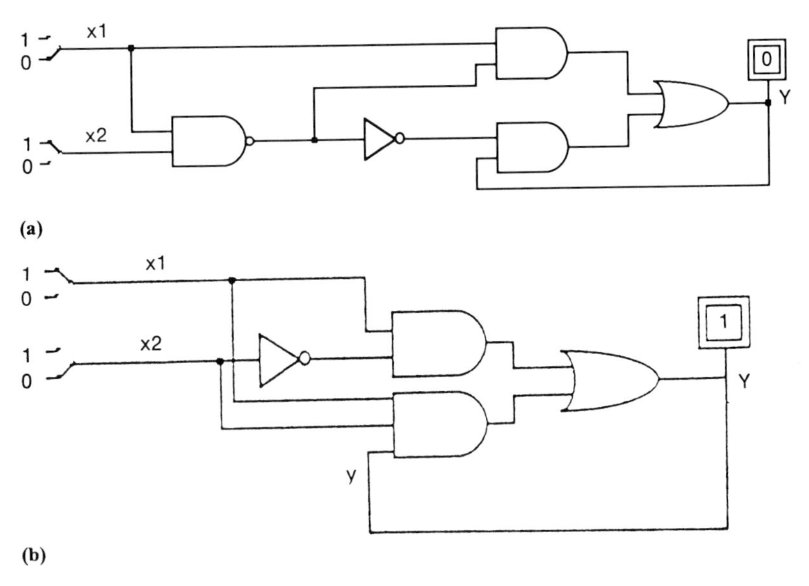

FIGURE 9.10
Asynchronous circuit. (a) With a dynamic hazard (b) With hazard eliminated

without its complement. For example, the foregoing equation can be replaced with the following logically equivalent equation:

$$Y = x_1x_2' + x_1x_2y$$

A circuit implemented using this equation has no dynamic hazard. ■

Essential Hazards

A type of hazard that is peculiar to fundamental-mode circuits is the **essential hazard**. Like static and dynamic hazards, essential hazards are caused by propagation delays in a combinational logic circuit. However, essential hazards may only be present in sequential logic circuits that contain 2 or more feedback paths.

■ **EXAMPLE 9.4**

A circuit with an essential hazard is illustrated in Figure 9.11.

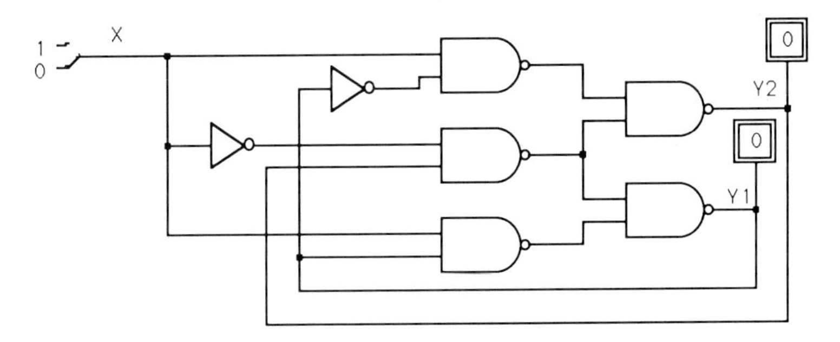

FIGURE 9.11
Logic diagram of a circuit with an essential hazard

The excitation equations and maps for the circuit in Figure 9.11 are shown in Figure 9.12.

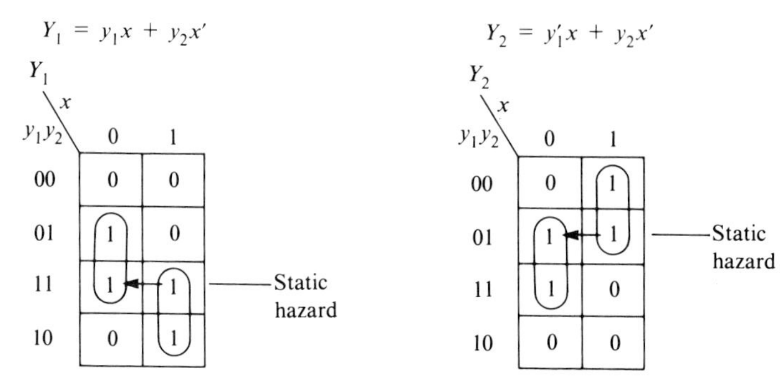

FIGURE 9.12
Excitation equations and maps

The individual excitation maps can be combined to form a circuit excitation map, as shown in Figure 9.13. Each cell of the excitation map contains the binary code for that state.

FIGURE 9.13
Excitation map for circuit with essential hazard

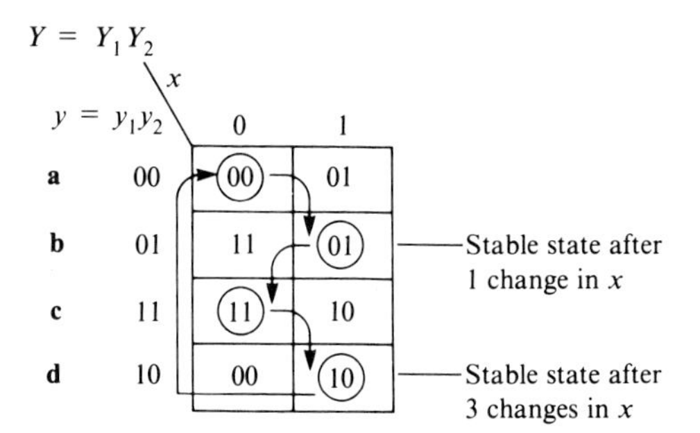

The presence of an essential hazard can be detected by an analysis of the combined excitation map for the circuit shown in Figure 9.13. If there is switching noise on input x, x may change several times when it is supposed to change only once. A circuit excitation map has an essential hazard if, and only if, for some initial stable state, the stable state reached after 1 change in an input x is different from the stable state reached after 3 changes in x. Note that an essential hazard exists in the excitation map shown in Figure 9.13.

A map of transitions can also be used to determine the presence of essential hazards. If, after 3 or more (odd) changes in an input, the number of alpha or beta transitions for a given state variable changes from odd to even (or even to odd), then the circuit contains an essential hazard. Note that the transitions for a stable state cell are the same as the cell entry in the excitation map. A map of transitions (alpha, beta, 1, 0) for the circuit is presented in Figure 9.14. Note that the transition code for a stable state is the same as the binary state code.

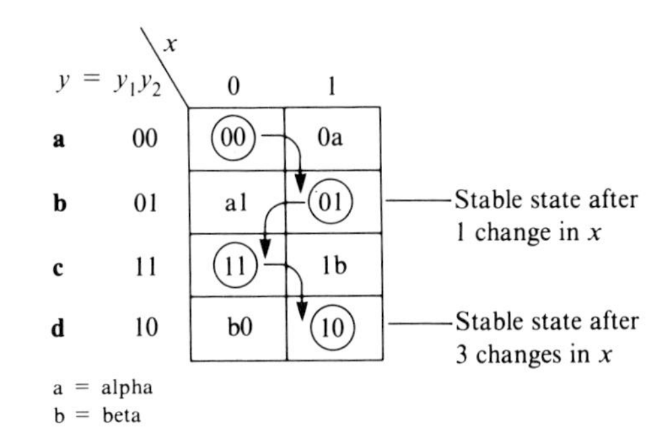

FIGURE 9.14
Map of transitions illustrating an essential hazard

■

Race Conditions in a Fundamental-Mode Circuit

An asynchronous sequential-logic circuit has no clock to synchronize changes in the state variables. A race condition exists when a transition from 1 stable state to another involves a change in 2 or more state (feedback) variables. If an input change can cause a transition to either of 2 stable states the race is a critical race, and the circuit may either oscillate or stabilize in the wrong stable state. Therefore, it is necessary to design asynchronous circuits in such a way that race conditions do not exist. Example 9.5 describes how a race condition can be detected in an asynchronous circuit.

■ **EXAMPLE 9.5**

A logic diagram of a fundamental-mode circuit with 1 input and 2 feedback lines is presented in Figure 9.15.

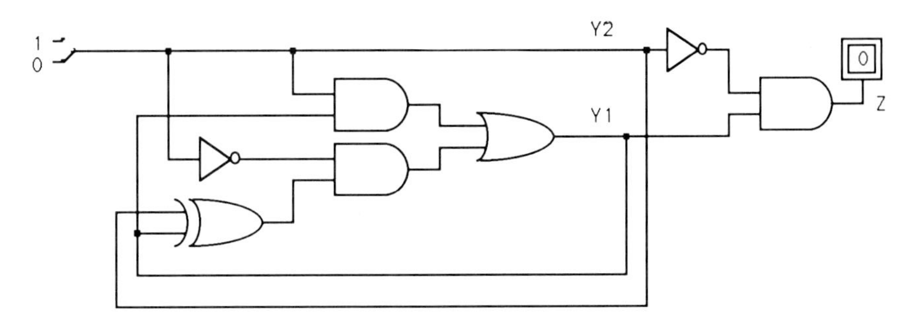

FIGURE 9.15
Fundamental-mode circuit with 2 feedback lines

The analysis of the fundamental-mode circuit can be accomplished using the following procedure:

1. Given a detailed logic diagram of a sequential logic circuit,
   a. Identify the following circuit elements: external input(s), output(s), and feedback line(s).
   b. Starting with the external input x and feedback variables y_1 and y_2, use the logic diagram to obtain an excitation equation for each output Y_1 and Y_2 of the underlying combinational logic circuit:

$$Y_1 = (y_1 \oplus y_2)x' + y_1 x$$
$$Y_2 = x$$

The output equation is

$$Z = Y_1 Y_2'$$

Y_1 and Y_2 can also be represented by the excitation maps in Figure 9.16.

FIGURE 9.16
Individual excitation maps

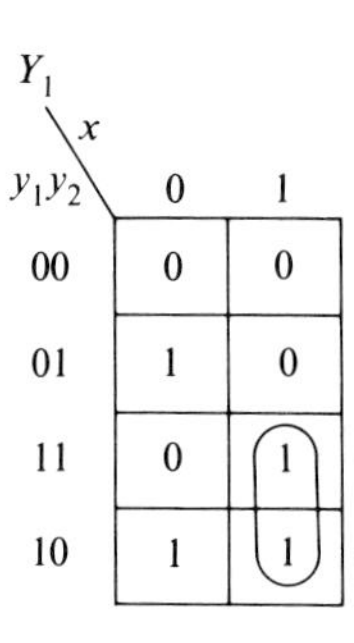

Y_2

y_1y_2 \ x	0	1
00	0	1
01	0	1
11	0	1
10	0	1

2. Combine the individual excitation maps to form the circuit's excitation map, shown in Figure 9.17, and analyze the excitation map to determine if a race condition exists.

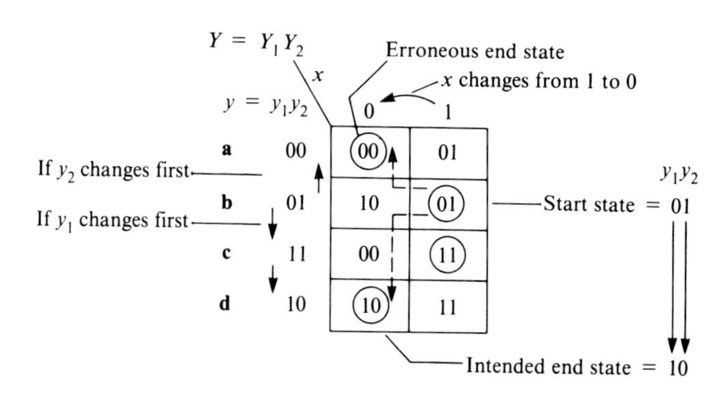

FIGURE 9.17
Circuit excitation map

Analysis of Figure 9.17 shows that the path between stable states 01 and 10 requires both state variables to change, indicating a race condition. Since there are 2 stable states in the destination column, the race is a **critical race**.

If the circuit is in stable start state 01 and x changes from 1 to 0, then an erroneous end state (00) is reached if state variable y_2 changes first; if y_1 changes first, the proper end state (10) is reached. Similarly, the path between stable states 11 and 00 contains a critical race condition.

A map of transitions (Figure 9.18) can be plotted from the excitation map. If any cell of the map of transitions contains aa, ab, ba, or bb (a = alpha, b = beta) then a race condition exists in the circuit. If the column containing an alpha-beta combination has 2 or more stable states, then a critical race condition exists.

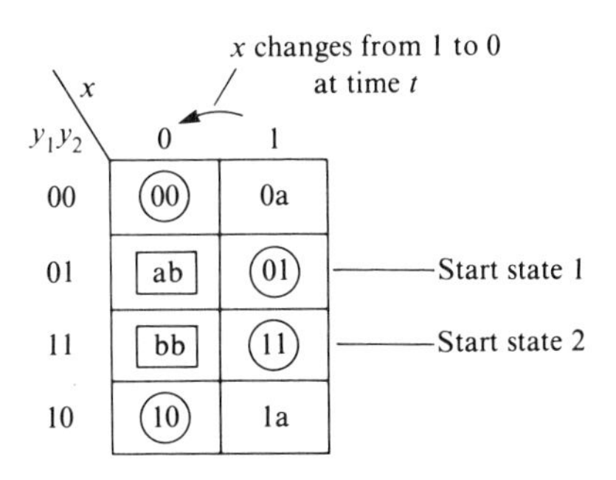

FIGURE 9.18
Transition map illustrating a critical race

A race condition is indicated when an intermediate cell in a path from a start state to an end state contains 2 letter transitions (aa, ab, ba, or bb), indicating that both state variables must change. Note that the transitions for a stable state cell are the same as the cell entry in the next-state map.

3. The excitation map (Figure 9.17) can be converted to a flow table and a corresponding state diagram (Figure 9.19) by replacing each cell's binary code with a symbol representing that state.

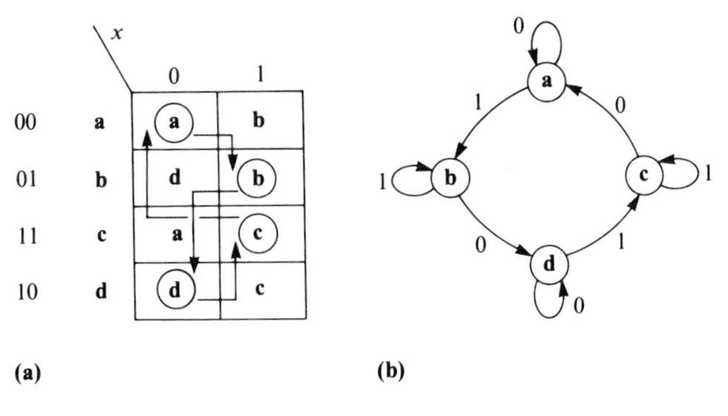

FIGURE 9.19
Illustration of a critical race. (a) Flow table (b) State diagram

Analysis Procedure for Asynchronous Sequential-Logic Circuits

An asynchronous sequential-logic circuit, represented by a logic diagram, can be analyzed using the step-by-step procedure illustrated in the preceding example. The steps in the analysis procedure are summarized as follows:

1. Given a detailed logic diagram of a sequential logic circuit,
   a. Identify the following circuit elements: external input(s), output(s), and feedback line(s).
   b. Starting with the external inputs and feedback variable(s), use the logic diagram to obtain an excitation equation for each output Y_i of the underlying combinational logic circuit. Determine the circuit output equation by tracing the circuit output back to the external inputs and feedback variable(s). Use the excitation equations to construct an excitation map for each output Y_i.
2. Construct a circuit excitation map by combining the individual excitation maps. Analyze the circuit excitation map to determine if any hazards or critical race conditions exist.
3. Construct a state diagram and flow table from the circuit excitation map.
4. Interpret the state diagram to determine the function of the circuit.

9.4 DESIGN OF FUNDAMENTAL-MODE CIRCUITS

The design of a fundamental-mode circuit is the process whereby the design specifications are translated into excitation and output equations that are in turn used to draw a logic diagram of the circuit. The design process is essentially the reverse of the analysis process described in Section 9.3.

Recall from Chapter 8 that 2 states of a completely specified state table are equivalent if for any input combination their next states are equivalent and their outputs are the same. If 2 or more states are equivalent, then all but 1 of the states are redundant and may be eliminated.

The state table of an asynchronous circuit is usually referred to as a flow table. For fundamental-mode operation the state (flow) table is an incompletely specified state table because input changes must satisfy a unit distance constraint (0–1 for 1 input; 00–01–11–10 for 2 inputs).

The state reduction of a flow table can be accomplished by the following 2-step process:

1. Eliminate the redundant stable states. Two states are equivalent if they are stable for the same input combination and if for any valid input combination their next states are equivalent and their outputs are the same. If 2 or more stable states are equivalent, then all but 1 of the states are redundant and can be eliminated.
2. Eliminate the intermediate unstable states by merging. Two rows of a flow table can be merged into a single row if there are no conflicts between their entries in any column.

Design Procedures for a Fundamental-Mode Circuit

This section describes the design of fundamental-mode circuits to realize sequential processes whose inputs are signals that may change from 0 to 1 or from 1 to 0 at arbitrary times.

As indicated in Chapter 7, functional sequential processes (with either synchronous or asynchronous inputs) can be represented by a finite state machine. However, the asynchronous nature of the inputs and the constraint of fundamental-mode operation require design procedures that are somewhat different than those used to design synchronous sequential-logic circuits.

We will again use the basic 4-step problem-solving procedure, modified as required, to model the asynchronous process and accomplish the state reduction and state assignment peculiar to fundamental-mode circuits. The resulting (modified) 4-step procedure for designing fundamental-mode circuits is as follows:

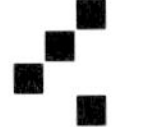

PSP

Step 1: Problem Statement. The design specifications include a functional description of the sequential process. It may also specify sequences of inputs and the corresponding sequences of outputs.

Step 2: Conceptualization. A logic model of the finite state machine with asynchronous inputs takes the form of a primitive state diagram and/or a primitive flow table; these can be constructed from the design specifications.

Step 3: Solution/Simplification.

a. State reduction.
  (1) Eliminate redundant stable states, thereby reducing the primitive flow table to an equivalent nonredundant flow table.
  (2) Eliminate intermediate unstable states by merging nonconflicting rows into a single row, thereby producing a minimal-row merged flow table.

b. State assignment. The machine states are assigned binary codes in such a way that the adjacencies required for a race-free circuit are satisfied. This state assignment is accomplished by the following substeps as appropriate for the particular merged flow table:
  (1) Make state assignments without using cycles or adding states.
  (2) Make state assignments using available paths for cycles.
  (3) Make state assignments by adding states for cycle paths.
  Convert the merged flow table to a circuit excitation (next-state) map by replacing each state label with its assigned binary code.

c. Separate the circuit excitation map into individual excitation maps, and derive the excitation equations and output equations.

Step 4: Realization. Use the excitation and output equations to construct a logic diagram of the fundamental-mode circuit. ■

Design Specification and Machine Conceptualization

This section describes the first 2 steps (problem statement and conceptualization) of the 4-step problem-solving procedure to design a fundamental-mode circuit that realizes a given sequential process that has asynchronous inputs. We will use the design of a sequence detector with 2 asynchronous inputs to illustrate the procedures involved.

The first step of the 4-step problem-solving design procedure is to write a problem statement that describes the operation of a particular sequential process and includes a set of specifications for the design of a sequential logic circuit that realizes the process.

Step 1: Problem Statement. Given 2 asynchronous inputs, describe a process that detects a specific sequence (00, 01, 11) of combinations of the 2 inputs. Then, design a fundamental-mode circuit that realizes the sequence detection process. The sequence-detector circuit has 2 asynchronous level-signal inputs, x_1 and x_2, and 1 output, Z, which is initially 0. When the circuit detects the input sequence $x_1x_2 = 00, 01, 11$, the output Z goes to 1 and remains at 1 as long as the inputs remain unchanged. If any other sequence is input, the circuit returns to its reset state (state **a**) on the next 00 input following an input not in the sequence 00, 01, 11.

The second step of the 4-step problem-solving design procedure is to conceptualize a sequential machine that represents the sequential process with asynchronous inputs. This conceptualization is accomplished by creating a logic model in the form of a primitive flow table or a primitive state diagram.

Step 2: Conceptualization. The operation of the sequence detector can be graphically described by a primitive state diagram, as shown in Figure 9.20.

FIGURE 9.20
Primitive state diagram for 00–01–11 sequence detector

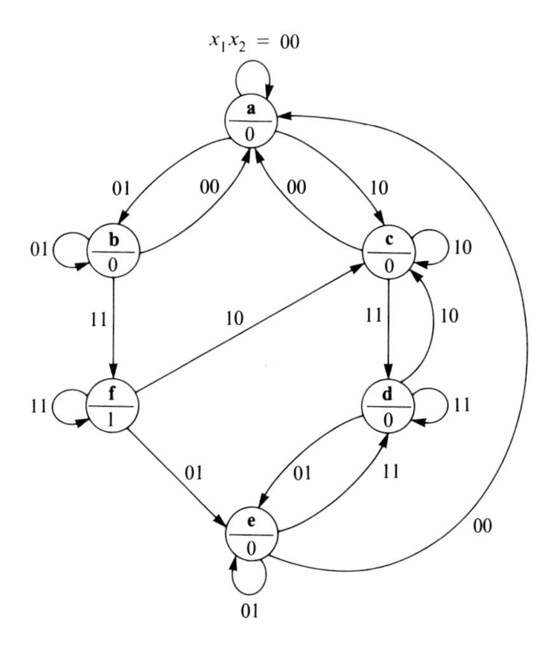

A corresponding primitive flow table for the circuit (Table 9.5) can be constructed using the primitive state diagram. A primitive flow table defines the operation of a sequential process by tabulating the transitions between pairs of stable states. Table 9.5 is a primitive flow table for the sequence detection process described in the problem statement. The flow table contains the same information as the primitive state diagram. However, the form of the flow table makes it a useful device to accomplish the state reduction required to design an optimal sequential machine.

TABLE 9.5
Primitive flow table for 00–01–11 sequence-detector

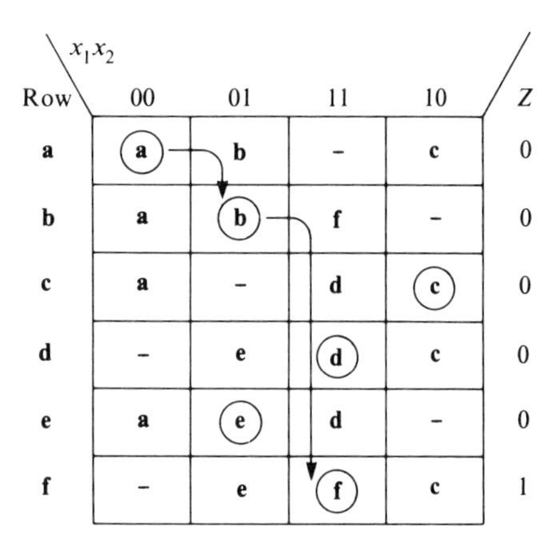

Row \ x_1x_2	00	01	11	10	Z
a	(a)	b	-	c	0
b	a	(b)	f	-	0
c	a	-	d	(c)	0
d	-	e	(d)	c	0
e	a	(e)	d	-	0
f	-	e	(f)	c	1

In the next section, we begin the solution/simplification step by describing the state reduction process required to design an optimal sequential machine.

State Reduction

The process of state reduction in finite state machines representing processes with asynchronous inputs can result in a significant simplification of asynchronous circuits. This state reduction process consists of two phases: (1) the elimination of any redundant stable states and (2) the merger of intermediate unstable states. If the asynchronous circuit is implemented using latches, the state reduction process may also reduce the number of latches required, since the number of states in a sequential circuit is related to the number of memory devices in the circuit.

Elimination of Redundant Stable States

Two states of a finite state machine are equivalent if they are stable for the same input combination, their next states are equivalent for any input combination, and their outputs are the same for any input combination. If 2 or more stable states are equivalent, then all but 1 of the states are redundant and may be eliminated. If a primitive flow table contains 1 or more sets of equivalent states, the primitive flow table can be reduced to an equivalent nonredundant flow table by eliminating these redundant stable states.

A fundamental-mode circuit is incompletely specified because the circuit's inputs are restricted to unit-distance changes. That is, only 1 input may change at a time and only when the circuit is in a stable state. Even though a fundamental-mode circuit is incompletely specified, its redundant stable states can be eliminated using the implication chart method or the equivalence-class partitioning method, described in Chapter 8, for completely specified state tables.

The translation of a set of design specifications into a primitive flow table may produce a primitive flow table that has 1 or more redundant stable states. Example 9.6 describes the reduction of a primitive flow table (Table 9.6) that contains redundant stable states.

EXAMPLE 9.6: State Reduction of a Flow Table with Redundant Stable States

This example begins with an arbitrary primitive flow table that has been developed from a set of design specifications. An examination of Table 9.6 reveals that 1 or more columns contain 2 or more stable states, indicating the possible existence of redundant stable states.

TABLE 9.6
Primitive flow table with redundant stable states

Row \ x_1x_2	00	01	11	10	Z
a	(a)	b	–	c	0
b	a	(b)	d	–	0
c	a	–	g	(c)	0
d	–	f	(d)	h	0
e	a	–	g	(e)	1
f	i	(f)	d	–	1
g	–	f	(g)	e	0
h	i	–	d	(h)	1
i	(i)	b	–	c	0

The sets of potentially equivalent stable states are found by grouping the sets of states that are stable in the same column and have the same outputs, as shown in Table 9.7. The sets of potentially equivalent stable states are

$$\{a, i\} \qquad \{d, g\} \qquad \{e, h\}$$

TABLE 9.7
Primitive flow table with potentially equivalent states

Row	00	01	11	10	Z
b	a	(b)	d	–	0
c	a	–	g	(c)	0
d	–	f	(d)	h	0
g	–	f	(g)	e	0
a	(a)	b	–	c	0
i	(i)	b	–	c	0
f	i	(f)	d	–	1
e	a	–	g	(e)	1
h	i	–	d	(h)	1

The sets of potentially equivalent states, together with their next states and outputs, are listed in Table 9.8.

TABLE 9.8
Sets of potentially equivalent states

Present State	Next State Inputs 00	01	11	10	Output Z
a	(**a**)	**b**	–	**c**	0
i	(**i**)	**b**	–	**c**	0
d	–	**f**	(**d**)	**h**	0
g	–	**f**	(**g**)	**e**	0
e	**a**	–	**g**	(**e**)	1
h	**i**	–	**d**	(**h**)	1

Stable states **a** and **i** in column 00 are equivalent since they have the same outputs and their next states are the same for each input combination. Since states **a** and **i** are equivalent, stable states **e** and **h** in column 10 are equivalent if states **d** and **g** are equivalent. Stable states **d** and **g** in column 11 are equivalent if states **e** and **h** are equivalent. Since the equivalence of **d** and **g** implies the equivalence of **e** and **h**, and conversely, then **d** and **g** are equivalent and **e** and **h** are equivalent.

Equivalent stable state pairs: **a** = **i**, **d** = **g**, **e** = **h**

It should be noted that an implication chart could have been used to determine these equivalent stable states.

The original primitive flow table (Table 9.6) with redundant stable states can be reduced to a nonredundant primitive flow table by deleting rows **g**, **h**, and **i** and replacing states **g**, **h**, and **i** with their equivalent states **d**, **e**, and **a**, respectively, in the remaining rows as shown in Table 9.9.

TABLE 9.9
Nonredundant primitive flow table

Row	x_1x_2 00	01	11	10	Z
a	(**a**)	**b**	–	**c**	0
b	**a**	(**b**)	**d**	–	0
c	**a**	–	**d**	(**c**)	0
d	–	**f**	(**d**)	**e**	0
e	**a**	–	**d**	(**e**)	1
f	**a**	(**f**)	**d**	–	1

■

Elimination of Intermediate Unstable States by Merging

The second step of the 2-step state reduction process involves the elimination of intermediate unstable states by a merging operation. Two rows of a primitive flow table can be merged into a single row if, for both rows,

- □ Each column contains the same state label or a state label and a dash (–).
- □ The outputs are the same.

Therefore, 2 rows of a primitive flow table can be merged into 1 row if there are no conflicts in any of their columns. Our next task is to search the primitive flow table for pairs of nonconflicting rows. These nonconflicting row pairs are then used to construct Table 9.10.

TABLE 9.10
Table of nonconflicting rows of the flow table

Row \ x_1x_2	00	01	11	10	Z
a	(**a**)	**b**	–	**c**	0
b	**a**	(**b**)	**d**	–	0
c	**a**	–	**d**	(**c**)	0
d	–	**f**	(**d**)	**e**	0
e	**a**	–	**d**	(**e**)	1
f	**a**	(**f**)	**d**	–	1

It should be noted that the merger process is not transitive. That is, if rows **r** and **s** have no conflict and rows **s** and **t** have no conflict, it is not necessarily true that rows **r** and **t** have no conflict. Stated another way, the fact that rows **r** and **s** can be merged and rows **s** and **t** can be merged does not imply that rows **r** and **t** can be merged.

In order to determine the sets of rows that can be merged, we must examine each collection of pairwise mergeable rows. For example, in Table 9.10, rows **a**, **b**, and **c** form a collection in which members of each of the pairs (**a**, **b**), (**b**, **c**), and (**c**, **a**) are nonconflicting and, thus, are pairwise mergeable. Such a collection is called a **merger group**.

A merger graph is a diagram in which each row is a node and members of each pair of nonconflicting row nodes are connected by a branch, as shown in Figure 9.21. If, in a group of 2 or more nodes, each node is connected to each of the other nodes in the group, a complete subgraph is formed. A maximal merger group consists of any group of nodes that forms a complete subgraph and cannot be enlarged. The merger graph in Figure 9.21 contains the maximal merger groups

$$\{\mathbf{a}, \mathbf{b}, \mathbf{c}\}, \quad \{\mathbf{e}, \mathbf{f}\}, \quad \{\mathbf{d}\}$$

since each of these groups forms a complete subgraph. If a single node is not connected to any other node, it forms a complete subgraph within itself. In this graph, **d** is such a node.

FIGURE 9.21
Merger graph for flow Table 9.9

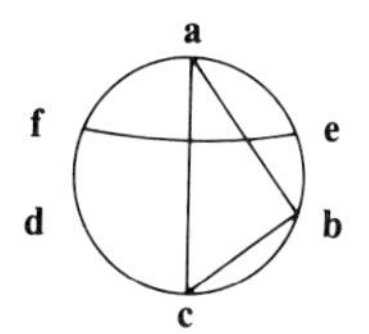

The nonredundant flow table (Table 9.9) can be reduced to a minimal-row merged flow table by selecting a set of mutually exclusive maximal merger groups that covers all the rows of the flow table. Thus, Table 9.9 can be reduced to a minimal-row flow table (Table 9.11) by merging the rows in each of the maximal merger groups. Note that a stable state and an unstable state with the same label in 2 nonconflicting (mergeable) rows are merged into a stable state (with the same label) in the merged flow table.

TABLE 9.11
Minimal-row flow table

	Row \ x_1x_2	00	01	11	10	Z
(abc)	A	(a)	(b)	d	(c)	0
d	B	–	f	(d)	e	0
(ef)	C	a	(f)	d	(e)	1

State Assignment for Critical Race-Free Circuits

The preceding section was concerned primarily with methods for state reduction. This section is concerned primarily with methods for state assignment that produce a critical race-free fundamental-mode circuit. For a given minimal-row flow table, the required adjacencies can be determined by listing the SS-to-SS paths of the flow table.

It can be shown that the required adjacencies for a given minimal-row flow table can be satisfied, and a race-free state assignment can be made, using the following algorithm:

- □ If the required adjacencies can be satisfied by logically adjacent state assignments without using cycles* or adding states (type 1), then make those logically adjacent state assignments.
- □ If the required adjacencies can be satisfied by available paths (type 2), then use available paths for cycles.
- □ If neither type 1 nor type 2 situations exist, then add additional states (rows) to provide paths for the cycles (type 3).

A flowchart for the state assignment algorithm is presented in Figure 9.22 (p. 426).

This section presents an example of each of the 3 ways in which a race-free state assignment is made, depending on the minimal-row flow table's number of required adjacencies and the cycle paths available.

---

*A cycle is a transition path that contains 2 or more consecutive unstable states.

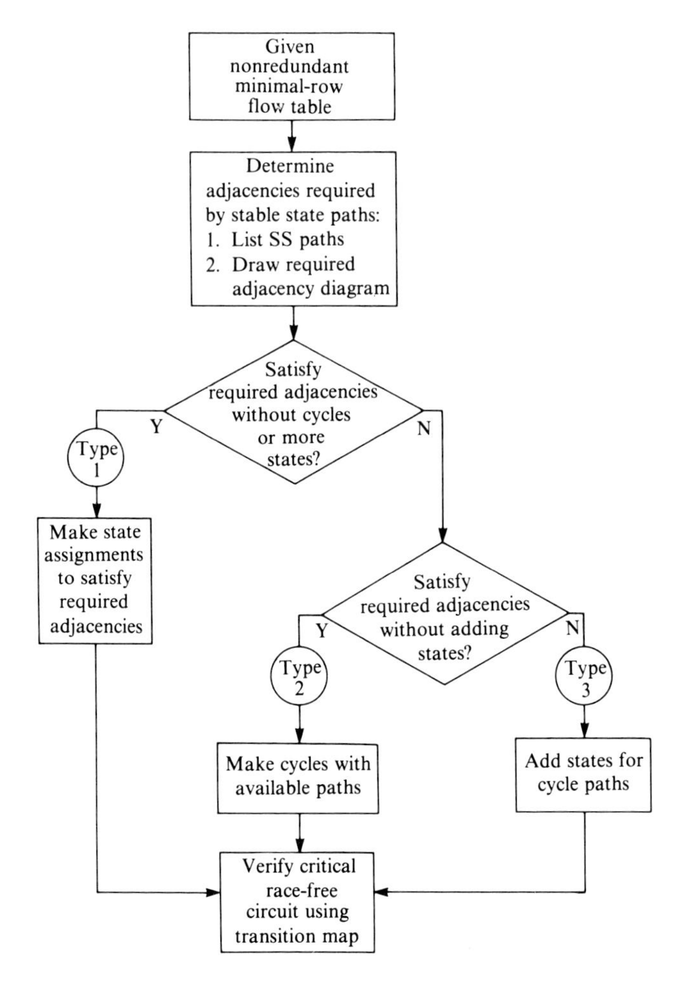

FIGURE 9.22
Flowchart for the state assignment algorithm

Required Adjacencies Satisfied Without Cycles or Additional States

A procedure for making a critical race-free state assignment for a type 1 minimal-row flow table is illustrated by the following example.

■ **EXAMPLE 9.7**

Given a minimal-row merged flow table, shown in Table 9.12, assign binary codes to the stable states.

TABLE 9.12
Minimal-row merged flow table

x_1x_2	00	01	11	10
a	(a)	(a)	d	(a)
b	(b)	(b)	c	(b)
c	b	a	(c)	(c)
d	a	b	(d)	(d)

State assignment is the process of assigning binary codes to the stable states in such a way that no race conditions exist in the circuit; that is, in any transition only 1 state variable is required to change. The circuit's transitions can be determined by listing the flow-table paths between stable states; these paths are shown in column order in Figure 9.23a.

A race-free assignment requires that every transition is a transition between logically adjacent states (so that only 1 state variable changes in any transition). The adjacencies required are determined by drawing an adjacency diagram (Figure 9.23b) that summarizes the distinct SS-to-SS paths of Figure 9.23a.

FIGURE 9.23
Determining required adjacencies. (a) Path table (b) Adjacency diagram

Column:	00	01	11	10
Paths:		**a–d**	**c–a**	**a–d**
		b–c	**d–b**	**b–c**

(a)

a b
c d

(b)

We see from Figure 9.23 that a race-free circuit requires that

State **a** must be logically adjacent to states **c** and **d**

State **b** must be logically adjacent to states **c** and **d**

State **d** must be logically adjacent to states **a** and **b**

State **c** must be logically adjacent to states **a** and **b**

Since the minimal-row flow table has 4 rows and there are only 4 required adjacencies, these adjacencies can be satisfied without cycles or additional states. Since there are 4 stable states, 2 state variables y_1 and y_2 with 4 distinct states are required to implement the circuit. The required adjacencies can be satisfied by using the state assignment matrix in Figure 9.24 to assign binary codes to the stable states as follows: **a** = 00, **c** = 01, **d** = 10, and **b** = 11.

FIGURE 9.24
State assignment matrix

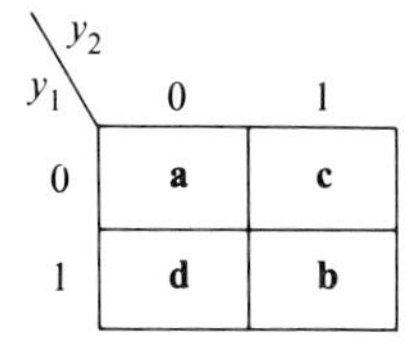

A race-free state assignment can, therefore, be accomplished by an effective reordering of the rows of the minimal-row flow table to satisfy the required adjacencies (**a–c–b–d–a**), as indicated in the state assignment matrix. The reordering of the rows of Table 9.12 produces a row-reordered table shown in Table 9.13.

TABLE 9.13
Row-reordered flow table

	x_1x_2	00	01	11	10
00	**a**	(a)	(a)	d	(a)
01	**c**	b	a	(c)	(c)
11	**b**	(b)	(b)	c	(b)
10	**d**	a	b	(d)	(d)

The properly ordered minimal-row flow table can then be converted to a circuit excitation map (Figure 9.25) by replacing each state label with its assigned binary code.

FIGURE 9.25
Circuit excitation (next-state) map

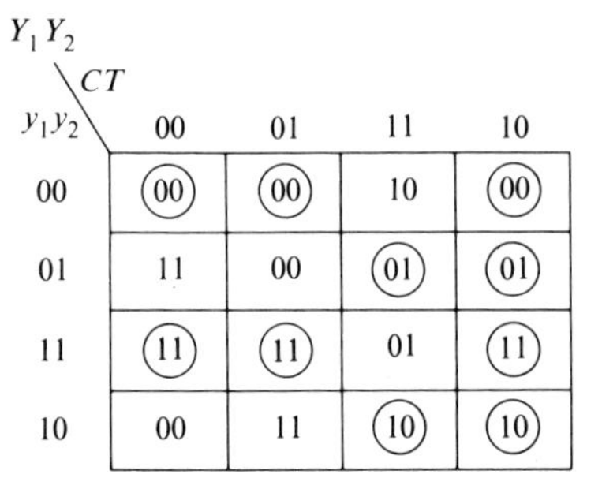

Y_1Y_2

y_1y_2 \ CT	00	01	11	10
00	(00)	(00)	10	(00)
01	11	00	(01)	(01)
11	(11)	(11)	01	(11)
10	00	11	(10)	(10)

A map of transitions can be plotted to verify that the state assignment is race free. For stable state cells ($Y_2Y_1 = y_2y_1$) only 00, 01, 10, or 11 transitions occur. Therefore, it is only necessary to plot transitions for the unstable state cells in each SS-to-SS path, as shown in Figure 9.26.

FIGURE 9.26
Map of circuit transitions

y_1y_2 \ CT	00	01	11	10
00			a0	
01	a1	0b		
11			b1	
10	b0	1a		

The map of transitions verifies that the state assignment is race free since no cell contains more than 1 alpha or 1 beta. ■

Required Adjacencies Satisfied Using Available Cycle Paths

A cycle is a transition path that contains 2 or more consecutive unstable states. A procedure for making a race-free state assignment for a type 2 minimal-row flow table is illustrated by the following example.

■ **EXAMPLE 9.8**

Given a minimal-row merged flow table, as shown in Table 9.14, assign binary codes to the stable states.

TABLE 9.14
Minimal-row merged flow table

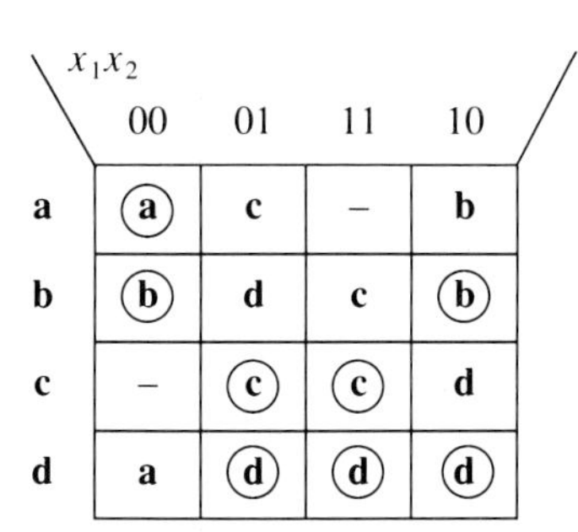

	x_1x_2 00	01	11	10
a	(a)	c	–	b
b	(b)	d	c	(b)
c	–	(c)	(c)	d
d	a	(d)	(d)	(d)

The circuit's transitions can be determined by listing the flow-table paths between stable states; these paths are shown in column order in Figure 9.27a.

A race-free assignment requires that every transition is a transition between logically adjacent states (so that only 1 state variable changes in any transition). The adjacencies required are determined by drawing an adjacency diagram (Figure 9.27b) that summarizes the distinct SS-to-SS paths of Figure 9.27a.

FIGURE 9.27
Determining required adjacencies. (a) Path table (b) Adjacency diagram

Column:	00	01	11	10
Paths:	**a–b**	**d–a**	**c–d**	**b–c**
	a–c			**d–a**
	b–d			

(a)

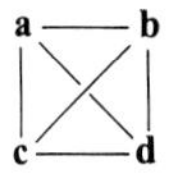

(b)

Table 9.14 shows that paths **a–c** and **b–d** have end points in nonadjacent rows. Further, column 01, which contains the end stable states for these paths, has no unspecified intermediate states. Therefore, if logically adjacent state assignments are made using the original row order **a**, **b**, **c**, **d**, then paths **a–c** and **b–d** would contain race conditions.

If the rows are effectively reordered in the order **a**, **b**, **d**, **c** (to make **a**, **c** adjacent and **b**, **d** adjacent) the race conditions in paths **a–c** and **b–d** are eliminated, but race conditions are introduced in paths **b–c** and **d–a**, as shown in Table 9.15.

TABLE 9.15
Row-reordered flow table

x_1x_2	00	01	11	10
a	(a)	c	–	b
b	(b)	d	c	(b)
d	a	(d)	(d)	(d)
c	–	(c)	(c)	d

Fortunately, in each of these paths, **b–c** and **d–a**, the column of the end stable state has a row with an unspecified state. Therefore, a cycle can be created for each path in which the start state and end state are not logically adjacent. A cycle-maker diagram (Figure 9.28b, p. 430) can be used to illustrate the required cycles. (Capital letters are used to denote the new unstable states in each cycle.)

The circuit's transitions can be determined by listing the flow-table paths between stable states; these paths are shown in column order in Figure 9.28a.

The result, obtained by effectively reordering rows (so **a**, **b**, **d**, **c** appear in Gray-code order) and by creating cycles (using unspecified and unstable intermediate states), produces Table 9.16.

The flow table (Table 9.16, p. 430) is then converted to a circuit excitation map by replacing each state letter symbol with its assigned binary code (**a** = 00, **b** = 01, **c** = 10, **d** = 11). The race-free excitation map and the corresponding map of transitions are shown in Figure 9.29a and b (p. 430).

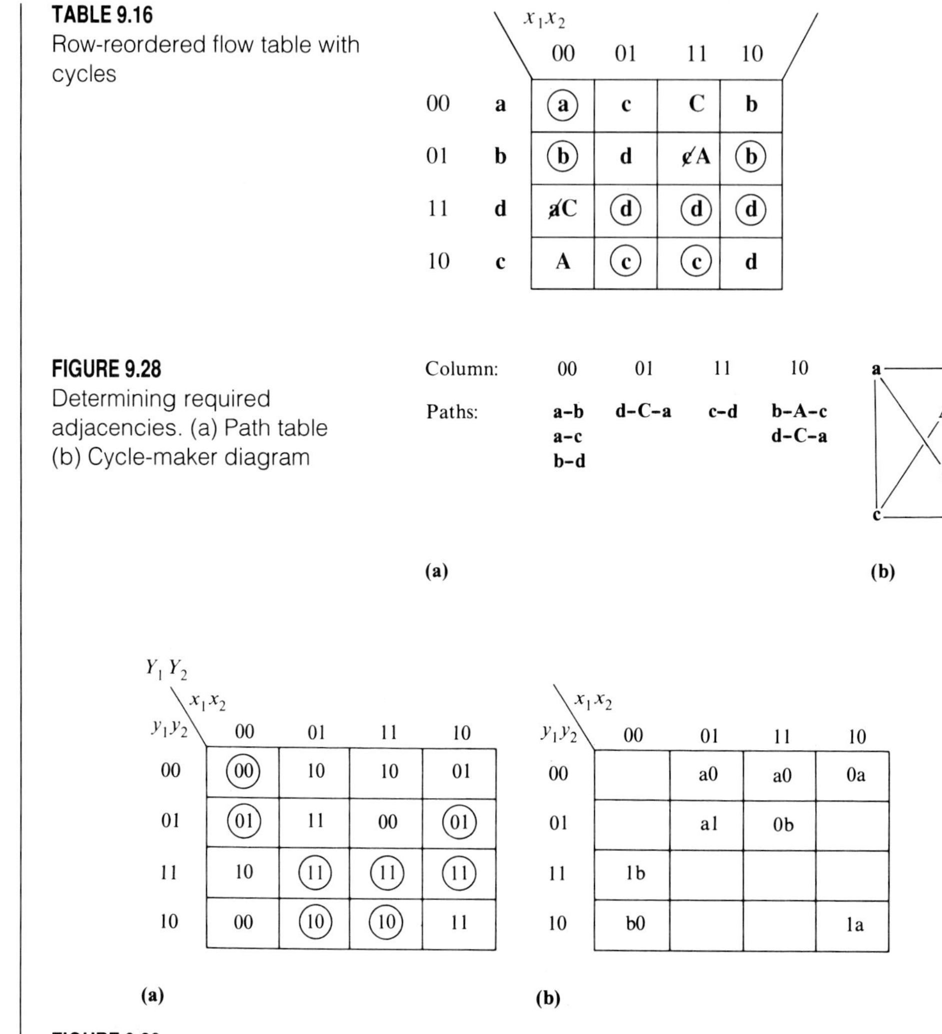

TABLE 9.16
Row-reordered flow table with cycles

		x_1x_2 00	01	11	10
00	a	(a)	c	C	b
01	b	(b)	d	~~c~~A	(b)
11	d	~~d~~C	(d)	(d)	(d)
10	c	A	(c)	(c)	d

Column:	00	01	11	10
Paths:	a–b a–c b–d	d–C–a	c–d	b–A–c d–C–a

(a)

(b)

FIGURE 9.28
Determining required adjacencies. (a) Path table (b) Cycle-maker diagram

Y_1Y_2

y_1y_2 \ x_1x_2	00	01	11	10
00	(00)	10	10	01
01	(01)	11	00	(01)
11	10	(11)	(11)	(11)
10	00	(10)	(10)	11

(a)

y_1y_2 \ x_1x_2	00	01	11	10
00		a0	a0	0a
01		a1	0b	
11	1b			
10	b0			1a

(b)

FIGURE 9.29
Verifying a race-free state assignment. (a) Circuit excitation map (b) Map of transitions

Since the map of transitions has no cell that contains two letter symbols (a = alpha, b = beta), we see that the circuit is race free. ■

Required Adjacencies Satisfied by Adding States for Cycle Paths

A procedure for making a race-free state assignment for a type 3 minimal-row flow table is illustrated by the following example.

■ **EXAMPLE 9.9**

A minimal-row merged flow table is shown in Table 9.17.

TABLE 9.17
Minimal-row merged flow table

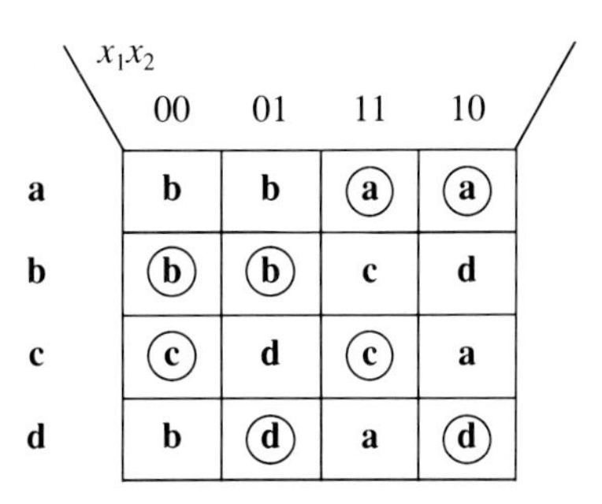

x_1x_2	00	01	11	10
a	b	b	(a)	(a)
b	(b)	(b)	c	d
c	(c)	d	(c)	a
d	b	(d)	a	(d)

The circuit's transitions can be determined by listing the flow-table paths between stable states; these paths are shown in column order in Figure 9.30a. The adjacencies required are determined by drawing an adjacency diagram (Figure 9.30b) that summarizes the distinct SS-to-SS paths of Figure 9.30a.

FIGURE 9.30
Determining required adjacencies. (a) Path table (b) Adjacency diagram

Column:	00	01	11	10
Paths:	b–d	b–c	a–b	a–b
	c–d	d–b	c–a	d–a
	c–a	d–a	c–d	d–b

(a)

a b d c

(b)

Since the number of required adjacencies is greater than the number of rows and insufficient cycle paths exist, a third state variable must be introduced to provide cycle paths to satisfy the required adjacencies. A state assignment matrix (Figure 9.31) can be used to assign binary codes to the states.

FIGURE 9.31
State assignment matrix

y_1 \ y_2y_3	00	01	11	10
0	a	b	G	d
1	c	F		E

A cycle can then be created for each path in which the start state and end state are not logically adjacent. A cycle-maker diagram (Figure 9.32) can be used to illustrate the required cycles.

FIGURE 9.32
Cycle-maker diagram

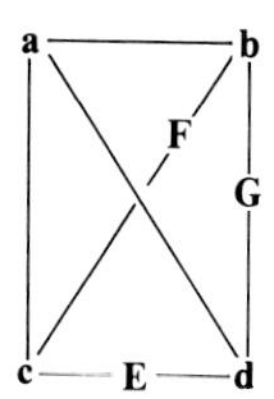

The original flow table is then augmented by rows **E**, **F**, and **G** to allow creation of the required cycles (Table 9.18, p. 432). Further, the original intermediate unstable state in each path requiring a cycle is replaced with the label of the cycle row (denoted by capital letters). For example, row **b** path **b–c–c** becomes path **b–F–c**, with a cycle from row **b** to row **F** and back to row **c**.

TABLE 9.18
Augmented-row flow table

x_1x_2	00	01	11	10
a	b	b	(a)	(a)
b	(b)	(b)	~~c~~F	~~d~~G
c	(c)	~~d~~E	(c)	a
d	~~b~~G	(d)	a	(d)
E	–	d	–	–
F	–	–	c	–
G	b	–	–	d

Note that row **G** is shared by intermediate states **b** and **d**. Hence, this method is called the **shared-row method.**

The rows of the augmented flow table (Table 9.17) are then reordered in accordance with the state assignment matrix, as shown in Table 9.19.

TABLE 9.19
Reordered augmented-row flow table

	x_1x_2	00	01	11	10
000	a	b	b	(a)	(a)
001	b	(b)	(b)	F	G
011	G	b	–	–	d
010	d	G	(d)	a	(d)
110	E	–	d	–	–
111	–	–	–	–	–
101	F	–	–	c	–
100	c	(c)	E	(c)	a

The flow table is then converted into an excitation map by replacing each label with its corresponding binary code. The excitation map and the associated map of transitions are presented in Figure 9.33.

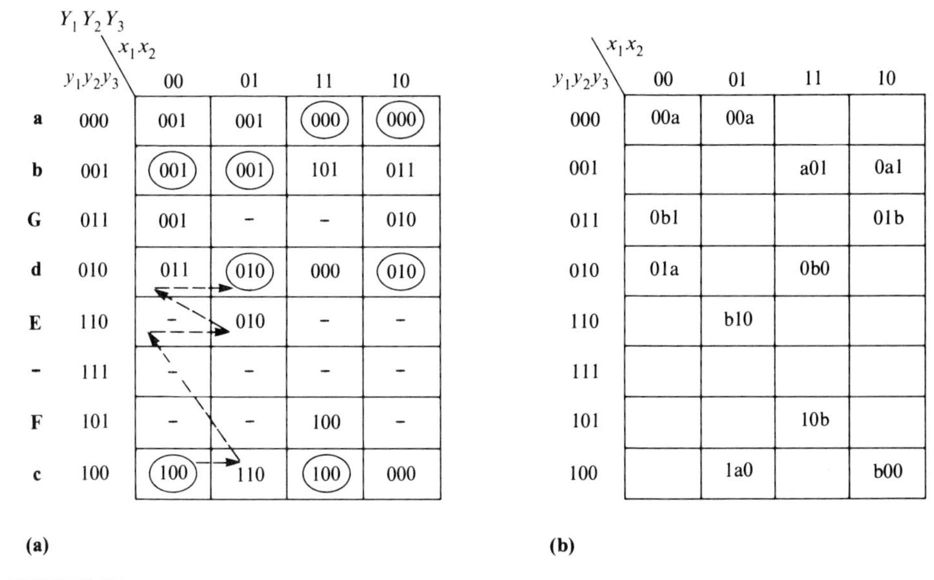

(a) $Y_1Y_2Y_3$

	$y_1y_2y_3$ \ x_1x_2	00	01	11	10
a	000	001	001	(000)	(000)
b	001	(001)	(001)	101	011
G	011	001	-	-	010
d	010	011	(010)	000	(010)
E	110	-	010	-	-
-	111	-	-	-	-
F	101	-	-	100	-
c	100	(100)	110	(100)	000

(b)

$y_1y_2y_3$ \ x_1x_2	00	01	11	10
000	00a	00a		
001			a01	0a1
011	0b1			01b
010	01a		0b0	
110		b10		
111				
101			10b	
100		1a0		b00

FIGURE 9.33
Verifying a race-free state assignment. (a) Circuit excitation map (b) Map of transitions

Since no cell of the map of transitions has more than one alpha or beta (a or b), it can be seen that a race-free state assignment has been accomplished. ■

Now that we understand how to accomplish each step of the 4-step problem-solving procedure for the design of fundamental-mode circuits, we will describe a complete design of a circuit in the next section.

A Complete Design of a Fundamental-Mode Circuit

The use of the 4-step problem-solving procedure to design a fundamental-mode circuit is described in Example 9.10.

■ **EXAMPLE 9.10: Design of a Sequence Detector with 2 Asynchronous Inputs**

Step 1: Problem Statement. Given 2 asynchronous inputs, describe a process that detects a specific sequence (00, 01, 11) of combinations of the 2 inputs. Then, design a fundamental-mode circuit that realizes the sequence detection process. The sequence-detector circuit has 2 asynchronous level-signal inputs, x_1 and x_2, and 1 output, Z, which is initially 0. When the circuit detects the input sequence $x_1x_2 = 00, 01, 11$ the output Z goes to 1 and remains at 1 as long as the inputs remain unchanged. If any other sequence is input, the circuit returns to its reset state (state **a**) on the next 00 input following an input not in the sequence 00, 01, 11.

Step 2: Conceptualization. The operation of the sequence detector can be graphically described by a primitive state diagram, as shown in Figure 9.34 (p. 434).

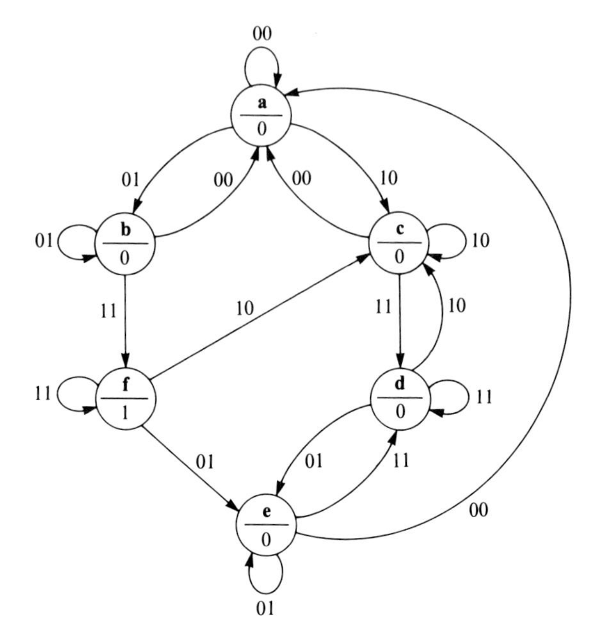

FIGURE 9.34
Primitive state diagram for Example 9.10

A corresponding primitive flow table for the circuit (Table 9.20) can be constructed using the primitive state diagram.

TABLE 9.20
Primitive flow table for Example 9.10

Row	x_1x_2 00	01	11	10	Z
a	(a)	b	–	c	0
b	a	(b)	f	–	0
c	a	–	d	(c)	0
d	–	e	(d)	c	0
e	a	(e)	d	–	0
f	–	e	(f)	c	1

Step 3: Solution/Simplification.

a. State reduction

(1) Eliminate redundant stable states. States **b** and **e**, both stable for input 01 and with output 0, are the only potentially equivalent states in Table 9.19. However, **b** and **e** are not equivalent because their next states **f** and **d** are not equivalent (**f** and **d** have different outputs). Therefore, there are no equivalent states in Table 9.19, so it is a nonredundant flow table.

(2) Eliminate intermediate unstable states. A nonredundant primitive flow table can be condensed into a minimal-row merged flow table by eliminating intermediate

unstable states; this condensation is accomplished by merging groups of nonconflicting (compatible) rows into a single row. An examination of the primitive flow table (Table 9.20) reveals the following groups of nonconflicting rows: {**a**, **b**}, {**c**, **d**, **e**}. By merging these rows in the primitive flow table, a minimal-row merged flow table is produced. Figure 9.35 presents the merged flow table and a corresponding reduced state diagram.

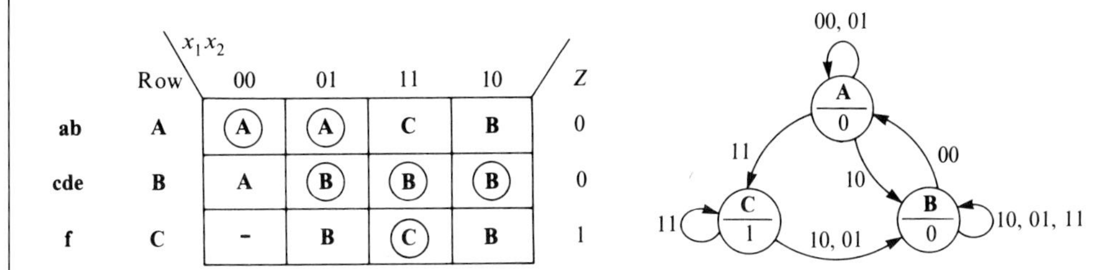

	Row \ x_1x_2	00	01	11	10	Z
ab	A	(A)	(A)	C	B	0
cde	B	A	(B)	(B)	(B)	0
f	C	-	B	(C)	B	1

FIGURE 9.35
Minimal-row merged flow table and reduced state diagram

b. State assignment. Remember that state assignment is the process of assigning binary codes to the stable states in such a way that no race conditions exist in the circuit; that is, in any transition, only 1 state variable is required to change. The circuit's transitions can be determined by listing the flow-table paths between stable states, as shown in Figure 9.36a. A race-free assignment requires that every transition is a transition between logically adjacent states (so that only one state variable changes in any transition). The adjacencies required are determined by drawing an adjacency diagram (Figure 9.36b) that summarizes the distinct SS-to-SS paths of Figure 9.36a.

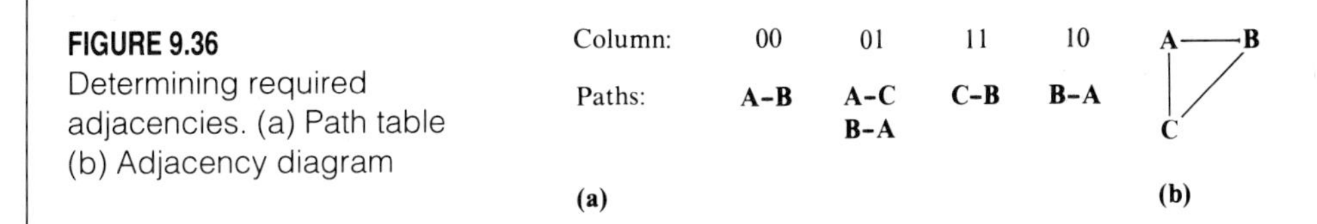

Column:	00	01	11	10
Paths:	**A–B**	**A–C** **B–A**	**C–B**	**B–A**

FIGURE 9.36
Determining required adjacencies. (a) Path table (b) Adjacency diagram

We see from Figure 9.36 that a race-free circuit requires that

State **A** must be logically adjacent to state **B**

State **B** must be logically adjacent to state **C**

State **C** must be logically adjacent to state **A**

The minimal-row merged flow table has 3 stable states. Therefore, 2 state variables, with 4 distinct states, are required to implement the circuit. Consequently, there are 4 choices (00, 01, 11, 10) to assign a binary code to the first stable state **A**. After state **A** has been assigned a code, there are 3 remaining code choices for state **B**, and finally there are 2 remaining code choices for state **C**. Thus, there are $4 \times 3 \times 2 = 12$ possible state assignments for 3 states. However, any one of these 12 possible state assignments will result in a race condition, as indicated in Figure 9.37.

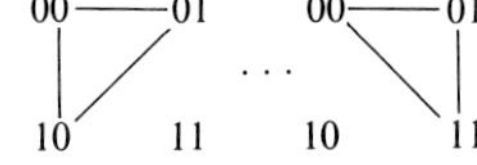

FIGURE 9.37
Race condition for a 3-row flow table

Therefore, for a race-free state assignment, a fourth row (state **D**) must be added to provide a cycle path: **C–D–B** = 10–11–01, where only 1 state variable changes in each transition; this eliminates the race condition in path **C–B** = 10–01. A state assignment matrix (Figure 9.38), with state variables denoted by y_1 and y_2, can be used to produce a race-free assignment: **A** = 00, **B** = 01, **C** = 10, **D** = 11.

You can see from the state assignment matrix that each SS-to-SS path in the adjacency diagram (Figure 9.36b) can be traversed by transitions in which only 1 state variable changes at a time.

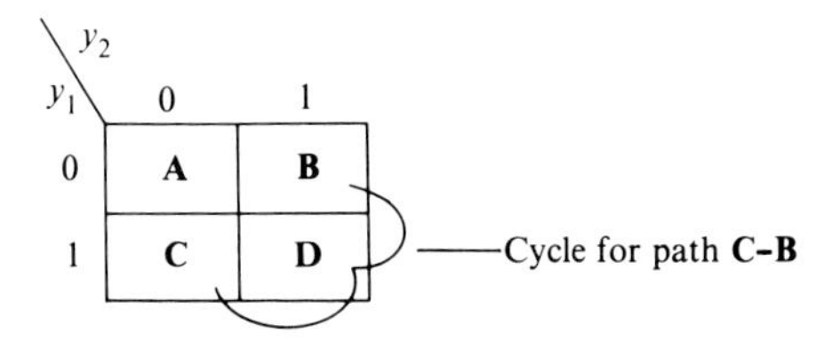

FIGURE 9.38
State assignment using a state assignment matrix

The resulting flow table, with row **D** added to provide cycles for paths **C** to **B**, is shown in Table 9.21.

TABLE 9.21
Flow table with state added for cycles

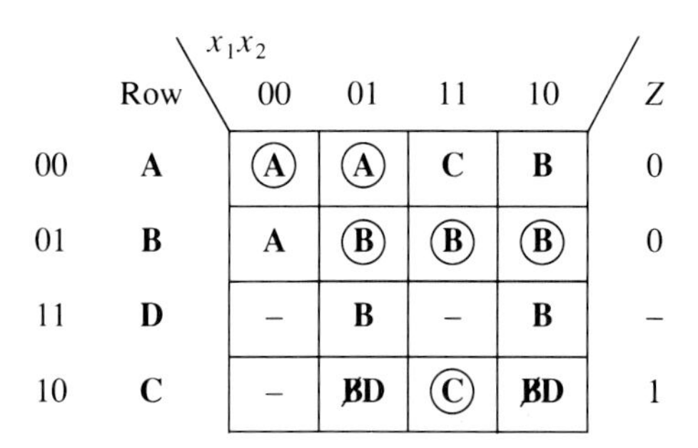

	Row \ x_1x_2	00	01	11	10	Z
00	A	(A)	(A)	C	B	0
01	B	A	(B)	(B)	(B)	0
11	D	–	B	–	B	–
10	C	–	~~B~~D	(C)	~~B~~D	1

The flow table in Table 9.21 is converted to a circuit excitation/output map (Figure 9.39) by replacing each state label with the binary code for that state. By tracing out each SS-to-SS path, we see that every transition can be made by changing only 1 state variable at a time. Therefore, the state assignment made using the state assignment matrix results in a race-free circuit.

FIGURE 9.39
Circuit excitation/output map

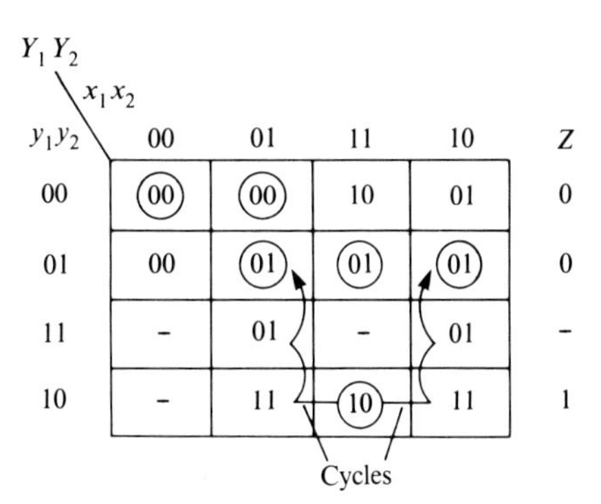

c. The circuit excitation map is then separated into individual excitation maps for the state variables, as shown in Figure 9.40.

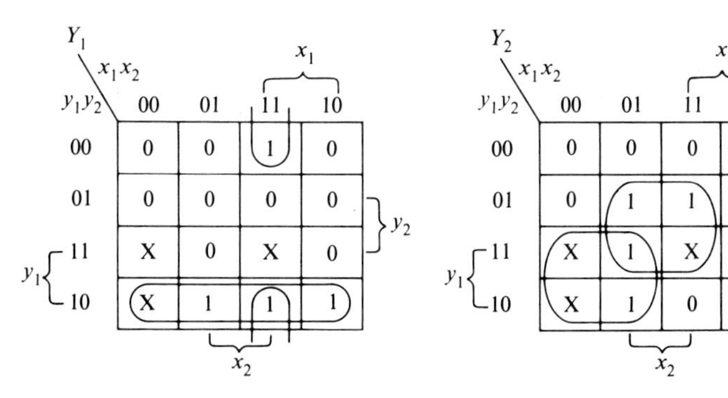

FIGURE 9.40
Excitation maps for state variables

The excitation (next-state) equations for the circuit, determined from the excitation maps, are

$$Y_1 = x_1\, x_2\, y_2' + y_1\, y_2'$$
$$Y_2 = x_1\, x_2' + x_1' y_1 + x_2\, y_2$$

and the output equation is

$$Z = Y_1$$

Step 4: Realization. The excitation equations and output equation are used to construct a logic diagram of the fundamental-mode circuit, as shown in Figure 9.41 (p. 438).

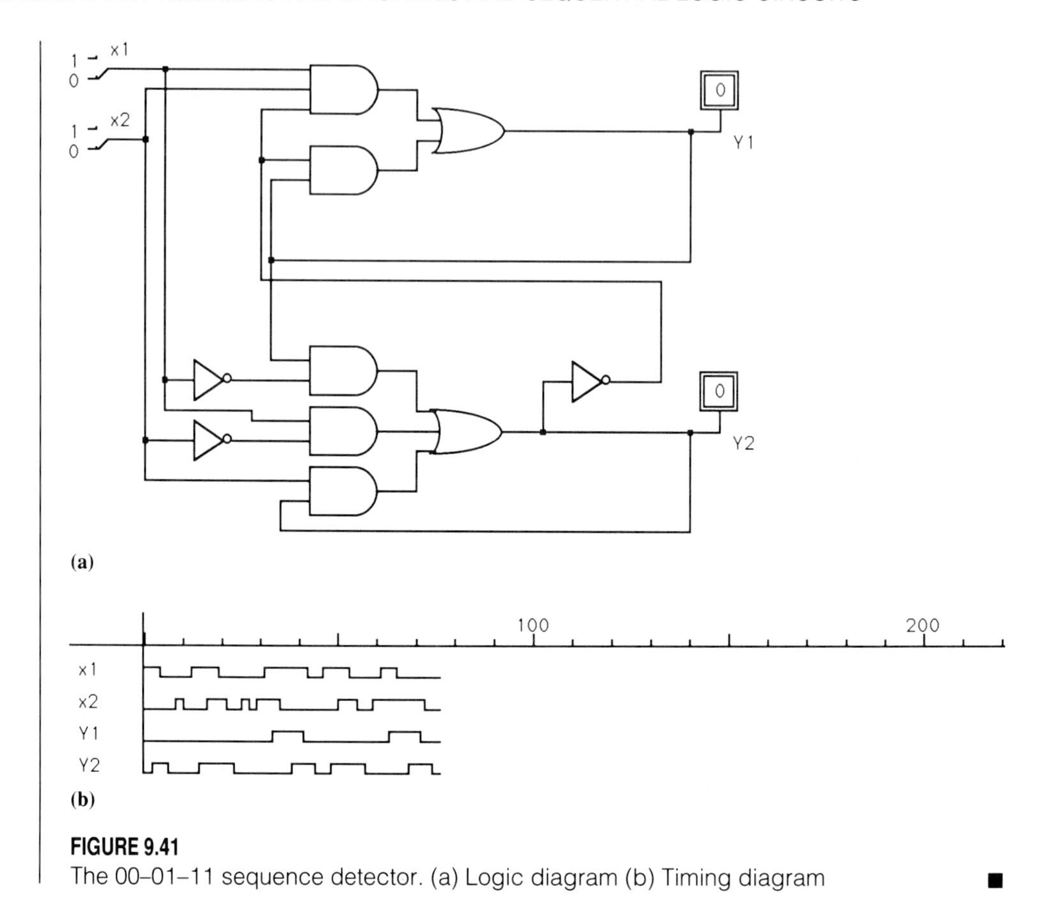

FIGURE 9.41
The 00–01–11 sequence detector. (a) Logic diagram (b) Timing diagram ■

This example design, with fairly straightforward state reduction and state assignment, was presented to illustrate the use of the 4-step problem-solving procedure to design a fundamental-mode circuit.

Section 9.5 describes the design of a number of representative fundamental-mode circuits using the 4-step problem-solving procedure illustrated in this section.

9.5 DESIGN OF REPRESENTATIVE FUNDAMENTAL-MODE CIRCUITS

Now that we have described and illustrated detailed procedures for designing fundamental-mode circuits, we will design several representative circuits, such as

- ***A negative edge-triggered* T *flip-flop***
- ***A pulse generator***
- ***A positive edge-triggered* D *flip-flop***

There is usually more than one approach that can be used to design a particular circuit, as we have shown in earlier chapters. This is also true for fundamental-mode circuits. In the following sections, you will learn two alternative approaches that can be used to design fundamental-mode circuits:

- A traditional design approach
- A modular design approach

Either of these approaches can be used within the framework of the 4-step problem-solving procedure for designing fundamental-mode circuits, as described and illustrated in Section 9.4.

A Traditional Design Approach

This section describes the design of representative fundamental-mode circuits using a traditional design approach in which the circuit is represented by a general asynchronous circuit model shown in Figure 9.42. Recall that this model includes the following components:

- A set of n external inputs, $x_1, \ldots, x_n$
- A combinational logic network consisting of a next-state logic section and an output logic section
- A set of m external outputs, $z_1, \ldots, z_m$
- A memory section that stores system state $y_1, \ldots, y_p$
- A set of feedback lines

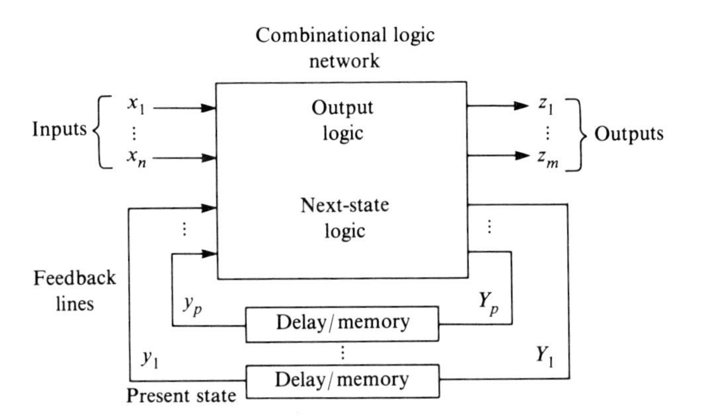

FIGURE 9.42
Model of an asynchronous sequential-logic circuit

INTERACTIVE DESIGN APPLICATION

Design of a Negative Edge-Triggered *T* Flip-Flop

This section describes the design of a negative edge-triggered (NET) T flip-flop using a traditional design approach within the framework of the 4-step problem-solving design procedure. It should be noted that one of the external inputs ($x_1, \ldots, x_n$) may be a clock signal provided that the inputs (including the clock) satisfy the fundamental-mode constraints described earlier.

Step 1: Problem Statement. Design a NET T flip-flop that has a control input T, a clock input C, and two complementary outputs Q and Q'. Input T operates asynchronously with respect to the clock C. The device is to operate as follows:

- If $T = 1$ and the clock makes a 1–0 transition, then the next state Q is the complement of present state q.
- Otherwise, the next state Q is the present state q.

Step 2: Conceptualization. A block diagram of the NET T flip-flop implemented using delay elements is shown in Figure 9.43.

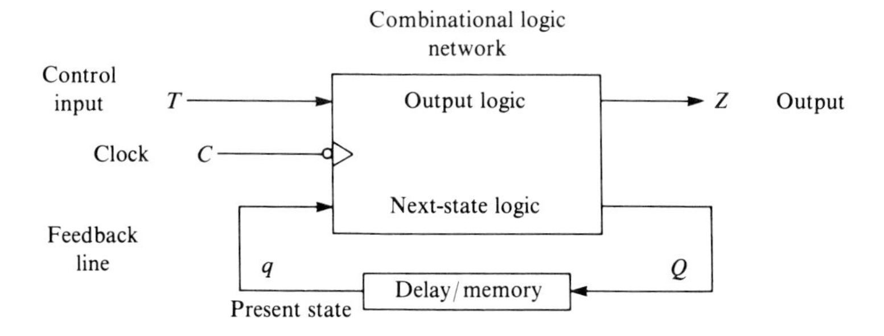

FIGURE 9.43
Block diagram of NET T flip-flop using delay element

A primitive state diagram (Moore form) can be developed using the design specifications described in the problem statement, as shown in Figure 9.44.

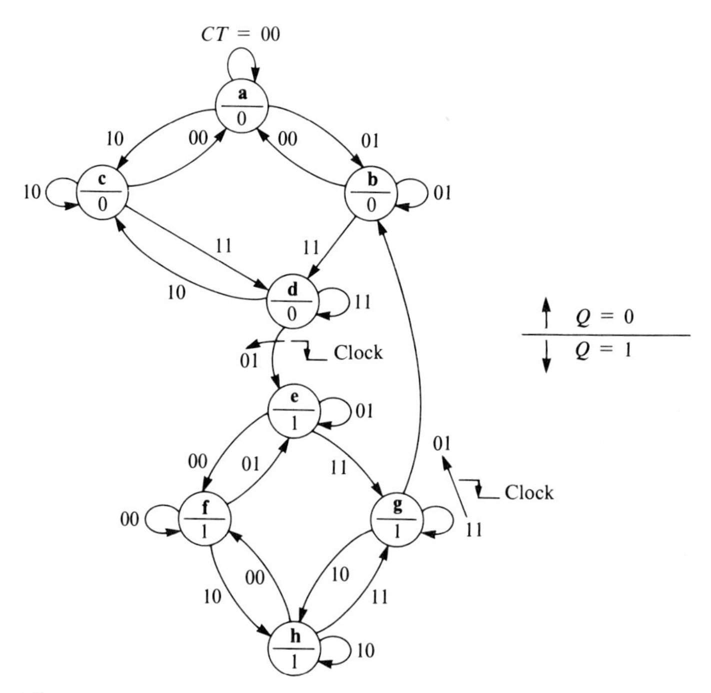

FIGURE 9.44
Primitive state diagram for NET T flip-flop

A primitive flow table (Table 9.22) can be constructed from the state diagram by tabulating the SS-to-SS transitions and outputs for each node.

TABLE 9.22
Primitive flow table for NET *T* flip-flop

Row	CT = 00	01	11	10	Q
a	(a)	b	–	c	0
b	a	(b)	d	–	0
c	a	–	d	(c)	0
d	–	e	(d)	c	0
e	f	(e)	g	–	1
f	(f)	e	–	h	1
g	–	b	(g)	h	1
h	f	–	g	(h)	1

Step 3: Solution/Simplification.

a. State reduction.

(1) Eliminate redundant stable states. An examination of Table 9.22 shows that there are no redundant stable states because no 2 states that are stable for the same input combination have the same output. Therefore, Table 9.22 is a nonredundant flow table.

(2) Eliminate intermediate unstable states by merging. The pairs of mergeable rows can be determined by searching the flow table for pairs of nonconflicting (compatible) rows and constructing a merger graph, as shown in Figure 9.45.

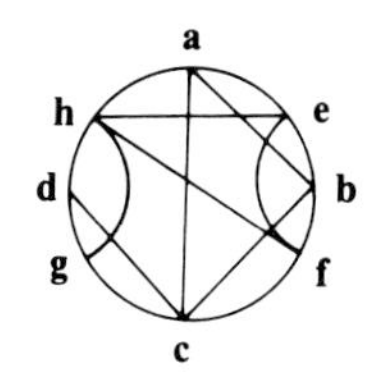

FIGURE 9.45
Merger graph for the NET *T* flip-flop

The maximal merger groups, as determined from the merger graph, are

{**a**, **b**, **c**} {**c**, **d**} {**e**, **f**, **h**} {**g**, **h**}

The nonredundant flow table (Table 9.22) can be reduced to a minimal-row merged flow table by selecting a set of mutually exclusive merger groups that covers all of the rows of the flow table. Thus, Table 9.22 can be reduced to a minimal-row table (Table 9.23, p. 442) by merging the rows in each of the mutually exclusive merger groups {**a**, **b**}, {**c**, **d**}, {**e**, **f**}, and {**g**, **h**}.

TABLE 9.23
Minimal-row flow table for NET *T* flip-flop

	Row \ CT	00	01	11	10	Q
(ab)	A	(a)	(b)	d	c	0
(cd)	B	a	e	(d)	(c)	0
(ef)	C	(f)	(e)	g	h	1
(gh)	D	f	b	(g)	(h)	1

A final merged flow table is obtained by replacing each primitive state symbol in Table 9.23 with its merged state symbol, as shown in Table 9.24.

TABLE 9.24
Final minimal-row flow table for NET *T* flip-flop

	Row \ CT	00	01	11	10	Q
(ab)	A	(A)	(A)	B	B	0
(cd)	B	A	C	(B)	(B)	0
(ef)	C	(C)	(C)	D	D	1
(gh)	D	C	A	(D)	(D)	1

b. State assignment. The circuit's transitions can then be determined by listing the flow-table paths between stable states, as shown in Figure 9.46a. The adjacencies required for a race-free state assignment are determined by drawing an adjacency diagram (Figure 9.46b) that summarizes the distinct SS-to-SS paths of Figure 9.46a.

Column:	00	01	11	10
Paths:	A–B	A–B	B–C	B–A
	C–D	C–D	D–A	D–C

(a)

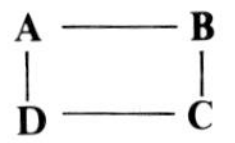

A — B
| |
D — C

(b)

FIGURE 9.46
Determining required adjacencies. (a) Path table (b) Adjacency diagram

The state assignment **A** = 00, **B** = 01, **C** = 11, **D** = 10 satisfies the requirements for a race-free circuit.

Finally, we can convert the final merged flow table (Table 9.24) to a circuit excitation map (Figure 9.47a) by replacing each state symbol with its assigned binary code. A merged state diagram, as shown in Figure 9.47b, is then constructed using the excitation map.

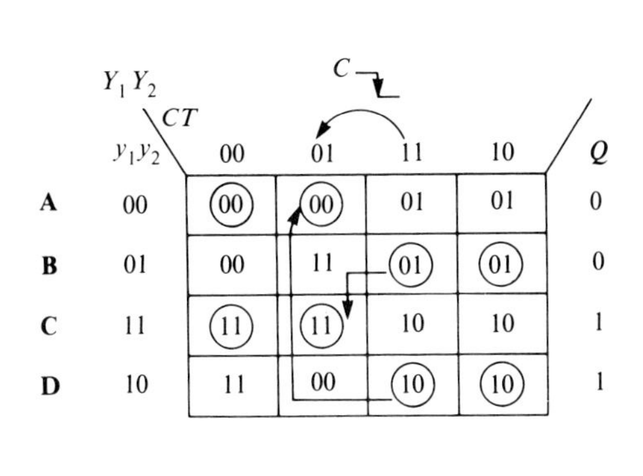

Y1 Y2
CT
y1y2
00 01 11 10
Q
A 00 00 00 01 01 0
B 01 00 11 01 01 0
C 11 11 11 10 10 1
D 10 11 00 10 10 1
C
(a)

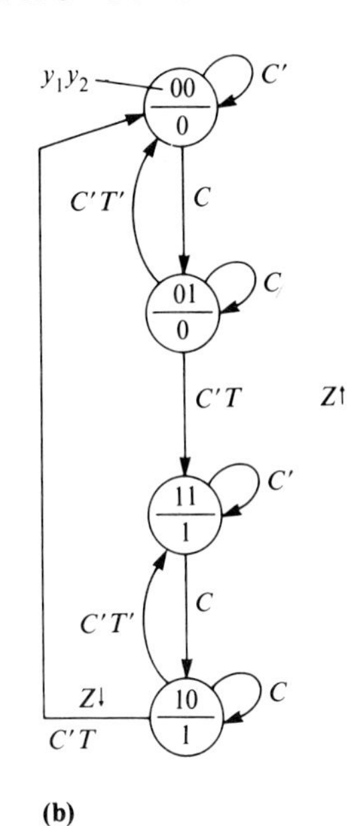

FIGURE 9.47
NET T flip-flop. (a) Circuit excitation map (b) Merged state diagram

c. The circuit excitation map can then be separated into individual excitation maps for Y_1 and Y_2, as shown in Figure 9.48.

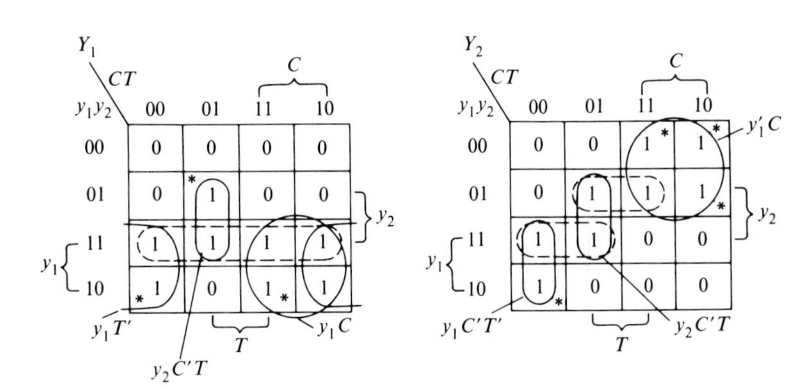

FIGURE 9.48
Excitation maps for a NET T flip-flop

The excitation equations, as determined from the excitation maps, are

$$Y_1 = y_1T' + y_1C + y_2C'T$$
$$Y_2 = y_1'C + y_1C'T' + y_2C'T$$

Entries in the output map (Figure 9.49, p. 444) are determined using the following rules:

1. For stable states, output Z is assigned the value of Q specified in the primitive flow table.
2. For unstable states,

a. If start state Q = end state Q, then $Z = Q$.
b. If start state $Q \neq$ end state Q, then Z is assigned a value such that the output makes only one 0-to-1 (or 1-to-0) transition in any SS-to-SS path.

The resulting output map is shown in Figure 9.49.

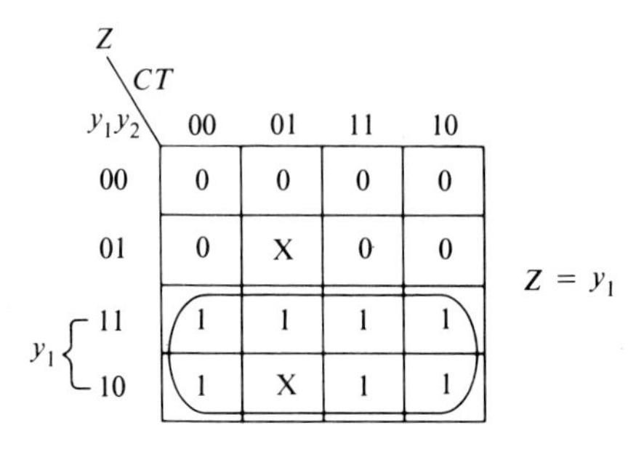

FIGURE 9.49
Output map for NET T flip-flop

The output equation, obtained from Figure 9.49, is

$$Z = y_1$$

Step 4: Realization. A logic diagram (Figure 9.50) of the NET T flip-flop can be constructed from the excitation equations and output equation.

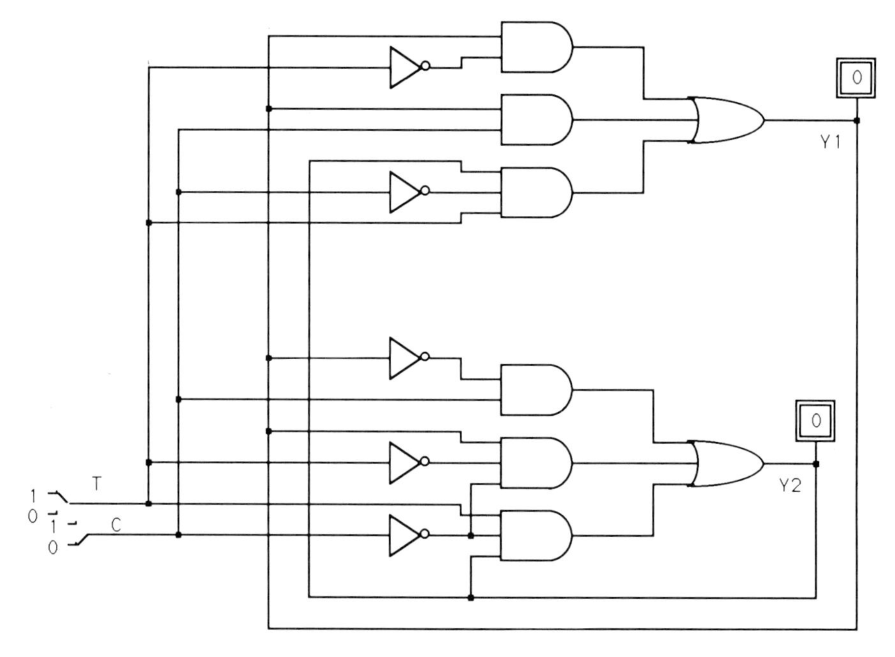

FIGURE 9.50
Logic diagram of a NET T flip-flop

INTERACTIVE DESIGN APPLICATION

Design of a Pulse Generator

This section describes the design of a fundamental-mode circuit that generates a pulse in response to a pushbutton control. Since the pushbutton can be depressed at any time, it is an asynchronous input. A traditional design approach is also used to design this device.

Step 1: Problem Statement. Design a pulse generator that produces an output pulse Z on the next 0–to–1 clock transition after a pushbutton P is depressed. Only 1 Z pulse is to be generated for each assertion of P, even if the duration of P is greater than a clock period. Short-duration (less than half the clock period) spikes and dropouts in P are to be ignored in generating Z.

Step 2: Conceptualization. A block diagram of the pulse generator implemented using delay elements is shown in Figure 9.51.

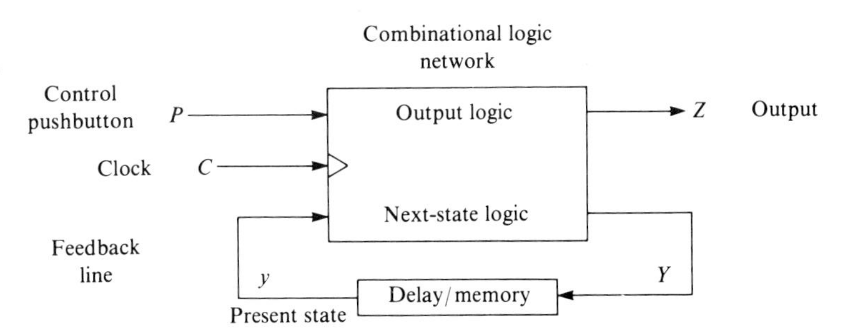

FIGURE 9.51
Block diagram of pulse generator using a delay element

A timing diagram can be constructed as follows: first, plot a number of cycles of the clock signal. Next, plot signal P so that all possible C, P combinations occur. Then, use the design specifications (if the pushbutton is depressed, an output pulse is generated on the next 0-to-1 clock transition) to plot the output signal Z. A representative timing diagram is shown in Figure 9.52.

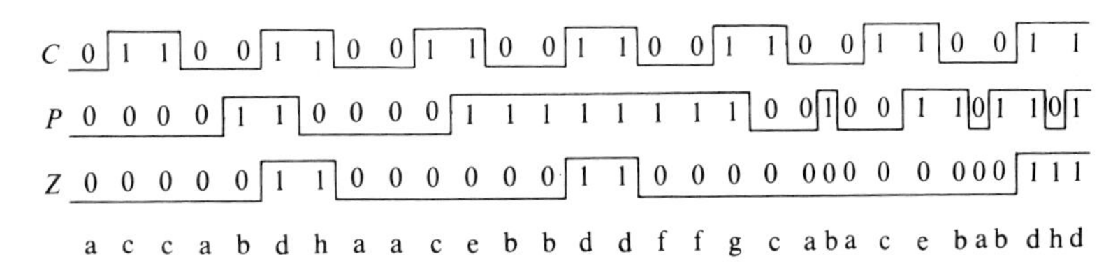

FIGURE 9.52
Timing diagram for pulse generator

Different stable states may contain the same input/output (CPZ) combination. Consequently a particular CPZ combination may have more than one label so that each stable state has a unique label. For example, $CPZ = 010$ occurs in both states b and f, and 110 occurs in both states e and g.

If the circuit is in a stable state, then valid sequences of input (C, P) changes, such as 00–01–11–10–00, produce a sequence of stable states. A primitive flow table may be generated as follows: First, construct an empty table with columns CP = 00, 01, 11, 10 and rows **a**, **b**, **c**, ... Then, use the timing diagram to fill in the table with each path between stable states, as shown in the primitive flow table in Table 9.25.

TABLE 9.25
Primitive flow table for pulse generator

Row	*CP* 00	01	11	10	*Z*
a	(a)	b	–	c	0
b	a	(b)	d	–	0
c	a	–	e	(c)	0
d	–	f	(d)	h	1
e	–	b	(e)	c	0
f	a	(f)	g	–	0
g	–	f	(g)	c	0
h	a	–	d	(h)	1

Step 3: Solution/Simplification.

a. State reduction.

(1) Eliminate redundant stable states. An examination of Table 9.25 shows that there are two pairs of potentially equivalent stable states: {**b**, **f**} and {**e**, **g**}. However, **b** and **f** are not equivalent because their next states **d** and **g** have different outputs. States **e** and **g** are not equivalent because their next states **b** and **f** are not equivalent. Therefore, Table 9.25 is a nonredundant flow table.

(2) Eliminate intermediate unstable states by merging. The pairs of mergeable rows can be determined by searching Table 9.25 for nonconflicting pairs of rows and constructing a merger graph, as shown in Figure 9.53.

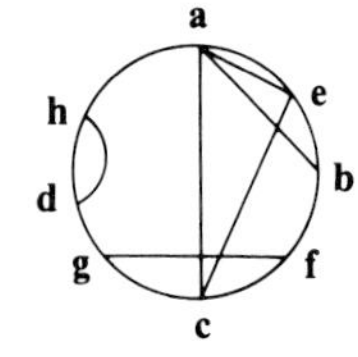

FIGURE 9.53
Merger graph for pulse generator

The maximal merger groups, as determined from the merger graph, are

{**a**, **c**, **e**} {**a**, **b**} {**f**, **g**} {**d**, **h**}

Rows **a** and **b** are not merged because the sequence **a**–**b**–**a** represents a *P* signal glitch (state **b**), which should not produce a *Z* pulse. Therefore, the circuit should return to state **a** ($CPZ = 000$) after **b**, so labels **a** and **b** should represent different stable states. The merger of rows within each of the mutually exclusive merger groups

{a, c, e} **{b}** **{f, g}** **{d, h}**

reduces the nonredundant primitive flow table to a minimal-row merged flow table as shown in Table 9.26.

TABLE 9.26
Minimal-row flow table for pulse generator

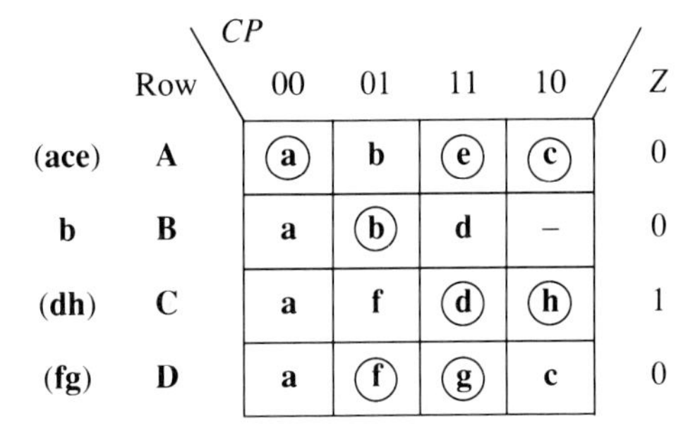

	Row	CP = 00	01	11	10	Z
(ace)	A	(a)	b	(e)	(c)	0
b	B	a	(b)	d	–	0
(dh)	C	a	f	(d)	(h)	1
(fg)	D	a	(f)	(g)	c	0

b. The circuit's transitions can then be determined by listing the flow table paths between stable states, as shown in Figure 9.54a. The adjacencies required for a race-free state assignment are determined by drawing an adjacency diagram (Figure 9.54b) that summarizes the distinct SS-to-SS paths of Figure 9.54a.

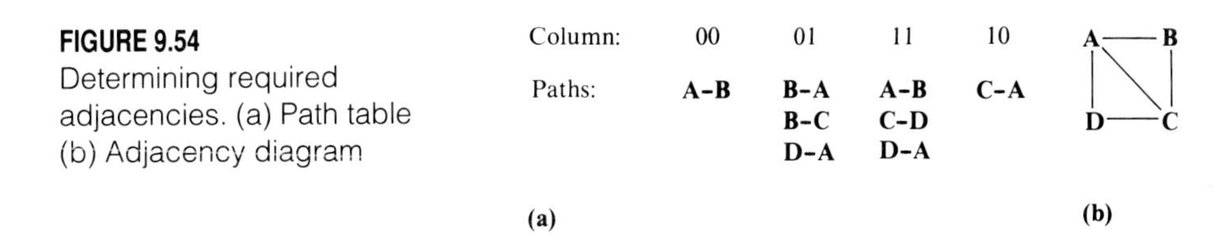

FIGURE 9.54
Determining required adjacencies. (a) Path table (b) Adjacency diagram

The **C–A** path requirement **h–a–a** can be ignored since stable state **h** followed by $CP = 00$ will eventually return to stable state **a** because each unstable state in column 00 is **a**. Consequently, no critical race occurs in path **C–A**. A state assignment matrix can be used to assign codes to the stable states to satisfy the required adjacencies, as shown in Figure 9.55.

FIGURE 9.55
State assignment matrix

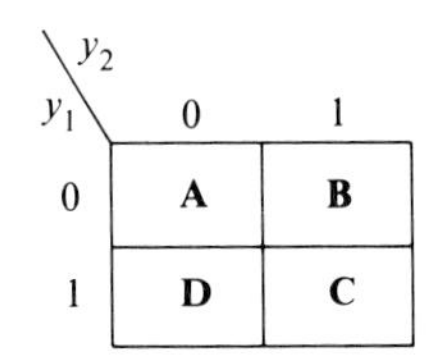

A final merged flow table (Figure 9.56a, p. 448) is obtained by replacing each primitive state symbol with its merged state symbol. The merged flow table can then be converted to a circuit excitation map (Figure 9.56b) by replacing each state symbol with its assigned binary code. A corresponding merged state diagram (in which each transition is documented by the condition for that transition) is shown in Figure 9.56c.

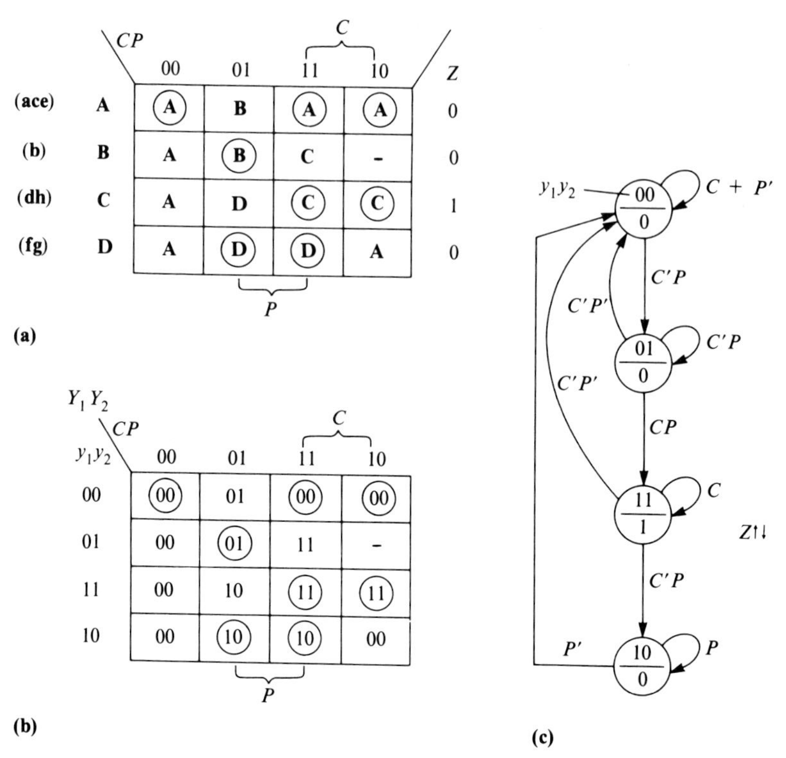

FIGURE 9.56
Pulse generator. (a) Minimal-row flow table (b) Circuit excitation map (c) Merged state diagram

c. The circuit map can be separated into individual excitation maps for Y_1 and Y_2, as shown in Figure 9.57.

FIGURE 9.57
Excitation maps for pulse generator

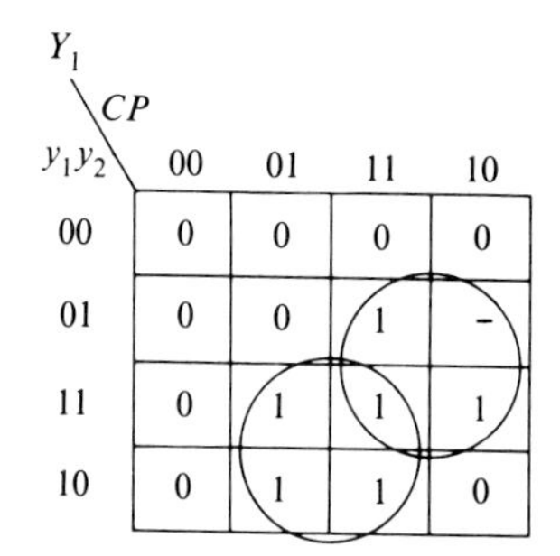

Y_2

y_1y_2 \ CP	00	01	11	10
00	0	1	0	0
01	0	1	1	-
11	0	0	1	1
10	0	0	0	0

The excitation equations, as determined from the excitation maps, are

$$Y_1 = y_1P + y_2C$$
$$Y_2 = y_1'C'P + y_2C + y_1'y_2P$$

The circuit is a Mealy circuit because the required output is a pulse signal Z that is generated concurrent with the first clock pulse following the assertion of P.

The values for the output Z (plotted in Figure 9.58) are determined by using the following rules:

1. For stable states, Z is assigned the value of the output specified in the primitive flow table. The output value is circled for each stable state cell.
2. For unstable states,
   a. If the start state output = the end state output, then the same value is used for the unstable state output and the output value is enclosed in parentheses.
   b. If the start state output ≠ the end state output, then the output in each cell in the transition path is assigned a value, enclosed in brackets, [ ], such that the output changes only once in the path between stable states.

The resulting output map is presented in Figure 9.58b, together with a merged flow table (Figure 9.58a), in which the transitions between stable states with the same output are indicated by solid vectors and the transitions between stable states with different outputs are indicated by dashed-line vectors.

FIGURE 9.58
Pulse generator. (a) Merged flow table (b) Output map

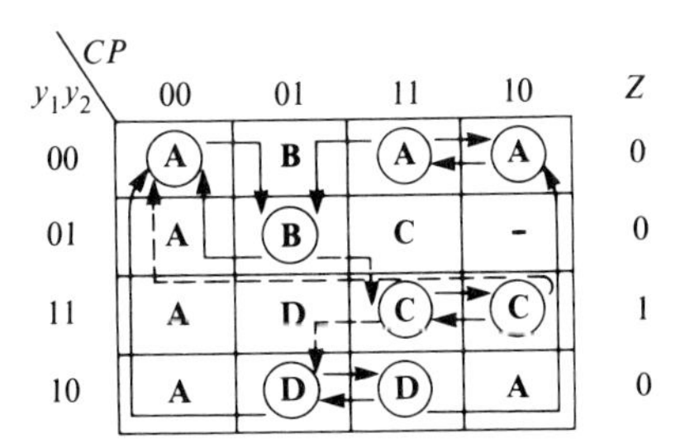

(a)

Z

y_1y_2 \ CP	00	01	11	10
00	0	(0)	0	0
01	(0)	0	[0]	-
11	[0]	[0]	1	1
10	(0)	0	0	(0)

$Z = y_1\, y_2\, C$

(b)

The output equation, obtained from Figure 9.58, is

$$Z = y_1\ y_2 C$$

Step 4: Realization. A logic diagram of the pulse generator (Figure 9.59a, p. 450) can be constructed using the excitation and output equations. A timing diagram for the pulse generator is also shown in Figure 9.59b.

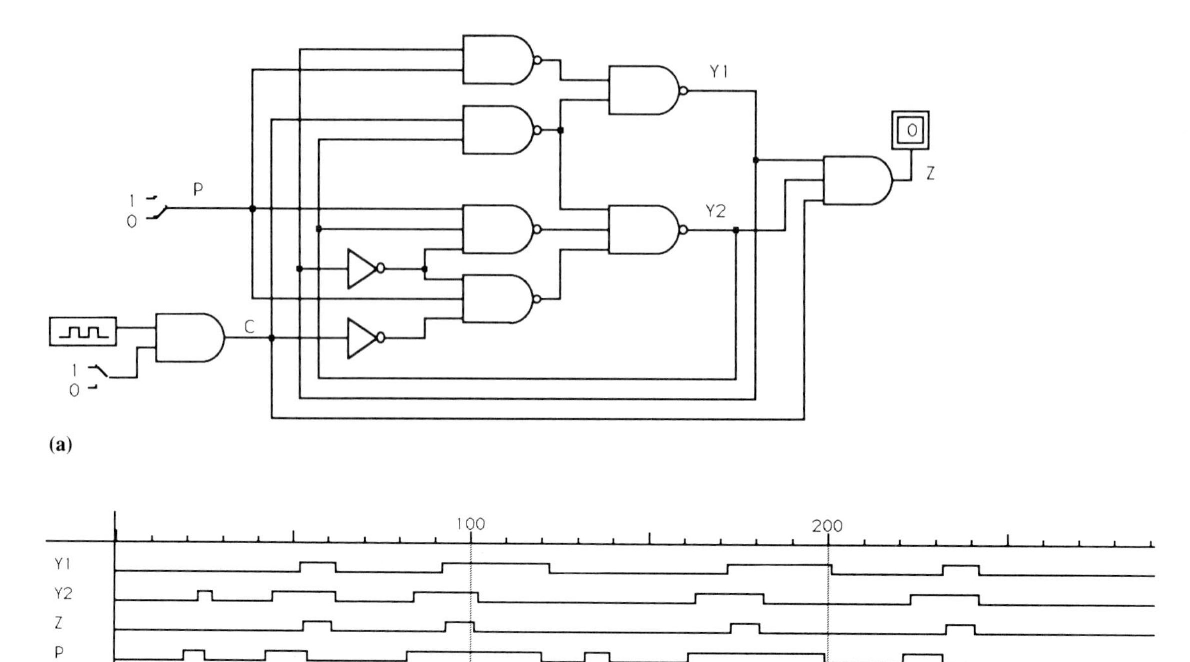

FIGURE 9.59
Pulse generator. (a) Logic diagram (b) Timing diagram

A Modular Design Approach

Up to this point we have used a traditional design approach to design fundamental-mode circuits. This section describes a modular design approach that uses *SR* latches as building blocks.

The block diagram presented in Figure 9.60 is used as a conceptual model of the fundamental-mode circuit. The block diagram is composed of an *SR* latch and a "driver" that is a fundamental-mode circuit whose outputs are fed back to the input side of the circuit. The driver outputs are also connected to the set and reset inputs of the latch.

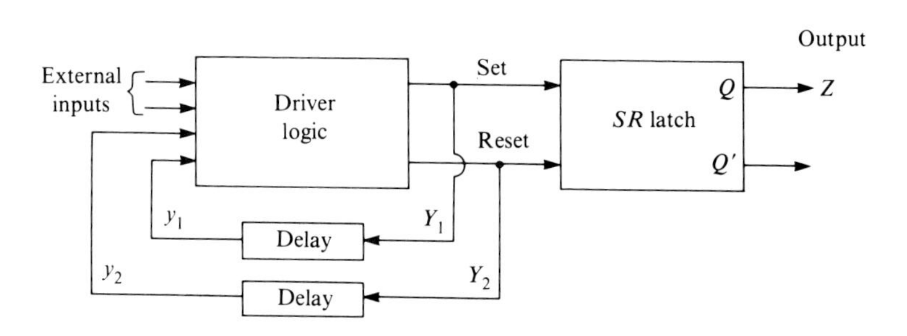

FIGURE 9.60
Block diagram of fundamental-mode circuit implemented with an *SR* latch

It should be noted that one of the external inputs of the circuit model in Figure 9.60 may be a clock signal. If the particular circuit to be designed has a clock signal as an external input, then that circuit is a synchronous device. However, the driver is a fundamental-mode circuit that operates under the constraint that only 1 driver external input (including the clock) can change at a time and only when the driver is in a stable state.

The design problem is to determine the latch inputs set and reset as functions of the driver's external inputs and the feedback variables.

INTERACTIVE DESIGN APPLICATION

Design of a Positive Edge-Triggered *D* Flip-Flop Using an *SR* Latch

We will design a positive edge-triggered (PET) *D* flip-flop to illustrate the modular design method. As before, the design is accomplished within the framework of the 4-step problem-solving design procedure.

Step 1: Problem Statement. Design a PET *D* flip-flop that has a data input *D*, a clock input *C*, and 2 complementary outputs *Q* and *Q*'. The device is to operate as follows:

- □ When the clock makes a 0-to-1 transition, the present *D* input becomes the next state *Q*
- □ Otherwise, the next state *Q* is the present state *q*

Step 2: Conceptualization. The PET *D* flip-flop is represented by a block diagram model that consists of a fundamental-mode circuit as a driver and an *SR* latch, as shown in Figure 9.61. The driver's outputs are connected to the set and reset inputs of the latch and also fed back to the input side of the driver.

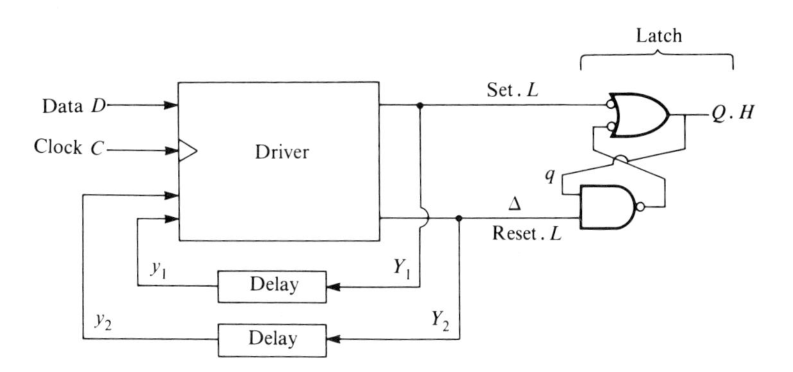

FIGURE 9.61 Block diagram of PET *D* flip-flop derived from an *SR* latch

The operation of the PET *D* flip-flop can be graphically described by a primitive state diagram (Figure 9.62, p. 452) in which each node is a stable state of the fundamental-mode driver. In states **a** and **b**, *q* can be either 0 or 1. The PET *D* flip-flop can change state only on the leading edge of the clock signal, that is, when **a** is the start state and **c** is the end state, or when **b** is the start state and **d** is the end state. If the circuit is in state **f** (with $q = 1$), a 00 input causes a transition to state **a** (still with $q = 1$); a 10 input then causes a transition to state **c** (with $q = 0$), so the flip-flop changes state from 1 to 0.

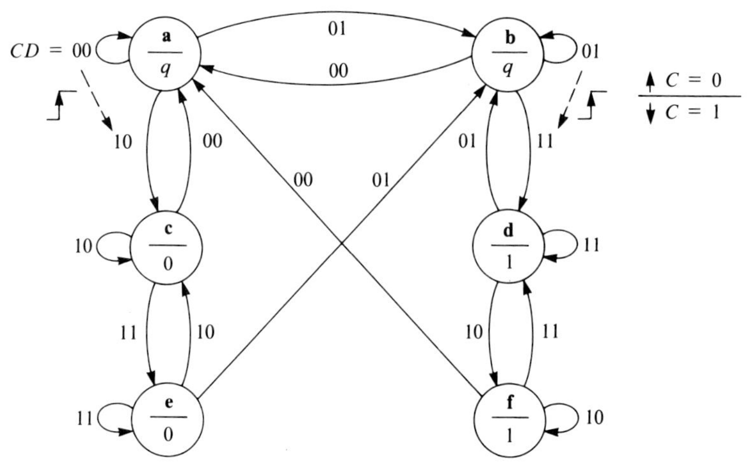

FIGURE 9.62
Primitive state diagram for PET *D* flip-flop

denotes the state changing (0-to-1) clock transition for the PET *D* flip-flop.

A primitive flow table for the PET *D* flip-flop can be constructed from the primitive state diagram, as shown in Table 9.27. Note that the PET *D* flip-flop can change state only on a 0-to-1 clock transition, either path **a**–**c** or **b**–**d**, as indicated by the vectors in Table 9.27.

TABLE 9.27
Primitive flow table for PET *D* flip-flop

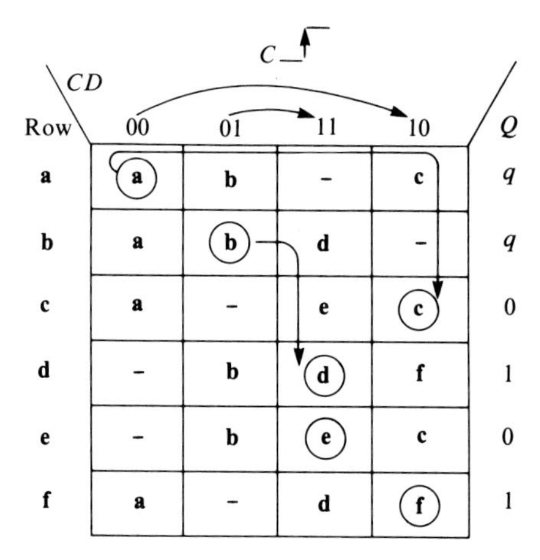

Row \ CD	00	01	11	10	Q
a	(a)	b	-	c	q
b	a	(b)	d	-	q
c	a	-	e	(c)	0
d	-	b	(d)	f	1
e	-	b	(e)	c	0
f	a	-	d	(f)	1

Step 3: Solution/Simplification.

a. State reduction

(1) Eliminate redundant stable states. An examination of Table 9.27 shows that there are no redundant stable states because no two states that are stable for the same input combination have the same output. Therefore, Table 9.27 is a nonredundant flow table.

(2) Eliminate intermediate unstable states by merging. The pairs of mergeable rows in the flow table, which can be determined by inspection of Table 9.27, are

$$\{\mathbf{a}, \mathbf{b}\} \qquad \{\mathbf{c}, \mathbf{e}\} \qquad \{\mathbf{d}, \mathbf{f}\}$$

The merger of these row pairs produces the minimal-row merged flow table shown in Table 9.28. The letters H, R, and S represent hold, reset, and set, respectively.

TABLE 9.28
Minimal-row merged flow table for PET D flip-flop

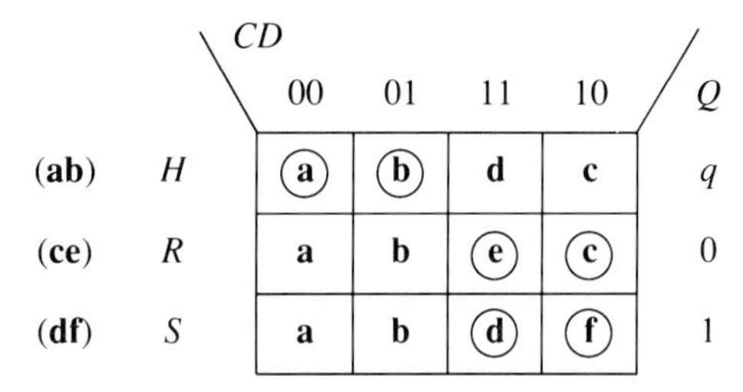

		CD 00	01	11	10	Q
(ab)	H	(a)	(b)	d	c	q
(ce)	R	a	b	(e)	(c)	0
(df)	S	a	b	(d)	(f)	1

The merged flow table can then be written in final form (Table 9.29) by replacing the primitive state symbols by their corresponding merged state symbols.

TABLE 9.29
Final minimal-row flow table for PET D flip-flop

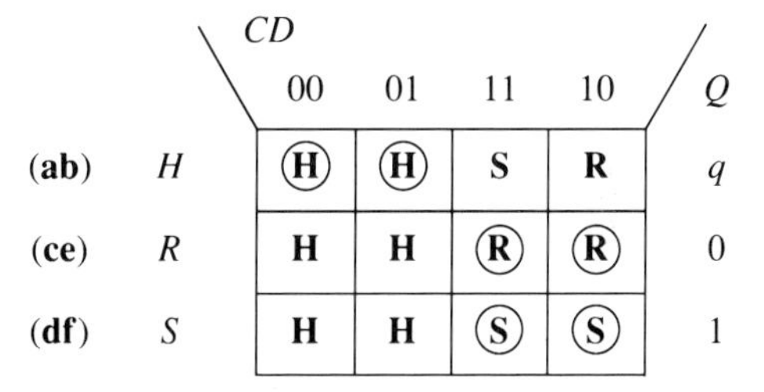

		CD 00	01	11	10	Q
(ab)	H	(H)	(H)	S	R	q
(ce)	R	H	H	(R)	(R)	0
(df)	S	H	H	(S)	(S)	1

b. State assignment. The letters H, R, and S correspond to the hold, reset, and set combinations 00, 01, and 10. Therefore, we assign the binary codes so that $H = 00$, $R = 01$, and $S = 10$ represent the hold, reset, and set functions of the latch, respectively.

Since the state variables Y_1 and Y_2 represent the set and reset controls of the SR latch, an additional state is added when the flow table is converted to an excitation map, as shown in Figure 9.63. Note that Y_1 and Y_2 are complementary when $C = 1$. Therefore, the requirement that $SR = 0$ is satisfied. Since no stable $Y_1Y_2 = 11$ exists in Figure 9.63, $y_1y_2 = 11$ does not occur.

FIGURE 9.63
Excitation map for PET D flip-flop

y_1y_2 \ CD (Y_1Y_2)	00	01	11	10	
00	(00)	(00)	10	01	q
01	00	00	(01)	(01)	0
11	XX	XX	XX	XX	
10	00	00	(10)	(10)	1

A map of transitions (Figure 9.64, p. 454) can be plotted to verify that a race-free state assignment has been accomplished.

FIGURE 9.64
Map of transitions for PET *D* flip-flop

y_1y_2 \ CD	00	01	11	10	q
00			a0	0a	q
01	0b	0b			0
11	XX	XX	XX	XX	
10	b0	b0			1

c. The circuit excitation map can then be separated into individual excitation maps for Y_1 and Y_2, as shown in Figure 9.65.

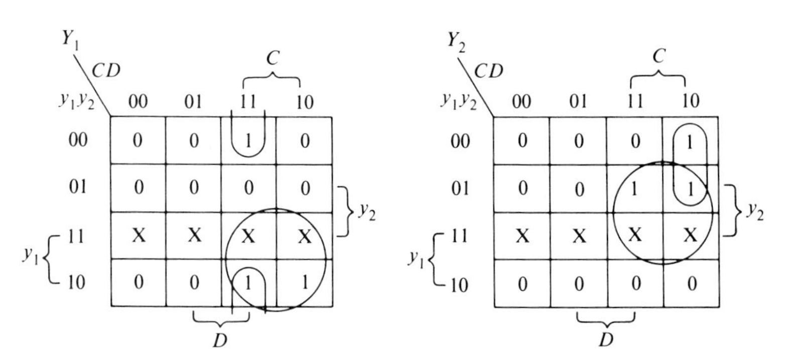

FIGURE 9.65
Excitation maps for PET *D* flip-flop

The excitation equations, as determined from the excitation maps, are

$$\text{Set} = Y_1 = y_1C + y_2'CD = C(y_1 + y_2'D)$$
$$\text{Reset} = Y_2 = y_2C + y_1'CD' = C(y_2 + y_1'D')$$

Therefore, the latch inputs set and reset can be expressed as functions of the external inputs (clock and data) and the feedback variables that are the set and reset inputs of the latch:

$$\text{Set} = \text{clock}(\text{set} + \text{reset}' \cdot \text{data})$$
$$\text{Reset} = \text{clock}(\text{reset} + \text{set}' \cdot \text{data}')$$

The output equation, as determined by the fact that the state variables Y_1 and Y_2 are the set/reset controls of the *SR* latch, is

$$Q.H = Y_1 + Y_2'q = \text{set} + \text{reset}' \cdot q$$

Step 4: Realization. A logic diagram (Figure 9.66) for the PET *D* flip-flop can be constructed directly from the excitation and output equations.

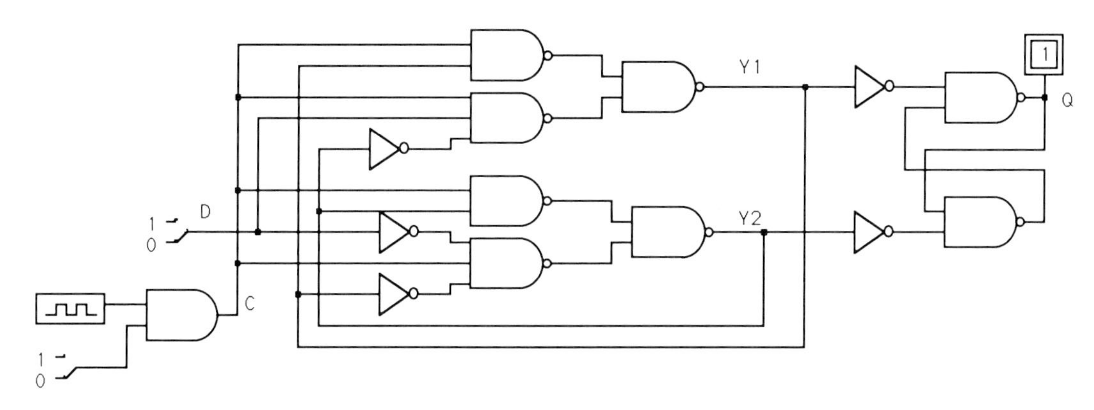

FIGURE 9.66
Logic diagram of PET *D* flip-flop

9.6 SUMMARY

The behavior of an asynchronous sequential-logic circuit, which has no clock signal to synchronize state changes, depends on the order in which its inputs change. Since its inputs may change at any instant of time, changes in the circuit state may occur at any time in response to these input changes. Consequently an asynchronous sequential-logic circuit is subject to a number of problems caused by unequal propagation delays in different signal paths. For example,

- A race condition exists if 2 or more state variables change in response to a change in an input variable. A critical race condition exists if the circuit can reach 2 or more stable states depending on the order in which the state variables change.
- An essential hazard (which can cause the circuit to malfunction) may exist due to unequal propagation delays along 2 or more signal paths from the same external input.
- Oscillation, where a circuit may repeatedly sequence through a number of unstable states without ever reaching a stable state, may occur.

We discussed three methods of identifying different types of hazards. These methods were

- An analysis of the circuit's next-state map to identify a static hazard
- An analysis of an excitation equation to identify a dynamic hazard
- An analysis of the flow table to identify an essential hazard

We then considered methods to eliminate the different types of hazards, such as

- The addition of a consensus term in an excitation equation to eliminate static or dynamic hazards
- The reformulation of an excitation equation to eliminate a dynamic hazard
- The redesign of a circuit to eliminate an essential hazard

Finally, we designed a number of fundamental-mode circuits using a traditional design approach that used gate network propagation delays as memory and a modular design approach that used a basic memory cell (*SR* latch) as a building block.

In this chapter, we modified the basic 4-step problem-solving procedure to help us solve the special problems associated with asynchronous sequential-logic circuits.

Chapter 9 concludes Part III (Sequential Processes and Machines and Sequential Logic Circuits). In Part IV, we will discuss systems-level design.

KEY TERMS

Lumped-delay model
Fundamental-mode circuit
Primary variables
Secondary variable
Excitation equation
Excitation table
Total state
Internal state
Primitive flow table
Hazard
Transient
Static hazard
Dynamic hazard
Essential hazard
Critical race
Merger group
Shared-row method

EXERCISES

General

1. Given the following primitive flow table,

a. List the SS-to-SS paths and construct the adjacency diagram.

b. If any race conditions exist in the flow table, determine if effective row reordering will eliminate race condition(s). If so, make state assignments for a race-free circuit and construct an excitation map.

Row \ x	0	1
a	(a)	b
b	d	(b)
c	a	(c)
d	(d)	c

2. Given the following primitive flow table,

a. Use an implication chart to determine sets of equivalent stable states.

b. Eliminate any redundant stable states and construct a nonredundant primitive flow table.

Row \ x_1x_2	00	01	11	10
a	(a)	b	–	g
b	a	(b)	c	–
c	–	d	(c)	g
d	e	(d)	i	–
e	(e)	h	–	f
f	a	–	i	(f)
g	a	–	i	(g)
h	a	(h)	i	–
i	–	h	(i)	g

3. Given the following nonredundant primitive flow table use an implication chart to determine nonconflicting (compatible) row pairs.

Row \ x_1x_2	00	01	11	10	Z
a	(a)	b	–	c	0
b	a	(b)	d	–	0
c	a	–	d	(c)	0
d	–	f	(d)	e	0
e	a	–	d	(e)	1
f	a	(f)	d	–	1

4. Given the following merged flow table,

a. List the SS-to-SS paths and construct an adjacency diagram.

b. If any race conditions exist in the flow table, determine whether effective row reordering will eliminate race condition(s). If so, construct a reordered-row flow table, make the state assignments for a race-free circuit, and construct a circuit excitation map.

Row \ x_1x_2	00	01	11	10
a	(a)	c	(a)	(a)
b	–	c	(b)	d
c	a	(c)	b	–
d	a	–	b	(d)

5. Given the following merged flow table,

a. List the SS-to-SS paths and construct an adjacency diagram.

b. If the number of required adjacencies > 4, use the following procedure to determine if cycle paths are available:

(1) Underline like unstable states in each column.

(2) Construct a flow table using the stable states, the like unstable states in each row, and the nonunderlined unstable states.

(3) For each SS-to-SS path, determine if a cycle path exists (by replacing deleted unstable state(s) with new unstable state(s)). If so, use a capital letter for the new unstable state.

(4) If unstable states in each row have same letter (ignoring capital and lower case), then sufficient cycles exist since *N* rows contain only *N* next-state rows. Make state assignments and construct a circuit excitation map.

Row \ x_1x_2	00	01	11	10
a	(a)	b	c	(a)
b	(b)	(b)	d	a
c	b	b	(c)	(c)
d	a	(d)	(d)	a

6. Given the following merged flow table,

a. List the SS-to-SS paths and construct an adjacency diagram.

b. If the number of required adjacencies > 4, determine if the (unspecified and specified) unstable states in the flow table can be used to create cycles required for a race-free circuit. Construct a flow table showing the cycle paths, make state assignments for a race-free circuit, and construct a circuit excitation map.

Row \ x_1x_2	00	01	11	10
a	(a)	b	–	c
b	a	(b)	d	(b)
c	(c)	d	(c)	(c)
d	a	(d)	(d)	–

7. Given the following merged flow table,

a. List the SS-to-SS paths and construct an adjacency diagram.

b. If the number of required adjacencies > 4, determine if the (unspecified and specified) unstable states in the flow table can be used to create cycles required for a race-free circuit. If so, construct a flow table showing the cycle paths, make state assignments for a race-free circuit, and construct a circuit excitation map. If not, construct an augmented flow table with additional states *E*, *F*, and *G* for cycles required to satisfy unit-distance assignments. Make state assignments for a race-free circuit and construct a circuit excitation map.

Note: There are no like unstable states in any column.

Row \ x_1x_2	00	01	11	10
a	(a)	(a)	b	c
b	c	(b)	(b)	d
c	(c)	b	(c)	(c)
d	a	(d)	c	(d)

8. Given the following merged flow table,

a. List the SS-to-SS paths and construct an adjacency diagram.

b. If the number of required adjacencies > 4, use the procedure given in Exercise 5 to determine if cycle paths are available.

Row \ x_1x_2	00	01	11	10
a	(a)	b	(a)	d
b	a	(b)	d	–
c	(c)	b	a	(c)
d	c	b	(d)	(d)

9. Given the following minimal-row merged flow table, determine row adjacencies required for a race-free state assignment. Construct a circuit excitation map, using state variables y_1 and y_2.

Row \ CT	00	01	11	10
a	(a)	(a)	d	(a)
b	(b)	(b)	c	(b)
c	b	a	(c)	(c)
d	a	b	(d)	(d)

10. Use the method of plotting a transition map to determine if any race conditions exist in the following next-state map.

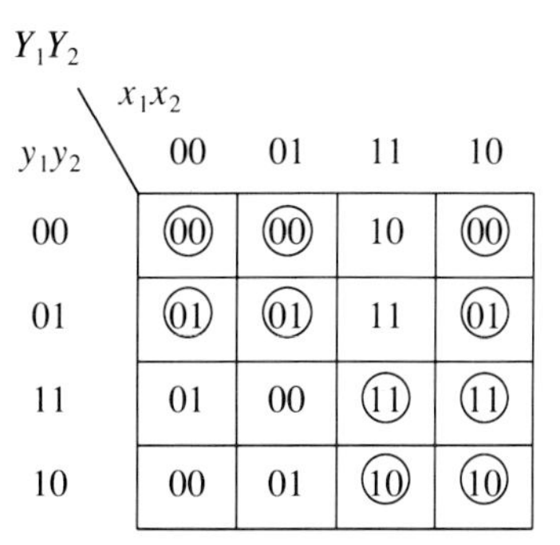

Y_1Y_2

y_1y_2 \ x_1x_2	00	01	11	10
00	(00)	(00)	10	(00)
01	(01)	(01)	11	(01)
11	01	00	(11)	(11)
10	00	01	(10)	(10)

11. Use the method of plotting a transition map to determine if any race conditions exist in the following next-state map.

Y_1Y_2

y_1y_2 \ x_1x_2	00	01	11	10
00	(00)	11	(00)	11
01	11	(01)	10	–
11	(11)	(11)	00	(11)
10	00	(10)	(10)	11

12. Use the method of plotting a transition map to determine if any race conditions exist in the following next-state map.

Y_1Y_2

y_1y_2 \ x_1x_2	00	01	11	10
00	(00)	(00)	10	(00)
01	(01)	(01)	11	(01)
11	01	(11)	(11)	00
10	00	(10)	(10)	01

13. Given the following minimal-row merged flow table, add a fourth state to construct a merged flow table with no race conditions.

Row \ x_1x_2	00	01	11	10	Z
a	(a)	(a)	b	c	0
b	–	a	(b)	c	0
c	a	–	b	(c)	1

14. Construct excitation and output maps for the resulting merged flow table of Exercise 13, using y_1 and y_2 as state variables.

15. Determine the excitation equations and output equation for Exercise 14.

16. Determine the excitation equations for Y_1 and Y_2 as defined by the following excitation maps.

Y_1

y_1y_2 \ x_1x_2	00	01	11	10
00	0	0	1	0
01	0	0	0	0
11	X	0	X	0
10	X	1	1	1

Y_2

y_1y_2 \ x_1x_2	00	01	11	10
00	0	0	0	1
01	0	1	1	1
11	X	1	X	1
10	X	1	0	1

Design/Implementation

1. Design a NET D flip-flop using the modular design approach.
2. Design a pulse generator, whose operation is defined by the following state diagram, using the traditional design approach.

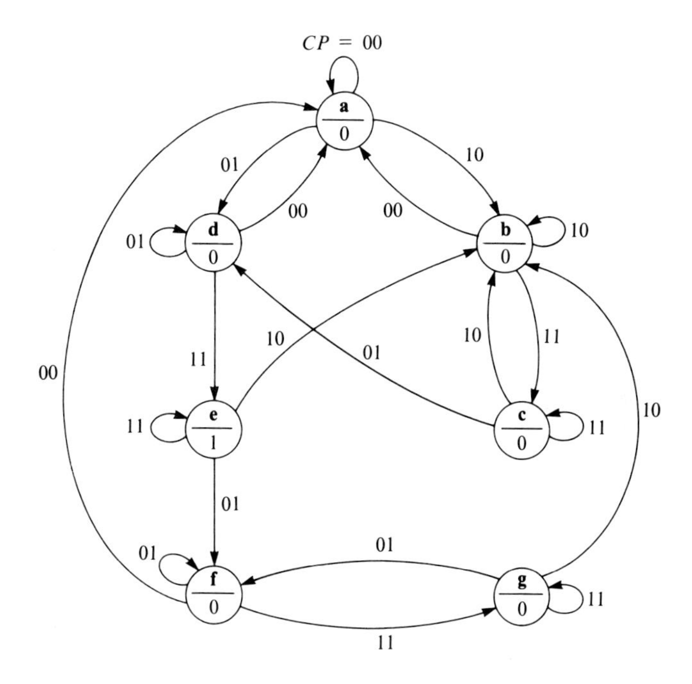

3. Design a PET D flip-flop using the traditional design approach.
4. Design a digital-logic combination lock that has 2 asynchronous inputs x_1x_2 and an output Z. When the input sequence 00, 10, 11, 01 is entered, output Z becomes 1 and the lock opens.
5. Design a digital logic circuit that has 2 asynchronous inputs x_1, x_2 and an output Z. When the input sequence 00, 01, 11, 01, 00, 10 is entered, Z becomes 1.

PART FOUR

Digital Systems Design—An Algorithmic Approach

10 Digital Systems Design 1—State Machines and Systems Design
11 Digital Systems Design 2—Multipliers, Dividers, and Control Units

Part IV (Chapters 10 and 11) uses an algorithmic approach to (1) describe sequential processes and their control, (2) develop process and process-control models, and (3) design digital systems to realize the models.

Chapter 10 presents basic concepts of systems, develops conceptual models and system-level design tools, and describes procedures for the design of digital systems. Chapter 10 uses finite-state machine concepts (developed in Chapter 7) to describe one-dimensional iterative processes such as serial addition, parity checking, and 2s complementation. It then shows how an iterative process can be realized by a digital system (sequential machine) with a *rudimentary* control mechanism. Next, Chapter 10 describes a conceptual model, called an algorithmic state machine (ASM), that is used to represent a sequential process whose control can be described by an algorithm.

Then, Chapter 10 describes a 4-step problem-solving procedure (PSP) for systems design. The serial addition process and its control algorithm, represented by an ASM model, are used to illustrate the use of the 4-step PSP to design a digital system with a *sophisticated* control mechanism.

In systems-level design, registers and register operations are the basic elements and operations of the system. A simple register transfer notation (RTN) is a design tool for describing these register operations. In conjunction with the 4-step PSP, Chapter 10 describes two other systems design tools:

- A system **structure diagram** that partitions the system into a process section and a control section, as shown in Figure A (p. 464).
- A **control flow diagram** of the process control algorithm.

Chapters	Coverage	
13–14	Design of various 8-bit computers	Various 8-bit computers; Computer; Processor; I/O
12	Building-block evolution of a computer	CPU; Memory; Data path; Control unit
11	Digital systems design 2—Algorithmic-process control	Multiplier; Divider; Control unit
10	Digital systems design 1—Algorithmic-process control	Algorithmic state machines
9	Asynchronous sequential logic	Asynchronous sequential logic circuits
8	Synchronous sequential logic	Synchronous sequential logic circuits
7	Finite-state machine concepts: Latches, flip-flops, counters, registers, memory	Counters; Registers; Memory; Latches; Flip-flops
6	Logical and decimal operations	Arithmetic logic unit; BCD arithmetic unit
5	Evolving combinational logic computing devices	2-bit adder; 4-bit adder; Adder/subtracter; Arithmetic unit
4	Combinational-logic function modules	Mux, Demux; Encoder, Decoder; Parity circuits; PLA, PAL, ROM; Shifters
3	Simplifying Boolean expressions	1-bit adder; Error detector; BCD divide; Inequality detector
2	Logic, switching circuits, Boolean algebra, logic gates	Gates: AND OR NOT XOR NAND NOR
		Simple control circuits and computing devices
1	Number systems Computer codes	Decimal, binary, octal, hexadecimal BCD, symmetric, reflected, unit distance

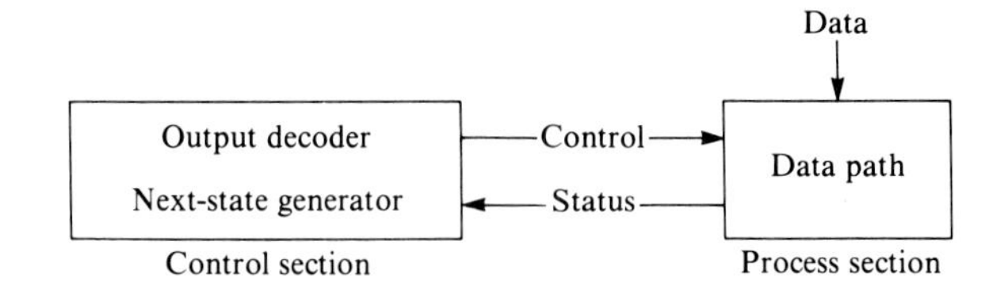

FIGURE A
Structure diagram of a digital logic system

A control flow diagram of the state sequencing defined by a control algorithm may be presented in any one of several conceptual forms. Two types of control flow diagrams are presented in Chapter 10: an **ASM** (algorithmic state machine) **chart** and a **DCAS** (documented control algorithm state) **diagram**.

The latter sections of Chapter 10 describe general processes (such as the game of dice or a blackjack dealer) governed by a set of rules (control algorithm). Such processes can be represented by an ASM model, whose organization and operation can be described using a structure diagram and a control flow diagram, respectively. Chapter 10 uses the 4-step PSP to design ASMs for the game of dice and for a blackjack dealer. The ASMs are then realized as digital logic systems and simulated using a logic simulation program.

Chapter 11 continues the algorithmic approach to systems design using realistic examples of sequential processes such as serial multiplication and division. For example, a process algorithm and process control algorithm for serial multiplication can be developed by:

- Analyzing the serial multiplication process to develop a process algorithm for generating a sequence of partial-product sums.
- Analyzing the control requirements (for accomplishing system initialization, loading of data, and control of the embedded process algorithm) to develop a process control algorithm.

Chapter 11 uses the 4-step PSP (developed in Chapter 10) to design a serial multiplier and a serial divider. Chapter 11 then describes elements of shared-data-path design and combined-control-unit design, common in developing multifunction devices, such as a combined multiplier/divider or a digital computer processor.

Finally, Chapter 11 describes the development of a multifunction processor by combining a serial multiplier and a serial divider. This requires (1) designing a data path that supports each function of the multifunction processor and (2) designing a control unit for the multifunction processor.

10

Digital Systems Design 1—State Machines and Systems Design

OBJECTIVES

After you complete this chapter, you will be able

- To use register transfer notation (RTN) to describe system operations
- To represent iterative processes in alternative forms: information distributed in space and information distributed in time
- To develop process algorithms that accomplish basic operations of a particular application
- To design systems with rudimentary control mechanisms to realize simple sequential processes
- To develop process control algorithms that accomplish system initialization, loading, and processing
- To develop algorithmic state machine (ASM) models to represent processes governed by a set of rules
- To use a problem-solving procedure for designing systems for the automation of processes that can be represented by ASMs
- To realize the ASM by a digital logic system, implemented in hardware or simulated using a logic simulation software program
- To describe systems-level design procedures

10.1 INTRODUCTION

As we discussed in Chapter 1, systems are an integral part of our lives. Also, systems vary in type, size, and complexity. For example, an accounting system may be as simple as a paycheck, a checkbook, and a set of rules for keeping track of our bank balance or as complex as a computerized accounting system that uses a database consisting of a number of files.

In 1854, George Boole wrote *An Investigation of the Laws of Thought*, where he described the elements of Boolean algebra, a mathematical discipline (system) consisting of logic variables, constants (0,1), operations (·, +, '), and rules (axioms, theorems). In 1938, Claude Shannon, a mathematician at Bell Laboratories, applied Boolean algebra to the analysis of relay switching networks.

A system can be represented by a model or an analog of the system. For example, the solar system can be represented by a model composed of a system of partial differential equations, whose solution requires rather complex numerical methods or algorithms. By contrast, a combinational logic network can be represented by a system of logic equations, whose solution can be effected by the relatively simple operations of Boolean algebra.

The study of digital logic encompasses the analysis and design of digital devices and systems, ranging from devices that perform simple logic functions to complex systems such as controllers and digital computers. In this chapter, we present the basic concepts of systems-level design, using a number of different processes for illustration. Finite-state machine (FSM) concepts (introduced in Chapter 7) are used to describe iterative processes and to develop simple sequential machines with rudimentary control mechanisms.

A process that is governed by a set of rules can be represented by an **algorithmic state machine** (ASM) model that, in turn, can be realized by hardware implementation of a digital logic system or by software simulation. For example, the rules for the game of dice can be translated into a control algorithm that, in turn, can be translated into a digital logic controller or into a computer program. The process of rolling the dice can be simulated in software by a random number generator or in hardware by a sequential machine such as a programmable logic sequencer.

ASMs (used to model process control algorithms) together with systems-level design tools are used to develop digital logic systems for a number of representative processes. These systems are then realized by simulation using a logic simulation program.

Registers, Register Operations, and Register Transfer Notation

In systems-level design, registers and register operations are the basic elements and operations of a system.

- An n-bit register $R = R < n - 1, 0 >$ is composed of n 1-bit memory cells, labeled $R_{n-1}, \ldots, R_1, R_0$, where R_0 is the LSB of the register.
- Registers are named using one or more capital letters. A register name may simply be a single letter (A, B, or C), or the name may indicate the function of the register. For example, AC indicates accumulator, MD indicates memory data, and MBR indicates memory buffer register.

- X and Y are used to represent n-bit register inputs, and Z denotes an n-bit register output.
- A simple **register transfer notation** (RTN), summarized in Table 10.1, is used to describe the operations of systems developed in the remainder of the text. The process of copying the contents of register B into register A is represented by the symbology $A \leftarrow B$.

TABLE 10.1
Register transfer notation operations

Function		Symbolic Representation	
Register Transfer Operations			
Copy (transfer) register B contents to register A		$A \leftarrow B$	
Load word X into register A		$A \leftarrow X$	
Reset (clear) register A		$A \leftarrow 0$	
Set (all bits of) register A		$A \leftarrow 1\text{s}$	
Increment a register's content		Inc(A)	
Decrement a register's content		Dec(A)	
Shift Register Operations			
Controls $S_1, S_0 =$	Hold	0, 0	
	Shift right	0, 1	
	Shift left	1, 0	
	Parallel load	1, 1	
Register Arithmetic Operations			
Addition		$C \leftarrow A + B$	
Subtraction		$C \leftarrow A - B$	
Multiplication		$C \leftarrow A \times B$	
Division		$C \leftarrow A/B$	
Register Logical Operations			
AND		$C \leftarrow A \wedge B$	$(C_i = A_i \wedge B_i)$
OR		$C \leftarrow A \vee B$	$(C_i = A_i \vee B_i)$
XOR		$C \leftarrow A \text{ XOR } B$	$(C_i = A_i \text{ XOR } B_i)$
Complement		$C \leftarrow A'$	$(C_i = A_i')$

10.2 ONE-DIMENSIONAL ITERATIVE PROCESSES

The operations of many digital logic systems are managed by sequential logic devices called **controllers.** ***In many cases, a controller (and the controlled process) can be represented by an ASM model. This section begins with a discussion of iterative processes and describes how they can be automated by digital logic systems that have rudimentary control mechanisms. After we present the basic concepts of process control, we will describe how simple iterative processes and general algorithmic processes can be automated by digital logic systems with more sophisticated controllers.***

An **iterative process** is one that exhibits both combinational and sequential characteristics. In an iterative process, a basic (combinational) operation is repeated a number of times (sequentially). Examples of one-dimensional and two-dimensional iterative processes include the addition and the multiplication of multibit numbers, respectively.

An iterative process can be accomplished by alternative combinational logic or sequential logic realizations that involve a trade-off between circuit complexity (space) and operating speed (time). For example, the binary addition of two n-bit numbers is an iterative process that can be accomplished using either a combinational logic adder (with information distributed in space), or a sequential logic adder (with information distributed in time). Other examples of one-dimensional iterative processes include

2s complement of an n-bit number

Parity check of an n-bit number

Comparison of two n-bit numbers

Distribution-in-space and distribution-in-time solutions of several one-dimensional iterative processes are described next.

Information Distributed in Space and Distributed in Time

This section describes the realization of some simple iterative processes using **information distributed in space** and **information distributed in time**. We will use these two approaches to describe, in parallel, alternative realizations of a number of representative processes. Since you are already familiar with the binary addition process, we will use this as our first example.

■ **EXAMPLE 10.1**

Recall from Chapter 5 that a 4-bit (parallel) ripple-carry adder can be implemented by cascading four 1-bit full adders (FA). Thus, the addition of two 4-bit unsigned binary numbers X and Y, and a 1-bit carry-in c_0, can be accomplished by a 4-bit parallel adder, which is realized by a linear cascade of four 1-bit full adders, as shown in Figure 10.1.

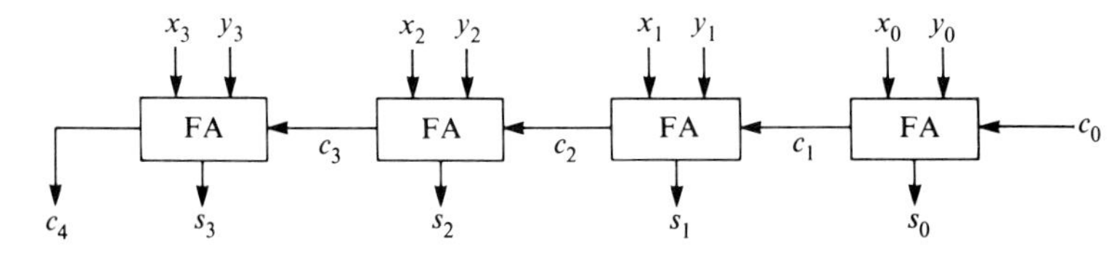

FIGURE 10.1
Binary adder with information distributed in space ■

The 4-bit cascaded adder illustrates the concept of information *distributed in space*, wherein a 4-bit adder is formed by iterating (repeating) a 1-bit FA module four times.

EXAMPLE 10.2

An iterative process can also be represented in a *distributed-in-time* form. In Chapter 7, we described how a functional sequential process can be represented by a model called a finite state machine (FSM). The pencil-and-paper method of adding two multibit numbers is a sequential process referred to as serial addition. We will use this as our example of an iterative process realized in information distributed-in-time form. ■

A one-dimensional iterative process is one in which a basic combinational operation is repeated a number of times. For example, the binary addition of two 4-bit numbers and a carry-in can be accomplished by computing the sum and carry-out for each column ($i = 0, 3$) of the addition as follows:

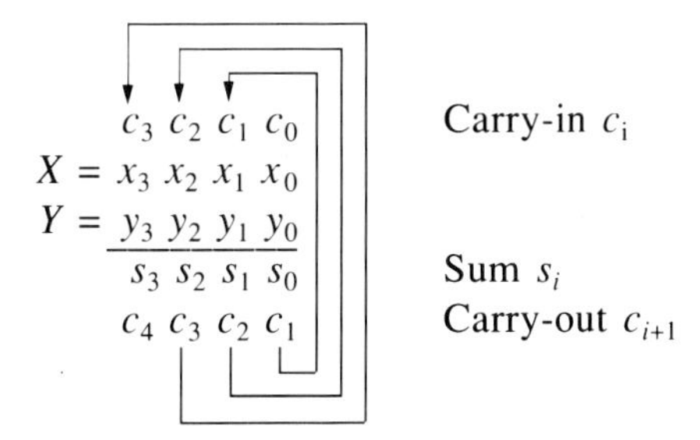

The basic operation, adding c_i, x_i, and y_i to produce a sum s_i and a carry-out c_{i+1}, is repeated for each column i, ($i = 0, 3$), where

$$s_i = c_i \oplus x_i \oplus y_i$$
$$c_{i+1} = x_i y_i + c_i x_i + c_i y_i$$

The basic operation can be accomplished by an FA module that has inputs c_i, x_i, and y_i and outputs s_i and c_{i+1}. The binary addition of two 4-bit numbers can, therefore, be accomplished by a synchronous sequential logic circuit of the form shown in Figure 10.2. This is the binary addition process in distributed-in-time form.

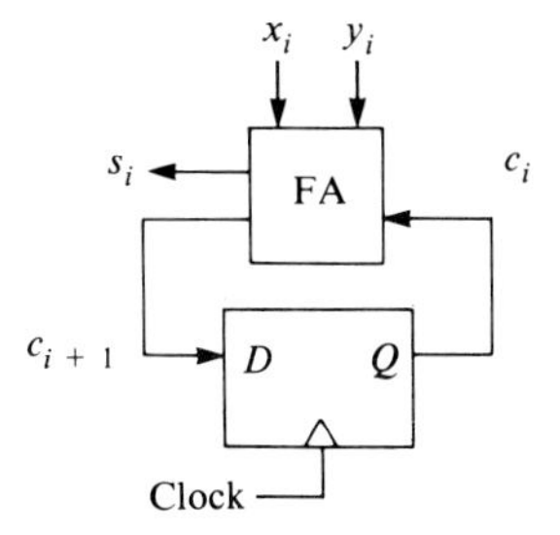

FIGURE 10.2
Binary addition with information distributed in time

The sequential logic adder uses an FA module to add (at time t_i) the column i elements x_i and y_i and present-state carry c_i, producing sum s_i and next-state carry c_{i+1}. When time advances from t_i to t_{i+1}, the next-state carry c_{i+1} is fed back as present-state carry c_i.

In order to study another example of an iterative process in both distributed-in-space and distributed-in-time forms, we will use the 2s complement process for the purpose of illustration. In Examples 10.3 and 10.4, we describe alternative realizations of a 2s complementer, a device that generates the 2s complement of an n-bit binary number.

EXAMPLE 10.3

Recall from Chapter 1 that the 2s complement of a binary number can be found by taking the 1s complement of the number and adding 1 to the result. The 1s complement of Y is obtained by complementing each bit of Y. For example,

$$\begin{array}{rrl} Y = & 00001010 & \\ & 11110101 & = Y' \text{ (1s complement)} \\ + & 1 & \\ \hline & 11110110 & = Y'' \text{ (2s complement)} \end{array}$$

■

The 1s complement y' of a 1-bit number y can be obtained by using an exclusive-OR gate that has a control input w and a data input y. The truth table (Table 10.2) shows that

$$\text{If } w = 0, \quad w \oplus y = y$$
$$\text{If } w = 1, \quad w \oplus y = y'$$

TABLE 10.2
XOR truth table

w	y	$w \oplus y$	
0	0	0	$0 \oplus y = y$
0	1	1	
1	0	1	$1 \oplus y = y'$
1	1	0	

The 2s complement of an 8-bit number Y can be generated by first complementing each bit y_i using an XOR gate with control $w = 1$ to obtain $Y' = y_7'y_6' \cdots y_0'$, and then adding $0 + Y' + 1$, as shown in the following example.

$$\begin{array}{rrl} Y = & 00001010 & \\ & 00000000 & \\ + & 11110101 & = Y' \text{ (1s complement)} \\ + & 1 & \\ \hline = & 11110110 & = Y'' \text{ (2s complement)} \end{array}$$

An 8-bit parallel 2s complementer is a combinational logic circuit that can be realized by iterating eight modules (FA with an XOR gate), as shown in Figure 10.3. The ith XOR gate with control $w = 1$ produces y_i', and the ith FA adds a 1-bit 0, y_i', and a carry-in. The rightmost carry-in equals 1. The parallel 2s complementer is in distributed-in-space form. An XOR gate is denoted by a + enclosed in a circle in Figure 10.3.

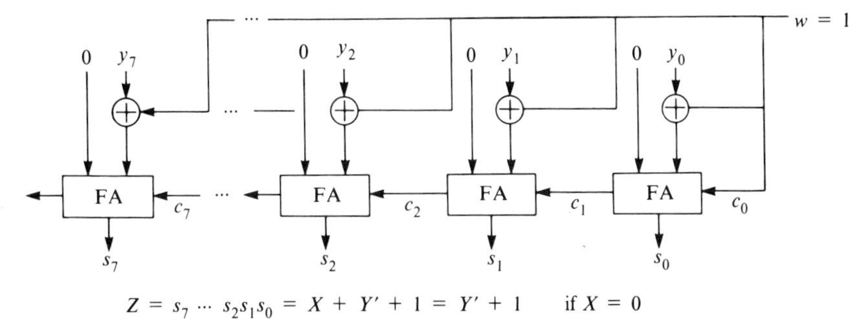

FIGURE 10.3
A 2s complementer in distributed-in-space form

EXAMPLE 10.4

We can see from Figure 10.3 that the parallel 2s complementer is realized by cascading the same module eight times. Each module consists of an FA and an XOR gate with control $w = 1$. A serial 2s complementer can be realized by a synchronous sequential-logic circuit formed by the full adder/XOR gate module and an edge-triggered D flip-flop, as shown in Figure 10.4. Note that the x_i input is 0, and the initial state of the D flip-flop is 1. This form of the 2s complementer is in distributed-in-time form.

FIGURE 10.4
A 2s complement in distributed-in-time form

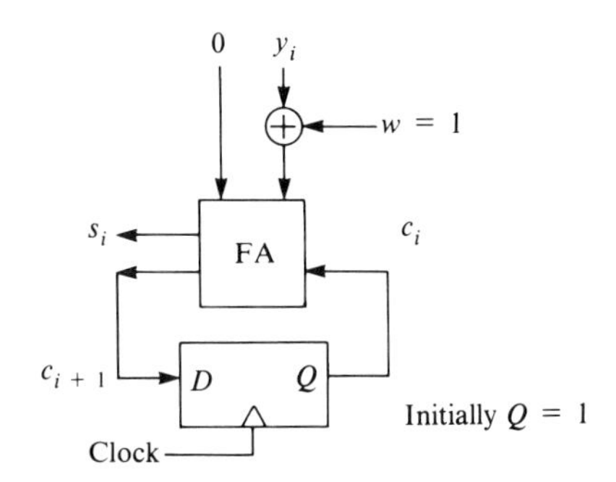

■

The examples in this section have been presented to provide a foundation for systems-level design. In Section 10.3, we will develop more of that foundation by designing a number of rudimentary control sequential machines that realize simple iterative processes. Then, we will be prepared to study systems-level design in Section 10.4.

10.3 SEQUENTIAL MACHINES WITH RUDIMENTARY CONTROL

In this section we will develop sequential machines that realize simple and familiar processes such as serial addition, parity checking, and 2s complementation. By using a simple process, such as serial addition, we can concentrate on the translation of the problem statement into a conceptual model, called a* sequential (state) machine. *A sequential machine is defined by a state diagram and a state table. Then, we can realize the sequential machine using a simple (rudimentary) control mechanism.

Later, after we understand how to develop sequential machines that realize simple processes, we will study more complicated processes. The 4-step problem-solving procedure for design will be used in this chapter to describe the design of sequential machines with a rudimentary control mechanism (Section 10.3) and with a sophisticated control mechanism (Section 10.4).

Design and Implementation of an Even-Parity Check Machine

In Chapter 7, we used the even-parity check process to illustrate state tables and state diagrams. Since we are already familiar with this process, it will be used as our first example to describe the development of a sequential machine with a *rudimentary* control mechanism.

EXAMPLE 10.5

Step 1: Problem Statement. Design an even-parity check machine that has the following specifications:

a. The machine reads an 8-bit word $w = ABCDEFGP$, which contains 7 data bits A, B, C, . . . , G and an even-parity bit P.

b. The machine generates an output function Z for which $Z = 1$ indicates an even-parity error (w has an odd number of 1s) and $Z = 0$ indicates no single error (w has an even number of 1s).

Step 2: Conceptualization. The first step in developing a finite-state machine model of the even-parity check process is to construct a state diagram, as shown in Figure 10.5. The even-parity check process states are represented by E = even number of 1s, and O = odd number of 1s. The initial (reset) state is indicated by a dashed line in the state diagram.

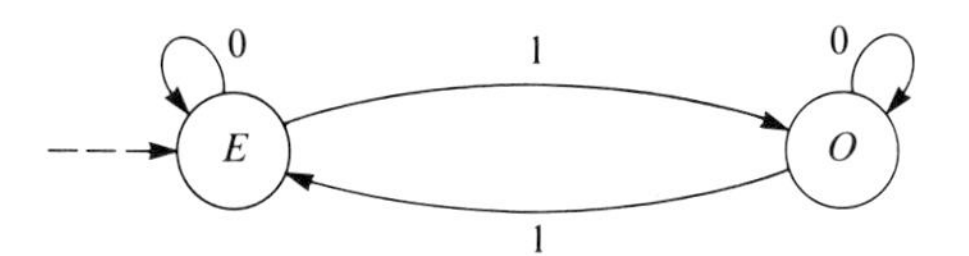

FIGURE 10.5
Even-parity check state diagram

The even-parity check process starts in initial state E (since no inputs have been read, the process has "seen" no 1s). If current state is E and if the next input is 0, then the process remains in state E; if the next input is 1, then the process goes to state O. If current state is O and if the next input is 0, then the process remains in state O; if the next input is 1, then the process goes to state E.

An even-parity check machine can be developed by examining a state table in map form (Table 10.3) in which the next state Y and output Z are tabulated for each possible combination of the input i and the present state y.

TABLE 10.3
Even-parity check state table

Present State y	Next State Y Input 0	Next State Y Input 1	Output Z Input 0	Output Z Input 1
E	E	O	0	1
O	O	E	1	0

Step 3: Solution/Simplification. We can use the state table to construct a next-state map and an output map for the even-parity check process (Figure 10.6).

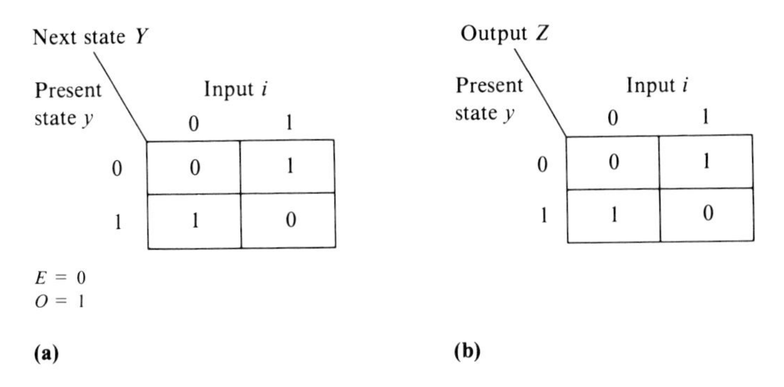

FIGURE 10.6
Even-parity checker. (a) Next-state map (b) Output map

Logic expressions for the next state Y and output Z, as determined from Figure 10.6, are

$$Y = yi' + y'i = y \oplus i$$
$$Z = yi' + y'i = y \oplus i$$

Step 4: Realization. An even-parity check FSM module can be implemented as a synchronous sequential-logic circuit consisting of a clock, a combinational logic network, and an edge-triggered D flip-flop, as shown in Figure 10.7. Note that the next-state function and the output function for the even-parity check process are the same function as shown in this logic diagram.

FIGURE 10.7
Even-parity check FSM module

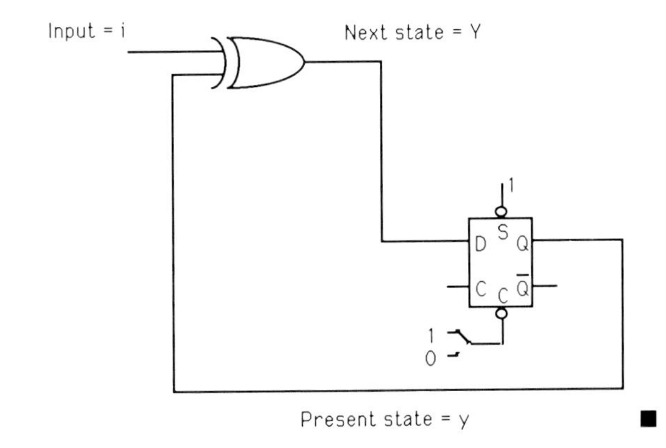

An 8-bit even-parity check machine with a rudimentary control mechanism can be implemented using the following components, as shown in Figure 10.8 (p. 474):

A system clock

Two hexadecimal keypads

An 8-bit parallel-load shift register (74198)

An even-parity check FSM module

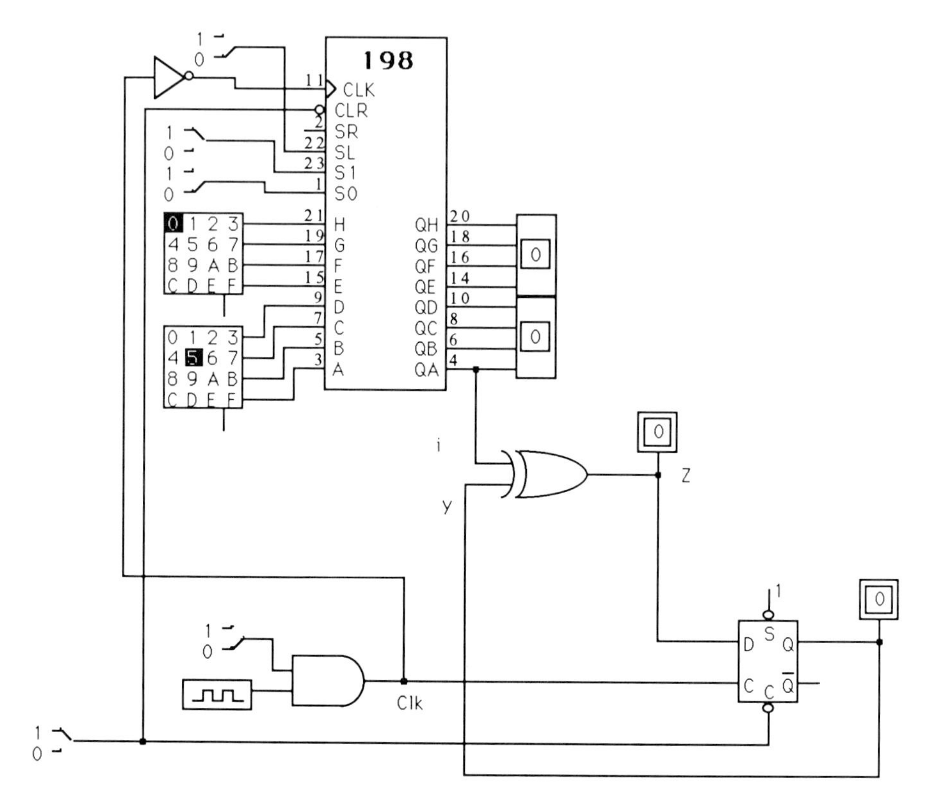

FIGURE 10.8
Complete even-parity check machine

Design and Implementation of a Serial 2s Complementer

Recall that another example of an iterative process is the generation of the 2s complement of an n-bit binary number. Since this is also a simple process, we will use it as our next example (Example 10.6) to illustrate the design of a sequential machine with a rudimentary control mechanism.

EXAMPLE 10.6

Step 1: Problem Statement. Design a 2s complement machine that has the following specifications:

a. The machine reads an 8-bit word $w = ABCDEFGH$, which represents a binary number.
b. The machine generates an output Z, which is the 2s complement of the input word w.

Step 2: Conceptualization. The first step in developing an FSM model of the 2s complement process is to construct a state diagram. In order to develop the state diagram, we need an algorithm for the process. A common algorithm is the copy-complement method for generating each output bit of the 2s complement process. In the copy-complement method, bits are processed from right to left as follows:

- Copy each bit (starting with the LSB) up to and including the first nonzero bit.
- Complement each of the remaining bits

The result is the 2s complement of the original number. The copy-complement method is illustrated as follows:

$w =$	00001010	
	10	Copy each bit including first nonzero bit
	111101	Complement the remaining bits
$Z =$	11110110	= 2s complement of original number w

The operation of the copy-complement method can be graphically described by the state diagram shown in Figure 10.9.

Copy — 0/0 — 1/1 — Comp — 0/1 — 1/0

FIGURE 10.9
Serial 2s complement state diagram

Step 3: Solution/Simplification. Let the copy state be denoted by 0 and the complement state be denoted by 1. The next state Y and output Z can be plotted for each possible combination of the input i and present state y, as shown in Figure 10.10.

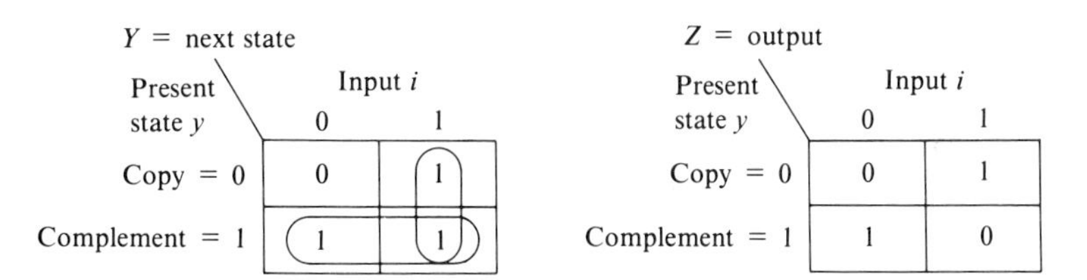

Y = next state
Present state y
Input i
0
1
Copy = 0
0
1
Complement = 1
1
1
Z = output
Present state y
Input i
0
1
Copy = 0
0
1
Complement = 1
1
0

FIGURE 10.10
Serial 2s complement next-state and output maps

We can use the next-state and output maps to determine logic expressions for the next state Y and output Z, as follows:

$$Y = i + y$$
$$Z = iy' + i'y = i \oplus y$$

Step 4: Realization. A 2s complement FSM module can be implemented as a synchronous sequential-logic circuit consisting of a clock, a combinational logic network, and an edge-triggered D flip-flop, as shown in Figure 10.11 (p. 476).

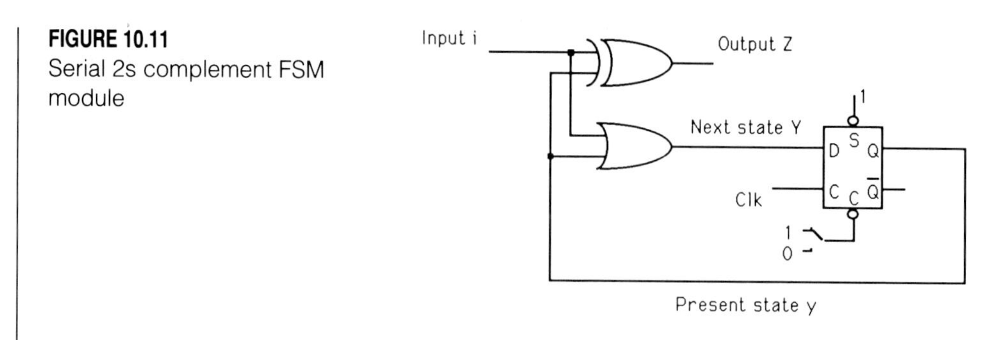

FIGURE 10.11
Serial 2s complement FSM module

A serial 2s complement machine with a rudimentary control mechanism can be implemented using the following components (see Figure 10.12):

A system clock

Two hexadecimal keypads

An 8-bit parallel-load shift register (74198)

A 2s complement FSM module

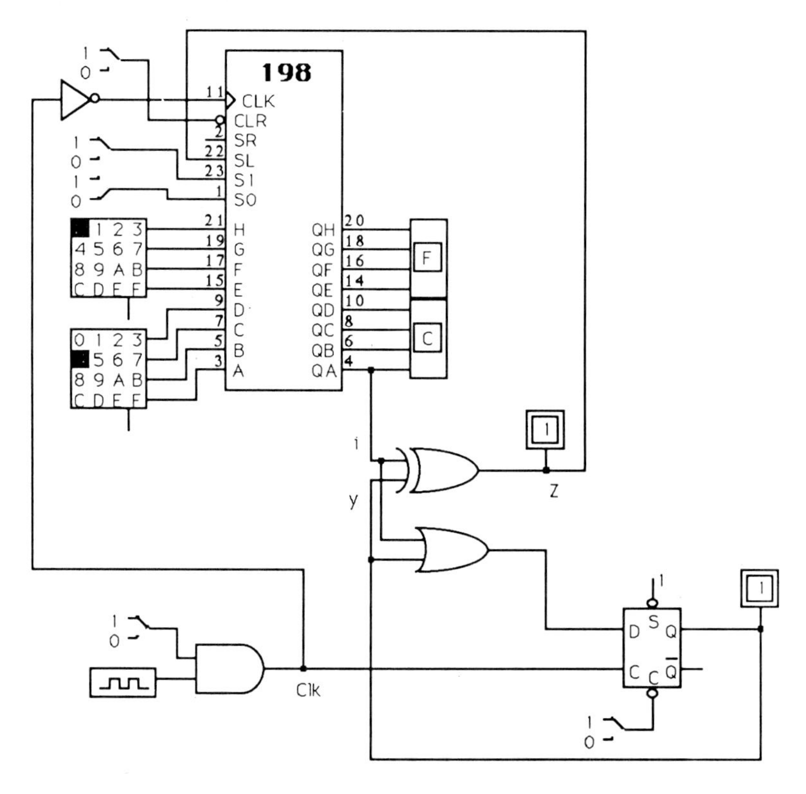

FIGURE 10.12
Serial 2s complement machine

Initially, word w is parallel loaded into the 8-bit shift register. Each bit Z of the 2s complement w'' is shifted into the register as the corresponding bit i of w is shifted out. The result is a serial 2s complement machine.

Design and Implementation of a Serial Addition Machine

We use the process of serial addition as our final example (Example 10.7) to illustrate the design of sequential machines with rudimentary control mechanisms. We saved this as the final example in this section so that we can use the same process in the next section to describe the design of digital logic systems with more sophisticated control mechanisms.

EXAMPLE 10.7

Step 1: Problem Statement. Design a serial addition machine that realizes the serial addition process for adding two unsigned 4-bit numbers $X = x_3x_2x_1x_0$ and $Y = y_3y_2y_1y_0$ and a carry-in c_0, as follows:

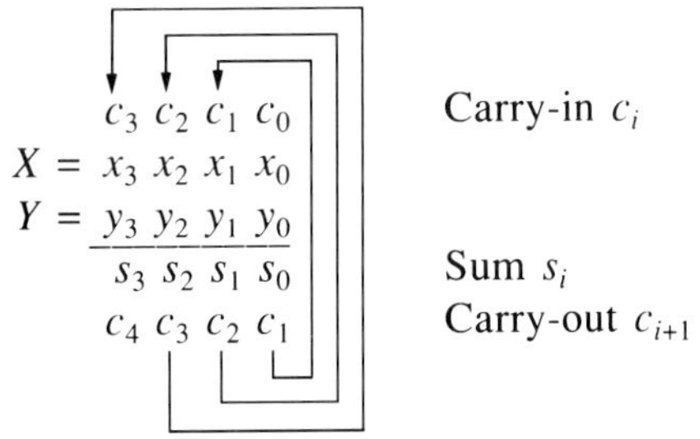

$$\begin{array}{rcccc} & c_3 & c_2 & c_1 & c_0 \\ X = & x_3 & x_2 & x_1 & x_0 \\ Y = & y_3 & y_2 & y_1 & y_0 \\ \hline & s_3 & s_2 & s_1 & s_0 \\ & c_4 & c_3 & c_2 & c_1 \end{array}$$

Carry-in c_i

Sum s_i
Carry-out c_{i+1}

Step 2: Conceptualization. The process of serial addition can be accomplished by computing the sum and carry-out for each column ($i = 0, 3$). The basic operation, adding c_i, x_i, and y_i to produce a sum s_i and a carry-out c_{i+1}, is repeated for each column i, ($i = 0, 3$). In each stage (column) of the process, inputs x and y and the present-state carry c are added to produce a sum S and next-state carry C. The process of serial addition can be defined by a state table (Table 10.4) that lists the next state C and output S for each combination of the present state C and the inputs x and y.

TABLE 10.4
State table for serial addition process

Present State	Next State C				Output S			
	Inputs xy				Inputs xy			
c	00	01	11	10	00	01	11	10
0	0	0	1	0	0	1	0	1
1	0	1	1	1	1	0	1	0

A state diagram for the process of serial addition is presented in Figure 10.13 (p. 478). For any input sequence, the state diagram can be used to determine the output sequence produced by the process (starting in a given state).

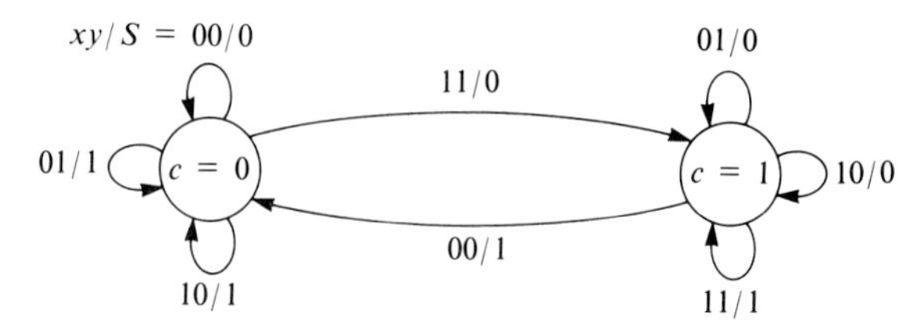

FIGURE 10.13
State diagram for the serial addition process

Step 3: Solution/Simplification. We can use the state table (Table 10.4) to plot the next state C and output S for each combination of the inputs x and y and present state c, as shown in Figure 10.14.

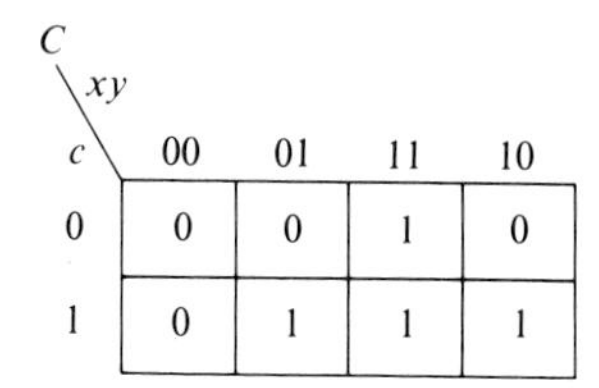

C

c \ xy	00	01	11	10
0	0	0	1	0
1	0	1	1	1

S

c \ xy	00	01	11	10
0	0	1	0	1
1	1	0	1	0

FIGURE 10.14
Serial addition. (a) Next-state map (b) Output map

We can use the next-state and output maps to obtain simplified expressions for C and S as follows:

$$C = xy + c(x + y)$$
$$S = c \oplus x \oplus y$$

Step 4: Realization. A serial-addition FSM module can be implemented as a synchronous sequential-logic circuit consisting of a clock, a combinational logic network, and an edge-triggered D flip-flop, as shown in Figure 10.15.

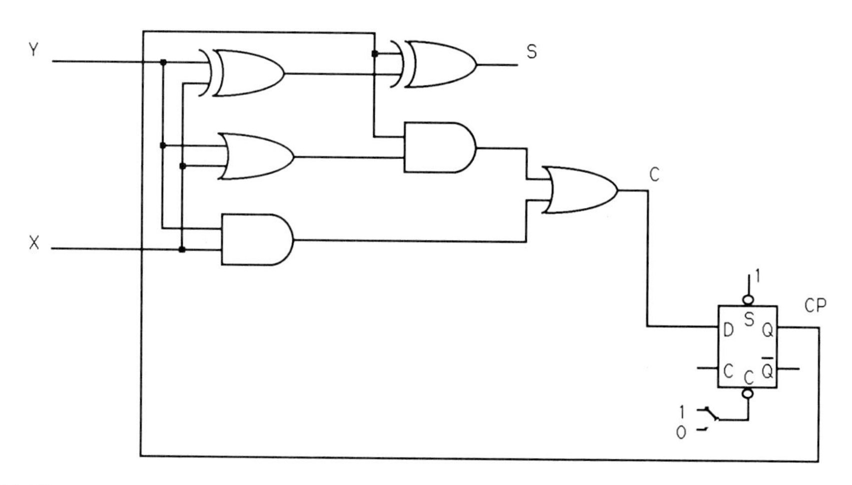

FIGURE 10.15
Serial addition FSM module

A building-block approach can then be used to implement a complete 4-bit serial adder. The following data path components are required.

Two hexadecimal keypads to enter X and Y

Two parallel-load shift registers (74194) (Initially, the X and Y shift registers will be loaded with the 4-bit numbers X and Y that are to be added.)

A serial addition FSM module as illustrated in Figure 10.15 (The FSM module is drawn in expanded form to emphasize that the output and next-state carry are generated by combinational logic circuits.)

A system clock to clock the shift registers and the carry state flip-flop

The components can be integrated into a rudimentary control serial addition machine as follows:

- Connect the outputs of each hexadecimal keypad to the parallel inputs of the corresponding shift registers X and Y.
- Connect the X and Y shift register outputs respectively to the x and y inputs of the serial-addition FSM module.
- Connect the S output of the FSM module to the Y shift register input so that for index i ($i = 0, 1, 2, 3$), s_i is shifted into the Y register and c_{i+1} is clocked into the carry state flip-flop.

Figure 10.16 (p. 480) presents logic and timing diagrams of the rudimentary control serial addition machine. Note in Figure 10.16 that the 74194 shift register has two controls, denoted S_0 and S_1. These controls are used to control the operation of the shift register as follows: $S_1, S_0 = 0, 0$ (hold), 1, 0 (shift left), 0, 1 (shift right), and 1, 1 (parallel load).

We have now completed the implementation of a rudimentary control machine that accomplishes the process of serial addition. In Section 10.4 we will design a more sophisticated mechanism to control the process of serial addition. In order to do this, we need to understand not only the serial addition process but also the process control mechanism of the rudimentary control machine that we have just completed.

The serial addition process produces a sum s_i and next-state carry c_{i+1} using the following process algorithm:

Inputs	$x_i, \ y_i$
Output	$s_i = c_i \oplus x_i \oplus y_i$
Next-state carry	$c_{i+1} = x_i y_i + c_i(x_i + y_i), \quad (i = 0, 3)$

A step-by-step procedure (algorithm) for control of the serial addition process is developed below. This algorithm will be used in Section 10.4 to design a controller (automated process-control mechanism) for a more sophisticated serial addition machine. The algorithm is as follows:

Step **a.** Initialize count (index i). Set the registers' controls to hold. Key X and Y into the respective hexadecimal keypad.

Step **b.** Set the registers' controls to load. Parallel load data into X and Y registers.

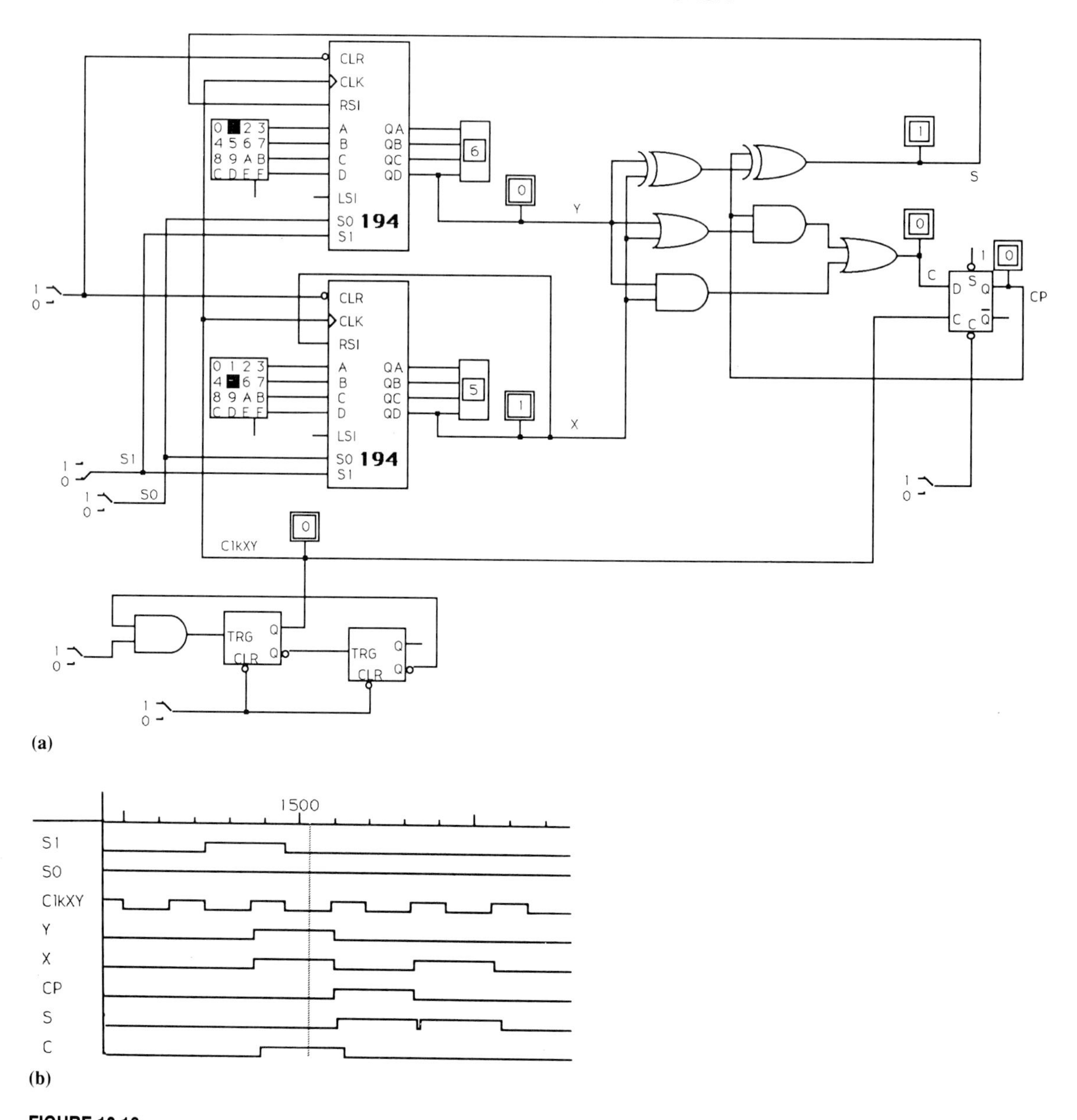

FIGURE 10.16
Rudimentary control serial addition machine. (a) Logic diagram (b) Timing diagram

Step **c**. Use the process algorithm to produce s_i and c_{i+1}. Set the registers' controls to shift right (S_1, $S_0 = 0, 1$). Shift right registers X and Y; increment i. If $i < 4$, then repeat Step c.

For index value i, ($i = 0, 1, 2, 3$), s_i is shifted into Y register and next-state carry c_{i+1} is clocked into the carry state flip-flop.

We see that the control of the serial addition process can be accomplished in three distinct steps. If we let each of these steps represent a state of the process control algorithm, we can rewrite this algorithm in a Pascal-like pseudocode as follows:

State **a**. Initialize index ($i := 0$)
Set registers' controls to hold ($S_1, S_0 := 0, 0$)

State **b**. Set registers' controls to load ($S_1, S_0 := 1, 1$)
Parallel load data into X and Y registers

State **c**. Use process algorithm to produce s_i and c_{i+1}
Set registers' controls to shift right ($S_1, S_0 := 0, 1$)
Shift right X and Y registers
Increment i
If $i < 4$ **then** {**goto c**} **else** {**goto a**}

10.4 ALGORITHMIC STATE MACHINES AND SYSTEMS-LEVEL DESIGN PROCEDURES

An algorithmic state machine (ASM) is a model that can be used to represent a sequential process whose control can be described by an algorithm. The terms ASM and FSM are quite similar. We choose to use the term ASM when referring to a model of a controller, because the name algorithmic state machine is derived from a control algorithm.

The example designs in Section 10.3 described the development of sequential machines with rudimentary control systems. In Example 10.7, we showed how the procedure for operating a control system can be described by a process control algorithm. This algorithm will be used in the first example of this section to design a more sophisticated control system for a serial addition machine.

A 4-step problem-solving procedure (PSP) is also used in this section to design digital logic systems that automate sequential processes. The procedure is as follows:

PSP

Step 1: Problem Statement. Organize the process in two parts (usually in pseudocode) and describe the system specifications.

a. Process (for example, parity check, 2s complement, and so forth).
b. Process control algorithm.

System Specifications. Specify requirements of the system that realize the process. These requirements are usually referred to as design specifications.

Step 2: Conceptualization. Develop a pictorial description of the system using the following systems-level design tools:

a. System structure diagram. This is a block diagram of system organization that partitions system into (1) a process section (data path), and (2) a control section (controller, composed of a next-state generator and an output decoder).

b. Control flow diagram. This is a diagrammatic representation of the operation of the process control algorithm that shows (1) the control states and conditions for state transitions and (2) the generation of signals to control data path components.

Step 3: Solution/Simplification. Develop a detailed description of the following:

a. Process data path. This takes the form of a detailed block diagram or a logic diagram.
b. Control unit.
(1) Next-state generator. This is developed from the process control algorithm next-state maps or tables.
(2) Output decoder. This is developed from the logic equations of data-path control signals.

Step 4: Realization. Develop the implementation of the following:

a. Process data path.
b. Control unit.
(1) Next-state generator.
(2) Output decoder.

The next-state generator can have alternative implementations; for example, it may be implemented using multiplexers, or *D* flip-flops or programmable logic sequencers. The digital system is then realized by integration of the control unit and the process data path. ■

We will devote the next four sections to describing these steps, together with the systems-level design tools. Each procedure step will be illustrated by the corresponding step of the design of an automatic control serial addition machine.

Problem Statement and System Specification

The first step in designing a system to realize a given process is to describe the process and develop a set of specifications for the system to be designed. This section develops the problem statement and design specifications for a serial addition machine.

Step 1: Problem Statement

Process. The addition of two 4-bit numbers $X = x_3x_2x_1x_0$ and $Y = y_3y_2y_1y_0$ and a carry-in c_0 can be accomplished by serial addition using the following **process algorithm**:

$$\left.\begin{array}{ll}\text{Process inputs} & x_i, \quad y_i \\ \text{Process output} & s_i = c_i \oplus (x_i \oplus y_i) \\ \text{Next-state carry} & c_{i+1} = x_i\, y_i + c_i(x_i + y_i)\end{array}\right\} \quad (i = 0, 3)$$

Process Control Algorithm. A procedure for system initialization, starting, loading, and processing [generating sequences s_i and c_{i+1} $(i = 0, 3)$] can be described by the following **process control algorithm**:

State **a**. Initialize index ($i := 0$)
Set registers' controls to hold ($S_1, S_0 := 0, 0$)
Data words X and Y are in hexadecimal keypads

State **b**. Set registers' controls to load ($S_1, S_0 := 1, 1$)
Parallel load data into X and Y registers

State **c**. Use process algorithm to produce s_i and c_{i+1}
Set registers' controls to shift right ($S_1, S_0 := 0, 1$)
Shift right X and Y registers
Increment i
If $i < 4$, **then** {**goto c**} **else** {**goto a**}

Note that a process algorithm is contained within this process control algorithm.

System Specifications. Design an ASM to accomplish the addition of two 4-bit numbers X and Y and a carry-in c_0. The problem can be expressed in RTN as follows:

$$X \leftarrow X + Y + c_0$$

System Conceptualization

The second step in designing a system to realize a given process is to develop a conceptual model of the system. If the process control can be described by an algorithm, we can use an ASM model to represent the process.

An ASM can be organized into two major subsystems or sections.

1. A process section (**data path**) composed of functional components, such as registers, multiplexers, demultiplexers, adders, encoders, and decoders. The data path components are organized to accomplish specified processing, as directed by control signals generated by the controller output decoder.
2. A control section (**controller**) that has two principal components:
   a. A next-state generator (realization of the control algorithm), which uses inputs and control algorithm present state to determine the control algorithm next state.
   b. An output decoder, which generates appropriately sequenced data-path control signals, at proper times, to cause the data path components to accomplish specified process functions.

Structure Diagrams and Control Flow Diagrams

Thus, the organization of an ASM system may be conceptualized using a structure diagram, which is organized into a process section (data path) and a control section (controller, consisting of a **next-state generator** and an **output decoder**), as shown in Figure 10.17 (p. 484).

While the organization of the system under consideration can be depicted by a structure diagram, the operation of the process control algorithm is graphically described by a control flow diagram. A control flow diagram shows the control states and conditions for state transitions, and the generation of control signals to data path components.

FIGURE 10.17
Structure diagram of an ASM

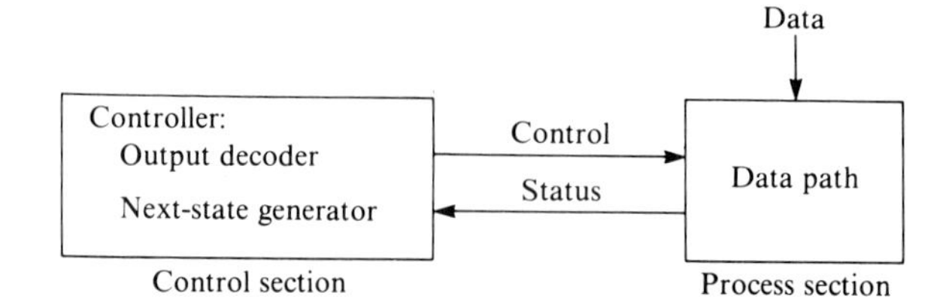

The control algorithm for an ASM defines all possible sequences of states of the machine; each next state of the machine is determined by the control algorithm, its present state, and the inputs at that state of the algorithm. A diagram of the state sequencing defined by a control algorithm may be presented in any one of several conceptual forms. Two types of control flow diagrams are presented in this text:

An ASM (algorithmic state machine) chart

A DCAS (documented control algorithm state) diagram

An **ASM chart** has the appearance of a standard flowchart used in computer programming. A **DCAS diagram** has the appearance of a state diagram in which the control algorithm state transitions are documented with the input conditions that cause each transition. The control algorithm language constructs for ASM charts and DCAS diagrams are defined in Table 10.5.

TABLE 10.5
Definition of constructs for ASM charts and DCAS diagrams

ASM Chart	Construct	DCAS Diagram
Rectangle	State	Circle
Diamond	Conditional branch	Half diamond
Oval	Conditional output	RTN (condition)
- - - -	Handshake	Linked circle pair

Examples of each construct are presented in Figure 10.18 and described in the next subsection.

A control algorithm for a sequential process is itself sequential in nature and is characterized by its state at any given time. The control algorithm next state is determined by its present state and its input(s) at that time.

ASM and DCAS Control-Flow Diagram Constructs

State. A control algorithm *state*, represented by a rectangle in ASM charts and a circle in DCAS diagrams has

An alphabetic state label

A binary state code

An unconditional (immediate) output represented by an action statement

The control algorithm makes a transition from its present state either

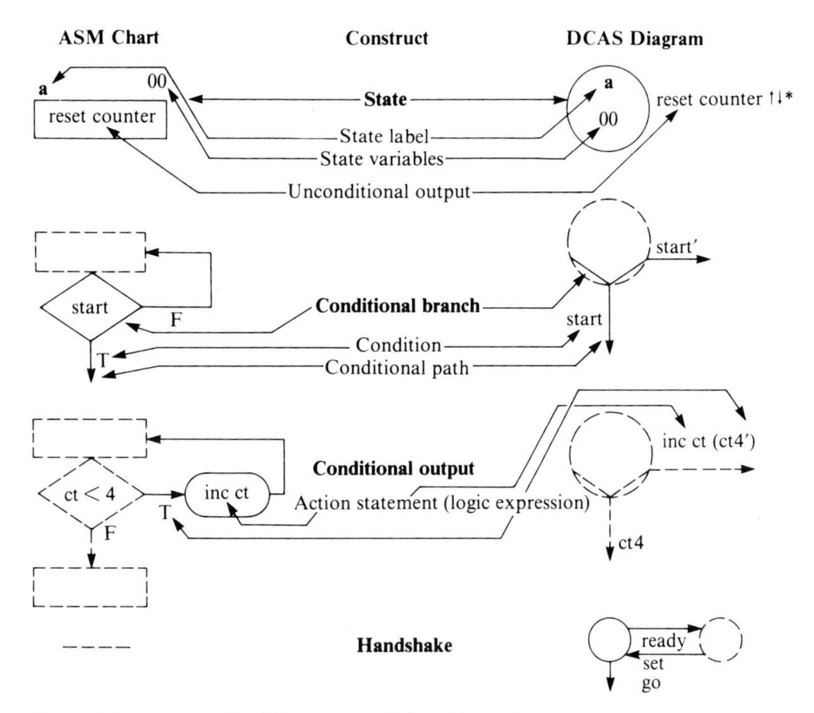

FIGURE 10.18
Examples of constructs in ASM charts and DCAS diagrams

Unconditionally to a next state of the control algorithm or

Conditionally to a next state, depending on the input(s) to the algorithm in its present state

Conditional Branch. A control algorithm *conditional branch* is represented by a diamond in ASM charts and a half diamond in the lower part of the state circle in DCAS diagrams. Each branch represents a conditional path from the algorithm's present state to a next state. Each branch is labeled with a logic expression denoting the input condition that causes that present-state/next-state transition.

Conditional Output. If a control algorithm state has an output associated with a conditional branch, that output is a *conditional* (input dependent) *output*. In an ASM chart a conditional output is represented by an oval containing an action statement. In a DCAS diagram a conditional output appears on its associated conditional branch and is denoted by an action statement with a logic expression in parentheses. The parenthetical logic expression indicates the condition under which the conditional output is asserted.

Handshake. A *handshake* (request/acknowledge) between systems or subsystems is accomplished by an "assert output" (request ready) and "wait for response" (acknowledge set). A request/acknowledge can be used with either unconditional or conditional outputs.

Note that for DCAS diagrams, assertion of an output is indicated by ↑ and de-assertion of an output is indicated by ↓. If a state has an unconditional output, that output is asserted on entry to the state and de-asserted on exit from the state. If a state has a conditional output, that output is usually asserted on the alternate clock transition and de-asserted on exit from the state. That is, if state variable changes occur on the 0-to-1 transition of the clock, conditional outputs occur on (or after) the 1-to-0 transition of the clock.

Also note the following exceptions for DCAS diagrams.

- Asynchronous inputs are not allowed in a DCAS diagram. If a system has an asynchronous input (denoted by *), that input, IN*, can be converted to a synchronous input (denoted IN·sync) by using IN* as input to a *D* flip-flop clocked by the alternate transition of the system clock; the flip-flop output would then be IN·sync.
- Voltage level assignments are not used in a DCAS diagram. The logic expressions used in a DCAS diagram represent input (logic) conditions and are used as map-entered variables in system next-state maps and tables.

Step 2: Conceptualization

Structure Diagram. The organization of an ASM is described by a structure diagram. The serial addition ASM consists of a controller and a process data path, as shown in Figure 10.19. Note that the controller has two principal components: a next-state generator and an output decoder.

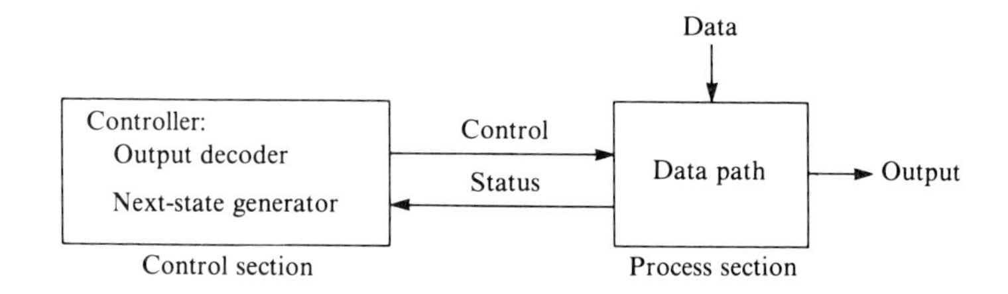

FIGURE 10.19
Structure diagram of a serial addition ASM

Control Flow Diagram. The control algorithm for an ASM defines all possible sequences of states of the machine; each next state of the machine is determined by the control algorithm, its present state and the inputs at that state of the algorithm. The operation of the process control algorithm is graphically described by a control flow diagram that shows the control states and conditions for state transitions and the generation of control signals to data path components.

The process control algorithm for the serial addition process (repeated below for convenience) is translated into a control flow diagram, which can be in the form of an ASM chart or a DCAS diagram, as shown in Figure 10.20. A system clock is used to synchronize the serial addition ASM. Signal eclk is an inversion of the system clock signal. That is, when signal clock makes a 0-to-1 transition, signal eclk makes a 1-to-0 transition. Note that the counter is incremented and the *X* and *Y* registers are loaded and shifted on the rising edge of signal eclk (trailing edge of signal clock).

State **a**. Initialize index (i := 0)
Set registers' controls to hold (S_1, S_0 := 0, 0)
Data words X and Y are in hexadecimal keypads

State **b**. Set registers' controls to load (S_1, S_0 := 1, 1)
Parallel load data into X and Y registers

State **c**. Use process algorithm to produce s_i and c_{i+1}
Set registers' controls to shift right (S_1, S_0 := 0, 1)
Shift right X and Y registers
Increment i
If $i < 4$, then {goto c} else {goto a}

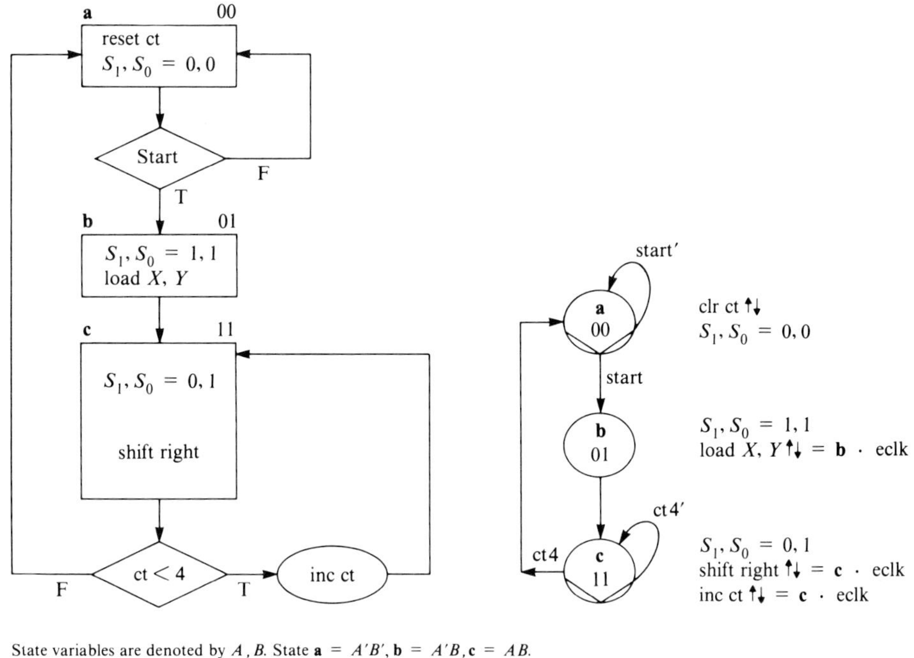

FIGURE 10.20
Control flow diagrams for a serial addition ASM. (a) ASM chart (b) DCAS diagram

System Design Solution and Simplification

The third step in designing a system to realize a given process is to develop a detailed description or a block diagram of the process data path, a table or map representation of the next-state generator operation, and logic equations for data-path control signals generated by the output decoder.

After analyzing the processing required by the ASM and developing a detailed description or a block diagram of the process data path, alternative representations of the next-state generator can be derived from the control flow diagram. The machine implementation form (using muxes or flip-flops) that you plan to use in Step 4 (Realization) dictates which representation you should use for the next-state generator. That is, a next-state table facilitates implementation using muxes, and a next-state map facilitates implementation using *D* flip-flops.

Finally, logic equations for the data-path control signals (generated by the output decoder) can be derived from the control flow diagram.

Step 3: Solution/Simplification

Process Data Path. The process data path is designed to accomplish the following process algorithm:

$$\left.\begin{array}{ll} \text{Process inputs} & x_i,\ \ y_i \\ \text{Process output} & s_i = c_i \oplus (x_i \oplus y_i) \\ \text{Next-state carry} & c_{i+1} = x_i\, y_i + c_i(x_i + y_i) \end{array}\right\} \quad (i = 0,\ 3)$$

The data path shown in the block diagram in Figure 10.15 (Section 10.3) satisfies the process requirements in this example.

Control Unit Next-State Generator. The DCAS diagram (Figure 10.20) can be used to tabulate the control algorithm next states for each present state, together with the control conditions for each present-state/next-state transition, as shown in Table 10.6. Note that an unused state, state **d**, must make a transition to the system reset state.

TABLE 10.6
Serial adder control algorithm table

Present State		Next State		Condition for Transition	Condition for	
AB	Name	Name	A^+B^+		$A^+ = 1$	$B^+ = 1$
00	**a**	**a**	00	start'	F	start
		b	01	start		
01	**b**	**c**	11	T	T	T
10	**d**	**a**	00	T	F	F
11	**c**	**c**	11	ct4'	ct4'	ct4'
		a	00	ct4		

As an alternative representation, for each present state *AB* of the serial adder control algorithm, the next state A^+B^+ can be determined using a next-state map in which the state variables *A* and *B* are map variables and the control conditions are map-entered variables. The circuit next-state map, derived from the control algorithm for the serial addition process, is shown in Figure 10.21a. The individual next-state maps for each state variable are shown in Figure 10.21b.

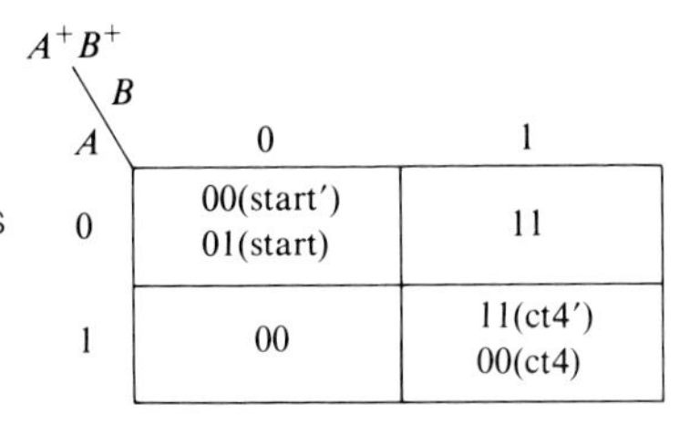

(a)

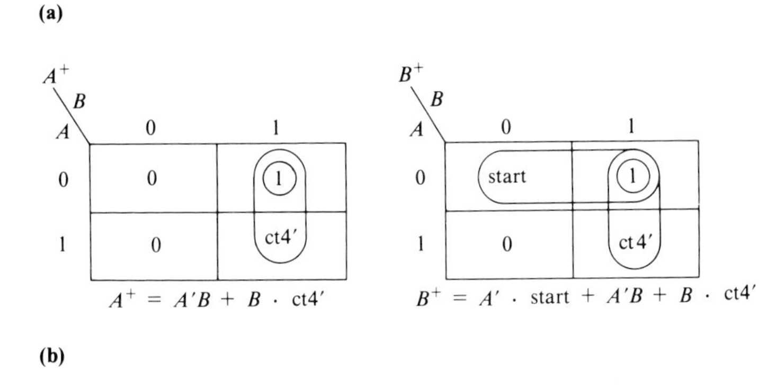

(b)

FIGURE 10.21
Serial adder control algorithm. (a) Next-state map (b) Next-state maps for individual state variables

Control Unit Output Decoder. Logic equations for the data-path control signals (generated by the output decoder), derived from the control flow diagram, are as follows:

$$S_1 = A'B$$
$$S_0 = B$$
$$\text{clr ctr} = A'B'$$
$$\text{clk-}X = (A'B + AB) \cdot \text{eclk} = B \cdot \text{eclk}$$
$$\text{clk-}Y = (A'B + AB) \cdot \text{eclk} = B \cdot \text{eclk}$$
$$\text{clk-}C = AB \cdot \text{eclk}$$
$$\text{inc ct} = AB \cdot \text{eclk}$$

Signal eclk is an inversion of the system clock signal. When signal clock makes a 0-to-1 transition, signal eclk makes a 1-to-0 transition, and vice versa. The state variable changes occur on the rising edge of signal clock. The counter is incremented and the X and Y registers and carry flip-flop C are strobed on the rising edge of signal eclk.

System Realization

The fourth step in designing a system to realize a given process is to realize the ASM system. The ASM is realized by a digital logic system. This system may be implemented in hardware or simulated using a logic simulation program.

After the system design has been completed, the machine can be implemented as follows:

Implement the process data path

Implement the output decoder

Use alternative implementations of the next-state generator; that is,

- Use multiplexers
- Use D flip-flops
- Use other implementations (for example, programmable logic sequencers)

Step 4: Realization

Process Data Path. The block diagram of Figure 10.19 can be used to construct a detailed logic diagram of the process data path, as shown in Figure 10.22.

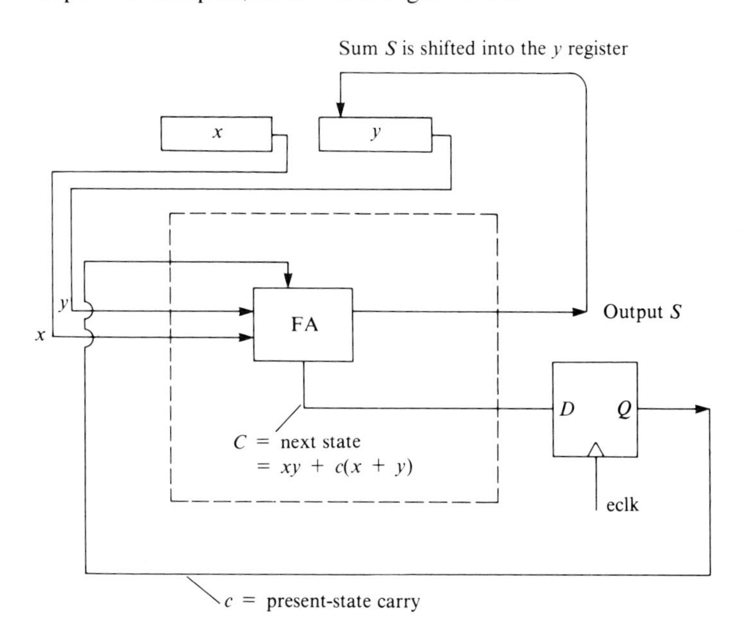

FIGURE 10.22
Logic diagram of the process data path

The process data path can be implemented using the following components:

Two hexadecimal keypads and a switch to enter X, Y, and c_0

Two parallel-load shift registers X and Y

An edge-triggered D flip-flop to store the present-state carry

A serial-addition FSM module to accomplish the process algorithm

At the completion of the addition, the 5-bit sum (X plus Y plus c_0) resides in the carry flip-flop (c_4) and the Y register ($s_3s_2s_1s_0$).

Control Unit Next-State Generator. First, we will implement the next-state generator using multiplexers. The serial adder control algorithm can be implemented directly from the control algorithm table using 4-to-1 multiplexers, where present state AB is used to select the input condition corresponding to the select value AB combination, as shown in Figure 10.23.

An alternative implementation uses D flip-flops. If the next-state generator is to be implemented using D flip-flops, the individual next-state maps shown in Figure 10.21b can be used to determine the D flip-flop excitation equations as follows:

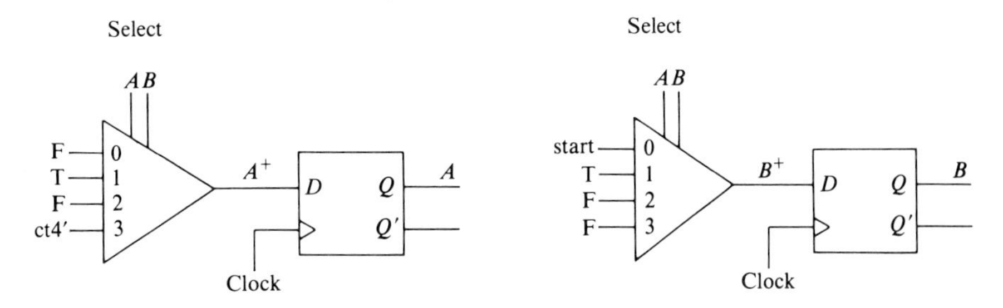

FIGURE 10.23
Multiplexer implementation of next-state generator

$$D_A = A'B + B \cdot \text{ct4}' \qquad D_B = A'B + A'\text{start} + B \cdot \text{ct4}'$$

Control Unit Output Decoder. The logic equations for the data-path control signals (generated by the output decoder) and derived in Step 3 are used to construct a logic diagram of the output decoder. Figure 10.24 presents a logic diagram of the controller output decoder and next-state generator (implemented using D flip-flops).

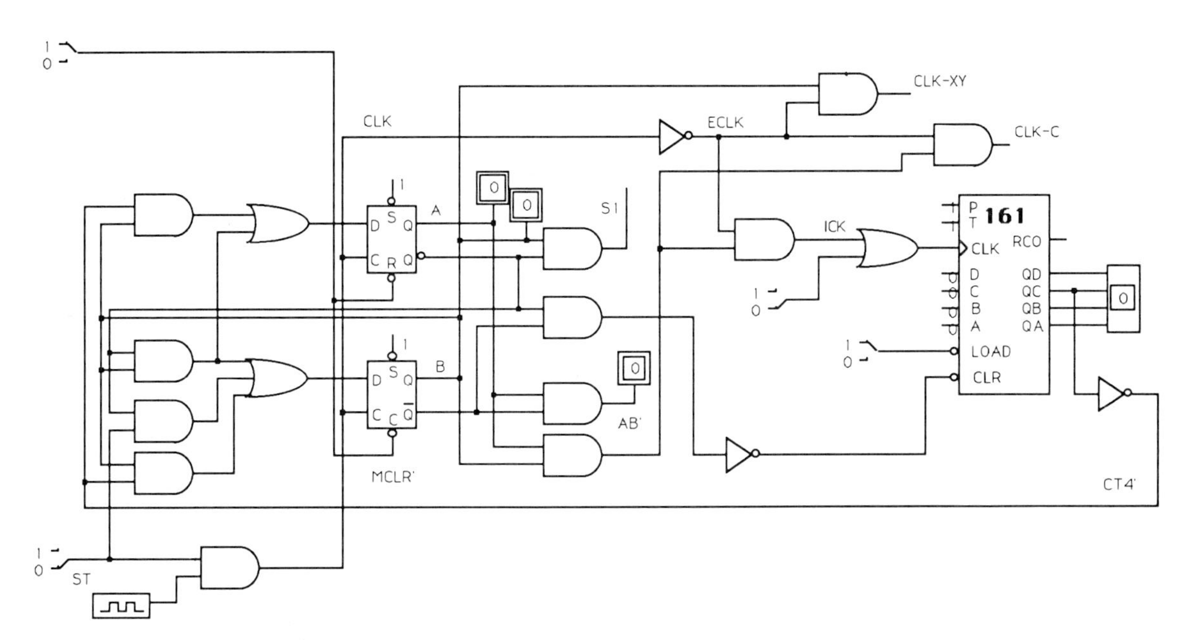

FIGURE 10.24
Controller output decoder and next-state generator logic diagram

Logic and timing diagrams of the complete serial addition ASM are presented in Figure 10.25 (p. 492).

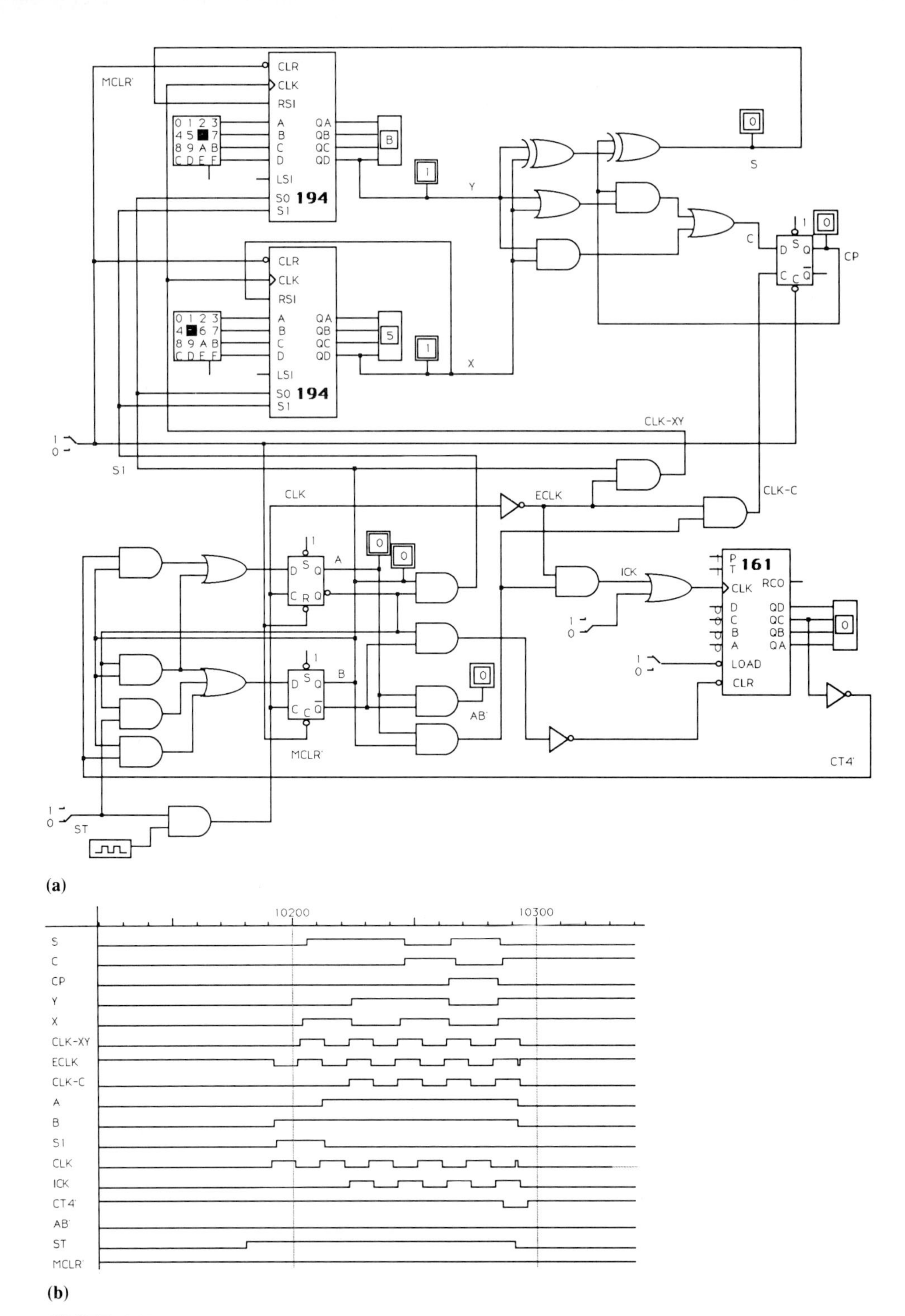

FIGURE 10.25
Complete serial addition ASM. (a) Logic diagram (b) Timing diagram

INTERACTIVE DESIGN APPLICATION

System Design of a Dice Game ASM

Now that we understand how to accomplish each step of a system design using systems-level tools, we will use the 4-step problem-solving procedure to design a complete system. This section describes the system design of a dice game ASM.

The *pyramid-dice game* is a simplification of the game of dice played with two 6-faced dice. This game is played using 2 pyramid-shaped dice with 4 faces, numbered 0, 1, 2, and 3. A player rolls the 2 dice, labeled X and Y. The outcome ($R = X + Y$) of a roll is one of 16 equally likely outcomes, as shown in Figure 10.26.

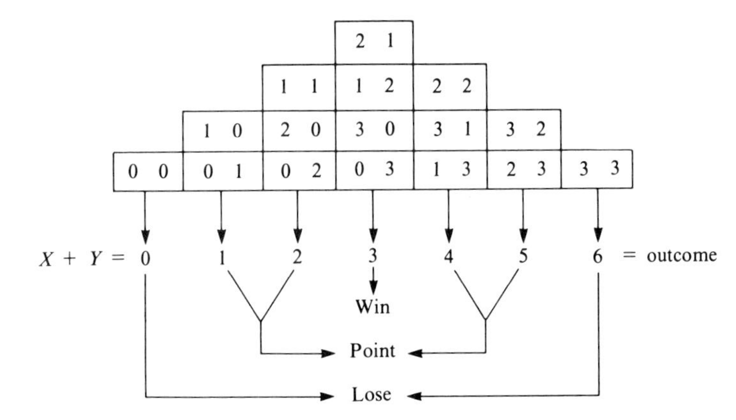

FIGURE 10.26
Possible outcomes of 2 pyramid dice

The player who is rolling the dice in a given game is referred to as the shooter. A game, consisting of a comeout (initial) roll and one or more point rolls, is governed by the following rules:

Step **a**. New shooter. Accepts dice and starts his first game
Initialize accumulated winnings ($T = 0$)

Step **b**. New game
Place bet B*
Comeout roll. In the comeout roll, a shooter can either
Win game (keep dice and go to **b**) if comeout roll $R = 3$
Lose game (lose dice and go to **a**) if comeout roll $R = 0$ or 6
Establish a point P (and go to **c**) if comeout roll $R = 1, 2, 4,$ or 5

Step **c**. Save point. Save comeout roll ($R = 1, 2, 4,$ or 5) as point and go to **d**.

Step **d**. Point roll. In a point roll, a shooter can either
Win game (keep dice and go to **b**) if point roll $R = P$
Lose game (lose dice and go to **a**) if point roll $R = 3$
Go to **d** (repeat point roll) if point roll R is not P or 3

*When the shooter wins a game, bet B is added to accumulated winnings T. When the shooter loses a game, bet B is subtracted from accumulated winnings T.

Now that you understand the rules governing the game of pyramid dice, we will use the 4-step problem-solving procedure to design an ASM that simulates this game. The rules of the game are translated into a process-control algorithm in Step 1. Each step (**a**, **b**, **c**, or **d**) becomes a state, with the same label, in the control algorithm. Comments are indicated by (* comment *).

Step 1: Problem Statement.

a. Process. The game of pyramid dice is played by 2 or more people using a pair of dice. Each die X (Y) has 4 faces labeled 0, 1, 2, 3. The outcome of a roll of the dice is $R = X + Y$, (* $0 \le R \le 6$ *).

b. Process control algorithm. The rules that govern the game of pyramid dice can be translated into an algorithm that has four distinct steps or states, denoted by **a**, **b**, **c**, and **d**.

State **a**. New shooter
    Accept dice and start first game
    Initialize accumulated winnings ($T = 0$)

State **b**. New game
    Place bet B
    Comeout roll. Shooter rolls dice with roll R ($0 \le R \le 6$)
      If $R = 3$, **then** {(* win *); $T := T + B$; **goto b**}
        else if ($R = 0$ or $R = 6$) **then** {(* lose *); $T := T - B$; **goto a**}
            else {(* save point *); **goto c**}

State **c**. Save point. (* If shooter does not win or lose on the comeout roll, the comeout roll R is used as a target "point" for the shooter *) Save comeout roll R as point P.

State **d**. Point roll. Shooter rolls dice with roll R (* $0 \le R \le 6$ *)
      If $R = P$ **then** {(* win on pt *); $T := T + B$; **goto b**}
        else if $R = 3$ **then** {(* lose on pt *); $T := T - B$; **goto a**}
            else {**goto d**}

System specifications. Design an ASM that simulates the game of pyramid dice.

Step 2: Conceptualization.

a. Structure diagram. Since the game of dice is an algorithmic process, it can be represented by an ASM. This ASM can be partitioned into a control section (controller) and a process section (data path). A structure diagram of the dice game ASM is presented in Figure 10.27.

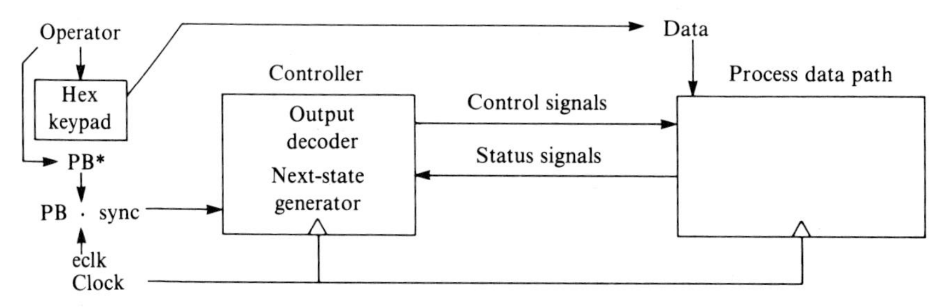

FIGURE 10.27
Structure diagram of dice game machine

Man–Machine Interface for the Dice Game ASM. A debounced pushbutton is used by the operator to signal that a roll ($R = X + Y$) of the dice has been entered into the hexadecimal keypad. (The * in Figure 10.28 denotes that output PB* is an asynchronous input to the controller.) The signal PB* is input to a D flip-flop, which is clocked by eclk (the inversion of the system clock); the flip-flop output PBS (denoting PB · sync) provides a synchronous input to the controller. Signal PBS is input to a second D flip-flop, which is also clocked by signal eclk, and has output PBD (denoting PB · delayed).

The debounced pushbutton PB* and the synchronizing and delaying flip-flops are shown in Figure 10.28.

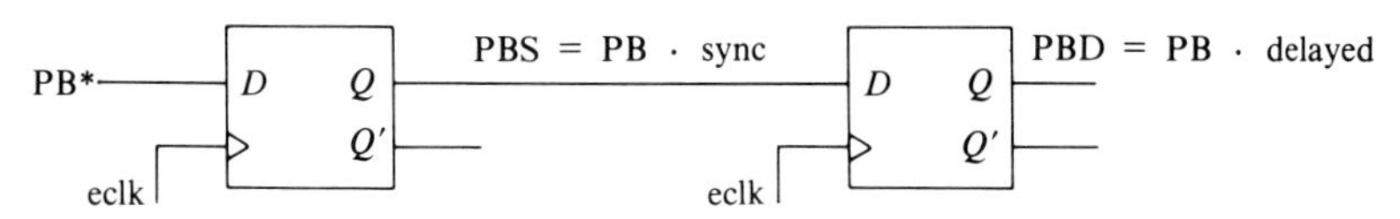

FIGURE 10.28
Synchronizing an asynchronous input PB*

A *ready-set-go* set of handshake signals is used to interface the dice game ASM controller with a debounced pushbutton PB* manipulated by an operator. The signals are as follows:

$$\text{rdy} = (A \oplus B)\ \text{PBS}' \cdot \text{PBD}'$$
$$\text{set} = (A \oplus B)\ \text{PBS}$$
$$\text{go} = \text{g} = (A \oplus B)\ \text{PBS}' \cdot \text{PBD}$$

Ready: When the dice game machine is in State **b** or **d** and signal rdy = 1, the dice game controller asserts "roll" to notify the operator (shooter) that the controller is ready for the dice to be rolled. The controller then waits for the operator to take the following actions:

1. The operator rolls a pair of pyramid dice X and Y, each with an outcome 0, 1, 2, or 3. The sum $R = X + Y$ is the current roll of the dice.
2. The result R ($0 \le R \le 6$) of a roll of the dice is keyed into a hexadecimal keypad.
3. The operator then presses debounced pushbutton PB*.

Set: When the controller detects set = 1, the controller de-asserts "roll."

Go: The controller waits for the operator to release PB*. When the controller detects go = 1, the roll result R is loaded into the R register.

b. Control flow diagram. The operation of the process control algorithm is graphically described by a control flow diagram that shows the control states and conditions for state transitions and the generation of control signals to data path components. The process control algorithm for the pyramid dice game can be translated into a control flow diagram (an ASM chart or a DCAS diagram), as shown in Figure 10.29 (p. 496).

Step 3: Solution/Simplification. Use the description of the process and the process control algorithm (control flow diagram) to develop the process data path, the next-state generator, and the output decoder.

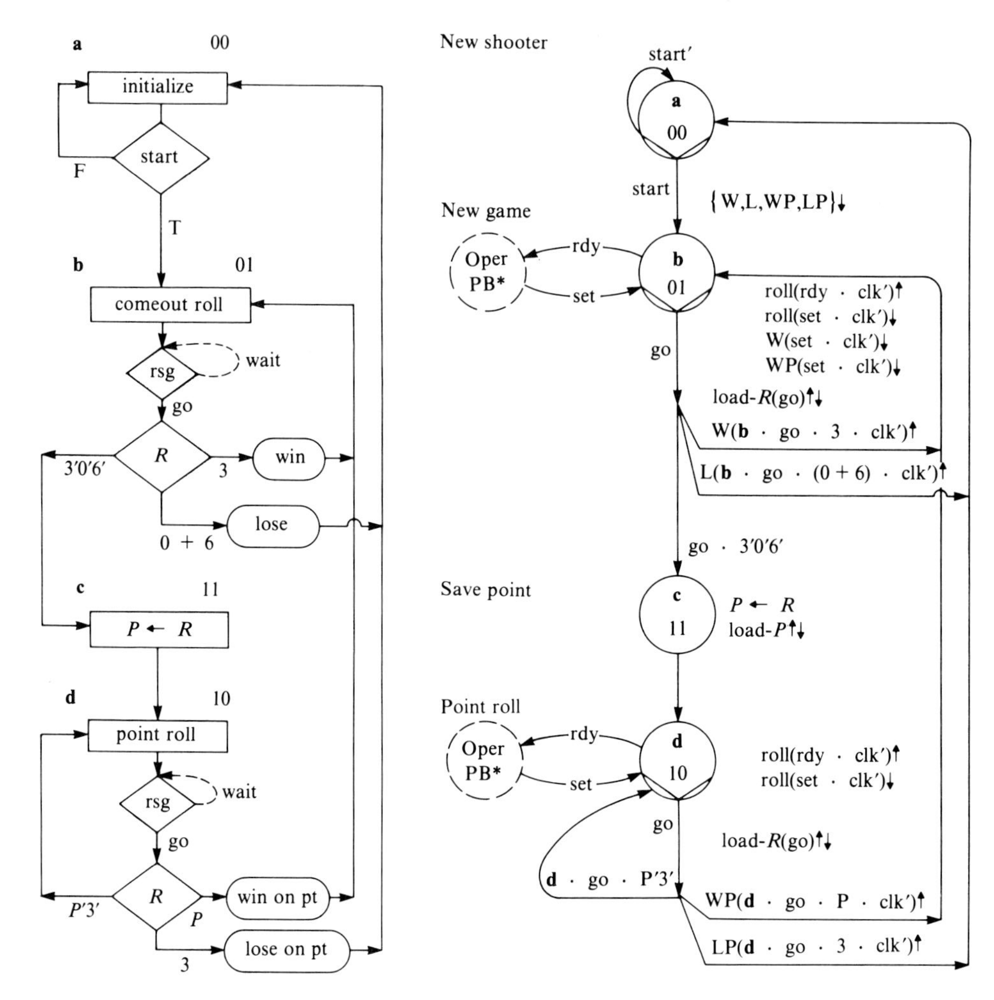

FIGURE 10.29
ASM chart and DCAS diagram for dice game ASM

a. Process data path. The roll of the dice can be simulated by entering a number between 0 and 6 into a hexadecimal keypad. Two registers R and P are required to store the comeout roll and the point roll, respectively. A combinational logic network is designed to determine if register R contains a 3, 0, or 6, or any other number. If a point is established, that point P is stored in register P. A comparator is required to determine if the established point P equals the point roll R.

b. Control unit.

(1) Next-state generator. Alternative representations of the next-state generator can be derived from the control flow diagram.

(a) The DCAS diagram (Figure 10.29) can be used to construct a **control algorithm table** (Table 10.7). This table tabulates the control algorithm next states for each present state, together with the control conditions for each present-state/next-state transition.

TABLE 10.7
Dice game control algorithm table

Present State		Next State		Condition for	Condition for	
AB	Name	Name	A^+B^+	Transition	$A^+ = 1$	$B^+ = 1$
00	**a**	**a**	00	start'	F	start
		b	01	start		
01	**b**	**a**	00	$g \cdot (0 + 6)$	$g \cdot 3'0'6'$	$g' + g \cdot (3 + 3'0'6')$
		b	01	$g \cdot 3 + g'$		
		c	11	$g \cdot 3'0'6'$		
10	**d**	**a**	00	$g \cdot 3$	$g \cdot P'3' + g'$	$g \cdot P$
		b	01	$g \cdot P$		
		d	10	$g \cdot P'3' + g'$		
11	**c**	**d**	10	Always true	T	F

Note: A completed handshake is denoted by g = go.

(b) For each present state AB of the dice game control algorithm, the next state A^+B^+ can be determined using a next-state map in which the state variables A and B are map variables and the input conditions are map-entered variables. The next-state map, derived from the control algorithm for the process (dice game), is shown in Figure 10.30.

FIGURE 10.30
Dice game control algorithm next-state map

A^+B^+

A \ B	0	1
0	00(start′) 01(start)	00(g(0 + 6)) 01(g · 3 + g′) 11(g · 3′0′6′)
1	00(g · 3) 01(g · P) 10(g · P′3′ + g′)	10

(2) Output decoder. The output decoder of the dice game controller generates the following data path control signals, as determined from the DCAS diagram in Figure 10.29. Note that clk' indicates that the JK flip-flops (for roll, W, WP, L, and LP) are clocked when signal clk makes a 1-to-0 transition. The logic equations are

$$\{W\downarrow, L\downarrow, WP\downarrow, LP\downarrow\} = A'B' \cdot \text{start} \cdot \text{clk}'$$
$$\text{rdy} = A \oplus B \cdot \text{PBS}' \cdot \text{PBD}'$$
$$\text{set} = A \oplus B \cdot \text{PBS}$$
$$\text{go} = g = A \oplus B \cdot \text{PBS}' \cdot \text{PBD}$$
$$\text{roll}\uparrow = \text{rdy} \cdot \text{clk}'$$
$$\text{roll}\downarrow = \text{set} \cdot \text{clk}'$$
$$W\downarrow = \text{set} \cdot \text{clk}'$$

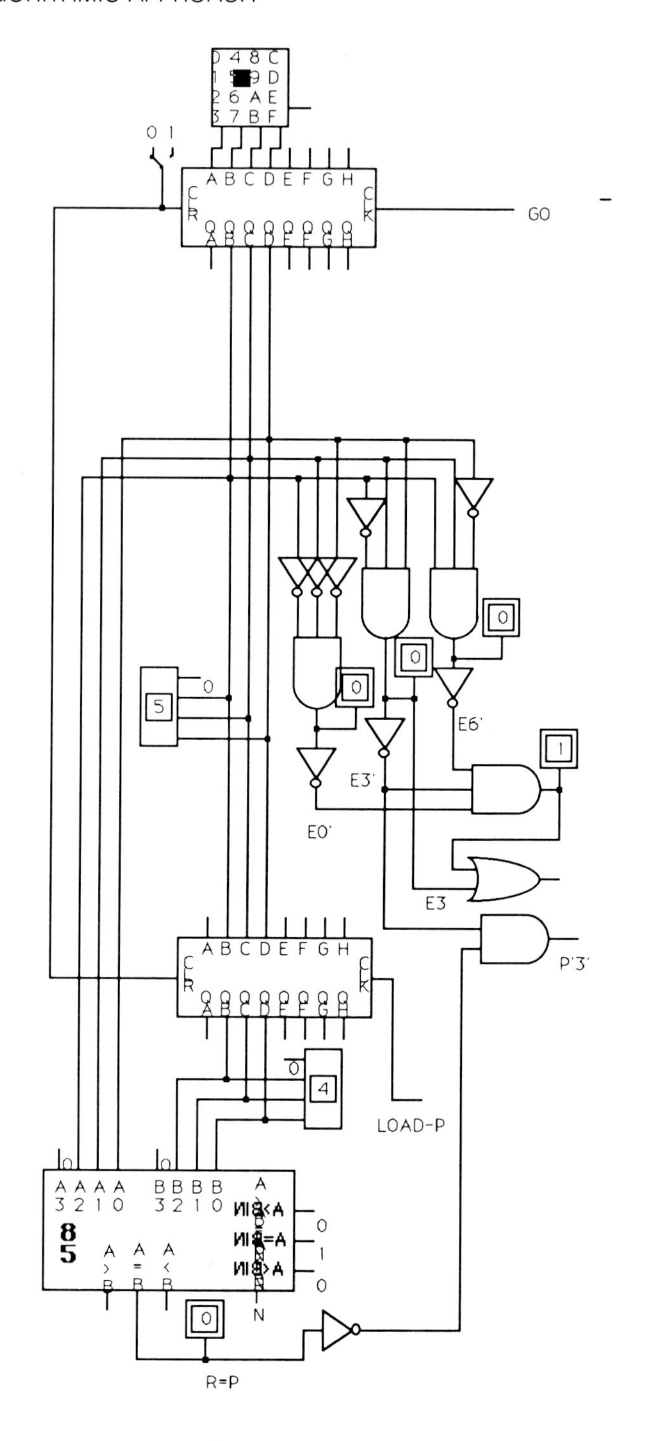

FIGURE 10.31
Logic diagram of the process data path

$$\text{WP}\downarrow = \text{set} \cdot \text{clk}'$$
$$\text{load-}R\uparrow\downarrow = \text{g}$$
$$\text{L}\uparrow = A'B\ (\text{g} \cdot (0 + 6) \cdot \text{clk}')$$
$$\text{W}\uparrow = A'B\ (\text{g} \cdot 3 \cdot \text{clk}')$$
$$\text{load-}P\uparrow\downarrow = AB$$
$$\text{LP}\uparrow = AB'\ (\text{g} \cdot 3 \cdot \text{clk}')$$
$$\text{WP}\uparrow = AB'\ (\text{g} \cdot P \cdot \text{clk}')$$

Step 4: Realization.

a. Process data path. A detailed logic diagram of the process data path (Figure 10.31) is developed from the description in Step 3.

b. Control unit.

(1) Next-state generator.

(a) Implemented using multiplexers. The dice game control algorithm can be implemented directly from the control algorithm table using 4-to-1 multiplexers, where present state AB is used to select the input condition corresponding to the select value AB combination, as shown in Figure 10.32.

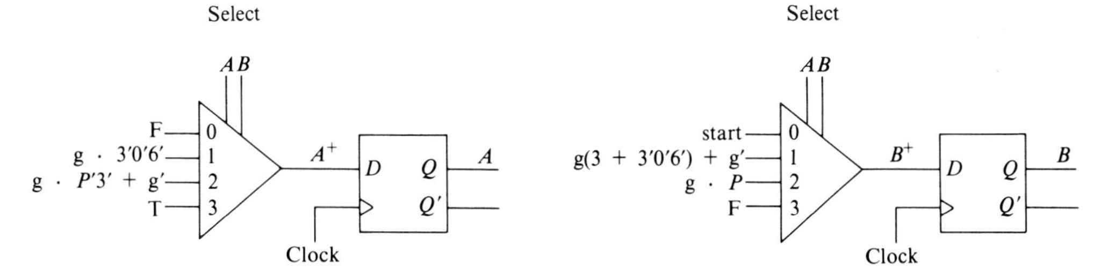

FIGURE 10.32
Multiplexer implementation of next-state generator

(b) Implemented using D flip flops. If the next state generator is to be implemented using D flip-flops, individual next-state maps for each state variable can be derived from the next-state map in Figure 10.30. The resulting next-state maps are shown in Figure 10.33.

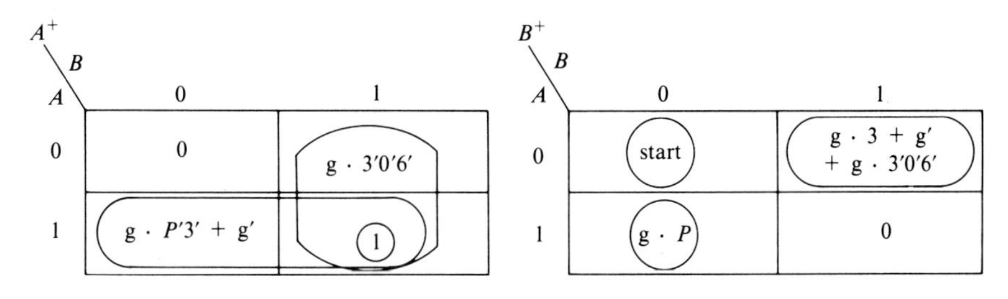

FIGURE 10.33
Next-state maps for controller state variables

The excitation equations for the D flip-flops, derived from the next-state maps in Figure 10.33, are

$$D_A = AB + A(g \cdot P'3' + g') + B \cdot g \cdot 3'0'6'$$
$$D_B = A'B'\text{start} + A'B(g[3 + 3'0'6'] + g') + AB'(g \cdot P)$$

(2) Output decoder. A logic diagram of the output decoder can be constructed using the logic equations developed from Figure 10.29 in Step 3. Figure 10.34 presents a timing diagram for the dice game ASM.

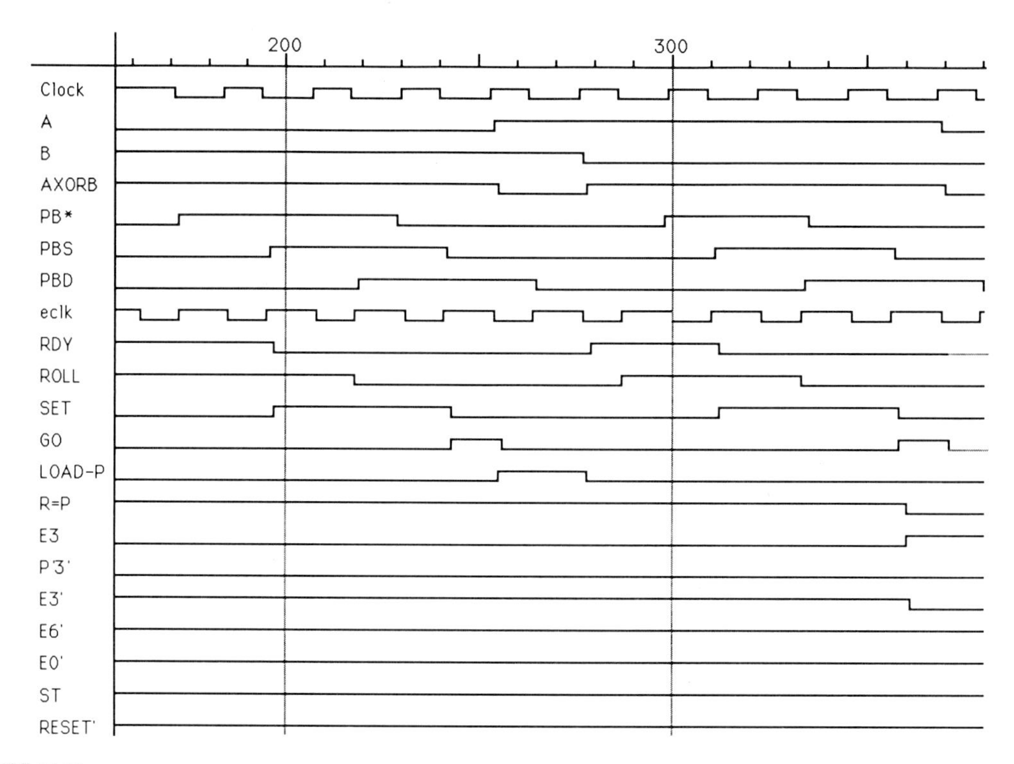

FIGURE 10.34
Timing diagram for dice game ASM

A complete logic diagram of the dice game ASM is presented in Figure 10.35. The data path section appears on the left side of the diagram and the control section on the right side.

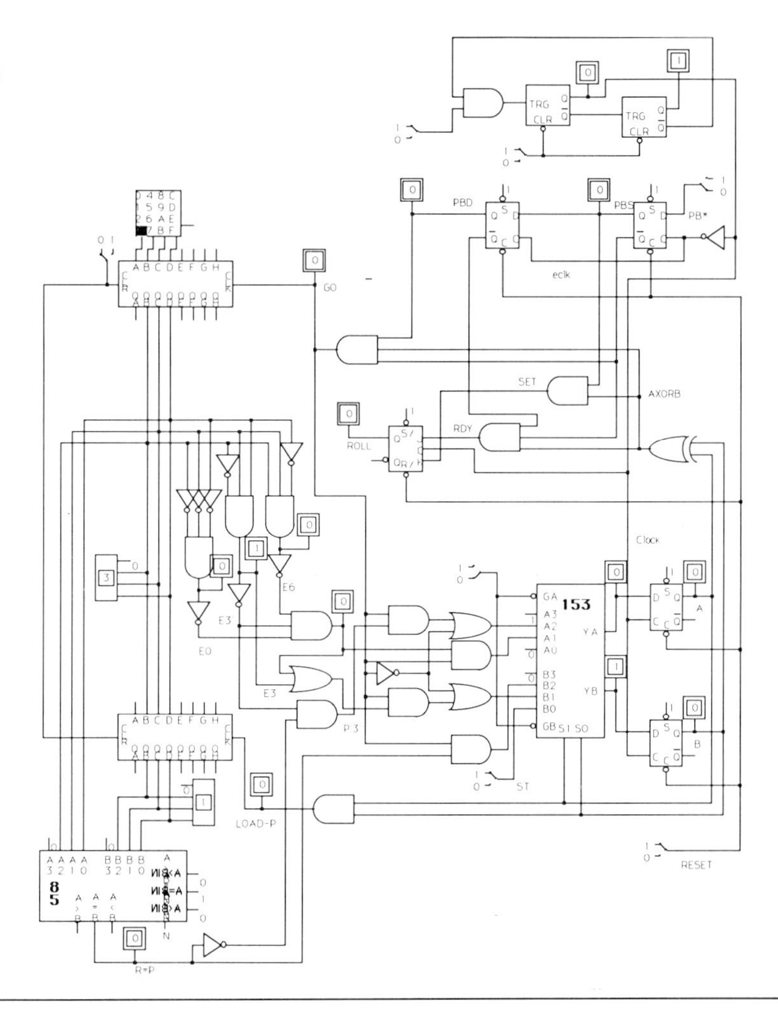

FIGURE 10.35
Complete dice game ASM logic diagram

INTERACTIVE DESIGN EXAMPLE

System Design of a Blackjack Dealer ASM

The game of blackjack (also called "21") is a card game played by two or more people. In a given game, one of the players is the dealer; the dealer deals cards to each of the other players and to himself (herself). Each numbered card has a value equal to its number, while the ten, jack, queen, and king each has value 10. An ace can be valued as 1 or 11. An ace and a value-10 card, together totaling 21, is referred to as a "blackjack."

Initially, each player is dealt 2 cards, the first face down and the second face up. After receiving these 2 cards, a player may elect to "stand pat" or to be dealt additional cards, up to a total of 5 cards. As a player receives a new card, one of the following actions take place:

1. The player is broke if his cards have total value more than 21.
2. The player wins if he has 5 cards with total value 21 or less.
3. The player may elect to stand on a total of 21 or less. The objective of each player is to get closer to 21 than the dealer does, since the dealer wins in the case of a tie.

The dealer may elect to stand only if his cards have total value between 17 and 21, inclusive. Thus, the dealer must continue to deal himself additional cards until the cards have a total value of 17 or more. If the total value is more than 21, the dealer is broke unless the last card was an ace (initial value 11), in which case the ace is re-valued as 1.

Now that you understand the rules governing the blackjack dealer, we will use the 4-step problem-solving procedure to design an ASM that simulates the dealer. These rules are translated into a process control algorithm in Step 1.

Step 1: Problem Statement.

a. Process. The dealer deals himself cards until the cards have a total value of 17 or more. If the total value is more than 21, the dealer is broke unless the last card was an ace (initial value 11), in which case the ace is re-valued as 1.

b. Process control algorithm. The rules that govern the blackjack dealer can be translated into an algorithm that has four distinct steps or states, denoted by **a**, **b**, **c**, and **d**:

State **a**. Idle
    If start = 1
      then {AC := 0; std := 0; brk := 0; **goto b**}
      else {**goto a**}

State **b**. Deal card *X*
   (* If dealer is "rdy" (ready-to-receive), notify operator *)
   (* Operator keys card value *X*, $2 \leq X \leq 11$, into hex keypad *)
   (* If operator is "set" (set-to-send), notify dealer *)
   (* "rdy" and "set" together imply "go" (handshake completed) *)
    If go = 1
      then {start := 0; mux-sel := 0}

State **c**. Add
    AC := AC + *X*; load AC
    If *X* = 11
      then {ace := 1; ace11 := 1}
      else {ace := 0; ace11 := 1}

State **d**. Analyze
    If AC ≤ 16
      then {**goto b**}
      else if AC ≤ 21
        then {std := 1; **goto a**}
        else if (ace · ace11)'
          then {brk := 1; **goto a**}
          else {ace11 := 0; mux-sel := 1; AC := AC – 10; **goto d**}

System specifications. Design an ASM that simulates the step-by-step procedures of a blackjack dealer.

Step 2: Conceptualization.

a. Structure diagram. The procedure followed by the dealer is an algorithmic process that can be represented by an ASM. This ASM can be partitioned into a control section (controller) and a process section (data path). A structure diagram of the blackjack dealer ASM is presented in Figure 10.36.

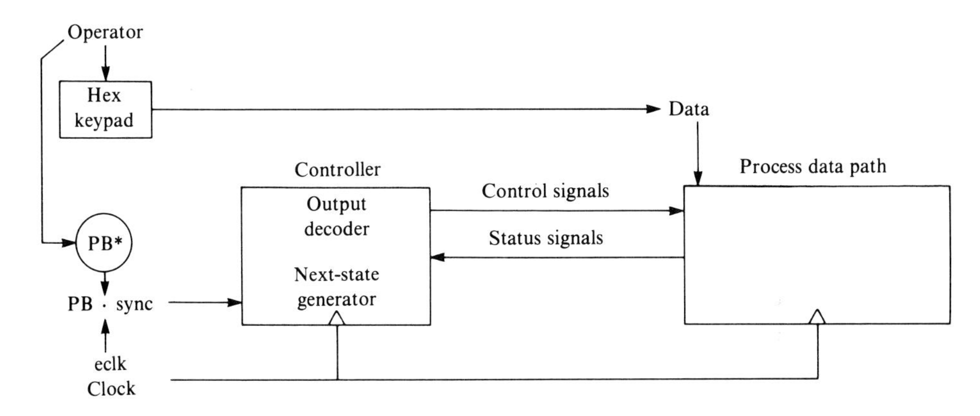

FIGURE 10.36
Structure diagram of blackjack dealer ASM

Man–Machine Interface for the Blackjack Dealer ASM A debounced pushbutton, PB*, is used by the operator to signal that a card value X has been entered into the hexadecimal keypad. The signal PB* is input to a D flip-flop, which is clocked with eclk (the inversion of the system clock). The flip-flop output PBS (PB · sync) provides a synchronous input to the controller. Signal PBS is input to a second D flip-flop, which is also clocked by signal eclk and has output PBD (PB · delayed). The debounced pushbutton PB* and the synchronizing and delaying flip-flops are shown in Figure 10.37.

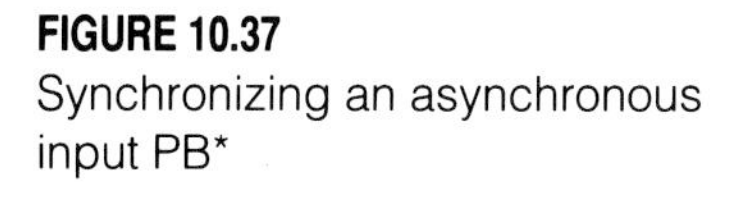
FIGURE 10.37
Synchronizing an asynchronous input PB*

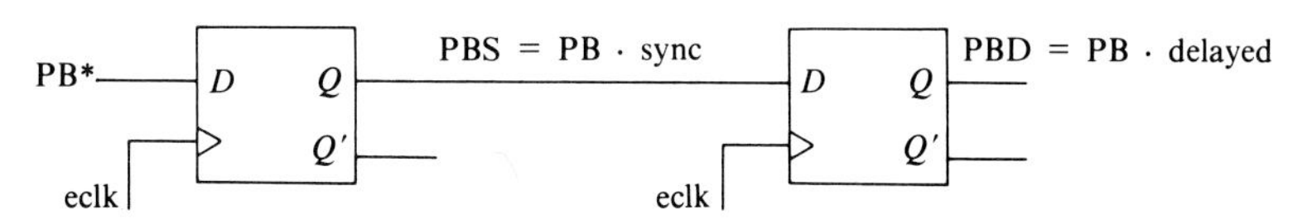

A *ready-set-go* set of handshake signals is used to interface the blackjack ASM controller with a debounced pushbutton PB* manipulated by the operator.

$$\text{rdy} = A'B \cdot \text{PBS}' \cdot \text{PBD}'$$
$$\text{set} = A'B \cdot \text{PBS}$$
$$\text{go} = \text{g} = A'B \cdot \text{PBS}' \cdot \text{PBD}$$

Ready: When the blackjack dealer ASM is in State **b** and signal rdy = 1, the blackjack controller asserts "deal" to notify the operator that the controller is ready for the card to be dealt.

The controller then waits for the operator to take the following actions:

1. The operator takes the next card from the deck with an outcome X, $(2 \leq X \leq 11)$.
2. Card value X is keyed into a hexadecimal keypad.
3. The operator then presses debounced pushbutton PB*.

Set: When the controller detects set = 1, the controller de-asserts "deal."

Go: The controller waits for the operator to release PB*. When the controller detects go = 1, the controller goes to State **c**, and the card value X is added to the accumulator AC.

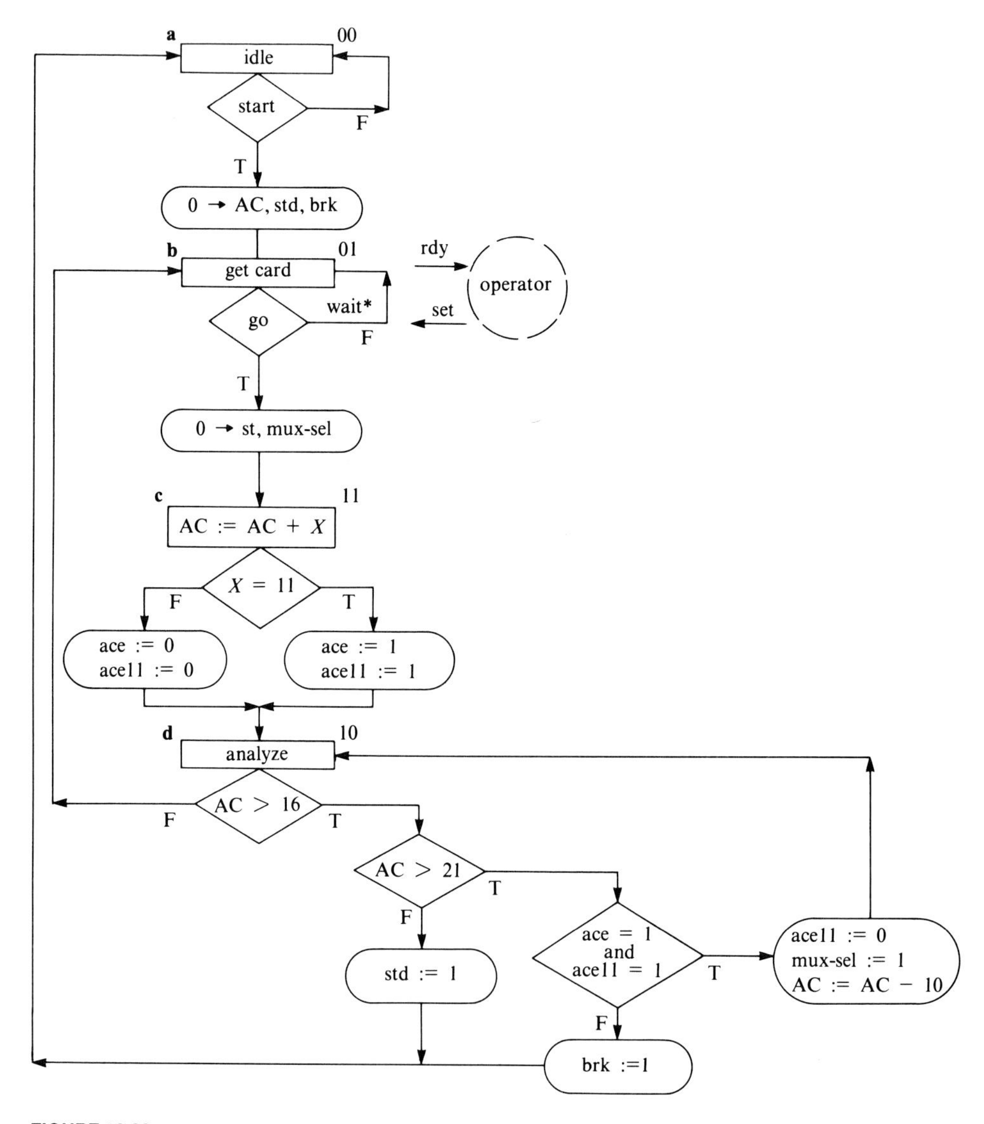

FIGURE 10.38
ASM chart for blackjack dealer ASM

b. Control flow diagram. The operation of the process control algorithm is graphically described by a control flow diagram that shows the control states and conditions for state transitions and the generation of control signals to data path components. The process control algorithm for the blackjack dealer can be translated into a control flow diagram (an ASM chart or a DCAS diagram), as shown in Figure 10.38.

Step 3: Solution/Simplification. Use the description of the process and the process control algorithm (control flow diagram) to develop the process data path and the control unit next-state generator and output decoder.

a. Process data path. The card that is dealt can be simulated by entering a number between 2 and 11 into a hexadecimal keypad. This number is stored in a register, and a comparator is used to set a flag if the number is an 11 (ace). An adder is used to add the number of the card dealt to a cumulative sum that is stored in an accumulator register. Two comparators are used to determine if the cumulative sum is either greater than 16 or greater than 21.

b. Control unit.

(1) Next-state generator. Alternative representations of the next-state generator can be derived from the control flow diagram.

(a) The ASM chart (Figure 10.38) can be used to construct a control algorithm table (Table 10.8). This table tabulates the control algorithm next states for each present state, together with the control conditions for each present-state/next-state transition.

TABLE 10.8
Blackjack dealer control algorithm table

Present State		Next State		Condition for Transition	Condition for	
AB	Name	Name	A^+B^+		$A^+ = 1$	$B^+ = 1$
00	**a**	**a**	00	start'	F	start
		b	01	start		
01	**b**	**b**	01	g'	g	T
		c	11	g		
10	**d**	**a**	00	G16 · G21' + G16 · G21 · (ace · a11)'	G16 · G21 · ace · a11	G16'
		b	01	G16'		
		d	10	G16 · G21 · ace · a11		
11	**c**	**d**	10	Always true	T	F

Notes: g denotes handshake completed (g = go).
ace denotes aceflag.
a11 denotes ace11.
G16 denotes total > 16.
G21 denotes total > 21.

(b) For each present state *AB* of the blackjack dealer control algorithm, the next state A^+B^+ can be determined using a next-state map in which the state variables *A* and B are map variables and the input conditions are map-entered variables. The next-state map, derived from the control algorithm for the process (blackjack dealer), is shown in Figure 10.39.

FIGURE 10.39
Blackjack dealer control algorithm next-state map

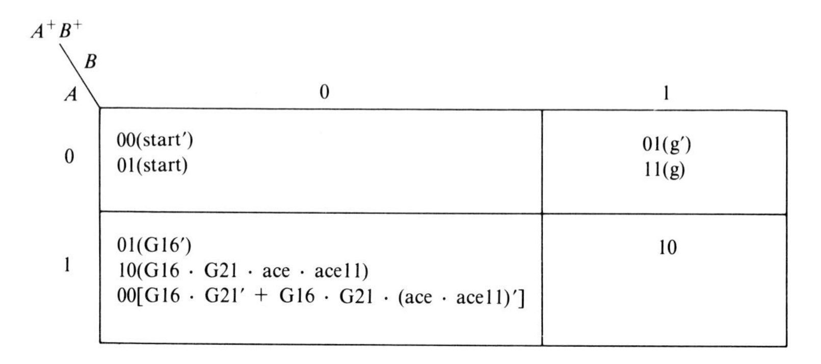

(2) Output decoder. The output decoder of the blackjack controller generates the following data path control signals as determined from the ASM chart shown in Figure 10.38. Note that g denotes handshake complete and clk indicates that the *JK* flip-flops (start, deal, std, brk, and a11) are clocked when signal clk makes a 1-to-0 transition. PBS and PBD are generated by PET *D* flip-flops, which are clocked by eclk.

$$\text{rdy} = A'B \cdot \text{PBS}' \cdot \text{PBD}'$$
$$\text{set} = A'B \cdot \text{PBS}$$
$$\text{go} = \text{g} = A'B \cdot \text{PBS}' \cdot \text{PBD}$$
$$\text{deal}\uparrow = \text{rdy} \cdot \text{clk}'$$
$$\text{deal}\downarrow = \text{set} \cdot \text{clk}'$$
$$\text{start}\downarrow = \text{g} \cdot \text{clk}'$$
$$\text{load-AC}\uparrow\downarrow = AB + AB' \cdot \text{G16} \cdot \text{G21} \cdot \text{ace} \cdot \text{a11} \cdot \text{eclk}$$
$$\text{a11}\uparrow = AB \cdot \text{ace} \cdot \text{clk}'$$
$$\text{a11}\downarrow = (AB \cdot \text{ace}' + AB' \cdot \text{G16} \cdot \text{G21} \cdot \text{ace} \cdot \text{a11})\ \text{clk}'$$
$$\text{mux-sel-1} = AB' \cdot \text{G16} \cdot \text{G21} \cdot \text{ace} \cdot \text{a11}$$
$$\text{std}\downarrow = A'B' \cdot \text{start} \cdot \text{clk}'$$
$$\text{brk}\downarrow = A'B' \cdot \text{start} \cdot \text{clk}'$$
$$\text{std}\uparrow = AB' \cdot \text{G16} \cdot \text{G21}' \cdot \text{clk}'$$
$$\text{brk}\uparrow = AB' \cdot \text{G16} \cdot \text{G21} \cdot (\text{ace} \cdot \text{a11})' \cdot \text{clk}'$$

For purposes of simplification, the accumulator AC, std, brk, and a11 can be reset by a master clear prior to start. ace = 1 (indicating aceflg) is set in data path by a comparator when a card value $X = 11$ is keyed into the hexadecimal keypad.

Step 4: Realization.

a. Process data path. A detailed logic diagram of the process data path (Figure 10.40) is developed from the description in Step 3.

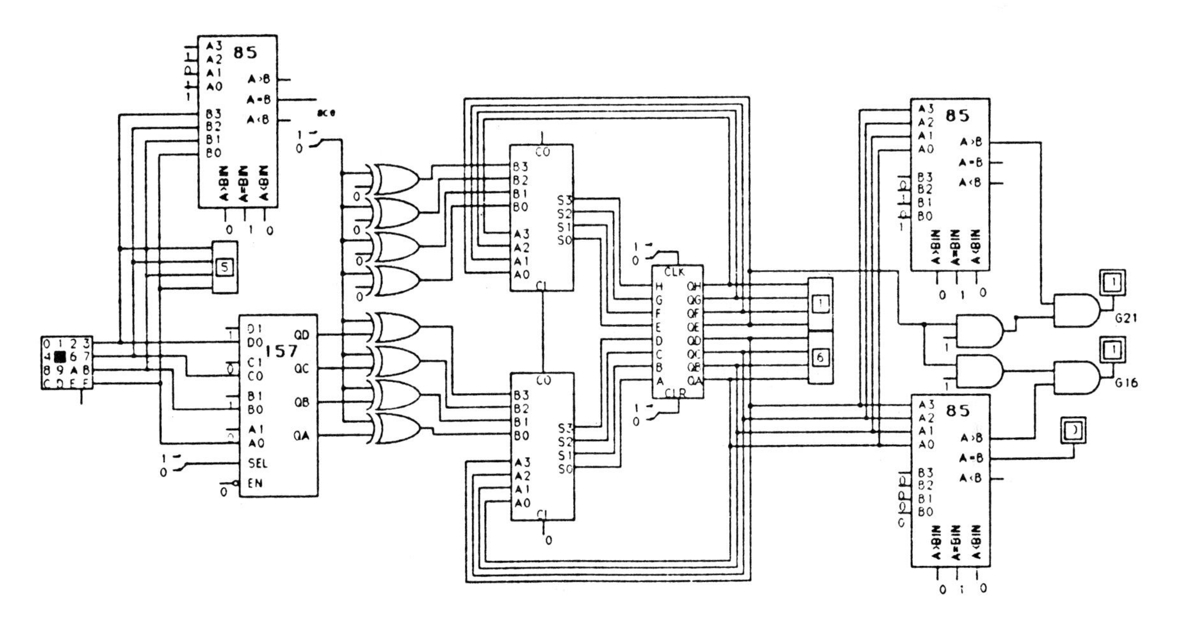

FIGURE 10.40
Logic diagram of the process data path

b. Control unit.

(1) Next-state generator.

(a) Implemented using multiplexers. The blackjack dealer control algorithm can be implemented directly from the control algorithm table (Table 10.8) using 4-to-1 multiplexers, where present state AB is used to select the input condition corresponding to the select value AB combination, as shown in Figure 10.41.

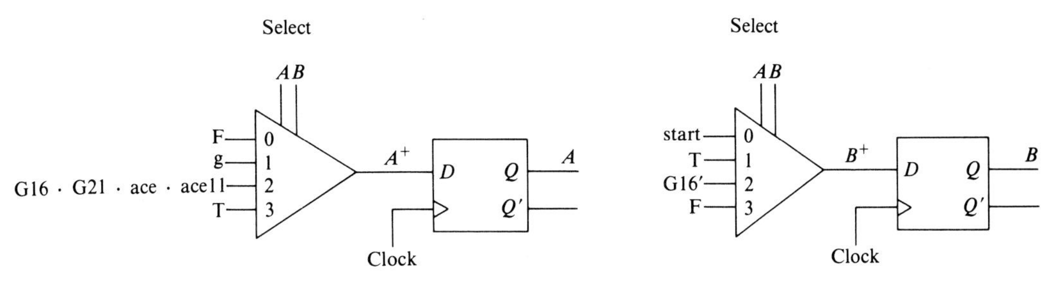

FIGURE 10.41
Mux implementation of blackjack next-state generator

(b) Implemented using *D* flip-flops. If the next-state generator is to be implemented using *D* flip-flops, individual next-state maps for each state variable can be derived from the next-state map in Figure 10.39. The resulting next-state maps are shown in Figure 10.42.

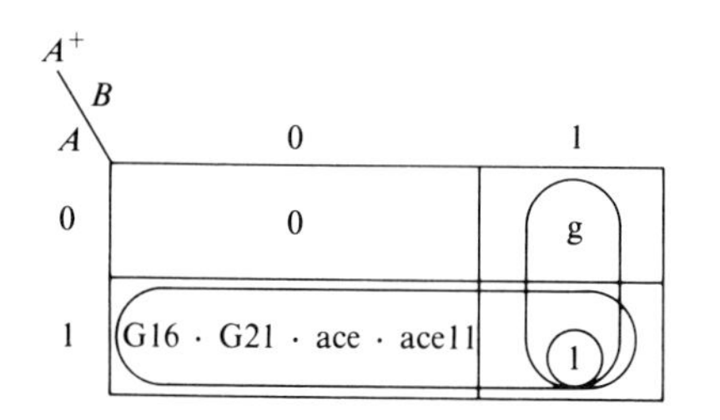

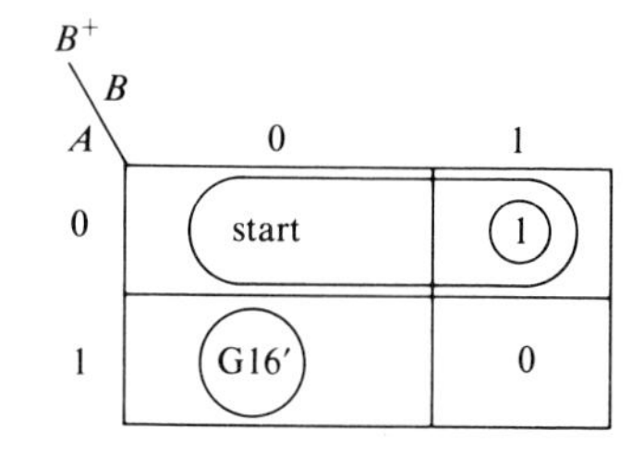

FIGURE 10.42
Next-state maps for controller state variables

The excitation equations for the PET *D* flip-flops, derived from the next-state maps in Figure 10.42, are

$$D_A = AB + A(\text{G16} \cdot \text{G21} \cdot \text{ace} \cdot \text{a11}) + B \cdot \text{g}$$
$$D_B = A'B + A'\text{start} + AB'(\text{G16}')$$

Note that control unit state changes occur on the 0-to-1 transition of the system clock.

(2) Output decoder. A logic diagram of the output decoder can be constructed using the logic equations from Step 3. Figure 10.43 presents a timing diagram for the blackjack dealer ASM.

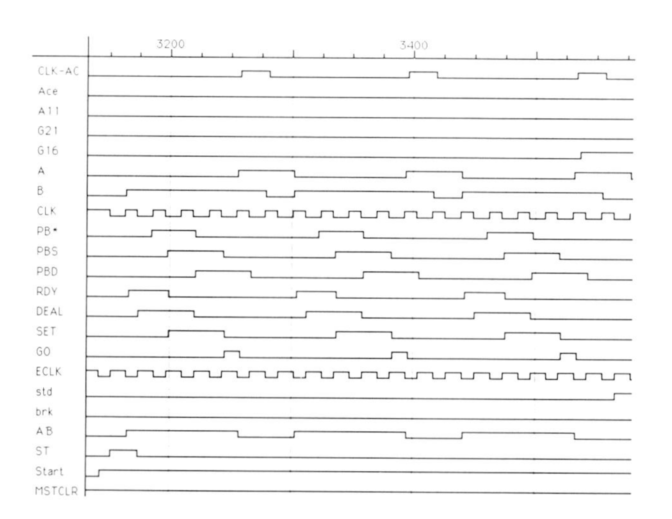

FIGURE 10.43
Timing diagram for blackjack ASM

A complete logic diagram of the blackjack dealer ASM is presented in Figure 10.44. The data path section appears on the right side of the diagram and the control section on the left side. Note that the next-state generator is implemented using a dual 4-line-to-1-line multiplexer (74153).

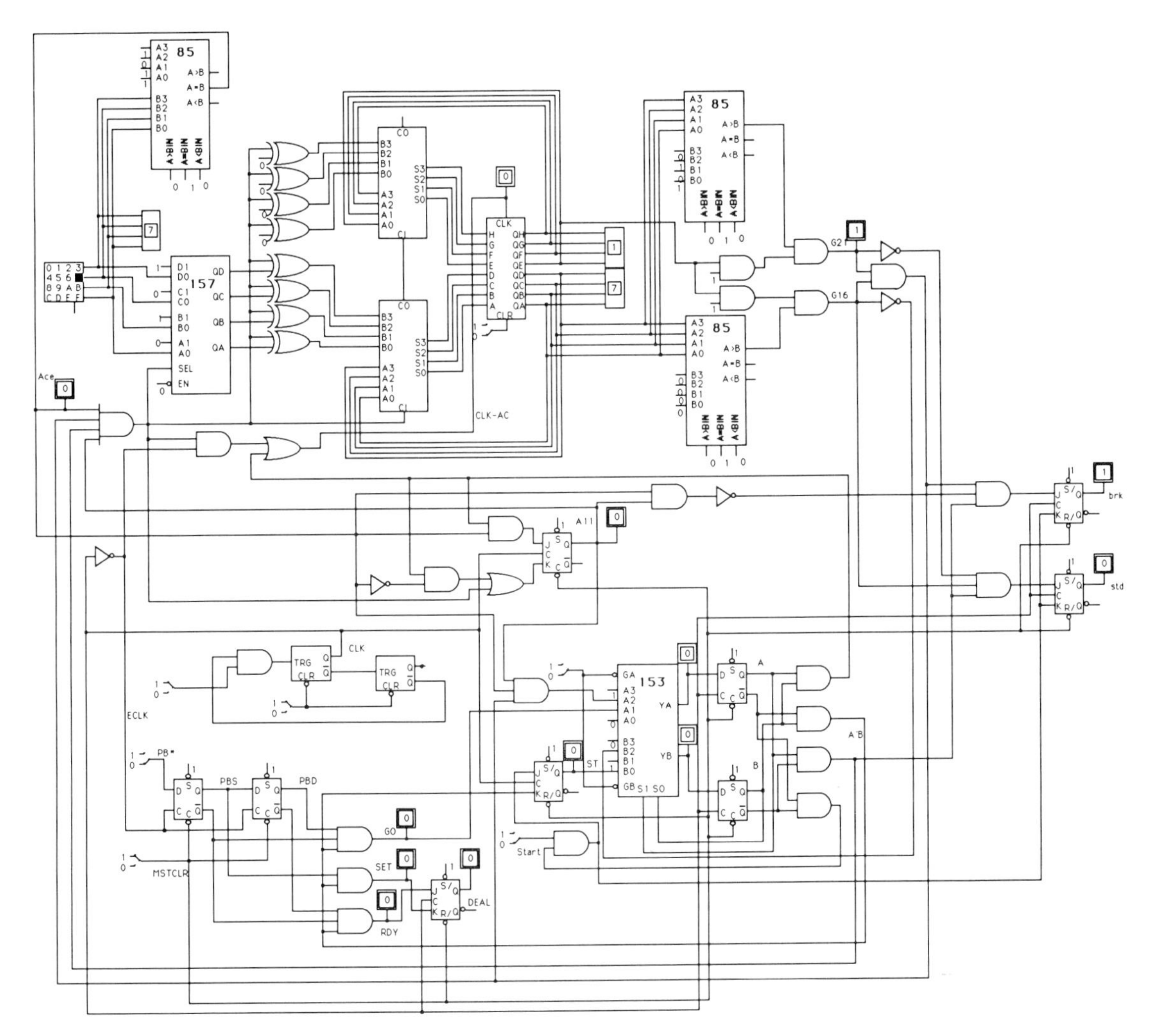

FIGURE 10.44
Complete blackjack dealer ASM logic diagram

10.5 SUMMARY

This chapter presented basic concepts of systems-level design, using a number of different processes for illustration. In systems-level design, registers and register operations are the basic elements and operations of a system. A simple register transfer notation (RTN), summarized in Table 10.1, is used to describe the operations of systems developed in the remainder of this chapter and the text.

The chapter began with a discussion of iterative processes (such as serial addition, parity checking, and 2s complementation) and described how they can be represented by

FSM models and realized by digital logic systems that have rudimentary control mechanisms. After learning how to develop sequential machines that realize simple processes, we used an algorithmic approach for modeling more complicated processes and for designing sophisticated process control mechanisms (controllers).

A process that is governed by a set of rules can be represented by an ASM model, which, in turn, can be realized by hardware implementation of a digital logic system or by software simulation. For example, the rules for the game of dice can be translated into a control algorithm, which in turn can be translated into a digital logic controller or into a computer program. The process of rolling the dice can be simulated in software by a random-number generator or in hardware by a sequential machine such as a programmable logic sequencer.

A 4-step problem-solving procedure for systems design is used in this and later chapters to design digital logic systems that realize sequential processes. In this chapter, we used the procedure to develop ASMs for a number of representative processes (serial addition, dice game, blackjack dealer). In the next chapter, we will develop ASMs for serial multiplication and serial division. These machines are then realized using a logic simulation program.

KEY TERMS

Algorithmic state machine (ASM)
Register transfer notation (RTN)
Controller
Iterative process
Information distributed in space
Information distributed in time
Sequential (state) machine
Process algorithm
Process control algorithm
Controller
Data path
Next-state generator
Output decoder
ASM chart
DCAS diagram
Control algorithm table

EXERCISES

General

1. Assume that a floating-point number $X = \pm F2^E$ is stored in a 12-bit word as follows:

S	$E^{\wedge}$	F
1	4	7

= seeee.fffffff

← Implied binary point

where $S = 0$ for positive X, and $S = 1$ for negative X
$E\wedge$ = base-2 exponent in excess-8 form. $E = E\wedge - 8$
F = fraction, normalized with leftmost bit = 1

a. Describe an algorithm for adding two positive floating-point numbers.
b. Draw a block diagram of a data path for the algorithm in part a.
c. Formulate a process control algorithm for a floating-point adder ASM.

2. A tour in a graph G is a closed path with no repeated edge. An Euler tour in a graph G is a tour in G that passes through each edge of G exactly once. The following algorithm generates a sequence of 4-bit numbers corresponding to the edges of an Euler tour (register $R = R_3R_2R_1R_0$):

Step **a.** A 4-bit shift register R initially contains 0001.
Step **b.** Shift R left 1 bit; set fill bit R_0 to 0.
If $R_3 = 0$ **then** {**goto b**}
else {add 1 to R, **goto c**}
Step **c.** Shift R left 1 bit; set fill bit R_0 to 0.
If $R_3 = 1$ **then** {**goto c**}
else {**goto d**}
Step **d.** **If** $R = 0000$ **then** {stop}
else {add 1 to R, **goto b**}.

a. Construct a flowchart of the algorithm.
b. The figure below presents the first 5 edges of a directed Euler tour generated by the algorithm. Complete the diagram.

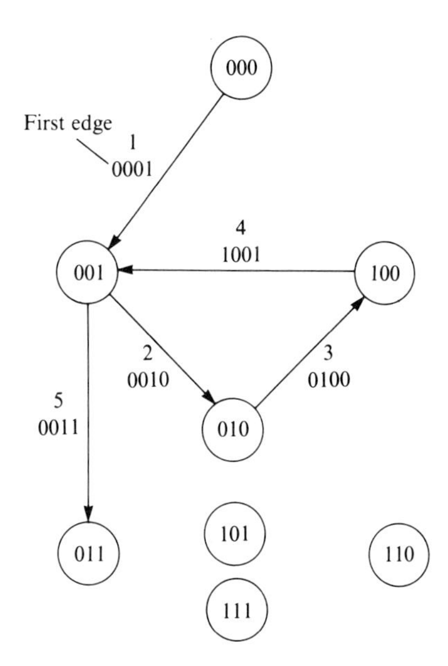

3. A control system for a railroad crossing signal light is pictured below. We assume switch i (or j) is closed when any part of the train is passing over the switch. Both switches cannot be closed at the same time. The controller is in State **a** prior to the approach of a train from either direction. The controller returns to State **a** when the train clears the region between the two switches. The length of the train is shorter than the switch-to-crossing distance.

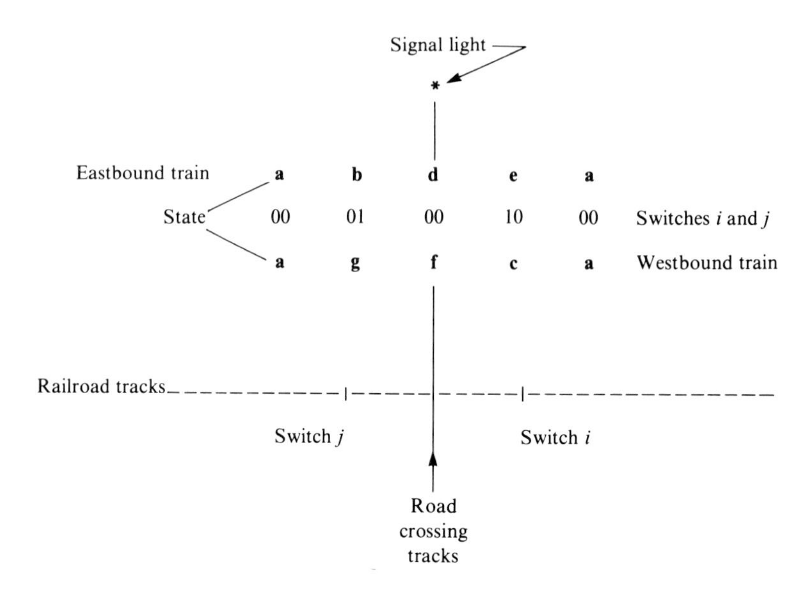

a. Complete the following primitive flow table for the railroad crossing signal control system.

Row \ ij	00	01	11	10	Z = light
a					0 (off)
b					1 (on)
c					1 (on)
d					1 (on)
e					1 (on)
f					1 (on)
g					1 (on)

b. Use an implication chart to determine compatible row pairs of the flow table in part a.

c. Reduce the nonredundant primitive flow table in part a to a minimal-row flow table by merging compatible rows.

d. Draw a state diagram corresponding to the minimal-row flow table in part c.

4. A state table for a simple vending machine is given below. The machine accepts nickels (5), dimes (10), and quarters (25) and has two select buttons, S (soda) and C (candy). When the machine has accumulated 25 cents or more, pressing S or C dispenses a soda or a candy bar and causes the machine to return to State **a**. Use the state table to draw a state diagram for the vending machine.

Accumulated Total Value	State	Input					Output				
		5	10	25	S	C					
0	a	b	c	f	a	a	0	0	0	0	0
5	b	c	d	f	b	b	0	0	0	0	0
10	c	d	e	f	c	c	0	0	0	0	0
15	d	e	f	f	d	d	0	0	0	0	0
20	e	f	f	f	e	e	0	0	0	0	0
25	f	f	f	f	a	a	0	0	0	soda	candy

5. A state diagram for a BCD recognizer FSM (automaton) is shown below. Use the state diagram to construct a state table for the BCD recognizer.

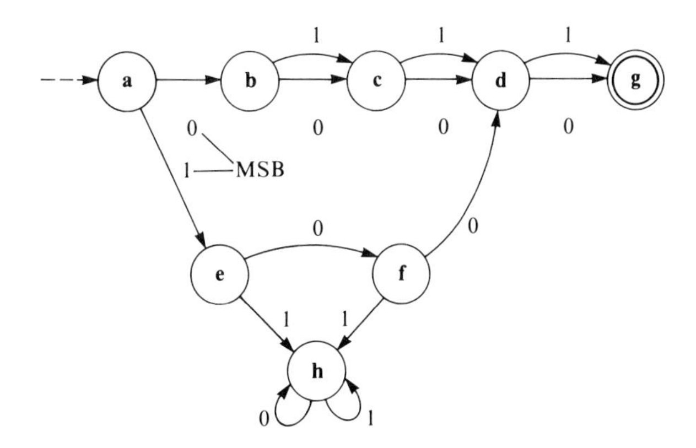

6. Use pseudocode to describe the rules of a 6-sided dice game.

7. Develop a control algorithm for a 6-sided dice game ASM.

Design/Implementation

Use the 4-step problem-solving procedure for systems design for each design exercise.

1. Design a floating-point adder ASM for positive numbers using the data path and control algorithm produced in General Exercise 1.

2. A deBruijn sequence is a circular arrangement $a_1a_2a_3 \ldots a_{2^n}$ of 0s and 1s such that every sequence of 0s and 1s of length n appears exactly once. An Euler tour in D_n can be used to generate a deBruijn sequence of length 2^n by using the first digit of each edge label in the tour to create the sequence. Design as ASM that generates a deBruijn sequence of length 16 (see General Exercise 2).

3. Design a railroad signal control system in accordance with the design specifications described in General Exercise 3.

4. Design an ASM to implement a 6-sided dice game.

5. The position of a rotating disk with 2^n sectors can be determined by n read heads that detect a 0 or 1, where each successive position is an edge of an Euler tour. Design a system to locate a specified sector on a rotating disk with 16 sectors.

6. Use a programmable logic sequencer to implement a 6-sided dice game.

11

Digital Systems Design 2—Multipliers, Dividers, and Control Units

OBJECTIVES

After you complete this chapter, you will be able

- ☐ To understand the concepts of two-dimensional iterative processes such as binary multiplication and division, BCD-to-binary and binary-to-BCD conversion
- ☐ To analyze general and specific sequential processes, with the idea of developing algorithms to represent these processes
- ☐ To translate word-description specifications into algorithms
- ☐ To represent a process and its control using as ASM model
- ☐ To design a control subsystem and a data path subsystem, and to integrate these subsystems into a functioning system
- ☐ To understand the elements of shared-data-path design and combined-control-unit design, common in developing multifunction devices such as a combined multiplier/divider or a digital computer
- ☐ To combine control units by combining control equations
- ☐ To evaluate trade-offs of alternative implementations—circuit complexity versus speed of execution
- ☐ To develop further skills in analyzing problems in digital logic design, developing algorithms for sequential processes, and translating control algorithms into control flow diagrams (ASM charts or DCAS diagrams)
- ☐ To organize a system into subsystems using a divide-and-conquer approach
- ☐ To compare alternative implementation forms, that is, combinational versus sequential logic and hardware implementation versus software implementation

11.1 INTRODUCTION—AN ALGORITHMIC APPROACH TO DESIGN

As you learned in Chapter 10, algorithms are useful for defining processes and process control. Algorithms can be used in any discipline to develop solutions for almost any conceivable type of problem.

Consider, for example, the following disciplines:

- Statistical methods, such as computing the mean, variance, and covariance, are accomplished by algorithms that involve computations on sets of elements $\{x(k)\}$, $\{y(k)\}$.
- Numerical methods, such as interpolation, curve fitting, numerical integration, and the solution of differential equations, are accomplished by algorithms that involve computations on sets of elements $\{x(k)\}$, $\{y(k)\}$.
- List processing of data $\{x(k)\}$ and $\{y(k)\}$ is accomplished by algorithms that involve adding, multiplying, comparing, sorting, and merging elements $x(k)$ and $y(k)$ in the given lists of data.

Thus, the solution of a large number of classes of problems for a variety of applications can be accomplished using appropriate algorithms.

An analysis of pencil-and-paper procedures for solving a given problem can lead to an algorithm for accomplishing the solution of that problem. Individual algorithm steps (add, subtract, multiply by a power of 2, and count) can be accomplished by functional modules such as adders, TC01s, shift registers, and counters. A properly organized algorithm can be translated into a digital logic device, which usually contains a data path (for processing data) and a control unit (for activating data path elements as required to accomplish the sequence of process steps involved in the algorithm).

The design of a computer involves determining a set of instructions that the computer must have to implement algorithms to be programmed. Obviously, a realistic computer must have the capability to multiply and divide, as well as add, subtract, and perform logical operations. Digital devices for adding, subtracting, and performing logical operations have been described in earlier chapters.

An analysis of the process of multiplying (or dividing) one integer by another leads to the development of a recursive algorithm that generates a sequence of intermediate results. Each iteration of the algorithm is a major cycle of the process. A number of microoperations are required to accomplish the processing within each iteration; each of these microoperations can be performed in a minor cycle of the overall (major) cycle. A composite of the microoperations determines a data path that supports each and every microoperation of the algorithm. The sequence of performing the microoperations determines the organization of the control unit that activates data path elements in the required order.

This chapter describes the design and implementation of sequential logic devices for multiplying two binary integers and for dividing one binary integer by another. We will use the 4-step problem-solving procedure for system design (described in Chapter 10) to design serial multiplication and serial division ASMs.

The individual functional modules (such as a serial multiplier and a serial divider) can be combined into a multifunction processor. This requires (1) designing a data path that supports each function of the multifunction processor and (2) designing a control unit for the multifunction processor.

11.2 DESIGN AND IMPLEMENTATION OF A SERIAL MULTIPLICATION ASM

There are several alternatives for implementing a multiply capability in a computer's CPU, including

- □ *Generating a sequence of* **partial-product sums** *(PPS), where successive partial-product sums are defined by a recursion formula. For example, a multiply function can be programmed as a subroutine that repeatedly invokes an addition instruction and a shift instruction.*
- □ ***A combinational-logic binary multiplier can be designed and implemented using an array of logic gates. It can also be implemented using a ROM table lookup.***
- □ ***A serial multiplication ASM can be designed and implemented using an n-bit adder and appropriate shift registers.***

A combinational logic multiplier has been described in Chapter 3. The design and implementation of a serial multiplication ASM is described in this section.

Analysis of Binary Integer Multiplication

For purposes of illustration, unsigned 4-bit numbers will be used for the multiplicand B and multiplier P. In binary positional notation, B and P can be expressed in the following form:

$$B = b_3\, 2^3 + b_2\, 2^2 + b_1\, 2^1 + b_0\, 2^0$$
$$P = p_3\, 2^3 + p_2\, 2^2 + p_1\, 2^1 + p_0\, 2^0$$

Algebraically, the product BP can be expressed as the sum of the bit products

$$\Sigma\, bp_i = \Sigma\, B(p_i\, 2^i)$$

or

$$BP = B(p_3 2^3) + B(p_2 2^2) + B(p_1 2^1) + B(p_0 2^0) \tag{11.1}$$

The procedure for multiplying two binary integers is analogous to the procedure for multiplying two decimal integers. The multiplicand B is multiplied by bit p_i of the multiplier ($i = 0, 1, 2, 3$). If $p_i = 1$, the bit product is formed by copying down the multiplicand B. If $p_i = 0$, the bit product = 0. Each succeeding bit product is shifted left one position relative to the preceding bit product. For example,

```
B =         1110
P =         1011
            ----
            1110  = B(p0 2^0) = bp0
           1110   = B(p1 2^1) = bp1
          0000    = B(p2 2^2) = bp2
         1110     = B(p3 2^3) = bp3
        --------
        10011010  = product BP = sum of bit products bp_i
```

An alternative to the standard pencil-and-paper multiplication is multiplication using a PPS method. The **PPS method**, starting with a zero cumulative sum, adds each successive bit product to the current cumulative sum to produce the next cumulative sum.

Starting with initial term $S_0 = 0$, a sequence of PPSs, S_{i+1} ($i = 0, 1, 2, 3$), are generated as follows:

		0000.		S_0
$i = 0$	Add	+ 1110.	+	Bp_0
		01110.	=	$Bp_0 + S_0$
	Shift right	0111.0		$S_1 = 2^{-1}(Bp_0 + S_0)$
$i = 1$	Add	+ 1110.	+	Bp_1
		10101.0	=	$Bp_1 + S_1$
	Shift right	1010.10		$S_2 = 2^{-1}(Bp_1 + S_1)$
$i = 2$	Add	+ 0000.	+	Bp_2
		01010.10	=	$Bp_2 + S_2$
	Shift right	0101.010		$S_3 = 2^{-1}(Bp_2 + S_2)$
$i = 3$	Add	+ 1110.	+	Bp_3
		10011.010	=	$Bp_3 + S_3$
	Shift right	1001.1010		$S_4 = 2^{-1}(Bp_3 + S_3)$
	$2^4 S_4$ =	10011010.	=	BP

An Algorithm for Binary Integer Multiplication

An algorithm for binary integer multiplication using the PPS method can be derived by factoring 2^4 from each term of Equation 11.1 and writing the product equation in the following form:

$$BP = 2^4 (Bp_3 2^{-1} + Bp_2 2^{-2} + Bp_1 2^{-3} + Bp_0 2^{-4}) \qquad \textbf{(11.2)}$$

Equation 11.2 can be written in **nested form** (with initial cumulative sum = 0) as follows:

$$BP = 2^4 [2^{-1}(Bp_3 + 2^{-1}\{Bp_2 + 2^{-1}[Bp_1 + 2^{-1}(Bp_0 + 0)]\})] \qquad \textbf{(11.3)}$$

The innermost term $(Bp_0 + 0)$ is obtained by adding Bp_0 and 0, where $0 = S_0$ is the initial cumulative sum. The 5-bit sum is stored in a register, and the register is shifted right 1 bit to accomplish multiplication by 2^{-1}. The product BP can be generated by four repetitions of adding and shifting right, followed by a left shift of 4 bit positions.

Starting with initial term $S_0 = 0$, the sequence of PPSs, representing successive terms of the nested expression, are

$$S_1 = 2^{-1}\,(Bp_0 + S_0) = 2^{-1}\,(Bp_0 + 0)$$
$$S_2 = 2^{-1}\,(Bp_1 + S_1) = 2^{-1}\,[Bp_1 + 2^{-1}\,(Bp_0 + 0)]$$
$$S_3 = 2^{-1}\,(Bp_2 + S_2) = 2^{-1}\,\{Bp_2 + 2^{-1}\,[\,Bp_1 + 2^{-1}\,(Bp_0 + 0)]\}$$
$$S_4 = 2^{-1}\,(Bp_3 + S_3) = 2^{-1}\,(Bp_3 + 2^{-1}\,\{Bp_2 + 2^{-1}\,[Bp_1 + 2^{-1}\,(Bp_0 + 0)]\})$$

PPS Algorithm of Binary Integer Multiplication

Each term of the binary integer multiplication sequence can be generated by the following **recursion formula**:

$$S_{i+1} = 2^{-1}\,(Bp_i + S_i), \quad (i = 0, 1, 2, 3), \quad \text{with starter } S_0 = 0$$

Add
Shift right

This recursion formula generates term S_{i+1} by computing the sum $Bp_i + S_i$, and shifting the sum right 1 bit. The desired product $BP = 2^4\,S_4$.

INTERACTIVE DESIGN APPLICATION

Design of a Serial-Multiplication ASM

An ASM that accomplishes the serial multiplication process can be designed using the 4-step problem-solving procedure.

Step 1: Problem Statement. Organize the process in two parts and describe the system specifications.

a. Process. Starting with initial term $S_0 = 0$, the sequence of PPSs can be generated by the following recursion formula:

$$S_{i+1} = 2^{-1}\,(Bp_i + S_i), \quad (i = 0, 1, 2, 3,), \quad \text{with starter } S_0 = 0$$

Add
Shift right

b. Process control algorithm. Develop a step-by-step procedure to accomplish the serial multiplication process, including initialization, loading data, processing data, and output of results. This can be accomplished by a control algorithm that has four distinct states **a**, **b**, **c**, and **d**, as follows:

State **a**. Wait: Wait for start

State **b**. Reset, Load: $0 \to i, S_0$
Multiplicand $\to B$
Multiplier $\to P$

State **c**. Add: $Z = Bp_i + S_i$ } Process algorithm
State **d**. Shift right: $Z2^{-1} \rightarrow S_{i+1}$
$i = i + 1$
If $i < 4$, repeat States **c** and **d**.
If $i = 4$, process of multiplying BP is completed.

System specification. Design an ASM to accomplish the multiplication of two 4-bit numbers X and Y. The problem can be expressed in register transfer notation (RTN), as follows:

$$X \leftarrow XY$$

Step 2: Conceptualization. Develop a pictorial description of the system.

a. System structure diagram. The organization of an ASM may be conceptualized using a structure diagram that is organized into a control section and a process data-path section, as shown in Figure 11.1.

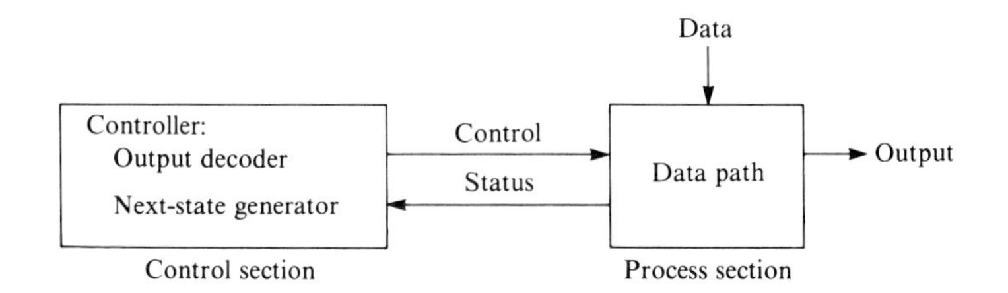

FIGURE 11.1
Structure diagram of an ASM

b. Control flow diagram. The operation of the process control algorithm is graphically described by a control flow diagram that shows the control states and conditions for state transitions and the generation of control signals to data path components. The control algorithm for the PPS method of binary integer multiplication can be translated into the ASM chart shown in Figure 11.2. Note that "st" is a control parameter to denote "start the process." When st = 0, the multiply process is in a wait state. When st = 1, the multiply process starts.

Step 3: Solution/Simplification.

a. Process data path. Design a data path that supports the microoperations required to implement the PPS integer multiplication algorithm.

A 4-bit adder can be used to add product Bp_0 to the accumulator register A, which is initially 0. The X and Y adder inputs, Bp_0 and A respectively, are added to form a 5-bit PPS Z, which is then stored back in registers E and A. After the addition, registers E, A, and P are shifted right 1 bit to align the binary points of the adder inputs.

Each succeeding bit product Bp_i is a 4-bit input to the adder's X inputs, and register A provides a 4-bit input to the adder's Y inputs. After each addition, registers E, A, and P must be shifted right 1 bit to align the binary points of the adder inputs. Thus, the LSB of the multiplier is discarded after its use, and the PPS in AP grows by 1 bit in each successive sum.

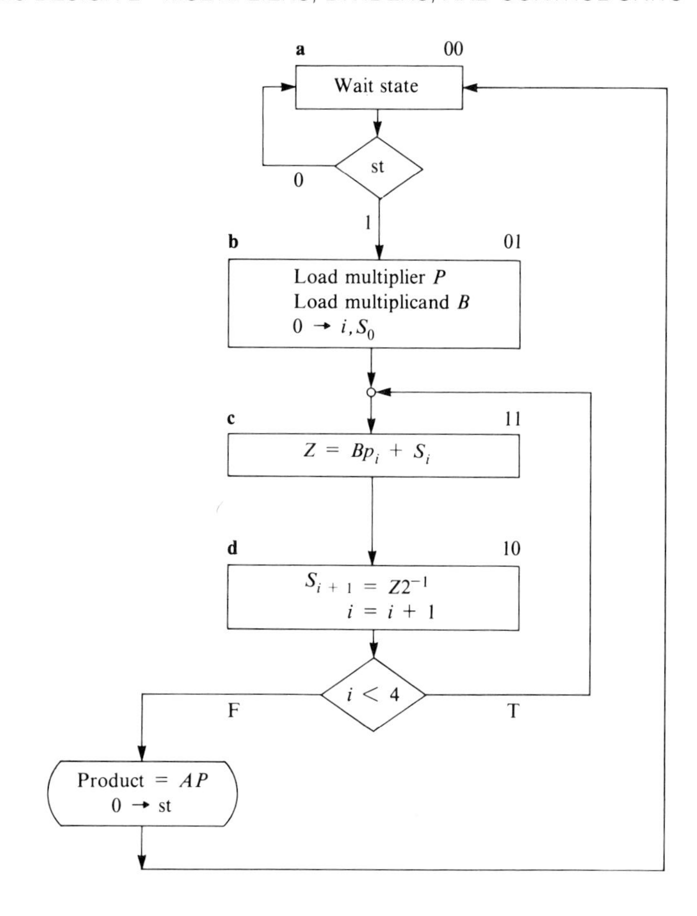

FIGURE 11.2
ASM chart for 4-bit serial multiplication

The PPS algorithm for binary integer multiplication can be implemented using a 4-bit adder, a 4-bit register B for the multiplicand, 4-bit parallel load shift registers A and P, a 1-bit extension flip-flop E. Initially, A is zero, and P holds multiplier P. E and A form a 5-bit register for storing the sum of two 4-bit numbers. Registers E, A, and P are combined to form a 9-bit shift right register. A block diagram of the binary multiplier data path is presented in Figure 11.3. Z denotes the 4-bit sum output by the adder, and z_4 denotes the carry-out of the adder.

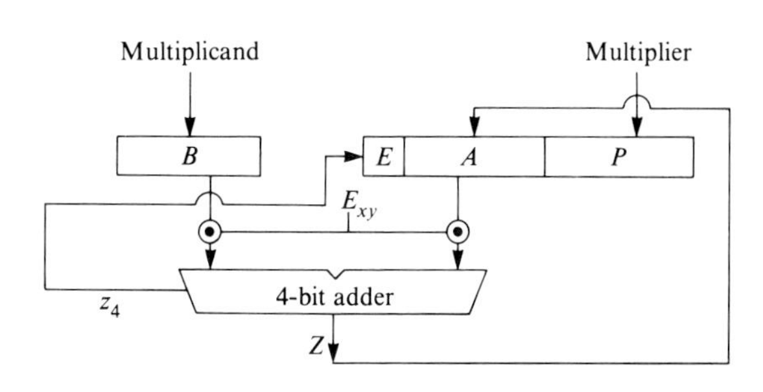

FIGURE 11.3
Block diagram of data path for serial multiplication ASM

b. Control unit. Design a control unit to generate control signals as required to activate data path elements at the appropriate times.

(1) Next-state generator.

(a) The ASM chart (Figure 11.2) can be used to tabulate the control algorithm next states for each present state, together with the control conditions for each present-state/next-state transition, as shown in Table 11.1. A binary counter is used for index i; ct4 denotes count $i = 4$.

TABLE 11.1
Serial multiplication control algorithm table

Present State		Next State		Condition for	Condition for	
AB	Name	Name	A^+B^+	Transition	$A^+ = 1$	$B^+ = 1$
00	**a**	**a**	00	st'	F	st
		b	01	st		
01	**b**	**c**	11	T	T	T
10	**d**	**c**	11	ct4'	ct4'	ct4'
		a	00	ct4		
11	**c**	**d**	10	T	T	F

(b) As an alternative, for each present state AB of the serial adder control algorithm, the next state A^+B^+ can be determined using a next-state map, in which the state variables A and B are map variables and the input conditions are map-entered variables. The next-state map, derived from the control algorithm for the serial multiplication process, is shown in Figure 11.4.

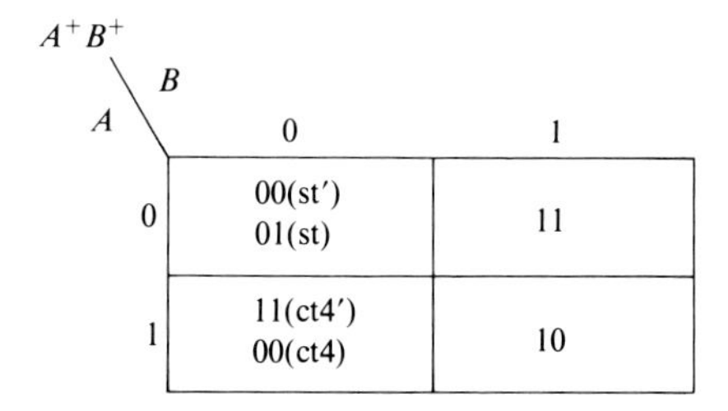

FIGURE 11.4
Serial multiplication control algorithm next-state map

(2) Output decoder. The control unit output decoder for the serial multiplication ASM can be designed by analyzing the ASM chart (Figure 11.2) and the data path diagram (Figure 11.3) to determine the sequence of microoperations required in each control state.

State **a** (00): Set register P controls for load ($S_1S_0 = 11$).

State **b** (01): Clear accumulator and counter. Load multiplier and multiplicand into registers P and B, respectively.

State **c** (11): Note that t_1 interval = State **c** = AB. Set A controls for load ($S_1S_0 = 11$). At time t_1, the contents of Bp_i and A appear at the X and Y adder inputs. After the adder gates' propagation delay, the sum appears at adder output Z, with carry-out = z_4. That is,

$$Z = Bp_i + S_i$$

At time $t_1 \cdot \text{eclk}\uparrow$ (midway through subinterval t_1), $z_4 \rightarrow E$ and $Z \rightarrow A$ (to store 5-bit sum in EA).

State **d** (10): Note that t_2 interval = State **d** = AB'. Set A, P controls for shift right ($S_1S_0 = 01$). At time $t_2 \cdot \text{eclk}\uparrow$ (midway through subinterval t_2), right shift EAP, and increment count i. That is,

$$S_{i+1} = 2^{-1} Z$$
$$i = i + 1$$

If count $i < 4$, repeat States **c** and **d**.

If count $i = 4$, $0 \rightarrow$ st.

Product BP is found in register AP with the binary point at the right end of register P.

This analysis leads to the timing diagram (for States **c** and **d**) shown in Figure 11.5. Note that t_1 and t_2 form a 2-phase clock system. The register controls S_1 and S_0 are set while

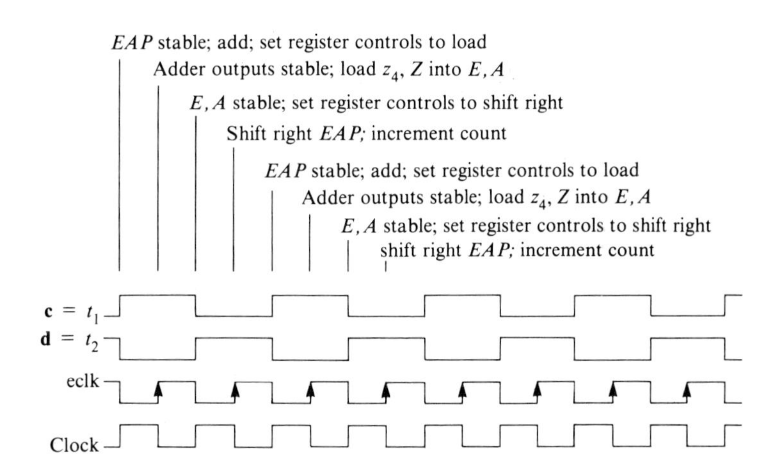

FIGURE 11.5
Timing diagram for serial multiplication states **c** and **d**

signal *clock* is high. The edge-triggered registers E, A, and P are loaded and shifted on the rising edge of signal *eclk*.

The States **a** and **b** microoperations and the timing diagram for States **c** and **d** can then be translated into a table (Table 11.2, p. 524) that lists the microoperations to be accomplished at specified times together with their respective control signals. States: **a** = 00 = $A'B'$; **b** = 01 = $A'B$; **c** = 11 = AB; **d** = 10 = AB'.

TABLE 11.2
Data-path control signals generated by output decoder

State/Time	Function	Microoperations	Control Signals	S_1S_0 P	S_1S_0 A
a (00)	Set S_1S_0 to load P			11	
b (01)	Clear accumulator		clr-$A = A'B$		
	Clear counter		clr-ct $= A'B$		
	Load multiplicand and multiplier	Load B	clk-$B = A'B \cdot$ eclk	11	
		Load P	clk-$P = A'B \cdot$ eclk		
c (11)	Add $Z = Bp_i + \text{A}$	Enable X, Y	$E_{xy} = AB \cdot p_0$		
		Carry-in = 0	Carry-in = 0		
	Set S_1S_0 to load A				11
	Store $z_4 \rightarrow E$	Load E^*	clk-$E = AB \cdot$ eclk		
	$z \rightarrow A$	Load A	clk-$A = AB \cdot p_0 \cdot$ eclk		
d (10)	Set S_1S_0 to shift right			01	01
	Shift right EAP	Shift right EAP	clk-$A = AB' \cdot$ eclk		
		$\rightarrow EAP$	clk-$P = AB' \cdot$ eclk		
	Increment count	Inc ct	inc-ct $= AB' \cdot$ ct4' $\cdot$ eclk		

S_1S_0 are shift register controls.

The control unit output decoder generates control signals to the data path components at appropriate times, as indicated in Table 11.2. These control signals are obtained by gathering the microoperations as follows:

$$\text{clr-}A = A'B$$
$$\text{clr-ct} = A'B$$
$$\text{clk-}B = A'B \cdot \text{eclk}$$
$$^*\text{clk-}E = AB \cdot \text{eclk}$$
$$\text{clk-}A = (AB \cdot p_0 + AB')\text{eclk}$$
$$\text{clk-}P = (A'B + AB')\text{eclk}$$
$$E_{xy} = AB \cdot p_0$$
$$\text{carry-in} = 0$$
$$S_{1A} = AB$$
$$S_{1P} = A'$$
$$S_0 = 1$$
$$\text{inc ct} = AB' \cdot \text{ct4}' \cdot \text{eclk}$$
$$\text{clr st} = AB' \cdot \text{ct4} \cdot \text{eclk}$$

*E is reset if current $p_0 = 0$ ($p_0 = 0$ disables X and Y adder inputs, resulting in carry-out = 0).

Step 4: Realization.

a. Process data path. Use the block diagram in Figure 11.3 to draw a logic diagram of the serial multiplication data path, as shown in Figure 11.6.

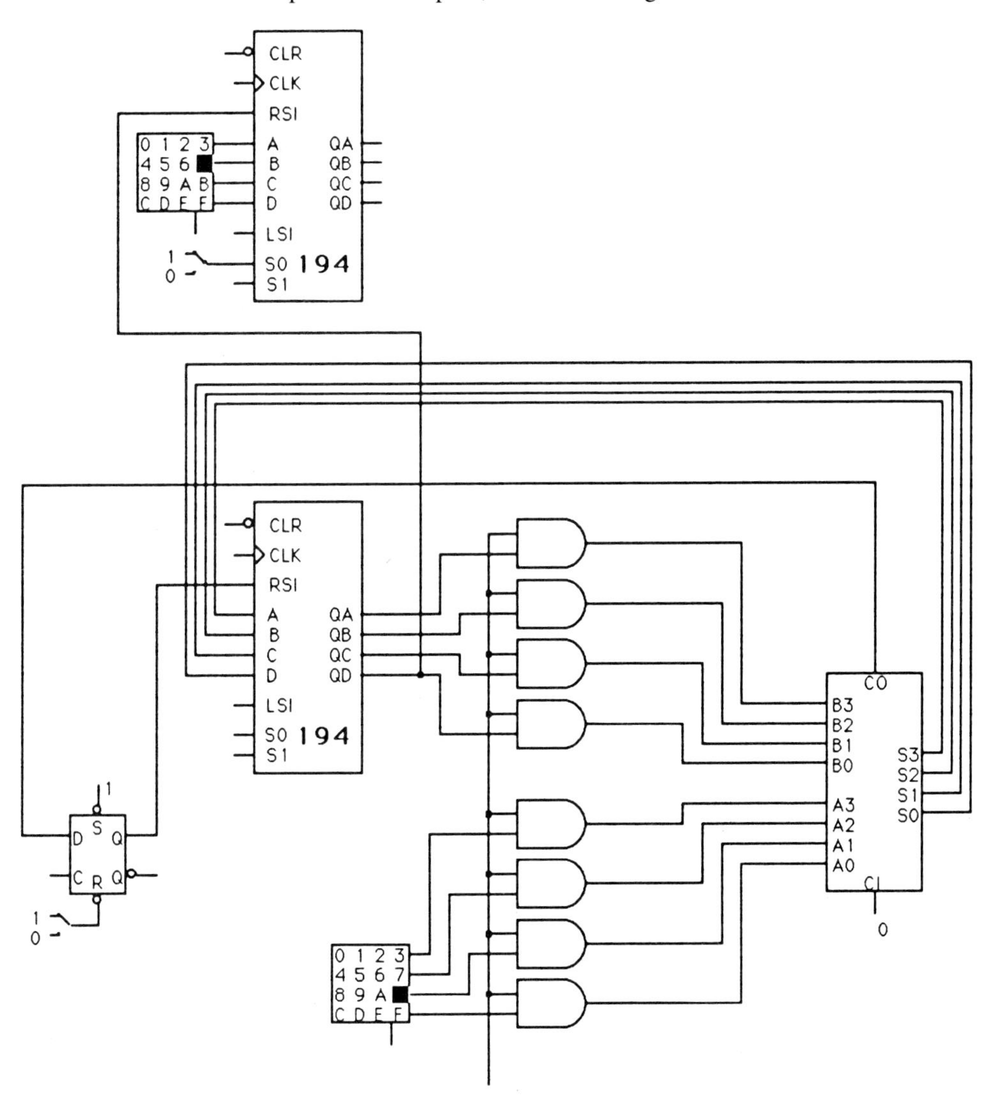

FIGURE 11.6
Logic diagram of serial multiplication data path

b. Control unit.

(1) Next-state generator.

(a) Implementation using multiplexers. The serial multiplication control algorithm can be implemented directly from the control algorithm table using 4-to-1 multiplexers, where present state *AB* is used to select the input condition corresponding to the select value *AB* combination, as shown in Figure 11.7 (p. 526). Note that changes in the state variables occur on the rising edge of signal *clock*.

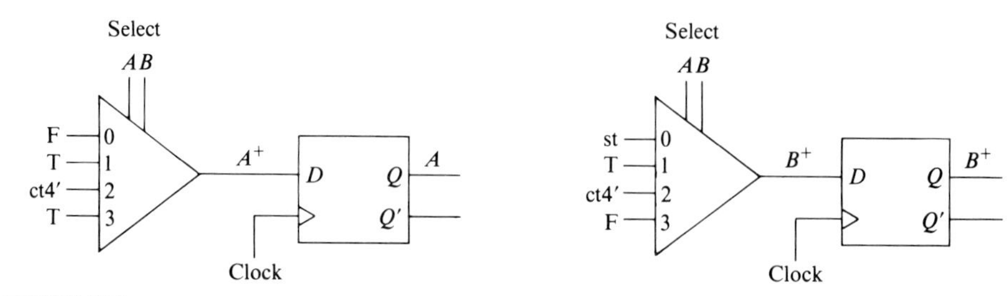

FIGURE 11.7
Multiplexer implementation of next-state generator

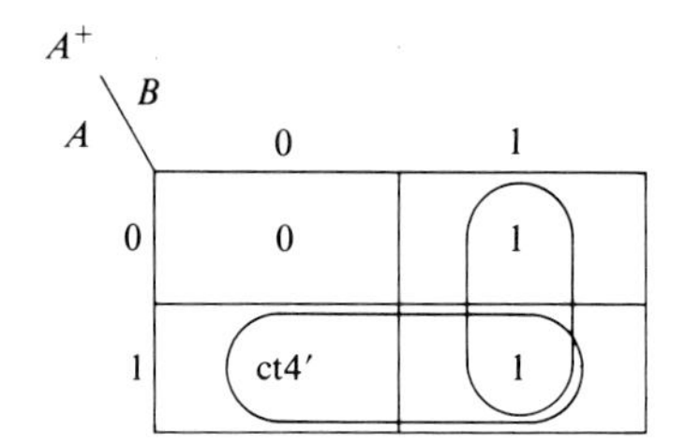

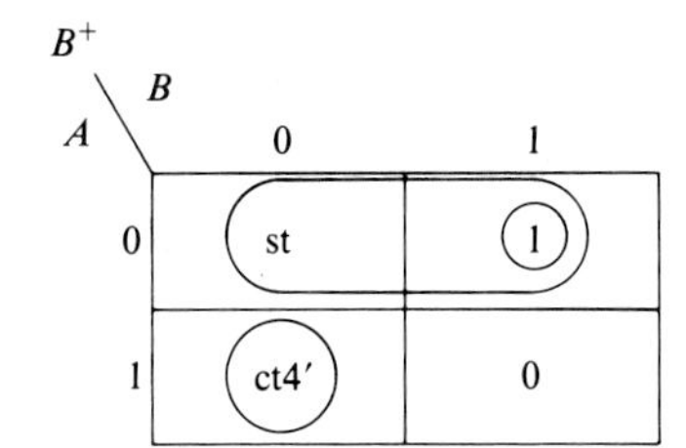

FIGURE 11.8
Next-state maps for controller state variables

(b) Implementation using D flip-flops. If the next-state generator is to be implemented using PET D flip-flops, individual next-state maps for each state variable can be derived from the next-state map in Figure 11.4. The individual next-state maps are shown in Figure 11.8.

The D flip-flop excitation equations, as determined from these maps, are

$$D_A = B + A \cdot \text{ct4'} \qquad D_B = A'B + A'\text{st} + AB' \cdot \text{ct4'}$$

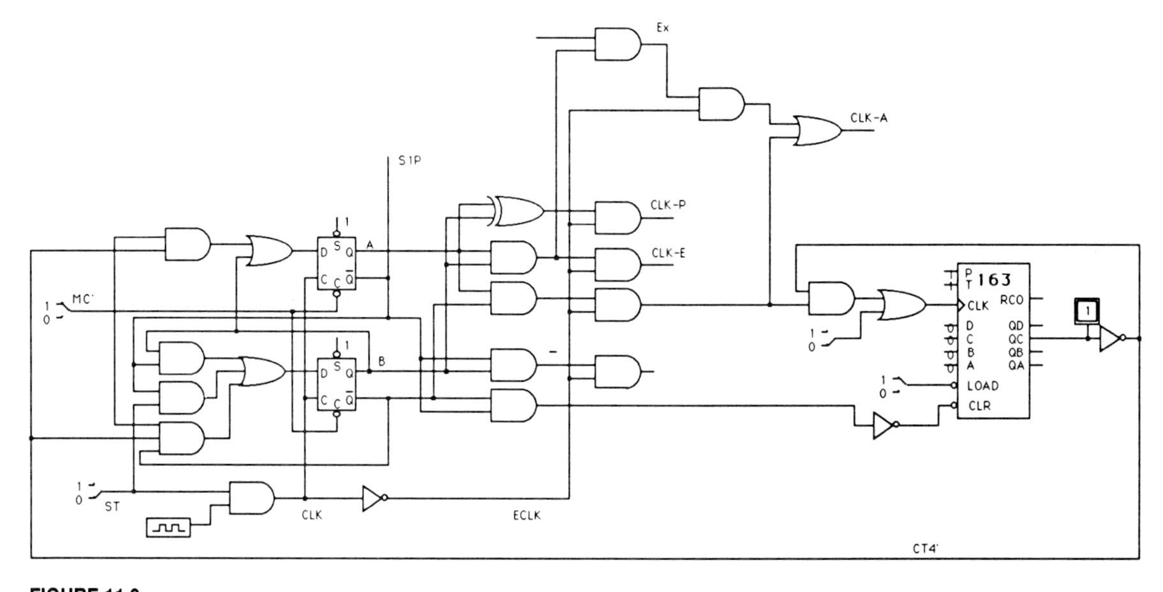

FIGURE 11.9
Logic diagram of output decoder and next-state generator

(2) Output decoder. Use the logic equations of the control signals generated by the output decoder to construct a logic diagram of the output decoder. Figure 11.9 presents a logic diagram of the output decoder and the next-state generator (implemented using PET D flip-flops).

A binary counter can be used to generate i as required for the major cycle counter. Each major cycle i is divided into 2 minor cycles, t_1 and t_2.

Combine the system control unit and the process data path by connecting each control unit output to the appropriate data-path activation point. The resulting configuration is a complete serial multiplication ASM. Logic and timing diagrams are presented in Figure 11.10.

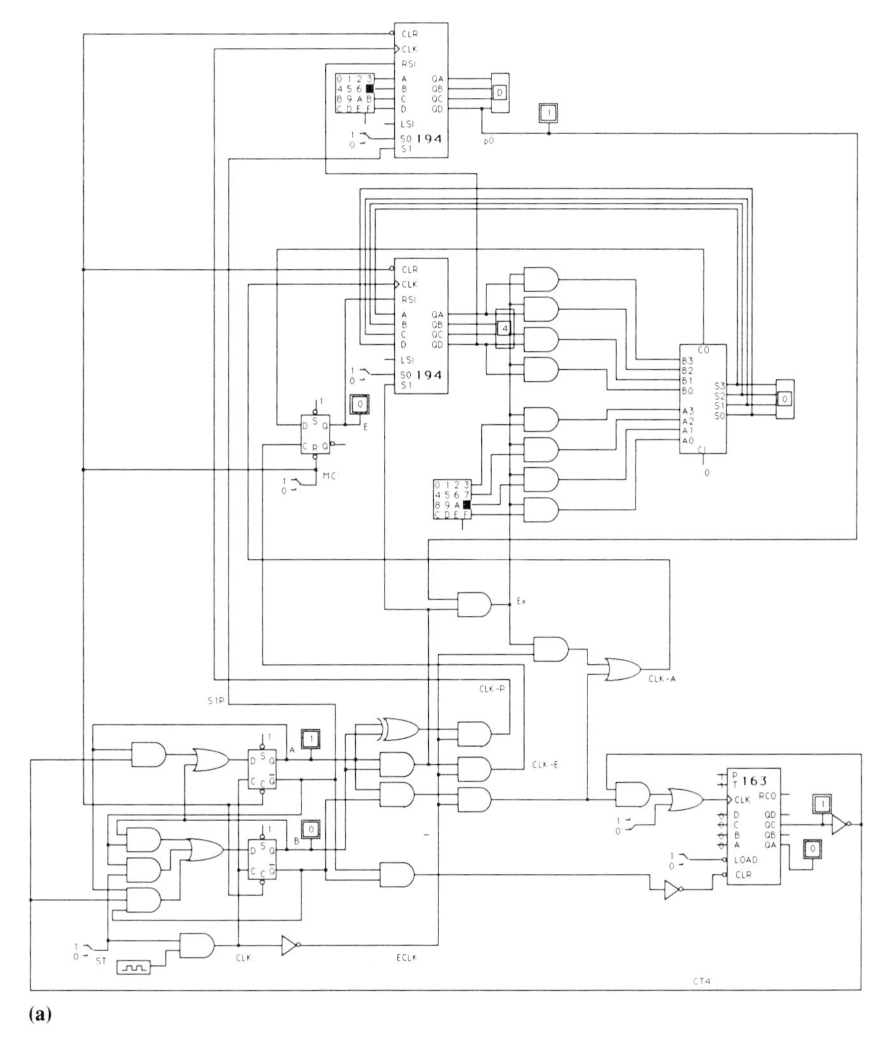

(a)

FIGURE 11.10
Complete serial multiplication ASM. (a) Logic diagram (b) Timing diagram

FIGURE 11.10
Continued

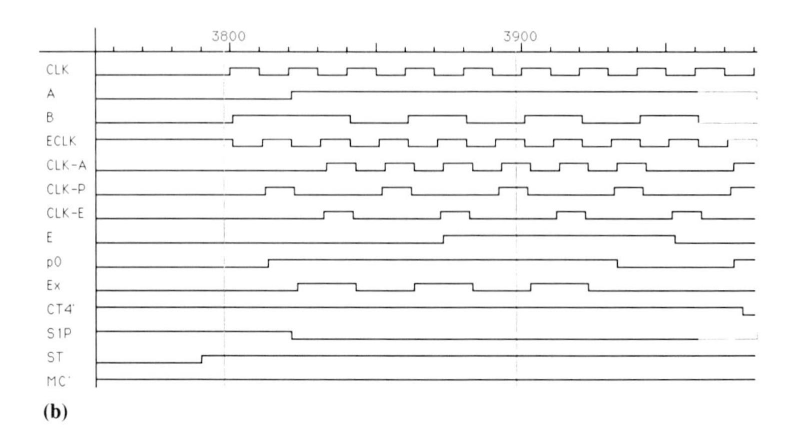

An example of binary integer multiplication using the PPS process algorithm is presented below, where $B = 1110$ is multiplied by $P = 1011$. The example shows the contents of the A and P registers, in both hexadecimal and binary form, at each stage ($i = 0, 1, 2, 3$) of the process.

Cycle i	Hex A P			E A P		
			$B = 1110$			
		Initial:	$S_0 = 0$	0 0000 1011	$p_3p_2p_1p_0$	
			Bp_0	1110	Add	
				0 1110 1011	$Z \rightarrow E, A$	t_1
0	7 5		$S_1 = 2^{-1}(Bp_0 + S_0)$	0 0111 0101	Shift right	t_2
			Bp_1	1110	Add	
				1 0101 0101	$Z \rightarrow E, A$	t_1
1	A A		$S_2 = 2^{-1}(Bp_1 + S_1)$	0 1010 1010	Shift right	t_2
			Bp_2	0000	Add	
				0 1010 1010	$Z \rightarrow E, A$	t_1
2	5 5		$S_3 = 2^{-1}(Bp_2 + S_2)$	0 0101 0101	Shift right	t_2
			Bp_3	1110	Add	
				1 0011 0101	$Z \rightarrow E, A$	t_1
3	9 A		$S_4 = 2^{-1}(Bp_3 + S_3)$	0 1001 1010	Shift right	t_2

The final product is

$$BP = 2^4 S_4 = 2^4 (Bp_3 2^{-1} + Bp_2 2^{-2} + Bp_1 2^{-3} + Bp_0 2^{-4})$$

This shifts the binary point 4 bits to the right (to the end of the P register).

11.3 DESIGN AND IMPLEMENTATION OF A SERIAL DIVISION ASM

There are several alternatives for implementing a divide capability in a computer's CPU, including

- ***Generating a sequence of partial remainders, where successive partial remainders are defined by a recursion formula. A division function can be programmed as a subroutine that repeatedly invokes a shift instruction and a subtraction instruction.***
- ***A serial division ASM can be designed and implemented using an $(n + 1)$-bit adder and appropriate shift registers.***

In the next section, we will analyze the pencil-and-paper procedures for dividing an n-bit dividend by an n-bit divisor to produce a quotient and remainder. Then, we will develop an algorithm for the division process, keeping in mind that the process will be implemented as a digital logic system consisting of a process data path and a process controller. Finally, we will use the 4-step problem-solving procedure for system design to design a serial division ASM.

Analysis of Binary Integer Division

The procedure for integer division can best be understood by examining numerical examples in both the decimal and binary number systems. As we will show, longhand division in the binary number system is simpler than decimal division because each coefficient of the binary number quotient is either 0 or 1.

In binary positional notation, P and Q can be expressed as

$$P = 2^3\,p_3 + 2^2\,p_2 + 2^1\,p_1 + 2^0\,p_0$$
$$Q = 2^3\,q_3 + 2^2\,q_2 + 2^1\,q_1 + 2^0\,q_0$$

where dividend $= P = p_3p_2p_1p_0$, quotient $= q_3q_2q_1q_0$, and divisor $= B$.

The **partial dividend** {PD}, indicated by { }, is that part of the dividend that is being divided in each step. In step j, the quotient digit or bit q_j indicates the number of times the divisor "goes into" the partial dividend. The true **partial remainder** $R_j = \{PD\} - Bq_j$.

Consider the following examples of binary and decimal division.

Decimal Division	*Binary Division*
Divide 2136 by 14	Divide 1101 by 11
0152. $= q_3q_2q_1q_0$	0100. $= q_3q_2q_1q_0$
14) {2}136	11) {1}101
– 0	– 0
{21}	{11}
– 14	– 11
{73}	{00}
– 70	– 0
{36}	{01}
– 28	– 0
8 = Remainder	1= Remainder

The following example uses algebra to illustrate the pencil-and-paper method of binary integer division. For each j, the true remainder $\{R_j\}$ is obtained by subtracting the divisor times quotient bit product (Bq_j) from partial dividend $\{PD\}$. That is, $\{R_j\} = \{PD\} - Bq_j$, for $j = 3, 2, 1, 0$.

$$
\begin{array}{rl}
 & \;\;\; q_3 \quad q_2 \quad q_1 \quad q_0 \\
B & \overline{\big)\; \{p_3\} \quad p_2 \quad p_1 \quad p_0}
\end{array}
$$

Index j		
	$- Bq_3$	**If** $B > p_3$, **then** $q_3 = 0$ **else** $q_3 = 1$
$3 =$	R_3	$R_3 = \{p_3\} - Bq_3$
	$\{2R_3 + p_2\}$	
	$- Bq_2$	**If** $B > 2R_3 + p_2$, **then** $q_2 = 0$ **else** $q_2 = 1$
$2 =$	R_2	$R_2 = \{2R_3 + p_2\} - Bq_2$
	$\{2R_2 + p_1\}$	
	$- Bq_1$	**If** $B > 2R_2 + p_1$, **then** $q_1 = 0$ **else** $q_1 = 1$
$1 =$	R_1	$R_1 = \{2R_2 + p_1\} - Bq_1$
	$\{2R_1 + p_0\}$	
	$- Bq_0$	**If** $B > 2R_1 + p_0$, **then** $q_0 = 0$ **else** $q_0 = 1$
$0 =$	R_0	$R_0 = \{2R_1 + p_0\} - Bq_0$

An Algorithm for Binary Integer Division

The notation used in the previous section can be used to develop an algorithm for binary integer division. For purposes of illustration, unsigned 4-bit numbers will be used for the dividend P and divisor B. The process of dividing P by B is expressed by the following equation:

$$P - BQ = R$$

where Q = quotient, R = remainder, and $R < B$.

Derivation of Formulas for Quotient Coefficients q_j

The division of P by B can be expressed in the following form:

$$2^3 p_3 + 2^2 p_2 + 2^1 p_1 + 2^0 p_0 - B(2^3 q_3 + 2^2 q_2 + 2^1 q_1 + 2^0 q_0) = R \tag{1}$$

The objective of the division process is to determine the quotient coefficients q_3, q_2, q_1, and q_0 and the remainder R. Coefficient q_j and true partial remainder R_j are determined for index j, ($j = 3, 2, 1, 0$), as follows:

Divide Equation (1) by 2^3 to isolate q_3:

$$\{p_3\} - Bq_3 = R_3 = \text{true partial remainder}$$

Divide Equation (1) by 2^2 to isolate q_2:

$$(2p_3 + p_2) - B(2q_3 + q_2) = R_2$$
$$2(p_3 - Bq_3) + p_2 - Bq_2 = R_2$$

Substituting R_3 for $p_3 - Bq_3$ produces

$$\{2R_3 + p_2\} - Bq_2 = R_2 = \text{true partial remainder}$$

Divide Equation (1) by 2^1 to isolate q_1:

$$(4p_3 + 2p_2 + p_1) - B(4q_3 + 2q_2 + q_1) = R_1$$
$$2[2(p_3 - Bq_3) + (p_2 - Bq_2)] + p_1 - Bq_1 = R_1$$
$$2[2R_3 + (p_2 - Bq_2)] + p_1 - Bq_1 = R_1$$

Substituting R_2 for $2R_3 + (p_2 - Bq_2)$ produces

$$\{2R_2 + p_1\} - Bq_1 = R_1 = \text{true partial remainder}$$

Divide Equation (1) by 2^0 to isolate q_0:

$$8p_3 + 4p_2 + 2p_1 + p_0 - B(8q_3 + 4q_2 + 2q_1 + q_0) = R_0$$
$$8(p_3 - Bq_3) + 4(p_2 - Bq_2) + 2(p_1 - Bq_1) + p_0 - Bq_0 = R_0$$
$$2[2\{2(p_3 - Bq_3) + (p_2 - Bq_2)\} + (p_1 - Bq_1)] + p_0 - Bq_0 = R_0$$
$$2[2\{2R_3 + (p_2 - Bq_2)\} + (p_1 - Bq_1)] + p_0 - Bq_0 = R_0$$
$$2[2R_2 + p_1 - Bq_1] + p_0 - Bq_0 = R_0$$

Substituting $R_1 = 2R_2 + p_1 - Bq_1$ produces

$$\{2R_1 + p_0\} - Bq_0 = R_0 = \text{true partial remainder}$$

Note that each equation for R_j ($j = 3, 2, 1, 0$) can be expressed in the following form:

$$R_j = \{PD\} - q_jB$$

The computation of q_j and R_j, using partial dividend = {PD}, for $j = 3, 2, 1, 0$, is accomplished as follows:

- For $j = 3$: $R_3 = \{p_3\} - q_3B$
  - Assume $q_3 = 1$; compute assumed remainder $Z_3 = \{p_3\} - 1B$.
    - If $Z_3 < 0$: division fails, $q_3 = 0$, $R_3 = \{p_3\} - 0B = Z_3 + B$
    - If $Z_3 \geq 0$: division succeeds, $q_3 = 1$, $R_3 = \{p_3\} - 1B = Z_3$
- For $j = 2$: $R_2 = \{2R_3 + p_2\} - q_2B$
  - Assume $q_2 = 1$; compute assumed remainder $Z_2 = \{2R_3 + p_2\} - 1B$.
    - If $Z_2 < 0$: division fails, $q_2 = 0$, $R_2 = \{2R_3 + p_2\} - 0B = Z_2 + B$.
    - If $Z_2 \geq 0$: division succeeds, $q_2 = 1$, $R_2 = \{2R_3 + p_2\} - 1B = Z_2$.
- For $j = 1$: $R_1 = \{2R_2 + p_1\} - q_1B$
  - Assume $q_1 = 1$; compute assumed remainder $Z_1 = \{2R_2 + p_1\} - 1B$.

If $Z_1 < 0$: division fails, $q_1 = 0$, $R_1 = \{2R_2 + p_1\} - 0B = Z_1 + B$
If $Z_1 \geq 0$: division succeeds, $q_1 = 1$, $R_1 = \{2R_2 + p_1\} - 1B = Z_1$.

- For $j = 0$: $R_0 = \{2R_1 + p_0\} - q_0B$.
  Assume $q_0 = 1$; compute assumed remainder $Z_0 = \{2R_1 + p_0\} - 1B$.
  If $Z_0 < 0$: division fails, $q_0 = 0$, $R_0 = \{2R_1 + p_0\} - 0B = Z_0 + B$.
  If $Z_0 \geq 0$: division succeeds, $q_0 = 1$, $R_0 = \{2R_1 + p_0\} - 1B = Z_0$.

Binary Division Algorithm

The formulas for computing the quotient coefficients q_j and the true partial remainders R_j can be summarized in the following algorithm:

- For $j = 3$: $R_3 = \{p_3\} - q_3B$
  Assume $q_3 = 1$; compute assumed remainder $Z_3 = \{p_3\} - 1B$.
  If $Z_3 < 0$: division fails, $q_3 = 0$, $R_3 = Z_3 + B$.
  If $Z_3 \geq 0$: division succeeds, $q_3 = 1$, $R_3 = Z_3$.
- For $j = 2, 1, 0$: $R_j = \{2(R_{j+1} + 2^{-1}p_j)\} - q_jB = \{PD\} - q_jB$
  Assume $q_j = 1$; compute assumed remainder $Z_j = \{2[R_{j+1} + 2^{-1}p_j]\} - 1B$.
  If $Z_j < 0$: division fails, $q_j = 0$, $R_j = Z_j + B$.
  If $Z_j \geq 0$: division succeeds, $q_j = 1$, $R_j = Z_j$.

Note that the formula $\{R_j\} = \{PD\} - q_jB$ is the same as in our initial example.

Restoring Method of Binary Integer Division

The assumption that $q_j = 1$ produces an assumed remainder $Z_j = \{PD\} - 1B$. If the quantity $Z_j < 0$, the division fails and $q_j = 0$. The **restoring method** of binary division restores the partial dividend by adding B back to Z_j. The true partial remainder R_j is obtained as follows:

- If $Z_j < 0$, then $q_j = 0$, and $R_j = \{PD\} - 0B = Z_j + B$.
- If $Z_j \geq 0$, then $q_j = 1$, and $R_j = \{PD\} - 1B = Z_j$.

The restoring method of binary integer division is accomplished by the following procedure:

Registers A and P are concatenated to form an 8-bit register.

Initial conditions: $R_4 = 0$; $A.P = 0000.p_3p_2p_1p_0$.

Let j denote the step in which q_j and R_j are determined.

For each j ($j = 3, 2, 1, 0$), the true partial remainder

$$R_j = \{2[R_{j+1} + 2^{-1}p_j]\} - q_jB$$

is computed by the following procedure:

1. Shift left AP: $A = 2[R_{j+1} + 2^{-1}p_j] = 2R_{j+1} + p_j$.
2. Subtract: (Assume $q_j = 1$) $Z_j = A - 1B = \{PD\} - 1B$.
3. If division fails, restore partial dividend by adding B to Z_j:

If $Z_j < 0$: division fails, $q_j = 0$, $\{PD\} - 0B = Z_j + B = R_j \rightarrow A$
If $Z_j \geq 0$: division succeeds, $q_j = 1$, $\{PD\} - 1B = Z_j = R_j \rightarrow A$

Note that

$$\text{If } Z_j < 0,\ (Z_j \text{ sign})' = (1)' = 0 \rightarrow q_j$$
$$\text{If } Z_j \geq 0,\ (Z_j \text{ sign})' = (0)' = 1 \rightarrow q_j$$

where sign denotes the sign bit of Z_j. Therefore, $q_j = (z_j \text{ sign})'$ in either case.

A more efficient method of integer binary division is the nonrestoring method presented next.

Nonrestoring Method of Binary Division

The **nonrestoring method** of binary integer division uses assumed partial remainder $Z_j = \{PD\} - 1B$, thereby reducing the partial dividend $\{PD\}$ by B for index j. If $Z_j < 0$, the division fails, so $q = 0$, and $2B$ is added to $\{PD\}$ in the next step to correct the partial dividend. The procedure for the nonrestoring method is illustrated in the following example. Note that if $Z_j < 0$, Z_j sign = 1 and $q_j = 0$; if $Z_j \geq 0$, Z_j sign = 0 and $q_j = 1$. Therefore, $q_j = (Z_j \text{ sign})'$. In the example, we divide $P = 1101$ by $B = 0011$.

			0100	= quotient	
			1111	= trial quotient	
	0011	)	0 0000 1101	$= 0\ 0000\ p_3p_2p_1p_0$	
Shift left	$\{p_3\}$		1		
Subtract	$-B$		-11		
	Z_3	=	-10	$Z_3 < 0$:	$q_3 = 0$ Add next
Shift left	$2Z_3 + p_2$		$-100 + 1$		
Add	$+B$		$+\ 11$		
	Z_2	=	0	$Z_2 = 0$:	$q_2 = 1$ Subtract next
Shift left	$2Z_2 + p_1$		$2(0) + 0$		
Subtract	$-B$		$-\ 11$		
	Z_1	=	-11	$Z_1 < 0$:	$q_1 = 0$ Add next
Shift left	$2Z_1 + p_0$		$-110 + 1$		
Add	$+B$		$+\ 11$		
	Z_0	=	-10	$Z_0 < 0$:	$q_0 = 0$ Add next
	$R = Z_0 + B$		-10		
			$+11$		
			$R = 1$	= remainder	

To summarize the preceding example, if the assumed partial remainder $Z_{j+1} < 0$, the division failed ($q_j = 0$), and we must *add* $2B$ to $\{PD\}$ in the next step to *correct* the partial dividend.

$$Z_j = \{2Z_{j+1} + 2B + p_j\} - 1B \leftrightarrow (2Z_{j+1} + p_j) + B$$

Shift left Add

If $Z_{j+1} \geq 0$, the division succeeded ($q_j = 1$), so the partial dividend *does not require correction* in the next step.

$$Z_j = \{2Z_{j+1} + 0B + p_j\} - 1B \leftrightarrow (2Z_{j+1} + p_j) - B$$

Shift left Subtract

$$(Z_j \text{ sign})' \rightarrow q_j$$

Let P be a 4-bit shift register that initially holds an unsigned dividend P. Let B be a 4-bit register that holds an unsigned divisor B. B_s is a flip-flop, appended to the left end of register B, to store a sign bit. Let A be a 4-bit shift register that initially is cleared. A_s is a flip-flop, appended to the left end of register A, to store a sign bit. Register A will be used to hold a partial dividend $\{PD\}$ at each stage of the process. Registers A and P are concatenated, with A to the left of P. Use 2s complement arithmetic to accomplish subtraction of the 5-bit number in B_sB from the 5-bit number in A_sA.

The nonrestoring method of division can be accomplished using the following algorithm:

- For $j = 3$, shift left AP: $\{PD\} = p_3$
  Assume $q_3 = 1$; compute $Z_3 = p_3 - 1B$; $Z_{3s}' \rightarrow q_3$.
- For $j = 2, 1, 0,$
  If $Z_{j+1} < 0$, add $2B$ to correct the partial dividend:

$$Z_j = \{2Z_{j+1} + 2B + p_j\} - 1B \leftrightarrow (2Z_{j+1} + p_j) + B$$

Shift left Add

If $Z_{j+1} \geq 0$,

$$Z_j = \{2Z_{j+1} + p_j\} - 1B \leftrightarrow (2Z_{j+1} + p_j) - B$$

Shift left Subtract

$$(Z_j \text{ sign})' \rightarrow q_j$$

The following is an example of nonrestoring binary division, where $P = 1101$ and $B = 0011$ (1 1101 is the 2s complement of B_sB).

j	Time	Function	B_s	B	A_s	A	P
			0	0011	0	0000	1101
3	t_1	Shift left			0	0001	101_
	t_2	Subtract			1	1101	
		$Z_s' \rightarrow q_3 = 0$			1	1110	1010

Add next since $Z_s = 1$

j	Time	Function	B_s B	A_s A P
2	t_1	Shift left		1 1101 010_
	t_2	Add		0 0011
		$Z_s' \rightarrow q_2 = 1$	Subtract next since $Z_s = 0$	0 0000 0101
1	t_1	Shift left		0 0000 101_
	t_2	Subtract		1 1101
		$Z_s' \rightarrow q_1 = 0$	Add next since $Z_s = 1$	1 1101 1010
0	t_1	Shift left		1 1011 010_
	t_2	Add		0 0011
		$Z_s' \rightarrow q_0 = 0$	Add next since $Z_s = 1$	1 1110 0100 → Quotient
	t_1	______		
	t_2	$R = A + B$		0 0011
				0 0001 = Remainder

INTERACTIVE DESIGN APPLICATION

Design of a Serial Division ASM

An ASM that accomplishes the serial division process (using the nonrestoring method) can be designed using the 4-step problem-solving procedure.

Step 1: Problem Statement. Organize the process in two parts and describe the system specifications.

a. Process.

□ For $j = 3$, shift left AP: $\{PD\} = p_3$
  Assume $q_3 = 1$; compute $Z_3 = p_3 - 1B$; $Z_{3s}' \rightarrow q_3$.

□ For $j = 2, 1, 0,$
  If $Z_{j+1} < 0$, add $2B$ to correct the partial dividend:

$$Z_j = \{2Z_{j+1} + 2B + p_j\} - 1B \leftrightarrow \underbrace{(2Z_{j+1} + p_j)}_{\text{Shift left}} \underbrace{+ B}_{\text{Add}}$$

If $Z_{j+1} \geq 0$,

$$Z_j = \{2Z_{j+1} + 0B + p_j\} - 1B \leftrightarrow \underbrace{(2Z_{j+1} + p_j)}_{\text{Shift left}} \underbrace{- B}_{\text{Subtract}}$$

$$(Z_j \text{ sign})' \rightarrow q_j$$

b. Process control algorithm. Develop a step-by-step procedure to accomplish the serial division process, including initialization, loading data, processing data, and output of results. This can be accomplished by a control algorithm that has five distinct states **a**, **b**, **c**, **d**, and **e**. In order to be compatible with the serial multiplication process (so that we can design a combined multiplier/divider later), a binary counter is used for index i. Note that index j counts down from 3 to 0 while index i counts up from 0 to 3.

State **a**. Wait: Wait for start.

State **b**. Reset, Load: $0 \rightarrow i, A$; $3 \rightarrow j$
Divisor $\rightarrow B$
Dividend $\rightarrow P$
Define $Z_4 = 0$

State **c**. Shift left AP: $Z = 2Z_{j+1} + p_j$

State **d**. Add/subtract: If $Z_{j+1} < 0$, $Z_j = Z + B$
If $Z_{j+1} \geq 0$, $Z_j = Z - B$
$i + 1 \rightarrow i, j - 1 \rightarrow j$
If $i < 4$, repeat States **c** and **d**;
If $i = 4$, process is completed; go to **e**.

State **e**. Restore remainder: If $Z_0 < 0$, remainder $R = Z_0 + B$

System specifications. Design an ASM to accomplish the integer division of X by Y, where X and Y are 4-bit unsigned integers. The problem can be expressed in register transfer notation as

$$X \leftarrow X/Y$$

Step 2: Conceptualization. Develop a pictorial description of the system.

a. System structure diagram. The organization of an ASM may be conceptualized using a structure diagram that is organized into a control section and a process data path section, as shown in Figure 11.11.

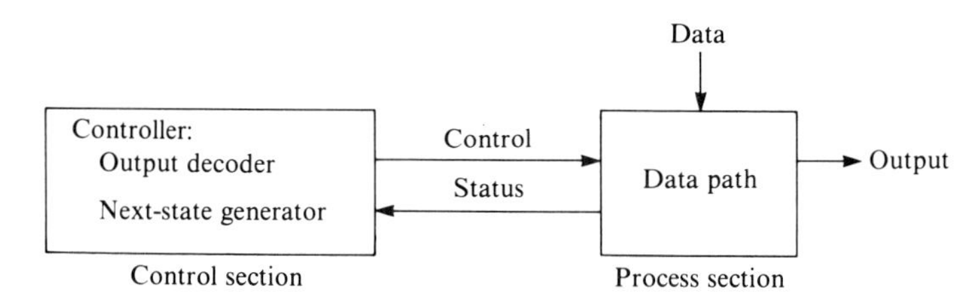

FIGURE 11.11
Structure diagram of a serial division ASM

b. Control flow diagram. The operation of the process control algorithm is graphically described by a control flow diagram that shows the control states and conditions for state transitions and the generation of control signals to data path components. The control algorithm for the nonrestoring method of binary integer division can be translated into the diagram shown in Figure 11.12. Note that for each j ($j = 2, 1, 0$):

$$\text{If } Z_{j+1} < 0,\ Z_j = 2Z_{j+1} + p_j + B;\ Z_{js}' \rightarrow q_j$$
$$\text{If } Z_{j+1} \geq 0,\ Z_j = 2Z_{j+1} + p_j - B;\ Z_{js}' \rightarrow q_j$$

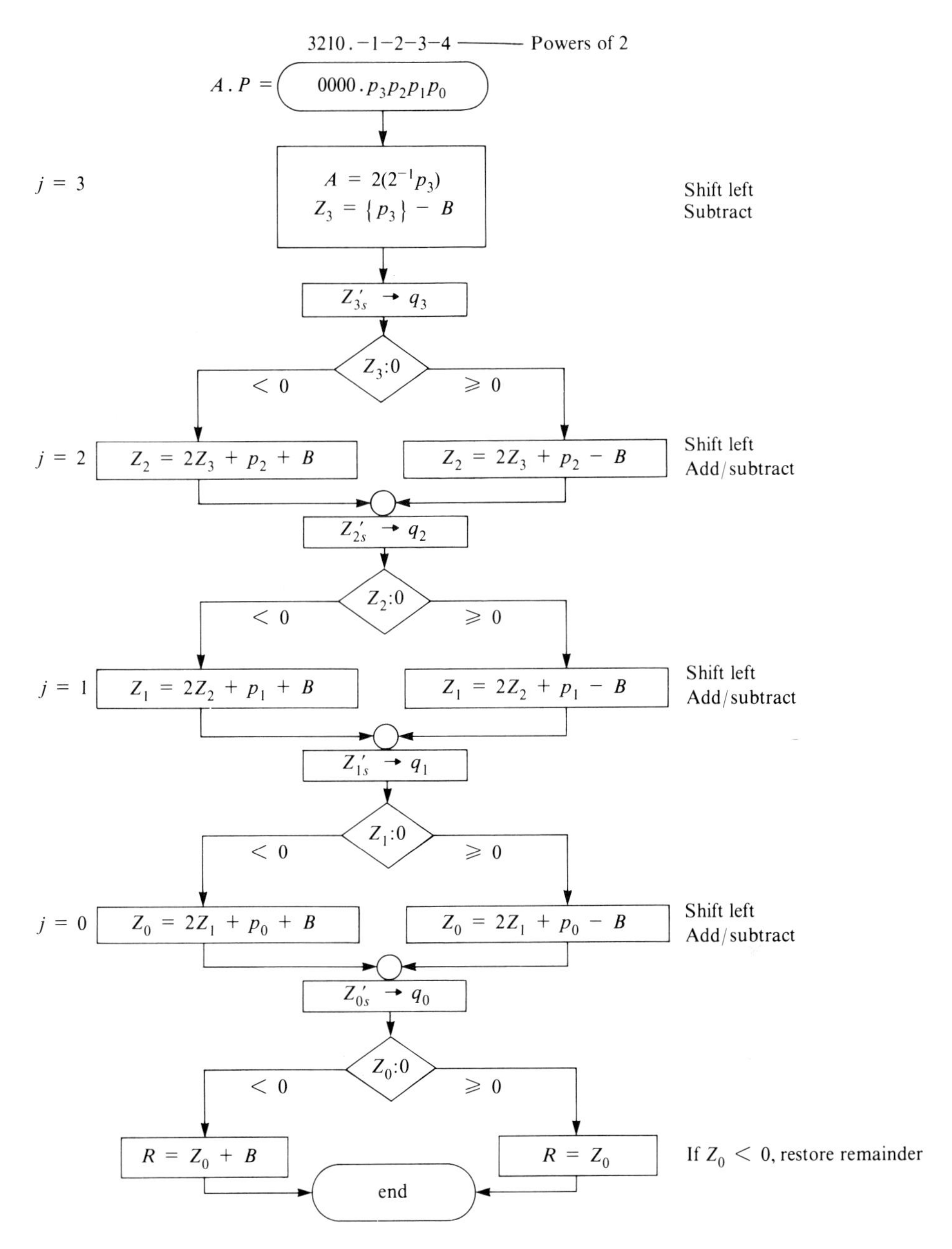

FIGURE 11.12
Diagram for nonrestoring division

The nonrestoring division algorithm can be translated into the ASM chart shown in Figure 11.13 (p. 538). Note that $Z_{sn} = (Z'_{so} \oplus A_s) \oplus C_o$. Z_{so} and Z_{sn} denote old and new values of the sign Z_s of the result Z_j, respectively.

FIGURE 11.13
ASM chart for nonrestoring division

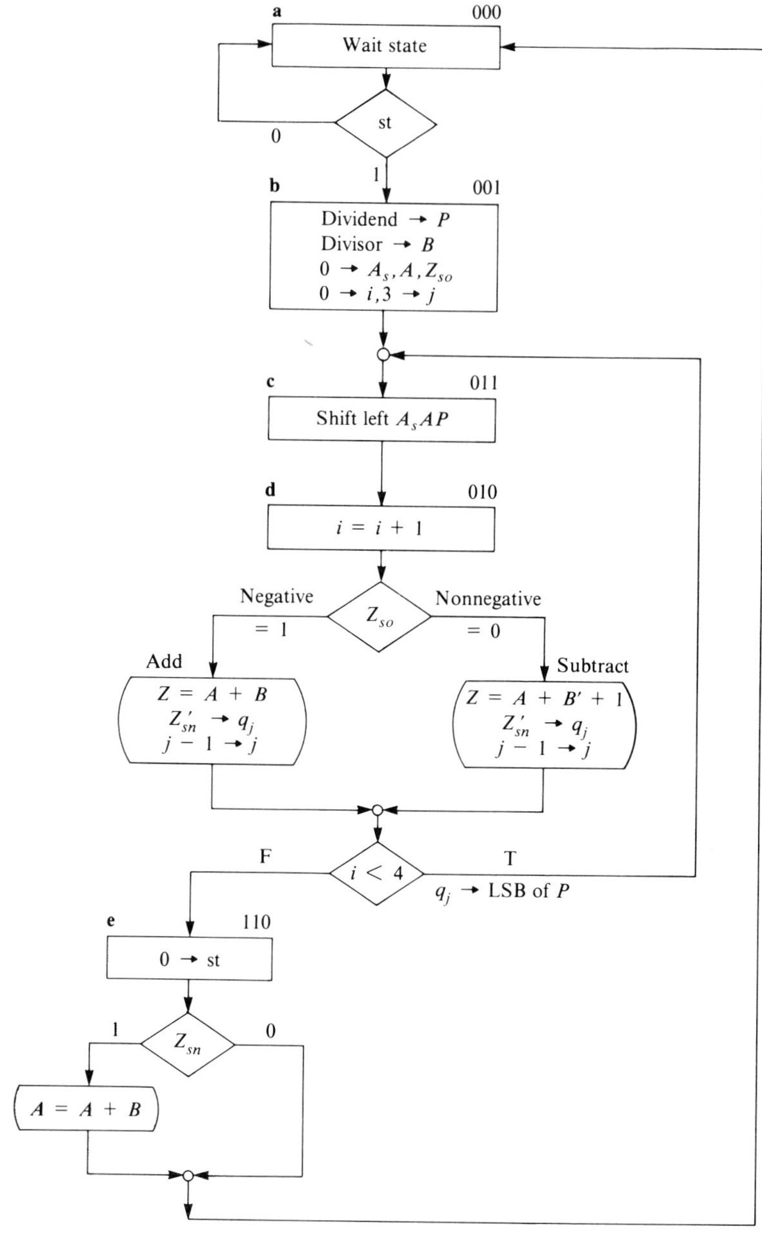

Step 3: Solution/Simplification.

a. Process data path. Design a data path that supports the microoperations required to implement the nonrestoring integer division algorithm. The nonrestoring division algorithm for binary integer division can be implemented using a 4-bit adder, a 4-bit register B for the divisor, 4-bit parallel load shift registers A and P, and 1-bit flip-flops A_s and Z_s. (Note that Z_s is a fifth bit used for the sign of B). Initially, A_s, A, and Z_s are zero. Initially P holds the dividend. In each major cycle j, q_j is stored in the LSB of P. When the division is completed, register P holds the quotient $q_3q_2q_1q_0$. Let Z denote the 4-bit sum output by the adder, and let z_4 denote the carry-out of the adder. A 2-to-1 mux is used to interface register P with both a hexadecimal keypad (for loading the dividend) and $p_3p_2p_1Z'_{sn}$. The 5-bit addition/subtraction is accomplished using

$$Z_{sn} = (Z'_{so} \oplus A_s) \oplus C_o$$

Figure 11.14 shows the block diagram for the data path.

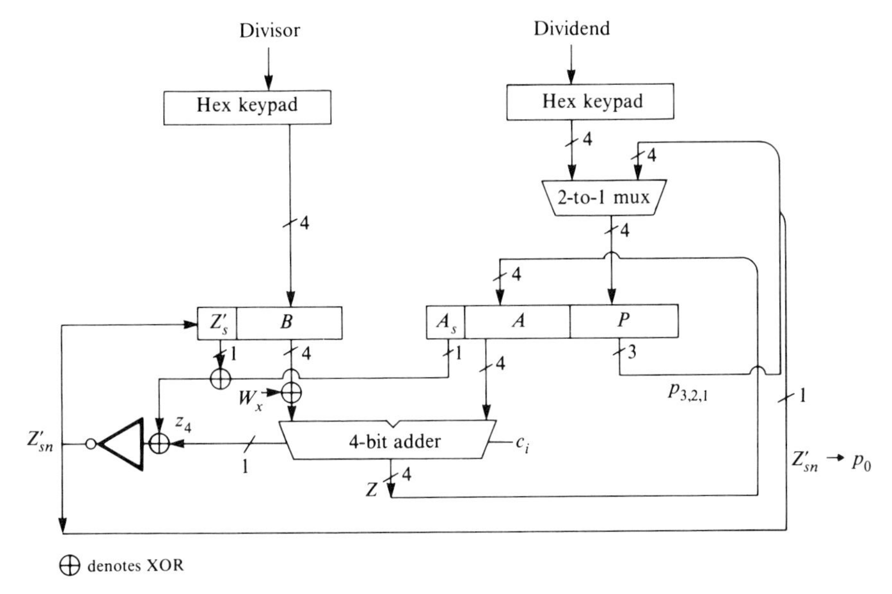

FIGURE 11.14
Block diagram of data path for serial division ASM

b. Control unit. Design a control unit to generate control signals as required to activate the data path elements at the appropriate times.

(1) Next-state generator.

(a) The ASM chart (Figure 11.13) can be used to tabulate the control algorithm next states for each present state, together with the control conditions for each present-state/next-state transition, as shown in Table 11.3.

TABLE 11.3
Serial division control algorithm table

Present State		Next State		Condition for Transition	Condition for		
ABC	Name	Name	$A^+B^+C^+$		$A^+ = 1$	$B^+ = 1$	$C^+ = 1$
000	a	a	000	st'	F	F	st
		b	001	st			
001	b	c	011	T	F	T	T
010	d	c	011	ct4'	ct4	T	ct4'
		e	110	ct4			
011	c	d	010	T	F	T	F
110	e	a	000	T	F	F	F

(b) As an alternative representation, for each present state ABC of the serial division control algorithm, the next state $A^+B^+C^+$ can be determined using a next-state map in which the state variables A, B, and C are map variables and the input conditions are map-entered variables. The next-state map, derived from the control algorithm for the serial division process, is shown in Figure 11.15.

FIGURE 11.15
Serial division control algorithm next-state map

$A^+B^+C^+$

A \ BC	00	01	11	10
0	000(st′) 001(st)	011	010	011(ct4′) 110(ct4)
1	XXX	XXX	XXX	000

(2) Output decoder. The control unit output decoder for the serial divider can be designed by analyzing the ASM chart (Figure 11.13) and the data path diagram (Figure 11.14) to determine the sequence of microoperations required in each control state.

State **a** (000): Set register controls for load ($S_1S_0 = 11$).

State **b** (001): Clear counter ($0 \rightarrow i$); $3 \rightarrow j$. Load dividend and divisor into registers P and B, respectively.

State **c** (011): Shift left.

$$Z = 2Z_{j+1} + p_j$$

State **d** (010): Add/subtract. (For $j = 3$, subtract since $Z(4) = 0$ by definition.)
If $Z_{j+1} < 0$, $Z_j = Z + B$.
If $Z_{j+1} \geq 0$, $Z_j = Z - B$.
Increment counter: $i = i + 1$; $j = j - 1$.
If $i < 4$, repeat States **c** and **d**.
If $i = 4$, process is completed; go to **e**.

State **e** (110): Restore remainder.
If $Z_0 < 0$, remainder $R = Z_0 + B$.
$0 \rightarrow$ st.

This analysis leads to the algorithm timing diagram for States **c** and **d**, as shown in Figure 11.16.

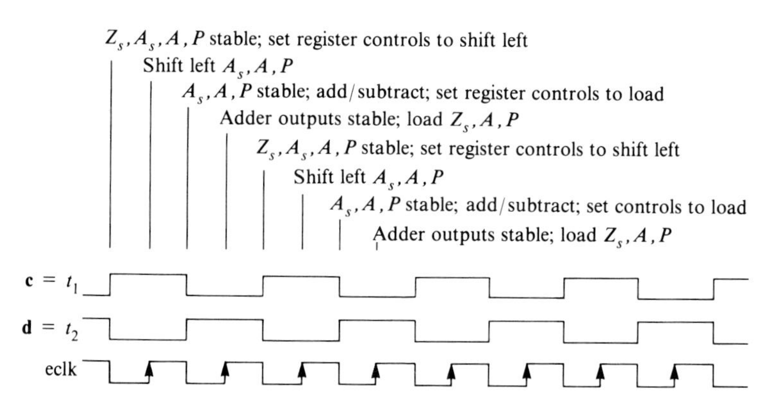

FIGURE 11.16
Timing diagram for states **c** and **d** of a serial division ASM

The timing diagram can then be translated into a table (Table 11.4) that lists the microoperations to be accomplished at specified times, together with their respective control signals.

TABLE 11.4
Data-path control signals generated by output decoder

State/Time	Function	Microoperations	Control Signal	S_1S_0	Mux Select
a (000)	Set S_1S_0 to load			11	
b (001)	Clear counter		clr-ct = $A'B'C$		
	Clear accumulator		clr-A = $A'B'C$		
	Load divisor and dividend	Load B	clk-B = $A'B'C \cdot$ eclk		
		Load P	clk-P = $A'B'C \cdot$ eclk	11	1
c (011)	Set S_1S_0 to shift left			10	
	Shift left A_sAP	S.L A_sAP	clk-A_s = $A'BC \cdot$ eclk		
		$\rightarrow A_sAP$	clk-A = $A'BC \cdot$ eclk		
			clk-P = $A'BC \cdot$ eclk		
d (010)	Set S_1S_0 to load			11	
	Add/Subtract ($Z = A \pm B$)	$A \pm B \rightarrow A$	$W_x = A'BC'\,Z'_{so}$		
			carry-in = $A'BC' \cdot Z'_{so}$		
	Increment counter	$i + 1 \rightarrow i$	inc-ct = $A'BC' \cdot$ eclk		
	Store Z_s	$Z_{sn} \rightarrow Z_s$	clk-Z_s = $A'BC' \cdot$ eclk		
	Store A	$Z \rightarrow A$	clk-A = $A'BC' \cdot$ eclk		
	Store P	$Z'_{sn} \rightarrow p_0$	clk-P = $A'BC' \cdot$ eclk		
e (110)	Restore remainder	$A + B \rightarrow A$	clk-A = $ABC'Z_{sn} \cdot$ eclk	11	
		$0 \rightarrow$ st	clr-st = $ABC' \cdot$ eclk		

Note: $Z_{sn} = (A_s \oplus Z_{so}') \oplus C_o$, where C_o = carry-out

The output decoder generates the following control signals to the data path components:

$$\text{clr-}A = A'B'C$$
$$\text{clr-ct} = A'B'C$$
$$\text{clk-}B = A'B'C \cdot \text{eclk}$$
$$\text{clk-}A = (A'BC + A'BC' + ABC'Z_{sn})\text{eclk}$$
$$\text{clk-}A_s = A'BC \cdot \text{eclk}$$
$$\text{clk-}P = (A'BC + A'BC' + A'B'C)\text{eclk}$$
$$\text{clk-}Z_s = A'BC' \cdot \text{eclk}$$
$$W_x = A'BC'Z'_{so}$$
$$\text{carry-in} = A'BC'Z'_{so}$$
$$S_1 = 1$$

$$S_0 = (A'BC)'$$
$$\text{mux-sel} = A'B'C$$
$$\text{inc-ct} = A'BC' \cdot \text{eclk}$$
$$\text{clr-st} = ABC' \cdot \text{eclk}$$

where initial conditions are $A_s = 0$, $Z_s = 0$ ($Z_s' = 1$), and $P = p_3p_2p_1p_0$.

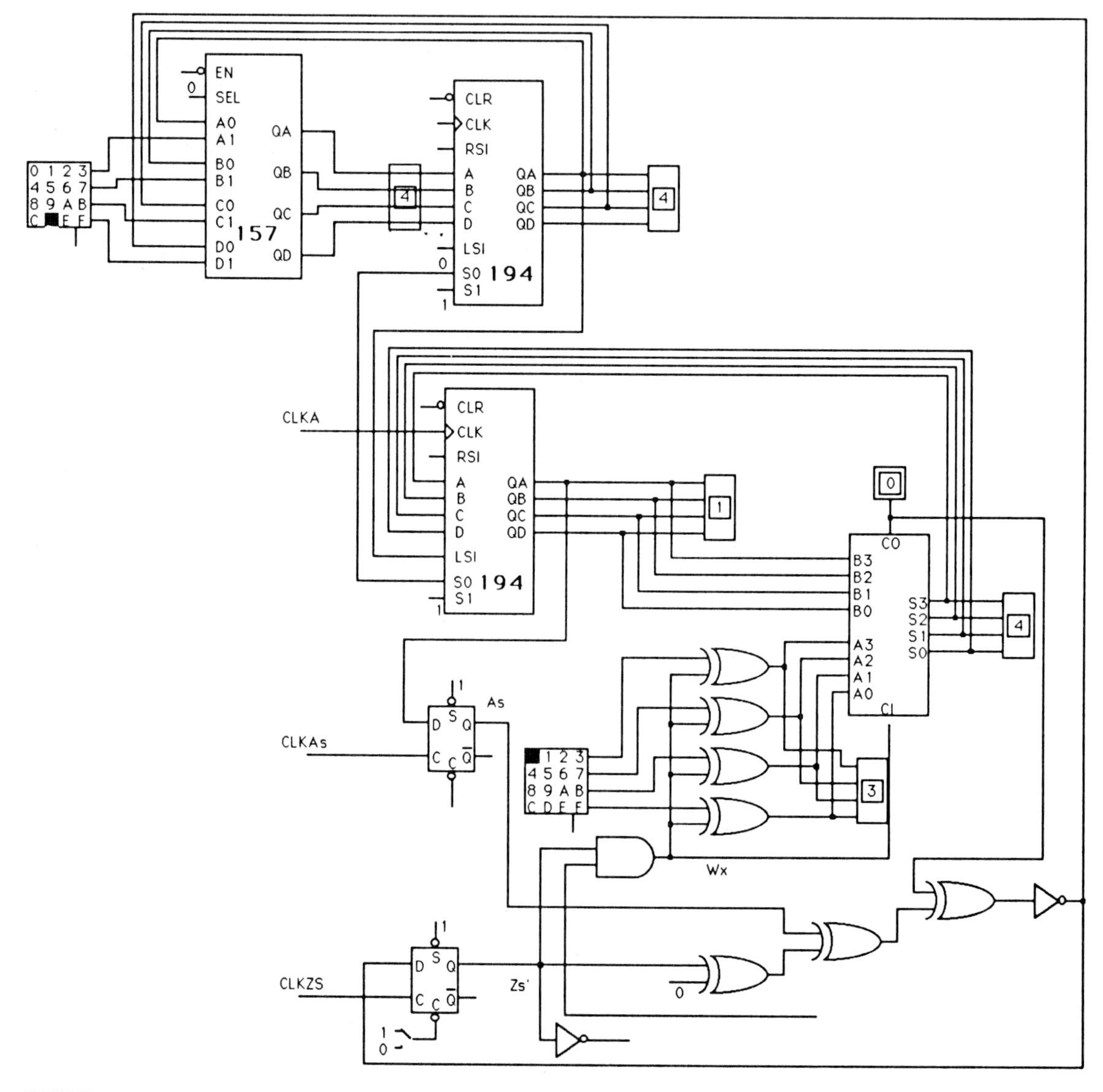

FIGURE 11.17
Logic diagram of the process data path

Step 4: Realization.

a. Process data path. Use the block diagram in Figure 11.14 to construct a logic diagram of the process data path, as shown in Figure 11.17.

(1) Next-state generator

(a) Implementation using multiplexers. The serial division control algorithm can be implemented directly from the control algorithm table using 8-to-1 multiplexers, where present state *ABC* is used to select the input condition corresponding to the select value *ABC* combination, as shown in Figure 11.18.

FIGURE 11.18
Multiplexer implementation of next-state generator

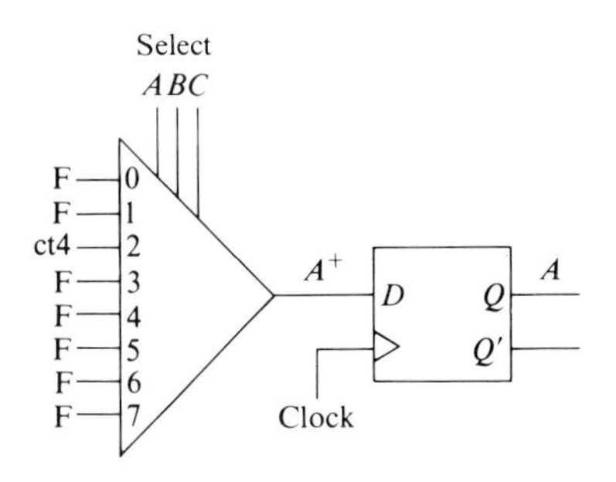

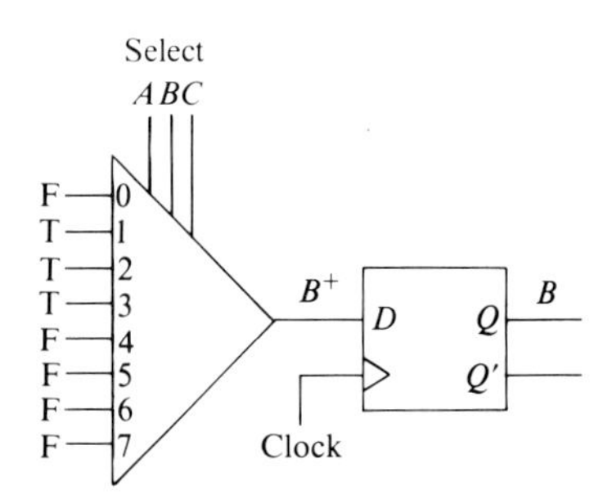

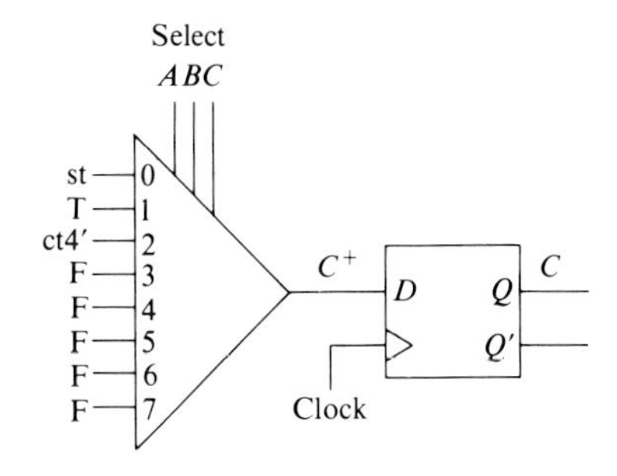

(b) Implementation using D flip-flops. If the next-state generator is to be implemented using D flip-flops, individual next-state maps for each state variable can be derived from the next-state map in Figure 11.15. The individual next-state maps are shown in Figure 11.19. Note that unused states (100, 101, 111) have next state 000.

FIGURE 11.19
Next-state maps for controller state variables

A \ BC	00	01	11	10
0	0	0	0	ct4
1	0	0	0	0

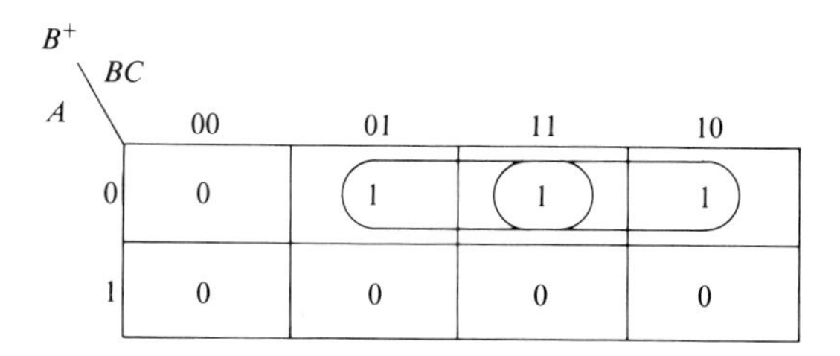

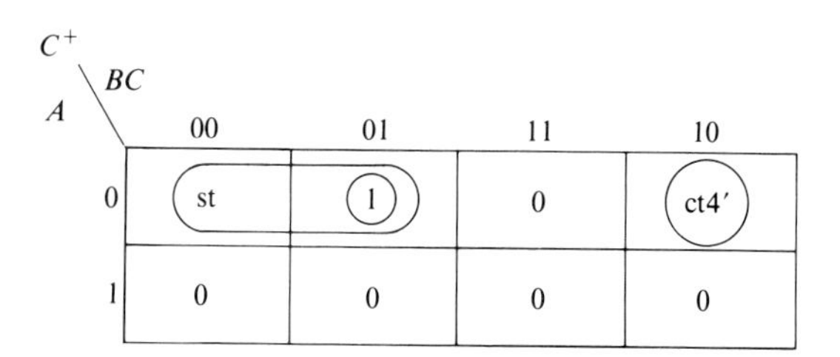

The D flip-flop excitation equations, as determined from these maps, are

$$D_A = A'BC' \cdot \text{ct4} \qquad D_B = A'C + A'B \qquad D_C = A'B'C + A'B' \cdot \text{st} + A'BC' \cdot \text{ct4}'$$

(2) Output decoder. Use the logic equations derived from Table 11.4 to construct a logic diagram of the output decoder. Figure 11.20 presents a logic diagram of the output decoder and next-state generator.

The logic diagrams of the process data path, output decoder, and next-state generator (implemented using PET D flip-flops) are combined to form a logic diagram of the complete serial division ASM, as shown in Figure 11.21a (p. 547). A timing diagram for the serial division ASM is presented in Figure 11.21b (p. 548).

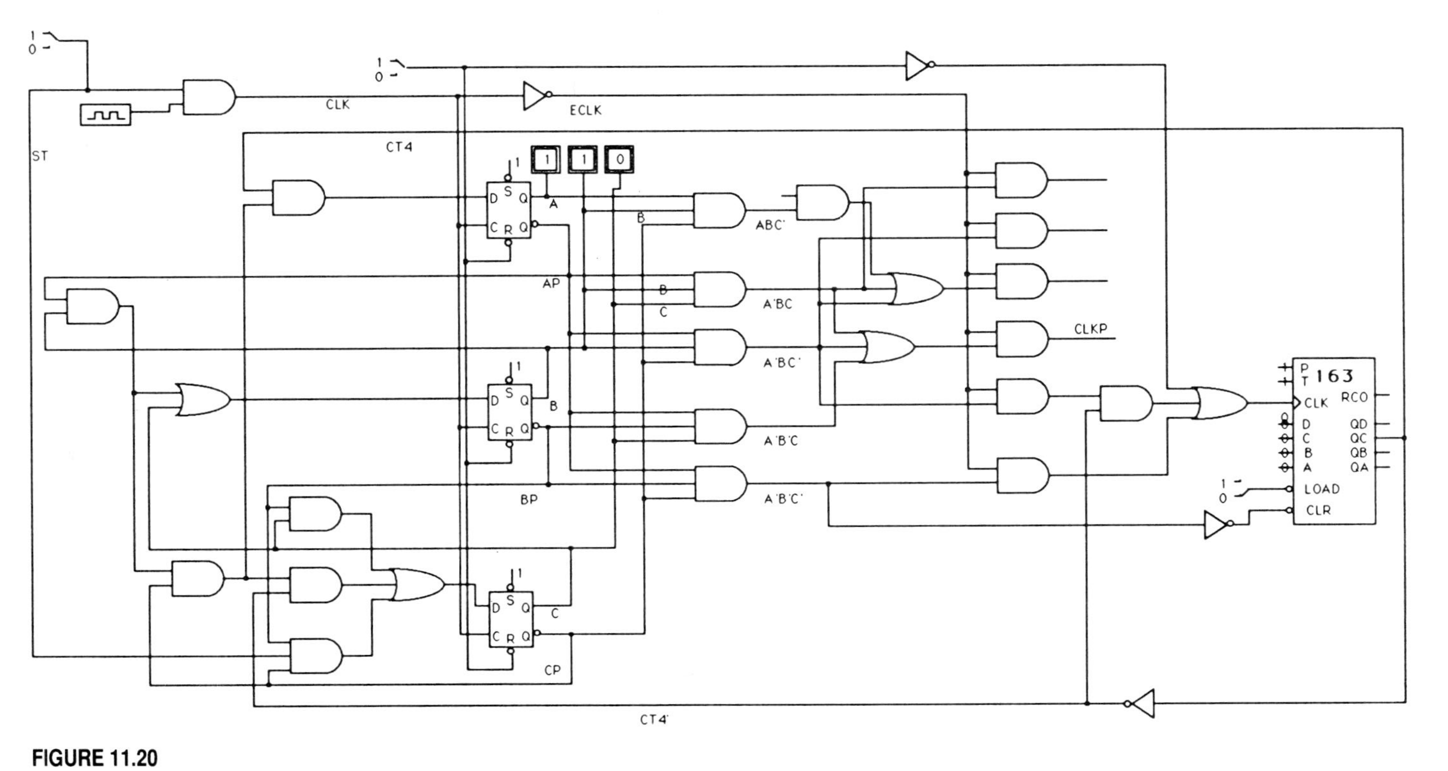

FIGURE 11.20
Logic diagram of the output decoder and next-state generator

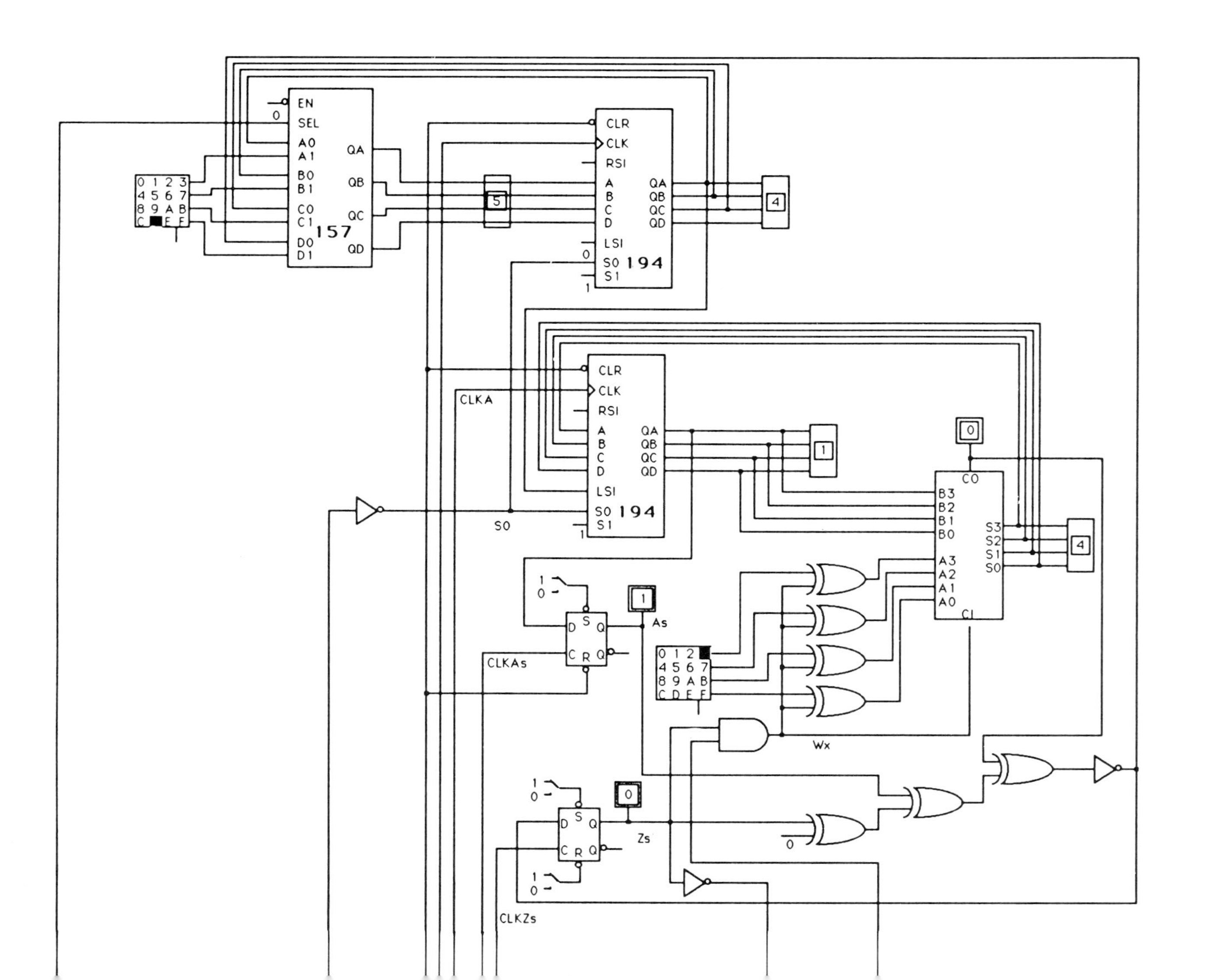
157
194
194
CLKA
CLKAs
CLKZs
S0
As
Zs
Wx

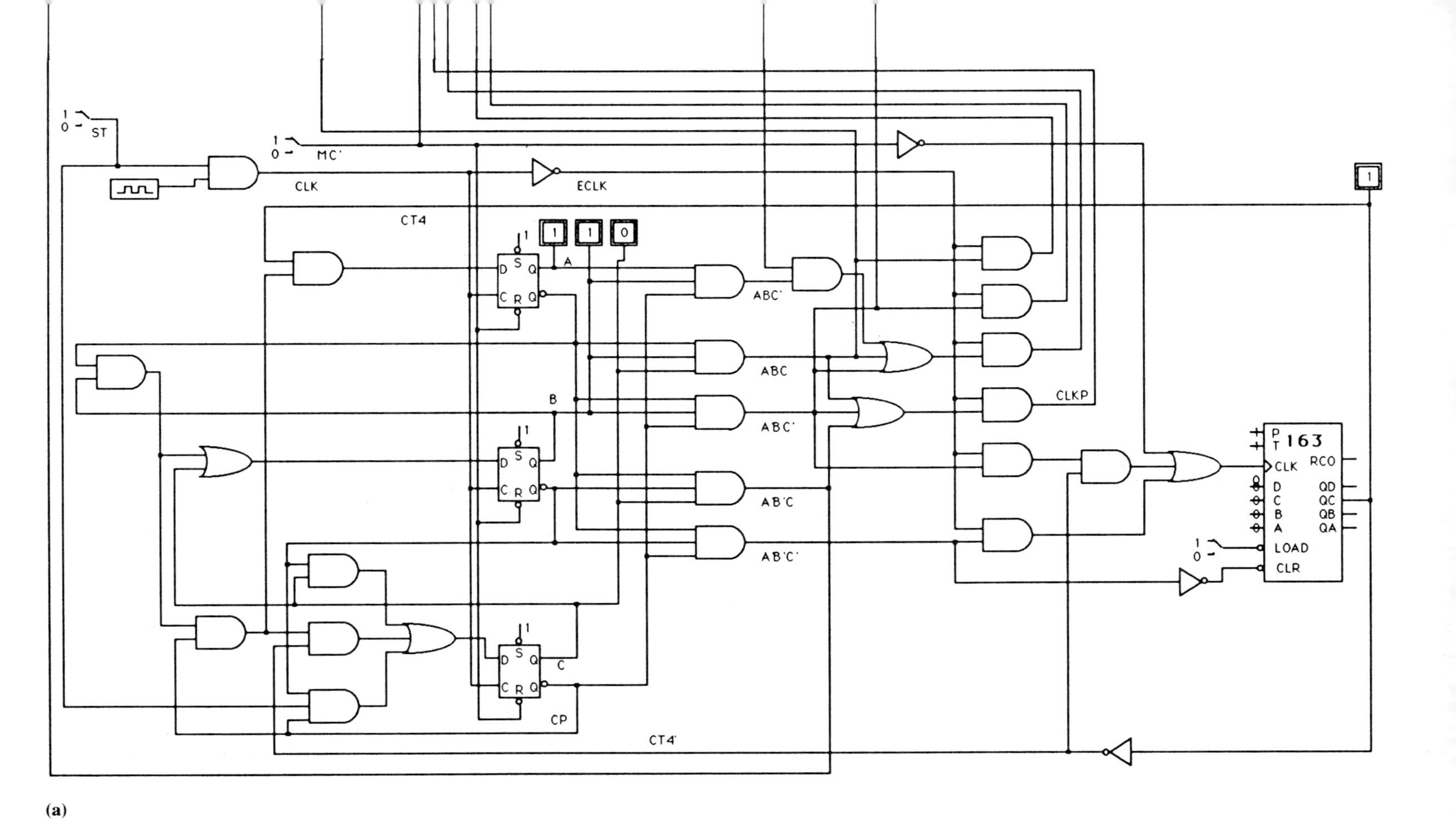

(a)

FIGURE 11.21 (pp. 546–548)
Complete serial division ASM. (a) Logic diagram (b) Timing diagram

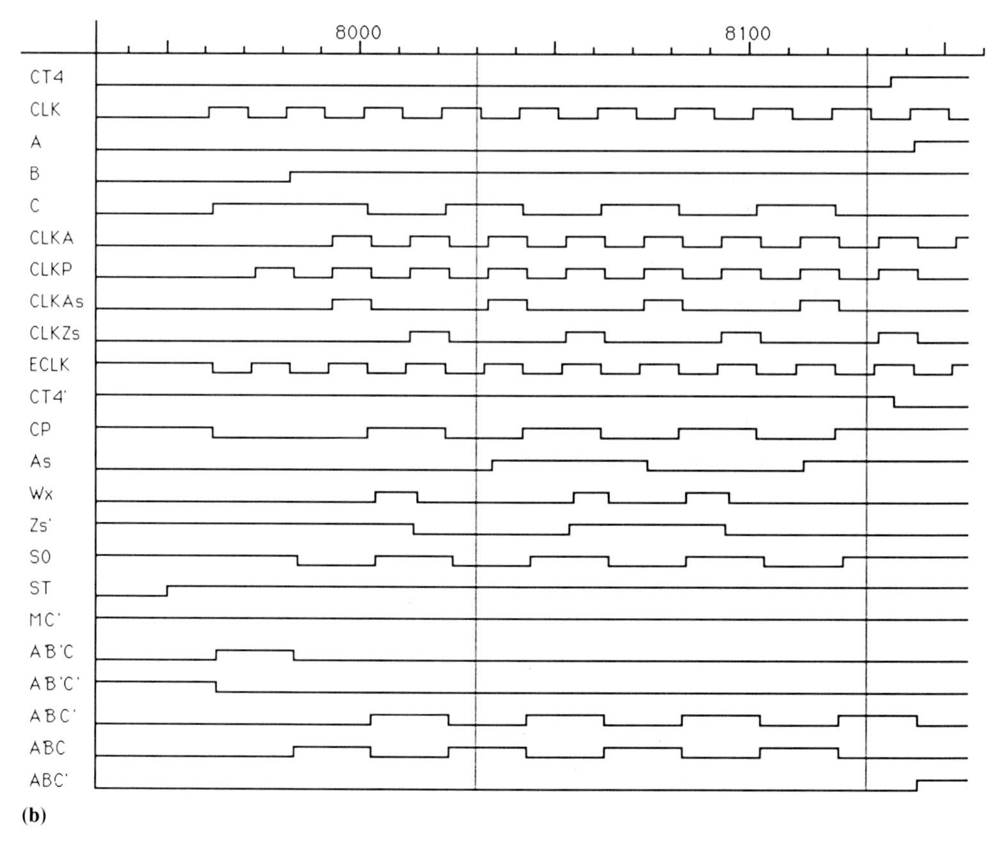

FIGURE 11.21
Continued

11.4 DESIGN AND IMPLEMENTATION OF A COMBINED SERIAL MULTIPLIER/DIVIDER

The design of a computer involves determining a set of instructions, which the computer must have to implement algorithms to be programmed. Obviously, a realistic computer must have the capability to multiply and divide, as well as add, subtract, and perform logical operations. This section describes the design of a combined multiplier/divider unit.

The individual functional modules (such as a serial multiplier and a serial divider) can be combined into a multifunction processor. This requires (1) designing a data path that supports each function of the multifunction processor and (2) designing a control unit for the multifunction processor.

In this section, you will learn the basics of **shared-data-path design** and **combined-control-unit design**, both common in developing multifunction devices, such as a combined multiplier/divider or a digital computer. A combined serial multiplier/divider can be designed since each (multiplier and divider) has essentially the same data path and each process is controlled by a 2-state algorithm:

1. For multiply, add and then shift right.
2. For divide, shift left and then add or subtract.

Analysis

Let's start by comparing the data paths of the serial multiplication and serial division machines, as shown in Figures 11.3 and 11.14, respectively. Each data path has a 4-bit adder and a composite 9-bit shift register (*EAP* for multiply, A_sAP for divide). Each has a 4-bit register *B*, but the divider has a fifth bit Z_s on the left of register *B*. Z_s is used for 5-bit addition/subtraction, which is accomplished by

$$Z_{sn} = (Z'_{so} \oplus A_s) \oplus C_o$$

Since the data path of the multiplier is a subset of the divider's data path, we can use Figure 11.14 as a block diagram for the combined multiplier/divider.

We can compare the next-state generators by examining Tables 11.1 and 11.3. The multiplier States **a**, **b**, **c**, and **d** can be made the same as the corresponding divider states by appending a 0 on the left of each multiplier state binary code. State **e** of the divider is used to restore the remainder, denoted by *R*. Finally, we can compare the output decoders by examining Tables 11.2 and 11.4.

It appears from this preliminary analysis that a combined multiplier/divider can be designed using the following approach:

- Develop a data path that supports both serial multiplication and serial division.
- Develop a control unit for the combined multiplier/divider.

The next section describes the design of a combined multiplier/divider using the 4-step problem-solving procedure for system design.

INTERACTIVE DESIGN APPLICATION

Design of a Combined Multiplier/Divider ASM

An ASM that accomplishes either the serial multiplication process or the serial division process (using the nonrestoring method) can be designed as follows. Note that we will use a 2-state control, *D*, to select the operation to be performed: $D = 0$ for multiply; $D = 1$ for divide.

Step 1: Problem Statement. Organize the process in two parts and describe the system specifications.

a. Process. If $D = 0$, accomplish the serial multiplication process as described in Section 11.2. If $D = 1$, accomplish the serial division process as described in Section 11.3.

b. Process control algorithm. Develop a control algorithm for the combined multiplier/divider ASM. The start, load, and processing operations required in the combined multiplier/divider can be accomplished by a control algorithm that has 5 distinct states:

a, **b**, **c**, **d**, and **e**. (The letters M and D in parenthetical remarks denote multiplication and division, respectively.) Index j is used in place of i in division formulas to be consistent with Section 11.3.

State **a**. Wait: Wait for start

State **b**. Reset, Load: $0 \rightarrow i, E, A_s, A$; $3 \rightarrow j$
Multiplicand or divisor $\rightarrow B$
Multiplier or dividend $\rightarrow P$
Define $Z_4 = 0$

State **c**. M—Add, store: $B\,p_i + A \rightarrow E, A$*
D—Shift left: $Z = 2Z_{j+1} + p_j$

State **d**. M—Shift right: Shift right EAP
D—Add/subtract: If $Z_{j+1} < 0$, $Z_j = Z + B$
If $Z_{j+1} \geq 0$, $Z_j = Z - B$
M or D: $i + 1 \rightarrow i$; $j - 1 \rightarrow j$
If $i < 4$, repeat States **c** and **d**.
If $i = 4$, process completed, go to **e**.

State **e**. Restore remainder: If $Z_0 < 0$, remainder $R = Z_0 + B$

System specification. Design an ASM to accomplish the serial multiplication process if $D = 0$ or the serial division process if $D = 1$. The problem can be expressed in register transfer notation as follows:

$$\text{If } D = 0,\ X \leftarrow XY$$
$$\text{If } D = 1,\ X \leftarrow X/Y$$

Step 2: Conceptualization. Develop a pictorial description of the combined multiplier/divider.

a. System structure diagram. The organization of a combined multiplier/divider ASM may be conceptualized using a structure diagram that is organized into a control section and a process data path section, as shown in Figure 11.22.

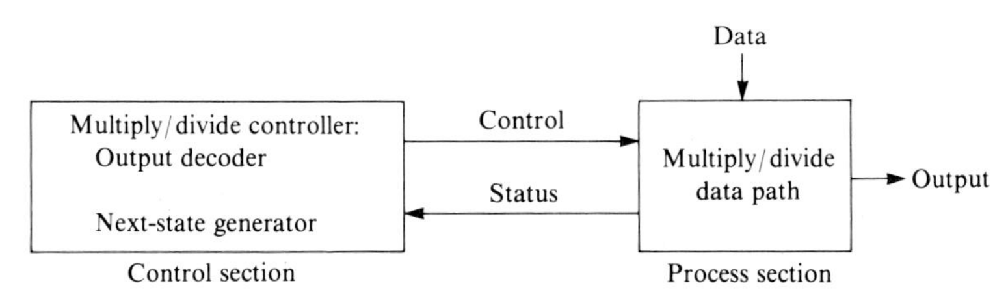

FIGURE 11.22
Structure diagram of a combined multiplier/divider

b. Control flow diagram. The control flow diagrams for the serial multiplier (Figure 11.2) and the serial divider (Figure 11.13) can be combined into a control flow diagram (in the form of an ASM chart) for the multiplier/divider, as shown in Figure 11.23.

*Reset E if $p_0 = 0$.

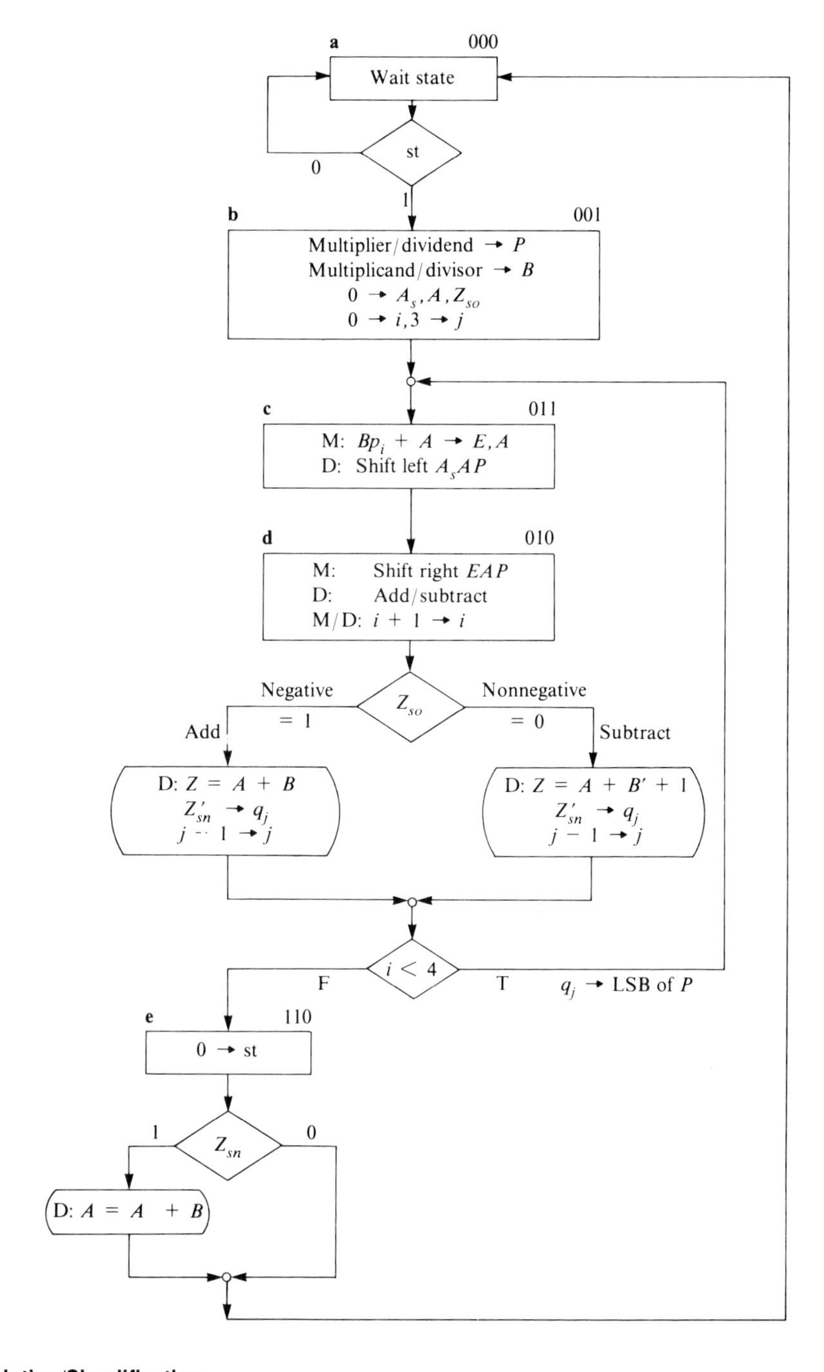

FIGURE 11.23
ASM chart for combined multiplier/divider

Step 3. Solution/Simplification.

a. Process data path. Block diagrams of the data paths of the serial multiplication and serial division machines, respectively, were presented in Figures 11.3 and 11.14. In the preliminary analysis, we noted that each data path has

A 4-bit adder and a composite 9-bit shift register (*EAP* for multiply, A_sAP for divide)

A 4-bit register *B*, but the divider has a fifth bit Z_s on the left of register *B*. Z_s is used for 5-bit addition/subtraction, which is accomplished by

$$Z_{sn} = (Z'_{so} \oplus A_s) \oplus C_o$$

Since the data path of the multiplier is a subset of the divider's data path, we can use Figure 11.14 (modified to show its dual use) as a block diagram for the combined multiplier/divider. The result, shown in Figure 11.24, is a block diagram of a combined multiplier/divider data path that supports both serial multiplication and serial division.

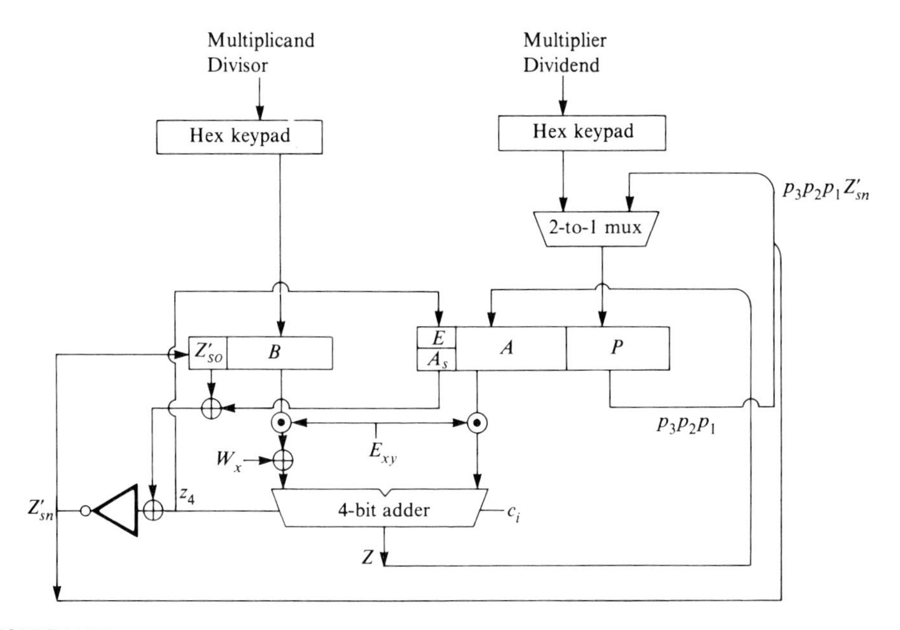

FIGURE 11.24
Data path for a combined multiplier/divider

b. Control unit.

(1) Next-state generator. We can compare the next-state generators by examining Tables 11.1 and 11.3. The multiplier States **a**, **b**, **c**, and **d** can be made the same as the corresponding divider states by appending a 0 on the left of each multiplier state binary code. State **e** of the divider is used to restore the remainder, denoted by *R*.

The design of an integrated multiply/divide control unit requires a next-state generator that will correctly generate the next states for the multiplier or for the divider, depending on which function is being executed at the time. We will use a control *D* to indicate divide or multiply: $D = 1$ indicates division; $D = 0$ indicates multiplication.

Since the serial division requires 5 states while the serial multiplication requires only 4 states, let's examine the control algorithm table for division first. A composite multiply/divide control algorithm table (Table 11.5) can be constructed from the control flow diagram (ASM chart) in Figure 11.23. It should be noted that Table 11.5 (combined multiply/divide) is identical to Table 11.3 (divide).

TABLE 11.5
Combined multiply/divide control algorithm table

Present State		Next State		Condition for Transition	Condition for		
ABC	Name	Name	$A^+B^+C^+$		$A^+ = 1$	$B^+ = 1$	$C^+ = 1$
000	**a**	**a**	000	st'	F	F	st
		b	001	st			
001	**b**	**c**	011	T	F	T	T
010	**d**	**c**	011	ct4'	ct4	T	ct4'
		e	110	ct4			
011	**c**	**d**	010	T	F	T	F
110	**e**	**a**	000	T	F	F	F

ct4' denotes i is < 4; ct4 denotes $i = 4$

To make the control algorithm tables compatible, we can modify the multiply control algorithm table (Table 11.1) by inserting a 0 on the left of each binary code for a multiply state and place an "F" in the "Condition for $A^+ = 1$" column (so that state variables A, B, and C are used instead for the multiplier). The result is Table 11.6.

TABLE 11.6
Serial multiplication control algorithm table

Present State		Next State		Condition for Transition	Condition for		
ABC	Name	Name	$A^+B^+C^+$		$A^+ = 1$	$B^+ = 1$	$C^+ = 1$
000	**a**	**a**	000	st'	F	F	st
		b	001	st			
001	**b**	**c**	011	T	F	T	T
010	**d**	**c**	011	ct4'	F	ct4'	ct4'
		a	000	ct4			
011	**c**	**d**	010	T	F	T	F

If the next state **a** of present state **d** is replaced by state **e** and a row with state **e** is added to Table 11.6, we see that Table 11.6 is identical to Table 11.5. Therefore, Table 11.6 is compatible with Table 11.5. Consequently Table 11.5 can be used either for multiplication or division.

(2) Output decoder. The design of an integrated multiply/divide control unit also requires an output decoder that will correctly generate the data path control signals for the multiplier or for the divider, depending on which function is being executed at the time. Logic equations for the control signals generated by the binary multiplier output decoder, derived in Section 11.2, are repeated here for convenience.

$$\text{clr-}A = A'B$$
$$\text{clr-ct} = A'B$$
$$\text{clk-}B = A'B \cdot \text{eclk}$$

$$\text{clk-}E = AB \cdot \text{eclk}$$
$$\text{clk-}A = (AB \cdot p_0 + AB')\text{eclk}$$
$$\text{clk-}P = (A'B + AB')\text{eclk}$$
$$E_{xy} = AB \cdot p_0$$
$$\text{carry-in} = 0$$
$$S_{1A} = AB$$
$$S_{1P} = A'$$
$$S_0 = 1$$
$$\text{inc-ct} = AB' \cdot \text{ct4}' \cdot \text{eclk}$$

In order to make these logic equations compatible with the divider equations, replace *A* and *B* with *B* and *C* and insert *A'* before each expression; clr st is done in State **e** instead of State **d**.

$$\text{clr-A} = A'B'C$$
$$\text{clr-ct} = A'B'C$$
$$\text{clk-B} = A'B'C \cdot \text{eclk}$$
$$\text{clk-E} = A'BC \cdot \text{eclk}$$
$$\text{clk-A} = (A'BC \cdot p_0 + A'BC')\text{eclk}$$
$$\text{clk-P} = A'(B'C + BC')\text{eclk}$$
$$E_{xy} = A'BC \cdot p_0$$
$$\text{carry-in} = 0$$
$$S_{1A} = A'BC$$
$$S_{1P} = A'B'$$
$$S_0 = 1$$
$$\text{inc-ct} = A'BC' \cdot \text{ct4}' \cdot \text{eclk}$$

Logic equations for the control signals generated by the binary divider output decoder, derived in Section 11.2, are repeated here for convenience.

$$\text{clr-}A = A'B'C$$
$$\text{clr-ct} = A'B'C$$
$$\text{clk-}B = A'B'C \cdot \text{eclk}$$
$$\text{clk-}A = (A'BC + A'BC' + ABC'Z_{sn})\text{eclk}$$
$$\text{clk-}A_s = A'BC \cdot \text{eclk}$$
$$\text{clk-}P = (A'BC + A'BC' + A'B'C)\text{eclk}$$
$$\text{clk-}Z_s = A'BC' \cdot \text{eclk}$$
$$W_x = A'BC'Z'_{so}$$
$$\text{carry-in} = A'BC'Z'_{so}$$
$$S_1 = 1$$
$$S_0 = (A'BC)'$$
$$\text{mux-sel} = A'B'C$$

$$\text{inc-ct} = A'BC' \cdot \text{ct4}' \cdot \text{eclk}$$
$$\text{clr-st} = ABC' \cdot \text{eclk}$$

where initial conditions are $A_s = 0$, $Z_s = 0$ ($Z_s' = 1$), and $P = p_3p_2p_1p_0$.

The combined output decoder generates control signals to the data path components at appropriate times, as indicated in Tables 11.2 and 11.4. The letter D indicates division ($D = 1$), and the letter M ($D = 0$) indicates multiplication. That is, $M = D'$. Initially, index i is set to 0 and is incremented in each pass. For each data path control point X, combine the control equations for X using the following logical equation (where $M = D'$):

$$X = D\cdot(\text{divide expression}) + M\cdot(\text{multiply expression})$$

The resulting logic equations for the control signals generated by a combined multiplier/divider output decoder are

$$\text{clk-}E = M \cdot A'BC \cdot \text{eclk} \quad (\text{Load } E \text{ if } p_0 = 1\text{; reset } E \text{ if } p_0 = 0)$$
$$\text{clk-}A_s = D \cdot A'BC \cdot \text{eclk}$$
$$\text{clk-}Z_s = D \cdot A'BC' \cdot \text{eclk}$$
$$\text{clr-}A = A'B'C$$
$$\text{clr-ct} = A'B'C$$
$$\text{inc-ct} = A'BC' \cdot \text{ct4}' \cdot \text{eclk}$$
$$\text{clk-}B = A'B'C \cdot \text{eclk}$$
$$\text{clr-st} = ABC' \cdot \text{eclk}$$
$$E_{xy} = M \cdot A'BC \cdot p_0 + D\ 1$$
$$W_x = D \cdot A'BC'Z_{so}'$$
$$\text{carry-in} = D \cdot A'BC'Z_{so}'$$
$$\text{mux-sel} = A'B'C$$
$$S_{1P} = M \cdot A'B' + D \cdot 1$$
$$S_{1A} = M \cdot A'BC + D \cdot 1$$
$$S_0 = M \cdot 1 + D(A'BC)'$$
$$\begin{aligned}\text{clk-}A &= M(A'BC \cdot p_0 + A'BC')\text{eclk} \\ &\quad + D(A'BC + A'BC' + ABC'Z_{sn})\text{eclk}\end{aligned}$$
$$\begin{aligned}\text{clk-}P &= M \cdot A'(B'C + BC')\text{eclk} \\ &\quad + D(A'BC + A'BC' + A'B'C)\text{eclk} \\ &= (A'[B'C + BC'] + D \cdot A'BC)\text{eclk}\end{aligned}$$

Step 4: Realization.

a. Process data path. Use Figure 11.24 to construct a logic diagram of the process data path for the combined multiplier/divider, as shown in Figure 11.25 (p. 556).

b. Control unit.

(1) Next-state generator. For an implementation using multiplexers, the multiply/divide control algorithm can be implemented directly from the control algorithm table using 8-to-1 multiplexers, where present state ABC is used to select the input condition corresponding to the select value ABC combination, as shown in Figure 11.26 (p. 556).

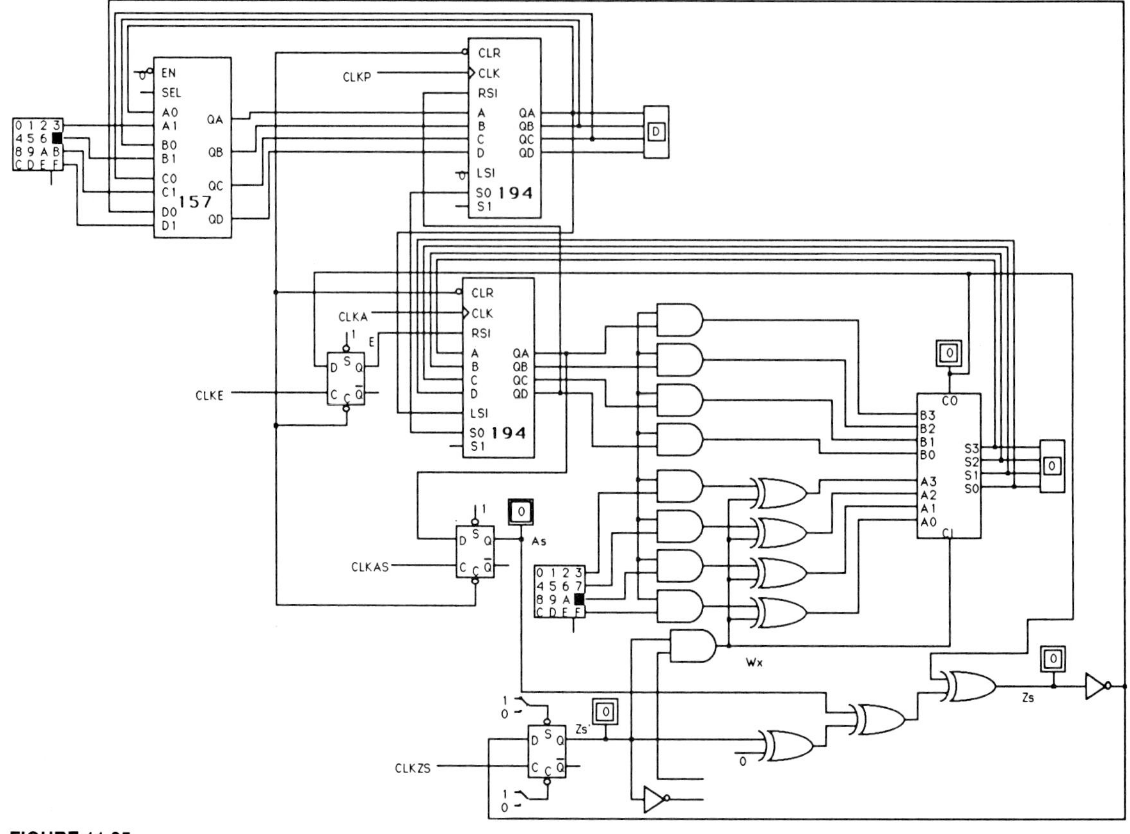

FIGURE 11.25
Logic diagram of the process data path

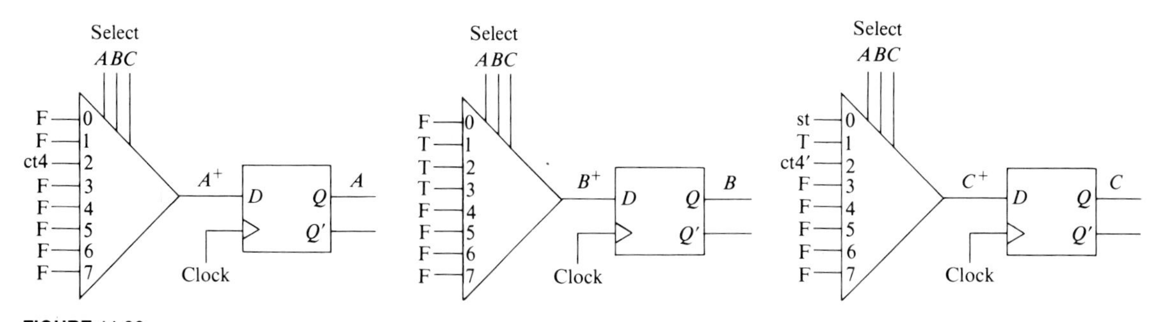

FIGURE 11.26
Multiplexer implementation of next-state generator

(2) Output decoder. Use the logic equations derived in Step 3 to construct a logic diagram of the output decoder. Figure 11.27 presents a logic diagram of the output decoder and next-state generator.

The logic diagrams of the process data path, output decoder, and next-state generator are combined to form a logic diagram of the complete multiply/divide ASM, as shown in Figure 11.28 (pp. 558–559).

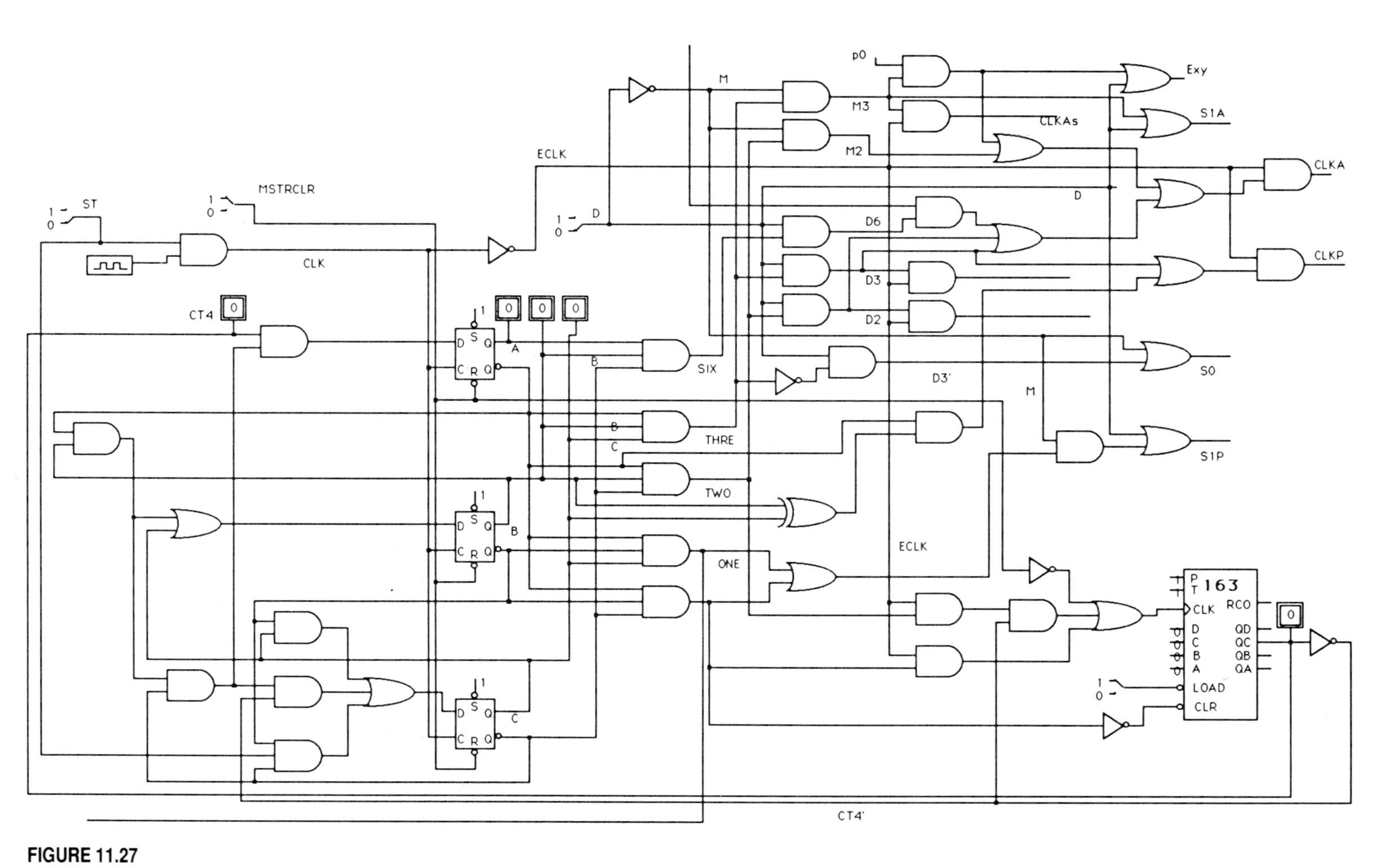

FIGURE 11.27
Logic diagram of the output decoder and next-state generator

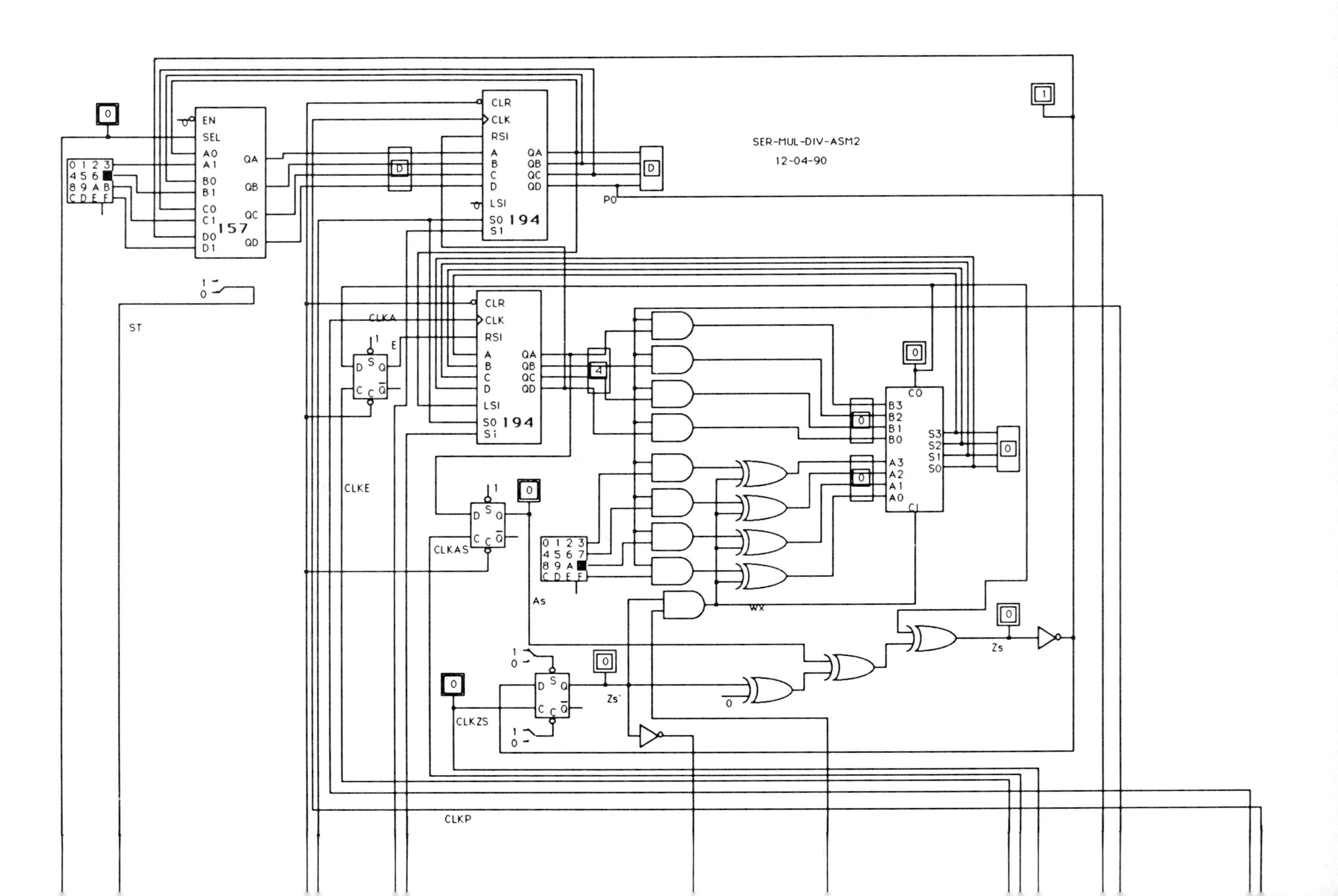
SER-MUL-DIV-ASM2
12-04-90
157
194
194
P0
ST
CLKA
CLKE
CLKAS
CLKZS
CLKP
As
Wx
Zs
Zs'

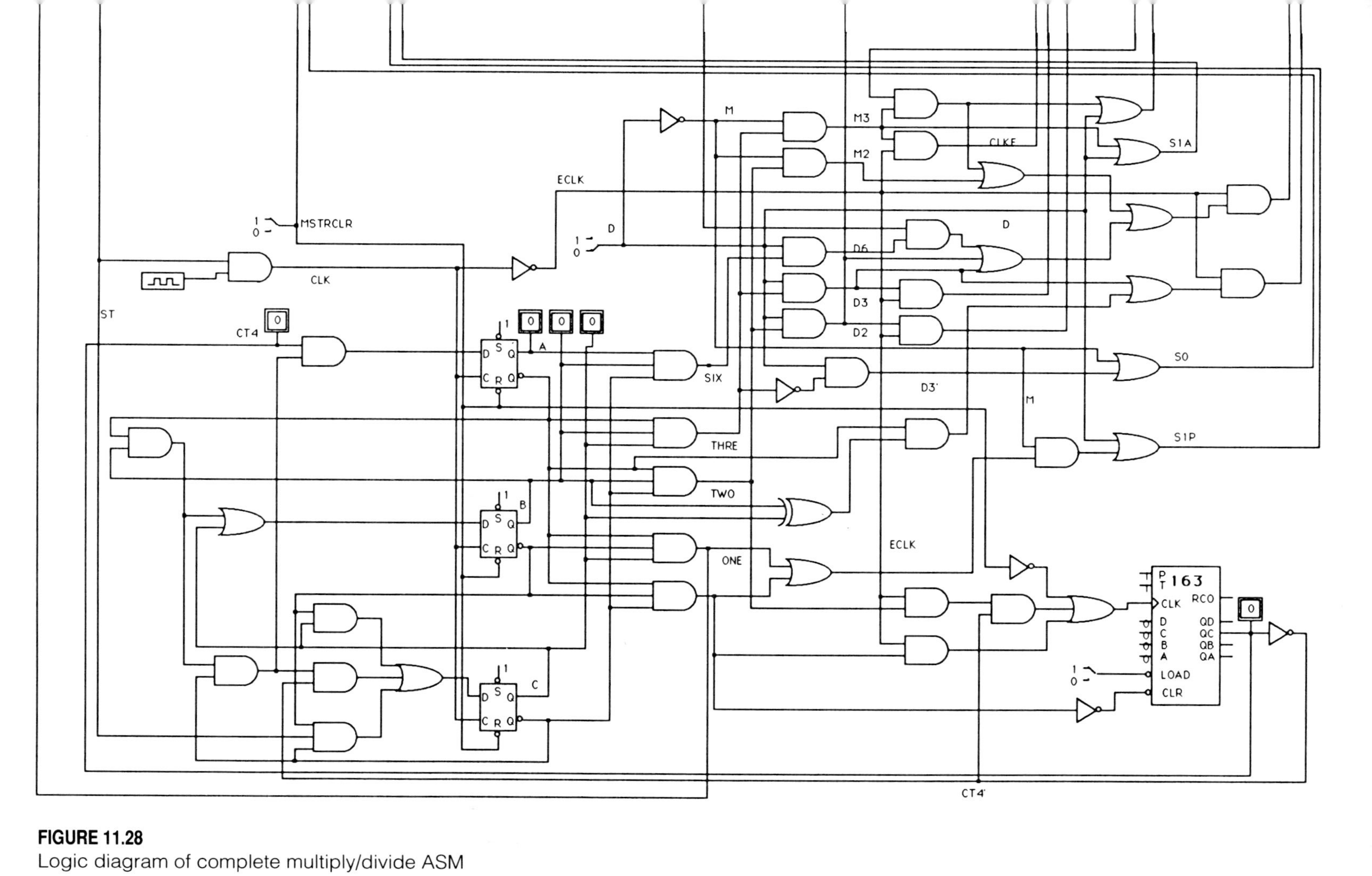

FIGURE 11.28
Logic diagram of complete multiply/divide ASM

11.5 BCD-TO-BINARY AND BINARY-TO-BCD CONVERSION

Before leaving Chapter 11, we consider two additional algorithms for iterative processes. Algorithms for converting from BCD to binary and from binary to BCD are presented next.

BCD-to-Binary Shift-and-Subtract Algorithm

Given a decimal integer N represented in BCD form, N can be converted from BCD to binary using the following algorithm (the binary register is initially empty):

1. Shift both registers right 1 bit position.
2. If any decade (after the shift) has MSB = 1, subtract 3 from that decade.
3. Repeat steps 1 and 2 until the BCD register contains only 0s.

The following table is an example of the BCD-to-binary shift-and-subtract algorithm. We will convert $N = 349 = 0011\ 0100\ 1001$ to a binary integer.

Decade:	2	1	0		
	BCD register				
	8421	8421	8421	Binary register	Description
349 =	0011	0100	1001		
	001	1010	0100	1	Shift right
	001	0111	0100		Subtract 3 from decade 1
	00	1011	1010	01	Shift right
	00	1000	0111		Subtract 3 from decades 0, 1
	0	0100	0011	101	Shift right
		0010	0001	1101	Shift right
		001	0000	11101	Shift right
		00	1000	011101	Shift right
		00	0101		Subtract 3 from decade 0
		0	0010	1011101	Shift right
			0001	01011101	Shift right
			000	101011101	Shift right

The result is $N = 349 = 0011\ 0100\ 1001 \rightarrow$ binary 101011101.

Binary-to-BCD Add-and-Shift Algorithm

Given an integer N in binary form, N can be converted from binary to BCD using the following algorithm. All BCD decades are initially empty.

1. Before any shift, add 3 to each decade that exceeds 0100.
2. Shift both registers left 1 bit position.
3. Repeat steps 1 and 2 until the binary register is empty.

The following table is an example of the binary-to-BCD add-and-shift algorithm. We will convert $N = 101011101$ to BCD form.

Decade: 2	1	0		
BCD register				
8421	8421	8421	Binary register	Description
			101011101	
		1	01011101	Shift left
		10	1011101	Shift left
		101	011101	Shift left
		1000	011101	Add 3 to decade 0
	1	0000	11101	Shift left
	10	0001	1101	Shift left
	100	0011	101	Shift left
	1000	0111	01	Shift left
	1011	1010	01	Add 3 to decades 0, 1
1	0111	0100	1	Shift left
1	1010	0100	1	Add 3 to decade 1
11	0100	1001		Shift left

The result is 3 4 9 $\leftrightarrow$ 101011101

11.6 SUMMARY

In this chapter, we have learned to analyze two-dimensional iterative processes (binary multiplication and division) and to develop recursive algorithms for these processes.

We began by representing a multibit number as a polynomial in powers of 2. The representation of a polynomial in nested form then led to a recursion formula for evaluating the polynomial. Then the recursion formulas were translated into sequential machines (algorithmic state machines).

Next, we designed and implemented two digital logic systems—a binary multiplier and a binary divider. These systems included designing a control unit subsystem and a data path subsystem and integrating these subsystems into a functioning system.

Finally, we designed and implemented a combined multiplier/divider. Elements of shared-data-path design and combined-control-unit design are common in developing multifunction devices, such as the combined multiplier/divider.

KEY TERMS

Partial product sums (PPS)

PPS method

Nested form

Recursive formula

Partial dividend

Partial remainder

Restoring method

Nonrestoring method

Shared-data-path design

Combined-control-unit design

EXERCISES

General

Use the BCD-to-binary and binary-to-BCD algorithms in Exercises 1 through 6.

1. Use the BCD-to-binary shift-and-subtract algorithm to convert $N = 245 = 0010\ 0100\ 0101$ to a binary integer.
2. Draw a block diagram of a process data path for a BCD-to-binary converter.
3. Write a pseudocode description of a process control algorithm for a BCD-to-binary converter. Include steps for initializing and loading the registers.
4. Use the binary-to-BCD add-and-shift algorithm to convert the binary integer $N = 11110101$ to BCD form.
5. Draw a block diagram of a process data path for a binary-to-BCD converter.
6. Write a pseudocode description of a process control algorithm for a binary-to-BCD converter. Include steps for initializing and loading the registers.
7. Draw a block diagram of a process data path for a parallel-to-serial converter.
8. Write a pseudocode description of a process control algorithm for a parallel-to-serial converter. Include steps for initializing and loading the registers.
9. Assume that a floating-point number $X = \pm F 2^E$ is stored in a 12-bit word

S	$E^\wedge$	F
1	4	7

= seeee.fffffff (Implied binary point)

where $S = 0$ for positive X, and $S = 1$ for negative X;
$E^\wedge$ = base-2 exponent in excess-8 form. $E = E^\wedge - 8$.
F = fraction, normalized with leftmost bit = 1.

Describe an algorithm for multiplying two positive floating-point numbers X and Y, where $X = F_x 2^{E_x}$ and $Y = F_y 2^{E_y}$.

10. Draw a block diagram of a data path for a floating-point multiplier.
11. Formulate a process control algorithm for a floating-point multiplier.

Design/Implementation

Use the 4-step problem-solving procedure for systems design for each design exercise.

1. Design a rudimentary control serial multiplier.
2. Design a rudimentary control serial divider (nonrestoring method).
3. Design an ASM to implement a BCD-to-binary converter that uses the shift-and-subtract algorithm.
4. Design an ASM to implement a binary-to-BCD converter that uses the add-and-shift algorithm.
5. Design an ASM to implement a floating-point multiplier.

PART FIVE

Computer Design, Simulation/ Implementation, and Programming

12 Building-Block Evolution of a Simple Computer
13 Design of Mux-Oriented 8-Bit Computers
14 Design of Bus-Oriented 8-Bit Computers

Part V (Chapters 12, 13, and 14) presents an introduction to computer design. In a systems context, a computer **processor** is a digital system composed of a **memory subsystem** and a **central processing unit** (CPU). The CPU can be represented by an algorithmic state machine (ASM) model consisting of a control section and a process section (data path). Figure A (p. 566) presents a structure diagram of a simple computer processor.

Part V describes the design of eight complete computers, ranging from a simple mux-oriented processor to a fairly sophisticated bus-oriented computer. A complete logic diagram for each computer allows the reader to implement the computer in hardware or simulate it using a logic-simulation program, such as Capilano Computing's Designworks (Macintosh). An **assembly language** (mnemonic code) for each computer allows the reader to program the computer and debug both the computer and the program. We use Designworks for this purpose.

Chapter 12 uses a building-block design approach to describe the evolution from a 4-bit adder to a functional stored-program computer processor in twelve steps. Each step describes a functional building block and its integration with the preceding components of its subsystem.

Chapter 13 describes in detail general procedures for the design of simple stored-program digital computers. These procedures are illustrated by describing the design evolution of a hierarchy of four 8-bit mux-oriented stored-program computers (MC–8X). Each succeeding computer is designed to implement an increasingly complex instruction set (architecture). For example, the MC–8A is an 8-bit computer designed and implemented with register

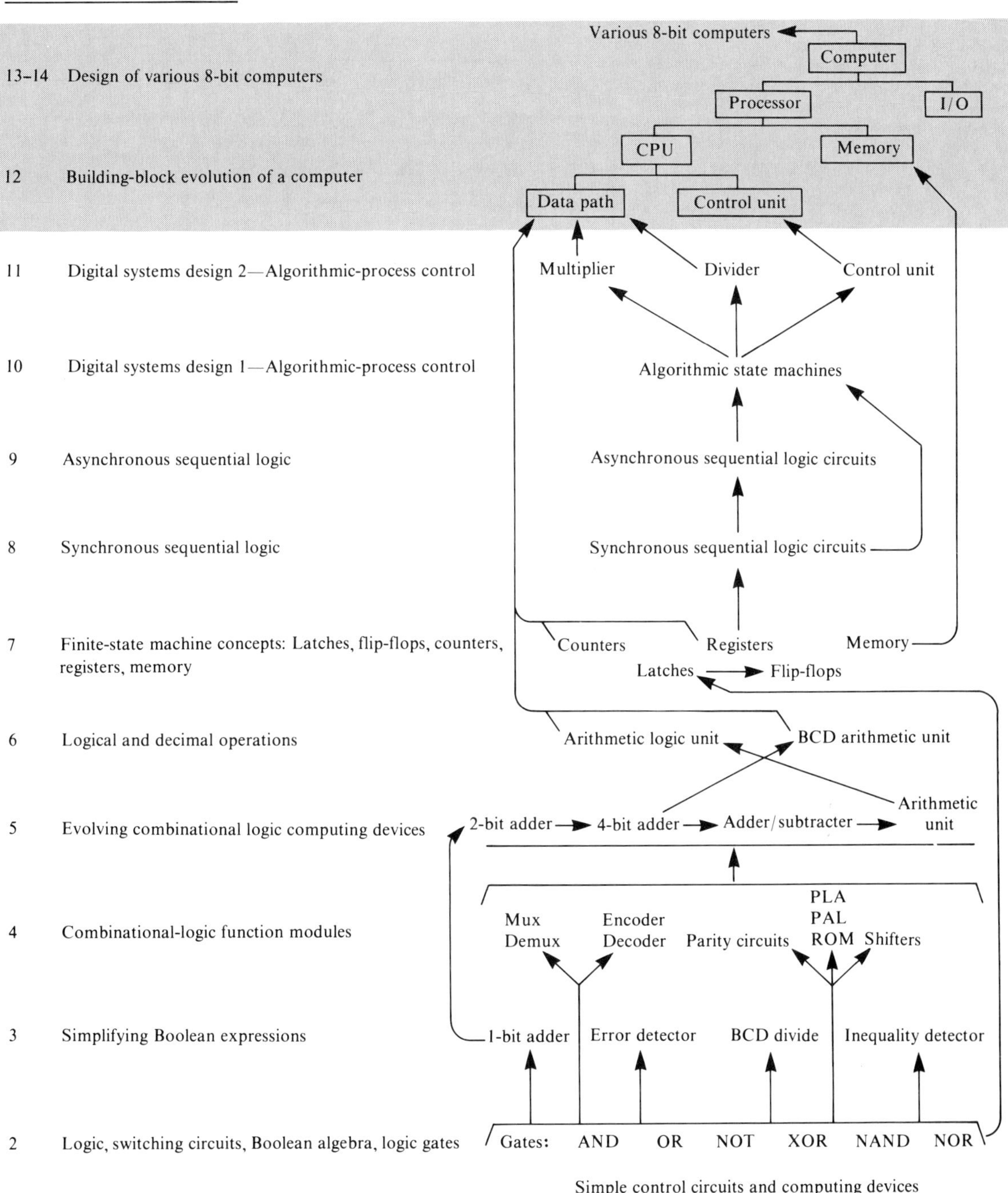
Chapters
Coverage
13–14 Design of various 8-bit computers
12 Building-block evolution of a computer
11 Digital systems design 2—Algorithmic-process control
10 Digital systems design 1—Algorithmic-process control
9 Asynchronous sequential logic
8 Synchronous sequential logic
7 Finite-state machine concepts: Latches, flip-flops, counters, registers, memory
6 Logical and decimal operations
5 Evolving combinational logic computing devices
4 Combinational-logic function modules
3 Simplifying Boolean expressions
2 Logic, switching circuits, Boolean algebra, logic gates
1 Number systems
Computer codes
Various 8-bit computers
Computer
Processor
I/O
CPU
Memory
Data path
Control unit
Multiplier
Divider
Control unit
Algorithmic state machines
Asynchronous sequential logic circuits
Synchronous sequential logic circuits
Counters
Registers
Memory
Latches
Flip-flops
Arithmetic logic unit
BCD arithmetic unit
Arithmetic unit
2-bit adder
4-bit adder
Adder/subtracter
Mux
Demux
Encoder
Decoder
Parity circuits
PLA
PAL
ROM
Shifters
1-bit adder
Error detector
BCD divide
Inequality detector
Gates: AND OR NOT XOR NAND NOR
Simple control circuits and computing devices
Decimal, binary, octal, hexadecimal
BCD, symmetric, reflected, unit distance

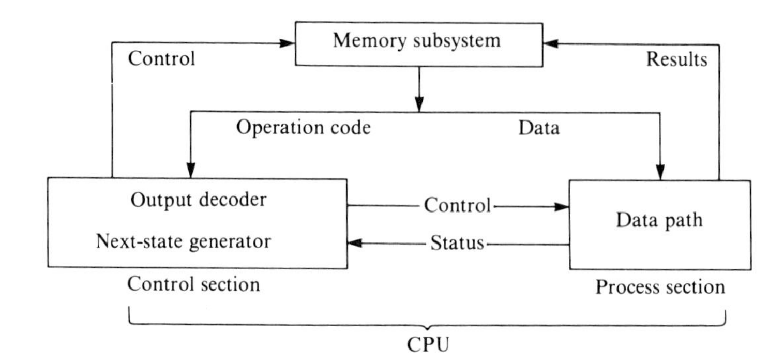

FIGURE A
A computer processor (digital system with memory)

reference, I/O, branch, and direct-address mode memory-reference instructions; the MC–8B incorporates compare, conditional branch, and indirect-address mode instructions into the computer's repertoire; the MC–8C incorporates subroutine jump, multiply, divide, and logic instructions; and the MC–8D is a simple, yet realistic, list-processing computer with indexed-address mode and instructions with 2 implied operands.

Chapter 14 describes the design evolution of a hierarchy of three 8-bit bus-oriented computers (BC–8X) with instruction sets corresponding to those of the computers designed in Chapter 13.

The design of each computer in Part V is accompanied by a description of its assembly language instruction set. The MC–8D assembly language can implement a wide range of algorithms, which include

- List processing (sorting, merging, comparing lists of data)
- Statistics (computing mean, variance, regression, correlation)
- Numerical methods (interpolation, curve fitting, numerical integration, solution of differential equations)

The chapter design/implementation exercises are structured to reinforce the design concepts in each chapter. The reader is encouraged to use a logic-simulation program to implement the digital logic devices, ranging from a 4-bit adder to a number of 8-bit computers and including the eight computers described in Chapters 12 through 14. Each function module, higher-level device, and computer is a complete, functional entity that can be simulated using a logic-simulation program or implemented using SSI/MSI integrated circuits.

12
Building-Block Evolution of a Simple Computer

OBJECTIVES

After you complete this chapter, you will be able

- To understand the concept of computer instructions and programs
- To apply systems-level design techniques to the design of a computer processor represented by a programmable ASM model
- To design and implement a **data path subsystem** that supports each instruction of a specified instruction set
- To design and implement a **control unit subsystem** that fetches, decodes, and executes each program instruction
- To design and implement a **memory subsystem** that stores program instructions, data, and program-generated results
- To develop proficiency in evolving digital logic devices
- To create a loop so that results of the current operation can be used as one of the inputs to the next operation
- To determine the time sequence in which to perform individual steps (microoperations) in an overall operation

12.1 INTRODUCTION

Systems are often designed using a* hierarchical approach*; that is, first, the overall system requirements are identified. Second, a divide-and-conquer technique is then used to identify subsystems and their requirements. If further refinement is required, subsystems are divided into functional units, each with its individual specifications. This hierarchical design process is referred to as a* top-down design. *As an example, consider the top-down design diagram for a BCD arithmetic unit (see Chapter 6) shown in Figure 12.1.

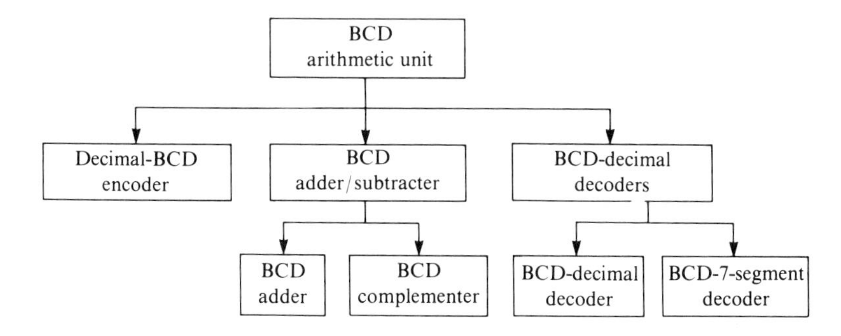

FIGURE 12.1
Top-down design of a BCD arithmetic unit

The implementation of a hierarchical system also can be accomplished using a bottom-up approach. That is, if a building block exists for each functional component in the lowest level of the hierarchy, the building blocks within a given subsystem can be integrated into a subsystem. After all of the subsystems have been implemented, they can, in turn, be integrated to form the overall system, as illustrated in Figure 12.2.

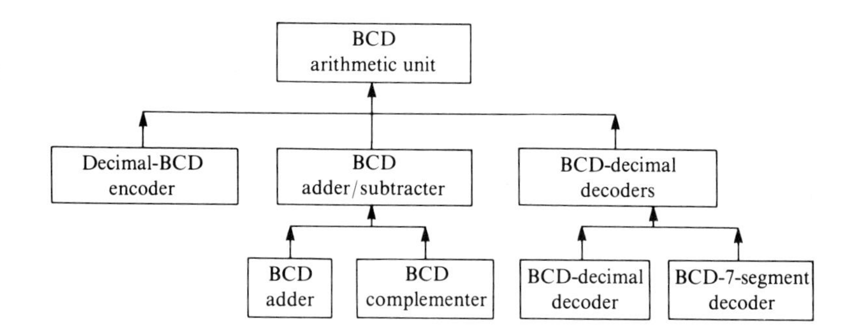

FIGURE 12.2
Bottom-up implementation of a system

Design of a Computer

A computer can be divided into two major components: (1) a processor and (2) an input-output subsystem. In turn, the processor can be divided into a CPU and a memory subsystem. Finally, the CPU can be divided into a data path subsystem and a control unit subsystem. Figure 12.3 illustrates the hierarchical organization of a simple digital computer.

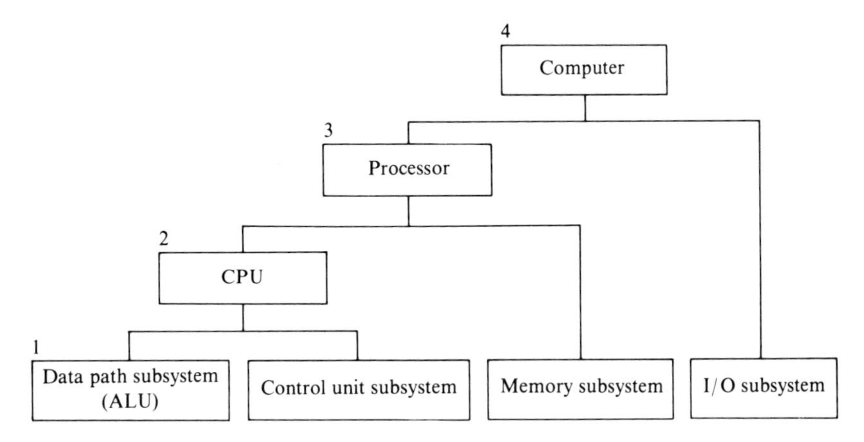

FIGURE 12.3
Hierarchical view of a computer system

Figure 12.3 illustrates a complete computer with input-output (I/O) organized as a hierarchical model containing a succession of "computers" of increasing sophistication and complexity (labeled 1, 2, 3, and 4), as follows:

1. A data path subsystem is essentially a **manual-control computer** that can perform 1 of 8 functions depending on the setting of three 2-state controls (or 1 of 2^n functions with n controls).
2. A data path subsystem combined with a control unit subsystem is a CPU (**semi-automatic computer**) that uses an instruction operation code (opcode) to control data processing in the data path.
3. A CPU combined with a memory subsystem is a processor that operates under control of program instructions stored in memory.
4. A processor combined with an I/O subsystem is a complete computer.

Thus, a simple, general-purpose digital computer can be viewed as a system that has four basic subsystems:

Data path subsystem

Control unit subsystem

Memory subsystem

Input-output subsystem

Each of these subsystems can be constructed by integrating an appropriate set of component building blocks, as described in this chapter. Assuming that each of these subsystems has been implemented, a complete computer can be assembled using bottom-up implementation; that is, a data path subsystem and a control unit subsystem are combined to form a CPU, the CPU is combined with a memory subsystem to form a processor, and finally, the processor is combined with an I/O subsystem to form a complete computer.

A Side-Trip Through a Railroad Switchyard

Before we launch into the evolution of a computer, let's take a brief side-trip to learn more about how a control unit "manipulates" the controls of data path components to accomplish a particular operation. This section uses an analogy to introduce some basic computer terminology and develop some concepts related to the organization and operation of a computer's control unit.

Manual Control of a Switchyard

Visualize a data path subsystem as a railroad switchyard for a grain cooperative warehouse, as illustrated in Figure 12.4. A member of the co-op may store wheat in the warehouse or withdraw wheat from the warehouse. Each member has a boxcar to transport wheat to and from the warehouse. The procedures for storing or withdrawing wheat are

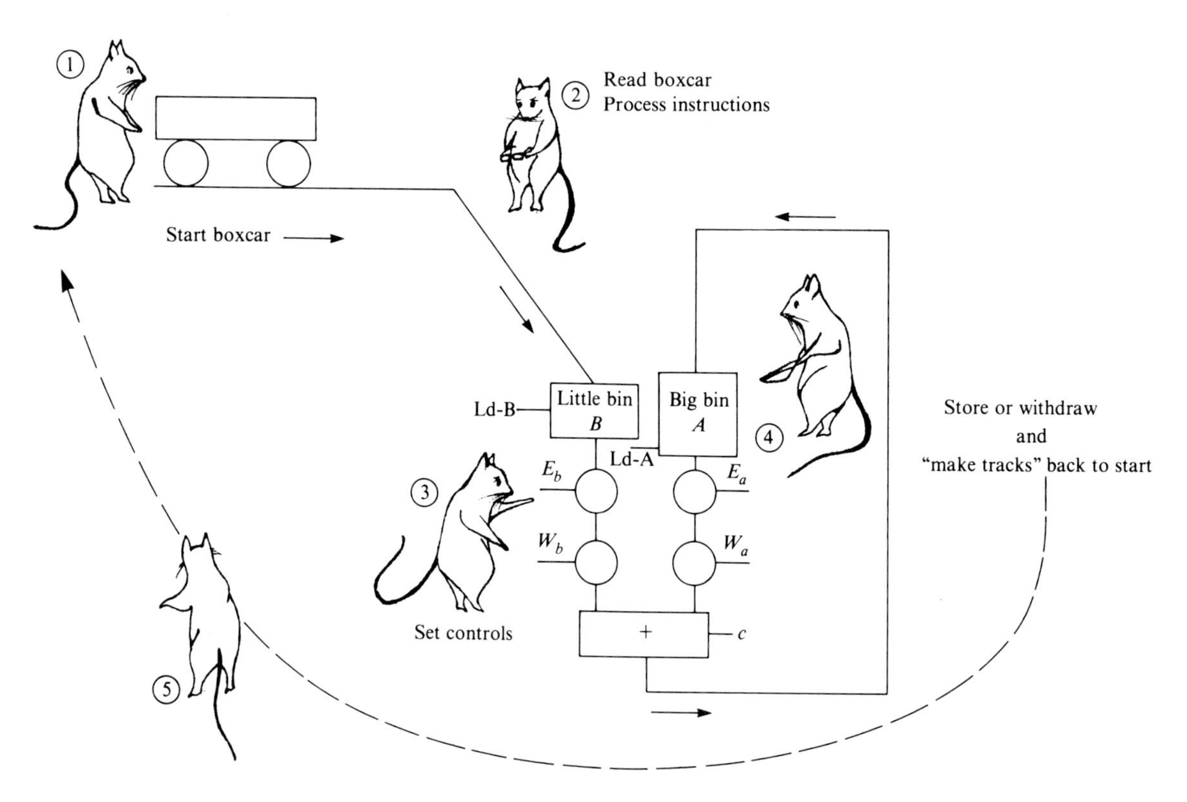

FIGURE 12.4
Diagram of manual control switchyard

1. Store transaction. A loaded boxcar, carrying X bushels of wheat to be stored, is routed through the switchyard to the warehouse and the cargo is unloaded (X is added to the accumulated amount A).
2. Withdraw transaction. An empty boxcar, to be loaded with X bushels of wheat being withdrawn from the co-op, is routed to the warehouse where it is loaded (X is subtracted from A).

Processing instructions that specify the transaction type (store/withdraw) and the transaction quantity (X = number of bushels of wheat) are affixed to the side of each boxcar.

In a manually controlled switchyard, each of the data path **controls** (Ld-B, Ld-A, E_a, W_a, E_b, W_b, C) is manually set at the appropriate time T_i to accomplish the transaction specified by the boxcar's processing instructions. The processing of each boxcar is accomplished by the following step-by-step procedure (algorithm):

1. At time T_0, the boxcar enters the switchyard.
2. At time T_1, the boxcar processing instructions are decoded.
3. At time T_2, the amount X (to be stored or withdrawn) is entered into a data register B.
4. At time T_3, the amount X is added to accumulator register A (for a store operation) or subtracted from A (for a withdraw operation).

Semiautomatic Control of Switchyard

The efficiency of the switchyard operation can be improved by automating the process of decoding each boxcar's processing instructions and activating switchyard (data path) control points at appropriate times. A complete cycle for processing a boxcar requires four time intervals: T_0, T_1, T_2, and T_3. A clock-ring counter (Chapter 7) can be used to generate these time intervals and then repeat them for the next boxcar's cycle. A semiautomatic control unit logic array then activates the switchyard controls at appropriate times depending on the boxcar's processing instructions.

The semiautomatic control unit uses the following step-by-step procedure to direct the processing operations of each boxcar:

1. At time T_0 (cycle start), the ith boxcar enters the switchyard; data X_i is keyed into a hexadecimal keypad, and the boxcar processing **instruction opcode** is entered into switches (assume opcodes are 101 = store; 110 = withdraw).
2. At time T_1, the opcode is loaded into instruction register IR; the opcode is decoded and activates the associated instruction line (add, subtract). The active instruction line, through a logic array, then activates data path components at appropriate times to accomplish the required processing.
3. At time T_2, the amount X_i (to be stored or withdrawn) is loaded into a memory data register B.
4. At time T_3, the amount X_i is added to A (for a store operation) or subtracted from A (for a withdraw operation); the boxcar then exits the switchyard (cycle end).

Figure 12.5 (p. 572) presents an illustration of a semiautomatic control switchyard.

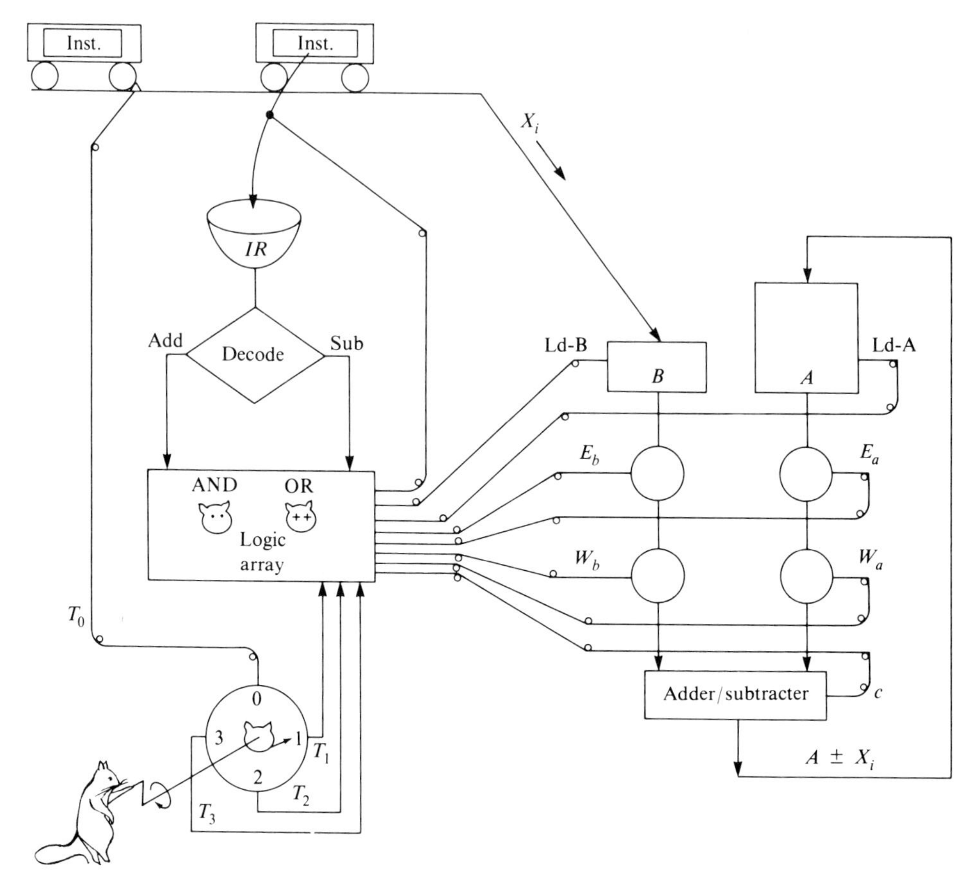

FIGURE 12.5
Diagram of a semiautomatic control switchyard

Automatic Control of Switchyard

The efficiency of the operation of the semiautomatic switchyard can be improved by grouping each day's processing instructions into a **program** for the day. The program and program data are stored in a memory device. As each program instruction is read, the control unit decodes the instruction opcode and generates control signals to the switchyard elements to accomplish the processing required by that instruction. Thus, the control unit automates the process of decoding each boxcar's processing instructions and activating data path control points at appropriate times.

Each day the switchyard is required to process a number of boxcars. The warehouse holding bin is empty at the start of each day (register *A* is cleared before the day's processing begins). The processing instructions (store/withdraw, amount *X*) are stored in the memory device. For simplicity, the boxcars to be processed on a given day are numbered sequentially 1, 2, 3, . . . , n.

A sample program can be written to process the instructions for the boxcars handled by the grain co-op on a given day. As an example, let's assume that today the odd-numbered boxcars contain wheat to be stored and the even-numbered boxcars are empties that will be used to withdraw wheat from the warehouse. We can write a program that updates the accumulated amount (A) as each boxcar is processed. For this example, assume that no withdrawal exceeds the accumulated amount at that time. Fortunately, the first boxcar (number 1) contains wheat to be stored. The program for today's processing can be summarized as follows:

Register A is cleared before the first boxcar is processed. Given a list of numbers X_i ($i = 1, n$), where X_i = amount for the ith boxcar:

If i is an odd number, add amount X_i to the accumulated total in register A; that is, add amount for an odd-numbered boxcar.

If i is an even number, subtract amount X_i from the accumulated total in register A; that is, subtract amount for an even-numbered boxcar.

The sample program can be written as follows:

Boxcar	Instruction	Processing Accomplished
—	CLA - -	Clear register A
1	ADD X_1	Add X_1 to A
2	SUB X_2	Subtract X_2 from A
3	ADD X_3	Add X_3 to A
4	SUB X_4	Subtract X_4 from A
5	ADD X_5	Add X_5 to A
⋮	. . .	. . .
	. . .	. . .

The execution of each program instruction (to process one boxcar) requires one complete **instruction cycle**. Each instruction is accomplished by performing a sequence of microoperations:

- Read instruction (opcode, data X_i).
- Activate Ld-IR to load opcode into instruction register IR and decode opcode.
- Activate data path controls to accomplish processing in accordance with decoded opcode; activate Ld-B to load data X_i into register B and activate Ld-A to load the result into A.

Figure 12.6 (p. 574) presents an illustration of the automatic control switchyard.

The preceding analogy illustrates an evolutionary building-block approach to the design of a completely automated system. This chapter describes such an approach to

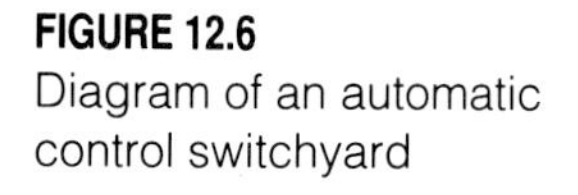

FIGURE 12.6
Diagram of an automatic control switchyard

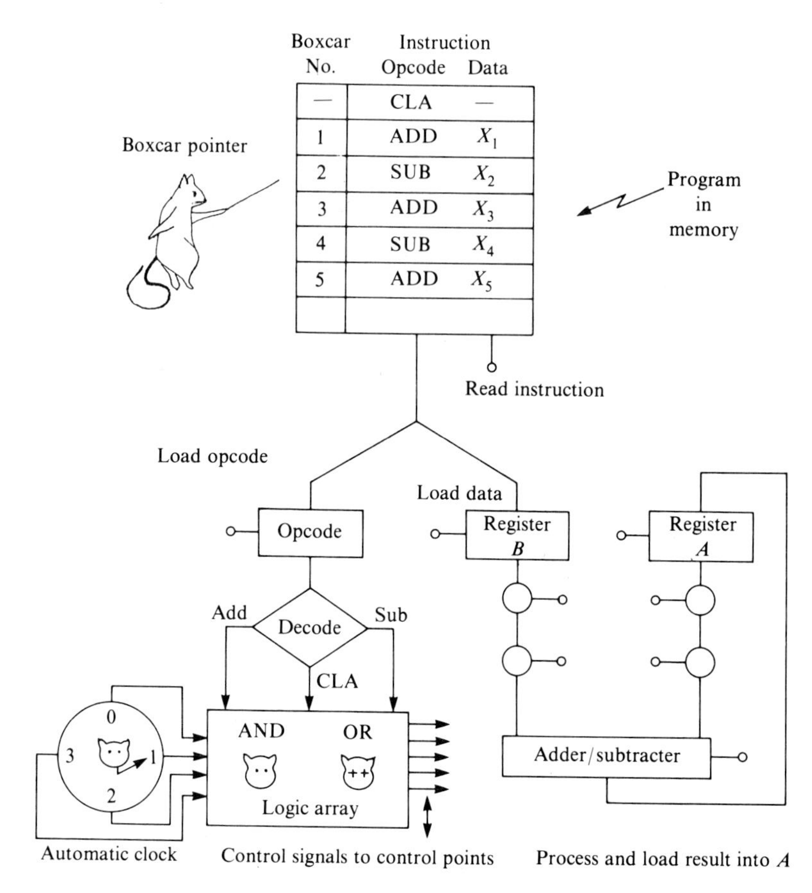

evolve from a 4-bit adder to a functional stored-program computer processor. At each step, another functional building block is described, and the building block is integrated with the preceding components of its subsystem as shown in Figure 12.7.

In Figure 12.7, steps 1–4 describe an evolution from a 4-bit adder to a manual control computer (data path subsystem). In steps 5–8, a 4-step problem-solving procedure for

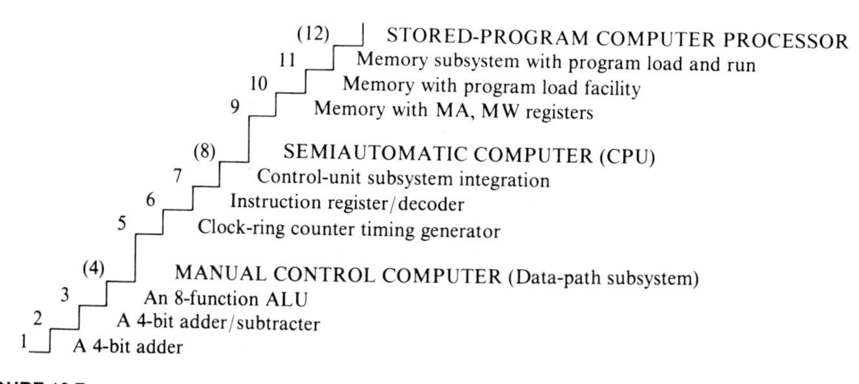

FIGURE 12.7
Design evolution of a computer processor

systems design (described in Chapters 10 and 11) is used to design a CPU consisting of a data path subsystem and a control unit subsystem. Steps 9–12 describe an evolution of a memory subsystem and integration of the memory subsystem and CPU to form a processor.

Sections 12.2, 12.3, and 12.4, respectively, describe the building-block evolution of a data path subsystem, a control unit subsystem, and a memory subsystem. The natural order of development of the subsystems is such that at the end of each section, you will have designed and implemented functional computers of increasing complexity (and of a functionally higher order), as shown in Table 12.1.

TABLE 12.1
Evolution of functional computers

Section	Design	Resulting Computer
12.2	Data path subsystem	Manual control computer
12.3	Control unit subsystem	Semiautomatic computer (CPU)
12.4	Memory subsystem	Stored-program computer processor

In general, an operation (such as add or subtract) performed by a computer is referred to as an **instruction**. A given instruction directs the computer to perform that operation. The set of instructions that a computer can execute is called the **computer's instruction repertoire**.

If the computer is used to solve an application problem, the sequence of instructions that effect the solution of the problem is called a **computer program**. If the instructions are written using the binary opcodes, the program is a **machine language program**. Since it is difficult to think in binary, languages were developed to represent each binary opcode by an easy-to-remember combination of letters (called a *mnemonic*). A program that is written using mnemonic codes must be translated into machine language (assembled) before it can be executed. Hence languages that use mnemonic codes are called **assembly languages**.

Each instruction in Table 12.2 is identified by its opcode and the corresponding mnemonic code. Registers A and B denote an accumulator register and a memory-data register, respectively.

TABLE 12.2
Computer instruction repertoire

Opcode	Mnemonic	Register Transfer	Description of Instruction
000	CLA	$A \leftarrow 0$	Clear A
001	CMP	$A \leftarrow A'$	Complement A
010	INC	$A \leftarrow A + 1$	Increment A
011	NEG	$A \leftarrow A' + 1$	Negate A = 2s complement A
100	LDA	$A \leftarrow B$	Transfer B to A
101	ADD	$A \leftarrow A + B$	Add B to A
110	SUB	$A \leftarrow A + B' + 1$	Subtract B from A
111	STA	Memory $\leftarrow A$	Store A in memory

Each instruction is composed of a 3-bit opcode and a 5-bit memory address.

Sections 12.2, 12.3, and 12.4 describe an evolutionary design process and implementation of a stored-program computer processor. In a systems context, a computer processor is a digital system composed of a memory subsystem and a CPU, where the CPU consists of a control unit subsystem and a data path subsystem.

We will use the 4-step problem-solving procedure for system design (described in Chapters 10 and 11) to design a computer processor.

Step 1: Problem Statement. Given an instruction set expressed in register transfer notation (Table 12.2), design a computer that has the capability of loading and storing program instructions and data and executing the program stored in memory.

The process of program execution is organized and described in two parts:

a. Process. Fetch (read from memory), decode, and execute each program instruction (and accomplish the required processing of data and storing of computer-generated results).

b. Process control algorithm. Use a step-by-step procedure to accomplish the fetch, decode, and execute of each program instruction.

Step 2: Conceptualization. Develop a pictorial description of the system:

a. System structure diagram. Draw a block diagram of the system organization that partitions the system into
(1) A data path subsystem
(2) A control unit subsystem composed of a next-state generator and an output decoder
(3) A memory subsystem

A computer processor can be represented by a **programmable algorithmic state machine** (PASM) model, as shown in Figure 12.8.

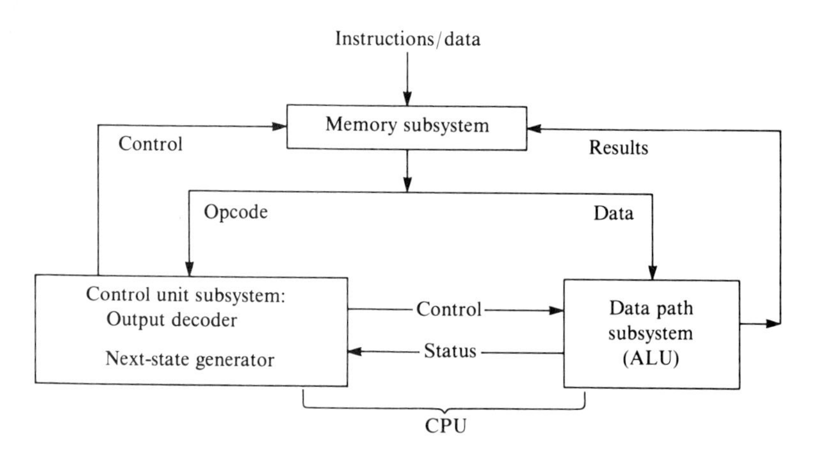

FIGURE 12.8
Structure diagram of a computer processor

b. Control flow diagram. Draw a diagrammatic representation of the operation of the process-control algorithm. A partial control flow diagram (showing instructions CLA, LDA, ADD, and SUB) is presented in Figure 12.9. Each instruction cycle is divided into 4 time subintervals: T_0, T_1, T_2, and T_3.

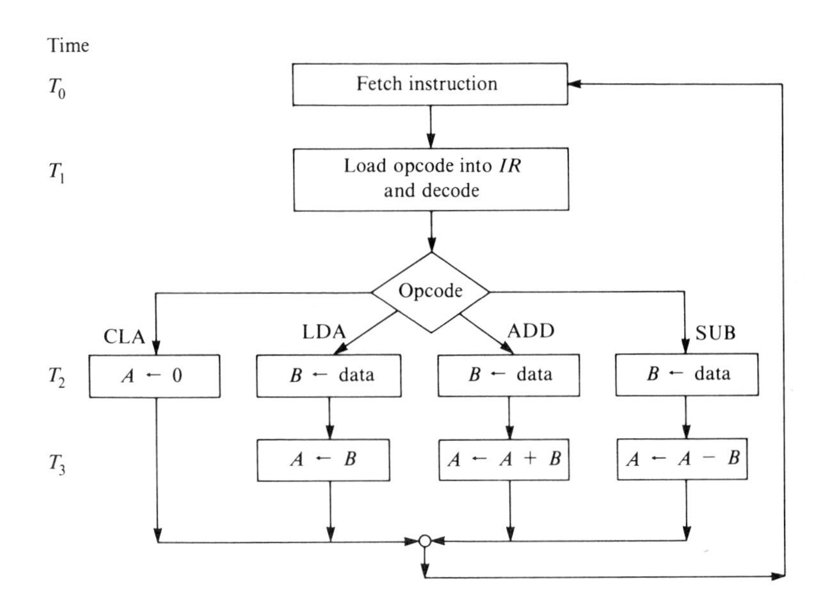

FIGURE 12.9
A process control algorithm control flow diagram (partial)

Step 3: Solution/Simplification.

a. Data path subsystem (Section 12.2). Design a data path that supports each instruction in the computer's repertoire.

b. Control unit subsystem (Section 12.3).

(1) Next-state generator. Develop a process control algorithm next-state table.

(2) Output decoder. Develop logic equations for the data path control signals.

c. Memory subsystem (Section 12.4). Design a memory subsystem to store program instructions/data and computer-generated results of the data processing.

Step 4: Realization.

a. Implement the data path subsystem (Section 12.2).

b. Implement the control unit subsystem (Section 12.3). Integrate data path and control unit to form the CPU.

c. Implement the memory subsystem. Integrate the CPU and memory subsystem to form the processor.

12.2 DESIGN AND IMPLEMENTATION OF A DATA PATH SUBSYSTEM

The evolution of a stored-program computer processor begins with the development of a data path subsystem that supports each of the eight instructions in the computer's repertoire (Table 12.2). The evolution of a digital-logic data-path subsystem is accomplished in four steps:

1. ***Implementation of a 4-bit MSI binary adder***
2. ***Implementation of a 4-bit adder/subtracter***

3. ***Implementation of a 4-bit arithmetic unit***
4. ***Implementation of a data path subsystem***

Implementation of a 4-Bit MSI Adder

A digital computer is a device that can execute a variety of arithmetic, logical, branch, store, and I/O instructions. Basic arithmetic instructions include the operations of addition, subtraction, multiplication, and division. A binary adder is a fundamental component of the CPU of many digital computers. Hence, the evolution of a computer naturally begins with a binary adder.

Addition is one of the most basic arithmetic operations performed by a digital computer. A 4-bit binary adder computes the sum of two 4-bit numbers, A and B, and a 1-bit carry-in C. The adder generates a 4-bit sum Z and a 1-bit carry-out C_o, as shown in Figure 12.10. Hexadecimal keypads may be used to input each of the 4-bit numbers A and B, and a switch is used to input C. A hexadecimal display and a probe may be used for the 4-bit sum Z and carry-out C_o, respectively.

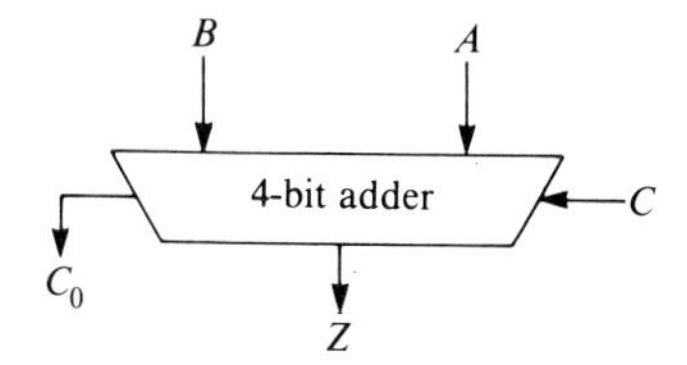

FIGURE 12.10
Block diagram of a 4-bit binary adder

Implementation of a 4-Bit Adder/Subtracter

The next step in evolving from a 4-bit adder to a computer is the development of an adder/subtracter. The adder circuit of Figure 12.10 can be modified to function as an adder/subtracter by inserting an XOR gate in each subtrahend bit path between the input device and the adder, as shown in Figure 12.11. A single 2-state control W is connected to the control line of the XOR gates. For addition, $W = 0$; for subtraction, $W = 1$ (to produce a 1s complement), and the carry-in $C = 1$ (to add 1 to produce the 2s complement).

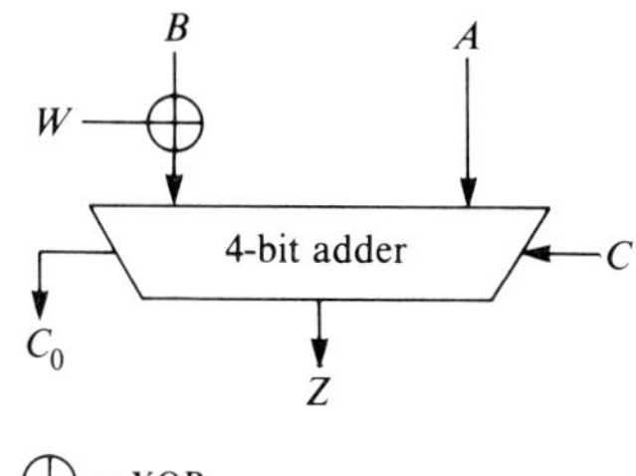

FIGURE 12.11
A 4-bit adder/subtracter circuit

The setting of control W determines whether the adder/subtracter performs an add operation ($W = 0$) or a subtract operation ($W = 1$). A function table for the device follows:

Control	Operation
W = 0	Add
W = 1	Subtract

Implementation of a 4-Bit ALU

The adder/subtracter developed in the preceding section has the ability to perform addition or subtraction depending on the setting of control W. A next logical step in evolving from a 4-bit adder to a computer is to develop an ALU that can perform a variety of arithmetic and logical operations. Such an ALU can be implemented by incorporating enabling (AND) gates and XOR gates in the adder input paths (see Section 6.2). The resulting 4-bit binary ALU can be represented by the block diagram shown in Figure 12.12.

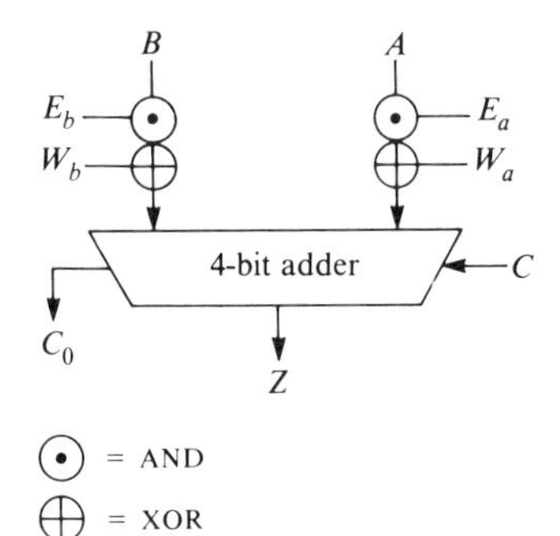

FIGURE 12.12
Block diagram of a binary ALU

Thirty-two possible operations could be defined using 5 independent controls E_b, W_b, E_a, W_a, and C. Table 12.3 lists a set of 8 operations (corresponding to the instruction set in Table 12.2) that can be performed by the ALU.

TABLE 12.3
Table of ALU operations

E_b	W_b	E_a	W_a	C	Register Transfer	Explanation
0	0	0	0	0	$Z \leftarrow 0$	Clear
0	0	1	1	0	$Z \leftarrow A'$	Complement A
0	0	1	0	1	$Z \leftarrow A + 1$	Increment A
0	0	1	1	1	$Z \leftarrow A' + 1$	Negate A (= 2s complement A)
1	0	0	0	0	$Z \leftarrow B$	Transfer B
1	0	1	0	0	$Z \leftarrow A + B$	Add B to A
1	1	1	0	1	$Z \leftarrow A + B' + 1$	Subtract B from A
0	0	1	0	0	$Z \leftarrow A$	Pass A

Implementation of a Data Path Subsystem

A simple data path subsystem can be formed by combining an ALU (Figure 12.12) with the following two data-storage registers:

A register (B) to hold the data value X

An accumulator register (A) to hold the result of successive arithmetic/logic operations

Registers A and B are loaded by strobing the load controls Ld-A and Ld-B, respectively. The output of the binary adder becomes an input to the accumulator. The accumulator output in turn becomes an input to the A enabling (AND) gates. The AND gate and XOR gate combination is represented as a true/complement/zero/one (TC01) function module in the resulting configuration (Figure 12.13) that forms a simple data path subsystem.

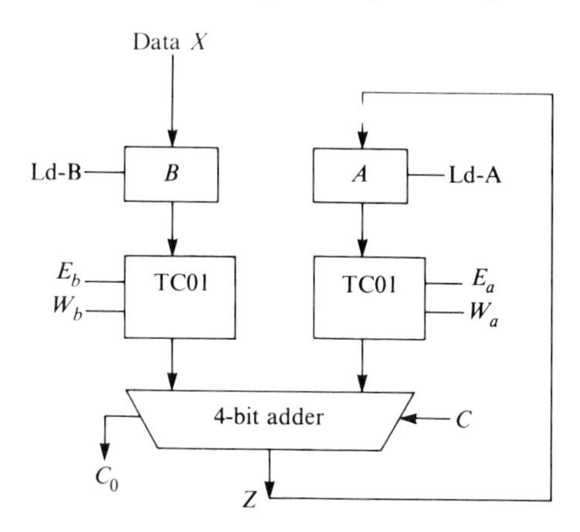

FIGURE 12.13
Block diagram of a data path subsystem

The data path subsystem has the ability to execute a computer instruction (Table 12.4) if

Controls E_b, W_b, E_a, W_a, and C are properly set

Adder inputs remain stable until the adder output is loaded into the accumulator register A

Registers B and A are strobed at the appropriate times

TABLE 12.4
Instruction execution table

Instruction Mnemonic	Controls					Register Transfers	
	E_b	W_b	E_a	W_a	C	T_2	T_3
CLA	0	0	0	0	0	$A \leftarrow 0$	
CMP	0	0	1	1	0	$A \leftarrow A'$	
INC	0	0	1	0	1	$A \leftarrow A + 1$	
NEG	0	0	1	1	1	$A \leftarrow A' + 1$	
LDA	1	0	0	0	0	$B \leftarrow$ data	$A \leftarrow B$
ADD	1	0	1	0	0	$B \leftarrow$ data	$A \leftarrow A + B$
SUB	1	1	1	0	1	$B \leftarrow$ data	$A \leftarrow A - B$
STA	0	0	1	0	0	$Z \leftarrow A$	

Note that 5 instructions (CLA, CMP, INC, NEG, and STA) in Table 12.4 do not require register B, while 3 instructions (LDA, ADD, and SUB) do. The instruction-execution table (Table 12.4) illustrates that some instructions (e.g., CLA and CMP)

require fewer microoperations, hence fewer time intervals, than other instructions (e.g., LDA and ADD).

The following example illustrates the operation of the data path subsystem. We will add a list of numbers (0001, 0010, 0011, . . .) using the partial-sums algorithm. Then, a sequence of instructions are executed to accomplish the following algorithm:

1. Key the first number (0001) into the X input device, and strobe Ld-B to load register B. Manually set controls E_b, W_b, E_a, W_a, and C to 1, 0, 0, 0, and 0. The control combination will cause the binary adder to pass B (= 0001) through the adder to the adder output lines. Then, when Ld-A is strobed, the adder output will be loaded into the accumulator register A.
2. To add the next number in the list, key that number (0010, 0011, . . .) into the X input device, and strobe Ld-B to load B. Set the controls E_b, W_b, E_a, W_a, and C to 1, 0, 1, 0, and 0. Strobe Ld-A to load the result, $A + B$, into A.
3. Repeat Step 2, keying in each successive number in the list into the X input device. The control combination will remain 1, 0, 1, 0, 0 to add A and B. At the end of the list of numbers, the accumulator A will contain the sum of the numbers in the list.

A manual control computer is defined here as a computing device that can perform a variety of arithmetic and logic operations depending on the settings of a number of manual controls. By this definition, an abacus is an example of an early manual control computer, and a calculator is an example of a more recent manual control computer. The data path subsystem developed in this section can be considered a manual control computer. The instruction repertoire (Table 12.2) of the stored-program processor is also the instruction repertoire of the manual control computer. Table 12.5 includes the opcode of each instruction.

TABLE 12.5
Instruction repertoire for manual control computer

Opcode	Mnemonic	Register Transfer	Description of Instruction
000	CLA	$A \leftarrow 0$	Clear A
001	CMP	$A \leftarrow A'$	Complement A
010	INC	$A \leftarrow A + 1$	Increment A
011	NEG	$A \leftarrow A' + 1$	Negate A = 2s complement A
100	LDA	$A \leftarrow B$	Transfer B to A
101	ADD	$A \leftarrow A + B$	Add B to A
110	SUB	$A \leftarrow A + B' + 1$	Subtract B from A
111	STA	$Z \leftarrow A$	Pass A

12.3 DESIGN AND IMPLEMENTATION OF A CONTROL UNIT SUBSYSTEM

The data path subsystem described in Section 12.2 supports each computer instruction listed in Table 12.2. We noted that the data-path subsystem can be interpreted as a manual computer that has the ability to execute each of these instructions. However,

the operation of the device is tedious and time consuming, requiring the human operator to determine and manually set the E_b, W_b, E_a, W_a, and C controls for each instruction and strobe the data path register load points in the order required for each instruction. These disadvantages serve as our motivation to design a more efficient device (control unit) that

- ***Fetches (reads) program instructions from memory***
- ***Decodes each instruction***
- ***Manipulates the data path control points at appropriate times to accomplish the instruction's register transfer operations***

This section describes the design and implementation of a control unit subsystem that performs the three functions listed above. The control unit subsystem and data path subsystem are then combined to form the CPU of a computer processor, as shown in Figure 12.14.

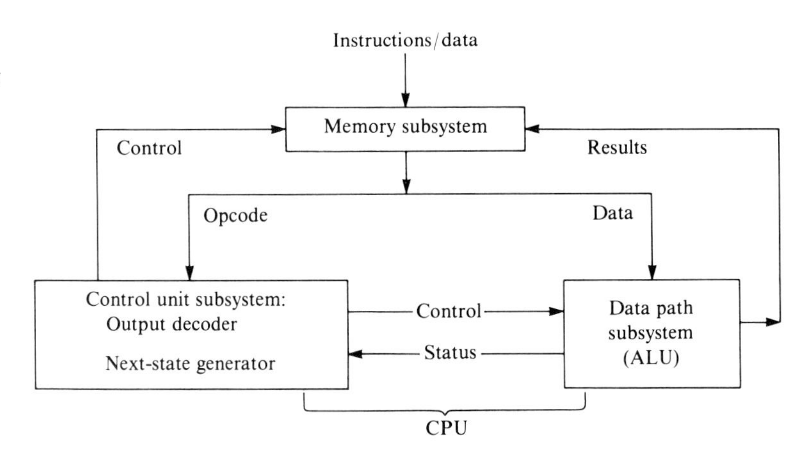

FIGURE 12.14
Structure diagram of a computer processor

Design of Control Unit Subsystem

The control unit subsystem is required to fetch each instruction from memory, decode the instruction opcode, and, at appropriate times, generate control signals to those data path components involved in the execution of that instruction. The control unit process control algorithm accomplishes fetching, decoding, and executing each computer instruction in 4 time subintervals T_0, T_1, T_2, and T_3 as follows (see Figure 12.15):

T_0, instruction fetched (read from memory)

T_1, opcode loaded into an instruction register *IR*, decoded by a 3-to-8 decoder and opcode-associated instruction line asserted

T_2–T_3, **microoperations** (register transfers) required for that instruction accomplished

The control unit subsystem is composed of a next-state generator and an output decoder. The operation of these function modules is described next.

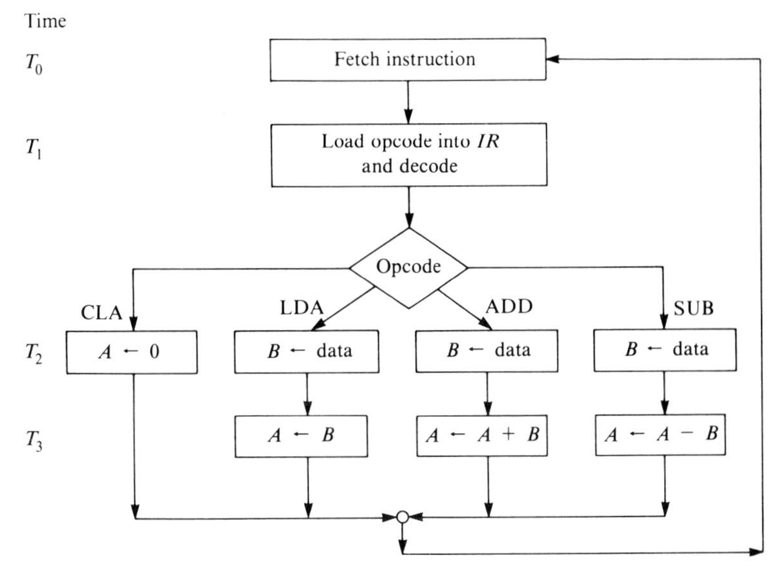

FIGURE 12.15
Process control algorithm control flow diagram (partial)

Control Unit Next-State Generator

We can see from the control flow diagram (Figure 12.15) that fetching, decoding, and executing any instruction can be accomplished in 4 states **a**, **b**, $\mathbf{c}_i$, and $\mathbf{d}_i$ (corresponding to times T_0, T_1, T_2, and T_3). A ring counter (see Chapter 7), which generates time pulses T_0, T_1, T_2, and T_3, can be used as a next-state generator for the control unit subsystem. A timing diagram of the ring counter outputs T_0 to T_3 is presented in Figure 12.16.

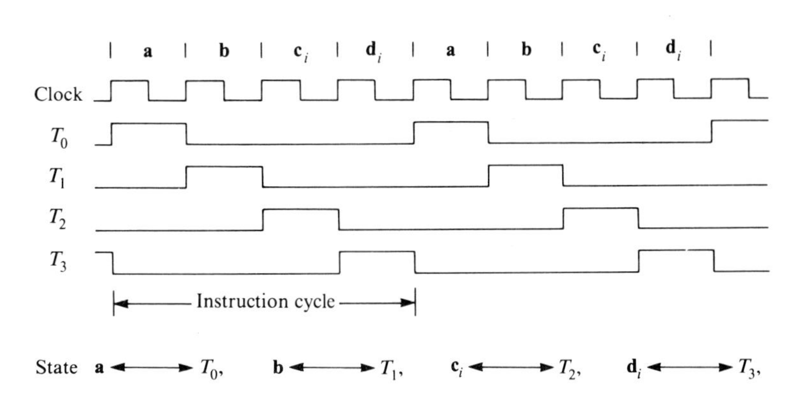

FIGURE 12.16
Timing diagram of a 4-bit ring counter

The organization of the control unit subsystem is presented in the form of a block diagram (Figure 12.17, p. 584).

Our next and final task in designing the control unit subsystem is to design an output decoder, which (for each instruction) generates control signals (at appropriate times) to those data path components involved in the execution of that instruction.

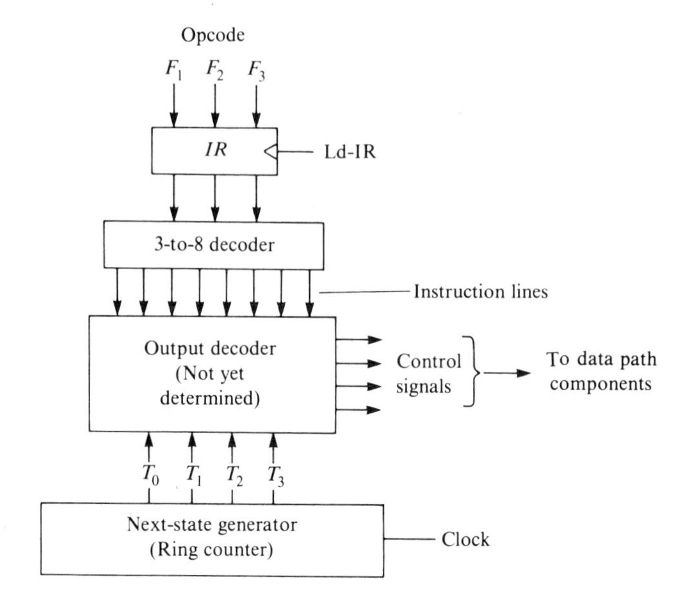

FIGURE 12.17
Block diagram of control unit subsystem

Control Unit Output Decoder

The control unit block diagram (Figure 12.17) shows that the output decoder has 8 instruction-line inputs (one for each decoded opcode) and 4 time-pulse inputs T_0, T_1, T_2, and T_3. The function of the output decoder is to manipulate the CPU controls to accomplish the following: Fetch each instruction, load the opcode into instruction register IR, and accomplish execution of the instruction by generating data path control signals at the times required for that instruction.

In states **b**, $\mathbf{c}_i$, and $\mathbf{d}_i$, the output decoder generates control signals by combining the signal on the active instruction line with the time-pulse signal (T_1, T_2, or T_3). These control signals activate the data path components at appropriate times to accomplish fetch, decode, and execute of that instruction. An **instruction-time event table** for instructions CLA (clear A), LDA (load A from B), ADD (add B to A), and SUB (subtract B from A) is as follows:

Instruction (Opcode) \ Time (Event)	T_0	T_1	T_2	T_3
CLA (000)	Fetch inst.	Ld-IR	[00000],Ld-A	—
LDA (100)	Fetch inst.	Ld-IR	Ld-B	[10000],Ld-A
ADD (101)	Fetch inst.	Ld-IR	Ld-B	[10100],Ld-A
SUB (110)	Fetch inst.	Ld-IR	Ld-B	[11101],Ld-A

where [#####] denotes setting of controls E_b, W_b, E_a, W_a, and C.

Note the following:

- Strobing Ld-IR accomplishes loading opcode into *IR*
- Strobing Ld-B accomplishes loading data word into *B*
- Strobing Ld-A accomplishes loading result into *A*
- Data path controls E_a, W_a, E_b, W_b, and C are level signals activated throughout interval T_i.
- Registers *IR*, *A*, and *B* are positive edge-triggered registers that are loaded midway through interval T_i.
- Each column-row intersection of the event table represents a Boolean algebra product term (p-term). Each **p-term** is a logical product of a time T_i *and* an instruction (CLA, LDA, ADD, SUB).
- A logic (control) equation for each data path control point can be derived by collecting all p-terms associated with that control point. For example, the control equation for loading register *A* is

$$\text{Ld-A} = T_2 \cdot \text{CLA} + T_3 \cdot \text{LDA} + T_3 \cdot \text{ADD} + T_3 \cdot \text{SUB}$$

The execution of instruction I_i (LDA,ADD,SUB, . . .) used to process data item X_i is accomplished by a sequence of operations:

T_0, fetch instruction

T_1, strobe Ld-IR to load instruction opcode into register *IR*, whose output lines are connected to the input lines of a 3-to-8 decoder. (The opcode is decoded, asserting instruction line *i* to the control-unit programmable logic array.)

T_2–T_3, the programmable logic array (PLA) consists of an array of AND gates and an array of OR gates that together implement the Boolean equations for the activation of data path components.

The activation of a given data path control point depends on the instruction being executed and the present state of the control algorithm. The control unit output decoder activations of data path control points are determined by tracing the execution of each instruction through the data path. Table 12.6 is a summary of these activations for 4 instructions (CLA, LDA, ADD, SUB).

TABLE 12.6
P-term/control point activation table

	Control Points								
Opcode · Time	**Fetch**	**Ld-IR**	**Ld-B**	E_b	W_b	E_a	W_a	C	**Ld-A**
ALL · T_0	1								
ALL · T_1		1							
CLA · T_2				0	0	0	0	0	1
LDA · T_2			1						
LDA · T_3				1	0	0	0	0	1
ADD · T_2			1						
ADD · T_3				1	0	1	0	0	1
SUB · T_2			1						
SUB · T_3				1	1	1	0	1	1

Table 12.6 can be interpreted as an AND-OR array; that is,

- The opcode and time columns are ANDed.
- All the rows under any control point are ORed.

The following logic equations for activating the control points are the result.

$$\text{Fetch} = T_0$$
$$\text{Ld-IR} = T_1$$
$$\text{Ld-B} = T_2 \cdot \text{LDA} + T_2 \cdot \text{ADD} + T_2 \cdot \text{SUB}$$
$$E_b = T_3 \cdot \text{LDA} + T_3 \cdot \text{ADD} + T_3 \cdot \text{SUB}$$
$$W_b = T_3 \cdot \text{SUB}$$
$$E_a = T_3 \cdot \text{ADD} + T_3 \cdot \text{SUB}$$
$$W_a = 0$$
$$C = T_3 \cdot \text{SUB}$$
$$\text{Ld-A} = T_2 \cdot \text{CLA} + T_3 \cdot \text{LDA} + T_3 \cdot \text{ADD} + T_3 \cdot \text{SUB}$$

The control equations implemented in the output decoder PLA for a given instruction repertoire are obtained by gathering the p-terms associated with each control point. The control point activations for the edge-triggered registers, loaded midway through interval T_i on the rising edge of a clock signal ECLK, are

$$\text{Ld-IR} = T_1 \cdot \text{ECLK}$$
$$\text{Ld-B} = (T_2 \cdot \text{LDA} + T_2 \cdot \text{ADD} + T_2 \cdot \text{SUB}) \cdot \text{ECLK}$$
$$\text{Ld-A} = (T_2 \cdot \text{CLA} + T_3 \cdot \text{LDA} + T_3 \cdot \text{ADD} + T_3 \cdot \text{SUB}) \cdot \text{ECLK}$$

Control points E_a, W_a, E_b, W_b, and C and fetch are activated by the following level signals:

$$E_b = T_3 \cdot \text{LDA} + T_3 \cdot \text{ADD} + T_3 \cdot \text{SUB}$$
$$W_b = T_3 \cdot \text{SUB}$$
$$E_a = T_3 \cdot \text{ADD} + T_3 \cdot \text{SUB}$$
$$W_a = 0$$
$$C = T_3 \cdot \text{SUB}$$
$$\text{Fetch} = T_0$$

The output decoder can be realized by a PLA in which the above logic equations are implemented in 2-level AND-OR form. This completes the solution of the control unit design.

The actual implementation of the control unit subsystem is described in the following subsections. The implementation of the CPU, formed by integrating the control unit and data path subsystems is accomplished in the final subsection.

Implementation of a Clock-Ring Counter Circuit

The ring counter generates 4 time pulses, $T_0, \ldots, T_3$, which are used in conjunction with each instruction opcode to cause the control unit to generate control signals (Ld-IR, Ld-A, E_a, E_b, etc.) required to activate the functional components (*IR*, *A*, *A*-enable, *B*-enable, etc.). A logic diagram of a clock-ring counter circuit is shown in Figure 12.18.

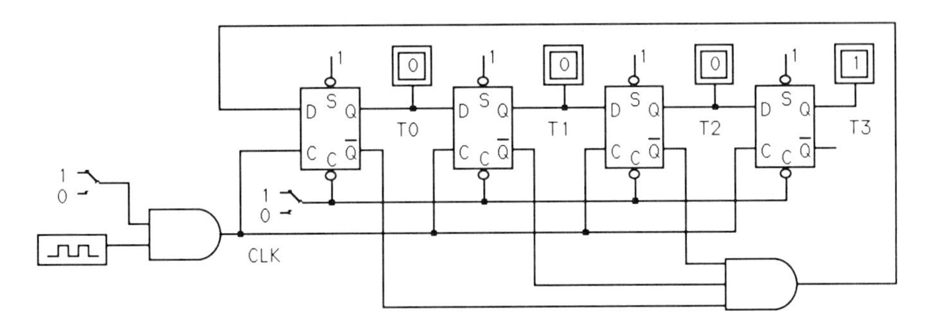

FIGURE 12.18
Logic diagram of clock-ring counter circuit

Implementation of an Instruction Register and Decoder

F_1, F_2, and F_3 denote the 3-bit opcode of the instruction to be executed. The opcode is loaded into an IR and decoded by a 3-to-8 decoder. Each output line of the decoder represents one of the 8 computer instructions (Table 12.2). The IR and decoder shown in the control unit block diagram (Figure 12.17) are implemented using a 4-bit edge-triggered *D* register and a 3-to-8 decoder, as shown in Figure 12.19. Note that the outputs of the decoder are active-low.

FIGURE 12.19
Implementation of instruction register and decoder

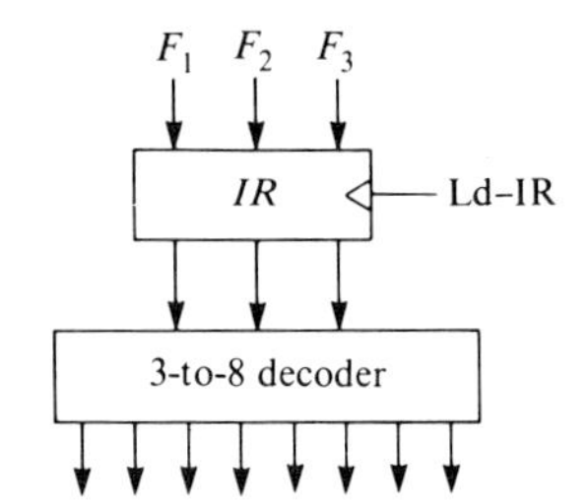

Implementation of a PLA and Its Integration into the Control Unit

The output decoder logic equations, derived earlier, are implemented in the form of a PLA.

Now, the complete control unit subsystem can be formed by integrating the following components (see Figure 12.20):

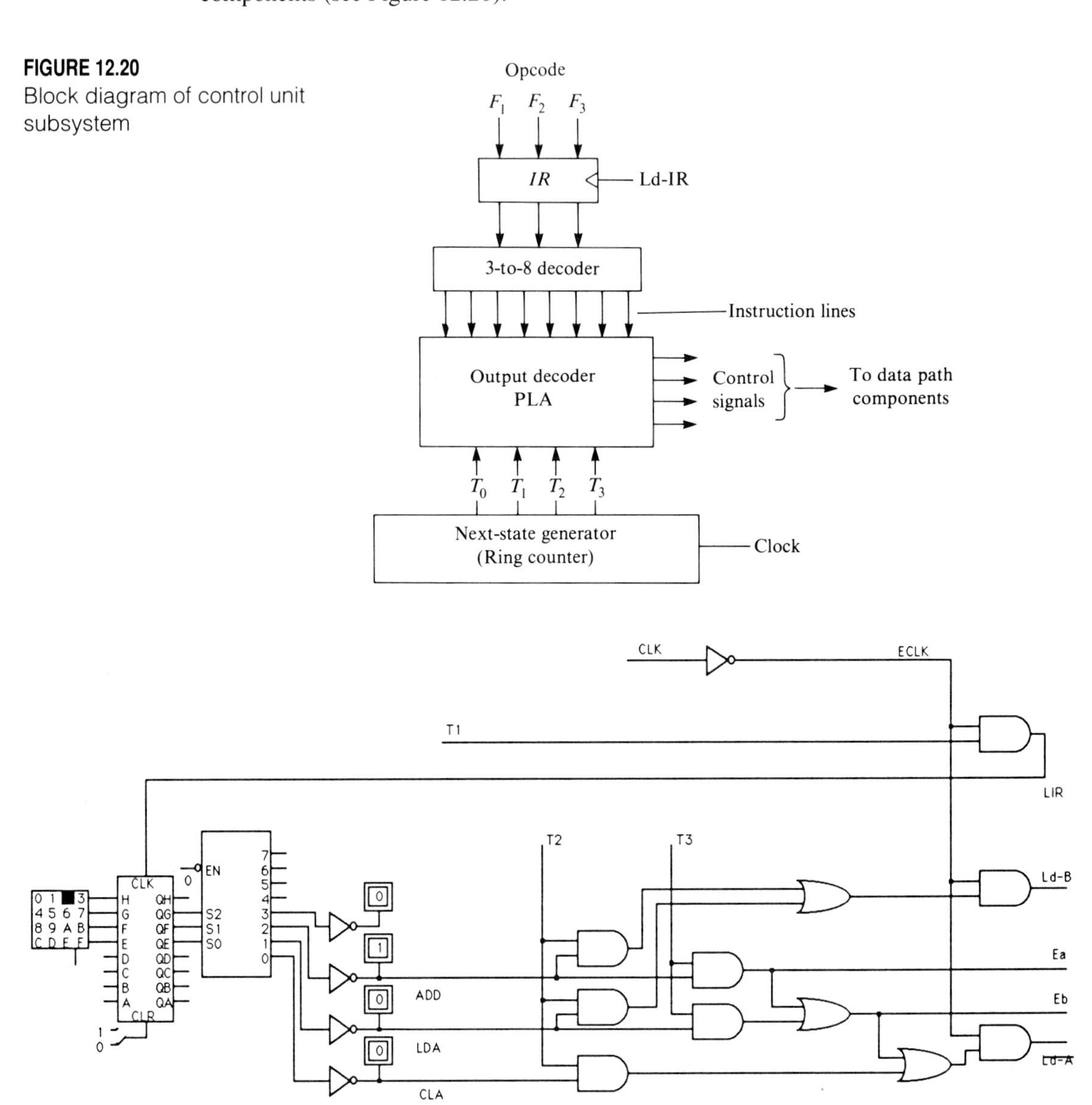

FIGURE 12.20
Block diagram of control unit subsystem

FIGURE 12.21
Logic diagram of control unit subsystem

A clock-ring counter to generate time pulses T_0, T_1, T_2, T_3

Opcode lines F_1, F_2, F_3

An IR and a 3-to-8 decoder

A PLA

A logic diagram of a control unit subsystem is presented in Figure 12.21.

Integration of Control Unit and Data Path to Form a CPU

The implementation of the CPU is accomplished by integrating the control unit and data path subsystems, as shown in the block diagram in Figure 12.22.

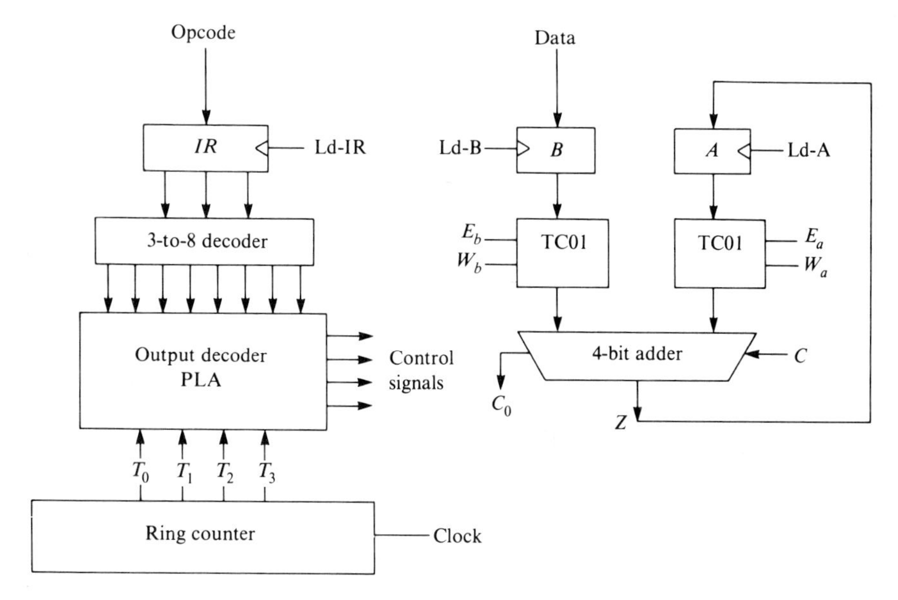

FIGURE 12.22
Block diagram of the CPU

A complete logic diagram and timing diagram of the CPU (with a 4-bit data path) is presented in Figure 12.23 (p. 590).

The CPU designed in this section can be interpreted as a semiautomatic computer that, given opcode and data inputs, can execute all of the instructions in Table 12.2, except for the STA instruction (which requires a memory subsystem).

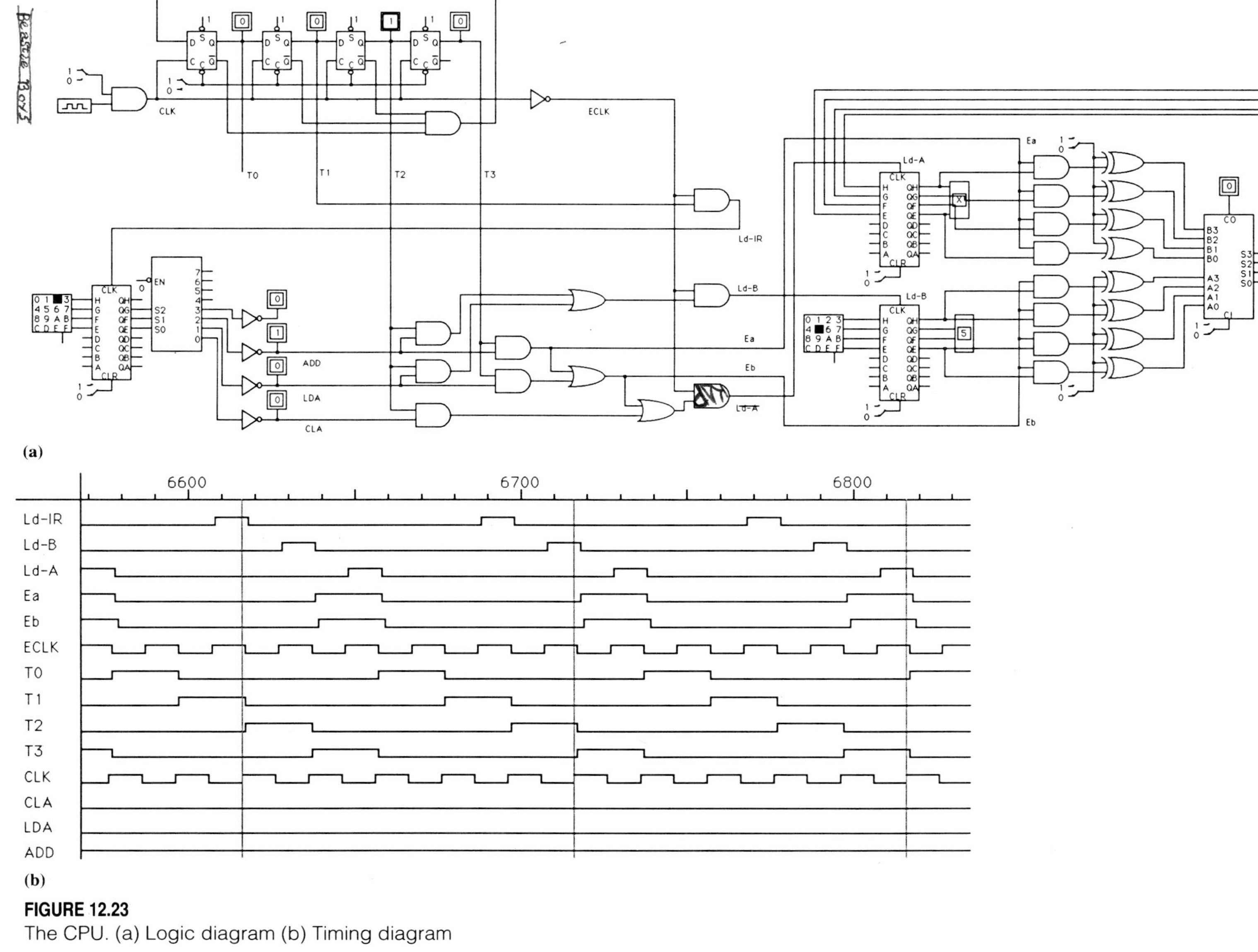

FIGURE 12.23
The CPU. (a) Logic diagram (b) Timing diagram

12.4 DESIGN AND IMPLEMENTATION OF A MEMORY SUBSYSTEM

This section completes the evolution of a stored-program processor. First, we describe the design and implementation of a memory subsystem and then the integration of the memory subsystem and an 8-bit CPU to form a complete stored-program processor. The 8-bit CPU, which processes 8-bit data words, is an extension of the 4-bit CPU described in Section 12.3.

The processor will have the ability to execute a set of 8-bit instructions as described in Table 12.7. Each instruction is composed of a 3-bit opcode field and a 5-bit address field. The first four instructions (CLA, CMP, INC, and NEG) use register A as an implied operand and do not require an address field.

TABLE 12.7
Processor instruction repertoire

Mnemonic	Binary	Description
CLA -----	000-----	Clear accumulator A
CMP -----	001-----	Complement accumulator A
INC -----	010-----	Increment accumulator A
NEG -----	011-----	Negate accumulator
LDA address	100xxxxx	Load A with content of memory address X
ADD address	101xxxxx	Add content of memory address X to A
SUB address	110xxxxx	Subtract content of memory address X from A
STA address	111xxxxx	Store accumulator in memory address X

Five instructions (CLA, CMP, INC, NEG, and STA) are register reference instructions (in which register A is an implied operand), and 3 instructions (LDA, ADD, and SUB) are memory reference instructions that process a data word stored in memory. A particular sequence of instructions constitute a program. Each program instruction will direct the CPU to accomplish a particular sequence of microoperations, which together accomplish that instruction. Program instructions and data words X_i are loaded in memory by a data-entry device.

Design of a Memory Subsystem

The processor can be created by adding a memory unit to a CPU (such as the one designed in Section 12.3). Thus, the processor is composed of the following subsystems:

A data path (Section 12.2) that processes memory data under control of the control unit

A control unit (Section 12.3) that interprets program instructions from memory and controls data path elements accordingly

A memory subsystem consisting of a memory unit and associated registers, a program-load facility, and a program-run facility

A memory subsystem of a computer must have the capability of

Loading program instructions and data into the memory prior to execution of the program

Providing both program instructions and data words to the CPU during program execution

Storing program-generated results in the memory

A computer read/write memory may be visualized as a set of mail boxes, the location of each being specified by an address. Each mail box can hold one instruction or one data item. A given location in memory is referenced by an address on the memory address lines. An address decoder decodes the address lines to determine the specified memory location to be read from or written to. Data and program instructions are transferred to or from addressed memory locations by way of data input lines and output lines, as illustrated in Figure 12.24.

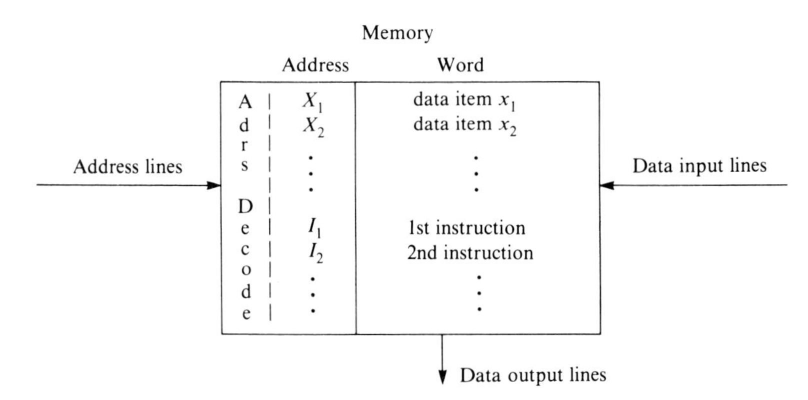

FIGURE 12.24
A functional view of a read/write memory

A read/write memory can store either a program instruction or a data word in any memory location. An 8-bit memory word is assumed for most of the computers designed in this text. Numeric data words are stored in 2s complement representation. A typical read/write memory is shown in symbolic form in Figure 12.25. In addition to the address lines, data input lines, and data output lines, the memory device usually has a number of control lines, such as

A **read enable** (OE') to enable the output buffer

A **write enable** (WE') to enable the input buffer

A **chip enable** (CE') to enable or select the memory chip

CE′ OE′ WE′
Memory
Address lines → Adrs
D_{in} ← Data input lines
D_{out}
Data output lines

FIGURE 12.25
Symbolic representation of a read/write memory

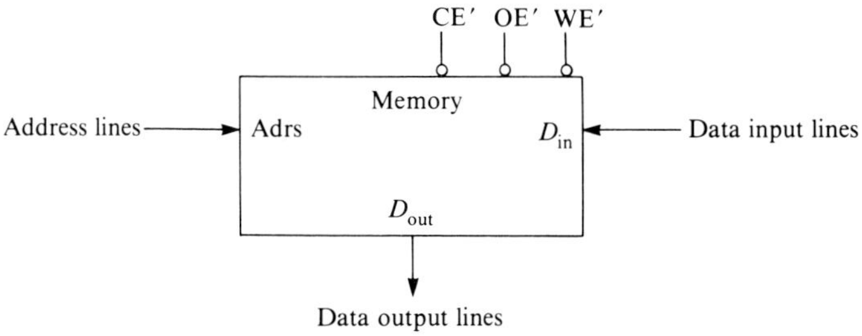

Some memories have bidirectional I/O lines rather than separate input lines and output lines.

A program instruction or data word can be written to a specified memory location by placing an 8-bit address on the memory address lines, asserting the chip enable line (CE') low, and asserting the write enable line (WE') low. A specific instruction or data word can be read from memory by placing the 8-bit address on the memory address lines, asserting the chip enable line (CE') low, and asserting the read enable line (OE') low.

Implementation of a Read/Write Memory and Associated Registers

A basic memory subsystem can be implemented by augmenting a **read/write memory** (R/WM) with a **memory address** (MA) register, and a **memory write** (MW) register, as shown in Figure 12.26.

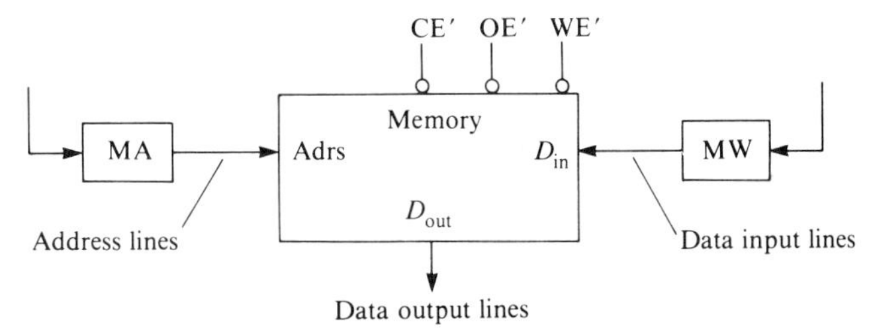

FIGURE 12.26
A basic memory subsystem with MA and MW registers

For a memory write, the MA register contains the memory address at which the word in the MW register is to be stored. For a memory read, the MA register contains the memory address of the word to be read out onto the data output lines.

Adding a Program-Load Facility to the Memory Subsystem

Prior to execution of a program, an input device is used to load the program instructions and data into the memory via the MW register. Each instruction/data word is stored in the memory location specified by the content of the MA register.

A multiplexer can be used to interface the MW register with a program load device (hexadecimal keypad) and the data path. A second multiplexer is used to interface the MA register with the load-address lines and the run-address lines. A block diagram of a memory subsystem with a load facility is presented in Figure 12.27 (p. 594).

The unused multiplexer inputs in Figure 12.27 will be connected to the data path and run-address lines at a later time (when the memory subsystem is interfaced with a CPU). Each program instruction or data word is loaded as follows:

- A load/run switch is set to 1 to select the load-instruction/data and load-address lines.
- Each instruction/data word address is keyed into the hexadecimal keypad(s), and the address is loaded into the MA register by strobing the PLd-MA control line.
- Each instruction/data word is keyed into the hexadecimal keypad(s), and the word is loaded into the MW register by strobing the PLd-MW control line.

- The word in the MW register is written into memory when the chip enable (CE') and write enable (WE') lines are asserted low.

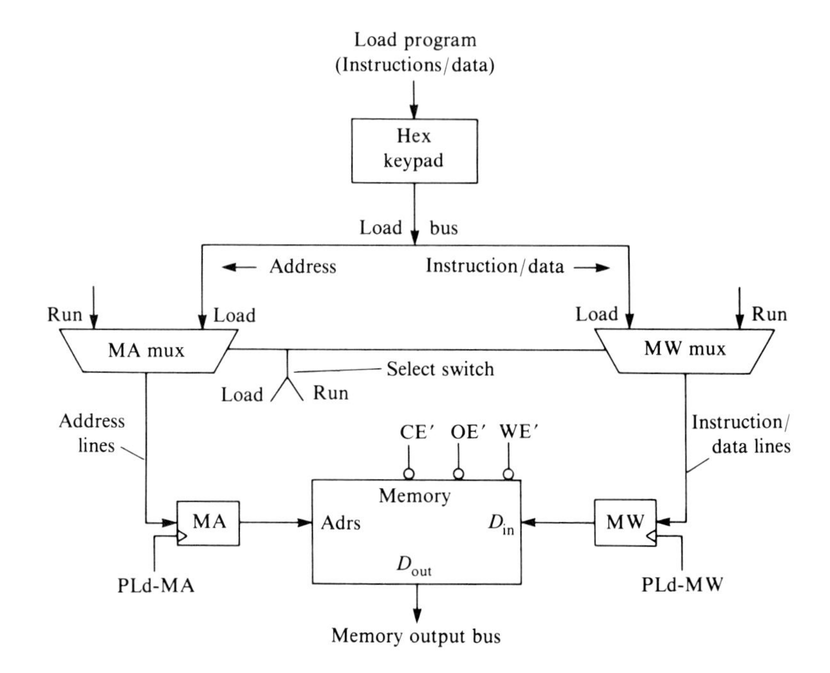

FIGURE 12.27
A memory subsystem with a program-load facility

Adding a Program-Run Facility to the Memory Subsystem

During program execution, the address of the instruction to be executed is contained in a **program counter** (PC). A third multiplexer (DA/PC mux) can be used to interface the program counter and data-address lines with the memory subsystem. The unused input of the MW mux is to be connected to the output of the accumulator when the memory subsystem is integrated with the CPU. Figure 12.28 presents a block diagram of a complete memory subsystem that uses multiplexers to interface a load facility and a run facility with the memory via the MA and MW registers.

After the program instructions and data have been loaded into memory, the address of the first program instruction is entered into the program counter and the load/run switch is set to the run position. During program execution, at the appropriate time within an instruction-execution cycle, the DA/PC mux selects either the instruction-address lines or data-address lines; thus instruction addresses (from the PC) and data addresses (contained in instructions) are routed to the MA register. The addressed instruction or data word is read from memory by asserting the chip enable (CE') and read enable (OE') low.

Program-generated results to be stored in memory, at an address specified by the MA register, are transferred from the accumulator to the MW register. The word is written to memory by asserting the chip enable (CE') and the write enable (WE') low.

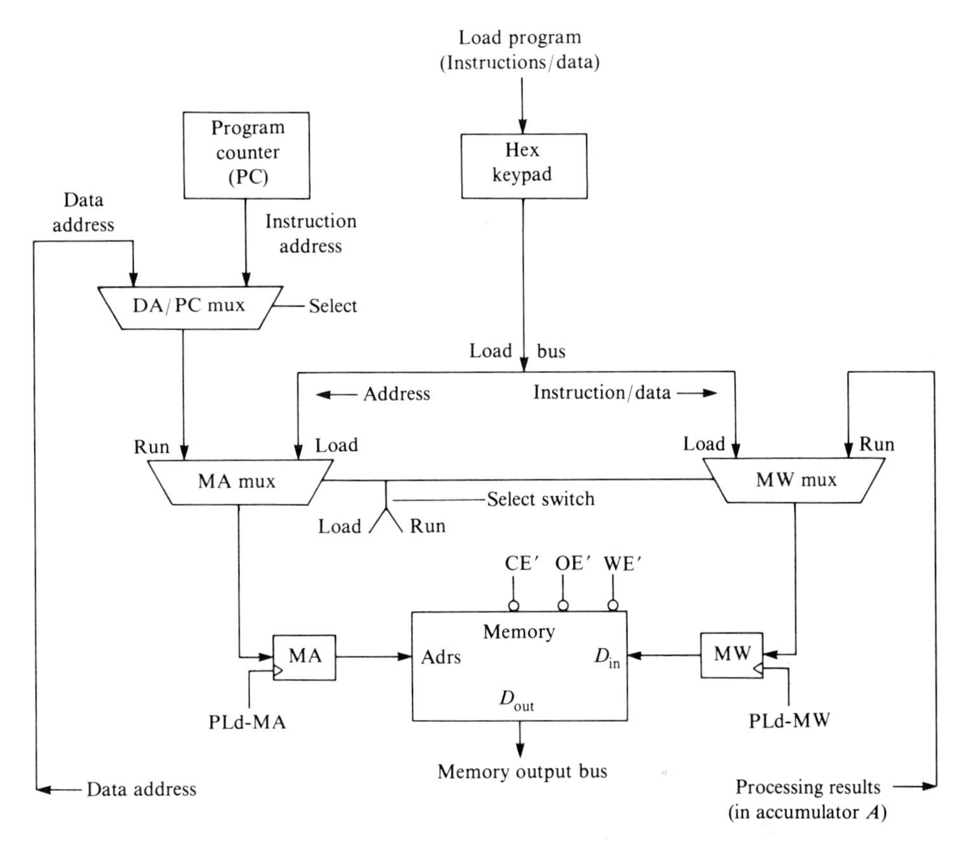

FIGURE 12.28
A memory subsystem with load and run facilities

Integration of Memory Subsystem and a CPU

The memory subsystem can be integrated with a CPU (with an 8-bit data path) to form a processor, as shown in Figure 12.29 (p. 596). The MA register is connected through a multiplexer to the PC for reading instruction addresses and to memory output data-address lines for reading data item addresses. The memory output data bus is also connected to the control unit (opcode lines) and to the data path (data-word lines).

Suppose that the memory has been loaded with the following data and instructions, as illustrated in Figure 12.29. Data items and instructions are represented in decimal equivalent and mnemonic form, respectively.

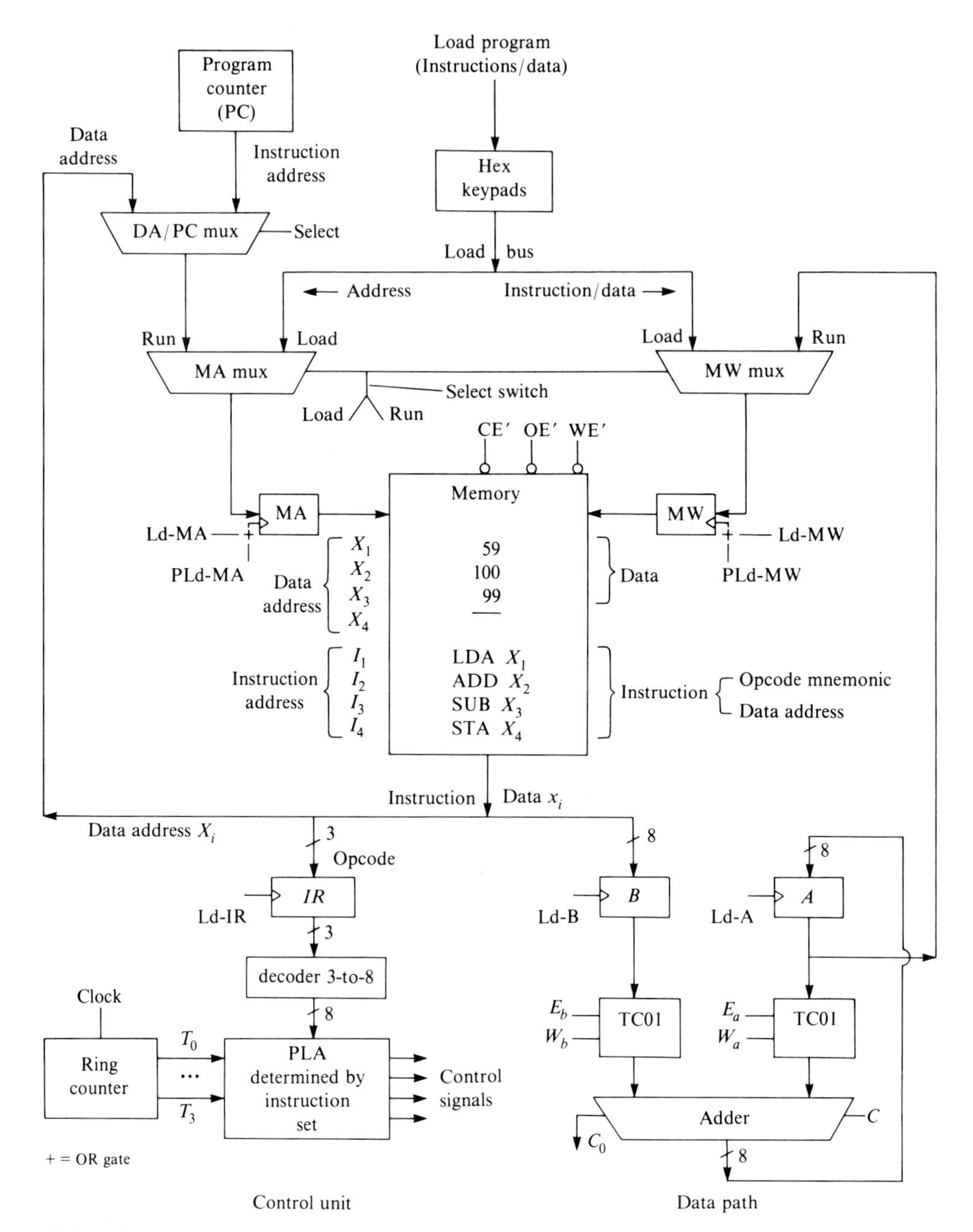

FIGURE 12.29
Block diagram of a simple processor

Basic Operation of a Stored-Program Processor

The processor accomplishes the execution of each program instruction by performing the following actions within the appropriate time subintervals of an instruction cycle:

- **Fetch** (read) the current instruction from memory.
- **Decode** the instruction to determine what specific actions are implied by that instruction. For memory reference instructions, this includes reading a data word (byte).
- **Execute** the decoded instruction by activating control signals of those data path components implied by that instruction.

The processing (fetch, decode, execute) of each instruction is accomplished in an instruction cycle that is divided into 4 clock cycles (states T_0, T_1, T_2, and T_3). Figure 12.30 illustrates a basic timing diagram for a simple processor. For simplicity, we assume that a memory read or write operation requires one-half of a clock cycle time. Note that signal ECLK is an inversion of signal clock.

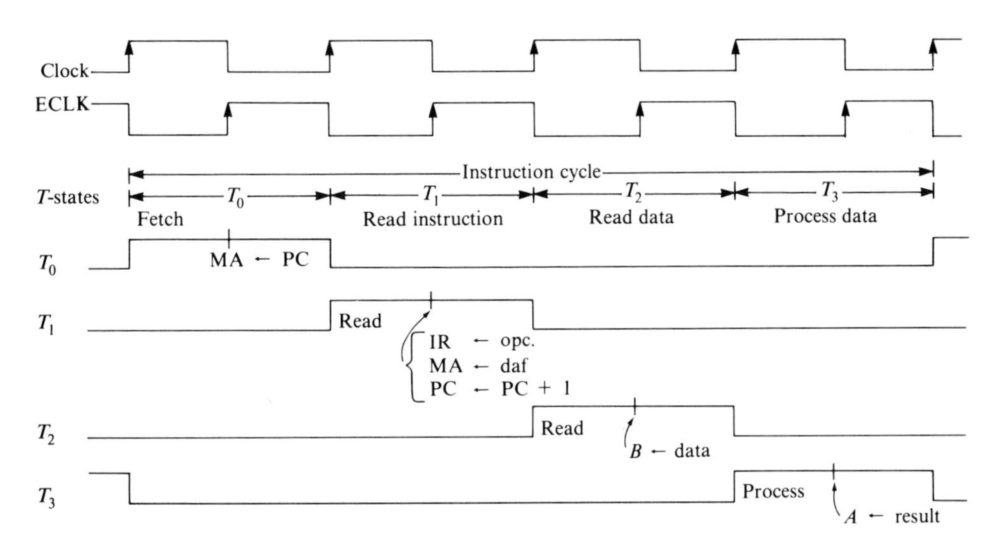

FIGURE 12.30
Timing diagram for a simple processor

The execution of each program instruction involves the following:

Instruction Fetch and Decode. The address of the instruction to be executed is contained in the PC. In state T_0 for all instructions, the PC is loaded into the MA register on the rising edge of signal ECLK midway through state T_0. By the end of state T_0, a valid instruction address is present in the memory address (MA) register.

At the beginning of state T_1 a memory read command is initiated to read the instruction. After delays for memory access time and memory bus interconnection, the instruction is read out onto the memory output bus. Midway through state T_1 the instruction opcode and **direct address field** (daf) are loaded into the IR and MA register, respectively, on the rising edge of ECLK, and the PC is incremented. By the end of state T_1, the IR contains the opcode, the MA contains the daf, and the PC contains the incremented PC (PC now points to the next instruction to be executed). The opcode in the IR is decoded by the 3-to-8 decoder, activating the appropriate instruction line to the control unit PLA.

Instruction Execution. If the instruction being executed requires reading a data word from memory, a memory read command is issued at the beginning of state T_2, and the data word is read from memory. At the midpoint of state T_2, the data word is loaded into the B register. In state T_3, the data word is processed by the data path components. At the midpoint of state T_3, after propagation delays in the data path and set-up time for the A register, the instruction result is loaded into the A register.

The execution of instruction I_i (LDA, ADD, SUB, STA) used to process data item X_i is accomplished by a sequence of microoperations. Table 12.8 summarizes the microoperations involved in fetching, decoding, and executing the LDA, ADD, SUB, and STA instructions.

TABLE 12.8
Instruction-time event table

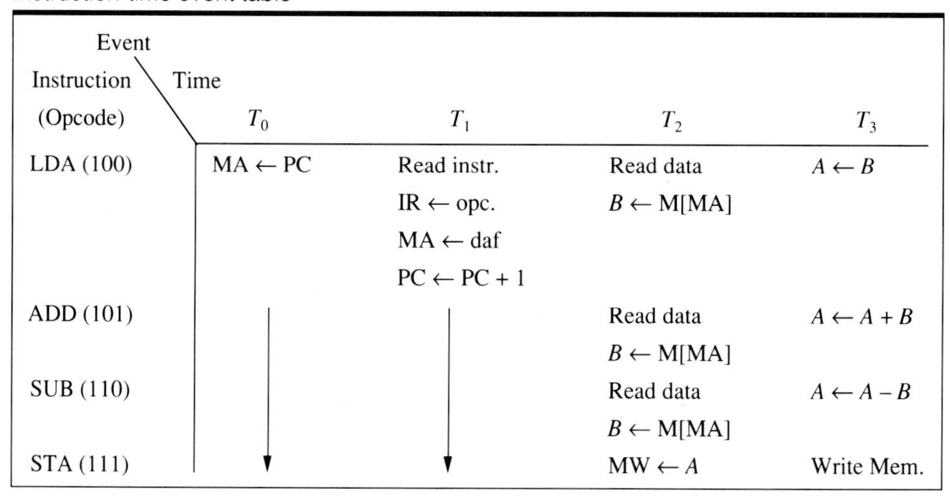

Event / Instruction (Opcode) \ Time	T_0	T_1	T_2	T_3
LDA (100)	MA ← PC	Read instr. IR ← opc. MA ← daf PC ← PC + 1	Read data B ← M[MA]	$A \leftarrow B$
ADD (101)	↓	↓	Read data B ← M[MA]	$A \leftarrow A + B$
SUB (110)	↓	↓	Read data B ← M[MA]	$A \leftarrow A - B$
STA (111)	↓	↓	MW ← A	Write Mem.

The activation of a given data path control point depends on the instruction being executed and the present state of the control algorithm.

The control unit output decoder activations of data-path control points are determined by tracing the execution of each instruction through the data path. Table 12.9 is a summary of these activations for 4 instructions (LDA, ADD, SUB, and STA).

This table can be interpreted as an AND-OR array; that is,

- The opcode *and* time columns are ANDed
- All the rows under any control signal are ORed
- Data path controls E_a, W_a, E_b, W_b, and C are level signals asserted throughout interval T_i.
- Registers MA, IR, A, B, and MW are positive edge-triggered registers that are loaded midway through interval T_i.
- Each column-row intersection of the event table represents a Boolean algebra product term (p-term). Each p-term is a logical product of a time T_i *and* an instruction (LDA, ADD, SUB, STA).

TABLE 12.9
P-term control point activation table

Opcode · Time	Control Points												
	Ld-MA	Ld-IR	RD	Ld-B	E_b	W_b	E_a	W_a	C	Pinc	Ld-A	Ld-MW	WR
ALL · T_0	1 (PC)												
ALL · T_1	1 (daf)	1 (opcode)	1 (inst.)							1 (PC + 1)			
LDA · T_2			1 (data)	1 (data)									
ADD · T_2			1 (data)	1 (data)									
SUB · T_2			1 (data)	1 (data)									
STA · T_2												1 (result)	
LDA · T_3					1	0	0	0	0		1 (data)		
ADD · T_3					1	0	1	0	0		1 (result)		
SUB · T_3					1	1	1	0	1		1 (result)		
STA · T_3													1 (result)

Note: The parenthetical entries indicate what is read, written, or loaded.

□ A logic equation for each data path control point can be derived by collecting all p-terms associated with that control point. For example, the control equation for loading the *A* register is

$$\text{Ld-A} = T_3 \cdot \text{LDA} + T_3 \cdot \text{ADD} + T_3 \cdot \text{SUB}$$

The PLA consists of an array of AND gates and an array of OR gates, which together implement Boolean equations for the activation of data path components. The following Boolean expressions for activating the control points are the result.

$$\text{Ld-MA} = T_0 \cdot \text{ALL} + T_1 \cdot \text{ALL}$$
$$\text{RD} = T_1 \cdot \text{ALL} + T_2 \cdot (\text{LDA} + \text{ADD} + \text{SUB})$$
$$\text{Ld-IR} = T_1 \cdot \text{ALL}$$
$$\text{Pinc} = T_1 \cdot \text{ALL}$$
$$\text{Ld-B} = T_2 \cdot (\text{LDA} + \text{ADD} + \text{SUB})$$

$$E_b = T_3 \cdot (\text{LDA} + \text{ADD} + \text{SUB})$$
$$W_b = T_3 \cdot \text{SUB}$$
$$E_a = T_3 \cdot (\text{ADD} + \text{SUB})$$
$$W_a = 0$$
$$C = T_3 \cdot \text{SUB}$$
$$\text{Ld-A} = T_3 \cdot (\text{LDA} + \text{ADD} + \text{SUB})$$
$$\text{Ld-MW} = T_2 \cdot \text{STA}$$
$$\text{WR} = T_3 \cdot \text{STA}$$

The control equations implemented in the output decoder PLA for a given instruction repertoire are obtained by gathering the p-terms associated with each control point. The control point activations for edge-triggered registers, loaded midway through interval T_i on the rising edge of signal ECLK, are

$$\text{Ld-MA} = (T_0 + T_1) \cdot \text{ECLK}$$
$$\text{Ld-IR} = T_1 \cdot \text{ECLK}$$
$$\text{Ld-}B = T_2 \cdot (\text{LDA} + \text{ADD} + \text{SUB}) \cdot \text{ECLK}$$
$$\text{Ld-}A = T_3 \cdot (\text{LDA} + \text{ADD} + \text{SUB}) \cdot \text{ECLK}$$
$$\text{Ld-MW} = T_2 \cdot \text{STA} \cdot \text{ECLK}$$

Control points E_a, W_a, E_b, W_b, C are activated by level signals. Thus,

$$E_a = T_3 \cdot (\text{ADD} + \text{SUB})$$
$$W_a = 0$$
$$E_b = T_3 \cdot (\text{LDA} + \text{ADD} + \text{SUB})$$
$$W_b = T_3 \cdot \text{SUB}$$
$$C = T_3 \cdot \text{SUB}$$

The PC, implemented using a binary counter, is incremented at time T_1 for each instruction. Thus,

$$\text{Pinc} = T_1$$

Memory read and write are accomplished by asserting OE' and WE' low while RD and WR, respectively, are asserted. Thus,

$$\text{RD} = T_1 + T_2 \cdot (\text{LDA} + \text{ADD} + \text{SUB})$$
$$\text{WR} = T_3 \cdot \text{STA}$$

Detailed Operation of the Processor

The control unit fetches, decodes, and executes the instruction specified in the PC. The fetch, decode, and execute phases together comprise a complete instruction cycle. Since each operation takes a finite amount of time, the instruction cycle is divided into time intervals denoted T_0, T_1 (fetch/decode), and T_2, T_3 (execute).

As each instruction is executed, the instruction control unit will send control signals to individual data path components at appropriate times, as required by the particular instruction (LDA, ADD, . . .). In this way, a data item x_i is processed (see Figure 12.31), that is, loaded into A, added to content of A, and so forth.

A logic diagram of simple 4-bit and 8-bit processors (which process 4-bit and 8-bit data words, respectively) are presented in Figures 12.32 (pp. 602–604) and 12.33 (pp. 606–

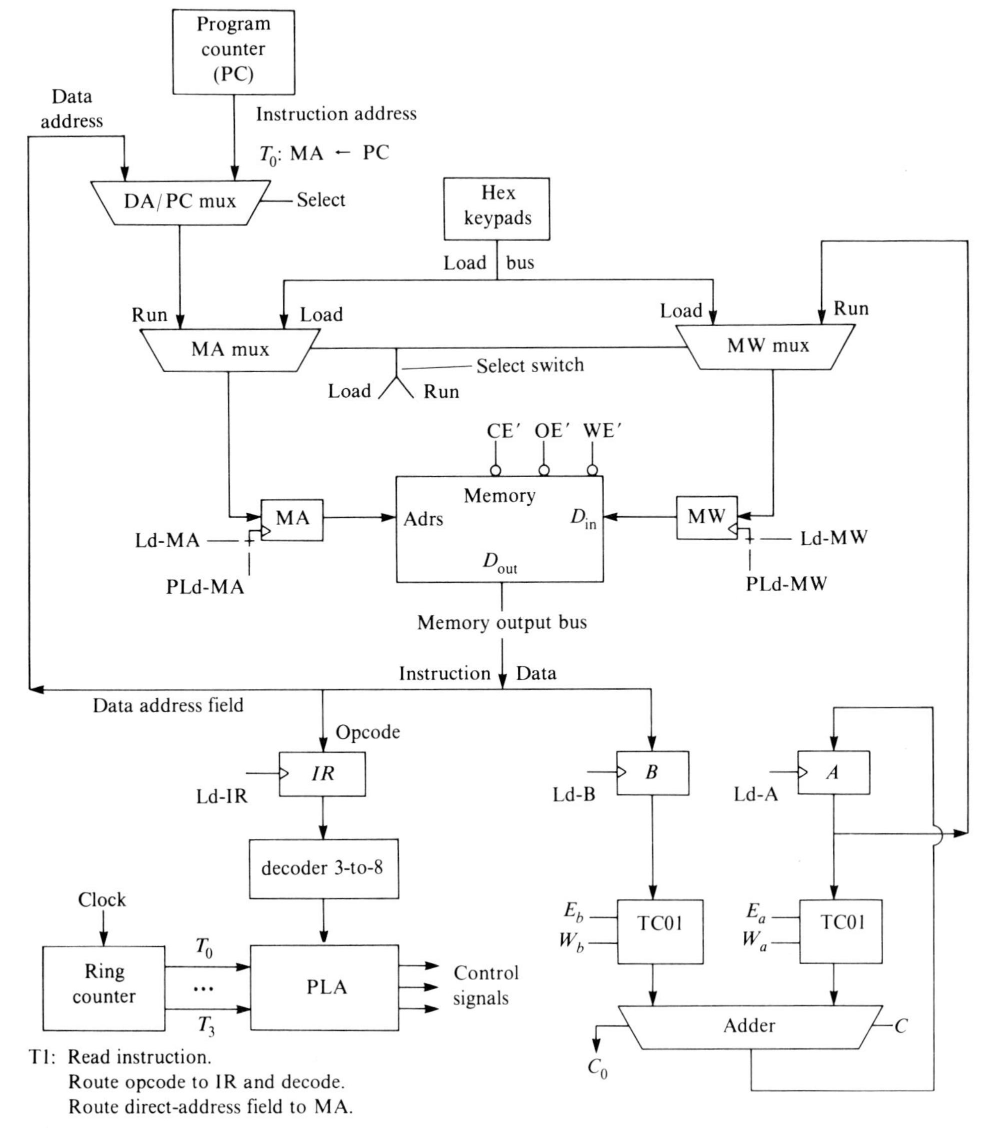

T1: Read instruction.
Route opcode to IR and decode.
Route direct-address field to MA.

T2, T3: Execute decoded opcode to process data item x_i.

FIGURE 12.31
Control and data flow during an instruction cycle

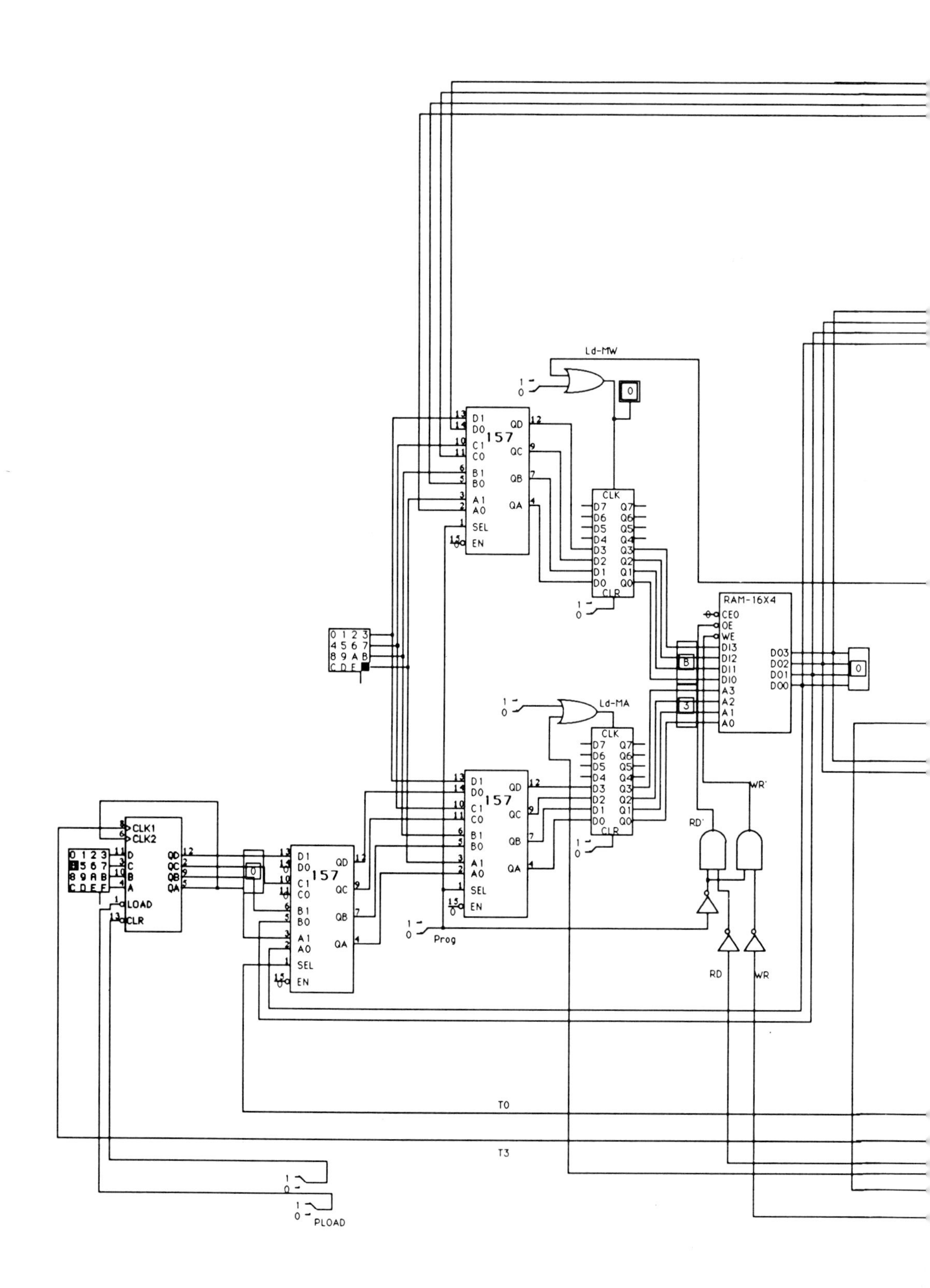

Ld-MW
157
CLK
RAM-16X4
Ld-MA
157
157
CLK1
CLK2
LOAD
CLR
Prog
RD'
WR'
RD
WR
T0
T3
PLOAD

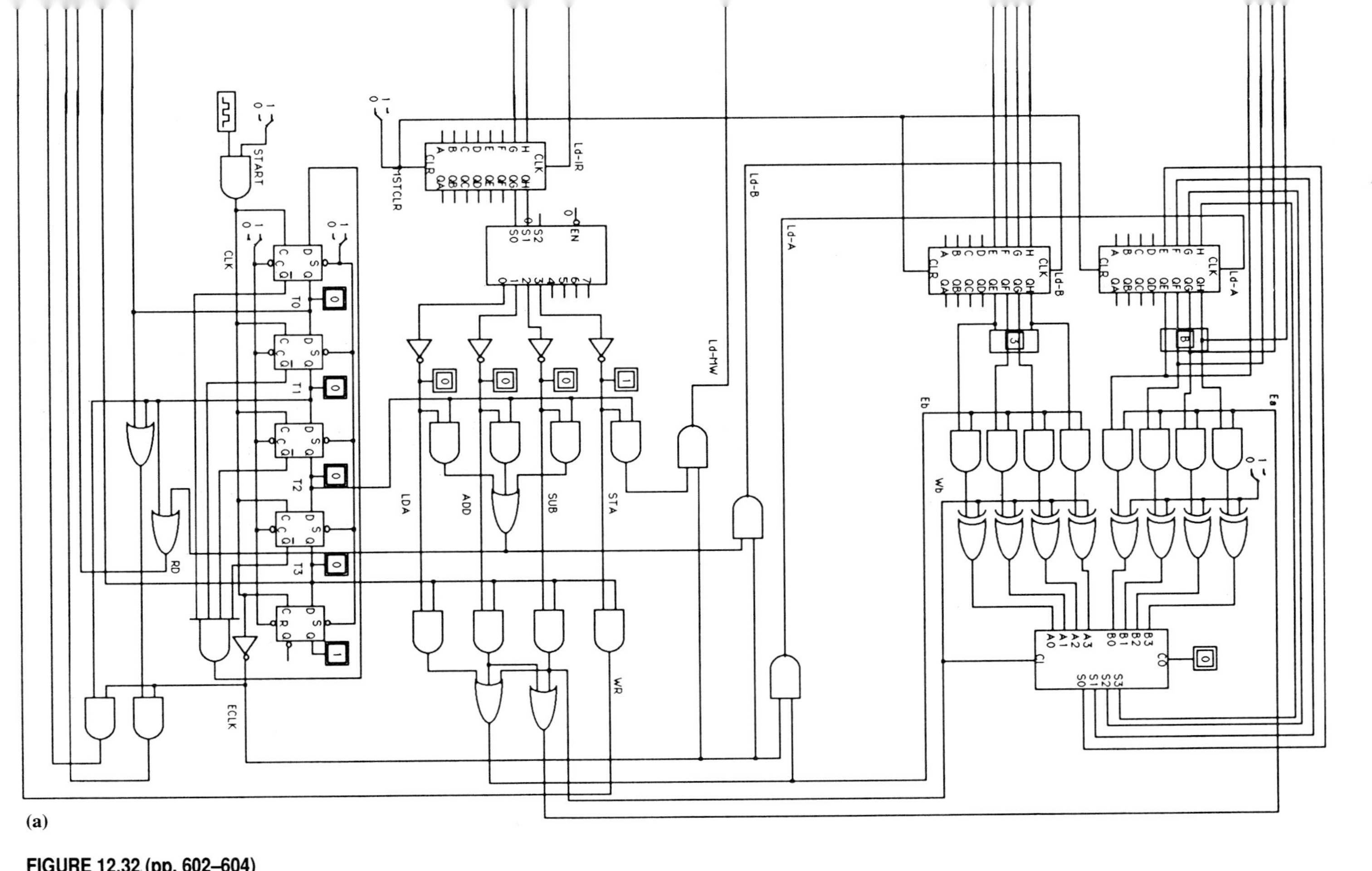

(a)

FIGURE 12.32 (pp. 602–604)
A 4-bit processor. (a) Logic diagram (b) Timing diagram

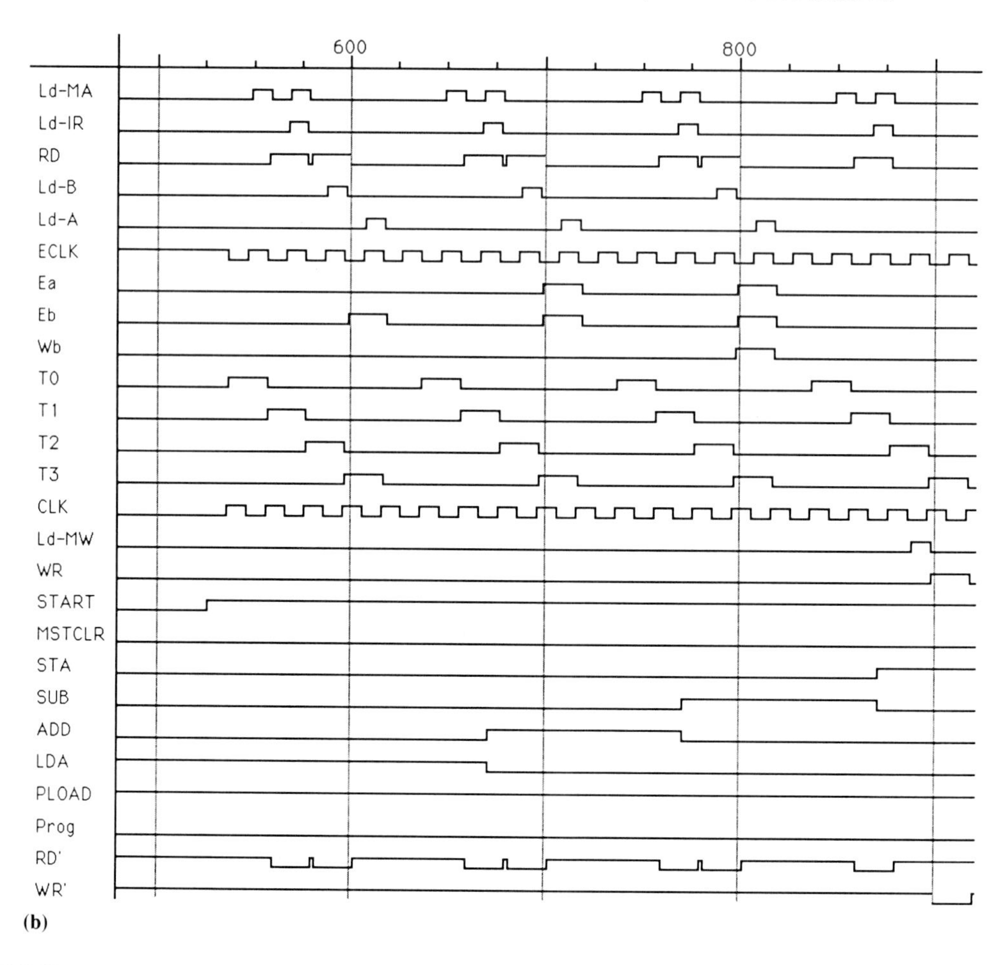

FIGURE 12.32
Continued

608). The 4-bit processor has a 16×4 memory and a 4-bit data path; each 4-bit instruction is composed of a 2-bit opcode (LDA = 00, ADD = 01, SUB = 10, STA = 11) and a 2-bit address field. The 8-bit processor has a 256×8 memory and an 8-bit data path; each 8-bit instruction is composed of a 3-bit opcode and a 5-bit address field (see Table 12.7). Only four instructions (LDA, ADD, SUB, STA) are implemented in the PLA of Figures 12.32 and 12.33.

12.5 SUMMARY

A stored-program digital computer with input-output (I/O) can be organized as a hierarchical model containing a succession of computers of increasing sophistication and complexity as follows:

1. A data path subsystem is essentially a manual control computer that can perform 1 of 2^n functions depending on the setting of n 2-state controls.
2. A data path subsystem combined with a control unit subsystem is a CPU (semiautomatic computer) that uses an instruction operation code (opcode) to control data processing in the data path.
3. A CPU combined with a memory subsystem is a processor that operates under control of program instructions stored in memory.
4. A processor combined with an I/O subsystem is a complete stored-program digital computer.

A building-block approach was used to evolve from a 4-bit adder to a functional stored-program digital computer processor in this chapter. At each step, another functional building block was described, and the building block was integrated with the preceding components of its subsystem. The steps in this evolution are as follows:

1. Implement a 4-bit MSI binary adder.
2. Implement a 4-bit adder/subtracter.
3. Implement a 4-bit 8-operation ALU.
4. Implement a manual control computer (data path subsystem).
5. Implement a clock-ring counter circuit.
6. Implement an instruction register and decoder.
7. Implement a PLA and integration of control unit components.
8. Integrate the control unit and data path subsystems to form a CPU.
9. Implement a read/write memory and associated registers.
10. Add a program load facility to the memory subsystem.
11. Add a program run facility to the memory subsystem.
12. Integrate the memory subsystem and CPU to form a processor.

In this chapter, we presented the basic concepts relating to the organization and operation of a stored-program computer. We also developed various types of computer program instructions, wrote machine language and assembly language programs, used a PASM model to represent a computer processor, and applied systems-level design techniques to the design of digital computers.

In Chapters 13 and 14 we will design a number of stored-program digital computers with increasingly complex instruction sets. In Chapter 13, multiplexers will be used to interface the subsystems of a computer. In Chapter 14, buses and devices with 3-state buffers will be used for interfacing.

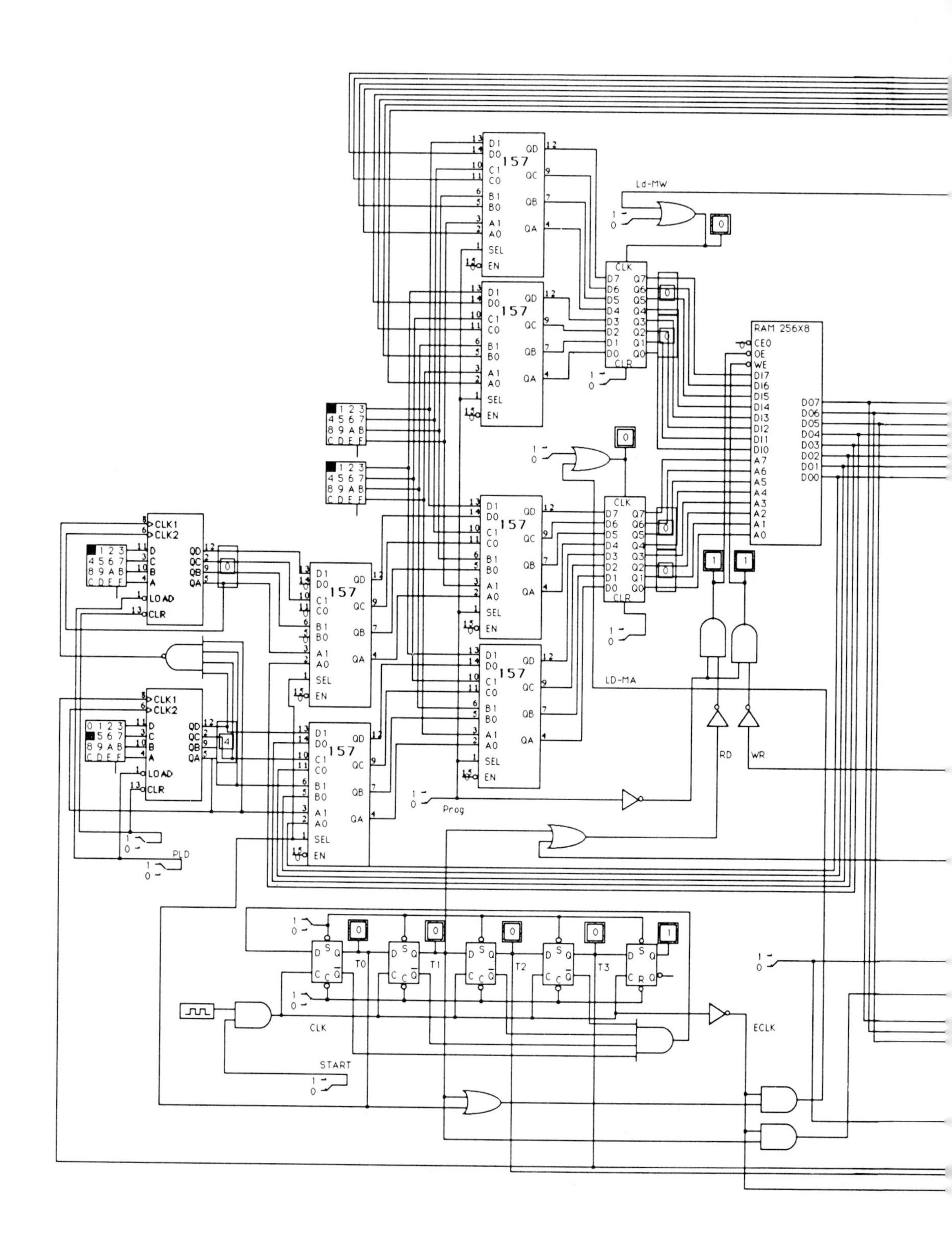

157
157
157
157
157
157
Ld-MW
CLK
CLR
RAM 256X8
LD-MA
RD
WR
Prog
PLD
CLK1
CLK2
LOAD
CLR
T0
T1
T2
T3
CLK
ECLK
START

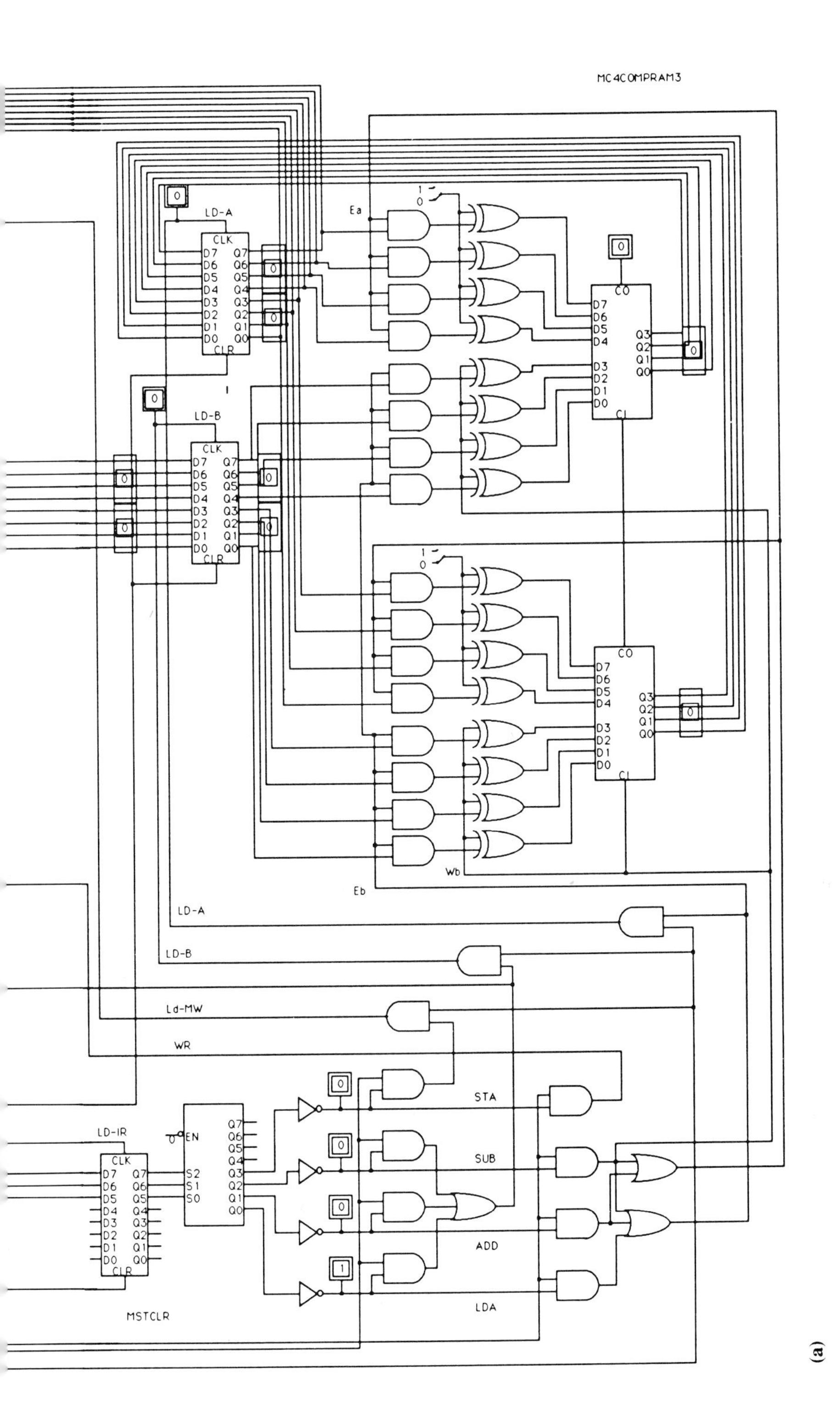

(a)

FIGURE 12.33 (pp. 606–608)
An 8-bit processor. (a) Logic diagram (b) Timing diagram

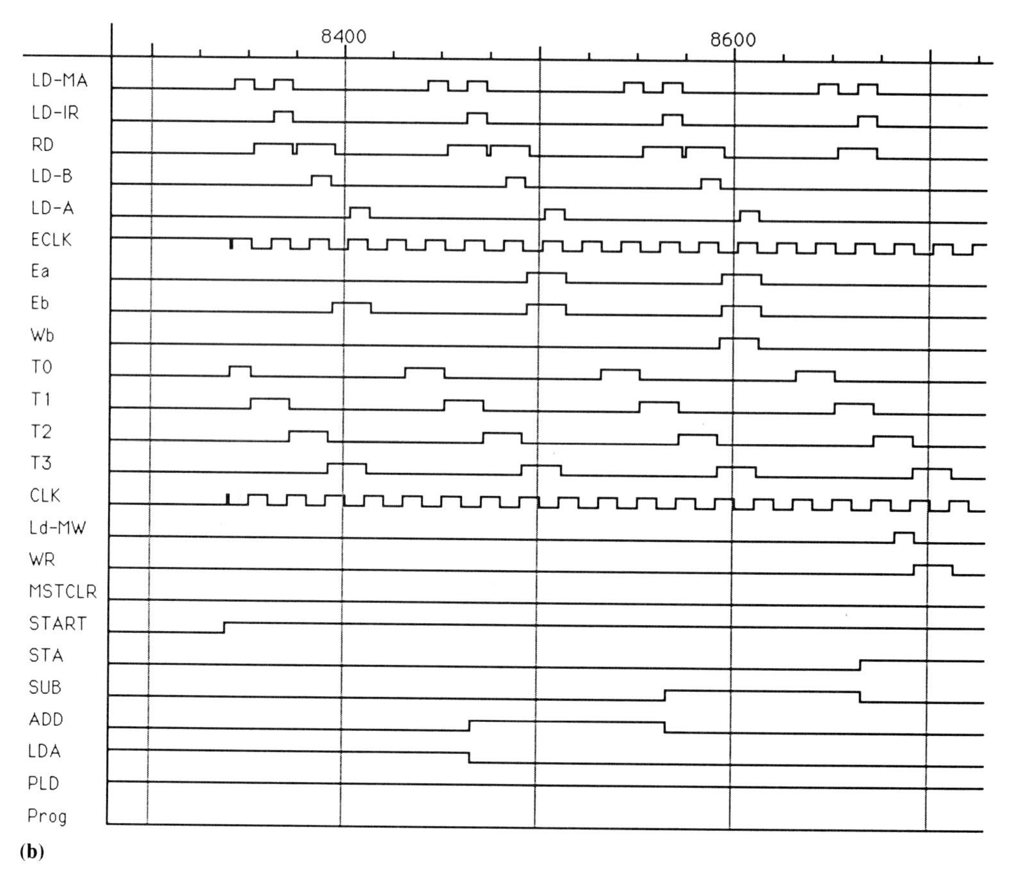

(b)

FIGURE 12.33
Continued

KEY TERMS

- Data path subsystem
- Control unit subsystem
- Memory subsystem
- Hierarchical approach
- Top-down design
- Manual control computer
- Semiautomatic computer
- Controls
- Instruction opcode
- Program
- Instruction cycle
- Computer's instruction repertoire
- Computer program
- Machine language program
- Assembly language
- Programmable algorithmic state machine (PASM)
- Microoperations

Instruction-time event table

p-term

Read enable (OE')

Write enable (WE')

Chip enable (CE')

Read/write memory (R/WM)

Memory address (MA)

Memory write (MW)

Program counter (PC)

Fetch

Decode

Execute

Direct address field (daf)

EXERCISES

General

Most exercises are related to the building-block approach used in this chapter to evolve from a 4-bit adder to a stored-program computer processor. Use 2s complement form for data words.

1. Verify the microoperations for each of the instructions in Table 12.4.
2. Write a simple machine language program using the following instructions: CLA (000), LDA (100), ADD (101), SUB (110), STA (111).
3. Draw a timing diagram for a 4-bit ring-counter circuit.

Design/Implementation

1. Implement an 8-bit adder using 7483 MSI adders, hexadecimal keypads for 8-bit inputs A and B, a switch for carry-in, a probe for carry-out, and hexadecimal displays for 8-bit output Z. Verify the operation of the circuit for the following:
   a. $A = 00000100$, $B = 00000011$
   b. $A = 00001010$, $B = 00000001$
   c. $A = 00001100$, $B = 00001111$
2. Implement an 8-bit adder/subtracter by incorporating eight XOR gates in the B-input path. Use a switch for control W ($W = 0$ for add, $W = 1$ for subtract). Use the data of Exercise 1 to verify the proper operation of the circuit for the following:
   a. Addition ($Z = A + B$)
   b. Subtraction ($Z = A - B$)
3. Design an overflow-detect circuit for the adder/subtracter circuit of Design/Implementation Exercise 2.
4. Design an AC-zero-detect circuit for the adder/subtracter circuit of Design/Implementation Exercise 2.
5. Implement an 8-bit 8-function ALU by incorporating AND and XOR gates in each input path of the adder. Verify the proper operation of the ALU for each of the operations in Table 12.3.
6. Implement a manual control computer (data path subsystem). Use hexadecimal keypads, connected to the input lines of the B register, to input a data byte X.
   a. Add a list of numbers using the manual-control computer.
   b. Use the manual control computer to compute the result of

$$N_1 - N_2 + N_3 - N_4$$

where N_1, N_2, N_3, and N_4 are suitable integers.

7. Implement a self-starting 4-bit ring-counter circuit as a T-state generator.
8. Implement an instruction register/decoder circuit using a 3-bit opcode. Note that the outputs of a 74138 decoder are active low. Verify the operation of the circuit for each 3-bit opcode.
9. Construct a control unit PLA, using AND and OR gates, to implement the following instruction set: LDA (100), ADD (101), SUB (110), STA (111). Use the logic equations accompanying Table 12.6 to implement the PLA. Integrate the T-state generator, instruction register/decoder, and PLA to form a control unit for a semiautomatic computer (CPU).
10. Construct a CPU by integrating the control unit (Design/Implementation Exercise 9) and the data path subsystem (Design/Implementation Exercise 6).
    a. Verify the proper operation of the semiautomatic computer for each of the instructions listed in Design/Implementation Exercise 9.
    b. Add a list of numbers using the semiautomatic computer.
    c. Use the semiautomatic computer to compute the result of

$$N_1 - N_2 + N_3 - N_4$$

where N_1, N_2, N_3, and N_4 are suitable integers.

11. Implement a 256×8 RAM memory with MA and MW registers.
12. Add load keypads and bus, MA mux, and MW mux to the circuit of Design/Implementation Exercise 11. Use a switch to select MA mux, MW mux inputs for program load or run mode (program load = 1, run = 0).
13. Add a PC and a DA/PC mux to the circuit of Design/Implementation Exercise 12 to form a complete memory subsystem.
14. Implement an 8-bit stored-program computer by integrating the memory subsystem and the CPU of Design/Implementation Exercise 10. Write a simple machine language program using each of the instructions LDA, ADD, SUB, and STA.
15. Verify the proper operation of the stored-program computer (see Figure 12.33a) using the program in Design/Implementation Exercise 14.

13 Design of Mux-Oriented 8-Bit Computers

OBJECTIVES

After you complete this chapter, you will be able

- To apply systems-design tools and techniques in the design of stored-program computers; that is,

  Use a programmable algorithmic state machine (PASM) model to represent a computer

  Integrate function modules (building blocks) to implement a digital system (computer)

  Describe basic categories of computer instructions and specific instructions within each category

- To use the 4-step problem-solving procedure (PSP) to design three simple stored-program computers:

  First, a computer that illustrates the basic categories of instructions: memory reference, register reference, input-output, and branch

  Second, a computer that is evolved from the first by adding indirect address mode, compare, and conditional branch instructions

  Third, a computer that is evolved from the second by adding multiply, divide, subroutine jump, and logic instructions

- To design an instruction set for a computer that accomplishes a variety of list-processing algorithms
- To describe an assembly language for each computer
- To accomplish the realization of complete, functional computers using a logic simulation program

13.1 INTRODUCTION

In Chapter 12, a building-block approach was used to evolve manual control, semiautomatic, and stored-program computers in a hierarchy. Thus, the emphasis was on* computer organization, *that is, the organization of components into subsystems and the integration of subsystems into a computer system. In Chapters 13 and 14, computer design will be approached more from the viewpoint of the programmer. That is, the set of instructions a computer must have to implement algorithms to be programmed will drive the design of the computer.

Different programmers may be interested in different applications areas. For example, programmer A may desire to write programs related to list processing (to sort or merge lists of data words). Programmer B may desire to write programs related to statistics (to compute the mean, variance, correlation, and so forth), and programmer C may wish to write programs related to numerical methods (to handle interpolation, curve fitting, numerical integration, or solution of differential equations).

It is possible to design a relatively simple computer that can satisfy the needs of all three programmers. Processing lists of data $\{X(k)\}$, $\{Y(k)\}$ involves adding, multiplying, comparing, sorting, and merging elements $x(k)$, $y(k)$ in the given lists of data. Statistical methods (computing mean, variance, and covariance) involve computations on the elements $x(k)$, $y(k)$. Numerical methods (interpolation, curve fitting, numerical integration, solution of differential equations) also involve computations on the elements $x(k)$, $y(k)$.

Given a **computer architecture** (a set of instructions to be implemented), there are four primary creative activities in the design of a simple computer:

1. Design of a data path subsystem that supports each and every instruction specified.
2. Design of a control unit subsystem that interprets memory instructions and generates control signals to data path components at appropriate times to accomplish the microoperations implied by each instruction.
3. Integration of a memory subsystem with the data path and control unit subsystems.
4. Integration of an I/O system with the data path, control unit, and memory subsystems.

In Chapter 13 most of the interfacing is accomplished using multiplexers. The designation MC–8X indicates mux computer. In Chapter 14 most of the interfacing will be accomplished using buses. The designation BC–8X indicates bus computer.

13.2 A PROGRAMMER'S VIEW OF A STORED-PROGRAM COMPUTER

From a programmer's viewpoint, a simple computer must have the ability to

- ***Load program instructions and data into memory***
- ***Read, interpret, and execute program instructions***
- ***Input data during program execution***
- ***Process data (original, derived, and input)***
- ***Store data-processing results in memory***
- ***Output data-processing results***

Thus, the focus of Chapter 13 is to describe representative instructions and develop procedures for designing computers to implement these instructions. Our approach is to show (1) how a computer, as a digital system, can be conceptualized as a programmable algorithmic state machine (PASM) model and (2) how to use systems-level design tools (RTN, system structure diagrams, and control flow diagrams) to design a series of computers with increasingly complex instruction sets.

Recall from Chapter 12 that a simple general-purpose digital computer can be viewed as a system that has four basic subsystems:

Data path subsystem

Control unit subsystem

Memory subsystem

Input-output subsystem

A data path subsystem and a control unit subsystem are combined to form a central processing unit (CPU); the CPU is combined with a memory subsystem to form a processor; and finally, the processor is combined with an I/O subsystem to form a complete computer.

The design of each computer in Chapters 13 and 14 is accomplished using the following 4-step problem-solving procedure for system design:

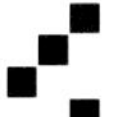

PSP

Step 1: Problem Statement. Given an instruction set, design a computer that implements that instruction set. The computer memory subsystem must be provided with a facility to load program instructions and data prior to program execution. The process of program execution is organized and described in two parts.

a. Process. Fetch, decode, and execute each program instruction (and accomplish the required processing of data and storing of computer-generated results).
b. Process control algorithm. A step-by-step procedure to accomplish the fetch, decode, and execute for each program instruction.

Step 2: Conceptualization. Develop a pictorial description of the system.

a. System structure diagram. Draw a block diagram that partitions the system into a memory subsystem, a data path subsystem, and a control unit subsystem.
b. Control flow diagram. Draw a diagrammatic representation of the operation of the process control algorithm that accomplishes the fetch, decode, and execute for each program instruction.

Step 3: Solution/Simplification.

a. Design a memory subsystem to store program instructions, data, and computer-generated results of the data processing.
b. Design a data path subsystem to support each instruction in the computer's repertoire.
c. Design a control unit subsystem to include the following subcomponents:

(1) Control unit output decoder.
  (a) Construct an instruction-time (p-term) event table.
  (b) Construct a p-term/control point activation table.
  (c) Determine an SOP expression for each output.
(2) Control unit next-state generator. Construct a next-state diagram.

Step 4: Realization.

a. Implement the memory subsystem.
b. Implement the data path subsystem.
c. Implement the control unit subsystem.
  (1) Integrate data path and control unit to form the CPU.
  (2) Integrate CPU and memory subsystem to form the processor. ■

These procedures are illustrated (throughout Chapters 13 and 14) by designing a series of computers that have increasingly complex instruction sets, where each computer is evolved from the one preceding it. That is, the first computer is designed to illustrate basic categories of instructions that are required in most computers (memory reference [direct address], register reference, I/O, and branch instructions). A second computer is evolved from the first by adding indirect address mode, compare, and conditional branch instructions. A third computer is evolved from the second by adding multiply, divide, subroutine jump, and logic instructions. Finally, an instruction set for a fourth computer is designed to accomplish a variety of list-processing algorithms (including statistics and numerical methods algorithms) to satisfy the needs of our hypothetical programmers A, B, and C.

The design of each computer will be approached from the viewpoint of architecture, that is, from the viewpoint of the instruction set specified by the programmer. The design of each computer is accompanied by a description of its assembly language instructions.

13.3 DIRECT ADDRESS MODE, REGISTER REFERENCE, I/O, AND BRANCH INSTRUCTIONS

This section describes the basic categories of computer instructions and uses the 4-step problem-solving procedure for system design to design a stored-program computer with a minimal instruction set.

The basic categories of computer instructions are

- Memory reference instructions that perform data transfer and arithmetic operations on data words (operands) stored in memory
- Register reference instructions that perform data transfer and arithmetic or logical operations on operands residing in special registers, such as an accumulator
- Input-output instructions that handle data transfer to or from I/O devices
- Compare instructions that provide a decision-making capability based on the comparison of two operands
- Branch instructions that provide a capability for a transfer of program control, such as a transfer to another instruction (not next in sequence) in a main program or a transfer from a main program to a subprogram

A data word referenced in an instruction may be addressed in a variety of ways, including the following address modes: immediate, direct, indirect, indexed, relative, page, and combinations thereof. In immediate mode, the data operand is contained within the instruction (see CMPIMM instruction in Section 13.4). Examples of specific instructions are described within each major category. These example instructions include the following:

- Memory reference instructions (direct address mode only)
  - LDADIR Load accumulator from specified memory location
  - ADDIR Add content of specified memory location to accumulator
  - SUBDIR Subtract content of specified memory location from accumulator
  - STADIR Store accumulator in specified memory location
- Register reference instructions
  - CLA Clear accumulator
  - CMP Complement accumulator
  - NEG Negate accumulator = 2s complement
- Input-output instructions
  - INP Input from a specified device
  - OUT Output to a specified device
- Compare instructions
  - COMPEQ Compare equal
  - COMPLT Compare less than
  - COMPGT Compare greater than
- Branch instructions
  - JMP Branch unconditionally to target address (tgtadrs)
  - SRJ Subroutine jump
  - BRLT Branch to target address on condition "less than"
  - BREQ Branch to target address on condition "equal to"
  - BRGT Branch to target address on condition "greater than"
- Logic instructions
  - AND For bit-wise masking
  - OR For bit-wise merging
  - XOR For bit-wise complementing

Direct-Address Mode Memory-Reference Instructions

Direct address mode (introduced in Chapter 12) is the simplest way to address a data word in memory. A direct-address memory-reference instruction consists of an operation code (opcode) field and a direct address field (daf). The opcode specifies the function of the instruction, and the direct address field contains the memory address of the data word (operand) referenced by the instruction. The number of bits allocated for the opcode is determined by the number of instructions in the computer's repertoire. If the computer has only 8 instructions, 3 bits are required to specify the opcode. Figure 13.1 presents a format

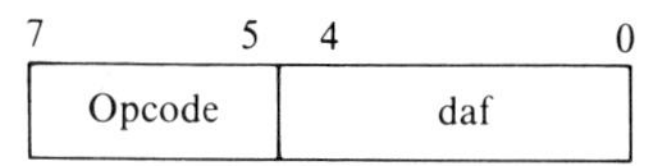

FIGURE 13.1
Format for a direct-address memory-reference instruction

for an 8-bit direct-address memory-reference instruction that has a 3-bit opcode. The opcode is contained in bits 7 through 5; the daf is contained in bits 4 through 0.

The following is an example program using direct-address memory-reference instructions. This program adds a list of numbers x, y, and z (stored at locations X, Y, Z) and stores the sum at location S.

LDADIR X	$(A \leftarrow x)$
ADDIR Y	$(A \leftarrow A + y = x + y)$
SUBDIR Z	$(A \leftarrow A - z = x + y - z)$
STADIR S	$(S \leftarrow \text{A})$

The following is the binary form of the program:

$000x \ldots x$	(LDADIR X)
$001y \ldots y$	(ADDIR Y)
$010z \ldots z$	(SUBDIR Z)
$011s \ldots s$	(STADIR S)

Figure 13.2 illustrates the direct address mode.

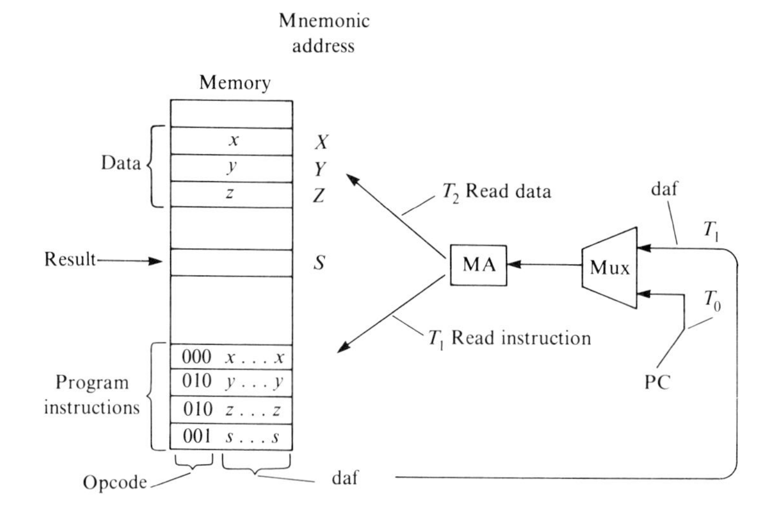

FIGURE 13.2
Illustration of direct address mode

INTERACTIVE DESIGN APPLICATION

Design of a Simple 8-Bit Computer, the MC–8A

The 4-step problem-solving procedure for systems design is used to design a simple 8-bit computer with a minimal instruction set.

Step 1: Problem Statement. Given a specified set of instructions (Table 13.1), design an 8-bit computer that implements the instruction set. The instruction set (specified without consideration of the computer's organization) includes instructions in each of the following basic categories:

Memory reference (direct address mode only)

Register reference

Input-output

Branch

TABLE 13.1
MC–8A computer instruction repertoire

Mnemonic	Binary	Description
LDADIR	000xxxxx	Load *A* with content of memory address *X*
ADDIR	001xxxxx	Add content of memory address *X* to *A*
SUBDIR	010xxxxx	Subtract content of memory address *X* from *A*
STADIR	011xxxxx	Store accumulator in memory address *X*
CLA	100– – – – –	Clear accumulator *A*
INP	101– – – – –	Load accumulator *A* from device
OUT	110– – – – –	Output accumulator *A* to device
JMP	111xxxxx	Branch unconditionally to target address *X*

Note: Each instruction is composed of a 3-bit opcode and a 5-bit memory address.

xxxxx denotes a 5-bit direct-address field *X*.

Computer I/O can be implemented using one of two basic techniques:

1. Memory-mapped I/O, in which all instructions that deal with the memory are available to handle data transfer to or from I/O devices
2. Isolated I/O, in which two instructions, *input* and *output*, are used to handle data transfer from and to I/O devices

Step 2: Conceptualization. Develop a pictorial description of the system.

a. System structure diagram. Draw a block diagram that partitions the system into a memory subsystem, a data path subsystem, and a control unit subsystem. In a systems context, a computer processor is a digital system composed of a memory subsystem and a CPU, where the CPU consists of a control unit subsystem and a data path subsystem. A computer processor can be represented by a PASM model, as shown in Figure 13.3 (p. 618).

b. Control flow diagram. Draw a diagrammatic representation of the operation of the process control algorithm that accomplishes the fetch, decode, and execute for each program instruction. The execution of each program instruction involves three principal actions:

□ Instruction fetch. The **program counter** (PC) is routed to the memory address (MA) register via the MA multiplexer, and the instruction specified by the PC is read from

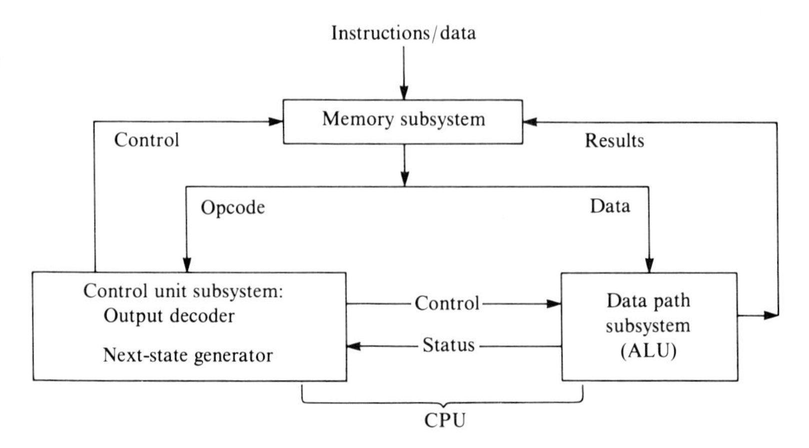

FIGURE 13.3
Structure diagram of a computer processor

memory (fetched). The instruction opcode is loaded into the **instruction register** (IR). The instruction direct address field (daf) is routed to the MA. The PC is incremented so that it "points to" the next instruction to be executed.

- Instruction decode. The instruction opcode in the IR is passed to a 3-to-8 decoder. The decoder interprets the opcode and activates the corresponding instruction line of the control unit PLA.
- Instruction execute. The control unit activates, at appropriate times, those data path components that carry out the register operations required to execute the instruction.

A partial control flow diagram (showing instructions CLA, LDADIR, ADDIR, STADIR) is presented in Figure 13.4. Each instruction cycle is divided into 4 time subintervals T_0, T_1, T_2, and T_3.

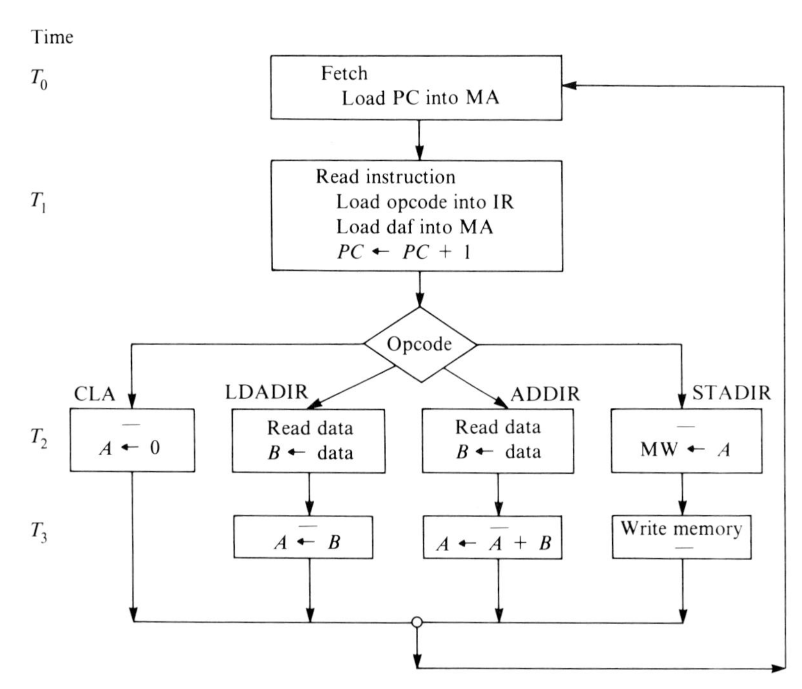

FIGURE 13.4
Process-control algorithm control-flow diagram (partial)

The individual operations involved in fetching, decoding, and executing an instruction are register transfers, usually referred to as **microoperations**. The accomplishment of each microoperation requires a finite amount of time. For example, if registers A and B are stable at time T_i, the sum S ($S = A + B$) is available for loading into A after a time that includes propagation delays through the TC01s and adder plus set-up time for the A register. An **instruction cycle** is the time required to complete the execution of an instruction (see Figure 13.5).

A **data flow cycle** is a time interval that allows each functional component in the data path to be used once. Each data path register may acquire a new value in a data flow cycle. The time duration of a data flow cycle is determined as the total time required to accomplish the microoperations in the longest-time path in the data path subsystem. This time includes gate propagation delays and register setup times.

A memory read operation, initiated at time T_i, causes an addressed memory location to be read out on the memory bus after delays for memory access time and memory bus interconnection. For simplicity it is assumed that a memory read or write operation requires one data flow cycle time.

The processing (fetch, decode, execute) of each instruction is accomplished in an instruction cycle that is divided into data flow cycles. The selection of a system clock is dependent on the data flow cycle time. In the system to be implemented, a clock cycle equal to two data flow cycles is selected, and an instruction cycle is divided into 4 clock cycles (t-states T_0, T_1, T_2, and T_3). Figure 13.5 illustrates a basic timing diagram for a simple computer. Note that a memory read is initiated at the start of interval T_i. Edge-triggered registers are loaded on the rising edge of signal ECLK, midway through the T_i interval. Also note that signal ECLK is an inversion of the system clock signal.

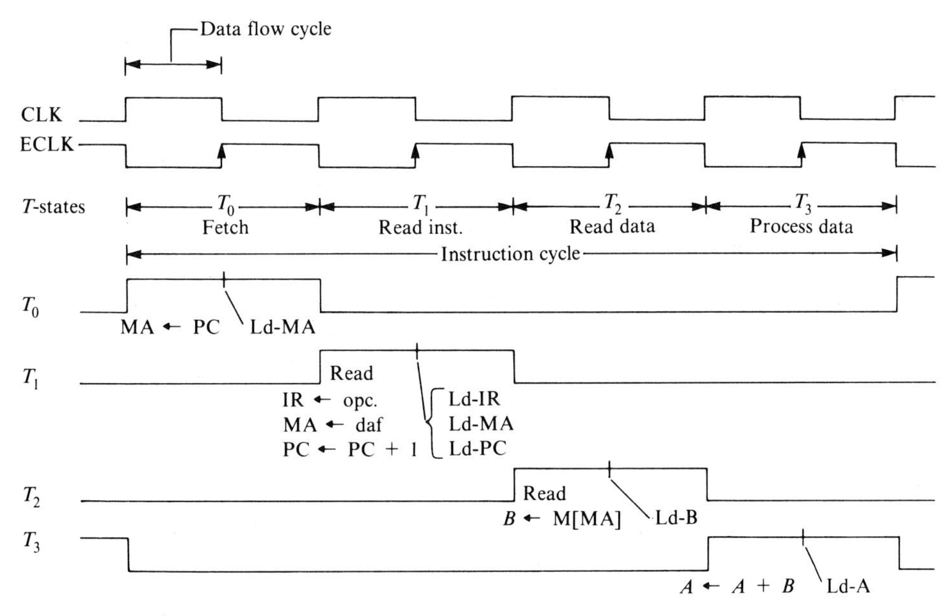

FIGURE 13.5
Timing diagram for a simple computer

Step 3: Solution/Simplification.

a. Design a memory subsystem. In order to concentrate on the design of a CPU that can execute a specified set of instructions, we will use the memory subsystem of Chapter 12 as the basic memory nucleus for the computers designed in this chapter. The facility to load instructions and data and store data-processing results can be realized by the memory subsystem shown in Figure 13.6.

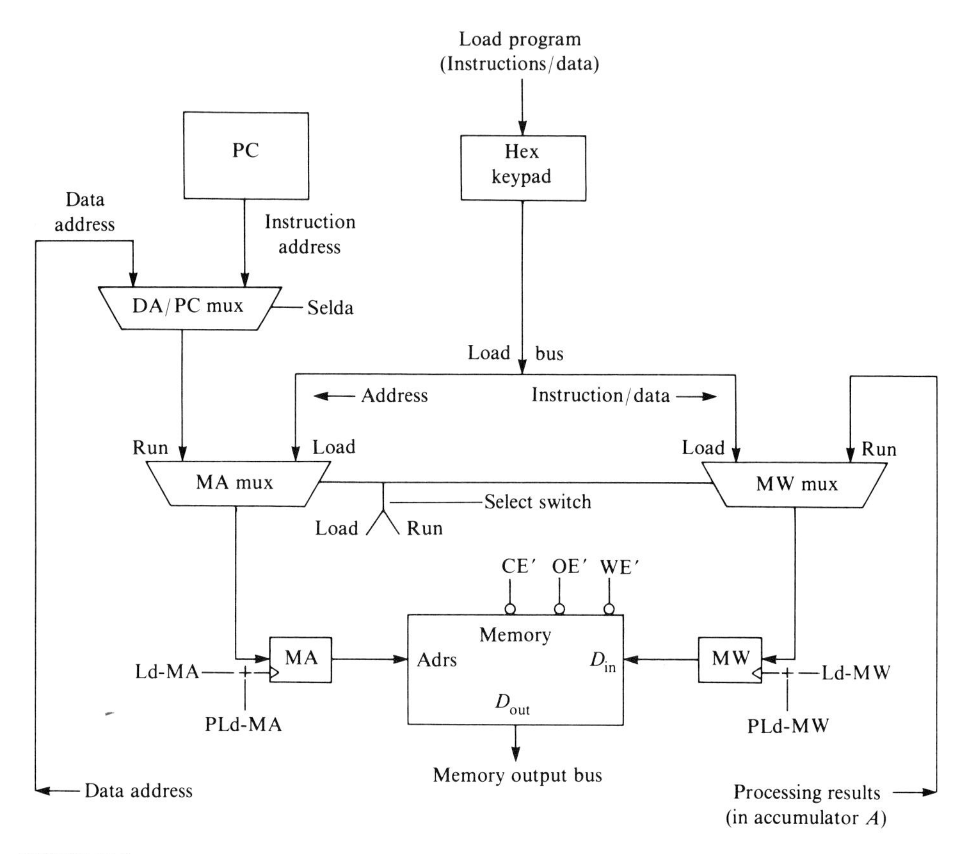

FIGURE 13.6
Memory subsystem with load and run facilities

b. Design a data path subsystem. A parallel ALU is used as a starting point (trial data path) for the data path subsystem of the MC–8A computer. The memory subsystem shown in Figure 13.6 is then interfaced with this trial data path by making the following connections (Figure 13.7):

□ The "run" input of the MW mux is connected to the accumulator output to provide for the STADIR instruction.

□ The memory output bus is connected to the input of the B register for transferring data from memory to the data path.

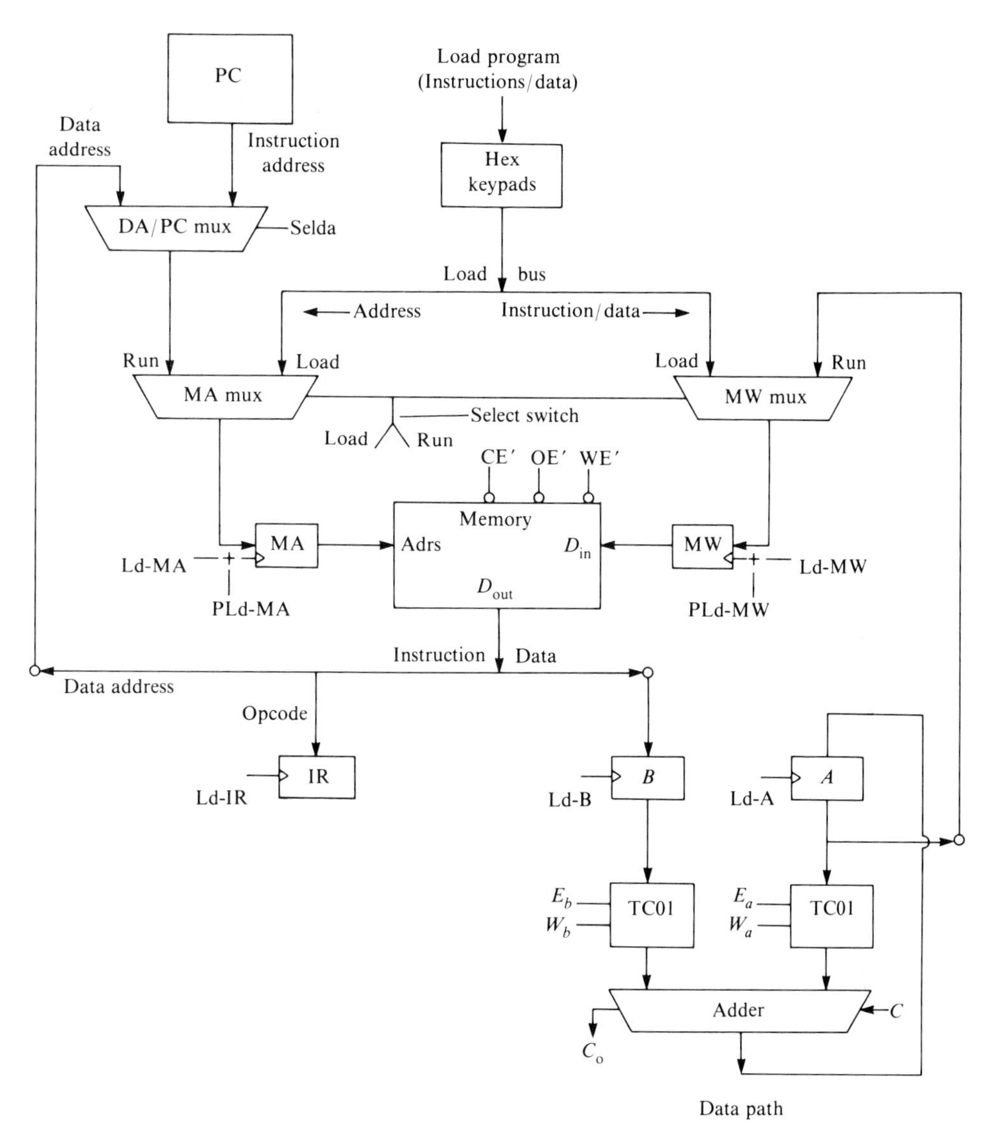

FIGURE 13.7
Memory subsystem and trial data path

□ The data address input of the DA/PC mux is connected to the 5 lower-order bits (daf) of the memory output bus. (The DA/PC mux is used to route either a data address (DA) or an instruction address (PC) to the run input of the MA register.) The details of the control unit remain unspecified at this time.

The trial data path is now modified as required until it supports all instructions specified in Table 13.1. After all instructions have been implemented, the data path can be refined by eliminating redundancy and increasing the efficiency of organization of the data path components. The trial data path of Figure 13.7 is modified by the following changes:

- An input register (IN) is multiplexed with the *B* register to provide a data path for an input instruction.
- An output register is connected to the *Z* bus to provide for an output instruction.
- The program counter in Figure 13.7 is replaced by a PC register that is multiplexed with the *A* register. The PC can then be incremented using the adder (with *X* input = 0, *Y* input = PC, *C* = 1). This also facilitates the JMP (jump to tgtadrs) instruction using the following microoperations:

$$\text{Read tgtadrs}, \quad B \leftarrow \text{tgtadrs}, \quad \text{PC} \leftarrow B$$

- By connecting the *Z* bus to DA/PC mux (whose output is connected to the run input of the MA mux), the PC is routed to the MA register (via the *A* mux, the adder, and the *Z* bus) on path PC2MA. This facilitates a memory fetch (involving microoperation MA ← PC).
- Multiplexers usually have an enable control. The mux enable control and a complementer can accomplish the functions of a TC01. Therefore, the TC01 can be replaced by a complementer and the mux enable.
- For a STA (store accumulator in memory) instruction, the accumulator output can be routed (via the *A* mux and the adder) to the *Z* bus which is connected to the run input of the MW mux (path Z2MW).

As a result of the foregoing changes, the trial data path is modified to become the MC–8A data path, illustrated in Figure 13.8.

The execution of each program instruction involves

- Instruction fetch. The address of the instruction to be executed is contained in the PC. In state T_0 for all instructions, the PC is moved to the MA and loaded into the MA on the rising edge of the ECLK signal (inverted clock) midway through state T_0. In register transfer notation, Ld-MA = $T_0 \cdot$ ALL $\cdot$ ECLK. By the end of state T_0, a valid instruction address is present in register MA. At the beginning of state T_1 a memory read command is initiated to read the instruction. After delays for memory access time and memory bus interconnection, the instruction is read out onto the memory bus. Midway through state T_1, the instruction opcode and daf are loaded into the IR and MA, respectively, on the rising edge of the ECLK and the PC is incremented. By the end of state T_1 the IR contains the opcode, the MA contains the daf, and the PC contains the incremented PC (PC now points to the next instruction to be executed).
- Instruction decode. The opcode in the IR is decoded by the 3-to-8 decoder, activating the appropriate instruction line to the control unit PLA.
- Instruction execute. If the instruction being executed requires reading a data byte from memory, a memory read command is issued at the beginning of state T_2, and the data byte is read from memory. At the midpoint of state T_2, the data word is loaded into the *B* register. In state T_3, the data byte is processed by the data path components. At the midpoint of state T_3, after propagation delays in the data path and set-up time for the *A* register, the result is loaded into the *A* register.

This process will be described in more detail in the remaining sections of this chapter.

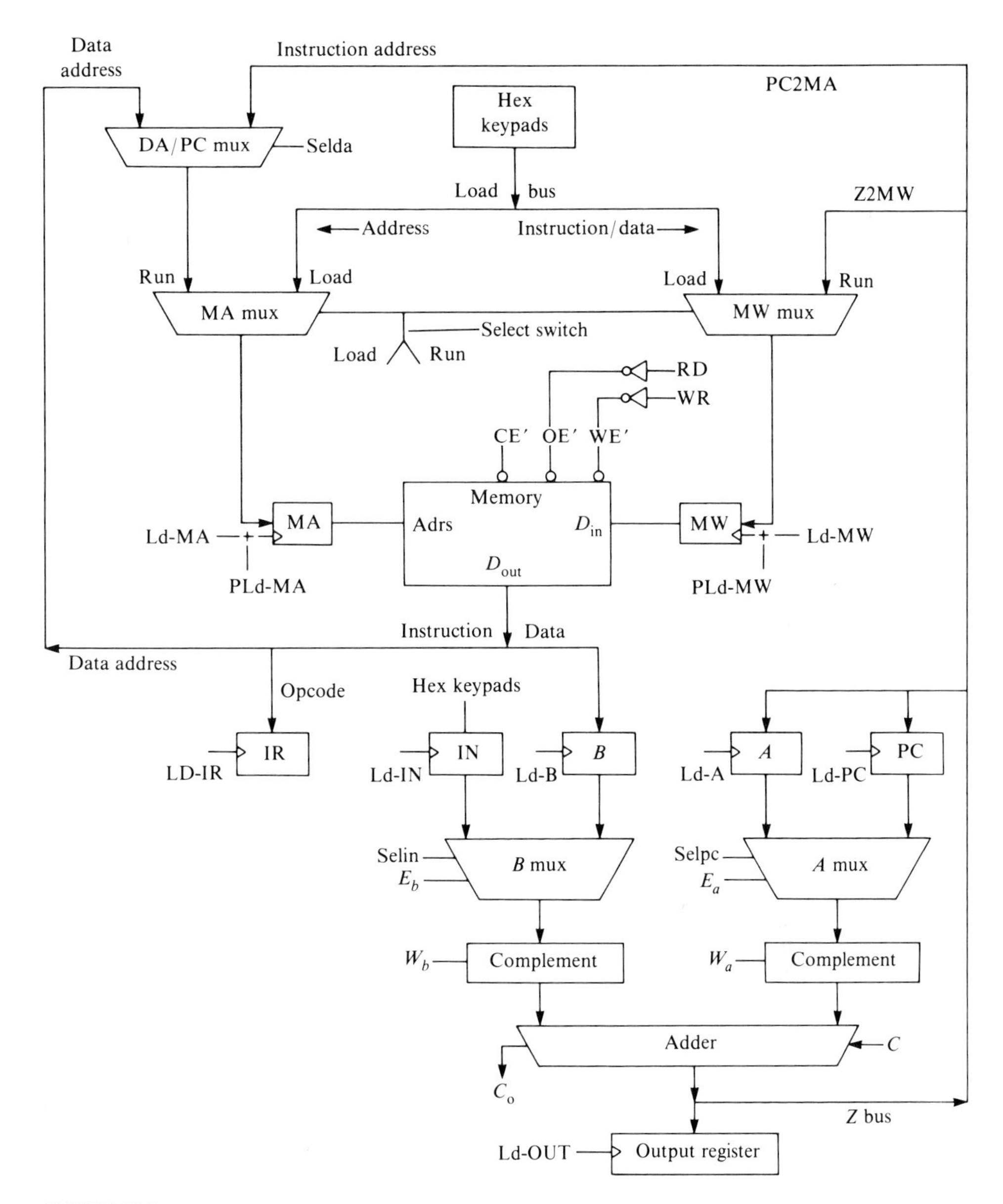

FIGURE 13.8
Memory and data path subsystems of computer MC–8A

A detailed control flow diagram can be constructed, as shown in Figure 13.9 (p. 624), by tracing the flow of each instruction through the revised data path for the MC-8A computer. Note that a memory read is initiated at the beginning of interval T_i. Edge-triggered registers are loaded midway through interval T_i on the rising edge of signal ECLK.

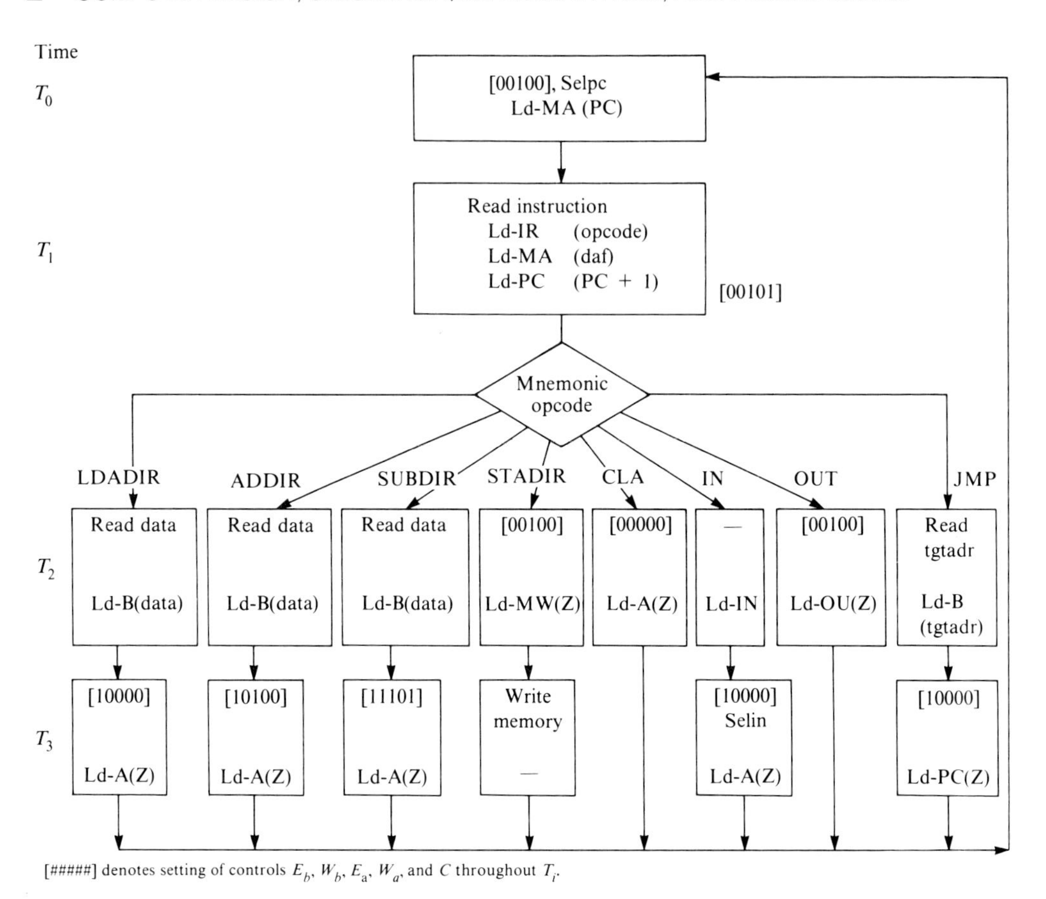

FIGURE 13.9
Detailed control flow diagram for MC–8A computer

c. Design a control unit subsystem

(1) Design the output decoder.

(a) Construct an instruction-time (p-term) event table. The execution of each instruction is accomplished by a sequence of microoperations at specific times within the overall instruction cycle. By tracing the control flow for each instruction, an instruction-time (p-term) event table can be constructed, as shown in Table 13.2.

(b) Construct a p-term/control point activation table. An examination of the instruction-time (p-term) event table (Table 13.2) to identify each control point leads to the construction of a p-term/control point activation table, as shown in Table 13.3 (p. 626). The reader can verify the entries by tracing each instruction's microoperations through the data path in Figure 13.8.

(c) Determine the output decoder control equations. Determine the Boolean SOP expressions for the output decoder outputs (to activate control points) by ORing the p-terms in each control point column of Table 13.3. The SOP expressions are

$$\text{PC2MA} = T_0 \cdot \text{ALL}$$

$$\text{Ld-MA} = T_0 \cdot \text{ALL} + T_1 \cdot \text{ALL}$$

TABLE 13.2
MC–8A instruction-time (p-term) event table

Instruction \ Event Time	T_0	T_1	T_2	T_3
000 LDADIR	Fetch MA $\leftarrow$ PC	Read inst. IR $\leftarrow$ opc. MA $\leftarrow$ daf PC $\leftarrow$ PC + 1	Read data $B \leftarrow$ M[MA]	$A \leftarrow B$
001 ADDIR	↓	↓	Read data $B \leftarrow$ M[MA]	$A \leftarrow A + B$
010 SUBDIR	↓	↓	Read data $B \leftarrow$ M[MA]	$A \leftarrow A - B$
011 STADIR	↓	↓	MW $\leftarrow A$	Write mem.
100 CLA	↓	↓	$A \leftarrow 0$	—
101 INP	↓	↓	Load IN reg.	$A \leftarrow$ IN
110 OUT	↓	↓	OUT $\leftarrow A$	—
111 JMP	↓	↓	Read tgtadrs $B \leftarrow$ M[MA]	PC $\leftarrow B$

M[MA] denotes memory location specified by MA.

$$\text{RD} = T_1 \cdot \text{ALL} + T_2 \cdot (\text{LDADIR} + \text{ADDIR} + \text{SUBDIR} + \text{JMP})$$
$$\text{Ld-IR} = T_1 \cdot \text{ALL}$$
$$\text{Ld-B} = T_2 \cdot (\text{LDADIR} + \text{ADDIR} + \text{SUBDIR} + \text{JMP})$$
$$E_b = T_3 \cdot (\text{LDADIR} + \text{ADDIR} + \text{SUBDIR} + \text{INP} + \text{JMP})$$
$$W_b = T_3 \cdot \text{SUBDIR}$$
$$E_a = T_0 \cdot \text{ALL} + T_1 \cdot \text{ALL} + T_2 \cdot (\text{STADIR} + \text{OUT}) + T_3 \cdot (\text{ADDIR} + \text{SUBDIR})$$
$$W_a = 0$$
$$C = T_1 \cdot \text{ALL} + T_3 \cdot \text{SUBDIR}$$
$$\text{Selin} = T_3 \cdot \text{INP}$$
$$\text{Selpc} = T_0 \cdot \text{ALL} + T_1 \cdot \text{ALL}$$
$$\text{Ld-IN} = T_2 \cdot \text{INP}$$
$$\text{Ld-PC} = T_1 \cdot \text{ALL} + T_3 \cdot \text{JMP}$$
$$\text{Ld-A} = T_2 \cdot \text{CLA} + T_3 \cdot (\text{LDADIR} + \text{ADDIR} + \text{SUBDIR} + \text{INP})$$
$$\text{Ld-OUT} = T_2 \cdot \text{OUT}$$
$$\text{Ld-MW} = T_2 \cdot \text{STADIR}$$
$$\text{WR} = T_3 \cdot \text{STADIR}$$

Note that the edge-triggered registers (PC, MA, IR, A, B, IN, OUT, and MW) are loaded on the rising edge of signal ECLK, midway through subinterval T_i. Memory read

TABLE 13.3
P-term/control point activation table

P-Term	Control Point																	
	PC2MA	Ld MA	RD	Ld IR	Ld B	Selin	Selpc	E_b	W_b	E_a	W_a	C	Ld IN	Ld PC	Ld A	Ld OU	Ld MW	WR
$T_0 \cdot$ ALL	1	1					1	0	0	1	0	0						
$T_1 \cdot$ ALL		1	1	1			1	0	0	1	0	1		1				
000 $T_2 \cdot$ LDADIR			1		1													
000 $T_3 \cdot$ LDADIR								1	0	0	0	0			1			
001 $T_2 \cdot$ ADDIR			1		1													
001 $T_3 \cdot$ ADDIR								1	0	1	0	0			1			
010 $T_2 \cdot$ SUBDIR			1		1													
010 $T_3 \cdot$ SUBDIR								1	1	1	0	1			1			
011 $T_2 \cdot$ STADIR								0	0	1	0	0					1	
011 $T_3 \cdot$ STADIR																		1
100 $T_2 \cdot$ CLA								0	0	0	0	0			1			
101 $T_2 \cdot$ INP													1					
101 $T_3 \cdot$ INP						1		1	0	0	0	0			1			
110 $T_2 \cdot$ OUT								0	0	1	0	0				1		
111 $T_2 \cdot$ JMP			1		1													
111 $T_3 \cdot$ JMP								1	0	0	0	0		1				

(RD) and write (WR) are initiated at the start of subinterval T_i. The remaining signals are level signals asserted throughout subinterval T_i.

(2) Design the next-state generator. Determine the control unit next-state functions and construct a state diagram (Figure 13.10) using the detailed control flow diagram (Figure 13.9) as a guide. The control functions are as follows:

$$T_1 = \mathrm{N}(T_0 \cdot \mathrm{ALL})$$
$$T_2 = \mathrm{N}(T_1 \cdot \mathrm{ALL})$$
$$T_3 = N(T_2 \cdot [\mathrm{LDADIR, ADDIR, SUBDIR, STADIR, INP, JMP}])$$
$$T_0 = \mathrm{N}(T_2 \cdot [\mathrm{CLA, OUT}] + T_3$$

Step 4: Realization. The control unit subsystem is realized by implementing the output decoder and the next-state generator. The control unit, data path, and memory subsystems are then integrated to form a complete MC–8A computer, as shown in Figure 13.11.

A logic diagram and timing diagram for the MC–8A computer are presented in Figures 13.12a and b (pp. 628–630), respectively.

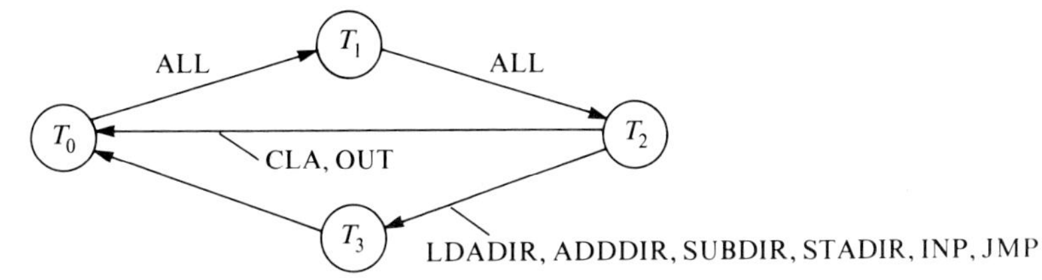

FIGURE 13.10
MC–8A control unit state diagram

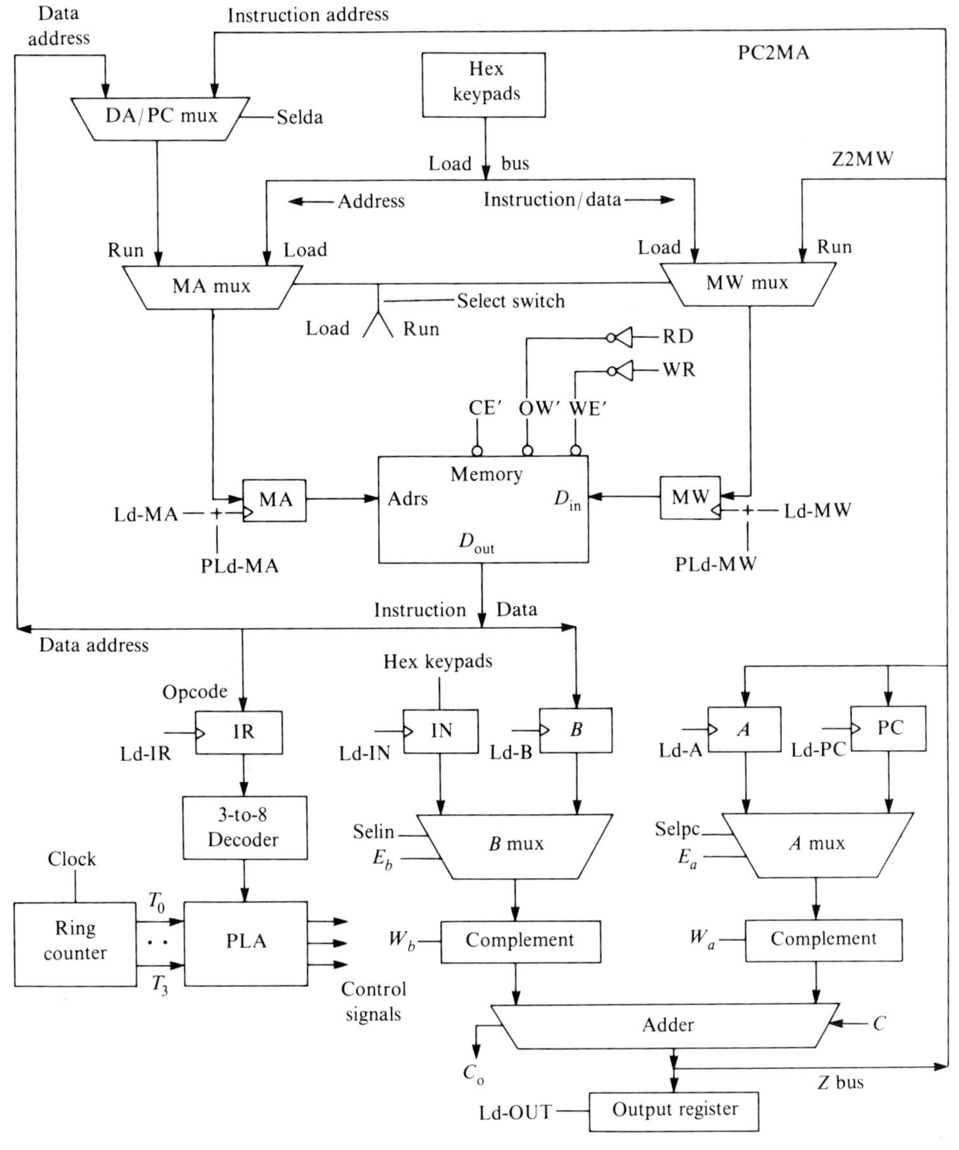

FIGURE 13.11
Block diagram of MC–8A computer

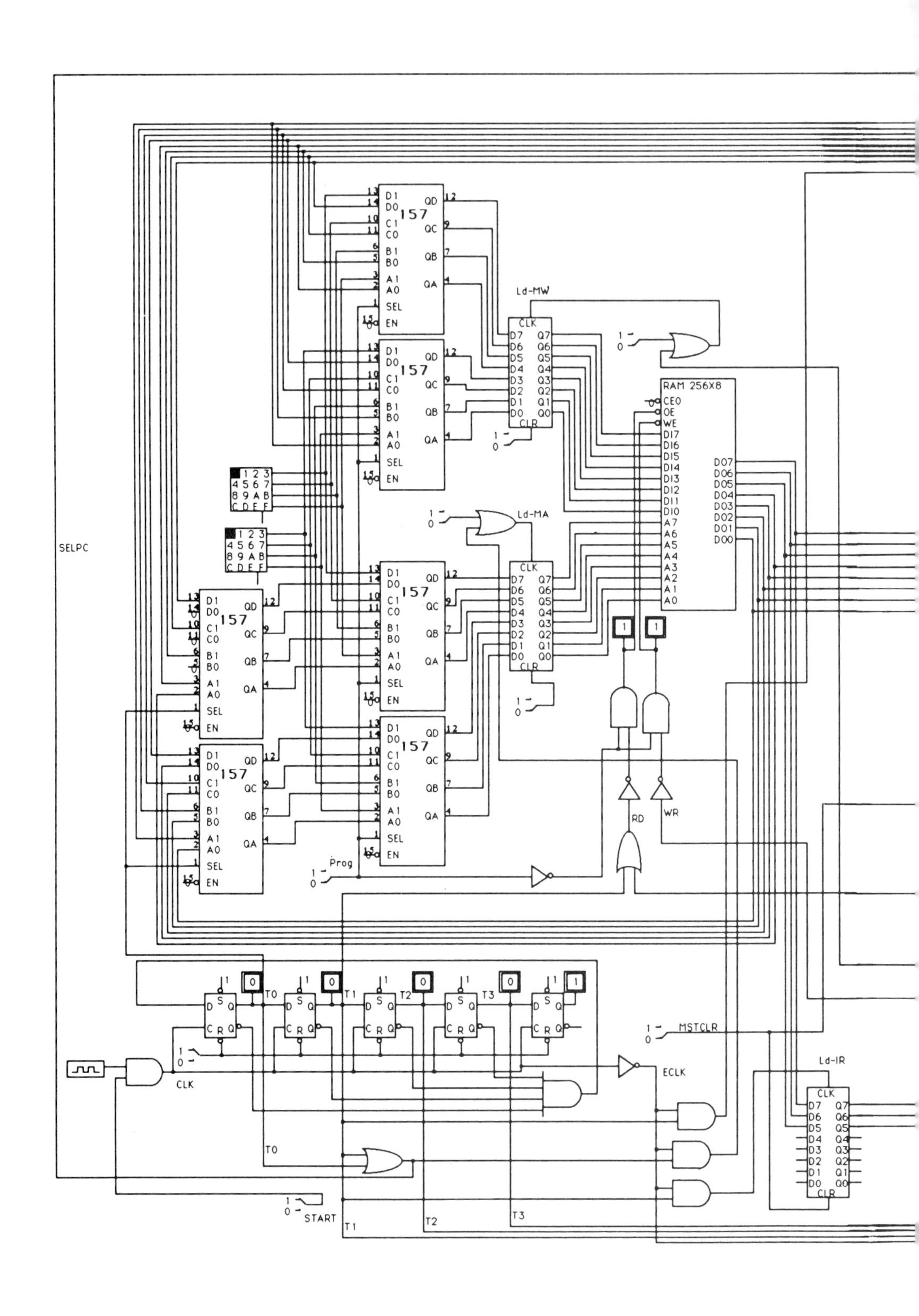

SELPC
157
Ld-MW
Ld-MA
RAM 256X8
CEO
OE
WE
RD
WR
Prog
T0
T1
T2
T3
MSTCLR
ECLK
CLK
START
Ld-IR

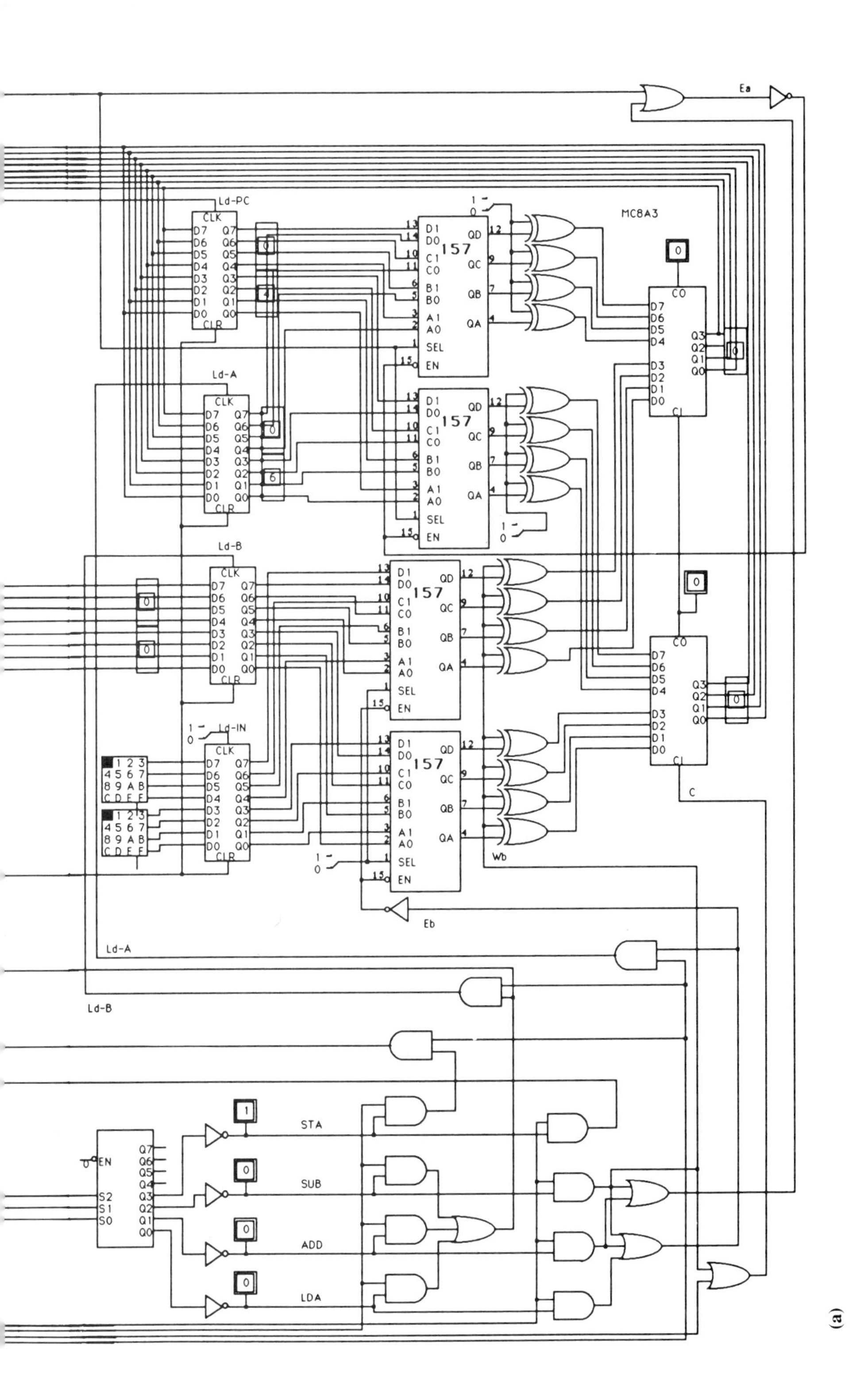

(a)

FIGURE 13.12 (pp. 628–630)
MC–8A computer. (a) Logic diagram (b) Timing diagram

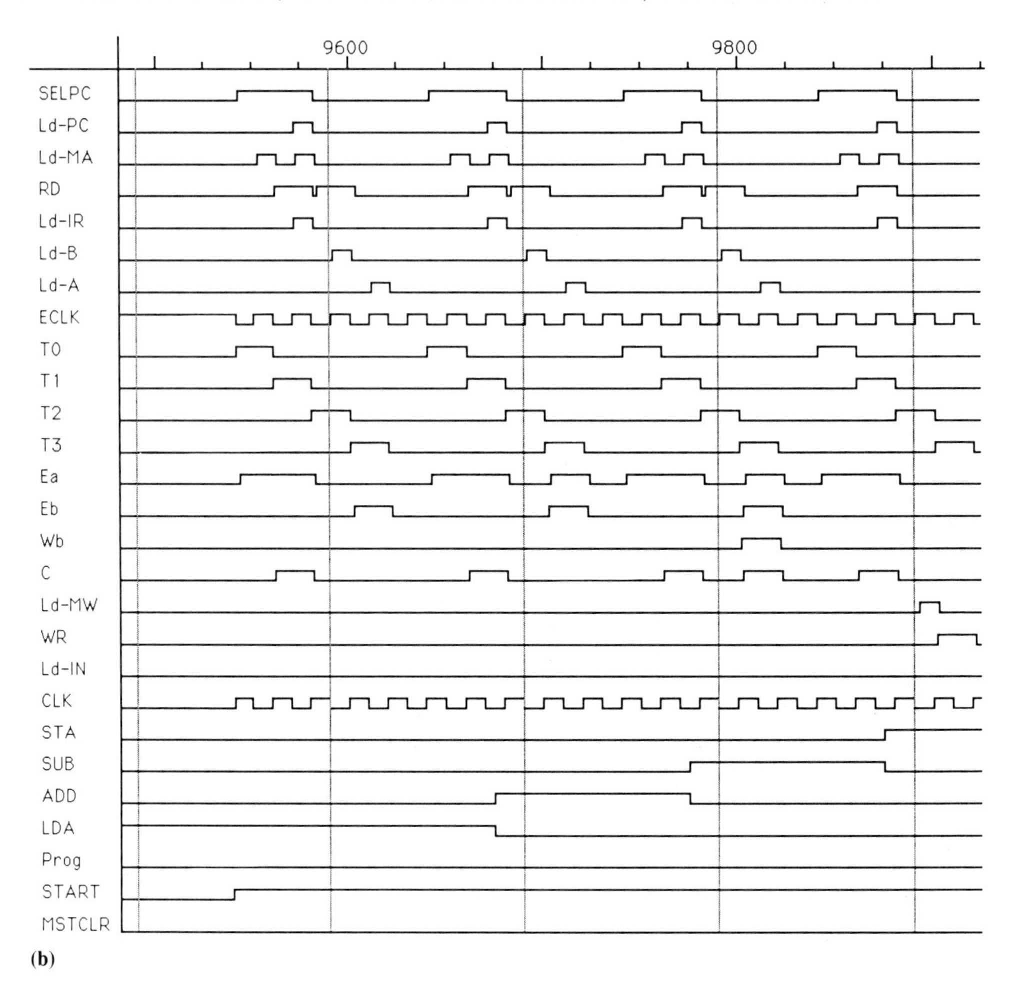

(b)

FIGURE 13.12
Continued

13.4 INDIRECT ADDRESS MODE, COMPARE, AND CONDITIONAL BRANCH INSTRUCTIONS

This section describes additional types of instructions in the basic categories, including the following:

- □ ***Indirect-address mode memory-reference instructions***
- □ ***Register reference instruction—complement* A**
- □ ***Compare instruction (using immediate address mode)***
- □ ***Conditional branch instructions.***

These additional instructions are incorporated into an expanded instruction set for our next computer, the MC–8B, which is designed in this section.

Indirect-Address Mode Memory-Reference Instructions

If the address of a data word to be read or written is greater than 00011111 (hex 1F = maximum address for 5-bit direct address mode), then that word must be located using address modes other than direct address mode. A memory reference instruction that uses **indirect address mode** locates the data word in memory as follows: The daf of the instruction comprises the 5 LSBs of an 8-bit memory address, 000daf. This location contains an 8-bit indirect address (iadrs). This address is the address of the data word. The indirect address mode is illustrated in Figure 13.13.

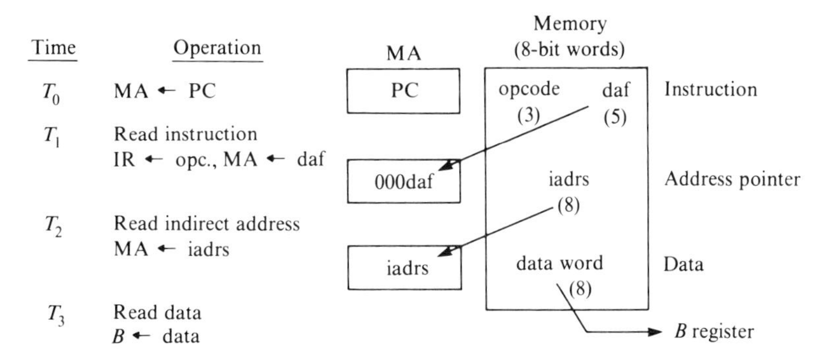

FIGURE 13.13
Illustration of indirect address mode

The execution of an indirect address mode instruction is accomplished as follows:

- Instruction fetch. The address of the instruction to be executed is contained in the PC. The content of the PC is moved to the MA register before the end of the T_0 state. At time T_1 the instruction is read from memory onto the memory bus. The opcode field is loaded into the instruction register (IR), and the daf is loaded into the MA register. The program counter is incremented so that it contains the address of the next instruction to be executed. The opcode in register IR is decoded by the 3-to-8 decoder, activating the appropriate instruction line to the control unit PLA.
- Indirect address read. If the decoded instruction is an indirect-address mode memory-reference instruction, the indirect address (iadrs) is read at time T_2, and iadrs is loaded into the MA register. At time T_3, the data word is read from memory.
- Instruction execute. Figure 13.14 illustrates the microoperations involved in an indirect-address mode memory-reference instruction (ADDIND).

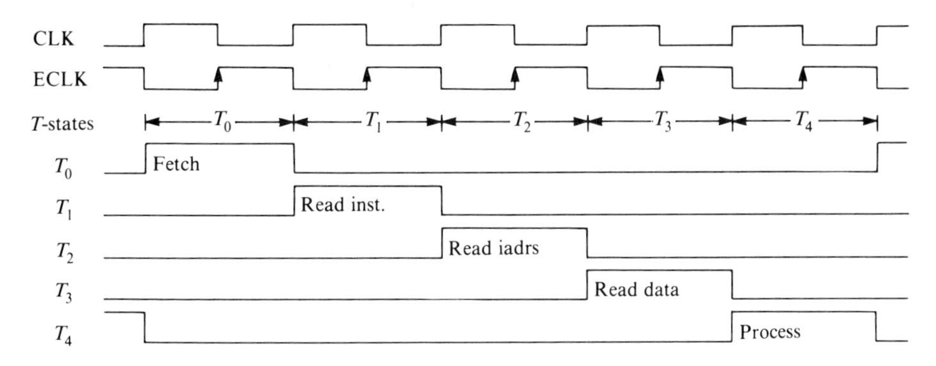

FIGURE 13.14
Microoperations involved in indirect address mode

An example of indirect address mode instruction microoperations follows

Opcode	Instruction	T_0	T_1	T_2	T_3	T_4
100	ADDIND	Fetch	Read inst.	Read iadrs	Read data	
		MA ← PC	IR ← opc.	MA ← M[MA]	B ← M[MA]	$A \leftarrow A + B$
			MA ← daf			
			PC ← PC + 1			

The following example program uses indirect-address mode memory-reference instructions. This program adds a list of numbers x, y, u, and v, and stores the sum at location S. Data words x and y are located using indirect address mode while u and v are located using direct address mode.

Binary	Assembly		Explanation
110	CLA		$A \leftarrow 0$
100x . . . x	ADDIND	X	$A \leftarrow A + x = 0 + x$
100y . . . y	ADDIND	Y	$A \leftarrow A + y = x + y$
001a . . . u	ADDIR	U	$A \leftarrow A + u = x + y + u$
001b . . . v	ADDIR	V	$A \leftarrow A + v = x + y + u + v$
011s . . . s	STADIR	S	$S \leftarrow A$

Figure 13.15 illustrates the indirect address mode.

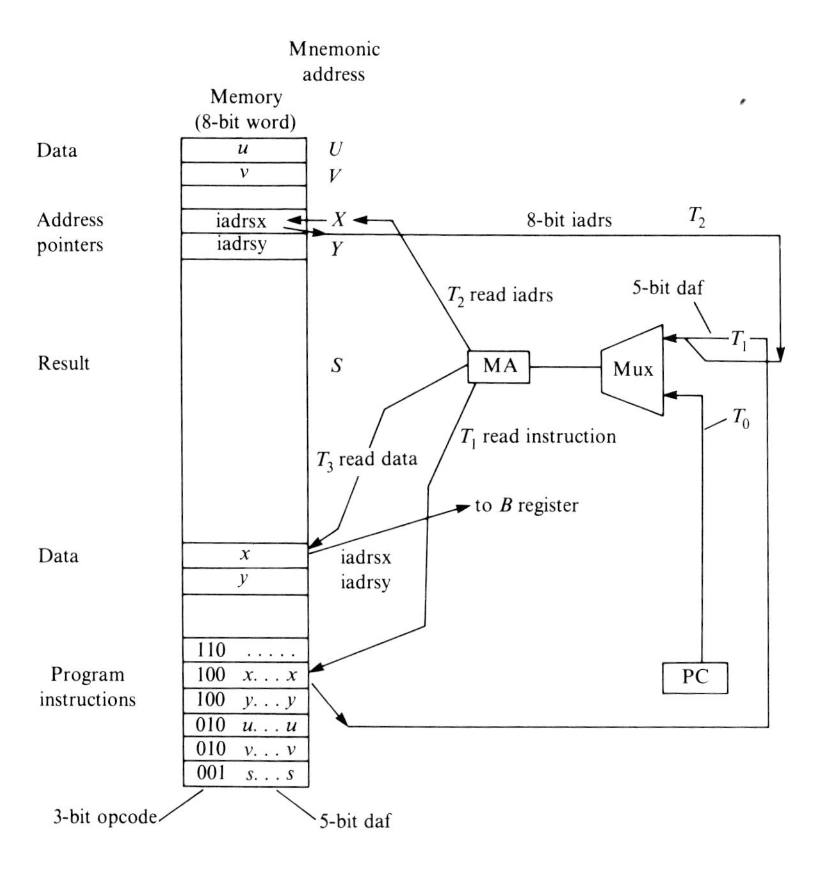

FIGURE 13.15
Illustration of indirect address mode

Compare Immediate Instruction

A computer should have the ability to compare two numbers and take appropriate action depending on the result of the compare operation. This capability can be implemented in a variety of ways. We choose to use a **status register** that contains flip-flops E, G, and L (for equal, greater than, and less than). These flip-flops are called **condition code** flip-flops.

The **compare** instruction compares a data word with the A register content and sets the E, G, and L flip-flops as follows:

$$\text{If } A = (A - \text{data}) \begin{cases} > 0, & E \leftarrow 0,\ G \leftarrow 1,\ L \leftarrow 0 \\ = 0, & E \leftarrow 1,\ G \leftarrow 0,\ L \leftarrow 0 \\ < 0, & E \leftarrow 0,\ G \leftarrow 0,\ L \leftarrow 1 \end{cases}$$

Note that $G = (E + L)'$.

Since data words are stored in 2s complement form, the result, $A - B$, in A is

< 0 if the sign of A is 1
$= 0$ if all bits of A are 0
> 0 if the sign of A is 0 and any other bit of A is 1

The **compare immediate** (CMPIMM) instruction is defined as a 2-byte instruction that has the following format:

Byte 1: Opcode (11100 . . .)

Byte 2: Data word to be compared to A register content

The CMPIMM instruction is accomplished by the following microoperations:

Opcode	Instruction	T_0	T_1	T_2	T_3	T_4
11100	CMPIMM	Fetch	Read inst.	MA ← PC	Read data	$A \leftarrow A - B$
	data	MA ← PC	IR ← opc.		B ← M[MA]	Set E, G, L
			MA ← daf		PC ← PC + 1	
			PC ← PC + 1			

Conditional Branch Instructions

A **condition code branch** is an instruction (2 bytes) that branches to a target address (tgtadrs) or does not branch depending on the setting of a condition code flip-flop (E, G, or L). A condition code branch instruction must be preceded by a compare instruction that sets the condition code flip-flops E, G, and L.

Because of the limited number of remaining opcodes, we choose to implement only one condition code branch instruction (BREQ) in the MC–8B computer. The BREQ (branch on equal) instruction is defined as a 2-byte instruction that has the following format:

Byte 1: Opcode (11111 . . .)

Byte 2: Target address (tgtadrs)

The BREQ instruction is accomplished by the following microoperations:

Opcode	Instruction	T_0	T_1	T_2	T_3	T_4
11111	BREQ	Fetch	Read inst.	MA ← PC	Read tgtadrs	If $E = 1$
	tgtadrs	MA ← PC	IR ← opc.		B ← M[MA]	PC ← B
			MA ← daf		PC ← PC + 1	
			PC ← PC + 1			

Some conditional branch instructions need not be preceded by a compare instruction. For example, a branch on $A = 0$ (BRA0) instruction either branches to a target address if $A = 0$, or it does not branch if $A \neq 0$. The BRA0 instruction requires a circuit that indicates the state of the accumulator (i.e., $A = 0$, or $A \neq 0$). The accumulator A is zero if all of its bits are zero. The output of this zero-detect circuit will be an input to the control unit. The BRA0 instruction is defined as a 2-byte instruction, which has the following format:

Byte 1: Opcode (11110 . . .)

Byte 2: Target address (tgtadrs)

The BRA0 instruction is accomplished by the following microoperations:

Opcode	Instruction	T_0	T_1	T_2	T_3	T_4
11110	BRA0	Fetch	Read inst.	MA ← PC	Read tgtadrs	If *Azero*
	tgtadrs	MA ← PC	IR ← opc.		B ← M[MA]	PC ← B
			MA ← daf		PC ← PC + 1	
			PC ← PC + 1			

INTERACTIVE DESIGN APPLICATION

Design of an 8-Bit Computer (MC–8B) with Additional Capability

The second computer (MC–8B) is evolved from the MC–8A by adding instructions that provide the following additional capabilities:

- □ Indirect address mode for memory reference instructions
- □ Additional register reference instruction (complement A)
- □ Compare instruction (using immediate address mode)
- □ Conditional branch instructions.

Step 1: Problem Statement. Define an instruction set (Table 13.4) that has instructions in each basic category (memory reference, register reference, I/O, and branch) and add instructions that provide the additional capabilities listed above.

TABLE 13.4
MC–8B computer instruction repertoire

Mnemonic	Binary	Description
LDADIR	000xxxxx	Load *A* with content of memory address *X*
ADDIR	001xxxxx	Add content of memory address *X* to *A*
SUBDIR	010xxxxx	Subtract content of memory address *X* from *A*
STADIR	011xxxxx	Store accumulator in memory address *X*
ADDIND	100xxxxx	Add content of indirect-address memory location to *A*
SUBIND	101xxxxx	Subtract content of indirect-address memory location from *A*
CLA	11000---	Clear *A*
COMP	11001---	Complement *A*
INP	11010---	Load *A* from an input device
OUT	11011---	Transfer *A* to an output device
CMPIMM	11100---	Compare immediate operand with *A* [2-byte instruction]
JMP	11101---	Branch unconditionally to target address
BRA0	11110---	Branch to target address on condition $A = 0$
BREQ	11111---	Branch to target address on condition code equal

Branch instructions: [2 bytes; byte-2 = target address]

Step 2: Conceptualization. Develop a pictorial description of the system.

a. System structure diagram. Draw a block diagram that partitions the system into a memory subsystem, a data path subsystem, and a control unit subsystem.

In a systems context, a computer processor is a digital system composed of a memory subsystem and a CPU, where the CPU consists of a control unit subsystem and a data path subsystem. A computer processor can be represented by a PASM model, as shown in Figure 13.16.

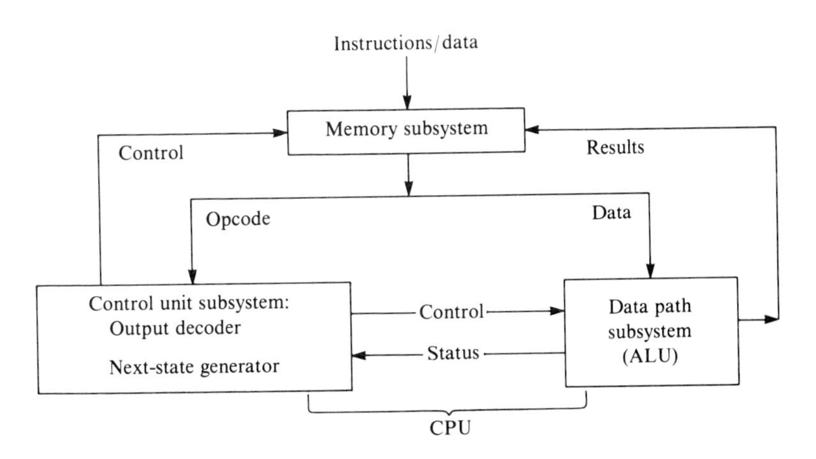

FIGURE 13.16
Structure diagram of a computer processor

b. Control flow diagram. Draw a diagrammatic representation of the operation of the process control algorithm that accomplishes the fetch, decode, and execute for each program instruction. A partial control flow diagram (showing instructions CLA, LDADIR, ADDIR, SUBIND) is presented in Figure 13.17. Each instruction cycle is divided into 5 time subintervals T_0, T_1, T_2, T_3, and T_4.

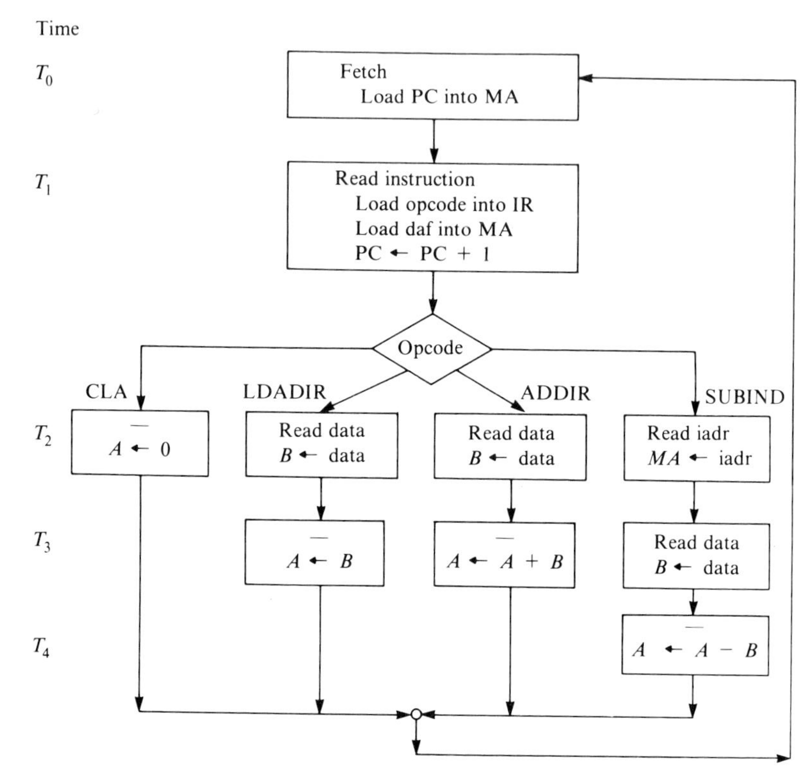

FIGURE 13.17
A process-control algorithm control-flow diagram for the MC–8B (partial)

Step 3: Solution/Simplification.

a. Design a memory subsystem. In order to concentrate on the design of a CPU that can execute a specified set of instructions, we will use the memory subsystem of the MC–8A computer as the memory subsystem of the MC–8B computer. The data path of the MC–8A computer will be used as a starting (trial) data path of the MC–8B computer.

b. Design a data path subsystem. Determine a data path for each MC–8B instruction (Table 13.4), using the data path of the MC–8A computer (Figure 13.18) as a starting point. Modify this data path as necessary to support all specified instructions.

Modification 1. Since there are more than 8 instructions in the MC–8B repertoire, it is necessary to use more than 3 bits for some opcodes, as indicated in Table 13.4.

a. The memory reference instructions are defined by 3-bit opcodes; direct-address mode memory-reference instructions (daf = 5 LSBs) require a mask to clear the 3 MSBs of the MA mux address lines.

b. The nonmemory reference instructions (CLA, COMP, INP, OUT, CMPIMM) and branch instructions are defined by 5-bit opcodes. In addition to the 3-to-8 decoder for the 3 MSBs, these 5-bit opcodes require a 2-to-4 decoder to decode the 2 LSBs of the opcode.

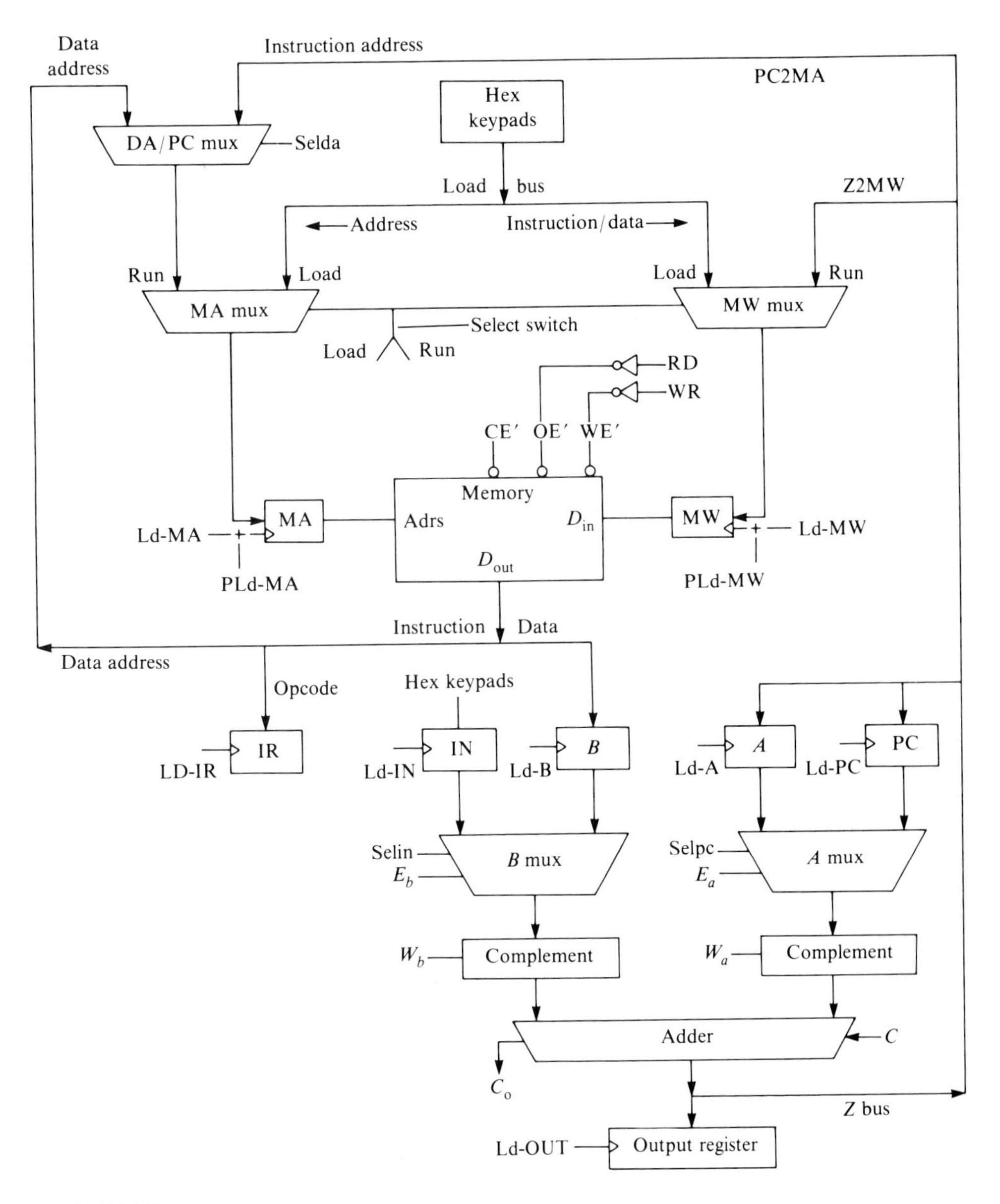

FIGURE 13.18
Memory and data path subsystems of the MC–8A

Modification 2. A detect-Azero circuit is used to determine if register $A = 0$. The output of this detect-Azero circuit will be an input to the control unit.

Modification 3. The compare immediate (CMPIMM) instruction compares a data word with the A register content and sets the E, G, and L flip-flops as follows:

$$\begin{aligned}\text{If } A = (A - \text{data}) &> 0,\ E \leftarrow 0,\ G \leftarrow 1,\ L \leftarrow 0\\ &= 0,\ E \leftarrow 1,\ G \leftarrow 0,\ L \leftarrow 0\\ &< 0,\ E \leftarrow 0,\ G \leftarrow 0,\ L \leftarrow 1\end{aligned}$$

Note that the MSB of A is the sign bit, denoted A_s. If $A_s = 1$, the content of A is negative.

Modifications 1, 2, and 3 convert the MC–8A data path and control unit to an MC–8B data path and control unit (decoder only). A block diagram of the MC–8B computer, presented in Figure 13.19, illustrates modifications 1, 2, and 3 (output decoder only).

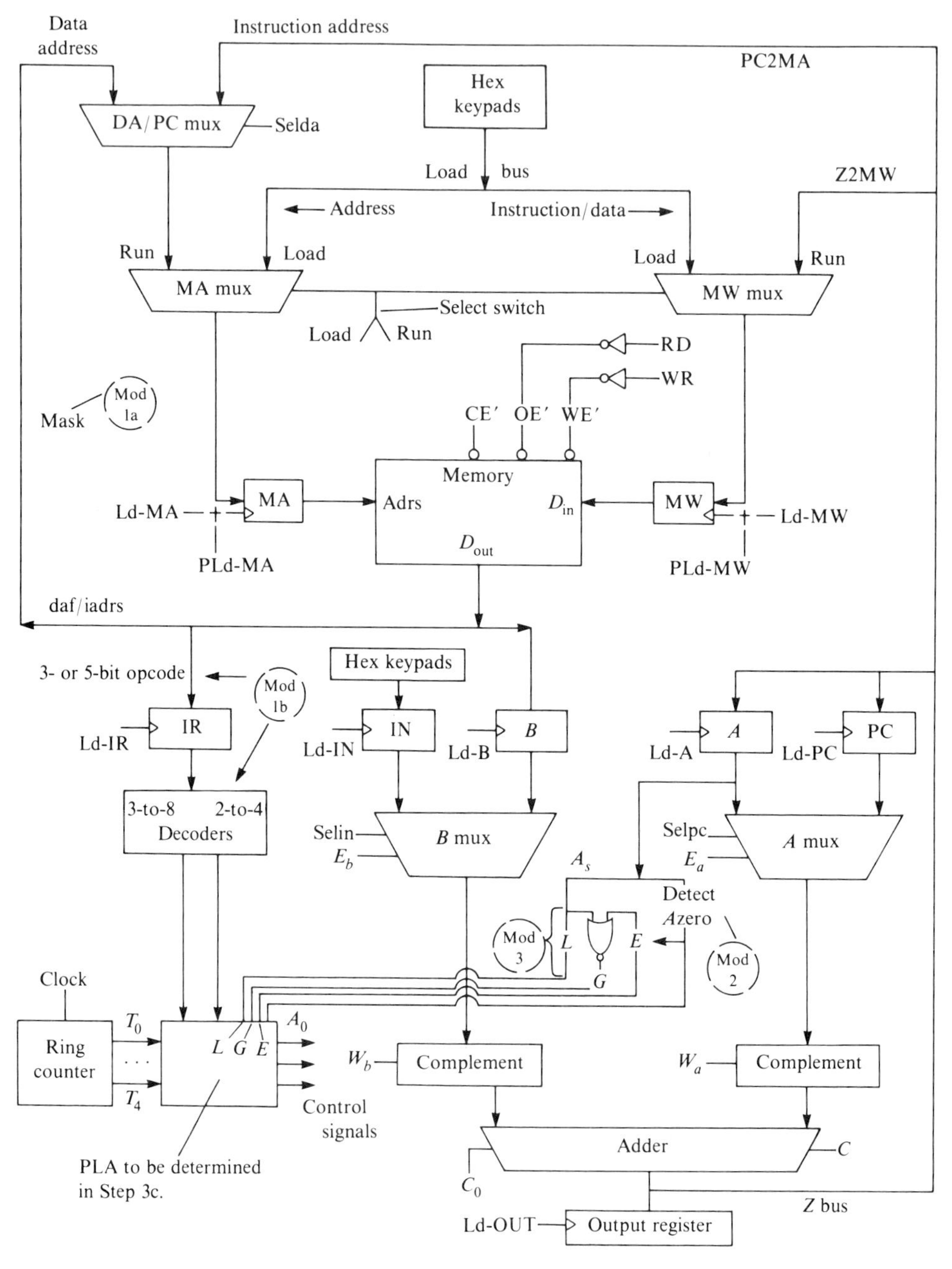

FIGURE 13.19
Block diagram of MC–8B computer with blank PLA

A detailed control flow diagram can be constructed, as shown in Figure 13.20, by tracing the flow of each instruction through the revised data path for the MC–8B computer.

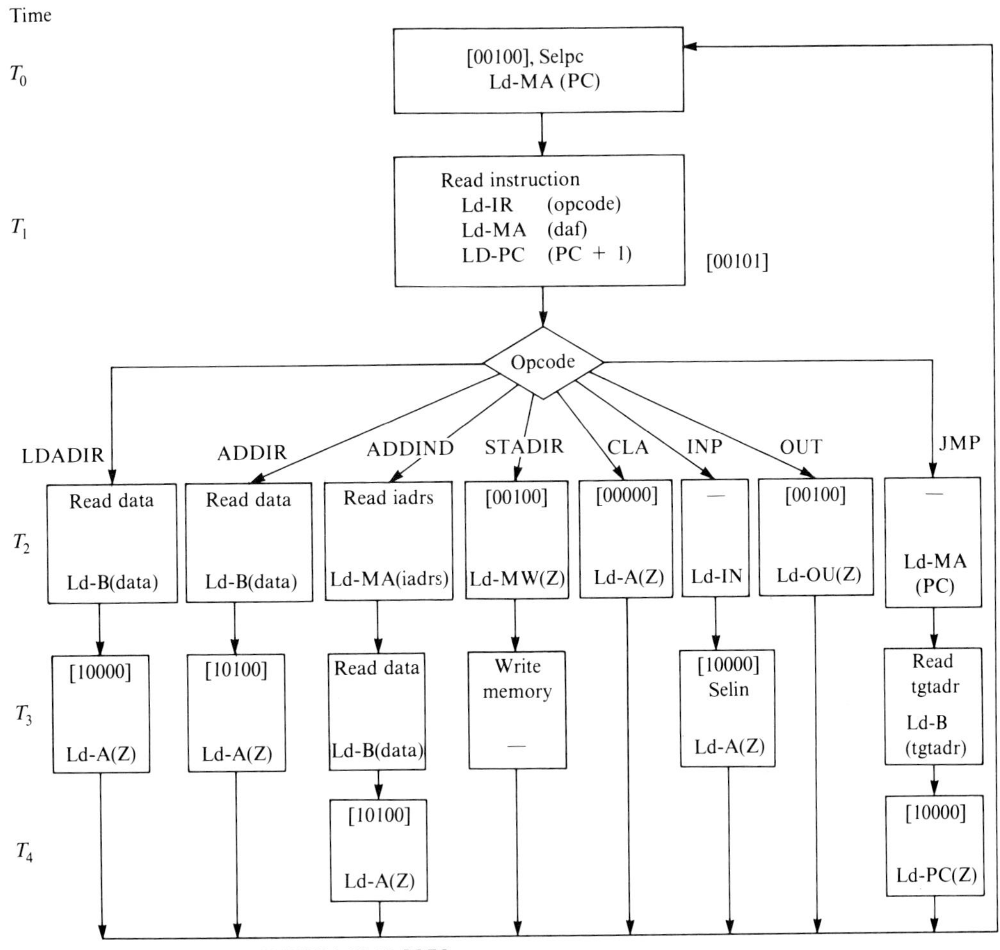

Not shown: SUBDIR, SUBIND, CMPIMM, BRA0, BREQ
[#####] denotes setting of controls E_b, W_b, E_a, W_a, C throughout T_i

FIGURE 13.20
Detailed control flow diagram for MC–8B computer (partial)

c. Design a control unit subsystem.

(1) Design the output decoder.

(a) Construct an instruction-time (p-term) event table. The execution of each instruction is accomplished by a sequence of microoperations at specific times within the overall instruction cycle. By tracing the data flow for each instruction, an instruction-time (p-term) event table can be constructed, as shown in Table 13.5 (p. 640). Note that a memory read or write is initiated at the beginning of interval T_i and edge-triggered registers are loaded midway through interval T_i.

TABLE 13.5
MC–8B instruction-time event table

Instruction		T_0	T_1	T_2	T_3	T_4
000	LDADIR	Fetch MA ← PC	Read inst. IR ← opc. MA ← daf PC ← PC + 1	Read data B ← M[MA]	$A \leftarrow B$	
001	ADDIR	↓	↓	Read data B ← M[MA]	$A \leftarrow A + B$	
010	SUBDIR	↓	↓	Read data B ← M[MA]	$A \leftarrow A - B$	
011	STADIR	↓	↓	MW ← A	Write mem.	
100	ADDIND	↓	↓	Read iadrs MA ← M[MA]	Read data B ← M[MA]	$A \leftarrow A + B$
101	SUBIND	↓	↓	Read iadrs MA ← M[MA]	Read data B ← M[MA]	$A \leftarrow A - B$
11000	CLA	↓	↓	$A \leftarrow 0$	—	
01	COMP	↓	↓	$A \leftarrow A'$	—	
10	INP	↓	↓	Load IN reg.	A ← IN	
11	OUT	↓	↓	OUT ← A	—	
11100	CMPIMM data	↓	↓	MA ← PC	Read data B ← M[MA] PC ← PC + 1	$A \leftarrow A - B$ Set E, G, L
01	JMP tgtadrs	↓	↓	MA ← PC	Read tgtadrs B ← M[MA]	PC ← B
10	BRA0 tgtadrs	↓	↓	MA ← PC	Read tgtadrs B ← M[MA] PC ← PC + 1	If $Azero$ PC ← B
11	BREQ tgtadrs	↓	↓	MA ← PC	Read tgtadrs B ← M[MA] PC ← PC + 1	If $E = 1$ PC ← B

Note: Precede BREQ instruction with CMPIMM instruction.

(b) Construct a p-term/control point activation table. An examination of the instruction-time event table to identify each control point leads to the construction of a p-term/control point activation table as shown in Table 13.6.

(c) Determine the output decoder control equations. Determine the Boolean SOP expressions for the output decoder outputs (to activate control points) by ORing the p-terms in each control point column in Table 13.6. The following are the SOP expressions for the MC–8B computer.

TABLE 13.6
MC–8B p-term/control point activation table

		Control Point																		
Opcode	**P-Term**	**Mask**	**Selda**	**Ld MA**	**RD**	**Ld IR**	**Ld B**	**Selin**	**Selpc**	E_b	W_b	E_a	W_a	C	**Ld IN**	**Ld PC**	**Ld A**	**Ld OU**	**Ld MW**	**WR**
	$T_0 \cdot$ ALL			1					1	0	0	1	0	0						
	$T_1 \cdot$ ALL	1	1	1	1	1			1	0	0	1	0	1		1				
000	$T_2 \cdot$ LDADIR				1		1													
000	$T_3 \cdot$ LDADIR									1	0	0	0	0			1			
001	$T_2 \cdot$ ADDIR				1		1													
001	$T_3 \cdot$ ADDIR									1	0	1	0	0			1			
010	$T_2 \cdot$ SUBDIR				1		1													
010	$T_3 \cdot$ SUBDIR									1	1	1	0	1			1			
011	$T_2 \cdot$ STADIR									0	0	1	0	0					1	
011	$T_3 \cdot$ STADIR																			1
100	$T_2 \cdot$ ADDIND		1	1	1															
100	$T_3 \cdot$ ADDIND				1		1													
100	$T_4 \cdot$ ADDIND									1	0	1	0	0			1			
101	$T_2 \cdot$ SUBIND		1	1	1															
101	$T_3 \cdot$ SUBIND				1		1													
101	$T_4 \cdot$ SUBIND									1	1	1	0	1			1			
10000	$T_2 \cdot$ CLA									0	0	0	0	0			1			
10001	$T_2 \cdot$ COMP									0	0	1	1	0			1			
10010	$T_2 \cdot$ INP														1					
11010	$T_3 \cdot$ INP							1		1	0	0	0	0			1			
10011	$T_2 \cdot$ OUT									0	0	1	0	0				1		
11100	$T_2 \cdot$ CMPIMM			1					1	0	0	1	0	0						
11100	$T_3 \cdot$ CMPIMM				1		1		1	0	0	1	0	1		1				
11000	$T_4 \cdot$ CMPIMM									1	1	1	0	1			1	Set E, G, L		
11101	$T_2 \cdot$ JMP			1					1	0	0	1	0	0						
11101	$T_3 \cdot$ JMP				1		1													
11001	$T_4 \cdot$ JMP									1	0	0	0	0		1				
11110	$T_2 \cdot$ BRA0			1					1	0	0	1	0	0						
11110	$T_3 \cdot$ BRA0				1		1		1	0	0	1	0	1		1				
11010	$T_4 \cdot$ BRA0									1	0	0	0	0		$A0$				
11111	$T_2 \cdot$ BREQ			1					1	0	0	1	0	0						
11111	$T_3 \cdot$ BREQ				1		1		1	0	0	1	0	1		1				
11011	$T_4 \cdot$ BREQ									1	0	0	0	0		E				

Selda = T_1 · ALL + T_2 · (ADDIND + SUBIND)

Mask = T_1 · ALL

Ld-MA = T_0 · ALL + T_1 · ALL
+ T_2 · (ADDIND + SUBIND + CMPIMM + JMP + BRA0 + BREQ)

RD = T_1 · ALL + T_2 · (LDADIR + ADDIR + SUBDIR + ADDIND + SUBIND)
+ T_3 · (ADDIND + SUBIND + CMPIMM + JMP + BRA0 + BREQ)

Ld-IR = T_1 · ALL

Ld-B = T_2 · (LDADIR + ADDIR + SUBDIR)
+ T_3 · (ADDIND + SUBIND + CMPIMM + JMP + BRA0 + BREQ)

Selin = T_3 · INP

Selpc = T_0 · ALL + T_1 · ALL + T_2 · (CMPIMM + JMP + BRA0 + BREQ)
+ T_3 · (CMPIMM + BRA0 + BREQ)

E_b = T_3 · (LDADIR + ADDIR + SUBDIR + INP)
+ T_4 · (ADDIND + SUBIND + CMPIMM + JMP + BRA0 + BREQ)

W_b = T_3 · SUBDIR + T_4 · (SUBIND + CMPIMM)

E_a = T_0 · ALL + T_1 · ALL + T_2 · (COMP + OUT + STADIR + CMPIMM + JMP
+ BRA0 + BREQ) + T_3 · (ADDIR + SUBDIR + CMPIMM + BRA0 + BREQ)
+ T_4 · (ADDIND + SUBIND + CMPIMM)

W_a = T_2 · COMP

C = T_1 · ALL + T_3 · (SUBDIR + CMPIMM + BRA0 + BREQ)
+ T_4 · (SUBIND + CMPIMM)

Ld-IN = T_2 · INP

Ld-PC = T_1 · ALL + T_3 · (CMPIMM + BRA0 + BREQ)
+ T_4 · (JMP + BRA0 · A0 + BREQ · E)

Ld-A = T_2 · (CLA + COMP) + T_3 · (LDADIR + ADDIR + SUBDIR + INP)
+ T_4 · (ADDIND + SUBIND + CMPIMM)

Ld-OUT = T_2 · OUT

Ld-MW = T_2 · STADIR

WR = T_3 · STADIR

(2) Design the next-state generator. Determine the control unit next-state functions and construct a state diagram (Figure 13.21) using the event table (Table 13.5) as a guide.

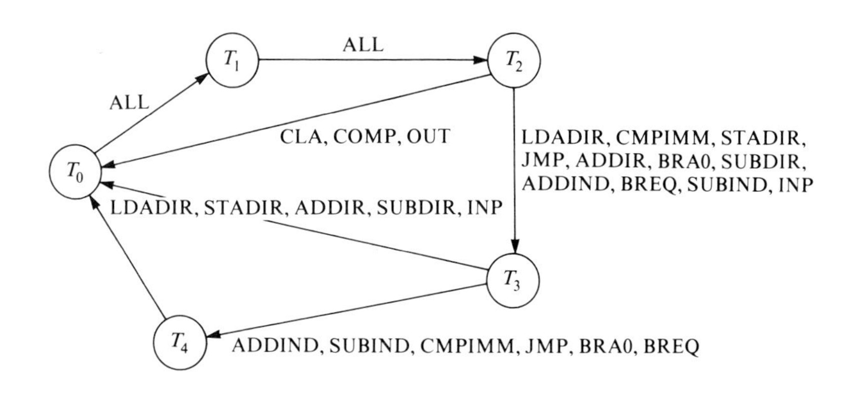

FIGURE 13.21
State diagram for MC–8B computer

Step 4: Realization. The control unit subsystem is realized by implementing the output decoder and the next-state generator. The control unit, data path, and memory subsystems are then integrated to form a complete MC–8B computer, as shown in Figure 13.22.

A logic diagram and timing diagram for the MC–8B computer are presented in Figures 13.23a and b (pp. 644–646), respectively.

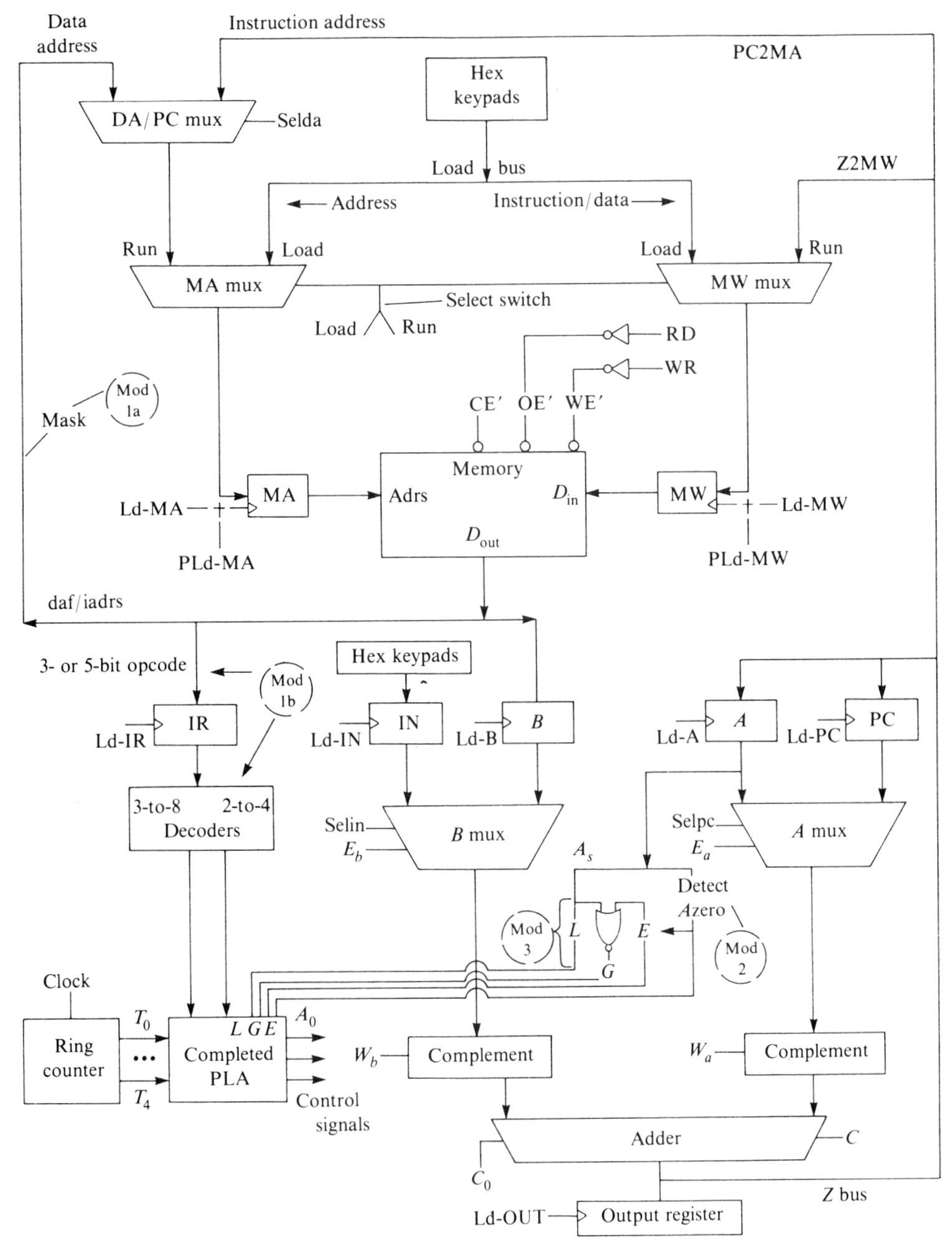

FIGURE 13.22
Block diagram of completed MC–8B computer

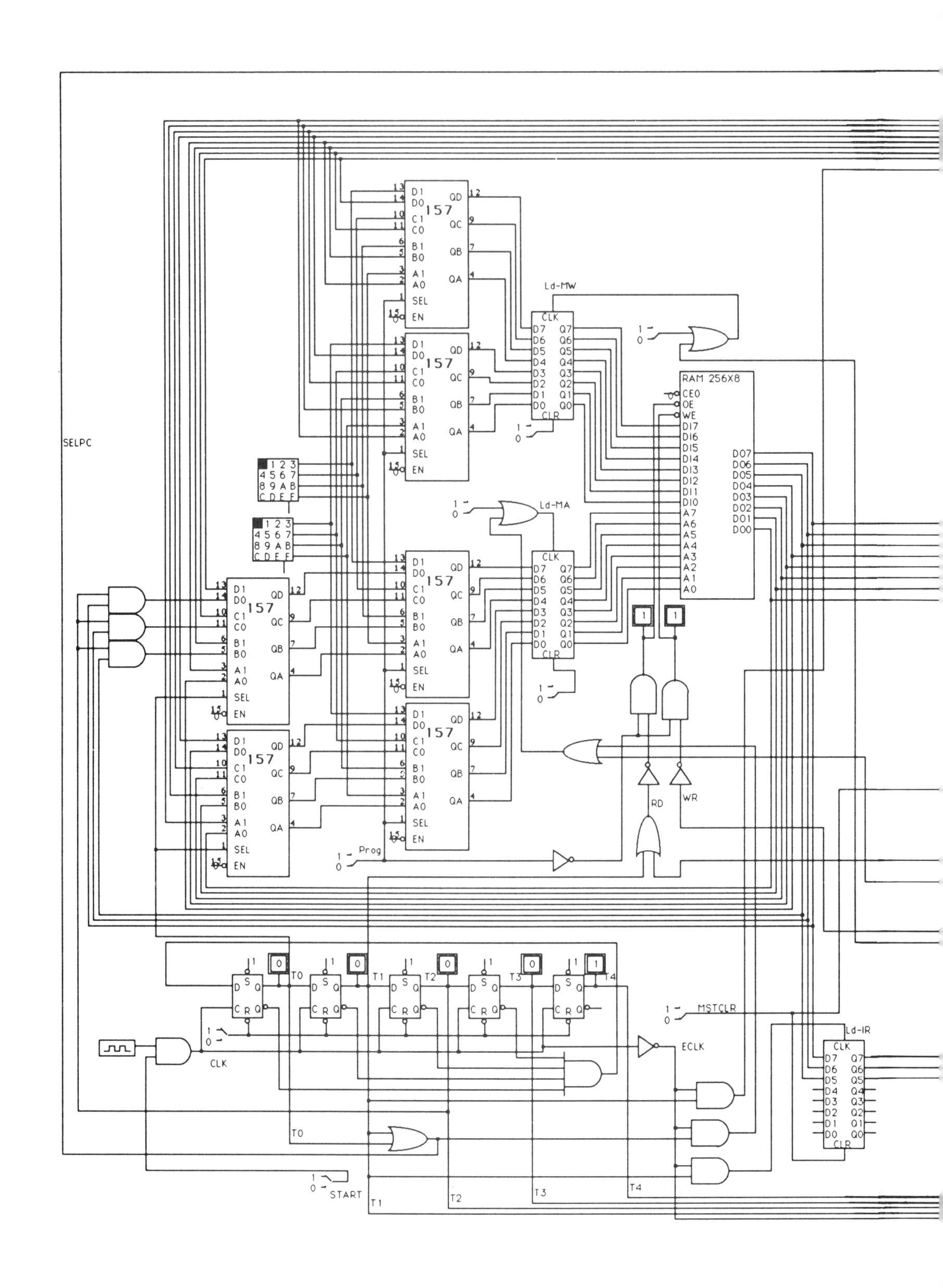

SELPC
157
Ld-MW
Ld-MA
RAM 256X8
RD
WR
Prog
MSTCLR
ECLK
CLK
Ld-IR
START
T0
T1
T2
T3
T4

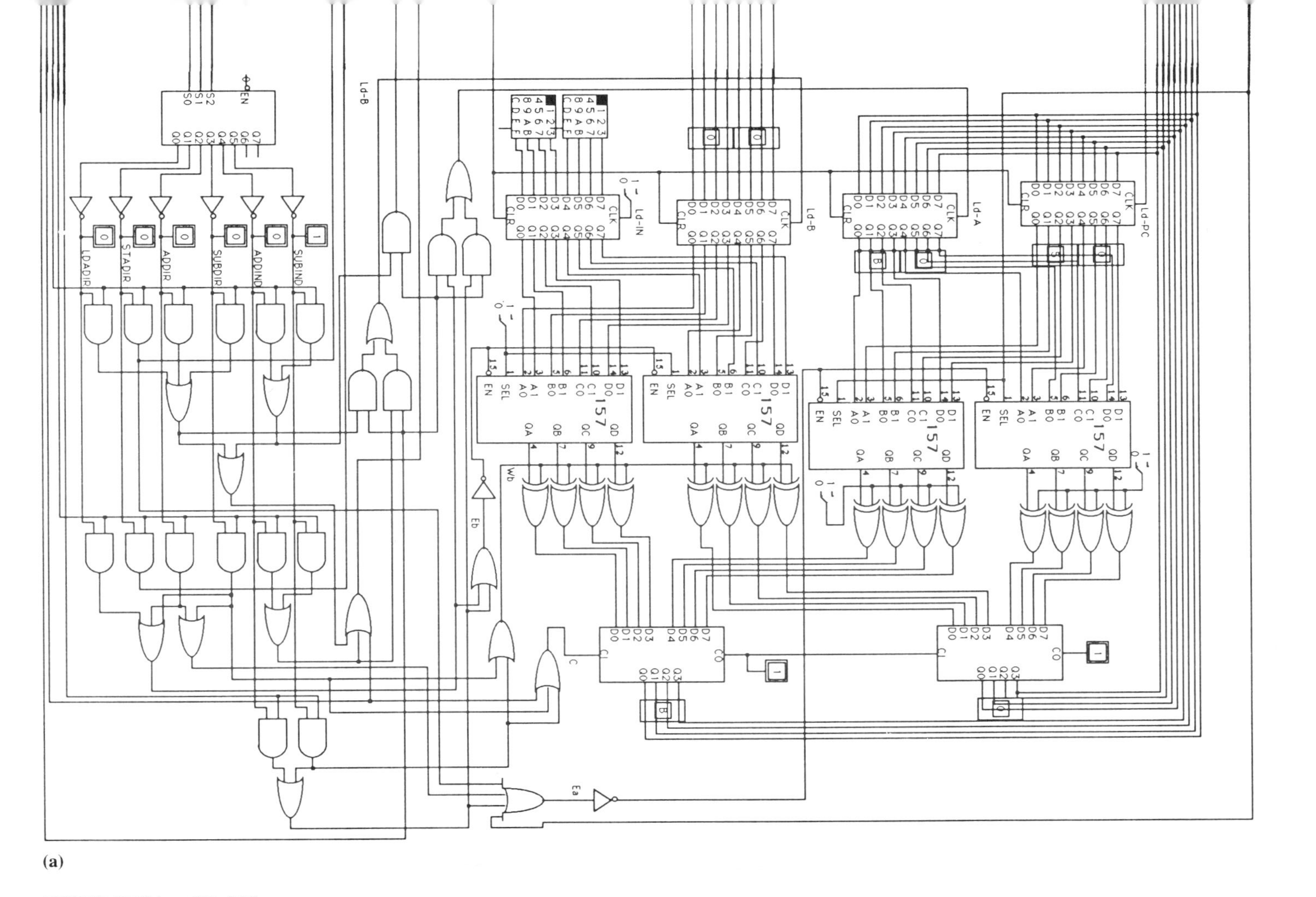

(a)

FIGURE 13.23 (pp. 644–646)
MC–8B computer. (a) Logic diagram (b) Timing diagram

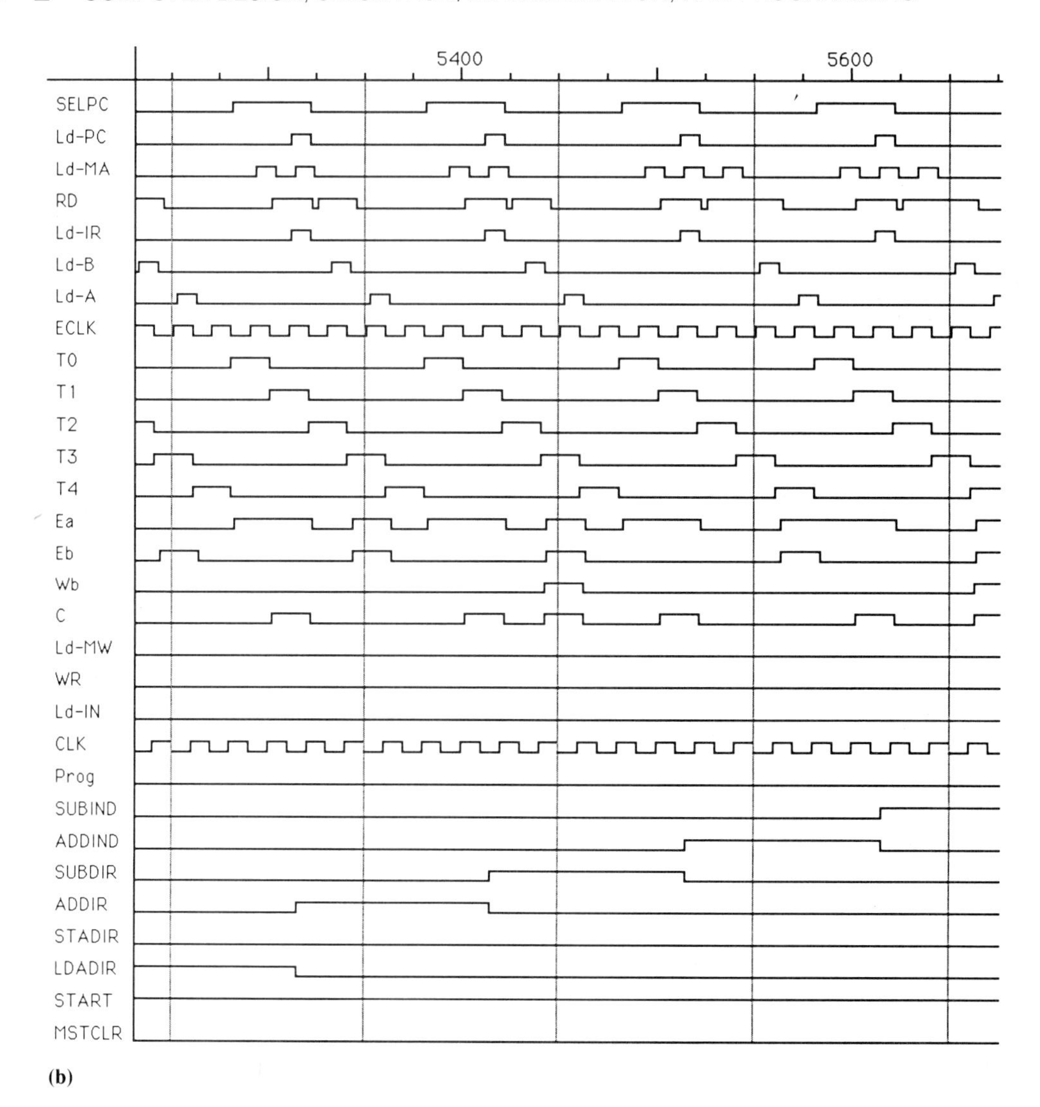

(b)

FIGURE 13.23
Continued

13.5 SUBROUTINE JUMP, JUMP INDIRECT, MULTIPLY, DIVIDE, AND LOGIC INSTRUCTIONS

The objective in this section is to introduce and describe various features found in many computers and design a useful 8-bit computer. Each succeeding computer in this chapter has a wider range of instructions and is, therefore, able to solve a wider range of problems.

According to structured-programming theory, any program can be written using three basic logic structures: sequence, selection, and repetition. Further, any computational algorithm can be accomplished using the arithmetic operations of addition, subtraction, multiplication, and division (since exponentiation can be accomplished by repeated multiplication).

The MC–8C computer designed in this section will augment the MC–8B instruction set with the following additional capabilities:

Subroutine jump, jump indirect instructions

Multiply, divide instructions

Logic (AND,OR) instructions

Instructions to implement selection and repetition structures

Instructions with two implied operands

Memory Reference Instructions

At a minimum, two memory reference instructions are required: an instruction to load a register (*B* or *A*) from a memory location and an instruction to store a register (*B* or *A*) in a memory location. Let's use the *B* register as a destination register for the load instruction. For example,

000 LDBDIR = load register *B* from memory using direct address mode

Since the accumulator *A* is used to hold the result of most instructions, let's use *A* as the source register for the store instruction. For example,

001 STADIR = store accumulator *A* in memory using direct-address mode

Alternatively, we could define opcode 000 and 001 as LDBIND and STAIND (using indirect address mode). The indirect address mode provides a greater range of accessible memory locations than direct address mode. However, each indirect address mode instruction requires an address pointer in the directly addressable portion of memory.

Instructions with 2 Implied Operands

If we restrict our computer to an 8-bit word size, a larger instruction set can be implemented by using more than 3 bits for the opcode (see MC–8B nonmemory reference instructions). If a total of 5 bits are used for the opcode, the remaining 3 bits of the instruction byte are insufficient to specify a useful direct address field. In these cases, 2 operand instructions (which require no memory reference) can be implemented using registers as the 2 implied operands. For example,

01000 = ADDAB = add register *B* to *A*; store result in *A*

01001 = SUBAB = subtract register *B* from *A*; store result in *A*

01010 = ANDAB = AND registers *B* and *A*; store result in *A*

01011 = ORAB = OR registers *B* and *A*; store result in *A*

The first data word to be processed can be brought into register *A* using the instruction sequence CLA, LDBDIR, and ADDAB. The second data word can then be loaded into the *B* register using the LDBDIR instruction. The sum, difference, AND, and OR of the two data words can then be produced using the ADDAB, SUBAB, ANDAB, or ORAB instructions, respectively.

The **multiply instruction** (MULBP) requires the multiplier in the P register and the multiplicand in the B register at the start of the multiply process. The **divide instruction** (DIVBP) requires the dividend in the P register and the divisor in the B register at the start of the divide process. The multiplier x (or dividend x) can be brought into the P register by the instruction sequence CLA, LDBDIR, ADDAB, and MOVA2P, where MOVA2P moves the content of A to register P. The multiplicand y (or divisor y) can be loaded into the B register using the LDBDIR instruction.

A multiply instruction (MULBP) produces a double length product in which the most significant byte is in A and the least significant byte is in the P register. The most significant byte can, therefore, be stored in memory using a STADIR instruction. The least significant byte can be stored in memory using the instruction sequence MOVP2A and STADIR.

The divide instruction (DIVBP) produces an 8-bit quotient in the P register and an 8-bit remainder in A. The remainder can be stored in memory using a STADIR instruction. The quotient can be stored in memory using the instruction sequence MOVP2A and STADIR.

Subroutines

The same algorithm may be needed more than once in solving a particular problem. For example, the computation of the covariance of two sets of data $\{X_k\}$ and $\{Y_k\}$ requires that one sample mean (x^) be computed for set $\{X_k\}$, and another sample mean (y^) be computed for set $\{Y_k\}$. In such cases, an algorithm that is invoked more than once in solving a problem can be coded as a **subroutine**. That subroutine can then be called (invoked) as required by the main program, which solves the overall problem.

It is, therefore, helpful if the computer's instruction set includes a subroutine jump instruction. There are various techniques for implementing a subroutine jump capability in a computer. We choose to use a technique first used in an early computer, the LGP–30. The LGP–30 subroutine jump technique is illustrated in Figure 13.24.

FIGURE 13.24
LGP–30 subroutine jump technique

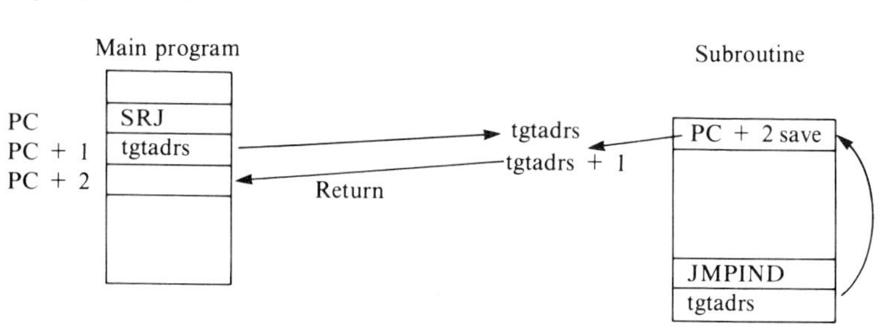

The LGP–30 subroutine jump technique is implemented by a 2-byte instruction with the following format:

Byte-1: Opcode

Byte-2: Subroutine target address (tgtadrs)

In the main program, the **subroutine jump instruction** (SRJ) must save a return address to which the program returns after executing the subroutine. In the subroutine, the first location is the save location for the return address (PC + 2). The second location is the first executable instruction of the subroutine, and so on. The last two locations are the jump indirect instruction via the save location of the return address.

Instructions for Indexing and Looping

The repetition structure in structured-programming theory involves the repeated execution of one or more instructions until some condition is satisfied. For example, the problem of adding a list of numbers $x(1), x(2), \ldots, x(n)$ can easily be solved using indexed variables $x(k)$ and a repetition structure. The solution of this simple problem (adding a list of numbers) suggests that we need at least 4 index-related instructions, as follows:

LDNDX value = Load an initial value into register NDX (NDX ← value)

LDBNDXBAS base = Load B from memory using indexed address mode
(B ← M[base + NDX])

DECNDX = Decrement register NDX (NDX ← NDX − 1)

BRNDXPOS tgtadrs = Branch on condition that index is positive

The following FORTRAN, Pascal, and assembly language examples illustrate the use of a loop in which the partial sum is computed, using indexed variable $x(k)$, in each repetition.

FORTRAN

```
SUM = 0.0
DO 20 K = 1,N
20 SUM = SUM + X(K)
```

Pascal

```
SUM:= 0
FOR K:= 1 TO N
DO SUM:= SUM + X(K)
```

MC–8C Assembly Language

```
       CLA                (A ← 0)
       LDNDX value        (NDX ← value)
start: LDBNDXBAS base     (B ← M[base + NDX])
       ADDAB              (A ← A + B)
       DECNDX             (NDX ← NDX - 1)
       BRNDXPOS start     (Branch to start if index value
                          positive)
```

INTERACTIVE DESIGN APPLICATION

Design of an 8-Bit Computer (MC–8C) with Expanded Capability

The third computer (MC–8C) is evolved from the MC–8B by adding instructions that provide the following additional capabilities:

- □ Subroutine jump, and jump indirect instructions
- □ Multiply and divide instructions

- Logic (AND, OR) instructions
- Instructions to implement selection and repetition logic structures
- Instructions with two implied operands

Step 1: Problem Statement. Define an instruction set (Table 13.7) that has instructions in

TABLE 13.7
MC–8C computer instruction repertoire

Mnemonic	Binary	Description
LDBDIR	000	Load *B* from specified memory location
STADIR	001	Store *A* in specified memory location
ADDAB	01000	Add *B* to *A*; store result in *A*
SUBAB	01001	Subtract *B* from *A*; store result in *A*
ANDAB	01010	AND *B* and *A*; store result in *A*
ORAB	01011	OR *B* and *A*; store result in A
MULBP	01100	Multiply *B* by *P* register; store product in *AP*
DIVBP	01101	Divide *P* register by *B*; store quotient in *P*; store remainder in *A*
MOVA2P	01110	Move *A* to *P* register
MOVP2A	01111	Move *P* register to *A*
CLA	10000	Clear *A*
COMP	10001	Complement *A*
INP	10010	Input
OUT	10011	Output
BRLT tgtadrs	10100	Branch on condition code *L* (Less Than)
BREQ tgtadrs	10101	Branch on condition code *E* (Equal)
BRGT tgtadrs	10110	Branch on condition code *G* (Greater Than)
BRA0 tgtadrs	10111	Branch on *A*zero
SRJ tgtadrs	11000	Subroutine jump
JMPIND iadrs	11001	Jump using indirect address mode
CMPIMM operand	11010	Compare immediate operand with *A*; set *E*, *G*, *L* flip-flops
CMPAB	11011	Compare *B* and *A*; set *E*, *G*, *L* flip-flops
LDNDX	11100	Load index register with initial value
LDBNDXBAS	11101	Load *B* from memory using indexed address mode
DECNDX	11110	Decrement index register
BRNDXPOS tgtadrs	11111	Branch on positive index register value

Branch instructions: [2 bytes; byte-2 = target address]

each basic category (memory reference, register reference, I/O, branch) and add the instructions that provide expanded capability.

Step 2: Conceptualization. Develop a pictorial description of the system:

a. System structure diagram. Draw a block diagram that partitions the system into a memory subsystem, a data path subsystem, and a control unit subsystem. In a systems context, a computer processor is a digital system composed of a memory subsystem and a CPU, where the CPU consists of a control unit subsystem and a data path subsystem. A computer processor can be represented by a PASM model, as shown in Figure 13.25.

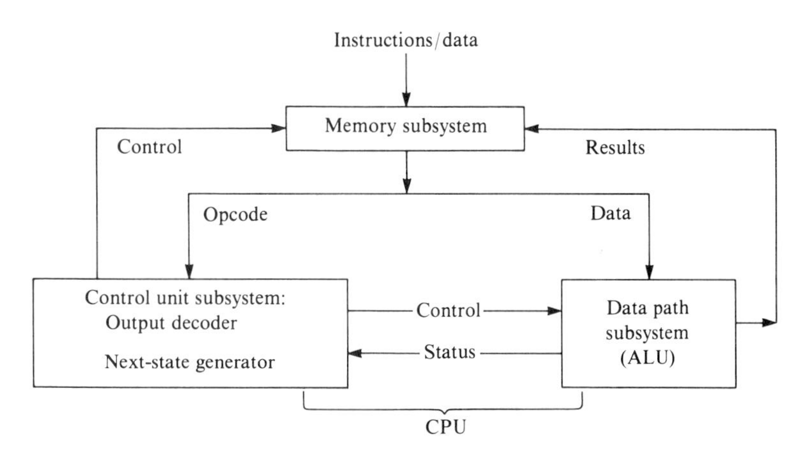

FIGURE 13.25
Structure diagram of a computer processor

b. Control flow diagram. Draw a diagrammatic representation of the operation of the process control algorithm that accomplishes the fetch, decode, and execute for each program instruction. A partial control flow diagram is presented in Figure 13.26 (p. 652). Each instruction cycle is divided into 6 time subintervals T_0, T_1, T_2, T_3, T_4, and T_5.

Step 3: Solution/Simplification.

a. Design a memory subsystem. In order to concentrate on the design of a CPU that can execute a specified set of instructions, we will use the memory subsystem of the MC–8B computer as the memory subsystem of the MC–8C computer.

b. Design a data path subsystem. The data path of the MC–8B computer will be used as a starting (trial) data path of the MC–8C computer. We will modify this data path as necessary to support all instructions specified in Table 13.7. Modifications 1, 2, and 3 (incorporated in MC–8B) are implied but are not explicitly shown in the block diagram of the MC–8C computer (Figure 13.27, p. 653). The additional modifications required are as follows:

Modification 4. This modification consists of a set of AND and OR gates inserted between the complementer outputs and the inputs of a 4-to-1 mux that is inserted between the adder and the *Z*-bus. The mux controls *R* and *S* select either add, ANDAB, or ORAB.

Modification 5. This modification consists of the addition of an index register (NDX) and a *P* shift register, the use of a shift register for *A*, and replacement of the 2-to-1 *A* mux

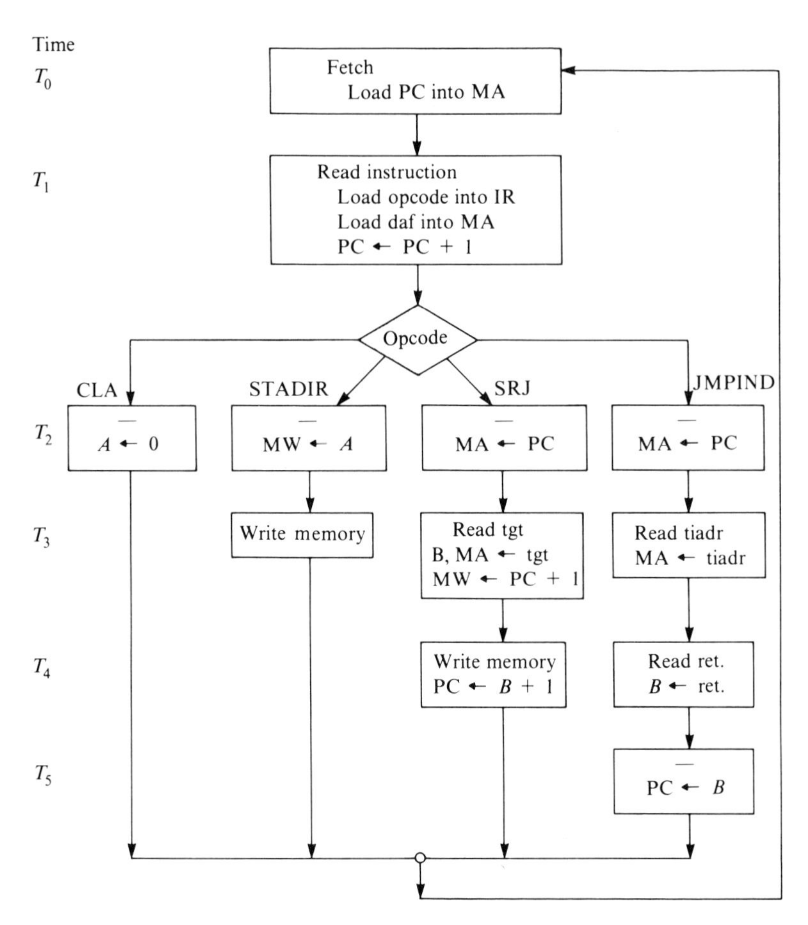

FIGURE 13.26
A process-control algorithm control-flow diagram (partial)

with a 4-to-1 A mux, with inputs from the A, P, PC, and NDX registers. A sign bit, A_s, is appended to the left of the 8-bit shift register A. Together A and P form a 16-bit shift register. An extension bit E, for storing the adder carry-out C_o, is implemented using a D flip-flop. See Figure 13.29 (p. 660) for a block diagram of the MC-8C computer.

c. Design a control unit subsystem.

(1) Design the output decoder.

(a) Construct an instruction-time (p-term) event table. The execution of each instruction is accomplished by a sequence of microoperations at time subintervals within the overall instruction cycle. By tracing the data flow for each instruction, we can construct an instruction-time event table as shown in Table 13.8.

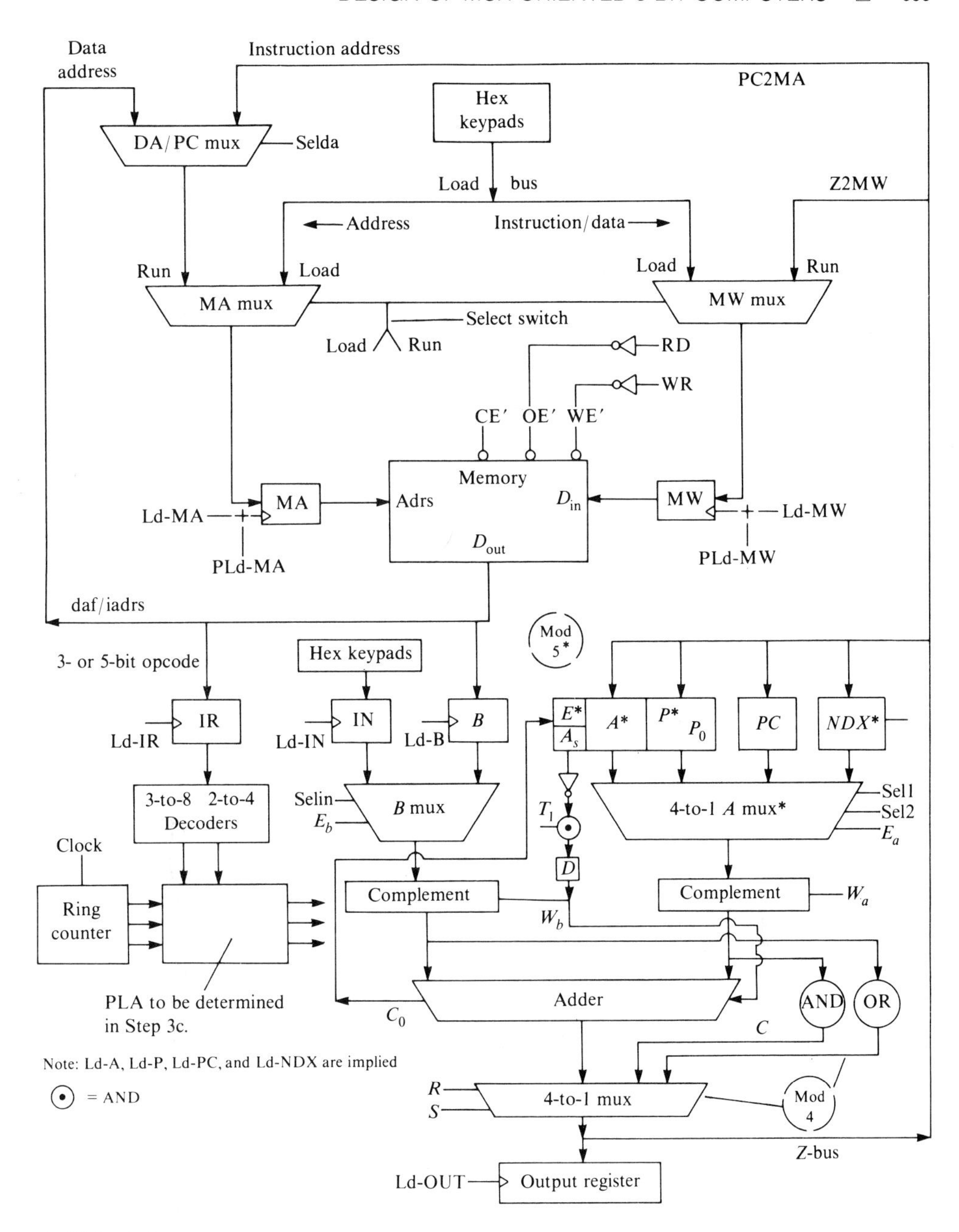

FIGURE 13.27
Block diagram of MC–8C computer with blank PLA

TABLE 13.8
MC–8C instruction-time (p-term) event table

Instruction \ Event / Time	T_0	T_1	T_2	T_3	T_4	T_5
000 LDBDIR	Fetch MA ← PC	Read inst. IR ← opc. MA ← daf PC ← PC + 1	Read data B ← M[MA]	—	—	—
001 STADIR			MW ← A	Write mem.	—	—
01000 ADDAB			$A \leftarrow A + B$	—	—	—
01 SUBAB			$A \leftarrow A - B$	—	—	—
10 ANDAB			$A \leftarrow A \wedge B$	—	—	—
11 ORAB			$A \leftarrow A \vee B$	—	—	—
01100 MULBP	*Note:* A_s, P_s not in *EAP* for shift		(t1)* Compute A+B $E \leftarrow C_o$ If $P_0 = 1$, Ld-A	(t2)* R.Shft *EAP*	Repeat t1, t2 steps 7 times	
01 DIVBP	P has no sign. A_s used in compute and in left shift *AP*		D FF ← A_s L.Shft *AP*	W_b, $C_i \leftarrow D$ Add/subtr. $P_0 \leftarrow Z_s'$ Ld-A	Repeat t1, t2 steps 7 times	
10 MOVA2P			$P \leftarrow A$			
11 MOVP2A			$A \leftarrow P$			
10000 CLA			$A \leftarrow 0$	—	—	—
01 COMP			$A \leftarrow A'$	—	—	—
10 INP			Ld-IN	A ← IN	—	—
11 OUT			OUT ← A	—	—	—
10100 BRLT tgtadrs			MA ← PC	Read tgtadrs B ← M[MA] PC ← PC + 1	If $L = 1$ PC ← B	
01 BREQ tgtadrs			MA ← PC	Read tgtadrs B ← M[MA] PC ← PC + 1	If $E = 1$ PC ← B	

Opcode	Instruction	T_0	T_1	T_2	T_3	T_4	T_5
10	BRGT tgtadrs			MA ← PC	Read tgtadrs B ← M[MA] PC ← PC + 1	If $G = 1$ PC ← B	
11	BRA0 tgtadrs			MA ← PC	Read tgtadrs B ← M[MA] PC ← PC + 1	If $A = 0$ PC ← B	
11000	SRJ tgtadrs			MA ← PC	Read tgtadrs B, MA ← M[MA] MW ← PC + 1	Write mem. PC ← B + 1	
01	JMPIND tiadrs			MA ← PC	Read tiadrs MA ← M[MA]	Read ret. B ← M[MA]	PC ← B
10	CMPIMM data			MA ← PC	Read data B ← M[MA] PC ← PC + 1	$A \leftarrow A - B$ Set E, G, L	
11	CMPAB			$A \leftarrow A - B$ Set E, G, L	—	—	
11100	LDNDX ndxval			MA ← PC	Read ndxval B ← M[MA] PC ← PC + 1	NDX ← B	
01	LDBNDXBAS base			MA ← PC	Read base B ← M[MA] PC ← PC + 1	MA ← B + NDX	Read B ← M[MA]
10	DECNDX			NEG NDX	NDX ← NDX + 1	NEG NDX	
11	BRNDXPOS tgtadrs			MA ← PC	Read tgtadrs B ← M[MA] PC ← PC + 1	If NDX > 0 PC ← B	

*For MULBP and DIVBP, cycles t_1, t_2 are executed 8 times in T_2 to T_5 interval.

T_0 T_1 T_2 T_3 T_4 T_5

t_1, t_2 t_1, t_2 t_1, t_2 t_1, t_2 t_1, t_2 t_1, t_2 t_1, t_2 t_1, t_2

(b) Construct a p-term/control point activation table. An examination of the instruction-time event table to identify each control point leads to the construction of a p-term/control point activation table as shown in Table 13.9.

TABLE 13.9
MC–8C p-term/control point activation table

		Control Point																			
Opcode	**P-Term**	**Mask**	**Selda**	**Ld MA**	**RD**	**Ld IR**	**Ld B**	**Selin**	**SS ee ll 12**	E_b	W_b	E_a	W_a	C	**Ld IN**	**Ld PC**	**Ld A**	**Ld OU**	**Ld MW**	**WR**	**LL dd XP**
	$T_0 \cdot$ ALL			1					PC	0	0	1	0	0							
	$T_1 \cdot$ ALL	1	1	1	1	1			PC	0	0	1	0	1		1					
000	$T_2 \cdot$ LDBDIR				1		1														
001	$T_2 \cdot$ STADIR								*A*	0	0	1	0	0					1		
	$T_3 \cdot$ STADIR																			1	
01000	$T_2 \cdot$ ADDAB								*A*	1	0	1	0	0			1				
01001	$T_2 \cdot$ SUBAB								*A*	1	1	1	0	1			1				
01010	$T_2 \cdot$ ANDAB								*A*	1	0	1	0	0			1				
01011	$T_2 \cdot$ ORAB								*A*	1	0	1	0	0			1				
01100	$T_2 \cdot$ MULBP								Perform 8 t1, t2 cycles												
	$T_3 \cdot$ MULBP																				
	$T_4 \cdot$ MULBP																				
	$T_5 \cdot$ MULBP																				
01101	$T_2 \cdot$ DIVBP								Perform 8 t_1, t_2 cycles												
	$T_3 \cdot$ DIVBP																				
	$T_4 \cdot$ DIVBP																				
	$T_5 \cdot$ DIVBP																				
01110	$T_2 \cdot$ MOVA2P								*A*	0	0	1	0	0							1
01111	$T_2 \cdot$ MOVP2A								*P*	0	0	1	0	0			1				
10000	$T_2 \cdot$ CLA								*A*	0	0	0	0	0			1				
10001	$T_2 \cdot$ COMP								*A*	0	0	1	1	0			1				
10010	$T_2 \cdot$ INP														1						
	$T_3 \cdot$ INP							1		1	0	0	0	0			1				
10011	$T_2 \cdot$ OUT								*A*	0	0	1	0	0				1			
10100	$T_2 \cdot$ BRLT			1					PC	0	0	1	0	0							
	$T_3 \cdot$ BRLT				1		1		PC	0	0	1	0	1		1					
	$T_4 \cdot$ BRLT									1	0	0	0	0		*L*					
10101	$T_2 \cdot$ BREQ			1					PC	0	0	1	0	0							
	$T_3 \cdot$ BREQ				1		1		PC	0	0	1	0	1		1					
	$T_4 \cdot$ BREQ									1	0	0	0	0		*E*					

TABLE 13.9
Continued

Opcode	P-Term	Control Point																			
		Mask	Sela	Ld MA	RD	Ld IR	Ld B	Selin	SSeell12	E_b	W_b	E_a	W_a	C	Ld IN	Ld PC	Ld A	Ld OU	Ld MW	W R	LLddXP
10110	$T_2 \cdot$ BRGT			1					PC	0	0	1	0	0							
	$T_3 \cdot$ BRGT				1		1		PC	0	0	1	0	1		1					
	$T_4 \cdot$ BRGT									1	0	0	0	0		G					
10111	$T_2 \cdot$ BRA0			1					PC	0	0	1	0	0							
	$T_3 \cdot$ BRA0				1		1		PC	0	0	1	0	1		1					
	$T_4 \cdot$ BRA0									1	0	0	0	0		$A0$					
11000	$T_2 \cdot$ SRJ			1					PC	0	0	1	0	0							
	$T_3 \cdot$ SRJ		1	1	1		1		PC	0	0	1	0	1					1		
	$T_4 \cdot$ SRJ									1	0	0	0	1		1				1	
11001	$T_2 \cdot$ JMPIND			1					PC	0	0	1	0	0							
	$T_3 \cdot$ JMPIND		1	1	1																
	$T_4 \cdot$ JMPIND				1		1														
	$T_5 \cdot$ JMPIND									1	0	0	0	0		1					
11010	$T_2 \cdot$ CMPIMM			1					PC	0	0	1	0	0							
	$T_3 \cdot$ CMPIMM				1		1		PC	0	0	1	0	1		1					
	$T_4 \cdot$ CMPIMM								A	1	1	1	0	1			1 set EGL				
11011	$T_2 \cdot$ CMPAB								A	1	1	1	0	1			1 set EGL				
11100	$T_2 \cdot$ LDNDX			1					PC	0	0	1	0	0							
	$T_3 \cdot$ LDNDX				1		1		PC	0	0	1	0	1		1					
	$T_4 \cdot$ LDNDX									1	0	0	0	0							1
11101	$T_2 \cdot$ LDBXBS			1					PC	0	0	1	0	0							
	$T_3 \cdot$ LDBXBS				1		1		PC	0	0	1	0	1		1					
	$T_4 \cdot$ LDBXBS			1					X	1	0	1	0	0							
	$T_5 \cdot$ LDBXBS				1		1														
11110	$T_2 \cdot$ DECNDX								X	0	0	1	1	1							1
	$T_3 \cdot$ DECNDX								X	0	0	1	0	1							1
	$T_4 \cdot$ DECNDX								X	0	0	1	1	1							1
11111	$T_2 \cdot$ BRNDXPS			1					PC	0	0	1	0	0							
	$T_3 \cdot$ BRNDXPS				1		1		PC	0	0	1	0	1		1					
	$T_4 \cdot$ BRNDXPS									1	0	0	0	0		X+					

(c) Determine the output decoder control equations. Determine the Boolean SOP expressions for the output decoder outputs (to activate control points) by ORing the p-terms in each control point column in Table 13.9. The MC–8C SOP expressions are as follows:

Selda = T_1 · ALL + T_3 · (SRJ + JMPIND)

Mask = T_1 · ALL

Ld-MA = T_0 · ALL + T_1 · ALL + T_2 · (BRLT + BREQ + BRGT + BRA0)
+ T_2 · (CMPIMM + JMPIND + SRJ + LDNDX + LDBNDXBAS
+ BRNDXPOS) + T_3 · (SRJ + JMPIND) + T_4 · LDBNDXBAS

RD = T_1 · ALL + T_2 · LDBDIR + T_3 · (BRLT + BREQ + BRGT + BREQ + SRJ)
+ T_3 · (JMPIND + CMPIMM + LDNDX + LDBNDXBAS + BRNDXPOS)
+ T_4 · JMPIND + T_5 · LDBNDXBAS

Ld-IR = T_1 · ALL

Ld-B = T_2 · LDBDIR + T_3 · (BRLT + BREQ + BRGT + BRA0 + SRJ + CMPIMM)
+ T_3 · (LDNDX + BRNDXPOS) + T_4 · JMPIND
+ T_5 · LDBNDXBAS

Selin = T_3 INP

E_b = T_2 · (ADDAB + SUBAB + ANDAB + ORAB + CMPAB) + T_3 · INP
+ T_4 · (BRLT + BREQ + BRGT + BRA0 + SRJ)
+ T_4 · (CMPIMM + LDNDX + LDBNDXBAS + BRNDXPOS) + T_5 · JMPIND

W_b = T_2 · (SUBAB + CMPAS) + T_4 · CMPIMM

E_a = T_0 · ALL + T_1 · ALL + T_2 · (STADIR + COMP + OUT + MOVA2P)
+ T_2 · (MOVP2A + ADDAB + SUBAB + ANDAB + ORAB + CMPAB)
+ T_2 · (BRLT + BREQ + BRGT + BRA0 + SRJ + JMPIND)
+ T_2 · (CMPIMM + LDNDX + LDBNDXBAS + DECNDX + BRNDXPOS)
+ T_3 · (BRLT + BREQ + BRGT + BRA0 + SRJ)
+ T_3 · (CMPIMM + LDNDX + LDBNDXBAS + DECNDX + BRNDXPOS)
+ T_4 · (CMPIMM + LDBNDXBAS + DECNDX)

W_a = T_2 · (COMP + DECNDX) + T_4 · DECNDX

C = T_1 · ALL + T_2 · (SUBAB + CMPAB + DECNDX)
+ T_3 · (BRLT + BREQ + BRGT + BRA0 + SRJ)
+ T_3 · (CMPIMM + LDNDX + LDBNDXBAS + DECNDX + BRNDXPOS)
+ T_4 · (SRJ + CMPIMM + DECNDX)

Ld-IN = T_2 · INP

Ld-PC = T_1 · ALL + T_3 · (BRLT + BREQ + BRGT + BRA0)
+ T_3 · (CMPIMM + LDNDX + LDBNDXBAS + BRNDXPOS)
+ T_4 · (SRJ + BRLT · L + BREQ · E + BRGT · G + BRA0 · A0)
+ T_4 · (BRNDXPOS · NDX $\geq$ 0) + T_5 · JMPIND

Ld-A = T_2 · (ADDAB + SUBAB + ANDAB + ORAB + CMPAB)
+ T_2 · (CLA + COMP + MOVP2A) + T_3 · INP + T_4 · CMPIMM

Ld-OUT = T_2 · OUT

Ld-MW = T_2 · STADIR + T_3 · SRJ

WR = T_3 · STADIR + T_4 · SRJ

Ld-X = T_2 · DECNDX + T_3 · DECNDX + T_4 · (DECNDX + LDNDX)

Ld-P = T_2 · MOVA2P

Set EGL = T_2 · CMPAB + T_4 · CMPIMM

(2) Design the next-state generator. Determine the FSM control unit next-state functions by drawing a state diagram (Figure 13.28). (Use event Table 13.8 as a guide.)

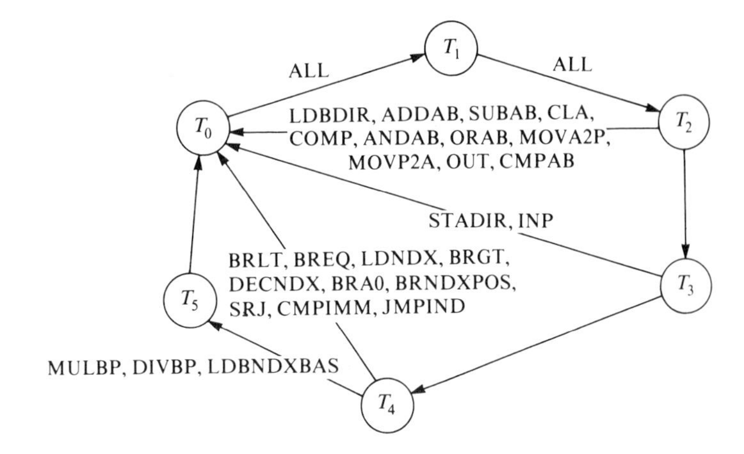

FIGURE 13.28
State diagram for MC–8C computer

Step 4: Realization. Implement the control unit output functions and next-state functions to realize the control unit subsystem. Then integrate the control unit, data path, and memory subsystems to form the MC–8C computer, as shown in Figure 13.29 (p. 660).

13.6 INDEXED ADDRESS MODE, COMPARE, SWAP, AND SKIP INSTRUCTIONS

There are many algorithms that involve the processing of lists of numeric, alphabetic, or alphanumeric data items. You are probably already familiar with a number of list-processing tasks, such as tasks related to one list of data items {X(k)}:

- □ ***Add a list of numbers: S = x(1) + x(2) + · · · + x(n).***
- □ ***Compute the average (x^) of a list of numbers.***
- □ ***Compute the variance of a list of numbers.***
- □ ***Find the largest (or smallest) number in list {X(k)}.***
- □ ***Sort the list {X(k)} in ascending or descending order.***

Other tasks related to two lists of data items {X(k)}, {Y(k)} are

- □ ***Add corresponding numbers in two lists: z(k) = x(k) + y(k).***
- □ ***Compare corresponding data items x(k),y(k) in two lists.***
- □ ***Merge two sorted lists into a single sorted list.***
- □ ***Multiply corresponding x(k) and y(k), and compute the sum of their products. (Some applications of this task include correlation, regression, and covariance computations.)***

There are also a wide range of algorithms in numerical methods that involve processing lists of numbers. Some examples include interpolation, curve fitting (least squares), numerical integration, and the solution of differential equations. This section describes an instruction set for a simple, yet powerful, list-processing computer, designated MC–8D, which is capable of solving any of the problems listed above.

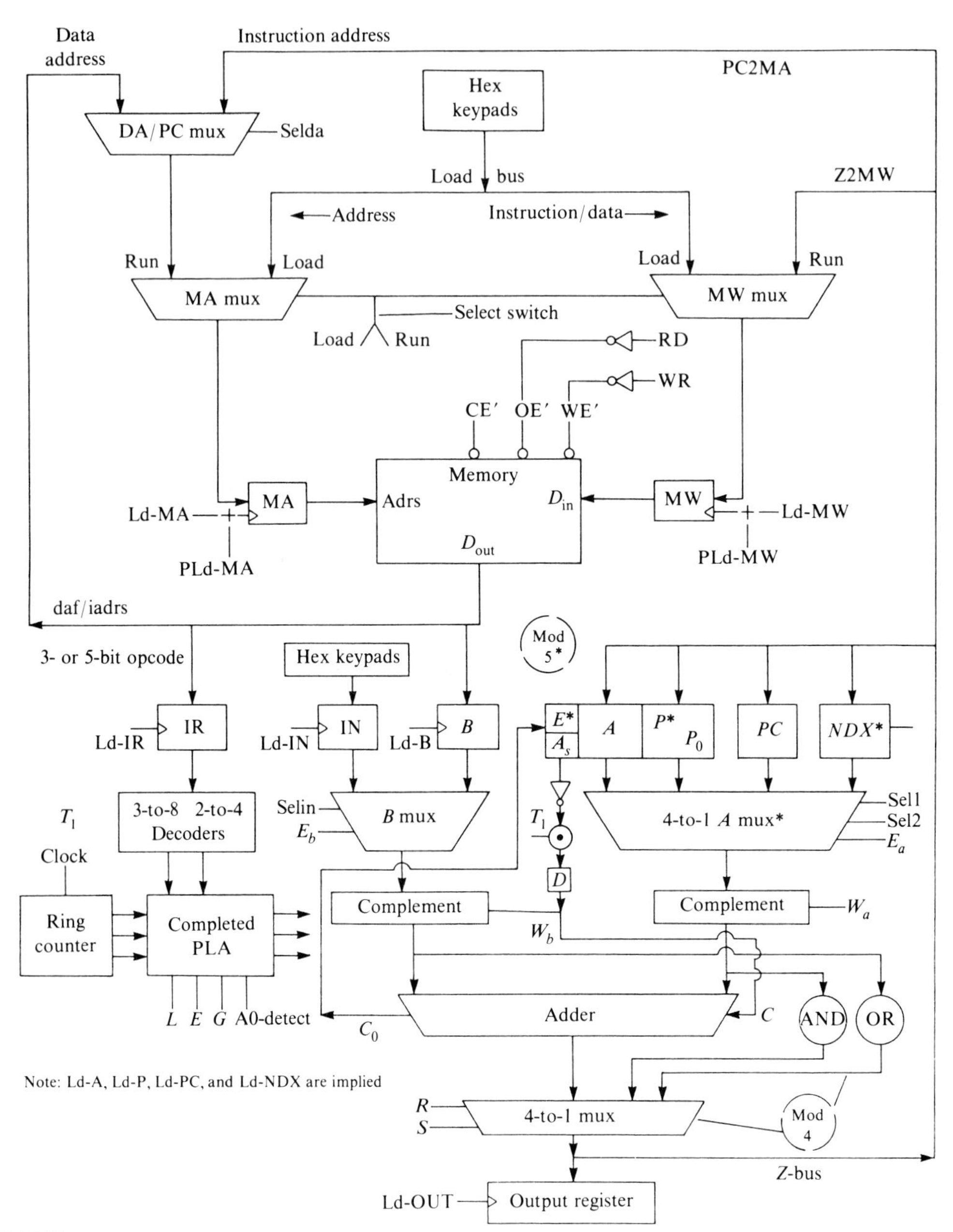

FIGURE 13.29
Block diagram of computer MC–8C

Description of an Instruction Set (MC–8D) for List Processing

The MC–8D instruction set has instructions in each basic category (memory reference, register reference, I/O, and branch), and additional instructions that are useful in performing list processing. The instruction set includes the following:

- Memory reference instructions using immediate mode (2 bytes; byte-2 = data operand)
  - LDBIMM Load *B* from memory using immediate mode
  - STAIMM Store *A* in memory using immediate mode
- Indexed address instructions
  - LDNDX Load index register with initial value
  - LDBNDXBAS Load *B* from memory using indexed-address mode
  - STBNDXBAS Store *B* in memory using indexed-address mode
  - LDANDXBAS Load *A* from memory using indexed-address mode
  - STANDXBAS Store *A* in memory using indexed-address mode
  - DECNDX Decrement index register
  - BRNDXPOS Branch on positive index register value
  - INCNDX Increment index register
  - BRNDX limit tgtadrs Branch if index value < limit
- Index-and-counter addressed instructions
  - STANDXCTR Store *A* using index and counter addressing
  - STBNDXCTR Store *B* using index and counter addressing
  - LDCTR Load counter using immediate mode
  - INCCTR Increment counter
  - MOVCT2NDX Move counter value to index register
  - SAVCTR Save counter in memory location 0
- Register reference instructions
  - CLA Clear *A*
  - COMP Complement *A*
  - ADDAB Add *B* to *A*; store result in *A*
  - SUBAB Subtract *B* and *A*; store result in *A*
  - SWAPAB Exchange contents of *B* and *A*
  - MULBP Multiply *B* by *P* register; store product in *AP*
  - DIVBP Divide *P* register by *B*; store quotient in *P*; store remainder in *A*
- Input-output instructions
  - INP Input
  - OUT Output
- Compare instruction
  - CMPAB Compare *B* and *A*; set flip-flops *E*, *G*, *L*
- Branch instructions
  - SRJ Subroutine jump
  - JMPIND Jump using indirect-address mode
  - BRLT Branch on condition code *L* (less than)
  - BREQ Branch on condition code *E* (equal)
  - BRGT Branch on condition code *G* (greater than)
  - BRA0 Branch on *A*zero

□ Skip instructions

SKPLT Skip next instruction if condition code $L = 1$
SKPEQ Skip next instruction if condition code $E = 1$
SKPGT Skip next instruction if condition code $G = 1$

The design of the MC–8D computer is not presented because of the lengthiness of the instruction-time event table, the p-term/control point activation table, and the list of Boolean expressions for the control unit output decoder.

13.7 SUMMARY

This chapter described the basic categories of instructions used in computers, such as memory reference, register reference, I/O, compare, branch, subroutine jump, multiply, divide, and logic instructions.

A 4-step problem-solving procedure (PSP) for system (computer) design was described. The 4-step PSP uses a programmable algorithmic state machine (PASM) model to represent a computer processor, composed of a memory subsystem and a central processing unit (CPU). The CPU is composed of a data path subsystem and control unit subsystem. This problem solving procedure for system (computer) design was used to design three computers, (MC–8A, B, and C), each with increasingly complex instruction sets. Each computer was evolved from the one preceding it. The design of each computer included a description of its assembly language instructions.

In order to concentrate on the design of a CPU that can execute a specified set of instructions, we used a memory subsystem with a program-load facility as a memory subsystem nucleus for each computer. The interfacing between the memory subsystem and the CPU was accomplished using multiplexers.

A parallel ALU was used as a starting point in the design of the data path of the first computer. This data path was then modified as required to support any and all of the specified instructions. The data path of each succeeding computer was evolved from the data path of the computer that preceded it.

The control unit next-state generator and output decoder were designed using the techniques described in Section 12.3. The control unit of each computer can be implemented either in hardwired form or microprogrammed form.

The MC–8A computer used a 3-bit opcode to define 8 instructions. This minimal instruction set illustrated the basic categories of instructions required in most computers: memory reference (direct address), register reference, I/O, and branch instructions. A second computer (MC–8B) used a 5-bit opcode to provide additional types of instructions, including compare, branch (unconditional and conditional), register reference, and indirect-address mode memory-reference instructions. The third computer (MC–8C) used a 5-bit opcode and instructions with 2 implied operands to define an expanded instruction set that included multiply, divide, subroutine jump, and logic (AND, OR) instructions.

Finally, an extensive instruction set, designed to accomplish a variety of list-processing algorithms, was described for a fourth computer (MC–8D). The instruction set included a variety of instructions of the following categories: memory reference (immediate mode, indexed address mode, index-and-counter address mode), register reference, I/O, compare, subroutine jump, branch, and skip.

Each MC–8X computer was implemented with a partial instruction set. Because of the PLA complexity for a full implementation, it was impractical to present a full instruction set here. The exercises will allow you to gain skill in implementing other partial instruction sets.

KEY TERMS

Computer organization
Computer architecture
Programmable algorithmic state machine (PASM)
Program counter (PC)
Instruction register (IR)
Microoperations
Instruction cycle
Data flow cycle
Indirect address mode
Status register
Condition code
Compare
Compare immediate (CMPIMM)
Condition code branch
Multiply instruction (MULBP)
Divide instruction (DIVBP)
Subroutine
Subroutine jump instruction (SRJ)

EXERCISES

General

1. Write an assembly language program for the MC–8A computer shown in Figure 13.12a. The program is to add the integers 1, 2, 3, 4, and 5. Convert the program to machine language (hexadecimal).
2. Write an assembly language program for the MC–8B computer shown in Figure 13.23a. The program is to add the integers 1, 2, 3, 4, and 5. Assume that the data is stored in memory at addresses 32, 33, 34, 35, and 36. Convert the program to machine language (hexadecimal).
3. Verify the PLA of the MC–8A computer shown in Figure 13.12a.
4. Verify the PLA of the MC–8B computer shown in Figure 13.23a.
5. Draw a logic diagram of an MC–8A computer control-unit PLA that implements the following instruction set: ADDIR (001), SUBDIR (010), STADIR (011), and CLA (100).
6. Draw a timing diagram of the microoperations involved in each of the instructions in Exercise 5.
7. Draw a logic diagram of an MC–8A computer control unit PLA that implements the following instruction set: ADDIR (001), SUBDIR (010), STADIR (011), CLA (100), and JMP (111).
8. Draw a timing diagram of the microoperations involved in each of the instructions in Exercise 7.
9. Draw a logic diagram of an MC–8B computer control unit PLA that implements the following instruction set: LDADIR (000), ADDIND (100), SUBIND (101), and STADIR (011).
10. Draw a timing diagram of the microoperations involved in each of the instructions in Exercise 9.

Design/Implementation

1. Implement the MC–8A computer with the following instruction set: LDADIR (000), ADDIR (001), SUBDIR (010), and STADIR (011). Load and run the following program:

Address	Instruction		Hexadecimal		
(Decimal)	Assembly	Binary	Address	Inst.	Remarks
0	LDADIR 4	000 00100	00	04	
1	ADDIR 5	001 00101	01	25	
2	SUBDIR 6	010 00110	02	46	
3	STADIR 7	011 00111	03	67	
4	4	0000 0100	04	04	Direct address data
5	5	0000 0101	05	05	
6	3	0000 0011	06	03	
7	—	0000 0000	07	—	Result

2. Implement the MC–8B computer with the following instruction set: LDADIR (000), STADIR (001), ADDIR (010), SUBDIR (011), ADDIND (100), and SUBIND (101). Load and run the following program:

Address	Instruction		Hexadecimal		
(Decimal)	Assembly	Binary	Address	Inst.	Remarks
0	LDADIR 8	000 01000	00	08	
1	ADDIR 9	010 01001	01	49	
2	SUBDIR 10	011 01010	02	6A	
3	ADDIND 11	100 01011	03	8B	
4	SUBIND 12	101 01100	04	AC	
5	STADIR 7	001 00111	05	27	
6	—		06	—	
7	—		07	—	Result
8	7	0000 0111	08	07	Direct address data
9	8	0000 1000	09	08	
10	5	0000 0101	0A	05	
11	32	0010 0000	0B	20	Iadrs pointers
12	33	0010 0001	0C	21	
—					
32	2	0000 0010	20	02	Iadrs data
33	1	0000 0001	21	01	

3. Implement the MC–8C computer with the following instruction set: LDBDIR (000), STADIR (001), ADDAB (01000), SUBAB (01001), SRJ (11000), and JMPIND (11001). Make up a program using these instructions; load and run the program.

4. Implement the MC–8C computer with the following instruction set: LDBDIR (000), STADIR (001), ADDAB (01000), SUBAB (01001), CMPIMM (11010), BRLT (10100), and BREQ (10101). Make up a program using these instructions; load and run the program.

14 Design of Bus-Oriented 8-Bit Computers

OBJECTIVES

After you complete this chapter, you will be able

- ☐ To design bus-oriented computers using the 4-step problem-solving procedure used in Chapter 13 to design mux-oriented computers
- ☐ To interface using buses and 3-state devices instead of interfacing using multiplexers
- ☐ To design the control unit output decoder so that no bus conflicts occur

14.1 INTRODUCTION

In Chapter 13, multiplexers were used to interface multiple sources to a single destination. A simpler alternative to multiplexer interfacing is to use buses and devices with 3-state outputs for interfacing. Figure 14.1 illustrates a memory subsystem and a data path that use a bus and buffers with 3-state outputs for interfacing multiple sources to multiple destinations. A memory with bidirectional data lines could further simplify implementation of the memory subsystem.

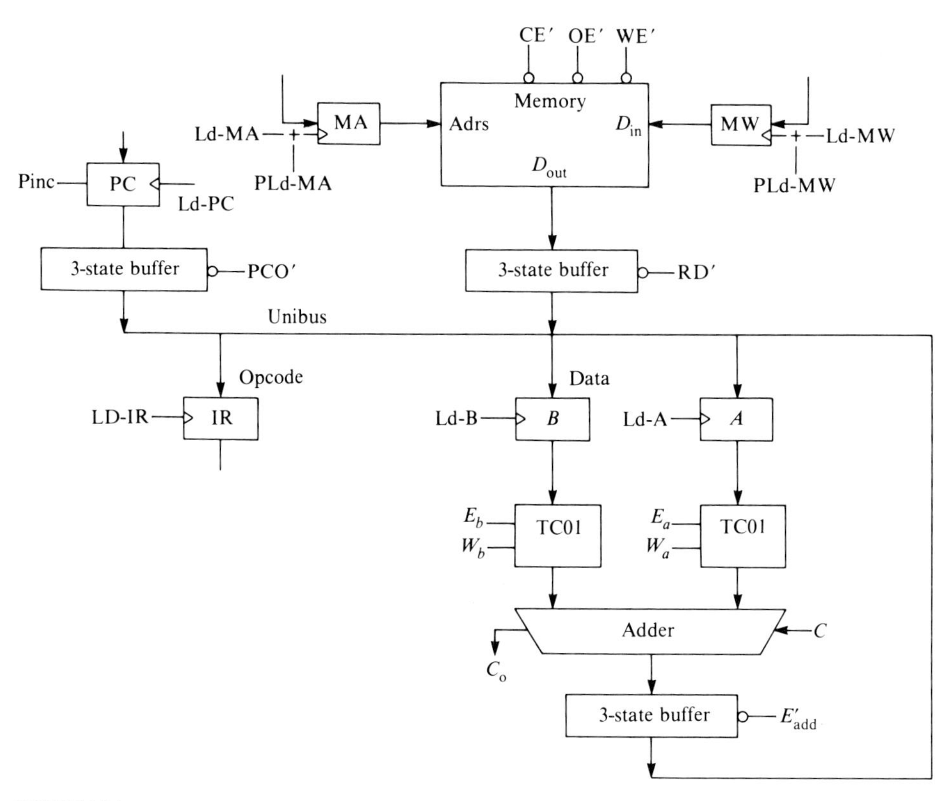

FIGURE 14.1
Memory subsystem and data path using a single bus (Unibus)

In general, the design of digital systems is simpler when using bus structures and devices with 3-state outputs than using multiplexer interfacing. The objective of this chapter is to illustrate the design of simple **bus-oriented computers** BC–8X. Each of the computers in this chapter has the same instruction set as the corresponding computer in Chapter 13.

In order to concentrate on the design of a CPU that can execute a specified set of instructions, we will use a memory subsystem with a program-load facility (from Chapter 12, Figure 12.26) as a nucleus of the memory subsystem of each computer designed in this

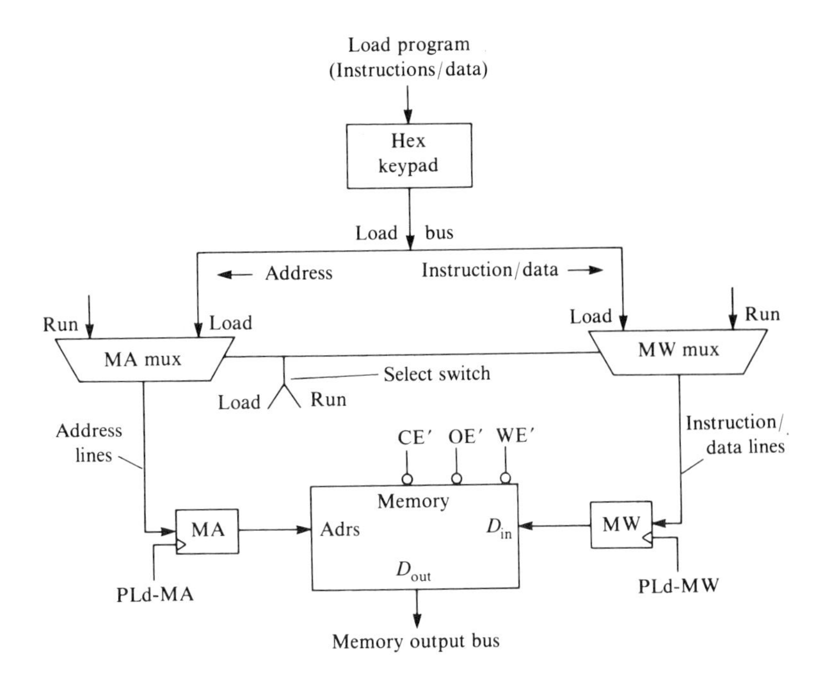

FIGURE 14.2
Memory subsystem with a program-load facility

chapter. For convenience, Figure 14.2 presents a block diagram of the memory subsystem with a program-load facility.

In the memory subsystem shown in Figure 14.2, input lines are available on the MA mux and MW mux for run-time addresses and data, respectively.

The design of each computer in this chapter is accomplished using the same 4-step problem-solving procedure for system design described in Chapter 13.

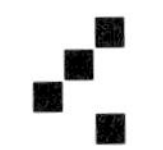

PSP

Step 1: Problem Statement. Given an instruction set, design a bus-oriented computer that implements that instruction set. The computer memory subsystem must be provided with a facility to load program instructions and data prior to program execution. The process of program execution is organized and described in two parts:

a. Process. Fetch, decode, and execute each program instruction (and accomplish the required processing of data and storing of computer-generated results).

b. Process control algorithm. A step-by-step procedure to accomplish the fetch, decode, and execute for each program instruction.

Step 2: Conceptualization. Develop a pictorial description of the system:

a. System structure diagram. Draw a block diagram that partitions the system into a memory subsystem, a data path subsystem, and a control unit subsystem.

b. Control flow diagram. Draw a diagrammatic representation of the operation of the process control algorithm that accomplishes the fetch, decode, and execute for each program instruction.

Step 3: Solution/Simplification.

a. Design a memory subsystem to store program instructions, data, and computer-generated results of the data processing.
b. Design a data path subsystem to support each instruction in the computer's repertoire.
c. Design a control unit subsystem to include the following subcomponents:
 (1) Control unit output decoder.
  (a) Construct an instruction-time (p-term) event table.
  (b) Construct a p-term/control point activation table.
  (c) Determine an SOP expression for each output.
 (2) Control unit next-state generator. Construct a next-state diagram.

Step 4: Realization.

a. Implement the memory subsystem.
b. Implement the data path subsystem.
c. Implement the control unit subsystem.
d. Integrate the control unit, data path, and memory subsystems to form a computer processor. ■

14.2 DIRECT ADDRESS MODE, REGISTER REFERENCE, I/O, AND BRANCH INSTRUCTIONS

This section describes the design of a bus-oriented computer whose instruction set includes instructions in each of the basic categories (memory reference [direct address mode only], register reference, I/O, and branch). A description of specific instructions within each of these categories was presented in Section 13.2.

INTERACTIVE DESIGN APPLICATION

Design of a Simple Bus-Oriented Computer, the BC–8A

This section describes the design of a simple 8-bit bus-oriented computer (BC–8A). The instruction set of the BC–8A computer is the same as that used in the MC–8A mux-oriented computer.

Step 1: Problem Statement. Given a specified set of instructions (Table 14.1), design a bus-oriented 8-bit computer that implements the instruction set in Table 14.1.

TABLE 14.1
BC–8A computer instruction repertoire

Mnemonic	Binary	Description
LDADIR	000xxxxx	Load *A* with content of memory address *X*
ADDIR	001xxxxx	Add content of memory address *X* to *A*
SUBDIR	010xxxxx	Subtract content of memory address *X* from *A*
STADIR	011xxxxx	Store accumulator in memory address *X*
CLA	100– – – – –	Clear accumulator *A*
INP	101– – – – –	Load accumulator *A* from device
OUT	110– – – – –	Output accumulator *A* to device
JMP	111xxxxx	Branch unconditionally to target address

Note: Each instruction is composed of a 3-bit opcode and a 5-bit memory address.

xxxxx denotes a 5-bit direct-address field *X*.

Step 2: Conceptualization. Develop a pictorial description of the system.

a. System structure diagram. Draw a block diagram that partitions the system into a memory subsystem, a data path subsystem, and a control unit subsystem.

In a systems context, a computer processor is a digital system composed of a memory subsystem and a central processing unit (CPU), where the CPU consists of a control unit subsystem and a data path subsystem. A computer processor can be represented by a programmable algorithmic state machine (PASM) model, as shown in Figure 14.3.

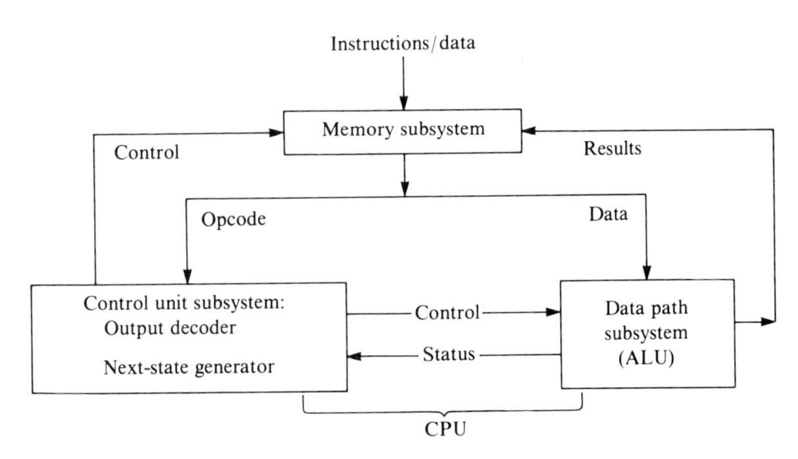

FIGURE 14.3
Structure diagram of a computer processor

b. Control flow diagram. Draw a diagrammatic representation of the operation of the process control algorithm that accomplishes the fetch, decode, and execute for each program instruction.

A partial control flow diagram (showing instructions CLA, LDADIR, ADDIR, and STADIR) is presented in Figure 14.4. Each instruction cycle is divided into four time subintervals T_0, T_1, T_2, and T_3.

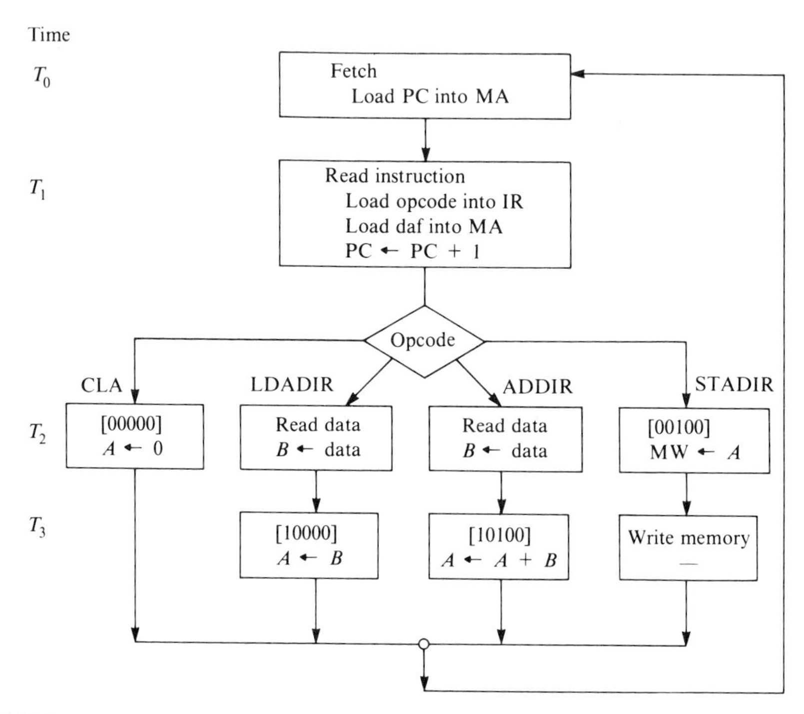

FIGURE 14.4
A process-control algorithm control-flow diagram (partial)

Step 3: Solution/Simplification.

a. Design a memory subsystem. We will describe the design of an 8-bit bus-oriented computer and illustrate interfacing using buses and 3-state devices (versus multiplexer interfacing). For this reason, we will use the memory and data path subsystems of the mux-oriented MC–8A computer (Figure 14.5) as a starting point for the design of a bus-oriented BC–8A computer. We will then replace multiplexer interfacing with buses and 3-state devices. However, the nucleus memory will be retained (as will the MA and MW muxes) so that we can concentrate on the design of the CPU.

b. Design a data path subsystem. The MC–8A data path (Figure 14.5), consisting of a basic parallel ALU data path augmented by IN and PC registers, is used as a starting point for

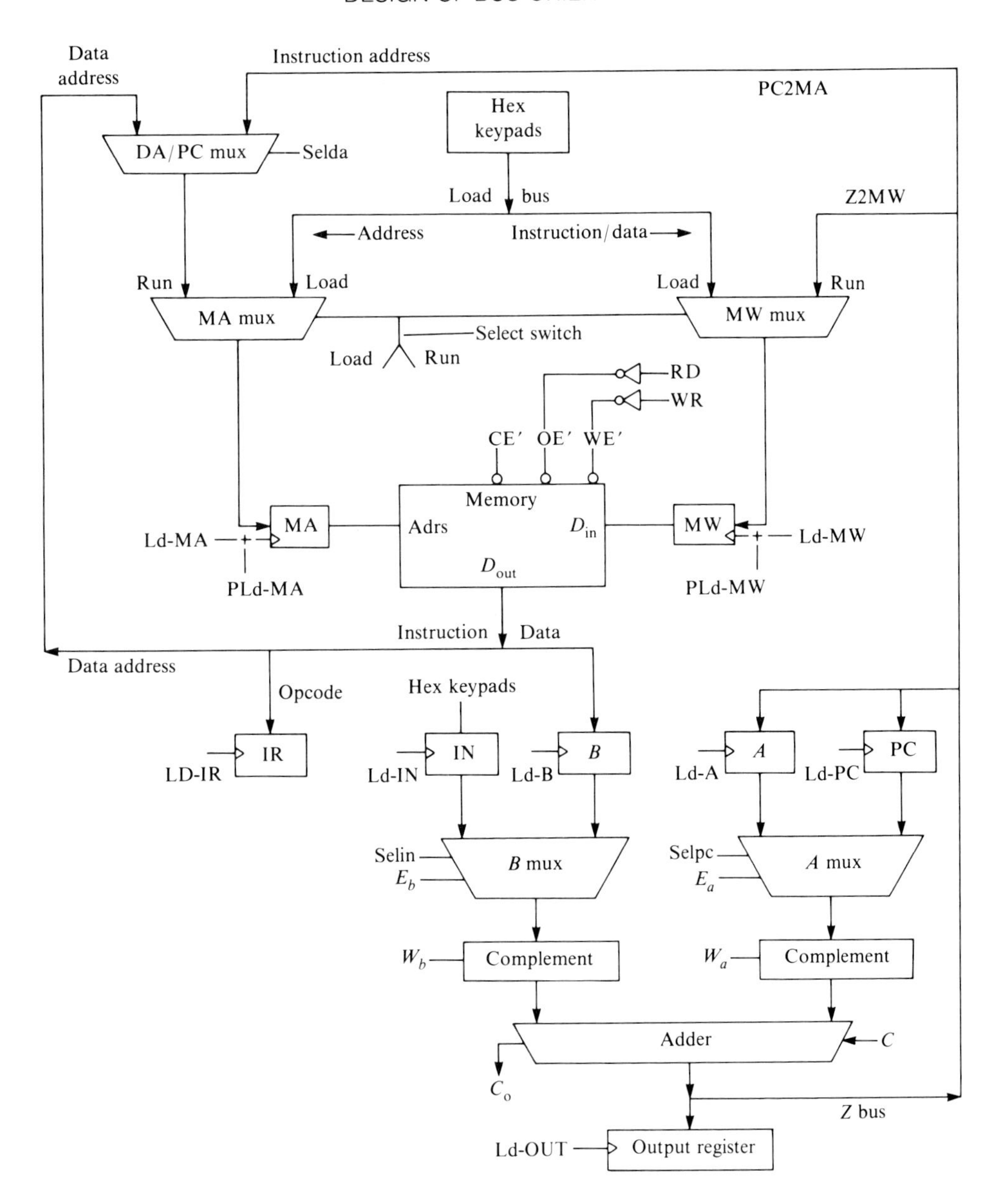

FIGURE 14.5
Memory and data path subsystems of the MC–8A computer

the design of the BC–8A data path. Since the objective is to eliminate as many muxes as possible, the IN and PC registers are temporarily removed, thereby obviating the need for the *A* mux and *B* mux. The data path that remains (see Figure 14.6) is a simple parallel ALU data path (first described in Section 6.2).

The simple parallel ALU data path in Figure 14.6 has a single bus, the Z-bus. Since the adder is the only device feeding the bus, no 3-state buffer is required for the parallel ALU data path.

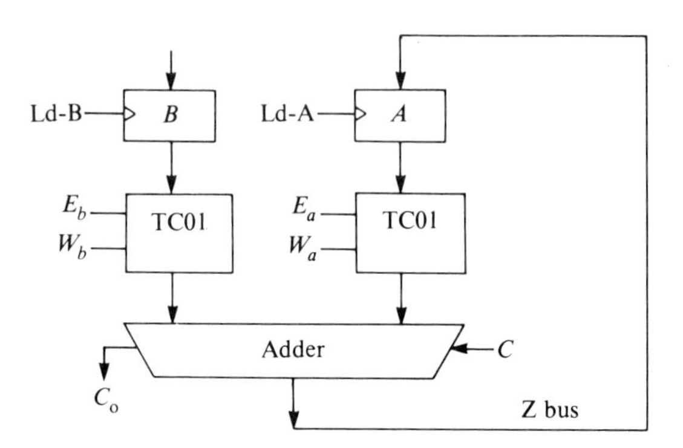

FIGURE 14.6
Basic parallel ALU data path

Now, we will add back the IN and PC registers using bus/3-state interfacing. If more than one device feeds the bus, each device feeding the bus must be interfaced by a 3-state buffer. We must also ensure that no **bus contention** takes place among devices feeding the bus. One way to avoid bus contention is by ensuring that no two devices feed the bus at the same time.

A 3-state buffer is placed between the adder and the *Z* bus to allow other devices to feed the bus. The re-insertion of the IN register and the PC register is accomplished by interfacing these registers with the *Z* bus using 3-state buffers, or by using 3-state buffer registers for the IN and PC registers. Note that the program counter (PC) can be implemented by a counter that increments itself when control PCO is activated to place the instruction address on the bus. Consequently, the microoperation PC ← PC + 1 could also be accomplished concurrent with the microoperation MA ← PC.

The next (and final) step is to interface the memory subsystem nucleus with the data path. This is accomplished by using a 3-state buffer to interface the memory output line with the *Z* bus and to connect the *Z* bus to the run input of the MA mux for instruction and data addresses and to the run input of the MW mux to provide for a "store" instruction. The resulting configuration, with the *Z* bus redesignated as the unibus, is shown in Figure 14.7.

The details of the control unit (unspecified at this time) are determined in Step 3c.

c. Design a control unit subsystem.

(1) Design the output decoder.

(a) Construct an instruction-time (p-term) event table. The execution of each instruction is accomplished by a sequence of microoperations at specific times within the overall instruction cycle. By tracing the control flow for each instruction, an instruction-time (p-term) event table can be constructed, as shown in Table 14.2 (p. 674).

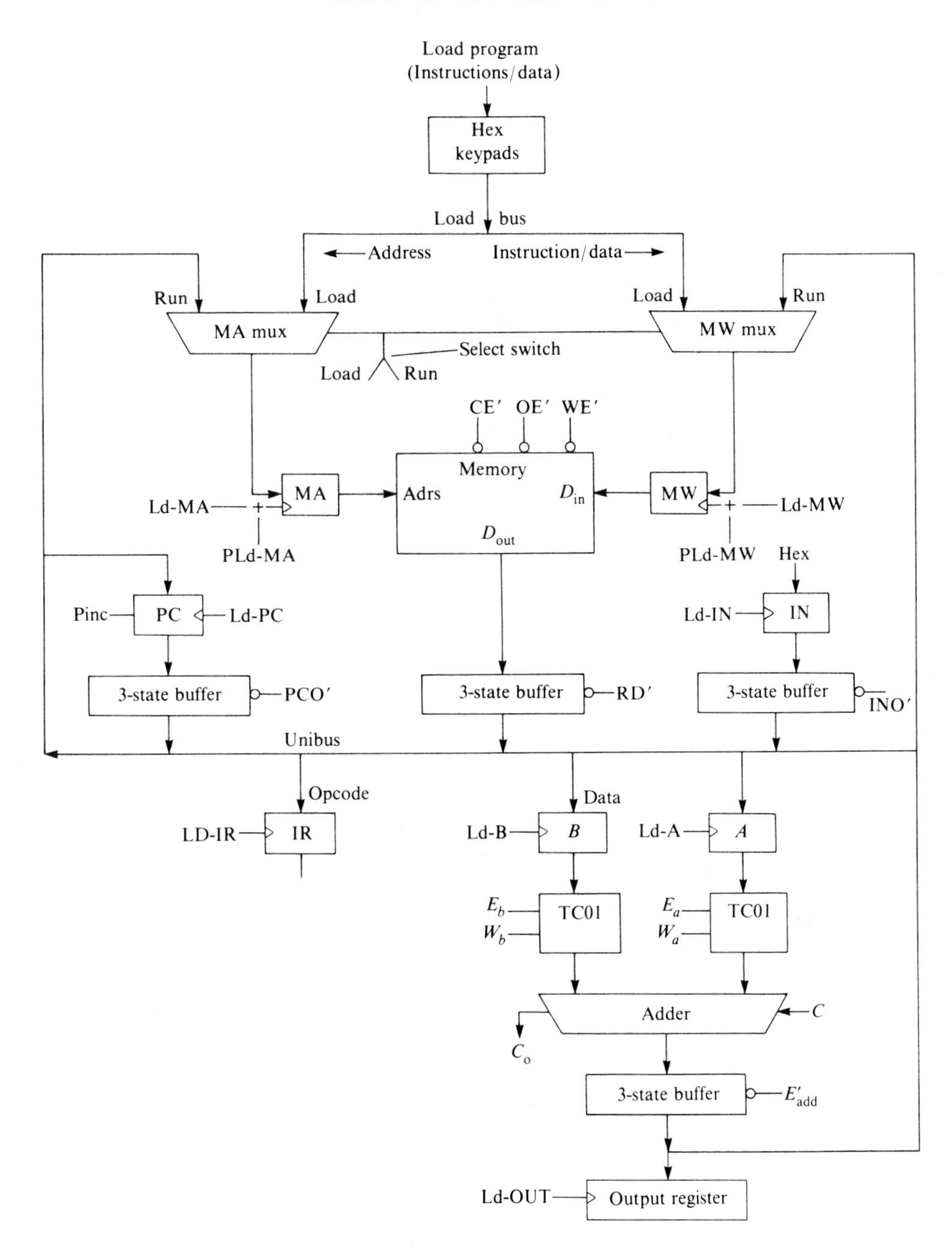

FIGURE 14.7
Memory and data path of a bus-oriented computer

TABLE 14.2
BC–8A instruction-time (p-term) event table

Instruction \ Event Time		T_0	T_1	T_2	T_3
000	LDADIR	Fetch MA ← PC	Read inst. IR ← opc. MA ← daf PC ← PC + 1	Read data A ← M[MA]	—
001	ADDIR	↓	↓	Read data B ← M[MA]	$A \leftarrow A + B$
010	SUBDIR	↓	↓	Read data B ← M[MA]	$A \leftarrow A - B$
011	STADIR	↓	↓	MW ← A	Write mem.
100	CLA	↓	↓	$A \leftarrow 0$	—
101	INP	↓	↓	Ld-IN	A ← IN
110	OUT	↓	↓	OUT ← A	—
111	JMP tgtadrs	↓	↓	MA ← PC	Read tgtadrs PC ← M[MA]

(b) Construct a p-term/control point activation table. An examination of the instruction-time (p-term) event table (Table 14.2) to identify each control point leads to the construction of a p-term/control point activation table, as shown in Table 14.3. The reader can verify the entries by tracing each instruction's microoperations through the data path in Figure 14.7.

(c) Determine the output decoder control equations. Determine the Boolean SOP expressions for the output decoder outputs (to activate control points) by ORing the p-terms in each control point column of Table 14.3.

$$\text{PCO} = T_0 \cdot \text{ALL} + T_2 \cdot \text{JMP}$$
$$\text{Ld-PC} = T_3 \cdot \text{JMP}$$
$$\text{Ld-MA} = T_0 \cdot \text{ALL} + T_1 \cdot \text{ALL} + T_2 \cdot \text{JMP}$$
$$\text{RD} = T_1 \cdot \text{ALL} + T_2 \cdot (\text{LDADIR} + \text{ADDIR} + \text{SUBDIR}) + T_3 \cdot \text{JMP}$$
$$\text{Ld-IR} = T_1 \cdot \text{ALL}$$
$$\text{Ld-B} = T_2 \cdot (\text{ADDIR} + \text{SUBDIR})$$
$$E_b = T_3 \cdot (\text{ADDIR} + \text{SUBDIR})$$
$$W_b = T_3 \cdot \text{SUBDIR}$$

TABLE 14.3
BC–8A p-term/control point activation table

P-Term	Control Point																		
	PCO	Ld PC	Ld MA	RD	Ld IR	Ld B	E_b	W_b	E_a	W_a	C	E Ad	Ld A	Ld MW	WR	Ld IN	INO	Ld OU	Pi nc
$T_0 \cdot$ ALL	1		1																
$T_1 \cdot$ ALL			1	1	1														1
$T_2 \cdot$ LDADIR				1									1						
$T_2 \cdot$ ADDIR				1		1													
$T_3 \cdot$ ADDIR							1	0	1	0	0	1	1						
$T_2 \cdot$ SUBDIR				1		1													
$T_3 \cdot$ SUBDIR							1	1	1	0	1	1	1						
$T_2 \cdot$ STADIR							0	0	1	0	0	1		1					
$T_3 \cdot$ STADIR															1				
$T_2 \cdot$ CLA							0	0	0	0	0	1	1						
$T_2 \cdot$ INP																1			
$T_3 \cdot$ INP													1				1		
$T_2 \cdot$ OUT							0	0	1	0	0	1						1	
$T_2 \cdot$ JMP	1		1																
$T_3 \cdot$ JMP		1		1															

$$E_a = T_2 \cdot (\text{OUT} + \text{STADIR}) + T_3 \cdot (\text{ADDIR} + \text{SUBDIR})$$
$$W_a = 0$$
$$C = T_3 \cdot \text{SUBDIR}$$
$$\text{E-Add} = T_2 \cdot (\text{STADIR} + \text{CLA} + \text{OUT}) + T_3 \cdot (\text{ADDIR} + \text{SUBDIR})$$
$$\text{Ld-A} = T_2 \cdot (\text{LDADIR} + \text{CLA}) + T_3 \cdot (\text{ADDIR} + \text{SUBDIR} + \text{INP})$$
$$\text{Ld-IN} = T_2 \cdot \text{INP}$$
$$\text{INO} = T_3 \cdot \text{INP}$$
$$\text{Ld-MW} = T_2 \cdot \text{STADIR}$$
$$\text{WR} = T_3 \cdot \text{STADIR}$$
$$\text{Ld-OUT} = T_2 \cdot \text{OUT}$$
$$\text{Pinc} = T_1 \cdot \text{ALL}$$

Note that the edge-triggered registers (PC, MA, IR, B, A, IN, OUT, MW) are loaded on the rising edge of signal ECLK, midway through subinterval T_i. Memory read (RD) and write (WR) are initiated at the start of subinterval T_i. The remaining signals are level signals asserted throughout subinterval T_i.

(2) Design the next-state generator. Use Figure 14.7 to determine the control unit next-state functions and construct a state diagram (Figure 14.8) for the CPU. The control functions are as follows:

$$T_1 = \mathrm{N}(T_0 \cdot \mathrm{ALL})$$
$$T_2 = \mathrm{N}(T_1 \cdot \mathrm{ALL})$$
$$T_3 = \mathrm{N}(T_2 \cdot [\mathrm{ADDIR} + \mathrm{SUBDIR} + \mathrm{INP} + \mathrm{JMP} + \mathrm{STADIR}])$$
$$T_0 = \mathrm{N}(T_2 \cdot [\mathrm{LDADIR} + \mathrm{CLA} + \mathrm{OUT}] + T_3)$$

FIGURE 14.8
State diagram for BC–8A computer

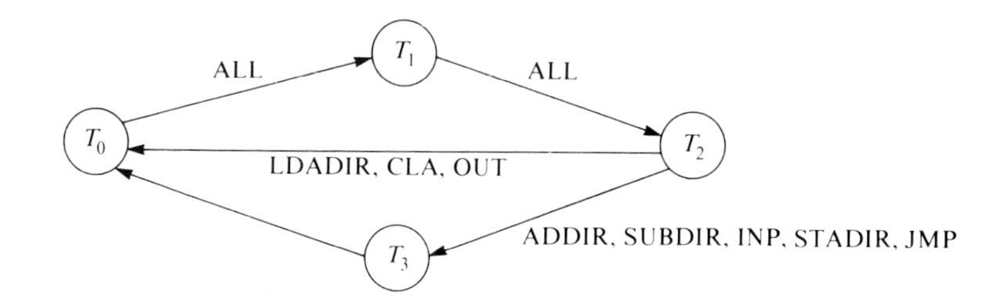

Step 4: Realization. Implement the control unit output functions and next-state functions to realize the control unit subsystem. The control unit, data path, and memory subsystems are then integrated to form a complete BC–8A computer, as shown in Figure 14.9.

A 74244 buffer with 3-state outputs was used in the BC-8X computers to interface the source devices (memory, program counter, and adder) with the unibus.

A logic diagram and timing diagram for the BC–8A computer are presented in Figure 14.10a and b (pp. 678–680), respectively.

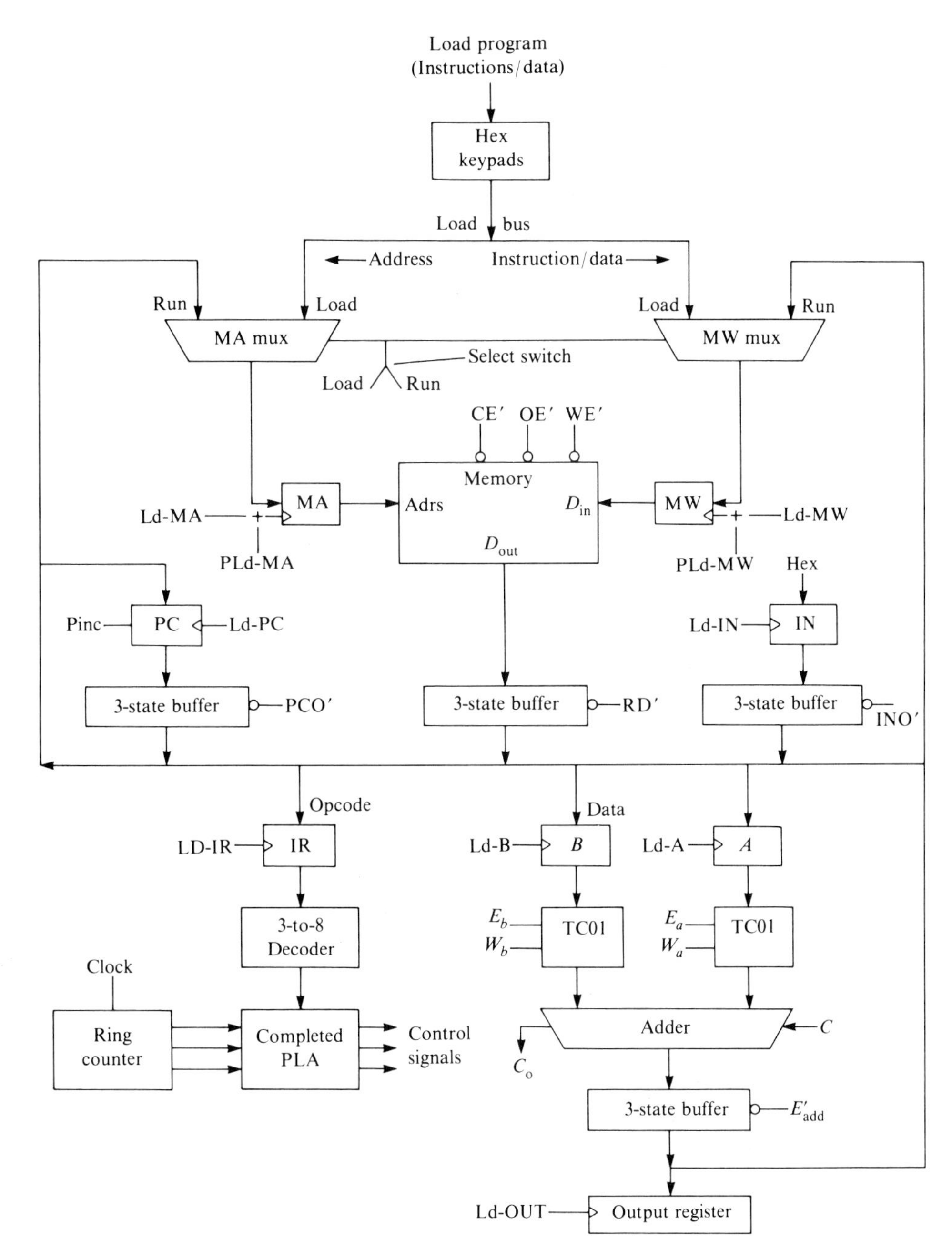

FIGURE 14.9
Block diagram of BC–8A bus-oriented computer

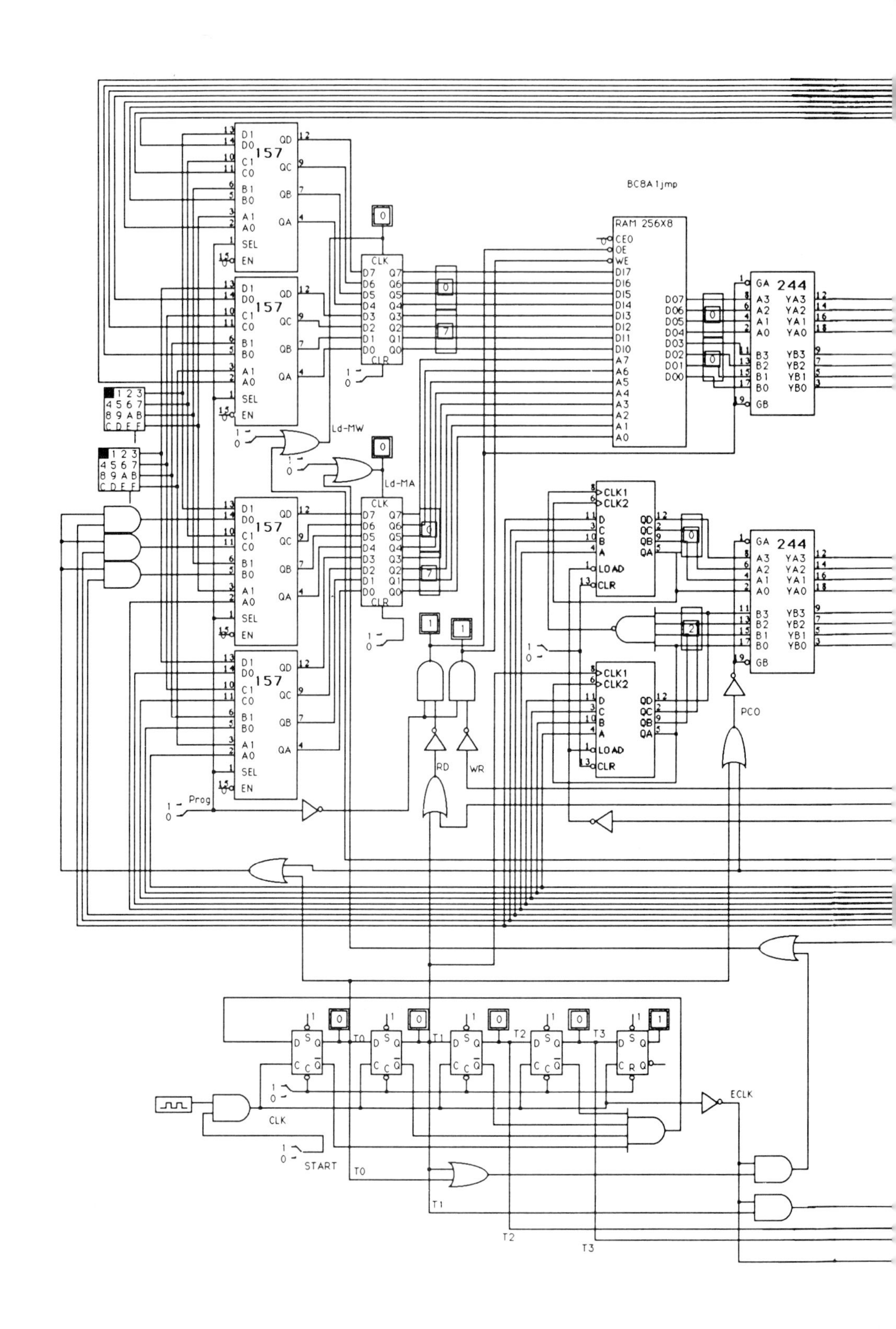

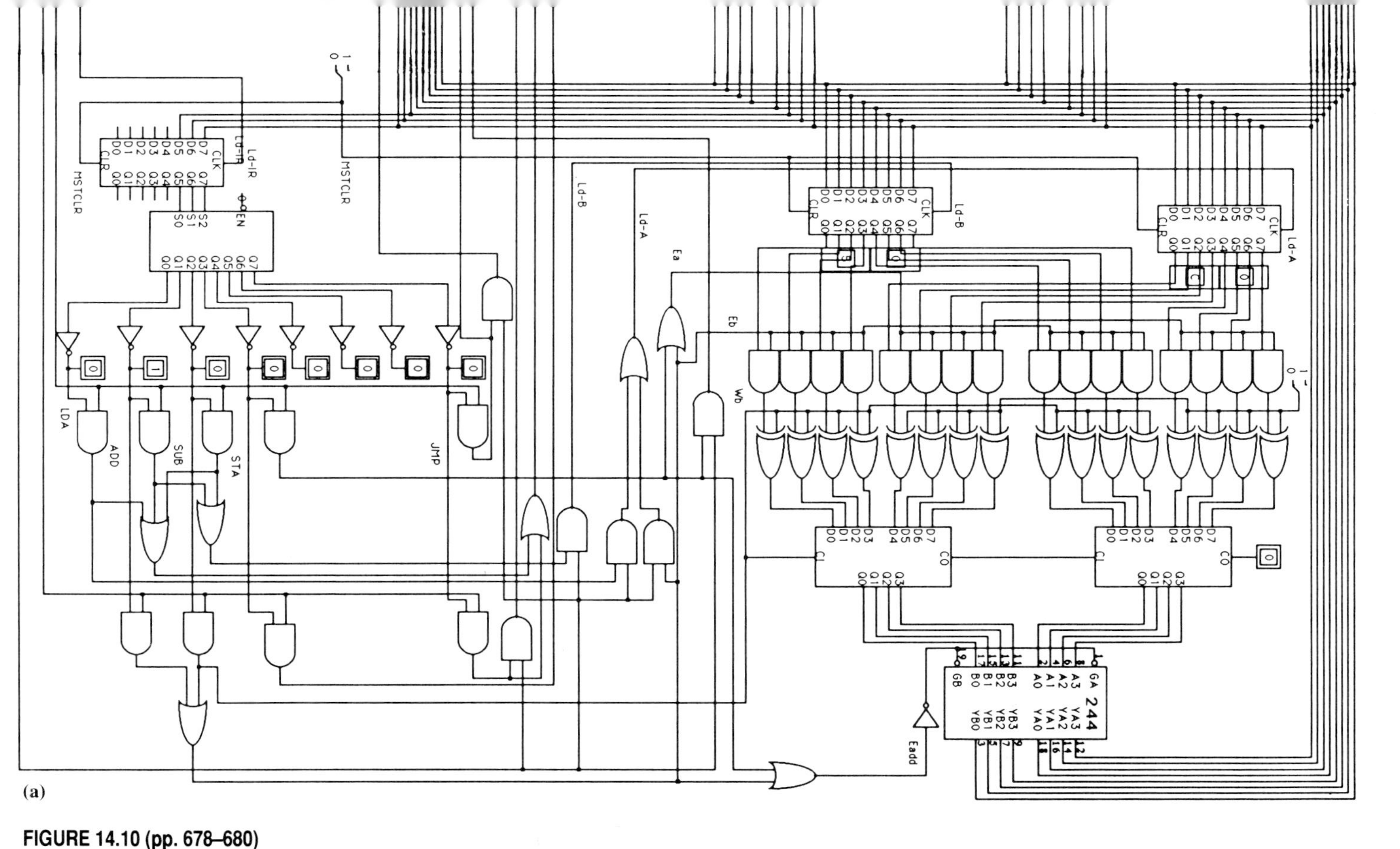

(a)

FIGURE 14.10 (pp. 678–680)
BC–8A bus-oriented computer. (a) Logic diagram (b) Timing diagram

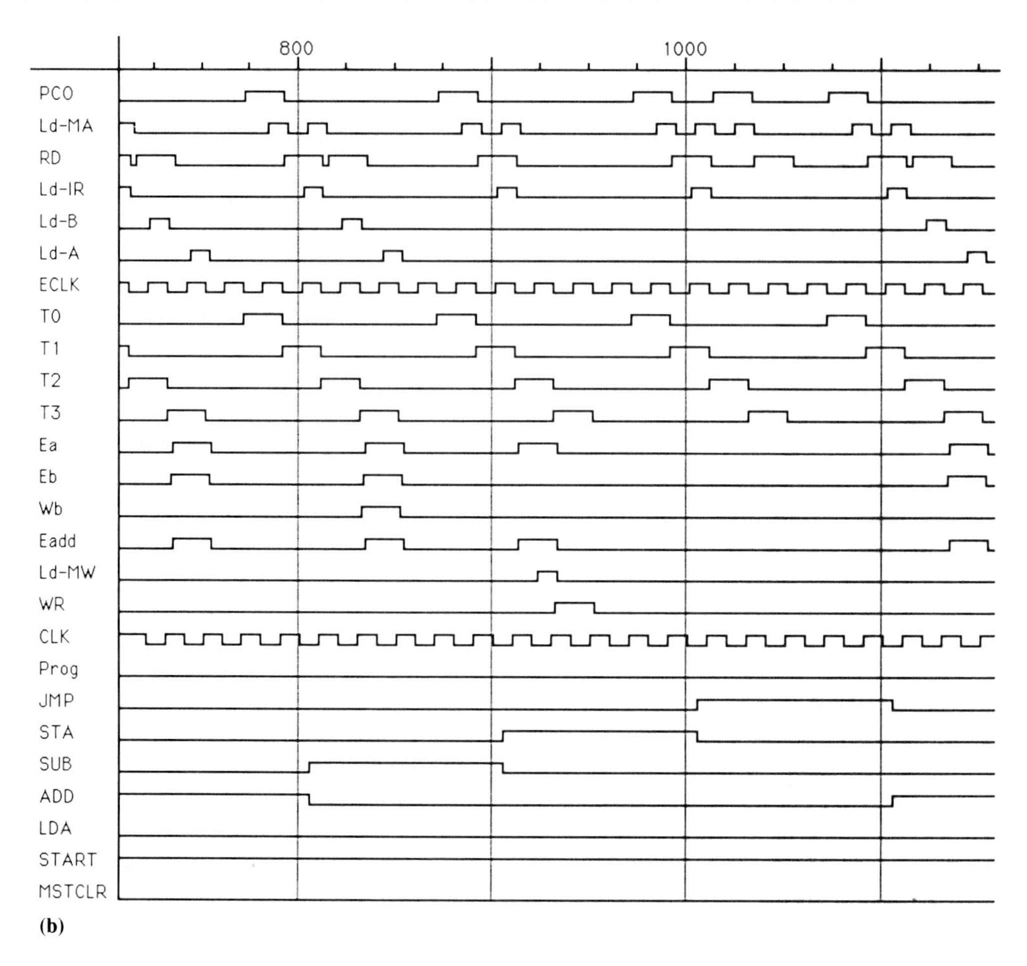

FIGURE 14.10
Continued

14.3 INDIRECT ADDRESS MODE, COMPARE, AND CONDITIONAL BRANCH INSTRUCTIONS

This section describes the design of a bus-oriented computer whose instruction set includes additional types of instructions in the basic categories, such as indirect-address mode memory-reference instructions, register reference instruction (complement A), compare instruction (using immediate address mode), and conditional branch instructions. A description of specific instructions within each of these categories was presented in Section 13.3.

INTERACTIVE DESIGN APPLICATION

Design a Bus-Oriented Computer (BC–8B) with Additional Capability

The design of an 8-bit bus-oriented computer (BC–8B) with additional capability is described in this section. The instruction set of the BC–8B computer is the same as that used in the MC–8B mux-oriented computer. This second computer is evolved from the BC–8A by adding instructions that provide the following additional capabilities:

- Indirect-address mode for memory-reference instructions
- Additional register reference instruction (complement *A*)
- Compare instruction (using immediate address mode)
- Conditional branch instructions

Step 1: Problem Statement. Given an instruction set that has instructions in each basic category (memory reference, register reference, I/O, and branch) and instructions that provide additional capability, design a bus-oriented computer that implements the instruction set in Table 14.4.

TABLE 14.4
BC–8B computer instruction repertoire

Mnemonic	Binary	Description
LDADIR	000xxxxx	Load *A* with content of memory address *X*
ADDIR	001xxxxx	Add content of memory address *X* to *A*
SUBDIR	010xxxxx	Subtract content of memory address *X* from *A*
STADIR	011xxxxx	Store accumulator in memory address *X*
ADDIND	100xxxxx	Add content of indirect-address memory location to *A*
SUBIND	101xxxxx	Subtract content of indirect-address memory location from *A*
CLA	11000	Clear *A*
COMP	11001	Complement *A*
INP	11010	Load *A* from an input device
OUT	11011	Transfer *A* to an output device
CMPIMM	11100	Compare immediate operand with *A* (2-byte instruction)
JMP	11101	Branch unconditionally to target address
BRA0	11110	Branch to target address on condition *A* = zero
BREQ	11111	Branch to target address on condition code equal

Branch instructions: [2 bytes; byte-2 = target address]

Step 2: Conceptualization. Develop a pictorial description of the system.

a. System structure diagram. Draw a block diagram that partitions the system into a memory subsystem, a data path subsystem, and a control unit subsystem.

In a systems context, a computer processor is a digital system composed of a memory subsystem and a central processing unit (CPU), where the CPU consists of a control unit subsystem and a data path subsystem. A computer processor can be represented by a PASM model (see Figure 14.3).

b. Control flow diagram. Draw a diagrammatic representation of the operation of the process control algorithm which accomplishes the fetch, decode, and execute for each program instruction. (See Figure 14.4 for an example.)

Step 3: Solution/Simplification.

a. Design a memory subsystem. In order to concentrate on the design of a CPU that can execute the specified set of instructions, we will use the memory subsystem of the BC–8A computer as the memory subsystem of the BC–8B computer.

b. Design a data path subsystem. The data path of the simple bus-oriented computer BC–8A (Figure 14.9) is used as a starting point for development of the BC–8B data path. We will modify this data path as necessary to support all specified instructions in Table 14.4. The required modifications are as follows:

Modification 1. Since there are more than 8 instructions in the MC–8B repertoire, it is necessary to use more than 3 bits for some opcodes.

a. The memory reference instructions are defined by 3-bit opcodes; direct-address mode memory-reference instructions require a mask enable to clear the 3 MSBs of the MA address lines.

b. The nonmemory reference instructions are defined by 5-bit opcodes. In addition to the 3-to-8 decoder for the 3 MSBs, these 5-bit opcodes require a 2-to-4 decoder to decode the 2 LSBs of the opcode.

Modification 2. A detect-*A*zero circuit is used to determine if register $A = 0$. The output of this detect-*A*zero circuit will be an input to the control unit.

Modification 3. The compare immediate (CMPIMM) instruction compares a data word with the A register content and sets the E, G, and L flip-flops as follows:

$$\text{If } A = (A - \text{data}) \begin{cases} > 0, & E \leftarrow 0,\ G \leftarrow 1,\ L \leftarrow 0 \\ = 0, & E \leftarrow 1,\ G \leftarrow 0,\ L \leftarrow 0 \\ < 0, & E \leftarrow 0,\ G \leftarrow 0,\ L \leftarrow 1 \end{cases}$$

The E, G, and L flip-flop outputs will be inputs to the control unit. Note that flip-flops E, G, and L will be reset by all instructions (other than CMPIMM) after register A is loaded.

Modifications 1, 2, and 3 convert the BC–8A data path and control unit to a BC–8B data path and control unit (decoder only). A block diagram of the BC–8B is presented in Figure 14.11.

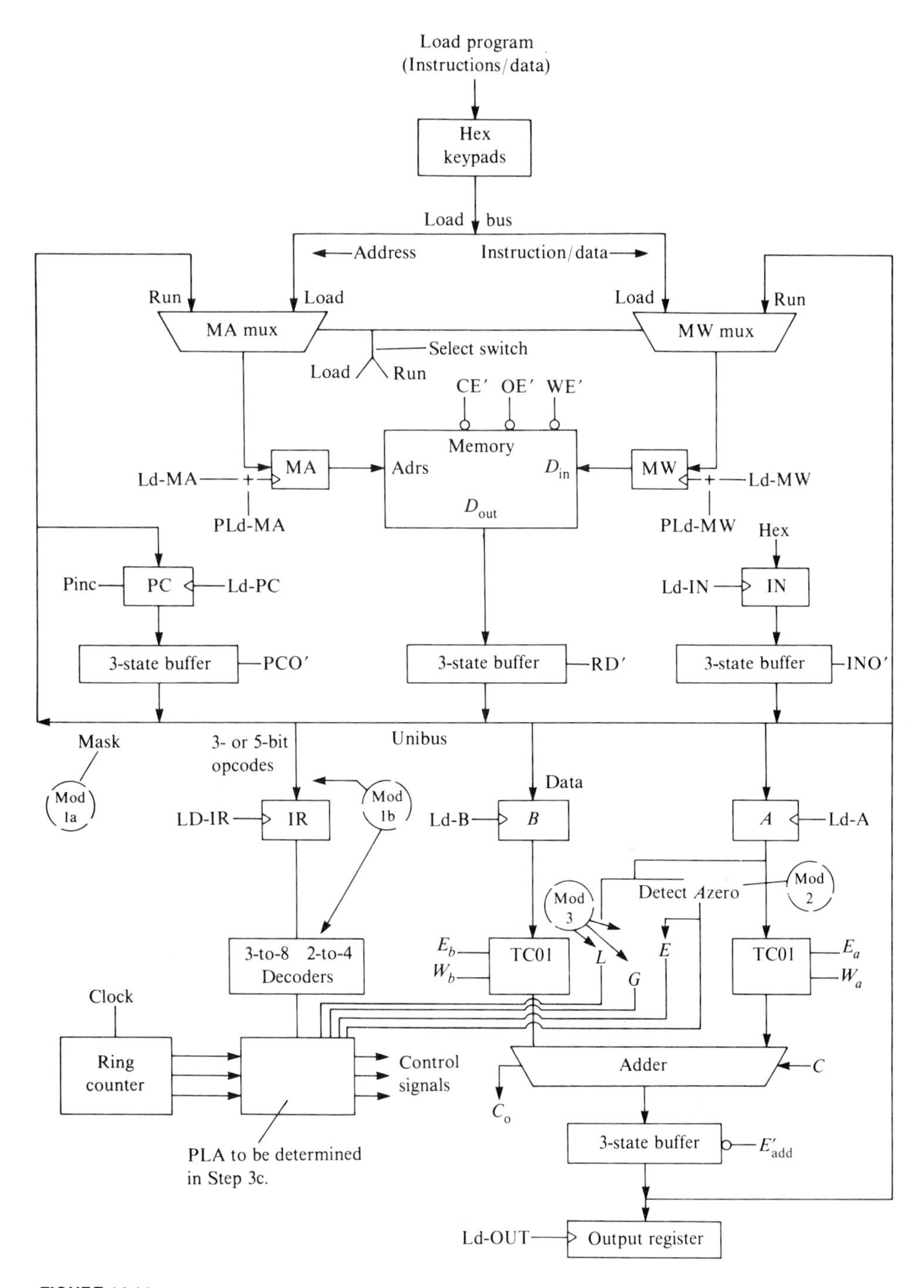

FIGURE 14.11
Block diagram of BC–8B bus-oriented computer with blank PLA

c. Design a control unit subsystem.

(1) Design the output decoder.

(a) Construct an instruction-time (p-term) event table. The execution of each instruction is accomplished by a sequence of microoperations at time subintervals within the overall instruction cycle. By tracing the data flow for each instruction, an instruction-time (p-term) event table can be constructed, as shown in Table 14.5.

TABLE 14.5
BC–8B instruction-time (p-term) event table

Instruction (Event / Time)		T_0	T_1	T_2	T_3	T_4
000	LDADIR	Fetch MA ← PC	Read inst. IR ← opc. MA ← daf PC ← PC + 1	Read data A ← M[MA]	—	—
001	ADDIR	↓	↓	Read data B ← M[MA]	$A \leftarrow A + B$	—
010	SUBDIR			Read data B ← M[MA]	$A \leftarrow A - B$	—
011	STADIR			MW ← A	Write mem.	—
100	ADDIND			Read iadrs* MA ← M[MA]	Read data B ← M[MA]	$A \leftarrow A + B$
101	SUBIND			Read iadrs* MA ← M[MA]	Read data B ← M[MA]	$A \leftarrow A - B$
11000	CLA			$A \leftarrow 0$	—	—
11001	COMP			$A \leftarrow A'$	—	—
10110	INP			Ld-IN	A ← IN	—
11011	OUT			OUT ← A	—	—
11100	CMPIMM data			MA ← PC* PC ← PC + 1	Read data B ← M[MA]	$A \leftarrow A - B$ $E \leftarrow 1$ if $A = 0$
11101	JMP tgtadrs			MA ← PC*	Read tgtadrs PC ← M[MA]	—
11110	BRA0 tgtadrs			MA ← PC* PC ← PC + 1	Read tgtadrs PC ← M[MA] if $A = 0$	—
11111	BREQ tgtadrs	↓	↓	MA ← PC* PC ← PC + 1	Read tgtadrs PC ← M[MA] if $E = 1$	—

Note: Precede BREQ instruction with CMPIMM instruction

*Enable bits 7–5 to MA register

(b) Construct a p-term/control point activation table. An examination of the instruction-time (p-term) event table to identify each control point leads to the construction of a p-term/control point activation table as shown in Table 14.6.

TABLE 14.6
BC–8B p-term/control point activation

	Control Point																		
P-Term	PCO	Ld PC	Ld MA	RD	Ld IR	Ld B	E_b	W_b	E_a	W_a	C	E Ad	Ld A	Ld MW	WR	Ld IN	INO	Ld OU	Pi nc
$T_0 \cdot$ ALL	1		1																
$T_1 \cdot$ ALL			1	1	1														1
$T_2 \cdot$ LDADIR				1									1						
$T_2 \cdot$ ADDIR				1		1													
$T_3 \cdot$ ADDIR							1	0	1	0	0	1	1						
$T_2 \cdot$ SUBDIR				1		1													
$T_3 \cdot$ SUBDIR							1	1	1	0	1	1	1						
$T_2 \cdot$ STADIR							0	0	1	0	0	1		1					
$T_3 \cdot$ STADIR															1				
$T_2 \cdot$ ADDIND			1	1															
$T_3 \cdot$ ADDIND				1		1													
$T_4 \cdot$ ADDIND							1	0	1	0	0	1	1						
$T_2 \cdot$ SUBIND			1	1															
$T_3 \cdot$ SUBIND				1		1													
$T_4 \cdot$ SUBIND							1	1	1	0	1	1	1						
$T_2 \cdot$ CLA							0	0	0	0	0	1	1						
$T_2 \cdot$ COMP							0	0	1	1	0	1	1						
$T_2 \cdot$ INP																1			
$T_3 \cdot$ INP													1				1		
$T_2 \cdot$ OUT							0	0	1	0	0	1						1	
$T_2 \cdot$ CMPIMM	1		1																1
$T_3 \cdot$ CMPIMM				1		1													
$T_4 \cdot$ CMPIMM							1	1	1	0	1	1	1*						
$T_2 \cdot$ JMP	1		1																
$T_3 \cdot$ JMP		1		1															
$T_2 \cdot$ BRA0	1		1																1
$T_3 \cdot$ BRA0		A0		1															
$T_2 \cdot$ BREQ	1		1																1
$T_3 \cdot$ BREQ		E		1															

* Set E, G, L.

(c) Determine the output decoder control equations. Determine the Boolean SOP expressions for the control unit output functions (to activate control points) by ORing the p-terms in each control point column in Table 14.6. The SOP expressions are as follows:

PCO = T_0 · ALL + T_2 · (CMPIMM + JMP + BRA0 + BREQ)
Ld-PC = T_3 · (JMP + BRA0 · A0 + BREQ · E)
Ld-MA = T_0 · ALL + T_1 · ALL + T_2 · (ADDIND + SUBIND + CMPIMM + JMP + BRA0 + BREQ)
RD = T_1 · ALL + T_2 · (LDADIR + ADDIR + SUBDIR + ADDIND + SUBIND) + T_3 · (ADDIND + SUBIND + CMPIMM + JMP + BRA0 + BREQ)
Ld-IR = T_1 · ALL
Ld-B = T_2 · (ADDIR + SUBDIR) + T_3 · (ADDIND + SUBIND + CMPIMM)
E_b = T_3 · (ADDIR + SUBDIR) + T_4 · (ADDIND + SUBIND + CMPIMM)
W_b = T_3 · SUBDIR + T_4 · (SUBIND + CMPIMM)
E_a = T_2 · (STADIR + COMP + OUT) + T_3 · (ADDIR + SUBDIR) + T_4 · (ADDIND + SUBIND + CMPIMM)
W_a = T_2 · COMP
C = T_3 · SUBDIR + T_4 · (SUBIND + CMPIMM)
E-Add = T_2 · (STADIR + CLA + COMP + OUT) + T_3 · (ADDIR + SUBDIR) + T_4 · (ADDIND + SUBIND + CMPIMM)
Ld-A = T_2 · (LDADIR + CLA + COMP) + T_3 · (ADDIR + SUBDIR + INP) + T_4 · (ADDIND + SUBIND + CMPIMM)
Ld-IN = T_2 · INP
INO = T_3 · INP
Ld-MW = T_2 · STADIR
WR = T_3 · STADIR
Ld-OUT = T_2 · OUT
E_b765 = T_2 · (ADDIND + SUBIND + JMP + BRA0 + BREQ + CMPIMM)
Pinc = T_1 · ALL + T_2 · (CMPIMM + BRA0 + BREQ)

(2) Design the next-state generator. Determine the control unit next-state functions and construct a state diagram (Figure 14.12) using the event table (Table 14.5) as a guide. The control functions are as follows:

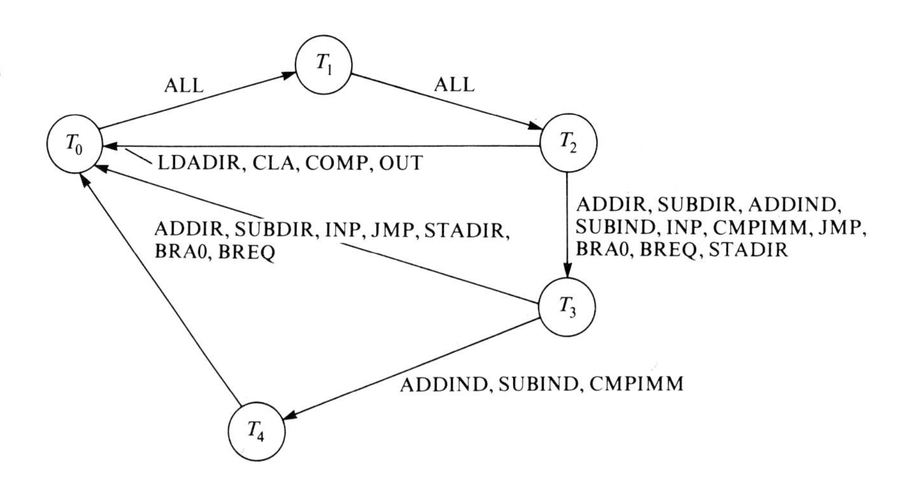

FIGURE 14.12
State diagram for BC–8B computer

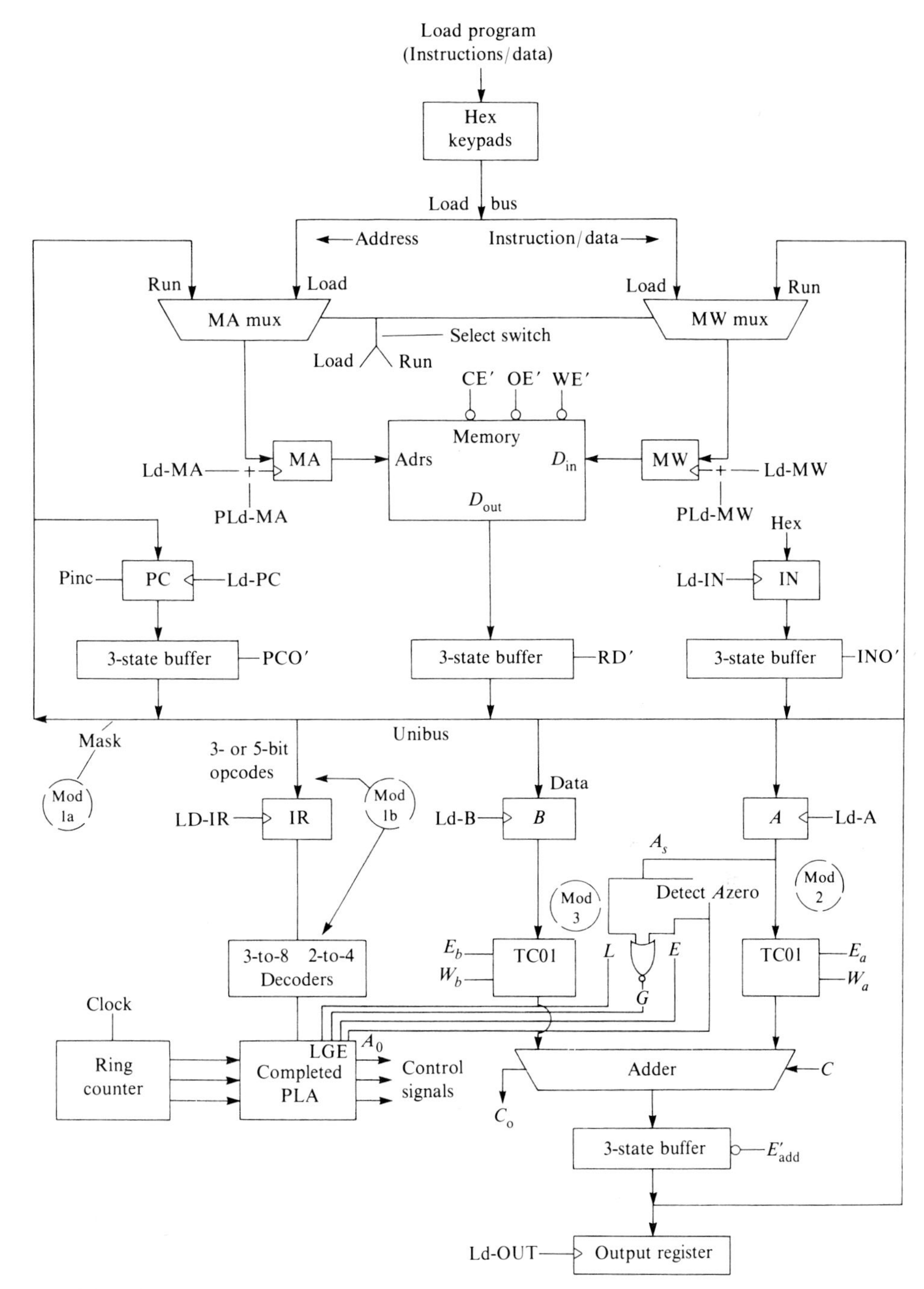

FIGURE 14.13
Block diagram of complete BC–8B bus-oriented computer

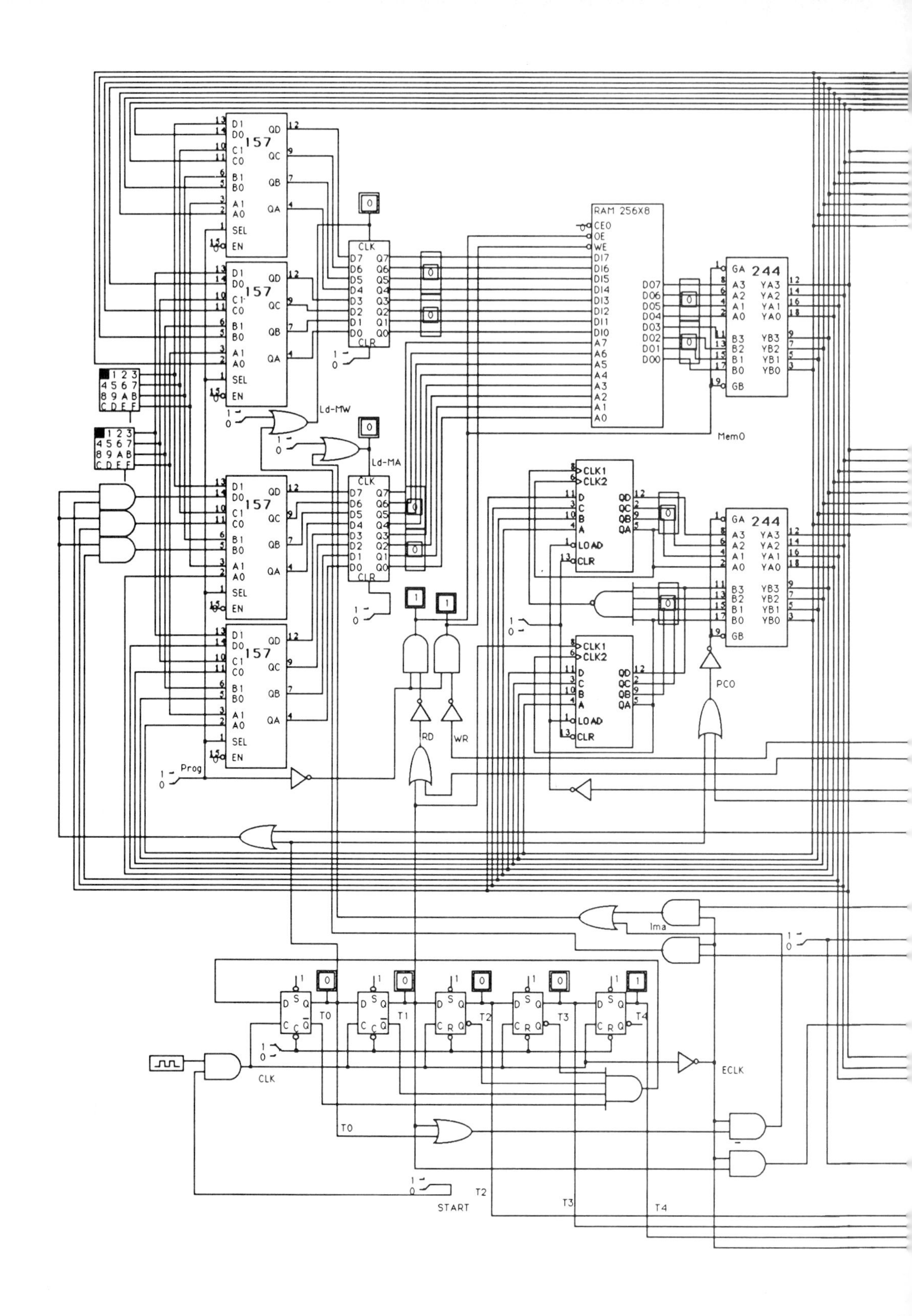

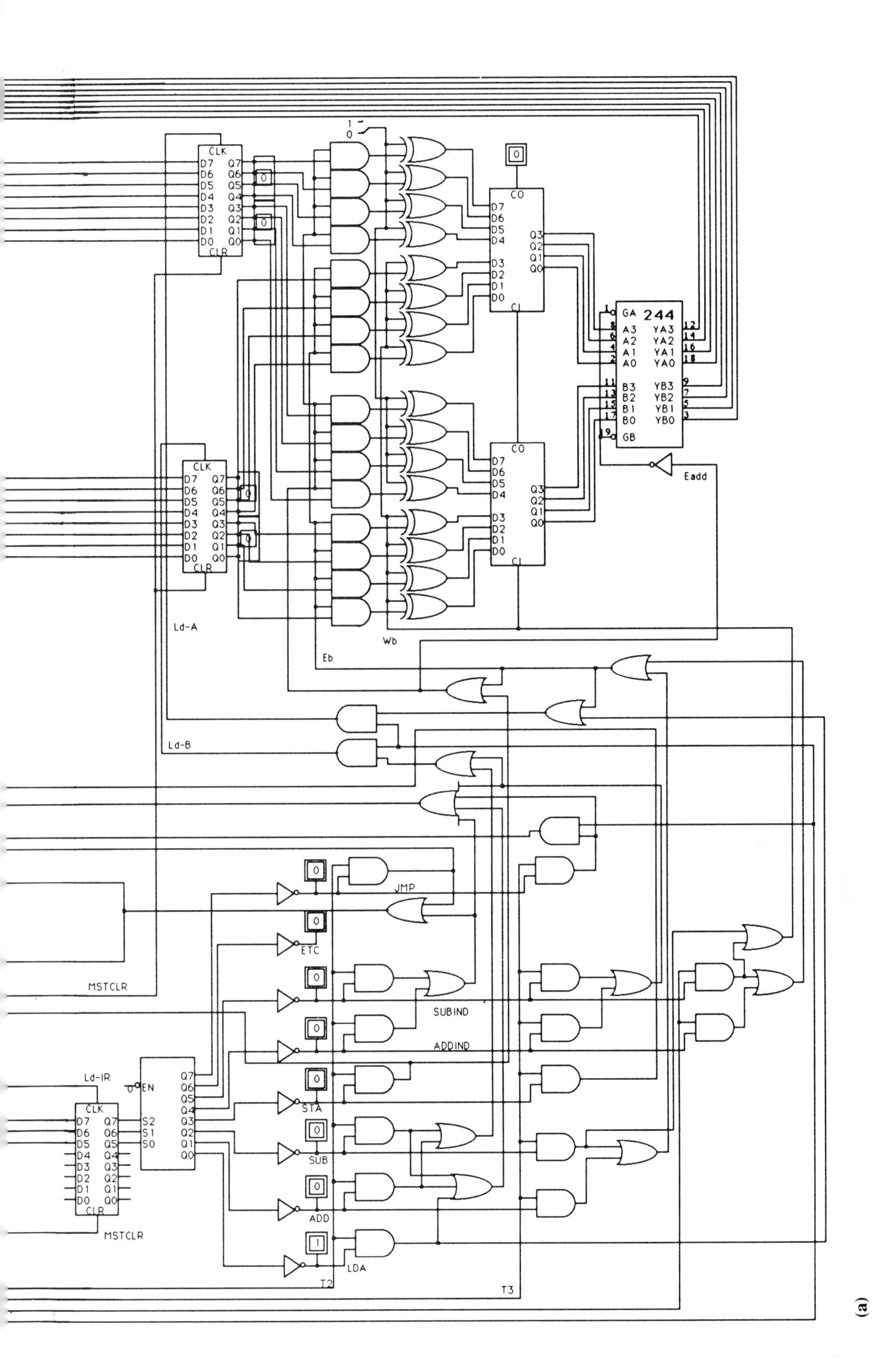

(a)

FIGURE 14.14 (pp. 688–690)
BC–8B bus-oriented computer. (a) Logic diagram (b) Timing diagram

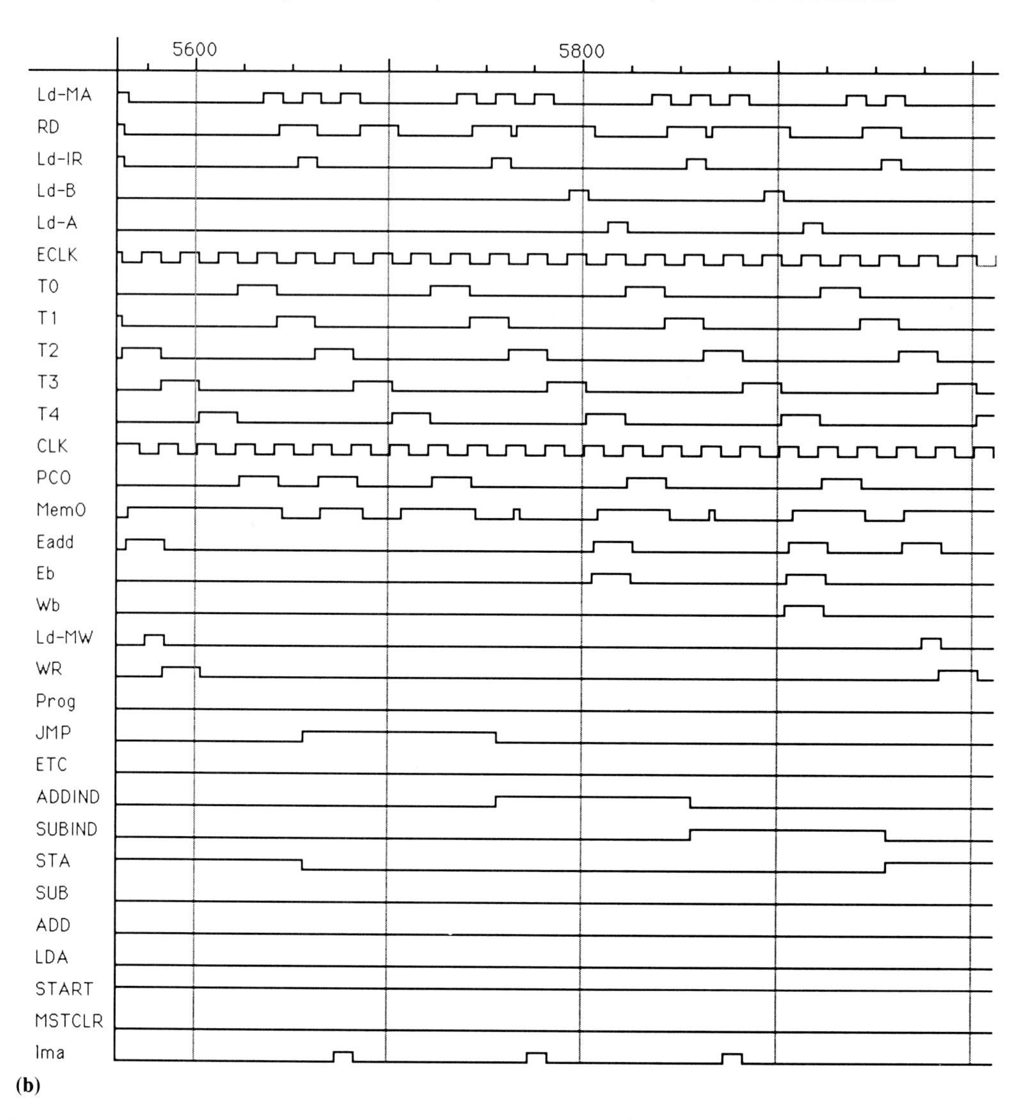

(b)

FIGURE 14.14
Continued

$$T_1 = \mathrm{N}(T_0 \cdot \mathrm{ALL})$$

$$T_2 = \mathrm{N}(T_1 \cdot \mathrm{ALL})$$

$$T_3 = \mathrm{N}(T_2 \cdot [\mathrm{ADDIR} + \mathrm{SUBDIR} + \mathrm{ADDIND} + \mathrm{SUBIND} + \mathrm{INP} + \mathrm{CMPIMM} + \mathrm{JMP} + \mathrm{BRA0} + \mathrm{BREQ} + \mathrm{STADIR}])$$

$$T_4 = \mathrm{N}(T_3 \cdot [\mathrm{ADDIND} + \mathrm{SUBIND} + \mathrm{CMPIMM}])$$

$$T_0 = \mathrm{N}(T_2 \cdot [\mathrm{LDADIR} + \mathrm{CLA} + \mathrm{COMP} + \mathrm{OUT}] + T_3 \cdot [\mathrm{ADDIR} + \mathrm{SUBDIR} + \mathrm{INP} + \mathrm{JMP} + \mathrm{BRA0} + \mathrm{BREQ} + \mathrm{STADIR}] + T_4)$$

Step 4: Realization. The control unit subsystem is realized by implementing the output decoder and the next-state generator. The control unit, data path, and memory subsystems are then integrated to form a complete BC–8B computer, as shown in Figure 14.13 (p. 687).

A logic diagram and timing diagram for the BC–8B computer are presented in Figure 14.14a and b (pp. 688–690), respectively.

14.4 SUBROUTINE JUMP, JUMP INDIRECT, MULTIPLY, DIVIDE, AND LOGIC INSTRUCTIONS

This section describes the design of a bus-oriented computer whose instruction set includes additional types of instructions in the basic categories. The additional instructions include subroutine jump, jump indirect instructions, multiply and divide instructions, logic (AND, OR) instructions, instructions to implement selection and repetition structures, and instructions with two implied operands. A description of the specific instructions within each of these categories was presented in Section 13.5.

INTERACTIVE DESIGN APPLICATION

Design of a Bus-Oriented Computer (BC–8C) with Expanded Capability

The design of an 8-bit bus-oriented computer (BC–8C) with expanded capability is described in this section. The instruction set of the BC–8C computer is the same as that used in the MC–8C mux-oriented computer. This third computer (BC–8C) is evolved from the BC–8B.

Step 1: Problem Statement. Given an instruction set that has instructions in each basic category (memory reference, register reference, I/O, and branch) and instructions that provide expanded capability (Table 14.7), design a bus-oriented computer that implements the instruction set.

TABLE 14.7
BC–8C computer instruction repertoire

Mnemonic	Binary	Description
LDBDIR	000	Load *B* from specified memory location
STADIR	001	Store *A* in specified memory location
ADDAB	01000	Add *B* to *A*; to store result in *A*
SUBAB	01001	Subtract *B* from *A*; store result in *A*
ANDAB	01010	AND *B* and *A*; store result in *A*
ORAB	01011	OR *B* and *A*; store result in *A*

continues

TABLE 14.7
Continued

Mnemonic	Binary	Description
MULBP	01100	Multiply *B* by *P* register; store product in *AP*
DIVBP	01101	Divide *P* register by *B*; store quotient in *P*; store remainder in *A*
MOVA2P	01110	Move *A* to *P* register
MOVP2A	01111	Move *P* register to *A*
CLA	10000	Clear *A*
COMP	10001	Complement *A*
INP	10010	Input
OUT	10011	Output
BRLT tgtadrs	10100	Branch on condition code *L* (less than)
BREQ tgtadrs	10101	Branch on condition code *E* (equal)
BRGT tgtadrs	10110	Branch on condition code *G* (greater than)
BRA0 tgtadrs	10111	Branch on *A*zero
SRJ tgtadrs	11000	Subroutine jump
JMPIND iadrs	11001	Jump using indirect-address mode
CMPIMM operand	11010	Compare immediate operand with *A*; set *E*, *G*, *L* flip-flops
CMPAB	11011	Compare *B* and *A*; set *E*, *G*, *L* flip-flops
LDNDX	11100	Load index register with initial value
LDBNDXBAS	11101	Load *B* from memory using indexed address mode
DECNDX	11110	Decrement index register
BRNDXPOS	11111	Branch on positive index register value

Branch instructions: [2 bytes; byte-2 = target address]

Step 2: Conceptualization. Develop a pictorial description of the system.

a. System structure diagram. Draw a block diagram (see Figure 14.3) that partitions the system into a memory subsystem, a data path subsystem, and a control unit subsystem.

b. Control flow diagram. Draw a diagram of the operation of the process control algorithm that accomplishes the fetch, decode, and execute for each program instruction.

Step 3: Solution/Simplification.

a. Design a memory subsystem. In order to concentrate on the design of a CPU that can execute a specified set of instructions, we will use the memory subsystem of the BC–8B computer as the memory subsystem of the BC–8C computer.

b. Design a data path subsystem. The data path of the BC–8B computer (Figure 14.13) will be used as a starting (trial) data path of the BC–8C computer. We will modify this data path to support all instructions specified in Table 14.7. Note that modifications 1, 2, and 3 (incorporated in BC–8B) are implied but are not explicitly shown in the block diagram of the BC–8C computer (Figure 14.15). The required modifications are as follows:

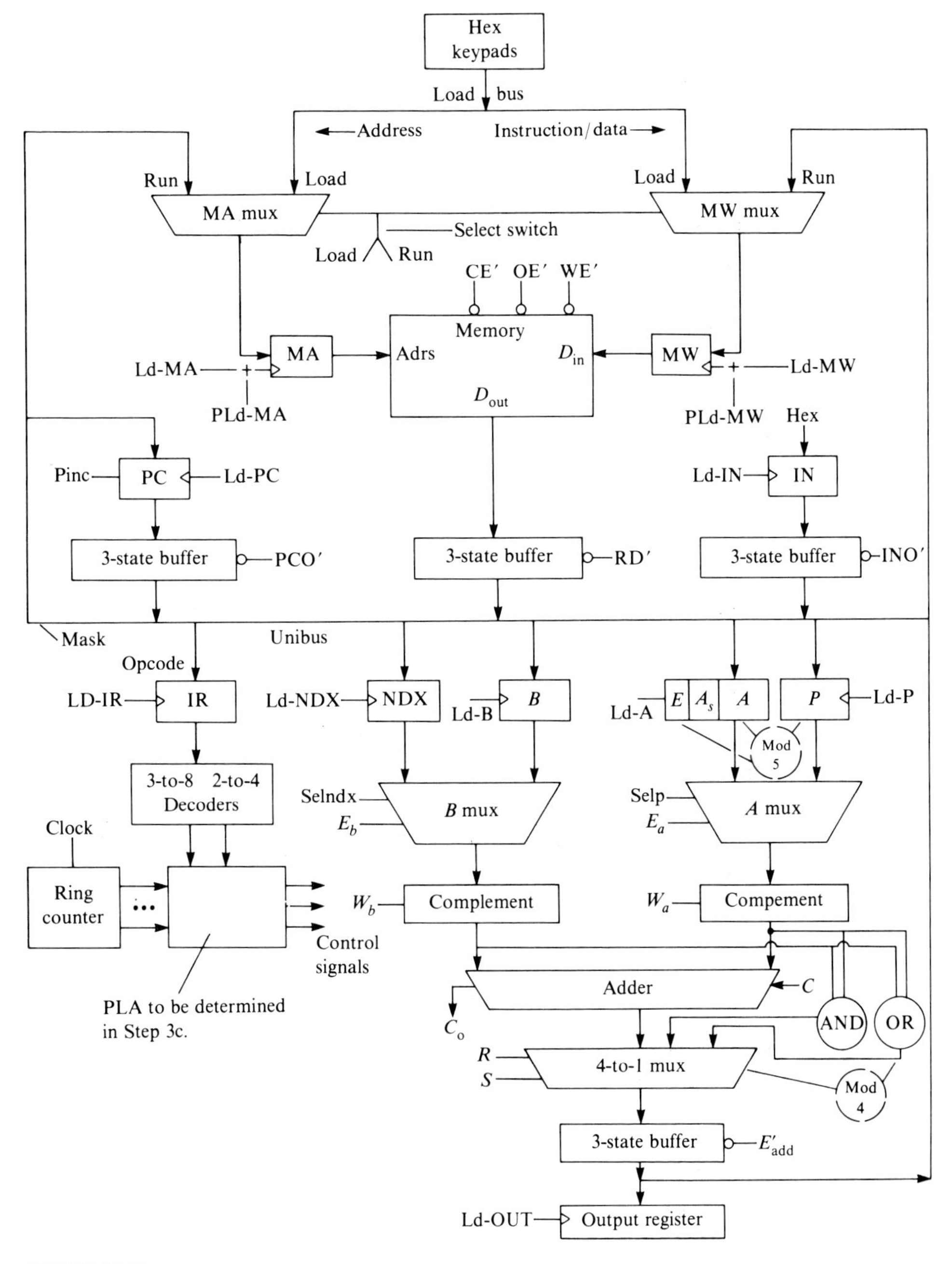

FIGURE 14.15
Block diagram of BC–8C computer with blank PLA

TABLE 14.8
BC–8C instruction-time (p-term) event table

Instruction (Event \ Time)		T_0	T_1	T_2	T_3	T_4	T_5
000	LDBDIR	Fetch MA ← PC	Read inst. IR ← opc. MA ← daf PC ← PC + 1	Read data $B \leftarrow$ M[MA]	—	—	—
011	STADIR			MW ← A	Write mem.	—	—
01000	ADDAB			$A \leftarrow A + B$	—	—	—
01	SUBAB			$A \leftarrow A - B$	—	—	—
10	ANDAB			$A \leftarrow A \wedge B$	—	—	—
11	ORAB			$A \leftarrow A \vee B$	—	—	—
01100	MULBP	*Note:* A_s, P_s not in *EAP* for shift		(t1)* Compute $A + B$ $E \leftarrow C_o$ If $P_0 = 1$, Ld-A	(t2)* R.Shf *EAP*	Repeat t1, t2 steps 7 times	
01	DIVBP	P has no sign. A_s used in compute and in left shift *AP*		D FF ← A_s	W_b, $C_i \leftarrow D$ L.Shf *AP* $P_0 \leftarrow Z_s'$ Ld-A	Repeat t1, t2 steps 7 times Add/subt.	
10	MOVA2P			$P \leftarrow A$	—	—	—
11	MOVP2A			$A \leftarrow P$	—	—	—
10000	CLA			$A \leftarrow 0$	—	—	—
01	COMP			$A \leftarrow A'$	—	—	—
10	INP			Ld-IN	$A \leftarrow$ IN	—	—
11	OUT			OUT ← A	—	—	—
10100	BRLT tgtadrs			MA ← PC**	Read target $B \leftarrow$ M[MA] PC ← PC + 1	If $L = 1$ PC ← B	—

Code	Instruction						
01	BREQ tgtadrs			MA ← PC**	Read target B ← M[MA] PC ← PC + 1	If $E = 1$ PC ← B	—
10	BRGT tgtadrs			MA ← PC**	Read tgtadrs B ← M[MA] PC ← PC + 1	If $G = 1$ PC ← B	—
11	BRA0 tgtadrs			MA ← PC**	Read tgtadrs B ← M[MA] PC ← PC + 1	If $A = 0$ PC ← B	—
11000	SRJ tgtadrs			MA ← PC**	Read tgtadrs B, MA ← M[MA]**	MW ← PC + 1	Wr mem. PC ← $B + 1$
01	JMPIND tiadrs			MA ← PC**	Read tiadrs MA ← M[MA]**	Read ret. B ← M[MA]	PC ← B
10	CMPIMM data			MA ← PC**	Read data B ← M[MA] PC ← PC + 1	$A \leftarrow A - B$ Set E, G, L	—
11	CMPAB			$A \leftarrow A - B$ Set E, G, L	—	—	—
11100	LDNDX ndxval			MA ← PC**	Read ndxval B ← M[MA] PC ← PC + 1	NDX ← B	—
01	LDBNDXBAS base			MA ← PC**	Read base P ← M[MA] PC ← PC + 1	MA ← P + NDX**	Read B ← M[MA]
10	DECNDX			NEG NDX	NDX ← NDX + 1	NEG NDX	—
11	BRNDXPOS tgtadrs			MA ← PC** PC ← PC + 1	Read tgtadrs B ← M[MA] PC ← PC + 1	If NDX > 0 PC ← B	—

*For MULBP, DIVBP cycles t_1 and t_2 are executed 8 times in T_2–T_5.

T_0	T_1	T_2	T_3	T_4	T_5
		t_1, t_2 t_1, t_2	t_1, t_2 t_1, t_2	t_1, t_2 t_1, t_2	t_1, t_2 t_1, t_2

**Enable bits 7–5 to MA register

Modification 4. This modification consists of a set of AND and OR gates between the complementer outputs and the inputs of a 4-to-1 mux that is inserted between the adder and the 3-state buffer. The mux controls R and S select either add, ANDAB, or ORAB.

Modification 5. This modification consists of the addition of an index register (NDX) and a P shift register and implementation of A using a shift register. Together A and P form a 16-bit shift register. A sign bit, A_s, is appended to the left of the 8-bit shift register A. An extension bit E, for storing the adder carry-out C_o, is implemented using a D flip-flop.

A block diagram of the BC–8C is shown in Figure 14.15.

c. Design a control unit subsystem.

(1) Design the output decoder.

(a) Construct an instruction-time (p-term) event table. The execution of each instruction is accomplished by a sequence of microoperations at time subintervals within the overall instruction cycle. By tracing the data flow for each instruction, we can construct an instruction-time event table as shown in Table 14.8 (pp. 694–695).

(b) Construct a p-term/control point activation table. An examination of the instruction-time event table to identify each control point leads to the construction of a p-term/control point activation table as shown in Table 14.9.

TABLE 14.9
BC–8C p-term/control point activation table

	Control Point																				
P-Term	PCO	Ld PC	Ld MA	RD	Ld IR	Ld B	E_b	W_b	E_a	W_a	C	E Ad	Ld A	Ld MW	WR	Ld IN	IN O	Ld OU	Pi nc	Ld X	Ld P
$T_0 \cdot$ ALL	1		1																		
$T_1 \cdot$ ALL			1	1	1														1		
$T_2 \cdot$ LDBDIR				1		1															
$T_2 \cdot$ STADIR							0	0	1	0	0	1		1							
$T_3 \cdot$ STADIR															1						
$T_2 \cdot$ ADDAB							1	0	1	0	0	1	1								
$T_2 \cdot$ SUBAB							1	1	1	0	1	1	1								
$T_2 \cdot$ ANDAB							1	0	1	0	0	1	1								
$T_2 \cdot$ ORAB							1	0	1	0	0	1	1								
$T_2 \cdot$ MULBP							Perform 8 t_1, t_2 cycles														
$T_3 \cdot$ MULBP																					
$T_4 \cdot$ MULBP																					
$T_5 \cdot$ MULBP																					
$T_2 \cdot$ DIVBP							Perform 8 t_1, t_2 cycles														
$T_3 \cdot$ DIVBP																					
$T_4 \cdot$ DIVBP																					
$T_5 \cdot$ DIVBP																					
$T_2 \cdot$ MOVA2P							0	0	1	0	0	1									1
$T_2 \cdot$ MOVP2A							0	0	1	0	0*	1	1								
$T_2 \cdot$ CLA							0	0	0	0	0	1	1								
$T_2 \cdot$ COMP							0	0	1	1	0	1	1								

TABLE 14.9
Continued

P-Term	Control Point																				
	PCO	Ld PC	Ld MA	RD	Ld IR	Ld B	E_b	W_b	E_a	W_a	C	E Ad	Ld A	Ld MW	WR	Ld IN	IN O	Ld OU	Pi nc	Ld X	Ld P
$T_2 \cdot$ INP																1					
$T_3 \cdot$ INP													1				1				
$T_2 \cdot$ OUT							0	0	1	0	0	1						1			
$T_2 \cdot$ BRLT	1		1																		
$T_3 \cdot$ BRLT				1		1													1		
$T_4 \cdot$ BRLT		L					1	0	0	0	0	1									
$T_2 \cdot$ BREQ	1		1																		
$T_3 \cdot$ BREQ				1		1													1		
$T_4 \cdot$ BREQ		E					1	0	0	0	0	1									
$T_2 \cdot$ BRGT	1		1																		
$T_3 \cdot$ BRGT				1		1													1		
$T_4 \cdot$ BRGT		G					1	0	0	0	0	1									
$T_2 \cdot$ BRA0	1		1																		
$T_3 \cdot$ BRA0				1		1													1		
$T_4 \cdot$ BRA0		$A0$					1	0	0	0	0	1									
$T_2 \cdot$ SRJ	1		1																		
$T_3 \cdot$ SRJ			1	1		1													1		
$T_4 \cdot$ SRJ	1													1							
$T_5 \cdot$ SRJ		1					1	0	0	0	1*	1			1						
$T_2 \cdot$ JMPIND	1		1																		
$T_3 \cdot$ JMPIND			1	1																	
$T_4 \cdot$ JMPIND				1		1															
$T_5 \cdot$ JMPIND		1					1	0	0	0	0	1									
$T_2 \cdot$ CMPIMM	1		1																		
$T_3 \cdot$ CMPIMM				1		1													1		
$T_4 \cdot$ CMPIMM							1	1	1	0	1	1	1 set E,G,L								
$T_2 \cdot$ CMPAB							1	1	1	0	1	1	1 set E,G,L								
$T_2 \cdot$ LDNDX	1		1																		
$T_3 \cdot$ LDNDX				1		1													1		
$T_4 \cdot$ LDNDX							1	0	0	0	0	1								1	
$T_2 \cdot$ LDBXBS	1		1																		
$T_3 \cdot$ LDBXBS				1															1		1
$T_4 \cdot$ LDBXBS			1				1	0	1	0	0	1*†									
$T_5 \cdot$ LDBXBS				1		1															
$T_2 \cdot$ DECNDX							1	1	0	0	1†	1								1	
$T_3 \cdot$ DECNDX							1	0	0	0	1†	1								1	
$T_4 \cdot$ DECNDX							1	1	0	0	1†	1								1	
$T_2 \cdot$ BRNDXPOS	1		1																1		
$T_3 \cdot$ BRNDXPOS				1		1													1		
$T_4 \cdot$ BRNDXPOS		X+					1	0	0	0	0	1									

*Selp †Selndx

(c) Determine the output decoder control equations. Determine the Boolean SOP expressions for the output decoder outputs (to activate control points) by ORing the p-terms in each control point column in Table 14.9. The SOP expressions are as follows:

PCO = T_0 · ALL + T_2 · (CMPIMM + JMPIND + SRJ + BRLT + BREQ + BRGT + BRA0) + T_2 · (LDNDX + LDBNDXBAS + BRNDXPOS) + T_4 · SRJ

Ld-PC = T_4 · (BRLT · L + BREQ · E + BRGT · G + BRA0 · A0 + BRNDXPOS · NDX $\geq$ 0) + T_5 · (SRJ + JMPIND)

Ld-MA = T_0 · ALL + T_1 · ALL + T_2 · (CMPIMM + JMPIND + SRJ + BRLT + BREQ + BRGT + BRA0) + T_2 · (LDNDX + LDBNDXBAS + BRNDXPOS) + T_3 · (SRJ + JMPIND) + T_4 · LDBNDXBAS

RD = T_1 · ALL + T_2 · LDBDIR + T_3 · (BRLT + BREQ + BRGT + BRA0 + SRJ + JMPIND) + T_3 · (CMPIMM + LDNDX + LDBNDXBAS + BRNDXPOS) + T_4 · JMPIND + T_5 · LDBNDXBAS

Ld-IR = T_1 · ALL

Ld-B = T_2 · LDBDIR + T_3 · (BRLT + BREQ + BRGT + BRA0 + SRJ + CMPIMM) + T_3 · (LDNDX + BRNDXPOS) + T_5 · LDBNDXBAS

E_b = T_2(ADDAB + SUBAB + ANDAB + ORAB + CMPAB + DECNDX) + T_3 · DECNDX + T_4 · (BRLT + BREQ + BRGT + BRA0) + T_4 · (CMPIMM + LDNDX + LDBNDXBAS + DECNDX + BRNDXPOS) + T_5 · (SRJ + JMPIND)

W_b = T_2 · (SUBAB + DECNDX) + T_4 · (CMPIMM + DECNDX)

E_a = T_2 · (STADIR + COMP + OUT + MOVA2P + MOVP2A) + T_2 · (ADDAB + SUBAB + ANDAB + ORAB + CMPAB) + T_4 · (CMPIMM + LDBNDXBAS)

W_a = T_2 · COMP

C = T_2 · (SUBAB + CMPAB + DECNDX) + T_3 · DECNDX + T_4 · DECNDX + T_5 · SRJ

Eadd = T_2 · (ADDAB + SUBAB + ANDAB + ORAB + CMPAB + DECNDX) + T_2 · (STADIR + COMP + OUT + MOVA2P + MOVP2A + CLA) + T_3 · DECNDX + T_4 · (BRLT + BREQ + BRGT + BRA0) + T_4 · (CMPIMM + LDNDX + LDBNDXBAS + DECNDX + BRNDXPOS) + T_5 · (SRJ + JMPIND)

Ld-A = T_2 · (ADDAB + SUBAB + ANDAB + ORAB + CMPAB) + T_2 · (CLA + COMP + MOVP2A) + T_3 · INP + T_4 · CMPIMM

Ld-IN = T_2 · INP

INO = T_3 · INP

Ld-MW = T_2 · STADIR + T_4 · SRJ

WR = T_3 · STADIR + T_5 · SRJ

Ld-OUT = T_2 · OUT

Eb765 = $T_2 \cdot$ (BRLT + BREQ + BRGT + BRA0 + SRJ + JMPIND + CMPIMM)
+ $T_2 \cdot$ (LDNDX + LDBNDXBAS + BRNDXPOS) + $T_3 \cdot$ (SRJ + JMPIND)
+ $T_4 \cdot$ LDBNDXBAS

Pinc = $T_1 \cdot$ ALL
+ $T_3 \cdot$ (BRLT + BREQ + BRGT + BRA0 + SRJ + CMPIMM + BRNDXPOS)
+ $T_3 \cdot$ (LDNDX + LDBNDXBAS + BRNDXPOS)

Ld-NDX = $T_2 \cdot$ DECNDX + $T_3 \cdot$ DECNDX + $T_4 \cdot$ (DECNDX + LDNDX)

Ld-P = $T_2 \cdot$ MOVA2P + $T_3 \cdot$ LDBNDXBAS

Set EGL = $T_2 \cdot$ CMPAB + $T_4 \cdot$ CMPIMM

(2) Design the next-state generator. Determine the control unit next-state functions by drawing a state diagram (use event Table 14.8 as a guide). See Figure 14.16.

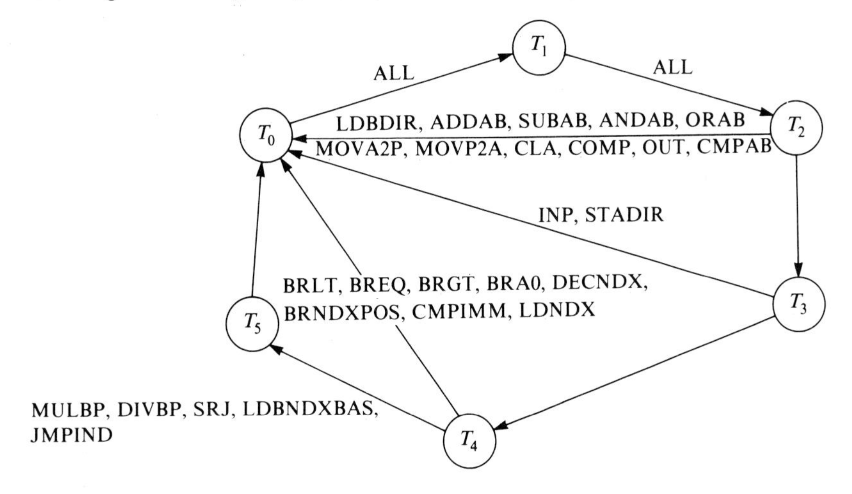

FIGURE 14.16
State diagram for BC–8C computer

Step 4: Realization. Implement the control unit output functions and next-state functions to realize the control unit subsystem. Then integrate the control unit, data path, and memory subsystems to form the BC–8C computer, as shown in Figure 14.17 (p. 700).

14.5 SUMMARY

The 4-step problem-solving procedure for system (computer) design, described in Chapter 13, is used in this chapter to design three bus-oriented computers (BC–8A, B, and C), each with increasingly complex instructions sets and each evolving from the one preceding it. Each computer has the same instruction set as the corresponding computer in Chapter 13.

The BC–8A computer uses a 3-bit opcode to define 8 instructions. This minimal instruction set illustrates basic categories of instructions required in most computers: memory reference (direct address), register reference, I/O, and branch instructions.

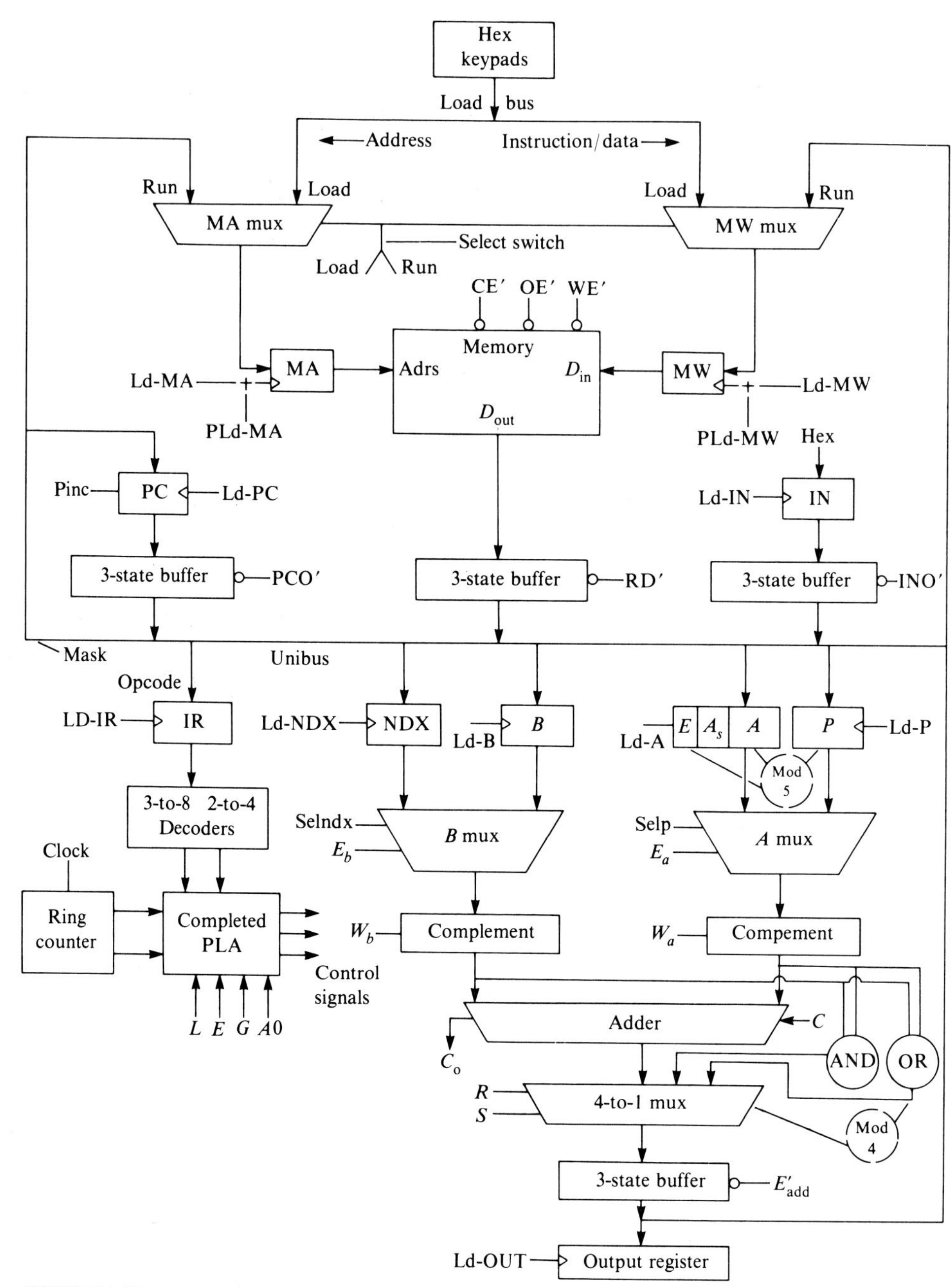

FIGURE 14.17
Block diagram of BC–8C computer

A second computer (BC–8B) uses a 5-bit opcode to provide additional types of instructions, including compare, branch (unconditional and conditional), register reference, and indirect-address mode memory-reference instructions.

A third computer (BC–8C) uses a 5-bit opcode and instructions with 2 implied operands to define an expanded instruction set that includes multiply, divide, subroutine jump, and logic (AND, OR) instructions.

The design of each computer is accompanied by a description of its assembly language instructions.

Each BC–8X computer uses a common memory subsystem nucleus (with a program-load facility) so that you can concentrate on the design of the CPU, which implements the specified instruction set. The interfacing between the memory subsystem and the CPU is accomplished using a single bus (unibus) and devices with 3-state buffers.

The design of each computer CPU begins with the development of a data path that supports each and every instruction in the computer's repertoire. A parallel ALU can be used as a starting point in the design of the data path of the first computer (BC–8A). This data path is then modified as required to support any and all of the specified instructions. The data path of each succeeding computer is evolved from the data path of the computer which preceded it.

The control unit next-state generator and output decoder can be designed using the Chapter 13 approach, with the additional requirement that no two sources feed the bus at any given time. The control unit of each computer can be implemented either in hardwired form or microprogrammed form.

If more than one device feeds the bus, each device feeding the bus must be interfaced by a 3-state buffer. We must also ensure that no bus contention takes place among devices feeding the bus. One way to avoid bus contention is by ensuring that no two devices feed the bus at the same time. The control unit output decoder must be designed so that only one source feeds the bus at any given time. An examination of the output decoder logic equations can be used to determine if bus contention exists. A timing diagram of the computer's operation can also be used to locate two or more sources that attempt to feed the bus at the same time.

Each BC–8X computer was implemented with a partial instruction set because of the PLA complexity for a full implementation. The exercises will provide additional skill in implementing other partial instruction sets.

KEY TERMS

Bus-oriented computers

Bus contention

EXERCISES

General

1. Write an assembly language program for the BC–8A computer shown in Figure 14.10a. The program is to add the integers 1, 2, 3, 4, and 5. Convert the program to machine language (hexadecimal).

2. Write an assembly language program for the BC–8B computer shown in Figure 14.14a. The program is to add the integers 1, 2, 3, 4, and 5. Assume that the data is stored in memory at addresses 32, 33, 34, 35, and 36. Convert the program to machine language (hexadecimal).
3. Verify the PLA of the BC–8A computer shown in Figure 14.10a.
4. Verify the PLA of the BC–8B computer shown in Figure 14.14a.
5. Draw a logic diagram of a BC–8A computer control unit PLA that implements the following instruction set: ADDIR (001), SUBDIR (010), STADIR (011), CLA (100).
6. Draw a timing diagram of the microoperations involved in each of the instructions in Exercise 5.
7. Draw a logic diagram of a BC–8A computer control unit PLA that implements the following instruction set: ADDIR (001), SUBDIR (010), STADIR (011), CLA (100), JMP (111).
8. Draw a timing diagram of the microoperations involved in each of the instructions in Exercise 7.
9. Draw a logic diagram of a BC–8B computer control unit PLA that implements the following instruction set: LDADIR (000), ADDIND (100), SUBIND (101), STADIR (011).
10. Draw a timing diagram of the microoperations involved in each of the instructions in Exercise 9.

Design/Implementation

1. Implement the BC–8A computer with the following instruction set: LDADIR (000), ADDIR (001), SUBDIR (010), STADIR (011), JMP (111). Load and run the following program:

Address (Decimal)	Instruction		Hexadecimal		Remarks
	Assembly	Binary	Address	Inst.	
0	LDADIR 6	000 00110	00	06	
1	ADDIR 7	001 00111	01	27	
2	SUBDIR 8	010 01000	02	48	
3	STADIR 9	011 01001	03	69	
4	JMP	111 00000	04	E0	
5	1	0000 0001	05	01	Tgtadrs
6	4	0000 0100	06	04	Direct address data
7	5	0000 0101	07	05	Direct address data
8	2	0000 0010	08	02	Direct address data
9	—	0000 0000	09	00	Result

2. Implement the BC–8B computer with the following instruction set: LDADIR (000), ADDIR (001), SUBDIR (010), STADIR (011), ADDIND (100), SUBIND (101), JMP (111). Load and run the following program:

Address (Decimal)	Instruction		Hexadecimal		Remarks
	Assembly	Binary	Address	Inst.	
0	LDADIR 6	000 00110	00	06	
1	ADDIR 7	001 00111	01	27	
2	SUBDIR 8	010 01000	02	48	
3	STADIR 9	011 01001	03	69	
4	JMP	111 00000	04	E0	
5	10	0000 1010	05	0A	Tgtadrs
6	4	0000 0100	06	04	Direct address data
7	5	0000 0101	07	05	
8	2	0000 0010	08	02	
9	—	0000 0000	09	00	—
10	ADDIND 14	100 01110	0A	8E	
11	SUBIND 15	101 01111	0B	AF	
12	STADIR 13	011 01101	0C	6D	
13	—	—			Result
14	32	0010 0000	0E	20	Iadrs pointers
15	33	0010 0001	0F	21	
—					
32	2	0000 0010	20	02	Iadrs data
33	1	0000 0001	21	01	

3. Implement the BC–8C computer with the following instruction set: LDBDIR (000), STADIR (001), ADDAB (01000), SUBAB (01001), SRJ (11000), JMPIND (11001). Make up a program using these instructions; load and run the program.

4. Implement the BC–8C computer with the following instruction set: LDBDIR (000), STADIR (001), ADDAB (01000), SUBAB (01001), CMPIMM (11010), BRLT (10100), and BREQ (10101). Make up a program using these instructions; load and run the program.

Appendices

A Integrated Circuit List
B ANSI/IEEE Standard 91–1984 for Logic Functions
C TTL User's Guide
D Introduction to Signetics Programmable Logic

A INTEGRATED CIRCUIT LIST

Number	Figure	Type
7408	2.11, 2.12	Quad 2-input AND
7432	2.13	Quad 2-input OR
7400	2.14	Quad 2-input NAND
7402	2.15	Quad 2-input NOR
7486	2.16	Quad 2-input XOR
7404	2.17	Hex inverter
74153	4.14	Dual 4-to-1 multiplexer
7407	4.19	Hex buffer (O.C.)
74148	ref 4.25	8-input priority encoder
74154	ref 4.27	4-to-16 decoder
74180	ref 4.34	9-bit parity generator/checker
74138	ref 4.37	3-to-8 decoder
PLS100	ref 4.46	Programmable AND-Array Logic (PAL)
74S182	ref 5.10	Carry look-ahead generator
7483	5.14, 5.20, 6.9, 6.17, 6.18, 6.21, 11.6, 11.10, 11.17, 11.21, 11.25, 11.28, 12.23, 12.32, 12.33, 13.12, 13.23, 14.10, 14.14	4-bit MSI adder
74181	ref 5.10	4-bit MSI ALU
7475	7.19	Quad *D* Latch (level-sensitive)
7474	7.25, 7.53, 7.60, 7.71, 7.76, 8.4, 8.8, 8.15, 8.22, 8.63, 10.7, 10.8, 10.12, 10.15, 10.16, 10.18, 10.24, 10.25, 10.35, 10.44, 11.9, 11.10, 11.17, 11.20, 11.21, 11.25, 11.27, 11.28, 12.18, 12.23, 12.32, 12.33, 13.12, 13.23, 14.10, 14.14	Dual positive edge-triggered *D* flip-flop
7476	7.38, 7.42, 7.49, 7.54, 7.55, 7.56, 7.65, 8.12, 8.44, 8.56, 8.71, 8.78, 10.44	Dual negative edge-triggered *JK* flip-flop
74LS163A	7.57	4-bit binary counter
74193	7.58	4-bit binary up/down counter
74165	8.8, 8.22, 8.56, 8.78	PISO 8-bit shift register
74198	10.8, 10.12	8-bit parallel-load shift register
74194	10.16, 10.25, 11.6, 11.10, 11.17, 11.25, 11.28	4-bit parallel-load shift register
74161	10.24, 10.25	4-bit binary counter
74163	11.9, 11.10, 11.20, 11.21, 11.27, 11.28	4-bit binary counter

7485	6.4, 10.31, 10.35, 10.40, 10.44	4-bit comparator
74--	10.31, 10.35, 10.40, 10.44, 12.21, 12.23, 12.32, 12.33, 13.12, 13.23, 14.10, 14.14	8-bit data register
74153	10.35, 10.44	Dual 4-to-1 multiplexer
74157	10.40, 10.44, 11.17, 11.21, 11.25, 11.28, 12.32, 12.33, 13.12, 13.23, 14.10, 14.14	Quad 2-to-1 multiplexer
74138	12.21, 12.23, 12.32, 12.33, 13.12, 13.23, 14.10, 14.14	1-to-8 decoder
--	12.32, 12.33, 13.12, 13.23, 14.10, 14.14	Read/write memory
74S244	14.10, 14.14	3-state buffer/driver

B ANSI/IEEE STANDARD 91–1984 FOR LOGIC FUNCTIONS

In 1984 the American National Standards Institute (ANSI) and the Institute for Electrical and Electronics Engineers (IEEE) developed a new standard graphic language for documenting logic circuits: *Graphic Symbols for Logic Functions* (ANSI/IEEE Std. 91–1984). The objective of this new standard is to provide a standard, consistent, and concise notation to describe the behavior of logic functions and circuits.

The ANSI/IEEE Std. 91–1984 uses a uniform shape symbol (the rectangle) with qualifying symbols to define common logic functions. The standard also employs a dependency notation that defines relationships among inputs, outputs, and inputs and outputs without showing all the elements and interconnections in the circuit.

Each type of basic logic gate can be represented by a distinctive shape symbol (used throughout this text) or by the ANSI/IEEE Std. 91–1984 uniform shape symbol containing a general qualifying symbol such as & for AND, ≥1 for OR, and =1 for XOR. These symbols are shown in Figure B.1.

FIGURE B.1
ANSI/IEEE Std. 91–1984 uniform shape and distinctive shape logic symbols for basic logic gates

Function	Logic Symbol	
	Uniform Shape	Distinctive Shape
AND gate	&	
OR gate	≥ 1	
Inverter	1	
XOR gate	= 1	
NAND gate	&	
NOR gate	≥ 1	

In the ANSI/IEEE standard, the half arrow (⊵) denotes the signal is asserted low (active-low voltage). This text uses a bubble (o) to denote a signal that is asserted low.

Large scale functions may be represented by a uniform shape symbol (rectangle) containing a general qualifying symbol, I/O qualifying symbols, and dependency notation symbols (if required), as shown in Figure B.2.

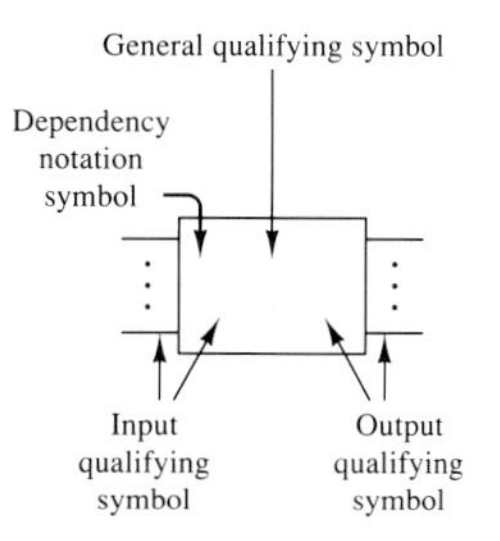

FIGURE B.2
Symbolic representation of a function

Table B.1 summarizes some of the more common general qualifying symbols (other than &, ≥1, and =1), I/O qualifying symbols, and dependency notation symbols defined by the ANSI/IEEE Std. 91–1984.

TABLE B.1
ANSI/IEEE Std. 91–1984 Symbols

General Qualifying Symbols		Input/Output Qualifying Symbols	
MUX	Multiplexer		Inverted input
DMUX	Demultiplexer		Inverted output
X/Y	Code converter		Asserted-low input
Σ	Adder		Asserted-low output
π	Multiplier		Dynamic input
▷	Buffer/driver	D	Data input
RGn	*n*-bit register	J	*J* input of *JK* flip-flop
SRGn	*n*-bit shift register	K	*K* input of *JK* flip-flop
CTRn	*n*-bit counter	T	Toggle input
RAM	Random access memory	→	Shift right
ROM	Read only memory	←	Shift left

Dependency Notation Symbols	
Type	**Function**
A	Address
C	Control*
G	AND (Selection)
V	OR
N	Negate
EN	Enable
S	Set
R	Reset
M	Mode
Z	Interconnection

*Used in general for clock or timing inputs

For example, a 2-to-1 multiplexer and a 1-to-4 demultiplexer can be represented, in the ANSI/IEEE Std. 91–1984 notation, as shown in Figure B.3 and Figure B.4, respectively.

FIGURE B.3
ANSI/IEEE Std. 91–1984 representation of a 2-to-1 MUX

MUX
Select line a — G1
Data inputs b — 1, c — 1'
d

Dependency symbol G1 indicates that the mux select line (a) determines which data input (b or c) is routed to the output line (d).

FIGURE B.4
ANSI/IEEE Std. 91–1984 representation of a 1-to-4 DEMUX

Dependency symbol $G\frac{0}{3}$ indicates the select lines (a and b) determine which output line is selected to be connected to the data input line (c).

Another example of the ANSI/IEEE Std. 91–1984 notation is presented in Figure B.5, which shows the pin configuration, logic symbol, and the logic symbol (IEEE/IEC) for a '107 dual *JK* flip-flop. The '107 is dual flip-flop with individual *J*, *K*, clock, and direct reset inputs. The 74107 is a pulse-triggered flip-flop. *JK* information is loaded into the master while the clock is high and transferred to the slave on the high-to-low clock transition. For these devices, the *J* and *K* inputs should be stable while the clock is high for conventional operation.

The 74LS107 is a negative edge-triggered flip-flop. The *J* and *K* inputs must be stable one set-up time prior to the high-to-low clock transition for predictable operation. The reset ($\overline{R}_D$) is an asynchronous active-low input. When low, it overrides the clock and data inputs, forcing the Q output low and the $\overline{Q}$ output high.

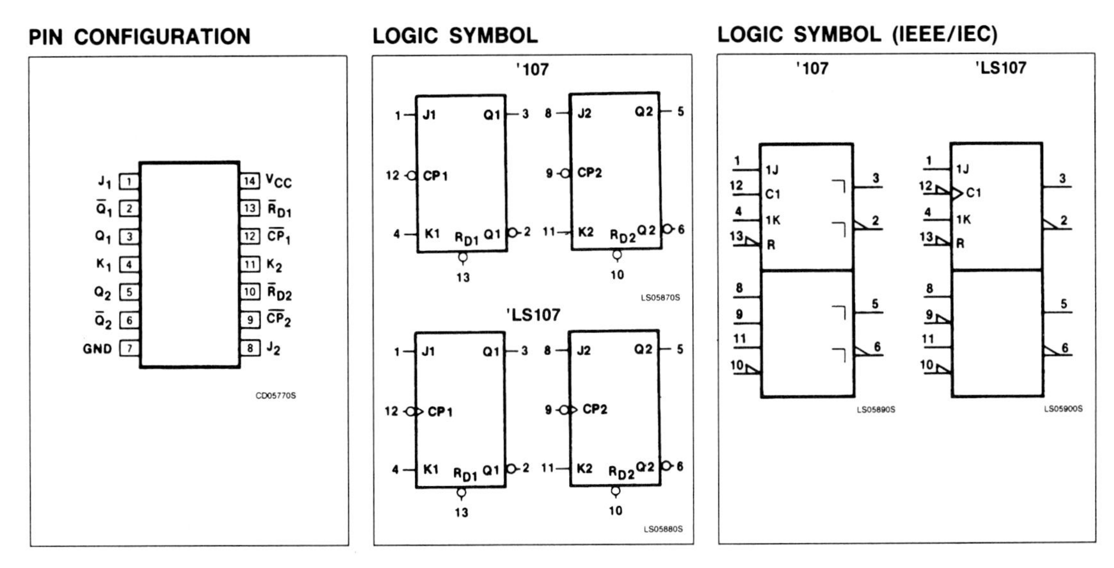

FIGURE B.5
Symbolic representations of a '107 *JK* flip-flop (Reprinted by courtesy of Philips Components/Signetics.)

A common control block (spade-shaped symbol) can be used in conjunction with the uniform shape function symbol (rectangle) to represent more complex functions that have a common control, as shown in Figure B.6.

FIGURE B.6
Symbolic representation of a complex function

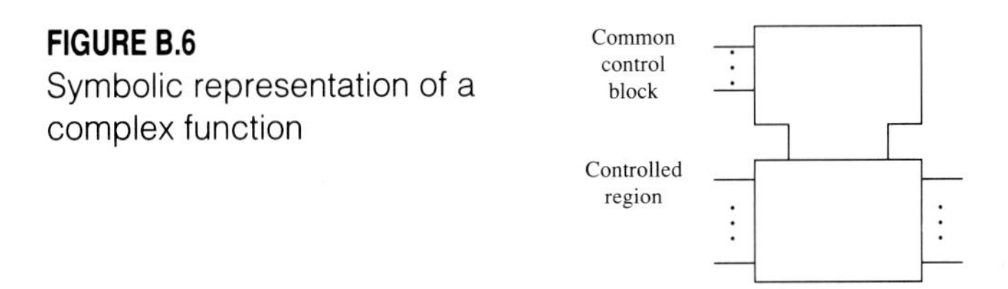

For example, a 7475 quad gated *D* latch can be represented, in the ANSI/IEEE Std. 91–1984 notation, as shown in Figure B.7. Note that the controlled region may be composed of a number of stacked rectangles.

In Figure B.7, C denotes *control* and 1 (or 2) indicates which data input (in the controlled region) is controlled. Therefore, the C1 in the common control block indicates that the number 1 controls data input 1D in the first two function blocks, and C2 indicates that the number 2 controls data input 2D in the last two function blocks.

A final example of the ANSI/IEEE Std. 91–1984 notation is presented in Figure B.8, which shows the pin configuration, logic symbol, and the logic symbol (IEEE/IEC) for a 7494 4-bit shift

FIGURE B.7
ANSI/IEEE logic symbol for 7475 gated latch

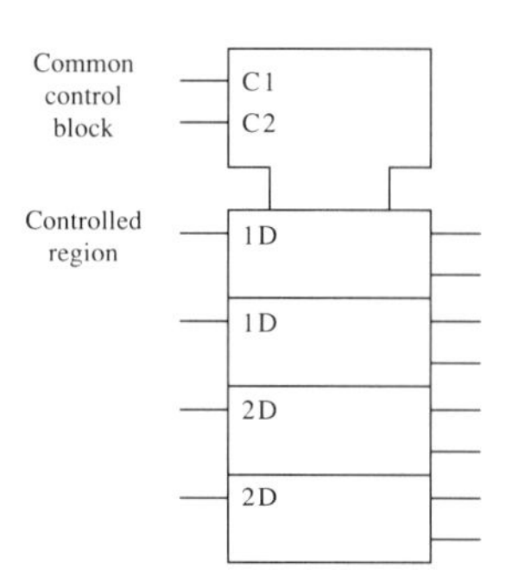

register. The 7494 is a 4-bit shift register with serial and parallel data entry. To facilitate parallel 1s transfer from two sources, two parallel-load inputs (PL_0 and PL_1) with associated parallel data inputs ($D_{0a} - D_{0d}$ and $D_{1a} - D_{1d}$) are provided. To accommodate these extra inputs, only the output of the last stage is available. The asynchronous master reset (MR) is active high. When MR is high, it overrides the clock and clears the register, forcing Q_d low. Four flip-flops are connected so that shifting is synchronous; they change state when the clock goes from low-to-high. Data is accepted at the serial D_s input prior to this clock transition.

The common control block for the 7494 shift register contains the following elements:

- A dynamic input for the clock signal
- Control symbol C1, which controls serial data input 1D
- Select symbols G2 and G3, which enable parallel load (set) inputs 2S and 3S, respectively
- Reset symbol R, which is an asynchronous reset for all four flip-flops

PIN CONFIGURATION

D_{0a} 1, D_{0b} 2, D_{0c} 3, D_{0d} 4, V_{CC} 5, PL_0 6, D_S 7, CP 8, 16 D_{1a}, 15 PL_1, 14 D_{1b}, 13 D_{1c}, 12 GND, 11 D_{1d}, 10 MR, 9 Q_d

CD05740S

LOGIC SYMBOL

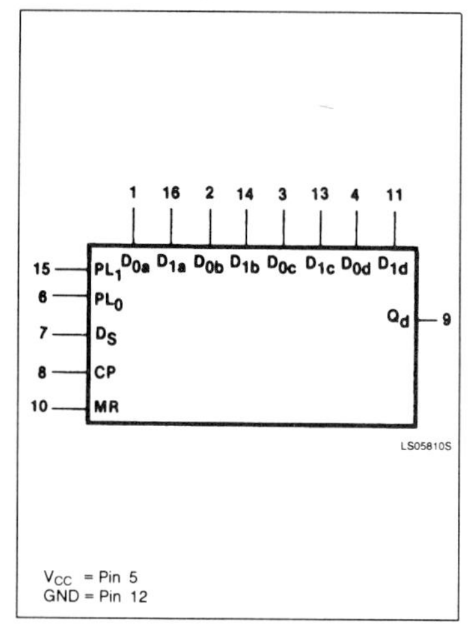

LOGIC SYMBOL (IEEE/IEC)

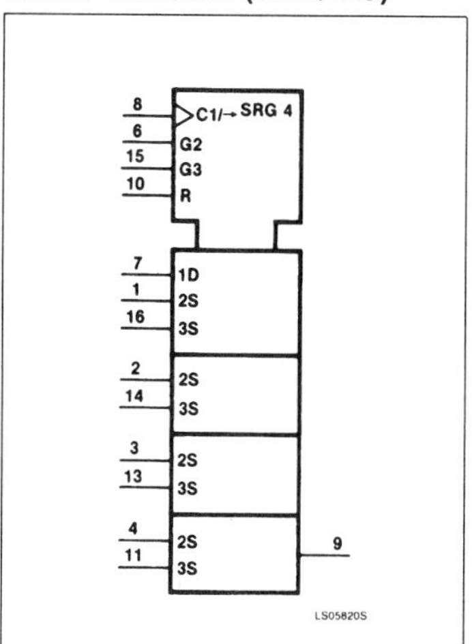

FIGURE B.8
Symbolic representations of a 7494 shift register (Reprinted by courtesy of Philips Components/Signetics.)

Logic Products

INTRODUCTION

The TTL Logic devices described in this data manual differ widely in function, complexity and performance, but their electrical input and output characteristics are very similar and are defined and tested to guarantee compatibility. The data sheets that make up this book cover four major categories of TTL circuits and a series of TTL compatible interface products.

The oldest TTL product category is the gold-doped double-diffused type which is made up of the 7400 family of devices. This family reflects the same performance ranges and differ only in functions and pin configuration.

The remaining two categories of products are fabricated with a non-saturating Schottky clamped transistor technique. The 74S00 family of TTL products are very high performance, high power devices. The most popular TTL category is the 74LS Low Power Schottky family. These products feature the performance of the 74 family at about 1/4 the power.

ABSOLUTE MAXIMUM RATINGS

The Absolute Maximum Ratings constitute limiting values above which serviceability of the device may be impaired. Provisions should be made in system design and testing to limit voltages and currents as shown below.

OPERATING TEMPERATURE AND VOLTAGE RANGES

The nominal supply voltage (V_{CC}) for all TTL circuits is +5.0 volts. Commercial grade parts are guaranteed to perform with a ±5% supply tolerance (±250mV) over an ambient temperature range of 0°C to 70°C.

The actual junction temperature can be calculated by multiplying the power dissipation of the device with the thermal resistance of the package and adding it to the measured ambient temperature T_A or package (case) temperature T_C. The thermal resistance for the various packages in which the TTL products are offered is specified with the Package Information in Section 9 of this manual.

GENERAL TTL CIRCUIT CHARACTERISTICS

All TTL products are derived from a common NAND logic structure. The NAND circuit is actually five subcircuits as shown in Figure 1 and each performs a separate function. The input circuit (1) is an AND gate usually fabricated with a multi-emitter transistor which characterizes TTL technology. Many Schottky processed circuits have been designed with PNP or diode inputs in order to optimize the speed/power performance of the circuits.

The phase splitter (2) provides the inversion and amplification in the circuit. It determines whether the outputs are active level HIGH or active level LOW. The level shifter (3) provides noise immunity between the HIGH and LOW output levels, and minimizes the possibility of having both HIGH level driver (4) and LOW level driver (5) on simultaneously.

NAND Gate Example

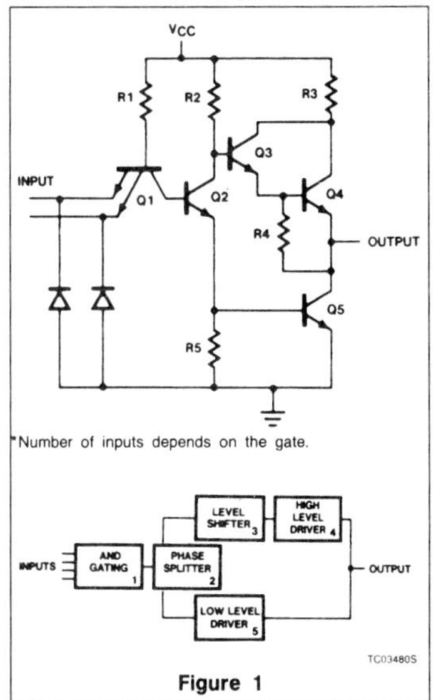

Figure 1

The level shifter (3) and HIGH level driver (4) combine to form an emitter follower circuit that tracks the voltage at the collector of the

ABSOLUTE MAXIMUM RATINGS

PARAMETER	74	74S	74LS
V_{CC} supply voltage, continuous (Note a)	7.0V	7.0V	7.0V
Input voltage, continuous (Notes a and b)	−0.5V to +5.5V	−0.5V to +5.5V	−0.5V to +7.0V[(b)]
Input current, continuous	−30mA to +5mA	−30mA to +5mA	−30mA to +1mA
Voltage applied to HIGH outputs (Note a)	−0.5V to V_{CC}	−0.5V to 7.0V	−0.5V to V_{CC}
Voltage applied to "off" Open Collector outputs (Notes a and c)	−0.5V to 7.0V	−0.5V to 7.0V	−0.5V to 7.0V
Current into LOW standard output, continuous	30mA	40mA	15mA
Current into LOW buffer output, continuous	80mA	100mA	50mA
Operating free air temperature range (Com'l)	0°C to +70°C		
Storage temperature range	−65°C to +150°C		

NOTES:

a. Voltages are referenced to device ground terminal.

b. LS devices are generally limited to 7.0V maximum input voltage. Exceptions are called out on individual product data sheets.

c. Some open collector devices are specially processed to handle higher output voltages of from 15V to 30V. The Absolute Maximum voltage for these devices is 10% over the specified V_{OUT} test condition.

TTL INPUT CONFIGURATIONS

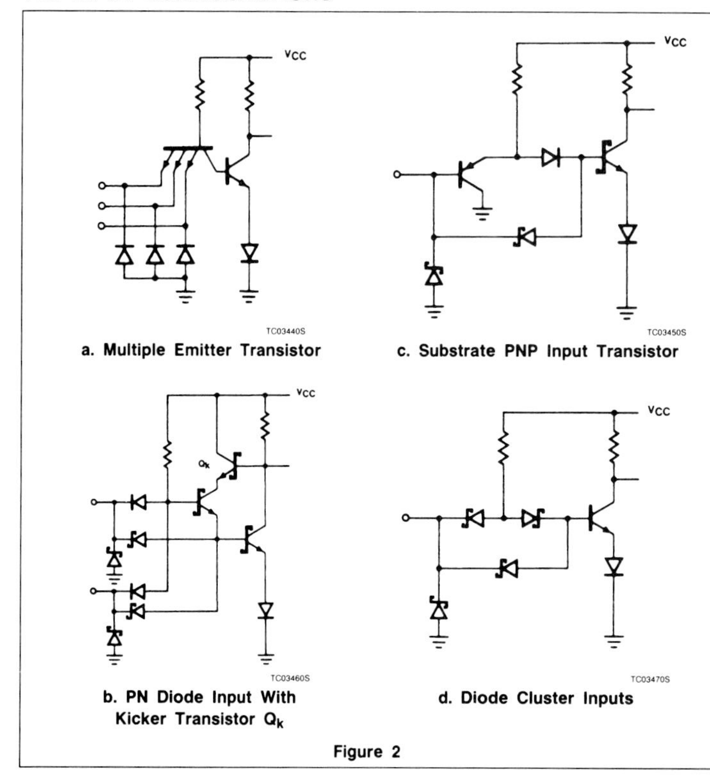

a. Multiple Emitter Transistor

c. Substrate PNP Input Transistor

b. PN Diode Input With Kicker Transistor Q_k

d. Diode Cluster Inputs

Figure 2

phase splitter. This circuit is usually designed to drive very heavy capacitive loads so that the initial rise time of the output is determined primarily by the rise time at the phase splitter collector. The LOW level driver (5) is usually a saturating transistor for the gold-doped process devices, or a Schottky diode clamped transistor for the Schottky processed devices. These output transistors are designed to sink the rated fan-out current which characterizes the various TTL families.

Input Circuits

The input circuits as described above are basically AND gate configurations designed with multiple-emitter NPN transistors (MET), substrate PNP transistors, or various junction and Schottky diodes as shown in Figure 2. All of the circuit configurations have very high impedance in the HIGH state. When the input voltage is above the circuit threshold vòltage, all of the inputs act like reversed biased diodes.

The MET transistors are actually operated in the inverse mode, but the gain is so low there is very little current flowing into the devices.

The LOW level input impedance of the MET and diode inputs is determined by the internal pull-up resistor. This resistor is nominally 2kΩ for 54S/74S inputs, and it is 16kΩ to 20kΩ for the 54LS/74LS inputs. Some 54LS/74LS buffer devices have substrate PNP inputs which exhibit very high impedance at both HIGH and LOW input logic levels. This is used to minimize the input load factor and produce better output drive and performance.

The inputs to all Signetics TTL devices have clamp diodes to ground to minimize negative ringing effects. These diodes are designed to operate in the ac mode and cannot handle heavy dc currents for long periods.

Output Circuits

The output circuit configurations used for the TTL products in this manual are shown in Figure 3. The basic advantages and disadvantages of each configuration are given for reference. The different circuits are used to optimize the functional and performance requirements of the various devices, and are not necessarily restricted to individual TTL families. The pull-down circuit (not shown) on the base of the LOW level driver is usually a resistor which provides a means of turning off the output transistor. The majority of the 54S/74S and 54LS/74LS devices use a resistor-transistor network which acts to square-up the $V_{IN} - V_{OUT}$ transfer characteristics of the device.

A resistive pull-up can be added to any TTL output circuit increasing V_{OH} to almost V_{CC}, but only circuits "c," "d," and "e" can be pulled higher than V_{CC}, e.g., to +7.0V for driving MOS circuits. Configurations "a" and "b" have a diode associated with the resistor at the output which clamps the output one diode drop above V_{CC}. This is an important consideration in large systems where sections might be powered down ($V_{CC} = 0$). In this state, the outputs of circuits "a" and "b" represent a very low impedance at a fairly low voltage (< 1.0V), while the outputs of circuits "c," "d," and "e" represent a high impedance and thus a logic HIGH, more appropriate for isolation from the rest of the system.

The output impedance of a typical TTL device in both the LOW and HIGH state is shown in Figure 4. In the LOW state, the output impedance is determined by a saturated transistor (about 8Ω to 10Ω). However, at very high sinking current, especially at low temperature, the output device is not able to stay in saturation and the output impedance rises as shown.

When switching from the LOW to the HIGH state, the totem-pole output structure provides a low output impedance capable of rapidly charging capacitive loads. However, charge and discharge currents must also flow through the V_{CC} and the ground distribution networks. The V_{CC} and ground lines should therefore be short and adequately decoupled.

3-State Outputs

Some of the buffers and registers have 3-state outputs designed for "busing." This type of output electrically performs as a totem-pole output with the additional feature that the output may be disabled, neither sinking nor sourcing current. The 3-state outputs are designed to be tied together, but they are not designed to be active simultaneously. In order to minimize noise and protect the outputs from excessive power dissipation, only one 3-state output should be active at any time.

DESIGN CONSIDERATIONS

The properties of high speed TTL logic circuits dictate that some care be used in the design and layout of a system. Some general "design considerations" are included in this section. This is not intended to be a thorough guideline for designing TTL systems, but a reference for some of the constraints and

TTL OUTPUT CONFIGURATIONS

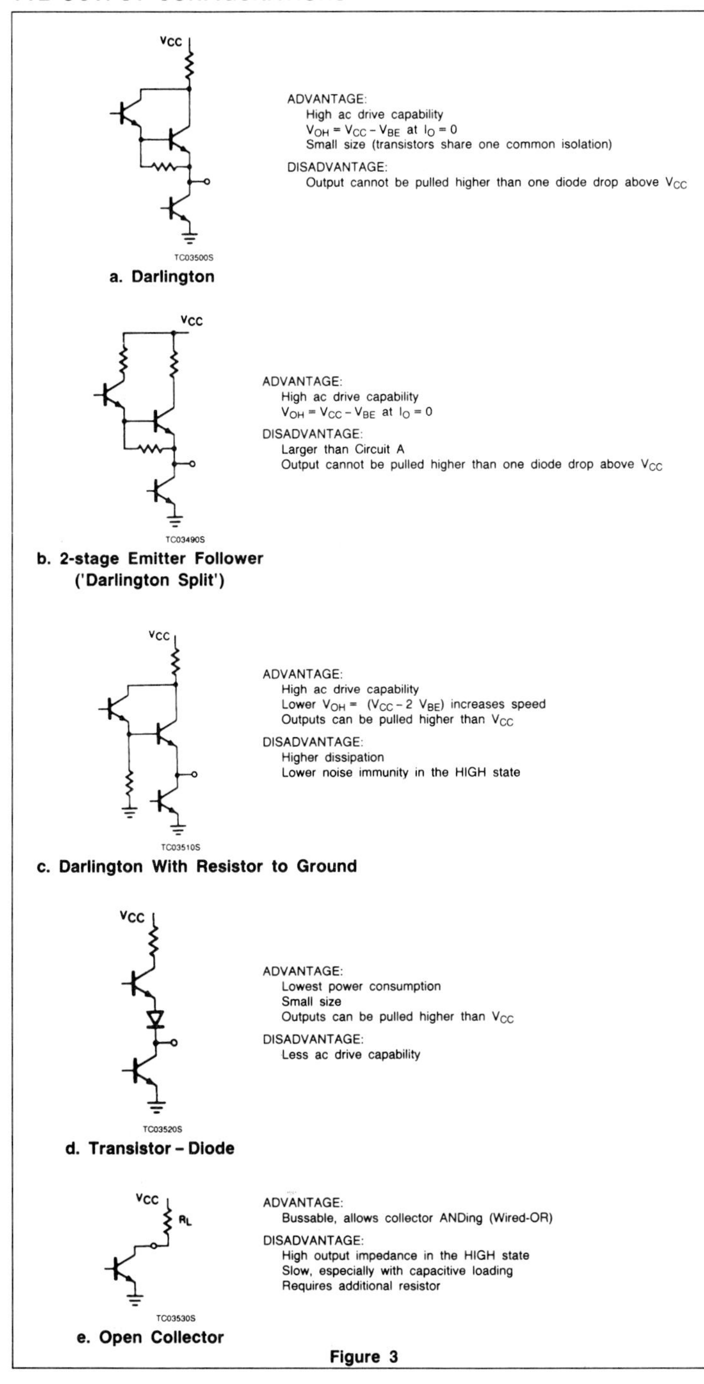

Figure 3

TYPICAL INPUT/OUTPUT CHARACTERISTICS

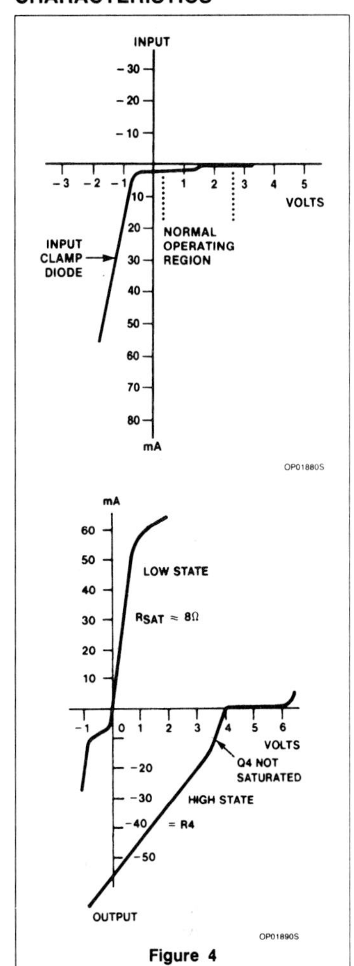

Figure 4

techniques to be considered when designing the system.

Mixing 74 and 74LS is less restrictive, and the overall system design need not be so elaborate. Standard two sided PC boards can be used with good, well decoupled power and ground grid systems. The signal transitions are slower and therefore generate less noise. However, good high speed design techniques are still required, especially when working with counters, registers, or other devices with memory.

Clock Pulse Requirements

Most TTL flip-flop circuits are master-slave devices which makes their clock inputs level sensitive. This is an improvement over ac coupled clock inputs, but it does not make the devices fully insensitive to clock edge rates. The dc level at which the data in the master (input section) is transferred to the slave (output section) is the normal threshold voltage for the devices. For most Signetics TTL devices this level is 1.4V at 25°C, and it changes at a rate of about −4mV/°C.

When the clock input reaches the threshold voltage, the internal gates and the changing outputs start to dump current into the ground lead of the device. If there are enough internal gates or loaded outputs changing at the same time, the chip ground reference level (and therefore the clock input reference level) can rise by as much as 500mV. This ground noise is the algebraic sum of the internal and external ground plane noise. If the clock input of a positive edge triggered device is at or near the threshold of the device during the ground noise transient period, it is quite possible for the internal device to receive multiple clock pulses.

For this reason the rise time on positive edge-triggered devices should be less than the nominal clock to output delay time measured between the 0.8V and 2.0V levels of the clock driver. This edge rate is obtainable from almost any Signetics TTL device of the same family, as long as it is driving no more than rated fan out and no more than 12 to 16 inches of line. When clock pulses are distributed on lines over 16 inches long, all of the clock inputs should be clustered at the receiving end of the line to avoid reflection problems at the driving end.

Special Note

Some of the Signetics Counters and registers have been designed with a special clock buffer that includes a small amount of hysteresis to minimize clock edge rate and noise problems. The LS160A, LS161A, LS162A, LS163A, LS364, and LS374 all have the special clock buffers to increase their tolerance of slow positive clock edges and heavy ground noise.

TTL OUTPUTS TIED TOGETHER

The only TTL outputs that are designed to be tied together are open collector and 3-state outputs. Standard TTL outputs should not be tied together unless their logic levels will always be the same; either all HIGH or all LOW. When connecting open collector or 3-state outputs together some general guidelines must be observed.

Open Collector

These devices must be used whenever two or more OR-tied outputs will be at opposite logic levels at the same time. These devices must have a pull-up resistor (or resistors) added between the OR-tie connector and V_{CC} to establish an active HIGH level. Only special high voltage buffers can be tied to a higher voltage than V_{CC}. The minimum and maximum size of the pull-up resistor is determined as follows:

$$R(Min) = \frac{V_{CC}(Max) - V_{OL}}{I_{OL} - N_2(I_{IL})}$$

$$R(Max) = \frac{V_{CC}(Min) - V_{OH}}{N_1(I_{OH}) + N_2(I_{IH})}$$

where: I_{OL} = Minimum I_{OL} guarantee or OR-tied elements.

$N_2(I_{IL})$ = Cumulative maximum input LOW current for all inputs tied to OR-tie connection.

$N_1(I_{OH})$ = Cumulative maximum output HIGH leakage current for all outputs tied to OR-tie connection.

$N_2(I_{IH})$ = Cumulative maximum input HIGH leakage current for all inputs tied to OR-tie connection.

If a resistor divider network is used to provide the HIGH level, the R(Max) must be decreased enough to provide the required (V_{OH}/R(pull-down)) current.

Minimum propagation delay results when the minimum value of external pull-up resistor is used in Load Circuit 1, Figure 5.

Diodes should be fast recovery 1N4376 or equivalent. External pull-up resistor Load Circuits 2 and 3 give progressively slower propagation delays.

3-STATE OUTPUTS

3-State Outputs are designed to be tied together, but they are not designed to be active simultaneously. In order to minimize noise and protect the outputs from excessive power dissipation, only one 3-state output should be active at any time. This generally requires that the Output Enable signals be non-overlapping. When TTL decoders are used to enable 3-state outputs, the decoder should be disabled while the address is being changed. Since all TTL decoder outputs are subject to decoding spikes, non-overlapping signals cannot normally be guaranteed when the address is changing.

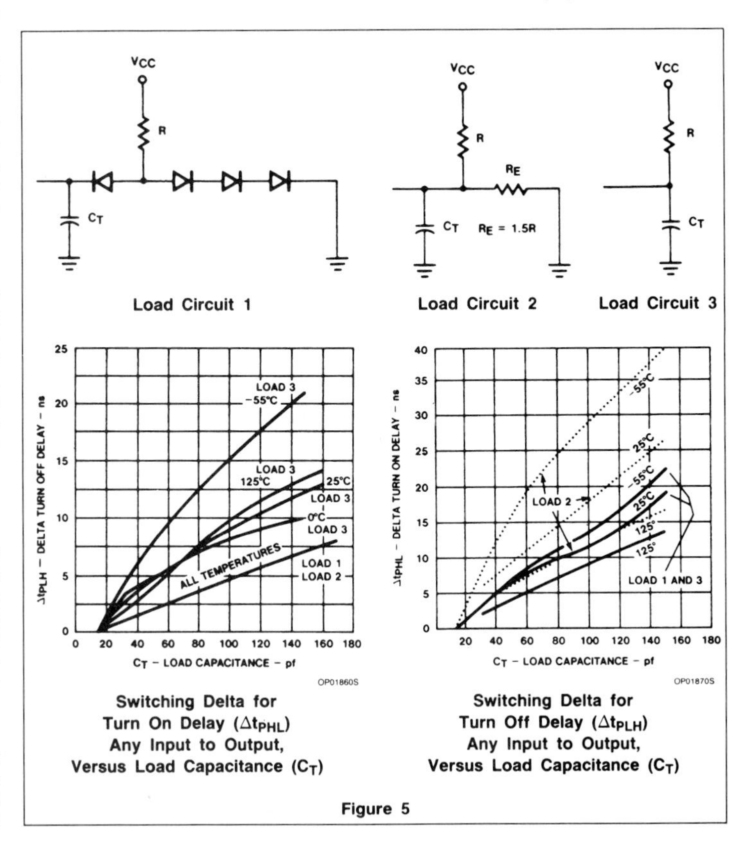

Switching Delta for Turn On Delay (Δt_{PHL}) Any Input to Output, Versus Load Capacitance (C_T)

Switching Delta for Turn Off Delay (Δt_{PLH}) Any Input to Output, Versus Load Capacitance (C_T)

Figure 5

Since most 3-state Output Enable signals are active LOW, shift registers or edge-triggered storage registers provide good Output Enable buffers. Shift registers with one circulating LOW bit, like the "164" or "194" are ideal for sequential enable signals. The "174" or "273" can be used to buffer enable signals from TTL decoders or microcode (ROM) devices. Since the outputs of these registers will change from LOW-to-HIGH faster than from HIGH-to-LOW, the selection of one device at a time is assured.

POWER SUPPLY DECOUPLING

Power supply capacitance decoupling is required for any TTL system. Generally 0.01μF per synchronously driven gate and at least 0.1μF for each 20 gates is required regardless of synchronization. Counters and shift registers are especially susceptible to power and ground line noise. They should be decoupled with a 0.1μF capacitor for each eight internal flip-flops, or one capacitor for each two devices put as close as possible to the devices. Buffers and line drivers should be heavily decoupled at the driver power pins, due to the large current transients needed to charge and discharge the lines.

On-Board Regulation

In most digital systems, there is a large current requirement, and the current supplied usually comes from a main supply. TTL logic tends to generate current spikes during switching due to the overlap in conduction of both upper and lower transistors, thus creating V_{CC} noise. An on-board regulator would not only regulate the power supplied to the circuits on-board, but also would isolate the noise otherwise propagated to the reset of the system. Systems designed using this technique would not need tight regulation on the main power supply.

LINE DRIVING AND RECEIVING

Open wire connections between TTL circuits should not be bundled, tied, or routed together. Instead, point-to-point wiring should be used, preferably above a ground plane which reduces coupling between conductors.

Single line wire interconnections should not exceed two feet; for wires longer than 15 inches, a ground plane is essential to provide adequate system performance. Over 2-foot twisted pairs or coaxial cable should be used. The characteristic impedance of an open wire over a ground plane is about 150Ω, while for twisted pairs of #26 wire the impedance is about 120Ω. For added protection against crosstalk, coaxial cables can be used but coaxial cables having very low characteristic impedances are difficult to drive. For best performance, coaxial cables with a characteristic impedance R_O of 100Ω should be used. Resistive pull-ups at the receiving end can be used to increase noise margins. If reflection effects are unacceptable, the line must be terminated in its characteristic impedance. One method is shown in Figure 6 where the output of the line is tied to V_{CC} through a resistor equivalent to the characteristic impedance of the line. Therefore, R_O is fairly small, and the driving gate must sink the current through it in addition to the current from the inputs being driven. Terminating the line in a voltage divider with two resistors, each twice the line impedance, reduces the extra sink current by 50%. It is preferable to dedicate gates solely for line driving if the line length is in excess of five feet.

Clamp Diode Effect on Negative Input Voltages

All Signetics TTL circuits are provided with clamp diodes on the device inputs to minimize negative ringing effects. These diodes should not be used to clamp negative dc voltages or long duration negative pulses especially for 74LS product. If the input voltage of an LS device is taken more than 0.5 volts negative (referenced to the device ground terminal) for more than 0.5 microseconds, it is possible to activate a parasitic circuit component which can cause the HIGH level output of that gate to degrade sufficiently to cause a logic error.

Disposition of Unused Inputs

Electrically open inputs degrade ac noise immunity as well as the switching speed of a circuit. To optimize performance, each input must be connected to a low impedance source. Unused active HIGH NOR or OR inputs must be returned to ground or a LOW level output. Unused active HIGH NAND or AND inputs should be maintained at a voltage greater than 2.7V, but not exceeding the Absolute Maximum Rating. This eliminates the distributed capacitance associated with the floating input, bond wire, and package lead, and ensures that no degradation will occur in the propagation delay times.

Possible ways of handling unused inputs are:

1. Connect the unused active LOW inputs of the TTL devices to ground. The active HIGH inputs should be tied through a resistor of from 1K to 10kΩ to V_{CC}. The unused active HIGH LS inputs can be tied directly to V_{CC}, as long as the leads are very short and the supply is adequately decoupled.
2. Connect the unused HIGH input to the output of an unused gate that is forced HIGH.
3. Tie unused NAND or AND inputs (multi-emitter inputs) of non-LS devices to a used input of the same gate, provided the HIGH level fan out of the driving circuit is not exceeded. Note that the LOW level fan out is not increased by this connection because the inputs share a common base pull-up resistor.

Unused Gates

It is recommended that the outputs of unused gates be forced HIGH by tying a NAND gate input or all NOR gate inputs to ground. This lowers the power dissipation and supplies a logic HIGH at the gate output which can be used at unused inputs to other gates.

Increasing Fan Out

To increase fan out, inputs and outputs of gates on the same package may be paralleled. It is advisable to limit the gates being paralleled to those in a single package to avoid large transient supply currents due to different switching times of the gates. This is not detrimental to the devices, but could cause logic problems if the gates are being used as clock drivers.

TTL DRIVING TWISTED PAIR

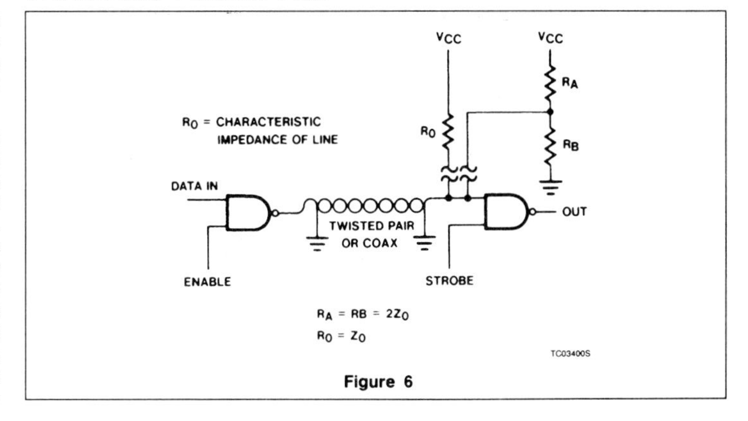

Figure 6

Isolation Diodes

NEVER REVERSE THE V_{CC} AND GROUND POTENTIALS. Catastrophic failure can occur if more than 100mA is conducted through a forward biased substrate (isolation) diode.

Input Loading and Output Drive Characteristics

The logic levels of all the TTL products are fully compatible with each other. However, the inputs loading and output drive characteristics of each of these families is different and must be taken into consideration when mixing the TTL families in a system. Table 1 shows the relative drive capabilities of each family for the Commercial temperature and voltage ranges. For Military ranges the 74LS drive capabilities must be cut in half. You will note that the 74LS Buffers have three times the drive capability of the standard 74LS devices; in fact, they can drive more loads than any other non-buffer TTL device.

Table 1

DRIVING DEVICE	NUMBER OF LOADS DRIVEN					
	74F	**74LS**	**74**	**74S**	**8200 and 9300**	**82500**
74F	33	50	12.5	10	12	50
74F Buffer	106	160	40	32	40	160
74LS	13	20	5	4	5	20
*74LS Buffer	40	60	15	12	15	60
74	26	40	10	8	10	40
74 Buffer	78	120	30	24	30	120
74S	33	50	12.5	10	12	50
74S Buffer	100	150	37.5	30	37	150
8200 & 9300	26	40	10	8	10	40
82500	33	50	12	10	12	50

*The 74LS Buffers include 3-state outputs except LS253 & LS670

NOTE:

For 74LS devices do not connect multiple inputs of a common gate together. This would increase the input coupling capacitance and reduce the ac noise immunity.

Mixing TTL Families

Most TTL families are intended to used together, but this cannot be done indiscriminately. Each family of TTL devices has unique input and output characteristics optimized to get the desired speed or power features. Fast devices like 74S and 74F are designed with relatively low input and output impedances. The speed of these devices is determined primarily by fast rise and fall times internally as well as at the input and output nodes. These fast transitions cause noise of various types in the system. Power and ground line noise is generated by the large currents needed to charge and discharge the circuit and load capacitances during the switching transitions. Signal line noise is generated by the fast output transitions and the relatively low output impedances, which tend to increase reflections.

The noise generated by these 74S and 74F devices can only be tolerated in systems designed with very short signal leads, elaborate ground planes, and good, well decoupled power distribution networks. Mixing the slower TTL families like 74 and 74LS with the higher speed families is also possible but must be done with caution. The slower speed families are more susceptible to induced noise than the higher speed families due to their higher input and output impedances. The low power Schottky 74LS family is especially sensitive to induced noise and must be isolated as much as possible from the 74S and 74F devices. Separate or isolated power and ground systems are recommended, and the LS input signal lines should not run adjacent to lines driven by 74S and 74F devices.

DC SYMBOLS AND DEFINITIONS

Voltages — All voltages are referenced to ground. Negative voltage limits are specified as absolute values (i.e., −10V is greater than −1.0V).

Currents — Positive current is defined as conventional current flow into a device. Negative current is defined as conventional current flow out of a device. All current limits are specified as absolute values.

V_{CC}	**Supply voltage:** The range of power supply voltage over which the device is guaranteed to operate within the specified limits.
V_{IK}	**Input clamp voltage:** The most negative voltage at an input when the specified current is forced out of that input terminal. This parameter guarantees the integrity of the input diode intended to clamp negative ringing at the input terminal.
V_{IH}	**Input HIGH voltage:** The range of input voltages recognized by the device as a logic HIGH.
V_{IH}(MIN)	**Minimum Input HIGH voltage:** This value is the guaranteed input HIGH threshold for the device. The minimum allowed input HIGH in a logic system.
V_{IL}	**Input LOW voltage:** The range of input voltages recognized by the device as a logic LOW.
V_{IL}(MAX)	**Maximum Input LOW voltage:** This value is the guaranteed input LOW threshold for the device. The maximum allowed input LOW in a logic system.
V_M	**Measurement voltage:** The reference voltage level on ac waveforms for determining ac performance. Usually specified as 1.5V for most TTL families, but 1.3V for the Low Power Schottky 74LS family.
V_{OH}(MIN)	**Output HIGH voltage:** The minimum guaranteed HIGH voltage at an output terminal for the specified output current I_{OH} and at the minimum V_{CC} value.
V_{OL}(MAX)	**Output LOW voltage:** The maximum guaranteed LOW voltage at an output terminal sinking the specified load current I_{OL}.
V_{T+}	**Positive-going threshold voltage:** The input voltage of a variable threshold device which causes operation according to specification as the input transition rises from below V_{T-}(MIN).
V_{T-}	**Negative-going threshold voltage:** The input voltage of a variable threshold device which causes operation according to specification as the input transition falls from above V_{T+}(MAX).

I_{CC}	**Supply current:** The current flowing into the V_{CC} supply terminal of the circuit with specified input conditions and open outputs. Input conditions are chosen to guarantee worst case operation unless specified.
I_I	**Input leakage current:** The current flowing into an input when the maximum allowed voltage is applied to the input. This parameter guarantees the minimum breakdown voltage for the input.
I_{IH}	**Input HIGH current:** The current flowing into an input when a specified HIGH level voltage is applied to that input.
I_{IL}	**Input LOW current:** The current flowing out of an input when a specified LOW level voltage is applied to that input.
I_{OH}	**Output HIGH current:** The leakage current flowing into a turned off open collector output with a specified HIGH output voltage applied. For devices with a pull-up circuit, the I_{OH} is the current flowing out of an output which is in the HIGH state.
I_{OL}	**Output LOW current:** The current flowing into an output which is in the LOW state.
I_{OS}	**Output short-circuit current:** The current flowing out of an output which is in the HIGH state when that output is short-circuit to ground.
I_{OZH}	**Output off current HIGH:** The current flowing into a disabled 3-state output with a specified HIGH output voltage applied.
I_{OZL}	**Output off current LOW:** The current flowing out of a disabled 3-state output with a specified LOW output voltage applied.

AC SWITCHING PARAMETERS AND DEFINITIONS

f_{MAX}	**The maximum clock frequency:** The maximum input frequency at a clock input for predictable performance. Above this frequency the device may cease to function.
t_{PLH}	**Propagation delay time:** The time between the specified reference points on the input and output waveforms with the output changing from the defined LOW level to the defined HIGH level.
t_{PHL}	**Propagation delay time:** The time between the specified reference points on the input and output waveforms with the output changing from the defined HIGH level to the defined LOW level.
t_{PHZ}	**Output disable time from HIGH level of a 3-state output:** The delay time between the specified reference points on the input and output voltage waveforms with the 3-state output changing from the HIGH level to a high impedance "off" state.
t_{PLZ}	**Output disable time from LOW level of 3-state output:** The delay time between the specified reference points on the input and output voltage waveforms with the 3-state output changing from the LOW level to a high impedance "off" state.
t_{PZH}	**Output enable time to a HIGH level of a 3-state output:** The delay time between the specified reference points on the input and output voltage waveforms with the 3-state output changing from a high impedance "off" state to the HIGH level.
t_{PZL}	**Output enable time to a LOW level of a 3-state output:** The delay time between the specified reference points on the input and output voltage waveforms with the 3-state output changing from a high impedance "off" state to the LOW level.

t_h	**Hold time:** The interval immediately following the active transition of the timing pulse (usually the clock pulse) or following the transition of the control input to its latching level, during which interval the data to be recognized must be maintained at the input to ensure its continued recognition. A negative hold time indicates that the correct logic level may be released prior to the active transition of the timing pulse and still be recognized.
t_s	**Set-up time:** The interval immediately preceding the active transition of the timing pulse (usually the clock pulse) or preceding the transition of the control input to its latching level, during which interval the data to be recognized must be maintained at the input to ensure its recognition. A negative set-up time indicates that the correct logic level may be initiated sometime after the active transition of the timing pulse and still be recognized.
t_w	**Pulse width:** The times between the specified reference points on the leading and trailing edges of a pulse.
t_{rec}	**Recovery time:** The time between the reference point on the trailing edge of an asynchronous input control pulse and the reference point on the activating edge of a synchronous (clock) pulse input such that the device will respond to the synchronous input.
t_{TLH}	**Transition time:** LOW to HIGH, the time between two specified reference points on a waveform, normally 10% and 90% points, that is changing from LOW to HIGH.
t_{THL}	**Transition time:** LOW to HIGH, the time between two specified reference points on a waveform, normally 90% and 10% points, that is changing from LOW to HIGH.
t_r, t_f	**Clock input rise and fall times:** 10% to 90% value.

D INTRODUCTION TO SIGNETICS PROGRAMMABLE LOGIC

Application Specific Products

Author: K. A. H. Noach

INTRODUCTION

Custom logic is expensive – too expensive if your production run is short. 'Random logic' is cheaper but occupies more sockets and board space. Signetics Programmable Logic bridges the gap. Using PLD, you can configure an off-the-shelf chip to perform just the logic functions you need. Design and development times are much shorter, and risk much lower than for custom logic. Connections are fewer than for random logic, and, for all but the simplest functions, propagation delay is usually shorter. Yet another advantage that PLD has over custom logic is that it allows you to redesign the functions without redesigning the chip – giving you an invaluable margin not only for cut-and-try during system development, but also for later revision of system design. You're not tied down by the need to recover capital invested in a custom chip.

A PLD chip is an array of logic elements – gates, inverters, and flip-flops, for instance. In the virgin state, everything is connected to everything else by nichrome fuses, and although the chip has the capacity to perform an extensive variety of logic functions, it doesn't have the ability to. What gives it that is programming: selectively blowing undesired fuses so that those that remain provide the interconnections necessary for the required functions.

Signetics Series 20 PLD, named for the number of pins, supplements the well-known Series 28. The package is smaller – little more than a third the size, in fact – but the improved architecture, with user-programmable shared I/O. compensates for the fewer pins. The series comprises the following members, in order of increasing complexity:

- **PLS151 – field-programmable gate array**
- **PLS153 – field-programmable logic array**
- **PLS155 – field-programmable logic sequencer**
- **PLS157 – field-programmable logic sequencer**
- **PLS159 – field-programmable logic sequencer**

Entry to all the devices is via a product matrix, an array of input and shared I/O lines fuse-connected to the multiple inputs of an array of AND gates (see Figures 1, 2 and 5). To exploit the capacity of any device, it is important to make the most economical use of the AND gates it has available. Application of de Morgan's theorem can help in this. For example, inputs for the function

$$F = A + B + C + D$$

would occupy four of the AND gates of the product matrix. However, the same function rewritten as

$$\overline{F} = \overline{A}\ \overline{B}\ \overline{C}\ \overline{D}$$

would occupy only one. Moreover, the second function could be done on the simplest of the Series 20 devices (*and* leave eleven gates over for other functions), whereas the first could not. The fact that all inputs of the Series 20 devices, including the shared ones, incorporate double buffers that make the true and complement forms of all input variables equally accessible, greatly facilitates the use of de Morgan's theorem for logic minimization.

To convert the minimized logic equations to the pattern of fuses to be blown, you can use either a programming sheet (see e.g. Table 1) or Boolean equation program-entry software that lets you enter the equations via the keyboard of a terminal. The direct programmability of logic equations makes system design with PLD simple and sure. Functional changes can be made by replacing one PLD chip by another differently programmed. In many cases you can even remove the original one, reprogram it on the spot, and re-insert it. Programming machines qualified for the Series 20 are at present available from DATA I/O, KONTRON, and STAG.

FPGA PLS151

The field-programmable gate array is the simplest of the Series 20 PLD devices; Figure 1 shows the functional diagram. The array can accept up to 18 inputs. There are six dedicated input pins (A) and twelve (A') that can be programmed as inputs, outputs, or bidirectional I/O. All input variables, whether on dedicated or programmed input pins, are available in both true and complement form in the product matrix (B), and both forms are buffered: either form can drive all 12 product lines if required. In the virgin state, all the input variables and their complements are connected to all the product lines via a diode and a fuse (C), and the product matrix is effectively inoperative. To enable it to generate the required functions, unrequired connections between individual input lines and product lines are severed by blowing the connecting fuses.

At the output of the product matrix are 12 NAND gates, each with 36 inputs to accommodate the 18 possible input variables and their complements. Each of the product terms is normally active-Low, but a unique feature of Signetics PLD is that any or all of them can be independently programmed active-High. This is done by means of an array of exclusive-OR gates (D) at the NAND-gate outputs; when the fuse that grounds the second input of each OR gate is blown, the output of that gate is inverted.

The product matrix and exclusive OR-gate connections shown in Figure 1 illustrate the flexibility conferred by having buffered complements of all input variables internally available, together with independently programmable output polarities. Output B_{11}, shown with its exclusive OR-gate fuse intact, is programmed

$$\overline{B_{11}} = I_0 I_1 \overline{I_5}$$

Reprinted by courtesy of Philips Components/Signetics.

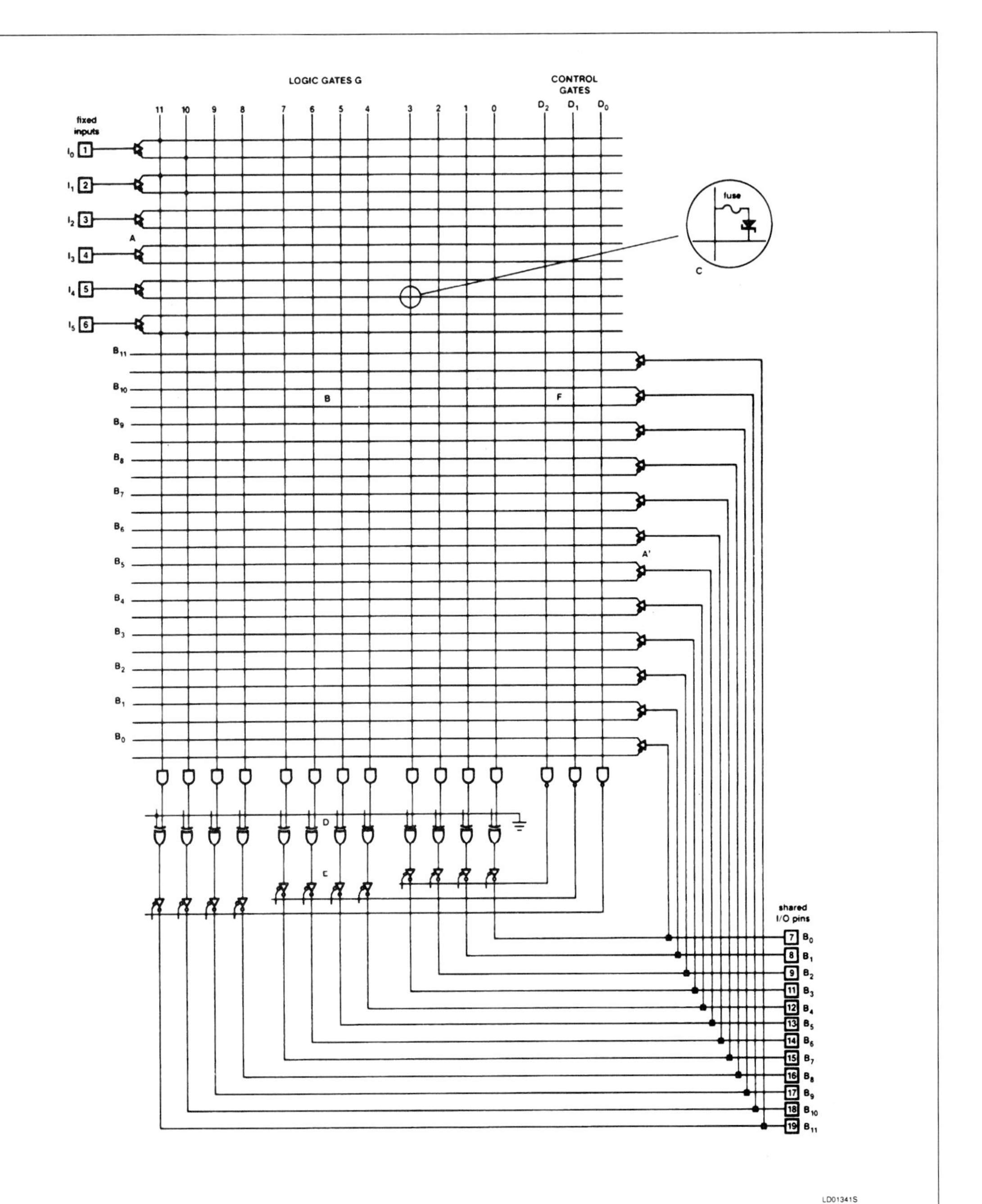

NOTE:
A, dedicated inputs; A', programmable I/O. B, product (NAND) matrix with fused connections C; each of the vertical lines in the matrix represents 36 inputs to the terminating NAND gated. D, exclusive-OR array with inputs grounded via fuses for polarity control. E, programmable Tri-state output buffers. F, fuse-programmable control matrix. Square dots (■) represent permanent ccnnections; round dots (●) intact fuse connections. Connected as shown, the array is programmed for the functions $\overline{B}_{11} = I_0\ I_1\ \overline{I_5}$ and $B_{10} = \overline{I_0}\ \overline{I_1}\ \overline{I_5}$.

Figure 1. Field-Programmable Gate Array PLS151A

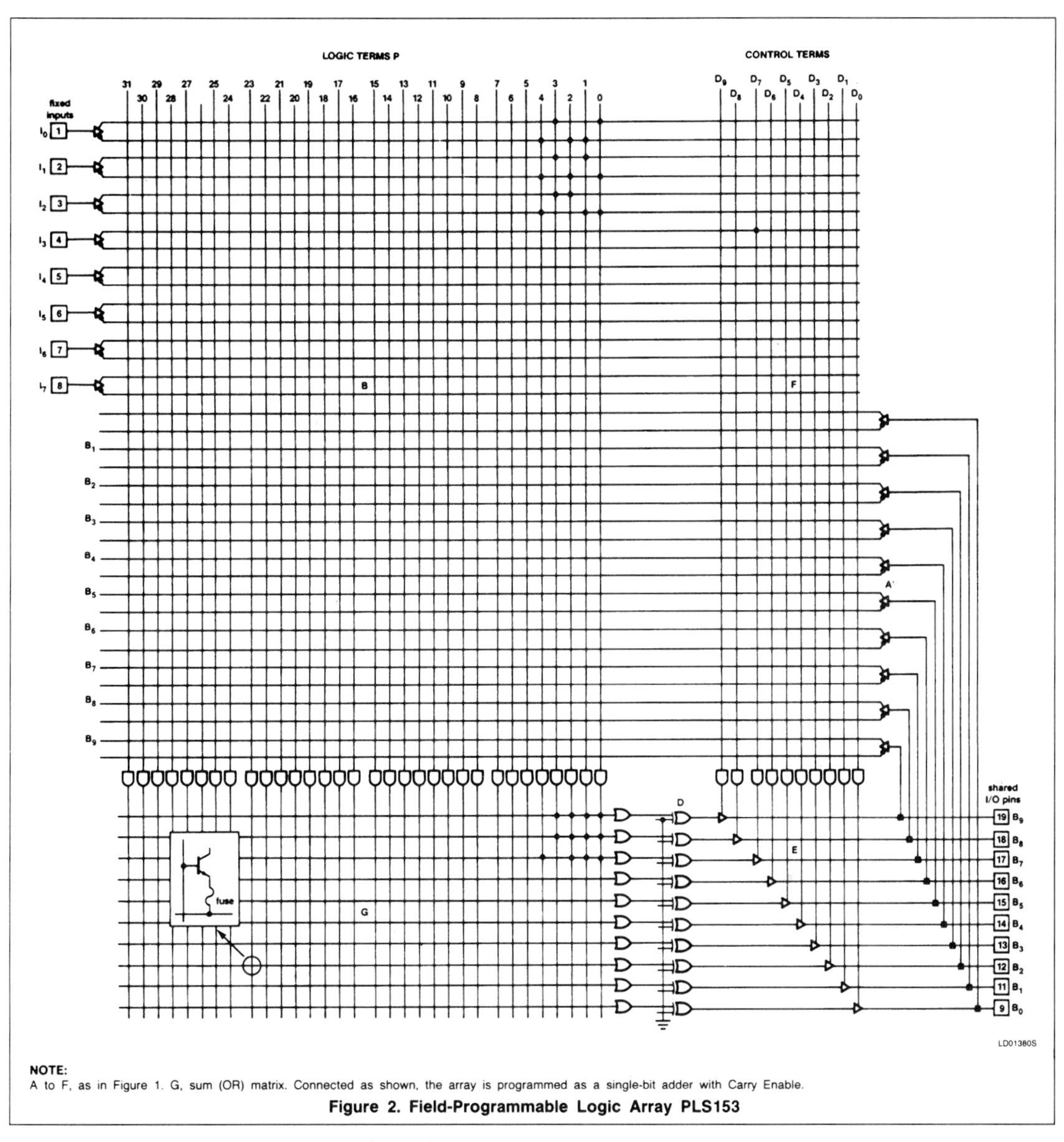

NOTE:
A to F, as in Figure 1. G, sum (OR) matrix. Connected as shown, the array is programmed as a single-bit adder with Carry Enable.

Figure 2. Field-Programmable Logic Array PLS153

At the same time, and without using any additional inputs, output B_{10} (fuse blown) is programmed

$$B_{10} = \overline{I_0}\,\overline{I_1}\,\overline{I_5}$$

Each of the exclusive-OR gates drives a three-state output buffer. In the virgin state all the buffers (E) are disabled and therefore in the high-impedance state. The function of the programmable I/O pins (A') is then determined by the I/O control matrix (F). The three AND gates at the control-matrix output are Active-High, and when one of them is in the High state, the four output buffers it controls are enabled; the corresponding I/O pins then act as outputs. conversely, when a control-matrix AND-gate output is Low and the control fuse for the corresponding Tri-state buffer is intact, the pins controlled by that gate act as inputs. Thus, these pins can be programmed in groups of up to four to act as inputs or outputs according to the state of selected input variables. If required, any of the programmable I/O pins can be made a dedicated output by blowing the control fuse of the output buffer associated with it.

The speed of the FPGA compares favorably with TTL, although its propagation delay is longer than the individual gate delay of TTL. When the number of inputs required is large, however, the FPGA more than makes up for this. When more than eight inputs are required, for example, the FPGA has a distinct advantage. Then, the overall propagation delay of TTL often amounts to two or three gate delays, but that of the FPGA to only one.

FPLA PLS153

Architecture

With two levels of logic embodied in a product matrix terminating in 32 AND gates coupled to a ten-output OR matrix (Figure 2), the FPLA is a step up in complexity from the FPGA. Again, there is provision for 18 input variables, internally complemented and buffered, but here divided between eight dedicated input pins and ten individually programmable I/O pins. As before, exclusive-OR gates grounded by fuses provide output polarity control, and any of the programmable I/O pins can be made a dedicated output by blowing the control fuse of the output buffer associated with it.

Programming

When the required functions have been defined, corresponding programming instructions are entered in a programming table, the layout of which reflects the FPLA architecture. (A Signetics computer program named AMAZE, which accepts Boolean equations as input and generates an FPLA programming table as output, is also available.) The programming machine blows the FPLA fuses in the pattern prescribed by the table.

As an illustration of FPLA programming, consider a full adder. Figure 3 shows a TTL version (74LS80) and the corresponding logic equations. Note that the feedback of $\overline{C}_{n+1}$ introduces a second propagation delay. In the FPLA this is eliminated by redefining Σ in terms of A, B, and C_n, as shown in Figure 4, and using the right side of the equation for $\overline{C}_{n+1}$ instead of the term itself. At first glance this would appear to require a minimum of three product terms for $\overline{C}_{n+1}$ plus four for Σ, or a total of seven. The Karnaugh maps, however, show considerable overlap between the two functions: the map for $\overline{C}_{n+1}$ differs from that for Σ only by having A B C_n instead of $\overline{A}\,\overline{B}\,\overline{C_n}$. Rewriting the equation for $\overline{C}_{n+1}$ to introduce $\overline{A}\,\overline{B}\,\overline{C_n}$ and eliminate A B C_n,

$$\overline{C}_{n+1} = A\,\overline{B}\,\overline{C_n} + \overline{A}\,B\,\overline{C_n} + \overline{A}\,\overline{B}\,C_n + \overline{A}\,\overline{B}\,\overline{C_n}$$

increases the number of product terms by one, but now $\overline{C}_{n+1}$ and Σ have three terms in common. Therefore, since the FPLA allows multiple use of product terms, it is sufficient to program each of the common terms only once; thus, the original seven product terms are effectively reduced to five.

To fill in the programming table (Table 1), first allocate inputs and outputs.

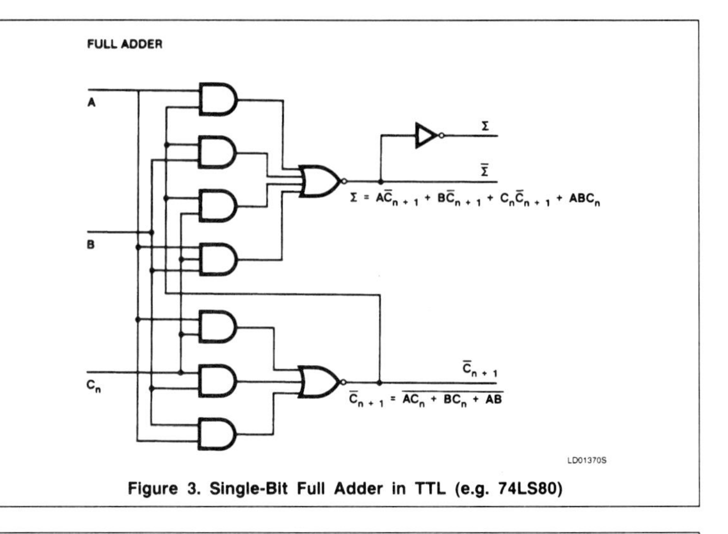

Figure 3. Single-Bit Full Adder in TTL (e.g. 74LS80)

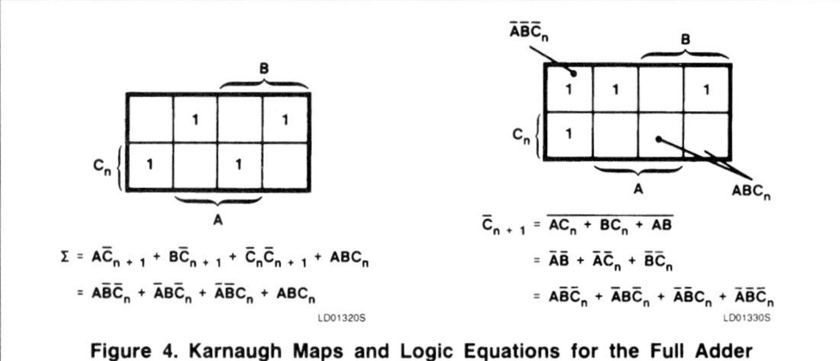

Figure 4. Karnaugh Maps and Logic Equations for the Full Adder of Figure 3, Illustrating how the Equations are Reduced for the FPLA Implementation Shown in Figure 2 and Table 1

Inputs:		Outputs:	
A	$= I_0$	$\overline{C}_{n+1}$	$= B_7$
B	$= I_1$	$\overline{\Sigma}$	$= B_8$
C_n	$= I_2$	Σ	$= B_9$

Next, enter the product terms of Σ in the product-matrix (AND) part of the table, using H to indicate a true input and L a false one.

- Term 0 is A $\overline{B}$ $\overline{C_n}$: mark H, L, L in columns I_0, I_1, I_2 of row 0
- Term 1 is $\overline{A}$ B $\overline{C_n}$: mark L, H, L in columns I_0, I_1, I_2 of row 1
- Term 2 is $\overline{A}$ $\overline{B}$ C_n: mark L, L, H in columns I_0, I_1, I_2 of row 2
- Term 3 is A B C_n: mark H, H, H in columns I_0, I_1, I_2 of row 3.

Fill the rest of rows 0, 1, 2, and 3 with dashes to indicate that all other inputs are to be disconnected from Terms 0, 1, 2, and 3 (fuses blown).

The product terms of Σ must be added to form the sum-of-products required at output B_9. Indicate the required addition by putting an A (for Attached, i.e. fuse unblown) in the Term 0, 1, 2, and 3 spaces of column $B(O)_9$; Term 4 is not required for Σ, so put a dot in the Term 4 space to indicate that it is to be disconnected (fuse blown). To indicate that the output is to be Active-High, put an H in the polarity square above the $B(O)_9$ column. Finally, fill row D_9 with dashes to indicate that all fuses on line D_9 of the control matrix are to be blown and B_9 is to be a dedicated output. This completes the programming of Σ.

The $\overline{\Sigma}$ output on B_8 is programmed in just the same way, except that the polarity square above the $B(O)_8$ column is marked L to indicate Active-Low. (Note that in the FPLA, the Σ and $\overline{\Sigma}$ outputs change simultaneously, because all output signals traverse the exclusive-OR array (D), whether they are Active-High or Active-Low. In the TTL full adder shown in Figure 3, the output inverter delays the change of Σ with respect to $\overline{\Sigma}$.)

Table 1. FPLA Programming Table Filled in for the Full Adder of Figure 2

PROGRAM TABLE ENTRIES:

I, B(I)	
INACTIVE	0
I, B	H
$\overline{I}$, $\overline{B}$	L
DON'T CARE	—

(AND)

B(O)	
ACTIVE	A
INACTIVE	•

B(O)	
HIGH	H
LOW	L

(POL.)

TERM	AND: I								AND: B(I)										OR: B(O)									
POLARITY																			H	L	L							
	7	6	5	4	3	2	1	0	9	8	7	6	5	4	3	2	1	0	9	8	7	6	5	4	3	2	1	0
0	—	—	—	—	—	L	L	H	—	—	—	—	—	—	—	—	—	—	A	A	A							
1	—	—	—	—	—	L	H	L	—	—	—	—	—	—	—	—	—	—	A	A	A							
2	—	—	—	—	—	H	L	L	—	—	—	—	—	—	—	—	—	—	A	A	A							
3	—	—	—	—	—	H	H	H	—	—	—	—	—	—	—	—	—	—	A	A	•							
4	—	—	—	—	—	L	L	L	—	—	—	—	—	—	—	—	—	—	•	•	A							
5																												
6																												
7																												
8																												
9																												
10																												
11																												
12																												
13																												
14																												
15																												
16																												
17																												
18																												
19																												
20																												
21																												
22																												
23																												
24																												
25																												
26																												
27																												
28																												
29																												
30																												
31																												
D9	—	—	—	—	—	—	—	—	—	—	—	—	—	—	—	—	—	—										
D8	—	—	—	—	—	—	—	—	—	—	—	—	—	—	—	—	—	—										
D7	—	—	—	—	H	—	—	—	—	—	—	—	—	—	—	—	—	—										
D6																												
D5																												
D4																												
D3																												
D2																												
D1																												
D0																												
PIN NO.	8	7	6	5	4	3	2	1	19	18	17	16	15	14	13	12	11	9										
VARIABLE NAME					ENABLE	C_n	B	A	Σ	$\overline{\Sigma}$	$\overline{C_{n+1}}$																	

TB01480S

The output C_{n+1} on B_7 contains three of the same terms as Σ, plus the term $\overline{A}\ \overline{B}\ \overline{C_n}$. Only this last term needs to be additionally programmed in the product matrix: mark L, L, L in columns I_0, I_1, I_2 of the Term 4 row. Indicate the addition

$$A\ \overline{B}\ \overline{C_n} + \overline{A}\ B\ \overline{C_n} + \overline{A}\ \overline{B}\ C_n + \overline{A}\ \overline{B}\ \overline{C_n}$$

by putting an A in rows 0, 1, 2, and 4 of column $B(O)_7$, and show that Term 3 (A B C_n) is not required by putting a dot in the Term 3 row to indicate disconnection (fuse blown). Put an L in the $B(O)_7$ polarity square to indicate Active-Low.

Identifying B_7 as a dedicated output by indicating that all the fuses to control term D_7 are to be blown, would now complete the programming of the full adder. However, a useful supplementary feature would be a Carry Enable function to keep the B_7 output buffer in the high-impedance state except when the enable input I_3 is true. The output buffer is enabled when both the fuses of a control term are blown, or when one is blown and the term that controls the output buffer is true. Thus, a Carry Enable can be provided via the I_3 input by leaving intact the fuse for Active-High operation of the enable signal to control term D_7. To indicate this, put an H in the I_3 column of row D_7 and fill the rest of the row with dashes.

The full adder with output Carry Enable uses only four of the eight dedicated inputs, three of the ten programmable I/O pins, and five of the 32 AND gates. The remaining capacity can be used for programming other functions which may, if required, also make use of AND-gate outputs already programmed for the full adder.

All fuses not indicated as blown in the programming table are normally left intact to preserve capacity for later program revisions or the addition of supplementary functions. If it is essential to minimize propagation delay, however, the finalized program should include instructions for blowing all unused fuses to minimize load capacitance.

FPLS PLS155 – PLS157 – PLS159

Architecture

The FPLS (Figure 5) is the most complex of the Series 20 PLD devices. Like the FPLA, it has a 32-term product matrix followed by an OR matrix. In the FPLS, however, the OR matrix is larger and comprises three distinct parts, with architecture differing in detail from type to type. In the PLS155, for instance, the first part consists of eight 32-input gates coupled, like those of the FPLA, to an output-polarity-controlling exclusive-OR array. The second consists of twelve additional gates which control four flip-flops. These are what give the FPLS its sequential character, enabling it to dictate its next state as a function of its present state. The third part is the deceptively simple Complement Array (I in Figure 5): a single OR gate with its output inverted and fed back into the product matrix. This enables a chosen sum-of-products to become a common factor of any or all the product terms and makes it possible to work factored sum-of-products equations. it is also useful for handshaking control when interfacing with a processor and for altering the sequence of a state machine without resorting to a large number of product terms.

PLS155 has four dedicated inputs and eight programmable I/O pins that can be allocated in the same way as in the FPLA. It also has four shared I/O pins (L) whereby the flip-flops can be interfaced with a bidirectional data bus. Two product terms, L_A and L_B in the control matrix F, control the loading of the flip-flops, in pairs, synchronized with the clock.

Figure 6 shows the architecture of the flip-flop circuitry in the PLS155. The flip-flops are positive-edge-triggered and can be dynamically changed to J-K, T, or D types according to the requirements of the function being performed; this considerably lessens the demands on the logic. The Tri-state inverter between the J and K inputs governs the mode of operation, under the control of the product term F:

- When the inverter is in the high-impedance state, the flip-flop is a J-K type, or a T type when J = K.
- When the inverter is active, $K = \overline{J}$ and the flip-flop is a D type; the K input must then be disconnected from the OR matrix.

All the product terms from the product matrix (T_0 to T_{31} in Figure 5) are fuse-connected to the J and K input OR gates. if both fuses of any one product term are left intact, J = K and the flip-flop is a T type.

The flip-flops of the PLS155 have asynchronous Preset and Reset controlled by terms in the OR matrix that take priority over the clock. Their three-state output buffers can be controlled from the enable pin OE or permanently enabled or disabled by blowing fuses or leaving them intact in the enable array (K in Figure 5).

The PLS157 and PLS159 sequencers have, respectively, six and eight flip-flops. The architecture differs in detail but is similar in principle to that of PLS155.

Programming

The FPLS is programmed in much the same way as the FPLA, using a table to instruct the machine that blows the undesired fuses. It is not necessary to work with a circuit diagram; in fact, it is even undesirable to do so, since applying the necessary logic reduction techniques would in most cases make the diagram difficult to read and more a hindrance than a help. An example of how to program the FPLS as a universal counter/shift-register is given in the Appendix.

DEVELOPMENT AND PRODUCTION ECONOMY WITH PLD

Underlying the design philosophy of the Signetics Series 20 PLD is the concept of programmable arrays whose architecture emulates logic equation formats rather than mere aggregations of gates. The unique combination of features which support this philosophy includes:

- double-buffered true and complement inputs
- programmable-polarity outputs
- programmable I/O for internal feedback and maximum freedom in allocating inputs and outputs
- truth-table programming format

These features are common to all the PLD devices. In the field-programmable logic sequencers they are further supported by:

- flip-flops with dynamically alterable operating modes
- a complement array for simplified handshaking control

From the development engineer's point of view an important advantage of PLD is that it eliminates breadboarding. Once the functions required in terms of minimized logic equations are worked out, a PLD can be programmed accordingly. Once programmed, it will perform those functions.

Loading the instructions into the programming machine usually takes no more than a couple of hours; after that, the machine can program the devices at a rate of 100 an hour. Moreover, since any PLD can be programmed in many different ways, PLD has considerable potential for simplifying purchasing and stock control. One type of device can be programmed to perform a diversity of tasks for which it would otherwise be necessary to purchase and stock many different devices.

Series 20 PLD is second-sourced by Harris Semiconductor.

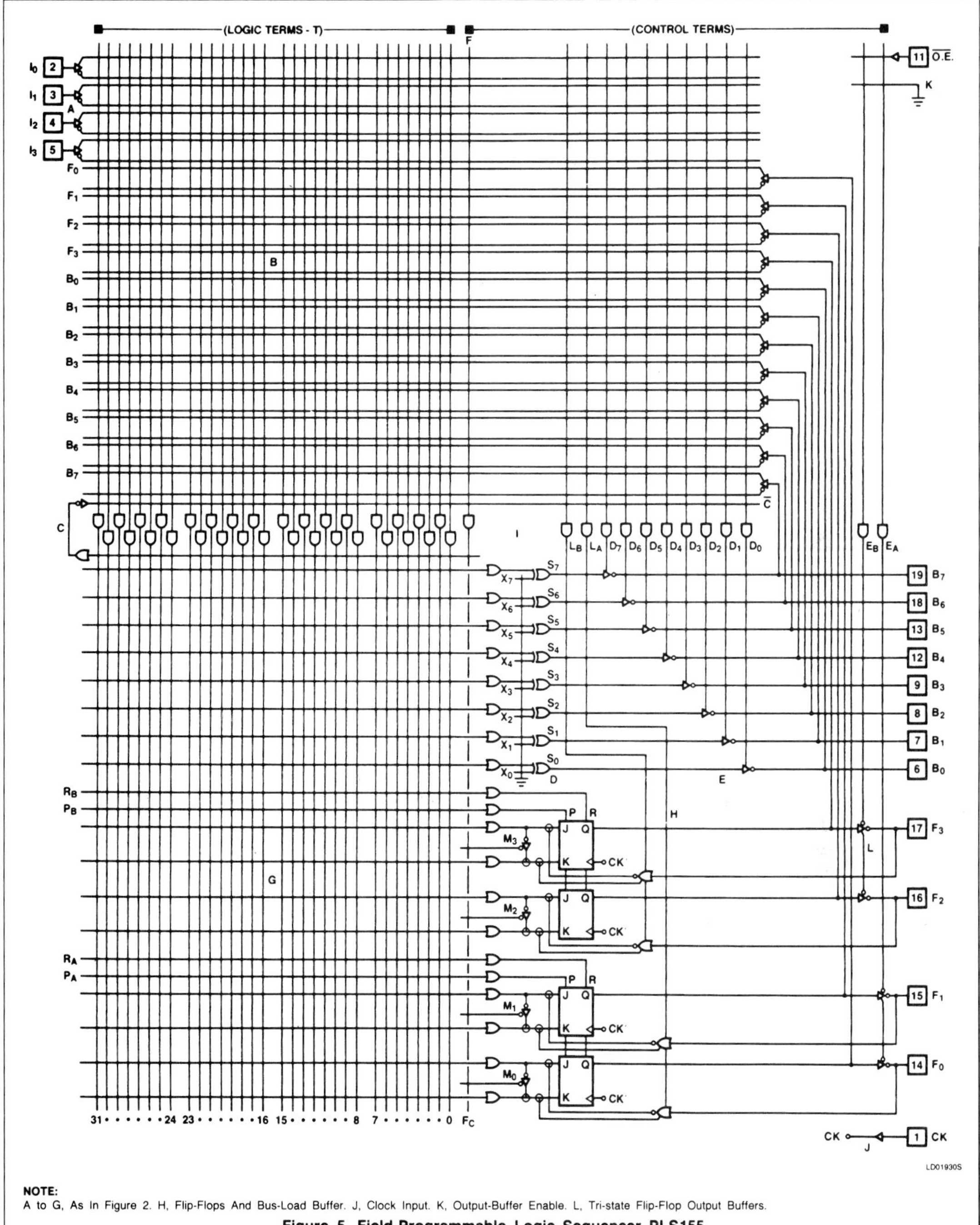

NOTE:
A to G, As In Figure 2. H, Flip-Flops And Bus-Load Buffer. J, Clock Input. K, Output-Buffer Enable. L, Tri-state Flip-Flop Output Buffers.

Figure 5. Field-Programmable Logic Sequencer PLS155

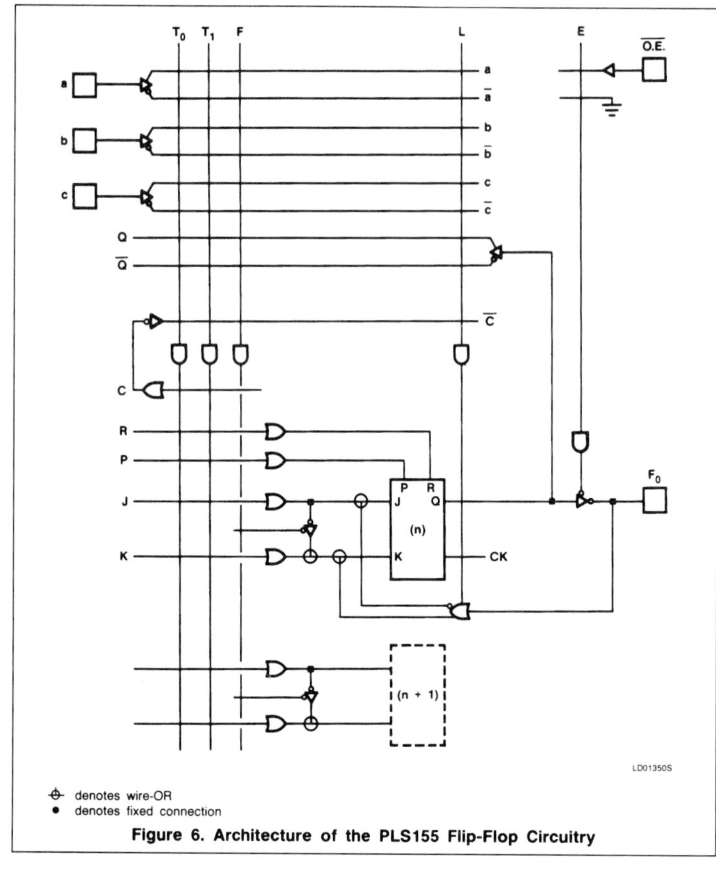

Figure 6. Architecture of the PLS155 Flip-Flop Circuitry

APPENDIX

Programming an FPLS as a Counter/Shift-Register

Objective: to program a PLS155 FPLS as a count-up, count-down, shift-right, shift-left machine governed by three control terms – COUNT/$\overline{\text{SHIFT}}$, RIGHT/UP, LEFT/DOWN. Direct implementation would result in a machine with 64 state transitions (see Table A-1), which is beyond the scope of the PLS155 or even the 28-pin PLS105. Logic reduction is therefore necessary.

As there are only four feedback variables (D, C, B, A), you can do the reduction by hand, one mode at a time; the control terms need not be included till the summary equations are written. Using the transition mapping method suggested here, you can examine the excitation equations for all types of flip-flops (R-S, J-K, D, T) and choose those types that will perform the required functions using the fewest product terms. Table A-2 summarizes the rules for flip-flop implementation using transition maps; the transition symbols used in the table mean:

PRESENT STATE	NEXT STATE	TRANSITION SYMBOL
0	0	0
0	1	α
1	0	β
1	1	1

Using these symbols, construct Table A-3 from Table A-1 to enable you to examine the excitation equations for all types of flip-flops. Proceeding one mode at a time, transfer the state conditions from Table A-3 to Karnaugh maps, as in Figure A-1. Following the rules in Table A-2, derive the excitation equations for the different types of flip-flops (the examples shown in Figure A-1 omit the T type because it is the same as the J-K type when J = K). In deciding which types of flip-flop to use, remember that logic minimization with PLD is different from logic minimization with 'random logic': with random logic you seek to reduce the number of standard packages required; with PLD you seek to reduce the number of product terms.

From Figure A-1 it is evident that you should choose J-K or T flip-flops for the counter mode and D flip-flops for the shift mode, for you then require only one product term per flip-flop per mode. Table A-4 summarizes the number of product terms per mode the various types of flip-flops would require.

Table A-5 shows the completed programming table for the counter/shift-register. The programming of Terms 0 to 15 reflects the flip-flop excitation equations and illustrates the value of being able to switch the flip-flops dynamically from one type of operation to another. Terms 16, 17 and 18, respectively, provide for INITIALIZE, asynchronous RESET, and STOP functions.

The programming of the two additional inputs $\overline{\text{HALT}}$ and $\overline{\text{BUSY}}$ illustrates the value of the complementary, which is made active when $\overline{\text{HALT}}$ and $\overline{\text{BUSY}}$ are Low (A in the Complement square of Term 18) and propagated into all the other terms (dot in the Complement squares of Terms 0 to 17). This means that unless the $\overline{\text{HALT}}$ and $\overline{\text{BUSY}}$ inputs are High, none of the product terms will be true and the state of the machine will not change. If the Complement Array were not used, twice the number of product terms would be required, even if one of the additional inputs were omitted.

As it is, the design uses only 19 of the 32 product terms available, so there is ample capacity for extending its capabilities. For example, the shift-left function can be augmented by a binary multiplication capability, using a D type flip-flop to make it shift one, two, or three places according to the state of two extra inputs, X and Y. Table A-6 shows the revised programming table. The binary multiplication function occupies nine additional product terms.

ACKNOWLEDGEMENT

Electronic Components and Applications; Vol. 4, No. 2, February 1982. Reprinted with the permission of PHILIPS.

Table A-1. Present-State/Next-State Table for Counter/Shift-Register

STATE NO.	PRESENT STATE				NEXT STATE															
					Count Down				Count Up				Shift Left				Shift Right			
	D	C	B	A	D	C	B	A	D	C	B	A	D	C	B	A	D	C	B	A
0	0	0	0	0	1	1	1	1	0	0	0	1	0	0	0	0	0	0	0	0
1	0	0	0	1	0	0	0	0	0	0	1	0	0	0	1	0	1	0	0	0
2	0	0	1	0	0	0	0	1	0	0	1	1	0	1	0	0	0	0	0	1
3	0	0	1	1	0	0	1	0	0	1	0	0	0	1	1	0	1	0	0	1
4	0	1	0	0	0	0	1	1	0	1	0	1	1	0	0	0	0	0	1	0
5	0	1	0	1	0	1	0	0	0	1	1	0	1	0	1	0	1	0	1	0
6	0	1	1	0	0	1	0	1	0	1	1	1	1	1	0	0	0	0	1	1
7	0	1	1	1	0	1	1	0	1	0	0	0	1	1	1	0	1	0	1	1
8	1	0	0	0	0	1	1	1	1	0	0	1	0	0	0	1	0	1	0	0
9	1	0	0	1	1	0	0	0	1	0	1	0	0	0	1	1	1	1	0	0
10	1	0	1	0	1	0	0	1	1	0	1	1	0	1	0	1	0	1	0	1
11	1	0	1	1	1	0	1	0	1	1	0	0	0	1	1	1	1	1	0	1
12	1	1	0	0	1	0	1	1	1	1	0	1	1	0	0	1	0	1	1	0
13	1	1	0	1	1	1	0	0	1	1	1	0	1	0	1	1	1	1	1	0
14	1	1	1	0	1	1	0	1	1	1	1	1	1	1	0	1	0	1	1	1
15	1	1	1	1	1	1	1	0	0	0	0	0	1	1	1	1	1	1	1	1
CONTROL TERMS																				
COUNT/$\overline{\text{SHIFT}}$					1				1				0				0			
RIGHT/UP					0				1				0				1			
LEFT/DOWN					1				0				1				0			

Table A-2. Rules for Flip-Flop Implementation Using Transition Maps

FLIP-FLOP TYPE	INPUT	MUST INCLUDE	MUST EXCLUDE	REDUNDANT
R-S	S	α	β,0	1,x
	R	β	α,1	0,x
D	D	α,1	β,0	x
T	T	α,β	0,1	x
J-K	J	α	0	1,β,x
	K	β	1	0,α,x

Table A-3. Transition Table for Counter/Shift-Register

STATE NO.	PRESENT STATE				TRANSITION															
					Count Down				Count Up				Shift Left				Shift Right			
	D	C	B	A	D	C	B	A	D	C	B	A	D	C	B	A	D	C	B	A
0	0	0	0	0	α	α	α	α	0	0	0	α	0	0	0	0	0	0	0	0
1	0	0	0	1	0	0	0	β	0	0	α	β	0	0	α	β	α	0	0	β
2	0	0	1	0	0	0	β	α	0	0	1	α	0	α	β	0	0	0	β	α
3	0	0	1	1	0	0	1	β	0	α	β	β	0	α	1	β	α	0	β	1
4	0	1	0	0	0	β	α	α	0	1	0	α	α	β	0	0	0	β	α	0
5	0	1	0	1	0	1	0	β	0	1	α	β	α	β	α	β	α	β	α	β
6	0	1	1	0	0	1	β	α	0	1	1	α	α	1	β	0	0	β	1	α
7	0	1	1	1	0	1	1	β	α	β	β	β	α	1	1	β	α	β	1	1
8	1	0	0	0	β	α	α	α	1	0	0	α	β	0	0	α	β	α	0	0
9	1	0	0	1	1	0	0	β	1	0	α	β	β	0	α	1	1	α	0	β
10	1	0	1	0	1	0	β	α	1	0	1	α	β	α	β	α	β	α	β	α
11	1	0	1	1	1	0	1	β	1	α	β	β	β	α	1	1	1	α	β	1
12	1	1	0	0	1	β	α	α	1	1	0	α	1	β	0	α	β	1	α	0
13	1	1	0	1	1	1	0	β	1	1	α	β	1	β	α	1	1	1	α	β
14	1	1	1	0	1	1	β	α	1	1	1	α	1	1	β	α	β	1	1	α
15	1	1	1	1	1	1	1	β	β	β	β	β	1	1	1	1	1	1	1	1

Table A-4. Number of Product Terms Required for Counter/Shift-Register Flip-Flop Excitation

FLIP-FLOP TYPE	COUNT UP	COUNT DOWN	SHIFT RIGHT	SHIFT LEFT	TOTAL
SR only	8	8	8	8	32
JK only	4	4	8	8	24
D only	10	10	4	4	28
FPLS	4(J-K)	4(J-K)	4(D)	4(D)	16

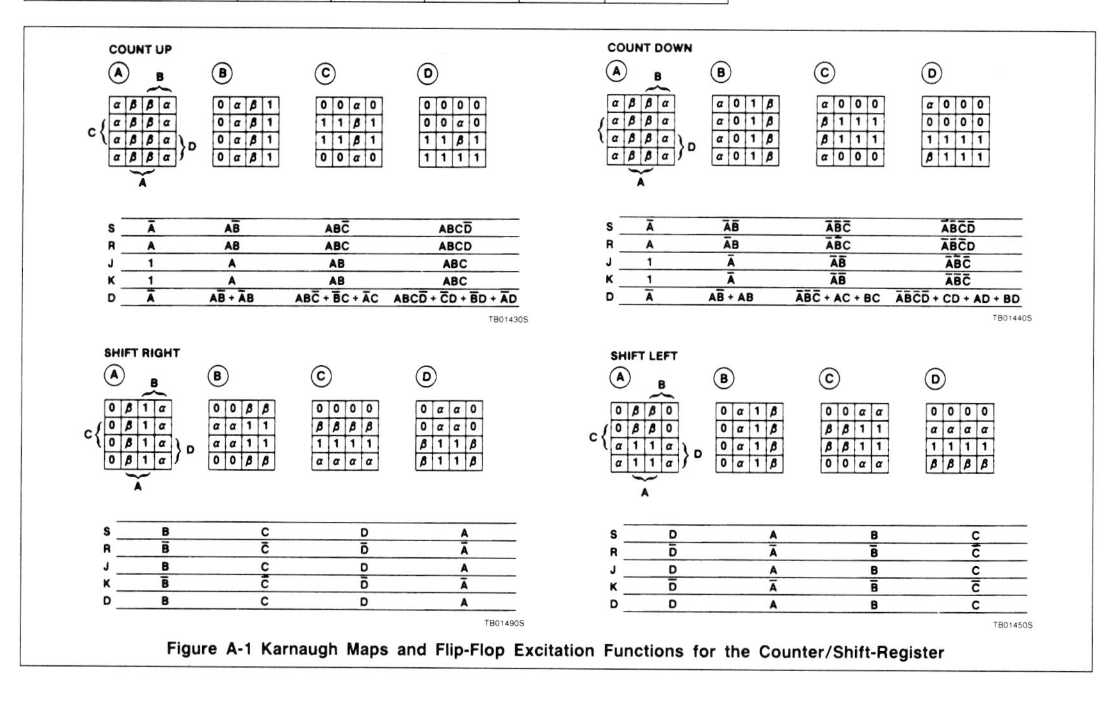

Figure A-1 Karnaugh Maps and Flip-Flop Excitation Functions for the Counter/Shift-Register

Table A-5. PLS155 FPLS Programming Table for the Counter/Shift-Register

PROGRAM TABLE ENTRIES

C	
INACTIVE	O
GENERATE	A
PROPAGATE	•
TRANSPARENT	—
AND	

$(Q = J/K)$	
TOGGLE	O
SET	H
RESET	L
HOLD	—
(OR)	

E_A, E_B	
IDLE	O
CONTROL	A
ENABLE	•
DISABLE	—

I, B(I), Q(P)	
INACTIVE	O
I, B, Q	H
$\bar{I}, \bar{B}, \bar{Q}$	L
DON'T CARE	—
(AND)	

P, R, B(O), (Q = D)	
ACTIVE	A
INACTIVE	•
(OR)	

(POL.)	
HIGH	H
LOW	L

F/F TYPE	
J/K	•
D OR J/K	A
(OR)	

F/F TYPE: A A A A

$E_B = A$, $E_A = A$: A A A A

POLARITY: (7 blank) H

AND

TERM	C	I 3	I 2	I 1	I 0	B(I) 7	B(I) 6	B(I) 5	B(I) 4	B(I) 3	B(I) 2	B(I) 1	B(I) 0	Q(P) 3	Q(P) 2	Q(P) 1	Q(P) 0
0	•	L	H	L	H	L		—	—	—	—	—	—	—	—	—	—
1	•	L	H	L	H	L		C	O	UN	T	—	—	—	—	—	L
2	•	L	H	L	H	L		—	D	O	W	N	—	—	—	L	L
3	•	L	H	L	H	L		—	—	—	—	—	—	—	L	L	L
4	•	L	L	H	H	L		—	—	—	—	—	—	—	—	—	—
5	•	L	L	H	H	L		CO	UN	T	—	—	—	—	—	—	H
6	•	L	L	H	H	L			U	P	—	—	—	—	—	H	H
7	•	L	L	H	H	L		—	—	—	—	—	—	—	H	H	H
8	•	L	H	L	L	L		—	—	—	—	—	—	H	—	—	—
9	•	L	H	L	L	L		SH	IF	T	—	—	—	—	—	—	H
10	•	L	H	L	L	L		—	LE	FT	—	—	—	—	—	H	—
11	•	L	H	L	L	L		—	—	—	—	—	—	—	H	—	—
12	•	L	L	H	L	L		—	—	—	—	—	—	—	—	H	—
13	•	L	L	H	L	L		SH	IF	T	—	—	—	—	H	—	—
14	•	L	L	H	L	L		—	RIG	HT	—	—	—	H	—	—	—
15	•	L	L	H	L	L		—	—	—	—	—	—	—	—	—	H
16	•	H	—	—	—	L	—	INI	TIA	LI	ZE	—	—	—	—	—	—
17	•	—	—	—	—	H	—	R	E	SE	T	—	—	—	—	—	—
18	A	—	—	—	—	—	L	L	—	S	TO	P	—	—	—	—	—
19																	
20																	
21																	
22																	
23																	
24																	
25																	
26																	
27																	
28																	
29																	
30																	
31																	
F_C	—	—	—	—	L	—	—	—	—	—	—	—	—	—	—	—	—
L_B																	
L_A																	
D_7	•	0	0	0	0	0	0	0	0	0	0	0	0	0	0	0	0
D_6	•	0	0	0	0	0	0	0	0	0	0	0	0	0	0	0	0
D_5	•	0	0	0	0	0	0	0	0	0	0	0	0	0	0	0	0
D_4																	
D_3																	
D_2																	
D_1																	
D_0	—	—	—	—	—	—	—	L	—	—	—	—	—	—	—	—	—
		5	4	3	2	19	18	13	12	9	8	7	6	17	16	15	14
VARIABLE NAME		INIT	LEFT/DOWN	RIGHT/UP	COUNT/$\overline{SHIFT}$	RESET	$\overline{HALT}$	$\overline{BUSY}$					ACTIVE	D	C	B	A

OR

TERM	Q(N) 3	Q(N) 2	Q(N) 1	Q(N) 0	P B	P A	B	A	B(O) 7	B(O) 6	B(O) 5	B(O) 4	B(O) 3	B(O) 2	B(O) 1	B(O) 0
0	—	—	—	O	—	—	—	—	—	—	—	—	—	—	—	A
1	—	—	O	—	—	—	—	—	—	—	—	—	—	—	—	A
2	—	O	—	—	—	—	—	—	—	—	—	—	—	—	—	A
3	O	—	—	—	—	—	—	—	—	—	—	—	—	—	—	A
4	—	—	—	O	—	—	—	—	—	—	—	—	—	—	—	A
5	—	—	O	—	—	—	—	—	—	—	—	—	—	—	—	A
6	—	O	—	—	—	—	—	—	—	—	—	—	—	—	—	A
7	O	—	—	—	—	—	—	—	—	—	—	—	—	—	—	A
8	—	—	—	H	—	—	—	—	—	—	—	—	—	—	—	A
9	—	—	H	—	—	—	—	—	—	—	—	—	—	—	—	A
10	—	H	—	—	—	—	—	—	—	—	—	—	—	—	—	A
11	H	—	—	—	—	—	—	—	—	—	—	—	—	—	—	A
12	—	—	—	H	—	—	—	—	—	—	—	—	—	—	—	A
13	—	—	H	—	—	—	—	—	—	—	—	—	—	—	—	A
14	—	H	—	—	—	—	—	—	—	—	—	—	—	—	—	A
15	H	—	—	—	—	—	—	—	—	—	—	—	—	—	—	A
16	L	L	L	L	—	—	—	—	—	—	—	—	—	—	—	•
17	—	—	—	—	—	—	A	A	—	—	—	—	—	—	—	•
18	—	—	—	—	—	—	—	—	—	—	—	—	—	—	—	•
19–31																

TB01470S

Table A-6. Modified PLS155 FPLS Programming Table for the Counter/Shift-Register With the Addition of a Binary Multiplier

PROGRAM TABLE ENTRIES

E_A, E_B	
IDLE	O
CONTROL	A
ENABLE	•
DISABLE	-

(Q = J/K)	
TOGGLE	O
SET	H
RESET	L
HOLD	-

(OR)

F/F TYPE	
J/K	•
D OR J/K	A

(OR)

C	
INACTIVE	O
GENERATE	A
PROPAGATE	•
TRANSPARENT	-

AND

(POL.)	
HIGH	H
LOW	L

P, R, B(O), (Q = D)	
ACTIVE	A
INACTIVE	•

(OR)

I, B(I), Q(P)	
INACTIVE	O
I, B, Q	H
Ī, B̄, Q̄	L
DON'T CARE	-

(AND)

														F/F TYPE				E_B = A		E_A = A						POLARITY							
														A	A	A	A	A	A	A	A												H
TERM	AND																	OR															
	C	I				B(I)								Q(P)				Q(N)				P				B(O)							
		3	2	1	0	7	6	5	4	3	2	1	0	3	2	1	0	3	2	1	0	B	A	B	A	7	6	5	4	3	2	1	0
0	•	L	H	L	H	L	-	-	-	-	-	-	-	-	-	-	-	-	-	-	O	-	-	-	-	-	-	-	-	-	-	-	A
1	•	L	H	L	H	L	-	-	-	-	-	-	-	-	-	-	L	-	-	O	-	-	-	-	-	-	-	-	-	-	-	-	A
2	•	L	H	L	H	L	-	-	-	-	-	-	-	-	-	L	L	-	O	-	-	-	-	-	-	-	-	-	-	-	-	-	A
3	•	L	H	L	H	L	-	-	-	-	-	-	-	-	L	L	L	O	-	-	-	-	-	-	-	-	-	-	-	-	-	-	A
4	•	L	L	H	H	L	-	-	-	-	-	-	-	-	-	-	-	-	-	-	O	-	-	-	-	-	-	-	-	-	-	-	A
5	•	L	L	H	H	L	-	-	-	-	-	-	-	-	-	-	H	-	-	O	-	-	-	-	-	-	-	-	-	-	-	-	A
6	•	L	L	H	H	L	-	-	-	-	-	-	-	-	-	H	H	-	O	-	-	-	-	-	-	-	-	-	-	-	-	-	A
7	•	L	L	H	H	L	-	-	-	-	-	-	-	-	H	H	H	O	-	-	-	-	-	-	-	-	-	-	-	-	-	-	A
8	•	L	H	L	L	L	-	-	L	H	-	-	-	H	-	-	-	-	-	-	H	-	-	-	-	-	-	-	-	-	-	-	A
9	•	L	H	L	L	L	-	-	H	L	-	-	-	-	H	-	-	-	-	-	H	-	-	-	-	-	-	-	-	-	-	-	A
10	•	L	H	L	L	L	-	-	H	H	-	-	-	-	-	H	-	-	-	-	H	-	-	-	-	-	-	-	-	-	-	-	A
11	•	L	H	L	L	L	-	-	L	H	-	-	-	-	-	-	H	-	-	H	-	-	-	-	-	-	-	-	-	-	-	-	A
12	•	L	H	L	L	L	-	-	H	L	-	-	-	H	-	-	-	-	-	H	-	-	-	-	-	-	-	-	-	-	-	-	A
13	•	L	H	L	L	L	-	-	H	H	-	-	-	-	H	-	-	-	-	H	-	-	-	-	-	-	-	-	-	-	-	-	A
14	•	L	H	L	L	L	-	-	L	H	-	-	-	-	-	H	-	-	H	-	-	-	-	-	-	-	-	-	-	-	-	-	A
15	•	L	H	L	L	L	-	-	H	L	-	-	-	-	-	-	H	-	H	-	-	-	-	-	-	-	-	-	-	-	-	-	A
16	•	L	H	L	L	L	-	-	H	H	-	-	-	H	-	-	-	-	H	-	-	-	-	-	-	-	-	-	-	-	-	-	A
17	•	L	H	L	L	L	-	-	L	H	-	-	-	-	H	-	-	H	-	-	-	-	-	-	-	-	-	-	-	-	-	-	A
18	•	L	H	L	L	L	-	-	H	L	-	-	-	-	-	H	-	H	-	-	-	-	-	-	-	-	-	-	-	-	-	-	A
19	•	L	H	L	L	L	-	-	H	H	-	-	-	-	-	-	H	H	-	-	-	-	-	-	-	-	-	-	-	-	-	-	A
20	•	L	L	H	L	L	-	-	-	-	-	-	-	-	-	H	-	-	-	-	H	-	-	-	-	-	-	-	-	-	-	-	A
21	•	L	L	H	L	L	-	-	-	-	-	-	-	-	H	-	-	-	-	H	-	-	-	-	-	-	-	-	-	-	-	-	A
22	•	L	L	H	L	L	-	-	-	-	-	-	-	H	-	-	-	-	H	-	-	-	-	-	-	-	-	-	-	-	-	-	A
23	•	L	L	H	L	L	-	-	-	-	-	-	-	-	-	-	H	H	-	-	-	-	-	-	-	-	-	-	-	-	-	-	A
24	•	H	-	-	-	L	-	-	-	-	-	-	-	-	-	-	-	-	-	-	-	-	-	-	-	-	-	-	-	-	-	-	•
25	•	-	-	-	-	H	-	-	-	-	-	-	-	-	-	-	-	-	-	-	-	-	-	-	-	-	-	-	-	-	-	-	•
26	A	-	-	-	-	-	L	L	-	-	-	-	-	-	-	-	-	-	-	-	-	-	-	-	-	-	-	-	-	-	-	-	•
27																																	
28																																	
29																																	
30																																	
31																																	
F_C	-	-	-	-	L	-	-	-	-	-	-	-	-	-	-	-	-																
L_B																																	
L_A																																	
D_7	•	O	O	O	O	O	O	O	O	O	O	O	O	O	O	O	O																
D_6	•	O	O	O	O	O	O	O	O	O	O	O	O	O	O	O	O																
D_5	•	O	O	O	O	O	O	O	O	O	O	O	O	O	O	O	O																
D_4	•	O	O	O	O	O	O	O	O	O	O	O	O	O	O	O	O																
D_3	•	O	O	O	O	O	O	O	O	O	O	O	O	O	O	O	O																
D_2																																	
D_1																																	
D_0	-	-	-	-	-	-	-	L	-	-	-	-	-	-	-	-	-																
		5	4	3	2	19	18	13	12	9	8	7	6	17	16	15	14																
VARIABLE NAME		INIT (overbarred)	LEFT/DOWN	RIGHT/UP	COUNT/SHIFT	RESET	HALT (overbarred)	BUSY (overbarred)					ACTIVE	D	C	B	A																

TB01460S

Solutions to Odd-Numbered Exercises

CHAPTER 1

General Exercises

1. **a.** 101011 **b.** 10110000 **c.** 1001100 **d.** 10010

3. **a.** 25 **b.** 42 **c.** 54 **d.** 29

5. **a.** A **b.** F **c.** 10 **d.** 1F **e.** 20

7. **a.** $1 = 1 \cdot 2^0$

b. $1010 = 1 \cdot 2^3 + 0 \cdot 2^2 + 1 \cdot 2^1 + 0 \cdot 2^0$

c. $10010 = 1 \cdot 2^4 + 0 \cdot 2^3 + 0 \cdot 2^2 + 1 \cdot 2^1 + 0 \cdot 2^0$

d. $100010 = 1 \cdot 2^5 + 0 \cdot 2^4 + 0 \cdot 2^3 + 0 \cdot 2^2 + 1 \cdot 2^1 + 0 \cdot 2^0$

9. **a.**
$$\begin{array}{r} 10001 \\ 11011 \\ \hline 101100 \end{array} = 44$$

b.
$$\begin{array}{r} 101 \\ 010 \\ \hline 111 \end{array} = 7$$

11. **a.** $1 \cdot 2^3 + 1 \cdot 2^2 + 1 \cdot 2^1 + 0 \cdot 2^0$ **b.** $1 \cdot 2^4 + 0 \cdot 2^3 + 1 \cdot 2^2 + 1 \cdot 2^1 + 1 \cdot 2^0$

13. **a.**
$$\begin{array}{r} 76 \\ -55 \\ \hline 21 \end{array} \qquad \begin{array}{r} 76 \\ +44 \\ \hline 120 \\ \rightarrow 1 \\ \hline 21 \end{array}$$

b.
$$\begin{array}{r} 5 \\ -4 \\ \hline 1 \end{array} \qquad \begin{array}{r} 5 \\ +5 \\ \hline 10 \\ \rightarrow 1 \\ \hline 1 \end{array}$$

c.
$$\begin{array}{r} 7 \\ -5 \\ \hline 2 \end{array} \qquad \begin{array}{r} 7 \\ +4 \\ \hline 11 \\ \rightarrow 1 \\ \hline 2 \end{array}$$

d.
$$\begin{array}{r} 171 \\ -028 \\ \hline 143 \end{array} \qquad \begin{array}{r} 171 \\ +971 \\ \hline 1142 \\ \rightarrow 1 \\ \hline 143 \end{array}$$

15.

1s Complement	2s Complement
1011	1100
1110	1111
1001	1010
1000	1001

17.

x_1x_0 \ y_1y_0	00	01	11	10
00	0000	0001 →	0011 →	0010
01	0100 →	(0101)	0111	0110
11	1100 ←	1101 ←	1111 ←	1110
10	1000	1001	1011	1010

0001 0011 0010 0110 1110

0101 0100 1100 1101 1111

CHAPTER 2

General Exercises

1. $F = A'C + B$

3. $A'C + AB$

5. $F = A(B + B')(C + C') + (A + A')BC$
$= ABC + ABC' + AB'C + AB'C' + ABC + A'BC$
$= ABC + ABC' + AB'C + AB'C' + A'BC$

7. $F = A'B'(C + C') + A'(B + B')C$
$= A'B'C + A'B'C' + A'BC + A'B'C$
$= A'B'C + A'B'C' + A'BC$

9. $G = (A + B')(B + C)$

11. Dual $G = (A' + B)(B' + C)$
$F' = (A + B')(B + C')$

13. $F = (F')' = [(A + BC + CD')']'$
$= [A' \cdot (BC)' \cdot (CD')']'$

15.

x	y	Z	$x + (y + Z)$	$(x + y) + Z$
0	0	0	$0 + 0 = 0$	$0 + 0 = 0$
0	0	1	$0 + 1 = 1$	$0 + 1 = 1$
0	1	0	$0 + 1 = 1$	$1 + 0 = 1$
0	1	1	$0 + 1 = 1$	$1 + 1 = 1$
1	0	0	$1 + 0 = 1$	$1 + 0 = 1$
1	0	1	$1 + 1 = 1$	$1 + 1 = 1$
1	1	0	$1 + 1 = 1$	$1 + 0 = 1$
1	1	1	$1 + 1 = 1$	$1 + 1 = 1$

x	y	Z	$x \cdot (y \cdot Z)$	$(x \cdot y) \cdot Z$
0	0	0	$0 \cdot 0 = 0$	$0 \cdot 0 = 0$
0	0	1	$0 \cdot 0 = 0$	$0 \cdot 1 = 0$
0	1	0	$0 \cdot 0 = 0$	$0 \cdot 0 = 0$
0	1	1	$0 \cdot 1 = 0$	$0 \cdot 1 = 0$
1	0	0	$1 \cdot 0 = 0$	$0 \cdot 0 = 0$
1	0	1	$1 \cdot 0 = 0$	$0 \cdot 1 = 0$
1	1	0	$1 \cdot 0 = 0$	$1 \cdot 0 = 0$
1	1	1	$1 \cdot 1 = 1$	$1 \cdot 1 = 1$

17. $AB + A'C = AB(C + C') + A'C(B + B')$
$= ABC + ABC' + A'BC + A'B'C$
$= ABC + ABC' + A'BC + A'B'C + A'BC + ABC$ Idempotent
$= AB(C + C') + A'C(B + B') + BC(A' + A)$
$= AB + A'C + BC$

Design/Implementation Exercises

1.

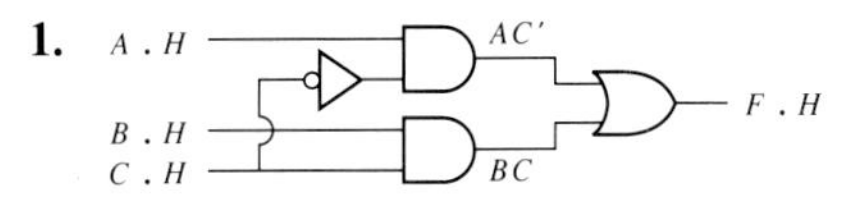

3. $F = A \cdot B + B \cdot C$

2 1 2

A . H
B . H
C . H
AB
BC
F . H = AB + BC

5. $F = (A \cdot C)' + B$

2 1

A . H
C . H
B . H
(AC)'
F . L

7.

B . H
A . H
C . H
F . H

9. $A \cdot B' + C \cdot D$

2 1 2

A . H
B . H
C . H
D . H
F . H

11. $G = ABC' + AB'C + A'BC + ABC$

13. $D = 1$ denotes door is open
$W = 1$ denotes window is open
$S = 1$ denotes skylight is open
$E = 1$ denotes switch is closed

E
D
W
S
A

15.
$$F = A'B'CD' + A'BCD + AB'C'D' + ABC'D$$
$$= A'C(B'D' + BD) + AC'(B'D' + BD)$$
$$= (A'C + AC') \cdot (B'D' + BD)$$

17.

F	R	C	Danger	
0	0	0	0	No one crosses
0	0	1	1	Corn crosses
0	1	0	0	Rooster crosses (Trip 2)
0	1	1	1	Rooster, corn cross
1	0	0	1	Fox crosses
1	0	1	0	Fox, corn cross (Trip 1)
1	1	0	1	Fox, rooster cross
1	1	1	1	All cross

2 trips

CHAPTER 3

General Exercises

1. **a.** $F = A'C + AC'$

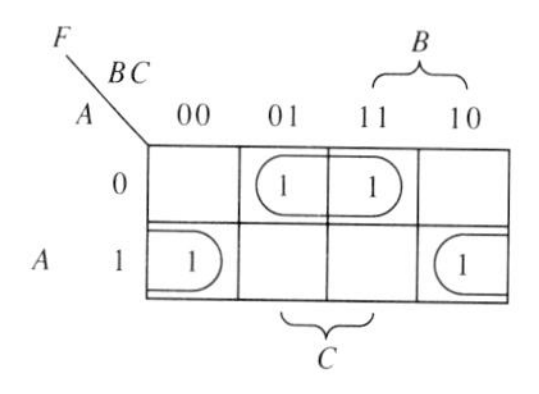

b. $F = A'B'C + A'BC + AB'C' + ABC'$
$= A'C(B' + B) + AC'(B' + B)$
$= A'C + AC'$

3. $F = A'B' + AC$

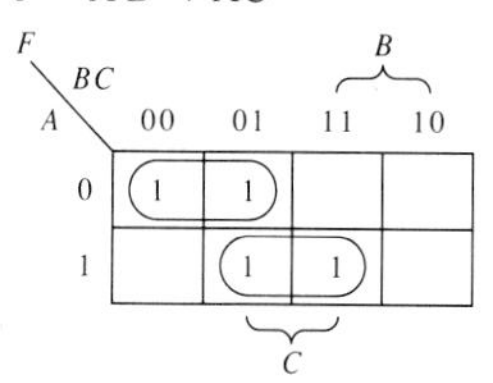

5. $F = A'B'C' + A'BC' + AB'C' + ABC + ABC'$

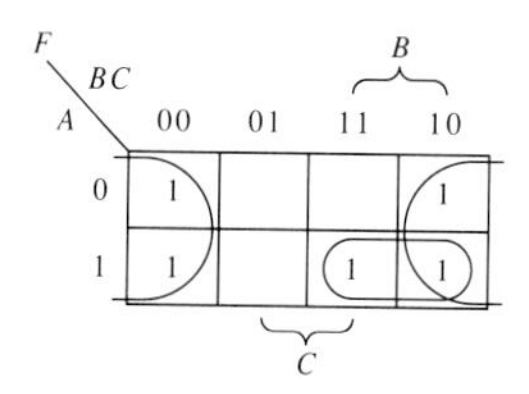

7. $F = m_5 + m_7 + m_6$
$F' = m_0 + m_1 + m_2 + m_4$
$F = (F')' = m_0' \cdot m_1' \cdot m_2' \cdot m_3' \cdot m_4'$
$= m_0 \cdot m_1 \cdot m_2 \cdot m_3 \cdot m_4$
$= (A + B + C)(A + B + C')(A + B' + C)(A + B' + C')(A' + B + C)$

A \ BC	00	01	11	10
0	0	0	0	0
1	0	1	1	1

9. $F = AB + B'C$

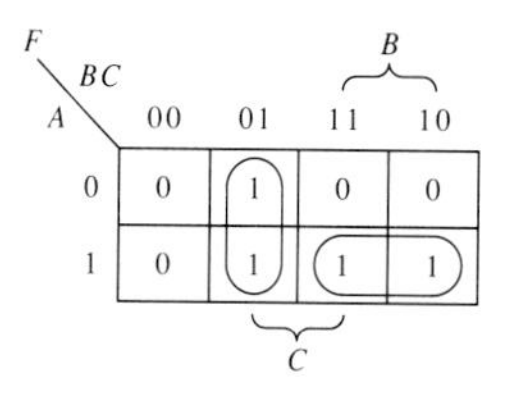

11.

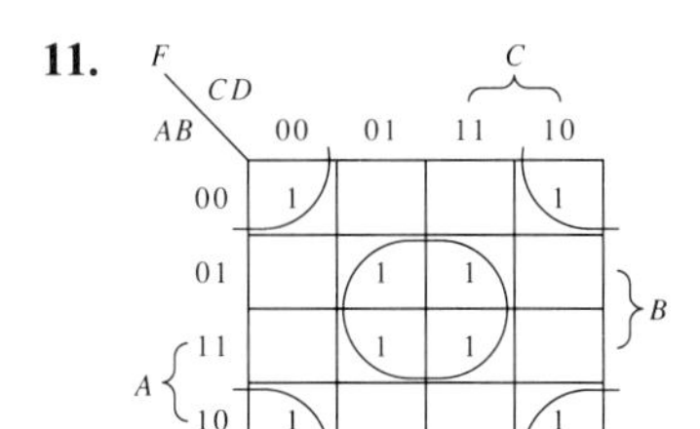

13. EPIs: ABD', $A'B'D$, BC

PIs: ABD', $A'B'D$, BC, $A'CD$

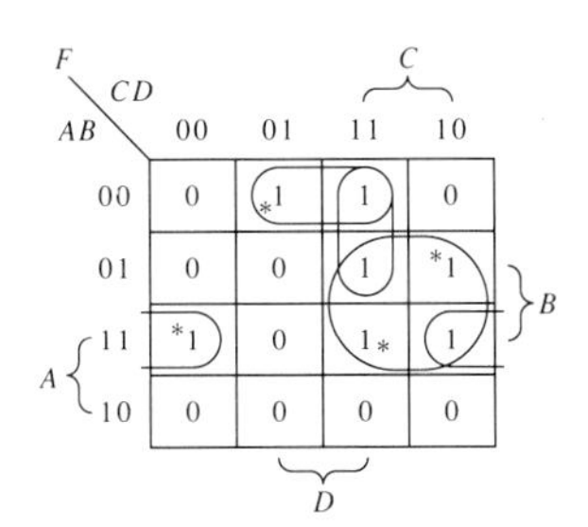

15. $F = A + B'C$

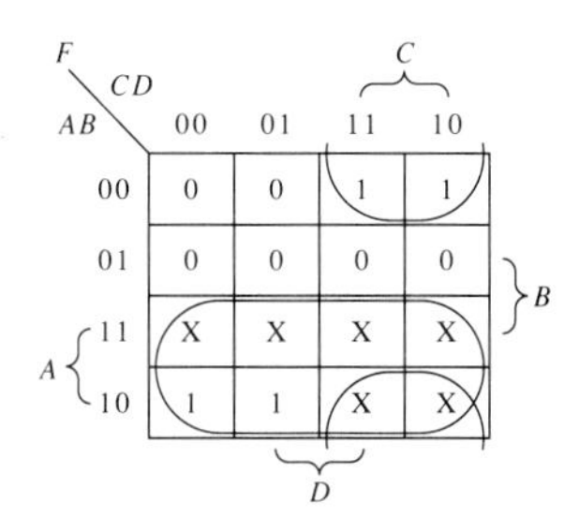

17. EPIs: $B'C'D'$, BCD', $A'CD$

Unchecked terms: 5, 9, 13

$$F = B'C'D' + BCD' + A'CD + \begin{cases} A'BD + AC'D \\ AB'C' + BC'D \end{cases}$$

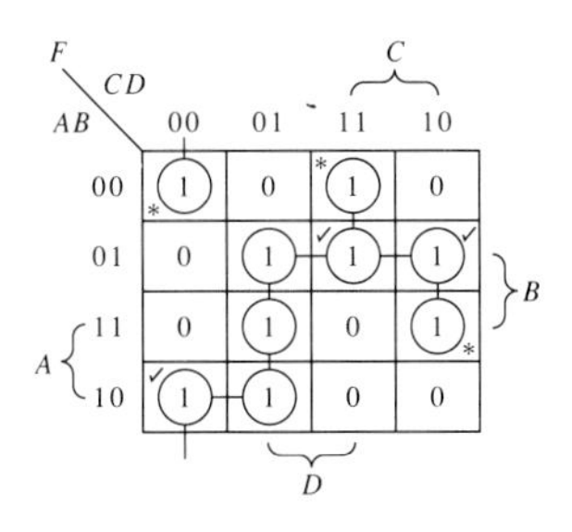

19. EPIs: BC', $AB'D$, $A'CD$

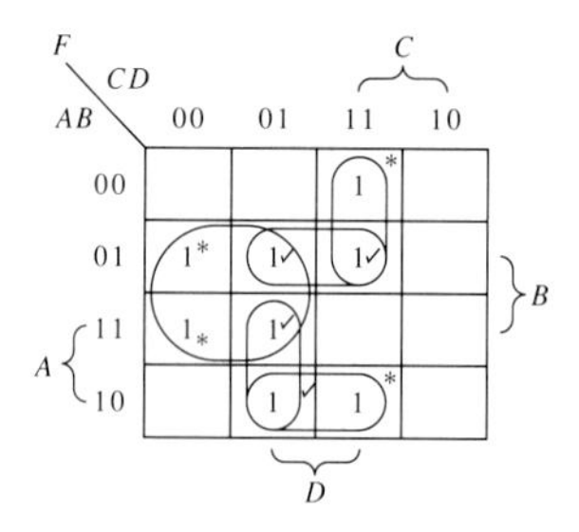

21.

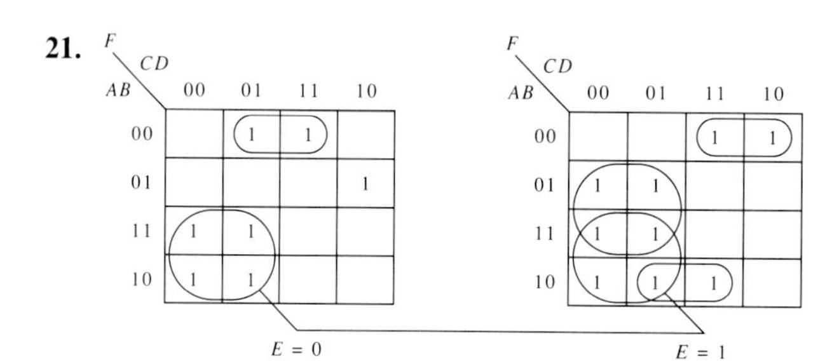

23.

Dec.	E	ABCD	f	MEV*
0	0	0000		
16	1	0000		
1	0	0001	1	E'
17	1	0001		
2	0	0010		
18	1	0010	1	E
3	0	0011	1	1
19	1	0011	1	
4	0	0100		
20	1	0100	1	E
5	0	0101		
21	1	0101	1	E
6	0	0110	1	E'
22	1	0110		
7	0	0111		
23	1	0111		
8	0	1000	1	1
24	1	1000	1	
9	0	1001	1	1
25	1	1001	1	
10	0	1010		
26	1	1010		
11	0	1011		
27	1	1011	1	E
12	0	1100	1	1
28	1	1100	1	
13	0	1101	1	1
29	1	1101	1	
14	0	1110		
30	1	1110		
15	0	1111		
31	1	1111		

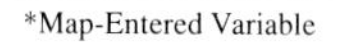
*Map-Entered Variable

Variable-entered map (alternate form)

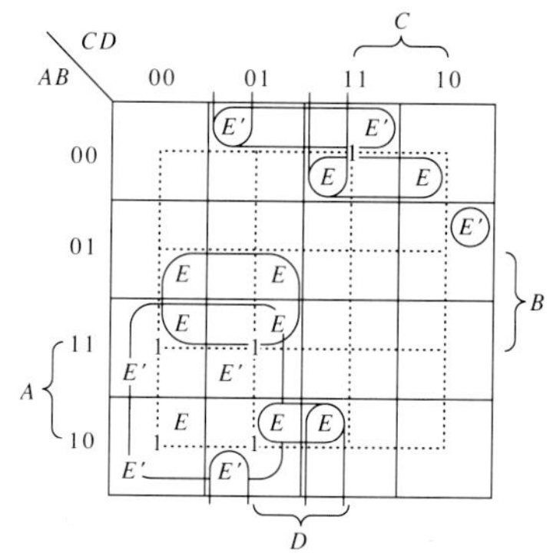

Note similarity to 5-variable minterm-ring map

Variable-entered map

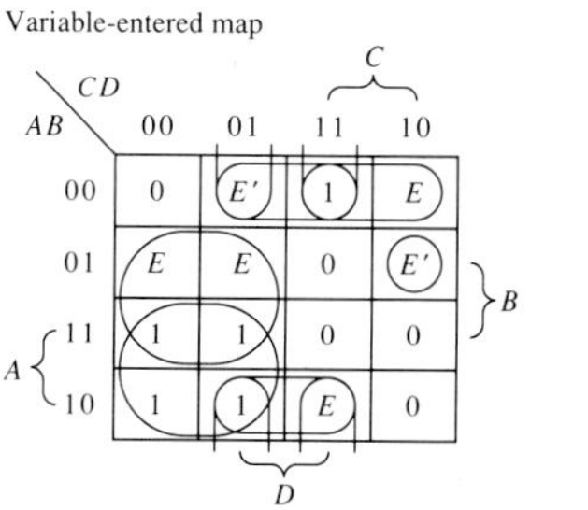

A 5-variable minterm-ring map

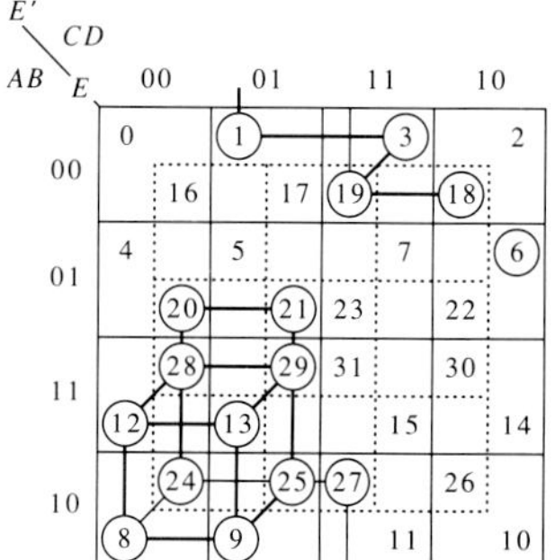

25.

1s	M	Pairs	Quartets	Octets	
5	31✓				
4	30✓	(31, 30)✓			
	23✓	(31, 23)✓			
	15✓	(31, 15)✓			
3	28✓	(30, 28)✓			
	22✓	(30, 22)✓			
	14✓	(30, 14)✓	(31, 30, 15, 14)✓		
	7✓	(23, 22)✓	(31, 30, 23, 22)✓		
		(23, 7)✓			
		(15, 7)✓	(31, 23, 15, 7)✓		
		(15, 14)✓			
2	18✓	(28, 12)✓	(30, 14, 28*, 12)		= ABD' (EPI)
	17✓	(22, 18)			
	12✓	(22, 6)✓			
	6✓	(14, 12)✓			
		(14, 6)✓	(30, 22, 14, 6)✓		EPIs
		(7, 6)✓	(23, 22, 7, 6)✓		
			(15, 14, 7, 6)✓	(31*, 30, 23*, 22, 15*, 14, 7*, 6)	= BC (EPI)
1	16✓	(18, 16)			
	4✓	(17*, 16)			= $EA'B'C'$ (EPI)
		(12, 4)✓			
		(6, 4)✓	(14, 12, 6, 4)		
0	0	(16, 0)			
		(4, 0)			

$$f(E, A, B, C, D) = \Sigma(0, 4, 6, 7, 12, 14, 15, 16, 17, 18, 22, 23, 28, 30, 31)$$

Prime implicant table

Minterms

			✓	*	✓	✓	*	✓	*		✓	*	*	✓	*	
PIs	0	4	6	7	12	14	15	16	17	18	22	23	28	30	31	EPIs
(31*, 30, 23*, 22 *15, 14, 7*, 6)			x	x		x	x				x	x		x	x	= BC
(30, 14, 28*, 12)					x	x							x	x		= ABD'
(14, 12, 6, 4)		x	x		x	x										
(17*, 16)								x	x							= $EA'B'C'$
(22, 18)											x	x				
(18, 16)									x		x					
(16, 0)	x							x								
(4, 0)	x	x														

Reduced PI table

	PIs	0	4	18	PIs	
Dominated by (4, 0)	(14, 12, 6, 4)		x			
Choice	(22, 18)			x	$= EA'CD'$	Choice
Choice	(18, 16)			x	$= EA'B'D'$	Choice
	(4, 0)	x	x		$= E'A'C'D'$	
Dominated by (4, 0)	(16, 0)	x				

Design/Implementation Exercises

1. $F = A'B'C' + ABC$

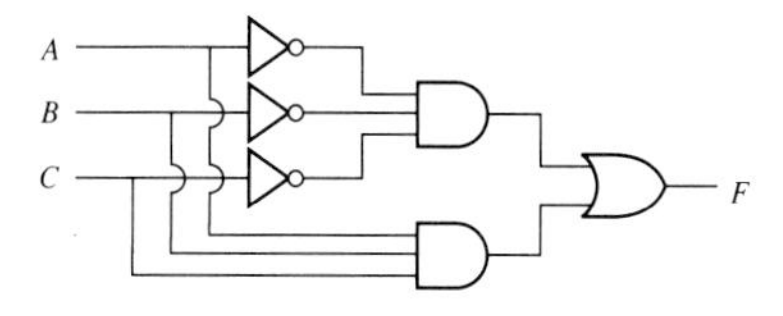

3. **a.** $F = A'B'C' + A'B'C + A'BC' + AB'C'$

A	B	C	F
0	0	0	1
0	0	1	1
0	1	0	1
0	1	1	0
1	0	0	1
1	0	1	0
1	1	0	0
1	1	1	0

b. $F = A'B' + A'C' + B'C'$

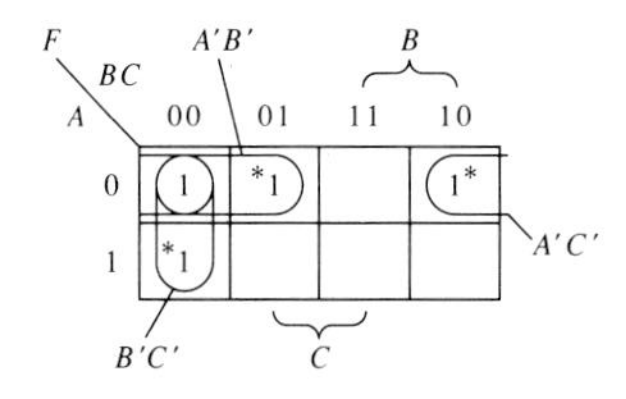

c.

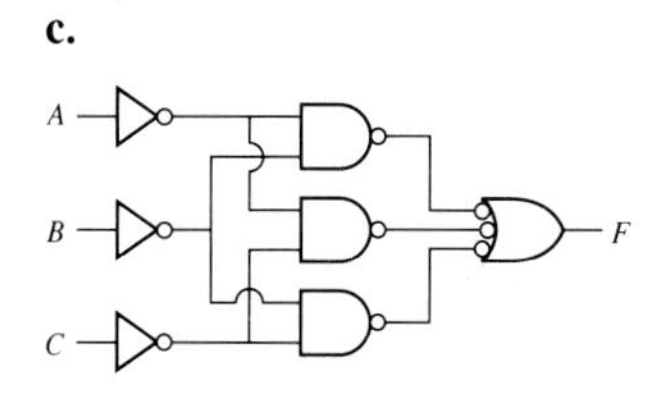

CHAPTER 4

General Exercises

1.

Gray				Binary			
A	*B*	*C*	*D*	*W*	*X*	*Y*	*Z*
0	0	0	0	0	0	0	0
0	0	0	1	0	0	0	1
0	0	1	1	0	0	1	0
0	0	1	0	0	0	1	1
0	1	1	0	0	1	0	0
0	1	1	1	0	1	0	1
0	1	0	1	0	1	1	0
0	1	0	0	0	1	1	1
1	1	0	0	1	0	0	0
1	1	0	1	1	0	0	1
1	1	1	1	1	0	1	0
1	1	1	0	1	0	1	1
1	0	1	0	1	1	0	0
1	0	1	1	1	1	0	1
1	0	0	1	1	1	1	0
1	0	0	0	1	1	1	1

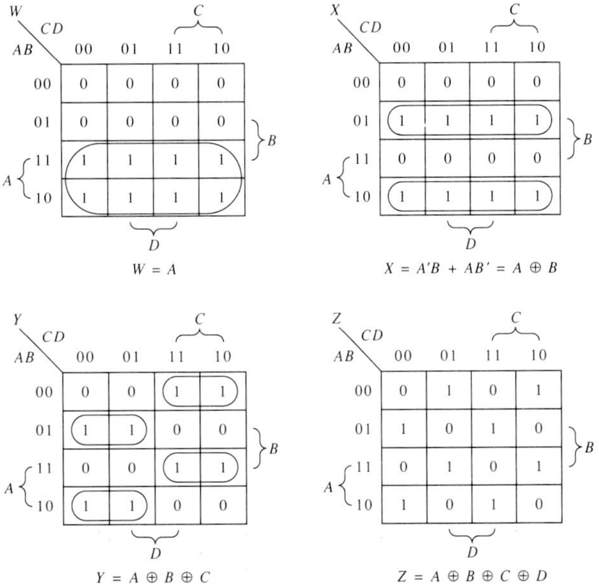

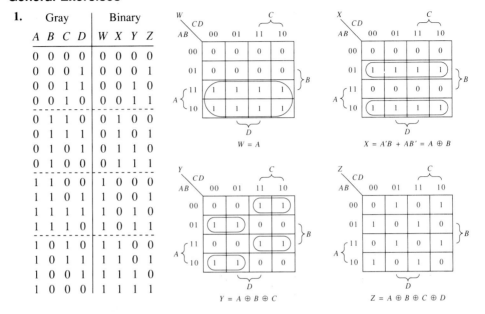

3. 3-bit multiplication: $Y = FEA$, $X = BCD$; Product $= p_5p_4p_3p_2p_1p_0$.

Truth table

Y Decimal	*X* Decimal	Minterm Number (Binary) *F*	*E*	*A*	*B*	*C*	*D*	Product Dec.	p_5	p_4	p_3	p_2	p_1	p_0	Minterm Number (Decimal)
0	0	0	0	0	0	0	0	0	0	0	0	0	0	0	0
	1				0	0	1	0	0	0	0	0	0	0	1
	2				0	1	0	0	0	0	0	0	0	0	2
	3				0	1	1	0	0	0	0	0	0	0	3
	4				1	0	0	0	0	0	0	0	0	0	4
	5				1	0	1	0	0	0	0	0	0	0	5
	6				1	1	0	0	0	0	0	0	0	0	6
	7				1	1	1	0	0	0	0	0	0	0	7
1	0	0	0	1	0	0	0	0	0	0	0	0	0	0	8
	1				0	0	1	1	0	0	0	0	0	1	9
	2				0	1	0	2	0	0	0	0	1	0	10
	3				0	1	1	3	0	0	0	0	1	1	11
	4				1	0	0	4	0	0	0	1	0	0	12
	5				1	0	1	5	0	0	0	1	0	1	13
	6				1	1	0	6	0	0	0	1	1	0	14
	7				1	1	1	7	0	0	0	1	1	1	15
2	0	0	1	0	0	0	0	0	0	0	0	0	0	0	16
	1				0	0	1	2	0	0	0	0	1	0	17

continues

Y Decimal	X Decimal	F	E	A	B	C	D	Product Dec.	p_5	p_4	p_3	p_2	p_1	p_0	Minterm Number (Decimal)
		Minterm Number (Binary)													
	2				0	1	0	4	0	0	0	1	0	0	18
	3				0	1	1	6	0	0	0	1	1	0	19
	4				1	0	0	8	0	0	1	0	0	0	20
	5				1	0	1	10	0	0	1	0	1	0	21
	6				1	1	0	12	0	0	1	1	0	0	22
	7				1	1	1	14	0	0	1	1	1	0	23
3	0	0	1	1	0	0	0	0	0	0	0	0	0	0	24
	1				0	0	1	3	0	0	0	0	1	1	25
	2				0	1	0	6	0	0	0	1	1	0	26
	3				0	1	1	9	0	0	1	0	0	1	27
	4				1	0	0	12	0	0	1	1	0	0	28
	5				1	0	1	15	0	0	1	1	1	1	29
	6				1	1	0	18	0	1	0	0	1	0	30
	7				1	1	1	21	0	1	0	1	0	1	31
4	0	1	0	0	0	0	0	0	0	0	0	0	0	0	32
	1				0	0	1	4	0	0	0	1	0	0	33
	2				0	1	0	8	0	0	1	0	0	0	34
	3				0	1	1	12	0	0	1	1	0	0	35
	4				1	0	0	16	0	1	0	0	0	0	36
	5				1	0	1	20	0	1	0	1	0	0	37
	6				1	1	0	24	0	1	1	0	0	0	38
	7				1	1	1	28	0	1	1	1	0	0	39
5	0	1	0	1	0	0	0	0	0	0	0	0	0	0	40
	1				0	0	1	5	0	0	0	1	0	1	41
	2				0	1	0	10	0	0	1	0	1	0	42
	3				0	1	1	15	0	0	1	1	1	1	43
	4				1	0	0	20	0	1	0	1	0	0	44
	5				1	0	1	25	0	1	1	0	0	1	45
	6				1	1	0	30	0	1	1	1	0	0	46
	7				1	1	1	35	1	0	0	0	1	1	47
6	0	1	1	0	0	0	0	0	0	0	0	0	0	0	48
	1				0	0	1	6	0	0	0	1	1	0	49
	2				0	1	0	12	0	0	1	1	0	0	50
	3				0	1	1	18	0	1	0	0	1	0	51
	4				1	0	0	24	0	1	1	0	0	0	52
	5				1	0	1	30	0	1	1	1	1	0	53
	6				1	1	0	36	1	0	0	1	0	0	54
	7				1	1	1	42	1	0	1	0	1	0	55
7	0	1	1	1	0	0	0	0	0	0	0	0	0	0	56
	1				0	0	1	7	0	0	0	1	1	1	57
	2				0	1	0	14	0	0	1	1	1	0	58
	3				0	1	1	21	0	1	0	1	0	1	59
	4				1	0	0	28	0	1	1	1	0	0	60
	5				1	0	1	35	1	0	0	0	1	1	61
	6				1	1	0	42	1	0	1	0	1	0	62
	7				1	1	1	49	1	1	0	0	0	1	63

$p_0 = \Sigma(9, 11, 13, 15, 25, 27, 29, 31, 41, 43, 45, 47, 57, 59, 61, 63)$
$p_1 = \Sigma(10, 11, 14, 15, 17, 19, 21, 23, 25, 26, 29, 30, 42, 43, 47, 49, 51, 53, 55, 57, 58, 61, 62)$
$p_2 = \Sigma(12, 13, 14, 15, 18, 19, 22, 23, 26, 28, 29, 31, 33, 35, 37, 39, 41, 43, 44, 46, 49, 50, 53, 54, 57, 58, 59, 60)$
$p_3 = \Sigma(20, 21, 22, 23, 27, 28, 29, 34, 35, 38, 39, 42, 43, 45, 46, 50, 52, 53, 55, 58, 60, 62)$
$p_4 = \Sigma(30, 31, 36, 37, 38, 39, 44, 45, 46, 51, 52, 53, 59, 60, 63)$
$p_5 = \Sigma(47, 54, 55, 61, 62, 63)$

Construct a 6-variable minterm ring map to simplify each of the above expressions:
$p_5, p_4, p_3, p_2, p_1, p_0$.

5.

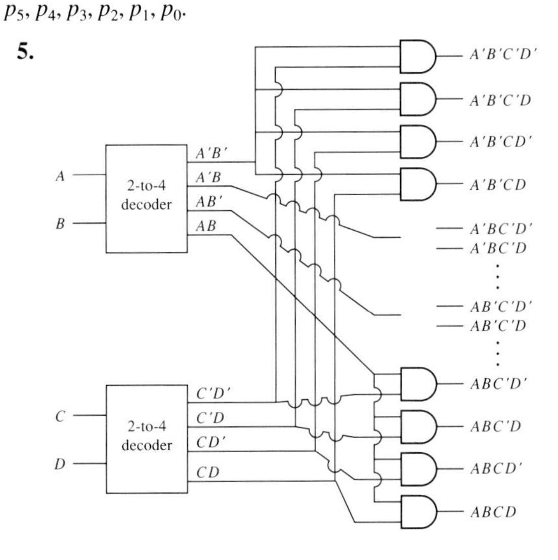

Design/Implementation Exercises

In Design Exercises that use the 4-step problem-solving procedure, the problem statement is Step 1 and will not be repeated here.

1. Step 2: Conceptualization. Truth table.

Inputs			Outputs							
A_2	A_1	A_0	O_0	O_1	O_2	O_3	O_4	O_5	O_6	O_7
0	0	0	1	0	0	0	0	0	0	0
0	0	1	0	1	0	0	0	0	0	0
0	1	0	0	0	1	0	0	0	0	0
0	1	1	0	0	0	1	0	0	0	0
1	0	0	0	0	0	0	1	0	0	0
1	0	1	0	0	0	0	0	1	0	0
1	1	0	0	0	0	0	0	0	1	0
1	1	1	0	0	0	0	0	0	0	1

Step 3: Solution/Simplification.

$O_0 = A'B'C'$ $\quad O_1 = A'B'C$

$O_2 = A'BC'$ $\quad O_3 = A'BC$

$O_4 = AB'C'$ $\quad O_5 = AB'C$

$O_6 = ABC'$ $\quad O_7 = ABC$

Step 4: Realization.

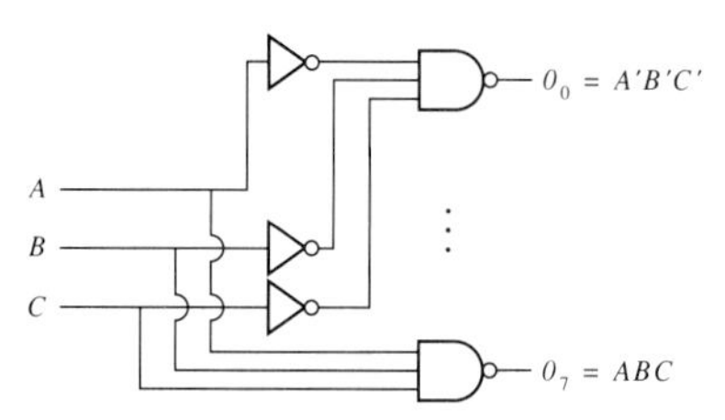

3. Step 2: Conceptualization. See General Exercise 1, truth table.

Step 3: Solution/Simplification. See General Exercise 1, Karnaugh maps.

$$W = A, \quad X = A \oplus B, \quad Y = A \oplus B \oplus C, \quad Z = A \oplus B \oplus C \oplus D$$

Step 4: Realization.

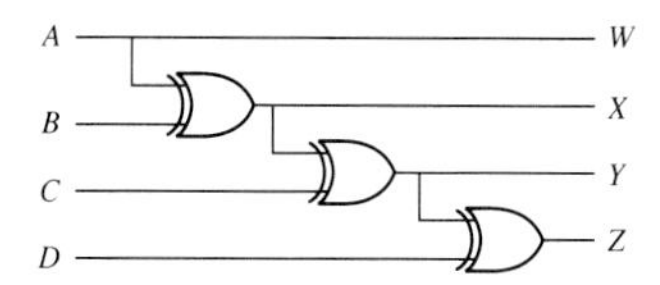

5. Step 2: Conceptualization. Truth table.

	Inputs				Outputs						
Dec	*A*	*B*	*C*	*D*	*a*	*b*	*c*	*d*	*e*	*f*	*g*
0	0	0	0	0	1	1	1	1	1	1	0
1	0	0	0	1	0	1	1	0	0	0	0
2	0	0	1	0	1	1	0	1	1	0	1
3	0	0	1	1	1	1	1	1	0	0	1
4	0	1	0	0	0	1	1	0	0	1	1
5	0	1	0	1	1	0	1	1	0	1	1
6	0	1	1	0	1	0	1	1	1	1	1
7	0	1	1	1	1	1	1	0	0	0	0
8	1	0	0	0	1	1	1	1	1	1	1
9	1	0	0	1	1	1	1	0	0	1	1
	1	0	1	0	} Don't Cares						
	⋮										

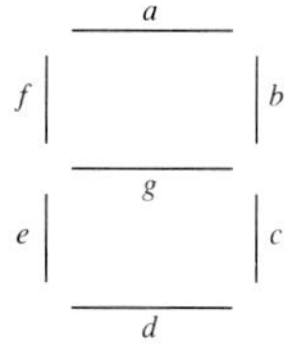

Step 3: Solution/Simplification. Write canonical SOP expressions for each output directly from truth table.

Step 4: Realization. Use a ROM to implement canonical SOP expressions determined in Step 3.

7. Step 2: Conceptualization. Truth table.

	Inputs				Outputs						
Hex	*A*	*B*	*C*	*D*	*a*	*b*	*c*	*d*	*e*	*f*	*g*
0	0	0	0	0	1	1	1	1	1	1	0
1	0	0	0	1	0	1	1	0	0	0	0
2	0	0	1	0	1	1	0	1	1	0	1
3	0	0	1	1	1	1	1	1	0	0	1
4	0	1	0	0	0	1	1	0	0	1	1
5	0	1	0	1	1	0	1	1	0	1	1
6	0	1	1	0	1	0	1	1	1	1	1
7	0	1	1	1	1	1	1	0	0	0	0
8	1	0	0	0	1	1	1	1	1	1	1
9	1	0	0	1	1	1	1	0	0	1	1
A	1	0	1	0	1	1	1	0	1	1	1
b	1	0	1	1	0	0	1	1	1	1	1
C	1	1	0	0	1	0	0	1	1	1	0
d	1	1	0	1	0	1	1	1	1	0	1
E	1	1	1	0	1	0	0	1	1	1	1
F	1	1	1	1	1	0	0	0	1	1	1

Step 3: Solution/Simplification. Write canonical SOP expression for each output directly from truth table.

Step 4: Realization. Use a ROM to implement canonical SOP expressions in Step 3.

9. Step 2: Conceptualization.

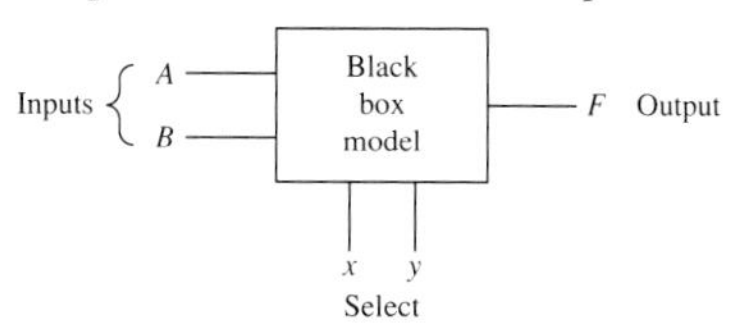

Step 3: Solution/Simplification.

$$F = x'y'A \cdot B + x'y(A + B) + xy'(A \oplus B) + xyA'$$

Step 4: Realization.

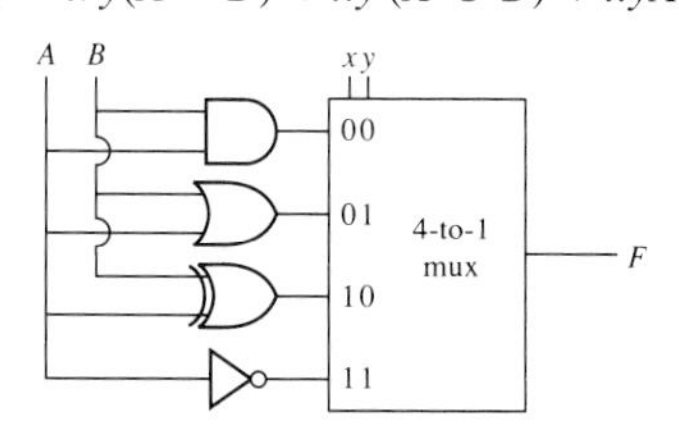

CHAPTER 5

General Exercises

1. **a.** 6 = 0110 **b.** 2 = 10010 (Drop)

3. **a.** 2 = 10010 (Drop) **b.** +8 = 1000 ↔ –8

Yes, subtraction [5 – (–3)] produces an erroneous negative result.

5.

```
            1000
   −5        1011
 + (−4)      1100
   −9        0111
 Positive ↗
```

7. Overflow = $c_i \oplus c_o$

9. See Section 5.2, Design of a 2-Bit Direct Binary Adder Without Cascading.

11. Speed is slow because the carry must ripple through each stage.

Design/Implementation Exercises

1.

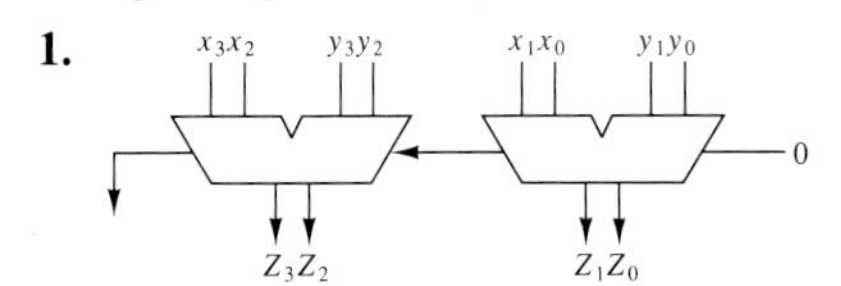

3. Use the mixed-logic algorithm with all inputs and outputs active high.

$C_1 = G_0 + C_0P_0$ (1, 2)

$C_2 = G_1 + G_0P_1 + C_0P_0P_1$ (1, 2, 1, 2)

5. Make up values for X and Y. MSB = sign bit.

CHAPTER 6

General Exercises

1. **a.** Truth table

a_1	a_0	b_1	b_0	E	G	L
0	0	0	0	1	0	0
0	0	0	1	0	0	1
0	0	1	0	0	0	1
0	0	1	1	0	0	1
0	1	0	0	0	1	0
0	1	0	1	1	0	0
0	1	1	0	0	0	1
0	1	1	1	0	0	1
1	0	0	0	0	1	0
1	0	0	1	0	1	0
1	0	1	0	1	0	0
1	0	1	1	0	0	1
1	1	0	0	0	1	0
1	1	0	1	0	1	0
1	1	1	0	0	1	0
1	1	1	1	1	0	0

b. $E = a_1'a_0'b_1'b_0' + a_1'a_0b_1'b_0 + a_1a_0b_1b_0 + a_1a_0'b_1b_0'$

$G = a_1'a_0b_1'b_0' + a_1a_0'b_1'b_0' + a_1a_0'b_1'b_0 + a_1a_0b_1'b_0' + a_1a_0b_1'b_0 + a_1a_0b_1b_0'$

$L = a_1'a_0'b_1'b_0 + a_1'a_0'b_1b_0' + a_1'a_0'b_1b_0 + a_1'a_0b_1b_0' + a_1'a_0b_1b_0 + a_1a_0'b_1b_0$

c.

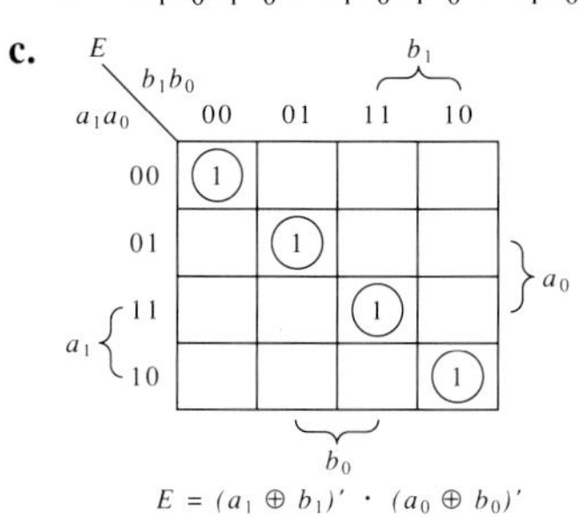

$E = (a_1 \oplus b_1)' \cdot (a_0 \oplus b_0)'$

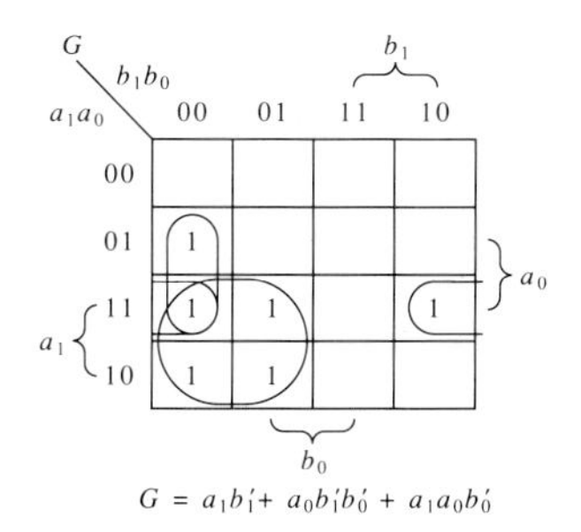

$G = a_1b_1' + a_0b_1'b_0' + a_1a_0b_0'$

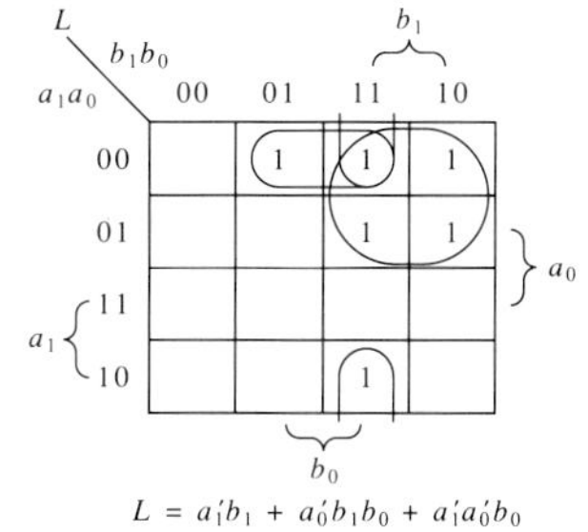

$L = a_1'b_1 + a_0'b_1b_0 + a_1'a_0'b_0$

3. a.

WX \ YZ (Decimal)	00	01	11	10
00	1	8	7	6
01	2	3	4	5
11	X	X	X	X
10	Start 0	End 9	X	X

A path to generate unit distance code

Truth table

BCD				Unit Distance				
A	B	C	D	W	X	Y	Z	Decimal
0	0	0	0	1	0	0	0	0
0	0	0	1	0	0	0	0	1
0	0	1	0	0	1	0	0	2
0	0	1	1	0	1	0	1	3
0	1	0	0	0	1	1	1	4
0	1	0	1	0	1	1	0	5
0	1	1	0	0	0	1	0	6
0	1	1	1	0	0	1	1	7
1	0	0	0	0	0	0	1	8
1	0	0	1	1	0	0	1	9
1	0	1	0	X	X	X	X	
1	0	1	1	X	X	X	X	
1	1	0	0	X	X	X	X	
1	1	0	1	X	X	X	X	
1	1	1	0	X	X	X	X	
1	1	1	1	X	X	X	X	

Note: Other solutions may be generated using a different starting point.

b. Canonical SOP expressions.

$$W = A'B'C'D' + AB'C'D$$
$$X = A'B'CD + A'B'CD' + A'BC'D' + A'BC'D$$
$$Y = A'BC'D' + A'BC'D + A'BCD + A'BCD'$$
$$Z = A'B'CD + A'BC'D' + A'BCD + AB'C'D' + AB'C'D$$

c.

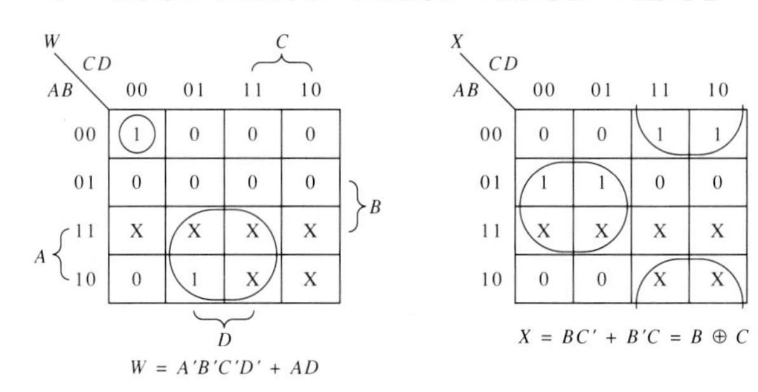

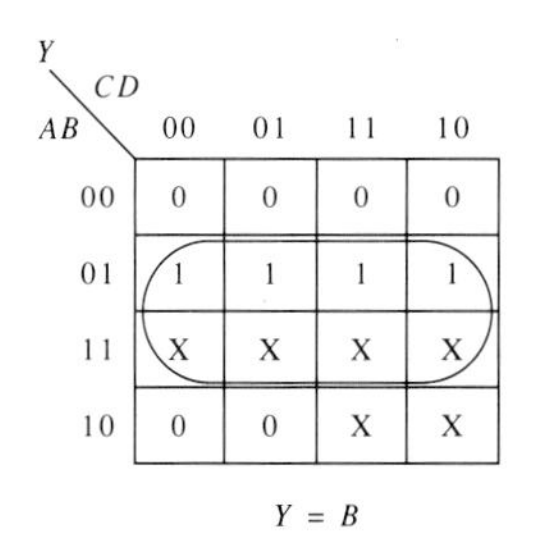

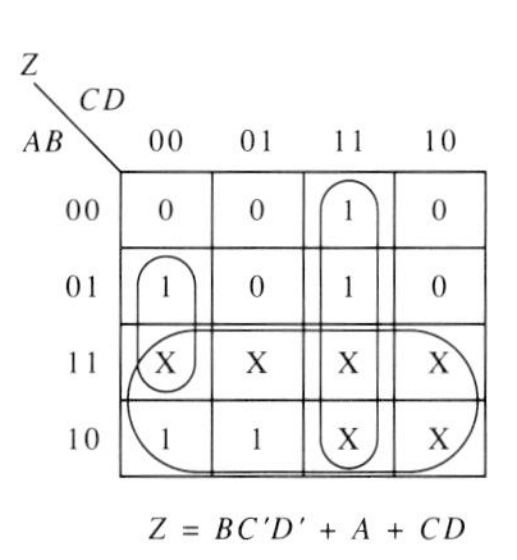

Design/Implementation Exercises

1. Step 2: Conceptualization. Use the truth table derived in General Exercise 2.

   Step 3: Solution/Simplification. Use the minimal SOP derived in General Exercise 2.

   Step 4: Realization.

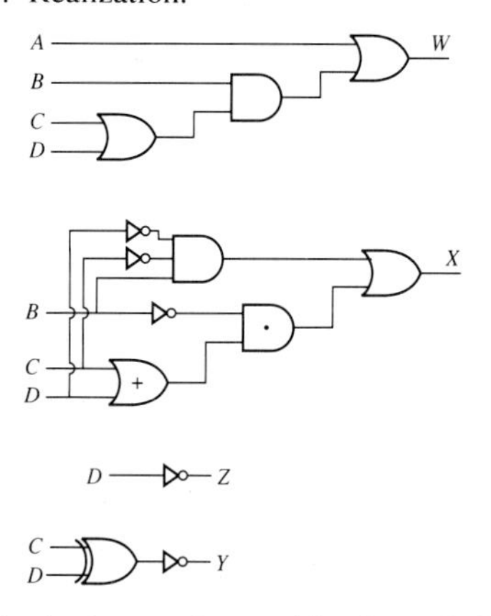

3. *WXYZ* = don't cares for combinations (5, 6, 7, 13, 14, 15).
   WXYZ = *ABCD* if N = 0, 1, 2, 3, or 4.
   WXYZ = *ABCD* – 0011 if N = 8, 9, 10, 11, or 12.

Step 2: Conceptualization. Truth table.

	Inputs				Outputs			
N	A	B	C	D	W	X	Y	Z
0	0	0	0	0	0	0	0	0
1	0	0	0	1	0	0	0	1
2	0	0	1	0	0	0	1	0
3	0	0	1	1	0	0	1	1
4	0	1	0	0	0	1	0	0
5	0	1	0	1	X	X	X	X
6	0	1	1	0	X	X	X	X
7	0	1	1	1	X	X	X	X
8	1	0	0	0	0	1	0	1
9	1	0	0	1	0	1	1	0
10	1	0	1	0	0	1	1	1
11	1	0	1	1	1	0	0	0
12	1	1	0	0	1	0	0	1
13	1	1	0	1	X	X	X	X
14	1	1	1	0	X	X	X	X
15	1	1	1	1	X	X	X	X
	BCD decade after right shift				corrected BCD			

Step 3: Solution/Simplification.

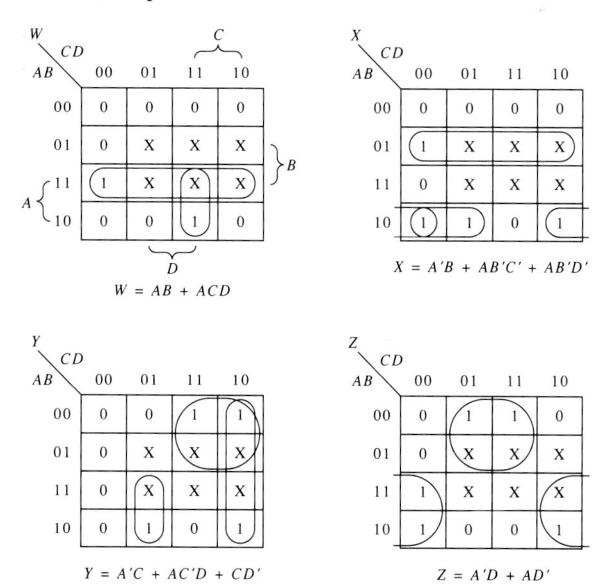

Step 4: Realization. Implement

$$W = AB + ACD$$
$$X = A'B + AB'C' + AB'D'$$
$$Y = A'C + AC'D + CD'$$
$$Z = A'D + AD' = A \oplus D$$

5. Use Figure 6.7.

7. Use Figure 6.17.

9. Use Figure 6.21.

11. Use Figure 6.24.

CHAPTER 7

General Exercises

1.

Function	S	R	Q	Description
Hold	0	0	q	next state = Present state
Reset	0	1	0	next state = 0
Set	1	0	1	next state = 1
Not allowed	1	1	x	next state = Indeterminate

3.

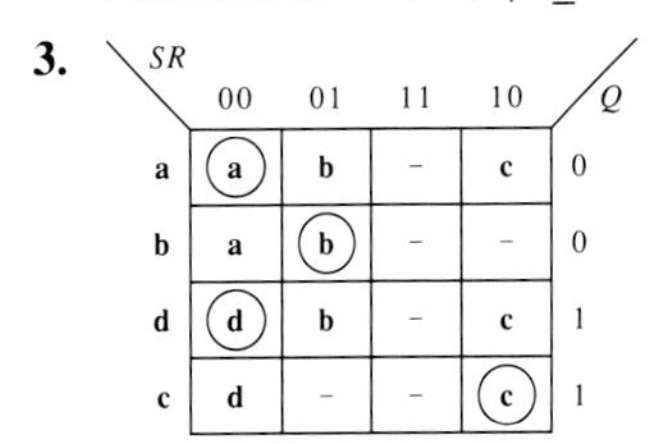

5.

Trailing edge

Period

1

0

Time

Leading edge

7.

Function	J	K	q		Q	transition
Hold	0	0	0	→	0	0
	0	0	1	→	1	1
Reset	0	1	0	→	0	0
	0	1	1	→	0	beta
Set	1	0	0	→	1	alpha
	1	0	1	→	1	1
Toggle	1	1	0	→	1	alpha
	1	1	1	→	0	beta

9. $Q = Jq' + K'q$

11. $Q = \mathrm{D}$

13. A T flip-flop has the following characteristic table:

T	q	Q
0	0	0
0	1	1
1	0	1
1	1	0

Conversion

Existing Excitation

Transition	q	Q	T	T	q	S	R
0	0	0	0	0	0	0	X
alpha	0	1	1	1	0	1	0
beta	1	0	1	1	1	0	1
1	1	1	0	0	1	X	0

Target Excitation

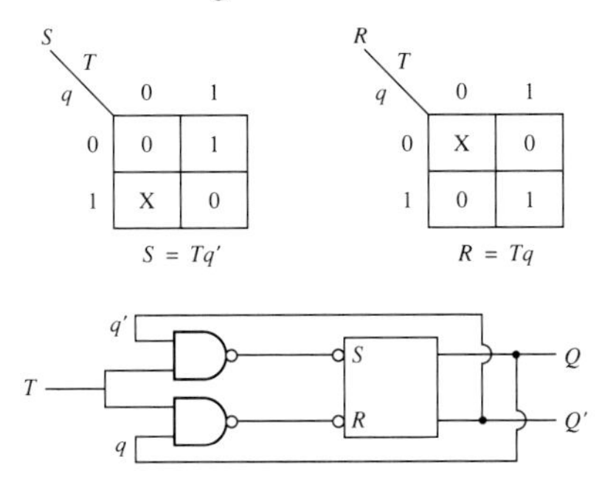

15.

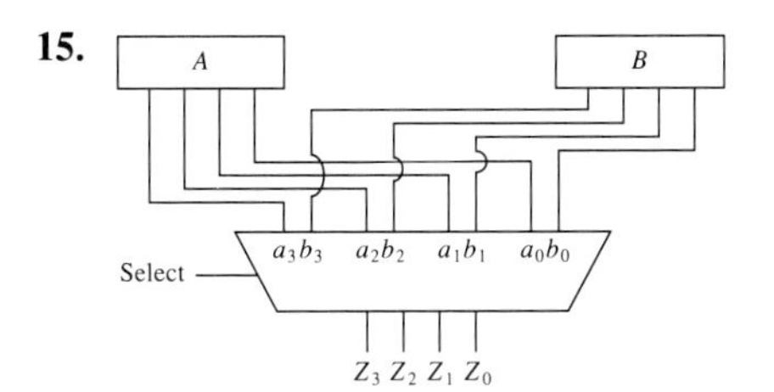

Design/Implementation Exercises

1. Step 2: Conceptualization.

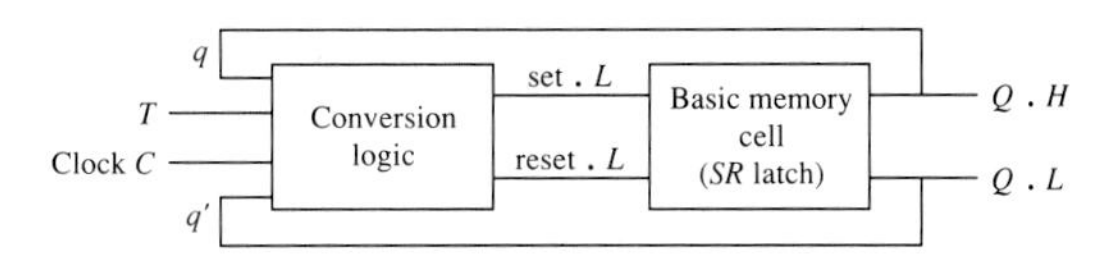

Step 3: Solution/Simplification. Clocked T flip-flop characteristic table.

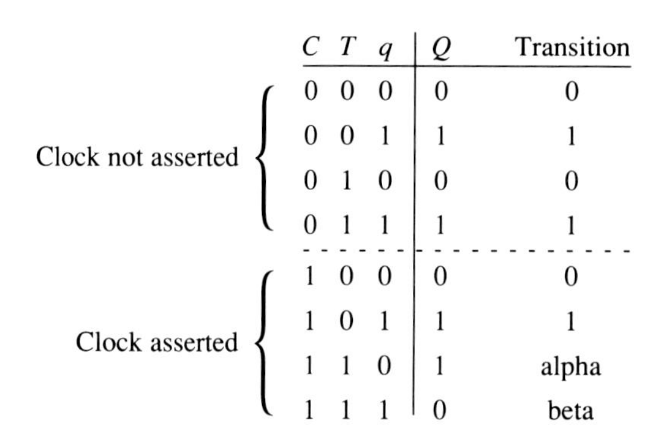

	C	T	q	Q	Transition
Clock not asserted	0	0	0	0	0
	0	0	1	1	1
	0	1	0	0	0
	0	1	1	1	1
Clock asserted	1	0	0	0	0
	1	0	1	1	1
	1	1	0	1	alpha
	1	1	1	0	beta

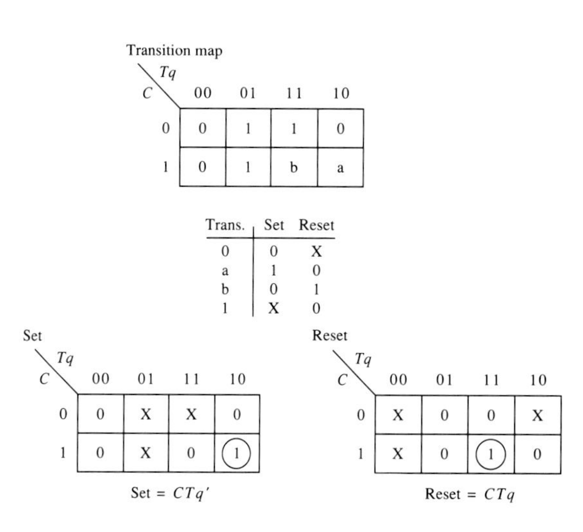

Trans.	Set	Reset
0	0	X
a	1	0
b	0	1
1	X	0

Step 4: Realization

3. $S.H = D + sU'$

5. Step 2: Conceptualization. Construct a count sequence/transition table and then a transition map for each flip-flop.

Count sequence/transition table.

A	B	C	D	Transition			
0	0	0	0	0	0	0	a
0	0	0	1	0	0	a	b
0	0	1	0	0	0	1	a
0	0	1	1	0	a	b	b
0	1	0	0	0	1	0	a
0	1	0	1	0	1	a	b
0	1	1	0	0	1	1	a
0	1	1	1	a	b	b	b
1	0	0	0	1	0	0	a
1	0	0	1	1	0	a	b
1	0	1	0	1	0	1	a
1	0	1	1	1	a	b	b
1	1	0	0	1	1	0	a
1	1	0	1	1	1	a	b
1	1	1	0	1	1	1	a
1	1	1	1	b	b	b	b

AB \ CD	00	01	11	10
00	0	0	0	0
01	0	0	a	0
11	1	1	b	1
10	1	1	1	1

Flip-flop A

AB \ CD	00	01	11	10
00	0	0	a	0
01	1	1	b	1
11	1	1	b	1
10	0	0	a	0

Flip-flop B

AB \ CD	00	01	11	10
00	0	a	b	1
01	0	a	b	1
11	0	a	b	1
10	0	a	b	1

Flip-flop C

AB \ CD	00	01	11	10
00	a	b	b	a
01	a	b	b	a
11	a	b	b	a
10	a	b	b	a

Flip-flop D

Step 3: Solution/Simplification.

A. Substitute the D flip-flop general input equations ($1_D = 1$, a; $0_D = 0$, b) into the transition maps to obtain flip-flop input maps.

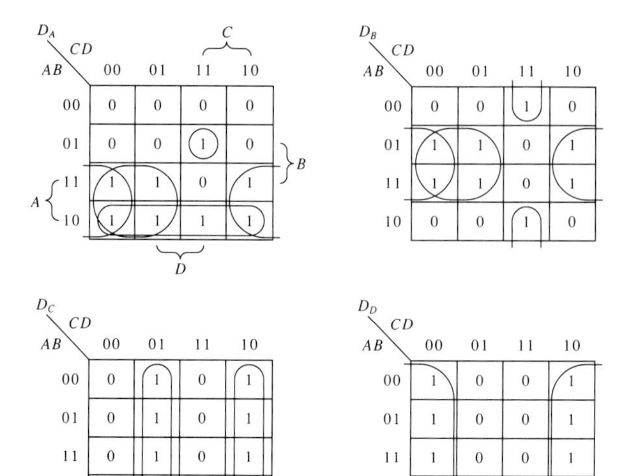

B. Use the flip-flop input maps to determine the application equations for each flip-flop.

$D_A = AB' + AC' + AD' + A'BCD = A(B' + C' + D') + A'(BCD) = A \oplus (BCD)$
$D_B = BC' + BD' + B'CD = B(C' + D') + B'(CD) = B \oplus (CD)$
$D_C = C'D + CD' = C \oplus D$
$D_D = D'$

Step 4: Realization. Use the application equations to construct a logic diagram of the counter.

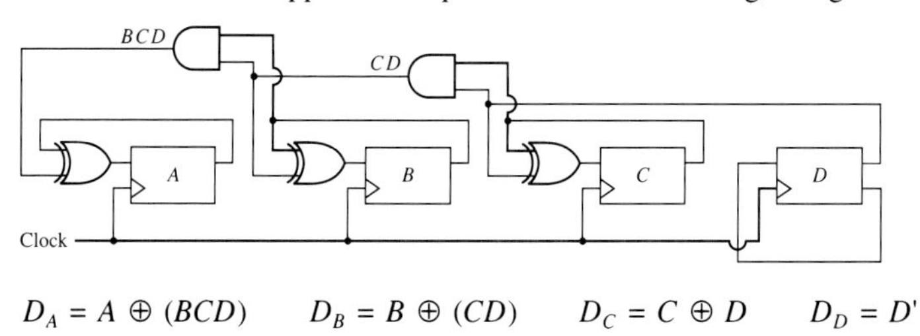

$D_A = A \oplus (BCD) \qquad D_B = B \oplus (CD) \qquad D_C = C \oplus D \qquad D_D = D'$

7. Step 2: Conceptualization. Construct a count sequence/transition table and then a transition map for each flip-flop.

Dec.	ABC	Transitions
0	000	
		0 a 0
2	010	
		a b 0
4	100	
		1 a 0
6	110	
		1 1 a
7	111	
		1 b 1
5	101	
		b a 1
3	011	
		0 b 1
1	001	
		0 0 b
0	000	

FF A

A \ BC	00	01	11	10
0	0	0	0	a
1	1	b	1	1

FF B

A \ BC	00	01	11	10
0	0	0	b	b
1	a	a	b	1

FF C

A \ BC	00	01	11	10
0	0	b	1	0
1	0	1	1	a

Step 3: Solution/Simplification.

A. Substitute the *JK* flip-flop general input equations into the transition maps to obtain flip-flop input maps.

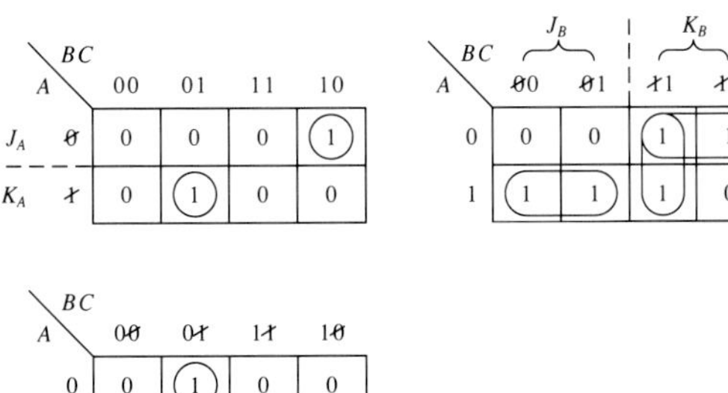

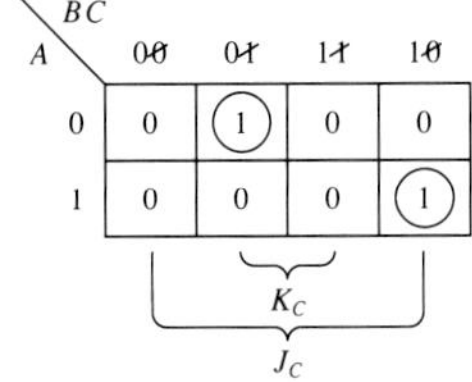

B. Use the flip-flop input maps to determine the application equations for each flip-flop.

$$J_A = BC' \qquad J_B = A \qquad J_C = AB$$
$$K_A = B'C \qquad K_B = A' + C \qquad K_C = A'B'$$

Step 4: Realization. Use the application equations to construct a logic diagram of the counter.

9. Step 2: Conceptualization. Construct a count sequence/transition table and then a transition map for each flip-flop.

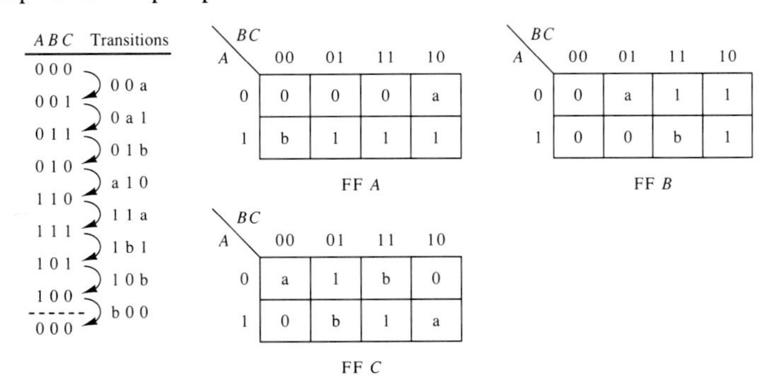

Step 3: Solution/Simplification.

A. Substitute the D flip-flop general input equations ($1_D = 1$, a; $0_D = 0$, b) into the transition maps to obtain flip-flop input maps.

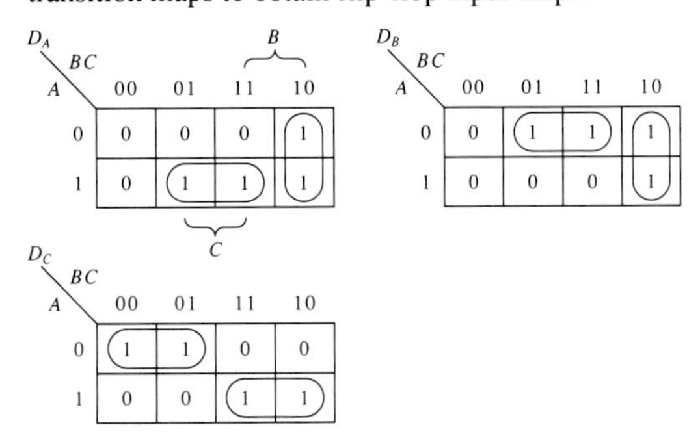

B. Use the flip-flop input maps to determine the application equations for each flip-flop.

$$D_A = AC + BC' \qquad D_B = A'C + BC' \qquad D_C = A'B' + AB$$

Step 4: Realization. Use the application equations to construct a logic diagram of the counter.

11. Step 2: Conceptualization. Construct a count sequence/transition table and then a transition map for each flip-flop.

Dec.	ABC	Transitions
0	000	
		a 0 0
4	100	
		1 a 0
6	110	
		1 1 a
7	111	
		b 1 1
3	011	
		0 b 1
1	001	
		a 0 1
5	101	
		b 0 b
0	000	

FF A

A \ BC	00	01	11	10
0	0	a	0	X
1	1	b	b	1

FF B

A \ BC	00	01	11	10
0	0	0	b	X
1	a	0	1	1

FF C

A \ BC	00	01	11	10
0	0	1	1	X
1	0	b	1	a

Step 3: Solution/Simplification.

A. Substitute the D flip-flop general input equations ($1_D = 1$, a; $0_D = 0$, b) into the transition maps to obtain flip-flop input maps.

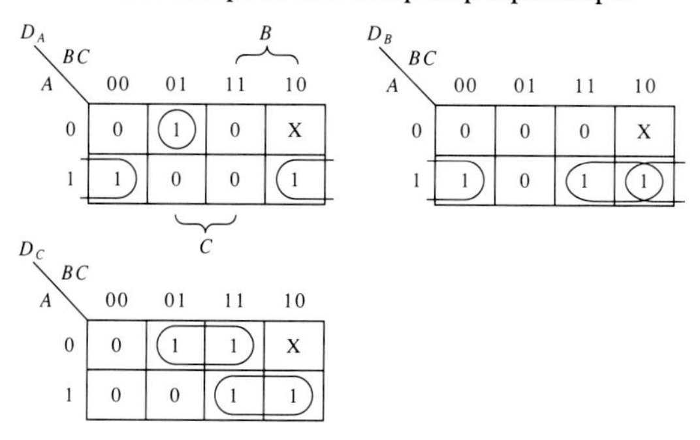

B. Use the flip-flop input maps to determine the application equations for each flip-flop.

$$D_A = AC' + A'B'C \qquad D_B = AB + AC' \qquad D_C = A'C + AB$$

Step 4: Realization. Use the application equations to construct a logic diagram of the counter.

13. Step 2: Conceptualization. Each cell must have the capability of being loaded in parallel, shifted right, and output serially.

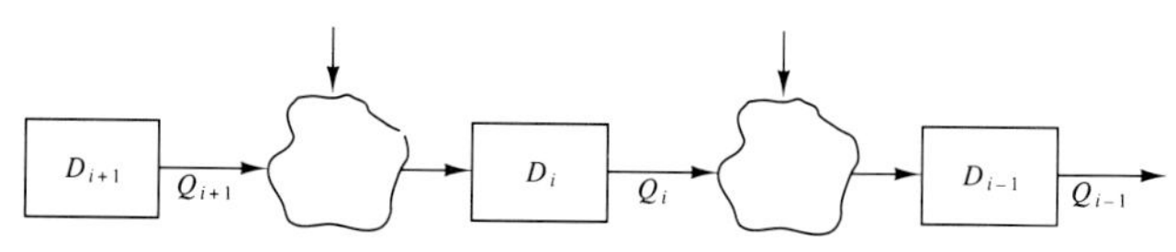

Let L = load, S = shift

Cell i input = (L and data$_i$) or (S and Q_{i+1})

Step 3: Solution/Simplification.

$$\begin{aligned} D_i &= L \cdot \text{data}_i + S \cdot Q_{i+1} \\ &= L \cdot \text{data}_i + L' \cdot Q_{i+1} \end{aligned} \qquad \text{where } S = L'$$

Step 4: Realization.
For $L = 1$, load register
For $L = 0$, shift register

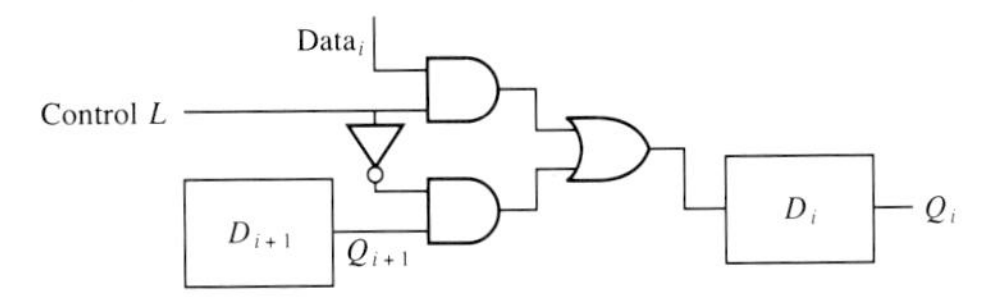

CHAPTER 8

General Exercises

1.

Present State	Next State $I = 0$	Next State $I = 1$	Output $I = 0$	Output $I = 1$
a	**b**	**c**	0	0
b	**d**	**e**	0	0
c	**f**	**g**	0	0
d	**d**	**a**	1	0
e	**b**	**a**	0	0
f	**d**	**a**	0	0
g	**b**	**a**	0	0

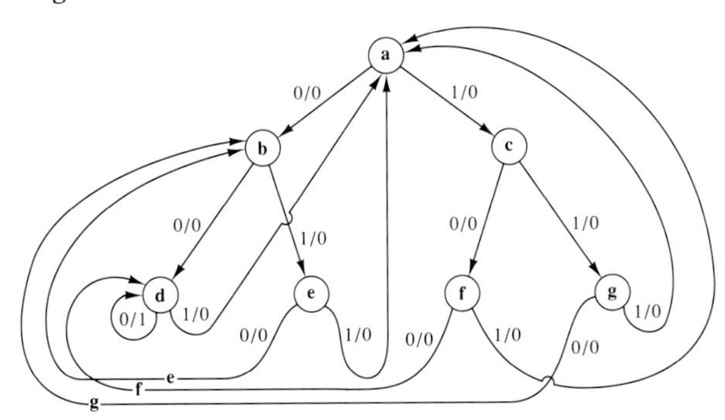

3.

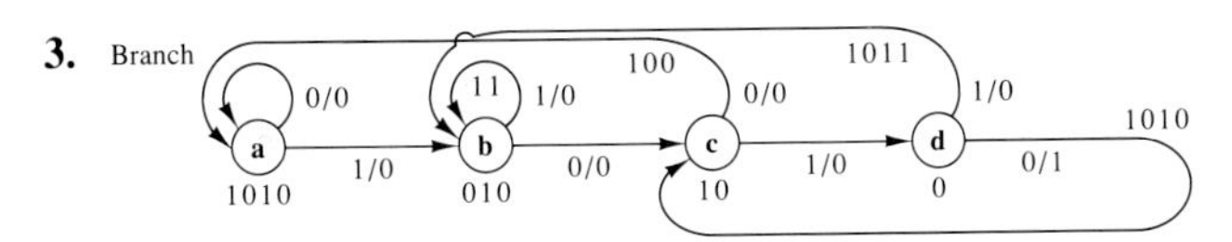

R-C Branch No.	Node No.	**a** 1010	**b** 010	**c** 10	**d** 0
11		↑	11–010	11–10	11–0
100			100–010	100–10	100–0
1011		Any	1011–010	1011–10	1011–0
1010		↓	1010–010	1010–10	1010–0

5. Cell R,C = N(C, 0) N(R, 0)
N(C, 1) N(R, 1)

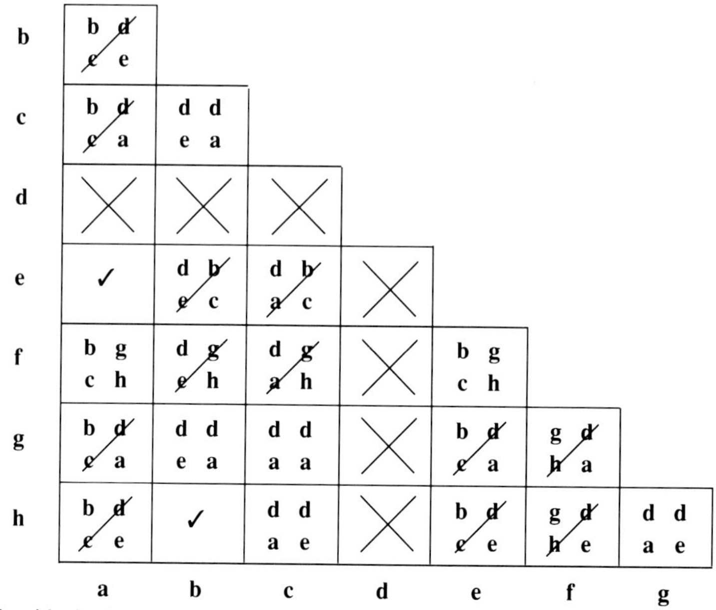

b	b ~~d~~ ~~e~~ e						
c	b ~~d~~ ~~e~~ a	d d e a					
d	X	X	X				
e	✓	d ~~b~~ ~~e~~ c	d ~~b~~ ~~a~~ c	X			
f	b g c h	d ~~g~~ ~~e~~ h	d ~~g~~ ~~a~~ h	X	b g c h		
g	b ~~d~~ ~~e~~ a	d d e a	d d a a	X	b ~~d~~ ~~e~~ a	g ~~d~~ ~~h~~ a	
h	b ~~d~~ ~~e~~ e	✓	d d a e	X	b ~~d~~ ~~e~~ e	g ~~d~~ ~~h~~ e	d d a e
	a	**b**	**c**	**d**	**e**	**f**	**g**

Cells with check mark represent unconditionally equivalent state pairs:

$$\mathbf{a = e} \qquad \mathbf{b = h}$$

Then,

$$\mathbf{a = e} \rightarrow \begin{cases} \mathbf{b = c} \\ \left.\begin{matrix}\mathbf{b = g} \\ \mathbf{c = h}\end{matrix}\right\} \rightarrow \left\{\begin{matrix}\mathbf{a = f} \\ \mathbf{e = f}\end{matrix}\right. \\ \mathbf{g = h} \end{cases}$$

Sets of equivalent states:

$$\mathbf{a = e = f} \qquad \mathbf{b = c = g = h} \qquad \mathbf{d}$$

Minimal-state table.

Present State	Next State $I = 0$	Next State $I = 1$	Output $I = 0$	Output $I = 1$
a	**b**	**b**	0	0
b	**d**	**a**	0	0
d	**a**	**b**	0	1

7.

Present State	Next State $I = 0$	Next State $I = 1$	Output $I = 0$	Output $I = 1$
a	**b**	**c**	0	0
b	**c**	**a**	0	0
c	**g**	**d**	0	0
d	**f**	**g**	0	1
e	**b**	**c**	0	0
f	**g**	**h**	0	0
g	**c**	**a**	0	0
h	**a**	**e**	0	0

a = e (rows a and e bracketed)
b = g (rows b and g bracketed)

Minimal-row table

Present State	Next State $I=0$	Next State $I=1$	Output $I=0$	Output $I=1$
a	**b**	**c**	0	0
b	**c**	**a**	0	0
c	**b**	**d**	0	0
d	**f**	**b**	0	1
f	**b**	**h**	0	0
h	**a**	**a**	0	0

9. Partition 1.

Present Class	Present State	Next State $I=0$	Next State $I=1$	Next Class	
S_1	**[d]**	**f**	**g**	S_2	S_2
S_2	**a**	**b**	**a**	S_2	S_2
	b	**b**	**c**	S_2	S_2
	<c>	**[d]**	**e**	S_1	S_2
	e	**h**	**a**	S_2	S_2
	f	**b**	**g**	S_2	S_2
	<g>	**[d]**	**a**	S_1	S_2
	h	**f**	**g**	S_2	S_2

Partition 2.

Present Class	Present State	Next State $I=0$	Next State $I=1$	Next Class	
S_1	**[d]**	**f**	**g**	S_{22}	S_{21}
S_{21}	**<c>**	**d**	**e**	S_1	S_{22}
	<g>	**d**	**a**	S_1	S_{22}
S_{22}	**a**	**b**	**a**	S_{22}	S_{22}
	(b)	**b**	**<c>**	S_{22}	S_{21}
	e	**h**	**a**	S_{22}	S_{22}
	(f)	**b**	**<g>**	S_{22}	S_{21}
	(h)	**f**	**<g>**	S_{22}	S_{21}

Partition 3.

Present Class	Present State	Next State $I=0$	Next State $I=1$	Next Class	
S_1	**d**	**f**	**g**	S_{31}	S_{21}
S_{21}	**c**	**d**	**e**	S_1	S_{32}
	g	**d**	**a**	S_1	S_{32}
S_{31}	**b**	**b**	**c**	S_{31}	S_{21}
	f	**b**	**g**	S_{31}	S_{21}
	h	**f**	**g**	S_{31}	S_{21}
S_{32}	**a**	**b**	**a**	S_{31}	S_{32}
	e	**h**	**a**	S_{31}	S_{32}

Since the 0-successors of each group are all in the same group and the 1-successors of each group are all in the same group, no further partition can be made. The state table has been partitioned into the following equivalence classes: (**a**, **e**), (**b**, **f**, **h**), (**c**, **g**), and (**d**).

11. Partition 1.

Present Class	Present State	Next State $I = 0$	Next State $I = 1$	Next Class	
S_1	[**d**]	**e**	**a**	S_2	S_2
S_2	**a**	**c**	**b**	S_2	S_2
	<**b**>	**c**	[**d**]	S_2	S_1
	c	**e**	**f**	S_2	S_2
	e	**a**	**b**	S_2	S_2
	<**f**>	**e**	[**d**]	S_2	S_1

Partition 2.

Present Class	Present State	Next State $I = 0$	Next State $I = 1$	Next Class	
S_1	**d**	**e**	**a**	S_{22}	S_{22}
S_{21}	<**b**>	**c**	[**d**]	S_{22}	S_1
	<**f**>	**e**	[**d**]	S_{22}	S_1
S_{22}	**a**	**c**	<**b**>	S_{22}	S_{21}
	c	**e**	<**f**>	S_{22}	S_{21}
	e	**a**	<**b**>	S_{22}	S_{21}

The state table has been partitioned into the following equivalence classes: (**a**, **c**, **e**), (**b**, **f**), and (**d**).

Design/Implementation Exercises

1. Step 2: Conceptualization. Construct a state table and then a next-state map for each flip-flop.

State diagram

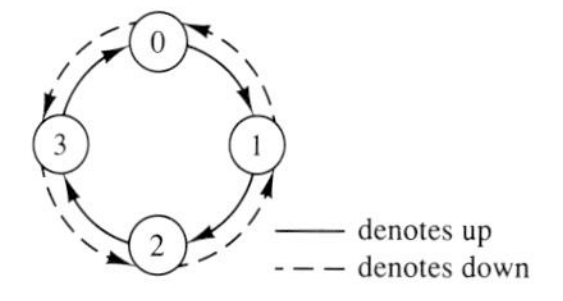

Dec.	Present State A B	Next State A^+B^+ $U = 0$	$U = 1$
0	0 0	11	01
1	0 1	00	10
2	1 0	01	11
3	1 1	10	00

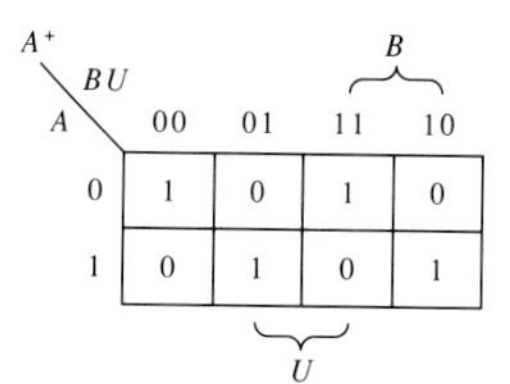

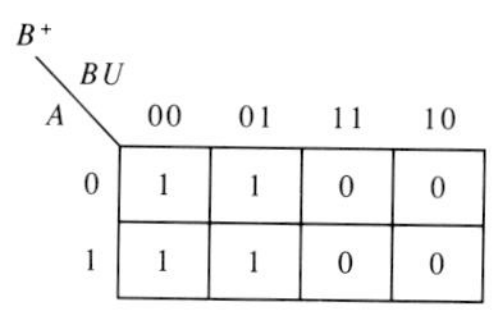

Step 3: Solution/Simplification.

A. Select the type of flip-flops that are to be used. For *JK* flip-flops use transition definitions to convert the next-state maps into transition maps.

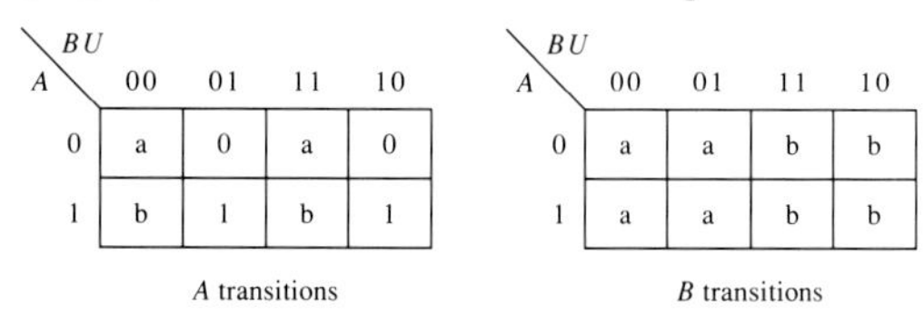

B. Substitute the flip-flop general input equations (1_J = a, 0_J = 0; 1_K = b, 0_K = 1) into transition maps to obtain flip-flop input maps.

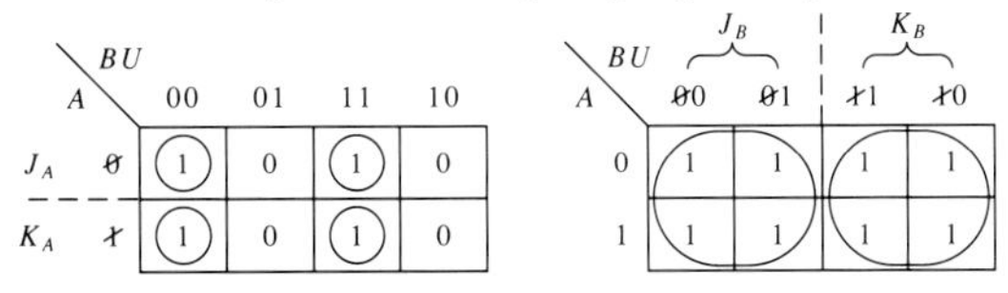

C. Use the flip-flop input maps to determine the application equations for each flip-flop.

$$J_A = B'U' + BU = (B \oplus U)' \qquad J_B = 1$$
$$K_A = B'U' + BU = (B \oplus U)' \qquad K_B = 1$$

Step 4: Realization. Use the application equations to construct a logic diagram of the counter.

3. Step 2: Conceptualization. Branch numbers are underlined.

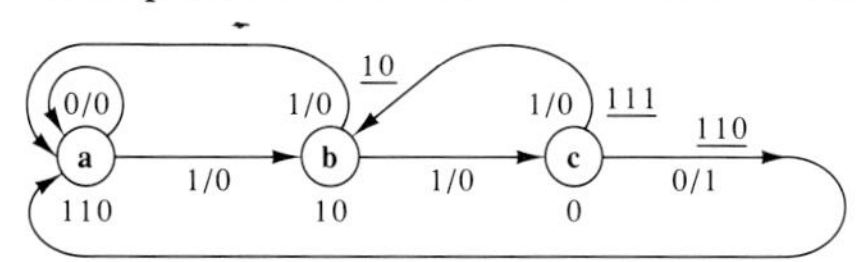

Step 3: Solution/Simplification.

State reduction. State diagram requires no reduction. Minimal state table.

Present State	Next State $I = 0$	Next State $I = 1$	Output $I = 0$	Output $I = 1$
a	**a**	**b**	0	0
b	**a**	**c**	0	0
c	**a**	**b**	1	0

State assignment.

By rule 1: **a–b** **a–c** **b–c**

A \ B	0	1
0	**a**	**b**
1	**c**	

By rule 2: **a–b** **a–c**

Circuit state table.

	AB	A^+B^+ $I = 0$	A^+B^+ $I = 1$	Output $I = 0$	Output $I = 1$
a	00	00	01	0	0
b	01	00	10	0	0
	11	XX	XX	–	–
c	10	00	01	1	0

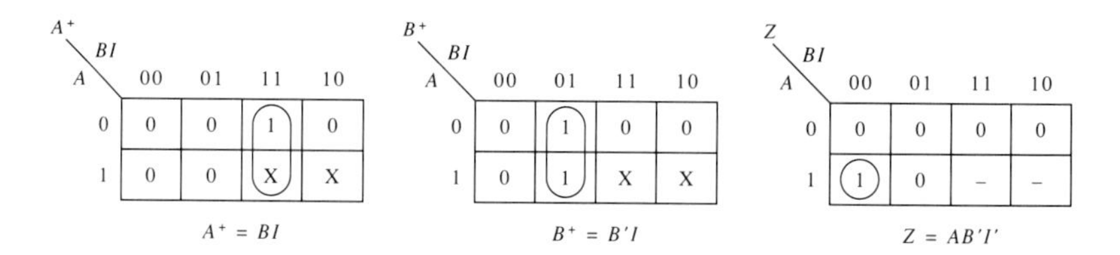

Step 4: Realization. Use the application equations and output equation (determined in Step 3) to construct a logic diagram of the sequence detector.

5. Step 2: Conceptualization.

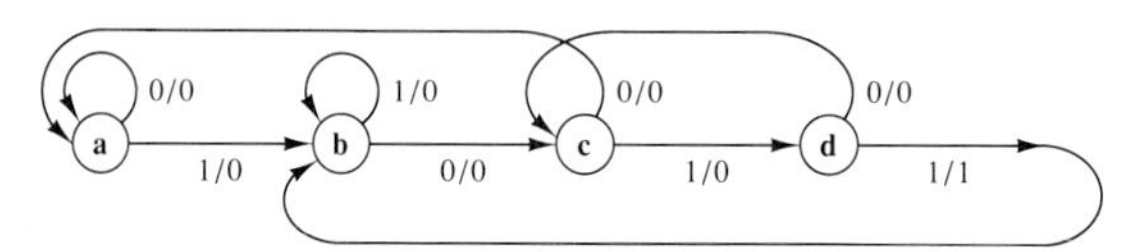

Present State	Next State $I=0$	Next State $I=1$	Output $I=0$	Output $I=1$
a	**a**	**b**	0	0
b	**c**	**b**	0	0
c	**a**	**d**	0	0
d	**c**	**b**	0	1

Step 3: Solution/Simplification.

By rule 1: **a–c b–d a–b a–d b–d**

$A \backslash B$	0	1
0	**a**	**b**
1	**d**	**c**

By rule 2: **a–b a–d c–b**

	AB	A^+B^+ $I=0$	A^+B^+ $I=1$	Output $I=0$	Output $I=1$
a	00	00	01	0	0
b	01	11	01	0	0
c	11	00	10	0	0
d	10	11	01	0	1

A^+

$A \backslash BI$	00	01	11	10
0	0	0	0	1
1	1	0	1	0

$A^+ = AB'I' + ABI + A'BI'$

B^+

$A \backslash BI$	00	01	11	10
0	0	1	1	1
1	1	1	0	0

$B^+ = AB' + A'B + B'I$

Z

$A \backslash BI$	00	01	11	10
0	0	0	0	0
1	0	1	0	0

$Z = AB'I$

$A^+ = AB'I' + ABI + A'BI'$
$B^+ = AB' + A'B + B'I$
$Z = AB'I$

7. Step 2: Conceptualization.

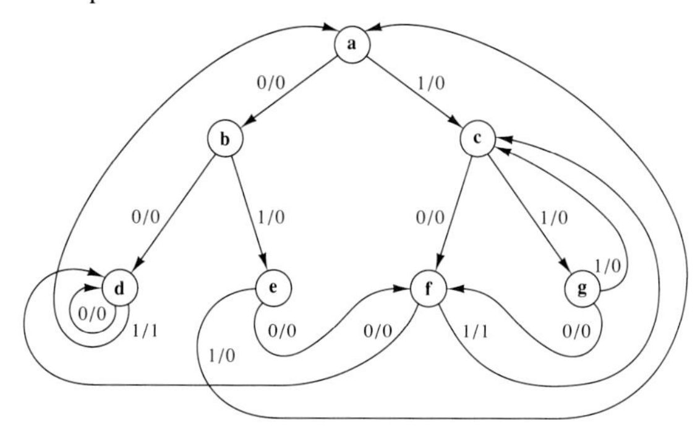

Present State	Next State $I = 0$	Next State $I = 1$	Output $I = 0$	Output $I = 1$
a	**b**	**c**	0	0
b	**d**	**e**	0	0
c	**f**	**g**	0	0
d	**d**	**a**	0	1
e	**f**	**a**	0	0
f	**d**	**c**	0	1
g	**f**	**c**	0	0

Step 3: Solution/Simplification.

By rule 1: **b–d b–f d–f c–e c–g e–g a–f a–g f–g**

AB \ C	0	1
00	**a**	**d**
01	**g**	**f**
11	**c**	**b**
10	**e**	

By rule 2: **b–c d–e d–a d–c f–c f–a f–g**

	ABC	$A^+B^+C^+$ $I = 0$	$A^+B^+C^+$ $I = 1$	Output $I = 0$	Output $I = 1$
a	000	111	110	0	0
d	001	001	000	0	1
f	011	001	110	0	1
g	010	011	110	0	0
c	110	011	010	0	0
b	111	001	100	0	0
–	101	xxx	xxx	–	–
e	100	011	000	0	0

$Z = A'CI$

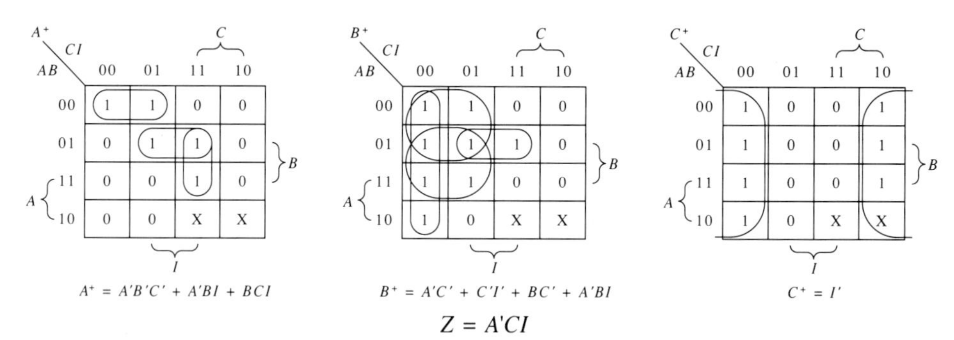

$$Z = A'CI$$

Step 4: Realization. Use the application equations and output equation to draw a logic diagram for the circuit.

CHAPTER 9
General Exercises

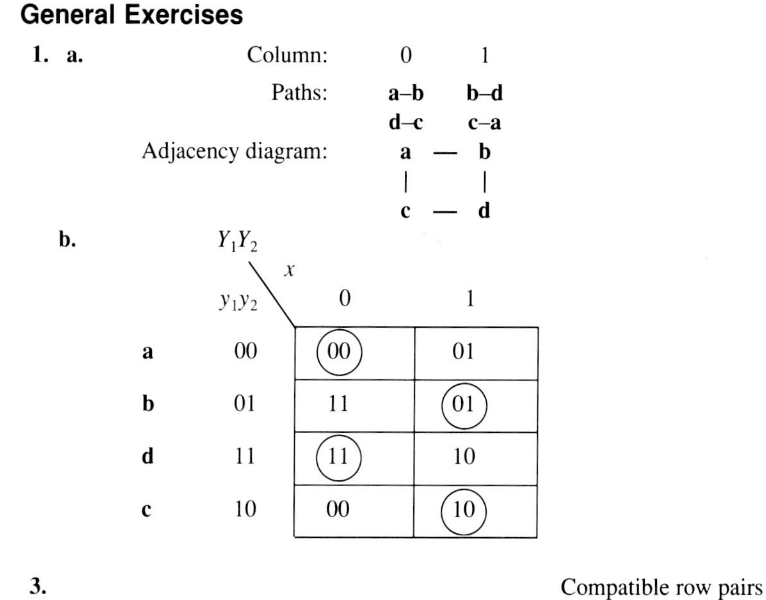

1. a.

Column:	0	1
Paths:	**a–b**	**b–d**
	d–c	**c–a**

Adjacency diagram:

```
a — b
|   |
c — d
```

b.

	y_1y_2 \ x (Y_1Y_2)	0	1
a	00	(00)	01
b	01	11	(01)
d	11	(11)	10
c	10	00	(10)

3.

						Compatible row pairs
b	✓					**(a, b)**
c	✓	✓				**(a, c), (b, c)**
d	b–f, c–e (crossed out)	b–f (crossed out)	c–e (crossed out)			
e	X	X	X	X		
f	X	X	X	X	✓	**(e, f)**
	a	**b**	**c**	**d**	**e**	

5. a.

Column:	00	01	11	10
Paths:	**a–b**	**b–d**	**c–b**	**a–c**
	b–a	**d–a**	**d–a**	**c–b**

Adjacency diagram:

```
a — b
| \/ |
| /\ |
c    d
```

b.

(1)

Row \ x_1x_2	00	01	11	10
a	(a)	b	c	(a)
b	(b)	(b)	d	a
c	b	b	(c)	(c)
d	a	(d)	(d)	a

(2)

Row \ x_1x_2	00	01	11	10
a	(a)		c	(a)
b	(b)	(b)	d	
c	b	b	(c)	(c)
d	a	(d)	(d)	a

(3)

Row \ x_1x_2	00	01	11	10
a	(a)	~~b~~C	c	(a)
b	(b)	(b)	d	~~a~~D
c	b	b	(c)	(c)
d	a	(d)	(d)	a

(4)

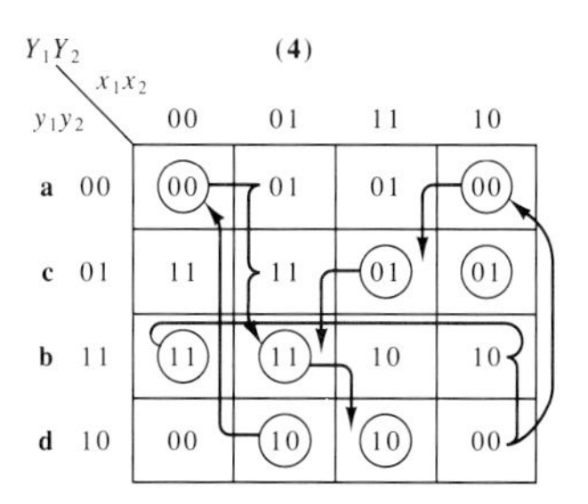

Note that reordered rows (a, c, b, d) and cycles are required for race-free circuit.

7. a.

Column:	00	01	11	10
Paths:	**a–c**	**a–b**	**b–d**	**d–c**
	c–b	**b–c**	**c–b**	**d–a**
		d–a		
		d–c		

Adjacency diagram:

```
a — b
| \/ |
| /\ |
d — c
```

b.

Row \ x_1x_2	00	01	11	10
a	(a)	(a)	b	c
b	~~c~~F	(b)	(b)	~~d~~G
G				d
d	a	(d)	~~c~~E	(d)
E			c	
F	c			
c	(c)	b	(c)	(c)

$y_1y_2y_3$ \ x_1x_2	00	01	11	10
a 000	(000)	(000)	001	100
b 001	101	(001)	(001)	011
G 011				010
d 010	000	(010)	110	(010)
E 110			100	
111				
F 101	100			
c 100	(100)	001	(100)	(100)

9.

Column:	00	01	11	10
Paths:		**a–d**	**c–a**	**a–d**
		b–c	**d–b**	**b–c**
				c–b
				d–a

Adjacency diagram:

a — **d**
| |
c — **b**

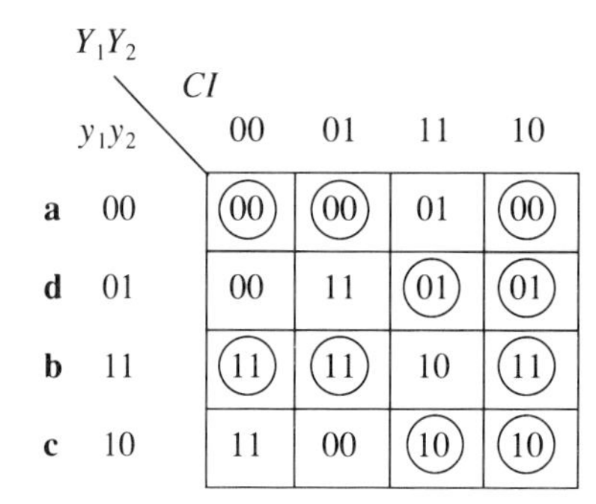

Y_1Y_2 / y_1y_2 \ CI		00	01	11	10
a	00	(00)	(00)	01	(00)
d	01	00	11	(01)	(01)
b	11	(11)	(11)	10	(11)
c	10	11	00	(10)	(10)

11. Transitions

y_1y_2 \ x_1x_2	00	01	11	10
00		aa		aa
01	a1		ab	
11			bb	
10	b0			1a

13.

Row \ x_1x_2	00	01	11	10	Z
a	(**a**)	(**a**)	**b**	**c**	0
b	–	**a**	(**b**)	**D**	0
D	–	–	**b**	**c**	
c	**a**	–	**D**	(**c**)	1

15. Y_1

y_1y_2 \ x_1x_2	00	01	11	10
00	0	0	0	1
01		0	0	1
11			0	1
10	0		1	1

Y_2

y_1y_2 \ x_1x_2	00	01	11	10
00	0	0	1	0
01		0	1	1
11			1	0
10	0		1	0

$$Y_1 = x_1x_2' + y_1y_2'x_1$$
$$Y_2 = x_1x_2 + y_1'y_2x_1$$
$$Z = y_1$$

Design/Implementation Exercises

1. Step 2: Conceptualization.

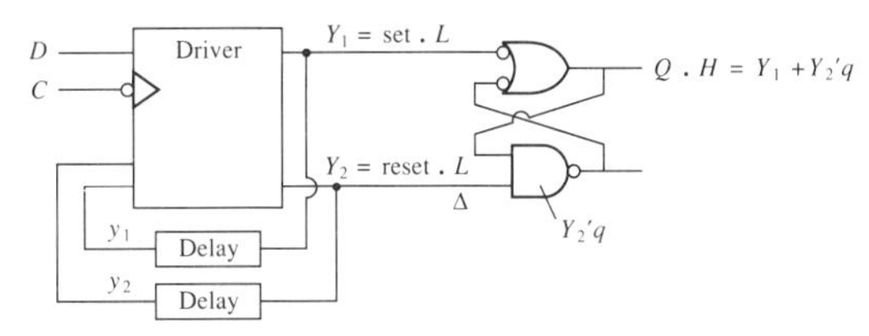

NET D flip-flop primitive state diagram.

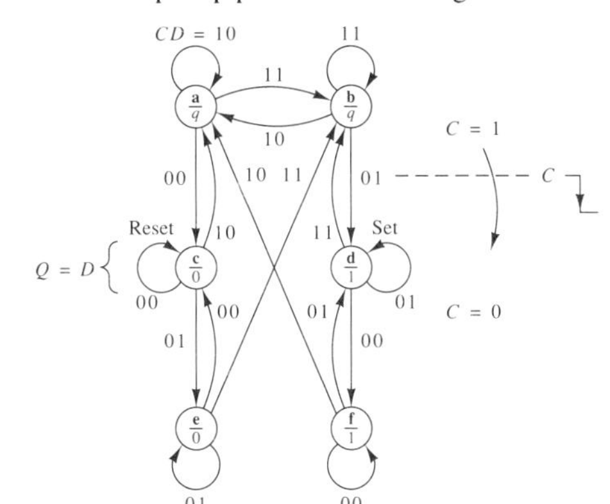

Primitive flow table

	CD 00	01	11	10	Q
a	c	–	b	(a)	q
b	–	d	(b)	a	q
c	(c)	e	–	a	0
d	f	(d)	b	–	1
e	c	(e)	b	–	0
f	(f)	d	–	a	1

Step 3: Solution/Simplification.

Merged flow table (by inspection)

CD	00	01	11	10	Q
(a, b) H	c	d	(b)	(a)	q
(c, e) R	(c)	(e)	b	a	0
(d, f) S	(f)	(d)	b	a	1

State table

y_1y_2 \ CD (Y_1Y_2)	00	01	11	10
00	01	10	(00)	(00)
01	(01)	(01)	00	00
11	X	X	X	X
10	(10)	(10)	00	00

Excitation maps

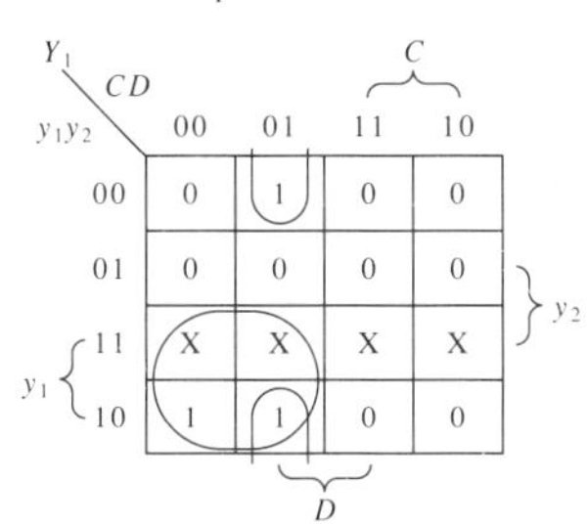

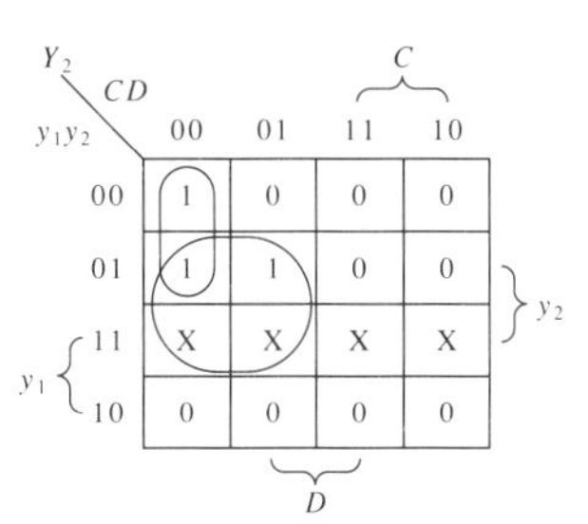

Application equations:

$$Y_1 = y_1C' + y_2'C'D$$
$$= C'(y_1 + y_2'D) = \text{set}.L$$

$$Y_2 = y_2C' + y_1'C'D'$$
$$= C'(y_2 + y_1'D') = \text{reset}.L$$

Output equation:

$$Q = Y_1 + Y_2'q$$

Step 4: Realization. Use the application equations and output equation (determined in Step 3) to draw a logic diagram for the circuit.

3. Step 2: Conceptualization.

Primitive flow diagram

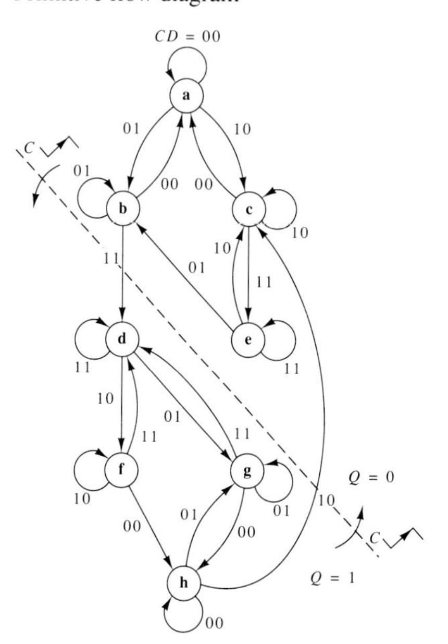

Primitive flow table

	CD 00	01	11	10	Q
a	(a)	b	–	c	0
b	a	(b)	d	–	0
c	a	–	e	(c)	0
d	–	g	(d)	f	1
e	–	b	(e)	c	0
f	h	–	d	(f)	1
g	h	(g)	d	–	1
h	(h)	g	–	c	1

Step 3: Solution/Simplification.

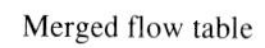
Merged flow table

		CD 00	01	11	10	Q
(a, c, e)	A	(a)	b	(e)	(c)	0
b	B	a	(b)	d	–	0
(d, f, g)	C	h	(g)	(d)	(f)	1
h	D	(h)	g	–	c	1

State table

	Y_1Y_2 / y_1y_2	CD 00	01	11	10
A	00	(00)	01	(00)	(00)
B	01	00	(01)	11	–
C	11	10	(11)	(11)	(11)
D	10	(10)	11	–	00

Excitation maps

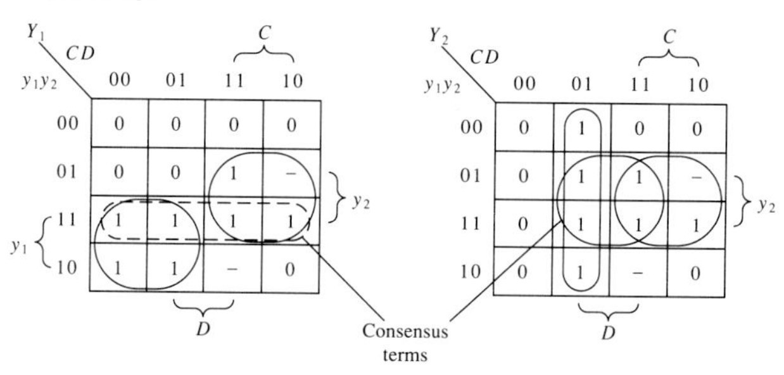

Excitation equations:

$$Y_1 = y_1C' + y_2C + y_1y_2 \qquad Y_2 = y_2C + C'D + y_2D$$
$$\therefore Y_1 = y_2C + y_1(y_2 + C') \qquad \therefore Y_2 = y_2C + D(y_2 + C')$$

Step 4: Realization.

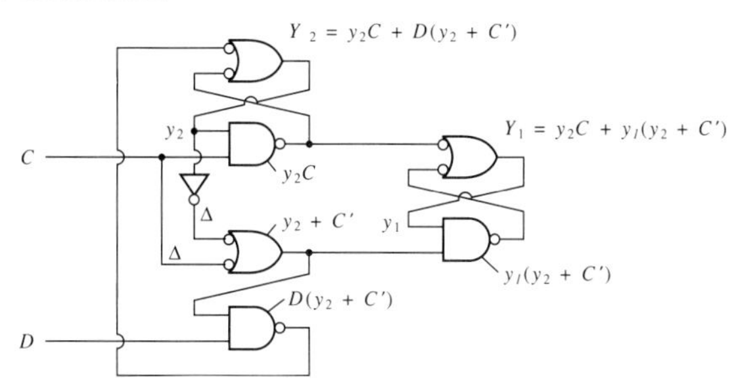

5. Step 2: Conceptualization.

Primitive flow table

		x_1x_2 00	01	11	10	Z
Reset	**a**	(**a**)	**b**	–	**g**	0
	b	**a**	(**b**)	**c**	–	0
	c	–	**d**	(**c**)	**g**	0
	d	**e**	(**d**)	**i**	–	0
	e	(**e**)	**h**	–	**f**	0
	f	**a**	–	**i**	(**f**)	1
Failure routes to reset	**g**	**a**	–	**i**	(**g**)	0
	h	**a**	(**h**)	**i**	–	0
	i	–	**h**	(**i**)	**g**	0

Step 3: Solution/Simplification.

Merged flow table

		00	01	11	10	Z
(a, b)	B	a	b	c	g	0
c	C	–	d	c	g	0
d	D	e	d	i	–	0
e	E	e	h	–	f	0
f	F	a	–	i	f	1
(g, h, i)	G	a	h	i	g	0

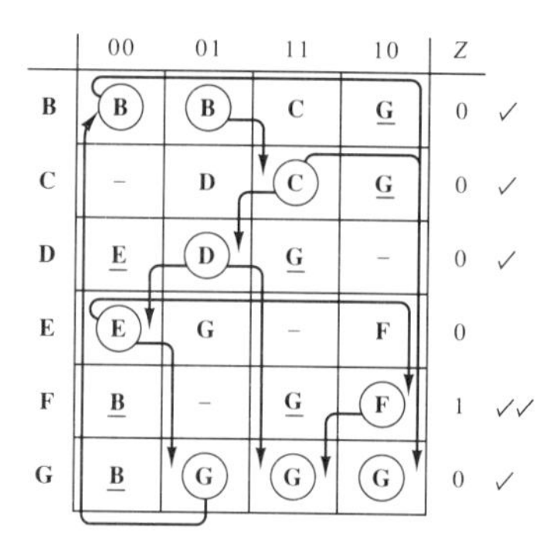

Check adjacency requirements

Columns:	00	01	11	10
Paths:	**B–G**	**B–C**	**C–D**	**F–G**
	E–G	**D–E**	**C–G**	**F–B**
	E–F	**D–G**		**G–B**
		G–B		

y_1 \ y_2y_3	00	01	11	10
0	B	C	D	E
1	G			F

Create cycle paths

	00	01	11	10	Z
B	B	B	C	G	0
C	–	D	C	~~C~~ B	0
D	E	D	~~G~~ E	–	0
E	E	~~G~~ F	F	F	0
F	~~B~~ G	G	G	F	1
G	B	G	G	G	0

Next-state map

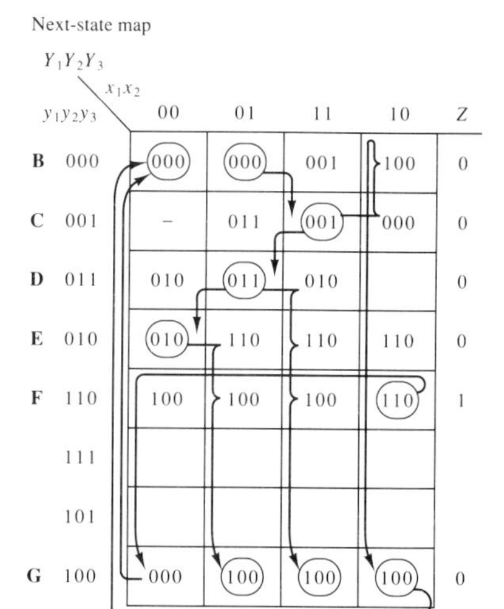

CHAPTER 10

General Exercises

1. a. Algebraic solution: $X + Y = F_x 2^{e_x} + F_y 2^{e_y} = F_x 2^{e_x} + F_y 2^{e_y - e_x} 2^{e_x} = (F_x + F_y 2^{e_y - e_x}) 2^{e_x}$

$X = F_x(2^{e_x})$ and $Y = F_y(2^{e_y})$, where $e_x \geq e_y$

Algorithm:

1. Determine difference in exponents.
2. Shift F_y right $e_x - e_y$ places $\rightarrow f_y$.
3. Add $F_x + f_y = F_Z$.

4. Normalize F_Z and adjust exponents.

b.

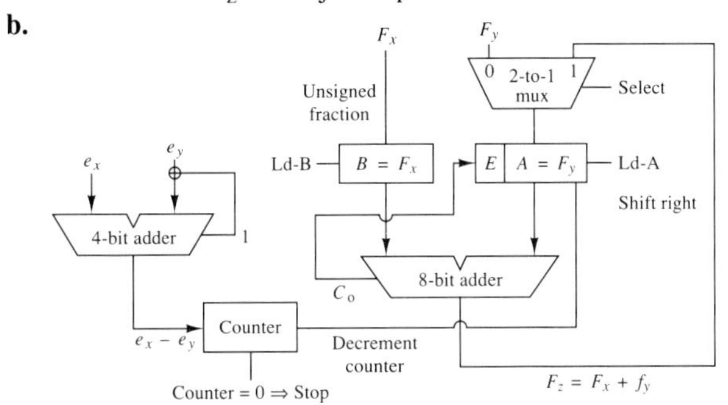

c. a. Load counter with $e_x - e_y$.
Load F_x and F_y into B and A registers.

b. Shift A right 1 bit and decrement counter
If counter = 0, **then** {**goto c**} **else** {**goto b**}

c. Select mux
Load $F_Z = F_x + f_y$ into A; load C_o into E

d. Normalize F_Z and adjust exponent.
If $C_o = 1$, EA shift right; $e_x + 1 \rightarrow e_x$

3. a.

Row \ ij	00	01	11	10	Z = Light
a	(a)	b	X	c	0 (off)
b	d	(b)	X	–	1 (on)
c	f	–	X	(c)	1 (on)
d	(d)	–	X	e	1 (on)
e	a	–	X	(e)	1 (on)
f	(f)	g	X	–	1 (on)
g	a	(g)	X	–	1 (on)

There are no redundant stable states (by inspection).

b.

Merger diagram

							Compatible row pairs
b	X						
c	X	*					
d	X	✓	*				**(b, d)**
e	X	*	*	*			
f	X	*	✓	*	*		**(c, f)**
g	X	*	*	*	✓	*	**(e, g)**
	a	**b**	**c**	**d**	**e**	**f**	

* denotes conflicting entries; X denotes conflicting outputs

c. Minimal-row flow table

	ij	00	01	11	10	Z
(a)	A	(a)	b	X	c	1
(b, d)	B	(d)	(b)	X	e	0
(c, f)	C	(f)	g	X	(c)	0
(e, g)	D	a	(g)	X	(e)	0

ij	00	01	11	10	Z
A	(A)	B	X	C	1
B	(B)	(B)	X	D	0
C	(C)	D	X	(C)	0
D	A	(D)	X	(D)	0

d.

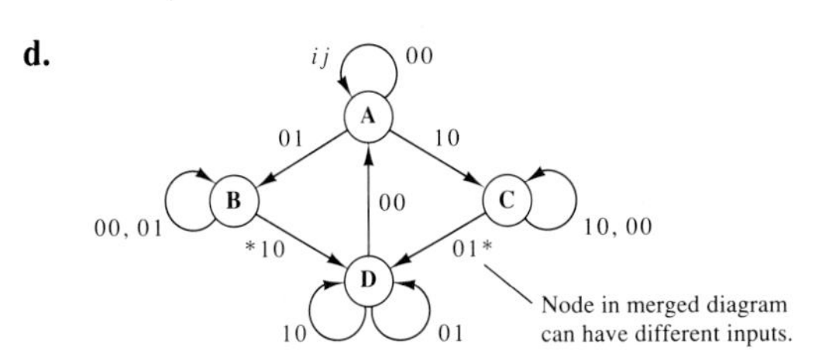

5.

	Present State	Next State $I = 0$	Next State $I = 1$	Output $I = 0$	Output $I = 1$	
Reset	**a**	**b**	**e**	0	0	
	b	**c**	**c**	0	0	
	c	**d**	**d**	0	0	
	d	**g**	**g**	0	0	
	e	**f**	**h**	0	0	
	f	**d**	**h**	0	0	
	g	–	–	1	1	} Reset to **a** after 4 bits
	h	**h**	**h**	0	0	

7. See Interactive Design Application, Dice Game ASM.

Design/Implementation Exercises

1. Step 2: Conceptualization.
Structure diagram

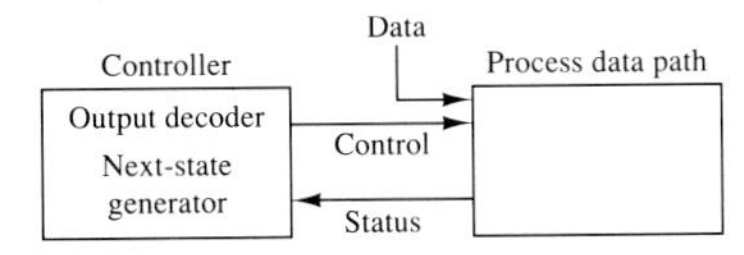

Control-flow diagram. Use control algorithm developed in General Exercise 1c.

Step 3: Solution/Simplification.
Process data path. Use logic diagram developed in General Exercise 1b.
Controller. Consists of manual controls of data path elements.

Step 4: Realization. Use a logic simulation program to realize a rudimentary-control floating-point adder.

3. Step 2: Conceptualization. Use the primitive flow table developed in General Exercise 3.

Step 3: Solution/Simplification.

State reduction. Use the minimal-row flow table developed in General Exercise 3.
State assignment. Path table.

Column:	00	01	11	10
Paths:	**A–B**	**D–A**	—	**D–A**
	A–C			
	B–D			
	C–D			

Adjacency diagram:

```
A — B
| \ |
C — D
```

State assignment matrix

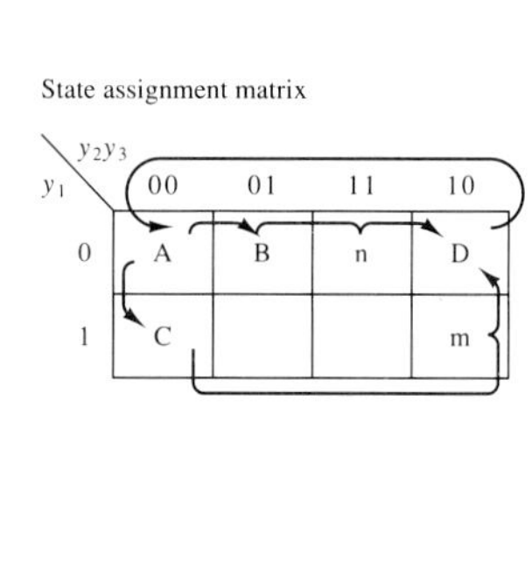

Flow table with cycles

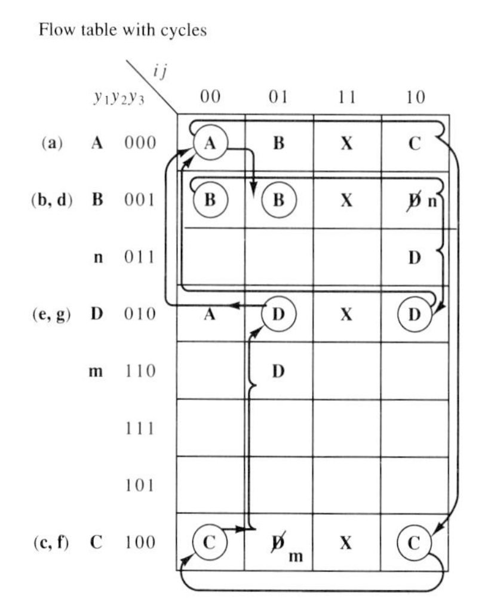

	$y_1y_2y_3$ \ ij	00	01	11	10
(a)	A 000	A	B	X	C
(b, d)	B 001	B	B	X	∅ n
	n 011				D
(e, g)	D 010	A	D	X	D
	m 110		D		
	111				
	101				
(c, f)	C 100	C	∅ m	X	C

Next-state map

$Y_1Y_2Y_3$

	$y_1y_2y_3$ \ ij	00	01	11	10
A	000	000	001	XXX	100
B	001	001	001	XXX	011
n	011				010
D	010	000	010	XXX	010
m	110		010		
	111				
	101				
C	100	100	110	XXX	100

Excitation maps

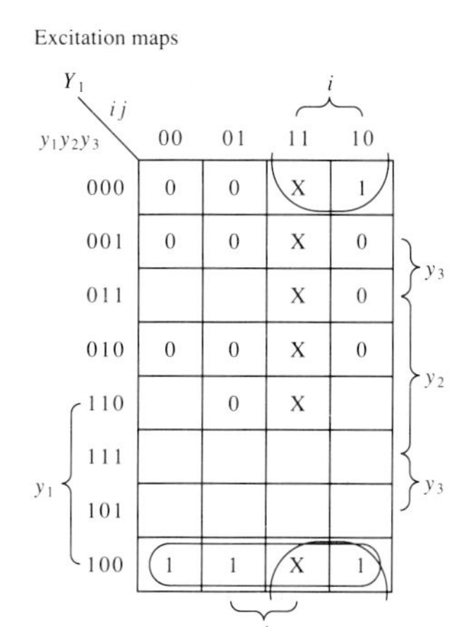

Y_1

$y_1y_2y_3$ \ ij	00	01	11	10
000	0	0	X	1
001	0	0	X	0
011			X	0
010	0	0	X	0
110		0	X	
111				
101				
100	1	1	X	1

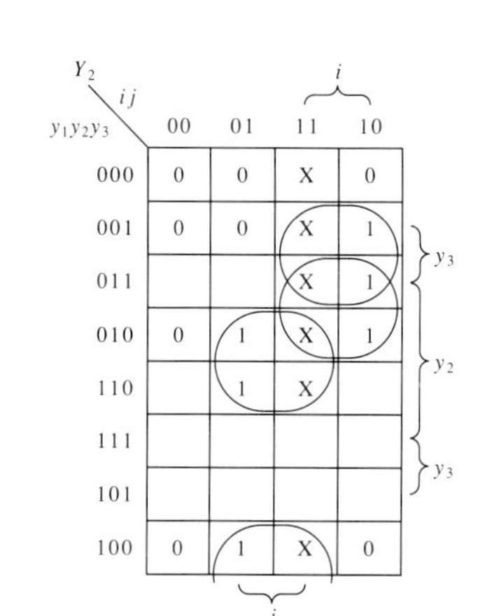

Y_2

$y_1y_2y_3$ \ ij	00	01	11	10
000	0	0	X	0
001	0	0	X	1
011			X	1
010	0	1	X	1
110		1	X	
111				
101				
100	0	1	X	0

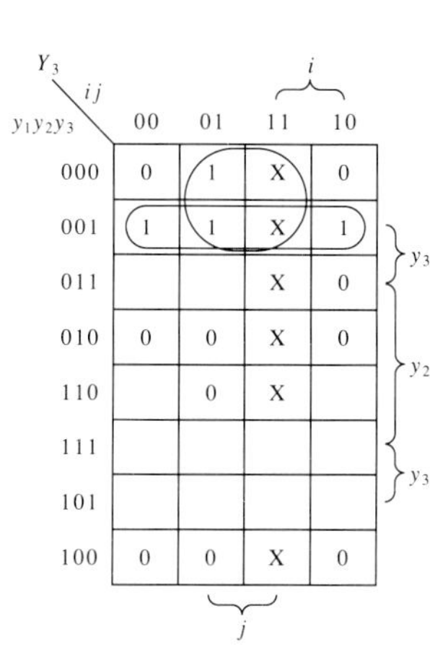

Y_3

$y_1y_2y_3$ \ ij	00	01	11	10
000	0	1	X	0
001	1	1	X	1
011			X	0
010	0	0	X	0
110		0	X	
111				
101				
100	0	0	X	0

Next-state generator:

$$Y_1 = y_1y_2'y_3' + y_2'y_3'i \qquad Y_2 = y_1'y_3i + y_1'y_2i + y_2y_3'j + y_2y_3'j \qquad Y_3 = y_1'y_2'y_3 + y_1'y_2'j$$

Output decoder: $Z = (y_1'y_2'y_3')'$

Step 4: Realization. Implement the next-state and output equations.

5. See Design/Implementation Exercise 2.

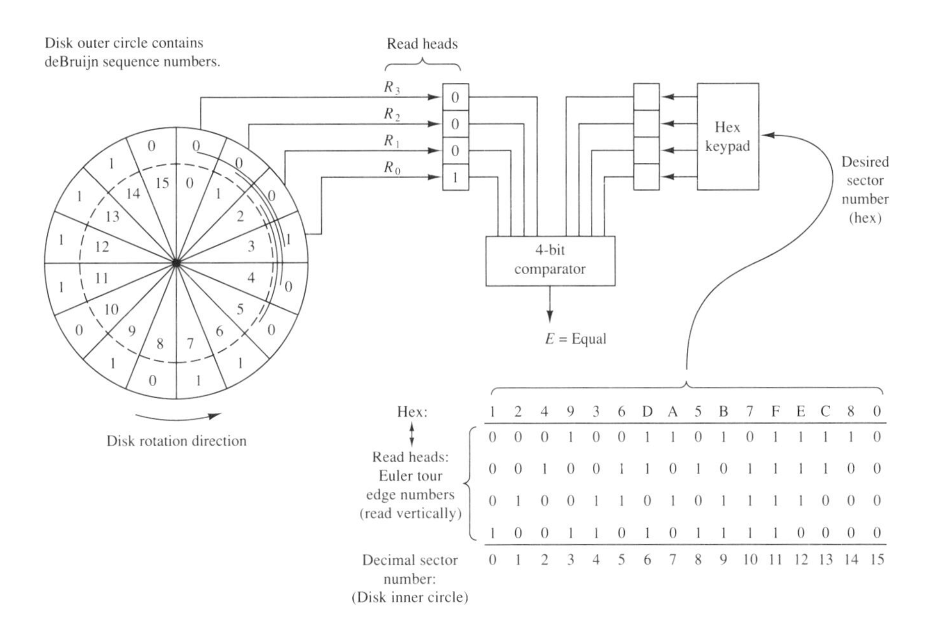

CHAPTER 11

General Exercises

1. Decade: 2 1 0

	BCD register				
	8421	8421	8421	Binary register	
245 =	0010	0100	0101		
	001	0010	0010	1	Shift right
	00	1001	0001	01	Shift right
	00	0110	0001		Subtract 3 from decade 1
	0	0011	0000	101	Shift right
		0001	1000	0101	Shift right
		0001	0101		Subtract 3 from decade 0
		000	1010	10101	Shift right
		000	0111		Subtract 3 from decade 0
		00	0011	110101	Shift right
		0	0001	1110101	Shift right
			0000	11110101	Shift right

Result: $N = 245 = 0010\ 0100\ 0101 \rightarrow$ binary 11110101.

3. Algorithm:

a. Load BCD register
Clear Binary register

b. Shift right; for each decade subtract 3 **if** MSB = 1

c. **If** BCD register = 0, **then** {stop}
else {**goto b**}

5. Binary–BCD converter data path.

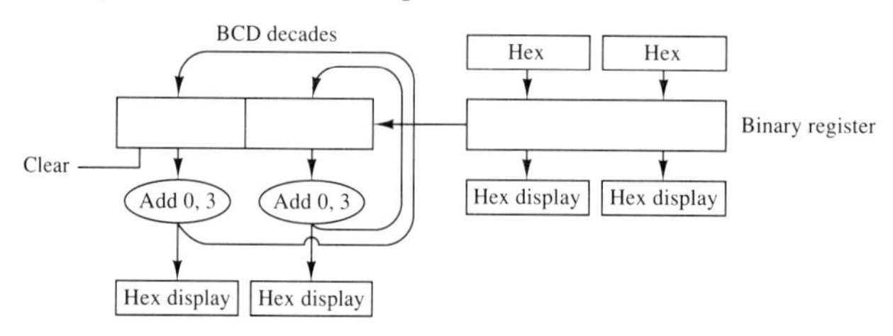

7.

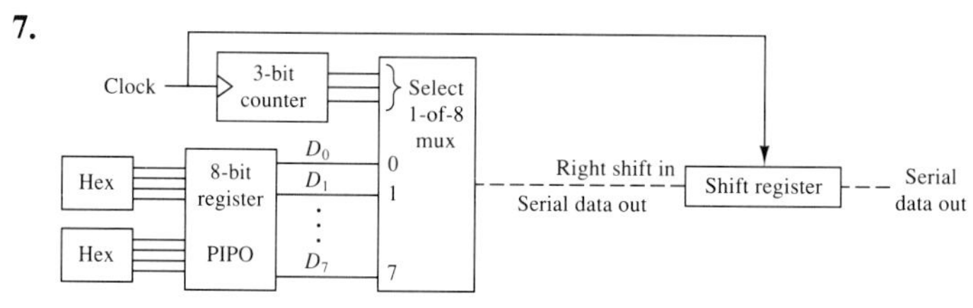

9. $X \cdot Y = F_x \cdot F_y\ 2^{(E_x + E_y)}$

Algorithm: 1. Multiply fractions
2. Add exponents
3. Normalize (shift fraction; correct exponent)

11.

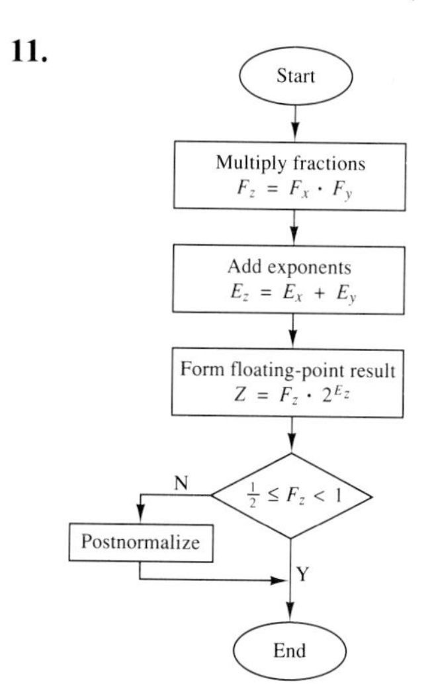

Design/Implementation Exercises

1. See text for discussion.
3. Use the data path and control algorithm developed in General Exercises 2 and 3.
5. Use the data path and control algorithm developed in General Exercises 10 and 11.

CHAPTER 12

General Exercises

1. Implement the data path in Figure 12.12 and use it to verify the instructions.
3. See text Figure 12.17.

Design/Implementation Exercises

1. See Figure 12.9 for the implementation.
3. For addition, $V = A_3'B_3'Z_3 + A_3B_3Z_3'$. For subtraction, $V = A_3B_3'Z_3' + A_3'B_3Z_3$. $Z_3 = 1$ denotes a negative result.
5. See text Figure 12.12 to implement the ALU.
7. See Figure 12.18.

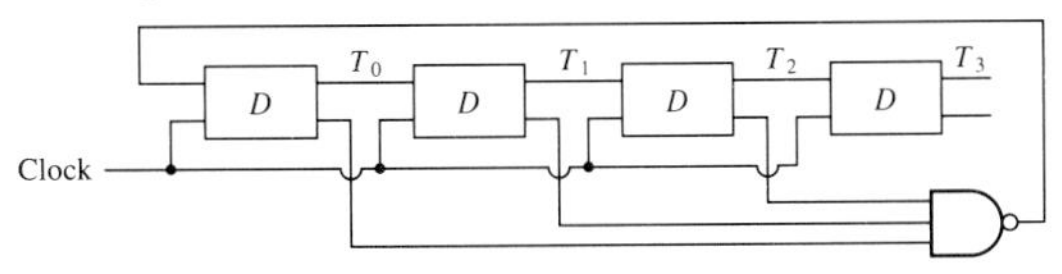

9. See Figure 12.22.
11. See Figure 12.25.
13. See Figure 12.28.
15. See Figure 12.31.

CHAPTER 13

General Exercises

1.

Decimal Address	Instruction		Hexadecimal Adrs.	Hexadecimal Inst.
0	LDADIR 6		00	06
1	ADDIR 7		01	27
2	ADDIR 8		02	28
3	ADDIR 9		03	29
4	ADDIR 10		04	2A
6	1	Data	06	01
7	2	Data	07	02
8	3	Data	08	03
9	4	Data	09	04
10	5	Data	0A	05

LDADIR = 000

ADDIR = 001

3. See Figure 13.12a.

5. Use a logic simulation program.
7. Use a logic simulation program.
9. Use a logic simulation program.

Design/Implementation Exercises

1. See Figure 13.12a.
3. See Figure 13.29.

CHAPTER 14

General Exercises

1. Same as answer to General Exercise 1 in Chapter 13.
3. See Figure 14.10a.
5. Use a logic simulation program.
7. Use a logic simulation program.
9. Use a logic simulation program.

Design/Implementation Exercises

1. See Figure 14.10a.
3. See Figure 14.17.

Index

Absorption law of Boolean algebra, 67
Accumulator register, 575, 580
Active high, 68–75
Active low, 68–75
Adder
  BCD, 236, 239–242
  carry look-ahead, 4-bit, 207–209
  cascaded, 2-bit, 198–199
  cascaded, 4-bit, 205–206
  direct, 2-bit, 200–205
  full (1-bit), 83–85
  half, 44, 81–83
  information distributed in space, 468
  information distributed in time, 469
  MSI cascaded 8-bit, 217
  MSI, 4-bit, 578
  serial, 266–267, 477–481, 482–483, 486–492
Adder/subtracter, 4-bit, 578
Adding multibit numbers, 266–267
Address
  field, 30–31, 597, 616
  pointer, 632
Address modes
  direct address, 615–616
  immediate, 633
  indexed-address, 659
  indirect address, 631–632, 686
Adjacency conditions, 371
  illustrated, 371, 373, 381, 385, 390
Algebra of sets, 94
Algorithm, 9, 516
  BCD-to-binary conversion, 560
  binary-to-BCD conversion, 560–561
  logic gate network design, 77
  serial addition, 479
  serial division, 530–532
  serial multiplication, 518–519
Algorithmic
  approach to design, 516
  process, 464
Al-Khowarizmi, 9
Alpha-beta test for critical race, 417
Alternate state transition method, 335
Alternative design approaches (of a 2-bit adder)
  formal solution, 200
  intuitive approach, 198
ALU (arithmetic logic unit), 222, 579
  bit-slice, 228–232
  MSI adder and TC01s, 232–235
  operations, 230, 232, 233, 579
Analog system, 6
Analog-to-digital converter, 7
Analysis
  of fundamental-mode circuits, 409–410
  of general synchronous sequential circuits, 335–348
  of mixed-logic gate network, 76–77
  procedure for (fundamental-mode circuits), 415
AND-OR form expression, 73
Application of switching algebra, 61
Architecture, computer, 612
Arithmetic logic unit (*See* ALU)
Arithmetic operations
  second 4-bit ALU, 233
  simple 4-bit ALU, 232
ASCII code, 6, 36

Assertion, 60–61, 71–77, 410
ASM (algorithmic state machine), 462, 466, 481
  chart, 464, 484–486, 487, 496, 504, 521, 538, 551
Assembly language, 564, 649
Assignment of adjacent codes
  application of, 371, 373, 381, 385
  rules for, 368
Astable multivibrator, 274
Asynchronous process, 402
Asynchronous sequential circuit design
  modular, 450–455
  traditional, 439–450
Augmented-row flow table, 432

Balance-beam problem, 43
Basic memory element, 268
  operation of, 268–272
BCD (binary coded decimal)
  adder, 236, 239–243
  adder/subtracter, 245–249
  complementer, 244–245
  definition, 6
  subtraction, 242–244
Binary codes assigned to machine states, 367
Binary digit, 6
Bit slice
  definition, 222
  design of a 4-bit ALU, 228–232
Black-box diagram
  of a half adder, 82
  of a full adder, 84
Blackjack dealer ASM, 501–509
Boole, George, 46
Boolean algebra (*See* Switching algebra)
Bubble in mixed logic, 76
Buffer, memory, 592
Building block, 81–85
  evolution of computer processor, 561–608
  implementation of higher-level devices
    full adder, 83
    pyramid-dice game, 186–187
    tiny calculator, 184–186
Bus
  memory load, 594–596
  memory output, 594–596
  unibus, 666, 672, 673
  Z bus, 671, 672
Bus design
  centralized-access bus, 167
  decentralized-access bus, 167–168
Bus implementation
  using a mux, 167
  using open-collector buffer/drivers, 168
  using 3-state buffers, 168
Bus-oriented computers, 665–701
Byte, 31

Canonical form of logic expressions
  POS (product-of-sums), 58–59
  SOP (sum-of-products), 58–59
Carry look-ahead
  carry generate, 207
  carry propagate, 207
  equation, 207
  group generate, 208
  group propagate, 208
  logic diagram of a carry look-ahead adder, 209
  recursion formula, 208
Cartesian
  coordinates, 49
  product, 50, 93–95, 101
Cascaded adder
  2-bit, 198–199
  4-bit, 205–206
Cell of Karnaugh map
  identified by minterm number, 96
  identified by product term, 95
  identified by row and column, 93, 95
Characteristic table, 273
  SR latch, 273
Characteristic equation
  D flip-flop, 286
  JK flip-flop, 287
  SR flip-flop, 288
Chip enable, 592
Clock-ring counter circuit, 587
Clock signal, 274
  clock period, 274
Clocked bistable memory element, 274
Code converters, 175–177
Coefficient of product term, 93
Combinational logic functions
  analysis of, 156–158
  from specification through realization, 156
Combined control unit design, 548
Combined serial multiplier/divider, 548
Combining logically adjacent cells
  Karnaugh maps, 97–100, 105–108
  minterm-ring maps, 123–135

Compare operations and comparators, 224–226
Complement forms
  1s complement, 19
  2s complement, 20
  9s complement, 12
  10s complement, 13
Complementer, 2s
  information distributed in space, 470
  information distributed in time, 471
Computer codes
  alphanumeric, 36
    ASCII, 36
    EBCDIC, 36
    flexowriter, 36
    Hollerith, 36
  decimal, 31–34
    nonweighted unit distance, 33
    nonweighted reflected, 34
    weighted nonsymmetric, 32
    weighted symmetric, 32–33
  Gray, 34–35
Computer, building-block evolution
  control unit subsystem, 582–590
  data path subsystem, 577–581
  integration of memory subsystem and CPU, 595–604
  memory subsystem, 591–595
Computers
  mux-oriented (MC-8A, B, C, D), 611–663
  bus-oriented (BC-8A, B, C, D), 665–701
Conceptual model, 42
  ASM model of controlled process, 483
  ASM model of CPU, 564
  FSM model of sequential process, 258, 265
  PASM model of computer, 613
  truth table model of combinational logic function, 155
Conceptualization
  in logic design, 44
  system, 483–487
Conceptualization step (*See* PSP)
Condition code flip-flops, 633
Conditional branch, 485
  instructions, 633, 680
Conditional output, 481
Configuration of switches, 53
Connectives, logic, 46
Consensus
  term, 410, 412
  theorem, 67
Constructs
  in ASM chart, 484–486
  in DCAS diagram, 484–486
Control algorithm, 479–481
  blackjack dealer ASM, 502
  dice game ASM, 494
  serial adder, 479–481, 482–483
  serial adder ASM, 486–487
  serial divider ASM, 536
  serial multiplier ASM, 519–520
Control algorithm table
  blackjack dealer, 505
  dice game ASM, 487
  serial divider ASM, 539–540
  serial multiplier ASM, 522
Control equations, output decoder, 624, 642, 658, 676, 686, 698
Control flow diagram, 462, 482
  BC-8A computer, 670
  blackjack dealer ASM, 504
  computer processor, 577, 583
  dice game ASM, 495–496
  MC-8A computer, 618, 624
  MC-8B computer, 636, 639
  MC-8C computer, 652
  serial adder ASM, 486–487
  serial (nonrestoring) division, 538
  serial multiplication, 521
Control flow diagram, types
  ASM chart, 484–486
  DCAS diagram, 484–486
Control system
  rudimentary, 471–480
  sophisticated, 481–509
Control unit, 222
  next-state generator, 488, 490, 496, 499, 507, 522, 539, 552, 583
  output decoder, 489, 491, 497, 506, 522, 539, 552, 584
Converting flip-flops
  D-to-*JK*, 289–290
  SR-to-*JK*, 290–292
Converting fractions
  binary to decimal, 27
  decimal to binary, 24–27
Converting integers
  binary to decimal, 24
  decimal to binary, 23
Converting logic expressions
  AND-OR to NAND-NAND, 65, 162

Converting logic expressions *(Continued)*
OR-AND to NOR-NOR, 66, 162
Copy-complement method, 474–475
Counter, 262
arbitrary count sequence, 303–304
asynchronous binary, 308–310
classifications, 292
design, 293–324
divide-by-*N*, 304–308
IC, 311–312
ring, 319–321
shift-register, 319–324
twisted-ring (Johnson, Moebius), 322–324
synchronous binary, 293–303
CPU (central processing unit), 222, 589
Critical race condition, 416
Critical race-free state assignment
type 1, 426–428
type 2, 428–430
type 3, 430–433, 436
Cycle
in fundamental-mode circuit, 428
maker diagram, 429, 430, 432
paths, 428, 437

Daf (direct address field), 597
format for, 616
Data flow cycle, 619
Data path
BC-8A computer, 677
BC-8B computer, 683, 687
BC-8C computer, 693, 700
combined serial multiplier/divider ASM, 556
MC-8A computer, 623, 637
MC-8B computer, 638, 643
MC-8C computer, 653
parallel ALU, 672
serial divider ASM, 539
serial multiplier ASM, 521, 525
subsystem of a simple computer, 580, 596
DC (direct current) voltage, 67–68
DCAS (documented control algorithm state) diagram, 464, 484, 496
Decimal number system, 8–16
Decoders, 173–174
Decoding a counter, 310
D flip-flop
characteristic table, 286
excitation table, 286
Delay element, 403
Design
1-bit comparator, 224–225
2-bit comparator, 225–226
2-bit inequality detector, 107–108
4-bit adder/subtracter, 210–213
4-bit ALU, 232–234
4-bit comparator, 226
4-bit 8-function arithmetic unit, 214–217
6311 error detector, 104–106
BCD adder, 236, 239–243
BCD adder/subtracter, 245–249
BCD arithmetic unit, 235–252
BCD complementer, 244–245
BCD divide-by-3 function, 108–109
BCD error detector, 252–254
BCD-to-decimal decoder, 249–250
BCD-to-7 segment decoder, 251–252
cascaded (bit-slice) ALU, 228–232
counters, 293–310
gated *SR* latch, 275–276
Mealy 111 sequence detector, 352–353
pulse-triggered (M/S) flip-flop, 283–284
shift-register counters, 319–324
SR latch, 268–273
Design (4-step PSP)
1-to-4 demultiplexer (demux), 165–166
2-to-1 multiplexer (mux, 163–164
2-to-4 decoder, 173–174
3-bit up/down counter, 374–377
3-input majority circuit, 161–162
4-to-1 multiplexer, 164–165
4-input priority encoder, 172–173
4-to-2 encoder, 170–171
BCD adder, 240–242
binary adders, 2-bit, 197–205
binary-to-Gray code converter, 176–177
clocked, level-sensitive *JK* flip-flop, 280–283
decimal-to-BCD encoder, 238–239
digital logic bus, 166–168
even-parity checker, 179, 472–474
even-parity generator, 177–179
filter circuit, 377–383
full adder, 83–85
gated *D* latch, 276–279
half adder, 81–83
logical shifter, 180–181
Moore 111 sequence detector, 348–351
pyramid-dice game machine, 186–189

sequence detector, 433–438
serial 2s complementer, 474–477
serial addition machine, 477–480
tiny calculator, 184–186
Design applications, interactive (IDA)
2-person game, 384–387
BC-8A computer, 668–680
BC-8B computer, 681–691
BC-8C computer, 691–700
blackjack dealer ASM, 501–509
combined serial multiplier/divider ASM, 549
dice game ASM, 493–501
MC-8A computer, 616–630
MC-8B computer, 634–646
MC-8C computer, 649–659
Mealy 0101 sequence detector, 393–396
NET *T* flip-flop, 439–444
PET *D* flip-flop, 451–455
pulse generator, 445–450
ring counter, 4-bit, 319–321
sequential-logic combination lock, 387–392
serial divider ASM, 535
serial multiplier ASM, 519–528
twisted-ring counter, 4-bit, 322–324
Design of a 4-bit adder
by cascading four 1-bit full adders, 205–206
using carry look-ahead, 206–210
Design evolution
1-bit adder to 8-bit arithmetic unit, 196–217
bus-oriented computers, 666–700
computer processor, 574–608
mux-oriented computers, 612–662
Design procedure
algorithmic state machines (ASM), 481–482
combinational logic circuits, 160
computers, bus-oriented, 667–668
computers, mux-oriented, 613–614
general synchronous sequential circuits, 348
fundamental-mode asynchronous circuits, 418–419
Design problems, model for solution of, 43–46
Design specifications, 83
machine conceptualization, 419
Designworks logic-simulation program, 86, 564
Detect-*A*zero circuit, 637, 682
Dice game ASM, design of, 493–501
Digital system, 6, 7, 222
Digital systems design
state machines and systems design, 462–510
multipliers, dividers, and control units, 515–561
Direct adder, 2-bit, 200–205
Division, binary integer
algorithm, 530–532
nonrestoring method, 533–535
restoring method, 532–533
Don't care conditions in Karnaugh maps, 108
Duty cycle, clock, 335

EAC (end-around carry), 12–14, 21–22
EBCDIC code, 6, 36
ECLK (inverted clock signal), 619, 622, 676
Edge-triggered
D flip-flop, PET, 284–286, 451–455
JK flip-flop, 286–287
registers, 619, 623, 625, 676
SR flip-flop, 286–287
T flip-flop, NET, 439–444
types, 274
E, *G*, *L* flip-flops, 637–638
EHBBA (evolutionary hierarchical building-block approach), 2
diagram, 3, 150, 259, 463, 565
Eliminating
hazards, 159–160
intermediate unstable states in asynchronous circuit, 424
redundant stable states in asynchronous circuit, 421–423
redundant states in sequential circuit, 379
Elimination law of Boolean algebra, 67
Encoders, 168–173
End-around carry (*See* EAC)
Equivalence class
definition, 360
partitioning method, 360–364
Equivalent stable states, 421, 423
Erratic circuit operation, 410
Erroneous end stable state, 416
Evolution
1-bit adder to an 8-bit arithmetic unit, 195–217
computer processor, 567–608
rudimentary to sophisticated control systems, 467–509
simple to more complex computers, 613–662, 666–701

Excitation in asynchronous sequential circuit, 404
  analysis, 409–410
  equation, 406, 412, 414, 416, 437, 448, 454, 526, 544
  map, 409, 414, 416, 428, 430, 433, 437, 443, 448, 453
Excitation map, *SR* latch, 270
Excitation network simplification, 368
Excitation table
  JK flip-flop, 287
  lumped delay *SR* latch, 270
  SR latch, 273
Exclusive-OR as a selective complementer, 211–212
Exponent, 9

Fetch, instruction, 582, 597
Fixed-logic systems, 68
Flip-flop, definition, 274
  characteristic and excitation, 286
  characteristic and excitation for ET *D* flip-flop, 286
  characteristic and excitation for ET *JK* flip-flop, 286–287
  general input equations for, 286–288
Floating-point numbers
  base-2 operations, 22–23
  base-2 representation, 22
  base-10 operations, 15–16
  base-10 representation, 14–16
Flow table of asynchronous circuit
  augmented-row, 432, 436
  minimal-row reduced, 425, 427, 435, 442, 447, 453
  nonredundant, 421, 423
  primitive, 407, 417, 422, 434, 441, 446
  row-reordered, 427, 429
  row-reordered with cycles, 430
Fractions, 9
  base-2 representation, 16
  base-10 representation, 10
FSM (finite state machine)
  concepts, 264–267
  modules
    even-parity check, 473
    serial 2s complementer, 475–476
    serial adder, 478
Full adder, design of, 83–85
Function, Boolean (switching), domain, mapping, range, 57
Functions in algebra and logic, 49
Fundamental-mode asynchronous sequential circuit, 404

GAIL (gate asserted input level), 77
GAOL (gate asserted output level), 77
Gate function tables, 71–75
Gate realization of logic functions, 70–75
Gated latches, 275–279
Gray code, 34
  wheel, 175

Half adder, 44
  design of, 81–83
Handshake, 485, 495, 503
Hazards
  combinational logic circuits, 158–160
  elimination of, 159–160, 411–415
  fundamental-mode circuits, 410–415
Hazardous transition, 411
Hexadecimal (hex) keypads, 473, 476
Hierarchical
  diagram of a computer, 569
  model, 7
Hindu-Arabic number system, 8

IC (integrated circuit), 70–75, 217, 226, 707–708
  pin configuration, 70
IC counters, 311–312
IDA (interactive design applications) (*See* Design applications)
Idealized gates, 404–405
Idempotent law of Boolean algebra, 67
Implementation of Boolean functions using MSI ICs
  full adder using multiplexers, 181–182
  full adder using a decoder, 182–183
Implicant pairs, quartets, and octets, 123–135
Implicants of a switching function, 109–115
  defined in terms of minterms, 122–123
  essential prime implicants (EPI), 109–115
  prime implicants (PI), 109–115
Implication chart method, 364–366
Index register, 651
Information
  distributed in space, 468, 470
  distributed in time, 469, 471
Inputs, 60
  alternate state transition method, 335
  asynchronous, 258

flip-flop, 280–288
gated latch, 275–279
process, 263–268
register, 312–314, 317–318
shift register, 315–316
SR latch, 268–273
synchronous, 258
Instruction
cycle, 573, 597, 618–619
decoder, 582, 587, 597
execution, 597
fetch, decode, execute, 618, 622, 631
mnemonic, 573
register, 582, 585, 587, 597
with 2 implied operands, 647
word, 593
Instruction set (repertoire), 575
BC-8A computer, 669
BC-8B computer, 681
BC-8C computer, 691–692
MC-8A computer, 617
MC-8B computer, 635
MC-8C computer, 650
MC-8D computer, 661
Instruction-time event table, 584, 598
for BC-8A, 674
for BC-8B, 684
for BC-8C, 694–695
for MC-8A, 625
for MC-8B, 640
for MC-8C, 654–655
Integration
data path and control unit, 589
memory and CPU, 595
Interactive design applications (*See* Design applications)
Interfacing
using bus, 666
using mux, 612
Internal state of fundamental-mode circuit, 406
I/O (input/output) instructions, 615
Iterative process, 263, 469

JK flip-flop, 287
Jump instruction, 622, 669
Jump indirect instruction, 647, 661

Karnaugh maps (K-maps), 92–117, 119–120
algorithm, 106
pairs, quartets, and octets, 97–98, 100, 102–115, 120
simplification techniques, 103–105
sophisticated algorithm, 110–111
sophisticated techniques, 109–110

Latch
gated *D*, 276–279
gated *SR*, 275–276
SR, 268–273
Level-sensitive memory element, 274
LGP-30 computer, 648
List processing, 612
Literal, 57
Load bus (*See* Bus)
Logic
constants, 46
diagram, 62
duals, 52–53
equation, 62
expressions, 46
functions, 44, 46–50, 56–60, 70, 709–712
gates, 54, 70–75
identifier, 68–70
model, 44–49
operations, 46–47, 59–60
product, 47
simulation programs, 42
sum, 47
symbols, 62
systems, 52–53, 68–70
variables, 46
Logic diagram, 377, 383, 387, 392, 396, 474, 476, 480, 492, 501, 509, 526, 546, 558, 628, 644, 678, 688
Logic functions, ANSI/IEEE, 709–712
Logic gate network design, 77–81
Logic instructions, 647, 650
Logic operations
of a 1-bit ALU, 229
of a 4-bit ALU, 230
AND, OR, XOR, 227–228
Logic structures, 646, 649
Logical adjacency theorem, 67
Logically equivalent
excitation equation, 412
expression, 92
Looping instructions, 649
LSB (least significant bit), 36
Lumped-delay model
asynchronous sequential circuit, 404
enabled *D* latch, 405
NOR gate memory cell, 270

MA (memory address) register, 593, 621, 631
Machine language instruction, 30–31
Machine language program, 575
Man-machine interface
  blackjack dealer ASM, 503–504
  dice game ASM, 495
Manual-control computer, 569
Map-entered variables (*See* Variable-entered map)
Map-form state table, 374, 381, 386, 390
Map of transitions
  illustrating essential hazard, 415
  illustrating race-free condition, 428, 430, 433, 454
Mapping a switching function, 49–50, 93–143
Mask, address, 635, 638
Maxterm, 58–59
McCalla minterm-ring algorithm, 126–130, 131–135
MCS-4 microcomputer, 222
Mealy, 265
Memory, 222
  asynchronous sequential circuit, 402
  element derived from *SR* latch, 274
  instruction fetch, 618
  load (*See* Bus)
  output (*See* Bus)
  read operation, 619
  synchronous sequential circuit, 402
Memory reference instructions, 615, 635, 647
Memory subsystem
  bus-oriented computer, 673
  design and implementation, 591, 620
  MC-8A computer, 623
Merged state diagram, 443, 448
Merger
  graph, 424–425, 441, 446
  group, 424
  intermediate unstable states, 421
  maximal merger groups, 425, 441, 446
Microoperations
  serial division, 538
  serial multiplication, 524
Minimal
  cover for a K-map, 106
  NOR-NOR form, 100
  POS using K-map, 100–101
  SOP using K-map, 96–100
  SOP in 2-level AND-OR form, 107, 202–203
Minimal state table
  Mealy 111 sequence detector, 360
  Moore 000/111 sequence detector, 372
Minterm, 57–59
  distinguished, 110, 125
  isolated, 110, 123
  links, 113–115
  pairs, quartets, octets, 103–105, 123–124
  rings in a K-map, 115–116
Minterm-maxterm relations, 58
Minterm-ring algorithm (McCalla), 126–130, 131–135
  procedure guide for, 129, 133
Minterm-ring maps
  in 5 variables, 123, 125, 128, 201–203
  in 6 variables, 130–132
Minterm-ring table
  for a 4-variable K-map, 119
  in 5 variables, 122
Missing-link theorems, 125–126
Mixed-logic system, 69–81
  analysis of gate network, 76–77
  diagrams, 75–81
  implementing logic equations in, 75–76
  table of symbols, 75
Mnemonic code, 31
Model
  combinational logic functions, 154–155
  finite state machine, 265
  general asynchronous sequential circuit, 403
  general synchronous sequential circuit, 334
Modulo-4 addition, 332
Moore, 265
  111 sequence detector, 333–334
MSB (most significant bit), 20, 36
MSI (medium scale integration) ICs, 85
MSI implementation
  8-bit computing devices, 217
  full adder (using decoder), 182–183
  full adder (using mux), 181–182
Multifunction processor, 548
Multiplication, binary integer, 528
Multiplier
  2-bit combinational, 189
  4-bit serial, 517–519
Multiply instruction (MULBP), 648, 661
Mux enable control, 622
Mux implementation of next-state generator
  serial divider ASM, 543

serial multiplier ASM, 526
serial multiplier/divider ASM, 556
Mux interfacing, 620
Mux-oriented computers, 611–663
MW (memory write) register, 593, 620

NAND-NAND form, 73
Negative logic, 52
Nested form of polynomial, 518
NET (negative edge-triggered) *T* flip-flop, 439
Next state, 264
maps, 473, 475, 478, 489, 497, 506
Nibble, 31
Nonconflicting rows of flow table, 424–425
Nonequivalent state assignments, 367
Nonunique minimal cover in K-map, 114–115
Number systems
binary, 16–23
decimal, 8–16
hexadecimal, 28–31
octal, 28
sexagesimal (base 60), 8
vigesimal (base 20), 9

Octet of logically adjacent minterms, 104–106, 111, 113, 119–120, 123–135, 201
Opcode (operation code), 30, 575, 585, 616
Operations in logic
AND, OR, NOT, 46, 59
exclusive-NOR, 60
exclusive-OR, 60
NAND, NOR, 59–60
ORAB instruction, 647
Output decoder (*See* Control unit)
Output instruction (OUT), 615, 661, 669
Output map generation, for fundamental-mode circuit, 443–444, 449
Overflow detection in binary addition and subtraction, 213

Pair of logically adjacent minterms, 94, 97, 100, 102–108, 111–115, 123–135, 202
Parity
even, 263
generation and checking, 177–179
Partial dividend, 529
Partial remainder, true, 529–530
PASM (programmable algorithmic state machine), 576, 613, 618, 635
Path table for state adjacencies, 427, 429, 430, 435, 442, 447, 452
PC (program counter), 594–601, 621
incrementing, 622
PDP-8 minicomputer, 222
Physical function table, 45, 48
Pin configuration of IC, 70
PIPO (parallel in/parallel out) shift register, 315, 316
PISO (parallel in/serial out) shift register, 315, 316
PLA (programmable logic array) (*See* Programmable logic devices)
POS (product-of-sums) expression, 100–101, 161–162
Positional notation, 8, 16–17, 28–29, 529
Positive logic, 52
Potentially equivalent states, 422
PPS (partial-product-sums) method, 517–518
Present state, 264
Prime implicants (PI) of switching function, 109–111, 124–127
essential prime implicant (EPI), 109–115
determined by counting minterm links, 120, 125–127
Prime implicant table, 136–138
reduced, 128, 134, 138
Primitive flow table, 407, 421, 434, 441, 446
Primitive state diagram, 408–409, 420, 434, 440
for *SR* latch, 271
Process algorithm
serial addition, 479, 482, 488
serial division, 532, 534
serial multiplication, 519–520
Process control algorithm
blackjack dealer ASM, 502, 504–506
dice game ASM, 494, 496–497
serial addition, 479–481, 482, 486
serial divider ASM, 536–540
serial multiplier ASM, 519–520
serial multiplier/divider ASM, 549–550
Process control flow diagram, 483–487, 496, 504, 521, 538, 551, 577, 583
Process section (data path)
blackjack dealer ASM, 507
combined serial multiplier/divider ASM, 550–552, 555–556
dice game ASM, 498
serial addition ASM, 490

Process section (data path) *(Continued)*
serial divider ASM, 536, 538–539, 542
serial multiplier ASM, 520–521, 525
Processor, 566, 568–569, 591–608, 613–614, 617–618, 635, 651, 669
block diagram, 596
data flow cycle, 619
logic diagram, 602–603
operation, 597, 600–601, 606–607
structure diagram, 566, 618, 635, 651, 669
timing diagram, 597, 604, 608, 630, 646
Product-of-sums expression (*See* POS)
Program, 573
load facility, 593–594, 667
run facility, 594–595
Programmable logic devices (PLD)
introduction to, 721–732
programmable AND-array logic (PAL), 190
programmable logic array (PLA), 188–189, 585, 589, 596, 618, 631, 653, 677, 683, 687
read-only memory, 190–191
Programmer's view of a computer, 612
Propagation delays, 402, 410, 412
PSP (problem-solving procedure) for design, 42
for computer design 576–577
model for 4-step PSP, 44
PSP design examples [*See* Design (4-step PSP) entries]
P-term
control point activation table, 585, 599, 626, 641, 656–657, 675, 685, 696–697
time/instruction, 585, 598
Pulse generator design, 445–450
Pulse-triggered (master slave) flip-flop, 274

Quartet of logically adjacent minterms, 103–108, 111–115, 123–135, 201, 203
Quine–McCluskey method, 135–140
Quotient formulas, 530–532

Race conditions, 415–417
critical, 415–417
test for, 417
for 3-row flow table, 436
Race-free state assignment, 425, 427
Railroad switchyard
automatic control, 572–574
manual control, 570–571
semiautomatic control, 571–572

Read enable, memory, 592, 594, 596
Ready-set-go (handshake)
blackjack dealer ASM, 503–504
dice game ASM, 495
Realization of sequential machine, 267–268
Recursion formula, 265
binary integer division, 531–532, 534
binary integer multiplication, 519
Reduced state table
1011 sequence detector, 364
0101 sequence detector, 366, 368
Redundant
stable states of fundamental-mode circuit, 421
states, 360
Register, 312
data, 313–314
reference instructions, 615, 634
transfer, 317, 582
transfer notation, table, 318, 467
Reliability, fundamental-mode circuit, 410
Repetition structure, 649
Required state adjacencies
satisfied by adding states for cycle paths, 430–433
satisfied by using available cycle paths, 428–430
satisfied without cycles or more states, 426–427
Right-shift register, 315
4-bit SIPO, 316
4-bit SISO, 316
Ring counter
design, 319–321
timing diagram, 319, 321, 583
RS latch (*See SR* latch)
RTN (register transfer notation), 318, 466–467
Rudimentary control machine
2s complement machine, 474–477
even-parity check, 472–474
serial addition machine, 477–480
R/W (read/write) memory, 592

Semiautomatic computer, 569
Sequence
detection process, 420
detector, 420–421, 433–438
generation process, 262
Sequential (finite state) machine
concepts, 264–265
with rudimentary control, 471

Sequential process, 263
Shared-data-path design, 548
Shared-row method, 432
Shift operations, 180
Shift register
  counters, 262
  types, 315
Shift register property, 314–315
Shortcut *JK* method, 296, 302, 305, 376, 382, 387, 391, 395
Signal, 7
Signed integers, base-2 representations
  1s complement form, 19
  2s complement form, 20
  sign-magnitude form, 19
Simplifying Boolean expressions
  algorithmic methods
    McCalla minterm-ring algorithm, 126–135
    Quine–McCluskey method, 135–140
  mapping techniques
    Karnaugh maps, 92–117, 131, 139
    minterm-ring maps, 119–135, 139
Simulation (logic) programs, 42
SIPO (serial in/parallel out), 315, 316
SISO (serial in/serial out), 315, 316
Skip instruction, 659
Solution/simplification step (*See* PSP)
SOP (sum-of-products) expression, 48, 93–135
  canonical, 93–94, 96, 161, 164, 182
  minimal, 96–100, 106, 110–114, 130, 135, 162, 177–178, 202–203
SR latch, 262, 268–272
SSI (small scale integration) ICs, 85
Stable state to stable state path, 427
State assignment (asynchronous)
  algorithm for race-free circuit, 419
  matrix, 427, 436, 447
  rules for race-free assignment, 425
State assignment (synchronous)
  matrix, 369–370
  state tables with unused states, 371–374
  synchronous sequential circuits, 367
  techniques, 368–371
State-changing transition, 315
State diagram generation
  binary tree procedure, 352–353
  multiple-sequence-path algorithm, 355–357
  single-sequence-path algorithm, 354–355, 393
State reduction (fundamental-mode)
  eliminating redundant stable states, 421–423
  examples, 434, 441, 446, 452
  merging intermediate unstable states, 424–425
State table and diagram, 264
  Mealy 111 sequence detector, 353
  Mealy 0101 sequence detector, 355
  Mealy 0000/1111 sequence detector, 357
  modulo-4 addition, 332
  Moore 111 sequence detector, 334
State table reduction methods
  equivalence class partitioning, 360–364
  implication chart, 364–366, 379
  row elimination, 358–360, 379
State transitions, 273
Store accumulator instruction, 622, 647, 669
Stored-program computer, 569, 572–573, 591–604
Structure diagram, 462, 464, 481, 484
  BC-8A computer, 669
  blackjack dealer ASM, 503
  computer processor, 576, 582
  dice game ASM, 494
  MC-8A computer, 618
  MC-8B computer, 635
  MC-8C computer, 651
  serial adder ASM, 486
  serial divider ASM, 536
  serial multiplier ASM, 520
  serial multiplier/divider ASM, 550
Subroutine, 648
Subroutine jump instruction, 647–648, 661
Subscript, 9
Subtraction
  1s complement method, 18–20
  2s complement method, 21–22
  9s complement method, 12–13
  10s complement method, 13–14
Sum-of-products expression (*See* SOP)
Swap instruction, 659
Switches in series or parallel, 50–54
Switch network algebra, 50–54
Switching algebra, 54–67
  algebraic techniques for simplifying expressions, 66–67
  axiomatic definition, 54–55
  properties of, 63–65
    complement of a Boolean expression, 64
    converting logic expressions, 65–66

Switching algebra, *(Continued)*
dual of a Boolean expression, 64
equivalence of expressions, 63–64
functional completeness, 63
theorems, 55–56, 67, 97
variables, expressions, and functions, 56–60
Switching functions, 56–60
and K-maps, 93–104
table of *f(x, y)*, 59
in 3 or more variables, 60
System design
solution and simplification, 487–489
realization, 489–492
Symbolic logic, 46

Target address, 634
TC01 (true/complement/zero/one), 214–217, 579, 589
Thermocouple, 7
Time subintervals of instruction cycle, 582
Timing diagram, 285, 528, 630, 646, 680, 690
Timing generator, 319
Top-down design
BCD arithmetic unit, 234–252
definition, 222
Total induction, 67
Total state (fundamental-mode), 406, 409, 411
Transient, 410
Transition between stable states, 404, 410
Transparency property, 315
Tri-state (3-state) buffer, 168, 666, 672
Truth table, 45–48, 50, 52, 57, 61, 65, 71–75, 82, 84, 93, 99, 107, 141, 155, 157, 161, 171, 172, 174, 176, 182, 200–201
and corresponding K-maps, 96–100, 105, 108, 117, 143, 178
T-states, 619
TTL (transistor transistor logic), 68
ICs, 70–75, 85
User's Guide, 712–720

Underlying combinational logic circuit, 406
Unibus, 666, 687 (*See also* Bus)
Unstable state
asynchronous circuit, 404, 409
unspecified, 429

Variable-entered map
in 3 variables, 140–142
in 4 variables, 142–144
serial divider next-state generator, 544
serial multiplier next-state generator, 522
Variables (fundamental-mode)
primary (input), 405
secondary (feedback), 405, 415–416
Venn diagram, 94
Voltage polarity, 68–69
Voltage-referenced signal, 68

Write enable (memory), 592

Z bus (*See* Bus)